Shigley's
Mechanical
Engineering
Design

Shigley's Mechanical Engineering Design

Tenth Edition

Richard G. Budynas
Professor Emeritus, Kate Gleason College of Engineering, Rochester Institute of Technology

J. Keith Nisbett
Associate Professor of Mechanical Engineering, Missouri University of Science and Technology

Adapted by

Kiatfa Tangchaichit
Khon Kaen University

Mc
Graw
Hill
Education

Dedication

To my wife, Joanne, my family, and my late brother, Bill, who advised me to enter the field of mechanical engineering. In many respects, Bill had considerable insight, skill, and inventiveness.

Richard G. Budynas

To my wife, Kim, for her unwavering support.

J. Keith Nisbett

Dedication to Joseph Edward Shigley

Joseph Edward Shigley (1909–1994) is undoubtedly one of the most well-known and respected contributors in machine design education. He authored or coauthored eight books, including *Theory of Machines and Mechanisms* (with John J. Uicker, Jr.), and *Applied Mechanics of Materials*. He was coeditor-in-chief of the well-known *Standard Handbook of Machine Design*. He began *Machine Design* as sole author in 1956, and it evolved into *Mechanical Engineering Design*, setting the model for such textbooks. He contributed to the first five editions of this text, along with coauthors Larry Mitchell and Charles Mischke. Uncounted numbers of students across the world got their first taste of machine design with Shigley's textbook, which has literally become a classic. Nearly every mechanical engineer for the past half century has referenced terminology, equations, or procedures as being from "Shigley." McGraw-Hill is honored to have worked with Professor Shigley for more than 40 years, and as a tribute to his lasting contribution to this textbook, its title officially reflects what many have already come to call it—*Shigley's Mechanical Engineering Design*.

Having received a bachelor's degree in Electrical and Mechanical Engineering from Purdue University and a master of science in Engineering Mechanics from the University of Michigan, Professor Shigley pursued an academic career at Clemson College from 1936 through 1954. This led to his position as professor and head of Mechanical Design and Drawing at Clemson College. He joined the faculty of the Department of Mechanical Engineering of the University of Michigan in 1956, where he remained for 22 years until his retirement in 1978.

Professor Shigley was granted the rank of Fellow of the American Society of Mechanical Engineers in 1968. He received the ASME Mechanisms Committee Award in 1974, the Worcester Reed Warner Medal for outstanding contribution to the permanent literature of engineering in 1977, and the ASME Machine Design Award in 1985.

Joseph Edward Shigley indeed made a difference. His legacy shall continue.

About the Authors

Richard G. Budynas is Professor Emeritus of the Kate Gleason College of Engineering at Rochester Institute of Technology. He has more than 50 years experience in teaching and practicing mechanical engineering design. He is the author of a McGraw-Hill textbook, *Advanced Strength and Applied Stress Analysis,* Second Edition; and coauthor of a McGraw-Hill reference book, *Roark's Formulas for Stress and Strain,* Eighth Edition. He was awarded the BME of Union College, MSME of the University of Rochester, and the PhD of the University of Massachusetts. He is a licensed Professional Engineer in the state of New York.

J. Keith Nisbett is an Associate Professor and Associate Chair of Mechanical Engineering at the Missouri University of Science and Technology. He has more than 30 years of experience with using and teaching from this classic textbook. As demonstrated by a steady stream of teaching awards, including the Governor's Award for Teaching Excellence, he is devoted to finding ways of communicating concepts to the students. He was awarded the BS, MS, and PhD of the University of Texas at Arlington.

Brief Contents

Contents

Objectives

This text is intended for students beginning the study of mechanical engineering design. The focus is on blending fundamental development of concepts with practical specification of components. Students of this text should find that it inherently directs them into familiarity with both the basis for decisions and the standards of industrial components. For this reason, as students transition to practicing engineers, they will find that this text is indispensable as a reference text. The objectives of the text are to:

- Cover the basics of machine design, including the design process, engineering mechanics and materials, failure prevention under static and variable loading, and characteristics of the principal types of mechanical elements.
- Offer a practical approach to the subject through a wide range of real-world applications and examples.
- Encourage readers to link design and analysis.
- Encourage readers to link fundamental concepts with practical component specification.

New to This Edition

Enhancements and modifications to the tenth edition are described in the following summaries:

- A new Chap. 20, *Geometric Dimensioning and Tolerancing,* has been added to introduce an important topic in machine design. Most of the major manufacturing companies utilize geometric dimensioning and tolerancing (GD&T) as a standardized means of accurately representing machine parts and assemblies for the purposes of design, manufacture, and quality control. Unfortunately, many mechanical engineers do not have sufficient exposure to the notation and concepts of GD&T to interpret the drawings.

  During the time when GD&T was becoming most prevalent in manufacturing, many engineering schools were phasing out comprehensive drafting courses in favor of computerized CAD instruction. This was followed by another transition to 3D solid modeling, where the part was drawn with ideal dimensions. Unfortunately, this ability to draw a perfect part in three dimensions is all too often accompanied by a neglect of focus on how to accurately and uniquely represent the part for manufacture and inspection.

  A full understanding of GD&T is usually obtained through an intensive course or training program. Some mechanical engineers will benefit from such a rigorous training. *All* mechanical engineers, however, should be familiar with the basic concepts and notation. The purpose of the coverage of GD&T in this new chapter is to provide this foundational exposure that is essential for all machine designers.

  It is always a challenge to find time to include additional material in a course. To facilitate this, the chapter is arranged and presented at a level appropriate for students to learn in an independent study format. The problems at the end of the chapter are more like quiz questions, and are focused on checking comprehension of the most fundamental concepts. Instructors are encouraged to consider using this chapter as a reading assignment, coupled with even a minimal lecture or online discussion. Of course, there is ample material for expanded presentation and discussion as well.

- Chapter 1, *Introduction to Mechanical Engineering Design,* has been expanded to provide more insight into design practices. Further discussion of the development of the *design factor* is presented, as well as the statistical relationships between *reliability* and the *probability of failure,* and reliability and the design factor. Statistical considerations are provided here rather than in a chapter at the end of the text as in past editions. The section on Dimensions and Tolerances has been expanded to emphasize the designer's role in specifying dimensions and tolerances as a critical part of machine design.

- The chapter of the previous edition, *Statistical Considerations,* has been eliminated. However, the material of that chapter pertinent to this edition has been integrated within the sections that utilize statistics. The stand-alone section on stochastic methods in Chap. 6, *Fatigue Failure Resulting from Variable Loading,* has also been eliminated. This is based on user input and the authors' convictions that the excessive amount of development and data provided in that section was far too involved for the simple class of problems that could be solved. For instructors who still want access to this material, it is available on McGraw-Hill's Online Learning Center at www.mhhe.com/shigley.

- In Chap. 11, *Rolling-Contact Bearings,* the Weibull probability distribution is defined and related to bearing life.

- In conjunction with the Connect Engineering resource, the end-of-chapter problems have been freshly examined to ensure they are clearly stated with less room for vague interpretations. Approximately 50 percent of the problems are targeted for Connect implementation. With the problem parameterization available in this Web-based platform, students can be assigned basic problems with minimal duplication from student to student and semester to semester. For a good balance, this edition maintains many end-of-chapter problems that are open-ended and suitable for exploration and design.

Connect Engineering

The tenth edition continues to feature McGraw-Hill Connect Engineering, a Web-based assignment and assessment platform that allows instructors to deliver assignments, quizzes, and tests easily online. Students can practice important skills at their own pace and on their own schedule.

McGraw-Hill LearnSmart®

McGraw-Hill LearnSmart is an adaptive learning system designed to help students learn faster, study more efficiently, and retain more knowledge for greater success. Through a series of adaptive questions, Learnsmart pinpoints concepts the student does not understand and maps out a personalized study plan for success. It also lets instructors see exactly what students have accomplished, and it features a built-in assessment tool for graded assignments. Ask your McGraw-Hill Representative for more information, and visit www.mhlearnsmart.com for a demonstration.

SMARTBOOK™

McGraw-Hill SmartBook™

Powered by the intelligent and adaptive LearnSmart engine, SmartBook is the first and only continuously adaptive reading experience available today. Distinguishing what students know from what they don't, and honing in on concepts they are most likely to forget, SmartBook personalizes content for each student. Reading is no longer a passive and linear experience but an engaging and dynamic one, where students are more likely to master and retain important concepts, coming to class better prepared. SmartBook includes powerful reports that identify specific topics and learning objectives students need to study. These valuable reports also provide instructors insight into how students are progressing through textbook content and are useful for identifying class trends, focusing precious class time, providing personalized feedback to students, and tailoring assessment.

    How does SmartBook work? Each SmartBook contains four components: Preview, Read, Practice, and Recharge. Starting with an initial preview of each chapter and key learning objectives, students read the material and are guided to topics for which they need the most practice based on their responses to a continuously adapting diagnostic. Read and practice continue until SmartBook directs students to recharge important material they are most likely to forget to ensure concept mastery and retention.

Electronic Textbooks

This text is available as an eBook at www.CourseSmart.com. At CourseSmart your students can take advantage of significant savings off the cost of a print textbook, reduce their impact on the environment, and gain access to powerful web tools for learning. CourseSmart eBooks can be viewed online or downloaded to a computer.

    The eBooks allow students to do full text searches, add highlighting and notes, and share notes with classmates. CourseSmart has the largest selection of eBooks available anywhere. Visit www.CourseSmart.com to learn more and to try a sample chapter.

McGraw-Hill Create™

With McGraw-Hill Create, you can easily rearrange chapters, combine material from other content sources, and quickly upload content you have written, like your course syllabus or teaching notes. Find the content you need in Create by searching through thousands of leading McGraw-Hill textbooks. Arrange your book to fit your teaching style. Create even allows you to personalize your book's appearance by selecting the cover and adding your name, school, and course information. Order a Create book and you'll receive a complimentary print review copy in 3–5 business days or a complimentary electronic review copy (eComp) via e-mail in minutes. Go to www.mcgrawhillcreate.com today and register to experience how McGraw-Hill Create empowers you to teach your students your way.

    Additional media offerings available at www.mhhe.com/shigley include instructor and student supplements.

Supplements

Instructor Supplements (provided to instructors using this SI edition)

http://highered.mheducation.com/sites/9814595284

- *Solutions manual.* The solutions manual contains solutions to most end-of-chapter nondesign problems.

- *PowerPoint slides.* Slides of important figures and tables from the text are provided in PowerPoint format for use in lectures.

The following are related to users of the U.S. edition

www.mhhe.com/shigley

- *Fundamentals of Engineering (FE) exam questions for machine design.* Interactive problems and solutions serve as effective, self-testing problems as well as excellent preparation for the FE exam.
- *C.O.S.M.O.S.* A complete online solutions manual organization system that allows instructors to create custom homework, quizzes, and tests using end-of-chapter problems from the text.

Rationale for Non-Conversion to SI Units

In some cases the U.S. customary units have been retained in this SI edition. The reasons are as follows:

- Trade/brand products commonly used in the industry adopt U.S. customary units. Examples and problems based on these products will therefore not have the units converted to SI. Examples of these can be found in Chapter 8.
- The variable diametral pitch is only used with U.S. customary units. Some examples found in Chapters 13 and 14, as well as Case Study Part 3 in Chapter 18, are not converted to SI units because they are based on diametral pitch.
- Empirical expressions like the AGMA equations (see Chapter 15) were derived from research done based on products in U.S. customary units. Examples that use these equations are not converted to SI units.

Acknowledgments

The authors would like to acknowledge those who have contributed to this text for over 50 years and nine editions. We are especially grateful to those who provided input to this tenth edition:

Expanded Connect Implementation

Peter J. Schuster, *California Polytechnic State University*

Drawings for GD&T Chapter

Glenn Traner, *Tech Manufacturing, LLC*

CAD Model Used in Cover Design

Jedrzej Galecki, *University of the West of England*

Reviewers

Kenneth Huebner, *Arizona State*
Gloria Starns, *Iowa State*
Tim Lee, *McGill University*
Robert Rizza, *MSOE*
Richard Patton, *Mississippi State University*
Stephen Boedo, *Rochester Institute of Technology*
Om Agrawal, *Southern Illinois University*
Arun Srinivasa, *Texas A&M*
Jason Carey, *University of Alberta*
Patrick Smolinski, *University of Pittsburgh*
Dennis Hong, *Virginia Tech*

List of Symbols

This is a list of common symbols used in machine design and in this book. Specialized use in a subject-matter area often attracts fore and post subscripts and superscripts. To make the table brief enough to be useful, the symbol kernels are listed. See Table 14–1, pp. 727–728 for spur and helical gearing symbols, and Table 15–1, pp. 781–782 for bevel-gear symbols.

A	Area, coefficient
a	Distance
B	Coefficient
Bhn	Brinell hardness
b	Distance, Weibull shape parameter, range number, width
C	Basic load rating, bolted-joint constant, center distance, coefficient of variation, column end condition, correction factor, specific heat capacity, spring index
c	Distance, viscous damping, velocity coefficient
COV	Coefficient of variation
D	Diameter, helix diameter
d	Diameter, distance
E	Modulus of elasticity, energy, error
e	Distance, eccentricity, efficiency, Naperian logarithmic base
F	Force, fundamental dimension force
f	Coefficient of friction, frequency, function
fom	Figure of merit
G	Torsional modulus of elasticity
g	Acceleration due to gravity, function
H	Heat, power
H_B	Brinell hardness
HRC	Rockwell C-scale hardness
h	Distance, film thickness
$\hbar_{CR}$	Combined overall coefficient of convection and radiation heat transfer
I	Integral, linear impulse, mass moment of inertia, second moment of area
i	Index
$\mathbf{i}$	Unit vector in x-direction
J	Mechanical equivalent of heat, polar second moment of area, geometry factor
$\mathbf{j}$	Unit vector in the y-direction
K	Service factor, stress-concentration factor, stress-augmentation factor, torque coefficient
k	Marin endurance limit modifying factor, spring rate
$\mathbf{k}$	Unit vector in the z-direction
L	Length, life, fundamental dimension length

$\mathscr{L}$	Life in hours
l	Length
M	Fundamental dimension mass, moment
$\mathbf{M}$	Moment vector
m	Mass, slope, strain-strengthening exponent
N	Normal force, number, rotational speed, number of cycles
n	Load factor, rotational speed, factor of safety
n_d	Design factor
P	Force, pressure, diametral pitch
PDF	Probability density function
p	Pitch, pressure, probability
Q	First moment of area, imaginary force, volume
q	Distributed load, notch sensitivity
R	Radius, reaction force, reliability, Rockwell hardness, stress ratio, reduction in area
$\mathbf{R}$	Vector reaction force
r	Radius
$\mathbf{r}$	Distance vector
S	Sommerfeld number, strength
s	Distance, sample standard deviation, stress
T	Temperature, tolerance, torque, fundamental dimension time
$\mathbf{T}$	Torque vector
t	Distance, time, tolerance
U	Strain energy
u	Strain energy per unit volume
V	Linear velocity, shear force
v	Linear velocity
W	Cold-work factor, load, weight
w	Distance, gap, load intensity
X	Coordinate, truncated number
x	Coordinate, true value of a number, Weibull parameter
Y	Coordinate
y	Coordinate, deflection
Z	Coordinate, section modulus, viscosity
z	Coordinate, dimensionless transform variable for normal distributions
α	Coefficient, coefficient of linear thermal expansion, end-condition for springs, thread angle
β	Bearing angle, coefficient
Δ	Change, deflection
δ	Deviation, elongation
ϵ	Eccentricity ratio, engineering (normal) strain
ε	True or logarithmic normal strain
Γ	Gamma function, pitch angle
γ	Pitch angle, shear strain, specific weight
λ	Slenderness ratio for springs
μ	Absolute viscosity, population mean
ν	Poisson ratio
ω	Angular velocity, circular frequency
ϕ	Angle, wave length

ψ	Slope integral
ρ	Radius of curvature, mass density
σ	Normal stress
σ'	Von Mises stress
$\hat{\sigma}$	Standard deviation
τ	Shear stress
θ	Angle, Weibull characteristic parameter
¢	Cost per unit weight
$	Cost

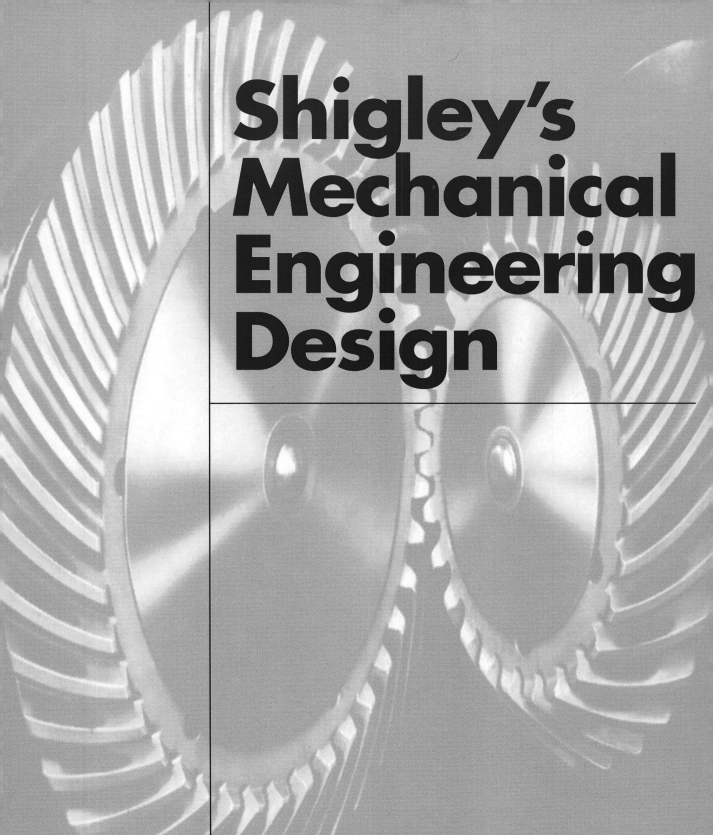

Shigley's
Mechanical
Engineering
Design

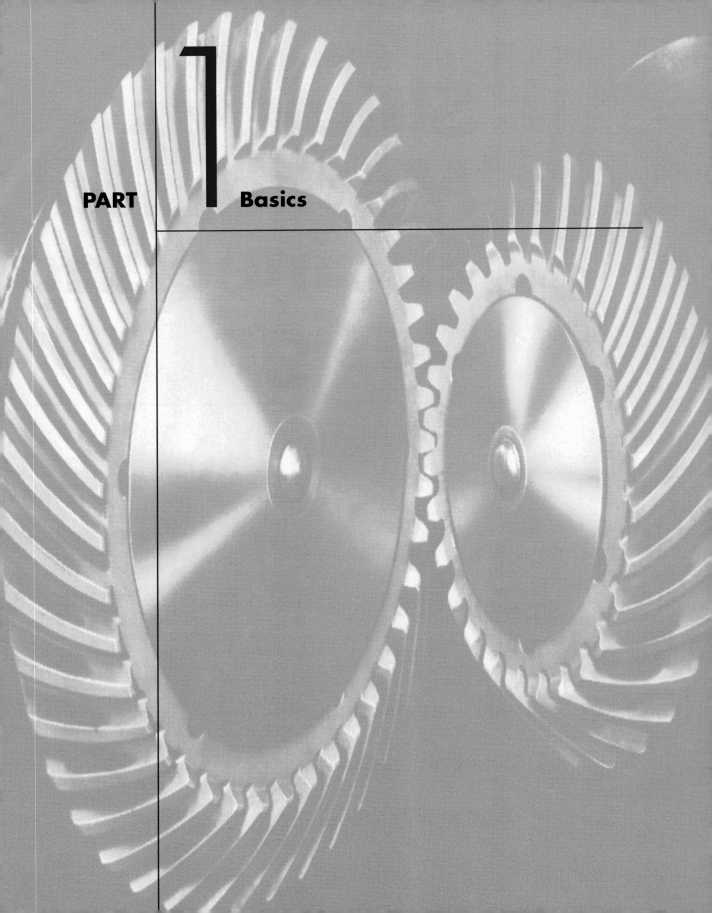

PART

1 Basics

1

Introduction to Mechanical Engineering Design

Mechanical design is a complex process, requiring many skills. Extensive relationships need to be subdivided into a series of simple tasks. The complexity of the process requires a sequence in which ideas are introduced and iterated.

We first address the nature of design in general, and then mechanical engineering design in particular. Design is an iterative process with many interactive phases. Many resources exist to support the designer, including many sources of information and an abundance of computational design tools. Design engineers need not only develop competence in their field but they must also cultivate a strong sense of responsibility and professional work ethic.

There are roles to be played by codes and standards, ever-present economics, safety, and considerations of product liability. The survival of a mechanical component is often related through stress and strength. Matters of uncertainty are ever-present in engineering design and are typically addressed by the design factor and factor of safety, either in the form of a deterministic (absolute) or statistical sense. The latter, statistical approach, deals with a design's *reliability* and requires good statistical data.

In mechanical design, other considerations include dimensions and tolerances, units, and calculations.

This book consists of four parts. Part 1, *Basics,* begins by explaining some differences between design and analysis and introducing some fundamental notions and approaches to design. It continues with three chapters reviewing material properties, stress analysis, and stiffness and deflection analysis, which are the principles necessary for the remainder of the book.

Part 2, *Failure Prevention,* consists of two chapters on the prevention of failure of mechanical parts. Why machine parts fail and how they can be designed to prevent failure are difficult questions, and so we take two chapters to answer them, one on preventing failure due to static loads, and the other on preventing fatigue failure due to time-varying, cyclic loads.

In Part 3, *Design of Mechanical Elements,* the concepts of Parts 1 and 2 are applied to the analysis, selection, and design of specific mechanical elements such as shafts, fasteners, weldments, springs, rolling contact bearings, film bearings, gears, belts, chains, and wire ropes.

Part 4, *Special Topics,* provides introductions to two important methods used in mechanical design, finite element analysis and geometric dimensioning and tolerancing. This is optional study material, but some sections and examples in Parts 1 to 3 demonstrate the use of these tools.

There are two appendixes at the end of the book. Appendix A contains many useful tables referenced throughout the book. Appendix B contains answers to selected end-of-chapter problems.

1–1 Design

To design is either to formulate a plan for the satisfaction of a specified need or to solve a specific problem. If the plan results in the creation of something having a physical reality, then the product must be functional, safe, reliable, competitive, usable, manufacturable, and marketable.

Design is an innovative and highly iterative process. It is also a decision-making process. Decisions sometimes have to be made with too little information, occasionally with just the right amount of information, or with an excess of partially contradictory information. Decisions are sometimes made tentatively, with the right reserved to adjust as more becomes known. The point is that the engineering designer has to be personally comfortable with a decision-making, problem-solving role.

Design is a communication-intensive activity in which both words and pictures are used, and written and oral forms are employed. Engineers have to communicate effectively and work with people of many disciplines. These are important skills, and an engineer's success depends on them.

A designer's personal resources of creativeness, communicative ability, and problem-solving skill are intertwined with the knowledge of technology and first principles. Engineering tools (such as mathematics, statistics, computers, graphics, and languages) are combined to produce a plan that, when carried out, produces a product that is *functional, safe, reliable, competitive, usable, manufacturable, and marketable,* regardless of who builds it or who uses it.

1–2   Mechanical Engineering Design

Mechanical engineers are associated with the production and processing of energy and with providing the means of production, the tools of transportation, and the techniques of automation. The skill and knowledge base are extensive. Among the disciplinary bases are mechanics of solids and fluids, mass and momentum transport, manufacturing processes, and electrical and information theory. Mechanical engineering design involves all the disciplines of mechanical engineering.

Real problems resist compartmentalization. A simple journal bearing involves fluid flow, heat transfer, friction, energy transport, material selection, thermomechanical treatments, statistical descriptions, and so on. A building is environmentally controlled. The heating, ventilation, and air-conditioning considerations are sufficiently specialized that some speak of heating, ventilating, and air-conditioning design as if these were separate and distinct from mechanical engineering design. Similarly, internal-combustion engine design, turbomachinery design, and jet-engine design are sometimes considered discrete entities. Here, the leading string of words preceding the word design is merely a product descriptor. Similarly, there are phrases such as machine design, machine-element design, machine-component design, systems design, and fluid-power design. All of these phrases are somewhat more focused *examples* of mechanical engineering design. They all draw on the same bodies of knowledge, are similarly organized, and require similar skills.

1–3   Phases and Interactions of the Design Process

What is the design process? How does it begin? Does the engineer simply sit down at a desk with a blank sheet of paper and jot down some ideas? What happens next? What factors influence or control the decisions that have to be made? Finally, how does the design process end?

The complete design process, from start to finish, is often outlined as in Fig. 1–1. The process begins with an identification of a need and a decision to do something about it. After many iterations, the process ends with the presentation of the plans for satisfying the need. Depending on the nature of the design task, several design phases may be repeated throughout the life of the product, from inception to termination. In the next several subsections, we shall examine these steps in the design process in detail.

Identification of need generally starts the design process. Recognition of the need and phrasing the need often constitute a highly creative act, because the need may be

Figure 1–1

The phases in design,
acknowledging the many
feedbacks and iterations.

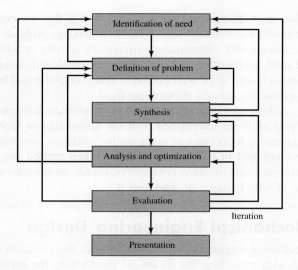

only a vague discontent, a feeling of uneasiness, or a sensing that something is not right. The need is often not evident at all; recognition can be triggered by a particular adverse circumstance or a set of random circumstances that arises almost simultaneously. For example, the need to do something about a food-packaging machine may be indicated by the noise level, by a variation in package weight, and by slight but perceptible variations in the quality of the packaging or wrap.

There is a distinct difference between the statement of the need and the definition of the problem. The *definition of problem* is more specific and must include all the specifications for the object that is to be designed. The specifications are the input and output quantities, the characteristics and dimensions of the space the object must occupy, and all the limitations on these quantities. We can regard the object to be designed as something in a black box. In this case we must specify the inputs and outputs of the box, together with their characteristics and limitations. The specifications define the cost, the number to be manufactured, the expected life, the range, the operating temperature, and the reliability. Specified characteristics can include the speeds, feeds, temperature limitations, maximum range, expected variations in the variables, dimensional and weight limitations, etc.

There are many implied specifications that result either from the designer's particular environment or from the nature of the problem itself. The manufacturing processes that are available, together with the facilities of a certain plant, constitute restrictions on a designer's freedom, and hence are a part of the implied specifications. It may be that a small plant, for instance, does not own cold-working machinery. Knowing this, the designer might select other metal-processing methods that can be performed in the plant. The labor skills available and the competitive situation also constitute implied constraints. Anything that limits the designer's freedom of choice is a constraint. Many materials and sizes are listed in supplier's catalogs, for instance, but these are not all easily available and shortages frequently occur. Furthermore, inventory economics requires that a manufacturer stock a minimum number of materials and sizes. An example of a specification is given in Sec. 1–18. This example is for a case study of a power transmission that is presented throughout this text.

The *synthesis* of a scheme connecting possible system elements is sometimes called the *invention of the concept* or *concept design*. This is the first and most important step in the synthesis task. Various schemes must be proposed, investigated, and quantified in terms of established metrics.[1] As the fleshing out of the scheme progresses, analyses must be performed to assess whether the system performance is satisfactory or better, and, if satisfactory, just how well it will perform. System schemes that do not survive analysis are revised, improved, or discarded. Those with potential are optimized to determine the best performance of which the scheme is capable. Competing schemes are compared so that the path leading to the most competitive product can be chosen. Figure 1–1 shows that synthesis and *analysis and optimization* are intimately and iteratively related.

We have noted, and we emphasize, that design is an iterative process in which we proceed through several steps, evaluate the results, and then return to an earlier phase of the procedure. Thus, we may synthesize several components of a system, analyze and optimize them, and return to synthesis to see what effect this has on the remaining parts of the system. For example, the design of a system to transmit power requires attention to the design and selection of individual components (e.g., gears, bearings, shaft). However, as is often the case in design, these components are not independent. In order to design the shaft for stress and deflection, it is necessary to know the applied forces. If the forces are transmitted through gears, it is necessary to know the gear specifications in order to determine the forces that will be transmitted to the shaft. But stock gears come with certain bore sizes, requiring knowledge of the necessary shaft diameter. Clearly, rough estimates will need to be made in order to proceed through the process, refining and iterating until a final design is obtained that is satisfactory for each individual component as well as for the overall design specifications. Throughout the text we will elaborate on this process for the case study of a power transmission design.

Both analysis and optimization require that we construct or devise abstract models of the system that will admit some form of mathematical analysis. We call these models mathematical models. In creating them it is our hope that we can find one that will simulate the real physical system very well. As indicated in Fig. 1–1, *evaluation* is a significant phase of the total design process. Evaluation is the final proof of a successful design and usually involves the testing of a prototype in the laboratory. Here we wish to discover if the design really satisfies the needs. Is it reliable? Will it compete successfully with similar products? Is it economical to manufacture and to use? Is it easily maintained and adjusted? Can a profit be made from its sale or use? How likely is it to result in product-liability lawsuits? And is insurance easily and cheaply obtained? Is it likely that recalls will be needed to replace defective parts or systems? The project designer or design team will need to address a myriad of engineering and non-engineering questions.

Communicating the design to others is the final, vital *presentation* step in the design process. Undoubtedly, many great designs, inventions, and creative works have been lost to posterity simply because the originators were unable or unwilling to properly explain their accomplishments to others. Presentation is a selling job. The engineer, when presenting a new solution to administrative, management, or supervisory persons, is attempting to sell or to prove to them that their solution is a better one. Unless this can be done successfully, the time and effort spent on obtaining the

---

[1]An excellent reference for this topic is presented by Stuart Pugh, *Total Design—Integrated Methods for Successful Product Engineering,* Addison-Wesley, 1991. A description of the *Pugh method* is also provided in Chap. 8, David G. Ullman, *The Mechanical Design Process,* 3rd ed., McGraw-Hill, 2003.

solution have been largely wasted. When designers sell a new idea, they also sell themselves. If they are repeatedly successful in selling ideas, designs, and new solutions to management, they begin to receive salary increases and promotions; in fact, this is how anyone succeeds in his or her profession.

Design Considerations

Sometimes the strength required of an element in a system is an important factor in the determination of the geometry and the dimensions of the element. In such a situation we say that strength is an important *design consideration*. When we use the expression design consideration, we are referring to some characteristic that influences the design of the element or, perhaps, the entire system. Usually quite a number of such characteristics must be considered and prioritized in a given design situation. Many of the important ones are as follows (not necessarily in order of importance):

1	Functionality	14	Noise
2	Strength/stress	15	Styling
3	Distortion/deflection/stiffness	16	Shape
4	Wear	17	Size
5	Corrosion	18	Control
6	Safety	19	Thermal properties
7	Reliability	20	Surface
8	Manufacturability	21	Lubrication
9	Utility	22	Marketability
10	Cost	23	Maintenance
11	Friction	24	Volume
12	Weight	25	Liability
13	Life	26	Remanufacturing/resource recovery

Some of these characteristics have to do directly with the dimensions, the material, the processing, and the joining of the elements of the system. Several characteristics may be interrelated, which affects the configuration of the total system.

1-4 Design Tools and Resources

Today, the engineer has a great variety of tools and resources available to assist in the solution of design problems. Inexpensive microcomputers and robust computer software packages provide tools of immense capability for the design, analysis, and simulation of mechanical components. In addition to these tools, the engineer always needs technical information, either in the form of basic science/engineering behavior or the characteristics of specific off-the-shelf components. Here, the resources can range from science/engineering textbooks to manufacturers' brochures or catalogs. Here too, the computer can play a major role in gathering information.[2]

Computational Tools

Computer-aided design (CAD) software allows the development of three-dimensional (3-D) designs from which conventional two-dimensional orthographic views with

---

[2]An excellent and comprehensive discussion of the process of "gathering information" can be found in Chap. 4, George E. Dieter, *Engineering Design, A Materials and Processing Approach,* 3rd ed., McGraw-Hill, New York, 2000.

automatic dimensioning can be produced. Manufacturing tool paths can be generated from the 3-D models, and in some cases, parts can be created directly from a 3-D database by using a rapid prototyping and manufacturing method (stereolithography)— *paperless manufacturing!* Another advantage of a 3-D database is that it allows rapid and accurate calculations of mass properties such as mass, location of the center of gravity, and mass moments of inertia. Other geometric properties such as areas and distances between points are likewise easily obtained. There are a great many CAD software packages available such as Aries, AutoCAD, CadKey, I-Deas, Unigraphics, Solid Works, and ProEngineer, to name a few.

The term *computer-aided engineering* (CAE) generally applies to all computer-related engineering applications. With this definition, CAD can be considered as a subset of CAE. Some computer software packages perform specific engineering analysis and/or simulation tasks that assist the designer, but they are not considered a tool for the creation of the design that CAD is. Such software fits into two categories: engineering-based and non-engineering-specific. Some examples of engineering-based software for mechanical engineering applications—software that might also be integrated within a CAD system—include finite-element analysis (FEA) programs for analysis of stress and deflection (see Chap. 19), vibration, and heat transfer (e.g., Algor, ANSYS, and MSC/NASTRAN); computational fluid dynamics (CFD) programs for fluid-flow analysis and simulation (e.g., CFD++, FIDAP, and Fluent); and programs for simulation of dynamic force and motion in mechanisms (e.g., ADAMS, DADS, and Working Model).

Examples of non-engineering-specific computer-aided applications include software for word processing, spreadsheet software (e.g., Excel, Lotus, and Quattro-Pro), and mathematical solvers (e.g., Maple, MathCad, MATLAB,[3] Mathematica, and TKsolver).

Your instructor is the best source of information about programs that may be available to you and can recommend those that are useful for specific tasks. One caution, however: Computer software is no substitute for the human thought process. *You are the driver here; the computer is the vehicle to assist you on your journey to a solution.* Numbers generated by a computer can be far from the truth if you entered incorrect input, if you misinterpreted the application or the output of the program, if the program contained bugs, etc. It is your responsibility to assure the validity of the results, so be careful to check the application and results carefully, perform benchmark testing by submitting problems with known solutions, and monitor the software company and user-group newsletters.

Acquiring Technical Information

We currently live in what is referred to as the *information age,* where information is generated at an astounding pace. It is difficult, but extremely important, to keep abreast of past and current developments in one's field of study and occupation. The reference in footnote 2 provides an excellent description of the informational resources available and is highly recommended reading for the serious design engineer. Some sources of information are:

- *Libraries (community, university, and private).* Engineering dictionaries and encyclopedias, textbooks, monographs, handbooks, indexing and abstract services, journals, translations, technical reports, patents, and business sources/brochures/catalogs.

---

[3]MATLAB is a registered trademark of The MathWorks, Inc.

- *Government sources.* Departments of Defense, Commerce, Energy, and Transportation; NASA; Government Printing Office; U.S. Patent and Trademark Office; National Technical Information Service; and National Institute for Standards and Technology.
- *Professional societies.* American Society of Mechanical Engineers, Society of Manufacturing Engineers, Society of Automotive Engineers, American Society for Testing and Materials, and American Welding Society.
- *Commercial vendors.* Catalogs, technical literature, test data, samples, and cost information.
- *Internet.* The computer network gateway to websites associated with most of the categories listed above.[4]

This list is not complete. The reader is urged to explore the various sources of information on a regular basis and keep records of the knowledge gained.

1-5 The Design Engineer's Professional Responsibilities

In general, the design engineer is required to satisfy the needs of *customers* (management, clients, consumers, etc.) and is expected to do so in a competent, responsible, ethical, and professional manner. Much of engineering course work and practical experience focuses on competence, but when does one begin to develop engineering responsibility and professionalism? To start on the road to success, you should start to develop these characteristics early in your educational program. You need to cultivate your professional work ethic and process skills before graduation, so that when you begin your formal engineering career, you will be prepared to meet the challenges.

It is not obvious to some students, but communication skills play a large role here, and it is the wise student who continuously works to improve these skills—*even if it is not a direct requirement of a course assignment!* Success in engineering (achievements, promotions, raises, etc.) may in large part be due to competence but if you cannot communicate your ideas clearly and concisely, your technical proficiency may be compromised.

You can start to develop your communication skills by keeping a neat and clear journal/logbook of your activities, entering dated entries frequently. (Many companies require their engineers to keep a journal for patent and liability concerns.) Separate journals should be used for each design project (or course subject). When starting a project or problem, in the definition stage, make journal entries quite frequently. Others, as well as yourself, may later question why you made certain decisions. Good chronological records will make it easier to explain your decisions at a later date.

Many engineering students see themselves after graduation as practicing engineers designing, developing, and analyzing products and processes and consider the need of good communication skills, either oral or writing, as secondary. This is far from the truth. Most practicing engineers spend a good deal of time communicating with others, writing proposals and technical reports, and giving presentations and interacting with engineering and nonengineering support personnel. You have the time now to sharpen your communication skills. When given an assignment to write or

---

[4]Some helpful Web resources, to name a few, include www.globalspec.com, www.engnetglobal.com, www.efunda.com, www.thomasnet.com, and www.uspto.gov.

make any presentation, technical *or* nontechnical, accept it enthusiastically, and work on improving your communication skills. It will be time well spent to learn the skills now rather than on the job.

When you are working on a design problem, it is important that you develop a systematic approach. Careful attention to the following action steps will help you to organize your solution processing technique.

- *Understand the problem.* Problem definition is probably the most significant step in the engineering design process. Carefully read, understand, and refine the problem statement.

- *Identify the knowns.* From the refined problem statement, describe concisely what information is known and relevant.

- *Identify the unknowns and formulate the solution strategy.* State what must be determined, in what order, to arrive at a solution to the problem. Sketch the component or system under investigation, identifying known and unknown parameters. Create a flowchart of the steps necessary to reach the final solution. The steps may require the use of free-body diagrams; material properties from tables; equations from first principles, textbooks, or handbooks relating the known and unknown parameters; experimentally or numerically based charts; specific computational tools as discussed in Sec. 1–4; etc.

- *State all assumptions and decisions.* Real design problems generally do not have unique, ideal, closed-form solutions. Selections, such as the choice of materials, and heat treatments, require decisions. Analyses require assumptions related to the modeling of the real components or system. All assumptions and decisions should be identified and recorded.

- *Analyze the problem.* Using your solution strategy in conjunction with your decisions and assumptions, execute the analysis of the problem. Reference the sources of all equations, tables, charts, software results, etc. Check the credibility of your results. Check the order of magnitude, dimensionality, trends, signs, etc.

- *Evaluate your solution.* Evaluate each step in the solution, noting how changes in strategy, decisions, assumptions, and execution might change the results, in positive or negative ways. Whenever possible, incorporate the positive changes in your final solution.

- *Present your solution.* Here is where your communication skills are important. At this point, you are selling yourself and your technical abilities. If you cannot skillfully explain what you have done, some or all of your work may be misunderstood and unaccepted. Know your audience.

As stated earlier, all design processes are interactive and iterative. Thus, it may be necessary to repeat some or all of the above steps more than once if less than satisfactory results are obtained.

In order to be effective, all professionals must keep current in their fields of endeavor. The design engineer can satisfy this in a number of ways by: being an active member of a professional society such as the American Society of Mechanical Engineers (ASME), the Society of Automotive Engineers (SAE), and the Society of Manufacturing Engineers (SME); attending meetings, conferences, and seminars of societies, manufacturers, universities, etc.; taking specific graduate courses or programs at universities; regularly reading technical and professional journals; etc. An engineer's education does not end at graduation.

The design engineer's professional obligations include conducting activities in an ethical manner. Reproduced here is the *Engineers' Creed* from the National Society of Professional Engineers (NSPE)[5]:

> *As a Professional Engineer I dedicate my professional knowledge and skill to the advancement and betterment of human welfare.*
>
> *I pledge:*
>
> *To give the utmost of performance;*
>
> *To participate in none but honest enterprise;*
>
> *To live and work according to the laws of man and the highest standards of professional conduct;*
>
> *To place service before profit, the honor and standing of the profession before personal advantage, and the public welfare above all other considerations.*
>
> *In humility and with need for Divine Guidance, I make this pledge.*

1-6   Standards and Codes

A *standard* is a set of specifications for parts, materials, or processes intended to achieve uniformity, efficiency, and a specified quality. One of the important purposes of a standard is to limit the multitude of variations that can arise from the arbitrary creation of a part, material, or process.

A *code* is a set of specifications for the analysis, design, manufacture, and construction of something. The purpose of a code is to achieve a specified degree of safety, efficiency, and performance or quality. It is important to observe that safety codes *do not* imply *absolute safety*. In fact, absolute safety is impossible to obtain. Sometimes the unexpected event really does happen. Designing a building to withstand a 190 km/h wind does not mean that the designers think a 220 km/h wind is impossible; it simply means that they think it is highly improbable.

All of the organizations and societies listed below have established specifications for standards and safety or design codes. The name of the organization provides a clue to the nature of the standard or code. Some of the standards and codes, as well as addresses, can be obtained in most technical libraries or on the Internet. The organizations of interest to mechanical engineers are:

Aluminum Association (AA)
American Bearing Manufacturers Association (ABMA)
American Gear Manufacturers Association (AGMA)
American Institute of Steel Construction (AISC)
American Iron and Steel Institute (AISI)
American National Standards Institute (ANSI)
American Society of Heating, Refrigerating and Air-Conditioning Engineers (ASHRAE)
American Society of Mechanical Engineers (ASME)
American Society of Testing and Materials (ASTM)
American Welding Society (AWS)

---

[5]Adopted by the National Society of Professional Engineers, June 1954. "The Engineer's Creed." Reprinted by permission of the National Society of Professional Engineers. NSPE also publishes a much more extensive *Code of Ethics for Engineers* with rules of practice and professional obligations. For the current revision, July 2007 (at the time of this book's printing), see the website www.nspe.org/Ethics/CodeofEthics/index.html.

ASM International
British Standards Institution (BSI)
Industrial Fasteners Institute (IFI)
Institute of Transportation Engineers (ITE)
Institution of Mechanical Engineers (IMechE)
International Bureau of Weights and Measures (BIPM)
International Federation of Robotics (IFR)
International Standards Organization (ISO)
National Association of Power Engineers (NAPE)
National Institute for Standards and Technology (NIST)
Society of Automotive Engineers (SAE)

1–7 Economics

The consideration of cost plays such an important role in the design decision process that we could easily spend as much time in studying the cost factor as in the study of the entire subject of design. Here we introduce only a few general concepts and simple rules.

First, observe that nothing can be said in an absolute sense concerning costs. Materials and labor usually show an increasing cost from year to year. But the costs of processing the materials can be expected to exhibit a decreasing trend because of the use of automated machine tools and robots. The cost of manufacturing a single product will vary from city to city and from one plant to another because of overhead, labor, taxes, and freight differentials and the inevitable slight manufacturing variations.

Standard Sizes

The use of standard or stock sizes is a first principle of cost reduction. An engineer who specifies an AISI 1020 bar of hot-rolled steel 53 mm square has added cost to the product, provided that a bar 50 or 60 mm square, both of which are preferred sizes, would do equally well. The 53-mm size can be obtained by special order or by rolling or machining a 60-mm square, but these approaches add cost to the product. To ensure that standard or preferred sizes are specified, designers must have access to stock lists of the materials they employ.

A further word of caution regarding the selection of preferred sizes is necessary. Although a great many sizes are usually listed in catalogs, they are not all readily available. Some sizes are used so infrequently that they are not stocked. A rush order for such sizes may add to the expense and delay. Thus you should also have access to a list such as those in Table A–17 for preferred inch and millimeter sizes.

There are many purchased parts, such as motors, pumps, bearings, and fasteners, that are specified by designers. In the case of these, too, you should make a special effort to specify parts that are readily available. Parts that are made and sold in large quantities usually cost somewhat less than the odd sizes. The cost of rolling bearings, for example, depends more on the quantity of production by the bearing manufacturer than on the size of the bearing.

Large Tolerances

Among the effects of design specifications on costs, tolerances are perhaps most significant. Tolerances, manufacturing processes, and surface finish are interrelated and influence the producibility of the end product in many ways. Close tolerances

Figure 1–2

Cost versus tolerance/machining process.
(From David G. Ullman, The Mechanical Design Process, *3rd ed., McGraw-Hill, New York, 2003.)*

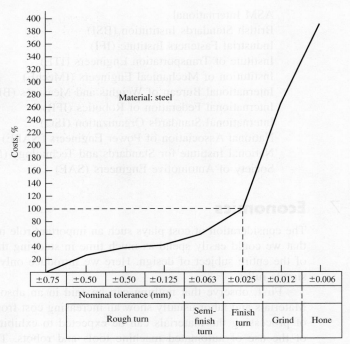

may necessitate additional steps in processing and inspection or even render a part completely impractical to produce economically. Tolerances cover dimensional variation and surface-roughness range and also the variation in mechanical properties resulting from heat treatment and other processing operations.

Since parts having large tolerances can often be produced by machines with higher production rates, costs will be significantly smaller. Also, fewer such parts will be rejected in the inspection process, and they are usually easier to assemble. A plot of cost versus tolerance/machining process is shown in Fig. 1–2, and illustrates the drastic increase in manufacturing cost as tolerance diminishes with finer machining processing.

Breakeven Points

Sometimes it happens that, when two or more design approaches are compared for cost, the choice between the two depends on a set of conditions such as the quantity of production, the speed of the assembly lines, or some other condition. There then occurs a point corresponding to equal cost, which is called the *breakeven point.*

As an example, consider a situation in which a certain part can be manufactured at the rate of 25 parts per hour on an automatic screw machine or 10 parts per hour on a hand screw machine. Let us suppose, too, that the setup time for the automatic is 3 h and that the labor cost for either machine is $20 per hour, including overhead. Figure 1–3 is a graph of cost versus production by the two methods. The breakeven point for this example corresponds to 50 parts. If the desired production is greater than 50 parts, the automatic machine should be used.

Figure 1–3

A breakeven point.

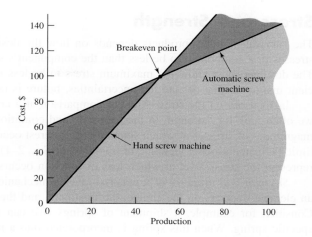

Cost Estimates

There are many ways of obtaining relative cost figures so that two or more designs can be roughly compared. A certain amount of judgment may be required in some instances. For example, we can compare the relative value of two automobiles by comparing the dollar cost per pound of weight. Another way to compare the cost of one design with another is simply to count the number of parts. The design having the smaller number of parts is likely to cost less. Many other cost estimators can be used, depending upon the application, such as area, volume, horsepower, torque, capacity, speed, and various performance ratios.[6]

1–8 Safety and Product Liability

The *strict liability* concept of product liability generally prevails in the United States. This concept states that the manufacturer of an article is liable for any damage or harm that results because of a defect. And it doesn't matter whether the manufacturer knew about the defect, or even could have known about it. For example, suppose an article was manufactured, say, 10 years ago. And suppose at that time the article could not have been considered defective on the basis of all technological knowledge then available. Ten years later, according to the concept of strict liability, the manufacturer is still liable. Thus, under this concept, the plaintiff needs only to prove that the article was defective and that the defect caused some damage or harm. Negligence of the manufacturer need not be proved.

The best approaches to the prevention of product liability are good engineering in analysis and design, quality control, and comprehensive testing procedures. Advertising managers often make glowing promises in the warranties and sales literature for a product. These statements should be reviewed carefully by the engineering staff to eliminate excessive promises and to insert adequate warnings and instructions for use.

---

[6]For an overview of estimating manufacturing costs, see Chap. 11, Karl T. Ulrich and Steven D. Eppinger, *Product Design and Development*, 3rd ed., McGraw-Hill, New York, 2004.

1–9   Stress and Strength

The survival of many products depends on how the designer adjusts the maximum stresses in a component to be less than the component's strength at critical locations. The designer must allow the maximum stress to be less than the strength by a sufficient margin so that despite the uncertainties, failure is rare.

In focusing on the stress-strength comparison at a critical (controlling) location, we often look for "strength in the geometry and condition of use." Strengths are the magnitudes of stresses at which something of interest occurs, such as the proportional limit, 0.2 percent-offset yielding, or fracture (see Sec. 2–1). In many cases, such events represent the stress level at which loss of function occurs.

Strength is a property of a material or of a mechanical element. The strength of an element depends on the choice, the treatment, and the processing of the material. Consider, for example, a shipment of springs. We can associate a strength with a specific spring. When this spring is incorporated into a machine, external forces are applied that result in load-induced stresses in the spring, the magnitudes of which depend on its geometry and are independent of the material and its processing. If the spring is removed from the machine undamaged, the stress due to the external forces will return to zero. But the strength remains as one of the properties of the spring. Remember, then, that *strength is an inherent property of a part,* a property built into the part because of the use of a particular material and process.

Various metalworking and heat-treating processes, such as forging, rolling, and cold forming, cause variations in the strength from point to point throughout a part. The spring cited above is quite likely to have a strength on the outside of the coils different from its strength on the inside because the spring has been formed by a cold winding process, and the two sides may not have been deformed by the same amount. Remember, too, therefore, that a strength value given for a part may apply to only a particular point or set of points on the part.

In this book we shall use the capital letter S to denote *strength,* with appropriate subscripts to denote the type of strength. Thus, S_y is a yield strength, S_u an ultimate strength, S_{sy} a shear yield strength, and S_e an endurance strength.

In accordance with accepted engineering practice, we shall employ the Greek letters σ (sigma) and τ (tau) to designate normal and shear *stresses,* respectively. Again, various subscripts will indicate some special characteristic. For example, σ_1 is a principal normal stress, σ_y a normal stress component in the y direction, and σ_r a normal stress component in the radial direction.

Stress is a state property at a *specific point* within a body, which is a function of load, geometry, temperature, and manufacturing processing. In an elementary course in mechanics of materials, stress related to load and geometry is emphasized with some discussion of thermal stresses. However, stresses due to heat treatments, molding, assembly, etc. are also important and are sometimes neglected. A review of stress analysis for basic load states and geometry is given in Chap. 3.

1–10   Uncertainty

Uncertainties in machinery design abound. Examples of uncertainties concerning stress and strength include:

- Composition of material and the effect of variation on properties.
- Variations in properties from place to place within a bar of stock.
- Effect of processing locally, or nearby, on properties.

- Effect of nearby assemblies such as weldments and shrink fits on stress conditions.
- Effect of thermomechanical treatment on properties.
- Intensity and distribution of loading.
- Validity of mathematical models used to represent reality.
- Intensity of stress concentrations.
- Influence of time on strength and geometry.
- Effect of corrosion.
- Effect of wear.
- Uncertainty as to the length of any list of uncertainties.

Engineers must accommodate uncertainty. Uncertainty always accompanies change. Material properties, load variability, fabrication fidelity, and validity of mathematical models are among concerns to designers.

There are mathematical methods to address uncertainties. The primary techniques are the deterministic and stochastic methods. The deterministic method establishes a *design factor* based on the absolute uncertainties of a loss-of-function parameter and a maximum allowable parameter. Here the parameter can be load, stress, deflection, etc. Thus, the design factor n_d is defined as

$$n_d = \frac{\text{loss-of-function parameter}}{\text{maximum allowable parameter}} \quad (1\text{-}1)$$

If the parameter is load (as would be the case for column buckling), then the maximum allowable load can be found from

$$\text{Maximum allowable load} = \frac{\text{loss-of-function load}}{n_d} \quad (1\text{-}2)$$

EXAMPLE 1-1

Consider that the maximum load on a structure is known with an uncertainty of ±20 percent, and the load causing failure is known within ±15 percent. If the load causing failure is *nominally* 10 kN, determine the design factor and the maximum allowable load that will offset the absolute uncertainties.

Solution

To account for its uncertainty, the loss-of-function load must increase to 1/0.85, whereas the maximum allowable load must decrease to 1/1.2. Thus to offset the absolute uncertainties the design factor, from Eq. (1–1), should be

Answer

$$n_d = \frac{1/0.85}{1/1.2} = 1.4$$

From Eq. (1–2), the maximum allowable load is found to be

Answer

$$\text{Maximum allowable load} = \frac{10}{1.4} = 7.1 \text{ kN}$$

Stochastic methods are based on the statistical nature of the design parameters and focus on the probability of survival of the design's function (that is, on reliability). This is discussed further in Secs. 1–12 and 1–13.

1–11 Design Factor and Factor of Safety

A general approach to the allowable load versus loss-of-function load problem is the deterministic design factor method, and sometimes called the classical method of design. The fundamental equation is Eq. (1–1) where n_d is called the *design factor*. All loss-of-function modes must be analyzed, and the mode leading to the smallest design factor governs. After the design is completed, the *actual* design factor may change as a result of changes such as rounding up to a standard size for a cross section or using off-the-shelf components with higher ratings instead of employing what is calculated by using the design factor. The factor is then referred to as the *factor of safety, n*. The factor of safety has the same definition as the design factor, but it generally differs numerically.

Since stress may not vary linearly with load (see Sec. 3–19), using load as the loss-of-function parameter may not be acceptable. It is more common then to express the design factor in terms of a stress and a relevant strength. Thus Eq. (1–1) can be rewritten as

$$n_d = \frac{\text{loss-of-function strength}}{\text{allowable stress}} = \frac{S}{\sigma(\text{or } \tau)} \qquad (1\text{–}3)$$

The stress and strength terms in Eq. (1–3) must be of the same type and units. Also, the stress and strength must apply to the same critical location in the part.

EXAMPLE 1–2

A solid circular rod of diameter d undergoes a bending moment $M = 100$ N · m inducing a stress $\sigma = 16M/(\pi d^3)$. Using a material strength of 170 MPa and a *design factor* of 2.5, determine the minimum diameter of the rod. Using Table A–17, select a preferred fractional diameter and determine the resulting *factor of safety*.

Solution

From Eq. (1–3), $\sigma = S/n_d$, then

$$\sigma = \frac{16M}{\pi d^3} = \frac{S}{n_d}$$

Solving for d yields

Answer

$$d = \left(\frac{16Mn_d}{S\pi}\right)^{1/3} = \left(\frac{16(100)2.5}{170(10)^6 2.5}\right)^{1/3} = 0.02111 \text{ m} = 21.11 \text{ mm}$$

From Table A–17, the next higher preferred size is 22 mm. Thus, when n_d is replaced with n in the equation developed above, the factor of safety n is

Answer

$$n = \frac{\pi S d^3}{16M} = \frac{\pi(170)(10^6) 0.022^3}{16(100)} = 3.55$$

It is tempting to offer some recommendations concerning the assignment of the design factor for a given application.[7] The problem in doing so is with the evaluation

---

[7]If the reader desires some examples of assigning design factor values see David G. Ullman, *The Mechanical Design Process*, 4th ed., McGraw-Hill, New York, 2010, App. C.

of the many uncertainties associated with the loss-of-function modes. The reality is, the designer must attempt to account for the variance of all the factors that will affect the results. Then, the designer must rely on experience, company policies, and the many codes that may pertain to the application (e.g. the ASME Boiler and Pressure Vessel Code) to arrive at an appropriate design factor. An example might help clarify the intricacy of assigning a design factor.

EXAMPLE 1–3

A vertical round rod is to be used to support a hanging weight. A person will place the weight on the end without dropping it. The diameter of the rod can be manufactured within ± 1 percent of its nominal dimension. The support ends can be centered within ± 1.5 percent of the nominal diameter dimension. The weight is known within ± 2 percent of the nominal weight. The strength of the material is known within ± 3.5 percent of the nominal strength value. If the designer is using nominal values and the nominal stress equation, $\sigma_{\text{nom}} = P/A$ (as in the previous example), determine what design factor should be used so that the stress does not exceed the strength.

Solution

There are two hidden factors to consider here. The first, due to the possibility of eccentric loading, the maximum stress *is not* $\sigma = P/A$ (review Chap. 3). Second, the person may not be placing the weight onto the rod support end *gradually,* and the load application would then be considered dynamic.

Consider the eccentricity first. With eccentricity, a bending moment will exist giving an additional stress of $\sigma = 32 M/(\pi d^3)$ (see Sec. 3–10). The bending moment is given by $M = Pe$, where e is the eccentricity. Thus, the maximum stress in the rod is given by

$$\sigma = \frac{P}{A} + \frac{32Pe}{\pi d^3} = \frac{P}{\pi d^2/4} + \frac{32Pe}{\pi d^3} \tag{1}$$

Since the eccentricity tolerance is expressed as a function of the diameter, we will write the eccentricity as a percentage of d. Let $e = k_e d$, where k_e is a constant. Thus, Eq. (1) is rewritten as

$$\sigma = \frac{4P}{\pi d^2} + \frac{32Pk_e d}{\pi d^3} = \frac{4P}{\pi d^2}(1 + 8k_e) \tag{2}$$

Applying the tolerances to achieve the maximum the stress can reach gives

$$\sigma_{\text{max}} = \frac{4P(1 + 0.02)}{\pi[d(1 - 0.01)]^2}[1 + 8(0.015)] = 1.166\left(\frac{4P}{\pi d^2}\right) \tag{3}$$

$$= 1.166\sigma_{\text{nom}}$$

Suddenly applied loading is covered in Sec. 4–17. If a weight is dropped from a height, h, from the support end, the maximum load in the rod is given by Eq. (4–59) which is

$$F = W + W\left(1 + \frac{hk}{W}\right)^{1/2}$$

where F is the force in the rod, W is the weight, and k is the rod's spring constant. Since the person is not dropping the weight, $h = 0$, and with $W = P$, then $F = 2P$. This assumes the person is *not* gradually placing the weight on, and there is no damping in the rod. Thus Eq. (3) is modified by substituting $2P$ for P and the maximum stress is

$$\sigma_{max} = 2(1.166)\,\sigma_{nom} = 2.332\,\sigma_{nom}$$

The minimum strength is

$$S_{min} = (1 - 0.035)\,S_{nom} = 0.965\,S_{nom}$$

Equating the maximum stress to the minimum strength gives

$$2.332\,\sigma_{nom} = 0.965\,S_{nom}$$

From Eq. (1–3), the design factor using nominal values should be

Answer

$$n_d = \frac{S_{nom}}{\sigma_{nom}} = \frac{2.332}{0.965} = 2.42$$

Obviously, if the designer takes into account all of the uncertainties in this example and accounts for all of the tolerances in the stress and strength in the calculations, a design factor of one would suffice. However, in practice, the designer would probably use the nominal geometric and strength values with the simple $\sigma = P/A$ calculation. The designer would probably not go through the calculations given in the example and would assign a design factor. This is where the experience factor comes in. The designer should make a list of the loss-of-function modes and estimate a factor, n_i, for each. For this example, the list would be

Loss-of-function	Estimated accuracy	n_i
Geometry dimensions	Good tolerances	1.05
Stress calculation		
Dynamic load	Not gradual loading	2.0*
Bending	Slight possibility	1.1
Strength data	Well known	1.05

*Minimum

Each term directly affects the results. Therefore, for an estimate, we evaluate the product of each term

$$n_d = \Pi n_i = 1.05(2.0)(1.1)(1.05) = 2.43$$

1–12   Reliability and Probability of Failure

In these days of greatly increasing numbers of liability lawsuits and the need to conform to regulations issued by governmental agencies such as EPA and OSHA, it is very important for the designer and the manufacturer to know the reliability of their product. The *reliability method* of design is one in which we obtain the distribution of stresses and the distribution of strengths and then relate these two in order to achieve an acceptable success rate. The statistical measure of the probability that a mechanical element will not fail in use is called the *reliability* of that element and as we will see, is related to the *probability of failure, p_f*.

Figure 1–4

The shape of the normal
distribution curve: (*a*) small $\hat{\sigma}$;
(*b*) large $\hat{\sigma}$.

Figure 1–5

Transformed normal
distribution function of
Table A–10.

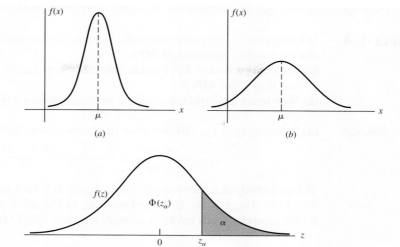

Probability of Failure

The probability of failure, p_f, is obtained from the *probability density function* (PDF), which represents the distribution of events within a given range of values. A number of standard discrete and continuous probability distributions are commonly applicable to engineering problems. The two most important continuous probability distributions for our use in this text are the *Gaussian (normal) distribution* and the *Weibull distribution*. We will describe the normal distribution in this section and in Sec. 2–2. The Weibull distribution is widely used in rolling-contact bearing design and will be described in Chap. 11.

The continuous Gaussian (normal) distribution is an important one whose *probability density function* (PDF) is expressed in terms of its mean, μ_x, and its standard deviation[8] as

$$f(x) = \frac{1}{\hat{\sigma}_x \sqrt{2\pi}} \exp\left[-\frac{1}{2}\left(\frac{x - \mu_x}{\hat{\sigma}_x}\right)^2\right] \tag{1–4}$$

Plots of Eq. (1–4) are shown in Fig. 1–4 for small and large standard deviations. The bell-shaped curve is taller and narrower for small values of $\hat{\sigma}$ and shorter and broader for large values of $\hat{\sigma}$. Note that the area under each curve is unity. That is, the probability of all events occurring is one (100 percent).

To obtain values of p_f, integration of Eq. (1–4) is necessary. This can come easily from a table if the variable x is placed in dimensionless form. This is done using the transform

$$z = \frac{x - \mu_x}{\hat{\sigma}_x} \tag{1–5}$$

The integral of the transformed normal distribution is tabulated in Table A–10, where α is defined, and is shown in Fig. 1–5. The value of the normal density function is used so often, and manipulated in so many equations, that it has its own particular symbol, $\Phi(z)$. The transform variant z has a mean value of zero and a standard deviation of unity. In Table A–10, the probability of an observation less than z is $\Phi(z)$ for negative values of z and $1 - \Phi(z)$ for positive values of z.

[8]The symbol σ is normally used for the standard deviation. However, in this text σ is used for stress. Consequently, we will use $\hat{\sigma}$ for the standard deviation.

EXAMPLE 1–4   In a shipment of 250 connecting rods, the mean tensile strength is found to be 315 MPa and the standard deviation 35 MPa.

(*a*) Assuming a normal distribution, how many rods can be expected to have a strength less than 276.5 MPa?

(*b*) How many are expected to have a strength between 276.5 and 416.5 MPa?

Solution   (*a*) Substituting in Eq. (20–16) gives the standardized z variable as

$$z_{276.5} = \frac{x - \mu_x}{\sigma_x} = \frac{S - \bar{S}}{\sigma_S} = \frac{276.5 - 315}{35} = -1.10$$

The probability that the strength is less than 276.5 MPa can be designated as $F(z) = \Phi(-1.10)$. Using Table A–10, and referring to Fig. 20–7, we find $\Phi(z_{276.5}) = 0.1357$. So the number of rods having a strength less than 276.5 MPa is,

| Figure 1–6

Answer $$N\Phi(z_{276.5}) = 250(0.1357) = 33.9 \approx 34$$

because $\Phi(z_{276.5})$ represents the proportion of the population N having a strength less than 276.5 MPa.

(*b*) Corresponding to $S = 416.5$ MPa, we have

$$z_{416.5} = \frac{416.5 - 315}{35} = 2.90$$

Referring again to Fig. 20–7, we see that the probability that the strength is less than 416.5 MPa is $F(z) = \Phi(z_{416.5})$. Since the z variable is positive, we need to find the value complementary to unity. Thus, from Table A–10,

$$\Phi(2.90) = 1 - \Phi(-2.90) = 1 - 0.001\,87 = 0.998\,13$$

The probability that the strength lies between 276.5 and 416.5 MPa is the area between the ordinates at $z_{276.5}$ and $z_{416.5}$ in Fig. 20–7. This probability is found to be

$$p = \Phi(z_{416.5}) - \Phi(z_{276.5}) = \Phi(2.90) - \Phi(-1.10)$$
$$= 0.998\,13 - 0.1357 = 0.862\,43$$

Therefore the number of rods expected to have strengths between 276.5 and 416.5 MPa is

Answer $$Np = 250(0.862) = 215.5 \approx 216$$

Events typically arise as *discrete distributions,* which can be approximated by continuous distributions. Consider N samples of events. Let x_i be the value of an event $(i = 1, 2, \ldots k)$ and f_i is the class frequency or number of times the event x_i occurs within the class frequency range. The *discrete* mean, $\bar{x}$, and standard deviation, defined as s_x, are given by

$$\bar{x} = \frac{1}{N} \sum_{i=1}^{k} f_i x_i \tag{1-6}$$

$$s_x = \sqrt{\frac{\sum_{i=1}^{k} f_i x_i^2 - N\bar{x}^2}{N - 1}} \tag{1-7}$$

EXAMPLE 1–5

Five tons of 51 mm round rods of 1030 hot-rolled steel have been received for work-piece stock. Nine standard-geometry tensile test specimens have been machined from random locations in various rods. In the test report, the ultimate tensile strength was given in MPa. The data in the ranges 427-448, 448-469, 469-490 and 490-510 MPa is given in histographic form as follows:

S_{ut} (MPa)	438	459	479	500
f	2	2	3	2

where the values of S_{ut} are the midpoints of each range. Find the mean and standard deviation of the data.

Solution

Table 1–1 provides a tabulation of the calculations for the solution.

| Table 1–1

Class Midpoint x, MPa	Class Frequency f	Extension fx	fx^2
438	2	876	383,688
459	2	918	421,362
479	3	1437	688,323
500	2	1000	500,000
$\sum$	9	4231	1,993,373

From Eq. (1–6),

Answer

$$\bar{x} = \frac{1}{N} \sum_{i=1}^{k} f_i x_i = \frac{1}{9}(4231) = 470.1111 = 470.1 \text{ MPa}$$

From Eq. (1–7),

Answer

$$s_x = \sqrt{\frac{\sum_{i=1}^{k} f_i x_i^2 - N\bar{x}^2}{N - 1}} = \sqrt{\frac{1,993,373 - 9(470.1111^2)}{9 - 1}}$$

$$= 23.27 \text{ MPa}$$

Reliability

The reliability R can be expressed by

$$R = 1 - p_f \qquad (1-8)$$

where p_f is the *probability of failure,* given by the number of instances of failures per total number of possible instances. The value of R falls in the range $0 \le R \le 1$. A reliability of $R = 0.90$ means that there is a 90 percent chance that the part will perform its proper function without failure. The failure of 6 parts out of every 1000 manufactured, $p_f = 6/1000$, might be considered an acceptable failure rate for a certain class of products. This represents a reliability of $R = 1 - 6/1000 = 0.994$ or 99.4 percent.

In the *reliability method of design,* the designer's task is to make a judicious selection of materials, processes, and geometry (size) to achieve a specific reliability goal. Thus, if the objective reliability is to be 99.4 percent, as above, what combination of materials, processing, and dimensions is needed to meet this goal?

If a mechanical system fails when any one component fails, the system is said to be a *series system.* If the reliability of component i is R_i in a series system of n components, then the reliability of the system is given by

$$R = \sum_{i=1}^{n} R_i \qquad (1-9)$$

For example, consider a shaft with two bearings having reliabilities of 95 percent and 98 percent. From Eq. (1–9), the overall reliability of the shaft system is then

$$R = R_1 R_2 = 0.95(0.98) = 0.93$$

or 93 percent.

Analyses that lead to an assessment of reliability address uncertainties, or their estimates, in parameters that describe the situation. Stochastic variables such as stress, strength, load, or size are described in terms of their means, standard deviations, and distributions. If bearing balls are produced by a manufacturing process in which a diameter distribution is created, we can say upon choosing a ball that there is uncertainty as to size. If we wish to consider weight or moment of inertia in rolling, this size uncertainty can be considered to be *propagated* to our knowledge of weight or inertia. There are ways of estimating the statistical parameters describing weight and inertia from those describing size and density. These methods are variously called *propagation of error, propagation of uncertainty,* or *propagation of dispersion.* These methods are integral parts of analysis or synthesis tasks when probability of failure is involved.

It is important to note that good statistical data and estimates are essential to perform an acceptable reliability analysis. This requires a good deal of testing and validation of the data. In many cases, this is not practical and a deterministic approach to the design must be undertaken.

1–13 Relating Design Factor to Reliability

Reliability is the statistical probability that machine systems and components will perform their intended function satisfactorily without failure. Stress and strength are statistical in nature and very much tied to the reliability of the stressed component. Consider the probability density functions for stress and strength, σ and S, shown in

Figure 1–7

Plots of density functions showing how the interference of S and σ is used to explain the stress margin m. (a) Stress and strength distributions. (b) Distribution of interference; the reliability R is the area of the density function for $m > 0$; the interference is the area $(1 - R)$.

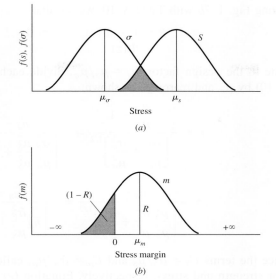

Fig. 1–7a. The mean values of stress and strength are $\overline{\sigma} = \mu_\sigma$ and $\overline{S} = \mu_S$, respectively. Here, the "average" design factor is

$$\overline{n}_d = \frac{\mu_S}{\mu_\sigma} \qquad (a)$$

The *margin of safety* for any value of stress σ and strength S is defined as

$$m = S - \sigma \qquad (b)$$

The average of the margin of safety is $\overline{m} = \mu_S - \mu_\sigma$. However, for the overlap of the distributions shown by the shaded area in Fig. 1–7a, the stress exceeds the strength. Here, the margin of safety is negative, and these parts are expected to fail. This shaded area is called the *interference* of σ and S.

Figure 1–7b shows the distribution of m, which obviously depends on the distributions of stress and strength. The reliability that a part will perform without failure, R, is the area of the margin of safety distribution for $m > 0$. The interference is the area, $1 - R$, where parts are expected to fail. Assuming that σ and S each have a normal distribution, the stress margin m will also have a normal distribution. Reliability is the probability p that $m > 0$. That is,

$$R = p(S > \sigma) = p(S - \sigma > 0) = p(m > 0) \qquad (c)$$

To find the probability that $m > 0$, we form the z variable of m and substitute $m = 0$. Noting that $\mu_m = \mu_S - \mu_\sigma$,[9] and $\hat{\sigma}_m = (\hat{\sigma}_S^2 + \hat{\sigma}_\sigma^2)^{1/2}$, use Eq. (1–5) to write

$$z = \frac{m - \mu_m}{\hat{\sigma}_m} = \frac{0 - \mu_m}{\hat{\sigma}_m} = -\frac{\mu_m}{\hat{\sigma}_m} = -\frac{\mu_S - \mu_\sigma}{(\hat{\sigma}_S^2 + \hat{\sigma}_\sigma^2)^{1/2}} \qquad (1\text{–}10)$$

---

[9]Note: If a and b are normal distributions, and $c = a \pm b$, then c is a normal distribution with a mean of $\mu_c = \mu_a \pm \mu_b$, and a standard deviation of $\hat{\sigma}_c = (\hat{\sigma}_a^2 + \hat{\sigma}_b^2)^{1/2}$. Tabular results for means and standard deviations for simple algebraic operations can be found in R. G. Budynas and J. K. Nisbett, *Shigley's Mechanical Engineering Design*, 9th ed., McGraw-Hill, New York, 2011, Table 20-6, p. 993.

Comparing Fig. 1–7b with Table A–10, we see that

$$
\begin{aligned}
R &= 1 - \Phi(z) \qquad z \le 0 \\
&= \Phi(z) \qquad\quad z > 0
\end{aligned}
\tag{d}
$$

To relate to the design factor, $\bar{n}_d = \mu_S/\mu_\sigma$, divide each term on the right side of Eq. (1–10) by μ_σ and rearrange as shown:

$$
z = -\dfrac{\dfrac{\mu_S}{\mu_\sigma} - 1}{\left[\dfrac{\hat{\sigma}_S^2}{\mu_\sigma^2} + \dfrac{\hat{\sigma}_\sigma^2}{\mu_\sigma^2}\right]^{1/2}} = -\dfrac{\bar{n}_d - 1}{\left[\dfrac{\hat{\sigma}_S^2}{\mu_\sigma^2}\dfrac{\mu_S^2}{\mu_S^2} + \dfrac{\hat{\sigma}_\sigma^2}{\mu_\sigma^2}\right]^{1/2}}
$$

$$
= -\dfrac{\bar{n}_d - 1}{\left[\dfrac{\mu_S^2}{\mu_\sigma^2}\dfrac{\hat{\sigma}_S^2}{\mu_S^2} + \dfrac{\hat{\sigma}_\sigma^2}{\mu_\sigma^2}\right]^{1/2}} = -\dfrac{\bar{n}_d - 1}{\left[\bar{n}_d^2\dfrac{\hat{\sigma}_S^2}{\mu_S^2} + \dfrac{\hat{\sigma}_\sigma^2}{\mu_\sigma^2}\right]^{1/2}}
\tag{e}
$$

Introduce the terms $C_S = \hat{\sigma}_S/\mu_S$ and $C_\sigma = \hat{\sigma}_\sigma/\mu_\sigma$, called the *coefficients of variance* for strength and stress, respectively. Equation (e) is then rewritten as

$$
z = -\dfrac{\bar{n}_d - 1}{\sqrt{\bar{n}_d^2 C_S^2 + C_\sigma^2}}
\tag{1–11}
$$

Squaring both sides of Eq. (1–11) and solving for $\bar{n}_d$ results in

$$
\bar{n}_d = \dfrac{1 \pm \sqrt{1 - (1 - z^2 C_S^2)(1 - z^2 C_\sigma^2)}}{1 - z^2 C_S^2}
\tag{1–12}
$$

The plus sign is associated with $R > 0.5$, and the minus sign with $R \le 0.5$.

Equation (1–12) is remarkable in that it relates the design factor $\bar{n}_d$ to the reliability goal R (through z) and the coefficients of variation of the strength and stress.

EXAMPLE 1–6

A round cold-drawn 1018 steel rod has 0.2 percent mean yield strength $\bar{S}_y = 540$ MPa with a standard deviation of 40 MPa. The rod is to be subjected to a mean static axial load of $\bar{P} = 222$ kN with a standard deviation of 18 kN. Assuming the strength and load have normal distributions, what value of the design factor $\bar{n}_d$ corresponds to a reliability of 0.999 against yielding? Determine the corresponding diameter of the rod.

Solution

For strength, $C_S = \hat{\sigma}_S/\mu_S = 40/540 = 0.074$. For stress,

$$
\sigma = \dfrac{P}{A} = \dfrac{4P}{\pi d^2}
$$

Since the tolerance on the diameter will be an order of magnitude less than that of the load or strength, the diameter will be treated deterministically. Thus, statistically, the stress is linearly proportional to the load, and $C_\sigma = C_P = \hat{\sigma}_P/\mu_P = 18/222 = 0.081$. From Table A–10, for $R = 0.999$, $z = -3.09$. Then, Eq. (1–12) gives

Answer

$$
\bar{n}_d = \dfrac{1 + \sqrt{1 - [1 - (-3.09)^2(0.074)^2][1 - (-3.09)^2(0.081)^2]}}{1 - (-3.09)^2(0.074)^2} = 1.408
$$

The diameter is found deterministically from

$$\bar{\sigma} = \frac{4\bar{P}}{\pi d^2} = \frac{\bar{S}_y}{\bar{n}_d}$$

Solving for d gives

Answer
$$d = \sqrt{\frac{4\bar{P}\bar{n}_d}{\pi \bar{S}_y}} = \sqrt{\frac{4(222 \times 10^3)(1.408)}{\pi(540 \times 10^6)}} = 27.14 \text{ mm}$$

1–14   Dimensions and Tolerances

Part of a machine designer's task is to specify the parts and components necessary for a machine to perform its desired function. Early in the design process, it is usually sufficient to work with nominal dimensions to determine function, stresses, deflections, and the like. However, eventually it is necessary to get to the point of specificity that every component can be purchased and every part can be manufactured. For a part to be manufactured, its essential shape, dimensions, and tolerances must be communicated to the manufacturers. This is usually done by means of a machine drawing, which may either be a multiview drawing on paper, or digital data from a CAD file. Either way, the drawing usually represents a legal document between the parties involved in the design and manufacture of the part. It is essential that the part be defined precisely and completely so that it can only be interpreted in one way. The designer's intent must be conveyed in such a way that any manufacturer can make the part and/or component to the satisfaction of any inspector.

Common Dimensioning Terminology

Before going further, we will define a few terms commonly used in dimensioning.

- *Nominal size.* The size we use in speaking of an element. For example, we may specify a 40-mm pipe or a 12-mm bolt. Either the theoretical size or the actual measured size may be quite different. The theoretical size of a 40-mm pipe is 47.5 mm for the outside diameter. And the diameter of the 12-mm bolt, say, may actually measure 11.8 mm.
- *Limits.* The stated maximum and minimum dimensions.
- *Tolerance.* The difference between the two limits.
- *Bilateral tolerance.* The variation in both directions from the basic dimension. That is, the basic size is between the two limits, for example, 25 ± 0.05 mm. The two parts of the tolerance need not be equal.
- *Unilateral tolerance.* The basic dimension is taken as one of the limits, and variation is permitted in only one direction, for example,

$$25 \, {}^{+0.05}_{-0.000} \text{ mm}$$

- *Clearance.* A general term that refers to the mating of cylindrical parts such as a bolt and a hole. The word clearance is used only when the internal member is smaller than the external member. The *diametral clearance* is the measured difference in the two diameters. The *radial clearance* is the difference in the two radii.
- *Interference.* The opposite of clearance, for mating cylindrical parts in which the internal member is larger than the external member (e.g., press-fits).

- *Allowance.* The minimum stated clearance or the maximum stated interference for mating parts.
- *Fit.* The amount of clearance or interference between mating parts. See Sec. 7–8 for a standardized method of specifying fits for cylindrical parts, such as gears and bearings onto a shaft.
- *GD&T.* Geometric Dimensioning and Tolerancing (GD&T) is a comprehensive system of symbols, rules, and definitions for defining the nominal (theoretically perfect) geometry of parts and assemblies, along with the allowable variation in size, location, orientation, and form of the features of a part. See Chap. 20 for an overview of GD&T.

Choice of Tolerances

The choice of tolerances is the designer's responsibility and should not be made arbitrarily. Tolerances should be selected based on a combination of considerations including functionality, fit, assembly, manufacturing process ability, quality control, and cost. While there is need for balancing these considerations, functionality must not be compromised. If the functionality of the part or assembly cannot be achieved with a reasonable balance of the other considerations, the entire design may need to be reconsidered. The relationship of tolerances to functionality is usually associated with the need to assemble multiple parts. For example, the diameter of a shaft does not generally need a tight tolerance, except for the portions that must fit with components like bearings or gears. The bearings need a particular press fit in order to function properly. Section 7–8 addresses this issue in detail.

Manufacturing methods evolve over time. The manufacturer is free to use any manufacturing process, as long as the final part meets the specifications. This allows the manufacturer to take advantage of available materials and tools, and to specify the least expensive manufacturing methods. Excessive precision on the part of the designer may seem like an easy way to achieve functionality, but it is actually a poor design choice in that it limits the manufacturing options and drives up the cost. In a competitive manufacturing environment, the designer must embrace the idea that less expensive manufacturing methods *should* be selected, even though the parts may be less than perfect. Since tight tolerances usually correlate to higher production costs, as shown in Fig. 1–2, the designer should generally be thinking in terms of loosening the tolerances as much as possible, while still achieving the desired functionality.

Choice of Dimensions

Dimensioning a part is a designer's responsibility, since the choice of which dimensions to specify can make a difference in the functionality of the part. A properly dimensioned part will include just enough information, with no extraneous information that can lead to confusion or multiple interpretations. For example, the part shown in Fig. 1–8a is over-specified in its length dimensions. Note, in machine drawings, the units for the dimensions are typically specified in an overall note on the drawing, and are not shown with the dimensions. If all the dimensions were theoretically perfect, there would be no inconsistency in the over-specified dimensions. But in reality every dimension can only be manufactured to some less-than-perfect level of accuracy. Suppose every dimension in Fig. 1–8a is specified with a tolerance of $+/- 1$. It would be possible to manufacture the part such that some dimensions were within the specified tolerance, while forcing related redundant dimensions to be out of tolerance.

Figure 1–8

Example of over-specified dimensions. (*a*) Five nominal dimensions specified. (*b*) With +/− 1 tolerances, two dimensions are incompatible.

(*a*)   (*b*)

Figure 1–9

Examples of choice of which dimensions to specify.

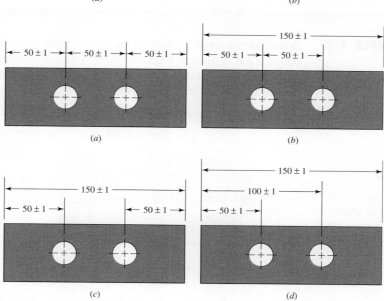

(*a*)   (*b*)

(*c*)   (*d*)

For example, in Fig. 1–8*b*, three of the dimensions are within the +/− 1 tolerance, but they force the other two dimensions to be out of tolerance. In this example, only three length dimensions should be specified. The designer should determine which three are most important to the functioning and assembly of the part.

Figure 1–9 shows four different choices of how the length dimensions might be specified for the same part. None of them are incorrect, but they are not all equivalent in terms of satisfying a particular function. For example, if the two holes are to mate with a pair of corresponding features from another part, the distance between the holes is critical. The choice of dimensions in Fig. 1–9*c* would not be a good choice in this case. Even if the part is manufactured within the specified tolerances of +/− 1, the distance between the holes could range anywhere from 47 to 53, an effective tolerance of +/− 3. Choosing dimensions as shown in Fig. 1–9*a* or 1–9*b* would serve the purpose better to limit the dimension between the holes to a tolerance of +/− 1. For a different application, the distance of the holes to one or both edges might be important, while the overall length might be critical for another application. The point is, the designer should make this determination, not the manufacturer.

Tolerance Stack-up

Note that while there are always choices of which dimensions to specify, the cumulative effect of the individual specified tolerances must be allowed to accumulate *somewhere*.

This is known as *tolerance stack-up*. Figure 1–9a shows an example of *chain dimensioning,* in which several dimensions are specified in series such that the tolerance stack-up can become large. In this example, even though the individual tolerances are all +/− 1, the total length of the part has an implied tolerance of +/− 3 due to the tolerance stack-up. A common method of minimizing a large tolerance stack-up is to dimension from a common *baseline,* as shown in Fig. 1–9d.

The tolerance stack-up issue is also pertinent when several parts are assembled. A gap or interference will occur, and will depend on the dimensions and tolerances of the individual parts. An example will demonstrate the point.

EXAMPLE 1-7

A shouldered screw contains three hollow right circular cylindrical parts on the screw before a nut is tightened against the shoulder. To sustain the function, the gap w must equal or exceed 0.08 mm. The parts in the assembly depicted in Fig. 1–10 have dimensions and tolerances as follows:

$$a = 44.50 \pm 0.08 \text{ mm} \qquad b = 19.05 \pm 0.02 \text{ mm}$$

$$c = 3.05 \pm 0.13 \text{ mm} \qquad d = 22.23 \pm 0.02 \text{ mm}$$

Figure 1–10

An assembly of three cylindrical sleeves of lengths b, c, and d on a shoulder bolt shank of length a. The gap w is of interest.

All parts except the part with the dimension d are supplied by vendors. The part containing the dimension d is made in-house.
(a) Estimate the mean and tolerance on the gap w.
(b) What basic value of d will assure that $w \geq 0.08$ mm?

Solution

(a) The mean value of w is given by

Answer

$$\bar{w} = \bar{a} - \bar{b} - \bar{c} - \bar{d} = 44.50 - 19.05 - 3.05 - 22.23 = 0.17 \text{ mm}$$

For equal bilateral tolerances, the tolerance of the gap is

Answer

$$t_w = \sum_{\text{all}} t = 0.08 + 0.02 + 0.13 + 0.02 = 0.25 \text{ mm}$$

Then, $w = 0.17 \pm 0.25$, and

$$w_{\max} = \bar{w} + t_w = 0.17 + 0.25 = 0.42 \text{ mm}$$

$$w_{\min} = \bar{w} - t_w = 0.17 - 0.25 = -0.08 \text{ mm}$$

Thus, both clearance and interference are possible.
(b) If $w_{\min}$ is to be 0.08 mm, then, $\bar{w} = w_{\min} + t_w = 0.08 + 0.25 = 0.33$ mm. Thus,

Answer

$$\bar{d} = \bar{a} - \bar{b} - \bar{c} - \bar{w} = 44.50 - 19.05 - 3.05 - 0.33 = 22.07 \text{ mm}$$

The previous example represented an *absolute tolerance system*. Statistically, gap dimensions near the gap limits are rare events. Using a *statistical tolerance system,* the probability that the gap falls within a given limit is determined. This probability deals with the statistical distributions of the individual dimensions. For example, if the distributions of the dimensions in the previous example were normal and the tolerances, t, were given in terms of standard deviations of the dimension distribution, the standard deviation of the gap $\overline{w}$ would be $t_w = \sqrt{\sum_{\text{all}} t^2}$. However, this assumes a normal distribution for the individual dimensions, a rare occurrence. To find the distribution of w and/or the probability of observing values of w within certain limits requires a computer simulation in most cases. *Monte Carlo* computer simulations can be used to determine the distribution of w by the following approach:

1 Generate an instance for each dimension in the problem by selecting the value of each dimension based on its probability distribution.

2 Calculate w using the values of the dimensions obtained in step 1.

3 Repeat steps 1 and 2 N times to generate the distribution of w. As the number of trials increases, the reliability of the distribution increases.

1–15 Units

In the symbolic units equation for Newton's second law, $F = ma$,

$$F = MLT^{-2} \tag{1–13}$$

F stands for force, M for mass, L for length, and T for time. Units chosen for *any* three of these quantities are called *base* units. The first three having been chosen, the fourth unit is called a *derived* unit. When force, length, and time are chosen as base units, the mass is the derived unit and the system that results is called a *gravitational system of units*. When mass, length, and time are chosen as base units, force is the derived unit and the system that results is called an *absolute system of units*.

In some English-speaking countries, the *U.S. customary foot-pound-second system* (fps) and the *inch-pound-second system* (ips) are the two standard gravitational systems most used by engineers. In the fps system the unit of mass is

$$M = \frac{FT^2}{L} = \frac{(\text{pound-force})(\text{second})^2}{\text{foot}} = \text{lbf} \cdot \text{s}^2/\text{ft} = \text{slug} \tag{1–14}$$

Thus, length, time, and force are the three base units in the fps gravitational system.

The unit of force in the fps system is the pound, more properly the *pound-force*. We shall often abbreviate this unit as lbf; the abbreviation lb is permissible however, since we shall be dealing only with the U.S. customary gravitational system. In some branches of engineering it is useful to represent 1000 lbf as a kilopound and to abbreviate it as kip. *Note:* In Eq. (1–14) the derived unit of mass in the fps gravitational system is the lbf · s²/ft and is called a *slug;* there is no abbreviation for slug.

The unit of mass in the ips gravitational system is

$$M = \frac{FT^2}{L} = \frac{(\text{pound-force})(\text{second})^2}{\text{inch}} = \text{lbf} \cdot \text{s}^2/\text{in} \tag{1–15}$$

The mass unit lbf · s²/in has no official name.

The *International System of Units* (SI) is an absolute system. The base units are the meter, the kilogram (for mass), and the second. The unit of force is derived by using Newton's second law and is called the *newton*. The units constituting the newton (N) are

$$F = \frac{ML}{T^2} = \frac{(\text{kilogram})(\text{meter})}{(\text{second})^2} = \text{kg} \cdot \text{m/s}^2 = \text{N} \qquad (1\text{-}16)$$

The weight of an object is the force exerted upon it by gravity. Designating the weight as W and the acceleration due to gravity as g, we have

$$W = mg \qquad (1\text{-}17)$$

In the fps system, standard gravity is $g = 32.1740$ ft/s^2. For most cases this is rounded off to 32.2. Thus the weight of a mass of 1 slug in the fps system is

$$W = mg = (1 \text{ slug})(32.2 \text{ ft/s}^2) = 32.2 \text{ lbf}$$

In the ips system, standard gravity is 386.088 or about 386 in/s^2. Thus, in this system, a unit mass weighs

$$W = (1 \text{ lbf} \cdot \text{s}^2/\text{in})(386 \text{ in/s}^2) = 386 \text{ lbf}$$

With SI units, standard gravity is 9.806 or about 9.81 m/s. Thus, the weight of a 1-kg mass is

$$W = (1 \text{ kg})(9.81 \text{ m/s}^2) = 9.81 \text{ N}$$

A series of names and symbols to form multiples and submultiples of SI units has been established to provide an alternative to the writing of powers of 10. Table A–1 includes these prefixes and symbols.

Numbers having four or more digits are placed in groups of three and separated by a space instead of a comma. However, the space may be omitted for the special case of numbers having four digits. A period is used as a decimal point. These recommendations avoid the confusion caused by certain European countries in which a comma is used as a decimal point, and by the English use of a centered period. Examples of correct and incorrect usage are as follows:

1924 or 1 924 but not 1,924
0.1924 or 0.192 4 but not 0.192,4
192 423.618 50 but not 192,423.61850

The decimal point should always be preceded by a zero for numbers less than unity.

1–16   Calculations and Significant Figures

The discussion in this section applies to real numbers, not integers. The accuracy of a real number depends on the number of significant figures describing the number. Usually, but not always, three or four significant figures are necessary for engineering accuracy. Unless otherwise stated, *no less* than three significant figures should be used in your calculations. The number of significant figures is usually inferred by the number of figures given (except for leading zeros). For example, 706, 3.14, and 0.002 19 are assumed to be numbers with three significant figures. For trailing zeros, a little more clarification is necessary. To display 706 to four significant figures insert a trailing zero and display either 706.0, 7.060×10^2, or 0.7060×10^3. Also, consider a number such as 91 600. Scientific notation should be used to clarify the accuracy. For three significant figures express the number as 91.6×10^3. For four significant figures express it as 91.60×10^3.

Computers and calculators display calculations to many significant figures. However, you should never report a number of significant figures of a calculation any greater than the smallest number of significant figures of the numbers used for the calculation. Of course, you should use the greatest accuracy possible when performing a calculation. For example, determine the circumference of a solid shaft with a diameter of $d = 0.40$ in. The circumference is given by $C = \pi d$. Since d is given with two significant figures, C should be reported with only two significant figures. Now if we used only two significant figures for π our calculator would give $C = 3.1 (11) = 34.1$ mm. This rounds off to two significant figures as $C = 34$ mm. However, using $\pi = 3.141\ 592\ 654$ as programmed in the calculator, $C = 3.141\ 592\ 654 (11) = 34.557\ 519\ 190$ mm. This rounds off to $C = 35$ mm, which is 8.3 percent higher than the first calculation. Note, however, since d is given with two significant figures, it is implied that the range of d is 11 ± 0.12. This means that the calculation of C is only accurate to within $\pm 0.12/11 = \pm 0.0109 = \pm 1.09\%$. The calculation could also be one in a series of calculations, and rounding each calculation separately may lead to an accumulation of greater inaccuracy. Thus, it is considered good engineering practice to make all calculations to the greatest accuracy possible and report the results within the accuracy of the given input.

1–17   Design Topic Interdependencies

One of the characteristics of machine design problems is the interdependencies of the various elements of a given mechanical system. For example, a change from a spur gear to a helical gear on a drive shaft would add axial components of force, which would have implications on the layout and size of the shaft, and the type and size of the bearings. Further, even within a single component, it is necessary to consider many different facets of mechanics and failure modes, such as excessive deflection, static yielding, fatigue failure, contact stress, and material characteristics. However, in order to provide significant attention to the details of each topic, most machine design textbooks focus on these topics separately and give end-of-chapter problems that relate only to that specific topic.

To help the reader see the interdependence between the various design topics, this textbook presents many ongoing and interdependent problems in the end-of-chapter problem sections. Each row of Table 1–2 shows the problem numbers that apply to the same mechanical system that is being analyzed according to the topics being presented in that particular chapter. For example, in the second row, Probs. 3–40, 5–65, and 5–66 correspond to a pin in a knuckle joint that is to be analyzed for stresses in Chap. 3 and then for static failure in Chap. 5. This is a simple example of interdependencies, but as can be seen in the table, other systems are analyzed with as many as 10 separate problems. It may be beneficial to work through some of these continuing sequences as the topics are covered to increase your awareness of the various interdependencies.

In addition to the problems given in Table 1–2, Sec. 1–18 describes a power transmission case study where various interdependent analyses are performed throughout the book, when appropriate in the presentation of the topics. The final results of the case study are then presented in Chap. 18.

Table 1–2

Problem Numbers for Linked End-of-Chapter Problems*

3–1	4–50	4–74											
3–40	5–65	5–66											
3–68	4–23	4–29	4–35	5–39	6–37	7–7	11–14						
3–69	4–24	4–30	4–36	5–40	6–38	7–8	11–15						
3–70	4–25	4–31	4–37	5–41	6–39	7–9	11–16						
3–71	4–26	4–32	4–38	5–42	6–40	7–10	11–17						
3–72	4–27	4–33	4–39	5–43	6–41	7–11	7–19	7–20	7–34	11–27	11–28	13–38	14–36
3–73	4–28	4–34	4–40	5–44	6–42	7–12	7–21	7–22	7–35	11–29	11–30	13–39	14–37
3–74	5–45	6–43	7–13	11–41	13–42								
3–76	5–46	6–44	7–14	11–42	13–42								
3–77	5–47	6–45	7–15	11–18	13–40	14–38							
3–79	5–48	6–46	7–16	11–19	13–41	14–39							
3–80	4–41	4–71	5–49	6–47									
3–81	5–50	6–48											
3–82	5–51	6–49											
3–83	5–52	6–50											
3–84	4–43	4–73	5–53	5–56	6–51								
3–85	5–54	6–52											
3–86	5–55	6–53											
3–87	5–56												

*Each row corresponds to the same mechanical component repeated for a different design concept.

1–18 Power Transmission Case Study Specifications

A case study incorporating the many facets of the design process for a power transmission speed reducer will be considered throughout this textbook. The problem will be introduced here with the definition and specification for the product to be designed. Further details and component analysis will be presented in subsequent chapters. Chapter 18 provides an overview of the entire process, focusing on the design sequence, the interaction between the component designs, and other details pertinent to transmission of power. It also contains a complete case study of the power transmission speed reducer introduced here.

Many industrial applications require machinery to be powered by engines or electric motors. The power source usually runs most efficiently at a narrow range of rotational speed. When the application requires power to be delivered at a slower speed than supplied by the motor, a speed reducer is introduced. The speed reducer should transmit the power from the motor to the application with as little energy loss as practical, while reducing the speed and consequently increasing the torque. For example, assume that a company wishes to provide off-the-shelf speed reducers in various capacities and speed ratios to sell to a wide variety of target applications. The marketing team has determined a need for one of these speed reducers to satisfy the following customer requirements.

Design Requirements

Power to be delivered: 15 kW
Input speed: 30 rev/s
Output speed: 2 rev/s
Targeted for uniformly loaded applications, such as conveyor belts, blowers, and generators
Output shaft and input shaft in-line
Base mounted with 4 bolts
Continuous operation
6-year life, with 8 hours/day, 5 days/wk
Low maintenance
Competitive cost
Nominal operating conditions of industrialized locations
Input and output shafts standard size for typical couplings

In reality, the company would likely design for a whole range of speed ratios for each power capacity, obtainable by interchanging gear sizes within the same overall design. For simplicity, in this case study only one speed ratio will be considered.

Notice that the list of customer requirements includes some numerical specifics, but also includes some generalized requirements, e.g., low maintenance and competitive cost. These general requirements give some guidance on what needs to be considered in the design process, but are difficult to achieve with any certainty. In order to pin down these nebulous requirements, it is best to further develop the customer requirements into a set of product specifications that are measurable. This task is usually achieved through the work of a team including engineering, marketing, management, and customers. Various tools may be used (see footnote 1, p. 7) to prioritize the requirements, determine suitable metrics to be achieved, and to establish target values for each metric. The goal of this process is to obtain a product specification that identifies precisely what the product must satisfy. The following product specifications provide an appropriate framework for this design task.

Design Specifications

Power to be delivered: 15 kW
Power efficiency: >95%
Steady state input speed: 30 rev/s
Maximum input speed: 40 rev/s
Steady-state output speed: 1.5–2 rev/s
Usually low shock levels, occasional moderate shock
Input and output shafts extend 4 in outside gearbox
Input and output shaft diameter tolerance: ±0.025 mm
Input and output shafts in-line: concentricity ±0.125 mm, alignment ±0.001 rad
Maximum allowable loads on input shaft: axial, 220 N; transverse, 440 N
Maximum allowable loads on output shaft: axial, 220 N; transverse, 2220 N
Maximum gearbox size: 350-mm × 350-mm base, 550-mm height
Base mounted with 4 bolts
Mounting orientation only with base on bottom
100% duty cycle

Maintenance schedule: lubrication check every 2000 hours; change of lubrication every 8000 hours of operation; gears and bearing life >12,000 hours; infinite shaft life; gears, bearings, and shafts replaceable

Access to check, drain, and refill lubrication without disassembly or opening of gasketed joints.

Manufacturing cost per unit: <$300

Production: 10,000 units per year

Operating temperature range: −23° to 65°C

Sealed against water and dust from typical weather

Noise: <85 dB from 1 meter

PROBLEMS

1–1 Select a mechanical component from Part 3 of this book (roller bearings, springs, etc.), go to your university's library or the appropriate internet website, and, using the *Thomas Register of American Manufacturers* (www.thomasnet.com), report on the information obtained on five manufacturers or suppliers.

1–2 Select a mechanical component from Part 3 of this book (roller bearings, springs, etc.), go to the Internet, and, using a search engine, report on the information obtained on five manufacturers or suppliers.

1–3 Select an organization listed in Sec. 1–6, go to the Internet, and list what information is available on the organization.

1–4 Go to the Internet and connect to the NSPE website (www.nspe.org/ethics). Read the history of the *Code of Ethics* and briefly discuss your reading.

1–5 Go to the Internet and connect to the NSPE website (www.nspe.org/ethics). Read the complete *NSPE Code of Ethics for Engineers* and briefly discuss your reading.

1–6 Go to the Internet and connect to the NSPE website (www.nspe.org/ethics). Go to *Ethics Resources* and review one or more of the topics given. A sample of some of the topics may be:
(*a*) Education Publications
(*b*) Ethics Case Search
(*c*) Ethics Exam
(*d*) FAQ
(*e*) Milton Lunch Contest
(*f*) Other Resources
(*g*) You Be the Judge
Briefly discuss your reading.

1–7 Estimate how many times more expensive it is to grind a steel part to a tolerance of ±0.0125 mm in versus turning it to a tolerance of ±0.075 mm.

1–8 The costs to manufacture a part using methods A and B are estimated by $C_A = 10 + 0.8 P$ and $C_B = 60 + 0.8 P - 0.005 P^2$ respectively, where the cost C is in dollars and P is the number of parts. Estimate the breakeven point.

1–9 A cylindrical part of diameter d is loaded by an axial force P. This causes a stress of P/A, where $A = \pi d^2/4$. If the load is known with an uncertainty of ±10 percent, the diameter is known within ±5 percent (tolerances), and the stress that causes failure (strength) is known within ±15 percent, determine the minimum design factor that will guarantee that the part will not fail.

1–10 When one knows the true values x_1 and x_2 and has approximations X_1 and X_2 at hand, one can see where errors may arise. By viewing error as something to be added to an approximation to attain a true value, it follows that the error e_i is related to X_i and x_i as $x_i = X_i + e_i$

(a) Show that the error in a sum $X_1 + X_2$ is

$$(x_1 + x_2) - (X_1 + X_2) = e_1 + e_2$$

(b) Show that the error in a difference $X_1 - X_2$ is

$$(x_1 - x_2) - (X_1 - X_2) = e_1 - e_2$$

(c) Show that the error in a product $X_1 X_2$ is

$$x_1 x_2 - X_1 X_2 \approx X_1 X_2 \left(\frac{e_1}{X_1} + \frac{e_2}{X_2} \right)$$

(d) Show that in a quotient X_1/X_2 the error is

$$\frac{x_1}{x_2} - \frac{X_1}{X_2} \approx \frac{X_1}{X_2} \left(\frac{e_1}{X_1} - \frac{e_2}{X_2} \right)$$

1–11 Use the true values $x_1 = \sqrt{7}$ and $x_2 = \sqrt{8}$

(a) Demonstrate the correctness of the error equation from Prob. 1–10 for addition if X_1 and X_2 are obtained by truncating x_1 and x_2 to three digits.

(b) Demonstrate the correctness of the error equation for addition if X_1 and X_2 are obtained by rounding x_1 and x_2 to three significant figures.

1–12 A solid circular rod of diameter d undergoes a bending moment $M = 113 \text{ N} \cdot \text{m}$ inducing a stress $\sigma = 32M/(\pi d^3)$. Using a material strength of 172.4 MPa and a *design factor* of 2.5, determine the minimum diameter of the rod. Using Table A–17 select a preferred fractional diameter and determine the resulting *factor of safety*.

1–13 A fatigue test is performed on rotating beam specimens where, for each rotation cycle, the specimens experience tensile and compressive stresses of equal magnitude. The cycles-to-failure experience with 69 specimens of 5160H steel from 31.75 mm hexagonal bar stock was as follows:

L	60	70	80	90	100	110	120	130	140	150	160	170	180	190	200	210
f	2	1	3	5	8	12	6	10	8	5	2	3	2	1	0	1

where L is the life in thousands of cycles, and f is the class frequency of failures.

(a) Estimate the mean and standard deviation of the life for the population from which the sample was drawn.

(b) Presuming the distribution is normal, how many specimens are predicted to fail at less than 115 kcycles?

1–14 Determinations of the ultimate tensile strength S_{ut} of stainless-steel sheet (17-7PH, condition TH 1050), in sizes from 0.41 mm to 1.57 mm, in 197 tests combined into seven classes were

S_{ut} GPa	1.2	1.25	1.31	1.37	1.42	1.48	1.53
f	6	9	44	67	53	12	6

where S_{ut} is the class midpoint and f is the class frequency. Estimate the mean and standard deviation.

1–15 The lives of parts are often expressed as the number of cycles of operation that a specified percentage of a population will exceed before experiencing failure. The symbol L is used to designate this definition of life. Thus we can speak of L_{10} life as the number of cycles to failure exceeded by 90 percent of a population of parts. Given a normal distribution model, with a mean of $\overline{L} = 122.9$ kilocycles and standard deviation of $s_L = 30.3$ kilocycles, estimate the corresponding L_{10} life.

1–16   The tensile 0.2 percent offset yield strength of AISI 1137 cold-drawn steel bars up to 25.4 mm in diameter from 0.05 mm and 25 heats is reported as follows:

S_y	0.64	0.66	0.64	0.67	0.70	0.71	0.72	0.74	0.75	0.77
f	19	25	38	17	12	10	5	4	4	2

where S_y is the class midpoint in Mpa and f is the number in each class. Presuming the distribution is normal, what is the yield strength exceeded by 99 percent of the population?

1–17   A mechanical system comprises three subsystems in series with reliabilities of 98, 96, and 94 percent. What is the overall reliability of the system?

1–18   From Sec. 3–12, the maximum shear stress in a solid round bar of diameter, d, due to an applied torque, T, is given by $\tau_{max} = 16\,T/(\pi d^3)$. A round, cold-drawn 1018 steel rod is subjected to a mean torsional load of $\overline{T} = 1.5$ kN · m with a standard deviation of 145 N · m. The rod material has a mean shear yield strength of $\overline{S}_{sy} = 312$ MPa with a standard deviation of 23.5 MPa. Assuming the strength and load have normal distributions, what value of the design factor $\overline{n}_d$ corresponds to a reliability of 0.99 against yielding? Determine the corresponding diameter of the rod.

1–19   A round cold-drawn 1045 steel rod has a mean strength $\overline{S}_y = 658.45$ MPa with a standard deviation of $\hat{\sigma}_{S_y} = 45.44$ MPa. The rod is to be subjected to a mean static axial load of $\overline{P} = 289.13$ kN with a standard deviation of $\hat{\sigma}_P = 22.24$ kN. Assuming the strength and load have normal distributions, determine the reliabilities corresponding to the design factors of (*a*) 1.2, (*b*) 1.5. Also, determine the diameter corresponding to each case.

1–20   A beam subjected to axial loading will experience an axial stress, σ_a. If, in addition, the beam is subjected to a bending load, a bending stress, σ_b, will also occur at the outer fibers of the beam. The maximum stress at the outer fibers of the beam will be $\sigma_{max} = \sigma_a + \sigma_b$. Assume that σ_a and σ_b are independent and $\overline{\sigma}_a = 90$ MPa, $\hat{\sigma}_{\sigma_a} = 8.4$ MPa, $\overline{\sigma}_b = 383$ MPa, $\hat{\sigma}_{\sigma_b} = 22.3$ MPa. The rod is made of a steel with $\overline{S}_y = 553$ MPa and $\hat{\sigma}_{S_y} = 42.7$ MPa. Assuming the strength and load have normal distributions, determine the design factor and the reliability guarding against yielding.

1–21   Three blocks A, B, and C and a grooved block D have dimensions a, b, c, and d as follows:

$$a = 37.5 \pm 0.025 \text{ mm} \qquad b = 50 \pm 0.075 \text{ mm}$$

$$c = 75 \pm 0.1 \text{ mm} \qquad d = 163 \pm 0.25 \text{ mm}$$

Problem 1–21

(*a*) Determine the mean gap $\overline{w}$ and its tolerance.
(*b*) Determine the mean size of d that will assure that $w \geq 0.010$ in.

1-22 The volume of a rectangular parallelepiped is given by $V = xyz$. If $x = a \pm \Delta a$, $y = b \pm \Delta b$, $z = c \pm \Delta c$, show that

$$\frac{\Delta V}{V} \approx \frac{\Delta a}{a} + \frac{\Delta b}{b} + \frac{\Delta c}{c}$$

Use this result to determine the bilateral tolerance on the volume of a rectangular parallelepiped with dimensions

$$a = 37.5 \pm 0.05 \text{ mm} \qquad b = 47 \pm 0.075 \text{ mm} \qquad c = 75 \pm 0.1 \text{ mm}$$

1-23 A pivot in a linkage has a pin in the figure whose dimension $a \pm t_a$ is to be established. The thickness of the link clevis is 37.5 ± 0.125 mm. The designer has concluded that a gap of between 0.1 and 1.25 mm will satisfactorily sustain the function of the linkage pivot. Determine the dimension a and its tolerance.

Problem 1–23
Dimensions in mm.

1-24 A circular cross section O ring has the dimensions shown in the figure. In particular, an AS 568A standard No. 240 O ring has an inside diameter D_i and a cross-section diameter d of

$$D_i = 84.4 \pm 0.7 \text{ mm} \qquad d_i = 3.5 \pm 0.1 \text{ mm}$$

Estimate the mean outside diameter $\overline{D}_o$ and its bilateral tolerance.

Problem 1–24

1-25 to 1-28 For the table given, repeat Prob. 1–24 for the following O rings, given the AS 568A standard number. *Note:* The solutions require research.

Problem number	1–25	1–26	1–27	1–28
AS 568A No.	110	220	160	320

1-29 Convert the following to appropriate ips units:
(a) A stress, $\sigma = 150$ MPa.
(b) A force, $F = 2$ kN.
(c) A moment, $M = 150$ N · m.
(d) An area, $A = 1\,500$ mm^2.
(e) A second moment of area, $I = 750$ cm^4.
(f) A modulus of elasticity, $E = 145$ GPa.
(g) A speed, $v = 75$ km/h.
(h) A volume, $V = 1$ liter.

1–30 Convert the following to appropriate SI units:

(a) A length, $l = 5$ ft.

(b) A stress, $\sigma = 90$ kpsi.

(c) A pressure, $p = 25$ psi.

(d) A section modulus, $Z = 12$ in^3.

(e) A unit weight, $w = 0.208$ lbf/in.

(f) A deflection, $\delta = 0.001\ 89$ in.

(g) A velocity, $v = 1\ 200$ ft/min.

(h) A unit strain, $\epsilon = 0.002\ 15$ in/in.

(i) A volume, $V = 1830$ in^3.

1–31 Generally, final design results are rounded to or fixed to three digits because the given data cannot justify a greater display. In addition, prefixes should be selected to limit number strings to no more than four digits to the left of the decimal point. Using these rules, as well as those for the choice of prefixes, solve the following relations:

(a) $\sigma = M/Z$, where $M = 200$ N · m and $Z = 15.3 \times 10^3$ mm^3.

(b) $\sigma = F/A$, where $F = 42$ kN and $A = 600$ mm^2.

(c) $y = Fl^3/3EI$, where $F = 1200$ N, $l = 800$ mm, $E = 207$ GPa, and $I = 64 \times 10^3$ mm^4.

(d) $\theta = Tl/GJ$, where $T = 1100$ N · m, $l = 250$ mm, $G = 79.3$ GPa, and $d = 25$ mm.

1–32 Repeat Prob. 1–31 for the following:

(a) $\sigma = F/wt$, where $F = 1$ kN, $w = 25$ mm, and $t = 5$ mm.

(b) $I = bh^3/12$, where $b = 10$ mm and $h = 25$ mm.

(c) $I = \pi d^4/64$, where $d = 25.4$ mm.

(d) $\tau = 16\ T/\pi d^3$, where $T = 25$ N · m, and $d = 12.7$ mm.

1–33 Repeat Prob. 1–31 for:

(a) $\tau = F/A$, where $A = \pi d^2/4$, $F = 120$ kN, and $d = 20$ mm.

(b) $\sigma = 32\ Fa/\pi d^3$, where $F = 800$ N, $a = 800$ mm, and $d = 32$ mm.

(c) $Z = \pi(d_o^4 - d_i^4)/(32\ d_o)$ for $d_o = 36$ mm and $d_i = 26$ mm.

(d) $k = (d^4\ G)/(8\ D^3\ N)$, where $d = 1.6$ mm, $G = 79.3$ GPa, $D = 19.2$ mm, and $N = 32$ (a dimensionless number).

2

Materials

Chapter Outline

The selection of a material for a machine part or a structural member is one of the most important decisions the designer is called on to make. The decision is usually made before the dimensions of the part are established. After choosing the process of creating the desired geometry and the material (the two cannot be divorced), the designer can proportion the member so that loss of function can be avoided or the chance of loss of function can be held to an acceptable risk.

In Chaps. 3 and 4, methods for estimating stresses and deflections of machine members are presented. These estimates are based on the properties of the material from which the member will be made. For deflections and stability evaluations, for example, the elastic (stiffness) properties of the material are required, and evaluations of stress at a critical location in a machine member require a comparison with the strength of the material at that location in the geometry and condition of use. This strength is a material property found by testing and is adjusted to the geometry and condition of use as necessary.

As important as stress and deflection are in the design of mechanical parts, the selection of a material is not always based on these factors. Many parts carry no loads on them whatever. Parts may be designed merely to fill up space or for aesthetic qualities. Members must frequently be designed to also resist corrosion. Sometimes temperature effects are more important in design than stress and strain. So many other factors besides stress and strain may govern the design of parts that the designer must have the versatility that comes only with a broad background in materials and processes.

2–1 Material Strength and Stiffness

The standard tensile test is used to obtain a variety of material characteristics and strengths that are used in design. Figure 2–1 illustrates a typical tension-test specimen and its characteristic dimensions.[1] The original diameter d_0 and the gauge length l_0, used to measure the deflections, are recorded before the test is begun. The specimen is then mounted in the test machine and slowly loaded in tension while the load P and deflection are observed. The load is converted to stress by the calculation

$$\sigma = \frac{P}{A_0} \tag{2-1}$$

where $A_0 = \frac{1}{4}\pi d_0^2$ is the original area of the specimen.

Figure 2–1

A typical tension-test specimen. Some of the standard dimensions used for d_0 are 2.5, 6.25, and 12.5 mm and 0.505 in, but other sections and sizes are in use. Common gauge lengths l_0 used are 10, 25, and 50 mm and 1 and 2 in.

---

[1] See ASTM standards E8 and E-8 m for standard dimensions.

Figure 2–2

Stress-strain diagram obtained from the standard tensile test (*a*) Ductile material; (*b*) brittle material.
pl marks the proportional limit; *el*, the elastic limit; *y*, the offset-yield strength as defined by offset strain *a*; *u*, the maximum or ultimate strength; and *f*, the fracture strength.

(*a*)

(*b*)

The deflection, or extension of the gauge length, is given by $l - l_0$ where l is the gauge length corresponding to the load P. The normal strain is calculated from

$$\epsilon = \frac{l - l_0}{l_0} \qquad (2\text{–}2)$$

The results are plotted as a *stress-strain diagram.* Figure 2–2 depicts typical stress-strain diagrams for ductile and brittle materials. Ductile materials deform much more than brittle materials.

Point *pl* in Fig. 2–2*a* is called the *proportional limit.* This is the point at which the curve first begins to deviate from a straight line. No permanent set will be observable in the specimen if the load is removed at this point. In the linear range, the uniaxial stress-strain relation is given by *Hooke's law* as

$$\sigma = E\epsilon \qquad (2\text{–}3)$$

where the constant of proportionality E, the slope of the linear part of the stress-strain curve, is called *Young's modulus* or the *modulus of elasticity.* E is a measure of the stiffness of a material, and since strain is dimensionless, the units of E are the same as stress. Steel, for example, has a modulus of elasticity of about 207 GPa (30 Mpsi) *regardless of heat treatment, carbon content, or alloying.* Stainless steel is about 190 GPa (27.5 Mpsi).

Point *el* in Fig. 2–2 is called the *elastic limit.* If the specimen is loaded beyond this point, the deformation is said to be plastic and the material will take on a permanent set when the load is removed. Between *pl* and *el* the diagram is not a perfectly straight line, even though the specimen is elastic.

During the tension test, many materials reach a point at which the strain begins to increase very rapidly without a corresponding increase in stress. This point is called the *yield point.* Not all materials have an obvious yield point, especially for brittle materials. For this reason, *yield strength* S_y is often defined by an *offset method* as shown in Fig. 2–2, where line *ay* is drawn at slope E. Point *a* corresponds to a definite or stated amount of permanent set, usually 0.2 percent of the original gauge length ($\epsilon = 0.002$), although 0.01, 0.1, and 0.5 percent are sometimes used.

The *ultimate,* or *tensile, strength* S_u or S_{ut} corresponds to point *u* in Fig. 2–2 and is the maximum stress reached on the stress-strain diagram.[2] As shown in Fig. 2–2*a*,

---

[2]Usage varies. For a long time engineers used the term *ultimate strength,* hence the subscript *u* in S_u or S_{ut}. However, in material science and metallurgy the term *tensile strength* is used.

some materials exhibit a downward trend after the maximum stress is reached and fracture at point f on the diagram. Others, such as some of the cast irons and high-strength steels, fracture while the stress-strain trace is still rising, as shown in Fig. 2–2b, where points u and f are identical.

As noted in Sec. 1–9, *strength,* as used in this book, is a built-in property of a material, or of a mechanical element, because of the selection of a particular material or process or both. The strength of a connecting rod at the critical location in the geometry and condition of use, for example, is the same no matter whether it is already an element in an operating machine or whether it is lying on a workbench awaiting assembly with other parts. On the other hand, *stress* is something that occurs in a part, usually as a result of its being assembled into a machine and loaded. However, stresses may be built into a part by processing or handling. For example, shot peening produces a compressive *stress* in the outer surface of a part, and also improves the fatigue strength of the part. Thus, in this book we will be very careful in distinguishing between *strength,* designated by S, and *stress,* designated by σ or τ.

The diagrams in Fig. 2–2 are called *engineering* stress-strain diagrams because the stresses and strains calculated in Eqs. (2–1) and (2–2) are not *true* values. The stress calculated in Eq. (2–1) is based on the original area *before* the load is applied. In reality, as the load is applied the area reduces so that the *actual* or *true stress* is larger than the *engineering stress.* To obtain the true stress for the diagram the load and the cross-sectional area must be measured simultaneously during the test. Figure 2–2a represents a ductile material where the stress appears to decrease from points u to f. Typically, beyond point u the specimen begins to "neck" at a location of weakness where the area reduces dramatically, as shown in Fig. 2–3. For this reason, the true stress is much higher than the engineering stress at the necked section.

The engineering strain given by Eq. (2–2) is based on net change in length from the *original* length. In plotting the *true stress-strain diagram,* it is customary to use a term called *true strain* or, sometimes, *logarithmic strain.* True strain is the sum of the incremental elongations divided by the *current* gauge length at load P, or

$$\varepsilon = \int_{l_0}^{l} \frac{dl}{l} = \ln\frac{l}{l_0} \qquad (2\text{–}4)$$

where the symbol ε is used to represent true strain. The most important characteristic of a true stress-strain diagram (Fig. 2–4) is that the true stress continually increases all the way to fracture. Thus, as shown in Fig. 2–4, the true fracture stress σ_f is greater than the true ultimate stress σ_u. Contrast this with Fig. 2–2a, where the engineering fracture strength S_f is less than the engineering ultimate strength S_u.

Compression tests are more difficult to conduct, and the geometry of the test specimens differs from the geometry of those used in tension tests. The reason for this is that the specimen may buckle during testing or it may be difficult to distribute the stresses evenly. Other difficulties occur because ductile materials will bulge after yielding. However, the results can be plotted on a stress-strain diagram also, and the same strength definitions can be applied as used in tensile testing. For most ductile materials the compressive strengths are about the same as the tensile strengths. When substantial differences occur between tensile and compressive strengths, however, as

Figure 2–3

Tension specimen after necking.

Figure 2–4

True stress-strain diagram plotted in Cartesian coordinates.

is the case with the cast irons, the tensile and compressive strengths should be stated separately, S_{ut}, S_{uc}, where S_{uc} is reported as a *positive* quantity.

Torsional strengths are found by twisting solid circular bars and recording the torque and the twist angle. The results are then plotted as a *torque-twist diagram*. The shear stresses in the specimen are linear with respect to radial location, being zero at the center of the specimen and maximum at the outer radius r (see Chap. 3). The maximum shear stress $\tau_{\max}$ is related to the angle of twist θ by

$$\tau_{\max} = \frac{Gr}{l_0}\theta \qquad (2\text{–}5)$$

where θ is in radians, r is the radius of the specimen, l_0 is the gauge length, and G is the material stiffness property called the *shear modulus* or the *modulus of rigidity*. The maximum shear stress is also related to the applied torque T as

$$\tau_{\max} = \frac{Tr}{J} \qquad (2\text{–}6)$$

where $J = \frac{1}{2}\pi r^4$ is the polar second moment of area of the cross section.

The torque-twist diagram will be similar to Fig. 2–2, and, using Eqs. (2–5) and (2–6), the modulus of rigidity can be found as well as the elastic limit and the *torsional yield strength* S_{sy}. The maximum point on a torque-twist diagram, corresponding to point u on Fig. 2–2, is T_u. The equation

$$S_{su} = \frac{T_u r}{J} \qquad (2\text{–}7)$$

defines the *modulus of rupture* for the torsion test. Note that it is incorrect to call S_{su} the ultimate torsional strength, as the outermost region of the bar is in a plastic state at the torque T_u and the stress distribution is no longer linear.

All of the stresses and strengths defined by the stress-strain diagram of Fig. 2–2 and similar diagrams are specifically known as *engineering stresses* and *strengths* or *nominal stresses* and *strengths*. These are the values normally used in all engineering design calculations. The adjectives *engineering* and *nominal* are used here to emphasize that the stresses are computed by using the *original* or *unstressed cross-sectional area* of the specimen. In this book we shall use these modifiers only when we specifically wish to call attention to this distinction.

In addition to providing strength values for a material, the stress-strain diagram provides insight into the energy-absorbing characteristics of a material. This is because the stress-strain diagram involves both loads and deflections, which are directly related to energy. The capacity of a material to absorb energy within its elastic range is called *resilience*. The *modulus of resilience* u_R of a material is defined as the energy absorbed per unit volume without permanent deformation, and is equal to the area under the stress-strain curve up to the elastic limit. The elastic limit is often approximated by the yield point, since it is more readily determined, giving

$$u_R \approx \int_0^{\epsilon_y} \sigma \, d\epsilon \tag{2-8}$$

where ϵ_y is the strain at the yield point. If the stress-strain is linear to the yield point, then the area under the curve is simply a triangular area; thus

$$u_R \approx \frac{1}{2} S_y \epsilon_y = \frac{1}{2} (S_y)(S_y/E) = \frac{S_y^2}{2E} \tag{2-9}$$

This relationship indicates that for two materials with the same yield strength, the less stiff material (lower E), will have a greater resilience, that is, an ability to absorb more energy without yielding.

The capacity of a material to absorb energy without fracture is called *toughness*. The *modulus of toughness* u_T of a material is defined as the energy absorbed per unit volume without fracture, which is equal to the total area under the stress-strain curve up to the fracture point, or

$$u_T = \int_0^{\epsilon_f} \sigma \, d\epsilon \tag{2-10}$$

where ϵ_f is the strain at the fracture point. This integration is often performed graphically from the stress-strain data, or a rough approximation can be obtained by using the average of the yield and ultimate strengths and the strain at fracture to calculate an area; that is,

$$u_T \approx \left(\frac{S_y + S_{ut}}{2} \right) \epsilon_f \tag{2-11}$$

The units of toughness and resilience are energy per unit volume (lbf · in/in^3 or J/m^3), which are numerically equivalent to psi or Pa. These definitions of toughness and resilience assume the low strain rates that are suitable for obtaining the stress-strain diagram. For higher strain rates, see Sec. 2–5 for impact properties.

2–2   The Statistical Significance of Material Properties

There is some subtlety in the ideas presented in the previous section that should be pondered carefully before continuing. Figure 2–2 depicts the result of a *single* tension test (*one* specimen, now fractured). It is common for engineers to consider these important *stress* values (at points *pl*, *el*, *y*, *u*, and *f*) as properties and to denote them as strengths with a special notation, uppercase *S*, in lieu of lowercase sigma σ, with subscripts added: S_{pl} for proportional limit, S_y for yield strength, S_u for ultimate tensile strength (S_{ut} or S_{uc}, if tensile or compressive sense is important).

If there were 1000 nominally identical specimens, the values of strength obtained would be distributed between some minimum and maximum values. It follows that the description of strength, a material property, is distributional and thus is statistical in nature.

Consider the following example.

EXAMPLE 2–1 One thousand specimens of 1020 steel were tested to rupture, and the ultimate tensile strengths were reported as follows:

Range midpoint S_{ut} (kpsi)	56.5	57.5	58.5	59.5	60.5	61.5	62.5	63.5	64.5	65.5	66.5	67.5	68.5	69.5	70.5	71.5
Frequency, f_i	2	18	23	31	83	109	138	151	139	130	82	49	28	11	4	2

where the *range* of stress for each midpoint entry is $w = 1$ kpsi.

Plot a histogram of the data (a bar chart of f versus S_{ut}). Presuming the distribution is normal, plot Eq. (1–4). Compare the plots.

Solution The worksheet for the data is

Range Midpoint x_i (kpsi)	Frequency f_i	$x_i f_i$	$x_i^2 f_i$	Observed PDF* $f_i/(Nw)^†$	Normal PDF* $f(x)$
56.5	2	113.0	6 384.50	0.002	0.0035
57.5	18	1 035.0	59 512.50	0.018	0.0095
58.5	23	1 345.5	78 711.75	0.023	0.0218
59.5	31	1 844.5	109 747.75	0.031	0.0434
60.5	83	5 021.5	303 800.75	0.083	0.0744
61.5	109	6 703.5	412 265.25	0.109	0.1100
62.5	138	8 625.0	539 062.50	0.138	0.1400
63.5	151	9 588.5	608 869.75	0.151	0.1536
64.5	139	8 965.5	578 274.75	0.139	0.1453
65.5	130	8 515.0	577 732.50	0.130	0.1184
66.5	82	5 453.0	362 624.50	0.082	0.0832
67.5	49	3 307.5	223 256.25	0.049	0.0504
68.5	28	1 918.0	131 382.00	0.028	0.0260
69.5	11	764.5	53 132.75	0.011	0.0118
70.5	4	282.0	19 881.00	0.004	0.0046
71.5	2	143.0	10 224.50	0.002	0.0015
Σ	1000	63 625	4 054 864	1.000	

*PDF refers to Probability Density Function (see Sec. 1–12).
†To compare discrete frequency data with continuous density functions, f_i must be divided by Nw. Here, N = sample size = 1000, and w = width of range interval = 1 kpsi.

From Eq. (1–6),

$$\bar{x} = \frac{1}{N}\sum_{i=1}^{k} f_i x_i = \frac{1}{1000}(63\ 625) = 63.625 \text{ kpsi}$$

From Eq. (1–7),

$$s_x = \sqrt{\frac{\sum_{i=1}^{k} f_i x_i^2 - N\bar{x}^2}{N-1}} = \sqrt{\frac{4\ 054\ 864 - 1000(63.625^2)}{1000 - 1}}$$

$$= 2.594\ 245 = 2.594 \text{ kpsi}$$

From Eq. (1–4), with $\mu_x = \bar{x}$ and $\hat{\sigma}_x = s_x$, the PDF for a normal density function is

$$f(x) = \frac{1}{\hat{\sigma}_x\sqrt{2\pi}}\exp\left[-\frac{1}{2}\left(\frac{x-\mu_x}{\hat{\sigma}_x}\right)\right]$$

$$= \frac{1}{2.594\ 245\sqrt{2\pi}}\exp\left[-\frac{1}{2}\left(\frac{x-63.625}{2.594\ 245}\right)^2\right]$$

For example, $f(63.5) = 0.1536$.

The bar chart shown in Fig. 2–5 depicts the histogram of the PDF of the discrete data. A plot of the continuous normal PDF, $f(x)$, is also included.

Figure 2–5

Histogram for 1000 tensile tests on a 1020 steel from a single heat.

Note, in Ex. 2–1, the test program described 1020 property S_{ut}, for only one heat of one supplier. Testing is an involved and expensive process. Tables of properties are often prepared to be helpful to other persons. A statistical quantity is described by its mean, standard deviation, and distribution type. Many tables display a single number, which is often the mean, minimum, or some percentile, such as the 99th percentile. Always read the foonotes to the table. If no qualification is made in a single-entry table, the table is subject to serious doubt.

Since it is no surprise that useful descriptions of a property are statistical in nature, engineers, when ordering property tests, should couch the instructions so the data generated are enough for them to observe the statistical parameters and to identify the distributional characteristic. The tensile test program on 1000 specimens of 1020 steel is a large one. If you were faced with putting something in a table of ultimate tensile strengths and constrained to a single number, what would it be and just how would your footnote read?

2–3 Strength and Cold Work

Cold working is the process of plastic straining below the recrystallization temperature in the plastic region of the stress-strain diagram. Materials can be deformed plastically by the application of heat, as in forging or hot rolling, but the resulting mechanical properties are quite different from those obtained by cold working. The purpose of this section is to explain what happens to the significant mechanical properties of a material when that material is cold worked.

Consider the stress-strain diagram of Fig. 2–6a. Here a material has been stressed beyond the yield strength at *y* to some point *i*, in the plastic region, and then the load removed. At this point the material has a permanent plastic deformation ϵ_p. If the load corresponding to point *i* is now reapplied, the material will be elastically deformed by the amount ϵ_e. Thus at point *i* the total unit strain consists of the two components ϵ_p and ϵ_e and is given by the equation

$$\epsilon = \epsilon_p + \epsilon_e \tag{a}$$

This material can be unloaded and reloaded any number of times from and to point *i*, and it is found that the action always occurs along the straight line that is approximately parallel to the initial elastic line *Oy*. Thus

$$\epsilon_e = \frac{\sigma_i}{E} \tag{b}$$

The material now has a higher yield point, is less ductile as a result of a reduction in strain capacity, and is said to be *strain-hardened*. If the process is continued, increasing ϵ_p, the material can become brittle and exhibit sudden fracture.

Figure 2–6

(*a*) Stress-strain diagram showing unloading and reloading at point *i* in the plastic region; (*b*) analogous load-deformation diagram.

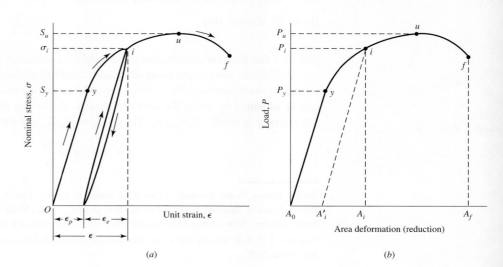

(*a*)

(*b*)

It is possible to construct a similar diagram, as in Fig. 2–6b, where the abscissa is the area deformation and the ordinate is the applied load. The *reduction in area* corresponding to the load P_f, at fracture, is defined as

$$R = \frac{A_0 - A_f}{A_0} = 1 - \frac{A_f}{A_0} \tag{2–12}$$

where A_0 is the original area. The quantity R in Eq. (2–12) is usually expressed in percent and tabulated in lists of mechanical properties as a measure of *ductility*. See Appendix Table A–20, for example. Ductility is an important property because it measures the ability of a material to absorb overloads and to be cold-worked. Thus such operations as bending, drawing, heading, and stretch forming are metal-processing operations that require ductile materials.

Figure 2–6b can also be used to define the quantity of cold work. The *cold-work factor W* is defined as

$$W = \frac{A_0 - A_i'}{A_0} \approx \frac{A_0 - A_i}{A_0} \tag{2–13}$$

where A_i' corresponds to the area after the load P_i has been released. The approximation in Eq. (2–13) results because of the difficulty of measuring the small diametral changes in the elastic region. If the amount of cold work is known, then Eq. (2–13) can be solved for the area A_i'. The result is

$$A_i' = A_0(1 - W) \tag{2–14}$$

Cold working a material produces a new set of values for the strengths, as can be seen from stress-strain diagrams. Datsko[3] describes the plastic region of the true stress–true strain diagram by the equation

$$\sigma = \sigma_0 \varepsilon^m \tag{2–15}$$

where σ = true stress

σ_0 = a strength coefficient, or strain-strengthening coefficient

ε = true plastic strain

m = strain-strengthening exponent

It can be shown[4] that

$$m = \varepsilon_u \tag{2–16}$$

provided that the load-deformation curve exhibits a stationary point (a place of zero slope).

Difficulties arise when using the gauge length to evaluate the true strain in the plastic range, since necking causes the strain to be nonuniform. A more satisfactory relation can be obtained by using the area at the neck. Assuming that the change in volume of the material is small, $Al = A_0 l_0$. Thus, $l/l_0 = A_0/A$, and the true strain is given by

$$\varepsilon = \ln \frac{l}{l_0} = \ln \frac{A_0}{A} \tag{2–17}$$

---

[3]Joseph Datsko, "Solid Materials," Chap. 32 in Joseph E. Shigley, Charles R. Mischke, and Thomas H. Brown, Jr. (eds.), *Standard Handbook of Machine Design,* 3rd ed., McGraw-Hill, New York, 2004. See also Joseph Datsko, "New Look at Material Strength," *Machine Design,* vol. 58, no. 3, Feb. 6, 1986, pp. 81–85.

[4]See Sec. 5–2, J. E. Shigley and C. R. Mischke, *Mechanical Engineering Design,* 6th ed., McGraw-Hill, New York, 2001.

Returning to Fig. 2–6b, if point i is to the left of point u, that is, $P_i < P_u$, then the new yield strength is

$$S_y' = \frac{P_i}{A_i'} = \sigma_0 \varepsilon_i^m \qquad P_i \leq P_u \qquad (2\text{–}18)$$

Because of the reduced area, that is, because $A_i' < A_0$, the ultimate strength also changes, and is

$$S_u' = \frac{P_u}{A_i'} \qquad (c)$$

Since $P_u = S_u A_0$, we find, with Eq. (2–14), that

$$S_u' = \frac{S_u A_0}{A_0 (1 - W)} = \frac{S_u}{1 - W} \qquad \varepsilon_i \leq \varepsilon_u \qquad (2\text{–}19)$$

which is valid only when point i is to the left of point u.

For points to the right of u, the yield strength is approaching the ultimate strength, and, with small loss in accuracy,

$$S_u' \approx S_y' \approx \sigma_0 \varepsilon_i^m \qquad \varepsilon_i > \varepsilon_u \qquad (2\text{–}20)$$

A little thought will reveal that a bar will have the same ultimate load in tension after being strain-strengthened in tension as it had before. The new strength is of interest to us not because the static ultimate load increases, but—since fatigue strengths are correlated with the local ultimate strengths—because the fatigue strength improves. Also the yield strength increases, giving a larger range of sustainable *elastic* loading.

EXAMPLE 2–2

An annealed AISI 1018 steel (see Table A–22) has $S_y = 220$ MPa, $S_u = 341$ MPa, $\sigma_f = 628$ MPa, $\sigma_0 = 620$ MPa, $m = 0.25$, and $\varepsilon_f = 1.05$ mm/mm. Find the new values of the strengths if the material is given 15 percent cold work.

Solution

From Eq. (2–16), we find the true strain corresponding to the ultimate strength to be

$$\varepsilon_u = m = 0.25$$

The ratio A_0/A_i is, from Eq. (2–13),

$$\frac{A_0}{A_i} = \frac{1}{1 - W} = \frac{1}{1 - 0.15} = 1.176$$

The true strain corresponding to 15 percent cold work is obtained from Eq. (2–17). Thus

$$\varepsilon_i = \ln \frac{A_0}{A_i} = \ln 1.176 = 0.1625$$

Since $\varepsilon_i < \varepsilon_u$, Eqs. (2–18) and (2–19) apply. Therefore,

Answer

$$S_y' = \sigma_0 \varepsilon_i^m = 620(0.1625)^{0.25} = 393.6 \text{ MPa}$$

Answer

$$S_u' = \frac{S_u}{1 - W} = \frac{341}{1 - 0.15} = 401.2 \text{ MPa}$$

2–4   **Hardness**

The resistance of a material to penetration by a pointed tool is called *hardness*. Though there are many hardness-measuring systems, we shall consider here only the two in greatest use.

Rockwell hardness tests are described by ASTM standard hardness method E–18 and measurements are quickly and easily made, they have good reproducibility, and the test machine for them is easy to use. In fact, the hardness number is read directly from a dial. Rockwell hardness scales are designated as *A, B, C, . . . ,* etc. The indenters are described as a diamond, a 1.6-mm-diameter ball, and a diamond for scales *A, B,* and *C,* respectively, where the load applied is either 60, 100, or 150 kg. Thus the Rockwell *B* scale, designated R_B, uses a 100-kg load and a No. 2 indenter, which is a 1.6-mm-diameter ball. The Rockwell *C* scale R_C uses a diamond cone, which is the No. 1 indenter, and a load of 150 kg. Hardness numbers so obtained are relative. Therefore a hardness $R_C = 50$ has meaning only in relation to another hardness number using the same scale.

The *Brinell hardness* is another test in very general use. In testing, the indenting tool through which force is applied is a ball and the hardness number H_B is found as a number equal to the applied load divided by the spherical surface area of the indentation. Thus the units of H_B are the same as those of stress, though they are seldom used. Brinell hardness testing takes more time, since H_B must be computed from the test data. The primary advantage of both methods is that they are nondestructive in most cases. Both are empirically and directly related to the ultimate strength of the material tested. This means that the strength of parts could, if desired, be tested part by part during manufacture.

Hardness testing provides a convenient and nondestructive means of estimating the strength properties of materials. The Brinell hardness test is particularly well known for this estimation, since for many materials the relationship between the minimum ultimate strength and the Brinell hardness number is roughly linear. The constant of proportionality varies between classes of materials, and is also dependent on the load used to determine the hardness. There is a wide scatter in the data, but for rough approximations for *steels,* the relationship is generally accepted as

$$S_u = \begin{cases} 0.5\,H_B & \text{kpsi} \\ 3.4\,H_B & \text{MPa} \end{cases} \tag{2–21}$$

Similar relationships for *cast iron* can be derived from data supplied by Krause.[5] The minimum strength, as defined by the ASTM, is found from these data to be

$$S_u = \begin{cases} 0.23\,H_B - 12.5 \text{ kpsi} \\ 1.58\,H_B - 86 \text{ MPa} \end{cases} \tag{2–22}$$

Walton[6] shows a chart from which the SAE minimum strength can be obtained, which is more conservative than the values obtained from Eq. (2–22).

---

[5]D. E. Krause, "Gray Iron—A Unique Engineering Material," ASTM Special Publication 455, 1969, pp. 3–29, as reported in Charles F. Walton (ed.), *Iron Castings Handbook,* Iron Founders Society, Inc., Cleveland, 1971, pp. 204, 205.

[6]Ibid.

EXAMPLE 2–3   It is necessary to ensure that a certain part supplied by a foundry always meets or exceeds ASTM No. 20 specifications for cast iron (see Table A–24). What hardness should be specified?

Solution   From Eq. (2–22), with $(S_u)_{min} = 138$ MPa, we have

Answer

$$H_B = \frac{S_u + 86}{1.58} = \frac{138 + 86}{1.58} = 142$$

If the foundry can control the hardness within 20 points, routinely, then specify $145 < H_B < 165$. This imposes no hardship on the foundry and assures the designer that ASTM grade 20 will always be supplied at a predictable cost.

2–5   Impact Properties

An external force applied to a structure or part is called an *impact load* if the time of application is less than one-third the lowest natural period of vibration of the part or structure. Otherwise it is called simply a *static load*.

The *Charpy* (commonly used) and *Izod* (rarely used) *notched-bar tests* utilize bars of specified geometries to determine brittleness and impact strength. These tests are helpful in comparing several materials and in the determination of low-temperature brittleness. In both tests the specimen is struck by a pendulum released from a fixed height, and the energy absorbed by the specimen, called the *impact value,* can be computed from the height of swing after fracture, but is read from a dial that essentially "computes" the result.

The effect of temperature on impact values is shown in Fig. 2–7 for a material showing a ductile-brittle transition. Not all materials show this transition. Notice the narrow region of critical temperatures where the impact value increases very rapidly. In the low-temperature region the fracture appears as a brittle, shattering type, whereas the appearance is a tough, tearing type above the critical-temperature region. The critical temperature seems to be dependent on both the material and the geometry of the notch. For this reason designers should not rely too heavily on the results of notched-bar tests.

Figure 2–7

A mean trace shows the effect of temperature on impact values. The result of interest is the brittle-ductile transition temperature, often defined as the temperature at which the mean trace passes through the 20 J level. The critical temperature *is* dependent on the geometry of the notch, which is why the Charpy V notch is closely defined.

Figure 2–8

Influence of strain rate on tensile properties.

The average strain rate used in obtaining the stress-strain diagram is about 0.025 mm/(mm · s) or less. When the strain rate is increased, as it is under impact conditions, the strengths increase, as shown in Fig. 2–8. In fact, at very high strain rates the yield strength seems to approach the ultimate strength as a limit. But note that the curves show little change in the elongation. This means that the ductility remains about the same. Also, in view of the sharp increase in yield strength, a mild steel could be expected to behave elastically throughout practically its entire strength range under impact conditions.

The Charpy and Izod tests really provide toughness data under dynamic, rather than static, conditions. It may well be that impact data obtained from these tests are as dependent on the notch geometry as they are on the strain rate. For these reasons it may be better to use the concepts of notch sensitivity, fracture toughness, and fracture mechanics, discussed in Chaps. 5 and 6, to assess the possibility of cracking or fracture.

2–6 Temperature Effects

Strength and ductility, or brittleness, are properties affected by the temperature of the operating environment.

The effect of temperature on the static properties of steels is typified by the strength versus temperature chart of Fig. 2–9. Note that the tensile strength changes only a small amount until a certain temperature is reached. At that point it falls off rapidly. The yield strength, however, decreases continuously as the environmental temperature is increased. There is a substantial increase in ductility, as might be expected, at the higher temperatures.

Many tests have been made of ferrous metals subjected to constant loads for long periods of time at elevated temperatures. The specimens were found to be permanently deformed during the tests, even though at times the actual stresses were less than the yield strength of the material obtained from short-time tests made at the same temperature. This continuous deformation under load is called *creep*.

Figure 2-9

A plot of the results of 145 tests of 21 carbon and alloy steels showing the effect of operating temperature on the yield strength S_y and the ultimate strength S_{ut}. The ordinate is the ratio of the strength at the operating temperature to the strength at room temperature. The standard deviations were $0.0442 \leq \hat{\sigma}_{Sy} \leq 0.152$ for S_y and $0.099 \leq \hat{\sigma}_{Sut} \leq 0.11$ for S_{ut}. (*Data source: E. A. Brandes (ed.),* Smithells Metal Reference Book, 6th ed., *Butterworth, London, 1983 pp. 22–128 to 22–131.*)

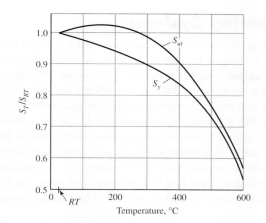

One of the most useful tests to have been devised is the long-time creep test under constant load. Figure 2–10 illustrates a curve that is typical of this kind of test. The curve is obtained at a constant stated temperature. A number of tests are usually run simultaneously at different stress intensities. The curve exhibits three distinct regions. In the first stage are included both the elastic and the plastic deformation. This stage shows a decreasing creep rate, which is due to the strain hardening. The second stage shows a constant minimum creep rate caused by the annealing effect. In the third stage the specimen shows a considerable reduction in area, the true stress is increased, and a higher creep eventually leads to fracture.

When the operating temperatures are lower than the transition temperature (Fig. 2–7), the possibility arises that a part could fail by a brittle fracture. This subject will be discussed in Chap. 5.

Of course, heat treatment, as will be shown, is used to make substantial changes in the mechanical properties of a material.

Heating due to electric and gas welding also changes the mechanical properties. Such changes may be due to clamping during the welding process, as well as heating; the resulting stresses then remain when the parts have cooled and the clamps have been removed. Hardness tests can be used to learn whether the strength has been changed by welding, but such tests will not reveal the presence of residual stresses.

Figure 2-10

Creep-time curve.

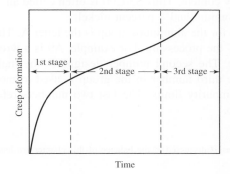

2–7 Numbering Systems

The Society of Automotive Engineers (SAE) was the first to recognize the need, and to adopt a system, for the numbering of steels. Later the American Iron and Steel Institute (AISI) adopted a similar system. In 1975 the SAE published the Unified Numbering System for Metals and Alloys (UNS); this system also contains cross-reference numbers for other material specifications.[7] The UNS uses a letter prefix to designate the material, as, for example, G for the carbon and alloy steels, A for the aluminum alloys, C for the copper-base alloys, and S for the stainless or corrosion-resistant steels. For some materials, not enough agreement has as yet developed in the industry to warrant the establishment of a designation.

For the steels, the first two numbers following the letter prefix indicate the composition, excluding the carbon content. The various compositions used are as follows:

G10	Plain carbon	G46	Nickel-molybdenum
G11	Free-cutting carbon steel with more sulfur or phosphorus	G48	Nickel-molybdenum
		G50	Chromium
G13	Manganese	G51	Chromium
G23	Nickel	G52	Chromium
G25	Nickel		
		G61	Chromium-vanadium
G31	Nickel-chromium	G86	Chromium-nickel-molybdenum
G33	Nickel-chromium	G87	Chromium-nickel-molybdenum
G40	Molybdenum	G92	Manganese-silicon
G41	Chromium-molybdenum	G94	Nickel-chromium-molybdenum
G43	Nickel-chromium-molybdenum		

The second number pair refers to the approximate carbon content. Thus, G10400 is a plain carbon steel with a nominal carbon content of 0.40 percent (0.37 to 0.44 percent). The fifth number following the prefix is used for special situations. For example, the old designation AISI 52100 represents a chromium alloy with about 100 points of carbon. The UNS designation is G52986.

The UNS designations for the stainless steels, prefix S, utilize the older AISI designations for the first three numbers following the prefix. The next two numbers are reserved for special purposes. The first number of the group indicates the approximate composition. Thus 2 is a chromium-nickel-manganese steel, 3 is a chromium-nickel steel, and 4 is a chromium alloy steel. Sometimes stainless steels are referred to by their alloy content. Thus S30200 is often called an 18-8 stainless steel, meaning 18 percent chromium and 8 percent nickel.

The prefix for the aluminum group is the letter A. The first number following the prefix indicates the processing. For example, A9 is a wrought aluminum, while A0 is a casting alloy. The second number designates the main alloy group as shown in Table 2–1. The third number in the group is used to modify the original alloy or to designate the impurity limits. The last two numbers refer to other alloys used with the basic group.

---

[7]Many of the materials discussed in the balance of this chapter are listed in the Appendix tables. Be sure to review these.

Table 2–1

Aluminum Alloy Designations

Aluminum 99.00% pure and greater	Ax1xxx
Copper alloys	Ax2xxx
Manganese alloys	Ax3xxx
Silicon alloys	Ax4xxx
Magnesium alloys	Ax5xxx
Magnesium-silicon alloys	Ax6xxx
Zinc alloys	Ax7xxx

The American Society for Testing and Materials (ASTM) numbering system for cast iron is in widespread use. This system is based on the tensile strength. Thus ASTM A18 speaks of classes; e.g., 30 cast iron has a minimum tensile strength of 138 MPa. Note from Appendix A-24, however, that the *typical* tensile strength is 152 MPa. You should be careful to designate which of the two values is used in design and problem work because of the significance of factor of safety.

2–8   Sand Casting

Sand casting is a basic low-cost process, and it lends itself to economical production in large quantities with practically no limit to the size, shape, or complexity of the part produced.

In sand casting, the casting is made by pouring molten metal into sand molds. A pattern, constructed of metal or wood, is used to form the cavity into which the molten metal is poured. Recesses or holes in the casting are produced by sand cores introduced into the mold. The designer should make an effort to visualize the pattern and casting in the mold. In this way the problems of core setting, pattern removal, draft, and solidification can be studied. Castings to be used as test bars of cast iron are cast separately and properties may vary.

Steel castings are the most difficult of all to produce, because steel has the highest melting temperature of all materials normally used for casting. This high temperature aggravates all casting problems.

The following rules will be found quite useful in the design of any sand casting:

1 All sections should be designed with a uniform thickness.

2 The casting should be designed to produce a gradual change from section to section where this is necessary.

3 Adjoining sections should be designed with generous fillets or radii.

4 A complicated part should be designed as two or more simple castings to be assembled by fasteners or by welding.

Steel, gray iron, brass, bronze, and aluminum are most often used in castings. The minimum wall thickness for any of these materials is about 5 mm, though with particular care, thinner sections can be obtained with some materials.

2–9   Shell Molding

The shell-molding process employs a heated metal pattern, usually made of cast iron, aluminum, or brass, which is placed in a shell-molding machine containing a mixture of dry sand and thermosetting resin. The hot pattern melts the plastic, which, together with the sand, forms a shell about 5 to 10 mm thick around the pattern. The shell is then baked at from 205 to 370°C for a short time while still on the pattern. It is then stripped from the pattern and placed in storage for use in casting.

In the next step the shells are assembled by clamping, bolting, or pasting; they are placed in a backup material, such as steel shot; and the molten metal is poured into the cavity. The thin shell permits the heat to be conducted away so that solidification takes place rapidly. As solidification takes place, the plastic bond is burned and the mold collapses. The permeability of the backup material allows the gases to escape and the casting to air-cool. All this aids in obtaining a fine-grain, stress-free casting.

Shell-mold castings feature a smooth surface, a draft that is quite small, and close tolerances. In general, the rules governing sand casting also apply to shell-mold casting.

2–10   Investment Casting

Investment casting uses a pattern that may be made from wax, plastic, or other material. After the mold is made, the pattern is melted out. Thus a mechanized method of casting a great many patterns is necessary. The mold material is dependent upon the melting point of the cast metal. Thus a plaster mold can be used for some materials while others would require a ceramic mold. After the pattern is melted out, the mold is baked or fired; when firing is completed, the molten metal may be poured into the hot mold and allowed to cool.

If a number of castings are to be made, then metal or permanent molds may be suitable. Such molds have the advantage that the surfaces are smooth, bright, and accurate, so that little, if any, machining is required. *Metal-mold castings* are also known as *die castings* and *centrifugal castings*.

2–11   Powder-Metallurgy Process

The powder-metallurgy process is a quantity-production process that uses powders from a single metal, several metals, or a mixture of metals and nonmetals. It consists essentially of mechanically mixing the powders, compacting them in dies at high pressures, and heating the compacted part at a temperature less than the melting point of the major ingredient. The particles are united into a single strong part similar to what would be obtained by melting the same ingredients together. The advantages are (1) the elimination of scrap or waste material, (2) the elimination of machining operations, (3) the low unit cost when mass-produced, and (4) the exact control of composition. Some of the disadvantages are (1) the high cost of dies, (2) the lower physical properties, (3) the higher cost of materials, (4) the limitations on the design, and (5) the limited range of materials that can be used. Parts commonly made by this process are oil-impregnated bearings, incandescent lamp filaments, cemented-carbide tips for tools, and permanent magnets. Some products can be made only by powder metallurgy: surgical implants, for example. The structure is different from what can be obtained by melting the same ingredients.

2–12   Hot-Working Processes

By *hot working* are meant such processes as rolling, forging, hot extrusion, and hot pressing, in which the metal is heated above its recrystallization temperature.

Figure 2–11

Common shapes available through hot rolling.

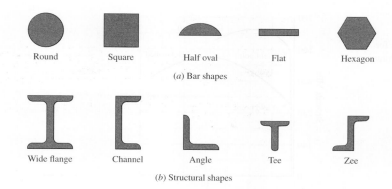

Round   Square   Half oval   Flat   Hexagon

(*a*) Bar shapes

Wide flange   Channel   Angle   Tee   Zee

(*b*) Structural shapes

Hot rolling is usually used to create a bar of material of a particular shape and dimension. Figure 2–11 shows some of the various shapes that are commonly produced by the hot-rolling process. All of them are available in many different sizes as well as in different materials. The materials most available in the hot-rolled bar sizes are steel, aluminum, magnesium, and copper alloys.

Tubing can be manufactured by hot-rolling strip or plate. The edges of the strip are rolled together, creating seams that are either butt-welded or lap-welded. Seamless tubing is manufactured by roll-piercing a solid heated rod with a piercing mandrel.

Extrusion is the process by which great pressure is applied to a heated metal billet or blank, causing it to flow through a restricted orifice. This process is more common with materials of low melting point, such as aluminum, copper, magnesium, lead, tin, and zinc. Stainless steel extrusions are available on a more limited basis.

Forging is the hot working of metal by hammers, presses, or forging machines. In common with other hot-working processes, forging produces a refined grain structure that results in increased strength and ductility. Compared with castings, forgings have greater strength for the same weight. In addition, drop forgings can be made smoother and more accurate than sand castings, so that less machining is necessary. However, the initial cost of the forging dies is usually greater than the cost of patterns for castings, although the greater unit strength rather than the cost is usually the deciding factor between these two processes.

2–13 Cold-Working Processes

By *cold working* is meant the forming of the metal while at a low temperature (usually room temperature). In contrast to parts produced by hot working, cold-worked parts have a bright new finish, are more accurate, and require less machining.

Cold-finished bars and shafts are produced by rolling, drawing, turning, grinding, and polishing. Of these methods, by far the largest percentage of products are made by the cold-rolling and cold-drawing processes. Cold rolling is now used mostly for the production of wide flats and sheets. Practically all cold-finished bars are made by cold drawing but even so are sometimes mistakenly called "cold-rolled bars." In the drawing process, the hot-rolled bars are first cleaned of scale and then drawn by pulling them through a die that reduces the size about 0.03 to 0.06 mm. This process does not remove material from the bar but reduces, or "draws" down, the size. Many different shapes of hot-rolled bars may be used for cold drawing.

Cold rolling and cold drawing have the same effect upon the mechanical properties. The cold-working process does not change the grain size but merely distorts

Figure 2–12

Stress-strain diagram for hot-rolled and cold-drawn UNS G10350 steel.

it. Cold working results in a large increase in yield strength, an increase in ultimate strength and hardness, and a decrease in ductility. In Fig. 2–12 the properties of a cold-drawn bar are compared with those of a hot-rolled bar of the same material.

Heading is a cold-working process in which the metal is gathered, or upset. This operation is commonly used to make screw and rivet heads and is capable of producing a wide variety of shapes. *Roll threading* is the process of rolling threads by squeezing and rolling a blank between two serrated dies. *Spinning* is the operation of working sheet material around a rotating form into a circular shape. *Stamping* is the term used to describe punch-press operations such as *blanking, coining, forming,* and *shallow drawing.*

2–14 The Heat Treatment of Steel

Heat treatment of steel refers to time- and temperature-controlled processes that relieve residual stresses and/or modifies material properties such as hardness (strength), ductility, and toughness. Other mechanical or chemical operations are sometimes grouped under the heading of heat treatment. The common heat-treating operations are annealing, quenching, tempering, and case hardening.

Annealing

When a material is cold- or hot-worked, residual stresses are built in, and, in addition, the material usually has a higher hardness as a result of these working operations. These operations change the structure of the material so that it is no longer represented by the equilibrium diagram. Full annealing and normalizing is a heating operation that permits the material to transform according to the equilibrium diagram. The material to be annealed is heated to a temperature that is approximately 38°C above the critical temperature. It is held at this temperature for a time that is sufficient for the carbon to become dissolved and diffused through the material. The object being treated is then allowed to cool slowly, usually in the furnace in which it was treated. If the transformation is complete, then it is said to have a full anneal. Annealing is used to soften a material and make it more ductile, to relieve residual stresses, and to refine the grain structure.

The term *annealing* includes the process called *normalizing.* Parts to be normalized may be heated to a slightly higher temperature than in full annealing. This produces a coarser grain structure, which is more easily machined if the material is a low-carbon steel. In the normalizing process the part is cooled in still air at room temperature. Since this cooling is more rapid than the slow cooling used in full anneal-ing, less time is available for equilibrium, and the material is harder than fully annealed steel. Normalizing is often used as the final treating operation for steel. The cooling in still air amounts to a slow quench.

Quenching

Eutectoid steel that is fully annealed consists entirely of pearlite, which is obtained from austenite under conditions of equilibrium. A fully annealed hypoeutectoid steel would consist of pearlite plus ferrite, while hypereutectoid steel in the fully annealed condition would consist of pearlite plus cementite. The hardness of steel of a given carbon content depends upon the structure that replaces the pearlite when full anneal-ing is not carried out.

The absence of full annealing indicates a more rapid rate of cooling. The rate of cooling is the factor that determines the hardness. A controlled cooling rate is called *quenching*. A mild quench is obtained by cooling in still air, which, as we have seen, is obtained by the normalizing process. The two most widely used media for quench-ing are water and oil. The oil quench is quite slow but prevents quenching cracks caused by rapid expansion of the object being treated. Quenching in water is used for carbon steels and for medium-carbon, low-alloy steels.

The effectiveness of quenching depends upon the fact that when austenite is cooled it does not transform into pearlite instantaneously but requires time to initiate and complete the process. Since the transformation ceases at about 425°C, it can be prevented by rapidly cooling the material to a lower temperature. When the material is cooled rapidly to 204°C or less, the austenite is transformed into a structure called *martensite*. Martensite is a supersaturated solid solution of carbon in ferrite and is the hardest and strongest form of steel.

If steel is rapidly cooled to a temperature between 204 and 425°C and held there for a sufficient length of time, the austenite is transformed into a material that is gen-erally called *bainite*. Bainite is a structure intermediate between pearlite and martens-ite. Although there are several structures that can be identified between the temperatures given, depending upon the temperature used, they are collectively known as bainite. By the choice of this transformation temperature, almost any variation of structure may be obtained. These range all the way from coarse pearlite to fine martensite.

Tempering

When a steel specimen has been fully hardened, it is very hard and brittle and has high residual stresses. The steel is unstable and tends to contract on aging. This ten-dency is increased when the specimen is subjected to externally applied loads, because the resultant stresses contribute still more to the instability. These internal stresses can be relieved by a modest heating process called *stress relieving,* or a combination of stress relieving and softening called *tempering* or *drawing*. After the specimen has been fully hardened by being quenched from above the critical temperature, it is reheated to some temperature below the critical temperature for a certain period of time and then allowed to cool in still air. The temperature to which it is reheated depends upon the composition and the degree of hardness or toughness desired.[8] This

---

[8]For the quantitative aspects of tempering in plain carbon and low-alloy steels, see Charles R. Mischke, "The Strength of Cold-Worked and Heat-Treated Steels," Chap. 33 in Joseph E. Shigley, Charles R. Mischke, and Thomas H. Brown, Jr. (eds.), *Standard Handbook of Machine Design,* 3rd ed., McGraw-Hill, New York, 2004.

Figure 2–13

The effect of thermal-mechanical history on the mechanical properties of AISI 4340 steel. *(Prepared by the International Nickel Company.)*

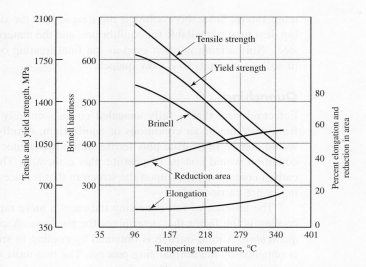

Condition	Tensile strength, MPa	Yield strength, MPa	Reduction in area, %	Elongation in 50 mm, %	Brinell hardness, Bhn
Normalized	1400	1029	20	10	410
As rolled	1330	1008	18	9	380
Annealed	840	693	43	18	228

reheating operation releases the carbon held in the martensite, forming carbide crystals. The structure obtained is called *tempered martensite*. It is now essentially a superfine dispersion of iron carbide(s) in fine-grained ferrite.

The effect of heat-treating operations upon the various mechanical properties of a low alloy steel is shown graphically in Fig. 2–13.

Case Hardening

The purpose of case hardening is to produce a hard outer surface on a specimen of low-carbon steel while at the same time retaining the ductility and toughness in the core. This is done by increasing the carbon content at the surface. Either solid, liquid, or gaseous carburizing materials may be used. The process consists of introducing the part to be carburized into the carburizing material for a stated time and at a stated temperature, depending upon the depth of case desired and the composition of the part. The part may then be quenched directly from the carburization temperature and tempered, or in some cases it must undergo a double heat treatment in order to ensure that both the core and the case are in proper condition. Some of the more useful case-hardening processes are pack carburizing, gas carburizing, nitriding, cyaniding, induction hardening, and flame hardening. In the last two cases carbon is not added to the steel in question, generally a medium carbon steel, for example SAE/AISI 1144.

Quantitative Estimation of Properties of Heat-Treated Steels

Courses in metallurgy (or material science) for mechanical engineers usually present the addition method of Crafts and Lamont for the prediction of heat-treated properties from the Jominy test for plain carbon steels.[9] If this has not been in your prerequisite experience, then refer to the *Standard Handbook of Machine Design,* where the addition method is covered with examples.[10] If this book is a textbook for a machine elements course, it is a good class project (many hands make light work) to study the method and report to the class.

For low-alloy steels, the multiplication method of Grossman[11] and Field[12] is explained in the *Standard Handbook of Machine Design* (Secs. 29.6 and 33.6).

Modern Steels and Their Properties Handbook explains how to predict the Jominy curve by the method of Grossman and Field from a ladle analysis and grain size.[13] Bethlehem Steel has developed a circular plastic slide rule that is convenient to the purpose.

2–15   Alloy Steels

Although a plain carbon steel is an alloy of iron and carbon with small amounts of manganese, silicon, sulfur, and phosphorus, the term *alloy steel* is applied when one or more elements other than carbon are introduced in sufficient quantities to modify its properties substantially. The alloy steels not only possess more desirable physical properties but also permit a greater latitude in the heat-treating process.

Chromium

The addition of chromium results in the formation of various carbides of chromium that are very hard, yet the resulting steel is more ductile than a steel of the same hardness produced by a simple increase in carbon content. Chromium also refines the grain structure so that these two combined effects result in both increased toughness and increased hardness. The addition of chromium increases the critical range of temperatures and moves the eutectoid point to the left. Chromium is thus a very useful alloying element.

Nickel

The addition of nickel to steel also causes the eutectoid point to move to the left and increases the critical range of temperatures. Nickel is soluble in ferrite and does not form carbides or oxides. This increases the strength without decreasing the ductility. Case hardening of nickel steels results in a better core than can be obtained with plain carbon steels. Chromium is frequently used in combination with nickel to obtain the toughness and ductility provided by the nickel and the wear resistance and hardness contributed by the chromium.

Manganese

Manganese is added to all steels as a deoxidizing and desulfurizing agent, but if the sulfur content is low and the manganese content is over 1 percent, the steel is classified as a manganese alloy. Manganese dissolves in the ferrite and also forms carbides.

---

[9]W. Crafts and J. L. Lamont, *Hardenability and Steel Selection,* Pitman and Sons, London, 1949.

[10]Charles R. Mischke, Chap. 33 in Joseph E. Shigley, Charles R. Mischke, and Thomas H. Brown, Jr. (eds.), *Standard Handbook of Machine Design,* 3rd ed., McGraw-Hill, New York, 2004, p. 33.9.

[11]M. A. Grossman, *AIME,* February 1942.

[12]J. Field, *Metals Progress,* March 1943.

[13]*Modern Steels and Their Properties,* 7th ed., Handbook 2757, Bethlehem Steel, 1972, pp. 46–50.

It causes the eutectoid point to move to the left and lowers the critical range of temperatures. It increases the time required for transformation so that oil quenching becomes practicable.

Silicon

Silicon is added to all steels as a deoxidizing agent. When added to very-low-carbon steels, it produces a brittle material with a low hysteresis loss and a high magnetic permeability. The principal use of silicon is with other alloying elements, such as manganese, chromium, and vanadium, to stabilize the carbides.

Molybdenum

While molybdenum is used alone in a few steels, it finds its greatest use when combined with other alloying elements, such as nickel, chromium, or both. Molybdenum forms carbides and also dissolves in ferrite to some extent, so that it adds both hardness and toughness. Molybdenum increases the critical range of temperatures and substantially lowers the transformation point. Because of this lowering of the transformation point, molybdenum is most effective in producing desirable oil-hardening and air-hardening properties. Except for carbon, it has the greatest hardening effect, and because it also contributes to a fine grain size, this results in the retention of a great deal of toughness.

Vanadium

Vanadium has a very strong tendency to form carbides; hence it is used only in small amounts. It is a strong deoxidizing agent and promotes a fine grain size. Since some vanadium is dissolved in the ferrite, it also toughens the steel. Vanadium gives a wide hardening range to steel, and the alloy can be hardened from a higher temperature. It is very difficult to soften vanadium steel by tempering; hence, it is widely used in tool steels.

Tungsten

Tungsten is widely used in tool steels because the tool will maintain its hardness even at red heat. Tungsten produces a fine, dense structure and adds both toughness and hardness. Its effect is similar to that of molybdenum, except that it must be added in greater quantities.

2–16   Corrosion-Resistant Steels

Iron-base alloys containing at least 12 percent chromium are called *stainless steels*. The most important characteristic of these steels is their resistance to many, but not all, corrosive conditions. The four types available are the ferritic chromium steels, the austenitic chromium-nickel steels, and the martensitic and precipitation-hardenable stainless steels.

The ferritic chromium steels have a chromium content ranging from 12 to 27 percent. Their corrosion resistance is a function of the chromium content, so that alloys containing less than 12 percent still exhibit some corrosion resistance, although they may rust. The quench-hardenability of these steels is a function of both the chromium and the carbon content. The very high carbon steels have good quench hardenability up to about 18 percent chromium, while in the lower carbon ranges it ceases at about 13 percent. If a little nickel is added, these steels retain some degree of hardenability up to 20 percent chromium. If the chromium content exceeds 18 percent, they become difficult to weld, and at the very high chromium levels the hardness becomes so great that very careful attention must be paid to the service conditions. Since chromium is expensive, the designer will choose the lowest chromium content consistent with the corrosive conditions.

The chromium-nickel stainless steels retain the austenitic structure at room temperature; hence, they are not amenable to heat treatment. The strength of these steels can be greatly improved by cold working. They are not magnetic unless cold-worked. Their work hardenability properties also cause them to be difficult to machine. All the chromium-nickel steels may be welded. They have greater corrosion-resistant properties than the plain chromium steels. When more chromium is added for greater corrosion resistance, more nickel must also be added if the austenitic properties are to be retained.

2-17   **Casting Materials**

Gray Cast Iron

Of all the cast materials, gray cast iron is the most widely used. This is because it has a very low cost, is easily cast in large quantities, and is easy to machine. The principal objections to the use of gray cast iron are that it is brittle and that it is weak in tension. In addition to a high carbon content (over 1.7 percent and usually greater than 2 percent), cast iron also has a high silicon content, with low percentages of sulfur, manganese, and phosphorus. The resultant alloy is composed of pearlite, ferrite, and graphite, and under certain conditions the pearlite may decompose into graphite and ferrite. The resulting product then contains all ferrite and graphite. The graphite, in the form of thin flakes distributed evenly throughout the structure, darkens it; hence, the name *gray cast iron*.

Gray cast iron is not readily welded, because it may crack, but this tendency may be reduced if the part is carefully preheated. Although the castings are generally used in the as-cast condition, a mild anneal reduces cooling stresses and improves the machinability. The tensile strength of gray cast iron varies from 100 to 400 MPa (15 to 60 kpsi), and the compressive strengths are 3 to 4 times the tensile strengths. The modulus of elasticity varies widely, with values extending all the way from 75 to 150 GPa (11 to 22 Mpsi).

Ductile and Nodular Cast Iron

Because of the lengthy heat treatment required to produce malleable cast iron, engineers have long desired a cast iron that would combine the ductile properties of malleable iron with the ease of casting and machining of gray iron and at the same time would possess these properties in the as-cast conditions. A process for producing such a material using magnesium-containing material seems to fulfill these requirements.

Ductile cast iron, or *nodular cast iron,* as it is sometimes called, is essentially the same as malleable cast iron, because both contain graphite in the form of spheroids. However, ductile cast iron in the as-cast condition exhibits properties very close to those of malleable iron, and if a simple 1-h anneal is given and is followed by a slow cool, it exhibits even more ductility than the malleable product. Ductile iron is made by adding MgFeSi to the melt; since magnesium boils at this temperature, it is necessary to alloy it with other elements before it is introduced.

Ductile iron has a high modulus of elasticity (172 GPa or 25 Mpsi) as compared with gray cast iron, and it is elastic in the sense that a portion of the stress-strain curve is a straight line. Gray cast iron, on the other hand, does not obey Hooke's law, because the modulus of elasticity steadily decreases with increase in stress. Like gray cast iron, however, nodular iron has a compressive strength that is higher than the tensile strength, although the difference is not as great. In 40 years it has become extensively used.

White Cast Iron

If all the carbon in cast iron is in the form of cementite and pearlite, with no graphite present, the resulting structure is white and is known as *white cast iron*. This may be produced in two ways. The composition may be adjusted by keeping the carbon and silicon content low, or the gray-cast-iron composition may be cast against chills in order to promote rapid cooling. By either method, a casting with large amounts of cementite is produced, and as a result the product is very brittle and hard to machine but also very resistant to wear. A chill is usually used in the production of gray-iron castings in order to provide a very hard surface within a particular area of the casting, while at the same time retaining the more desirable gray structure within the remaining portion. This produces a relatively tough casting with a wear-resistant area.

Malleable Cast Iron

If white cast iron within a certain composition range is annealed, a product called *malleable cast iron* is formed. The annealing process frees the carbon so that it is present as graphite, just as in gray cast iron but in a different form. In gray cast iron the graphite is present in a thin flake form, while in malleable cast iron it has a nodular form and is known as *temper carbon*. A good grade of malleable cast iron may have a tensile strength of over 350 MPa (50 kpsi), with an elongation of as much as 18 percent. The percentage elongation of a gray cast iron, on the other hand, is seldom over 1 percent. Because of the time required for annealing (up to 6 days for large and heavy castings), malleable iron is necessarily somewhat more expensive than gray cast iron.

Alloy Cast Irons

Nickel, chromium, and molybdenum are the most common alloying elements used in cast iron. Nickel is a general-purpose alloying element, usually added in amounts up to 5 percent. Nickel increases the strength and density, improves the wearing qualities, and raises the machinability. If the nickel content is raised to 10 to 18 percent, an austenitic structure with valuable heat- and corrosion-resistant properties results. Chromium increases the hardness and wear resistance and, when used with a chill, increases the tendency to form white iron. When chromium and nickel are both added, the hardness and strength are improved without a reduction in the machinability rating. Molybdenum added in quantities up to 1.25 percent increases the stiffness, hardness, tensile strength, and impact resistance. It is a widely used alloying element.

Cast Steels

The advantage of the casting process is that parts having complex shapes can be manufactured at costs less than fabrication by other means, such as welding. Thus the choice of steel castings is logical when the part is complex and when it must also have a high strength. The higher melting temperatures for steels do aggravate the casting problems and require closer attention to such details as core design, section thicknesses, fillets, and the progress of cooling. The same alloying elements used for the wrought steels can be used for cast steels to improve the strength and other mechanical properties. Cast-steel parts can also be heat-treated to alter the mechanical properties, and, unlike the cast irons, they can be welded.

2–18 Nonferrous Metals

Aluminum

The outstanding characteristics of aluminum and its alloys are their strength-weight ratio, their resistance to corrosion, and their high thermal and electrical conductivity. The density of aluminum is about 2770 kg/m^3, compared with 7750 kg/m^3 for steel. Pure aluminum has a tensile strength of about 90 MPa, but this can be improved considerably by cold working and also by alloying with other materials. The modulus of elasticity of aluminum, as well as of its alloys, is 71.7 GPa, which means that it has about one-third the stiffness of steel.

Considering the cost and strength of aluminum and its alloys, they are among the most versatile materials from the standpoint of fabrication. Aluminum can be processed by sand casting, die casting, hot or cold working, or extruding. Its alloys can be machined, press-worked, soldered, brazed, or welded. Pure aluminum melts at 660°C, which makes it very desirable for the production of either permanent or sand-mold castings. It is commercially available in the form of plate, bar, sheet, foil, rod, and tube and in structural and extruded shapes. Certain precautions must be taken in joining aluminum by soldering, brazing, or welding; these joining methods are not recommended for all alloys.

The corrosion resistance of the aluminum alloys depends upon the formation of a thin oxide coating. This film forms spontaneously because aluminum is inherently very reactive. Constant erosion or abrasion removes this film and allows corrosion to take place. An extra-heavy oxide film may be produced by the process called *anodizing*. In this process the specimen is made to become the anode in an electrolyte, which may be chromic acid, oxalic acid, or sulfuric acid. It is possible in this process to control the color of the resulting film very accurately.

The most useful alloying elements for aluminum are copper, silicon, manganese, magnesium, and zinc. Aluminum alloys are classified as *casting alloys* or *wrought alloys*. The casting alloys have greater percentages of alloying elements to facilitate casting, but this makes cold working difficult. Many of the casting alloys, and some of the wrought alloys, cannot be hardened by heat treatment. The alloys that are heat-treatable use an alloying element that dissolves in the aluminum. The heat treatment consists of heating the specimen to a temperature that permits the alloying element to pass into solution, then quenching so rapidly that the alloying element is not precipitated. The aging process may be accelerated by heating slightly, which results in even greater hardness and strength. One of the better-known heat-treatable alloys is duraluminum, or 2017 (4 percent Cu, 0.5 percent Mg, 0.5 percent Mn). This alloy hardens in 4 days at room temperature. Because of this rapid aging, the alloy must be stored under refrigeration after quenching and before forming, or it must be formed immediately after quenching. Other alloys (such as 5053) have been developed that age-harden much more slowly, so that only mild refrigeration is required before forming. After forming, they are artificially aged in a furnace and possess approximately the same strength and hardness as the 2024 alloys. Those alloys of aluminum that cannot be heat-treated can be hardened only by cold working. Both work hardening and the hardening produced by heat treatment may be removed by an annealing process.

Magnesium

The density of magnesium is about 1800 kg/m^3, which is two-thirds that of aluminum and one-fourth that of steel. Since it is the lightest of all commercial metals, its greatest use is in the aircraft and automotive industries, but other uses are now being found for it. Although the magnesium alloys do not have great strength, because of their light weight the strength-weight ratio compares favorably with the stronger aluminum and steel alloys. Even so, magnesium alloys find their greatest use in applications where strength is not an important consideration. Magnesium will not withstand elevated temperatures; the yield point is definitely reduced when the temperature is raised to that of boiling water.

Magnesium and its alloys have a modulus of elasticity of 45 GPa (6.5 Mpsi) in tension and in compression, although some alloys are not as strong in compression as in tension. Curiously enough, cold working reduces the modulus of elasticity. A range of cast magnesium alloys are also available.

Titanium

Titanium and its alloys are similar in strength to moderate-strength steel but weigh half as much as steel. The material exhibits very good resistence to corrosion, has low thermal conductivity, is nonmagnetic, and has high-temperature strength. Its modulus of elasticity is between those of steel and aluminum at 114 GPa (16.4 Mpsi). Because of its many advantages over steel and aluminum, applications include: aerospace and military aircraft structures and components, marine hardware, chemical tanks and processing equipment, fluid handling systems, and human internal replacement devices. The disadvantages of titanium are its high cost compared to steel and aluminum and the difficulty of machining it.

Copper-Base Alloys

When copper is alloyed with zinc, it is usually called *brass*. If it is alloyed with another element, it is often called *bronze*. Sometimes the other element is specified too, as, for example, *tin bronze* or *phosphor bronze*. There are hundreds of variations in each category.

Brass with 5 to 15 Percent Zinc

The low-zinc brasses are easy to cold work, especially those with the higher zinc content. They are ductile but often hard to machine. The corrosion resistance is good. Alloys included in this group are *gilding brass* (5 percent Zn), *commercial bronze* (10 percent Zn), and *red brass* (15 percent Zn). Gilding brass is used mostly for jewelry and articles to be gold-plated; it has the same ductility as copper but greater strength, accompanied by poor machining characteristics. Commercial bronze is used for jewelry and for forgings and stampings, because of its ductility. Its machining properties are poor, but it has excellent cold-working properties. Red brass has good corrosion resistance as well as high-temperature strength. Because of this it is used a great deal in the form of tubing or piping to carry hot water in such applications as radiators or condensers.

Brass with 20 to 36 Percent Zinc

Included in the intermediate-zinc group are *low brass* (20 percent Zn), *cartridge brass* (30 percent Zn), and *yellow brass* (35 percent Zn). Since zinc is cheaper than copper, these alloys cost less than those with more copper and less zinc. They also have better machinability and slightly greater strength; this is offset, however, by poor corrosion resistance and the possibility of cracking at points of residual stresses. Low brass is very similar to red brass and is used for articles requiring deep-drawing operations. Of the copper-zinc alloys, cartridge brass has the best combination of ductility and strength. Cartridge cases were originally manufactured entirely by cold working; the process consisted of a series of deep draws, each draw being followed by an anneal to place the material in condition for the next draw, hence the name cartridge brass. Although the hot-working ability of yellow brass is poor, it can be used in practically any other fabricating process and is therefore employed in a large variety of products.

When small amounts of lead are added to the brasses, their machinability is greatly improved and there is some improvement in their abilities to be hot-worked. The addition of lead impairs both the cold-working and welding properties. In this group are *low-leaded brass* ($32\frac{1}{2}$ percent Zn, $\frac{1}{2}$ percent Pb), *high-leaded brass* (34 percent Zn, 2 percent Pb), and *free-cutting brass* ($35\frac{1}{2}$ percent Zn, 3 percent Pb). The low-leaded brass is not only easy to machine but has good cold-working properties. It is used for various screw-machine parts. High-leaded brass, sometimes called *engraver's brass,* is used for instrument, lock, and watch parts. Free-cutting brass is also used for screw-machine parts and has good corrosion resistance with excellent mechanical properties.

Admiralty metal (28 percent Zn) contains 1 percent tin, which imparts excellent corrosion resistance, especially to saltwater. It has good strength and ductility but only fair machining and working characteristics. Because of its corrosion resistance it is used in power-plant and chemical equipment. *Aluminum brass* (22 percent Zn) contains 2 percent aluminum and is used for the same purposes as admiralty metal, because it has nearly the same properties and characteristics. In the form of tubing or piping, it is favored over admiralty metal, because it has better resistance to erosion caused by high-velocity water.

Brass with 36 to 40 Percent Zinc

Brasses with more than 38 percent zinc are less ductile than cartridge brass and cannot be cold-worked as severely. They are frequently hot-worked and extruded. *Muntz metal* (40 percent Zn) is low in cost and mildly corrosion-resistant. *Naval brass* has the same composition as Muntz metal except for the addition of 0.75 percent tin, which contributes to the corrosion resistance.

Bronze

Silicon bronze, containing 3 percent silicon and 1 percent manganese in addition to the copper, has mechanical properties equal to those of mild steel, as well as good corrosion resistance. It can be hot- or cold-worked, machined, or welded. It is useful wherever corrosion resistance combined with strength is required.

Phosphor bronze, made with up to 11 percent tin and containing small amounts of phosphorus, is especially resistant to fatigue and corrosion. It has a high tensile strength and a high capacity to absorb energy, and it is also resistant to wear. These properties make it very useful as a spring material.

Aluminum bronze is a heat-treatable alloy containing up to 12 percent aluminum. This alloy has strength and corrosion-resistance properties that are better than those of brass, and in addition, its properties may be varied over a wide range by cold working, heat treating, or changing the composition. When iron is added in amounts up to 4 percent, the alloy has a high endurance limit, a high shock resistance, and excellent wear resistance.

Beryllium bronze is another heat-treatable alloy, containing about 2 percent beryllium. This alloy is very corrosion resistant and has high strength, hardness, and resistance to wear. Although it is expensive, it is used for springs and other parts subjected to fatigue loading where corrosion resistance is required.

With slight modification most copper-based alloys are available in cast form.

2-19 Plastics

The term *thermoplastics* is used to mean any plastic that flows or is moldable when heat is applied to it; the term is sometimes applied to plastics moldable under pressure. Such plastics can be remolded when heated.

A *thermoset* is a plastic for which the polymerization process is finished in a hot molding press where the plastic is liquefied under pressure. Thermoset plastics cannot be remolded.

Table 2–2 lists some of the most widely used thermoplastics, together with some of their characteristics and the range of their properties. Table 2–3, listing some of the thermosets, is similar. These tables are presented for information only and should not be used to make a final design decision. The range of properties and characteristics

Table 2-2

The Thermoplastics *Source:* These data have been obtained from the *Machine Design Materials Reference Issue,* published by Penton/IPC, Cleveland. These reference issues are published about every 2 years and constitute an excellent source of data on a great variety of materials.

Name	S_u MPa	E GPa	Hardness Rockwell	Elongation %	Dimensional Stability	Heat Resistance	Chemical Resistance	Processing
ABS group	14–55	0.69–2.55	60–110R	3–50	Good	*	Fair	EMST
Acetal group	55–69	2.83–3.59	80–94M	40–60	Excellent	Good	High	M
Acrylic	34–69	1.38–3.24	92–110M	3–75	High	*	Fair	EMS
Fluoroplastic group	3.4–48	. . .	50–80D	100–300	High	Excellent	Excellent	MPR[†]
Nylon	55–97	1.24–3.10	112–120R	10–200	Poor	Poor	Good	CEM
Phenylene oxide	48–124	2.41–6.34	115R, 106L	5–60	Excellent	Good	Fair	EFM
Polycarbonate	55–110	2.34–5.93	62–91M	10–125	Excellent	Excellent	Fair	EMS
Polyester	55–124	1.93–11.03	65–90M	1–300	Excellent	Poor	Excellent	CLMR
Polyimide	41–345	. . .	88–120M	Very low	Excellent	Excellent	Excellent[†]	CLMP
Polyphenylene sulfide	97–131	0.76	122R	1.0	Good	Excellent	Excellent	M
Polystyrene group	10.3–83	0.97–4.14	10–90M	0.5–60	. . .	Poor	Poor	EM
Polysulfone	69	2.48	120R	50–100	Excellent	Excellent	Excellent[†]	EFM
Polyvinyl chloride	10.3–52	2.41–4.14	65–85D	40–450	. . .	Poor	Poor	EFM

*Heat-resistant grades available.

[†]With exceptions.

C Coatings   L Laminates   R Resins   E Extrusions   M Moldings   S Sheet   F Foams   P Press and sinter methods   T Tubing

Table 2-3

The Thermosets *Source:* These data have been obtained from the *Machine Design Materials Reference Issue,* published by Penton/IPC, Cleveland. These reference issues are published about every 2 years and constitute an excellent source of data on a great variety of materials.

Name	S_u MPa	E GPa	Hardness Rockwell	Elongation %	Dimensional Stability	Heat Resistance	Chemical Resistance	Processing
Alkyd	20–62	0.34–2.07	99M*	. . .	Excellent	Good	Fair	M
Allylic	28–69	. . .	105–120M	. . .	Excellent	Excellent	Excellent	CM
Amino group	34–55	0.90–1.65	110–120M	0.30–0.90	Good	Excellent*	Excellent*	LR
Epoxy	34–138	0.21–2.07*	80–120M	1–10	Excellent	Excellent	Excellent	CMR
Phenolics	34–62	0.69–1.72	70–95E	. . .	Excellent	Excellent	Good	EMR
Silicones	34–41	. . .	80–90M	. . .	. . .	Excellent	Excellent	CLMR

*With exceptions.

C Coatings   L Laminates   R Resins   E Extrusions   M Moldings   S Sheet   F Foams   P Press and sinter methods   T Tubing

that can be obtained with plastics is very great. The influence of many factors, such as cost, moldability, coefficient of friction, weathering, impact strength, and the effect of fillers and reinforcements, must be considered. Manufacturers' catalogs will be found quite helpful in making possible selections.

2-20   Composite Materials[14]

Composite materials are formed from two or more dissimilar materials, each of which contributes to the final properties. Unlike metallic alloys, the materials in a composite remain distinct from each other at the macroscopic level.

Most engineering composites consist of two materials: a reinforcement called a *filler* and a *matrix.* The filler provides stiffness and strength; the matrix holds the material together and serves to transfer load among the discontinuous reinforcements. The most common reinforcements, illustrated in Fig. 2–14, are continuous fibers, either straight or woven, short chopped fibers, and particulates. The most common matrices are various plastic resins although other materials including metals are used.

Figure 2-14

Composites categorized by type of reinforcement.

| Particulate composite | Randomly oriented short fiber composite | Unidirectional continuous fiber composite | Woven fabric composite |

[14]For references see I. M. Daniel and O. Ishai, *Engineering Mechanics of Composite Materials,* Oxford University Press, 1994, and *ASM Engineered Materials Handbook: Composites,* ASM International, Materials Park, OH, 1988.

Metals and other traditional engineering materials are uniform, or isotropic, in nature. This means that material properties, such as strength, stiffness, and thermal conductivity, are independent of both position within the material and the choice of coordinate system. The discontinuous nature of composite reinforcements, though, means that material properties can vary with both position and direction. For example, an epoxy resin reinforced with continuous graphite fibers will have very high strength and stiffness in the direction of the fibers, but very low properties normal or transverse to the fibers. For this reason, structures of composite materials are normally constructed of multiple plies (laminates) where each ply is oriented to achieve optimal structural stiffness and strength performance.

High strength-to-weight ratios, up to 5 times greater than those of high-strength steels, can be achieved. High stiffness-to-weight ratios can also be obtained, as much as 8 times greater than those of structural metals. For this reason, composite materials are becoming very popular in automotive, marine, aircraft, and spacecraft applications where weight is a premium.

The directionality of properties of composite materials increases the complexity of structural analyses. Isotropic materials are fully defined by two engineering constants: Young's modulus E and Poisson's ratio v. A single ply of a composite material, however, requires four constants, defined with respect to the ply coordinate system. The constants are two Young's moduli (the longitudinal modulus in the direction of the fibers, E_1, and the transverse modulus normal to the fibers, E_2), one Poisson's ratio (v_{12}, called the major Poisson's ratio), and one shear modulus (G_{12}). A fifth constant, the minor Poisson's ratio, v_{21}, is determined through the reciprocity relation, $v_{21}/E_2 = v_{12}/E_1$. Combining this with multiple plies oriented at different angles makes structural analysis of complex structures unapproachable by manual techniques. For this reason, computer software is available to calculate the properties of a laminated composite construction.[15]

2–21   Materials Selection

As stated earlier, the selection of a material for a machine part or structural member is one of the most important decisions the designer is called on to make. Up to this point in this chapter we have discussed many important material physical properties, various characteristics of typical engineering materials, and various material production processes. The actual selection of a material for a particular design application can be an easy one, say, based on previous applications (1020 steel is always a good candidate because of its many positive attributes), or the selection process can be as involved and daunting as any design problem with the evaluation of the many material physical, economical, and processing parameters. There are systematic and optimizing approaches to material selection. Here, for illustration, we will only look at how to approach some material properties. One basic technique is to list all the important material properties associated with the design, e.g., strength, stiffness, and cost. This can be prioritized by using a weighting measure depending on what properties are more important than others. Next, for each property, list all available materials and rank them in order beginning with the best material; e.g., for strength, high-strength steel such as 4340 steel should be near the top of the list. For completeness of available materials, this might require a large source of material data. Once the lists are

---

[15]About Composite Materials Software listing, http://composite.about.com/cs/software/index.htm.

formed, select a manageable amount of materials from the top of each list. From each reduced list select the materials that are contained within every list for further review. The materials in the reduced lists can be graded within the list and then weighted according to the importance of each property.

M. F. Ashby has developed a powerful systematic method using *materials selection charts.*[16] This method has also been implemented in a software package called CES Edupack.[17] The charts display data of various properties for the families and classes of materials listed in Table 2–4. For example, considering material stiffness

Table 2–4

Material Families and Classes

Family	Classes	Short Name
Metals (the metals and alloys of engineering)	Aluminum alloys Copper alloys Lead alloys Magnesium alloys Nickel alloys Carbon steels Stainless steels Tin alloys Titanium alloys Tungsten alloys Lead alloys Zinc alloys	Al alloys Cu alloys Lead alloys Mg alloys Ni alloys Steels Stainless steels Tin alloys Ti alloys W alloys Pb alloys Zn alloys
Ceramics Technical ceramics (fine ceramics capable of load-bearing application)	Alumina Aluminum nitride Boron carbide Silicon carbide Silicon nitride Tungsten carbide	Al_2O_3 AlN B_4C SiC Si_3N_4 WC
Nontechnical ceramics (porous ceramics of construction)	Brick Concrete Stone	Brick Concrete Stone
Glasses	Soda-lime glass Borosilicate glass Silica glass Glass ceramic	Soda-lime glass Borosilicate glass Silica glass Glass ceramic
Polymers (the thermoplastics and thermosets of engineering)	Acrylonitrile butadiene styrene Cellulose polymers Ionomers Epoxies Phenolics Polyamides (nylons) Polycarbonate	ABS CA Ionomers Epoxy Phenolics PA PC

(Continued)

---

[16]M. F. Ashby, *Materials Selection in Mechanical Design,* 3rd ed., Elsevier Butterworth-Heinemann, Oxford, 2005.

[17]Produced by Granta Design Limited. See www.grantadesign.com.

Table 2–4

(Continued)

Family	Classes	Short Name
Polymers (*continued*)	Polyesters	Polyester
	Polyetheretherkeytone	PEEK
	Polyethylene	PE
	Polyethylene terephalate	PET or PETE
	Polymethylmethacrylate	PMMA
	Polyoxymethylene(Acetal)	POM
	Polypropylene	PP
	Polystyrene	PS
	Polytetrafluorethylene	PTFE
	Polyvinylchloride	PVC
Elastomers	Butyl rubber	Butyl rubber
(engineering rubbers,	EVA	EVA
natural and synthetic)	Isoprene	Isoprene
	Natural rubber	Natural rubber
	Polychloroprene (Neoprene)	Neoprene
	Polyurethane	PU
	Silicon elastomers	Silicones
Hybrids	Carbon-fiber reinforced polymers	CFRP
Composites	Glass-fiber reinforced polymers	GFRP
	SiC reinforced aluminum	Al-SiC
Foams	Flexible polymer foams	Flexible foams
	Rigid polymer foams	Rigid foams
Natural materials	Cork	Cork
	Bamboo	Bamboo
	Wood	Wood

From M. F. Ashby, *Materials Selection in Mechanical Design,* 3rd ed., Elsevier Butterworth-Heinemann, Oxford, 2005. Table 4–1, pp. 49–50.

properties, a simple bar chart plotting Young's modulus E on the y axis is shown in Fig. 2–15. Each vertical line represents the range of values of E for a particular material. Only some of the materials are labeled. Now, more material information can be displayed if the x axis represents another material property, say density. Figure 2–16, called a "bubble" chart, represents Young's modulus E plotted against density ρ. The line ranges for each material property plotted two-dimensionally now form ellipses, or bubbles. Groups of bubbles outlined according to the material families of Table 2–4 are also shown. This plot is more useful than the two separate bar charts of each property. Now, we also see how stiffness/weight for various materials relate. The ratio of Young's modulus to density, E/ρ, is known as the *specific modulus,* or *specific stiffness.* This ratio is of particular interest when it is desired to minimize weight where the primary design limitation is deflection, stiffness, or natural frequency, rather than strength. Machine parts made from materials with higher specific modulus will exhibit lower deflection, higher stiffness, and higher natural frequency.

In the lower right corner of the chart in Figure 2–16, dotted lines indicate ratios of E^{β}/ρ. Several parallel dotted lines are shown for $\beta = 1$ that represent different

Figure 2–15

Young's modulus E for various materials. *(Figure courtesy of Prof. Mike Ashby, Granta Design, Cambridge, U.K.)*

values of the specific modulus E/ρ. This allows simple comparison of the specific modulus between materials. It can be seen, for example, that some woods and aluminum alloys have about the same specific modulus as steels. Different values of β allow comparisons for various relationships between stiffness and weight, such as in different loading conditions. The relationship is linear ($\beta = 1$) for axial loading, but nonlinear ($\beta = 1/2$) for bending loading [see Eq. (2–31) and its development]. Since the plot is on a log-log scale, the exponential functions still plot as straight lines. The $\beta = 1$ lines can also be used to represent constant values of the speed of sound in a material, since the relationship between E and ρ is linear in the equation for the speed of sound in a material, $c = (E/\rho)^{1/2}$. The same can be shown for natural frequency, which is a function of the ratio of stiffness to mass.

To see how β fits into the mix, consider the following. The performance metric P of a structural element depends on (1) the functional requirements, (2) the geometry, and (3) the material properties of the structure. That is,

$$P = \left[\begin{pmatrix} \text{functional} \\ \text{requirements } F \end{pmatrix}, \begin{pmatrix} \text{geometric} \\ \text{parameters } G \end{pmatrix}, \begin{pmatrix} \text{material} \\ \text{properties } M \end{pmatrix} \right]$$

Figure 2–16

Young's modulus E versus density ρ for various materials. *(Figure courtesy of Prof. Mike Ashby, Granta Design, Cambridge, U.K.)*

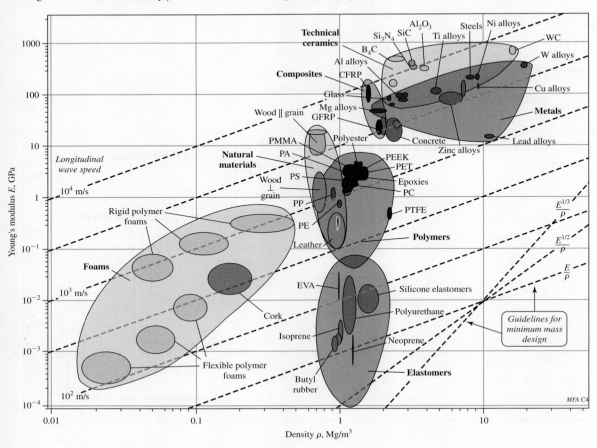

or, symbolically,

$$P = f(F, G, M) \tag{2-23}$$

If the function is *separable*, which it often is, we can write Eq. (2–23) as

$$P = f_1(F) \cdot f_2(G) \cdot f_3(M) \tag{2-24}$$

For optimum design, we desire to maximize or minimize P. With regard to material properties alone, this is done by maximizing or minimizing $f_3(M)$, called the *material efficiency coefficient*.

For illustration, say we want to design a light, stiff, end-loaded cantilever beam with a circular cross section. For this we will use the mass m of the beam for the performance metric to minimize. The stiffness of the beam is related to its material and geometry. The stiffness of a beam is given by $k = F/\delta$, where F and δ are the end load and deflection, respectively (see Chap. 4). The end deflection of an end-loaded cantilever beam is given in Table A–9, beam 1, as $\delta = y_{max} = (Fl^3)/(3EI)$, where E is Young's modulus, I the second moment of the area, and l the length of the beam. Thus, the stiffness is given by

$$k = \frac{F}{\delta} = \frac{3EI}{l^3} \tag{2–25}$$

From Table A–18, the second moment of the area of a circular cross section is

$$I = \frac{\pi D^4}{64} = \frac{A^2}{4\pi} \tag{2–26}$$

where D and A are the diameter and area of the cross section, respectively. Substituting Eq. (2–26) in (2–25) and solving for A, we obtain

$$A = \left(\frac{4\pi kl^3}{3E}\right)^{1/2} \tag{2–27}$$

The mass of the beam is given by

$$m = Al\rho \tag{2–28}$$

Substituting Eq. (2–27) into (2–28) and rearranging yields

$$m = 2\sqrt{\frac{\pi}{3}}(k^{1/2})(l^{5/2})\left(\frac{\rho}{E^{1/2}}\right) \tag{2–29}$$

Equation (2–29) is of the form of Eq. (2–24). The term $2\sqrt{\pi/3}$ is simply a constant and can be associated with any function, say $f_1(F)$. Thus, $f_1(F) = 2\sqrt{\pi/3}(k^{1/2})$ is the functional requirement, stiffness; $f_2(G) = (l^{5/2})$, the geometric parameter, length; and the material efficiency coefficient

$$f_3(M) = \frac{\rho}{E^{1/2}} \tag{2–30}$$

is the material property in terms of density and Young's modulus. To minimize m we want to minimize $f_3(M)$, or maximize

$$M = \frac{E^{1/2}}{\rho} \tag{2–31}$$

where M is called the *material index,* and $\beta = \frac{1}{2}$. Returning to Fig. 2–16, draw lines of various values of $E^{1/2}/\rho$ as shown in Fig. 2–17. Lines of increasing M move up and to the left as shown. Thus, we see that good candidates for a light, stiff, end-loaded cantilever beam with a circular cross section are certain woods, composites, and ceramics.

Other limits/constraints may warrant further investigation. Say, for further illustration, the design requirements indicate that we need a Young's modulus greater than 50 GPa. Figure 2–18 shows how this further restricts the search region. This eliminates woods as a possible material.

Another commonly useful chart, shown in Fig. 2–19, represents strength versus density for the material families. The ratio of strength to density is known as *specific strength,* and is particularly useful when it is desired to minimize weight where the primary design limitation is strength, rather than deflection. The guidelines in the

Figure 2–17

A schematic E versus ρ chart showing a grid of lines for various values the material index $M = E^{1/2}/\rho$. (From M. F. Ashby, Materials Selection in Mechanical Design, 3rd ed., Elsevier Butterworth-Heinemann, Oxford, 2005.)

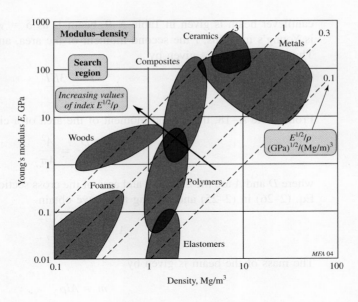

Figure 2–18

The search region of Fig. 2–16 further reduced by restricting $E \geq 50$ GPa, (From M. F. Ashby, Materials Selection in Mechanical Design, 3rd ed., Elsevier Butterworth-Heinemann, Oxford, 2005.)

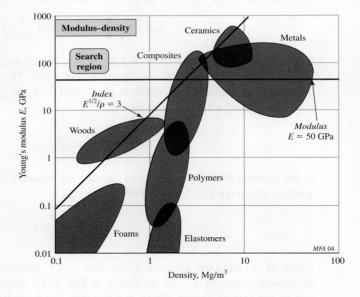

lower right corner represent different relationships between strength and density, in the form of S^β/ρ. Following an approach similar to that used before, it can be shown that for axial loading, $\beta = 1$, and for bending loading, $\beta = 2/3$.

Certainly, in a given design exercise, there will be other considerations such as environment, cost, availability, and machinability, and other charts may be necessary to investigate. Also, we have not brought in the material process selection part of the picture. If done properly, material selection can result in a good deal of bookkeeping. This is where software packages such as CES Edupack become very effective.

Figure 2–19

Strength S versus density ρ for various materials. For *metals*, S is the 0.2 percent offset yield strength. For *polymers*, S is the 1 percent yield strength. For *ceramics and glasses*, S is the compressive crushing strength. For *composites*, S is the tensile strength. For *elastomers*, S is the tear strength. (*Figure courtesy of Prof. Mike Ashby, Granta Design, Cambridge, U.K.*)

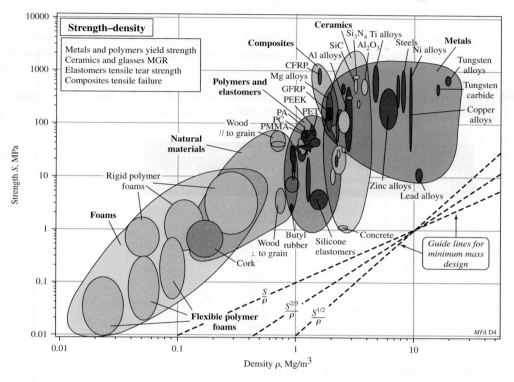

PROBLEMS

2-1 Determine the tensile and yield strengths for the following materials:
(*a*) UNS G10200 hot-rolled steel.
(*b*) SAE 1050 cold-drawn steel.
(*c*) AISI 1141 steel quenched and tempered at 540°C.
(*d*) 2024-T4 aluminum alloy.
(*e*) Ti-6Al-4V annealed titanium alloy.

2-2 Assume you were specifying an AISI 1060 steel for an application. Using Table A–21,
(*a*) how would you specify it if you desired to maximize the yield strength?
(*b*) how would you specify it if you desired to maximize the ductility?

2-3 Determine the yield strength-to-density ratios (specific strength) in units of kN · m/kg for AISI 1018 CD steel, 2011-T6 aluminum, Ti-6Al-4V titanium alloy, and ASTM No. 40 gray cast iron.

2-4 Determine the stiffness-to-weight density ratios (specific modulus) in units of meters for AISI 1018 CD steel, 2011-T6 aluminum, Ti-6Al-4V titanium alloy, and ASTM No. 40 gray cast iron.

2-5 Poisson's ratio v is a material property and is the ratio of the lateral strain and the longitudinal strain for a member in tension. For a homogeneous, isotropic material, the modulus of rigidity G is related to Young's modulus as

$$G = \frac{E}{2(1 + v)}$$

Using the tabulated values of G and E in Table A–5, calculate Poisson's ratio for steel, aluminum, beryllium copper, and gray cast iron. Determine the percent difference between the calculated values and the values tabulated in Table A–5.

2-6 A specimen of steel having an initial diameter of 12.8 mm was tested in tension using a gauge length of 50 mm. The following data were obtained for the elastic and plastic states:

Elastic State		Plastic State	
Load P kN	Elongation mm	Load P kN	Area A_i mm^2
4.5	0.01	39.6	128
9.0	0.015	41.4	127
13.5	0.025	41.0	126
18.0	0.0325	59.4	124
31.5	0.0575	68.4	121
37.8	0.07	76.5	101
39.6	0.09	73.8	84
41.4	0.2225	66.6	70

Note that there is some overlap in the data.
(a) Plot the engineering or nominal stress-strain diagram using two scales for the unit strain ϵ, one scale from zero to about 0.02 mm/mm and the other scale from zero to maximum strain.
(b) From this diagram find the modulus of elasticity, the 0.2 percent offset yield strength, the ultimate strength, and the percent reduction in area.
(c) Characterize the material as ductile or brittle. Explain your reasoning.
(d) Identify a material specification from Table A–20 that has a reasonable match to the data.

2-7 Compute the true stress and the logarithmic strain using the data of Prob. 2–6 and plot the results on log-log paper. Then find the plastic strength coefficient σ_0 and the strain-strengthening exponent m. Find also the yield strength and the ultimate strength after the specimen has had 20 percent cold work.

2-8 The stress-strain data from a tensile test on a cast-iron specimen are

Engineering stress, MPa	35	70	112	133	182	224	280	332	343	378
Engineering strain, $\epsilon \cdot 10^{-3}$ mm/mm	0.20	0.44	0.80	1.0	1.5	2.0	2.8	3.4	4.0	5.0

Plot the stress-strain locus and find the 0.1 percent offset yield strength, and the tangent modulus of elasticity at zero stress and at 140 MPa.

2-9 A part made from annealed AISI 1018 steel undergoes a 20 percent cold-work operation.
(a) Obtain the yield strength and ultimate strength before and after the cold-work operation. Determine the percent increase in each strength.
(b) Determine the ratios of ultimate strength to yield strength before and after the cold-work operation. What does the result indicate about the change of ductility of the part?

2-10 Repeat Prob. 2–9 for a part made from hot-rolled AISI 1212 steel.

2-11 Repeat Prob. 2–9 for a part made from 2024-T4 aluminum alloy.

2-12 A steel member has a Brinell of $H_B = 275$. Estimate the ultimate strength of the steel in MPa.

2-13 A gray cast iron part has a Brinell hardness number of $H_B = 200$. Estimate the ultimate strength of the part in MPa. Make a reasonable assessment of the likely grade of cast iron by comparing both hardness and strength to material options in Table A–24.

2-14 A part made from 1040 hot-rolled steel is to be heat treated to increase its strength to approximately 700 MPa. What Brinell hardness number should be expected from the heat-treated part?

2-15 Brinell hardness tests were made on a random sample of 10 steel parts during processing. The results were H_B values of 230, 232(2), 234, 235(3), 236(2), and 239. Estimate the mean and standard deviation of the ultimate strength in MPa.

2-16 Repeat Prob. 2–15 assuming the material to be cast iron.

2-17 For the material in Prob. 2–6: (a) Determine the modulus of resilience, and (b) Estimate the modulus of toughness, assuming that the last data point corresponds to fracture.

2-18 Some commonly used plain carbon steels are AISI 1010, 1018, and 1040. Research these steels and provide a comparative summary of their characteristics, focusing on aspects that make each one unique for certain types of application. Product application guides provided on the Internet by steel manufacturers and distributors are one source of information.

2-19 Repeat Prob. 2–18 for the commonly used alloy steels, AISI 4130 and 4340.

2-20 An application requires the support of an axial load of 400 kN with a round rod without exceeding the yield strength of the material. Assume the current cost per pound for round stock is given in the table below for several materials that are being considered. Material properties are available in Tables A–5, A–20, A–21, and A–24. Select one of the materials for each of the following additional design goals.
(a) Minimize diameter.
(b) Minimize weight.
(c) Minimize cost.
(d) Minimize axial deflection.

Material	Cost/N
1020 HR	$0.06
1020 CD	$0.07
1040 Q&T @800°F	$0.08
4140 Q&T @800°F	$0.18
Wrought Al 2024 T3	$0.24
Titanium alloy (Ti-6Al-4V)	$1.56

2–21 to
2–23

A 25-mm-diameter rod, 1 m long, of unknown material is found in a machine shop. A variety of inexpensive nondestructive tests are readily available to help determine the material, as described below:

(*a*) Visual inspection.

(*b*) Scratch test: Scratch the surface with a file; observe color of underlying material and depth of scratch.

(*c*) Check if it is attracted to a magnet.

(*d*) Measure weight (± 0.25 N).

(*e*) Inexpensive bending deflection test: Clamp one end in a vise, leaving 0.6 m cantilevered. Apply a force of 450 N (± 4.5 N). Measure deflection of the free end (within ± 0.8 mm).

(*f*) Brinell hardness test.

Choose which tests you would actually perform, and in what sequence, to minimize time and cost, but to determine the material with a reasonable level of confidence. The table below provides results that would be available to you if you choose to perform a given test. Explain your process, and include any calculations. You may assume the material is one listed in Table A–5. If it is carbon steel, try to determine an approximate specification from Table A–20.

Test	Results if test were made		
	Prob. 2–21	**Prob. 2–22**	**Prob. 2–23**
(*a*)	Dark gray, rough surface finish, moderate scale	Silvery gray, smooth surface finish, slightly tarnished	Reddish-brown, tarnished, smooth surface finish
(*b*)	Metallic gray, moderate scratch	Silvery gray, deep scratch	Shiny brassy color, deep scratch
(*c*)	Magnetic	Not magnetic	Not magnetic
(*d*)	$W = 36$ N	$W = 14$ N	$W = 40$ N
(*e*)	$\delta = 8$ mm	$\delta = 22$ mm	$\delta = 14$ mm
(*f*)	$H_B = 200$	$H_B = 95$	$H_B = 70$

2–24 Search the website noted in Sec. 2–20 (http://composite.about.com/cs/software/) and report your findings. Your instructor may wish to elaborate on the level of this report. The website contains a large variety of resources. The activity for this problem can be divided among the class.

2–25 Research the material Inconel, briefly described in Table A–5. Compare it to various carbon and alloy steels in stiffness, strength, ductility, and toughness. What makes this material so special?

2–26 Consider a rod transmitting a tensile force. The following materials are being considered: tungsten carbide, zinc alloy, polycarbonate polymer, and aluminum alloy. Using the Ashby charts, recommend the best material for a design situation in which failure is by exceeding the strength of the material, and it is desired to minimize the weight.

2–27 Repeat Prob. 2–26, except that the design situation is failure by excessive deflection, and it is desired to minimize the weight.

2–28 Consider a cantilever beam that is loaded with a transverse force at its tip. The following materials are being considered: tungsten carbide, high-carbon heat-treated steel, polycarbonate polymer, and aluminum alloy. Using the Ashby charts, recommend the best material for a design situation in which failure is by exceeding the strength of the material and it is desired to minimize the weight.

2–29   Repeat Prob. 2–28, except that the design situation is failure by excessive deflection, and it is desired to minimize the weight.

2–30   For an axially loaded rod, prove that $\beta = 1$ for the E^{β}/ρ guidelines in Fig. 2–16.

2–31   For an axially loaded rod, prove that $\beta = 1$ for the S^{β}/ρ guidelines in Fig. 2–19.

2–32   For a cantilever beam loaded in bending, prove that $\beta = 1/2$ for the E^{β}/ρ guidelines in Fig. 2–16.

2–33   For a cantilever beam loaded in bending, prove that $\beta = 2/3$ for the S^{β}/ρ guidelines in Fig. 2–19.

2–34   Consider a tie rod transmitting a tensile force F. The corresponding tensile stress is given by $\sigma = F/A$, where A is the area of the cross section. The deflection of the rod is given by Eq. (4–3), which is $\delta = (Fl)/(AE)$, where l is the length of the rod. Using the Ashby charts of Figs. 2–16 and 2–19, explore what ductile materials are best suited for a light, stiff, *and* strong tie rod. *Hint:* Consider stiffness and strength separately.

2–35   Repeat Prob. 1–13. Does the data reflect the number found in part (*b*)? If not, why? Plot a histogram of the data. Presuming the distribution is normal, plot Eq. (1–4) and compare it with the histogram.

3

Load and Stress Analysis

Chapter Outline

One of the main objectives of this book is to describe how specific machine components function and how to design or specify them so that they function safely without failing structurally. Although earlier discussion has described structural strength in terms of load or stress versus strength, failure of function for structural reasons may arise from other factors such as excessive deformations or deflections.

Here it is assumed that the reader has completed basic courses in statics of rigid bodies and mechanics of materials and is quite familiar with the analysis of loads, and the stresses and deformations associated with the basic load states of simple prismatic elements. In this chapter and Chap. 4 we will review and extend these topics briefly. Complete derivations will not be presented here, and the reader is urged to return to basic textbooks and notes on these subjects.

This chapter begins with a review of equilibrium and free-body diagrams associated with load-carrying components. One must understand the nature of forces before attempting to perform an extensive stress or deflection analysis of a mechanical component. An extremely useful tool in handling discontinuous loading of structures employs *Macaulay* or *singularity functions*. Singularity functions are described in Sec. 3–3 as applied to the shear forces and bending moments in beams. In Chap. 4, the use of singularity functions will be expanded to show their real power in handling deflections of complex geometry and statically indeterminate problems.

Machine components transmit forces and motion from one point to another. The transmission of force can be envisioned as a flow or force distribution that can be further visualized by isolating internal surfaces within the component. Force distributed over a surface leads to the concept of stress, stress components, and stress transformations (Mohr's circle) for all possible surfaces at a point.

The remainder of the chapter is devoted to the stresses associated with the basic loading of prismatic elements, such as uniform loading, bending, and torsion, and topics with major design ramifications such as stress concentrations, thin- and thick-walled pressurized cylinders, rotating rings, press and shrink fits, thermal stresses, curved beams, and contact stresses.

3–1 Equilibrium and Free-Body Diagrams

Equilibrium

The word *system* will be used to denote any *isolated* part or portion of a machine or structure—including all of it if desired—that we wish to study. A system, under this definition, may consist of a particle, several particles, a part of a rigid body, an entire rigid body, or even several rigid bodies.

If we assume that the system to be studied is motionless or, at most, has constant velocity, then the system has zero acceleration. Under this condition the system is said to be in *equilibrium*. The phrase *static equilibrium* is also used to imply that the system is *at rest*. For equilibrium, the forces and moments acting on the system balance such that

$$\sum \mathbf{F} = 0 \qquad\qquad (3\text{–}1)$$

$$\sum \mathbf{M} = 0 \qquad\qquad (3\text{–}2)$$

which states that *the sum of all forces* and the *sum of all moment vectors* acting upon a system in equilibrium is zero.

Free-Body Diagrams

We can greatly simplify the analysis of a very complex structure or machine by successively isolating each element and studying and analyzing it by the use of *free-body diagrams*. When all the members have been treated in this manner, the knowledge obtained can be assembled to yield information concerning the behavior of the total system. Thus, free-body diagramming is essentially a means of breaking a complicated problem into manageable segments, analyzing these simple problems, and then, usually, putting the information together again.

Using free-body diagrams for force analysis serves the following important purposes:

- The diagram establishes the directions of reference axes, provides a place to record the dimensions of the subsystem and the magnitudes and directions of the known forces, and helps in assuming the directions of unknown forces.

- The diagram simplifies your thinking because it provides a place to store one thought while proceeding to the next.

- The diagram provides a means of communicating your thoughts clearly and unambiguously to other people.

- Careful and complete construction of the diagram clarifies fuzzy thinking by bringing out various points that are not always apparent in the statement or in the geometry of the total problem. Thus, the diagram aids in understanding all facets of the problem.

- The diagram helps in the planning of a logical attack on the problem and in setting up the mathematical relations.

- The diagram helps in recording progress in the solution and in illustrating the methods used.

- The diagram allows others to follow your reasoning, showing *all* forces.

EXAMPLE 3–1

Figure 3–1*a* shows a simplified rendition of a gear reducer where the input and output shafts *AB* and *CD* are rotating at constant speeds ω_i and ω_o, respectively. The input and output torques (torsional moments) are $T_i = 28$ N · m and T_o, respectively. The shafts are supported in the housing by bearings at *A*, *B*, *C*, and *D*. The pitch radii of gears G_1 and G_2 are $r_1 = 20$ mm and $r_2 = 40$ mm, respectively. Draw the free-body diagrams of each member and determine the net reaction forces and moments at all points.

Solution

First, we will list all simplifying assumptions.

1   Gears G_1 and G_2 are simple spur gears with a standard pressure angle $\phi = 20°$ (see Sec. 13–5).

2   The bearings are self-aligning and the shafts can be considered to be simply supported.

3   The weight of each member is negligible.

4   Friction is negligible.

5   The mounting bolts at *E*, *F*, *H*, and *I* are the same size.

The separate free-body diagrams of the members are shown in Figs. 3–1*b–d*. Note that Newton's third law, called *the law of action and reaction*, is used extensively where each member mates. The force transmitted between the spur gears is not tangential but at the pressure angle ϕ. Thus, $N = F \tan \phi$.

(a) Gear reducer

(b) Gear box

(c) Input shaft

(d) Output shaft

Figure 3–1

(a) Gear reducer; (b–d) free-body diagrams. Diagrams are not drawn to scale.

Summing moments about the x axis of shaft AB in Fig. 3–1d gives

$$\sum M_x = F(0.02) - 28 = 0$$

$$F = 1400 \text{ N}$$

The normal force is $N = 1400 \text{ N} \tan 20° = 509.6 \text{ N}$.

Using the equilibrium equations for Figs. 3–1c and d, the reader should verify that: $R_{Ay} = 861.5$ N, $R_{Az} = 313.6$ N, $R_{By} = 538.5$ N, $R_{Bz} = 196$ N, $R_{Cy} = 861.5$ N, $R_{Cz} = 313.6$ N, $R_{Dy} = 538.5$ N, $R_{Dz} = 196$ N, and $T_o = 56$ N · m. The direction of the output torque T_o is opposite ω_o because it is the resistive load on the system opposing the motion ω_o.

Note in Fig. 3–1b the net force from the bearing reactions is zero whereas the net moment about the x axis is 0.06 (861.5) + 0.06 (538.5) = 84 N · m. This value is the same as $T_i + T_o = 28 + 56 = 84$ N · m, as shown in Fig. 3–1a. The reaction forces R_E, R_F, R_H, and R_I, from the mounting bolts cannot be determined from the equilibrium equations as there are too many unknowns. Only three equations are available, $\sum F_y = \sum F_z = \sum M_x = 0$. In case you were wondering about assumption 5, here is where we will use it (see Sec. 8–12). The gear box tends to rotate about the x axis because of a pure torsional moment of 84 N · m. The bolt forces must provide an equal but opposite torsional moment. The center of rotation relative to the bolts lies at the centroid of the bolt cross-sectional areas. Thus if the bolt areas are equal: the center of rotation is at the center of the four bolts, a distance of $\sqrt{(100/2)^2 + (125/2)^2} = 80$ mm from each bolt; the bolt forces are equal ($R_E = R_F = R_H = R_I = R$), and each bolt force is perpendicular to the line from the bolt to the center of rotation. This gives a net torque from the four bolts of $4R\,(0.08) = 84$. Thus, $R_E = R_F = R_H = R_I = 262.5$ N.

3–2   Shear Force and Bending Moments in Beams

Figure 3–2a *shows a beam supported by* reactions R_1 and R_2 and loaded by the concentrated forces F_1, F_2, and F_3. If the beam is cut at some section located at $x = x_1$ and the left-hand portion is removed as a free body, an *internal shear force V* and *bending moment M* must act on the cut surface to ensure equilibrium (see Fig. 3–2b). The shear force is obtained by summing the forces on the isolated section. The bending moment is the sum of the moments of the forces to the left of the section taken about an axis through the isolated section. The sign conventions used for bending moment and shear force in this book are shown in Fig. 3–3. Shear force and bending moment are related by the equation

$$V = \frac{dM}{dx} \tag{3–3}$$

Sometimes the bending is caused by a distributed load $q(x)$, as shown in Fig. 3–4; $q(x)$ is called the *load intensity* with units of force per unit length and is

Figure 3–2

Free-body diagram of simply-supported beam with V and M shown in positive directions (established by the conventions shown in Fig. 3–3).

Figure 3–3

Sign conventions for bending and shear.

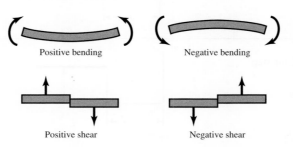

Figure 3–4

Distributed load on beam.

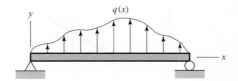

positive in the positive y direction. It can be shown that differentiating Eq. (3–3) results in

$$\frac{dV}{dx} = \frac{d^2M}{dx^2} = q \tag{3–4}$$

Normally the applied distributed load is directed downward and labeled w (e.g., see Fig. 3–6). In this case, $w = -q$.

Equations (3–3) and (3–4) reveal additional relations if they are integrated. Thus, if we integrate between, say, x_A and x_B, we obtain

$$\int_{V_A}^{V_B} dV = V_B - V_A = \int_{x_A}^{x_B} q \, dx \tag{3–5}$$

which states that *the change in shear force from A to B is equal to the area of the loading diagram between x_A and x_B.*

In a similar manner,

$$\int_{M_A}^{M_B} dM = M_B - M_A = \int_{x_A}^{x_B} V \, dx \tag{3–6}$$

which states that *the change in moment from A to B is equal to the area of the shear-force diagram between x_A and x_B.*

Table 3–1

Singularity (Macaulay[†]) Functions

Function	Graph of $f_n(x)$	Meaning
Concentrated moment (unit doublet)	$\langle x - a \rangle^{-2}$	$\langle x - a \rangle^{-2} = 0 \quad x \neq a$ $\langle x - a \rangle^{-2} = \pm \infty \quad x = a$ $\int \langle x - a \rangle^{-2} dx = \langle x - a \rangle^{-1}$
Concentrated force (unit impulse)	$\langle x - a \rangle^{-1}$	$\langle x - a \rangle^{-1} = 0 \quad x \neq a$ $\langle x - a \rangle^{-1} = +\infty \quad x = a$ $\int \langle x - a \rangle^{-1} dx = \langle x - a \rangle^{0}$
Unit step	$\langle x - a \rangle^{0}$	$\langle x - a \rangle^{0} = \begin{cases} 0 & x < a \\ 1 & x \geq a \end{cases}$ $\int \langle x - a \rangle^{0} dx = \langle x - a \rangle^{1}$
Ramp	$\langle x - a \rangle^{1}$	$\langle x - a \rangle^{1} = \begin{cases} 0 & x < a \\ x - a & x \geq a \end{cases}$ $\int \langle x - a \rangle^{1} dx = \dfrac{\langle x - a \rangle^{2}}{2}$

[†]W. H. Macaulay, "Note on the deflection of beams," *Messenger of Mathematics,* vol. 48, pp. 129–130, 1919.

3–3   Singularity Functions

The four singularity functions defined in Table 3–1, using the *angle brackets* $\langle\ \rangle$, constitute a useful and easy means of integrating across discontinuities. By their use, general expressions for shear force and bending moment in beams can be written when the beam is loaded by concentrated moments or forces. As shown in the table, the concentrated moment and force functions are zero for all values of x not equal to a. The functions are undefined for values of $x = a$. Note that the unit step and ramp functions are zero only for values of x that are less than a. The integration properties shown in the table constitute a part of the mathematical definition, too. The first two integrations of $q(x)$ for $V(x)$ and $M(x)$ do not require constants of integration provided *all* loads on the beam are accounted for in $q(x)$. The examples that follow show how these functions are used.

EXAMPLE 3–2   Derive the loading, shear-force, and bending-moment relations for the beam of Fig. 3–5a.

Figure 3–5

(*a*) Loading diagram for a simply-supported beam.
(*b*) Shear-force diagram.
(*c*) Bending-moment diagram.

Solution   Using Table 3–1 and $q(x)$ for the loading function, we find

Answer
$$q = R_1\langle x\rangle^{-1} - 800\langle x - 0.1\rangle^{-1} - 400\langle x - 0.25\rangle^{-1} + R_2\langle x - 0.5\rangle^{-1} \quad (1)$$

Integrating successively gives

Answer
$$V = \int q\,dx = R_1\langle x\rangle^0 - 800\langle x - 0.1\rangle^0 - 400\langle x - 0.25\rangle^0 + R_2\langle x - 0.5\rangle^0 \quad (2)$$

Answer
$$M = \int V\,dx = R_1\langle x\rangle^1 - 800\langle x - 0.1\rangle^1 - 400\langle x - 0.25\rangle^1 + R_2\langle x - 0.5\rangle^1 \quad (3)$$

Note that $V = M = 0$ at $x = 0$.

The reactions R_1 and R_2 can be found by taking a summation of moments and forces as usual, *or* they can be found by noting that the shear force and bending moment must be zero everywhere except in the region $0 \le x \le 0.5$ m. This means that Eq. (2)

should give $V = 0$ at x slightly larger than 0.5 m. Thus

$$R_1 - 800 - 400 + R_2 = 0 \qquad (4)$$

Since the bending moment should also be zero in the same region, we have, from Eq. (3),

$$R_1(0.5) - 800(0.5 - 0.1) - 400(0.5 - 0.25) = 0 \qquad (5)$$

Equations (4) and (5) yield the reactions $R_1 = 840$ N and $R_2 = 360$ N.

The reader should verify that substitution of the values of R_1 and R_2 into Eqs. (2) and (3) yield Figs. 3–5b and c.

EXAMPLE 3–3

Figure 3–6a shows the loading diagram for a beam cantilevered at A with a uniform load of 3.5 kN/m acting on the portion 75 mm $\leq x \leq$ 175 mm, and a concentrated counterclockwise moment of 1 kN·m at $x = 250$ mm. Derive the shear-force and bending-moment relations, and the support reactions M_1 and R_1.

Solution

Following the procedure of Example 3–2, we find the load intensity function to be

$$q = -M_1 \langle x \rangle^{-2} + R_1 \langle x \rangle^{-1} - 3.5 \langle x - 0.075 \rangle^0 + 3.5 \langle x - 0.175 \rangle^0 - 1 \langle x - 0.25 \rangle^{-2} \quad (1)$$

Note that the $3.5 \langle x - 0.175 \rangle^0$ term was necessary to "turn off" the uniform load at C. Integrating successively gives

Answers

$$V = -M_1 \langle x \rangle^{-1} + R_1 \langle x \rangle^0 - 3.5 \langle x - 0.075 \rangle^1 + 3.5 \langle x - 0.175 \rangle^1 - 1 \langle x - 0.25 \rangle^{-1} \quad (2)$$

$$M = -M_1 \langle x \rangle^0 + R_1 \langle x \rangle^1 - 1.75 \langle x - 0.075 \rangle^2 + 1.75 \langle x - 0.175 \rangle^2 - 1 \langle x - 0.25 \rangle^0 \quad (3)$$

The reactions are found by making x slightly larger than 250 mm, where both V and M are zero in this region. Equation (2) will then give

$$-M_1 (0.25)^{-1} + R_1 (1) - 3.5(0.25 - 0.075) + 3.5(0.25 - 0.175) - 1(0) = 0$$

Figure 3–6

(a) Loading diagram for a beam cantilevered at A.
(b) Shear-force diagram.
(c) Bending-moment diagram.

Answer which yields $R_1 = 350$ N.
From Eq. (3) we get

$$-M_1(1) + 0.35(0.25) - 1.75(0.25 - 0.075)^2 + 1.75(0.25 - 0.175)^2 - 1(0) = 0$$

Answer which yields $M_1 = 0.0438$ N·m.

Figures 3–6b and c show the shear-force and bending-moment diagrams. Note that the impulse terms in Eq. (2), $-M_1\langle x\rangle^{-1}$ and $-1\langle x - 0.25\rangle^{-1}$, are physically not forces and are not shown in the V diagram. Also note that both the M_1 and 1 kN·m moments are counterclockwise and negative singularity functions; however, by the convention shown in Fig. 3–2 the M_1 and 1 kN·m are negative and positive bending moments, respectively, which is reflected in Fig. 3–6c.

3–4 Stress

When an internal surface is isolated as in Fig. 3–2b, the net force and moment acting on the surface manifest themselves as force distributions across the entire area. The force distribution acting at a point on the surface is unique and will have components in the normal and tangential directions called *normal stress* and *tangential shear stress*, respectively. Normal and shear stresses are labeled by the Greek symbols σ and τ, respectively. If the direction of σ is outward from the surface it is considered to be a *tensile stress* and is a positive normal stress. If σ is into the surface it is a *compressive stress* and commonly considered to be a negative quantity. The units of stress in U.S. Customary units are pounds per square inch (psi). For SI units, stress is in newtons per square meter (N/m²); 1 N/m² = 1 pascal (Pa).

3–5 Cartesian Stress Components

The Cartesian stress components are established by defining three mutually orthogonal surfaces at a point within the body. The normals to each surface will establish the x, y, z Cartesian axes. In general, each surface will have a normal and shear stress. The shear stress may have components along two Cartesian axes. For example, Fig. 3–7 shows an infinitesimal surface area isolation at a point Q within a body where the surface normal is the x direction. The normal stress is labeled σ_x. The symbol σ indicates a normal stress and the subscript x indicates the direction of the surface normal. The net shear stress acting on the surface is $(\tau_x)_{net}$ which can be resolved into components in the y and z directions, labeled as τ_{xy} and τ_{xz}, respectively (see Fig. 3–7).

Figure 3–7

Stress components on surface normal to x direction.

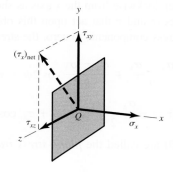

Figure 3–8

(*a*) General three-dimensional stress. (*b*) Plane stress with "cross-shears" equal.

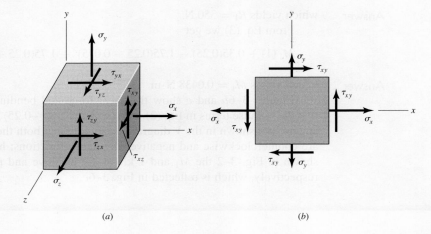

(*a*)   (*b*)

Note that double subscripts are necessary for the shear. The first subscript indicates the direction of the surface normal whereas the second subscript is the direction of the shear stress.

The state of stress at a point described by three mutually perpendicular surfaces is shown in Fig. 3–8*a*. It can be shown through coordinate transformation that this is sufficient to determine the state of stress on *any* surface intersecting the point. As the dimensions of the cube in Fig. 3–8*a* approach zero, the stresses on the hidden faces become equal and opposite to those on the opposing visible faces. Thus, in general, a complete state of stress is defined by nine stress components, σ_x, σ_y, σ_z, τ_{xy}, τ_{xz}, τ_{yx}, τ_{yz}, τ_{zx}, and τ_{zy}.

For equilibrium, in most cases, "cross-shears" are equal, hence

$$\tau_{yx} = \tau_{xy} \qquad \tau_{zy} = \tau_{yz} \qquad \tau_{xz} = \tau_{zx} \tag{3–7}$$

This reduces the number of stress components for most three-dimensional states of stress from nine to six quantities, σ_x, σ_y, σ_z, τ_{xy}, τ_{yz}, and τ_{zx}.

A very common state of stress occurs when the stresses on one surface are zero. When this occurs the state of stress is called *plane stress*. Figure 3–8*b* shows a state of plane stress, arbitrarily assuming that the normal for the stress-free surface is the z direction such that $\sigma_z = \tau_{zx} = \tau_{zy} = 0$. It is important to note that the element in Fig. 3–8*b* is still a three-dimensional cube. Also, here it is assumed that the cross-shears are equal such that $\tau_{yx} = \tau_{xy}$, and $\tau_{yz} = \tau_{zy} = \tau_{xz} = \tau_{zx} = 0$.

3–6 Mohr's Circle for Plane Stress

Suppose the $dx\,dy\,dz$ element of Fig. 3–8*b* is cut by an oblique plane with a normal n at an arbitrary angle ϕ counterclockwise from the x axis as shown in Fig. 3–9. Here, we are concerned with the stresses σ and τ that act upon this oblique plane. By summing the forces caused by all the stress components to zero, the stresses σ and τ are found to be

$$\sigma = \frac{\sigma_x + \sigma_y}{2} + \frac{\sigma_x - \sigma_y}{2} \cos 2\phi + \tau_{xy} \sin 2\phi \tag{3–8}$$

$$\tau = -\frac{\sigma_x - \sigma_y}{2} \sin 2\phi + \tau_{xy} \cos 2\phi \tag{3–9}$$

Equations (3–8) and (3–9) are called the *plane-stress transformation equations*.

| **Figure 3–9**

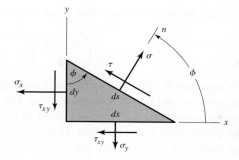

Differentiating Eq. (3–8) with respect to ϕ and setting the result equal to zero maximizes σ and gives

$$\tan 2\phi_p = \frac{2\tau_{xy}}{\sigma_x - \sigma_y} \qquad (3\text{–}10)$$

Equation (3–10) defines two particular values for the angle $2\phi_p$, one of which defines the maximum normal stress σ_1 and the other, the minimum normal stress σ_2. These two stresses are called the *principal stresses*, and their corresponding directions, the *principal directions*. The angle between the two principal directions is 90°. It is important to note that Eq. (3–10) can be written in the form

$$\frac{\sigma_x - \sigma_y}{2} \sin 2\phi_p - \tau_{xy} \cos 2\phi_p = 0 \qquad (a)$$

Comparing this with Eq. (3–9), we see that $\tau = 0$, meaning that the perpendicular *surfaces containing principal stresses have zero shear stresses.*

In a similar manner, we differentiate Eq. (3–9), set the result equal to zero, and obtain

$$\tan 2\phi_s = -\frac{\sigma_x - \sigma_y}{2\tau_{xy}} \qquad (3\text{–}11)$$

Equation (3–11) defines the two values of $2\phi_s$ at which the shear stress τ reaches an extreme value. The angle between the two surfaces containing the maximum shear stresses is 90°. Equation (3–11) can also be written as

$$\frac{\sigma_x - \sigma_y}{2} \cos 2\phi_p + \tau_{xy} \sin 2\phi_p = 0 \qquad (b)$$

Substituting this into Eq. (3–8) yields

$$\sigma = \frac{\sigma_x + \sigma_y}{2} \qquad (3\text{–}12)$$

Equation (3–12) tells us that the two surfaces containing the maximum shear stresses also contain equal normal stresses of $(\sigma_x + \sigma_y)/2$.

Comparing Eqs. (3–10) and (3–11), we see that $\tan 2\phi_s$ is the negative reciprocal of $\tan 2\phi_p$. This means that $2\phi_s$ and $2\phi_p$ are angles 90° apart, and thus the angles between the surfaces containing the maximum shear stresses and the surfaces containing the principal stresses are ±45°.

Formulas for the two principal stresses can be obtained by substituting the angle $2\phi_p$ from Eq. (3–10) in Eq. (3–8). The result is

$$\sigma_1, \sigma_2 = \frac{\sigma_x + \sigma_y}{2} \pm \sqrt{\left(\frac{\sigma_x - \sigma_y}{2}\right)^2 + \tau_{xy}^2} \qquad (3\text{–}13)$$

In a similar manner the two extreme-value shear stresses are found to be

$$\tau_1, \tau_2 = \pm\sqrt{\left(\frac{\sigma_x - \sigma_y}{2}\right)^2 + \tau_{xy}^2} \tag{3-14}$$

Your particular attention is called to the fact that an extreme value of the shear stress *may not be the same as the actual maximum value.* See Sec. 3–7.

It is important to note that the equations given to this point are quite sufficient for performing any plane stress transformation. However, extreme care must be exercised when applying them. For example, say you are attempting to determine the principal state of stress for a problem where $\sigma_x = 14$ MPa, $\sigma_y = -10$ MPa, and $\tau_{xy} = -16$ MPa. Equation (3–10) yields $\phi_p = -26.57°$ and $63.43°$, which locate the principal stress surfaces, whereas, Eq. (3–13) gives $\sigma_1 = 22$ MPa and $\sigma_2 = -18$ MPa for the principal stresses. If all we wanted was the principal stresses, we would be finished. However, what if we wanted to draw the element containing the principal stresses properly oriented relative to the x, y axes? Well, we have two values of ϕ_p and two values for the principal stresses. How do we know which value of ϕ_p corresponds to which value of the principal stress? To clear this up we would need to substitute one of the values of ϕ_p into Eq. (3–8) to determine the normal stress corresponding to that angle.

A graphical method for expressing the relations developed in this section, called *Mohr's circle diagram,* is a very effective means of visualizing the stress state at a point and keeping track of the directions of the various components associated with plane stress. Equations (3–8) and (3–9) can be shown to be a set of parametric equations for σ and τ, where the parameter is 2ϕ. The parametric relationship between σ and τ is that of a circle plotted in the σ, τ plane, where the center of the circle is located at $C = (\sigma, \tau) = [(\sigma_x + \sigma_y)/2, 0]$ and has a radius of $R = \sqrt{[(\sigma_x - \sigma_y)/2]^2 + \tau_{xy}^2}$. A problem arises in the sign of the shear stress. The transformation equations are based on a positive ϕ being counterclockwise, as shown in Fig. 3–9. If a positive τ were plotted above the σ axis, points would rotate clockwise on the circle 2ϕ in the opposite direction of rotation on the element. It would be convenient if the rotations were in the same direction. One could solve the problem easily by plotting positive τ below the axis. However, the classical approach to Mohr's circle uses a different convention for the shear stress.

Mohr's Circle Shear Convention

This convention is followed in drawing Mohr's circle:

- Shear stresses tending to rotate the element clockwise (cw) are plotted *above* the σ axis.

- Shear stresses tending to rotate the element counterclockwise (ccw) are plotted *below* the σ axis.

For example, consider the right face of the element in Fig. 3–8b. By Mohr's circle convention the shear stress shown is plotted *below* the σ axis because it tends to rotate the element counterclockwise. The shear stress on the top face of the element is plotted *above* the σ axis because it tends to rotate the element clockwise.

In Fig. 3–10 we create a coordinate system with normal stresses plotted along the abscissa and shear stresses plotted as the ordinates. On the abscissa, tensile (positive) normal stresses are plotted to the right of the origin O and compressive (negative) normal stresses to the left. On the ordinate, clockwise (cw) shear stresses are plotted up; counterclockwise (ccw) shear stresses are plotted down.

Figure 3–10

Mohr's circle diagram.

Using the stress state of Fig. 3–8b, we plot Mohr's circle, Fig. 3–10, by first look-ing at the right surface of the element containing σ_x to establish the sign of σ_x and the cw or ccw direction of the shear stress. The right face is called the *x face* where $\phi = 0°$. If σ_x is positive and the shear stress τ_{xy} is ccw as shown in Fig. 3–8b, we can establish point A with coordinates $(\sigma_x, \tau_{xy}^{ccw})$ in Fig. 3–10. Next, we look at the top *y face,* where $\phi = 90°$, which contains σ_y, and repeat the process to obtain point B with coordinates $(\sigma_y, \tau_{xy}^{cw})$ as shown in Fig. 3–10. The two states of stress for the element are $\Delta\phi = 90°$ from each other on the element so they will be $2\Delta\phi = 180°$ from each other on Mohr's circle. Points A and B are the same vertical distance from the σ axis. Thus, AB must be on the diameter of the circle, and the center of the circle C is where AB intersects the σ axis. With points A and B on the circle, and center C, the complete circle can then be drawn. Note that the extended ends of line AB are labeled x and y as references to the normals to the surfaces for which points A and B represent the stresses.

The entire Mohr's circle represents the state of stress at a *single* point in a struc-ture. Each point on the circle represents the stress state for a *specific* surface intersect-ing the point in the structure. Each pair of points on the circle 180° apart represent the state of stress on an element whose surfaces are 90° apart. Once the circle is drawn, the states of stress can be visualized for various surfaces intersecting the point being analyzed. For example, the principal stresses σ_1 and σ_2 are points D and E, respectively, and their values obviously agree with Eq. (3–13). We also see that the shear stresses are zero on the surfaces containing σ_1 and σ_2. The two extreme-value shear stresses, one clockwise and one counterclockwise, occur at F and G with mag-nitudes equal to the radius of the circle. The surfaces at F and G each also contain normal stresses of $(\sigma_x + \sigma_y)/2$ as noted earlier in Eq. (3–12). Finally, the state of stress on an arbitrary surface located at an angle ϕ counterclockwise from the x face is point H.

At one time, Mohr's circle was used graphically where it was drawn to scale very accurately and values were measured by using a scale and protractor. Here, we are strictly using Mohr's circle as a visualization aid and will use a semigraphical approach, calculating values from the properties of the circle. This is illustrated by the following example.

EXAMPLE 3–4

A plane stress element has $\sigma_x = 80$ MPa, $\sigma_y = 0$ MPa, and $\tau_{xy} = 50$ MPa cw, as shown in Fig. 3–11a.

(a) Using Mohr's circle, find the principal stresses and directions, and show these on a stress element correctly aligned with respect to the xy coordinates. Draw another stress element to show τ_1 and τ_2, find the corresponding normal stresses, and label the drawing completely.

(b) Repeat part a using the transformation equations only.

Solution

(a) In the semigraphical approach used here, we first make an approximate freehand sketch of Mohr's circle and then use the geometry of the figure to obtain the desired information.

Draw the σ and τ axes first (Fig. 3–11b) and from the x face locate $\sigma_x = 80$ MPa along the σ axis. On the x face of the element, we see that the shear stress is 50 MPa in the cw direction. Thus, for the x face, this establishes point A (80, 50^{cw}) MPa. Corresponding to the y face, the stress is $\sigma = 0$ and $\tau = 50$ MPa in the ccw direction. This locates point B (0, 50^{ccw}) MPa. The line AB forms the diameter of the required circle, which can now be drawn. The intersection of the circle with the σ axis defines σ_1 and σ_2 as shown. Now, noting the triangle ACD, indicate on the sketch the length of the legs AD and CD as 50 and 40 MPa, respectively. The length of the hypotenuse AC is

Answer

$$\tau_1 = \sqrt{(50)^2 + (40)^2} = 64.0 \text{ MPa}$$

and this should be labeled on the sketch too. Since intersection C is 40 MPa from the origin, the principal stresses are now found to be

Answer

$$\sigma_1 = 40 + 64 = 104 \text{ MPa} \quad \text{and} \quad \sigma_2 = 40 - 64 = -24 \text{ MPa}$$

The angle 2ϕ from the x axis cw to σ_1 is

Answer

$$2\phi_p = \tan^{-1}\tfrac{50}{40} = 51.3°$$

To draw the principal stress element (Fig. 3–11c), sketch the x and y axes parallel to the original axes. The angle ϕ_p on the stress element must be measured in the *same* direction as is the angle $2\phi_p$ on the Mohr circle. Thus, from x measure 25.7° (half of 51.3°) clockwise to locate the σ_1 axis. The σ_2 axis is 90° from the σ_1 axis and the stress element can now be completed and labeled as shown. Note that there are *no* shear stresses on this element.

The two maximum shear stresses occur at points E and F in Fig. 3–11b. The two normal stresses corresponding to these shear stresses are each 40 MPa, as indicated. Point E is 38.7° ccw from point A on Mohr's circle. Therefore, in Fig. 3–11d, draw a stress element oriented 19.3° (half of 38.7°) ccw from x. The element should then be labeled with magnitudes and directions as shown.

In constructing these stress elements it is important to indicate the x and y directions of the original reference system. This completes the link between the original machine element and the orientation of its principal stresses.

Figure 3-11

All stresses in MPa.

(a)

(b)

(c)

(d)

Answer

(b) The transformation equations are programmable. From Eq. (3–10),

$$\phi_p = \frac{1}{2} \tan^{-1}\left(\frac{2\tau_{xy}}{\sigma_x - \sigma_y}\right) = \frac{1}{2} \tan^{-1}\left(\frac{2(-50)}{80}\right) = -25.7°, \, 64.3°$$

From Eq. (3–8), for the first angle $\phi_p = -25.7°$,

$$\sigma = \frac{80 + 0}{2} + \frac{80 - 0}{2} \cos[2(-25.7)] + (-50) \sin[2(-25.7)] = 104.03 \text{ MPa}$$

The shear on this surface is obtained from Eq. (3–9) as

$$\tau = -\frac{80 - 0}{2} \sin[2(-25.7)] + (-50) \cos[2(-25.7)] = 0 \text{ MPa}$$

which confirms that 104.03 MPa is a principal stress. From Eq. (3–8), for $\phi_p = 64.3°$,

$$\sigma = \frac{80 + 0}{2} + \frac{80 - 0}{2} \cos[2(64.3)] + (-50) \sin[2(64.3)] = -24.03 \text{ MPa}$$

Answer

Substituting $\phi_p = 64.3°$ into Eq. (3–9) again yields $\tau = 0$, indicating that -24.03 MPa is also a principal stress. Once the principal stresses are calculated they can be ordered such that $\sigma_1 \geq \sigma_2$. Thus, $\sigma_1 = 104.03$ MPa and $\sigma_2 = -24.03$ MPa.

Since for $\sigma_1 = 104.03$ MPa, $\phi_p = -25.7°$, and since ϕ is defined positive ccw in the transformation equations, we rotate *clockwise* 25.7° for the surface containing σ_1. We see in Fig. 3–11c that this totally agrees with the semigraphical method.

To determine τ_1 and τ_2, we first use Eq. (3–11) to calculate ϕ_s:

$$\phi_s = \frac{1}{2}\tan^{-1}\left(-\frac{\sigma_x - \sigma_y}{2\tau_{xy}}\right) = \frac{1}{2}\tan^{-1}\left(-\frac{80}{2(-50)}\right) = 19.3°, 109.3°$$

For $\phi_s = 19.3°$, Eqs. (3–8) and (3–9) yield

Answer

$$\sigma = \frac{80 + 0}{2} + \frac{80 - 0}{2}\cos[2(19.3)] + (-50)\sin[2(19.3)] = 40.0 \text{ MPa}$$

$$\tau = -\frac{80 - 0}{2}\sin[2(19.3)] + (-50)\cos[2(19.3)] = -64.0 \text{ MPa}$$

Remember that Eqs. (3–8) and (3–9) are *coordinate* transformation equations. Imagine that we are rotating the x, y axes 19.3° counterclockwise and y will now point up and to the left. So a negative shear stress on the rotated x face will point down and to the right as shown in Fig. 3–11d. Thus again, results agree with the semigraphical method.

For $\phi_s = 109.3°$, Eqs. (3–8) and (3–9) give $\sigma = 40.0$ MPa and $\tau = +64.0$ MPa. Using the same logic for the coordinate transformation we find that results again agree with Fig. 3–11d.

3–7 General Three-Dimensional Stress

As in the case of plane stress, a particular orientation of a stress element occurs in space for which all shear-stress components are zero. When an element has this particular orientation, the normals to the faces are mutually orthogonal and correspond to the principal directions, and the normal stresses associated with these faces are the principal stresses. Since there are three faces, there are three principal directions and three principal stresses σ_1, σ_2, and σ_3. For plane stress, the stress-free surface contains the third principal stress which is zero.

In our studies of plane stress we were able to specify any stress state σ_x, σ_y, and τ_{xy} and find the principal stresses and principal directions. But six components of stress are required to specify a general state of stress in three dimensions, and the problem of determining the principal stresses and directions is more difficult. In design, three-dimensional transformations are rarely performed since most maximum stress states occur under plane stress conditions. One notable exception is contact stress, which is not a case of plane stress, where the three principal stresses are given in Sec. 3–19. In fact, *all* states of stress are truly three-dimensional, where they might be described one- or two-dimensionally with respect to *specific* coordinate axes. Here it is most important to understand the relationship among the *three*

Figure 3-12

Mohr's circles for three-dimensional stress.

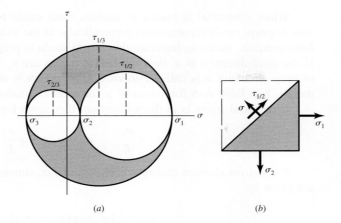

(a)   (b)

principal stresses. The process in finding the three principal stresses from the six stress components σ_x, σ_y, σ_z, τ_{xy}, τ_{yz}, and τ_{zx}, involves finding the roots of the cubic equation[1]

$$\sigma^3 - (\sigma_x + \sigma_y + \sigma_z)\sigma^2 + (\sigma_x\sigma_y + \sigma_x\sigma_z + \sigma_y\sigma_z - \tau_{xy}^2 - \tau_{yz}^2 - \tau_{zx}^2)\sigma$$
$$- (\sigma_x\sigma_y\sigma_z + 2\tau_{xy}\tau_{yz}\tau_{zx} - \sigma_x\tau_{yz}^2 - \sigma_y\tau_{zx}^2 - \sigma_z\tau_{xy}^2) = 0 \quad (3\text{--}15)$$

In plotting Mohr's circles for three-dimensional stress, the principal normal stresses are ordered so that $\sigma_1 \geq \sigma_2 \geq \sigma_3$. Then the result appears as in Fig. 3–12a. The stress coordinates σ, τ for any arbitrarily located plane will always lie on the boundaries or within the shaded area.

Figure 3–12a also shows the three *principal shear stresses* $\tau_{1/2}$, $\tau_{2/3}$, and $\tau_{1/3}$.[2] Each of these occurs on the two planes, one of which is shown in Fig. 3–12b. The figure shows that the principal shear stresses are given by the equations

$$\tau_{1/2} = \frac{\sigma_1 - \sigma_2}{2} \qquad \tau_{2/3} = \frac{\sigma_2 - \sigma_3}{2} \qquad \tau_{1/3} = \frac{\sigma_1 - \sigma_3}{2} \quad (3\text{--}16)$$

Of course, $\tau_{max} = \tau_{1/3}$ when the normal principal stresses are ordered ($\sigma_1 > \sigma_2 > \sigma_3$), so always order your principal stresses. Do this in any computer code you generate and you'll always generate τ_{max}.

3-8   **Elastic Strain**

Normal strain ϵ is defined and discussed in Sec. 2–1 for the tensile specimen and is given by Eq. (2–2) as $\epsilon = \delta/l$, where δ is the total elongation of the bar within the length l. Hooke's law for the tensile specimen is given by Eq. (2–3) as

$$\sigma = E\epsilon \quad (3\text{--}17)$$

where the constant E is called *Young's modulus* or the *modulus of elasticity*.

---

[1]For development of this equation and further elaboration of three-dimensional stress transformations see: Richard G. Budynas, *Advanced Strength and Applied Stress Analysis,* 2nd ed., McGraw-Hill, New York, 1999, pp. 46–78.

[2]Note the difference between this notation and that for a shear stress, say, τ_{xy}. The use of the shilling mark is not accepted practice, but it is used here to emphasize the distinction.

When a material is placed in tension, there exists not only an axial strain, but also negative strain (contraction) perpendicular to the axial strain. Assuming a linear, homogeneous, isotropic material, this lateral strain is proportional to the axial strain. If the axial direction is x, then the lateral strains are $\epsilon_y = \epsilon_z = -\nu\epsilon_x$. The constant of proportionality ν is called *Poisson's ratio*, which is about 0.3 for most structural metals. See Table A–5 for values of ν for common materials.

If the axial stress is in the x direction, then from Eq. (3–17)

$$\epsilon_x = \frac{\sigma_x}{E} \qquad \epsilon_y = \epsilon_z = -\nu\frac{\sigma_x}{E} \qquad (3\text{--}18)$$

For a stress element undergoing σ_x, σ_y, and σ_z simultaneously, the normal strains are given by

$$\epsilon_x = \frac{1}{E}[\sigma_x - \nu(\sigma_y + \sigma_z)]$$

$$\epsilon_y = \frac{1}{E}[\sigma_y - \nu(\sigma_x + \sigma_z)] \qquad (3\text{--}19)$$

$$\epsilon_z = \frac{1}{E}[\sigma_z - \nu(\sigma_x + \sigma_y)]$$

Shear strain γ is the change in a right angle of a stress element when subjected to pure shear stress, and Hooke's law for shear is given by

$$\tau = G\gamma \qquad (3\text{--}20)$$

where the constant G is the *shear modulus of elasticity* or *modulus of rigidity*.

It can be shown for a linear, isotropic, homogeneous material, the three elastic constants are related to each other by

$$E = 2G(1 + \nu) \qquad (3\text{--}21)$$

3–9 Uniformly Distributed Stresses

The assumption of a uniform distribution of stress is frequently made in design. The result is then often called *pure tension, pure compression,* or *pure shear,* depending upon how the external load is applied to the body under study. The word *simple* is sometimes used instead of *pure* to indicate that there are no other complicating effects. The tension rod is typical. Here a tension load F is applied through pins at the ends of the bar. The assumption of uniform stress means that if we cut the bar at a section remote from the ends and remove one piece, we can replace its effect by applying a uniformly distributed force of magnitude σA to the cut end. So the stress σ is said to be uniformly distributed. It is calculated from the equation

$$\sigma = \frac{F}{A} \qquad (3\text{--}22)$$

This assumption of uniform stress distribution requires that:

- The bar be straight and of a homogeneous material
- The line of action of the force contains the centroid of the section
- The section be taken remote from the ends and from any discontinuity or abrupt change in cross section

For simple compression, Eq. (3–22) is applicable with F normally being considered a negative quantity. Also, a slender bar in compression may fail by buckling, and this possibility must be eliminated from consideration before Eq. (3–22) is used.[3]

Another type of loading that assumes a uniformly distributed stress is known as *direct shear.* This occurs when there is a shearing action with no bending. An example is the action on a piece of sheet metal caused by the two blades of tin snips. Bolts and pins that are loaded in shear often have direct shear. Think of a cantilever beam with a force pushing down on it. Now move the force all the way up to the wall so there is no bending moment, just a force trying to shear the beam off the wall. This is direct shear. Direct shear is usually assumed to be uniform across the cross section, and is given by

$$\tau = \frac{V}{A} \tag{3–23}$$

where V is the shear force and A is the area of the cross section that is being sheared. The assumption of uniform stress is not accurate, particularly in the vicinity where the force is applied, but the assumption generally gives acceptable results.

3–10   Normal Stresses for Beams in Bending

The equations for the normal bending stresses in straight beams are based on the following assumptions:

- The beam is subjected to pure bending. This means that the shear force is zero, and that no torsion or axial loads are present (for most engineering applications it is assumed that these loads affect the bending stresses minimally).
- The material is isotropic and homogeneous.
- The material obeys Hooke's law.
- The beam is initially straight with a cross section that is constant throughout the beam length.
- The beam has an axis of symmetry in the plane of bending.
- The proportions of the beam are such that it would fail by bending rather than by crushing, wrinkling, or sidewise buckling.
- Plane cross sections of the beam remain plane during bending.

In Fig. 3–13 we visualize a portion of a straight beam acted upon by a positive bending moment M shown by the curved arrow showing the physical action of the moment together with a straight, double-headed, arrow indicating the moment vector. The x axis is coincident with the *neutral axis* of the section, and the xz plane, which

Figure 3–13

Straight beam in positive bending.

[3]See Sec. 4–11.

Figure 3–14

Bending stresses according to Eq. (3–24).

contains the neutral axes of all cross sections, is called the *neutral plane*. Elements of the beam coincident with this plane have zero stress. The location of the neutral axis with respect to the cross section is coincident with the *centroidal axis* of the cross section.

The bending stress varies linearly with the distance from the neutral axis, y, and is given by

$$\sigma_x = -\frac{My}{I} \tag{3–24}$$

where I is the *second-area moment* about the z axis. That is,

$$I = \int y^2 \, dA \tag{3–25}$$

The stress distribution given by Eq. (3–24) is shown in Fig. 3–14. The maximum magnitude of the bending stress will occur where y has the greatest magnitude. Designating σ_{max} as the maximum *magnitude* of the bending stress, and c as the maximum *magnitude* of y

$$\sigma_{max} = \frac{Mc}{I} \tag{3–26a}$$

Equation (3–24) can still be used to ascertain whether σ_{max} is tensile or compressive. Equation (3–26a) is often written as

$$\sigma_{max} = \frac{M}{Z} \tag{3–26b}$$

where $Z = I/c$ is called the *section modulus*.

EXAMPLE 3–5

A beam having a T section with the dimensions shown in Fig. 3–15 is subjected to a bending moment of 1600 N · m, about the negative z axis, that causes tension at the top surface. Locate the neutral axis and find the maximum tensile and compressive bending stresses.

Solution

Dividing the T section into two rectangles, numbered 1 and 2, the total area is $A = 12(75) + 12(88) = 1956 \text{ mm}^2$. Summing the area moments of these rectangles about the top edge, where the moment arms of areas 1 and 2 are 6 mm and $(12 + 88/2) = 56$ mm respectively, we have

$$1956c_1 = 12(75)(6) + 12(88)(56)$$

and hence $c_1 = 32.99$ mm. Therefore $c_2 = 100 - 32.99 = 67.01$ mm.

Next we calculate the second moment of area of each rectangle about its own centroidal axis. Using Table A–18, we find for the top rectangle

$$I_1 = \frac{1}{12}bh^3 = \frac{1}{12}(75)12^3 = 1.080 \times 10^4 \text{ mm}^4$$

Figure 3–15

Dimensions in millimeters.

For the bottom rectangle, we have

$$I_2 = \frac{1}{12}(12)88^3 = 6.815 \times 10^5 \text{ mm}^4$$

We now employ the *parallel-axis theorem* to obtain the second moment of area of the composite figure about its own centroidal axis. This theorem states

$$I_z = I_{ca} + Ad^2$$

where I_{ca} is the second moment of area about its own centroidal axis and I_z is the second moment of area about any parallel axis a distance d removed. For the top rectangle, the distance is

$$d_1 = 32.99 - 6 = 26.99 \text{ mm}$$

and for the bottom rectangle,

$$d_2 = 67.01 - \frac{88}{2} = 23.01 \text{ mm}$$

Using the parallel-axis theorem for both rectangles, we now find that

$$I = [1.080 \times 10^4 + 12(75)26.99^2] + [6.815 \times 10^5 + 12(88)23.01^2]$$
$$= 1.907 \times 10^6 \text{ mm}^4$$

Finally, the maximum tensile stress, which occurs at the top surface, is found to be

Answer
$$\sigma = \frac{Mc_1}{I} = \frac{1600(32.99)10^{-3}}{1.907(10^{-6})} = 27.68(10^6) \text{ Pa} = 27.68 \text{ MPa}$$

Similarly, the maximum compressive stress at the lower surface is found to be

Answer
$$\sigma = -\frac{Mc_2}{I} = -\frac{1600(67.01)10^{-3}}{1.907(10^{-6})} = -56.22(10^6) \text{ Pa} = -56.22 \text{ MPa}$$

Two-Plane Bending

Quite often, in mechanical design, bending occurs in both xy and xz planes. Considering cross sections with one or two planes of symmetry only, the bending stresses are given by

$$\sigma_x = -\frac{M_z y}{I_z} + \frac{M_y z}{I_y} \qquad (3\text{--}27)$$

where the first term on the right side of the equation is identical to Eq. (3–24), M_y is the bending moment in the xz plane (moment vector in y direction), z is the distance from the neutral y axis, and I_y is the second area moment about the y axis.

For *noncircular* cross sections, Eq. (3–27) is the superposition of stresses caused by the two bending moment components. The maximum tensile and compressive bending stresses occur where the summation gives the greatest positive and negative stresses, respectively. For solid *circular* cross sections, all lateral axes are the same and the plane containing the moment corresponding to the vector sum of M_z and M_y contains the maximum bending stresses. For a beam of diameter d the maximum distance from the neutral axis is $d/2$, and from Table A–18, $I = \pi d^4/64$. The maximum bending stress for a solid circular cross section is then

$$\sigma_m = \frac{Mc}{I} = \frac{(M_y^2 + M_z^2)^{1/2}(d/2)}{\pi d^4/64} = \frac{32}{\pi d^3}(M_y^2 + M_z^2)^{1/2} \qquad (3\text{--}28)$$

EXAMPLE 3–6

As shown in Fig. 3–16a, beam OC is loaded in the xy plane by a uniform load of 9 kN/m, and in the xz plane by a concentrated force of 0.4 kN at end C. The beam is 0.2 m long.

Figure 3–16

(a) Beam loaded in two planes; (b) loading and bending-moment diagrams in xy plane; (c) loading and bending-moment diagrams in xz plane.

(a) For the cross section shown determine the maximum tensile and compressive bending stresses and where they act.

(b) If the cross section was a solid circular rod of diameter, $d = 30$ mm, determine the magnitude of the maximum bending stress.

Solution

(a) The reactions at O and the bending-moment diagrams in the xy and xz planes are shown in Figs. 3–16b and c, respectively. The maximum moments in both planes occur at O where

$$(M_z)_O = -\frac{1}{2}(9)0.2^2 = -0.18 \text{ kN·m} \qquad (M_y)_O = 0.4(0.2) = 0.08 \text{ kN·m}$$

The second moments of area in both planes are

$$I_z = \frac{1}{12}(0.02)0.04^3 = 106.7 \times 10^{-9} \text{ m}^4 \qquad I_y = \frac{1}{12}(0.04)0.02^3 = 26.7 \times 10^{-9} \text{ m}^4$$

The maximum tensile stress occurs at point A, shown in Fig. 3–16a, where the maximum tensile stress is due to both moments. At A, $y_A = 0.02$ m and $z_A = 0.01$ m. Thus, from Eq. (3–27)

Answer

$$(\sigma_x)_A = -\frac{-0.18(0.02)}{106.7 \times 10^{-9}} + \frac{0.08(0.01)}{26.7 \times 10^{-9}} = 63\,702 \text{ kPa} = 63.7 \text{ MPa}$$

The maximum compressive bending stress occurs at point B where, $y_B = -0.02$ m and $z_B = -0.01$ m. Thus

Answer

$$(\sigma_x)_B = -\frac{-0.18(-0.02)}{106.7 \times 10^{-9}} + \frac{0.08(-0.01)}{26.7 \times 10^{-9}} = -63\,702 \text{ kPa} = -63.7 \text{ MPa}$$

(b) For a solid circular cross section of diameter, $d = 30$ mm, the maximum bending stress at end O is given by Eq. (3–28) as

Answer

$$\sigma_m = \frac{32}{\pi(0.03)^3}\left[0.08^2 + (-0.18)^2\right]^{1/2} = 74310.8 \text{ kPa} = 74.31 \text{ MPa}$$

Beams with Asymmetrical Sections[4]

The bending stress equations, given by Eqs. (3–24) and (3–27), can also be applied to beams having asymmetrical cross sections, provided the planes of bending coincide with the *area principal axes* of the section. The method for determining the orientation of the area principal axes and the values of the corresponding *principal second-area moments* can be found in any statics book. If a section has an axis of symmetry, that axis and its perpendicular axis are the area principal axes.

For example, consider a beam in bending, using an equal leg angle as shown in Table A–6. Equation (3–27) cannot be used if the bending moments are resolved about axis 1–1 and/or axis 2–2. However, Eq. (3–27) can be used if the moments are resolved about axis 3–3 and its perpendicular axis (let us call it, say, axis 4–4). Note, for this

[4]For further discussion, see Sec. 5.3, Richard G. Budynas, *Advanced Strength and Applied Stress Analysis,* 2nd ed., McGraw-Hill, New York, 1999.

cross section, axis 4–4 is an axis of symmetry. Table A–6 is a standard table, and for brevity, does not directly give all the information needed to use it. The orientation of the area principal axes and the values of I_{2-2}, I_{3-3}, and I_{4-4} are not given because they can be determined as follows: Since the legs are equal, the principal axes are oriented $\pm45°$ from axis 1–1, and $I_{2-2} = I_{1-1}$. The second-area moment I_{3-3} is given by

$$I_{3-3} = A(k_{3-3})^2 \tag{a}$$

where k_{3-3} is called the *radius of gyration*. The sum of the second-area moments for a cross section is invariant, so $I_{1-1} + I_{2-2} = I_{3-3} + I_{4-4}$. Thus, I_{4-4} is given by

$$I_{4-4} = 2\,I_{1-1} - I_{3-3} \tag{b}$$

where $I_{2-2} = I_{1-1}$. For example, consider a $80 \times 80 \times 6$ angle. Using Table A–6 and Eqs. (a) and (b), $I_{3-3} = 9.35\,(1.57)^2 = 23.047$ mm^4, and $I_{4-4} = 2\,(55.8) - 23.047 = 88.553$ mm^4.

3–11 Shear Stresses for Beams in Bending

Most beams have both shear forces and bending moments present. It is only occasionally that we encounter beams subjected to pure bending, that is to say, beams having zero shear force. The flexure formula is developed on the assumption of pure bending. This is done, however, to eliminate the complicating effects of shear force in the development of the formula. For engineering purposes, the flexure formula is valid no matter whether a shear force is present or not. For this reason, we shall utilize the same normal bending-stress distribution [Eqs. (3–24) and (3–26)] when shear forces are also present.

In Fig. 3–17a we show a beam segment of constant cross section subjected to a shear force V and a bending moment M at x. Because of external loading and V, the shear force and bending moment change with respect to x. At $x + dx$ the shear force and bending moment are $V + dV$ and $M + dM$, respectively. Considering forces in the x direction only, Fig. 3–17b shows the stress distribution σ_x due to the bending moments. If dM is positive, with the bending moment increasing, the stresses on the right face, for a given value of y, are larger in magnitude than the stresses on the left face. If we further isolate the element by making a slice at $y = y_1$ (see Fig. 3–17b), the net *force* in the x direction will be directed to the left with a value of

$$\int_{y_1}^{c} \frac{(dM)y}{I}\,dA$$

as shown in the rotated view of Fig. 3–17c. For equilibrium, a shear force on the bottom face, directed to the right, is required. This shear force gives rise to a shear stress τ, where, if assumed uniform, the force is $\tau b\,dx$. Thus

$$\tau b\,dx = \int_{y_1}^{c} \frac{(dM)y}{I}\,dA \tag{a}$$

The term dM/I can be removed from within the integral and $b\,dx$ placed on the right side of the equation; then, from Eq. (3–3) with $V = dM/dx$, Eq. (a) becomes

$$\tau = \frac{V}{Ib} \int_{y_1}^{c} y\,dA \tag{3–29}$$

In this equation, the integral is the first moment of the area A' with respect to the neutral axis (see Fig. 3–17c). This integral is usually designated as Q. Thus

$$Q = \int_{y_1}^{c} y\,dA = \bar{y}'A' \tag{3–30}$$

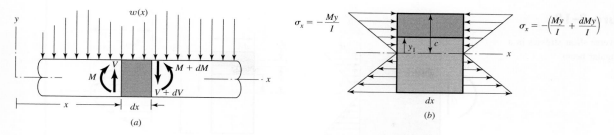

Figure 3–17

Beam section isolation.
Note: Only forces shown in x direction on dx element in (b).

where, for the isolated area y_1 to c, $\bar{y}'$ is the distance in the y direction from the neutral plane to the centroid of the area A'. With this, Eq. (3–29) can be written as

$$\tau = \frac{VQ}{Ib} \qquad (3\text{–}31)$$

This stress is known as the *transverse shear stress*. It is always accompanied with bending stress.

In using this equation, note that b is the width of the section at $y = y_1$. Also, I is the second moment of area of the entire section about the neutral axis.

Because cross shears are equal, and area A' is *finite*, the shear stress τ given by Eq. (3–31) and shown on area A' in Fig. 3–17c occurs only at $y = y_1$. The shear stress on the lateral area varies with y, normally maximum at $y = 0$ (where $\bar{y}'A'$ is maximum) and zero at the outer fibers of the beam where $A' = 0$.

The shear stress distribution in a beam depends on how Q/b varies as a function of y_1. Here we will show how to determine the shear stress distribution for a beam with a rectangular cross section and provide results of maximum values of shear stress for other standard cross sections. Figure 3–18 shows a portion of a beam with a rectangular cross section, subjected to a shear force V and a bending moment M. As a result of the bending moment, a normal stress σ is developed on a cross section such as A–A, which is in compression above the neutral axis and in tension below. To investigate the shear stress at a distance y_1 above the neutral axis, we select an element of area dA at a distance y above the neutral axis. Then, $dA = b\, dy$, and so Eq. (3–30) becomes

$$Q = \int_{y_1}^{c} y\, dA = b \int_{y_1}^{c} y\, dy = \frac{by^2}{2}\bigg|_{y_1}^{c} = \frac{b}{2}(c^2 - y_1^2) \qquad (b)$$

Substituting this value for Q into Eq. (3–31) gives

$$\tau = \frac{V}{2I}(c^2 - y_1^2) \qquad (3\text{–}32)$$

This is the general equation for shear stress in a rectangular beam. To learn something about it, let us make some substitutions. From Table A–18, the second moment of area for a rectangular section is $I = bh^3/12$; substituting $h = 2c$ and $A = bh = 2bc$ gives

$$I = \frac{Ac^2}{3} \qquad (c)$$

Figure 3–18

Transverse shear stresses in a rectangular beam.

(a) (b) (c)

(d)

If we now use this value of I for Eq. (3–32) and rearrange, we get

$$\tau = \frac{3V}{2A}\left(1 - \frac{y_1^2}{c^2}\right)$$

(3–33)

We note that the maximum shear stress exists when $y_1 = 0$, which is at the bending neutral axis. Thus

$$\tau_{max} = \frac{3V}{2A}$$

(3–34)

for a rectangular section. As we move away from the neutral axis, the shear stress decreases parabolically until it is zero at the outer surfaces where $y_1 = \pm c$, as shown in Fig. 3–18c. Horizontal shear stress is always accompanied by vertical shear stress of the same magnitude, and so the distribution can be diagrammed as shown in Fig. 3–18d. Figure 3–18c shows that the shear τ on the vertical surfaces varies with y. We are almost always interested in the horizontal shear, τ in Fig. 3–18d, which is nearly uniform over dx with constant $y = y_1$. The maximum horizontal shear occurs where the vertical shear is largest. This is usually at the neutral axis but may not be if the width b is smaller somewhere else. Furthermore, if the section is such that b can be minimized on a plane not horizontal, then the horizontal shear stress occurs on an inclined plane. For example, with tubing, the horizontal shear stress occurs on a radial plane and the corresponding "vertical shear" is not vertical, but tangential.

The distributions of transverse shear stresses for several commonly used cross sections are shown in Table 3–2. The profiles represent the VQ/Ib relationship, which is a function of the distance y from the neutral axis. For each profile, the formula for the maximum value at the neutral axis is given. Note that the expression given for the I beam is a commonly used approximation that is reasonable for a standard I beam with a thin web. Also, the profile for the I beam is idealized. In reality the transition from the web to the flange is quite complex locally, and not simply a step change.

Table 3–2

Formulas for Maximum
Transverse Shear Stress
from VQ/Ib

Beam Shape	Formula	Beam Shape	Formula
Rectangular $\tau_{avg} = \dfrac{V}{A}$	$\tau_{max} = \dfrac{3V}{2A}$	Hollow, thin-walled round $\tau_{avg} = \dfrac{V}{A}$	$\tau_{max} = \dfrac{2V}{A}$
Circular $\tau_{avg} = \dfrac{V}{A}$	$\tau_{max} = \dfrac{4V}{3A}$	Structural I beam (thin-walled) A_{web}	$\tau_{max} \approx \dfrac{V}{A_{web}}$

It is significant to observe that the transverse shear stress in each of these common cross sections is maximum on the neutral axis, and zero on the outer surfaces. Since this is exactly the opposite of where the bending and torsional stresses have their maximum and minimum values, the transverse shear stress is often not critical from a design perspective.

Let us examine the significance of the transverse shear stress, using as an example a cantilever beam of length L, with rectangular cross section $b \times h$, loaded at the free end with a transverse force F. At the wall, where the bending moment is the largest, at a distance y from the neutral axis, a stress element will include both bending stress and transverse shear stress. In Sec. 5–4 it will be shown that a good measure of the combined effects of multiple stresses on a stress element is the maximum shear stress. Inserting the bending stress (My/I) and the transverse shear stress (VQ/Ib) into the maximum shear stress equation, Eq. (3–14), we obtain a general equation for the maximum shear stress in a cantilever beam with a rectangular cross section. This equation can then be normalized with respect to L/h and y/c, where c is the distance from the neutral axis to the outer surface $(h/2)$, to give

$$\tau_{max} = \sqrt{\left(\frac{\sigma}{2}\right)^2 + \tau^2} = \frac{3F}{2bh}\sqrt{4(L/h)^2(y/c)^2 + [1 - (y/c)^2]^2} \qquad (d)$$

To investigate the significance of transverse shear stress, we plot τ_{max} as a function of L/h for several values of y/c, as shown in Fig. 3–19. Since F and b appear only as linear multipliers outside the radical, they will only serve to scale the plot in the vertical direction without changing any of the relationships. Notice that at the neutral axis where $y/c = 0$, τ_{max} is constant for any length beam, since the bending stress is zero at the neutral axis and the transverse shear stress is independent of L. On the other hand, on the outer surface where $y/c = 1$, τ_{max} increases linearly with L/h because of the bending moment. For y/c between zero and one, τ_{max} is nonlinear for low values of L/h, but behaves linearly as L/h increases, displaying the dominance of the bending stress as the moment arm increases. We can see from the graph that the critical stress element (the largest value of τ_{max}) will always be either on the outer surface $(y/c = 1)$ or at the neutral axis $(y/c = 0)$, and never between. Thus, for the rectangular cross section, the transition between these two locations occurs at $L/h = 0.5$ where the line for $y/c = 1$ crosses the horizontal line for $y/c = 0$. The critical stress element is either on the outer surface where the transverse shear is zero, or if L/h is small enough, it is on the neutral axis where the bending stress is zero.

Figure 3–19

Plot of maximum shear stress for a cantilever beam, combining the effects of bending and transverse shear stresses.

The conclusions drawn from Fig. 3–19 are generally similar for any cross section that does not increase in width farther away from the neutral axis. This notably includes solid round cross sections, but not I beams or channels. Care must be taken with I beams and channels that have thin webs that extend far enough from the neutral axis that the bending and shear may both be significant on the same stress element (See Ex. 3–7). For any common cross section beam, if the beam length to height ratio is greater than 10, the transverse shear stress is generally considered negligible compared to the bending stress at any point within the cross section.

EXAMPLE 3–7

A beam 0.3 m long is to support a load of 2 kN acting 80 mm from the left support, as shown in Fig. 3–20a. The beam is an I beam with the cross-sectional dimensions shown. To simplify the calculations, assume a cross section with square corners, as shown in Fig. 3–20c. Points of interest are labeled (a, b, c, and d) at distances y from the neutral axis of 0 mm, 32^- mm, 32^+ mm, and 38 mm (Fig. 3–20c). At the critical axial location along the beam, find the following information.

(a) Determine the profile of the distribution of the transverse shear stress, obtaining values at each of the points of interest.

(b) Determine the bending stresses at the points of interest.

(c) Determine the maximum shear stresses at the points of interest, and compare them.

Solution

First, we note that the transverse shear stress is not likely to be negligible in this case since the beam length to height ratio is much less than 10, and since the thin web and wide flange will allow the transverse shear to be large. The loading, shear-force, and bending-moment diagrams are shown in Fig. 3–20b. The critical axial location is at $x = 0.08$ m where the shear force and the bending moment are both maximum.

(a) We obtain the area moment of inertia I by evaluating I for a solid 76-mm × 58-mm rectangular area, and then subtracting the two rectangular areas that are not part of the cross section.

$$I = \frac{(58)(76)^3}{12} - 2\left[\frac{(27)(64)^3}{12}\right] = 942\,069 \text{ mm}^4$$

| **Figure 3–20**

Finding Q at each point of interest using Eq. (3–30) gives

$$Q_a = \left(32 + \frac{6}{2}\right)[(58)(6)] + \left(\frac{32}{2}\right)[(32)(4)] = 14\ 228 \text{ mm}^3$$

$$Q_b = Q_c = \left(32 + \frac{6}{2}\right)[(58)(6)] = 12\ 180 \text{ mm}^3$$

$$Q_d = (38)(0) = 0 \text{ mm}^3$$

Applying Eq. (3–31) at each point of interest, with V and I constant for each point, and b equal to the width of the cross section at each point, shows that the magnitudes of the transverse shear stresses are

$$\tau_a = \frac{VQ_a}{Ib_a} = \frac{(1470)(14\ 228 \times 10^{-9})}{(942\ 069 \times 10^{-12})(0.004)} = 5.55 \text{ MPa}$$

Answer

$$\tau_b = \frac{VQ_b}{Ib_b} = \frac{(1470)(12\ 180 \times 10^{-9})}{(942\ 069 \times 10^{-12})(0.004)} = 4.75 \text{ MPa}$$

$$\tau_c = \frac{VQ_c}{Ib_c} = \frac{(1470)(12\ 180 \times 10^{-9})}{(942\ 069 \times 10^{-12})(0.058)} = 0.33 \text{ MPa}$$

$$\tau_d = \frac{VQ_d}{Ib_d} = \frac{(1470)(0)}{(942\ 069 \times 10^{-12})(0.058)} = 0 \text{ MPa}$$

The magnitude of the idealized transverse shear stress profile through the beam depth will be as shown in Fig. 3–20d.

(b) The bending stresses at each point of interest are

Answer $$\sigma_a = \frac{My_a}{I} = \frac{(117.6)(0)}{942\ 069 \times 10^{-12}} = 0\ \text{MPa}$$

$$\sigma_b = \sigma_c = -\frac{My_b}{I} = -\frac{(117.6)(0.032)}{942\ 069 \times 10^{-12}} = -3.99\ \text{MPa}$$

$$\sigma_d = -\frac{My_d}{I} = -\frac{(117.6)(0.038)}{942\ 069 \times 10^{-12}} = -4.74\ \text{MPa}$$

(c) Now at each point of interest, consider a stress element that includes the bending stress and the transverse shear stress. The maximum shear stress for each stress element can be determined by Mohr's circle, or analytically by Eq. (3–14) with $\sigma_y = 0$,

$$\tau_{\text{max}} = \sqrt{\left(\frac{\sigma}{2}\right)^2 + \tau^2}$$

Thus, at each point

$$\tau_{\text{max},a} = \sqrt{0 + (5.55)^2} = 5.55\ \text{MPa}$$

$$\tau_{\text{max},b} = \sqrt{\left(\frac{-3.99}{2}\right)^2 + (4.75)^2} = 5.15\ \text{MPa}$$

$$\tau_{\text{max},c} = \sqrt{\left(\frac{-3.99}{2}\right)^2 + (0.33)^2} = 2.02\ \text{MPa}$$

$$\tau_{\text{max},d} = \sqrt{\left(\frac{-4.74}{2}\right)^2 + 0} = 2.37\ \text{MPa}$$

Answer Interestingly, the critical location is at point a where the maximum shear stress is the largest, even though the bending stress is zero. The next critical location is at point b in the web, where the thin web thickness dramatically increases the transverse shear stress compared to points c or d. These results are counterintuitive, since both points a and b turn out to be more critical than point d, even though the bending stress is maximum at point d. The thin web and wide flange increase the impact of the transverse shear stress. If the beam length to height ratio were increased, the critical point would move from point a to point b, since the transverse shear stress at point a would remain constant, but the bending stress at point b would increase. The designer should be particularly alert to the possibility of the critical stress element not being on the outer surface with cross sections that get wider farther from the neutral axis, particularly in cases with thin web sections and wide flanges. For rectangular and circular cross sections, however, the maximum bending stresses at the outer surfaces will dominate, as was shown in Fig. 3–19.

3-12 **Torsion**

Any moment vector that is collinear with an axis of a mechanical part is called a *torque vector,* because the moment causes the part to be twisted about that axis. A bar subjected to such a moment is said to be in *torsion.*

As shown in Fig. 3–21, the torque T applied to a bar can be designated by drawing arrows on the surface of the bar to indicate direction or by drawing torque-vector arrows along the axes of twist of the bar. Torque vectors are the hollow arrows shown on the x axis in Fig. 3–21. Note that they conform to the right-hand rule for vectors.

The *angle of twist,* in radians, for a solid round bar is

$$\theta = \frac{Tl}{GJ} \tag{3-35}$$

where T = torque
l = length
G = modulus of rigidity
J = polar second moment of area

Shear stresses develop throughout the cross section. For a round bar in torsion, these stresses are proportional to the radius ρ and are given by

$$\tau = \frac{T\rho}{J} \tag{3-36}$$

Designating r as the radius to the outer surface, we have

$$\tau_{\max} = \frac{Tr}{J} \tag{3-37}$$

The assumptions used in the analysis are:

• The bar is acted upon by a pure torque, and the sections under consideration are remote from the point of application of the load and from a change in diameter.
• The material obeys Hooke's law.
• Adjacent cross sections originally plane and parallel remain plane and parallel after twisting, and any radial line remains straight.

| **Figure 3–21**

The last assumption depends upon the axisymmetry of the member, so it does not hold true for noncircular cross sections. Consequently, Eqs. (3–35) through (3–37) apply *only* to circular sections. For a solid round section,

$$J = \frac{\pi d^4}{32} \tag{3-38}$$

where d is the diameter of the bar. For a hollow round section,

$$J = \frac{\pi}{32}(d_o^4 - d_i^4) \tag{3-39}$$

where the subscripts o and i refer to the outside and inside diameters, respectively.

There are some applications in machinery for noncircular cross section members and shafts where a regular polygonal cross section is useful in transmitting torque to a gear or pulley that can have an axial change in position. Because no key or keyway is needed, the possibility of a lost key is avoided. The development of equations for stress and deflection for torsional loading of noncircular cross sections can be obtained from the mathematical theory of elasticity. In general, the shear stress does not vary linearly with the distance from the axis, and depends on the specific cross section. In fact, for a rectangular section bar the shear stress is zero at the corners where the distance from the axis is the largest. The maximum shearing stress in a rectangular $b \times c$ section bar occurs in the middle of the *longest* side b and is of the magnitude

$$\tau_{\max} = \frac{T}{\alpha b c^2} \approx \frac{T}{bc^2}\left(3 + \frac{1.8}{b/c}\right) \tag{3-40}$$

where b is the width (longer side) and c is the thickness (shorter side). They can *not* be interchanged. The parameter α is a factor that is a function of the ratio b/c as shown in the following table.[5] The angle of twist is given by

$$\theta = \frac{Tl}{\beta bc^3 G} \tag{3-41}$$

where β is a function of b/c, as shown in the table.

b/c	1.00	1.50	1.75	2.00	2.50	3.00	4.00	6.00	8.00	10	∞
α	0.208	0.231	0.239	0.246	0.258	0.267	0.282	0.299	0.307	0.313	0.333
β	0.141	0.196	0.214	0.228	0.249	0.263	0.281	0.299	0.307	0.313	0.333

Equation (3–40) is also approximately valid for equal-sided angles; these can be considered as two rectangles, each of which is capable of carrying half the torque.[6]

It is often necessary to obtain the torque T from a consideration of the power and speed of a rotating shaft. For convenience when U. S. Customary units are used, three forms of this relation are

$$H = \frac{FV}{33\,000} = \frac{2\pi Tn}{33\,000(12)} = \frac{Tn}{63\,025} \tag{3-42}$$

---

[5]S. Timoshenko, *Strength of Materials,* Part I, 3rd ed., D. Van Nostrand Company, New York, 1955, p. 290.

[6]For other sections see W. C. Young, R. G. Budynas, and A. M. Sadegh, *Roark's Formulas for Stress and Strain,* 8th ed., McGraw-Hill, New York, 2012.

where H = power, hp

   T = torque, lbf · in

   n = shaft speed, rev/min

   F = force, lbf

   V = velocity, ft/min

When SI units are used, the equation is

$$H = T\omega \qquad (3\text{--}43)$$

where H = power, W

   T = torque, N · m

   ω = angular velocity, rad/s

The torque T corresponding to the power in watts is given approximately by

$$T = 9.55\frac{H}{n} \qquad (3\text{--}44)$$

where n is in revolutions per minute.

EXAMPLE 3–8   Figure 3–22 shows a crank loaded by a force $F = 1.3$ kN that causes twisting and bending of a 20-mm-diameter shaft fixed to a support at the origin of the reference system. In actuality, the support may be an inertia that we wish to rotate, but for the purposes of a stress analysis we can consider this a statics problem.

(a) Draw separate free-body diagrams of the shaft AB and the arm BC, and compute the values of all forces, moments, and torques that act. Label the directions of the coordinate axes on these diagrams.

(b) Compute the maxima of the torsional stress and the bending stress in the arm BC and indicate where these act.

(c) Locate a stress element on the top surface of the shaft at A, and calculate all the stress components that act upon this element.

(d) Determine the maximum normal and shear stresses at A.

| Figure 3–22

Solution   (a) The two free-body diagrams are shown in Fig. 3–23. The results are

At end C of arm BC: $\mathbf{F} = -1.3\mathbf{j}$ kN, $\mathbf{T}_C = -0.05\mathbf{k}$ kN · m
At end B of arm BC: $\mathbf{F} = 1.3\mathbf{j}$ kN, $\mathbf{M}_1 = 0.13\mathbf{i}$ kN · m, $\mathbf{T}_1 = 0.05\mathbf{k}$ kN · m
At end B of shaft AB: $\mathbf{F} = -1.3\mathbf{j}$ kN, $\mathbf{T}_2 = -0.13\mathbf{i}$ kN · m, $\mathbf{M}_2 = -0.05\mathbf{k}$ kN · m
At end A of shaft AB: $\mathbf{F} = 1.3\mathbf{j}$ kN, $\mathbf{M}_A = 0.66\mathbf{k}$ kN · m, $\mathbf{T}_A = 0.13\mathbf{i}$ kN · m

(b) For arm BC, the bending moment will reach a maximum near the shaft at B. If we assume this is 0.13 kN · m, then the bending stress for a rectangular section will be

Answer
$$\sigma = \frac{M}{I/c} = \frac{6M}{bh^2} = \frac{6(130)}{0.006(0.03)^2} = 144.4 \text{ MPa}$$

Of course, this is not exactly correct, because at B the moment is actually being transferred into the shaft, probably through a weldment.
For the torsional stress, use Eq. (3–43). Thus

Answer
$$\tau_{\max} = \frac{T}{bc^2}\left(3 + \frac{1.8}{b/c}\right) = \frac{50}{0.03(0.006)^2}\left(3 + \frac{1.8}{0.03/0.006}\right) = 155.6 \text{ MPa}$$

This stress occurs at the middle of the 30-mm side.
(c) For a stress element at A, the bending stress is tensile and is

Answer
$$\sigma_x = \frac{M}{I/c} = \frac{32M}{\pi d^3} = \frac{32(660)}{\pi(0.02)^3} = 840.3 \text{ MPa}$$

| Figure 3–23

The torsional stress is

Answer

$$\tau_{xz} = \frac{-T}{J/c} = \frac{-16T}{\pi d^3} = \frac{-16(130)}{\pi (0.02)^3} = -82.8 \text{ MPa}$$

where the reader should verify that the negative sign accounts for the direction of τ_{xz}.

(d) Point A is in a state of plane stress where the stresses are in the xz plane. Thus the principal stresses are given by Eq. (3–13) with subscripts corresponding to the x, z axes.

Answer

The maximum normal stress is then given by

$$\sigma_1 = \frac{\sigma_x + \sigma_z}{2} + \sqrt{\left(\frac{\sigma_x - \sigma_z}{2}\right)^2 + \tau_{xz}^2}$$

$$= \frac{840.3 + 0}{2} + \sqrt{\left(\frac{840.3 - 0}{2}\right)^2 + (-82.8)^2} = 848.4 \text{ MPa}$$

Answer

The maximum shear stress at A occurs on surfaces different than the surfaces containing the principal stresses or the surfaces containing the bending and torsional shear stresses. The maximum shear stress is given by Eq. (3–14), again with modified subscripts, and is given by

$$\tau_1 = \sqrt{\left(\frac{\sigma_x - \sigma_z}{2}\right)^2 + \tau_{xz}^2} = \sqrt{\left(\frac{840.3 - 0}{2}\right)^2 + (-82.8)^2} = 428.2 \text{ MPa}$$

EXAMPLE 3–9

The 40-mm-diameter solid steel shaft shown in Fig. 3–24a is simply supported at the ends. Two pulleys are keyed to the shaft where pulley B is of diameter 100 mm and pulley C is of diameter 200 mm. Considering bending and torsional stresses only, determine the locations and magnitudes of the greatest tensile, compressive, and shear stresses in the shaft.

Solution

Figure 3–24b shows the net forces, reactions, and torsional moments on the shaft. Although this is a three-dimensional problem and vectors might seem appropriate, we will look at the components of the moment vector by performing a two-plane analysis. Figure 3–24c shows the loading in the xy plane, as viewed down the z axis, where bending moments are actually vectors in the z direction. Thus we label the moment diagram as M_z versus x. For the xz plane, we look down the y axis, and the moment diagram is M_y versus x as shown in Fig. 3–24d.

The net moment on a section is the vector sum of the components. That is,

$$M = \sqrt{M_y^2 + M_z^2} \qquad (1)$$

At point B,

$$M_B = \sqrt{225^2 + 900^2} = 928 \text{ N} \cdot \text{m}$$

At point C,

$$M_C = \sqrt{450^2 + 450^2} = 636 \text{ N} \cdot \text{m}$$

(a)

(b)

(c)   (d)

Location: at B ($x = 250$ mm$^+$)

$$\beta = \tan^{-1}\frac{900}{225} = 76$$

(e)

| **Figure 3–24**

Thus the maximum bending moment is 928 N·m and the maximum bending stress at pulley B is

$$\sigma = \frac{M\,d/2}{\pi d^4/64} = \frac{32M}{\pi d^3} = \frac{32(928)}{\pi(0.04^3)} = 147.7 \text{ MPa}$$

The maximum torsional shear stress occurs between B and C and is

$$\tau = \frac{T\,d/2}{\pi d^4/32} = \frac{16T}{\pi d^3} = \frac{16(180)}{\pi(0.04^3)} = 14.3 \text{ MPa}$$

The maximum bending and torsional shear stresses occur just to the right of pulley B at points E and F as shown in Fig. 3–24e. At point E, the maximum tensile stress will be σ_1 given by

Answer
$$\sigma_1 = \frac{\sigma}{2} + \sqrt{\left(\frac{\sigma}{2}\right)^2 + \tau^2} = \frac{147.7}{2} + \sqrt{\left(\frac{147.7}{2}\right)^2 + 14.3^2} = 149.1 \text{ MPa}$$

At point F, the maximum compressive stress will be σ_2 given by

Answer
$$\sigma_2 = \frac{\sigma}{2} - \sqrt{\left(\frac{-\sigma}{2}\right)^2 + \tau^2} = \frac{147.7}{2} - \sqrt{\left(\frac{-147.7}{2}\right)^2 + 14.3^2} = -1.4 \text{ MPa}$$

The extreme shear stress also occurs at E and F and is

Answer
$$\tau_1 = \sqrt{\left(\frac{\pm\sigma}{2}\right)^2 + \tau^2} = \sqrt{\left(\frac{\pm147.7}{2}\right)^2 + 14.3^2} = 75.2 \text{ MPa}$$

Closed Thin-Walled Tubes $(t \ll r)$[7]

In closed thin-walled tubes, it can be shown that the product of shear stress times thickness of the wall τt is constant, meaning that the shear stress τ is inversely proportional to the wall thickness t. The total torque T on a tube such as depicted in Fig. 3–25 is given by

$$T = \int \tau t r\, ds = (\tau t)\int r\, ds = \tau t(2A_m) = 2A_m t \tau$$

where A_m is the *area enclosed by the section median line*. Solving for τ gives

$$\tau = \frac{T}{2A_m t} \qquad (3\text{–}45)$$

For constant wall thickness t, the angular twist (radians) per unit of length of the tube θ_1 is given by

$$\theta_1 = \frac{TL_m}{4GA_m^2 t} \qquad (3\text{–}46)$$

---

[7]See Sec. 3–13, F. P. Beer, E. R. Johnston, and J. T. De Wolf, *Mechanics of Materials*, 5th ed., McGraw-Hill, New York, 2009.

Figure 3–25

The depicted cross section is elliptical, but the section need not be symmetrical nor of constant thickness.

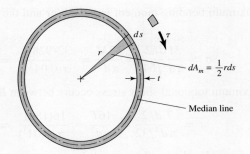

where L_m is the *length of the section median line.* These equations presume the buckling of the tube is prevented by ribs, stiffeners, bulkheads, and so on, and that the stresses are below the proportional limit.

EXAMPLE 3–10

A welded steel tube is 1 m long, has a 5-mm wall thickness, and a 45 mm by 70 mm rectangular cross section as shown in Fig. 3–26. Assume an allowable shear stress of 90 MPa and a shear modulus of 100 GPa.

(a) Estimate the allowable torque T.

(b) Estimate the angle of twist due to the torque.

Solution

(a) Within the section median line, the area enclosed is

$$A_m = (45 - 5)(70 - 5) = 2600 \text{ mm}^2$$

and the length of the median perimeter is

$$L_m = 2[(45 - 5) + (70 - 5)] = 210 \text{ mm}$$

From Eq. (3–45) the torque T is

Answer

$$T = 2A_m t\tau = 2(2600 \times 10^{-6})(0.005)(90 \times 10^6) = 2340 \text{ N·m}$$

(b) The angle of twist θ from Eq. (3–46) is

Answer

$$\theta = \theta_1 l = \frac{T L_m}{4 G A_m^2 t} l = \frac{2340(0.210)}{4(90 \times 10^9)(2600 \times 10^{-6})^2 (0.005)}(1) = 0.036 \text{ rad} = 2.08°$$

Figure 3–26

A rectangular steel tube produced by welding.

EXAMPLE 3–11

Compare the shear stress on a circular cylindrical tube with an outside diameter of 37 mm and an inside diameter of 32 mm, predicted by Eq. (3–37), to that estimated by Eq. (3–45).

Solution

From Eq. (3–37),

$$\tau_{\max} = \frac{Tr}{J} = \frac{Tr}{(\pi/32)\left(d_o^4 - d_i^4\right)} = \frac{T(0.0185)}{(\pi/32)(0.037^4 - 0.032^4)} = 228249.6\,T$$

From Eq. (3–45),

$$\tau = \frac{T}{2A_m t} = \frac{T}{2(\pi 0.0345^2/4)0.0025} = 213944.89\,T$$

Taking Eq. (3–37) as correct, the error in the thin-wall estimate is −6.27 percent.

Open Thin-Walled Sections

When the median wall line is not closed, the section is said to be an *open section*. Figure 3–27 presents some examples. Open sections in torsion, where the wall is thin, have relations derived from the membrane analogy theory[8] resulting in:

$$\tau = G\theta_1 c = \frac{3T}{Lc^2} \tag{3–47}$$

where τ is the shear stress, G is the shear modulus, θ_1 is the angle of twist per unit length, T is torque, and L is the length of the median line. The wall thickness is designated c (rather than t) to remind you that you are in open sections. By studying the table that follows Eq. (3–41) you will discover that membrane theory presumes $b/c \to \infty$. Note that open thin-walled sections in torsion should be avoided in design. As indicated in Eq. (3–47), the shear stress and the angle of twist are inversely proportional to c^2 and c^3, respectively. Thus, for small wall thickness, stress and twist can become quite large. For example, consider the thin round tube with a slit in Fig. 3–27. For a ratio of wall thickness of outside diameter of $c/d_o = 0.1$, the open section has greater magnitudes of stress and angle of twist by factors of 12.3 and 61.5, respectively, compared to a closed section of the same dimensions.

Figure 3–27

Some open thin-wall sections.

[8]See S. P. Timoshenko and J. N. Goodier, *Theory of Elasticity,* 3rd ed., McGraw-Hill, New York, 1970, Sec. 109.

EXAMPLE 3–12

A 0.3-m-long strip of steel is 3 mm thick and 25 mm wide, as shown in Fig. 3–28. If the allowable shear stress is 80 MPa and the shear modulus is 80 GPa, find the torque corresponding to the allowable shear stress and the angle of twist, in degrees, (a) using Eq. (3–47) and (b) using Eqs. (3–40) and (3–41).

Solution

(a) The length of the median line is 25 mm. From Eq. (3–47),

$$T = \frac{Lc^2\tau}{3} = \frac{(0.025)(0.003)^2 \, 80 \times 10^6}{3} = 6 \, \text{N} \cdot \text{m}$$

$$\theta = \theta_1 l = \frac{\tau l}{Gc} = \frac{80 \times 10^6 \, (0.3)}{80 \times 10^9 \, (0.003)} = 0.1 \, \text{rad} = 5.7°$$

A torsional spring rate k_t can be expressed as T/θ:

$$k_t = 6/0.1 = 60 \, \text{N} \cdot \text{m/rad}$$

(b) From Eq. (3–40),

$$T = \frac{\tau_{\text{max}} bc^2}{3 + 1.8/(b/c)} = \frac{80 \times 10^6 \, (0.025)(0.003)^2}{3 + 1.8/(25/3)} = 5.6 \, \text{N} \cdot \text{m}$$

From Eq. (3–41), with $b/c = 25/3 = 8.3$,

$$\theta = \frac{Tl}{\beta bc^3 G} = \frac{5.6 \, (0.3)}{0.307 \, (0.025)(0.003)^3 \, 80 \times 10^9} = 0.1011 \, \text{rad} = 5.8°$$

$$k_t = 5.6/0.1011 = 55.4 \, \text{N} \cdot \text{m/rad}$$

The cross section is not thin, where b should be greater than c by at least a factor of 10. In estimating the torque, Eq. (3–47) provides a value of 7.1 percent higher than Eq. (3–40), and is 8.0 percent higher than when the table on page 102 is used.

25 mm

3 mm

Figure 3–28

The cross-section of a thin strip of steel subjected to a torsional moment T.

$3\text{–}13$ **Stress Concentration**

In the development of the basic stress equations for tension, compression, bending, and torsion, it was assumed that no geometric irregularities occurred in the member under consideration. But it is quite difficult to design a machine without permitting some changes in the cross sections of the members. Rotating shafts must have shoulders designed on them so that the bearings can be properly seated and so that they will take thrust loads; and the shafts must have key slots machined into them for securing pulleys and gears. A bolt has a head on one end and screw threads on the other end, both of which account for abrupt changes in the cross section. Other parts require holes, oil grooves, and notches of various kinds. Any discontinuity in a machine part alters the stress distribution in the neighborhood of the discontinuity so that the elementary stress equations no longer describe the state of stress in the part at these locations. Such discontinuities are called *stress raisers,* and the regions in which they occur are called areas of *stress concentration.* Stress concentrations can also arise from some irregularity not inherent in the member, such as tool marks, holes, notches, grooves, or threads.

A *theoretical,* or *geometric, stress-concentration factor* K_t or K_{ts} is used to relate the actual maximum stress at the discontinuity to the *nominal stress.* The factors are defined by the equations

$$K_t = \frac{\sigma_{max}}{\sigma_0} \qquad K_{ts} = \frac{\tau_{max}}{\tau_0} \qquad (3\text{–}48)$$

where K_t is used for normal stresses and K_{ts} for shear stresses. The nominal stress σ_0 or τ_0 is the stress calculated by using the elementary stress equations and the net area, or net cross section. Sometimes the gross cross section is used instead, and so it is always wise to double check the source of K_t or K_{ts} before calculating the maximum stress.

The stress-concentration factor depends for its value only on the *geometry* of the part. That is, the particular material used has no effect on the value of K_t. This is why it is called a *theoretical* stress-concentration factor.

The analysis of geometric shapes to determine stress-concentration factors is a difficult problem, and not many solutions can be found. Most stress-concentration factors are found by using experimental techniques.[9] Though the finite-element method has been used, the fact that the elements are indeed finite prevents finding the exact maximum stress. Experimental approaches generally used include photoelasticity, grid methods, brittle-coating methods, and electrical strain-gauge methods. Of course, the grid and strain-gauge methods both suffer from the same drawback as the finite-element method.

Stress-concentration factors for a variety of geometries may be found in Tables A–15 and A–16.

An example is shown in Fig. 3–29, that of a thin plate loaded in tension where the plate contains a centrally located hole.

In *static loading,* stress-concentration factors are applied as follows. In *ductile materials* ($\epsilon_f \geq 0.05$), the stress-concentration factor *is not* usually applied to predict the critical stress, because plastic strain in the region of the stress concentration is localized and has a strengthening effect. In *brittle materials* ($\epsilon_f < 0.05$), the geometric stress-concentration factor K_t is applied to the nominal stress before comparing it with strength. Gray cast iron has so many inherent stress raisers that the stress raisers introduced by the designer have only a modest (but additive) effect.

Figure 3–29

Thin plate in tension or simple compression with a transverse central hole. The net tensile force is $F = \sigma w t$, where t is the thickness of the plate. The nominal stress is given by

$$\sigma_0 = \frac{F}{(w-d)t} = \frac{w}{(w-d)}\sigma$$

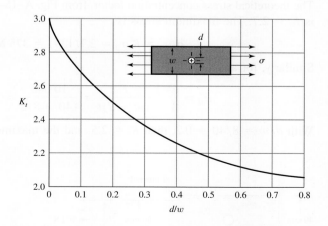

---

[9]The best source book is W. D. Pilkey and D. F. Pilkey, *Peterson's Stress Concentration Factors,* 3rd ed., John Wiley & Sons, New York, 2008.

Consider a part made of a ductile material and loaded by a gradually applied static load such that the stress in an area of a stress concentration goes beyond the yield strength. The yielding will be restricted to a very small region, and the permanent deformation as well as the residual stresses after the load is released will be insignificant and normally can be tolerated. If yielding does occur, the stress distribution changes and tends toward a more uniform distribution. In the region where yielding occurs, there is little danger of fracture of a ductile material, but if the possibility of a brittle fracture exists, the stress concentration must be taken seriously. Brittle fracture is not just limited to brittle materials. Materials often thought of as being ductile can fail in a brittle manner under certain conditions, e.g., any single application or combination of cyclic loading, rapid application of static loads, loading at low temperatures, and parts containing defects in their material structures (see Sec. 5–12). The effects on a ductile material of processing, such as hardening, hydrogen embrittlement, and welding, may also accelerate failure. Thus, care should always be exercised when dealing with stress concentrations.

For *dynamic loading,* the stress concentration effect is significant for *both* ductile and brittle materials and must always be taken into account (see Sec. 6–10).

EXAMPLE 3–13

The 2-mm-thick bar shown in Fig. 3–30 is loaded axially with a constant force of 10 kN. The bar material has been heat treated and quenched to raise its strength, but as a consequence it has lost most of its ductility. It is desired to drill a hole through the center of the 40-mm face of the plate to allow a cable to pass through it. A 4-mm hole is sufficient for the cable to fit, but an 8-mm drill is readily available. Will a crack be more likely to initiate at the larger hole, the smaller hole, or at the fillet?

Solution

Since the material is brittle, the effect of stress concentrations near the discontinuities must be considered. Dealing with the hole first, for a 4-mm hole, the nominal stress is

$$\sigma_0 = \frac{F}{A} = \frac{F}{(w - d)t} = \frac{10\,000}{(40 - 4)2} = 139 \text{ MPa}$$

The theoretical stress concentration factor, from Fig. A–15–1, with $d/w = 4/40 = 0.1$, is $K_t = 2.7$. The maximum stress is

Answer

$$\sigma_{\max} = K_t \sigma_0 = 2.7(139) = 375 \text{ MPa}$$

Similarly, for an 8-mm hole,

$$\sigma_0 = \frac{F}{A} = \frac{F}{(w - d)t} = \frac{10\,000}{(40 - 8)2} = 156 \text{ MPa}$$

With $d/w = 8/40 = 0.2$, then $K_t = 2.5$, and the maximum stress is

| **Figure 3–30**

40 mm   1 mm rad   34 mm   10 kN

Answer
$$\sigma_{max} = K_t\sigma_0 = 2.5(156) = 390 \text{ MPa}$$

Though the stress concentration is higher with the 4-mm hole, in this case the increased nominal stress with the 8-mm hole has more effect on the maximum stress.
For the fillet,

$$\sigma_0 = \frac{F}{A} = \frac{10\,000}{(34)2} = 147 \text{ MPa}$$

From Table A–15–5, $D/d = 40/34 = 1.18$, and $r/d = 1/34 = 0.026$. Then $K_t = 2.5$.

Answer
$$\sigma_{max} = K_t\sigma_0 = 2.5(147) = 368 \text{ MPa}$$

Answer
The crack will most likely occur with the 8-mm hole, next likely would be the 4-mm hole, and least likely at the fillet.

3–14 **Stresses in Pressurized Cylinders**

Cylindrical pressure vessels, hydraulic cylinders, gun barrels, and pipes carrying fluids at high pressures develop both radial and tangential stresses with values that depend upon the radius of the element under consideration. In determining the radial stress σ_r and the tangential stress σ_t, we make use of the assumption that the longitudinal elongation is constant around the circumference of the cylinder. In other words, a right section of the cylinder remains plane after stressing.

Referring to Fig. 3–31, we designate the inside radius of the cylinder by r_i, the outside radius by r_o, the internal pressure by p_i, and the external pressure by p_o. Then it can be shown that the tangential and radial stresses are[10]

$$\sigma_t = \frac{p_i r_i^2 - p_o r_o^2 - r_i^2 r_o^2 (p_o - p_i)/r^2}{r_o^2 - r_i^2}$$

$$\sigma_r = \frac{p_i r_i^2 - p_o r_o^2 + r_i^2 r_o^2 (p_o - p_i)/r^2}{r_o^2 - r_i^2}$$

(3–49)

As usual, positive values indicate tension and negative values, compression.
For the special case of $p_o = 0$, Eq. (3–49) gives

$$\sigma_t = \frac{r_i^2 p_i}{r_o^2 - r_i^2}\left(1 + \frac{r_o^2}{r^2}\right)$$

$$\sigma_r = \frac{r_i^2 p_i}{r_o^2 - r_i^2}\left(1 - \frac{r_o^2}{r^2}\right)$$

(3–50)

Figure 3–31

A cylinder subjected to both internal and external pressure.

[10]See Richard G. Budynas, *Advanced Strength and Applied Stress Analysis*, 2nd ed., McGraw-Hill, New York, 1999, pp. 348–352.

Figure 3-32

Distribution of stresses in a thick-walled cylinder subjected to internal pressure.

(a) Tangential stress distribution

(b) Radial stress distribution

The equations of set (3–50) are plotted in Fig. 3–32 to show the distribution of stresses over the wall thickness. It should be realized that longitudinal stresses exist when the end reactions to the internal pressure are taken by the pressure vessel itself. This stress is found to be

$$\sigma_l = \frac{p_i r_i^2}{r_o^2 - r_i^2} \tag{3-51}$$

We further note that Eqs. (3–49), (3–50), and (3–51) apply only to sections taken a significant distance from the ends and away from any areas of stress concentration.

Thin-Walled Vessels

When the wall thickness of a cylindrical pressure vessel is about one-tenth, or less, of its radius, the radial stress that results from pressurizing the vessel is quite small compared with the tangential stress. Under these conditions the tangential stress, called the *hoop stress,* can be obtained as follows: From Eq. (3–50), the average tangential stress is given by

$$(\sigma_t)_{\text{av}} = \frac{\int_{r_i}^{r_o} \sigma_t \, dr}{r_o - r_i} = \frac{\int_{r_i}^{r_o} \frac{r_i^2 p_i}{r_o^2 - r_i^2} \left(1 + \frac{r_o^2}{r_i^2}\right) dr}{r_o - r_i} = \frac{p_i r_i}{r_o - r_i} = \frac{p_i d_i}{2t} \tag{3-52}$$

where d_i is the inside diameter.

For a thin-walled pressure vessel, an approximation to the maximum tangential stress is

$$(\sigma_t)_{\text{max}} = \frac{p(d_i + t)}{2t} \tag{3-53}$$

In a closed cylinder, the longitudinal stress σ_l exists because of the pressure upon the ends of the vessel. If we assume this stress is also distributed uniformly over the wall thickness, we can easily find it to be

$$\sigma_l = \frac{p d_i}{4t} \tag{3-54}$$

EXAMPLE 3–14   An aluminum-alloy pressure vessel is made of tubing having an outside diameter of 200 mm and a wall thickness of 6 mm.

(a) What pressure can the cylinder carry if the permissible tangential stress is 82 MPa and the theory for thin-walled vessels is assumed to apply?

(b) On the basis of the pressure found in part (a), compute all of the stress components using the theory for thick-walled cylinders.

Solution   (a) Here $d_i = 200 - 2(6) = 188$ mm, $r_i = 188/2 = 94$ mm, and $r_o = 200/2 = 100$ mm. Then $t/r_i = 6/94 = 0.064$. Since this ratio is greater than $\frac{1}{20}$, the theory for thin-walled vessels may not yield safe results.

We first solve Eq. (3–53) to obtain the allowable pressure. This gives

Answer
$$p = \frac{2t(\sigma_t)_{max}}{d_i + t} = \frac{2(0.006)(82)}{0.188 + 0.006} = 5.07 \text{ MPa}$$

Then, from Eq. (3–54), we find the average longitudinal stress to be

$$\sigma_l = \frac{pd_i}{4t} = \frac{2.54(0.188)}{4(0.006)} = 19.9 \text{ MPa}$$

(b) The maximum tangential stress will occur at the inside radius, and so we use $r = r_i$ in the first equation of Eq. (3–49). This gives

Answer
$$(\sigma_t)_{max} = \frac{r_i^2 p_i}{r_o^2 - r_i^2}\left(1 + \frac{r_o^2}{r_i^2}\right) = p_i\frac{r_o^2 + r_i^2}{r_o^2 - r_i^2} = 2.54\frac{0.1^2 + 0.094^2}{0.1^2 - 0.094^2} = 41.1 \text{ MPa}$$

Similarly, the maximum radial stress is found, from the second equation of Eq. (3–50) to be

Answer
$$\sigma_r = -p_i = -2.54 \text{ MPa}$$

Equation (3–51) gives the longitudinal stress as

Answer
$$\sigma_l = \frac{p_i r_i^2}{r_o^2 - r_i^2} = \frac{2.54(0.094)^2}{0.1^2 - 0.094^2} = 19.28 \text{ MPa}$$

These three stresses, σ_t, σ_r, and σ_l, are principal stresses, since there is no shear on these surfaces. Note that there is no significant difference in the tangential stresses in parts (a) and (b), and so the thin-wall theory can be considered satisfactory.

3–15   **Stresses in Rotating Rings**

Many rotating elements, such as flywheels and blowers, can be simplified to a rotating ring to determine the stresses. When this is done it is found that the same tangential and radial stresses exist as in the theory for thick-walled cylinders except that they are caused by inertial forces acting on all the particles of the ring. The tangential and radial stresses so found are subject to the following restrictions:

- The outside radius of the ring, or disk, is large compared with the thickness $r_o \geq 10t$.
- The thickness of the ring or disk is constant.
- The stresses are constant over the thickness.

---

[11]Ibid, pp. 348–357.

The stresses are[11]

$$\sigma_t = \rho\omega^2\left(\frac{3+\nu}{8}\right)\left(r_i^2 + r_o^2 + \frac{r_i^2 r_o^2}{r^2} - \frac{1+3\nu}{3+\nu}r^2\right)$$

$$\sigma_r = \rho\omega^2\left(\frac{3+\nu}{8}\right)\left(r_i^2 + r_o^2 - \frac{r_i^2 r_o^2}{r^2} - r^2\right)$$

(3–55)

where r is the radius to the stress element under consideration, ρ is the mass density, and ω is the angular velocity of the ring in radians per second. For a rotating disk, use $r_i = 0$ in these equations.

3–16 Press and Shrink Fits

When two cylindrical parts are assembled by shrinking or press fitting one part upon another, a contact pressure is created between the two parts. The stresses resulting from this pressure may easily be determined with the equations of the preceding sections.

Figure 3–33 shows two cylindrical members that have been assembled with a shrink fit. Prior to assembly, the outer radius of the inner member was larger than the inner radius of the outer member by the *radial interference* δ. After assembly, an interference contact pressure p develops between the members at the nominal radius R, causing radial stresses $\sigma_r = -p$ in each member at the contacting surfaces. This pressure is given by[12]

$$p = \frac{\delta}{R\left[\frac{1}{E_o}\left(\frac{r_o^2 + R^2}{r_o^2 - R^2} + \nu_o\right) + \frac{1}{E_i}\left(\frac{R^2 + r_i^2}{R^2 - r_i^2} - \nu_i\right)\right]}$$

(3–56)

where the subscripts o and i on the material properties correspond to the outer and inner members, respectively. If the two members are of the same material with $E_o = E_i = E$, and $\nu_o = \nu_i$, the relation simplifies to

$$p = \frac{E\delta}{2R^3}\left[\frac{(r_o^2 - R^2)(R^2 - r_i^2)}{r_o^2 - r_i^2}\right]$$

(3–57)

Figure 3–33

Notation for press and shrink fits. (a) Unassembled parts; (b) after assembly.

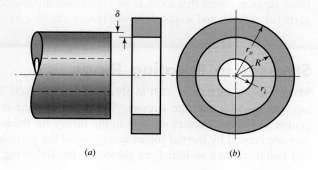

(a)   (b)

---

[12]Ibid, pp. 348–354.

For Eqs. (3–56) or (3–57), diameters can be used in place of R, r_i, and r_o, provided δ is the diametral interference (twice the radial interference).

With p, Eq. (3–49) can be used to determine the radial and tangential stresses in each member. For the inner member, $p_o = p$ and $p_i = 0$. For the outer member, $p_o = 0$ and $p_i = p$. For example, the magnitudes of the tangential stresses at the transition radius R are maximum for both members. For the inner member

$$(\sigma_t)_i \Big|_{r=R} = -p \frac{R^2 + r_i^2}{R^2 - r_i^2} \tag{3–58}$$

and, for the outer member

$$(\sigma_t)_o \Big|_{r=R} = p \frac{r_o^2 + R^2}{r_o^2 - R^2} \tag{3–59}$$

Assumptions

It is assumed that both members have the same length. In the case of a hub that has been press-fitted onto a shaft, this assumption would not be true, and there would be an increased pressure at each end of the hub. It is customary to allow for this condition by employing a stress-concentration factor. The value of this factor depends upon the contact pressure and the design of the female member, but its theoretical value is seldom greater than 2.

3–17 Temperature Effects

When the temperature of an unrestrained body is uniformly increased, the body expands, and the normal strain is

$$\epsilon_x = \epsilon_y = \epsilon_z = \alpha(\Delta T) \tag{3–60}$$

where α is the coefficient of thermal expansion and ΔT is the temperature change, in degrees. In this action the body experiences a simple volume increase with the components of shear strain all zero.

If a straight bar is restrained at the ends so as to prevent lengthwise expansion and then is subjected to a uniform increase in temperature, a compressive stress will develop because of the axial constraint. The stress is

$$\sigma = -\epsilon E = -\alpha(\Delta T)E \tag{3–61}$$

In a similar manner, if a uniform flat plate is restrained at the edges and also subjected to a uniform temperature rise, the compressive stress developed is given by the equation

$$\sigma = -\frac{\alpha(\Delta T)E}{1 - \nu} \tag{3–62}$$

Table 3–3

Coefficients of Thermal Expansion (Linear Mean Coefficients for the Temperature Range 0–100°C)

Material	Celsius Scale (°C^{-1})	Fahrenheit Scale (°F^{-1})
Aluminum	$23.9(10)^{-6}$	$13.3(10)^{-6}$
Brass, cast	$18.7(10)^{-6}$	$10.4(10)^{-6}$
Carbon steel	$10.8(10)^{-6}$	$6.0(10)^{-6}$
Cast iron	$10.6(10)^{-6}$	$5.9(10)^{-6}$
Magnesium	$25.2(10)^{-6}$	$14.0(10)^{-6}$
Nickel steel	$13.1(10)^{-6}$	$7.3(10)^{-6}$
Stainless steel	$17.3(10)^{-6}$	$9.6(10)^{-6}$
Tungsten	$4.3(10)^{-6}$	$2.4(10)^{-6}$

The stresses expressed by Eqs. (3–61) and (3–62) are called *thermal stresses*. They arise because of a temperature change in a clamped or restrained member. Such stresses, for example, occur during welding, since parts to be welded must be clamped before welding. Table 3–3 lists approximate values of the coefficients of thermal expansion.

3-18   Curved Beams in Bending[13]

The distribution of stress in a curved flexural member is determined by using the following assumptions:

- The cross section has an axis of symmetry in the plane of bending.
- Plane cross sections remain plane after bending.
- The modulus of elasticity is the same in tension as in compression.

We shall find that the neutral axis and the centroidal axis of a curved beam, unlike the axes of a straight beam, are not coincident and also that the stress does not vary linearly from the neutral axis. The notation shown in Fig. 3–34 is defined as follows:

$$r_o = \text{radius of outer fiber}$$
$$r_i = \text{radius of inner fiber}$$
$$h = \text{depth of section}$$
$$c_o = \text{distance from neutral axis to outer fiber}$$
$$c_i = \text{distance from neutral axis to inner fiber}$$
$$r_n = \text{radius of neutral axis}$$
$$r_c = \text{radius of centroidal axis}$$
$$e = r_c - r_n, \text{ distance from centroidal axis to neutral axis}$$
$$M = \text{bending moment; positive } M \text{ decreases curvature}$$

Figure 3–34 shows that the neutral and centroidal axes are not coincident. The location of the neutral axis with respect to the center of curvature O is given by the equation

$$r_n = \frac{A}{\int \dfrac{dA}{r}} \tag{3–63}$$

Figure 3–34

Note that y is positive in the direction toward the center of curvature, point O.

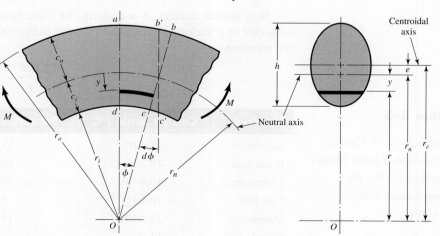

[13]For a complete development of the relations in this section, see Richard G. Budynas, *Advanced Strength and Applied Stress Analysis,* 2nd ed., McGraw-Hill, New York, 1999, pp. 309–317.

Furthermore, it can be shown that the stress distribution is given by

$$\sigma = \frac{My}{Ae(r_n - y)} \tag{3-64}$$

where M is positive in the direction shown in Fig. 3–34. The stress distribution given by Eq. (3–64) is *hyperbolic* and not linear as is the case for straight beams. The critical stresses occur at the inner and outer surfaces where $y = c_i$ and $y = -c_o$, respectively, and are

$$\sigma_i = \frac{Mc_i}{Aer_i} \qquad \sigma_o = -\frac{Mc_o}{Aer_o} \tag{3-65}$$

These equations are valid for pure bending. In the usual and more general case, such as a crane hook, the U frame of a press, or the frame of a C clamp, the bending moment is due to a force acting at a distance from the cross section under consideration. Thus, the cross section transmits a bending moment *and* an axial force. The axial force is located at *the centroidal axis* of the section and the bending moment is then computed at this location. The tensile or compressive stress due to the axial force, from Eq. (3–22), is then added to the bending stresses given by Eqs. (3–64) and (3–65) to obtain the resultant stresses acting on the section.

EXAMPLE 3–15

Plot the distribution of stresses across section *A-A* of the crane hook shown in Fig.3–35a. The cross section is rectangular, with $b = 18$ mm and $h = 100$ mm, and the load is $F = 22$ kN.

Solution

Since $A = bh$, we have $dA = b\,dr$ and, from Eq. (3–63),

$$r_n = \frac{A}{\displaystyle\int \frac{dA}{r}} = \frac{bh}{\displaystyle\int_{r_i}^{r_o} \frac{b}{r}\,dr} = \frac{h}{\ln \dfrac{r_o}{r_i}} \tag{1}$$

From Fig. 3–35b, we see that $r_i = 50$ mm, $r_o = 150$ mm, $r_c = 100$ mm, and $A = 1800$ mm^2. Thus, from Eq. (1)

$$r_n = \frac{h}{\ln(r_o/r_i)} = \frac{100}{\ln \frac{150}{50}} = 91 \text{ mm}$$

and so the eccentricity is $e = r_c - r_n = 100 - 91 = 9$ mm. The moment M is positive and is $M = Fr_c = 22(0.1) = 2.2$ kN·m. Adding the axial component of stress to Eq. (3–64) gives

$$\sigma = \frac{F}{A} + \frac{My}{Ae(r_n - y)} = \frac{22 \times 10^3}{1800 \times 10^{-6}} + \frac{(2.2 \times 10^3)(0.091 - r)}{1800 \times 10^{-6}\,(0.009)r} \tag{2}$$

Substituting values of r from 50 to 150 mm results in the stress distribution shown in Fig. 3–35c. The stresses at the inner and outer radii are found to be 123.6 and −41.2 MPa, respectively, as shown.

Figure 3–35

(a) Plan view of crane hook;
(b) cross section and notation;
(c) resulting stress distribution.
There is no stress
concentration.

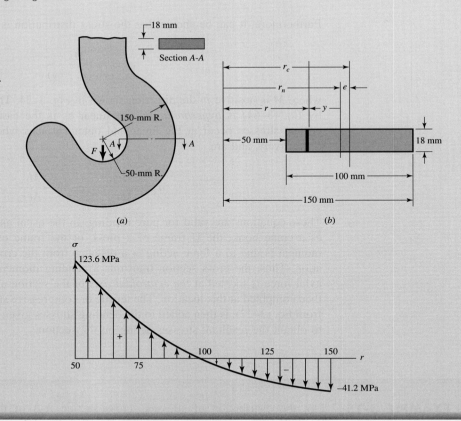

Note in the hook example, the symmetrical rectangular cross section causes the maximum tensile stress to be 3 times greater than the maximum compressive stress. If we wanted to design the hook to use material more effectively we would use more material at the inner radius and less material at the outer radius. For this reason, trapezoidal, T, or unsymmetric I, cross sections are commonly used. Sections most frequently encountered in the stress analysis of curved beams are shown in Table 3–4.

Alternative Calculations for e

Calculating r_n and r_c mathematically and subtracting the difference can lead to large errors if not done carefully, since r_n and r_c are typically large values compared to e. Since e is in the denominator of Eqs. (3–64) and (3–65), a large error in e can lead to an inaccurate stress calculation. Furthermore, if you have a complex cross section that the tables do not handle, alternative methods for determining e are needed. For a quick and simple approximation of e, it can be shown that[14]

$$e \approx \frac{I}{r_c A} \qquad (3\text{–}66)$$

This approximation is good for a large curvature where e is small with $r_n \approx r_c$. Substituting Eq. (3–66) into Eq. (3–64), with $r_n - y = r$, gives

$$\sigma \approx \frac{My}{I} \frac{r_c}{r} \qquad (3\text{–}67)$$

---

[14]Ibid., pp. 317–321. Also presents a numerical method.

Table 3–4

Formulas for Sections of Curved Beams

$$r_c = r_i + \frac{h}{2}$$

$$r_n = \frac{h}{\ln(r_o/r_i)}$$

$$r_c = r_i + \frac{h}{3}\frac{b_i + 2b_o}{b_i + b_o}$$

$$r_n = \frac{A}{b_o - b_i + [(b_i r_o - b_o r_i)/h]\ln(r_o/r_i)}$$

$$r_c = r_i + \frac{b_i c_1^2 + 2b_o c_1 c_2 + b_o c_2^2}{2(b_o c_2 + b_i c_1)}$$

$$r_n = \frac{b_i c_1 + b_o c_2}{b_i \ln[(r_i + c_1)/r_i] + b_o \ln[r_o/(r_i + c_1)]}$$

$$r_c = r_i + R$$

$$r_n = \frac{R^2}{2(r_c - \sqrt{r_c^2 - R^2})}$$

$$r_c = r_i + \frac{\frac{1}{2}h^2 t + \frac{1}{2}t_i^2(b_i - t) + t_o(b_o - t)(h - t_o/2)}{t_i(b_i - t) + t_o(b_o - t) + ht}$$

$$r_n = \frac{t_i(b_i - t) + t_o(b_o - t) + ht_o}{b_i \ln\dfrac{r_i + t}{r_i} + t\ln\dfrac{r_o - t_o}{r_i + t_i} + b_o \ln\dfrac{r_o}{r_o - t_o}}$$

$$r_c = r_i + \frac{\frac{1}{2}h^2 t + \frac{1}{2}t_i^2(b - t) + t_o(b - t)(h - t_o/2)}{ht + (b - t)(t_i + t_o)}$$

$$r_n = \frac{(b - t)(t_i + t_o) + ht}{b\left(\ln\dfrac{r_i + t_i}{r_i} + \ln\dfrac{r_o}{r_o - t_o}\right) + t\ln\dfrac{r_o - t_o}{r_i + t_i}}$$

If $r_n \approx r_c$, which it should be to use Eq. (3–67), then it is only necessary to calculate r_c, and to measure y from this axis. Determining r_c for a complex cross section can be done easily by most CAD programs or numerically as shown in the before-mentioned reference. Observe that as the curvature increases, $r \rightarrow r_c$, and Eq. (3–67) becomes the straight-beam formulation, Eq. (3–24). Note that the negative sign is missing because y in Fig. 3–34 is vertically downward, opposite that for the straight-beam equation.

EXAMPLE 3–16

A utility hook was formed from a round rod of diameter $d = 25$ mm into the geometry shown in the figure. What are the stresses at the inner and outer surfaces at section A–A if $F = 4$ kN, $L = 250$ mm, and $D_i = 80$ mm?

Solution

$d = 25$ mm, $r_i = 40$ mm, $r_o = 65$ mm.
From Table 3–4, for $R = 12.5$ mm,

$$r_c = 40 + 12.5 = 52.5 \text{ mm}$$

$$r_n = \frac{12.5^2}{2(52.5 - \sqrt{52.5^2 - 12.5^2})} = 51.7 \text{ mm}$$

$$e = r_c - r_n = 52.5 - 51.7 = 0.8 \text{ mm}$$
$$c_i = r_n - r_i = 51.7 - 40 = 11.7 \text{ mm}$$
$$c_o = r_o - r_n = 65 - 51.7 = 13.3 \text{ mm}$$
$$A = \pi d^2/4 = \pi(25)^2/4 = 491 \text{ mm}^2$$
$$M = Fr_c = 4000(52.5) = 210\,000 \text{ N} \cdot \text{mm}$$

Using Eq. (3–64)

Answer

$$\sigma_i = \frac{F}{A} + \frac{Mc_i}{Aer_i} = \frac{4000}{491} + \frac{210\,000(11.7)}{491(0.8)(40)} = 164.5 \text{ MPa}$$

Answer

$$\sigma_o = \frac{F}{A} - \frac{Mc_o}{Aer_o} = \frac{4000}{491} - \frac{210\,000(13.3)}{491(0.8)(65)} = -117.5 \text{ MPa}$$

Figure 3–36

3–19 Contact Stresses

When two bodies having curved surfaces are pressed together, point or line contact changes to area contact, and the stresses developed in the two bodies are three-dimensional. Contact-stress problems arise in the contact of a wheel and a rail, in automotive valve cams and tappets, in mating gear teeth, and in the action of rolling bearings. Typical failures are seen as cracks, pits, or flaking in the surface material.

The most general case of contact stress occurs when each contacting body has a double radius of curvature; that is, when the radius in the plane of rolling is different from the radius in a perpendicular plane, both planes taken through the axis of the contacting force. Here we shall consider only the two special cases of contacting spheres and contacting cylinders.[15] The results presented here are due to H. Hertz and so are frequently known as *Hertzian stresses*.

Spherical Contact

Figure 3–36 shows two solid spheres of diameters d_1 and d_2 pressed together with a force F. Specifying E_1, ν_1 and E_2, ν_2 as the respective elastic constants of the two spheres, the radius a of the circular contact area is given by the equation

$$a = \sqrt[3]{\frac{3F}{8} \frac{(1 - \nu_1^2)/E_1 + (1 - \nu_2^2)/E_2}{1/d_1 + 1/d_2}} \qquad (3\text{–}68)$$

The pressure distribution within the contact area of each sphere is hemispherical, as shown in Fig. 3–36b. The maximum pressure occurs at the center of the contact area and is

$$p_{max} = \frac{3F}{2\pi a^2} \qquad (3\text{–}69)$$

Equations (3–68) and (3–69) are perfectly general and also apply to the contact of a sphere and a plane surface or of a sphere and an internal spherical surface. For a plane surface, use $d = \infty$. For an internal surface, the diameter is expressed as a negative quantity.

The maximum stresses occur on the z axis, and these are principal stresses. Their values are

$$\sigma_1 = \sigma_2 = \sigma_x = \sigma_y = -p_{max}\left[\left(1 - \left|\frac{z}{a}\right|\tan^{-1}\frac{1}{|z/a|}\right)(1 + \nu) - \frac{1}{2\left(1 + \dfrac{z^2}{a^2}\right)}\right] \qquad (3\text{–}70)$$

$$\sigma_3 = \sigma_z = \frac{-p_{max}}{1 + \dfrac{z^2}{a^2}} \qquad (3\text{–}71)$$

Figure 3–36

(a) Two spheres held in contact by force F; (b) contact stress has a hemispherical distribution across contact zone of diameter $2a$.

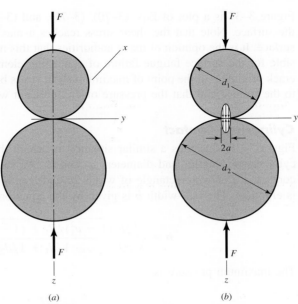

(a)   (b)

---

[15]A more comprehensive presentation of contact stresses may be found in Arthur P. Boresi and Richard J. Schmidt, *Advanced Mechanics of Materials,* 6th ed., Wiley, New York, 2003, pp. 589–623.

Figure 3–37

Magnitude of the stress components below the surface as a function of the maximum pressure of contacting spheres. Note that the maximum shear stress is slightly below the surface at $z = 0.48a$ and is approximately $0.3p_{max}$. The chart is based on a Poisson ratio of 0.30. Note that the normal stresses are all compressive stresses.

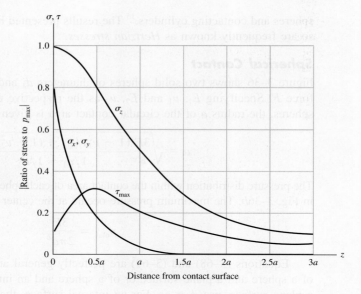

These equations are valid for either sphere, but the value used for Poisson's ratio must correspond with the sphere under consideration. The equations are even more complicated when stress states off the z axis are to be determined, because here the x and y coordinates must also be included. But these are not required for design purposes, because the maxima occur on the z axis.

Mohr's circles for the stress state described by Eqs. (3–70) and (3–71) are a point and two coincident circles. Since $\sigma_1 = \sigma_2$, we have $\tau_{1/2} = 0$ and

$$\tau_{max} = \tau_{1/3} = \tau_{2/3} = \frac{\sigma_1 - \sigma_3}{2} = \frac{\sigma_2 - \sigma_3}{2} \tag{3–72}$$

Figure 3–37 is a plot of Eqs. (3–70), (3–71), and (3–72) for a distance to $3a$ below the surface. Note that the shear stress reaches a maximum value slightly below the surface. It is the opinion of many authorities that this maximum shear stress is responsible for the surface fatigue failure of contacting elements. The explanation is that a crack originates at the point of maximum shear stress below the surface and progresses to the surface and that the pressure of the lubricant wedges the chip loose.

Cylindrical Contact

Figure 3–38 illustrates a similar situation in which the contacting elements are two cylinders of length l and diameters d_1 and d_2. As shown in Fig. 3–38b, the area of contact is a narrow rectangle of width $2b$ and length l, and the pressure distribution is elliptical. The half-width b is given by the equation

$$b = \sqrt{\frac{2F}{\pi l} \frac{(1 - \nu_1^2)/E_1 + (1 - \nu_2^2)/E_2}{1/d_1 + 1/d_2}} \tag{3–73}$$

The maximum pressure is

$$p_{max} = \frac{2F}{\pi bl} \tag{3–74}$$

Figure 3–38

(a) Two right circular cylinders held in contact by forces F uniformly distributed along cylinder length l.
(b) Contact stress has an elliptical distribution across the contact zone of width $2b$.

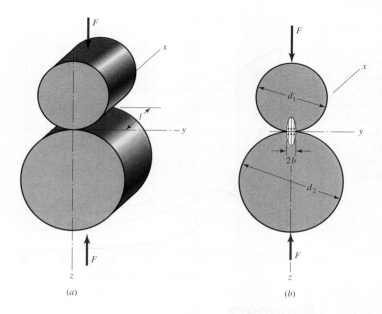

(a)

(b)

Equations (3–73) and (3–74) apply to a cylinder and a plane surface, such as a rail, by making $d = \infty$ for the plane surface. The equations also apply to the contact of a cylinder and an internal cylindrical surface; in this case d is made negative for the internal surface.

The stress state along the z axis is given by the equations

$$\sigma_x = -2\nu p_{max}\left(\sqrt{1 + \frac{z^2}{b^2}} - \left|\frac{z}{b}\right|\right) \tag{3–75}$$

$$\sigma_y = -p_{max}\left(\frac{1 + 2\dfrac{z^2}{b^2}}{\sqrt{1 + \dfrac{z^2}{b^2}}} - 2\left|\frac{z}{b}\right|\right) \tag{3–76}$$

$$\sigma_3 = \sigma_z = \frac{-p_{max}}{\sqrt{1 + z^2/b^2}} \tag{3–77}$$

These three equations are plotted in Fig. 3–39 up to a distance of $3b$ below the surface. For $0 \leq z \leq 0.436b$, $\sigma_1 = \sigma_x$, and $\tau_{max} = (\sigma_1 - \sigma_3)/2 = (\sigma_x - \sigma_z)/2$. For $z \geq 0.436b$, $\sigma_1 = \sigma_y$, and $\tau_{max} = (\sigma_y - \sigma_z)/2$. A plot of τ_{max} is also included in Fig. 3–39, where the greatest value occurs at $z/b = 0.786$ with a value of $0.300\, p_{max}$.

Hertz (1881) provided the preceding mathematical models of the stress field when the contact zone is free of shear stress. Another important contact stress case is *line of contact* with friction providing the shearing stress on the contact zone. Such shearing stresses are small with cams and rollers, but in cams with flat-faced followers, wheel-rail contact, and gear teeth, the stresses are elevated above the Hertzian field. Investigations of the effect on the stress field due to normal and shear stresses in the contact zone were begun theoretically by Lundberg (1939), and continued by Mindlin (1949), Smith-Liu (1949), and Poritsky (1949) independently. For further detail, see the reference cited in footnote 15, p. 137.

Figure 3–39

Magnitude of the stress components below the surface as a function of the maximum pressure for contacting cylinders. The largest value of τ_{max} occurs at $z/b = 0.786$. Its maximum value is $0.30p_{max}$. The chart is based on a Poisson ratio of 0.30. Note that all normal stresses are compressive stresses.

Distance from contact surface

3–20 Summary

The ability to quantify the stress condition at a critical location in a machine element is an important skill of the engineer. Why? Whether the member fails or not is assessed by comparing the (damaging) stress at a critical location with the corresponding material strength at this location. This chapter has addressed the description of stress.

Stresses can be estimated with great precision where the geometry is sufficiently simple that theory easily provides the necessary quantitative relationships. In other cases, approximations are used. There are numerical approximations such as finite element analysis (FEA, see Chap. 19), whose results tend to converge on the true values. There are experimental measurements, strain gauging, for example, allowing *inference* of stresses from the measured strain conditions. Whatever the method(s), the goal is a robust description of the stress condition at a critical location.

The nature of research results and understanding in any field is that the longer we work on it, the more involved things seem to be, and new approaches are sought to help with the complications. As newer schemes are introduced, engineers, hungry for the improvement the new approach *promises,* begin to use the approach. Optimism usually recedes, as further experience adds concerns. Tasks that promised to extend the capabilities of the nonexpert eventually show that expertise is not optional.

In stress analysis, the computer can be helpful if the necessary equations are available. Spreadsheet analysis can quickly reduce complicated calculations for parametric studies, easily handling "what if" questions relating trade-offs (e.g., less of a costly material or more of a cheaper material). It can even give insight into optimization opportunities.

When the necessary equations are not available, then methods such as FEA are attractive, but cautions are in order. Even when you have access to a powerful FEA code, you should be near an expert while you are learning. There are nagging questions of convergence at discontinuities. Elastic analysis is much easier than elastic-plastic analysis. The results are no better than the modeling of reality that was used to formulate the problem. Chapter 19 provides an idea of what finite-element analysis is and how it can be used in design. The chapter is by no means comprehensive in finite-element theory and the application of finite elements in practice. Both skill sets require much exposure and experience to be adept.

PROBLEMS

Problems marked with an asterisk (*) are linked with problems in other chapters, as summarized in Table 1–2 of Sec. 1–17, p. 34.

3–1* to
3–4

Sketch a free-body diagram of each element in the figure. Compute the magnitude and direction of each force using an algebraic or vector method, as specified.

Problem 3–1*   Problem 3–2

Problem 3–3   Problem 3–4

3–5 to
3–8

For the beam shown, find the reactions at the supports and plot the shear-force and bending-moment diagrams. Label the diagrams properly and provide values at all key points.

Problem 3–5

Dimensions in millimeters

Problem 3–6

Problem 3–7   Problem 3–8

3–9   Repeat Prob. 3–5 using singularity functions exclusively (including reactions).

3–10   Repeat Prob. 3–6 using singularity functions exclusively (including reactions).

3–11   Repeat Prob. 3–7 using singularity functions exclusively (including reactions).

3–12   Repeat Prob. 3–8 using singularity functions exclusively (including reactions).

3–13   For a beam from Table A–9, as specified by your instructor, find general expressions for the loading, shear-force, bending-moment, and support reactions. Use the method specified by your instructor.

3–14   A beam carrying a uniform load is simply supported with the supports set back a distance a from the ends as shown in the figure. The bending moment at x can be found from summing moments to zero at section x:

$$\sum M = M + \frac{1}{2}w(a + x)^2 - \frac{1}{2}wlx = 0$$

or

$$M = \frac{w}{2}[lx - (a + x)^2]$$

where w is the loading intensity in kN/m. The designer wishes to minimize the necessary weight of the supporting beam by choosing a setback resulting in the smallest possible maximum bending stress.

(a) If the beam is configured with $a = 0.06$ m, $l = 0.25$ m, and $w = 18$ kN/m, find the magnitude of the severest bending moment in the beam.

(b) Since the configuration in part (a) is not optimal, find the optimal setback a that will result in the lightest-weight beam.

Problem 3–14

3–15   For each of the plane stress states listed below, draw a Mohr's circle diagram properly labeled; find the principal normal and shear stresses, and determine the angle from the x axis to σ_1. Draw stress elements as in Fig. 3–11c and d and label all details.

(a) $\sigma_x = 20$ MPa, $\sigma_y = -10$ MPa, $\tau_{xy} = 8$ MPa cw
(b) $\sigma_x = 16$ MPa, $\sigma_y = 9$ MPa, $\tau_{xy} = 5$ MPa ccw
(c) $\sigma_x = 10$ MPa, $\sigma_y = 24$ MPa, $\tau_{xy} = 6$ MPa ccw
(d) $\sigma_x = -12$ MPa, $\sigma_y = 22$ MPa, $\tau_{xy} = 12$ MPa cw

3–16   Repeat Prob. 3–15 for:

(a) $\sigma_x = -8$ MPa, $\sigma_y = 7$ MPa, $\tau_{xy} = 6$ MPa cw
(b) $\sigma_x = 9$ MPa, $\sigma_y = -6$ MPa, $\tau_{xy} = 3$ MPa cw
(c) $\sigma_x = -4$ MPa, $\sigma_y = 12$ MPa, $\tau_{xy} = 7$ MPa ccw
(d) $\sigma_x = 6$ MPa, $\sigma_y = -5$ MPa, $\tau_{xy} = 8$ MPa ccw

3–17 Repeat Prob. 3–15 for:
(a) $\sigma_x = 12$ MPa, $\sigma_y = 6$ MPa, $\tau_{xy} = 4$ MPa, cw
(b) $\sigma_x = 30$ MPa, $\sigma_y = -10$ MPa, $\tau_{xy} = 10$ MPa ccw
(c) $\sigma_x = -10$ MPa, $\sigma_y = 18$ MPa, $\tau_{xy} = 9$ MPa cw
(d) $\sigma_x = 9$ MPa, $\sigma_y = 19$ MPa, $\tau_{xy} = 8$ MPa cw

3–18 For each of the stress states listed below, find all three principal normal and shear stresses. Draw a complete Mohr's three-circle diagram and label all points of interest.
(a) $\sigma_x = -80$ MPa, $\sigma_y = -30$ MPa, $\tau_{xy} = 20$ MPa cw
(b) $\sigma_x = 30$ MPa, $\sigma_y = -60$ MPa, $\tau_{xy} = 30$ MPa cw
(c) $\sigma_x = 40$ MPa, $\sigma_z = -30$ MPa, $\tau_{xy} = 20$ MPa ccw
(d) $\sigma_x = 50$ MPa, $\sigma_z = -20$ MPa, $\tau_{xy} = 30$ MPa cw

3–19 Repeat Prob. 3–18 for:
(a) $\sigma_x = 10$ MPa, $\sigma_y = -4$ MPa
(b) $\sigma_x = 10$ MPa, $\tau_{xy} = 4$ MPa ccw
(c) $\sigma_x = -2$ MPa, $\sigma_y = -8$ MPa, $\tau_{xy} = 4$ MPa cw
(d) $\sigma_x = 10$ MPa, $\sigma_y = -30$ MPa, $\tau_{xy} = 10$ MPa ccw

3–20 The state of stress at a point is $\sigma_x = -6$, $\sigma_y = 18$, $\sigma_z = -12$, $\tau_{xy} = 9$, $\tau_{yz} = 6$, and $\tau_{zx} = -15$ MPa. Determine the principal stresses, draw a complete Mohr's three-circle diagram, labeling all points of interest, and report the maximum shear stress for this case.

3–21 Repeat Prob. 3–20 with $\sigma_x = 20$, $\sigma_y = 0$, $\sigma_z = 20$, $\tau_{xy} = 40$, $\tau_{yz} = -20\sqrt{2}$, and $\tau_{zx} = 0$ MPa.

3–22 Repeat Prob. 3–20 with $\sigma_x = 10$, $\sigma_y = 40$, $\sigma_z = 40$, $\tau_{xy} = 20$, $\tau_{yz} = -40$, and $\tau_{zx} = -20$ MPa.

3–23 A 12-mm-diameter steel tension rod is 1.8 m long and carries a load of 9 kN. Find the tensile stress, the total deformation, the unit strains, and the change in the rod diameter.

3–24 Repeat Prob. 3–23 except change the rod to aluminum and the load to 1.8 kN.

3–25 A 30-mm-diameter copper rod is 1 m long with a yield strength of 70 MPa. Determine the axial force necessary to cause the diameter of the rod to reduce by 0.01 percent, assuming elastic deformation. Check that the elastic deformation assumption is valid by comparing the axial stress to the yield strength.

3–26 A diagonal aluminum alloy tension rod of diameter d and initial length l is used in a rectangular frame to prevent collapse. The rod can safely support a tensile stress of σ_{allow}. If $d = 15$ mm, $l = 3$ m, and $\sigma_{\text{allow}} = 135$ MPa, determine how much the rod must be stretched to develop this allowable stress.

3–27 Repeat Prob. 3–26 with $d = 16$ mm, $l = 3$ m, and $\sigma_{\text{allow}} = 140$ MPa.

3–28 Repeat Prob. 3–26 with $d = 15$ mm, $l = 3$ m, and $\sigma_{\text{allow}} = 105$ MPa.

3–29 Electrical strain gauges were applied to a notched specimen to determine the stresses in the notch. The results were $\epsilon_x = 0.0019$ and $\epsilon_y = -0.00072$. Find σ_x and σ_y if the material is carbon steel.

3–30 Repeat Prob. 3–29 for a material of aluminum.

3–31 The Roman method for addressing uncertainty in design was to build a copy of a design that was satisfactory and had proven durable. Although the early Romans did not have the intellectual tools to deal with scaling size up or down, you do. Consider a simply supported, rectangular-cross-section beam with a concentrated load F, as depicted in the figure.
(a) Show that the stress-to-load equation is

$$F = \frac{\sigma b h^2 l}{6ac}$$

(*b*) Subscript every parameter with *m* (for model) and divide into the above equation. Introduce a scale factor, $s = a_m/a = b_m/b = c_m/c$ etc. Since the Roman method was to not "lean on" the material any more than the proven design, set $\sigma_m/\sigma = 1$. Express F_m in terms of the scale factors and *F*, and comment on what you have learned.

Problem 3–31

3-32   Using our experience with concentrated loading on a simple beam, Prob. 3–31, consider a uniformly loaded simple beam (Table A–9–7).

(*a*) Show that the stress-to-load equation for a rectangular-cross-section beam is given by

$$W = \frac{4}{3}\frac{\sigma b h^2}{l}$$

where $W = wl$.

(*b*) Subscript every parameter with *m* (for model) and divide the model equation into the prototype equation. Introduce the scale factor *s* as in Prob. 3–31, setting $\sigma_m/\sigma = 1$. Express W_m and w_m in terms of the scale factor, and comment on what you have learned.

3-33   The Chicago North Shore & Milwaukee Railroad was an electric railway running between the cities in its corporate title. It had passenger cars as shown in the figure, which weighed 460 kN, had 10-m truck centers, 2-m-wheelbase trucks, and a coupled length of 17 m. Consider the case of a single car on a 30-m-long, simply supported deck plate girder bridge.

(*a*) What was the largest bending moment in the bridge?

(*b*) Where on the bridge was the moment located?

(*c*) What was the position of the car on the bridge?

(*d*) Under which axle is the bending moment?

Problem 3–33

3–34   For each section illustrated, find the second moment of area, the location of the neutral axis, and the distances from the neutral axis to the top and bottom surfaces. Consider that the section is transmitting a positive bending moment about the z axis, M_z, where $M_z = 50$ kN $\cdot$ m if the dimensions of the section are given in ips units, or $M_z = 1.13$ kN $\cdot$ m if the dimensions are in SI units. Determine the resulting stresses at the top and bottom surfaces and at every abrupt change in the cross section.

Problem 3–34

(a)

(b)

(c) Dimensions in mm

(d)

**3–35 to
3–38**   For the beam illustrated in the figure, find the locations and magnitudes of the maximum tensile bending stress due to M and the maximum shear stress due to V.

Problem 3–35

Problem 3–36

Problem 3–37

Problem 3–38

3–39 The figure illustrates a number of beam sections. Use an allowable bending stress of 80 MPa for steel and find the maximum safe uniformly distributed load that each beam can carry if the given lengths are between simple supports.

(a) Standard 50 mm × 5-mm tube, 3.6 m long

(b) Hollow steel tube 75 by 50 mm, outside dimensions, formed from 5-mm material and welded, 1.2 m long

(c) Steel angles 80 × 80 × 6 mm and 1.8 m long

(d) A 102 × 51-mm steel channel, 1.8 m long

Problem 3–39

(a)   (b)   (c)   (d)

3–40* A pin in a knuckle joint carrying a tensile load F deflects somewhat on account of this loading, making the distribution of reaction and load as shown in part (b) of the figure. A common simplification is to assume uniform load distributions, as shown in part (c). To further simplify, designers may consider replacing the distributed loads with point loads, such as in the two models shown in parts d and e. If $a = 12$ mm, $b = 18$ mm, $d = 12$ mm, and $F = 4$ kN, estimate the maximum bending stress and the maximum shear stress due to V for the three simplified models. Compare the three models from a designer's perspective in terms of accuracy, safety, and modeling time.

Problem 3–40*

(a)   (b)   (c)   (d)   (e)

3–41 Repeat Prob. 3–40 for $a = 6$ mm, $b = 18$ mm, $d = 12$ mm, and $F = 4$ kN.

3–42 For the knuckle joint described in Prob. 3–40, assume the maximum allowable tensile stress in the pin is 200 MPa and the maximum allowable shearing stress in the pin is 100 MPa. Use the model shown in part *c* of the figure to determine a minimum pin diameter for each of the following potential failure modes.

(*a*) Consider failure based on bending at the point of maximum bending stress in the pin.

(*b*) Consider failure based on the average shear stress on the pin cross section at the interface plane of the knuckle and clevis.

(*c*) Consider failure based on shear at the point of the maximum transverse shear stress in the pin.

3–43 The figure illustrates a pin tightly fitted into a hole of a substantial member. A usual analysis is one that assumes concentrated reactions *R* and *M* at distance *l* from *F*. Suppose the reaction is distributed linearly along distance *a*. Is the resulting moment reaction larger or smaller than the concentrated reaction? What is the loading intensity *q*? What do you think of using the usual assumption?

Problem 3–43

3–44 For the beam shown, determine (*a*) the maximum tensile and compressive bending stresses, (*b*) the maximum shear stress due to *V*, and (*c*) the maximum shear stress in the beam.

Problem 3–44

3–45 A cantilever beam with a 25-mm-diameter round cross section is loaded at the tip with a transverse force of 5 kN, as shown in the figure. The cross section at the wall is also shown, with labeled points *A* at the top, *B* at the center, and *C* at the midpoint between *A* and *B*. Study the

Problem 3–45

significance of the transverse shear stress in combination with bending by performing the following steps.

(a) Assume $L = 250$ mm. For points A, B, and C, sketch three-dimensional stress elements, labeling the coordinate directions and showing all stresses. Calculate magnitudes of the stresses on the stress elements. Do not neglect transverse shear stress. Calculate the maximum shear stress for each stress element.

(b) For each stress element in part (a), calculate the maximum shear stress if the transverse shear stress is neglected. Determine the percent error for each stress element from neglecting the transverse shear stress.

(c) Repeat the problem for $L = 100$, 25, and 2.5 mm. Compare the results and state any conclusions regarding the significance of the transverse shear stress in combination with bending.

3–46   Consider a simply supported beam of rectangular cross section of constant width b and variable depth h, so proportioned that the maximum stress σ_x at the outer surface due to bending is constant, when subjected to a load F at a distance a from the left support and a distance c from the right support. Show that the depth h at location x is given by

$$h = \sqrt{\frac{6Fcx}{lb\sigma_{\max}}} \qquad 0 \le x \le a$$

3–47   In Prob. 3–46, $h \to 0$ as $x \to 0$, which cannot occur. If the maximum shear stress $\tau_{\max}$ due to direct shear is to be constant in this region, show that the depth h at location x is given by

$$h = \frac{3}{2}\frac{Fc}{lb\tau_{\max}} \qquad 0 \le x \le \frac{3}{8}\frac{Fc\sigma_{\max}}{lb\tau_{\max}^2}$$

**3–48 and
3–49**   The beam shown is loaded in the xy and xz planes.

(a) Find the yz components of the reactions at the supports.

(b) Plot the shear-force and bending-moment diagrams for the xy and xz planes. Label the diagrams properly and provide the values at key points.

(c) Determine the net shear-force and bending-moment at the key points of part (b).

(d) Determine the maximum tensile bending stress. For Prob. 3–48, use the cross section given in Prob. 3–34, part (a). For Prob. 3–49, use the cross section given in Prob. 3–39, part (b).

Problem 3–48   Problem 3–49

3–50   Two steel thin-wall tubes in torsion of equal length are to be compared. The first is of square cross section, side length b, and wall thickness t. The second is a round of diameter b and wall thickness t. The largest allowable shear stress is τ_{all} and is to be the same in both cases.

(a) Determine the ratio of maximum torque for the square tube versus the round tube.

(b) Determine the ratio of the angle of twist per unit length for the square tube versus the round tube.

3–51   Consider a 25-mm-square steel thin-walled tube loaded in torsion. The tube has a wall thickness $t = 2$ mm, is 1 m long, and has a maximum allowable shear stress of 80 MPa. Determine the maximum torque that can be applied and the corresponding angle of twist of the tube.
(a) Assume that the internal radius at the corners $r_i = 0$.
(b) Assume that the internal radius at the corners is more realistically 3 mm.

Problem 3–51

3–52   The thin-walled open cross-section shown is transmitting torque T. The angle of twist per unit length of each leg can be determined separately using Eq. (3–47) and is given by

$$\theta_1 = \frac{3T_i}{GL_ic_i^3}$$

where for this case, $i = 1, 2, 3$, and T_i represents the torque in leg i. Assuming that the angle of twist per unit length for each leg is the same, show that

$$T = \frac{G\theta_1}{3}\sum_{i=1}^{3}L_ic_i^3 \qquad \text{and} \qquad \tau_{\max} = G\theta_1c_{\max}$$

Problem 3–52

3–53 to 3–55   Using the results from Prob. 3–52, consider a steel section with $\tau_{\text{allow}} = 80$ MPa.
(a) Determine the torque transmitted by each leg and the torque transmitted by the entire section.
(b) Determine the angle of twist per unit length.

Problem Number	c_1	L_1	c_2	L_2	c_3	L_3
3–53	2 mm	20 mm	3 mm	30 mm	0	0
3–54	1.5 mm	18 mm	3 mm	25 mm	1.5 mm	15 mm
3–55	2 mm	20 mm	3 mm	30 mm	2 mm	25 mm

3–56   Two 300-mm-long rectangular steel strips are placed together as shown. Using a maximum allowable shear stress of 80 MPa, determine the maximum torque and angular twist, and the torsional spring rate. Compare these with a single strip of cross section 30 mm by 4 mm. Solve the problem two ways: (*a*) using Eqs. (3–40) and (3–41), and (*b*) using Eq. (3–47). Compare and discuss your results

Problem 3–56

3–57   Using a maximum allowable shear stress of 70 MPa, find the shaft diameter needed to transmit 40 kW when
(*a*) The shaft speed is 2500 rev/min.
(*b*) The shaft speed is 250 rev/min.

3–58   Repeat Prob. 3–57 with an allowable shear stress of 140 MPa and a power of 37 kW.

3–59   Using an allowable shear stress of 50 MPa, determine the power that can be transmitted at 2000 rpm through a shaft with a 30-mm diameter.

3–60   A 20-mm-diameter steel bar is to be used as a torsion spring. If the torsional stress in the bar is not to exceed 110 MPa when one end is twisted through an angle of 15°, what must be the length of the bar?

3–61   A 0.6-m-long steel bar with a 20-mm-diameter is to be used as a torsion spring. If the torsional stress in the bar is not to exceed 210 MPa, what is the maximum angle of twist of the bar?

3–62   A 40-mm-diameter solid steel shaft, used as a torque transmitter, is replaced with a hollow shaft having a 40-mm OD and a 36-mm ID. If both materials have the same strength, what is the percentage reduction in torque transmission? What is the percentage reduction in shaft weight?

3–63   Generalize Prob. 3–62 for a solid shaft of diameter d replaced with a hollow shaft of the same material with an outside diameter d, and an inside diameter that is a fraction of the outside diameter, $x \times d$, where x is any value between zero and one. Obtain expressions for percentage reduction in torque transmission and percentage reduction in weight in terms of only x. Notice that the length and diameter of the shaft, and the material, are not needed for this comparison. Plot both results on the same axis for the range $0 < x < 1$. From the plot, what is the approximate value of x to obtain the greatest difference between the percent decrease in weight and the percent decrease in torque?

3–64   A hollow steel shaft is to transmit 4200 N · m of torque and is to be sized so that the torsional stress does not exceed 120 MPa.
(*a*) If the inside diameter is 70 percent of the outside diameter, what size shaft should be used? Use preferred sizes.
(*b*) What is the stress on the inside of the shaft when full torque is applied?

3–65   The figure shows an endless-belt conveyor drive roll. The roll has a diameter 120 mm and is driven at 10 rev/min by a geared-motor source rated at 1.5 kW. Determine a suitable shaft diameter d_C for an allowable torsional stress of 80 MPa.
(*a*) What would be the stress in the shaft you have sized if the motor starting torque is twice the running torque?
(*b*) Is bending stress likely to be a problem? What is the effect of different roll lengths B on bending?

Problem 3–65

(a)   (b)

3–66 The conveyer drive roll in the figure for Prob. 3–65 is 150 mm in diameter and is driven at 8 rev/min by a geared-motor source rated at 1 kW. Find a suitable shaft diameter d_C from the preferred decimal sizes in Table A–17, based on an allowable torsional stress of 75 MPa.

3–67 Consider two shafts in torsion, each of the same material, length, and cross-sectional area. One shaft has a solid square cross section and the other shaft has a solid circular section.
(a) Which shaft has the greater maximum shear stress and by what percentage?
(b) Which shaft has the greater angular twist θ and by what percentage?

3–68* to
3–71* A countershaft carrying two V-belt pulleys is shown in the figure. Pulley A receives power from a motor through a belt with the belt tensions shown. The power is transmitted through the shaft and delivered to the belt on pulley B. Assume the belt tension on the loose side at B is 15 percent of the tension on the tight side.
(a) Determine the tensions in the belt on pulley B, assuming the shaft is running at a constant speed.
(b) Find the magnitudes of the bearing reaction forces, assuming the bearings act as simple supports.
(c) Draw shear-force and bending-moment diagrams for the shaft. If needed, make one set for the horizontal plane and another set for the vertical plane.
(d) At the point of maximum bending moment, determine the bending stress and the torsional shear stress.
(e) At the point of maximum bending moment, determine the principal stresses and the maximum shear stress.

Problem 3–68*   *Problem 3–69**

Problem 3–70*
Dimensions in millimeters.

Problem 3–71*
Dimensions in millimeters.

3-72* to
3-73*

A gear reduction unit uses the countershaft shown in the figure. Gear A receives power from another gear with the transmitted force F_A applied at the 20° pressure angle as shown. The power is transmitted through the shaft and delivered through gear B through a transmitted force F_B at the pressure angle shown.

(*a*) Determine the force F_B, assuming the shaft is running at a constant speed.

(*b*) Find the bearing reaction forces, assuming the bearings act as simple supports.

(*c*) Draw shear-force and bending-moment diagrams for the shaft. If needed, make one set for the horizontal plane and another set for the vertical plane.

(*d*) At the point of maximum bending moment, determine the bending stress and the torsional shear stress.

(*e*) At the point of maximum bending moment, determine the principal stresses and the maximum shear stress.

Problem 3–72*

Problem 3–73*

3-74*

In the figure, shaft AB transmits power to shaft CD through a set of bevel gears contacting at point E. The contact force at E on the gear of shaft CD is determined to be $(\mathbf{F}_E)_{CD} = -400\mathbf{i} - 1600\mathbf{j} + 3600\mathbf{k}$ N. For shaft CD: (*a*) draw a free-body diagram and determine the reactions at C and D assuming simple supports (assume also that bearing C carries the thrust load), (b) draw the shear-force and bending-moment diagrams, (c) for the critical stress element, determine the torsional shear stress, the bending stress, and the axial stress, and (d) for the critical stress element, determine the principal stresses and the maximum shear stress.

Problem 3–74*

3–75 Repeat Prob. 3–74 except for a contact force at E of $(\mathbf{F}_E)_{CD} = -200\mathbf{i} - 600\mathbf{j} + 1800\mathbf{k}$ N and a shaft diameter of 25 mm.

3–76* Repeat the analysis of Prob. 3–74 for shaft AB. Assume that bearing A carries the thrust load.

3–77* A torque $T = 100$ N · m is applied to the shaft EFG, which is running at constant speed and contains gear F. Gear F transmits torque to shaft $ABCD$ through gear C, which drives the chain sprocket at B, transmitting a force P as shown. Sprocket B, gear C, and gear F have pitch diameters of $a = 150$, $b = 250$, and $c = 125$ mm, respectively. The contact force between the gears is transmitted through the pressure angle $\phi = 20°$. Assuming no frictional losses and considering the bearings at A, D, E, and G to be simple supports, locate the point on shaft $ABCD$ that contains the maximum tensile bending and maximum torsional shear stresses. Combine these stresses and determine the maximum principal normal and shear stresses in the shaft.

Problem 3–77*

View a–a

3–78 Repeat Prob. 3–77 with the chain parallel to the z axis with P in the positive z direction.

3–79* Repeat Prob. 3–77 with $T = 100$ N · m, $a = 150$ mm, $b = 125$ mm, $c = 250$ mm, $d = 35$ mm, $e = 100$ mm, $f = 250$ mm, and $g = 150$ mm.

3–80* The cantilevered bar in the figure is made from a ductile material and is statically loaded with $F_y = 800$ N and $F_x = F_z = 0$. Analyze the stress situation in rod AB by obtaining the following information.

(a) Determine the precise location of the critical stress element.

(b) Sketch the critical stress element and determine magnitudes and directions for all stresses acting on it. (Transverse shear may only be neglected if you can justify this decision.)

(c) For the critical stress element, determine the principal stresses and the maximum shear stress.

Problem 3–80*

3–81* Repeat Prob. 3–80 with $F_x = 0$, $F_y = 700$ N, and $F_z = 400$ N.

3–82* Repeat Prob. 3–80 with $F_x = 300$ N, $F_y = -800$ N, and $F_z = 400$ N.

3–83* For the handle in Prob. 3–80, one potential failure mode is twisting of the flat plate BC. Determine the maximum value of the shear stress due to torsion in the main section of the plate, ignoring the complexities of the interfaces at B and C.

3–84* The cantilevered bar in the figure is made from a ductile material and is statically loaded with $F_y = 1$ kN and $F_x = F_z = 0$. Analyze the stress situation in the small diameter at the shoulder at A by obtaining the following information:

Problem 3–84*

(a) Determine the precise location of the critical stress element at the cross section at A.

(b) Sketch the critical stress element and determine magnitudes and directions for all stresses acting on it. (Transverse shear may be neglected if you can justify this decision.)

(c) For the critical stress element, determine the principal stresses and the maximum shear stress.

3–85* Repeat Prob. 3–84 with $F_x = 1.2$ kN, $F_y = 1$ kN, and $F_z = 0$.

3–86* Repeat Prob. 3–84 with $F_x = 1.2$ kN, $F_y = 1$ kN, and $F_z = -0.4$ kN.

3–87* Repeat Prob. 3–84 for a brittle material, requiring the inclusion of stress concentration in the fillet radius.

3–88 Repeat Prob. 3–84 with $F_x = 1.2$ kN, $F_y = 1$ kN, and $F_z = 0$, and for a brittle material, requiring the inclusion of stress concentration in the fillet radius.

3–89 Repeat Prob. 3–84 with $F_x = 1.2$ kN, $F_y = 1$ kN, and $F_z = -0.4$ kN, and for a brittle material, requiring the inclusion of stress concentration in the fillet radius.

3–90 The figure shows a simple model of the loading of a square thread of a power screw transmitting an axial load F with an application of torque T. The torque is balanced by the frictional force F_f acting along the top surface of the thread. The forces on the thread are considered to be distributed along the circumference of the *mean diameter* d_m over the number of engaged threads, n_t. From the figure, $d_m = d_r + p/2$, where d_r is the *root diameter* of the thread and p is the *pitch* of the thread.

(a) Considering the thread to be a cantilever beam as shown in the cutaway view, show that the nominal bending stress at the root of the thread can be approximated by

$$\sigma_b = \pm \frac{6F}{\pi d_r n_t p}$$

(b) Show that the axial and maximum torsional shear stresses in the body of the shaft can be approximated by

$$\sigma_a = -\frac{4F}{\pi d_r^2} \qquad \text{and} \qquad \tau_t = \frac{16T}{\pi d_r^3}$$

(c) For the stresses of parts (a) and (b) show a three-dimensional representation of the state of stress on an element located at the intersection of the lower thread root base and the

Problem 3–90

thread body. Using the given coordinate system label the stresses using the notation given in Fig. 3–8a.

(d) A square-thread power screw has an outside diameter $d = 38$ mm, pitch $p = 6$ mm, and transmits a load $F = 6$ kN through the application of a torque $T = 26$ N · m. If $n_t = 2$, determine the key stresses and the corresponding *principal stresses* (normal and shear).

3–91   Develop the formulas for the maximum radial and tangential stresses in a thick-walled cylinder due to internal pressure only.

3–92   Repeat Prob. 3–91 where the cylinder is subject to external pressure only. At what radii do the maximum stresses occur?

3–93   Develop the equations for the principal stresses in a thin-walled spherical pressure vessel of inside diameter d_i, thickness t, and with an internal pressure p_i. You may wish to follow a process similar to that used for a thin-walled cylindrical pressure vessel on p. 128.

3–94 to   A pressure cylinder has an outer diameter d_o, wall thickness t, internal pressure p_i, and maxi-
3–96   mum allowable shear stress τ_{max}. In the table given, determine the appropriate value of x.

Problem Number	d_o	t	p_i	τ_{max}
3–94	150 mm	6 mm	x_{max}	70 MPa
3–95	200 mm	x_{min}	4 MPa	25 MPa
3–96	200 mm	6 mm	3 MPa	x

3–97 to   A pressure cylinder has an outer diameter d_o, wall thickness t, external pressure p_o, and maxi-
3–99   mum allowable shear stress τ_{max}. In the table given, determine the appropriate value of x.

Problem Number	d_o	t	p_o	τ_{max}
3–97	150 mm	6 mm	x_{max}	70 MPa
3–98	200 mm	x_{min}	4 MPa	25 MPa
3–99	200 mm	6 mm	3 MPa	x

3–100   An AISI 1040 cold-drawn steel tube has an OD = 50 mm and wall thickness 6 mm. What maximum external pressure can this tube withstand if the largest principal normal stress is not to exceed 80 percent of the minimum yield strength of the material?

3–101   Repeat Prob. 3–100 with an OD of 48 mm and wall thickness of 6 mm.

3–102   Repeat Prob. 3–100 with an internal pressure.

3–103   Repeat Prob. 3–101 with an internal pressure.

3–104   A thin-walled cylindrical steel water storage tank 9 m in diameter and 18 m long is oriented with its longitudinal axis vertical. The tank is topped with a hemispherical steel dome. The wall thickness of the tank and dome is 18 mm. If the tank is unpressurized and contains water 16 m above its base, and considering the weight of the tank, determine the maximum state of stress in the tank and the corresponding principal stresses (normal and shear). The weight density of water is 9810 N/m^3.

3–105   Repeat Prob. 3–104 with the tank being pressurized to 350 kPa.

3–106   Find the maximum shear stress in a 132-mm-diameter circular saw blade if it runs idle at 5000 rev/min. The saw is 14 gauge (1.9 mm) steel and is used on a 15-mm-diameter arbor. The thickness is uniform. What is the maximum radial component of stress?

3–107   The maximum recommended speed for a 250-mm-diameter abrasive grinding wheel is 2000 rev/min. Assume that the material is isotropic; use a bore of 20 mm, $\nu = 0.24$, and a mass density of 3320 kg/m^3, and find the maximum tensile stress at this speed.

3–108   An abrasive cutoff wheel has a diameter of 120 mm, is 1.5 mm thick, and has a 18-mm bore. It weighs 140 g and is designed to run at 12 000 rev/min. If the material is isotropic and $\nu = 0.20$, find the maximum shear stress at the design speed.

3–109   A rotary lawnmower blade rotates at 3500 rev/min. The steel blade has a uniform cross section 3 mm thick by 30 mm wide, and has a 12-mm-diameter hole in the center as shown in the figure. Estimate the nominal tensile stress at the central section due to rotation.

Problem 3–109

**3–110 to
3–115**   The table lists the maximum and minimum hole and shaft dimensions for a variety of standard press and shrink fits. The materials are both hot-rolled steel. Find the maximum and minimum values of the radial interference and the corresponding interface pressure. Use a collar diameter of 100 mm and 80 mm for even and odd numbered problems, respectively.

Problem Number	Fit Designation†	Basic Size	Hole		Shaft	
			D_{max}	D_{min}	d_{max}	d_{min}
3–110	50H7/p6	50 mm	50.025	50.000	50.042	50.026
3–111	40H7/p6	40 mm	40.025	40	40.042	40.026
3–112	50H7/s6	50 mm	50.025	50.000	50.059	50.043
3–113	40H7/s6	40 mm	40.025	40	40.059	40.043
3–114	50H7/u6	50 mm	50.025	50.000	50.086	50.070
3–115	40H7/u6	40 mm	40.025	40	40.076	40.060

†Note: See Table 7–9 for description of fits.

**3–116 to
3–119**   The table gives data concerning the shrink fit of two cylinders of differing materials and dimensional specification in inches. Elastic constants for different materials may be found in Table A–5. Identify the radial interference δ, then find the interference pressure p, and the tangential normal stress on both sides of the fit surface. If dimensional tolerances are given at fit surfaces, repeat the problem for the highest and lowest stress levels.

Problem Number	Inner Cylinder			Outer Cylinder		
	Material	d_i	d_o	Material	D_i	D_o
3–116	Steel	0	50.05	Steel	50	75
3–117	Steel	0	50.05	Cast iron	50	75
3–118	Steel	0	25.48	Steel	25.4/25.43	50
3–119	Aluminum	0	50.88/50.93	Steel	50.8/50.85	75

3–120   A utility hook was formed from a round rod of diameter $d = 20$ mm into the geometry shown in the figure. What are the stresses at the inner and outer surfaces at section A–A if $F = 4$ kN, $L = 250$ mm, and $D_i = 75$ mm?

Problem 3–120

3–121   Repeat Prob. 3–120 with $d = 18$ mm, $F = 3$ kN, $L = 240$ mm, and $D_i = 60$ mm.

3–122   The steel eyebolt shown in the figure is loaded with a force $F = 300$ N. The bolt is formed from wire of diameter $d = 6$ mm to a radius $R_i = 10$ mm in the eye and at the shank. Estimate the stresses at the inner and outer surfaces at section A–A.

Problem 3–122

3–123   For Prob. 3–122 estimate the stresses at the inner and outer surfaces at section B–B, located along the line between the radius centers.

3–124   Repeat Prob. 3–122 with $d = 6$ mm, $R_i = 12$ mm, and $F = 300$ N.

3–125   Repeat Prob. 3–123 with $d = 6$ mm, $R_i = 12$ mm, and $F = 300$ N.

3–126   Shown in the figure is a 2.8-mm by 20-mm latching spring that supports a load of $F = 100$ N. The inside radius of the bend is 3 mm.

    (a) Using straight-beam theory, determine the stresses at the top and bottom surfaces immediately to the right of the bend.

    (b) Using curved-beam theory, determine the stresses at the inner and outer surfaces at the bend.

    (c) By comparing the stresses at the bend with the nominal stresses before the bend, estimate effective stress concentration factors for the inner and outer surfaces.

Problem 3–126

Section A–A

3–127   Repeat Prob. 3–126 with a 3.6-mm material thickness.

3–128   Repeat Prob. 3–126 with a bend radius of 6 mm.

3–129   The cast-iron bell-crank lever depicted in the figure is acted upon by forces F_1 of 2.4 kN and F_2 of 3.2 kN. The section A–A at the central pivot has a curved inner surface with a radius of $r_i = 25$ mm. Estimate the stresses at the inner and outer surfaces of the curved portion of the lever.

Problem 3–129

Dimensions in mm

Section A–A

3–130   The crane hook depicted in Fig. 3–35 has a 18-mm-diameter hole in the center of the critical section. For a load of 30 kN, estimate the bending stresses at the inner and outer surfaces at the critical section.

3–131   An offset tensile link is shaped to clear an obstruction with a geometry as shown in the figure. The cross section at the critical location is elliptical, with a major axis of 100 mm and a minor axis of 50 mm. For a load of 90 kN, estimate the stresses at the inner and outer surfaces of the critical section.

Problem 3–131

250 mm R.

200 mm

3–132   A cast-steel C frame as shown in the figure has a rectangular cross section of 25 mm by 40 mm, with a 10-mm-radius semicircular notch on both sides that forms midflank fluting as shown. Estimate A, r_c, r_n, and e, and for a load of 13 kN, estimate the inner and outer surface stresses at the throat C. *Note:* Table 3–4 can be used to determine r_n for this section. From the table, the integral $\int dA/r$ can be evaluated for a rectangle and a circle by evaluating A/r_n for each shape [see Eq. (3–64)]. Subtracting A/r_n of the circle from that of the rectangle yields for the C frame, and r_n can then be evaluated.

Problem 3–132

3–133 Two carbon steel balls, each 30 mm in diameter, are pressed together by a force F. In terms of the force F in Newtons, find the maximum values of the principal stress, and the maximum shear stress, in MPa.

3–134 A carbon steel ball with 25-mm diameter is pressed together with an aluminum ball with a 40-mm diameter by a force of 10 N. Determine the maximum shear stress, and the depth at which it will occur for the aluminum ball. Assume Fig. 3–37, which is based on a typical Poisson's ratio of 0.3, is applicable to estimate the depth at which the maximum shear stress occurs for these materials.

3–135 Repeat Prob. 3–134 but determine the maximum shear stress and depth for the steel ball.

3–136 A carbon steel ball with a 30-mm diameter is pressed against a flat carbon steel plate with a force of 20 N. Determine the maximum shear stress, and the depth in the plate at which it will occur.

3–137 An AISI 1018 steel ball with 25-mm-diameter is used as a roller between a flat plate made from 2024 T3 aluminum and a flat table surface made from ASTM No. 30 gray cast iron. Determine the maximum amount of weight that can be stacked on the aluminum plate without exceeding a maximum shear stress of 140 MPa in any of the three pieces. Assume Fig. 3–37, which is based on a typical Poisson's ratio of 0.3, is applicable to estimate the depth at which the maximum shear stress occurs for these materials.

3–138 An aluminum alloy cylindrical roller with diameter 25 mm and length 50 mm rolls on the inside of a cast-iron ring having an inside radius of 100 mm, which is 50 mm thick. Find the maximum contact force F that can be used if the shear stress is not to exceed 28 MPa.

3–139 A pair of mating steel spur gears with a 18-mm face width transmits a load of 200 N. For estimating the contact stresses, make the simplifying assumption that the teeth profiles can be treated as cylindrical with instantaneous radii at the contact point of interest of 11 mm and 15 mm, respectively. Estimate the maximum contact pressure and the maximum shear stress experienced by either gear.

3–140 to 3–142 A wheel of diameter d and width w carrying a load F rolls on a flat rail.
Assume that Fig. 3–39, which is based on a Poisson's ratio of 0.3, is applicable to estimate the depth at which the maximum shear stress occurs for these materials. At this critical depth, calculate the Hertzian stresses σ_x, σ_y, σ_z, and τ_{max} for the wheel.

Problem Number	d	w	F	Wheel Material	Rail Material
3–140	125 mm	50 mm	3 kN	Steel	Steel
3–141	150 mm	40 mm	2 kN	Steel	Cast iron
3–142	72 mm	30 mm	1 kN	Cast iron	Cast iron

4

Deflection and Stiffness

Chapter Outline

All real bodies deform under load, either elastically or plastically. A body can be sufficiently insensitive to deformation that a presumption of rigidity does not affect an analysis enough to warrant a nonrigid treatment. If the body deformation later proves to be not negligible, then declaring rigidity was a poor decision, not a poor assumption. A wire rope is flexible, but in tension it can be robustly rigid and it distorts enormously under attempts at compressive loading. The same body can be both rigid and nonrigid.

Deflection analysis enters into design situations in many ways. A snap ring, or retaining ring, must be flexible enough to be bent without permanent deformation and assembled with other parts, and then it must be rigid enough to hold the assembled parts together. In a transmission, the gears must be supported by a rigid shaft. If the shaft bends too much, that is, if it is too flexible, the teeth will not mesh properly, and the result will be excessive impact, noise, wear, and early failure. In rolling sheet or strip steel to prescribed thicknesses, the rolls must be crowned, that is, curved, so that the finished product will be of uniform thickness. Thus, to design the rolls it is necessary to know exactly how much they will bend when a sheet of steel is rolled between them. Sometimes mechanical elements must be designed to have a particular force-deflection characteristic. The suspension system of an automobile, for example, must be designed within a very narrow range to achieve an optimum vibration frequency for all conditions of vehicle loading, because the human body is comfortable only within a limited range of frequencies.

The size of a load-bearing component is often determined on deflections, rather than limits on stress.

This chapter considers distortion of single bodies due to geometry (shape) and loading, then, briefly, the behavior of groups of bodies.

4–1 Spring Rates

Elasticity is that property of a material that enables it to regain its original configuration after having been deformed. A *spring* is a mechanical element that exerts a force when deformed. Figure 4–1a shows a straight beam of length l simply supported at the ends and loaded by the transverse force F. The deflection y is linearly related to the force, as long as the elastic limit of the material is not exceeded, as indicated by the graph. This beam can be described as a *linear spring*.

In Fig. 4–1b a straight beam is supported on two cylinders such that the length between supports decreases as the beam is deflected by the force F. A larger force is required to deflect a short beam than a long one, and hence the more this beam is deflected, the stiffer it becomes. Also, the force is not linearly related to the deflection, and hence this beam can be described as a *nonlinear stiffening spring*.

Figure 4–1c is an edge-view of a dish-shaped round disk. The force necessary to flatten the disk increases at first and then decreases as the disk approaches a flat

Figure 4–1

(a) A linear spring;
(b) a stiffening spring;
(c) a softening spring.

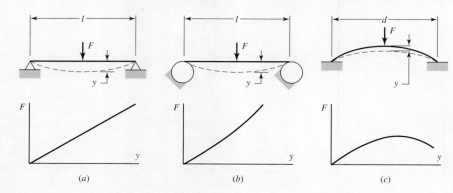

(a)   (b)   (c)

configuration, as shown by the graph. Any mechanical element having such a characteristic is called a *nonlinear softening spring.*

If we designate the general relationship between force and deflection by the equation

$$F = F(y) \tag{a}$$

then *spring rate* is defined as

$$k(y) = \lim_{\Delta y \to 0} \frac{\Delta F}{\Delta y} = \frac{dF}{dy} \tag{4-1}$$

where y must be measured in the direction of F and at the point of application of F. Most of the force-deflection problems encountered in this book are linear, as in Fig. 4–1a. For these, k is a constant, also called the *spring constant;* consequently Eq. (4–1) is written

$$k = \frac{F}{y} \tag{4-2}$$

We might note that Eqs. (4–1) and (4–2) are quite general and apply equally well for torques and moments, provided angular measurements are used for y. For linear displacements, the units of k are often pounds per inch or newtons per meter, and for angular displacements, pound-inches per radian or newton-meters per radian.

4–2   Tension, Compression, and Torsion

The total extension or contraction of a uniform bar in pure tension or compression, respectively, is given by

$$\delta = \frac{Fl}{AE} \tag{4-3}$$

This equation does not apply to a *long* bar loaded in compression if there is a possibility of buckling (see Secs. 4–11 to 4–15). Using Eqs. (4–2) and (4–3) with $\delta = y$, we see that the spring constant of an axially loaded bar is

$$k = \frac{AE}{l} \tag{4-4}$$

The angular deflection of a uniform solid or hollow round bar subjected to a twisting moment T was given in Eq. (3–35), and is

$$\theta = \frac{Tl}{GJ} \tag{4-5}$$

where θ is in radians. If we multiply Eq. (4–5) by $180/\pi$ and substitute $J = \pi d^4/32$ for a solid round bar, we obtain

$$\theta = \frac{583.6Tl}{Gd^4} \tag{4-6}$$

where θ is in degrees.

Equation (4–5) can be rearranged to give the torsional spring rate as

$$k = \frac{T}{\theta} = \frac{GJ}{l} \tag{4-7}$$

Equations (4–5), (4–6), and (4–7) apply *only* to circular cross sections. Torsional loading for bars with noncircular cross sections is discussed in Sec. 3–12 (p. 115). For the angular twist of rectangular cross sections, closed thin-walled tubes, and open thin-walled sections, refer to Eqs. (3–41), (3–46), and (3–47), respectively.

4–3 Deflection Due to Bending

The problem of bending of beams probably occurs more often than any other loading problem in mechanical design. Shafts, axles, cranks, levers, springs, brackets, and wheels, as well as many other elements, must often be treated as beams in the design and analysis of mechanical structures and systems. The subject of bending, however, is one that you should have studied as preparation for reading this book. It is for this reason that we include here only a brief review to establish the nomenclature and conventions to be used throughout this book.

The curvature of a beam subjected to a bending moment M is given by

$$\frac{1}{\rho} = \frac{M}{EI} \tag{4–8}$$

where ρ is the radius of curvature. From studies in mathematics we also learn that the curvature of a plane curve is given by the equation

$$\frac{1}{\rho} = \frac{d^2y/dx^2}{[1 + (dy/dx)^2]^{3/2}} \tag{4–9}$$

where the interpretation here is that y is the lateral deflection of the centroidal axis of the beam at any point x along its length. The slope of the beam at any point x is

$$\theta = \frac{dy}{dx} \tag{a}$$

For many problems in bending, the slope is very small, and for these the denominator of Eq. (4–9) can be taken as unity. Equation (4–8) can then be written

$$\frac{M}{EI} = \frac{d^2y}{dx^2} \tag{b}$$

Noting Eqs. (3–3) and (3–4) and successively differentiating Eq. (b) yields

$$\frac{V}{EI} = \frac{d^3y}{dx^3} \tag{c}$$

$$\frac{q}{EI} = \frac{d^4y}{dx^4} \tag{d}$$

It is convenient to display these relations in a group as follows:

$$\frac{q}{EI} = \frac{d^4y}{dx^4} \tag{4–10}$$

$$\frac{V}{EI} = \frac{d^3y}{dx^3} \tag{4–11}$$

$$\frac{M}{EI} = \frac{d^2y}{dx^2} \tag{4–12}$$

$$\theta = \frac{dy}{dx} \tag{4–13}$$

$$y = f(x) \tag{4–14}$$

| Figure 4–2

The nomenclature and conventions are illustrated by the beam of Fig. 4–2. Here, a beam of length $l = 0.5$ m is loaded by the uniform load $w = 14$ kN per inch of beam length. The x axis is positive to the right, and the y axis positive upward. All quantities—loading, shear, moment, slope, and deflection—have the same sense as y; they are positive if upward, negative if downward.

The reactions $R_1 = R_2 = +3.5$ kN and the shear forces $V_0 = +3.5$ kN and $V_l = -3.5$ kN are easily computed by using the methods of Chap. 3. The bending moment is zero at each end because the beam is simply supported. For a simply-supported beam, the deflections are also zero at each end.

EXAMPLE 4–1

For the beam in Fig. 4–2, the bending moment equation, for $0 \leq x \leq l$, is

$$M = \frac{wl}{2}x - \frac{w}{2}x^2$$

Using Eq. (4–12), determine the equations for the slope and deflection of the beam, the slopes at the ends, and the maximum deflection.

Solution

Integrating Eq. (4–12) as an indefinite integral we have

$$EI\frac{dy}{dx} = \int M \, dx = \frac{wl}{4}x^2 - \frac{w}{6}x^3 + C_1 \tag{1}$$

where C_1 is a constant of integration that is evaluated from geometric boundary conditions. We could impose that the slope is zero at the midspan of the beam, since the beam and loading are symmetric relative to the midspan. However, we will use the given boundary conditions of the problem and verify that the slope is zero at the midspan. Integrating Eq. (1) gives

$$EI \, y = \iint M \, dx = \frac{wl}{12}x^3 - \frac{w}{24}x^4 + C_1 x + C_2 \tag{2}$$

The boundary conditions for the simply supported beam are $y = 0$ at $x = 0$ and l. Applying the first condition, $y = 0$ at $x = 0$, to Eq. (2) results in $C_2 = 0$. Applying the second condition to Eq. (2) with $C_2 = 0$,

$$EI \, y(l) = \frac{wl}{12}l^3 - \frac{w}{24}l^4 + C_1 l = 0$$

Solving for C_1 yields $C_1 = -wl^3/24$. Substituting the constants back into Eqs. (1) and (2) and solving for the deflection and slope results in

$$y = \frac{wx}{24EI}(2lx^2 - x^3 - l^3) \tag{3}$$

$$\theta = \frac{dy}{dx} = \frac{w}{24EI}(6lx^2 - 4x^3 - l^3) \tag{4}$$

Comparing Eq. (3) with that given in Table A–9, beam 7, we see complete agreement. For the slope at the left end, substituting $x = 0$ into Eq. (4) yields

$$\theta|_{x=0} = -\frac{wl^3}{24EI}$$

and at $x = l$,

$$\theta|_{x=l} = \frac{wl^3}{24EI}$$

At the midspan, substituting $x = l/2$ gives $dy/dx = 0$, as earlier suspected. The maximum deflection occurs where $dy/dx = 0$. Substituting $x = l/2$ into Eq. (3) yields

$$y_{max} = -\frac{5wl^4}{384EI}$$

which again agrees with Table A–9–7.

The approach used in the example is fine for simple beams with continuous loading. However, for beams with discontinuous loading and/or geometry such as a step shaft with multiple gears, flywheels, pulleys, etc., the approach becomes unwieldy. The following section discusses bending deflections in general and the techniques that are provided in this chapter.

4–4 Beam Deflection Methods

Equations (4–10) through (4–14) are the basis for relating the intensity of loading q, vertical shear V, bending moment M, slope of the neutral surface θ, and the transverse deflection y. Beams have intensities of loading that range from $q = $ constant (uniform loading), variable intensity $q(x)$, to Dirac delta functions (concentrated loads).

The intensity of loading usually consists of piecewise contiguous zones, the expressions for which are integrated through Eqs. (4–10) to (4–14) with varying degrees of difficulty. Another approach is to represent the deflection $y(x)$ as a Fourier series, which is capable of representing single-valued functions with a finite number of finite discontinuities, then differentiating through Eqs. (4–14) to (4–10), and stopping at some level where the Fourier coefficients can be evaluated. A complication is the piecewise continuous nature of some beams (shafts) that are stepped-diameter bodies.

All of the above constitute, in one form or another, formal integration methods, which, with properly selected problems, result in solutions for q, V, M, θ, and y. These solutions may be

1 Closed-form, or

2 Represented by infinite series, which amount to closed form if the series are rapidly convergent, or

3 Approximations obtained by evaluating the first or the first and second terms.

The series solutions can be made equivalent to the closed-form solution by the use of a computer. Roark's[1] formulas are committed to commercial software and can be used on a personal computer.

There are many techniques employed to solve the integration problem for beam deflection. Some of the popular methods include:

- Superposition (see Sec. 4–5)
- The moment-area method[2]
- Singularity functions (see Sec. 4–6)
- Numerical integration[3]

The two methods described in this chapter are easy to implement and can handle a large array of problems.

There are methods that do not deal with Eqs. (4–10) to (4–14) directly. An energy method, based on Castigliano's theorem, is quite powerful for problems not suitable for the methods mentioned earlier and is discussed in Secs. 4–7 to 4–10. Finite element programs are also quite useful for determining beam deflections.

4–5   Beam Deflections by Superposition

The results of many simple load cases and boundary conditions have been solved and are available. Table A–9 provides a limited number of cases. Roark's[4] provides a much more comprehensive listing. *Superposition* resolves the effect of combined loading on a structure by determining the effects of each load separately and adding the results

---

[1]Warren C. Young, Richard G. Budynas, and Ali M. Sadegh, *Roark's Formulas for Stress and Strain,* 8th ed., McGraw-Hill, New York, 2012.

[2]See Chap. 9, F. P. Beer, E. R. Johnston Jr., and J. T. DeWolf, *Mechanics of Materials,* 5th ed., McGraw-Hill, New York, 2009.

[3]See Sec. 4–4, J. E. Shigley and C. R. Mischke, *Mechanical Engineering Design,* 6th ed., McGraw-Hill, New York, 2001.

[4]Warren C. Young, Richard G. Budynas, and Ali M. Sadegh, *Roark's Formulas for Stress and Strain,* 8th ed., McGraw-Hill, New York, 2012.

algebraically. Superposition may be applied provided: (1) each effect is linearly related to the load that produces it, (2) a load does not create a condition that affects the result of another load, and (3) the deformations resulting from any specific load are not large enough to appreciably alter the geometric relations of the parts of the structural system. The following examples are illustrations of the use of superposition.

EXAMPLE 4–2

Consider the uniformly loaded beam with a concentrated force as shown in Fig. 4–3. Using superposition, determine the reactions and the deflection as a function of x.

Solution

Considering each load state separately, we can superpose beams 6 and 7 of Table A–9. For the reactions we find

Answer
$$R_1 = \frac{Fb}{l} + \frac{wl}{2}$$

Answer
$$R_2 = \frac{Fa}{l} + \frac{wl}{2}$$

The loading of beam 6 is discontinuous and separate deflection equations are given for regions AB and BC. Beam 7 loading is not discontinuous so there is only one equation. Superposition yields

Answer
$$y_{AB} = \frac{Fbx}{6EIl}(x^2 + b^2 - l^2) + \frac{wx}{24EI}(2lx^2 - x^3 - l^3)$$

Answer
$$y_{BC} = \frac{Fa(l - x)}{6EIl}(x^2 + a^2 - 2lx) + \frac{wx}{24EI}(2lx^2 - x^3 - l^3)$$

| Figure 4–3

If the maximum deflection of a beam is desired, it will occur either where the slope is zero or at the end of the overhang if the beam has a free end. In the previous example, there is no overhang, so setting $dy/dx = 0$ will yield the equation for x that locates where the maximum deflection occurs. In the example there are two equations for y where only one will yield a solution. If $a = l/2$, the maximum deflection would obviously occur at $x = l/2$ because of symmetry. However, if $a < l/2$, where would the maximum deflection occur? It can be shown that as F moves toward the left support, the maximum deflection moves toward the left support also, but not as much as F (see Prob. 4–55). Thus, we would set $dy_{BC}/dx = 0$ and solve for x. If $a > l/2$, then we would set $dy_{AB}/dx = 0$. For more complicated problems, plotting the equations using numerical data is the simplest approach to finding the maximum deflection.

Sometimes it may not be obvious that we can use superposition with the tables at hand, as demonstrated in the next example.

EXAMPLE 4–3   Consider the beam in Fig. 4–4a and determine the deflection equations using super-position.

Solution   For region AB we can superpose beams 7 and 10 of Table A–9 to obtain

Answer
$$y_{AB} = \frac{wx}{24EI}(2lx^2 - x^3 - l^3) + \frac{Fax}{6EIl}(l^2 - x^2)$$

For region BC, how do we represent the uniform load? Considering the uniform load *only*, the beam deflects as shown in Fig. 4–4b. Region BC is straight since there is no bending moment due to w. The slope of the beam at B is θ_B and is obtained by taking the derivative of y given in the table with respect to x and setting $x = l$. Thus,

$$\frac{dy}{dx} = \frac{d}{dx}\left[\frac{wx}{24EI}(2lx^2 - x^3 - l^3)\right] = \frac{w}{24EI}(6lx^2 - 4x^3 - l^3)$$

Substituting $x = l$ gives

$$\theta_B = \frac{w}{24EI}(6ll^2 - 4l^3 - l^3) = \frac{wl^3}{24EI}$$

The deflection in region BC due to w is $\theta_B(x - l)$, and adding this to the deflection due to F, in BC, yields

Answer
$$y_{BC} = \frac{wl^3}{24EI}(x - l) + \frac{F(x - l)}{6EI}[(x - l)^2 - a(3x - l)]$$

Figure 4–4

(*a*) Beam with uniformly distributed load and overhang force; (*b*) deflections due to uniform load only.

(*a*)   (*b*)

EXAMPLE 4–4   Figure 4–5a shows a cantilever beam with an end load. Normally we model this problem by considering the left support as rigid. After testing the rigidity of the wall it was found that the translational stiffness of the wall was k_t force per unit vertical deflection, and the rotational stiffness was k_r moment per unit angular (radian) deflection (see Fig. 4–5b). Determine the deflection equation for the beam under the load F.

Solution

Here we will superpose the *modes* of deflection. They are: (1) translation due to the compression of spring k_t, (2) rotation of the spring k_r, and (3) the elastic deformation of beam 1 given in Table A–9. The force in spring k_t is $R_1 = F$, giving a deflection from Eq. (4–2) of

$$y_1 = -\frac{F}{k_t} \tag{1}$$

The moment in spring k_r is $M_1 = Fl$. This gives a clockwise rotation of $\theta = Fl/k_r$. Considering this mode of deflection only, the beam rotates rigidly clockwise, leading to a deflection equation of

$$y_2 = -\frac{Fl}{k_r}x \tag{2}$$

Finally, the elastic deformation of beam 1 from Table A–9 is

$$y_3 = \frac{Fx^2}{6EI}(x - 3l) \tag{3}$$

Adding the deflections from each mode yields

Answer

$$y = \frac{Fx^2}{6EI}(x - 3l) - \frac{F}{k_t} - \frac{Fl}{k_r}x$$

| Figure 4–5

(a)

(b)

4–6 Beam Deflections by Singularity Functions

Introduced in Sec. 3–3, singularity functions are excellent for managing discontinuities, and their application to beam deflection is a simple extension of what was presented in the earlier section. They are easy to program, and as will be seen later, they can greatly simplify the solution of statically indeterminate problems. The following examples illustrate the use of singularity functions to evaluate deflections of statically determinate beam problems.

EXAMPLE 4–5   Consider beam 6 of Table A–9, which is a simply supported beam having a concentrated load F not in the center. Develop the deflection equations using singularity functions.

Solution   First, write the load intensity equation from the free-body diagram,

$$q = R_1 \langle x \rangle^{-1} - F \langle x - a \rangle^{-1} + R_2 \langle x - l \rangle^{-1} \tag{1}$$

Integrating Eq. (1) twice results in

$$V = R_1 \langle x \rangle^0 - F \langle x - a \rangle^0 + R_2 \langle x - l \rangle^0 \tag{2}$$

$$M = R_1 \langle x \rangle^1 - F \langle x - a \rangle^1 + R_2 \langle x - l \rangle^1 \tag{3}$$

Recall that as long as the q equation is complete, integration constants are unnecessary for V and M; therefore, they are not included up to this point. From statics, setting $V = M = 0$ for x slightly greater than l yields $R_1 = Fb/l$ and $R_2 = Fa/l$. Thus Eq. (3) becomes

$$M = \frac{Fb}{l} \langle x \rangle^1 - F \langle x - a \rangle^1 + \frac{Fa}{l} \langle x - l \rangle^1$$

Integrating Eqs. (4–12) and (4–13) as indefinite integrals gives

$$EI \frac{dy}{dx} = \frac{Fb}{2l} \langle x \rangle^2 - \frac{F}{2} \langle x - a \rangle^2 + \frac{Fa}{2l} \langle x - l \rangle^2 + C_1$$

$$EI y = \frac{Fb}{6l} \langle x \rangle^3 - \frac{F}{6} \langle x - a \rangle^3 + \frac{Fa}{6l} \langle x - l \rangle^3 + C_1 x + C_2$$

Note that the first singularity term in both equations always exists, so $\langle x \rangle^2 = x^2$ and $\langle x \rangle^3 = x^3$. Also, the last singularity term in both equations does not exist until $x = l$, where it is zero, and since there is no beam for $x > l$ we can drop the last term. Thus

$$EI \frac{dy}{dx} = \frac{Fb}{2l} x^2 - \frac{F}{2} \langle x - a \rangle^2 + C_1 \tag{4}$$

$$EI y = \frac{Fb}{6l} x^3 - \frac{F}{6} \langle x - a \rangle^3 + C_1 x + C_2 \tag{5}$$

The constants of integration C_1 and C_2 are evaluated by using the two boundary conditions $y = 0$ at $x = 0$ and $y = 0$ at $x = l$. The first condition, substituted into Eq. (5), gives $C_2 = 0$ (recall that $\langle 0 - a \rangle^3 = 0$). The second condition, substituted into Eq. (5), yields

$$0 = \frac{Fb}{6l} l^3 - \frac{F}{6} (l - a)^3 + C_1 l = \frac{Fbl^2}{6} - \frac{Fb^3}{6} + C_1 l$$

Solving for C_1 gives

$$C_1 = -\frac{Fb}{6l} (l^2 - b^2)$$

Finally, substituting C_1 and C_2 in Eq. (5) and simplifying produces

$$y = \frac{F}{6EIl}[bx(x^2 + b^2 - l^2) - l\langle x - a\rangle^3] \qquad (6)$$

Comparing Eq. (6) with the two deflection equations for beam 6 in Table A–9, we note that the use of singularity functions enables us to express the deflection equation with a single discontinuous equation.

EXAMPLE 4–6

Determine the deflection equation for the simply supported beam with the load distribution shown in Fig. 4–6.

Solution

This is a good beam to add to our table for later use with superposition. The load intensity equation for the beam is

$$q = R_1\langle x\rangle^{-1} - w\langle x\rangle^0 + w\langle x - a\rangle^0 + R_2\langle x - l\rangle^{-1} \qquad (1)$$

where the $w\langle x - a\rangle^0$ is necessary to "turn off" the uniform load at $x = a$.

From statics, the reactions are

$$R_1 = \frac{wa}{2l}(2l - a) \qquad R_2 = \frac{wa^2}{2l} \qquad (2)$$

For simplicity, we will retain the form of Eq. (1) for integration and substitute the values of the reactions later.

Two integrations of Eq. (1) reveal

$$V = R_1\langle x\rangle^0 - w\langle x\rangle^1 + w\langle x - a\rangle^1 + R_2\langle x - l\rangle^0 \qquad (3)$$

$$M = R_1\langle x\rangle^1 - \frac{w}{2}\langle x\rangle^2 + \frac{w}{2}\langle x - a\rangle^2 + R_2\langle x - l\rangle^1 \qquad (4)$$

As in the previous example, singularity functions of order zero or greater starting at $x = 0$ can be replaced by normal polynomial functions. Also, once the reactions are determined, singularity functions starting at the extreme right end of the beam can be omitted. Thus, Eq. (4) can be rewritten as

$$M = R_1 x - \frac{w}{2}x^2 + \frac{w}{2}\langle x - a\rangle^2 \qquad (5)$$

| Figure 4–6

Integrating two more times for slope and deflection gives

$$EI\frac{dy}{dx} = \frac{R_1}{2}x^2 - \frac{w}{6}x^3 + \frac{w}{6}\langle x - a \rangle^3 + C_1 \tag{6}$$

$$EI\,y = \frac{R_1}{6}x^3 - \frac{w}{24}x^4 + \frac{w}{24}\langle x - a \rangle^4 + C_1 x + C_2 \tag{7}$$

The boundary conditions are $y = 0$ at $x = 0$ and $y = 0$ at $x = l$. Substituting the first condition in Eq. (7) shows $C_2 = 0$. For the second condition

$$0 = \frac{R_1}{6}l^3 - \frac{w}{24}l^4 + \frac{w}{24}(l - a)^4 + C_1 l$$

Solving for C_1 and substituting into Eq. (7) yields

$$EI\,y = \frac{R_1}{6}x(x^2 - l^2) - \frac{w}{24}x(x^3 - l^3) - \frac{w}{24l}x(l - a)^4 + \frac{w}{24}\langle x - a \rangle^4$$

Finally, substitution of R_1 from Eq. (2) and simplifying results gives

Answer
$$y = \frac{w}{24EI\,l}[2ax(2l - a)(x^2 - l^2) - xl(x^3 - l^3) - x(l - a)^4 + l\langle x - a \rangle^4]$$

As stated earlier, singularity functions are relatively simple to program, as they are omitted when their arguments are negative, and the $\langle \; \rangle$ brackets are replaced with $(\;)$ parentheses when the arguments are positive.

EXAMPLE 4–7   The steel step shaft shown in Fig. 4–7a is mounted in bearings at A and F. A pulley is centered at C where a total radial force of 2.6 kN is applied. Using singularity functions evaluate the shaft displacements at 12-mm increments. Assume the shaft is simply supported.

Solution   The reactions are found to be $R_1 = 1.56$ kN and $R_2 = 1.04$ kN. Ignoring R_2, using singularity functions, the moment equation is

$$M = 1560x - 2600\langle x - 0.2 \rangle \tag{1}$$

This is plotted in Fig. 4–7b.

For simplification, we will consider only the step at D. That is, we will assume section AB has the same diameter as BC and section EF has the same diameter as DE. Since these sections are short and at the supports, the size reduction will not add much to the deformation. We will examine this simplification later. The second area moments for BC and DE are

$$I_{BC} = \frac{\pi}{64}0.038^4 = 102.35 \times 10^{-9}\ \text{m}^4 \qquad I_{DE} = \frac{\pi}{64}0.045^4 = 2147 \times 10^{-9}\ \text{m}^4$$

Figure 4–7

Dimensions in millimeters.

(a)

(b)

(c)

A plot of M/I is shown in Fig. 4–7c. The values at points b and c, and the step change are

$$\left(\frac{M}{I}\right)_b = \frac{302}{102.35 \times 10^{-9}} = 2.95 \times 10^9 \text{ N/m}^3$$

$$\left(\frac{M}{I}\right)_c = \frac{302}{2147 \times 10^{-9}} = 0.141 \times 10^9 \text{ N/m}^3$$

$$\Delta\left(\frac{M}{I}\right) = \langle 0.141 - 2.95\rangle 10^9 = -2.81 \times 10^9 \text{ N/m}^3$$

The slopes for ab and cd, and the change are

$$m_{ab} = \frac{1560 - 2600}{102.39 \times 10^{-9}} = -10.16 \times 10^9 \text{ N/m}^4$$

$$m_{cd} = \frac{-0.141 \times 10^9}{0.29} = -0.49 \times 10^9 \text{ N/m}^4$$

$$\Delta m = -0.49 \times 10^9 - \langle -10.16 \times 10^9\rangle = 9.67 \times 10^9 \text{ N/m}^4$$

Dividing Eq. (1) by I_{BC} and, at $x = 0.21$ m, adding a step of -2.81×10^9 N/m^3 and a ramp of slope 9.67×10^9 N/m^4, gives

$$\frac{M}{I} = 15.24 \times 10^9 x - 25.4 \times 10^9 \langle x - 0.2\rangle^1 - 2.81 \times 10^9 \langle x - 0.21\rangle^0 \qquad (2)$$
$$+ 9.67 \times 10^9 \langle x - 0.21\rangle^1$$

Integrating twice gives

$$E\frac{dy}{dx} = 10^9 [7.62x^2 - 12.7\langle x - 0.2\rangle^2 - 2.81\langle x - 0.21\rangle^1 \qquad (3)$$
$$+ 4.84\langle x - 0.21\rangle^2] + C_1$$

and

$$Ey = 10^9 [2.54x^3 - 4.23\langle x - 0.2\rangle^3 - 1.41\langle x - 0.21\rangle^2 + 1.61\langle x - 0.21\rangle^5] + C_1 x + C_2$$
$$(4)$$

At $x = 0$, $y = 0$. This gives $C_2 = 0$ (remember, singularity functions do not exist until the argument is positive). At $x = 0.5$ m, $y = 0$, and

$$0 = 10^9[2.54\langle 0.5 \rangle^3 - 4.23\langle 0.5 - 0.2 \rangle^3 - 1.41\langle 0.5 - 0.21 \rangle^2 + 1.61\langle 0.5 - 0.21 \rangle^3] + 0.5C_1$$

Solving, gives $C_1 = -0.248 \times 10^9$ N/m^2. Thus, Eq. (4) becomes, with $E = 200$ GPa,

$$y = \frac{1}{200}[2.54x^3 - 4.23\langle x - 0.2 \rangle^3 - 1.41\langle x - 0.21 \rangle^2$$

$$+ 1.61\langle x - 0.21 \rangle^3 - 0.248x] \tag{5}$$

When using a spreadsheet, program the following equations:

$$y = \frac{1}{200}(2.54x^3 - 0.248x) \qquad\qquad 0 \le x \le 0.2 \text{ m}$$

$$y = \frac{1}{200}[2.54x^3 - 4.23(x - 0.2)^3 - 0.248x] \qquad\qquad 0.2 \le x \le 0.21 \text{ m}$$

$$y = \frac{1}{200}[2.54x^3 - 4.23(x - 0.2)^3 - 1.41(x - 0.21)^2$$

$$+ 1.61(x - 0.21)^3 - 0.248x] \qquad\qquad 0.21 \le x \le 0.5 \text{ m}$$

The following table results.

x (m)	y (mm)
0	0.0000
0.1	−0.1113
0.2	−0.0292
0.21	−0.0311
0.4	−0.0203
0.5	0.0000

where x is in meters and y is in millimeters. We see that the greatest deflection is at $x = 0.21$ m, where $y = -0.0311$ mm.

Substituting C_1 into Eq. (3) the slopes at the supports are found to be $\theta_A = 1.686(10^{-3})$ rad $= 0.09657$ deg, and $\theta_F = 1.198(10^{-3})$ rad $= 0.06864$ deg. You might think these to be insignificant deflections, but as you will see in Chap. 7, on shafts, they are not.

A finite-element analysis was performed for the same model and resulted in

$$y|_{x = 0.21 \text{ m}} = -0.0307 \text{ mm} \qquad \theta_A = -0.09653° \qquad \theta_F = 0.06868°$$

Virtually the same answer save some round-off error in the equations.

If the steps of the bearings were incorporated into the model, more equations result, but the process is the same. The solution to this model is

$$y|_{x = 0.21 \text{ m}} = -0.0315 \text{ mm} \qquad \theta_A = -0.09763° \qquad \theta_F = 0.06973°$$

The largest difference between the models is of the order of 1.3 percent. Thus the simplification was justified.

In Sec. 4–9, we will demonstrate the usefulness of singularity functions in solving statically indeterminate problems.

4–7 Strain Energy

The external work done on an elastic member in deforming it is transformed into *strain,* or *potential, energy.* If the member is deformed a distance y, and if the force-deflection relationship is linear, this energy is equal to the product of the average force and the deflection, or

$$U = \frac{F}{2}y = \frac{F^2}{2k} \tag{4-15}$$

This equation is general in the sense that the force F can also mean torque, or moment, provided, of course, that consistent units are used for k. By substituting appropriate expressions for k, strain-energy formulas for various simple loadings may be obtained. For tension and compression, for example, we employ Eq. (4–4) and obtain

$$\left.\begin{array}{r} U = \dfrac{F^2 l}{2AE} \\[4mm] \text{or} \\[4mm] U = \displaystyle\int \dfrac{F^2}{2AE} dx \end{array}\right\} \text{tension and compression}$$

$$\begin{array}{r}(4-16)\\[8mm](4-17)\end{array}$$

where the first equation applies when all the terms are constant throughout the length, and the more general integral equation allows for any of the terms to vary through the length.

Similarly, from Eq. (4–7), the strain energy for torsion is given by

$$\left.\begin{array}{r} U = \dfrac{T^2 l}{2GJ} \\[4mm] \text{or} \\[4mm] U = \displaystyle\int \dfrac{T^2}{2GJ} dx \end{array}\right\} \text{torsion}$$

$$\begin{array}{r}(4-18)\\[8mm](4-19)\end{array}$$

To obtain an expression for the strain energy due to direct shear, consider the element with one side fixed in Fig. 4–8a. The force F places the element in pure shear, and the work done is $U = F\delta/2$. Since the shear strain is $\gamma = \delta/l = \tau/G = F/AG$, we have

$$\left.\begin{array}{r} U = \dfrac{F^2 l}{2AG} \\[4mm] \text{or} \\[4mm] U = \displaystyle\int \dfrac{F^2}{2AG} dx \end{array}\right\} \text{direct shear}$$

$$\begin{array}{r}(4-20)\\[8mm](4-21)\end{array}$$

| Figure 4–8

(a) Pure shear element   (b) Beam bending element

The strain energy stored in a beam or lever by bending may be obtained by referring to Fig. 4–8b. Here AB is a section of the elastic curve of length ds having a radius of curvature ρ. The strain energy stored in this element of the beam is $dU = (M/2)d\theta$. Since $\rho\, d\theta = ds$, we have

$$dU = \frac{M\, ds}{2\rho} \tag{a}$$

We can eliminate ρ by using Eq. (4–8), $\rho = EI/M$. Thus

$$dU = \frac{M^2\, ds}{2EI} \tag{b}$$

For small deflections, $ds \approx dx$. Then, for the entire beam

$$U = \int dU = \int \frac{M^2}{2EI}dx \tag{c}$$

The integral equation is commonly needed for bending, where the moment is typically a function of x. Summarized to include both the integral and nonintegral form, the strain energy for bending is

$$\left. \begin{aligned} U &= \frac{M^2 l}{2EI} \\[2mm] \text{or} \\[2mm] U &= \int \frac{M^2}{2EI}dx \end{aligned} \right\} \text{bending} \qquad \begin{aligned} &(4\text{–}22) \\[6mm] &(4\text{–}23) \end{aligned}$$

Equations (4–22) and (4–23) are exact only when a beam is subject to pure bending. Even when transverse shear is present, these equations continue to give quite good results, except for very short beams. The strain energy due to shear loading of a beam is a complicated problem. An approximate solution can be obtained by using Eq. (4–20) with a correction factor whose value depends upon the shape of the cross section. If we use C for the correction factor and V for the shear force, then the strain energy due to shear in bending is

$$\left. \begin{aligned} U &= \frac{CV^2 l}{2AG} \\[2mm] \text{or} \\[2mm] U &= \int \frac{CV^2}{2AG}dx \end{aligned} \right\} \text{transverse shear} \qquad \begin{aligned} &(4\text{–}24) \\[6mm] &(4\text{–}25) \end{aligned}$$

Values of the factor C are listed in Table 4–1.

Table 4–1

Strain-Energy Correction Factors for Transverse Shear

Source: Richard G. Budynas, *Advanced Strength and Applied Stress Analysis,* 2nd ed., McGraw-Hill, New York, 1999. Copyright © 1999 The McGraw-Hill Companies.

Beam Cross-Sectional Shape	Factor C
Rectangular	1.2
Circular	1.11
Thin-walled tubular, round	2.00
Box sections[†]	1.00
Structural sections[†]	1.00

[†]Use area of web only.

EXAMPLE 4–8

A cantilever beam with a round cross section has a concentrated load F at the end, as shown in Fig. 4–9a. Find the strain energy in the beam.

Figure 4–9

(a) (b)

Solution

To determine what forms of strain energy are involved with the deflection of the beam, we break into the beam and draw a free-body diagram to see the forces and moments being carried within the beam. Figure 4–9b shows such a diagram in which the transverse shear is $V = -F$, and the bending moment is $M = -Fx$. The variable x is simply a variable of integration and can be defined to be measured from any convenient point. The same results will be obtained from a free-body diagram of the right-hand portion of the beam with x measured from the wall. Using the free end of the beam usually results in reduced effort since the ground reaction forces do not need to be determined.

For the transverse shear, using Eq. (4–24) with the correction factor $C = 1.11$ from Table 4–2, and noting that V is constant through the length of the beam,

$$U_{shear} = \frac{CV^2 l}{2AG} = \frac{1.11 F^2 l}{2AG}$$

For the bending, since M is a function of x, Eq. (4–23) gives

$$U_{bend} = \int \frac{M^2 dx}{2EI} = \frac{1}{2EI} \int_0^l (-Fx)^2 dx = \frac{F^2 l^3}{6EI}$$

The total strain energy is

Answer

$$U = U_{bend} + U_{shear} = \frac{F^2 l^3}{6EI} + \frac{1.11 F^2 l}{2AG}$$

Note, except for very short beams, the shear term (of order l) is typically small compared to the bending term (of order l^3). This will be demonstrated in the next example.

4–8 **Castigliano's Theorem**

A most unusual, powerful, and often surprisingly simple approach to deflection analysis is afforded by an energy method called *Castigliano's theorem*. It is a unique way of analyzing deflections and is even useful for finding the reactions of indeterminate structures. Castigliano's theorem states that *when forces act on elastic systems subject to small displacements, the displacement corresponding to any force, in the direction of the force, is equal to the partial derivative of the total strain energy with respect to that force.* The terms *force* and *displacement* in this statement are broadly interpreted to apply equally to moments and angular displacements. Mathematically, the theorem of Castigliano is

$$\delta_i = \frac{\partial U}{\partial F_i} \qquad (4\text{--}26)$$

where δ_i is the displacement of the point of application of the force F_i in the direction of F_i. For rotational displacement Eq. (4–26) can be written as

$$\theta_i = \frac{\partial U}{\partial M_i} \qquad (4\text{--}27)$$

where θ_i is the rotational displacement, in radians, of the beam where the moment M_i exists and in the direction of M_i.

As an example, apply Castigliano's theorem using Eqs. (4–16) and (4–18) to get the axial and torsional deflections. The results are

$$\delta = \frac{\partial}{\partial F}\left(\frac{F^2 l}{2AE}\right) = \frac{Fl}{AE} \qquad (a)$$

$$\theta = \frac{\partial}{\partial T}\left(\frac{T^2 l}{2GJ}\right) = \frac{Tl}{GJ} \qquad (b)$$

Compare Eqs. (*a*) and (*b*) with Eqs. (4–3) and (4–5).

EXAMPLE 4–9

The cantilever of Ex. 4–8 is a carbon steel bar 440 mm long with a 34-mm diameter and is loaded by a force $F = 580$ N.

(*a*) Find the maximum deflection using Castigliano's theorem, including that due to shear.

(*b*) What error is introduced if shear is neglected?

Solution

(*a*) From Ex. 4–8, the total energy of the beam is

$$U = \frac{F^2 l^3}{6EI} + \frac{1.11 F^2 l}{2AG} \qquad (1)$$

Then, according to Castigliano's theorem, the deflection of the end is

$$y_{\text{max}} = \frac{\partial U}{\partial F} = \frac{Fl^3}{3EI} + \frac{1.11 Fl}{AG} \qquad (2)$$

We also find that

$$I = \frac{\pi d^4}{64} = \frac{\pi (34)^4}{64} = 65597.24 \text{ mm}^4$$

$$A = \frac{\pi d^2}{4} = \frac{\pi (34)^2}{4} = 907.92 \text{ mm}^2$$

Substituting these values, together with $F = 580$ N, $l = 0.44$ m, $E = 209$ GPa, and $G = 79$ GPa, in Eq. (2) gives

Answer

$$y_{\text{max}} = 1.20 + 0.004 = 1.205 \text{ mm}$$

Note that the result is positive because it is in the *same* direction as the force F.

Answer

(*b*) The error in neglecting shear for this problem is $(1.205 - 1.201)/1.205 = 0.0033 = 0.33$ percent.

The relative contribution of transverse shear to beam deflection decreases as the length-to-height ratio of the beam increases, and is generally considered negligible for $l/d > 10$. Note that the deflection equations for the beams in Table A–9 do not include the effects of transverse shear.

Castigliano's theorem can be used to find the deflection at a point even though no force or moment acts there. The procedure is:

1. Set up the equation for the total strain energy U by including the energy due to a fictitious force or moment Q acting at the point whose deflection is to be found.

2. Find an expression for the desired deflection δ, in the direction of Q, by taking the derivative of the total strain energy with respect to Q.

3. Since Q is a fictitious force, solve the expression obtained in step 2 by setting Q equal to zero. Thus, the displacement at the point of application of the fictitious force Q is

$$\delta = \left. \frac{\partial U}{\partial Q} \right|_{Q=0} \tag{4–28}$$

In cases where integration is necessary to obtain the strain energy, it is more efficient to obtain the deflection directly without explicitly finding the strain energy, by moving the partial derivative inside the integral. For the example of the bending case,

$$\delta_i = \frac{\partial U}{\partial F_i} = \frac{\partial}{\partial F_i}\left(\int \frac{M^2}{2EI}dx\right) = \int \frac{\partial}{\partial F_i}\left(\frac{M^2}{2EI}\right)dx = \int \frac{2M\frac{\partial M}{\partial F_i}}{2EI}dx = \int \frac{1}{EI}\left(M\frac{\partial M}{\partial F_i}\right)dx$$

This allows the derivative to be taken before integration, simplifying the mathematics. This method is especially helpful if the force is a fictitious force Q, since it can be set to zero as soon as the derivative is taken. The expressions for the common cases in Eqs. (4–17), (4–19), and (4–23) are rewritten as

$$\delta_i = \frac{\partial U}{\partial F_i} = \int \frac{1}{AE}\left(F\frac{\partial F}{\partial F_i}\right)dx \qquad \text{tension and compression} \tag{4–29}$$

$$\theta_i = \frac{\partial U}{\partial M_i} = \int \frac{1}{GJ}\left(T\frac{\partial T}{\partial M_i}\right)dx \qquad \text{torsion} \tag{4–30}$$

$$\delta_i = \frac{\partial U}{\partial F_i} = \int \frac{1}{EI}\left(M\frac{\partial M}{\partial F_i}\right)dx \qquad \text{bending} \tag{4–31}$$

EXAMPLE 4–10

Using Castigliano's method, determine the deflections of points A and B due to the force F applied at the end of the step shaft shown in Fig. 4–10. The second area moments for sections AB and BC are I_1 and $2I_1$, respectively.

Solution

To avoid the need to determine the support reaction forces, define the origin of x at the left end of the beam as shown. For $0 \le x \le l$, the bending moment is

$$M = -Fx \tag{1}$$

| Figure 4–10

Since F is at A and in the direction of the desired deflection, the deflection at A from Eq. (4–31) is

$$\delta_A = \frac{\partial U}{\partial F} = \int_0^l \frac{1}{EI}\left(M\frac{\partial M}{\partial F}\right)dx \tag{2}$$

Substituting Eq. (1) into Eq. (2), noting that $I = I_1$ for $0 \leq x \leq l/2$, and $I = 2I_1$ for $l/2 \leq x \leq l$, we get

Answer

$$\delta_A = \frac{1}{E}\left[\int_0^{l/2}\frac{1}{I_1}(-Fx)(-x)\,dx + \int_{l/2}^l\frac{1}{2I_1}(-Fx)(-x)\,dx\right]$$

$$= \frac{1}{E}\left[\frac{Fl^3}{24I_1} + \frac{7Fl^3}{48I_1}\right] = \frac{3}{16}\frac{Fl^3}{EI_1}$$

which is positive, as it is in the direction of F.

For B, a fictitious force Q is necessary at the point. Assuming Q acts down at B, and x is as before, the moment equation is

$$\begin{aligned} M &= -Fx & 0 \leq x \leq l/2 \\ M &= -Fx - Q\left(x - \frac{l}{2}\right) & l/2 \leq x \leq l \end{aligned} \tag{3}$$

For Eq. (4–31), we need $\partial M/\partial Q$. From Eq. (3),

$$\begin{aligned} \frac{\partial M}{\partial Q} &= 0 & 0 \leq x \leq l/2 \\ \frac{\partial M}{\partial Q} &= -\left(x - \frac{l}{2}\right) & l/2 \leq x \leq l \end{aligned} \tag{4}$$

Once the derivative is taken, Q can be set to zero, so Eq. (4–31) becomes

$$\delta_B = \left[\int_0^l \frac{1}{EI}\left(M\frac{\partial M}{\partial Q}\right)dx\right]_{Q=0}$$

$$= \frac{1}{EI_1}\int_0^{l/2}(-Fx)(0)\,dx + \frac{1}{E(2I_1)}\int_{l/2}^l(-Fx)\left[-\left(x - \frac{l}{2}\right)\right]dx$$

Evaluating the last integral gives

Answer

$$\delta_B = \frac{F}{2EI_1}\left(\frac{x^3}{3} - \frac{lx^2}{4}\right)\Bigg|_{l/2}^l = \frac{5}{96}\frac{Fl^3}{EI_1}$$

which again is positive, in the direction of Q.

EXAMPLE 4–11   For the wire form of diameter d shown in Fig. 4–11a, determine the deflection of point B in the direction of the applied force F (neglect the effect of transverse shear).

Solution   Figure 4–11b shows free body diagrams where the body has been broken in each section, and internal balancing forces and moments are shown. The sign convention for the force and moment variables is positive in the directions shown. With energy methods, sign conventions are arbitrary, so use a convenient one. In each section, the variable x is defined with its origin as shown. The variable x is used as a variable of integration for each section independently, so it is acceptable to reuse the same variable for each section. For completeness, the transverse shear forces are included, but the effects of transverse shear on the strain energy (and deflection) will be neglected.

Element BC is in bending only so from Eq. (4–31),[5]

$$\frac{\partial U_{BC}}{\partial F} = \frac{1}{EI} \int_0^a (Fx)(x)\, dx = \frac{Fa^3}{3EI} \tag{1}$$

Element CD is in bending and in torsion. The torsion is constant so Eq. (4–30) can be written as

$$\frac{\partial U}{\partial F_i} = \left(T \frac{\partial T}{\partial F_i} \right) \frac{l}{GJ}$$

| Figure 4–11

(a)

(b)

---

[5]It is very tempting to mix techniques and try to use superposition also, for example. However, some subtle things can occur that you may visually miss. It is highly recommended that if you are using Castigliano's theorem on a problem, you use it for all parts of the problem.

where l is the length of the member. So for the torsion in member CD, $F_i = F$, $T = Fa$, and $l = b$. Thus,

$$\left(\frac{\partial U_{CD}}{\partial F}\right)_{\text{torsion}} = (Fa)(a)\frac{b}{GJ} = \frac{Fa^2b}{GJ} \tag{2}$$

For the bending in CD,

$$\left(\frac{\partial U_{CD}}{\partial F}\right)_{\text{bending}} = \frac{1}{EI}\int_0^b (Fx)(x)\,dx = \frac{Fb^3}{3EI} \tag{3}$$

Member DG is axially loaded and is bending in two planes. The axial loading is constant, so Eq. (4–29) can be written as

$$\frac{\partial U}{\partial F_i} = \left(F\frac{\partial F}{\partial F_i}\right)\frac{l}{AE}$$

where l is the length of the member. Thus, for the axial loading of DG, $F_i = F$, $l = c$, and

$$\left(\frac{\partial U_{DG}}{\partial F}\right)_{\text{axial}} = \frac{Fc}{AE} \tag{4}$$

The bending moments in each plane of DG are constant along the length, with $M_{DG2} = Fb$ and $M_{DG1} = Fa$. Considering each one separately in the form of Eq. (4–31) gives

$$\left(\frac{\partial U_{DG}}{\partial F}\right)_{\text{bending}} = \frac{1}{EI}\int_0^c (Fb)(b)\,dx + \frac{1}{EI}\int_0^c (Fa)(a)\,dx$$

$$= \frac{Fc(a^2 + b^2)}{EI} \tag{5}$$

Adding Eqs. (1) to (5), noting that $I = \pi d^4/64$, $J = 2I$, $A = \pi d^2/4$, and $G = E/[2(1 + \nu)]$, we find that the deflection of B in the direction of F is

Answer

$$(\delta_B)_F = \frac{4F}{3\pi E d^4}[16(a^3 + b^3) + 48c(a^2 + b^2) + 48(1 + \nu)a^2b + 3cd^2]$$

Now that we have completed the solution, see if you can physically account for each term in the result using an independent method such as superposition.

4–9   **Deflection of Curved Members**

Machine frames, springs, clips, fasteners, and the like frequently occur as curved shapes. The determination of stresses in curved members has already been described in Sec. 3–18. Castigliano's theorem is particularly useful for the analysis of deflections in curved parts too.[6] Consider, for example, the curved frame of Fig. 4–12a. We are

---

[6]For more solutions than are included here, see Joseph E. Shigley, "Curved Beams and Rings," Chap. 38 in Joseph E. Shigley, Charles R. Mischke, and Thomas H. Brown, Jr. (eds.), *Standard Handbook of Machine Design*, 3rd ed., McGraw-Hill, New York, 2004.

Figure 4–12

(a) Curved bar loaded by force F. R = radius to centroidal axis of section; h = section thickness. (b) Diagram showing forces acting on section taken at angle θ. $F_r = V$ = shear component of F; F_θ is component of F normal to section; M is moment caused by force F.

interested in finding the deflection of the frame due to F and in the direction of F. Unlike straight beams, the bending moment and axial force are coupled for curved beams, creating an additional energy term[7]. The energy due to the moment alone is

$$U_1 = \int \frac{M^2 \, d\theta}{2AeE} \tag{4-32}$$

In this equation, the eccentricity e is

$$e = R - r_n \tag{4-33}$$

where r_n is the radius of the neutral axis as defined in Sec. 3–18 and shown in Fig. 3–34.

Analogous to Eq. (4–17), the strain energy component due to the axial force F_θ alone is

$$U_2 = \int \frac{F_\theta^2 R \, d\theta}{2AE} \tag{4-34}$$

The additional coupling term between M and F_θ is

$$U_3 = -\int \frac{M F_\theta \, d\theta}{AE} \tag{4-35}$$

The negative sign of Eq. (4–35) can be appreciated by referring to both parts of Fig. 4–12. Note that the moment M tends to decrease the angle dθ. On the other hand, F_θ tends to increase dθ. Thus U_3 is negative. If F_θ had been acting in the opposite direction, then both M and F_θ would tend to decrease the angle dθ.

The fourth and last term is the transverse shear energy due to F_r. Adapting Eq. (4–25) gives

$$U_4 = \int \frac{C F_r^2 R \, d\theta}{2AG} \tag{4-36}$$

where C is the correction factor of Table 4–1.

---

[7]See Richard G. Budynas, *Advanced Strength and Applied Stress Analysis,* 2nd ed., Sec. 6.7, McGraw-Hill, New York, 1999.

Combining the four terms gives the total strain energy

$$U = \int \frac{M^2 d\theta}{2AeE} + \int \frac{F_\theta^2 R \, d\theta}{2AE} - \int \frac{MF_\theta \, d\theta}{AE} + \int \frac{CF_r^2 R \, d\theta}{2AG} \qquad (4\text{--}37)$$

The deflection produced by the force F can now be found. It is

$$\delta = \frac{\partial U}{\partial F} = \int \frac{M}{AeE}\left(\frac{\partial M}{\partial F}\right) d\theta + \int \frac{F_\theta R}{AE}\left(\frac{\partial F_\theta}{\partial F}\right) d\theta$$
$$- \int \frac{1}{AE}\frac{\partial(MF_\theta)}{\partial F} d\theta + \int \frac{CF_r R}{AG}\left(\frac{\partial F_r}{\partial F}\right) d\theta \qquad (4\text{--}38)$$

This equation is general and may be applied to any section of a thick-walled circular curved beam with application of appropriate limits of integration.

For the specific curved beam in Fig. 4–12b, the integrals are evaluated from 0 to π. Also, for this case we find

$$M = FR \sin\theta \qquad\qquad \frac{\partial M}{\partial F} = R\sin\theta$$

$$F_\theta = F\sin\theta \qquad\qquad \frac{\partial F_\theta}{\partial F} = \sin\theta$$

$$MF_\theta = F^2 R \sin^2\theta \qquad\qquad \frac{\partial(MF_\theta)}{\partial F} = 2FR\sin^2\theta$$

$$F_r = F\cos\theta \qquad\qquad \frac{\partial F_r}{\partial F} = \cos\theta$$

Substituting these into Eq. (4–38) and factoring yields

$$\delta = \frac{FR^2}{AeE}\int_0^\pi \sin^2\theta \, d\theta + \frac{FR}{AE}\int_0^\pi \sin^2\theta \, d\theta - \frac{2FR}{AE}\int_0^\pi \sin^2\theta \, d\theta$$
$$+ \frac{CFR}{AG}\int_0^\pi \cos^2\theta \, d\theta \qquad (4\text{--}39)$$

$$= \frac{\pi FR^2}{2AeE} + \frac{\pi FR}{2AE} - \frac{\pi FR}{AE} + \frac{\pi CFR}{2AG} = \frac{\pi FR^2}{2AeE} - \frac{\pi FR}{2AE} + \frac{\pi CFR}{2AG}$$

Because the first term contains the square of the radius, the second two terms will be small if the frame has a large radius.

For curved sections in which the radius is significantly larger than the thickness, say $R/h > 10$, the effect of the eccentricity is negligible, so that the strain energies can be approximated directly from Eqs. (4–17), (4–23), and (4–25) with a substitution of $R \, d\theta$ for dx. Further, as R increases, the contributions to deflection from the normal force and tangential force becomes negligibly small compared to the bending component. Therefore, an approximate result can be obtained for a thin circular curved member as

$$U \approx \int \frac{M^2}{2EI} R \, d\theta \qquad R/h > 10 \qquad (4\text{--}40)$$

$$\delta = \frac{\partial U}{\partial F} \approx \int \frac{1}{EI}\left(M\frac{\partial M}{\partial F}\right) R \, d\theta \qquad R/h > 10 \qquad (4\text{--}41)$$

EXAMPLE 4–12

The cantilevered hook shown in Fig. 4–13a is formed from a round steel wire with a diameter of 3.5 mm. The hook dimensions are $l = 30$ and $R = 70$ mm. A force P of 2 N is applied at point C. Use Castigliano's theorem to estimate the deflection at point D at the tip.

Solution

Since l/d and R/d are significantly greater than 10, only the contributions due to bending will be considered. To obtain the vertical deflection at D, a fictitious force Q will be applied there. Free-body diagrams are shown in Figs. 4–13b, c, and d, with breaks in sections AB, BC, and CD, respectively. The normal and shear forces, N and V respectively, are shown but are considered negligible in the deflection analysis.

For section AB, with the variable of integration x defined as shown in Fig. 4–13b, summing moments about the break gives an equation for the moment in section AB,

$$M_{AB} = P(R + x) + Q(2R + x) \tag{1}$$

$$\partial M_{AB}/\partial Q = 2R + x \tag{2}$$

Since the derivative with respect to Q has been taken, we can set Q equal to zero. From Eq. (4–31), inserting Eqs. (1) and (2),

$$(\delta_D)_{AB} = \left[\int_0^l \frac{1}{EI}\left(M_{AB}\frac{\partial M_{AB}}{\partial Q}\right) dx \right]_{Q=0} = \frac{1}{EI}\int_0^l P(R + x)(2R + x)\,dx$$

$$\tag{3}$$

$$= \frac{P}{EI}\int_0^l (2R^2 + 3Rx + x^2)\,dx = \frac{P}{EI}\left(2R^2 l + \frac{3}{2}l^2 R + \frac{1}{3}l^3\right)$$

Figure 4–13

(a)

(b)   (c)   (d)

For section BC, with the variable of integration θ defined as shown in Fig. 4–13c, summing moments about the break gives the moment equation for section BC.

$$M_{BC} = Q(R + R\sin\theta) + PR\sin\theta \tag{4}$$

$$\partial M_{BC}/\partial Q = R(1 + \sin\theta) \tag{5}$$

From Eq. (4–41), inserting Eqs. (4) and (5) and setting $Q = 0$, we get

$$(\delta_D)_{BC} = \left[\int_0^{\pi/2} \frac{1}{EI}\left(M_{BC}\frac{\partial M_{BC}}{\partial Q}\right)R\,d\theta\right]_{Q=0} = \frac{R}{EI}\int_0^{\pi/2}(PR\sin\theta)[R(1+\sin\theta)]\,d\theta$$

$$= \frac{PR^3}{EI}\left(1 + \frac{\pi}{4}\right) \tag{6}$$

Noting that the break in section CD contains nothing but Q, and after setting $Q = 0$, we can conclude that there is no actual strain energy contribution in this section. Combining terms from Eqs. (3) and (6) to get the total vertical deflection at D,

$$\delta_D = (\delta_D)_{AB} + (\delta_D)_{BC} = \frac{P}{EI}\left(2R^2l + \frac{3}{2}l^2R + \frac{1}{3}l^3\right) + \frac{PR^3}{EI}\left(1 + \frac{\pi}{4}\right)$$

$$= \frac{P}{EI}(1.785R^3 + 2R^2l + 1.5\,Rl^2 + 0.333l^3) \tag{7}$$

Substituting values, and noting $I = \pi d^4/64$, and $E = 207$ GPa for steel, we get

Answer
$$\delta_D = \frac{1}{207(10^9)[\pi(0.0035^4)/64]}[1.785(0.07^3) + 2(0.07^2)0.03$$

$$+ 1.5(0.07)0.03^2 + 0.333(0.03^3)]$$

$$= 1.32(10^{-3})\text{ m} = 1.32\text{ mm}$$

EXAMPLE 4–13 Deflection in a Variable-Cross-Section Punch-Press Frame

The general result expressed in Eq. (4–39),

$$\delta = \frac{\pi FR^2}{2AeE} - \frac{\pi FR}{2AE} + \frac{\pi CFR}{2AG}$$

is useful in sections that are uniform and in which the centroidal locus is circular. The bending moment is largest where the material is farthest from the load axis. Strengthening requires a larger second area moment I. A variable-depth cross section is attractive, but it makes the integration to a closed form very difficult. However, if you are seeking results, numerical integration with computer assistance is helpful.

Consider the steel C frame depicted in Fig. 4–14a in which the centroidal radius is 800 mm, the cross section at the ends is 50 mm × 50 mm, and the depth varies sinu-

Figure 4–14

(*a*) A steel punch press has a C frame with a varying-depth rectangular cross section depicted. The cross section varies sinusoidally from 50 mm × 50 mm at $\theta = 0°$ to 50 mm × 150 mm at $\theta = 90°$, and back to 50 mm × 50 mm at $\theta = 180°$. Of immediate interest to the designer is the deflection in the load axis direction under the load.
(*b*) Finite-element model.

soidally with an amplitude of 50 mm. The load is 4 kN. It follows that $C = 1.2$, $G = 80$ GPa, $E = 210$ GPa. The outer and inner radii are

$$R_{\text{out}} = 0.825 + 0.05\sin\theta \qquad R_{\text{in}} = 0.775 - 0.05\sin\theta$$

The remaining geometrical terms are

$$h = R_{\text{out}} - R_{\text{in}} = 0.05(1 + 0.05\sin\theta)$$

$$A = bh = 0.1(1 + 0.05\sin\theta)$$

$$r_n = \frac{h}{\ln[(R + h/2)/(R - h/2)]} = \frac{0.05(1 + 0.05\sin\theta)}{\ln[(0.825 + 0.05\sin\theta)/(0.775 - 0.05\sin\theta)]}$$

$$e = R - r_n = 0.8 - r_n$$

Note that

$$M = FR\sin\theta \qquad \partial M/\partial F = R\sin\theta$$

$$F_\theta = F\sin\theta \qquad \partial F_\theta/\partial F = \sin\theta$$

$$MF_\theta = F^2 R\sin^2\theta \qquad \partial MF_\theta/\partial F = 2FR\sin^2\theta$$

$$F_r = F\cos\theta \qquad \partial F_r/\partial F = \cos\theta$$

Substitution of the terms into Eq. (4–38) yields three integrals

$$\delta = I_1 + I_2 + I_3 \tag{1}$$

where the integrals are

$$I_1 = 213.33\,(10^{-3})\int_0^\pi \frac{\sin^2\theta\,d\theta}{(1 + 0.05\sin\theta)\left[0.8 - \dfrac{0.05(1 + 0.05\sin\theta)}{\ln\left(\dfrac{0.825 + 0.05\sin\theta}{0.775 - 0.05\sin\theta}\right)}\right]} \tag{2}$$

$$I_2 = -66.67\,(10^{-4})\int_0^\pi \frac{\sin^2\theta\,d\theta}{1 + 0.05\sin\theta} \tag{3}$$

$$I_3 = 208.70\,(10^{-4})\int_0^\pi \frac{\cos^2\theta\,d\theta}{1 + 0.05\sin\theta} \tag{4}$$

The integrals may be evaluated in a number of ways: by a program using Simpson's rule integration,[8] by a program using a spreadsheet, or by mathematics software. Using MathCad and checking the results with Excel gives the integrals as $I_1 = 1.915375$, $I_2 = -0.003975$, and $I_3 = 0.019325$. Substituting these into Eq. (1) gives

Answer

$$\delta = 1.93 \text{ mm}$$

Finite-element (FE) programs are also very accessible. Figure 4–14b shows a simple half-model, using symmetry, of the press consisting of 216 plane-stress (2-D) elements. Creating the model and analyzing it to obtain a solution took minutes. Doubling the results from the FE analysis yielded $\delta = 1.95$ mm, about 1 percent variation from the results of the numerical integration.

4–10 Statically Indeterminate Problems

A system is *overconstrained* when it has more unknown support (reaction) forces and/ or moments than static equilibrium equations. Such a system is said to be *statically indeterminate* and the extra constraint supports are called *redundant supports*. In addition to the static equilibrium equations, a deflection equation is required for *each* redundant support reaction in order to obtain a solution. For example, consider a beam in bending with a wall support on one end and a simple support on the other, such as beam 12 of Table A–9. There are three support reactions and only two static equilibrium equations are available. This beam has *one* redundant support. To solve for the three unknown support reactions we use the two equilibrium equations and *one* additional deflection equation. For another example, consider beam 15 of Table A–9. This beam has a wall on both ends, giving rise to *two* redundant supports requiring *two* deflection equations in addition to the equations from statics. The purpose of redundant supports is to provide additional safety and reduce deflection.

A simple example of a statically indeterminate problem is furnished by the nested helical springs in Fig. 4–15a. When this assembly is loaded by the compressive force F, it deforms through the distance δ. What is the compressive force in each spring?

Only one equation of static equilibrium can be written. It is

$$\sum F = F - F_1 - F_2 = 0 \qquad (a)$$

which simply says that the total force F is resisted by a force F_1 in spring 1 plus the force F_2 in spring 2. Since there are two unknowns and only one static equilibrium equation, the system is statically indeterminate.

To write another equation, note the deformation relation in Fig. 4–15b. The two springs have the same deformation. Thus, we obtain the second equation as

$$\delta_1 = \delta_2 = \delta \qquad (b)$$

If we now substitute Eq. (4–2) in Eq. (b), we have

$$\frac{F_1}{k_1} = \frac{F_2}{k_2} \qquad (c)$$

---

[8]See Case Study 4, p. 203, J. E. Shigley and C. R. Mischke, *Mechanical Engineering Design*, 6th ed., McGraw-Hill, New York, 2001.

| **Figure 4–15**

(a)

(b)

Now we solve Eq. (c) for F_1 and substitute the result in Eq. (a). This gives

$$F - \frac{k_1}{k_2}F_2 - F_2 = 0 \qquad \text{or} \qquad F_2 = \frac{k_2 F}{k_1 + k_2} \qquad (d)$$

Substituting F_2 into Eq. (c) gives $F_1 = k_1 F/(k_1 + k_2)$ and so $\delta = \delta_1 = \delta_2 = F/(k_1 + k_2)$. Thus, for two springs in parallel, the overall spring constant is $k = F/\delta = k_1 + k_2$.

In the spring example, obtaining the necessary deformation equation was very straightforward. However, for other situations, the deformation relations may not be as easy. A more structured approach may be necessary. Here we will show two basic procedures for general statically indeterminate problems.

Procedure 1

1 Choose the redundant reaction(s). There may be alternative choices (See Example 4–14).

2 Write the equations of static equilibrium for the remaining reactions in terms of the applied loads and the redundant reaction(s) of step 1.

3 Write the deflection equation(s) for the point(s) at the locations of the redundant reaction(s) of step 1 in terms of the applied loads and the redundant reaction(s) of step 1. Normally the deflection(s) is (are) zero. If a redundant reaction is a moment, the corresponding deflection equation is a rotational deflection equation.

4 The equations from steps 2 and 3 can now be solved to determine the reactions.

In step 3 the deflection equations can be solved in any of the standard ways. Here we will demonstrate the use of superposition and Castigliano's theorem on a beam problem.

EXAMPLE 4–14   The indeterminate beam 11 of Appendix Table A–9 is reproduced in Fig. 4–16. Determine the reactions using procedure 1.

Solution   The reactions are shown in Fig. 4–16b. Without R_2 the beam is a statically determinate cantilever beam. Without M_1 the beam is a statically determinate simply supported beam. In either case, the beam has only *one* redundant support. We will first solve this problem using superposition, choosing R_2 as the redundant reaction. For the second solution, we will use Castigliano's theorem with M_1 as the redundant reaction.

Solution 1   1   Choose R_2 at B to be the redundant reaction.

2   Using static equilibrium equations solve for R_1 and M_1 in terms of F and R_2. This results in

$$R_1 = F - R_2 \qquad M_1 = \frac{Fl}{2} - R_2 l \qquad (1)$$

3   Write the deflection equation for point B in terms of F and R_2. Using superposition of beam 1 of Table A–9 with $F = -R_2$, and beam 2 of Table A–9 with $a = l/2$, the deflection of B, at $x = l$, is

$$\delta_B = -\frac{R_2 l^2}{6EI}(l - 3l) + \frac{F(l/2)^2}{6EI}\left(\frac{l}{2} - 3l\right) = \frac{R_2 l^3}{3EI} - \frac{5Fl^3}{48EI} = 0 \qquad (2)$$

4   Equation (2) can be solved for R_2 directly. This yields

Answer
$$R_2 = \frac{5F}{16} \qquad (3)$$

Next, substituting R_2 into Eqs. (1) completes the solution, giving

Answer
$$R_1 = \frac{11F}{16} \qquad (4)$$

Note that the solution agrees with what is given for beam 11 in Table A–9.

Solution 2   1   Choose M_1 at O to be the redundant reaction.

2   Using static equilibrium equations solve for R_1 and R_2 in terms of F and M_1. This results in

$$R_1 = \frac{F}{2} + \frac{M_1}{l} \qquad R_2 = \frac{F}{2} - \frac{M_1}{l} \qquad (5)$$

| Figure 4–16

(a)   (b)

3 Since M_1 is the redundant reaction at O, write the equation for the angular deflection at point O. From Castigliano's theorem this is

$$\theta_O = \frac{\partial U}{\partial M_1} \tag{6}$$

We can apply Eq. (4–31), using the variable x as shown in Fig. 4–16b. However, simpler terms can be found by using a variable $\hat{x}$ that starts at B and is positive to the left. With this and the expression for R_2 from Eq. (5) the moment equations are

$$M = \left(\frac{F}{2} - \frac{M_1}{l}\right)\hat{x} \qquad 0 \le \hat{x} \le \frac{l}{2} \tag{7}$$

$$M = \left(\frac{F}{2} - \frac{M_1}{l}\right)\hat{x} - F\left(\hat{x} - \frac{l}{2}\right) \qquad \frac{l}{2} \le \hat{x} \le l \tag{8}$$

For both equations

$$\frac{\partial M}{\partial M_1} = -\frac{\hat{x}}{l} \tag{9}$$

Substituting Eqs. (7) to (9) in Eq. (6), using the form of Eq. (4–31) where $F_i = M_1$, gives

$$\theta_O = \frac{\partial U}{\partial M_1} = \frac{1}{EI}\left\{ \int_0^{l/2} \left(\frac{F}{2} - \frac{M_1}{l}\right)\hat{x}\left(-\frac{\hat{x}}{l}\right) d\hat{x} + \int_{l/2}^l \left[\left(\frac{F}{2} - \frac{M_1}{l}\right)\hat{x} \right.\right.$$
$$\left.\left. - F\left(\hat{x} - \frac{l}{2}\right)\right]\left(-\frac{\hat{x}}{l}\right) d\hat{x} \right\} = 0$$

Canceling $1/EIl$, and combining the first two integrals, simplifies this quite readily to

$$\left(\frac{F}{2} - \frac{M_1}{l}\right)\int_0^l \hat{x}^2\, d\hat{x} - F\int_{l/2}^l \left(\hat{x} - \frac{l}{2}\right)\hat{x}\, d\hat{x} = 0$$

Integrating gives

$$\left(\frac{F}{2} - \frac{M_1}{l}\right)\frac{l^3}{3} - \frac{F}{3}\left[l^3 - \left(\frac{l}{2}\right)^3\right] + \frac{Fl}{4}\left[l^2 - \left(\frac{l}{2}\right)^2\right] = 0$$

which reduces to

$$M_1 = \frac{3Fl}{16} \tag{10}$$

4 Substituting Eq. (10) into (5) results in

$$R_1 = \frac{11F}{16} \qquad R_2 = \frac{5F}{16} \tag{11}$$

which again agrees with beam 11 of Table A–9.

For some problems even procedure 1 can be a task. Procedure 2 eliminates some tricky geometric problems that would complicate procedure 1. We will describe the procedure for a beam problem.

Procedure 2

1 Write the equations of static equilibrium for the beam in terms of the applied loads and unknown restraint reactions.

2 Write the deflection equation for the beam in terms of the applied loads and unknown restraint reactions.

3 Apply boundary conditions to the deflection equation of step 2 consistent with the restraints.

4 Solve the equations from steps 1 and 3.

EXAMPLE 4–15

The steel beam $ABCD$ shown in Figure 4–17 is supported at C as shown and supported at B and D by shoulder steel bolts, each having a diameter of 8 mm. The lengths of BE and DF are 50 mm and 62 mm, respectively. The beam has a second area moment of $20.8(10^3)$ mm⁴. Prior to loading, the members are stress-free. A force of 2 kN is then applied at point A. Using procedure 2 of Sec. 4–10, determine the stresses in the bolts and the deflections of points A, B, and D. For steel, $E = 207$ GPa.

Figure 4–17

Dimensions in mm.

Solution

$$EI = 207(10^9)20.8(10^{-9}) = 4305.6 \text{ N} \cdot \text{m}^2$$

1
$$R_C + F_{BE} - F_{FD} = 2000 \tag{a}$$
$$0.075R_C + 0.15F_{BE} = 0.225(2000) = 450 \tag{b}$$

2
$$M = -2000x + F_{BE}\langle x - 0.075 \rangle^1 + R_C\langle x - 0.15 \rangle^1$$

$$EI\frac{dy}{dx} = -1000x^2 + \frac{F_{BE}}{2}\langle x - 0.075 \rangle^2 + \frac{R_C}{2}\langle x - 0.15 \rangle^2 + C_1$$

$$EIy = -\frac{1000}{3}x^3 + \frac{F_{BE}}{6}\langle x - 0.075 \rangle^3 + \frac{R_C}{6}\langle x - 0.15 \rangle^3 + C_1 x + C_2$$

$$y_B = \left(\frac{Fl}{AE}\right)_{BE} = -\frac{F_{BE}(0.05)}{(\pi/4)(0.008)^2(207)10^9} = -4.805(10^{-9})F_{BE}$$

Substituting and evaluating at $x = 0.075$ m

$$EIy_B = 4305.6(-4.805)10^{-9}F_{BE} = -\frac{1000}{3}(0.075^3) + 0.075\,C_1 + C_2$$

$$2.069(10^{-5})\,F_{BE} + 0.075\,C_1 + C_2 = 0.1406 \qquad\qquad (c)$$

Since $y = 0$ at $x = 0.15$ m

$$EIy|_{=0} = -\frac{1000}{3}(0.15^3) + \frac{F_{BE}}{6}(0.15 - 0.075)^3 + 0.15\,C_1 + C_2$$

$$7.03(10^{-5})\,F_{BE} + 0.15\,C_1 + C_2 = 1.125 \qquad\qquad (d)$$

$$y_D = \left(\frac{Fl}{AE}\right)_{DF} = \frac{F_{DF}(0.062)}{(\pi/4)(0.008)^2 207(10^9)} = 5.96(10^{-9})\,F_{DF}$$

Substituting and evaluating at $x = 0.225$ m

$$EIy_D = 4305.6(5.96)10^{-9}F_{DF} = -\frac{1000}{3}(0.225^3) + \frac{F_{BE}}{6}(0.225 - 0.075)^3$$

$$+ \frac{R_C}{6}(0.225 - 0.15)^3 + 0.225\,C_1 + C_2$$

$$7.03(10^{-5})\,R_C + 5.625(10^{-4})\,F_{BE} - 2.566(10^{-5})\,F_{DF} + 0.225\,C_1 + C_2 = 3.8 \quad (e)$$

$$\begin{bmatrix} 1 & 1 & -1 & 0 & 0 \\ 0.075 & 0.15 & 0 & 0 & 0 \\ 0 & 2.069(10^{-5}) & 0 & 0.075 & 1 \\ 0 & 7.03(10^{-5}) & 0 & 0.15 & 1 \\ 7.03(10^{-5}) & 5.625(10^{-4}) & -2.566(10^{-5}) & 0.225 & 1 \end{bmatrix} \begin{Bmatrix} R_C \\ F_{BE} \\ F_{DF} \\ C_1 \\ C_2 \end{Bmatrix} = \begin{Bmatrix} 2000 \\ 450 \\ 0.1406 \\ 1.125 \\ 3.8 \end{Bmatrix}$$

$$R_C = -2.657 \text{ kN}, \quad F_{BE} = 3783 \text{ kN}, \quad F_{DF} = -0.203 \text{ kN}$$

$$C_1 = 12 \text{ N} \cdot \text{m}^2, \quad C_2 = -0.85 \text{ N} \cdot \text{m}^3$$

Answer $\qquad \sigma_{BE} = \dfrac{4703}{(\pi/4)(8)^2} = 93.6$ MPa

Answer $\qquad \sigma_{DF} = -\dfrac{203}{(\pi/4)(8)^2} = -4.04$ MPa

Answer $\qquad y_A = \dfrac{1}{4305.6}(-0.85) = -0.000\,197$ m $= -0.197$ mm

Answer $\qquad y_B = \dfrac{1}{4305.6}\left[-\dfrac{1000}{3}(0.075^3) + 12(0.075) - 0.85\right] = -2.1(10^{-5})$ m $= 0.021$ mm

Answer $\qquad y_D = \dfrac{1}{4305.6}\left[-\dfrac{1000}{3}(0.225^3) + \dfrac{3783}{6}(0.225 - 0.075)^3\right.$

$$\left. + \dfrac{-2675}{6}(0.225 - 0.15)^3 + 12(0.225) - 0.85\right] = -1.34(10^{-6}) \text{ m}$$

Note that we could have easily incorporated the stiffness of the support at *B* if we were given a spring constant.

4-11   **Compression Members—General**

The analysis and design of compression members can differ significantly from that of members loaded in tension or in torsion. If you were to take a long rod or pole, such as a meterstick, and apply gradually increasing compressive forces at each end, very small axial deflections would happen at first, but then the stick would bend (buckle), and very quickly bend so much as to possibly fracture. Try it. The other extreme would occur if you were to saw off, say, a 5-mm length of the meterstick and perform the same experiment on the short piece. You would then observe that the failure exhibits itself as a mashing of the specimen, that is, a simple compressive failure. For these reasons it is convenient to classify compression members according to their length and according to whether the loading is central or eccentric. The term *column* is applied to all such members except those in which failure would be by simple or pure compression. Columns can be categorized then as:

1   Long columns with central loading
2   Intermediate-length columns with central loading
3   Columns with eccentric loading
4   Struts or short columns with eccentric loading

Classifying columns as above makes it possible to develop methods of analysis and design specific to each category. Furthermore, these methods will also reveal whether or not you have selected the category appropriate to your particular problem. The four sections that follow correspond, respectively, to the four categories of columns listed above.

4-12   **Long Columns with Central Loading**

Figure 4–18 shows long columns with differing end (boundary) conditions. If the axial force *P* shown acts along the centroidal axis of the column, simple compression of the member occurs for low values of the force. However, under certain conditions, when *P* reaches a specific value, the column becomes *unstable* and bending as shown in Fig. 4–18 develops rapidly. This force is determined by writing the bending deflection equation for the column, resulting in a differential equation where when the boundary conditions are applied, results in the *critical load* for unstable bending.[9] The critical force for the pin-ended column of Fig. 4–18a is given by

$$P_{cr} = \frac{\pi^2 EI}{l^2} \tag{4-42}$$

which is called the *Euler column formula*. Equation (4–42) can be extended to apply to other end-conditions by writing

$$P_{cr} = \frac{C\pi^2 EI}{l^2} \tag{4-43}$$

where the constant *C* depends on the end conditions as shown in Fig. 4–18.

---

[9]See F. P. Beer, E. R. Johnston, Jr., and J. T. DeWolf, *Mechanics of Materials,* 5th ed., McGraw-Hill, New York, 2009, pp. 610–613.

Figure 4–18

(*a*) Both ends rounded or pivoted; (*b*) both ends fixed; (*c*) one end free and one end fixed; (*d*) one end rounded and pivoted, and one end fixed.

(*a*) $C = 1$ (*b*) $C = 4$ (*c*) $C = \frac{1}{4}$ (*d*) $C = 2$

Using the relation $I = Ak^2$, where A is the area and k the *radius of gyration,* enables us to rearrange Eq. (4–43) into the more convenient form

$$\frac{P_{cr}}{A} = \frac{C\pi^2 E}{(l/k)^2}$$

(4–44)

where l/k is called the *slenderness ratio.* This ratio, rather than the actual column length, will be used in classifying columns according to length categories.

The quantity P_{cr}/A in Eq. (4–44) is the *critical unit load.* It is the load per unit area necessary to place the column in a condition of *unstable equilibrium.* In this state any small crookedness of the member, or slight movement of the support or load, will cause the column to begin to collapse. The unit load has the same units as strength, but this is the strength of a specific column, not of the column material. Doubling the length of a member, for example, will have a drastic effect on the value of P_{cr}/A but no effect at all on, say, the yield strength S_y of the column material itself.

Equation (4–44) shows that the critical unit load depends only upon the end conditions, the modulus of elasticity, and the slenderness ratio. Thus a column obeying the Euler formula made of high-strength alloy steel is *no stronger* than one made of low-carbon steel, since E is the same for both.

The factor C is called the *end-condition constant,* and it may have any one of the theoretical values $\frac{1}{4}$, 1, 2, and 4, depending upon the manner in which the load is applied. In practice it is difficult, if not impossible, to fix the column ends so that the factor $C = 2$ or $C = 4$ would apply. Even if the ends are welded, some deflection will occur. Because of this, some designers never use a value of C greater than unity. However, if liberal factors of safety are employed, and if the column load is accurately known, then a value of C not exceeding 1.2 for both ends fixed, or for one end rounded and one end fixed, is not unreasonable, since it supposes only partial fixation. Of course, the value $C = \frac{1}{4}$ must always be used for a column having one end fixed and one end free. These recommendations are summarized in Table 4–2.

When Eq. (4–44) is solved for various values of the unit load P_{cr}/A in terms of the slenderness ratio l/k, we obtain the curve *PQR* shown in Fig. 4–19. Since the yield strength of the material has the same units as the unit load, the horizontal line through S_y and Q has been added to the figure. This would appear to make the figure $S_y QR$ cover the entire range of compression problems from the shortest to the longest

Table 4–2

End-Condition Constants
for Euler Columns
[to Be Used with
Eq. (4–43)]

Column End Conditions	End-Condition Constant C		
	Theoretical Value	**Conservative Value**	**Recommended Value***
Fixed-free	$\frac{1}{4}$	$\frac{1}{4}$	$\frac{1}{4}$
Rounded-rounded	1	1	1
Fixed-rounded	2	1	1.2
Fixed-fixed	4	1	1.2

*To be used only with liberal factors of safety when the column load is accurately known.

Figure 4–19

Euler curve plotted using
Eq. (4–43) with $C = 1$.

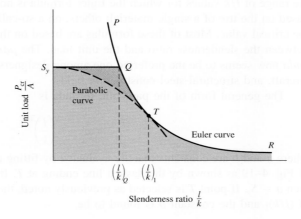

compression member. Thus it would appear that any compression member having an l/k value less than $(l/k)_Q$ should be treated as a pure compression member while all others are to be treated as Euler columns. Unfortunately, this is not true.

In the actual design of a member that functions as a column, the designer will be aware of the end conditions shown in Fig. 4–18, and will endeavor to configure the ends, using bolts, welds, or pins, for example, so as to achieve the required ideal end conditions. In spite of these precautions, the result, following manufacture, is likely to contain defects such as initial crookedness or load eccentricities. The existence of such defects and the methods of accounting for them will usually involve a factor-of-safety approach or a stochastic analysis. These methods work well for long columns and for simple compression members. However, tests show numerous failures for columns with slenderness ratios below and in the vicinity of point Q, as shown in the shaded area in Fig. 4–19. These have been reported as occurring even when near-perfect geometric specimens were used in the testing procedure.

A column failure is always sudden, total, unexpected, and hence dangerous. There is no advance warning. A beam will bend and give visual warning that it is overloaded, but not so for a column. For this reason neither simple compression methods nor the Euler column equation should be used when the slenderness ratio is near $(l/k)_Q$. Then what should we do? The usual approach is to choose some point T on the Euler curve of Fig. 4–19. If the slenderness ratio is specified as $(l/k)_1$ corresponding to point T, then use the Euler equation only when the actual slenderness ratio is greater than $(l/k)_1$. Otherwise, use one of the methods in the sections that follow. See Examples 4–17 and 4–18.

Most designers select point T such that $P_{cr}/A = S_y/2$. Using Eq. (4–43), we find the corresponding value of $(l/k)_1$ to be

$$\left(\frac{l}{k}\right)_1 = \left(\frac{2\pi^2 CE}{S_y}\right)^{1/2} \tag{4–45}$$

4–13 Intermediate-Length Columns with Central Loading

Over the years there have been a number of column formulas proposed and used for the range of l/k values for which the Euler formula is not suitable. Many of these are based on the use of a single material; others, on a so-called safe unit load rather than the critical value. Most of these formulas are based on the use of a linear relationship between the slenderness ratio and the unit load. The *parabolic* or *J. B. Johnson formula* now seems to be the preferred one among designers in the machine, automotive, aircraft, and structural-steel construction fields.

The general form of the parabolic formula is

$$\frac{P_{cr}}{A} = a - b\left(\frac{l}{k}\right)^2 \tag{a}$$

where a and b are constants that are evaluated by fitting a parabola to the Euler curve of Fig. 4–19 as shown by the dashed line ending at T. If the parabola is begun at S_y, then $a = S_y$. If point T is selected as previously noted, then Eq. (4–45) gives the value of $(l/k)_1$ and the constant b is found to be

$$b = \left(\frac{S_y}{2\pi}\right)^2 \frac{1}{CE} \tag{b}$$

Upon substituting the known values of a and b into Eq. (*a*), we obtain, for the parabolic equation,

$$\frac{P_{cr}}{A} = S_y - \left(\frac{S_y}{2\pi}\frac{l}{k}\right)^2 \frac{1}{CE} \qquad \frac{l}{k} \le \left(\frac{l}{k}\right)_1 \tag{4–46}$$

4–14 Columns with Eccentric Loading

We have noted before that deviations from an ideal column, such as load eccentricities or crookedness, are likely to occur during manufacture and assembly. Though these deviations are often quite small, it is still convenient to have a method of dealing with them. Frequently, too, problems occur in which load eccentricities are unavoidable.

Figure 4–20a shows a column in which the line of action of the column forces is separated from the centroidal axis of the column by the eccentricity e. From Fig. 4–20b, $M = -P(e + y)$. Substituting this into Eq. (4–12), $d^2y/dx^2 = M/EI$, results in the differential equation

$$\frac{d^2y}{dx^2} + \frac{P}{EI}y = -\frac{Pe}{EI} \tag{a}$$

Figure 4–20

Notation for an eccentrically loaded column.

The solution of Eq. (a), for the boundary conditions that $y = 0$ at $x = 0, l$ is

$$y = e\left[\tan\left(\frac{l}{2}\sqrt{\frac{P}{EI}}\right)\sin\left(\sqrt{\frac{P}{EI}}x\right) + \cos\left(\sqrt{\frac{P}{EI}}x\right) - 1\right] \qquad (b)$$

By substituting $x = l/2$ in Eq. (b) and using a trigonometric identity, we obtain

$$\delta = e\left[\sec\left(\sqrt{\frac{P}{EI}}\frac{l}{2}\right) - 1\right] \qquad (4–47)$$

The magnitude of the maximum bending moment also occurs at midspan and is

$$M_{\max} = P(e + \delta) = Pe\sec\left(\frac{l}{2}\sqrt{\frac{P}{EI}}\right) \qquad (4–48)$$

The magnitude of the maximum *compressive* stress at midspan is found by superposing the axial component and the bending component. This gives

$$\sigma_c = \frac{P}{A} + \frac{Mc}{I} = \frac{P}{A} + \frac{Mc}{Ak^2} \qquad (c)$$

Substituting $M_{\max}$ from Eq. (4–48) yields

$$\sigma_c = \frac{P}{A}\left[1 + \frac{ec}{k^2}\sec\left(\frac{l}{2k}\sqrt{\frac{P}{EA}}\right)\right] \qquad (4–49)$$

By imposing the compressive yield strength S_{yc} as the maximum value of σ_c, we can write Eq. (4–49) in the form

$$\frac{P}{A} = \frac{S_{yc}}{1 + (ec/k^2)\sec[(l/2k)\sqrt{P/AE}]} \qquad (4–50)$$

This is called the *secant column formula*. The term ec/k^2 is called the *eccentricity ratio*. Figure 4–21 is a plot of Eq. (4–50) for a steel having a compressive (and tensile) yield strength of 280 MPa. Note how the P/A contours asymptotically approach the Euler curve as l/k increases.

Figure 4–21

Comparison of secant and Euler equations for steel with $S_y = 280$ MPa.

Equation (4–50) cannot be solved explicitly for the load P. Design charts, in the fashion of Fig. 4–21, can be prepared for a single material if much column design is to be done. Otherwise, a root-finding technique using numerical methods must be used.

4–15 Struts or Short Compression Members

A short bar loaded in pure compression by a force P acting along the centroidal axis will shorten in accordance with Hooke's law, until the stress reaches the elastic limit of the material. At this point, permanent set is introduced and usefulness as a machine member may be at an end. If the force P is increased still more, the material either becomes "barrel-like" or fractures. When there is eccentricity in the loading, the elastic limit is encountered at smaller loads.

EXAMPLE 4–16

Develop specific Euler equations for the sizes of columns having
(a) Round cross sections
(b) Rectangular cross sections

Solution

(a) Using $A = \pi d^2/4$ and $k = \sqrt{I/A} = [(\pi d^4/64)/(\pi d^2/4)]^{1/2} = d/4$ with Eq. (4–44) gives

Answer

$$d = \left(\frac{64 P_{cr} l^2}{\pi^3 CE}\right)^{1/4} \tag{4–51}$$

(b) For the rectangular column, we specify a cross section $h \times b$ with the restriction that $h \leq b$. If the end conditions are the *same* for buckling in both directions, then buckling will occur in the direction of the least thickness. Therefore

$$I = \frac{bh^3}{12} \qquad A = bh \qquad k^2 = I/A = \frac{h^2}{12}$$

Substituting these in Eq. (4–44) gives

Answer

$$b = \frac{12 P_{cr} l^2}{\pi^2 CEh^3} \qquad h \leq b \tag{4–52}$$

Note, however, that rectangular columns do not generally have the same end conditions in both directions.

EXAMPLE 4–17   Specify the diameter of a round column 1.5 m long that is to carry a maximum load estimated to be 22 kN. Use a design factor $n_d = 4$ and consider the ends as pinned (rounded). The column material selected has a minimum yield strength of 500 MPa and a modulus of elasticity of 207 GPa.

Solution   We shall design the column for a critical load of

$$P_{cr} = n_d P = 4(22) = 88 \text{ kN}$$

Then, using Eq. (4–51) with $C = 1$ (see Table 4–2) gives

$$d = \left(\frac{64 P_{cr} l^2}{\pi^3 C E}\right)^{1/4} = \left[\frac{64(88)(1.5)^2}{\pi^3 (1)(207)}\right]^{1/4} \left(\frac{10^3}{10^9}\right)^{1/4} (10^3) = 37.48 \text{ mm}$$

Table A–17 shows that the preferred size is 40 mm. The slenderness ratio for this size is

$$\frac{l}{k} = \frac{l}{d/4} = \frac{1.5(10^3)}{40/4} = 150$$

To be sure that this is an Euler column, we use Eq. (5–51) and obtain

$$\left(\frac{l}{k}\right)_1 = \left(\frac{2\pi^2 C E}{S_y}\right)^{1/2} = \left[\frac{2\pi^2 (1)(207)}{500}\right]^{1/2} \left(\frac{10^9}{10^6}\right)^{1/2} = 90.4$$

where $l/k > (l/k)_1$ indicates that it is indeed an Euler column. So select

Answer $$d = 40 \text{ mm}$$

EXAMPLE 4–18   Repeat Ex. 4–16 for J. B. Johnson columns.

Solution   (*a*) For round columns, Eq. (4–46) yields

Answer $$d = 2\left(\frac{P_{cr}}{\pi S_y} + \frac{S_y l^2}{\pi^2 C E}\right)^{1/2} \qquad (4\text{–}53)$$

(*b*) For a rectangular section with dimensions $h \le b$, we find

Answer $$b = \frac{P_{cr}}{h S_y \left(1 - \dfrac{3l^2 S_y}{\pi^2 C E h^2}\right)} \qquad h \le b \qquad (4\text{–}54)$$

EXAMPLE 4–19   The hydraulic cylinder shown in Figure 4–22 has an 80-mm bore and is to operate at a pressure of 5.6 MPa. With the clevis mount shown, the piston rod should be sized as a column with both ends rounded for any plane of buckling. The rod is to be made of forged AISI 1030 steel without further heat treatment.

(*a*) Use a design factor $n_d = 3$ and select a preferred size for the rod diameter if the column length is 1.5 m.

(*b*) Repeat part (*a*) but for a column length of 0.5 m.

(*c*) What factor of safety actually results for each of the cases above?

| Figure 4–22

Solution

$$F = 5.6 \left(\frac{\pi}{4}\right)(80^2) = 28\ 149\ \text{N}, \quad S_y = 260\ \text{MPa}$$

$$P_{cr} = n_d F = 3(28\ 149) = 84\ 447\ \text{N}$$

(*a*) Assume Euler with $C = 1$

$$I = \frac{\pi}{64}d^4 = \frac{P_{cr}l^2}{C\pi^2 E} \quad \Rightarrow \quad d = \left[\frac{64 P_{cr} l^2}{\pi^3 C E}\right]^{1/4} = \left[\frac{64(84\ 447)(1.5)^2}{\pi^3(1)(207)10^9}\right]^{1/4} = 0.0371\ \text{m}$$

Use $d = 40$ mm; $k = d/4 = 10$

$$\frac{l}{k} = \frac{1500}{10} = 150$$

$$\left(\frac{l}{k}\right)_1 = \left(\frac{2\pi^2(1)(207)10^9}{260(10)^6}\right)^{1/2} = 125.4 \quad \therefore \text{use Euler}$$

$$P_{cr} = \frac{\pi^2(207)10^9(\pi/64)(0.04^4)}{1.5^2} = 114.1\ \text{kN}$$

Solution

$d = 40$ mm   is satisfactory.

(*b*) $\quad d = \left[\frac{64(84\ 447)(0.5)^2}{\pi^3(1)(207)10^9}\right]^{1/4} = 0.0214\ \text{m}, \quad$ so use 22 mm

$$k = \frac{22}{4} = 5.5\ \text{mm}$$

$$l/k = \frac{500}{5.5} = 90.9 \quad \text{try Johnson}$$

$$P_{cr} = \frac{\pi}{4}(0.022^2)\left[260(10^6) - \left(\frac{260(10^6)}{2\pi}90.9\right)^2 \frac{1}{1(207)(10^9)}\right] = 98\ 831\ \text{N}$$

Answer

Use $d = 22$ mm

Answer

(*c*) $\quad n_{(a)} = \frac{114\ 100}{28\ 149} = 4.05$

Answer

$$n_{(b)} = \frac{98\ 831}{28\ 149} = 3.51$$

A *strut* is a *short compression member* such as the one shown in Fig. 4–22. The magnitude of the maximum compressive stress in the x direction at point B in an intermediate section is the sum of a simple component P/A and a flexural component Mc/I; that is,

$$\sigma_c = \frac{P}{A} + \frac{Mc}{I} = \frac{P}{A} + \frac{PecA}{IA} = \frac{P}{A}\left(1 + \frac{ec}{k^2}\right) \tag{4–55}$$

Figure 4–22

Eccentrically loaded strut.

where $k = (I/A)^{1/2}$ and is the radius of gyration, c is the coordinate of point B, and e is the eccentricity of loading.

Note that the length of the strut does not appear in Eq. (4–55). In order to use the equation for design or analysis, we ought, therefore, to know the range of lengths for which the equation is valid. In other words, how long is a short member?

The difference between the secant formula Eq. (4–50) and Eq. (4–55) is that the secant equation, unlike Eq. (4–55), accounts for an increased bending moment due to bending deflection. Thus the secant equation shows the eccentricity to be magnified by the bending deflection. This difference between the two formulas suggests that one way of differentiating between a "secant column" and a strut, or short compression member, is to say that in a strut, the effect of bending deflection must be limited to a certain small percentage of the eccentricity. If we decide that the limiting percentage is to be 1 percent of e, then, from Eq. (4–44), the limiting slenderness ratio turns out to be

$$\left(\frac{l}{k}\right)_2 = 0.282 \left(\frac{AE}{P}\right)^{1/2} \qquad (4\text{–}56)$$

This equation then gives the limiting slenderness ratio for using Eq. (4–55). If the actual slenderness ratio is greater than $(l/k)_2$, then use the secant formula; otherwise, use Eq. (4–55).

EXAMPLE 4–20

Figure 4–23a shows a workpiece clamped to a milling machine table by a bolt tightened to a tension of 8.9 kN. The clamp contact is offset from the centroidal axis of the strut by a distance $e = 2.5$ mm, as shown in part b of the figure. The strut, or block, is steel, 25 mm square and 0.1 m long, as shown. Determine the maximum compressive stress in the block.

Solution

First we find $A = bh = 0.025(0.025) = 625 \times 10^{-6}$ m^2, $I = bh^3/12 = 0.025(0.025)^3/12$ $= 32.55 \times 10^{-9}$ m^4, $k^2 = I/A = 32.55 \times 10^{-9}/625 \times 10^{-6} = 52.1 \times 10^{-6}$ m^2, and $l/k =$ $0.1/(52.1 \times 10^{-6})^{1/2} = 13.9$. Equation (4–56) gives the limiting slenderness ratio as

$$\left(\frac{l}{k}\right)_2 = 0.282 \left(\frac{AE}{P}\right)^{1/2} = 0.282 \left[\frac{625 \times 10^{-6}\,(210 \times 10^9)}{8900}\right]^{1/2} = 34.2$$

Thus the block could be as long as

$$l = 34.2k = 34.2(52.1 \times 10^{-6})^{1/2} = 0.24 \text{ m}$$

before it need be treated by using the secant formula. So Eq. (4–55) applies and the maximum compressive stress is

Answer

$$\sigma_c = \frac{P}{A}\left(1 + \frac{ec}{k^2}\right) = \frac{8900}{625 \times 10^{-6}}\left[1 + \frac{0.0025(0.0125)}{52.1 \times 10^{-6}}\right] = 22.8 \text{ MPa}$$

Figure 4–23

A strut that is part of a workpiece clamping assembly.

(a) (b)

4–16   **Elastic Stability**

Section 4–12 presented the conditions for the unstable behavior of long, slender columns. *Elastic instability* can also occur in structural members other than columns. *Compressive loads/stresses within any long, thin structure can cause structural instabilities* (buckling). The compressive stress may be elastic or inelastic and the instability may be global or local. Global instabilities can cause *catastrophic* failure, whereas local instabilities may cause permanent deformation and function failure but not a catastrophic failure. The buckling discussed in Sec. 4–12 was global instability. However, consider a wide flange beam in bending. One flange will be in compression, and if thin enough, can develop localized buckling in a region where the bending moment is a maximum. Localized buckling can also occur in the web of the beam, where transverse shear stresses are present at the beam centroid. Recall, for the case of pure shear stress τ, a stress transformation will show that at 45°, a *compressive stress* of $\sigma = -\tau$ exists. If the web is sufficiently thin where the shear force V is a maximum, localized buckling of the web can occur. For this reason, additional support in the form of bracing is typically applied at locations of high shear forces.[10]

Thin-walled beams in bending can buckle in a torsional mode as illustrated in Fig. 4–24. Here a cantilever beam is loaded with a lateral force, F. As F is increased from zero, the end of the beam will deflect in the negative y direction normally according to the bending equation, $y = -FL^3/(3EI)$. However, if the beam is long enough and the ratio of b/h is sufficiently small, there is a critical value of F for which the beam will collapse in a twisting mode as shown. This is due to the *compression* in the bottom fibers of the beam which cause the fibers to buckle sideways (z direction).

There are a great many other examples of unstable structural behavior, such as thin-walled pressure vessels in compression or with outer pressure or inner vacuum, thin-walled open or closed members in torsion, thin arches in compression, frames in compression, and shear panels. Because of the vast array of applications and the complexity of their analyses, further elaboration is beyond the scope of this book. The intent of this section is to make the reader aware of the possibilities and potential safety issues. The key issue is that the designer should be aware that if any *unbraced* part of a structural member is *thin*, and/or *long*, and in *compression* (directly or *indirectly*), the possibility of buckling should be investigated.[11]

For unique applications, the designer may need to revert to a numerical solution such as using finite elements. Depending on the application and the finite-element code available, an analysis can be performed to determine the critical loading (see Fig. 4–25).

Figure 4–24

Torsional buckling of a thin-walled beam in bending.

Figure 4–25

Finite-element representation of flange buckling of a channel in compression.

---

[10]See C. G. Salmon, J. E. Johnson, and F. A. Malhas, *Steel Structures: Design and Behavior*, 5th ed., Prentice Hall, Upper Saddle River, NJ, 2009.

[11]See S. P. Timoshenko and J. M. Gere, *Theory of Elastic Stability*, 2nd ed., McGraw-Hill, New York, 1961. See also, Z. P. Bazant and L. Cedolin, *Stability of Structures*, Oxford University Press, New York, 1991.

$4-17$ **Shock and Impact**

Impact refers to the collision of two masses with initial relative velocity. In some cases it is desirable to achieve a known impact in design; for example, this is the case in the design of coining, stamping, and forming presses. In other cases, impact occurs because of excessive deflections, or because of clearances between parts, and in these cases it is desirable to minimize the effects. The rattling of mating gear teeth in their tooth spaces is an impact problem caused by shaft deflection and the clearance between the teeth. This impact causes gear noise and fatigue failure of the tooth surfaces. The clearance space between a cam and follower or between a journal and its bearing may result in crossover impact and also cause excessive noise and rapid fatigue failure.

Shock is a more general term that is used to describe any suddenly applied force or disturbance. Thus the study of shock includes impact as a special case.

Figure 4–26 represents a highly simplified mathematical model of an automobile in collision with a rigid obstruction. Here m_1 is the lumped mass of the engine. The displacement, velocity, and acceleration are described by the coordinate x_1 and its time derivatives. The lumped mass of the vehicle less the engine is denoted by m_2, and its motion by the coordinate x_2 and its derivatives. Springs k_1, k_2, and k_3 represent the linear and nonlinear stiffnesses of the various structural elements that compose the vehicle. Friction and damping can and should be included, but is not shown in this model. The determination of the spring rates for such a complex structure will almost certainly have to be performed experimentally. Once these values—the k's, m's, damping and frictional coefficients—are obtained, a set of nonlinear differential equations can be written and a computer solution obtained for any impact velocity. For sake of illustration, assuming the springs to be linear, isolate each mass and write their equations of motion. This results in

$$m\ddot{x}_1 + k_1 x_1 + k_2(x_1 - x_2) = 0$$
$$m\ddot{x}_2 + k_3 x_2 - k_2(x_1 - x_2) = 0$$

$(4-57)$

The analytical solution of the Eq. (4–57) pair is harmonic and is studied in a course on mechanical vibrations.[12] If the values of the m's and k's are known, the solution can be obtained easily using a program such as MATLAB.

Suddenly Applied Loading

A simple case of impact is illustrated in Fig. 4–27a. Here a weight W falls a distance h and impacts a cantilever of stiffness EI and length l. We want to find the maximum deflection and the maximum force exerted on the beam due to the impact.

Figure 4–26

Two-degree-of-freedom mathematical model of an automobile in collision with a rigid obstruction.

---

[12]See William T. Thomson and Marie Dillon Dahleh, *Theory of Vibrations with Applications*, 5th ed., Prentice Hall, Upper Saddle River, NJ, 1998.

Figure 4–27

(a) A weight free to fall a distance h to free end of a beam. (b) Equivalent spring model.

Figure 4–27b shows an abstract model of the system considering the beam as a simple spring. For beam 1 of Table A–9, we find the spring rate to be $k = F/y = 3EI/l^3$. The beam mass and damping can be accounted for, but for this example will be considered negligible. If the beam is considered massless, there is no momentum transfer, only energy. If the maximum deflection of the spring (beam) is considered to be δ, the drop of the weight is $h + \delta$, and the loss of potential energy is $W(h + \delta)$. The resulting increase in potential (strain) energy of the spring is $\frac{1}{2}k\delta^2$. Thus, for energy conservation $\frac{1}{2}k\delta^2 = W(h + \delta)$. Rearranging this gives

$$\delta^2 - 2\frac{W}{k}\delta - 2\frac{W}{k}h = 0 \tag{a}$$

Solving for δ yields

$$\delta = \frac{W}{k} \pm \frac{W}{k}\left(1 + \frac{2hk}{W}\right)^{1/2} \tag{b}$$

The negative solution is possible only if the weight "sticks" to the beam and vibrates between the limits of Eq. (b). Thus, the maximum deflection is

$$\delta = \frac{W}{k} + \frac{W}{k}\left(1 + \frac{2hk}{W}\right)^{1/2} \tag{4–58}$$

The maximum force acting on the beam is now found to be

$$F = k\delta = W + W\left(1 + \frac{2hk}{W}\right)^{1/2} \tag{4–59}$$

Note, in this equation, that if $h = 0$, then $F = 2W$. This says that when the weight is released while in contact with the spring but is not exerting any force on the spring, the largest force is double the weight.

Most systems are not as ideal as those explored here, so be wary about using these relations for nonideal systems.

PROBLEMS

Problems marked with an asterisk (*) are linked to problems in other chapters, as summarized in Table 1–2 of Sec. 1–17, p. 34.

4-1 The figure shows a torsion bar OA fixed at O, simply supported at A, and connected to a cantilever AB. The spring rate of the torsion bar is k_T, in newton-meters per radian, and that of the cantilever is k_l, in newtons per meter. What is the overall spring rate based on the deflection y at point B?

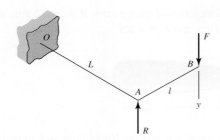

Problem 4–1

4–2 For Prob. 4–1, if the simple support at point A were eliminated and the cantilever spring rate of OA is given by k_L, determine the overall spring rate of the bar based on the deflection of point B.

4–3 A torsion-bar spring consists of a prismatic bar, usually of round cross section, that is twisted at one end and held fast at the other to form a stiff spring. An engineer needs a stiffer one than usual and so considers building in both ends and applying the torque somewhere in the central portion of the span, as shown in the figure. This effectively creates two springs in parallel. If the bar is uniform in diameter, that is, if $d = d_1 = d_2$, (a) determine how the spring rate and the end reactions depend on the location x at which the torque is applied, (b) determine the spring rate, the end reactions, and the maximum shear stress, if $d = 12$ mm, $x = 120$ mm, $l = 240$ mm, $T = 150$ N · m, and $G = 79$ GPa.

Problem 4–3

4–4 An engineer is forced by geometric considerations to apply the torque on the spring of Prob. 4–3 at the location $x = 0.4l$. For a uniform-diameter spring, this would cause one leg of the span to be underutilized when both legs have the same diameter. For optimal design the diameter of each leg should be designed such that the maximum shear stress in each leg is the same. This problem is to redesign the spring of part (b) of Prob. 4–3. Using $x = 0.4l$, $l = 240$ mm, $T = 150$ N · m, and $G = 79$ GPa, design the spring such that the maximum shear stresses in each leg are equal and the spring has the same spring rate (angle of twist) as part (b) of Prob. 4–3. Specify d_1, d_2, the spring rate k, and the torque and the maximum shear stress in each leg.

4–5 A bar in tension has a circular cross section and includes a tapered portion of length l, as shown.

(a) For the tapered portion, use Eq. (4–3) in the form of $\delta = \int_0^l [F/(AE)]\, dx$ to show that

$$\delta = \frac{4}{\pi}\frac{Fl}{d_1 d_2 E}$$

(b) Determine the elongation of each portion if $d_1 = 12$ mm, $d_2 = 18$ mm, $l = l_1 = l_2 = 48$ mm, $E = 207$ GPa, and $F = 4$ kN.

Problem 4–5

4–6 Instead of a tensile force, consider the bar in Prob. 4–5 to be loaded by a torque T.

(a) Use Eq. (4–5) in the form of $\theta = \int_0^l [T/(GJ)]\, dx$ to show that the angle of twist of the tapered portion is

$$\theta = \frac{32}{3\pi} \frac{Tl(d_1^2 + d_1 d_2 + d_2^2)}{Gd_1^3 d_2^3}$$

(b) Using the same geometry as in Prob. 4–5b with $T = 150$ N · m and $G = 79$ GPa, determine the angle of twist in degrees for each portion.

4–7 When a vertically suspended hoisting cable is long, the weight of the cable itself contributes to the elongation. If a 150-m steel cable has an effective diameter of 12 mm and lifts a load of 20 kN, determine the total elongation and the percent of the total elongation due to the cable's own weight.

4–8 Derive the equations given for beam 2 in Table A–9 using statics and the double-integration method.

4–9 Derive the equations given for beam 5 in Table A–9 using statics and the double-integration method.

4–10 The figure shows a cantilever consisting of steel angles size $100 \times 100 \times 12$ mm mounted back to back. Using superposition, find the deflection at B and the maximum stress in the beam.

Problem 4–10

4–11 A simply supported beam loaded by two forces is shown in the figure. Select a pair of structural steel channels mounted back to back to support the loads in such a way that the deflection at midspan will not exceed 1.6 mm and the maximum stress will not exceed 40 MPa. Use superposition.

Problem 4–11

4–12   Using superposition, find the deflection of the steel shaft at *A* in the figure. Find the deflection at midspan. By what percentage do these two values differ?

Problem 4–12

4–13   A rectangular steel bar supports the two overhanging loads shown in the figure. Using superposition, find the deflection at the ends and at the center.

Problem 4–13
Dimensions in millimeters.

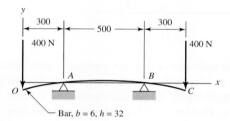

4–14   An aluminum tube with outside diameter of 50 mm and inside diameter of 38 mm is cantilevered and loaded as shown. Using the formulas in Appendix Table A–9 and superposition, find the deflection at *B*.

Problem 4–14

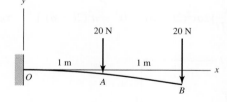

4–15   The cantilever shown in the figure consists of two structural-steel channels size 76 × 38 mm. Using superposition, find the deflection at *A*. Include the weight of the channels.

Problem 4–15

4-16 Using superposition for the bar shown, determine the minimum diameter of a steel shaft for which the maximum deflection is 2 mm.

Problem 4-16

Dimensions in millimeters.

4-17 A simply supported beam has a concentrated moment M_A applied at the left support and a concentrated force F applied at the free end of the overhang on the right. Using superposition, determine the deflection equations in regions AB and BC.

Problem 4-17

4-18 Calculating beam deflections using superposition is quite convenient provided you have a comprehensive table to refer to. Because of space limitations, this book provides a table that covers a great deal of applications, but not all possibilities. Take for example, Prob. 4-19, which follows this problem. Problem 4-19 is not directly solvable from Table A-9, but with the addition of the results of this problem, it is. For the beam shown, using statics and double integration, show that

$$R_1 = \frac{wa}{2l}(2l - a) \qquad R_2 = \frac{wa^2}{2l} \qquad V_{AB} = \frac{w}{2l}[2l(a - x) - a^2] \qquad V_{BC} = -\frac{wa^2}{2l}$$

$$M_{AB} = \frac{wx}{2l}(2al - a^2 - lx) \qquad M_{BC} = \frac{wa^2}{2l}(l - x)$$

$$y_{AB} = \frac{wx}{24EIl}[2ax^2(2l - a) - lx^3 - a^2(2l - a)^2] \qquad y_{BC} = y_{AB} + \frac{w}{24EI}(x - a)^4$$

Problem 4-18

4-19 Using the results of Prob. 4-18, use superposition to determine the deflection equations for the three regions of the beam shown.

Problem 4-19

4–20 Like Prob. 4–18, this problem provides another beam to add to Table A–9. For the simply supported beam shown with an overhanging uniform load, use statics and double integration to show that

$$R_1 = \frac{wa^2}{2l} \qquad R_2 = \frac{wa}{2l}(2l + a) \qquad V_{AB} = -\frac{wa^2}{2l} \qquad V_{BC} = w(l + a - x)$$

$$M_{AB} = -\frac{wa^2}{2l}x \qquad M_{BC} = -\frac{w}{2}(l + a - x)^2$$

$$y_{AB} = \frac{wa^2x}{12EIl}(l^2 - x^2) \qquad y_{BC} = -\frac{w}{24EI}[(l + a - x)^4 - 4a^2(l - x)(l + a) - a^4]$$

Problem 4–20

4–21 Consider the uniformly loaded simply supported steel beam with an overhang as shown. The second-area moment of the beam is $I = 19\ 500\ \text{mm}^4$. Use superposition (with Table A–9 and the results of Prob. 4–20) to determine the reactions and the deflection equations of the beam. Plot the deflections.

Problem 4–21

4–22 Illustrated is a rectangular steel bar with simple supports at the ends and loaded by a force F at the middle; the bar is to act as a spring. The ratio of the width to the thickness is to be about $b = 10h$, and the desired spring scale is 300 kN/m.

(*a*) Find a set of cross-section dimensions, using preferred fractional sizes from Table A–17.

(*b*) What deflection would cause a permanent set in the spring if this is estimated to occur at a normal stress of 400 MPa?

Problem 4–22

4–23* to 4–28* For the steel countershaft specified in the table, find the deflection and slope of the shaft at point A. Use superposition with the deflection equations in Table A–9. Assume the bearings constitute simple supports.

Problem Number	Problem, Page Number Defining Shaft
4–23*	3–68, 151
4–24*	3–69, 151
4–25*	3–70, 152
4–26*	3–71, 152
4–27*	3–72, 152
4–28*	3–73, 152

4–29* to 4–34* For the steel countershaft specified in the table, find the slope of the shaft at each bearing. Use superposition with the deflection equations in Table A–9. Assume the bearings constitute simple supports.

Problem Number	Problem, Page Number Defining Shaft
4–29*	3–68, 151
4–30*	3–69, 151
4–31*	3–70, 152
4–32*	3–71, 152
4–33*	3–72, 152
4–34*	3–73, 152

4–35* to 4–40* For the steel countershaft specified in the table, assume the bearings have a maximum slope specification of 0.06° for good bearing life. Determine the minimum shaft diameter.

Problem Number	Problem, Page Number Defining Shaft
4–35*	3–68, 151
4–36*	3–69, 151
4–37*	3–70, 152
4–38*	3–71, 152
4–39*	3–72, 152
4–40*	3–73, 152

4–41* The cantilevered handle in the figure is made from mild steel that has been welded at the joints. For $F_y = 800$ N, $F_x = F_z = 0$, determine the vertical deflection (along the y axis) at the tip. Use superposition. See the discussion on p. 116 for the twist in the rectangular cross section in section b/c.

Problem 4–41

4-42   For the cantilevered handle in Prob. 4–41, let $F_x = -600$ N, $F_y = 0$ N, $F_z = -400$ N. Find the deflection at the tip along the x axis.

4-43*   The cantilevered handle in Prob. 3–84, p. 154, is made from mild steel. Let $F_y = 1$ kN, $F_x = F_z = 0$. Determine the angle of twist in bar OC, ignoring the fillets but including the changes in diameter along the 400-mm effective length. Compare the angle of twist if the bar OC is simplified to be all of uniform 24-mm diameter. Use superposition to determine the vertical deflection (along the y axis) at the tip, using the simplified bar OC.

4-44   A flat-bed trailer is to be designed with a curvature such that when loaded to capacity the trailer bed is flat. The load capacity is to be 45 kN/m between the axles, which are 7 m apart, and the second-area moment of the steel structure of the bed is $I = 190 \times 10^6$ mm^4. Determine the equation for the curvature of the unloaded bed and the maximum height of the bed relative to the axles.

4-45   The designer of a shaft usually has a slope constraint imposed by the bearings used. This limit will be denoted as ξ. If the shaft shown in the figure is to have a uniform diameter d except in the locality of the bearing mounting, it can be approximated as a uniform beam with simple supports. Show that the minimum diameters to meet the slope constraints at the left and right bearings are, respectively,

$$d_L = \left| \frac{32Fb(l^2 - b^2)}{3\pi \, EI\xi} \right|^{1/4} \qquad d_R = \left| \frac{32Fa(l^2 - a^2)}{3\pi \, EI\xi} \right|^{1/4}$$

Problem 4–45

4-46   A steel shaft is to be designed so that it is supported by roller bearings. The basic geometry is shown in the figure from Prob. 4–45, with $l = 300$ mm, $a = 100$ mm, and $F = 3$ kN. The allowable slope at the bearings is 0.001 mm/mm without bearing life penalty. For a design factor of 1.28, what uniform-diameter shaft will support the load without penalty? Determine the maximum deflection of the shaft.

4–47   If the diameter of the steel beam shown is 30 mm, determine the deflection of the beam at $x = 160$ mm.

Problem 4–47
Dimensions in millimeters

4–48   For the beam of Prob. 4–47, plot the *magnitude* of the displacement of the beam in 2-mm increments. Approximate the maximum displacement and the value of x where it occurs.

4–49   Shown in the figure is a uniform-diameter shaft with bearing shoulders at the ends; the shaft is subjected to a concentrated moment $M = 130$ N $\cdot$ m. The shaft is of carbon steel and has $a = 80$ mm and $l = 200$ mm. The slope at the ends must be limited to 0.002 rad. Find a suitable diameter d.

Problem 4–49

4–50* and 4–51   The figure shows a rectangular member OB, made from 6-mm-thick aluminum plate, pinned to the ground at one end and supported by a 12-mm-diameter round steel rod with hooks formed on the ends. A load of 400 N is applied as shown. Use superposition to determine the vertical deflection at point B.

Problem 4–50*   *Problem 4–51*

4–52   The figure illustrates a stepped torsion-bar spring OA with an actuating cantilever AB. Both parts are of carbon steel. Use superposition and find the spring rate k corresponding to a force F acting at B.

Problem 4–52

4-53 Consider the simply supported beam 5 with a center load in Appendix A–9. Determine the deflection equation if the stiffness of the left and right supports are k_1 and k_2, respectively.

4-54 Consider the simply supported beam 10 with an overhanging load in Appendix A–9. Determine the deflection equation if the stiffness of the left and right supports are k_1 and k_2, respectively.

4-55 Prove that for a uniform-cross-section beam with simple supports at the ends loaded by a single concentrated load, the location of the maximum deflection will never be outside the range of $0.423l \leq x \leq 0.577l$ regardless of the location of the load along the beam. The importance of this is that you can always get a quick estimate of y_{max} by using $x = l/2$.

4-56 Solve Prob. 4–10 using singularity functions. Use statics to determine the reactions.

4-57 Solve Prob. 4–11 using singularity functions. Use statics to determine the reactions.

4-58 Solve Prob. 4–12 using singularity functions. Use statics to determine the reactions.

4-59 Solve Prob. 4–21 using singularity functions to determine the deflection equation of the beam. Use statics to determine the reactions.

4-60 Solve Prob. 4–13 using singularity functions. Since the beam is symmetric, only write the equation for half the beam and use the slope at the beam center as a boundary condition. Use statics to determine the reactions.

4-61 Solve Prob. 4–17 using singularity functions. Use statics to determine the reactions.

4-62 Solve Prob. 4–19 using singularity functions to determine the deflection equation of the beam. Use statics to determine the reactions.

4-63 Using singularity functions, write the deflection equation for the steel beam shown. Since the beam is symmetric, write the equation for only half the beam and use the slope at the beam center as a boundary condition. Plot your results and determine the maximum deflection.

Problem 4–63

4-64 Determine the deflection equation for the cantilever beam shown using singularity functions. Evaluate the deflections at B and C and compare your results with Example 4–10.

Problem 4–64

4–65 Use Castigliano's theorem to verify the maximum deflection for the uniformly loaded beam 7 of Appendix Table A–9. Neglect shear.

4–66 Use Castigliano's theorem to verify the maximum deflection for the uniformly loaded cantilever beam 3 of Appendix Table A–9. Neglect shear.

4–67 Solve Prob. 4–15 using Castigliano's theorem.

4–68 Solve Prob. 4–52 using Castigliano's theorem.

4–69 Determine the deflection at midspan for the beam of Prob. 4–63 using Castigliano's theorem.

4–70 Using Castigliano's theorem, determine the deflection of point B in the direction of the force F for the steel bar shown.

Problem 4–70

4–71* Solve Prob. 4–41 using Castigliano's theorem. Since Eq. (4–18) for torsional strain energy was derived from the angular displacement for circular cross sections, it is not applicable for section BC. You will need to obtain a new strain energy equation for the rectangular cross section from Eqs. (4–18) and (3–41).

4–72 Solve Prob. 4–42 using Castigliano's theorem.

4–73* The cantilevered handle in Prob. 3–84 is made from mild steel. Let $F_y = 1$ kN and $F_x = F_z = 0$. Using Castigliano's theorem, determine the vertical deflection (along the y axis) at the tip. Repeat the problem with shaft OC simplified to a uniform diameter of 24 mm for its entire length. What is the percent error from this simplification?

4–74* Solve Prob. 4–50 using Castigliano's theorem.

4–75 Solve Prob. 4–51 using Castigliano's theorem.

4–76 The steel curved bar shown has a rectangular cross section with a radial height $h = 6$ mm, and a thickness $b = 4$ mm. The radius of the centroidal axis is $R = 40$ mm. A force $P = 10$ N is applied as shown. Find the vertical deflection at B. Use Castigliano's method for a curved flexural member, and since $R/h < 10$, do not neglect any of the terms.

Problem 4–76

4–77 Repeat Prob. 4–76 to find the vertical deflection at *A*.

4–78 For the curved steel beam shown, *F* = 30 kN. Determine the relative deflection of the applied forces.

Problem 4–78

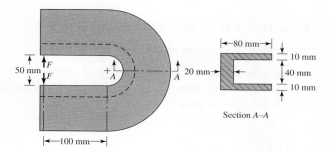

4–79 A steel piston ring has a mean diameter of 70 mm, a radial height *h* = 4.5 mm, and a thickness *b* = 3 mm. The ring is assembled using an expansion tool that separates the split ends a distance δ by applying a force *F* as shown. Use Castigliano's theorem and determine the force *F* needed to expand the split ends a distance δ = 1 mm.

Problem 4–79

4–80 For the steel wire form shown, use Castigliano's method to determine the horizontal reaction forces at *A* and *B* and the deflection at *C*.

Problem 4–80

4–81 and 4–82
The part shown is formed from a 3-mm diameter steel wire, with $R = 125$ mm and $l = 100$ mm. A force is applied with $P = 5$ N. Use Castigliano's method to estimate the horizontal deflection at point A. Justify any components of strain energy that you choose to neglect.

Problem 4–81   Problem 4–82

4–83 Repeat Prob. 4–81 for the vertical deflection at point A.

4–84 Repeat Prob. 4–82 for the vertical deflection at point A.

4–85 A hook is formed from a 2-mm-diameter steel wire and fixed firmly into the ceiling as shown. A 1-kg mass is hung from the hook at point D. Use Castigliano's theorem to determine the vertical deflection of point D.

Problem 4–85

4–86 The figure shows a rectangular member OB, made from 6-mm-thick aluminum plate, pinned to the ground at one end, and supported by a 12-mm-diameter round steel rod that is formed into an arc and pinned to the ground at C. A load of 500 N is applied at B. Use Castigliano's theorem to determine the vertical deflection at point B. Justify any choices to neglect any components of strain energy.

Problem 4–86

4–87 Repeat Prob. 4–86 for the vertical deflection at point A.

4–88   For the wire form shown, determine the deflection of point A in the y direction. Assume $R/h > 10$ and consider the effects of bending and torsion only. The wire is steel with $E = 200$ GPa, $\nu = 0.29$, and has a diameter of 6 mm. Before application of the 250-N force the wire form is in the xz plane where the radius R is 80 mm.

Problem 4–88

4–89   A 30-m cable is made using a 1.6-mm steel wire and three strands of 2-mm copper wire. Find the deflection of the cable and the stress in each wire if the cable is subjected to a tension of 2 kN.

4–90   The figure shows a steel pressure cylinder of diameter 100 mm that uses six SAE grade 5 steel bolts having a grip of 300 mm. These bolts have a proof strength (see Chap. 8) of 580 MPa. Suppose the bolts are tightened to 90 percent of this strength.

(a) Find the tensile stress in the bolts and the compressive stress in the cylinder walls.

(b) Repeat part (a), but assume now that a fluid under a pressure of 4 MPa is introduced into the cylinder.

Problem 4–90

4–91   A torsion bar of length L consists of a round core of stiffness $(GJ)_c$ and a shell of stiffness $(GJ)_s$. If a torque T is applied to this composite bar, what percentage of the total torque is carried by the shell?

4-92   A rectangular aluminum bar 10 mm thick and 60 mm wide is welded to fixed supports at the ends, and the bar supports a load $W = 4$ kN, acting through a pin as shown. Find the reactions at the supports and the deflection of point A.

Problem 4-92

4-93   Solve Prob. 4–92 using Castigliano's method and procedure 1 from Sec. 4–10.

4-94   An aluminum step bar is loaded as shown. (*a*) Verify that end *C* deflects to the rigid wall, and (*b*) determine the wall reaction forces, the stresses in each member, and the deflection of *B*.

Problem 4-94
(Not drawn to scale)

4-95   The steel shaft shown in the figure is subjected to a torque of 5 N · m applied at point *A*. Find the torque reactions at *O* and *B*; the angle of twist at *A*, in degrees; and the shear stress in sections *OA* and *AB*.

Problem 4-95

4-96   Repeat Prob. 4–95 with the diameters of section *OA* being 12 mm and section *AB* being 18 mm.

4-97   The figure shows a 10 by 40-mm rectangular steel bar welded to fixed supports at each end. The bar is axially loaded by the forces $F_A = 60$ kN and $F_B = 30$ kN acting on pins at *A* and *B*. Assuming that the bar will not buckle laterally, find the reactions at the fixed supports, the stress in section *AB*, and the deflection of point *A*. Use procedure 1 from Sec. 4–10.

Problem 4-97

4-98 For the beam shown, determine the support reactions using superposition and procedure 1 from Sec. 4–10.

Problem 4-98

4-99 Solve Prob. 4–98 using Castigliano's theorem and procedure 1 from Sec. 4–10.

4-100 Consider beam 13 in Table A–9, but with flexible supports. Let $w = 7$ kN/m, $l = 0.6$ m, $E = 207$ GPa, and $I = 330 \times 10^3$ mm^4. The support at the left end has a translational spring constant of $k_1 = 255(10^6)$ N/m and a rotational spring constant of $k_2 = 280$ kN · m. The right support has a translational spring constant of $k_3 = 340(10^6)$ N/m. Using procedure 2 of Sec. 4–10, determine the reactions at the supports and the deflection at the midpoint of the beam.

4-101 The steel beam $ABCD$ shown is simply supported at A and supported at B and D by steel cables, each having an effective diameter of 12 mm. The second area moment of the beam is $I = 8(10^5)$ mm^4. A force of 20 kN is applied at point C. Using procedure 2 of Sec. 4–10 determine the stresses in the cables and the deflections of B, C, and D.

Problem 4-101

4-102 The steel beam $ABCD$ shown is simply supported at C as shown and supported at B and D by shoulder steel bolts, each having a diameter of 8 mm. The lengths of BE and DF are 50 mm and 65 mm, respectively. The beam has a second area moment of $21(10^3)$ mm^4. Prior to loading, the members are stress-free. A force of 2 kN is then applied at point A. Using procedure 2 of Sec. 4–10, determine the stresses in the bolts and the deflections of points A, B, and D.

Problem 4-102

4-103 A thin ring is loaded by two equal and opposite forces F in part a of the figure. A free-body diagram of one quadrant is shown in part b. This is a statically indeterminate problem, because the moment M_A cannot be found by statics. (a) Find the maximum bending moment in the ring due to the forces F, and (b) find the increase in the diameter of the ring along the y axis. Assume that the radius of the ring is large so that Eq. (4–41) can be used.

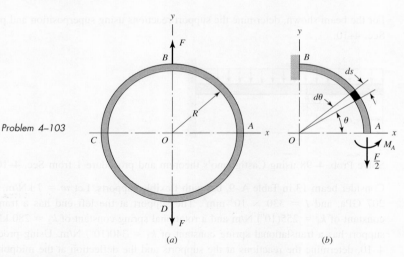

Problem 4–103

(a)   (b)

4–104   A round tubular column has outside and inside diameters of D and d, respectively, and a diametral ratio of $K = d/D$. Show that buckling will occur when the outside diameter is

$$D = \left[\frac{64 P_{cr} l^2}{\pi^3 CE(1 - K^4)} \right]^{1/4}$$

4–105   For the conditions of Prob. 4–104, show that buckling according to the parabolic formula will occur when the outside diameter is

$$D = 2 \left[\frac{P_{cr}}{\pi S_y (1 - K^2)} + \frac{S_y l^2}{\pi^2 CE(1 + K^2)} \right]^{1/2}$$

4–106   Link 2, shown in the figure, is 25 mm wide, has 12-mm-diameter bearings at the ends, and is cut from low-carbon steel bar stock having a minimum yield strength of 165 MPa. The end-condition constants are $C = 1$ and $C = 1.2$ for buckling in and out of the plane of the drawing, respectively.
(a) Using a design factor $n_d = 4$, find a suitable thickness for the link.
(b) Are the bearing stresses at O and B of any significance?

Problem 4–106

900 mm   750 mm   500 mm   800 N

4–107   Link 3, shown schematically in the figure, acts as a brace to support the 1.2-kN load. For buckling in the plane of the figure, the link may be regarded as pinned at both ends. For out-of-plane buckling, the ends are fixed. Select a suitable material and a method of manufacture, such as forging, casting, stamping, or machining, for casual applications of the brace in oil-field machinery. Specify the dimensions of the cross section as well as the ends so as to obtain a strong, safe, well-made, and economical brace.

Problem 4–107

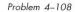

4–108

The hydraulic cylinder shown in the figure has a 50.8 mm bore and is to operate at a pressure of 5.6 MPa. With the clevis mount shown, the piston rod should be sized as a column with both ends rounded for any plane of buckling. The rod is to be made of forged AISI 1030 steel without further heat treatment.

Problem 4–108

(a) Use a design factor $n_d = 2.5$ and select a preferred size for the rod diameter if the column length is 1270 mm.
(b) Repeat part (a) but for a column length of 406.4 mm.
(c) What factor of safety actually results for each of the cases above?

4–109

The figure shows a schematic drawing of a vehicular jack that is to be designed to support a maximum mass of 300 kg based on the use of a design factor $n_d = 3.50$. The opposite-handed threads on the two ends of the screw are cut to allow the link angle θ to vary from 15 to 70°. The links are to be machined from AISI 1010 hot-rolled steel bars. Each of the four links is to consist of two bars, one on each side of the central bearings. The bars are to be 350 mm long and have a bar width of $w = 30$ mm. The pinned ends are to be designed to secure an end-condition constant of at least $C = 1.4$ for out-of-plane buckling. Find a suitable preferred thickness and the resulting factor of safety for this thickness.

Problem 4–109

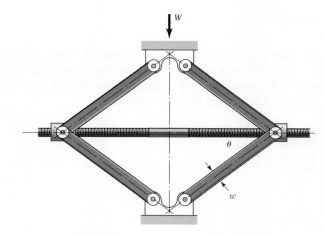

4–110 If drawn, a figure for this problem would resemble that for Prob. 4–90. A strut that is a standard hollow right circular cylinder has an outside diameter of 100 mm and a wall thickness of 10 mm and is compressed between two circular end plates held by four bolts equally spaced on a bolt circle of 140-mm diameter. All four bolts are hand-tightened, and then bolt A is tightened to a tension of 10 kN and bolt C, diagonally opposite, is tightened to a tension of 50 kN. The strut axis of symmetry is coincident with the center of the bolt circles. Find the maximum compressive load, the eccentricity of loading, and the largest compressive stress in the strut.

4–111 Design link CD of the hand-operated toggle press shown in the figure. Specify the cross-section dimensions, the bearing size and rod-end dimensions, the material, and the method of processing.

Problem 4–111
$L = 300$ mm, $l = 100$ mm,
$\theta_{\min} = 0°$.

4–112 Find the maximum values of the spring force and deflection of the impact system shown in the figure if $W = 150$ N, $k = 17$ kN/m, and $h = 50$ mm. Ignore the mass of the spring and solve using energy conservation.

Problem 4–112

4–113 As shown in the figure, the weight W_1 strikes W_2 from a height h. If $W_1 = 40$ N, $W_2 = 400$ N, $h = 200$ mm, and $k = 32$ kN/m, find the maximum values of the spring force and the deflection of W_2. Assume that the impact between W_1 and W_2 is *inelastic,* ignore the mass of the spring, and solve using energy conservation.

Problem 4–113

4–114   Part *a* of the figure shows a weight *W* mounted between two springs. If the free end of spring k_1 is suddenly displaced through the distance $x = a$, as shown in part *b*, determine the maximum displacement *y* of the weight. Let $W = 20$ N, $k_1 = 1.7$ kN/m, $k_2 = 3.4$ kN/m, and $a = 6$ mm. Ignore the mass of each spring and solve using energy conservation.

Problem 4–114

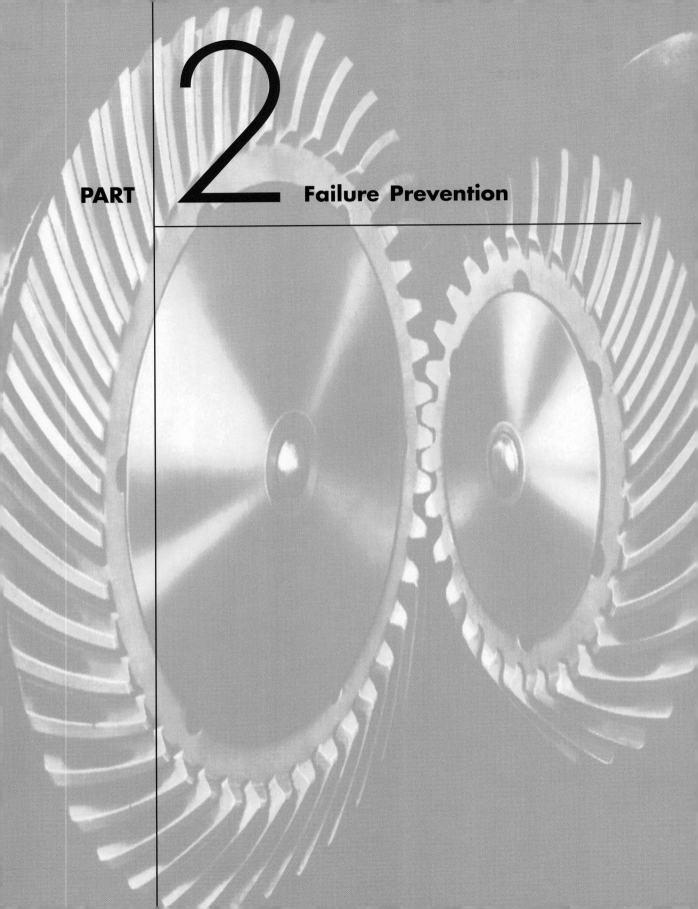

PART 2 Failure Prevention

5

Failures Resulting from Static Loading

In Chap. 1 we learned that *strength is a property or characteristic of a mechanical element*. This property results from the material identity, the treatment and processing incidental to creating its geometry, and the loading, and it is at the controlling or critical location.

In addition to considering the strength of a single part, we must be cognizant that the strengths of the mass-produced parts will all be somewhat different from the others in the collection or ensemble because of variations in dimensions, machining, forming, and composition. Descriptors of strength are necessarily statistical in nature, involving parameters such as mean, standard deviations, and distributional identification.

A *static load* is a stationary force or couple applied to a member. To be stationary, the force or couple must be unchanging in magnitude, point or points of application, and direction. A static load can produce axial tension or compression, a shear load, a bending load, a torsional load, or any combination of these. To be considered static, the load cannot change in any manner.

In this chapter we consider the relations between strength and static loading in order to make the decisions concerning material and its treatment, fabrication, and geometry for satisfying the requirements of functionality, safety, reliability, competitiveness, usability, manufacturability, and marketability. How far we go down this list is related to the scope of the examples.

"Failure" is the first word in the chapter title. Failure can mean a part has separated into two or more pieces; has become permanently distorted, thus ruining its geometry; has had its reliability downgraded; or has had its function compromised, whatever the reason. A designer speaking of failure can mean any or all of these possibilities. In this chapter our attention is focused on the predictability of permanent distortion or separation. In strength-sensitive situations the designer must separate mean stress and mean strength at the critical location sufficiently to accomplish his or her purposes.

Figures 5–1 to 5–5 are photographs of several failed parts. The photographs exemplify the need of the designer to be well-versed in failure prevention. Toward this end we shall consider one-, two-, and three-dimensional stress states, with and without stress concentrations, for both ductile and brittle materials.

(a)   (b)

Figure 5–1

(a) Failure of a truck drive-shaft spline due to corrosion fatigue. Note that it was necessary to use clear tape to hold the pieces in place. (b) Direct end view of failure. *(For permission to reprint Figs. 5–1 through 5–5, the authors are grateful for the personal photographs of Larry D. Mitchell, co-author of Mechanical Engineering Design, 4th ed., McGraw-Hill, New York, 1983.)*

Figure 5–2

Impact failure of a lawn-mower blade driver hub. The blade impacted a surveying pipe marker.

Figure 5–3

Failure of an overhead-pulley retaining bolt on a weightlifting machine. A manufacturing error caused a gap that forced the bolt to take the entire moment load.

(a)

(b)

Figure 5–4

Chain test fixture that failed in one cycle. To alleviate complaints of excessive wear, the manufacturer decided to case-harden the material. (a) Two halves showing fracture; this is an excellent example of brittle fracture initiated by stress concentration. (b) Enlarged view of one portion to show cracks induced by stress concentration at the support-pin holes.

Figure 5–5

Valve-spring failure caused by spring surge in an oversped engine. The fractures exhibit the classic 45° shear failure.

5–1 Static Strength

Ideally, in designing any machine element, the engineer should have available the results of a great many strength tests of the particular material chosen. These tests should be made on specimens having the same heat treatment, surface finish, and size as the element the engineer proposes to design; and the tests should be made under exactly the same loading conditions as the part will experience in service. This means that if the part is to experience a bending load, it should be tested with a bending load. If it is to be subjected to combined bending and torsion, it should be tested under combined bending and torsion. If it is made of heat-treated AISI 1040 steel drawn at 500°C with a ground finish, the specimens tested should be of the same material prepared in the same manner. Such tests will provide very useful and precise information. Whenever such data are available for design purposes, the engineer can be assured of doing the best possible job of engineering.

The cost of gathering such extensive data prior to design is justified if failure of the part may endanger human life or if the part is manufactured in sufficiently large quantities. Refrigerators and other appliances, for example, have very good reliabilities because the parts are made in such large quantities that they can be thoroughly tested in advance of manufacture. The cost of making these tests is very low when it is divided by the total number of parts manufactured.

You can now appreciate the following four design categories:

1 Failure of the part would endanger human life, or the part is made in extremely large quantities; consequently, an elaborate testing program is justified during design.

2 The part is made in large enough quantities that a moderate series of tests is feasible.

3 The part is made in such small quantities that testing is not justified at all; or the design must be completed so rapidly that there is not enough time for testing.

4 The part has already been designed, manufactured, and tested and found to be unsatisfactory. Analysis is required to understand why the part is unsatisfactory and what to do to improve it.

More often than not it is necessary to design using only published values of yield strength, ultimate strength, percentage reduction in area, and percentage elongation, such as those listed in Appendix A. How can one use such meager data to design against both static and dynamic loads, two- and three-dimensional stress states, high and low temperatures, and very large and very small parts? These and similar questions will be addressed in this chapter and those to follow, but think how much better it would be to have data available that duplicate the actual design situation.

5–2   Stress Concentration

Stress concentration (see Sec. 3–13) is a highly localized effect. In some instances it may be due to a surface scratch. If the material is ductile and the load static, the design load may cause yielding in the critical location in the notch. This yielding can involve strain strengthening of the material and an increase in yield strength at the small critical notch location. Since the loads are static and the material is ductile, that part can carry the loads satisfactorily with no general yielding. In these cases the designer sets the geometric (theoretical) stress-concentration factor K_t to unity.

The rationale can be expressed as follows. The worst-case scenario is that of an idealized non–strain-strengthening material shown in Fig. 5–6. The stress-strain curve rises linearly to the yield strength S_y, then proceeds at constant stress, which is equal to S_y. Consider a filleted rectangular bar as depicted in Fig. A–15–5, where the cross-section area of the small shank is 645 mm². If the material is ductile, with a yield point of 280 MPa, and the theoretical *stress-concentration factor* (SCF) K_t is 2,

- A load of 90 kN induces a nominal tensile stress of 140 MPa in the shank as depicted at point A in Fig. 5–6. At the critical location in the fillet the stress is 280 MPA, and the SCF is $K = \sigma_{max}/\sigma_{nom} = 40/20 = 2$.

- A load of 135 kN induces a nominal tensile stress of 210 MPa in the shank at point B. The fillet stress is still 280 MPa (point D), and the SCF $K = \sigma_{max}/\sigma_{nom} = S_y/\sigma = 280/210 = 1.33$.

- At a load of 180 kN the induced tensile stress (point C) is 280 MPa in the shank. At the critical location in the fillet, the stress (at point E) is 280 MPa. The SCF $K = \sigma_{max}/\sigma_{nom} = S_y/\sigma = 280/280 = 1$.

Figure 5–6

An idealized stress-strain curve. The dashed line depicts a strain-strengthening material.

For materials that strain-strengthen, the critical location in the notch has a higher S_y. The shank area is at a stress level a little below 280 MPa, is carrying load, and is very near its failure-by-general-yielding condition. This is the reason designers do not apply K_t in *static loading* of a *ductile material* loaded elastically, instead setting $K_t = 1$.

When using this rule for ductile materials with static loads, be careful to assure yourself that the material is not susceptible to brittle fracture (see Sec. 5–12) in the environment of use. The usual definition of geometric (theoretical) stress-concentration factor for normal stress K_t and shear stress K_{ts} is given by Eq. pair (3–48) as

$$\sigma_{\max} = K_t \sigma_{\text{nom}} \qquad\qquad (a)$$

$$\tau_{\max} = K_{ts} \tau_{\text{nom}} \qquad\qquad (b)$$

Since your attention is on the stress-concentration factor, and the definition of σ_{nom} or τ_{nom} is given in the graph caption or from a computer program, be sure the value of nominal stress is appropriate for the section carrying the load.

As shown in Fig. 2–2b, p. 43, brittle materials do not exhibit a plastic range. The stress-concentration factor given by Eq. (a) or (b) could raise the stress to a level to cause fracture to initiate at the stress raiser, and initiate a catastrophic failure of the member.

An exception to this rule is a brittle material that inherently contains microdiscontinuity stress concentration, worse than the macrodiscontinuity that the designer had in mind. Sand molding introduces sand particles, air, and water vapor bubbles. The grain structure of cast iron contains graphite flakes (with little strength), which are literally cracks introduced during the solidification process. When a tensile test on a cast iron is performed, the strength reported in the literature *includes* this stress concentration. In such cases K_t or K_{ts} need not be applied.

An important source of stress-concentration factors is R. E. Peterson, who compiled them from his own work and that of others.[1] Peterson developed the style of presentation in which the stress-concentration factor K_t is multiplied by the nominal stress σ_{nom} to estimate the magnitude of the largest stress in the locality. His approximations were based on photoelastic studies of two-dimensional strips (Hartman and Levan, 1951; Wilson and White, 1973), with some limited data from three-dimensional photoelastic tests of Hartman and Levan. A contoured graph was included in the presentation of each case. Filleted shafts in tension were based on two-dimensional strips. Table A–15 provides many charts for the theoretical stress-concentration factors for several fundamental load conditions and geometry. Additional charts are also available from Peterson.[2]

Finite element analysis (FEA) can also be applied to obtain stress-concentration factors. Improvements on K_t and K_{ts} for filleted shafts were reported by Tipton, Sorem, and Rolovic.[3]

---

[1] R. E. Peterson, "Design Factors for Stress Concentration," *Machine Design,* vol. 23, no. 2, February 1951; no. 3, March 1951; no. 5, May 1951; no. 6, June 1951; no. 7, July 1951.

[2] Walter D. Pilkey and Deborah Pilkey, *Peterson's Stress-Concentration Factors,* 3rd ed, John Wiley & Sons, New York, 2008.

[3] S. M. Tipton, J. R. Sorem Jr., and R. D. Rolovic, "Updated Stress-Concentration Factors for Filleted Shafts in Bending and Tension," *Trans. ASME, Journal of Mechanical Design,* vol. 118, September 1996, pp. 321–327.

5–3 **Failure Theories**

Section 5–1 illustrated some ways that loss of function is manifested. Events such as distortion, permanent set, cracking, and rupturing are among the ways that a machine element fails. Testing machines appeared in the 1700s, and specimens were pulled, bent, and twisted in simple loading processes.

If the failure mechanism is simple, then simple tests can give clues. Just what is simple? The tension test is uniaxial (that's simple) and elongations are largest in the axial direction, so strains can be measured and stresses inferred up to "failure." Just what is important: a critical stress, a critical strain, a critical energy? In the next several sections, we shall show failure theories that have helped answer some of these questions.

Unfortunately, there is no universal theory of failure for the general case of material properties and stress state. Instead, over the years several hypotheses have been formulated and tested, leading to today's accepted practices. Being accepted, we will characterize these "practices" as *theories* as most designers do.

Structural metal behavior is typically classified as being ductile or brittle, although under special situations, a material normally considered ductile can fail in a brittle manner (see Sec. 5–12). Ductile materials are normally classified such that $\varepsilon_f \geq 0.05$ and have an identifiable yield strength that is often the same in compression as in tension ($S_{yt} = S_{yc} = S_y$). Brittle materials, $\varepsilon_f < 0.05$, do not exhibit an identifiable yield strength, and are typically classified by ultimate tensile and compressive strengths, S_{ut} and S_{uc}, respectively (where S_{uc} is given as a positive quantity). The generally accepted theories are:

Ductile materials (yield criteria)

- Maximum shear stress (MSS), Sec. 5–4
- Distortion energy (DE), Sec. 5–5
- Ductile Coulomb-Mohr (DCM), Sec. 5–6

Brittle materials (fracture criteria)

- Maximum normal stress (MNS), Sec. 5–8
- Brittle Coulomb-Mohr (BCM), Sec. 5–9
- Modified Mohr (MM), Sec. 5–9

It would be inviting if we had one universally accepted theory for each material type, but for one reason or another, they are all used. Later, we will provide rationales for selecting a particular theory. First, we will describe the bases of these theories and apply them to some examples.

5–4 **Maximum-Shear-Stress Theory for Ductile Materials**

The *maximum-shear-stress* (MSS) *theory* predicts that *yielding begins whenever the maximum shear stress in any element equals or exceeds the maximum shear stress in a tension-test specimen of the same material when that specimen begins to yield.* The MSS theory is also referred to as the *Tresca* or *Guest theory*.

Many theories are postulated on the basis of the consequences seen from tensile tests. As a strip of a ductile material is subjected to tension, slip lines (called *Lüder lines*) form at approximately 45° with the axis of the strip. These slip lines are the beginning of yield, and when loaded to fracture, fracture lines are also seen at angles approximately 45° with the axis of tension. Since the shear stress is maximum at 45° from the axis of tension, it makes sense to think that this is the mechanism of failure. It will be shown in the next section, that there is a little more going on than this. However, it turns out the MSS theory is an acceptable but conservative predictor of failure; and since engineers are conservative by nature, it is quite often used.

Recall that for simple tensile stress, $\sigma = P/A$, and the maximum shear stress occurs on a surface 45° from the tensile surface with a magnitude of $\tau_{max} = \sigma/2$. So the maximum shear stress at yield is $\tau_{max} = S_y/2$. For a general state of stress, three principal stresses can be determined and ordered such that $\sigma_1 \geq \sigma_2 \geq \sigma_3$. The maximum shear stress is then $\tau_{max} = (\sigma_1 - \sigma_3)/2$ (see Fig. 3–12). Thus, for a general state of stress, the maximum-shear-stress theory predicts yielding when

$$\tau_{max} = \frac{\sigma_1 - \sigma_3}{2} \geq \frac{S_y}{2} \qquad \text{or} \qquad \sigma_1 - \sigma_3 \geq S_y \qquad (5\text{–}1)$$

Note that this implies that the yield strength in shear is given by

$$S_{sy} = 0.5S_y \qquad (5\text{–}2)$$

which, as we will see later is about 15 percent low (conservative).

For design purposes, Eq. (5–1) can be modified to incorporate a factor of safety, n. Thus,

$$\tau_{max} = \frac{S_y}{2n} \qquad \text{or} \qquad \sigma_1 - \sigma_3 = \frac{S_y}{n} \qquad (5\text{–}3)$$

Plane stress is a very common state of stress in design. However, it is extremely important to realize that plane stress is a *three-dimensional* state of stress. Plane stress transformations in Sec. 3–6 are restricted to the in-plane stresses only, where the in-plane principal stresses are given by Eq. (3–13) and labeled as σ_1 and σ_2. It is true that these are the principal stresses in the *plane of analysis,* but out of plane there is a third principal stress and it is *always zero* for plane stress. This means that if we are going to use the convention of ordering $\sigma_1 \geq \sigma_2 \geq \sigma_3$ for three-dimensional analysis, upon which Eq. (5–1) is based, we cannot arbitrarily call the in-plane principal stresses σ_1 and σ_2 until we relate them with the third principal stress of zero. To illustrate the MSS theory graphically for plane stress, we will first label the principal stresses given by Eq. (3–13) as σ_A and σ_B, and then order them with the zero principal stress according to the convention $\sigma_1 \geq \sigma_2 \geq \sigma_3$. Assuming that $\sigma_A \geq \sigma_B$, there are three cases to consider when using Eq. (5–1) for plane stress:

Case 1: $\sigma_A \geq \sigma_B \geq 0$. For this case, $\sigma_1 = \sigma_A$ and $\sigma_3 = 0$. Equation (5–1) reduces to a yield condition of

$$\sigma_A \geq S_y \qquad (5\text{–}4)$$

Case 2: $\sigma_A \geq 0 \geq \sigma_B$. Here, $\sigma_1 = \sigma_A$ and $\sigma_3 = \sigma_B$, and Eq. (5–1) becomes

$$\sigma_A - \sigma_B \geq S_y \qquad (5\text{–}5)$$

Case 3: $0 \geq \sigma_A \geq \sigma_B$. For this case, $\sigma_1 = 0$ and $\sigma_3 = \sigma_B$, and Eq. (5–1) gives

$$\sigma_B \leq -S_y \qquad (5\text{–}6)$$

Equations (5–4) to (5–6) are represented in Fig. 5–7 by the three lines indicated in the σ_A, σ_B plane. The remaining unmarked lines are cases for $\sigma_B \geq \sigma_A$, which completes the *stress yield envelope* but are not normally used. The maximum-shear-stress theory predicts yield if a stress state is outside the shaded region bordered by the

Figure 5–7

The maximum-shear-stress (MSS) theory yield envelope for plane stress, where σ_A and σ_B are the two nonzero principal stresses.

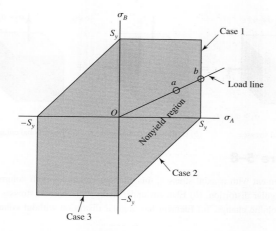

stress yield envelope. In Fig. 5–7, suppose point a represents the stress state of a critical stress element of a member. If the load is increased, it is typical to assume that the principal stresses will increase proportionally along the line from the origin through point a. Such a *load line* is shown. If the stress situation increases along the load line until it crosses the stress failure envelope, such as at point b, the MSS theory predicts that the stress element will yield. The factor of safety guarding against yield at point a is given by the ratio of strength (distance to failure at point b) to stress (distance to stress at point a), that is $n = Ob/Oa$.

Note that the first part of Eq. (5–3), $\tau_{\max} = S_y/2n$, is sufficient for design purposes provided the designer is careful in determining $\tau_{\max}$. For plane stress, Eq. (3–14) *does not always predict* $\tau_{\max}$. However, consider the special case when one normal stress is zero in the plane, say σ_x and τ_{xy} have values and $\sigma_y = 0$. It can be easily shown that this is a Case 2 problem, and the shear stress determined by Eq. (3–14) *is* $\tau_{\max}$. Shaft design problems typically fall into this category where a normal stress exists from bending and/or axial loading, and a shear stress arises from torsion.

5–5 Distortion-Energy Theory for Ductile Materials

The *distortion-energy theory* predicts that *yielding occurs when the distortion strain energy per unit volume reaches or exceeds the distortion strain energy per unit volume for yield in simple tension or compression of the same material.*

The distortion-energy (DE) theory originated from the observation that ductile materials stressed hydrostatically (equal principal stresses) exhibited yield strengths greatly in excess of the values given by the simple tension test. Therefore it was postulated that yielding was not a simple tensile or compressive phenomenon at all, but, rather, that it was related somehow to the angular distortion of the stressed element. To develop the theory, note, in Fig. 5–8a, the unit volume subjected to any three-dimensional stress state designated by the stresses σ_1, σ_2, and σ_3. The stress state shown in Fig. 5–8b is one of hydrostatic normal stresses due to the stresses σ_{av} acting in each of the same principal directions as in Fig. 5–8a. The formula for σ_{av} is simply

$$\sigma_{av} = \frac{\sigma_1 + \sigma_2 + \sigma_3}{3}$$ (a)

Thus the element in Fig. 5–8b undergoes pure volume change, that is, no angular distortion. If we regard σ_{av} as a component of σ_1, σ_2, and σ_3, then this component can be subtracted from them, resulting in the stress state shown in Fig. 5–8c. This element is subjected to pure angular distortion, that is, no volume change.

(a) Triaxial stresses   (b) Hydrostatic component   (c) Distortional component

Figure 5-8

(a) Element with triaxial stresses; this element undergoes both volume change and angular distortion. (b) Element under hydrostatic normal stresses undergoes only volume change. (c) Element has angular distortion without volume change.

The strain energy per unit volume for simple tension is $u = \frac{1}{2}\epsilon\sigma$. For the element of Fig. 5-8a the strain energy per unit volume is $u = \frac{1}{2}[\epsilon_1\sigma_1 + \epsilon_2\sigma_2 + \epsilon_3\sigma_3]$. Substituting Eq. (3-19) for the principal strains gives

$$u = \frac{1}{2E}[\sigma_1^2 + \sigma_2^2 + \sigma_3^2 - 2\nu(\sigma_1\sigma_2 + \sigma_2\sigma_3 + \sigma_3\sigma_1)] \tag{b}$$

The strain energy for producing only volume change u_v can be obtained by substituting σ_{av} for σ_1, σ_2, and σ_3 in Eq. (b). The result is

$$u_v = \frac{3\sigma_{av}^2}{2E}(1 - 2\nu) \tag{c}$$

If we now substitute the square of Eq. (a) in Eq. (c) and simplify the expression, we get

$$u_v = \frac{1 - 2\nu}{6E}(\sigma_1^2 + \sigma_2^2 + \sigma_3^2 + 2\sigma_1\sigma_2 + 2\sigma_2\sigma_3 + 2\sigma_3\sigma_1) \tag{5-7}$$

Then the distortion energy is obtained by subtracting Eq. (5-7) from Eq. (b). This gives

$$u_d = u - u_v = \frac{1 + \nu}{3E}\left[\frac{(\sigma_1 - \sigma_2)^2 + (\sigma_2 - \sigma_3)^2 + (\sigma_3 - \sigma_1)^2}{2}\right] \tag{5-8}$$

Note that the distortion energy is zero if $\sigma_1 = \sigma_2 = \sigma_3$.

For the simple tensile test, at yield, $\sigma_1 = S_y$ and $\sigma_2 = \sigma_3 = 0$, and from Eq. (5-8) the distortion energy is

$$u_d = \frac{1 + \nu}{3E}S_y^2 \tag{5-9}$$

So for the general state of stress given by Eq. (5-8), yield is predicted if Eq. (5-8) equals or exceeds Eq. (5-9). This gives

$$\left[\frac{(\sigma_1 - \sigma_2)^2 + (\sigma_2 - \sigma_3)^2 + (\sigma_3 - \sigma_1)^2}{2}\right]^{1/2} \geq S_y \tag{5-10}$$

If we had a simple case of tension σ, then yield would occur when $\sigma \geq S_y$. Thus, the left of Eq. (5-10) can be thought of as a *single, equivalent,* or *effective stress* for the entire general state of stress given by σ_1, σ_2, and σ_3. This effective stress is usually

Figure 5–9

The distortion-energy (DE) theory yield envelope for plane stress states. This is a plot of points obtained from Eq. (5–13) with $\sigma' = S_y$.

Pure shear load line $(\sigma_A = -\sigma_B = \tau)$

— DE
--- MSS

called the *von Mises stress*, σ', named after Dr. R. von Mises, who contributed to the theory. Thus Eq. (5–10), for yield, can be written as

$$\sigma' \geq S_y \tag{5–11}$$

where the von Mises stress is

$$\sigma' = \left[\frac{(\sigma_1 - \sigma_2)^2 + (\sigma_2 - \sigma_3)^2 + (\sigma_3 - \sigma_1)^2}{2} \right]^{1/2} \tag{5–12}$$

For plane stress, the von Mises stress can be represented by the principal stresses σ_A, σ_B, and zero. Then from Eq. (5–12), we get

$$\sigma' = (\sigma_A^2 - \sigma_A \sigma_B + \sigma_B^2)^{1/2} \tag{5–13}$$

Equation (5–13) is a rotated ellipse in the σ_A, σ_B plane, as shown in Fig. 5–9 with $\sigma' = S_y$. The dotted lines in the figure represent the MSS theory, which can be seen to be more restrictive, hence, more conservative.[4]

Using *xyz* components of three-dimensional stress, the von Mises stress can be written as

$$\sigma' = \frac{1}{\sqrt{2}} [(\sigma_x - \sigma_y)^2 + (\sigma_y - \sigma_z)^2 + (\sigma_z - \sigma_x)^2 + 6(\tau_{xy}^2 + \tau_{yz}^2 + \tau_{zx}^2)]^{1/2} \tag{5–14}$$

and for plane stress,

$$\sigma' = (\sigma_x^2 - \sigma_x \sigma_y + \sigma_y^2 + 3\tau_{xy}^2)^{1/2} \tag{5–15}$$

The distortion-energy theory is also called:

- The von Mises or von Mises–Hencky theory
- The shear-energy theory
- The octahedral-shear-stress theory

Understanding octahedral shear stress will shed some light on why the MSS is conservative. Consider an isolated element in which the normal stresses on each surface

---

[4]The three-dimensional equations for DE and MSS can be plotted relative to three-dimensional σ_1, σ_2, σ_3, coordinate axes. The failure surface for DE is a circular cylinder with an axis inclined at 45° from each principal stress axis, whereas the surface for MSS is a hexagon inscribed within the cylinder. See Arthur P. Boresi and Richard J. Schmidt, *Advanced Mechanics of Materials,* 6th ed., John Wiley & Sons, New York, 2003, Sec. 4.4.

Figure 5–10

Octahedral surfaces.

are equal to the hydrostatic stress σ_{av}. There are eight surfaces symmetric to the principal directions that contain this stress. This forms an octahedron as shown in Fig. 5–10. The shear stresses on these surfaces are equal and are called the *octahedral shear stresses* (Fig. 5–10 shows only one of the octahedral surfaces labeled). Through coordinate transformations the octahedral shear stress is given by[5]

$$\tau_{oct} = \frac{1}{3}[(\sigma_1 - \sigma_2)^2 + (\sigma_2 - \sigma_3)^2 + (\sigma_3 - \sigma_1)^2]^{1/2} \qquad (5\text{--}16)$$

Under the name of the octahedral-shear-stress theory, *failure is assumed to occur whenever the octahedral shear stress for any stress state equals or exceeds the octahedral shear stress for the simple tension-test specimen at failure.*

As before, on the basis of the tensile test results, yield occurs when $\sigma_1 = S_y$ and $\sigma_2 = \sigma_3 = 0$. From Eq. (5–16) the octahedral shear stress under this condition is

$$\tau_{oct} = \frac{\sqrt{2}}{3}S_y \qquad (5\text{--}17)$$

When, for the general stress case, Eq. (5–16) is equal or greater than Eq. (5–17), yield is predicted. This reduces to

$$\left[\frac{(\sigma_1 - \sigma_2)^2 + (\sigma_2 - \sigma_3)^2 + (\sigma_3 - \sigma_1)^2}{2}\right]^{1/2} \geq S_y \qquad (5\text{--}18)$$

which is identical to Eq. (5–10), verifying that the maximum-octahedral-shear-stress theory is equivalent to the distortion-energy theory.

The model for the MSS theory ignores the contribution of the normal stresses on the 45° surfaces of the tensile specimen. However, these stresses are $P/2A$, and *not* the hydrostatic stresses which are $P/3A$. Herein lies the difference between the MSS and DE theories.

The mathematical manipulation involved in describing the DE theory might tend to obscure the real value and usefulness of the result. The equations given allow the most complicated stress situation to be represented by a single quantity, the von Mises stress, which then can be compared against the yield strength of the material through Eq. (5–11). This equation can be expressed as a design equation by

$$\sigma' = \frac{S_y}{n} \qquad (5\text{--}19)$$

---

[5]For a derivation, see Arthur P. Boresi, op. cit., pp. 36–37.

The distortion-energy theory predicts no failure under hydrostatic stress and agrees well with all data for ductile behavior. Hence, it is the most widely used theory for ductile materials and is recommended for design problems unless otherwise specified.

One final note concerns the shear yield strength. Consider a case of pure shear τ_{xy}, where for plane stress $\sigma_x = \sigma_y = 0$. For yield, Eq. (5–11) with Eq. (5–15) gives

$$(3\tau_{xy}^2)^{1/2} = S_y \quad \text{or} \quad \tau_{xy} = \frac{S_y}{\sqrt{3}} = 0.577 S_y \qquad (5\text{–}20)$$

Thus, the shear yield strength predicted by the distortion-energy theory is

$$S_{sy} = 0.577 S_y \qquad (5\text{–}21)$$

which as stated earlier, is about 15 percent greater than the $0.5\, S_y$ predicted by the MSS theory. For pure shear, τ_{xy} the principal stresses from Eq. (3–13) are $\sigma_A = -\sigma_B = \tau_{xy}$. The load line for this case is in the third quadrant at an angle of 45° from the σ_A, σ_B axes shown in Fig. 5–9.

EXAMPLE 5–1

A hot-rolled steel has a yield strength of $S_{yt} = S_{yc} = 700$ MPa and a true strain at fracture of $\varepsilon_f = 0.55$. Estimate the factor of safety for the following principal stress states:

(a) 490, 490, 0 MPa.
(b) 210, 490, 0 MPa.
(c) 0, 490, −210 MPa.
(d) 0, −210, −490 MPa.
(e) 210, 210, 210 MPa.

Solution

Since $\varepsilon_f > 0.05$ and S_{yc} and S_{yt} are equal, the material is ductile and the distortion-energy (DE) theory applies. The maximum-shear-stress (MSS) theory will also be applied and compared to the DE results. Note that cases *a* to *d* are plane stress states.

(a) The ordered principal stresses are $\sigma_A = \sigma_1 = 490$, $\sigma_B = \sigma_2 = 490$, $\sigma_3 = 0$ MPa.

DE From Eq. (5–13),

$$\sigma' = [490^2 - 490(490) + 490^2]^{1/2} = 490 \text{ MPa}$$

Answer

$$n = \frac{S_y}{\sigma'} = \frac{700}{490} = 1.43$$

MSS Case 1, using Eq. (5–4) with a factor of safety,

Answer

$$n = \frac{S_y}{\sigma_A} = \frac{700}{490} = 1.43$$

(b) The ordered principal stresses are $\sigma_A = \sigma_1 = 490$, $\sigma_B = \sigma_2 = 210$, $\sigma_3 = 0$ MPa.

DE
$$\sigma' = [490^2 - 490(210) + 210^2]^{1/2} = 426 \text{ MPa}$$

Answer

$$n = \frac{S_y}{\sigma'} = \frac{700}{426} = 1.64$$

MSS   Case 1, using Eq. (5–4),

Answer
$$n = \frac{S_y}{\sigma_A} = \frac{700}{490} = 1.43$$

(c) The ordered principal stresses are $\sigma_A = \sigma_1 = 490$, $\sigma_2 = 0$, $\sigma_B = \sigma_3 = -210$ MPa.

DE
$$\sigma' = [490^2 - 490(-210) + (-210)^2]^{1/2} = 622 \text{ MPa}$$

Answer
$$n = \frac{S_y}{\sigma'} = \frac{700}{622} = 1.13$$

MSS   Case 2, using Eq. (5–5),

Answer
$$n = \frac{S_y}{\sigma_A - \sigma_B} = \frac{700}{490 - (-210)} = 1.00$$

(d) The ordered principal stresses are $\sigma_1 = 0$, $\sigma_A = \sigma_2 = -210$, $\sigma_B = \sigma_3 = -490$ MPa.

DE
$$\sigma' = [(-490)^2 - (-490)(-210) + (-210)^2]^{1/2} = 426 \text{ MPa}$$

Answer
$$n = \frac{S_y}{\sigma'} = \frac{700}{426} = 1.64$$

MSS   Case 3, using Eq. (5–6),

Answer
$$n = -\frac{S_y}{\sigma_B} = -\frac{700}{-490} = 1.43$$

(e) The ordered principal stresses are $\sigma_1 = 210$, $\sigma_2 = 210$, $\sigma_3 = 210$ MPa

DE   From Eq. (5–12),

$$\sigma' = \left[\frac{(210 - 210)^2 + (210 - 210)^2 + (210 - 210)^2}{2} \right]^{1/2} = 0 \text{ MPa}$$

Answer
$$n = \frac{S_y}{\sigma'} = \frac{700}{0} \rightarrow \infty$$

MSS   From Eq. (5–3),

Answer
$$n = \frac{S_y}{\sigma_1 - \sigma_3} = \frac{700}{210 - 210} \rightarrow \infty$$

A tabular summary of the factors of safety is included for comparisons.

	(a)	(b)	(c)	(d)	(e)
DE	1.43	1.64	1.13	1.64	∞
MSS	1.43	1.43	1.00	1.43	∞

Since the MSS theory is on or within the boundary of the DE theory, it will always predict a factor of safety equal to or less than the DE theory, as can be seen in the table. For each case, except case (e), the coordinates and load lines in the σ_A, σ_B plane are shown in Fig. 5–11. Case (e) is not plane stress. Note that the load line for case (a) is the only plane stress case given in which the two theories agree, thus giving the same factor of safety.

Figure 5–11

Load lines for Example 5–1.

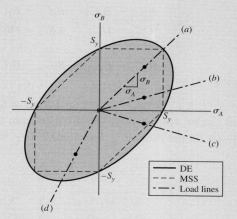

5–6 Coulomb-Mohr Theory for Ductile Materials

Not all materials have compressive strengths equal to their corresponding tensile values. For example, the yield strength of magnesium alloys in compression may be as little as 50 percent of their yield strength in tension. The ultimate strength of gray cast irons in compression varies from 3 to 4 times greater than the ultimate tensile strength. So, in this section, we are primarily interested in those theories that can be used to predict failure for materials whose strengths in tension and compression are not equal.

Historically, the Mohr theory of failure dates to 1900, a date that is relevant to its presentation. There were no computers, just slide rules, compasses, and French curves. Graphical procedures, common then, are still useful today for visualization. The idea of Mohr is based on three "simple" tests: tension, compression, and shear, to yielding if the material can yield, or to rupture. It is easier to define shear yield strength as S_{sy} than it is to test for it.

The practical difficulties aside, Mohr's hypothesis was to use the results of tensile, compressive, and torsional shear tests to construct the three circles of Fig. 5–12 defining a failure envelope tangent to the three circles, depicted as curve *ABCDE* in the figure. The argument amounted to the three Mohr circles describing the stress state in a body (see Fig. 3–12) growing during loading until one of them became tangent to the failure envelope, thereby defining failure. Was the form of the failure envelope straight, circular, or quadratic? A compass or a French curve defined the failure envelope.

A variation of Mohr's theory, called the *Coulomb-Mohr theory* or the *internal-friction theory,* assumes that the boundary *BCD* in Fig. 5–12 is straight. With this assumption only the tensile and compressive strengths are necessary. Consider the conventional ordering of the principal stresses such that $\sigma_1 \geq \sigma_2 \geq \sigma_3$. The largest circle connects σ_1 and σ_3, as shown in Fig. 5–13. The centers of the circles in Fig. 5–13 are C_1, C_2, and C_3. Triangles OB_iC_i are similar, therefore

$$\frac{B_2C_2 - B_1C_1}{OC_2 - OC_1} = \frac{B_3C_3 - B_1C_1}{OC_3 - OC_1}$$

or,

$$\frac{B_2C_2 - B_1C_1}{C_1C_2} = \frac{B_3C_3 - B_1C_1}{C_1C_3}$$

where $B_1C_1 = S_t/2$, $B_2C_2 = (\sigma_1 - \sigma_3)/2$, and $B_3C_3 = S_c/2$, are the radii of the right, center, and left circles, respectively. The distance from the origin to C_1 is $S_t/2$, to C_3 is $S_c/2$, and to C_2 (in the *positive* σ direction) is $(\sigma_1 + \sigma_3)/2$. Thus

$$\frac{\dfrac{\sigma_1 - \sigma_3}{2} - \dfrac{S_t}{2}}{\dfrac{S_t}{2} - \dfrac{\sigma_1 + \sigma_3}{2}} = \frac{\dfrac{S_c}{2} - \dfrac{S_t}{2}}{\dfrac{S_t}{2} + \dfrac{S_c}{2}}$$

Canceling the 2 in each term, cross-multiplying, and simplifying reduces this equation to

$$\frac{\sigma_1}{S_t} - \frac{\sigma_3}{S_c} = 1 \tag{5–22}$$

where either yield strength or ultimate strength can be used.

Figure 5–12

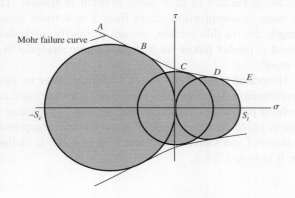

Three Mohr circles, one for the uniaxial compression test, one for the test in pure shear, and one for the uniaxial tension test, are used to define failure by the Mohr hypothesis. The strengths S_c and S_t are the compressive and tensile strengths, respectively; they can be used for yield or ultimate strength.

Figure 5–13

Mohr's largest circle for a general state of stress.

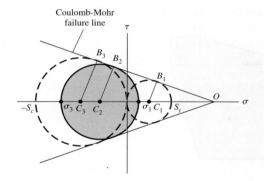

For plane stress, when the two nonzero principal stresses are $\sigma_A \geq \sigma_B$, we have a situation similar to the three cases given for the MSS theory, Eqs. (5–4) to (5–6). That is, the failure conditions are

Case 1: $\sigma_A \geq \sigma_B \geq 0$. For this case, $\sigma_1 = \sigma_A$ and $\sigma_3 = 0$. Equation (5–22) reduces to

$$\sigma_A \geq S_t \qquad (5\text{–}23)$$

Case 2: $\sigma_A \geq 0 \geq \sigma_B$. Here, $\sigma_1 = \sigma_A$ and $\sigma_3 = \sigma_B$, and Eq. (5–22) becomes

$$\frac{\sigma_A}{S_t} - \frac{\sigma_B}{S_c} \geq 1 \qquad (5\text{–}24)$$

Case 3: $0 \geq \sigma_A \geq \sigma_B$. For this case, $\sigma_1 = 0$ and $\sigma_3 = \sigma_B$, and Eq. (5–22) gives

$$\sigma_B \leq -S_c \qquad (5\text{–}25)$$

A plot of these cases, together with the normally unused cases corresponding to $\sigma_B \geq \sigma_A$, is shown in Fig. 5–14.

For design equations, incorporating the factor of safety n, divide all strengths by n. For example, Eq. (5–22) as a design equation can be written as

$$\frac{\sigma_1}{S_t} - \frac{\sigma_3}{S_c} = \frac{1}{n} \qquad (5\text{–}26)$$

Since for the Coulomb-Mohr theory we do not need the torsional shear strength circle we can deduce it from Eq. (5–22). For pure shear τ, $\sigma_1 = -\sigma_3 = \tau$. The torsional yield strength occurs when $\tau_{\max} = S_{sy}$. Substituting $\sigma_1 = -\sigma_3 = S_{sy}$ into Eq. (5–22) and simplifying gives

$$S_{sy} = \frac{S_{yt} S_{yc}}{S_{yt} + S_{yc}} \qquad (5\text{–}27)$$

Figure 5–14

Plot of the Coulomb-Mohr theory failure envelope for plane stress states.

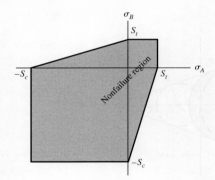

EXAMPLE 5–2   A 52-mm-diameter shaft is statically torqued to 935 N·m. It is made of Gray Cast Iron 30, with a yield strength in tension of 213.73 MPa and a yield strength in compression of 751.52 MPa. It is machined to final diameter. Estimate the factor of safety of the shaft.

Solution   The maximum shear stress is given by

$$\tau = \frac{16T}{\pi d^3} = \frac{16(935)}{\pi [52(10^{-3})]^3} = 33.87(10^6) \text{ N/m}^2 = 33.87 \text{ MPa}$$

The two nonzero principal stresses are 33.87 and −33.87 MPa, making the ordered principal stresses $\sigma_1 = 33.87$, $\sigma_2 = 0$, and $\sigma_3 = -33.87$ MPa. From Eq. (5–26), for yield,

Answer $$n = \frac{1}{\sigma_1/S_{yt} - \sigma_3/S_{yc}} = \frac{1}{33.87/213.73 - (-33.87)/751.52} = 4.91$$

Alternatively, from Eq. (5–27),

$$S_{sy} = \frac{S_{yt}S_{yc}}{S_{yt} + S_{yc}} = \frac{213.73(751.52)}{213.73 + 751.52} = 166.40 \text{ MPa}$$

and $\tau_{\text{max}} = 33.87$ MPa. Thus,

Answer $$n = \frac{S_{sy}}{\tau_{\text{max}}} = \frac{166.40}{33.87} = 4.91$$

5–7   Failure of Ductile Materials Summary

Having studied some of the various theories of failure, we shall now evaluate them and show how they are applied in design and analysis. In this section we limit our studies to materials and parts that are known to fail in a ductile manner. Materials that fail in a brittle manner will be considered separately because these require different failure theories.

To help decide on appropriate and workable theories of failure, Marin[6] collected data from many sources. Some of the data points used to select failure theories for ductile materials are shown in Fig. 5–15.[7] Marin also collected much data for copper and nickel alloys; if shown, the data points for these would be mingled with those already diagrammed. Figure 5–15 shows that either the maximum-shear-stress theory or the distortion-energy theory is acceptable for design and analysis of materials that would fail in a ductile manner.

The selection of one or the other of these two theories is something that you, the engineer, must decide. For design purposes the maximum-shear-stress theory is easy, quick to use, and conservative. If the problem is to learn *why* a part failed, then the distortion-energy theory may be the best to use; Fig. 5–15 shows that the plot of the distortion-energy theory passes closer to the central area of the data points, and thus is generally a better predictor of failure. However, keep in mind that though a failure curve passing through the center of the experimental data is *typical* of the data, its *reliability* from a statistical standpoint is about 50 percent. For design purposes, a larger factor of safety may be warranted when using such a failure theory.

Figure 5–15

Experimental data superposed on failure theories. *(From Fig. 7.11, p. 257, Mechanical Behavior of Materials, 2nd ed., N. E. Dowling, Prentice Hall, Englewood Cliffs, N.J., 1999. Modified to show only ductile failures.)*

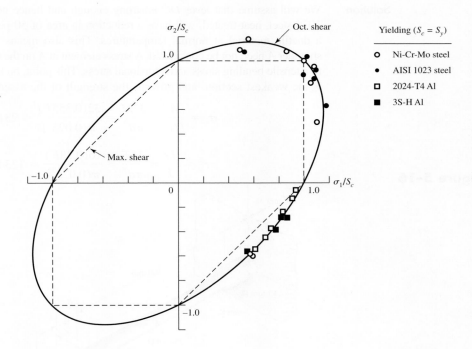

---

[6]Joseph Marin was one of the pioneers in the collection, development, and dissemination of material on the failure of engineering elements. He has published many books and papers on the subject. Here the reference used is Joseph Marin, *Engineering Materials*, Prentice-Hall, Englewood Cliffs, N.J., 1952. (See pp. 156 and 157 for some data points used here.)

[7]Note that some data in Fig. 5–15 are displayed along the top horizontal boundary where $\sigma_B \geq \sigma_A$. This is often done with failure data to thin out congested data points by plotting on the mirror image of the line $\sigma_B = \sigma_A$.

For ductile materials with unequal yield strengths, S_{yt} in tension and S_{yc} in compression, the Mohr theory is the best available. However, the theory requires the results from three separate modes of tests, graphical construction of the failure locus, and fitting the largest Mohr's circle to the failure locus. The alternative to this is to use the Coulomb-Mohr theory, which requires only the tensile and compressive yield strengths and is easily dealt with in equation form.

EXAMPLE 5–3

This example illustrates the use of a failure theory to determine the strength of a mechanical element or component. The example may also clear up any confusion existing between the phrases *strength of a machine part, strength of a material,* and *strength of a part at a point.*

A certain force F applied at D near the end of the 380-mm lever shown in Fig. 5–16, which is quite similar to a socket wrench, results in certain stresses in the cantilevered bar *OABC*. This bar (*OABC*) is of AISI 1035 steel, forged and heat-treated so that it has a minimum (ASTM) yield strength of 560 MPa. We presume that this component would be of no value after yielding. Thus the force F required to initiate yielding can be regarded as the strength of the component part. Find this force.

Solution

We will assume that lever *DC* is strong enough and hence not a part of the problem. A 1035 steel, heat-treated, will have a reduction in area of 50 percent or more and hence is a ductile material at normal temperatures. This also means that stress concentration at shoulder *A* need not be considered. A stress element at *A* on the top surface will be subjected to a tensile bending stress and a torsional stress. This point, on the 25-mm-diameter section, is the weakest section, and governs the strength of the assembly. The two stresses are

$$\sigma_x = \frac{M}{I/c} = \frac{32M}{\pi d^3} = \frac{32(0.355F)}{\pi(0.025^3)} = 231\,424\,F$$

$$\tau_{zx} = \frac{Tr}{J} = \frac{16T}{\pi d^3} = \frac{16(0.38F)}{\pi(0.025^3)} = 123\,860\,F$$

| Figure 5–16

Employing the distortion-energy theory, we find, from Eq. (5–15), that

$$\sigma' = \left(\sigma_x^2 + 3\tau_{zx}^2\right)^{1/2} = [(231\ 424F)^2 + 3(123\ 860F)^2]^{1/2} = 315\ 564\ F$$

Equating the von Mises stress to S_y, we solve for F and get

Answer

$$F = \frac{S_y}{315\ 564} = \frac{560 \times 10^6}{315\ 564} = 1.77\ \text{kN}$$

In this example the strength of the material at point A is $S_y = 560$ MPa. The strength of the assembly or component is $F = 1.8$ kN.

Let us see how to apply the MSS theory. For a point undergoing plane stress with only one non-zero normal stress and one shear stress, the two nonzero principal stresses σ_A and σ_B will have opposite signs and hence fit case 2 for the MSS theory. From Eq. (3–14),

$$\sigma_A - \sigma_B = 2\left[\left(\frac{\sigma_x}{2}\right)^2 + \tau_{zx}^2\right]^{1/2} = \left(\sigma_x^2 + 4\tau_{zx}^2\right)^{1/2}$$

For case 2 of the MSS theory, Eq. (5–3) applies and hence

$$\left(\sigma_x^2 + 4\tau_{zx}^2\right)^{1/2} = S_y$$

$$[(231\ 424F)^2 + 4(123\ 860F)^2]^{1/2} = 339\ 002\ F = 560 \times 10^6$$

$$F = 1.65\ \text{kN}$$

which is about 7 percent less than found for the DE theory. As stated earlier, the MSS theory is more conservative than the DE theory.

EXAMPLE 5–4

The cantilevered tube shown in Fig. 5–17 is to be made of 2014 aluminum alloy treated to obtain a specified minimum yield strength of 276 MPa. We wish to select a stock-size tube from Table A–8 using a design factor $n_d = 4$. The bending load is $F = 1.75$ kN, the axial tension is $P = 9.0$ kN, and the torsion is $T = 72$ N · m. What is the realized factor of safety?

Solution

The critical stress element is at point A on the top surface at the wall, where the bending moment is the largest, and the bending and torsional stresses are at their maximum values. The critical stress element is shown in Fig. 5–17b. Since the axial stress and bending stress are both in tension along the x axis, they are additive for the normal stress, giving

$$\sigma_x = \frac{P}{A} + \frac{Mc}{I} = \frac{9}{A} + \frac{120(1.75)(d_o/2)}{I} = \frac{9}{A} + \frac{105d_o}{I} \tag{1}$$

where, if millimeters are used for the area properties, the stress is in gigapascals.

The torsional stress at the same point is

$$\tau_{zx} = \frac{Tr}{J} = \frac{72(d_o/2)}{J} = \frac{36d_o}{J} \tag{2}$$

| Figure 5–17

(a)

(b)

For accuracy, we choose the distortion-energy theory as the design basis. The von Mises stress from Eq. (5–15), is

$$\sigma' = (\sigma_x^2 + 3\tau_{zx}^2)^{1/2} \tag{3}$$

On the basis of the given design factor, the goal for σ' is

$$\sigma' \le \frac{S_y}{n_d} = \frac{0.276}{4} = 0.0690 \text{ GPa} \tag{4}$$

where we have used gigapascals in this relation to agree with Eqs. (1) and (2).

Programming Eqs. (1) to (3) on a spreadsheet and entering metric sizes from Table A–8 reveals that a 42 × 5-mm tube is satisfactory. The von Mises stress is found to be $\sigma' = 0.06043$ GPa for this size. Thus the realized factor of safety is

Answer

$$n = \frac{S_y}{\sigma'} = \frac{0.276}{0.06043} = 4.57$$

For the next size smaller, a 42 × 4-mm tube, $\sigma' = 0.07105$ GPa giving a factor of safety of

$$n = \frac{S_y}{\sigma'} = \frac{0.276}{0.07105} = 3.88$$

Figure 5–18

Graph of maximum-normal-stress (MNS) theory failure envelope for plane stress states.

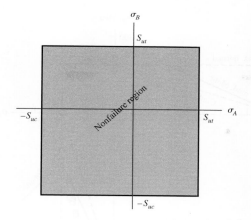

5–8 Maximum-Normal-Stress Theory for Brittle Materials

The maximum-normal-stress (MNS) theory states that *failure occurs whenever one of the three principal stresses equals or exceeds the strength*. Again we arrange the principal stresses for a general stress state in the ordered form $\sigma_1 \geq \sigma_2 \geq \sigma_3$. This theory then predicts that failure occurs whenever

$$\sigma_1 \geq S_{ut} \quad \text{or} \quad \sigma_3 \leq -S_{uc} \tag{5–28}$$

where S_{ut} and S_{uc} are the ultimate tensile and compressive strengths, respectively, given as positive quantities.

For plane stress, with the principal stresses given by Eq. (3–13), with $\sigma_A \geq \sigma_B$, Eq. (5–28) can be written as

$$\sigma_A \geq S_{ut} \quad \text{or} \quad \sigma_B \leq -S_{uc} \tag{5–29}$$

which is plotted in Fig. 5–18.

As before, the failure criteria equations can be converted to design equations. We can consider two sets of equations where $\sigma_A \geq \sigma_B$ as

$$\sigma_A = \frac{S_{ut}}{n} \quad \text{or} \quad \sigma_B = -\frac{S_{uc}}{n} \tag{5–30}$$

As will be seen later, the maximum-normal-stress theory is not very good at predicting failure in the fourth quadrant of the σ_A, σ_B plane. Thus, we will not recommend the theory for use. It has been included here mainly for historical reasons.

5–9 Modifications of the Mohr Theory for Brittle Materials

We will discuss two modifications of the Mohr theory for brittle materials: the Brittle-Coulomb-Mohr (BCM) theory and the modified Mohr (MM) theory. The equations provided for the theories will be restricted to plane stress and be of the design type incorporating the factor of safety.

Figure 5–19

Biaxial fracture data of gray
cast iron compared with
various failure criteria.
(*Dowling, N. E., Mechanical
Behavior of Materials, 2nd ed.,
1999, p. 261. Reprinted by
permission of Pearson
Education, Inc., Upper
Saddle River, New Jersey.*)

The Coulomb-Mohr theory was discussed earlier in Sec. 5–6 with Eqs. (5–23) to
(5–25). Written as design equations for a brittle material, they are:

Brittle-Coulomb-Mohr

$$\sigma_A = \frac{S_{ut}}{n} \qquad \sigma_A \geq \sigma_B \geq 0 \tag{5–31a}$$

$$\frac{\sigma_A}{S_{ut}} - \frac{\sigma_B}{S_{uc}} = \frac{1}{n} \qquad \sigma_A \geq 0 \geq \sigma_B \tag{5–31b}$$

$$\sigma_B = -\frac{S_{uc}}{n} \qquad 0 \geq \sigma_A \geq \sigma_B \tag{5–31c}$$

On the basis of observed data for the fourth quadrant, the modified Mohr theory
expands the fourth quadrant with the solid lines shown in the second and fourth
quadrants of Fig. 5–19 (where the factor of safety, n, is set to one).

Modified Mohr

$$\sigma_A = \frac{S_{ut}}{n} \qquad \sigma_A \geq \sigma_B \geq 0$$

$$\sigma_A \geq 0 \geq \sigma_B \quad \text{and} \quad \left|\frac{\sigma_B}{\sigma_A}\right| \leq 1 \tag{5–32a}$$

$$\frac{(S_{uc} - S_{ut})\sigma_A}{S_{uc}S_{ut}} - \frac{\sigma_B}{S_{uc}} = \frac{1}{n} \qquad \sigma_A \geq 0 \geq \sigma_B \quad \text{and} \quad \left|\frac{\sigma_B}{\sigma_A}\right| > 1 \tag{5–32b}$$

$$\sigma_B = -\frac{S_{uc}}{n} \qquad 0 \geq \sigma_A \geq \sigma_B \tag{5–32c}$$

Data are still outside this extended region. The straight line introduced by the modi-
fied Mohr theory, for $\sigma_A \geq 0 \geq \sigma_B$ and $|\sigma_B/\sigma_A| > 1$, can be replaced by a parabolic

relation which can more closely represent some of the data.[8] However, this introduces a nonlinear equation for the sake of a minor correction, and will not be presented here.

EXAMPLE 5–5	Consider the wrench in Ex. 5–3, Fig. 5–16, as made of cast iron, machined to dimension. The force F required to fracture this part can be regarded as the strength of the component part. If the material is ASTM grade 30 cast iron, find the force F with (*a*) Coulomb-Mohr failure model. (*b*) Modified Mohr failure model.
Solution	We assume that the lever DC is strong enough, and not part of the problem. Since grade 30 cast iron is a brittle material *and* cast iron, the stress-concentration factors K_t and K_{ts} are set to unity. From Table A–24, the tensile ultimate strength is 210 MPa and the compressive ultimate strength is 750 MPa. The stress element at A on the top surface will be subjected to a tensile bending stress and a torsional stress. This location, on the 1-in-diameter section fillet, is the weakest location, and it governs the strength of the assembly. The normal stress σ_x and the shear stress at A are given by

$$\sigma_x = K_t \frac{M}{I/c} = K_t \frac{32M}{\pi d^3} = (1)\frac{32(0.355)}{\pi(0.025)^3} = 231\ 424F$$

$$\tau_{xy} = K_{ts} \frac{Tr}{J} = K_{ts} \frac{16T}{\pi d^3} = (1)\frac{16(0.38)}{\pi(0.025)^3} = 123\ 860F$$

From Eq. (3–13) the nonzero principal stresses σ_A and σ_B are

$$\sigma_A, \sigma_B = \frac{231\ 424F + 0}{2} \pm \sqrt{\left(\frac{231\ 424F - 0}{2}\right)^2 + (123\ 860F)^2} = 285\ 213F,\ -53\ 789F$$

This puts us in the fourth-quadrant of the σ_A, σ_B plane.
(*a*) For BCM, Eq. (5–31*b*) applies with $n = 1$ for failure.

$$\frac{\sigma_A}{S_{ut}} - \frac{\sigma_B}{S_{uc}} = \frac{285\ 213F}{210 \times 10^6} - \frac{(-53\ 789F)}{750 \times 10^6} = 1$$

Solving for F yields

Answer
$$F = 699\ \text{N}$$

(*b*) For MM, the slope of the load line is $|\sigma_B/\sigma_A| = 53\ 789/285\ 213 = 0.189 < 1$. Obviously, Eq. (5–32*a*) applies.

Answer
$$\frac{\sigma_A}{S_{ut}} = \frac{285\ 213F}{210 \times 10^6} = 1$$

$$F = 736\ \text{N}$$

As one would expect from inspection of Fig. 5–19, Coulomb-Mohr is more conservative.

---

[8]See J. E. Shigley, C. R. Mischke, R. G. Budynas, *Mechanical Engineering Design,* 7th ed., McGraw-Hill, New York, 2004, p. 275.

Figure 5–20

A plot of experimental data points obtained from tests on cast iron. Shown also are the graphs of three failure theories of possible usefulness for brittle materials. Note points A, B, C, and D. To avoid congestion in the first quadrant, points have been plotted for $\sigma_A > \sigma_B$ as well as for the opposite sense. (Source of data: Charles F. Walton (ed.), Iron Castings Handbook, Iron Founders' Society, 1971, pp. 215, 216, Cleveland, Ohio.)

5–10 Failure of Brittle Materials Summary

We have identified failure or strength of brittle materials that conform to the usual meaning of the word *brittle*, relating to those materials whose true strain at fracture is 0.05 or less. We also have to be aware of normally ductile materials that for some reason may develop a brittle fracture or crack if used below the transition temperature. Figure 5–20 shows data for a nominal grade 30 cast iron taken under biaxial stress conditions, with several brittle failure hypotheses shown, superposed. We note the following:

- In the first quadrant the data appear on both sides and along the failure curves of maximum-normal-stress, Coulomb-Mohr, and modified Mohr. All failure curves are the same, and data fit well.

- In the fourth quadrant the modified Mohr theory represents the data best, whereas the maximum-normal-stress theory does not.

- In the third quadrant the points A, B, C, and D are too few to make any suggestion concerning a fracture locus.

5–11 Selection of Failure Criteria

For ductile behavior the preferred criterion is the distortion-energy theory, although some designers also apply the maximum-shear-stress theory because of its simplicity and conservative nature. In the rare case when $S_{yt} \neq S_{yc}$, the ductile Coulomb-Mohr method is employed.

For brittle behavior, the original Mohr hypothesis, constructed with tensile, compression, and torsion tests, with a curved failure locus is the best hypothesis we have. However, the difficulty of applying it without a computer leads engineers to choose modifications, namely, Coulomb Mohr, or modified Mohr. Figure 5–21 provides a

Figure 5–21

Failure theory selection flowchart.

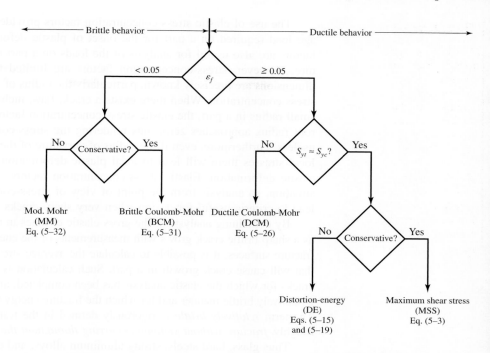

summary flowchart for the selection of an effective procedure for analyzing or predicting failures from static loading for brittle or ductile behavior. Note that the maximum-normal-stress theory is excluded from Fig. 5–21 as the other theories better represent the experimental data.

5–12 Introduction to Fracture Mechanics

The idea that cracks exist in parts even before service begins, and that cracks can grow during service, has led to the descriptive phrase "damage-tolerant design." The focus of this philosophy is on crack growth until it becomes critical, and the part is removed from service. The analysis tool is *linear elastic fracture mechanics* (LEFM). Inspection and maintenance are essential in the decision to retire parts before cracks reach catastrophic size. Where human safety is concerned, periodic inspections for cracks are mandated by codes and government ordinance.

We shall now briefly examine some of the basic ideas and vocabulary needed for the potential of the approach to be appreciated. The intent here is to make the reader aware of the dangers associated with the sudden brittle fracture of so-called ductile materials. The topic is much too extensive to include in detail here and the reader is urged to read further on this complex subject.[9]

---

[9]References on brittle fracture include:

H. Tada, P. C. Paris, and G. R. Irwin, *The Stress Analysis of Cracks Handbook,* 3rd ed., ASME Press, New York, 2000.

D. Broek, *Elementary Engineering Fracture Mechanics,* 4th ed., Martinus Nijhoff, London, 1985.

D. Broek, *The Practical Use of Fracture Mechanics,* Kluwar Academic Pub., London, 1988.

David K. Felbeck and Anthony G. Atkins, *Strength and Fracture of Engineering Solids,* 2nd ed., Prentice-Hall, Englewood Cliffs, N.J., 1995.

Kåre Hellan, *Introduction to Fracture Mechanics,* McGraw-Hill, New York, 1984.

The use of elastic stress-concentration factors provides an indication of the average load required on a part for the onset of plastic deformation, or yielding. These factors are also useful for analysis of the loads on a part that will cause fatigue fracture. However, stress-concentration factors are limited to structures for which all dimensions are precisely known, particularly the radius of curvature in regions of high stress concentration. When there exists a crack, flaw, inclusion, or defect of unknown small radius in a part, the elastic stress-concentration factor approaches infinity as the root radius approaches zero, thus rendering the stress-concentration factor approach useless. Furthermore, even if the radius of curvature of the flaw tip is known, the high local stresses there will lead to local plastic deformation surrounded by a region of elastic deformation. Elastic stress-concentration factors are no longer valid for this situation, so analysis from the point of view of stress-concentration factors does not lead to criteria useful for design when very sharp cracks are present.

By combining analysis of the gross elastic changes in a structure or part that occur as a sharp brittle crack grows with measurements of the energy required to produce new fracture surfaces, it is possible to calculate the average stress (if no crack were present) that will cause crack growth in a part. Such calculation is possible only for parts with cracks for which the elastic analysis has been completed, and for materials that crack in a relatively brittle manner and for which the fracture energy has been carefully measured. The term *relatively brittle* is rigorously defined in the test procedures,[10] but it means, roughly, *fracture without yielding occurring throughout the fractured cross section.*

Thus glass, hard steels, strong aluminum alloys, and even low-carbon steel below the ductile-to-brittle transition temperature can be analyzed in this way. Fortunately, ductile materials blunt sharp cracks, as we have previously discovered, so that fracture occurs at average stresses of the order of the yield strength, and the designer is prepared for this condition. The middle ground of materials that lie between "relatively brittle" and "ductile" is now being actively analyzed, but exact design criteria for these materials are not yet available.

Quasi-Static Fracture

Many of us have had the experience of observing brittle fracture, whether it is the breaking of a cast-iron specimen in a tensile test or the twist fracture of a piece of blackboard chalk. It happens so rapidly that we think of it as instantaneous, that is, the cross section simply parting. Fewer of us have skated on a frozen pond in the spring, with no one near us, heard a cracking noise, and stopped to observe. The noise is due to cracking. The cracks move slowly enough for us to see them run. The phenomenon is not instantaneous, since some time is necessary to feed the crack energy from the stress field to the crack for propagation. Quantifying these things is important to understanding the phenomenon "in the small." In the large, a static crack may be stable and will not propagate. Some level of loading can render the crack unstable, and the crack propagates to fracture.

The foundation of fracture mechanics was first established by Griffith in 1921 using the stress field calculations for an elliptical flaw in a plate developed by Inglis in 1913. For the infinite plate loaded by an applied uniaxial stress σ in Fig. 5–22, the maximum stress occurs at $(\pm a, 0)$ and is given by

$$(\sigma_y)_{\max} = \left(1 + 2\frac{a}{b}\right)\sigma \tag{5–33}$$

---

[10]BS 5447:1977 and ASTM E399-78.

| **Figure 5–22**

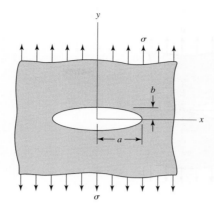

Note that when $a = b$, the ellipse becomes a circle and Eq. (5–33) gives a stress-concentration factor of 3. This agrees with the well-known result for an infinite plate with a circular hole (see Table A–15–1). For a fine crack, $b/a \to 0$, and Eq. (5–33) predicts that $(\sigma_y)_{max} \to \infty$. However, on a microscopic level, an infinitely sharp crack is a hypothetical abstraction that is physically impossible, and when plastic deformation occurs, the stress will be finite at the crack tip.

Griffith showed that the crack growth occurs when the energy release rate from applied loading is greater than the rate of energy for crack growth. Crack growth can be stable or unstable. Unstable crack growth occurs when the *rate* of change of the energy release rate relative to the crack length is equal to or greater than the *rate* of change of the crack growth rate of energy. Griffith's experimental work was restricted to brittle materials, namely glass, which pretty much confirmed his surface energy hypothesis. However, for ductile materials, the energy needed to perform plastic work at the crack tip is found to be much more crucial than surface energy.

Crack Modes and the Stress Intensity Factor

Three distinct modes of crack propagation exist, as shown in Fig. 5–23. A tensile stress field gives rise to mode I, the *opening crack propagation mode,* as shown in Fig. 5–23a. This mode is the most common in practice. Mode II is the *sliding mode,* is due to in-plane shear, and can be seen in Fig. 5–23b. Mode III is the *tearing mode,* which arises from out-of-plane shear, as shown in Fig. 5–23c. Combinations of these modes can also occur. Since mode I is the most common and important mode, the remainder of this section will consider only this mode.

| **Figure 5–23**

| Crack propagation modes.

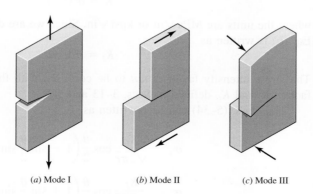

(*a*) Mode I   (*b*) Mode II   (*c*) Mode III

Figure 5–24

Mode I crack model.

Consider a mode I crack of length $2a$ in the infinite plate of Fig. 5–24. By using complex stress functions, it has been shown that the stress field on a $dx\,dy$ element in the vicinity of the crack tip is given by

$$\sigma_x = \sigma\sqrt{\frac{a}{2r}}\cos\frac{\theta}{2}\left(1 - \sin\frac{\theta}{2}\sin\frac{3\theta}{2}\right) \qquad (5\text{–}34a)$$

$$\sigma_y = \sigma\sqrt{\frac{a}{2r}}\cos\frac{\theta}{2}\left(1 + \sin\frac{\theta}{2}\sin\frac{3\theta}{2}\right) \qquad (5\text{–}34b)$$

$$\tau_{xy} = \sigma\sqrt{\frac{a}{2r}}\sin\frac{\theta}{2}\cos\frac{\theta}{2}\cos\frac{3\theta}{2} \qquad (5\text{–}34c)$$

$$\sigma_z = \begin{cases} 0 & \text{(for plane stress)} \\ \nu(\sigma_x + \sigma_y) & \text{(for plane strain)} \end{cases} \qquad (5\text{–}34d)$$

The stress σ_y near the tip, with $\theta = 0$, is

$$\sigma_y\big|_{\theta=0} = \sigma\sqrt{\frac{a}{2r}} \qquad (a)$$

As with the elliptical crack, we see that $\sigma_y\big|_{\theta=0} \to \infty$ as $r \to 0$, and again the concept of an infinite stress concentration at the crack tip is inappropriate. The quantity $\sigma_y\big|_{\theta=0}\sqrt{2r} = \sigma\sqrt{a}$, however, does remain constant as $r \to 0$. It is common practice to define a factor K called the *stress intensity factor* given by

$$K = \sigma\sqrt{\pi a} \qquad (b)$$

where the units are MPa$\sqrt{\text{m}}$ or kpsi$\sqrt{\text{in}}$. Since we are dealing with a mode I crack, Eq. (b) is written as

$$K_I = \sigma\sqrt{\pi a} \qquad (5\text{–}35)$$

The stress intensity factor is *not* to be confused with the static stress-concentration factors K_t and K_{ts} defined in Secs. 3–13 and 5–2.

Thus Eqs. (5–34) can be rewritten as

$$\sigma_x = \frac{K_I}{\sqrt{2\pi r}}\cos\frac{\theta}{2}\left(1 - \sin\frac{\theta}{2}\sin\frac{3\theta}{2}\right) \qquad (5\text{–}36a)$$

$$\sigma_y = \frac{K_I}{\sqrt{2\pi r}}\cos\frac{\theta}{2}\left(1 + \sin\frac{\theta}{2}\sin\frac{3\theta}{2}\right) \qquad (5\text{–}36b)$$

Figure 5-25

Off-center crack in a plate in longitudinal tension; solid curves are for the crack tip at A; dashed curves are for the tip at B.

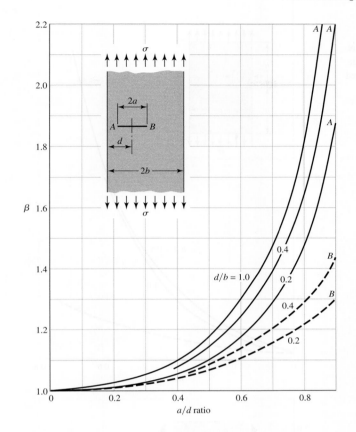

$$\tau_{xy} = \frac{K_I}{\sqrt{2\pi r}} \sin\frac{\theta}{2} \cos\frac{\theta}{2} \cos\frac{3\theta}{2} \qquad (5\text{-}36c)$$

$$\sigma_z = \begin{cases} 0 & \text{(for plane stress)} \\ \nu(\sigma_x + \sigma_y) & \text{(for plane strain)} \end{cases} \qquad (5\text{-}36d)$$

The stress intensity factor is a function of geometry, size and shape of the crack, and the type of loading. For various load and geometric configurations, Eq. (5–35) can be written as

$$K_I = \beta\sigma\sqrt{\pi a} \qquad (5\text{-}37)$$

where β is the *stress intensity modification factor*. Tables for β are available in the literature for basic configurations.[11] Figures 5–25 to 5–30 present a few examples of β for mode I crack propagation.

---

[11]See, for example:

H. Tada, P. C. Paris, and G. R. Irwin, *The Stress Analysis of Cracks Handbook,* 3rd ed., ASME Press, New York, 2000.

G. C. Sib, *Handbook of Stress Intensity Factors for Researchers and Engineers,* Institute of Fracture and Solid Mechanics, Lehigh University, Bethlehem, Pa., 1973.

Y. Murakami, ed., *Stress Intensity Factors Handbook,* Pergamon Press, Oxford, U.K., 1987.

W. D. Pilkey, *Formulas for Stress, Strain, and Structural Matrices,* 2nd ed. John Wiley & Sons, New York, 2005.

Figure 5-26

Plate loaded in longitudinal tension with a crack at the edge; for the solid curve there are no constraints to bending; the dashed curve was obtained with bending constraints added.

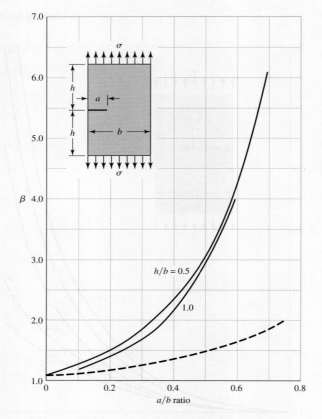

Figure 5-27

Beams of rectangular cross section having an edge crack.

Figure 5–28

Plate in tension containing a circular hole with two cracks.

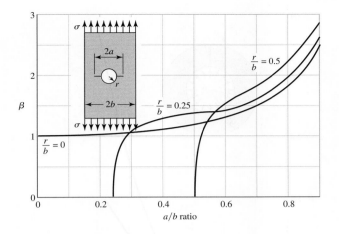

Figure 5–29

A cylinder loading in axial tension having a radial crack of depth a extending completely around the circumference of the cylinder.

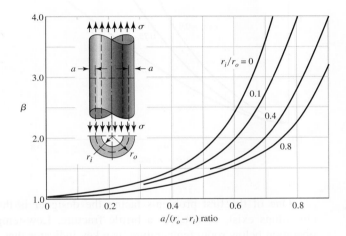

Fracture Toughness

When the magnitude of the mode I stress intensity factor reaches a critical value, K_{Ic}, crack propagation initiates. The *critical stress intensity factor* K_{Ic} is a material property that depends on the material, crack mode, processing of the material, temperature, loading rate, and the state of stress at the crack site (such as plane stress versus plane strain). The critical stress intensity factor K_{Ic} is also called the *fracture toughness* of the material. The fracture toughness for plane strain is normally lower than that for plane stress. For this reason, the term K_{Ic} is typically defined as the *mode I, plane strain fracture toughness*. Fracture toughness K_{Ic} for engineering metals lies in the range $20 \leq K_{Ic} \leq 200$ MPa $\cdot$ $\sqrt{m}$; for engineering polymers and ceramics, $1 \leq K_{Ic} \leq 5$ MPa $\cdot$ $\sqrt{m}$. For a 4340 steel, where the yield strength due to heat treatment ranges from 800 to 1600 MPa, K_{Ic} *decreases* from 190 to 40 MPa $\cdot$ $\sqrt{m}$.

Table 5–1 gives some approximate typical room-temperature values of K_{Ic} for several materials. As previously noted, the fracture toughness depends on many factors and the table is meant only to convey some typical magnitudes of K_{Ic}. For an actual application, it is recommended that the material specified for the application be certified using standard test procedures [see the American Society for Testing and Materials (ASTM) standard E399].

Figure 5–30

Cylinder subjected to internal pressure p_i, having a radial crack in the longitudinal direction of depth a. Use Eq. (3–50) for the tangential stress at $r = r_0$.

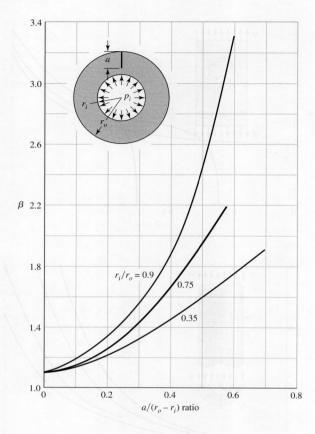

One of the first problems facing the designer is that of deciding whether the conditions exist, or not, for a brittle fracture. Low-temperature operation, that is, operation below room temperature, is a key indicator that brittle fracture is a possible failure mode. Tables of transition temperatures for various materials have not been published, possibly because of the wide variation in values, even for a single material. Thus, in many situations, laboratory testing may give the only clue to the possibility of a brittle fracture. Another key indicator of the possibility of fracture is the ratio of

Table 5–1

Values of K_{Ic} for Some Engineering Materials at Room Temperature

Material	K_{Ic}, MPa $\sqrt{m}$	S_y, MPa
Aluminum		
2024	26	455
7075	24	495
7178	33	490
Titanium		
Ti-6AL-4V	115	910
Ti-6AL-4V	55	1035
Steel		
4340	99	860
4340	60	1515
52100	14	2070

the yield strength to the ultimate strength. A high ratio of S_y/S_u indicates there is only a small ability to absorb energy in the plastic region and hence there is a likelihood of brittle fracture.

The strength-to-stress ratio K_{Ic}/K_I can be used as a factor of safety as

$$n = \frac{K_{Ic}}{K_I} \tag{5–38}$$

EXAMPLE 5–6   A steel ship deck plate is 30 mm thick and 12 m wide. It is loaded with a nominal uniaxial tensile stress of 50 MPa. It is operated below its ductile-to-brittle transition temperature with K_{Ic} equal to 28.3 MPa. If a 65-mm-long central transverse crack is present, estimate the tensile stress at which catastrophic failure will occur. Compare this stress with the yield strength of 240 MPa for this steel.

Solution   For Fig. 5–25, with $d = b$, $2a = 65$ mm and $2b = 12$ m, so that $d/b = 1$ and $a/d = 65/12(10^3) = 0.00542$. Since a/d is so small, $\beta = 1$, so that

$$K_I = \sigma\sqrt{\pi a} = 50\sqrt{\pi(32.5 \times 10^{-3})} = 16.0 \text{ MPa }\sqrt{m}$$

From Eq. (5–38),

$$n = \frac{K_{Ic}}{K_I} = \frac{28.3}{16.0} = 1.77$$

The stress at which catastrophic failure occurs is

Answer $$\sigma_c = \frac{K_{Ic}}{K_I}\sigma = \frac{28.3}{16.0}(50) = 88.4 \text{ MPa}$$

The yield strength is 240 MPa, and catastrophic failure occurs at $88.4/240 = 0.37$, or at 37 percent of yield. The factor of safety in this circumstance is $K_{Ic}/K_I = 28.3/16 = 1.77$ and *not* $240/50 = 4.8$.

EXAMPLE 5–7   A plate of width 1.4 m and length 2.8 m is required to support a tensile force in the 2.8-m direction of 4.0 MN. Inspection procedures will detect only through-thickness edge cracks larger than 2.7 mm. The two Ti-6AL-4V alloys in Table 5–1 are being considered for this application, for which the safety factor must be 1.3 and minimum weight is important. Which alloy should be used?

Solution   (*a*) We elect first to estimate the thickness required to resist yielding. Since $\sigma = P/wt$, we have $t = P/w\sigma$. For the weaker alloy, we have, from Table 5–1, $S_y = 910$ MPa. Thus,

$$\sigma_{all} = \frac{S_y}{n} = \frac{910}{1.3} = 700 \text{ MPa}$$

Thus

$$t = \frac{P}{w\sigma_{\text{all}}} = \frac{4.0(10)^3}{1.4(700)} = 4.08 \text{ mm or greater}$$

For the stronger alloy, we have, from Table 5–1,

$$\sigma_{\text{all}} = \frac{1035}{1.3} = 796 \text{ MPa}$$

and so the thickness is

Answer

$$t = \frac{P}{w\sigma_{\text{all}}} = \frac{4.0(10)^3}{1.4(796)} = 3.59 \text{ mm or greater}$$

(b) Now let us find the thickness required to prevent crack growth. Using Fig. 5–26, we have

$$\frac{h}{b} = \frac{2.8/2}{1.4} = 1 \qquad \frac{a}{b} = \frac{2.7}{1.4(10^3)} = 0.001\ 93$$

Corresponding to these ratios we find from Fig. 5–26 that $\beta \approx 1.1$, and $K_I = 1.1\sigma\sqrt{\pi a}$.

$$n = \frac{K_{Ic}}{K_I} = \frac{115\sqrt{10^3}}{1.1\sigma\sqrt{\pi a}}, \qquad \sigma = \frac{K_{Ic}}{1.1n\sqrt{\pi a}}$$

From Table 5–1, $K_{Ic} = 115 \text{ MPa}\sqrt{\text{m}}$ for the weaker of the two alloys. Solving for σ with $n = 1$ gives the fracture stress

$$\sigma = \frac{115}{1.1\sqrt{\pi(2.7 \times 10^{-3})}} = 1135 \text{ MPa}$$

which is greater than the yield strength of 910 MPa, and so yield strength is the basis for the geometry decision. For the stronger alloy $S_y = 1035$ MPa, with $n = 1$ the fracture stress is

$$\sigma = \frac{K_{Ic}}{nK_I} = \frac{55}{1(1.1)\sqrt{\pi(2.7 \times 10^{-3})}} = 542.9 \text{ MPa}$$

which is less than the yield strength of 1035 MPa. The thickness t is

$$t = \frac{P}{w\sigma_{\text{all}}} = \frac{4.0(10^3)}{1.4(542.9/1.3)} = 6.84 \text{ mm or greater}$$

This example shows that the fracture toughness K_{Ic} limits the geometry when the stronger alloy is used, and so a thickness of 6.84 mm or larger is required. When the weaker alloy is used the geometry is limited by the yield strength, giving a thickness of only 4.08 mm or greater. Thus the weaker alloy leads to a thinner and lighter weight choice since the failure modes differ.

5–13 Important Design Equations

The following equations and their locations are provided as a summary. *Note for plane stress:* The principal stresses in the following equations that are labeled σ_A and σ_B represent the principal stresses determined from the *two-dimensional* Eq. (3–13).

Maximum Shear Theory

p. 234
$$\tau_{\max} = \frac{\sigma_1 - \sigma_3}{2} = \frac{S_y}{2n}$$
(5–3)

Distortion-Energy Theory

Von Mises stress, p. 237

$$\sigma' = \left[\frac{(\sigma_1 - \sigma_2)^2 + (\sigma_2 - \sigma_3)^2 + (\sigma_3 - \sigma_1)^2}{2} \right]^{1/2}$$
(5–12)

p. 237 $\quad \sigma' = \frac{1}{\sqrt{2}}[(\sigma_x - \sigma_y)^2 + (\sigma_y - \sigma_z)^2 + (\sigma_z - \sigma_x)^2 + 6(\tau_{xy}^2 + \tau_{yz}^2 + \tau_{zx}^2)]^{1/2}$

(5–14)

Plane stress, p. 237

$$\sigma' = (\sigma_A^2 - \sigma_A\sigma_B + \sigma_B^2)^{1/2}$$
(5–13)

p. 237
$$\sigma' = (\sigma_x^2 - \sigma_x\sigma_y + \sigma_y^2 + 3\tau_{xy}^2)^{1/2}$$
(5–15)

Yield design equation, p. 238

$$\sigma' = \frac{S_y}{n}$$
(5–19)

Shear yield strength, p. 239

$$S_{sy} = 0.577\, S_y$$
(5–21)

Coulomb-Mohr Theory

p. 243
$$\frac{\sigma_1}{S_t} - \frac{\sigma_3}{S_c} = \frac{1}{n}$$
(5–26)

where S_t is tensile yield (ductile) or ultimate tensile (brittle), and S_t is compressive yield (ductile) or ultimate compressive (brittle) strengths.

Modified Mohr (Plane Stress)

$$\sigma_A = \frac{S_{ut}}{n} \qquad \sigma_A \geq \sigma_B \geq 0$$
(5–32a)

$$\sigma_A \geq 0 \geq \sigma_B \quad \text{and} \quad \left| \frac{\sigma_B}{\sigma_A} \right| \leq 1$$

p. 250
$$\frac{(S_{uc} - S_{ut})\sigma_A}{S_{uc}S_{ut}} - \frac{\sigma_B}{S_{uc}} = \frac{1}{n} \qquad \sigma_A \geq 0 \geq \sigma_B \quad \text{and} \quad \left| \frac{\sigma_B}{\sigma_A} \right| > 1$$
(5–32b)

$$\sigma_B = -\frac{S_{uc}}{n} \qquad 0 \geq \sigma_A \geq \sigma_B$$
(5–32c)

Failure Theory Flowchart

Fig. 5–21, p. 253

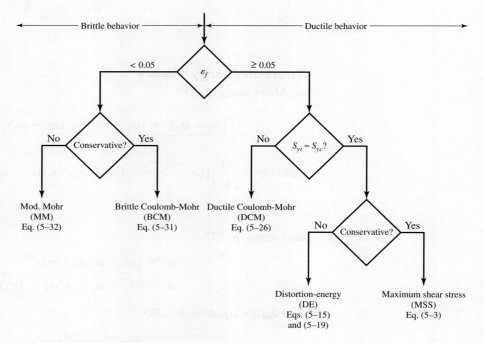

Fracture Mechanics

p. 257

$$K_I = \beta\sigma\sqrt{\pi a} \qquad (5\text{--}37)$$

where β is found in Figs. 5–25 to 5–30 (pp. 257 to 260)

p. 261

$$n = \frac{K_{Ic}}{K_I} \qquad (5\text{--}38)$$

where K_{Ic} is found in Table 5–1 (p. 260)

PROBLEMS

Problems marked with an asterisk (*) are linked to problems in other chapters, as summarized in Table 1–2 of Sec. 1–17, p. 34.

5–1   A ductile hot-rolled steel bar has a minimum yield strength in tension and compression of 350 MPa. Using the distortion-energy and maximum-shear-stress theories determine the factors of safety for the following plane stress states:

(a) $\sigma_x = 100$ MPa, $\sigma_y = 100$ MPa

(b) $\sigma_x = 100$ MPa, $\sigma_y = 50$ MPa

(c) $\sigma_x = 100$ MPa, $\tau_{xy} = -75$ MPa

(d) $\sigma_x = -50$ MPa, $\sigma_y = -75$ MPa, $\tau_{xy} = -50$ MPa

(e) $\sigma_x = 100$ MPa, $\sigma_y = 20$ MPa, $\tau_{xy} = -20$ MPa

5–2   Repeat Prob. 5–1 with the following principal stresses obtained from Eq. (3–13):

(a) $\sigma_A = 100$ MPa, $\sigma_B = 100$ MPa

(b) $\sigma_A = 100$ MPa, $\sigma_B = -100$ MPa

(c) $\sigma_A = 100$ MPa, $\sigma_B = 50$ MPa

(d) $\sigma_A = 100$ MPa, $\sigma_B = -50$ MPa

(e) $\sigma_A = -50$ MPa, $\sigma_B = -100$ MPa

5-3 Repeat Prob. 5–1 for a bar of AISI 1030 hot-rolled steel and:

(a) $\sigma_x = 175$ MPa, $\sigma_y = 105$ MPa

(b) $\sigma_x = 105$ MPa, $\sigma_y = -105$ MPa

(c) $\sigma_x = 140$ MPa, $\tau_{xy} = -70$ MPa

(d) $\sigma_x = -84$ MPa, $\sigma_y = 105$ MPa, $\tau_{xy} = -63$ MPa

(e) $\sigma_x = -168$ MPa, $\sigma_y = -168$ MPa, $\tau_{xy} = -105$ MPa

5-4 Repeat Prob. 5–1 for a bar of AISI 1015 cold-drawn steel with the following principal stresses obtained from Eq. (3–13):

(a) $\sigma_A = 210$ MPa, $\sigma_B = 210$ MPa

(b) $\sigma_A = 210$ MPa, $\sigma_B = -210$ MPa

(c) $\sigma_A = 210$ MPa, $\sigma_B = 105$ MPa

(d) $\sigma_A = -210$ MPa, $\sigma_B = -105$ MPa

(e) $\sigma_A = -350$ MPa, $\sigma_B = 70$ MPa

5-5 Repeat Prob. 5–1 by first plotting the failure loci in the σ_A, σ_B plane to scale; then, for each stress state, plot the load line and by graphical measurement estimate the factors of safety.

5-6 Repeat Prob. 5–3 by first plotting the failure loci in the σ_A, σ_B plane to scale; then, for each stress state, plot the load line and by graphical measurement estimate the factors of safety.

5-7 to 5-11 An AISI 1018 steel has a yield strength, $S_y = 295$ MPa. Using the distortion-energy theory for the given state of plane stress, (a) determine the factor of safety, (b) plot the failure locus, the load line, and estimate the factor of safety by graphical measurement.

Problem Number	σ_x (MPa)	σ_y (MPa)	τ_{xy} (MPa)
5–7	75	−35	0
5–8	−100	30	0
5–9	100	0	−25
5–10	−30	−65	40
5–11	−80	30	−10

5-12 A ductile material has the properties $S_{yt} = 420$ MPA and $S_{yc} = 525$ MPa. Using the ductile Coulomb-Mohr theory, determine the factor of safety for the states of plane stress given in Prob. 5–3.

5-13 Repeat Prob. 5–12 by first plotting the failure loci in the σ_A, σ_B plane to scale; then for each stress state, plot the load line and by graphical measurement estimate the factor of safety.

5-14 to 5-18 An AISI 4142 steel Q&T at 427°C exhibits $S_{yt} = 235$ MPa, $S_{yc} = 285$ MPa, and $\varepsilon_f = 0.07$. For the given state of plane stress, (a) determine the factor of safety, (b) plot the failure locus and the load line, and estimate the factor of safety by graphical measurement.

Problem Number	σ_x (MPa)	σ_y (MPa)	τ_{xy} (MPa)
5–14	150	−50	0
5–15	−150	50	0
5–16	125	0	−75
5–17	−80	−125	50
5–18	125	80	−75

5-19 A brittle material has the properties $S_{ut} = 210$ MPa and $S_{uc} = 630$ MPa. Using the brittle Coulomb-Mohr and modified-Mohr theories, determine the factor of safety for the following states of plane stress.

(a) $\sigma_x = 175$ MPa, $\sigma_y = 105$ MPa
(b) $\sigma_x = 105$ MPa, $\sigma_y = -105$ MPa
(c) $\sigma_x = 140$ MPa, $\tau_{xy} = -70$ MPa
(d) $\sigma_x = -105$ MPa, $\sigma_y = -70$ MPA, $\tau_{xy} = -105$ MPa
(e) $\sigma_x = -84$ MPa, $\tau_{xy} = -105$ MPa

5-20 Repeat Prob. 5-19 by first plotting the failure loci in the σ_A, σ_B plane to scale; then for each stress state, plot the load line and by graphical measurement estimate the factor of safety.

5-21 to
5-25 For an ASTM 30 cast iron, (a) find the factors of safety using the BCM and MM theories, (b) plot the failure diagrams in the σ_A, σ_B plane to scale and locate the coordinates of the stress state, and (c) estimate the factors of safety from the two theories by graphical measurements along the load line.

Problem Number	σ_x (MPa)	σ_y (MPa)	τ_{xy} (MPa)
5-21	105	70	0
5-22	105	-350	0
5-23	105	0	-70
5-24	-70	-175	-70
5-25	-245	91	-70

5-26 to
5-30 A cast aluminum 195-T6 exhibits $S_{ut} = 252$ MPa, $S_{uc} = 245$ MPa, and $\varepsilon_f = 0.045$. For the given state of plane stress, (a) using the Coulomb-Mohr theory, determine the factor of safety, (b) plot the failure locus and the load line, and estimate the factor of safety by graphical measurement.

Problem Number	σ_x (MPa)	σ_y (MPa)	τ_{xy} (MPa)
5-26	105	-70	0
5-27	-105	70	0
5-28	84	0	-56
5-29	-70	-105	70
5-30	105	56	-56

5-31 to
5-35 Repeat Probs. 5-26 to 5-30 using the modified-Mohr theory.

Problem number	5-31	5-32	5-33	5-34	5-35
Repeat problem	5-26	5-27	5-28	5-29	5-30

5-36 This problem illustrates that the factor of safety for a machine element depends on the particular point selected for analysis. Here you are to compute factors of safety, based upon the distortion-energy theory, for stress elements at A and B of the member shown in the figure. This bar is made of AISI 1006 cold-drawn steel and is loaded by the forces $F = 0.55$ kN, $P = 4.0$ kN, and $T = 25$ N · m.

Problem 5–36

100 mm

15-mm D.

5–37   For the beam in Prob. 3–44, p. 147, determine the minimum yield strength that should be considered to obtain a minimum factor of safety of 2 based on the distortion-energy theory.

5–38   A 1020 CD steel shaft is to transmit 15 kW while rotating at 1750 rpm. Determine the minimum diameter for the shaft to provide a minimum factor of safety of 3 based on the maximum-shear-stress theory.

5–39* to   For the problem specified in the table, build upon the results of the original problem to deter-
5–55*   mine the minimum factor of safety for yielding. Use both the maximum-shear-stress theory and the distortion-energy theory, and compare the results. The material is 1018 CD steel.

Problem Number	Original Problem, Page Number
5–39*	3–68, 151
5–40*	3–69, 151
5–41*	3–70, 151
5–42*	3–71, 151
5–43*	3–72, 152
5–44*	3–73, 152
5–45*	3–74, 152
5–46*	3–76, 153
5–47*	3–77, 153
5–48*	3–79, 153
5–49*	3–80, 153
5–50*	3–81, 154
5–51*	3–82, 154
5–52*	3–83, 154
5–53*	3–84, 154
5–54*	3–85, 155
5–55*	3–86, 155

5-56* Build upon the results of Probs. 3–84 and 3–87 to compare the use of a low-strength, ductile material (1018 CD) in which the stress-concentration factor can be ignored to a high-strength but more brittle material (4140 Q&T @ 400°F) in which the stress-concentration factor should be included. For each case, determine the factor of safety for yielding using the distortion-energy theory.

5-57 Using $F = 416$ MPa, design the lever arm CD of Fig. 5–16 by specifying a suitable size and material.

5-58 A spherical pressure vessel is formed of 1.25-mm-thick cold-drawn AISI 1020 sheet steel. If the vessel has a diameter of 200 mm, use the distortion-energy theory to estimate the pressure necessary to initiate yielding. What is the estimated bursting pressure?

5-59 This problem illustrates that the strength of a machine part can sometimes be measured in units other than those of force or moment. For example, the maximum speed that a flywheel can reach without yielding or fracturing is a measure of its strength. In this problem you have a rotating ring made of hot-forged AISI 1020 steel; the ring has a 150-mm inside diameter and a 250-mm outside diameter and is 40-mm in thick. Using the distortion-energy theory, determine the speed in revolutions per minute that would cause the ring to yield. At what radius would yielding begin? [*Note:* The maximum radial stress occurs at $r = (r_o r_i)^{1/2}$; see Eq. (3–55).]

5-60 A light pressure vessel is made of 2024-T3 aluminum alloy tubing with suitable end closures. This cylinder has a 100-mm in OD, a 1.5-mm wall thickness, and $\nu = 0.334$. The purchase order specifies a minimum yield strength of 320 MPa. Using the distortion-energy theory, determine the factor of safety if the pressure-release valve is set at 3.5 MPa.

5-61 A cold-drawn AISI 1015 steel tube is 300 mm OD by 200 mm ID and is to be subjected to an external pressure caused by a shrink fit. Using the distortion-energy theory, determine the maximum pressure that would cause the material of the tube to yield.

5-62 What speed would cause fracture of the ring of Prob. 5–59 if it were made of grade 30 cast iron?

5-63 The figure shows a shaft mounted in bearings at A and D and having pulleys at B and C. The forces shown acting on the pulley surfaces represent the belt tensions. The shaft is to be made of AISI 1035 CD steel. Using a conservative failure theory with a design factor of 2, determine the minimum shaft diameter to avoid yielding.

Problem 5–63

5-64   By modern standards, the shaft design of Prob. 5–63 is poor because it is so long. Suppose it is redesigned by halving the length dimensions. Using the same material and design factor as in Prob. 5–63, find the new shaft diameter.

5-65*   Build upon the results of Prob. 3–40, p. 146, to determine the factor of safety for yielding based on the distortion-energy theory for each of the simplified models in parts *c*, *d*, and *e* of the figure for Prob. 3–40. The pin is machined from AISI 1018 hot-rolled steel. Compare the three models from a designer's perspective in terms of accuracy, safety, and modeling time.

5-66*   For the clevis pin of Prob. 3–40, p. 146, redesign the pin diameter to provide a factor of safety of 2.5 based on a conservative yielding failure theory, and the most conservative loading model from parts *c*, *d*, and *e* of the figure for Prob. 3–40. The pin is machined from AISI 1018 hot-rolled steel.

5-67   A split-ring clamp-type shaft collar is shown in the figure. The collar is 50 mm OD by 25 mm ID by 12 mm wide. The screw is designated as M 6 × 1. The relation between the screw tightening torque T, the nominal screw diameter d, and the tension in the screw F_i is approximately $T = 0.2\,F_i d$. The shaft is sized to obtain a close running fit. Find the axial holding force F_x of the collar as a function of the coefficient of friction and the screw torque.

Problem 5–67

5-68   Suppose the collar of Prob. 5–67 is tightened by using a screw torque of 20 N · m. The collar material is AISI 1035 steel heat-treated to a minimum tensile yield strength of 450 MPa.
(*a*) Estimate the tension in the screw.
(*b*) By relating the tangential stress to the hoop tension, find the internal pressure of the shaft on the ring.
(*c*) Find the tangential and radial stresses in the ring at the inner surface.
(*d*) Determine the maximum shear stress and the von Mises stress.
(*e*) What are the factors of safety based on the maximum-shear-stress and the distortion-energy theories?

5-69   In Prob. 5–67, the role of the screw was to induce the hoop tension that produces the clamping. The screw should be placed so that no moment is induced in the ring. Just where should the screw be located?

5-70   A tube has another tube shrunk over it. The specifications are:

	Inner Member	**Outer Member**
ID	25 ± 0.05 mm	49.98 ± 0.01 mm
OD	50 ± 0.01 mm	75 ± 0.1 mm

Both tubes are made of a plain carbon steel.
(*a*) Find the nominal shrink-fit pressure and the von Mises stresses at the fit surface.
(*b*) If the inner tube is changed to solid shafting with the same outside dimensions, find the nominal shrink-fit pressure and the von Mises stresses at the fit surface.

5-71   Two steel tubes have the specifications:

	Inner Tube	Outer Tube
ID	20 ± 0.050 mm	39.98 ± 0.008 mm
OD	40 ± 0.008 mm	65 ± 0.10 mm

These are shrink-fitted together. Find the nominal shrink-fit pressure and the von Mises stress in each body at the fit surface.

5-72   Repeat Prob. 5–71 for maximum shrink-fit conditions.

5-73   A 50-mm diameter solid steel shaft has a gear with ASTM grade 20 cast-iron hub (E = 100 GPa) shrink-fitted to it. The specifications for the shaft are

$$50 \quad {}^{+0.0000}_{-0.01} \text{ mm}$$

The hole in the hub is sized at 49 mm ± 0.01 mm with an OD of 100 mm ± 0.8 mm. Using the midrange values and the modified Mohr theory, estimate the factor of safety guarding against fracture in the gear hub due to the shrink fit.

5-74   Two steel tubes are shrink-fitted together where the nominal diameters are 40, 45, and 50 mm. Careful measurement before fitting determined the diametral interference between the tubes to be 0.062 mm. After the fit, the assembly is subjected to a torque of 900 N · m and a bending-moment of 675 N · m. Assuming no slipping between the cylinders, analyze the outer cylinder at the inner and outer radius. Determine the factor of safety using distortion energy with S_y = 415 MPa.

5-75   Repeat Prob. 5–74 for the inner tube.

5-76 to   For the problem given in the table, the specifications for the press fit of two cylinders are given
5-81   in the original problem from Chap. 3. If both cylinders are hot-rolled AISI 1040 steel, determine the minimum factor of safety for the outer cylinder based on the distortion-energy theory.

Problem Number	Original Problem, Page Number
5–76	3–110, 157
5–77	3–111, 157
5–78	3–112, 157
5–79	3–113, 157
5–80	3–114, 157
5–81	3–115, 157

5-82   For Eqs. (5–36) show that the principal stresses are given by

$$\sigma_1 = \frac{K_I}{\sqrt{2\pi r}} \cos\frac{\theta}{2}\left(1 + \sin\frac{\theta}{2}\right)$$

$$\sigma_2 = \frac{K_I}{\sqrt{2\pi r}} \cos\frac{\theta}{2}\left(1 - \sin\frac{\theta}{2}\right)$$

$$\sigma_3 = \begin{cases} 0 & \text{(plane stress)} \\ \sqrt{\dfrac{2}{\pi r}}\, \nu K_I \cos\dfrac{\theta}{2} & \text{(plane strain)} \end{cases}$$

5–83   Use the results of Prob. 5–82 for plane strain near the tip with $\theta = 0$ and $\nu = \frac{1}{3}$. If the yield strength of the plate is S_y, what is σ_1 when yield occurs?

(a) Use the distortion-energy theory.

(b) Use the maximum-shear-stress theory. Using Mohr's circles, explain your answer.

5–84   A plate 150 mm wide, 300 mm long, and 8 mm thick is loaded in tension in the direction of the length. The plate contains a crack as shown in Fig. 5–26 with the crack length of 15.65 mm. The material is steel with $K_{Ic} = 490$ MPa $\cdot \sqrt{m}$ and $S_y = 1.1$ GPa. Determine the maximum possible load that can be applied before the plate (a) yields, and (b) has uncontrollable crack growth.

5–85   A cylinder subjected to internal pressure p_i has an outer diameter of 350 mm and a 25 mm wall thickness. For the cylinder material, $K_{Ic} = 80$ MPa $\cdot \sqrt{m}$, $S_y = 1200$ MPa, and $S_{ut} = 1350$ MPa. If the cylinder contains a radial crack in the longitudinal direction of depth 0.5 in determine the pressure that will cause uncontrollable crack growth.

6

Fatigue Failure Resulting from Variable Loading

In Chap. 5 we considered the analysis and design of parts subjected to static loading. The behavior of machine parts is entirely different when they are subjected to time-varying loading. In this chapter we shall examine how parts fail under variable loading and how to proportion them to successfully resist such conditions.

6–1   Introduction to Fatigue in Metals

In most testing of those properties of materials that relate to the stress-strain diagram, the load is applied gradually, to give sufficient time for the strain to fully develop. Furthermore, the specimen is tested to destruction, and so the stresses are applied only once. Testing of this kind is applicable, to what are known as *static conditions;* such conditions closely approximate the actual conditions to which many structural and machine members are subjected.

The condition frequently arises, however, in which the stresses vary with time or they fluctuate between different levels. For example, a particular fiber on the surface of a rotating shaft subjected to the action of bending loads undergoes both tension and compression for each revolution of the shaft. If the shaft is part of an electric motor rotating at 1725 rev/min, the fiber is stressed in tension and compression 1725 times each minute. If, in addition, the shaft is also axially loaded (as it would be, for example, by a helical or worm gear), an axial component of stress is superposed upon the bending component. In this case, some stress is always present in any one fiber, but now the *level* of stress is fluctuating. These and other kinds of loading occurring in machine members produce stresses that are called *variable, repeated, alternating,* or *fluctuating* stresses.

Often, machine members are found to have failed under the action of repeated or fluctuating stresses; yet the most careful analysis reveals that the actual maximum stresses were well below the ultimate strength of the material, and quite frequently even below the yield strength. The most distinguishing characteristic of these failures is that the stresses have been repeated a very large number of times. Hence the failure is called a *fatigue failure*.

When machine parts fail statically, they usually develop a very large deflection, because the stress has exceeded the yield strength, and the part is replaced before fracture actually occurs. Thus many static failures give visible warning in advance. But a fatigue failure gives no warning! It is sudden and total, and hence dangerous. It is relatively simple to design against a static failure, because our knowledge is comprehensive. Fatigue is a much more complicated phenomenon, only partially understood, and the engineer seeking competence must acquire as much knowledge of the subject as possible.

A fatigue failure has an appearance similar to a brittle fracture, as the fracture surfaces are flat and perpendicular to the stress axis with the absence of necking. The fracture features of a fatigue failure, however, are quite different from a static brittle fracture arising from three stages of development. *Stage I* is the initiation of one or more microcracks due to cyclic plastic deformation followed by crystallographic propagation extending from two to five grains about the origin. Stage I cracks are not normally discernible to the naked eye. *Stage II* progresses from microcracks to macrocracks forming parallel plateau-like fracture surfaces separated by longitudinal ridges. The plateaus are generally smooth and normal to the direction of maximum tensile stress. These surfaces can be wavy dark and light bands referred to as *beach marks* or *clamshell marks,* as seen in Fig. 6–1. During cyclic loading, these cracked surfaces open and close, rubbing together, and the beach mark appearance depends

Figure 6–1

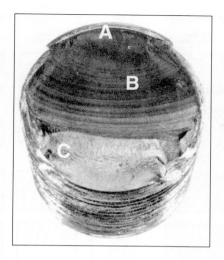

Fatigue failure of a bolt due to repeated unidirectional bending. The failure started at the thread root at *A*, propagated across most of the cross section shown by the beach marks at *B*, before final fast fracture at *C*. *(From ASM Handbook, Vol. 12:* Fractography, *2nd printing, 1992, ASM International, Materials Park, OH 44073-0002, fig 50, p. 120. Reprinted by permission of ASM International®, www.asminternational.org.)*

on the changes in the level or frequency of loading and the corrosive nature of the environment. *Stage III* occurs during the final stress cycle when the remaining material cannot support the loads, resulting in a sudden, fast fracture. A stage III fracture can be brittle, ductile, or a combination of both. Quite often the beach marks, if they exist, and possible patterns in the stage III fracture called *chevron lines,* point toward the origins of the initial cracks.

There is a good deal to be learned from the fracture patterns of a fatigue failure. Figure 6–2 shows representations of failure surfaces of various part geometries under differing load conditions and levels of stress concentration. Note that, in the case of rotational bending, even the direction of rotation influences the failure pattern.

Fatigue failure is due to crack formation and propagation. A fatigue crack will typically initiate at a discontinuity in the material where the cyclic stress is a maximum. Discontinuities can arise because of:

- Design of rapid changes in cross section, keyways, holes, etc. where stress concentrations occur as discussed in Secs. 3–13 and 5–2.
- Elements that roll and/or slide against each other (bearings, gears, cams, etc.) under high contact pressure, developing concentrated subsurface contact stresses (Sec. 3–19) that can cause surface pitting or spalling after many cycles of the load.
- Carelessness in locations of stamp marks, tool marks, scratches, and burrs; poor joint design; improper assembly; and other fabrication faults.
- Composition of the material itself as processed by rolling, forging, casting, extrusion, drawing, heat treatment, etc. Microscopic and submicroscopic surface and subsurface discontinuities arise, such as inclusions of foreign material, alloy segregation, voids, hard precipitated particles, and crystal discontinuities.

Various conditions that can accelerate crack initiation include residual tensile stresses, elevated temperatures, temperature cycling, a corrosive environment, and high-frequency cycling.

The rate and direction of fatigue crack propagation is primarily controlled by localized stresses and by the structure of the material at the crack. However, as with crack formation, other factors may exert a significant influence, such as environment, temperature, and frequency. As stated earlier, cracks will grow along planes normal

Figure 6–2

Schematics of fatigue fracture surfaces produced in smooth and notched components with round and rectangular cross sections under various loading conditions and nominal stress levels. *(From ASM Metals Handbook, Vol. 11: Failure Analysis and Prevention, 1986, ASM International, Materials Park, OH 44073-0002, fig 18, p. 111. Reprinted by permission of ASM International®, www.asminternational.org.)*

to the maximum tensile stresses. The crack growth process can be explained by fracture mechanics (see Sec. 6–6).

A major reference source in the study of fatigue failure is the 21-volume ASM *Metals Handbook*. Figures 6–1 to 6–8, reproduced with permission from ASM International, are but a minuscule sample of examples of fatigue failures for a great variety of conditions included in the handbook. Comparing Fig. 6–3 with Fig. 6–2, we see that failure occurred by rotating bending stresses, with the direction of rotation being clockwise with respect to the view and with a mild stress concentration and low nominal stress.

Figure 6–3

Fatigue fracture of an AISI 4320 drive shaft. The fatigue failure initiated at the end of the keyway at points *B* and progressed to final rupture at *C*. The final rupture zone is small, indicating that loads were low. *(From ASM Handbook, Vol. 12:* Fractography, *2nd printing, 1992, ASM International, Materials Park, OH 44073-0002, fig 51, p. 120. Reprinted by permission of ASM International®, www.asminternational.org.)*

Figure 6–4

Fatigue fracture surface of an AISI 8640 pin. Sharp corners of the mismatched grease holes provided stress concentrations that initiated two fatigue cracks indicated by the arrows. *(From ASM Handbook, Vol. 12:* Fractography, *2nd printing, 1992, ASM International, Materials Park, OH 44073-0002, fig 520, p. 331. Reprinted by permission of ASM International®, www.asminternational.org.)*

Figure 6–5

Fatigue fracture surface of a forged connecting rod of AISI 8640 steel. The fatigue crack origin is at the left edge, at the flash line of the forging, but no unusual roughness of the flash trim was indicated. The fatigue crack progressed halfway around the oil hole at the left, indicated by the beach marks, before final fast fracture occurred. Note the pronounced shear lip in the final fracture at the right edge. *(From ASM Handbook, Vol. 12:* Fractography, *2nd printing, 1992, ASM International, Materials Park, OH 44073-0002, fig 523, p. 332. Reprinted by permission of ASM International®, www.asminternational.org.)*

Figure 6–6

Fatigue fracture surface of a 200-mm (8-in) diameter piston rod of an alloy steel steam hammer used for forging. This is an example of a fatigue fracture caused by pure tension where surface stress concentrations are absent and a crack may initiate anywhere in the cross section. In this instance, the initial crack formed at a forging flake slightly below center, grew outward symmetrically, and ultimately produced a brittle fracture without warning. *(From ASM Handbook, Vol. 12:* Fractography, *2nd printing, 1992, ASM International, Materials Park, OH 44073-0002, fig 570, p. 342. Reprinted by permission of ASM International®, www.asminternational.org.)*

Figure 6–7

Fatigue failure of an ASTM A186 steel double-flange trailer wheel caused by stamp marks. (*a*) Coke-oven car wheel showing position of stamp marks and fractures in the rib and web. (*b*) Stamp mark showing heavy impression and fracture extending along the base of the lower row of numbers. (*c*) Notches, indicated by arrows, created from the heavily indented stamp marks from which cracks initiated along the top at the fracture surface. (*From ASM Metals Handbook, Vol. 11:* Failure Analysis and Prevention, *1986, ASM International, Materials Park, OH 44073-0002, fig 51, p. 130. Reprinted by permission of ASM International®, www.asminternational.org.*)

Figure 6–8

Aluminum alloy 7075-T73 landing-gear torque-arm assembly redesign to eliminate fatigue fracture at a lubrication hole. (*a*) Arm configuration, original and improved design (dimensions given in inches). (*b*) Fracture surface where arrows indicate multiple crack origins. (*From ASM Metals Handbook, Vol. 11:* Failure Analysis and Prevention, *1986, ASM International, Materials Park, OH 44073-0002, fig 23, p. 114. Reprinted by permission of ASM International®, www.asminternational.org.*)

6-2   Approach to Fatigue Failure in Analysis and Design

As noted in the previous section, there are a great many factors to be considered, even for very simple load cases. The methods of fatigue failure analysis represent a combination of engineering and science. Often science fails to provide the complete answers that are needed. But the airplane must still be made to fly—safely. And the automobile must be manufactured with a reliability that will ensure a long and troublefree life and at the same time produce profits for the stockholders of the industry. Thus, while science has not yet completely explained the complete mechanism of fatigue, the engineer must still design things that will not fail. In a sense this is a classic example of the true meaning of engineering as contrasted with science. Engineers use science to solve their problems if the science is available. But available or not, the problem must be solved, and whatever form the solution takes under these conditions is called *engineering*.

In this chapter, we will take a structured approach in the design against fatigue failure. As with static failure, we will attempt to relate to test results performed on simply loaded specimens. However, because of the complex nature of fatigue, there is much more to account for. From this point, we will proceed methodically, and in stages. In an attempt to provide some insight as to what follows in this chapter, a brief description of the remaining sections will be given here.

Fatigue-Life Methods (Secs. 6–3 to 6–6)

Three major approaches used in design and analysis to predict when, if ever, a cyclically loaded machine component will fail in fatigue over a period of time are presented. The premises of each approach are quite different but each adds to our understanding of the mechanisms associated with fatigue. The application, advantages, and disadvantages of each method are indicated. Beyond Sec. 6–6, only one of the methods, the stress-life method, will be pursued for further design applications.

Fatigue Strength and the Endurance Limit (Secs. 6–7 and 6–8)

The strength-life (S-N) diagram provides the fatigue strength S_f versus cycle life N of a material. The results are generated from tests using a simple loading of standard laboratory-controlled specimens. The loading often is that of sinusoidally reversing pure bending. The laboratory-controlled specimens are polished without geometric stress concentration at the region of minimum area.

For steel and iron, the S-N diagram becomes horizontal at some point. The strength at this point is called the *endurance limit* S'_e and occurs somewhere between 10^6 and 10^7 cycles. The prime mark on S'_e refers to the endurance limit of the *controlled laboratory specimen*. For nonferrous materials that do not exhibit an endurance limit, a fatigue strength at a specific number of cycles, S'_f, may be given, where again, the prime denotes the fatigue strength of the laboratory-controlled specimen.

The strength data are based on many controlled conditions that will not be the same as that for an actual machine part. What follows are practices used to account for the differences between the loading and physical conditions of the specimen and the actual machine part.

Endurance Limit Modifying Factors (Sec. 6–9)

Modifying factors are defined and used to account for differences between the specimen and the actual machine part with regard to surface conditions, size, loading, temperature, reliability, and miscellaneous factors. Loading is still considered to be simple and reversing.

Stress Concentration and Notch Sensitivity (Sec. 6–10)

The actual part may have a geometric stress concentration by which the fatigue behavior depends on the static stress-concentration factor and the component material's sensitivity to fatigue damage.

Fluctuating Stresses (Secs. 6–11 to 6–13)

These sections account for simple stress states from fluctuating load conditions that are not purely sinusoidally reversing axial, bending, or torsional stresses.

Combinations of Loading Modes (Sec. 6–14)

Here a procedure based on the distortion-energy theory is presented for analyzing combined fluctuating stress states, such as combined bending and torsion. It is assumed that the levels of the fluctuating stresses are in phase and not time varying.

Varying, Fluctuating Stresses; Cumulative Fatigue Damage (Sec. 6–15)

The fluctuating stress levels on a machine part may be time varying. Methods are provided to assess the fatigue damage on a cumulative basis.

Remaining Sections (Sec. 6–16 and 6–17)

The remaining two sections of the chapter pertain to surface fatigue strength and road maps with important equations.

6–3 Fatigue-Life Methods

The three major fatigue life methods used in design and analysis are the *stress-life method,* the *strain-life method,* and the *linear-elastic fracture mechanics method.* These methods attempt to predict the life in number of cycles to failure, N, for a specific level of loading. Life of $1 \leq N \leq 10^3$ cycles is generally classified as *low-cycle fatigue,* whereas *high-cycle fatigue* is considered to be $N > 10^3$ cycles.

The stress-life method, based on stress levels only, is the least accurate approach, especially for low-cycle applications. However, it is the most traditional method, since it is the easiest to implement for a wide range of design applications, has ample supporting data, and represents high-cycle applications adequately.

The strain-life method involves more detailed analysis of the plastic deformation at localized regions where the stresses and strains are considered for life estimates. This method is especially good for low-cycle fatigue applications. In applying this method, several idealizations must be compounded, and so some uncertainties will exist in the results. For this reason, it will be discussed only because of its value in adding to the understanding of the nature of fatigue.

The fracture mechanics method assumes a crack is already present and detected. It is then employed to predict crack growth with respect to stress intensity. It is most practical when applied to large structures in conjunction with computer codes and a periodic inspection program.

6–4 The Stress-Life Method

To determine the strength of materials under the action of fatigue loads, specimens are subjected to repeated or varying forces of specified magnitudes while the cycles or stress reversals are counted to destruction. The most widely used fatigue-testing device is the R. R. Moore high-speed rotating-beam machine. This machine subjects the specimen to pure bending (no transverse shear) by means of weights. The test specimen, shown in Fig. 6–9, is very carefully machined and polished, with a final polishing in an axial direction to avoid circumferential scratches. Other fatigue-testing machines are available for applying fluctuating or reversed axial stresses, torsional stresses, or combined stresses to the test specimens.

To establish the fatigue strength of a material, quite a number of tests are necessary because of the statistical nature of fatigue. For the rotating-beam test, a constant bending load is applied, and the number of revolutions (stress reversals) of the beam required for failure is recorded. The first test is made at a stress that is somewhat under the ultimate strength of the material. The second test is made at a stress that is less than that used in the first. This process is continued, and the results are plotted as an *S-N* diagram (Fig. 6–10). This chart may be plotted on semilog paper or on log-log paper. In the case of ferrous metals and alloys, the graph becomes horizontal after the material has been stressed for a certain number of cycles. Plotting on log paper emphasizes the bend in the curve, which might not be apparent if the results were plotted by using Cartesian coordinates.

The ordinate of the *S-N* diagram is called the *fatigue strength* S_f; a statement of this strength value must always be accompanied by a statement of the number of cycles *N* to which it corresponds.

Figure 6–9

Test-specimen geometry for the R. R. Moore rotating-beam machine. The bending moment is uniform, $M = Fa$, over the curved length and at the highest-stressed section at the mid-point of the beam.

Figure 6–10

An *S-N* diagram plotted from the results of completely reversed axial fatigue tests. Material: UNS G41300 steel, normalized; $S_{ut} = 810$ MPa; maximum $S_{ut} = 105$ MPa. *(Data from NACA Tech. Note 3866, December 1966.)*

Figure 6–11

S-N bands for representative aluminum alloys, excluding wrought alloys with $S_{ut} = 260$ MPa. *(From R. C. Juvinall,* Engineering Considerations of Stress, Strain and Strength. *Copyright © 1967 by The McGraw-Hill Companies, Inc. Reprinted by permission.)*

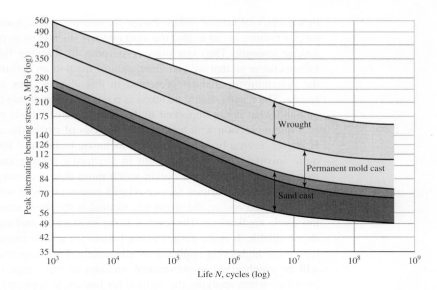

Soon we shall learn that *S-N* diagrams can be determined either for a test specimen or for an actual mechanical element. Even when the material of the test specimen and that of the mechanical element are identical, there will be significant differences between the diagrams for the two.

In the case of the steels, a knee occurs in the graph, and beyond this knee failure will not occur, no matter how great the number of cycles. The strength corresponding to the knee is called the *endurance limit* S_e, or the fatigue limit. The graph of Fig. 6–10 never does become horizontal for nonferrous metals and alloys, and hence these materials do not have an endurance limit. Figure 6–11 shows scatter bands indicating the *S-N* curves for most common aluminum alloys excluding wrought alloys having a tensile strength below 260 MPa. Since aluminum does not have an endurance limit, normally the fatigue strength S_f is reported at a specific number of cycles, normally $N = 5(10^8)$ cycles of reversed stress (see Table A–24).

The *S-N* diagram is usually obtained by *completely reversed* stress cycles, in which the stress level alternates between equal magnitudes of tension and compression. We note that a stress cycle ($N = 1$) constitutes a single application and removal of a load and then another application and removal of the load in the opposite direction. Thus $N = \frac{1}{2}$ means the load is applied once and then removed, which is the case with the simple tension test.

The body of knowledge available on fatigue failure from $N = 1$ to $N = 1000$ cycles is generally classified as *low-cycle fatigue,* as indicated in Fig. 6–10. *High-cycle fatigue,* then, is concerned with failure corresponding to stress cycles greater than 10^3 cycles.

We also distinguish a *finite-life region* and an *infinite-life region* in Fig. 6–10. The boundary between these regions cannot be clearly defined except for a specific material; but it lies somewhere between 10^6 and 10^7 cycles for steels, as shown in Fig. 6–10.

As noted previously, it is always good engineering practice to conduct a testing program on the materials to be employed in design and manufacture. This, in fact, is a requirement, not an option, in guarding against the possibility of a fatigue failure. *Because of this necessity for testing, it would really be unnecessary for us to proceed any further in the study of fatigue failure except for one important reason: the desire to know why fatigue failures occur so that the most effective method or methods can be used to improve fatigue strength.* Thus our primary purpose in studying fatigue is to understand why failures occur so that we can guard against them in an optimum manner. For this reason, the deterministic analysis presented in this chapter does not yield absolutely precise results. The results should be taken as a guide, as something that indicates what is important and what is not important in designing against fatigue failure.

In conjunction with testing, the methods of stochastic analysis must be employed. However, even a brief exposure to stochastic techniques for fatigue analysis would be quite involved.[1] To be proficient in stochastic fatigue analysis, it is highly recommended that the designer obtain much more than a cursory background in stochastic analysis and the various fatigue-related issues involved with each parameter.

As stated earlier, the stress-life method is the least accurate approach especially for low-cycle applications. However, it is the most traditional method, with much published data available. It is the easiest to implement for a wide range of design applications and represents high-cycle applications adequately. For these reasons the stress-life method will be emphasized in subsequent sections of this chapter. However, care should be exercised when applying the method for low-cycle applications, as the method does not account for the true stress-strain behavior when localized yielding occurs.

6–5   The Strain-Life Method

The best approach yet advanced to explain the nature of fatigue failure is called by some the *strain-life* method. The approach can be used to estimate fatigue strengths, but when it is so used it is necessary to compound several idealizations, and so some uncertainties will exist in the results. For this reason, the method is presented here only because of its value in explaining the nature of fatigue.

A fatigue failure almost always begins at a local discontinuity such as a notch, crack, or other area of stress concentration. When the stress at the discontinuity exceeds the elastic limit, plastic strain occurs. If a fatigue fracture is to occur, there must exist cyclic plastic strains. Thus we shall need to investigate the behavior of materials subject to cyclic deformation.

In 1910, Bairstow verified by experiment Bauschinger's theory that the elastic limits of iron and steel can be changed, either up or down, by the cyclic variations of stress.[2] In general, the elastic limits of annealed steels are likely to increase when subjected to cycles of stress reversals, while cold-drawn steels exhibit a decreasing elastic limit.

R. W. Landgraf investigated the low-cycle fatigue behavior of a large number of very high-strength steels, and during his research he made many cyclic stress-strain plots.[3] Figure 6–12 has been constructed to show the general appearance of these plots for the first few cycles of controlled cyclic strain. In this case the strength decreases with stress repetitions, as evidenced by the fact that the reversals occur at ever-smaller stress levels. As previously noted, other materials may be strengthened, instead, by cyclic stress reversals.

[1]A *brief* exposure can be found in previous editions of this book. See, for example, R. G. Budynas and J. K. Nisbett, *Shigley's Mechanical Engineering Design,* 9th ed., McGraw-Hill, New York, 2011, Sec. 6-17.

[2]L. Bairstow, "The Elastic Limits of Iron and Steel under Cyclic Variations of Stress," *Philosophical Transactions,* Series A, vol. 210, Royal Society of London, 1910, pp. 35–55.

[3]R. W. Landgraf, *Cyclic Deformation and Fatigue Behavior of Hardened Steels,* Report no. 320, Department of Theoretical and Applied Mechanics, University of Illinois, Urbana, 1968, pp. 84–90.

Figure 6–12

True stress–true strain hysteresis loops showing the first five stress reversals of a cyclic-softening material. The graph is slightly exaggerated for clarity. Note that the slope of the line AB is the modulus of elasticity E. The stress range is $\Delta\sigma$, $\Delta\varepsilon_p$ is the plastic-strain range, and $\Delta\varepsilon_e$ is the elastic strain range. The total-strain range is $\Delta\varepsilon = \Delta\varepsilon_p + \Delta\varepsilon_e$.

Figure 6–13

A log-log plot showing how the fatigue life is related to the true-strain amplitude for hot-rolled SAE 1020 steel. *(Reprinted with permission from SAE J1099_200208 © 2002 SAE International.)*

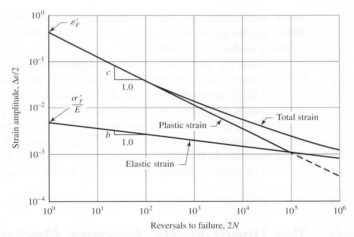

The SAE Fatigue Design and Evaluation Steering Committee released a report in 1975 in which the life in reversals to failure is related to the strain amplitude $\Delta\varepsilon/2$.[4] The report contains a plot of this relationship for SAE 1020 hot-rolled steel; the graph has been reproduced as Fig. 6–13. To explain the graph, we first define the following terms:

- *Fatigue ductility coefficient* ε'_F is the true strain corresponding to fracture in one reversal (point A in Fig. 6–12). The plastic-strain line begins at this point in Fig. 6–13.

- *Fatigue strength coefficient* σ'_F is the true stress corresponding to fracture in one reversal (point A in Fig. 6–12). Note in Fig. 6–13 that the elastic-strain line begins at σ'_F/E.

- *Fatigue ductility exponent* c is the slope of the plastic-strain line in Fig. 6–13 and is the power to which the life $2N$ must be raised to be proportional to the true plastic-strain amplitude. If the number of stress reversals is $2N$, then N is the number of cycles.

- *Fatigue strength exponent* b is the slope of the elastic-strain line, and is the power to which the life $2N$ must be raised to be proportional to the true-stress amplitude.

---

[4]*Technical Report on Fatigue Properties,* SAE J1099, 1975.

Now, from Fig. 6–12, we see that the total strain is the sum of the elastic and plastic components. Therefore the total strain amplitude is half the total strain range

$$\frac{\Delta \varepsilon}{2} = \frac{\Delta \varepsilon_e}{2} + \frac{\Delta \varepsilon_p}{2}$$ (a)

The equation of the plastic-strain line in Fig. 6–13 is

$$\frac{\Delta \varepsilon_p}{2} = \varepsilon'_F (2N)^c$$ (6–1)

The equation of the elastic strain line is

$$\frac{\Delta \varepsilon_e}{2} = \frac{\sigma'_F}{E} (2N)^b$$ (6–2)

Therefore, from Eq. (a), we have for the total-strain amplitude

$$\frac{\Delta \varepsilon}{2} = \frac{\sigma'_F}{E} (2N)^b + \varepsilon'_F (2N)^c$$ (6–3)

which is the Manson-Coffin relationship between fatigue life and total strain.[5] Some values of the coefficients and exponents are listed in Table A–23. Many more are included in the SAE J1099 report.[6]

Though Eq. (6–3) is a perfectly legitimate equation for obtaining the fatigue life of a part when the strain and other cyclic characteristics are given, it appears to be of little use to the designer. The question of how to determine the total strain at the bottom of a notch or discontinuity has not been answered. There are no tables or charts of strain-concentration factors in the literature. It is possible that strain-concentration factors will become available in research literature very soon because of the increase in the use of finite-element analysis. Moreover, finite element analysis can of itself approximate the strains that will occur at all points in the subject structure.[7]

6–6 The Linear-Elastic Fracture Mechanics Method

The first phase of fatigue cracking is designated as stage I fatigue. Crystal slip that extends through several contiguous grains, inclusions, and surface imperfections is presumed to play a role. Since most of this is invisible to the observer, we just say that stage I involves several grains. The second phase, that of crack extension, is called stage II fatigue. The advance of the crack (that is, new crack area is created) does produce evidence that can be observed on micrographs

---

[5]J. F. Tavernelli and L. F. Coffin, Jr., "Experimental Support for Generalized Equation Predicting Low Cycle Fatigue," and S. S. Manson, discussion, *Trans. ASME, J. Basic Eng.,* vol. 84, no. 4, pp. 533–537.

[6]See also, Landgraf, Ibid.

[7]For further discussion of the strain-life method see N. E. Dowling, *Mechanical Behavior of Materials,* 3rd ed., Prentice Hall, Upper Saddle River, N.J., 2007, Chap. 14.

from an electron microscope. The growth of the crack is orderly. Final fracture occurs during stage III fatigue, although fatigue is not involved. When the crack is sufficiently long that $K_I = K_{Ic}$ for the stress amplitude involved, where K_{Ic} is the critical stress intensity for the undamaged metal, then there is sudden, catastrophic failure of the remaining cross section in tensile overload (see Sec. 5–12). Stage III fatigue is associated with rapid acceleration of crack growth then fracture.

Crack Growth

Fatigue cracks nucleate and grow when stresses vary and there is some tension in each stress cycle. Consider the stress to be fluctuating between the limits of σ_{min} and σ_{max}, where the stress range is defined as $\Delta\sigma = \sigma_{max} - \sigma_{min}$. From Eq. (5–37) the stress intensity is given by $K_I = \beta\sigma\sqrt{\pi a}$. Thus, for $\Delta\sigma$, the stress intensity range per cycle is

$$\Delta K_I = \beta(\sigma_{max} - \sigma_{min})\sqrt{\pi a} = \beta\Delta\sigma\sqrt{\pi a} \tag{6–4}$$

To develop fatigue strength data, a number of specimens of the same material are tested at various levels of $\Delta\sigma$. Cracks nucleate at or very near a free surface or large discontinuity. Assuming an initial crack length of a_i, crack growth as a function of the number of stress cycles N will depend on $\Delta\sigma$, that is, ΔK_I. For ΔK_I below some threshold value $(\Delta K_I)_{th}$ a crack will not grow. Figure 6–14 represents the crack length a as a function of N for three stress levels $(\Delta\sigma)_3 > (\Delta\sigma)_2 > (\Delta\sigma)_1$, where $(\Delta K_I)_3 > (\Delta K_I)_2 > (\Delta K_I)_1$ for a given crack size. Notice the effect of the higher stress range in Fig. 6–14 in the production of longer cracks at a particular cycle count.

When the rate of crack growth per cycle, da/dN in Fig. 6–14, is plotted as shown in Fig. 6–15, the data from all three stress range levels superpose to give a sigmoidal curve. The three stages of crack development are observable, and the stage II data are linear on log-log coordinates, within the domain of linear elastic fracture mechanics (LEFM) validity. A group of similar curves can be generated by changing the stress ratio $R = \sigma_{min}/\sigma_{max}$ of the experiment.

Here we present a simplified procedure for estimating the remaining life of a cyclically stressed part after the discovery of a crack. This requires the assumption

Figure 6–14

The increase in crack length a from an initial length of a_i as a function of cycle count for three stress ranges, $(\Delta\sigma)_3 > (\Delta\sigma)_2 > (\Delta\sigma)_1$.

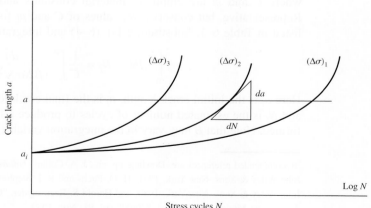

Figure 6–15

When da/dN is measured in Fig. 6–14 and plotted on log-log coordinates, the data for different stress ranges *superpose*, giving rise to a sigmoid curve as shown. $(\Delta K_\mathrm{I})_\mathrm{th}$ is the threshold value of ΔK_I, below which a crack does not grow. From threshold to rupture an aluminum alloy will spend 85–90 percent of life in region I, 5–8 percent in region II, and 1–2 percent in region III.

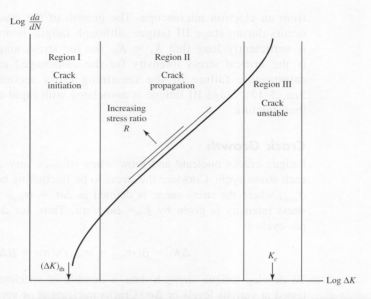

Table 6–1

Conservative Values of Factor C and Exponent m in Eq. (6–5) for Various Forms of Steel ($R = \sigma_\mathrm{max}/\sigma_\mathrm{min} \approx 0$)

Material	C, $\dfrac{\text{m/cycle}}{(\text{MPa}\sqrt{\text{m}})^m}$	C, $\dfrac{\text{in/cycle}}{(\text{kpsi}\sqrt{\text{in}})^m}$	m
Ferritic-pearlitic steels	$6.89(10^{-12})$	$3.60(10^{-10})$	3.00
Martensitic steels	$1.36(10^{-10})$	$6.60(10^{-9})$	2.25
Austenitic stainless steels	$5.61(10^{-12})$	$3.00(10^{-10})$	3.25

From J. M. Barsom and S. T. Rolfe, *Fatigue and Fracture Control in Structures, 2nd ed.,* Prentice Hall, Upper Saddle River, NJ, 1987, pp. 288–291, Copyright ASTM International. Reprinted with permission.

that plane strain conditions prevail.[8] Assuming a crack is discovered early in stage II, the crack growth in region II of Fig. 6–15 can be approximated by the *Paris equation,* which is of the form

$$\frac{da}{dN} = C(\Delta K_\mathrm{I})^m \tag{6–5}$$

where C and m are empirical material constants and ΔK_I is given by Eq. (6–4). Representative, but conservative, values of C and m for various classes of steels are listed in Table 6–1. Substituting Eq. (6–4) and integrating gives

$$\int_0^{N_f} dN = N_f = \frac{1}{C}\int_{a_i}^{a_f} \frac{da}{(\beta\,\Delta\sigma\sqrt{\pi a})^m} \tag{6–6}$$

Here a_i is the initial crack length, a_f is the final crack length corresponding to failure, and N_f is the estimated number of cycles to produce a failure *after* the initial crack is formed. Note that β may vary in the integration variable (e.g., see Figs. 5–25 to 5–30).

---

[8]Recommended references are: Dowling, op. cit.; J. A. Collins, *Failure of Materials in Mechanical Design,* John Wiley & Sons, New York, 1981; H. O. Fuchs and R. I. Stephens, *Metal Fatigue in Engineering,* John Wiley & Sons, New York, 1980; and Harold S. Reemsnyder, "Constant Amplitude Fatigue Life Assessment Models," *SAE Trans. 820688,* vol. 91, Nov. 1983.

If this should happen, then Reemsnyder[9] suggests the use of numerical integration employing the algorithm

$$\delta a_j = C(\Delta K_I)_j^m (\delta N)_j$$
$$a_{j+1} = a_j + \delta a_j$$
$$N_{j+1} = N_j + \delta N_j \qquad (6\text{-}7)$$
$$N_f = \sum \delta N_j$$

Here δa_j and δN_j are increments of the crack length and the number of cycles. The procedure is to select a value of δN_j, using a_i determine β and compute ΔK_I, determine δa_j, and then find the next value of a. Repeat the procedure until $a = a_f$.

The following example is highly simplified with β constant in order to give some understanding of the procedure. Normally, one uses fatigue crack growth computer programs such as NASA/FLAGRO 2.0 with more comprehensive theoretical models to solve these problems.

EXAMPLE 6-1

The bar shown in Fig. 6–16 is subjected to a repeated moment $0 \leq M \leq 135$ N·m. The bar is AISI 4430 steel with $S_{ut} = 1.28$ GPa, $S_y = 1.17$ GPa, and $K_{Ic} = 81$ MPa $\sqrt{m}$. Material tests on various specimens of this material with identical heat treatment indicate worst-case constants of $C = 114 \times 10^{-15}$ (m/cycle)/(MPa $\sqrt{m}$)^m and $m = 3.0$. As shown, a nick of size 0.1 mm has been discovered on the bottom of the bar. Estimate the number of cycles of life remaining.

Solution

The stress range $\Delta\sigma$ is always computed by using the nominal (uncracked) area. Thus

$$\frac{I}{c} = \frac{bh^2}{6} = \frac{0.006(0.012)^2}{6} = 144 \times 10^{-9} \text{ m}^3$$

Therefore, before the crack initiates, the stress range is

$$\Delta\sigma = \frac{\Delta M}{I/c} = \frac{135}{144 \times 10^{-9}} = 937.5 \text{ MPa}$$

which is below the yield strength. As the crack grows, it will eventually become long enough such that the bar will completely yield or undergo a brittle fracture. For the ratio of S_y/S_{ut} it is highly unlikely that the bar will reach complete yield. For brittle fracture, designate the crack length as a_f. If $\beta = 1$, then from Eq. (5–37) with $K_I = K_{Ic}$, we approximate a_f as

$$a_f = \frac{1}{\pi}\left(\frac{K_{Ic}}{\beta\sigma_{max}}\right)^2 \doteq \frac{1}{\pi}\left(\frac{81}{937.5}\right)^2 = 0.0024 \text{ m}$$

| Figure 6-16

M   M   6 mm   12 mm   Nick

[9]Op. cit.

From Fig. 5–27, we compute the ratio a_f/h as

$$\frac{a_f}{h} = \frac{0.0024}{0.012} = 0.2$$

Thus a_f/h varies from near zero to approximately 0.2. From Fig. 5–27, for this range β is nearly constant at approximately 1.05. We will assume it to be so, and re-evaluate a_f as

$$a_f = \frac{1}{\pi}\left(\frac{81}{1.05(937.5)}\right)^2 = 0.00216 \text{ m}$$

Thus, from Eq. (6–6), the estimated remaining life is

$$N_f = \frac{1}{C}\int_{a_i}^{a_f} \frac{da}{(\beta\Delta\sigma\sqrt{\pi a})^m} = \frac{1}{114 \times 10^{-15}}\int_{0.0001}^{0.00216} \frac{da}{[1.05(937.5)\sqrt{\pi a}]^3}$$

$$= -\frac{825.8}{\sqrt{a}}\Bigg|_{0.0001}^{0.00216} = 64.8(10^3) \text{ cycles}$$

6–7 The Endurance Limit

The determination of endurance limits by fatigue testing is now routine, though a lengthy procedure. Generally, stress testing is preferred to strain testing for endurance limits.

For preliminary and prototype design and for some failure analysis as well, a quick method of estimating endurance limits is needed. There are great quantities of data in the literature on the results of rotating-beam tests and simple tension tests of specimens taken from the same bar or ingot. By plotting these as in Fig. 6–17, it is possible to see whether there is any correlation between the two sets of results. The graph appears to suggest that the endurance limit ranges from about 40 to 60 percent of the tensile strength for steels up to about 210 kpsi (1450 MPa). Beginning at about $S_{ut} = 210$ kpsi (1450 MPa), the scatter appears to increase, but the trend seems to level off, as suggested by the dashed horizontal line at $S'_e = 105$ kpsi (735 MPa).

We wish now to present a method for estimating endurance limits. Note that estimates obtained from quantities of data obtained from many sources probably have a large spread and might deviate significantly from the results of actual laboratory tests of the mechanical properties of specimens obtained through strict purchase-order specifications. Since the area of uncertainty is greater, compensation must be made by employing larger design factors than would be used for static design.

For steels, simplifying our observation of Fig. 6–17, we will estimate the endurance limit as

$$S'_e = \begin{cases} 0.5S_{ut} & S_{ut} \le 200 \text{ kpsi (1400 MPa)} \\ 100 \text{ kpsi} & S_{ut} > 200 \text{ kpsi} \\ 700 \text{ MPa} & S_{ut} > 1400 \text{ MPa} \end{cases} \qquad (6\text{–}8)$$

where S_{ut} is the *minimum* tensile strength. The prime mark on S'_e in this equation refers to the *rotating-beam specimen* itself. We wish to reserve the unprimed symbol S_e for the endurance limit of an actual machine element subjected to any kind of loading. Soon we shall learn that the two strengths may be quite different.

Figure 6–17

Graph of endurance limits versus tensile strengths from actual test results for a large number of wrought irons and steels. Ratios of S'_e/S_{ut} of 0.60, 0.50, and 0.40 are shown by the solid and dashed lines. Note also the horizontal dashed line for S'_e = 735 MPa. Points shown having a tensile strength greater than 1470 MPa have a mean endurance limit of S'_e = 735 MPa. and a standard deviation of 95 MPa.
(*Collated from data compiled by H. J. Grover, S. A. Gordon, and L. R. Jackson in* Fatigue of Metals and Structures, *Bureau of Naval Weapons Document NAVWEPS 00-25-534, 1960; and from* Fatigue Design Handbook, *SAE, 1968, p. 42.)*

Steels treated to give different microstructures have different S'_e/S_{ut} ratios. It appears that the more ductile microstructures have a higher ratio. Martensite has a very brittle nature and is highly susceptible to fatigue-induced cracking; thus the ratio is low. When designs include detailed heat-treating specifications to obtain specific micro structures, it is possible to use an estimate of the endurance limit based on test data for the particular microstructure; such estimates are much more reliable and indeed should be used.

The endurance limits for various classes of cast irons, polished or machined, are given in Table A–24. Aluminum alloys do not have an endurance limit. The fatigue strengths of some aluminum alloys at $5(10^8)$ cycles of reversed stress are given in Table A–24.

6–8 Fatigue Strength

As shown in Fig. 6–10, a region of low-cycle fatigue extends from $N = 1$ to about 10^3 cycles. In this region the fatigue strength S_f is only slightly smaller than the tensile strength S_{ut}. An analytical approach has been given by Shigley, Mischke, and Brown[10]

---

[10]J. E. Shigley, C. R. Mischke, and T. H. Brown, Jr., *Standard Handbook of Machine Design,* 3rd ed., McGraw-Hill, New York, 2004, pp. 29.25–29.27.

for both high-cycle and low-cycle regions, requiring the parameters of the Manson-Coffin equation plus the strain-strengthening exponent m. Engineers often have to work with less information.

Figure 6–10 indicates that the high-cycle fatigue domain extends from 10^3 cycles for steels to the endurance limit life N_e, which is about 10^6 to 10^7 cycles. The purpose of this section is to develop methods of approximation of the S-N diagram in the high-cycle region, when information may be as sparse as the results of a simple tension test. Experience has shown high-cycle fatigue data are rectified by a logarithmic transform to both stress and cycles-to-failure. Equation (6–2) can be used to determine the fatigue strength at 10^3 cycles. Defining the specimen fatigue strength at a specific number of cycles as $(S'_f)_N = E\Delta\varepsilon_e/2$, write Eq. (6–2) as

$$(S'_f)_N = \sigma'_F(2N)^b \tag{6–9}$$

At 10^3 cycles,

$$(S'_f)_{10^3} = \sigma'_F(2 \cdot 10^3)^b = f S_{ut}$$

where f is the fraction of S_{ut} represented by $(S'_f)_{10^3}$ cycles. Solving for f gives

$$f = \frac{\sigma'_F}{S_{ut}}(2 \cdot 10^3)^b \tag{6–10}$$

Now, from Eq. (2–15), $\sigma'_F = \sigma_0\varepsilon^m$, with $\varepsilon = \varepsilon'_F$. If this true-stress–true-strain equation is not known, the SAE approximation[11] for steels with $H_B \leq 500$ may be used:

$$\sigma'_F = S_{ut} + 50 \text{ kpsi} \qquad \text{or} \qquad \sigma'_F = S_{ut} + 345 \text{ MPa} \tag{6–11}$$

To find b, substitute the endurance strength and corresponding cycles, S'_e and N_e, respectively into Eq. (6–9) and solving for b

$$b = -\frac{\log(\sigma'_F/S'_e)}{\log(2N_e)} \tag{6–12}$$

Thus, the equation $S'_f = \sigma'_F(2N)^b$ is known. For example, if $S_{ut} = 735$ MPa and $S'_e = 366$ MPa with $N_e = 10^6$ cycles,

Eq. (6–11) $\qquad\qquad \sigma'_F = 735 + 345 = 1080$ MPa

Eq. (6–12) $\qquad\qquad b = -\dfrac{\log(1080/366)}{\log(2 \cdot 10^6)} = -0.0746$

Eq. (6–10) $\qquad\qquad f = \dfrac{1080}{735}(2 \cdot 10^3)^{-0.0746} = 0.833$

and for Eq. (6–9), with $S'_f = (S'_f)_N$,

$$S'_f = 1080(2N)^{-0.0746} = 1026\,N^{-0.0746} \tag{a}$$

---

[11]*Fatigue Design Handbook,* vol. 4, Society of Automotive Engineers, New York, 1958, p. 27.

Figure 6-18

Fatigue strength fraction, f, of S_{ut} at 10^3 cycles for $S_e = S'_e = 0.5S_{ut}$ at 10^6 cycles.

The process given for finding f can be repeated for various ultimate strengths. Figure 6–18 is a plot of f for $490 \leq S_{ut} \leq 1400$ MPa. To be conservative, for $S_{ut} < 350$ MPa, let $f = 0.9$.

For an *actual* mechanical component, S'_e is reduced to S_e (see Sec. 6–9) which is less than $0.5 \, S_{ut}$. However, unless actual data is available, we recommend using the value of f found from Fig. 6–18. Equation (a), for the actual mechanical component, can be written in the form

$$S_f = a N^b \tag{6-13}$$

where N is cycles to failure and the constants a and b are defined by the points 10^3, $(S_f)_{10^3}$ and 10^6, S_e with $(S_f)_{10^3} = f S_{ut}$. Substituting these two points in Eq. (6–13) gives

$$a = \frac{(f S_{ut})^2}{S_e} \tag{6-14}$$

$$b = -\frac{1}{3} \log \left(\frac{f S_{ut}}{S_e} \right) \tag{6-15}$$

If a *completely reversed* stress σ_{rev} is given, setting $S_f = \sigma_{\text{rev}}$ in Eq. (6–13), the number of cycles-to-failure can be expressed as

$$N = \left(\frac{\sigma_{\text{rev}}}{a} \right)^{1/b} \tag{6-16}$$

Note that the typical *S-N* diagram, and thus Eq. (6–16), is only applicable for completely reversed loading. For general fluctuating loading situations, it is necessary to obtain an equivalent, completely reversing, stress that is considered to be equally as damaging as the actual fluctuating stress (see Ex. 6–12, p. 321).

Low-cycle fatigue is often defined (see Fig. 6–10) as failure that occurs in a range of $1 \leq N \leq 10^3$ cycles. On a log-log plot such as Fig. 6–10 the failure locus in this range is nearly linear below 10^3 cycles. A straight line between 10^3, $f S_{ut}$ and 1, S_{ut} (transformed) is conservative, and it is given by

$$S_f \geq S_{ut} N^{(\log f)/3} \qquad 1 \leq N \leq 10^3 \tag{6-17}$$

EXAMPLE 6–2   Given a 1050 HR steel, *estimate*

(*a*) the rotating-beam endurance limit at 10^6 cycles.

(*b*) the endurance strength of a polished rotating-beam specimen corresponding to 10^4 cycles to failure

(*c*) the expected life of a polished rotating-beam specimen under a completely reversed stress of 385 MPa.

Solution   (*a*) From Table A–20, $S_{ut} = 630$ MPa. From Eq. (6–8),

Answer
$$S_e' = 0.5(630) = 315 \text{ MPa}$$

(*b*) From Fig. 6–18, for $S_{ut} = 630$ MPa, $f \doteq 0.86$. From Eq. (6–14),

$$a = \frac{[0.86(630)^2]}{315} = 1084 \text{ MPa}$$

From Eq. (6–15),

$$b = -\frac{1}{3} \log \left[\frac{0.86(630)}{315} \right] = -0.0785$$

Thus, Eq. (6–13) is

$$S_f' = 1084 \, N^{-0.0785}$$

Answer   For 10^4 cycles to failure, $S_f' = 1084(10^4)^{-0.0785} = 526$ MPa

(*c*) From Eq. (6–16), with $\sigma_a = 385$ MPa,

Answer
$$N = \left(\frac{385}{1084} \right)^{1/-0.0785} = 53.3(10^4) \text{cycles}$$

Keep in mind that these are only *estimates*. So expressing the answers using three-place accuracy is a little misleading.

6–9   **Endurance Limit Modifying Factors**

We have seen that the rotating-beam specimen used in the laboratory to determine endurance limits is prepared very carefully and tested under closely controlled conditions. It is unrealistic to expect the endurance limit of a mechanical or structural member to match the values obtained in the laboratory. Some differences include

- *Material:* composition, basis of failure, variability
- *Manufacturing:* method, heat treatment, fretting corrosion, surface condition, stress concentration
- *Environment:* corrosion, temperature, stress state, relaxation times
- *Design:* size, shape, life, stress state, speed, fretting, galling

Marin[12] identified factors that quantified the effects of surface condition, size, loading, temperature, and miscellaneous items. The question of whether to adjust the endurance limit by subtractive corrections or multiplicative corrections was resolved by an extensive statistical analysis of a 4340 (electric furnace, aircraft quality) steel, in which a correlation coefficient of 0.85 was found for the multiplicative form and 0.40 for the additive form. A Marin equation is therefore written as

$$S_e = k_a k_b k_c k_d k_e k_f S'_e \tag{6-18}$$

where k_a = surface condition modification factor

k_b = size modification factor

k_c = load modification factor

k_d = temperature modification factor

k_e = reliability factor[13]

k_f = miscellaneous-effects modification factor

S'_e = rotary-beam test specimen endurance limit

S_e = endurance limit at the critical location of a machine part in the geometry and condition of use

When endurance tests of parts are not available, estimations are made by applying Marin factors to the endurance limit.

Surface Factor k_a

The surface of a rotating-beam specimen is highly polished, with a final polishing in the axial direction to smooth out any circumferential scratches. The surface modification factor depends on the quality of the finish of the actual part surface and on the tensile strength of the part material. To find quantitative expressions for common finishes of machine parts (ground, machined, or cold-drawn, hot-rolled, and as-forged), the coordinates of data points were recaptured from a plot of endurance limit versus ultimate tensile strength of data gathered by Lipson and Noll and reproduced by Horger.[14] The data can be represented by

$$k_a = a S_{ut}^b \tag{6-19}$$

where S_{ut} is the minimum tensile strength and a and b are to be found in Table 6–2.

---

[12]Joseph Marin, *Mechanical Behavior of Engineering Materials,* Prentice Hall, Englewood Cliffs, N.J., 1962, p. 224.

[13]The reliability factor given here is to account for the scatter in the fatigue data only and is not part of a complete stochastic analysis. The presentation in this text is strictly of a deterministic nature. For a more in-depth stochastic discussion see R. G. Budynas and J. K. Nisbett, op. cit.

[14]C. J. Noll and C. Lipson, "Allowable Working Stresses," *Society for Experimental Stress Analysis,* vol. 3, no. 2, 1946, p. 29. Reproduced by O. J. Horger (ed.), *Metals Engineering Design ASME Handbook,* McGraw-Hill, New York, 1953, p. 102.

Table 6–2

Parameters for Marin Surface Modification Factor, Eq. (6–19)

Surface Finish	Factor a S_{ut}, kpsi	Factor a S_{ut}, MPa	Exponent b
Ground	1.34	1.58	−0.085
Machined or cold-drawn	2.70	4.51	−0.265
Hot-rolled	14.4	57.7	−0.718
As-forged	39.9	272.	−0.995

From C. J. Noll and C. Lipson, "Allowable Working Stresses," *Society for Experimental Stress Analysis,* vol. 3, no. 2, 1946 p. 29. Reproduced by O.J. Horger (ed.) *Metals Engineering Design ASME Handbook,* McGraw-Hill, New York. Copyright © 1953 by The McGraw-Hill Companies, Inc. Reprinted by permission.

EXAMPLE 6–3

A steel has a minimum ultimate strength of 520 MPa and a machined surface. Estimate k_a.

Solution

From Table 6–2, $a = 4.51$ and $b = -0.265$. Then, from Eq. (6–19)

Answer

$$k_a = 4.51(520)^{-0.265} = 0.860$$

Again, it is important to note that this is an approximation as the data is typically quite scattered. Furthermore, this is not a correction to take lightly. For example, if in the previous example the steel was forged, the correction factor would be 0.540, a significant reduction of strength.

Size Factor k_b

The size factor has been evaluated using 133 sets of data points.[15] The results for bending and torsion may be expressed as

$$k_b = \begin{cases} (d/0.3)^{-0.107} = 0.879d^{-0.107} & 0.11 \le d \le 2 \text{ in} \\ 0.91d^{-0.157} & 2 < d \le 10 \text{ in} \\ (d/7.62)^{-0.107} = 1.24d^{-0.107} & 2.79 \le d \le 51 \text{ mm} \\ 1.51d^{-0.157} & 51 < d \le 254 \text{ mm} \end{cases} \quad (6\text{–}20)$$

For axial loading there is no size effect, so

$$k_b = 1 \quad (6\text{–}21)$$

but see k_c.

One of the problems that arises in using Eq. (6–20) is what to do when a round bar in bending is not rotating, or when a noncircular cross section is used. For example, what is the size factor for a bar 6 mm thick and 40 mm wide? The approach to be

[15]Charles R. Mischke, "Prediction of Stochastic Endurance Strength," *Trans. of ASME, Journal of Vibration, Acoustics, Stress, and Reliability in Design,* vol. 109, no. 1, January 1987, Table 3.

used here employs an *equivalent diameter* d_e obtained by equating the volume of material stressed at and above 95 percent of the maximum stress to the same volume in the rotating-beam specimen.[16] It turns out that when these two volumes are equated, the lengths cancel, and so we need only consider the areas. For a rotating round section, the 95 percent stress area is the area in a ring having an outside diameter d and an inside diameter of $0.95d$. So, designating the 95 percent stress area $A_{0.95\sigma}$, we have

$$A_{0.95\sigma} = \frac{\pi}{4}[d^2 - (0.95d)^2] = 0.0766d^2 \tag{6-22}$$

This equation is also valid for a rotating hollow round. For nonrotating solid or hollow rounds, the 95 percent stress area is twice the area outside of two parallel chords having a spacing of $0.95d$, where d is the diameter. Using an exact computation, this is

$$A_{0.95\sigma} = 0.01046d^2 \tag{6-23}$$

With d_e in Eq. (6–22), setting Eqs. (6–22) and (6–23) equal to each other enables us to solve for the effective diameter. This gives

$$d_e = 0.370d \tag{6-24}$$

as the effective size of a round corresponding to a nonrotating solid or hollow round.

A rectangular section of dimensions $h \times b$ has $A_{0.95\sigma} = 0.05hb$. Using the same approach as before,

$$d_e = 0.808(hb)^{1/2} \tag{6-25}$$

Table 6–3 provides $A_{0.95\sigma}$ areas of common structural shapes undergoing nonrotating bending.

EXAMPLE 6–4   A steel shaft loaded in bending is 32 mm in diameter, abutting a filleted shoulder 38 mm in diameter. The shaft material has a mean ultimate tensile strength of 690 MPa. Estimate the Marin size factor k_b if the shaft is used in

(*a*) A rotating mode.

(*b*) A nonrotating mode.

Solution   (*a*) From Eq. (6–20)

Answer $$k_b = \left(\frac{d}{7.62}\right)^{-0.107} = \left(\frac{32}{7.62}\right)^{-0.107} = 0.858$$

(*b*) From Table 6–3,

$$d_e = 0.37d = 0.37(32) = 11.84 \text{ mm}$$

From Eq. (6–20),

Answer $$k_b = \left(\frac{11.84}{7.62}\right)^{-0.107} = 0.954$$

---

[16]See R. Kuguel, "A Relation between Theoretical Stress-Concentration Factor and Fatigue Notch Factor Deduced from the Concept of Highly Stressed Volume," *Proc. ASTM*, vol. 61, 1961, pp. 732–748.

Table 6-3

$A_{0.95\sigma}$ Areas of Common Nonrotating Structural Shapes

$$A_{0.95\sigma} = 0.01046d^2$$
$$d_e = 0.370d$$

$$A_{0.95\sigma} = 0.05hb$$
$$d_e = 0.808\sqrt{hb}$$

$$A_{0.95\sigma} = \begin{cases} 0.10at_f & \text{axis 1-1} \\ 0.05ba & t_f > 0.025a \quad \text{axis 2-2} \end{cases}$$

$$A_{0.95\sigma} = \begin{cases} 0.05ab & \text{axis 1-1} \\ 0.052xa + 0.1t_f(b-x) & \text{axis 2-2} \end{cases}$$

Loading Factor k_c

Estimates of endurance limit, such as that given in Eq. (6–8), are typically obtained from testing with completely reversed bending. With axial or torsional loading, fatigue tests indicate different relationships between the endurance limit and the ultimate strength for each type of loading.[17] These differences can be accounted for with a load factor to adjust the endurance limit obtained from bending. Though the load factor is actually a function of the ultimate strength, the variation is minor, so it is appropriate to specify average values of the load factor as

$$k_c = \begin{cases} 1 & \text{bending} \\ 0.85 & \text{axial} \\ 0.59 & \text{torsion} \end{cases} \tag{6-26}$$

Note that the load factor for torsion is very close to the prediction from the distortion energy theory, Eq. (5–21), where for ductile materials the shear strength is 0.577 times the normal strength. This implies that the load factor for torsion is mainly accounting for the difference in shear strength versus normal strength. Therefore, use the torsion load factor *only* for pure torsional fatigue loading. When torsion is combined with

---

[17]H. J. Grover, S. A. Gordon, and L. R. Jackson, *Fatigue of Metals and Structures*, Bureau of Naval Weapons, Document NAVWEPS 00-2500435, 1960; R. G. Budynas and J. K. Nisbett, op. cit., pp. 332–333.

Table 6–4

Effect of Operating Temperature on the Tensile Strength of Steel.* (S_T = tensile strength at operating temperature; S_{RT} = tensile strength at room temperature; $0.099 \leq \hat{\sigma} \leq 0.110$)

Temperature, °C	S_T/S_{RT}	Temperature, °F	S_T/S_{RT}
20	1.000	70	1.000
50	1.010	100	1.008
100	1.020	200	1.020
150	1.025	300	1.024
200	1.020	400	1.018
250	1.000	500	0.995
300	0.975	600	0.963
350	0.943	700	0.927
400	0.900	800	0.872
450	0.843	900	0.797
500	0.768	1000	0.698
550	0.672	1100	0.567
600	0.549		

*Data source: Fig. 2–9.

other loading, such as bending, set $k_c = 1$, and the combined loading is managed by using the effective von Mises stress, as described in Sec. 6–14.

Temperature Factor k_d

When operating temperatures are below room temperature, brittle fracture is a strong possibility and should be investigated first. When the operating temperatures are higher than room temperature, yielding should be investigated first because the yield strength drops off so rapidly with temperature; see Fig. 2–9, p. 55. Any stress will induce creep in a material operating at high temperatures; so this factor must be considered too. Finally, it may be true that there is no fatigue limit for materials operating at high temperatures. Because of the reduced fatigue resistance, the failure process is, to some extent, dependent on time.

The limited amount of data available show that the endurance limit for steels increases slightly as the temperature rises and then begins to fall off in the 400 to 700°F range, not unlike the behavior of the tensile strength shown in Fig. 2–9. For this reason it is probably true that the endurance limit is related to tensile strength at elevated temperatures in the same manner as at room temperature.[18] It seems quite logical, therefore, to employ the same relations to predict endurance limit at elevated temperatures as are used at room temperature, at least until more comprehensive data become available. At the very least, this practice will provide a useful standard against which the performance of various materials can be compared.

Table 6–4 has been obtained from Fig. 2–9 by using only the tensile-strength data. Note that the table represents 145 tests of 21 different carbon and alloy steels. A fourth-order polynomial curve fit to the data underlying Fig. 2–9 gives

$$k_d = 0.975 + 0.432(10^{-3})T_F - 0.115(10^{-5})T_F^2$$
$$+ 0.104(10^{-8})T_F^3 - 0.595(10^{-12})T_F^4 \tag{6-27}$$

where $70 \leq T_F \leq 1000°F$.

---

[18]For more, see Table 2 of ANSI/ASME B106. 1M-1985 shaft standard, and E. A. Brandes (ed.), *Smithell's Metals Reference Book,* 6th ed., Butterworth, London, 1983, pp. 22–134 to 22–136, where endurance limits from 100 to 650°C are tabulated.

Two types of problems arise when temperature is a consideration. If the rotating-beam endurance limit is known at room temperature, then use

$$k_d = \frac{S_T}{S_{RT}} \qquad (6\text{--}28)$$

from Table 6–4 or Eq. (6–27) and proceed as usual. If the rotating-beam endurance limit is not given, then compute it using Eq. (6–8) and the temperature-corrected tensile strength obtained by using the factor from Table 6–4. Then use $k_d = 1$.

EXAMPLE 6–5

A 1035 steel has a tensile strength of 490 MPa and is to be used for a part that sees 230°C in service. Estimate the Marin temperature modification factor and $(S_e)_{230°}$ if
(a) The room-temperature endurance limit by test is $(S_e')_{37°} = 270$ MPa.
(b) Only the tensile strength at room temperature is known.

Solution

(a) First, from Eq. (6–27),

$$k_d = 0.9877 + 0.6507(10^{-3})(230) - 0.3414(10^{-5})(230^2)$$
$$+ 0.5621(10^{-8})(230^3) - 6.246(10^{-12})(230^4) = 1.00767$$

Thus,

Answer

$$(S_e)_{230°} = k_d\,(S_e')_{37°} = 1.00767(270) = 272.07 \text{ MPa}$$

(b) Interpolating from Table 6–4 gives

$$(S_T/S_{RT})_{230°} = 1.02 + (1.0 - 1.02)\frac{230 - 200}{250 - 200} = 1.0197$$

Thus, the tensile strength at 230°C is estimated as

$$(S_{ut})_{230°} = (S_T / S_{RT})_{230°}\,(S_{ut})_{37°} = 1.0197(490) = 499.7 \text{ MPa}$$

From Eq. (6–8) then,

Answer

$$(S_e)_{230°} = 0.5(S_{ut})_{230°} = 0.5(499.7) = 249.9 \text{ MPa}$$

Part *a* gives the better estimate due to actual testing of the particular material.

Reliability Factor k_e

The discussion presented here accounts for the scatter of data such as shown in Fig. 6–17 where the mean endurance limit is shown to be $S_e'/S_{ut} \approx 0.5$, or as given by Eq. (6–8). Most endurance strength data are reported as mean values. Data presented by Haugen and Wirching[19] show standard deviations of endurance strengths of less than 8 percent. Thus the reliability modification factor to account for this can be written as

$$k_e = 1 - 0.08\,z_a \qquad (6\text{--}29)$$

---

[19]E. B. Haugen and P. H. Wirsching, "Probabilistic Design," *Machine Design,* vol. 47, no. 12, 1975, pp. 10–14.

Table 6–5

Reliability Factors k_e
Corresponding to
8 Percent Standard
Deviation of the
Endurance Limit

Reliability, %	Transformation Variate z_a	Reliability Factor k_e
50	0	1.000
90	1.288	0.897
95	1.645	0.868
99	2.326	0.814
99.9	3.091	0.753
99.99	3.719	0.702
99.999	4.265	0.659
99.9999	4.753	0.620

Figure 6–19

The failure of a case-hardened
part in bending or torsion. In
this example, failure occurs
in the core.

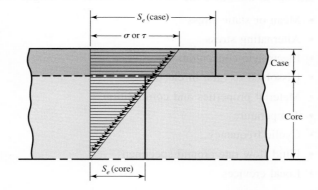

where z_a is defined by Eq. (1–5) and values for any desired reliability can be determined from Table A–10. Table 6–5 gives reliability factors for some standard specified reliabilities.

Miscellaneous-Effects Factor k_f

Though the factor k_f is intended to account for the reduction in endurance limit due to all other effects, it is really intended as a reminder that these must be accounted for, because actual values of k_f are not always available.

Residual stresses may either improve the endurance limit or affect it adversely. Generally, if the residual stress in the surface of the part is compression, the endurance limit is improved. Fatigue failures appear to be tensile failures, or at least to be caused by tensile stress, and so anything that reduces tensile stress will also reduce the possibility of a fatigue failure. Operations such as shot peening, hammering, and cold rolling build compressive stresses into the surface of the part and improve the endurance limit significantly. Of course, the material must not be worked to exhaustion.

The endurance limits of parts that are made from rolled or drawn sheets or bars, as well as parts that are forged, may be affected by the so-called *directional characteristics* of the operation. Rolled or drawn parts, for example, have an endurance limit in the transverse direction that may be 10 to 20 percent less than the endurance limit in the longitudinal direction.

Parts that are case-hardened may fail at the surface or at the maximum core radius, depending upon the stress gradient. Figure 6–19 shows the typical triangular stress distribution of a bar under bending or torsion. Also plotted as a heavy line in this figure are the endurance limits S_e for the case and core. For this example the endurance limit of the core rules the design because the figure shows that the stress σ or τ, whichever applies, at the outer core radius, is appreciably larger than the core endurance limit.

Corrosion

It is to be expected that parts that operate in a corrosive atmosphere will have a lowered fatigue resistance. This is, of course, true, and it is due to the roughening or pitting of the surface by the corrosive material. But the problem is not so simple as the one of finding the endurance limit of a specimen that has been corroded. The reason for this is that the corrosion and the stressing occur at the same time. Basically, this means that in time any part will fail when subjected to repeated stressing in a corrosive atmosphere. There is no fatigue limit. Thus the designer's problem is to attempt to minimize the factors that affect the fatigue life; these are:

- Mean or static stress
- Alternating stress
- Electrolyte concentration
- Dissolved oxygen in electrolyte
- Material properties and composition
- Temperature
- Cyclic frequency
- Fluid flow rate around specimen
- Local crevices

Electrolytic Plating

Metallic coatings, such as chromium plating, nickel plating, or cadmium plating, reduce the endurance limit by as much as 50 percent. In some cases the reduction by coatings has been so severe that it has been necessary to eliminate the plating process. Zinc plating does not affect the fatigue strength. Anodic oxidation of light alloys reduces bending endurance limits by as much as 39 percent but has no effect on the torsional endurance limit.

Metal Spraying

Metal spraying results in surface imperfections that can initiate cracks. Limited tests show reductions of 14 percent in the fatigue strength.

Cyclic Frequency

If, for any reason, the fatigue process becomes time-dependent, then it also becomes frequency-dependent. Under normal conditions, fatigue failure is independent of frequency. But when corrosion or high temperatures, or both, are encountered, the cyclic rate becomes important. The slower the frequency and the higher the temperature, the higher the crack propagation rate and the shorter the life at a given stress level.

Frettage Corrosion

The phenomenon of frettage corrosion is the result of microscopic motions of tightly fitting parts or structures. Bolted joints, bearing-race fits, wheel hubs, and any set of tightly fitted parts are examples. The process involves surface discoloration, pitting, and eventual fatigue. The frettage factor k_f depends upon the material of the mating pairs and ranges from 0.24 to 0.90.

6–10 Stress Concentration and Notch Sensitivity

In Sec. 3–13 it was pointed out that the existence of irregularities or discontinuities, such as holes, grooves, or notches, in a part increases the theoretical stresses significantly in the immediate vicinity of the discontinuity. Equation (3–48) defined a stress-concentration factor K_t (or K_{ts}), which is used with the nominal stress to obtain the maximum resulting stress due to the irregularity or defect. It turns out that some materials are not fully sensitive to the presence of notches and hence, for these, a reduced value of K_t can be used. For these materials, the effective maximum stress in fatigue is,

$$\sigma_{\max} = K_f \sigma_0 \qquad \text{or} \qquad \tau_{\max} = K_{fs} \tau_0 \qquad (6\text{–}30)$$

where K_f is a reduced value of K_t and σ_0 is the nominal stress. The factor K_f is commonly called a *fatigue stress-concentration factor,* and hence the subscript f. So it is convenient to think of K_f as a stress-concentration factor reduced from K_t because of lessened sensitivity to notches. The resulting factor is defined by the equation

$$K_f = \frac{\text{maximum stress in notched specimen}}{\text{stress in notch-free specimen}} \qquad (a)$$

Notch sensitivity q is defined by the equation

$$q = \frac{K_f - 1}{K_t - 1} \qquad \text{or} \qquad q_{\text{shear}} = \frac{K_{fs} - 1}{K_{ts} - 1} \qquad (6\text{–}31)$$

where q is usually between zero and unity. Equation (6–31) shows that if $q = 0$, then $K_f = 1$, and the material has no sensitivity to notches at all. On the other hand, if $q = 1$, then $K_f = K_t$, and the material has full notch sensitivity. In analysis or design work, find K_t first, from the geometry of the part. Then specify the material, find q, and solve for K_f from the equation

$$K_f = 1 + q(K_t - 1) \qquad \text{or} \qquad K_{fs} = 1 + q_{\text{shear}}(K_{ts} - 1) \qquad (6\text{–}32)$$

Notch sensitivities for specific materials are obtained experimentally. Published experimental values are limited, but some values are available for steels and aluminum. Trends for notch sensitivity as a function of notch radius and ultimate strength are shown in Fig. 6–20 for reversed bending or axial loading, and Fig. 6–21 for reversed torsion. In using these charts it is well to know that the actual test results

Figure 6–20

Notch-sensitivity charts for steels and UNS A92024-T wrought aluminum alloys subjected to reversed bending or reversed axial loads. For larger notch radii, use the values of q corresponding to the $r = 0.16$-in (4-mm) ordinate. *(From George Sines and J. L. Waisman (eds.),* Metal Fatigue, *McGraw-Hill, New York. Copyright © 1969 by The McGraw-Hill Companies, Inc. Reprinted by permission.)*

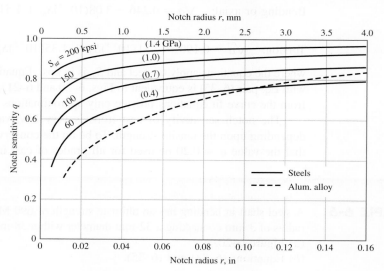

Figure 6–21

Notch-sensitivity curves for materials in reversed torsion. For larger notch radii, use the values of q_{shear} corresponding to $r = 0.16$ in (4 mm).

from which the curves were derived exhibit a large amount of scatter. Because of this scatter it is always safe to use $K_f = K_t$ if there is any doubt about the true value of q. Also, note that q is not far from unity for large notch radii.

Figure 6–20 has as its basis the *Neuber equation,* which is given by

$$K_f = 1 + \frac{K_t - 1}{1 + \sqrt{a/r}} \tag{6–33}$$

where $\sqrt{a}$ is defined as the *Neuber constant* and is a material constant. Equating Eqs. (6–31) and (6–33) yields the notch sensitivity equation

$$q = \frac{1}{1 + \dfrac{\sqrt{a}}{\sqrt{r}}} \tag{6–34}$$

correlating with Figs. 6–20 and 6–21 as

Bending or axial: $\sqrt{a} = 0.246 - 3.08(10^{-3})S_{ut} + 1.51(10^{-5})S_{ut}^2 - 2.67(10^{-8})S_{ut}^3$
$$\tag{6–35a}$$

Torsion: $\sqrt{a} = 0.190 - 2.51(10^{-3})S_{ut} + 1.35(10^{-5})S_{ut}^2 - 2.67(10^{-8})S_{ut}^3$ (6–35b)

where the equations apply to steel and S_{ut} is in kpsi. Equation (6–34) used in conjunction with Eq. pair (6–35) is equivalent to Figs. 6–20 and 6–21. As with the graphs, the results from the curve fit equations provide only approximations to the experimental data.

The notch sensitivity of cast irons is very low, varying from 0 to about 0.20, depending upon the tensile strength. To be on the conservative side, it is recommended that the value $q = 0.20$ be used for all grades of cast iron.

EXAMPLE 6–6   A steel shaft in bending has an ultimate strength of 690 MPa and a shoulder with a fillet radius of 3 mm connecting a 32-mm diameter with a 38-mm diameter. Estimate K_f using:
(a) Figure 6–20.
(b) Equations (6–33) and (6–35).

Solution From Fig. A–15–9, using $D/d = 38/32 = 1.1875$, $r/d = 3/32 = 0.093\,75$, we read the graph to find $K_t = 1.65$.

(a) From Fig. 6–20, for $S_{ut} = 690$ MPa and $r = 3$ mm, $q = 0.84$. Thus, from Eq. (6–32)

Answer
$$K_f = 1 + q(K_t - 1) = 1 + 0.84(1.65 - 1) = 1.55$$

(b) From Eq. (6–35a) with $S_{ut} = 690$ MPa $= 100$ kpsi, $\sqrt{a} = 0.0622\sqrt{\text{in}} = 0.313\sqrt{\text{mm}}$. Substituting this into Eq. (6–33) with $r = 3$ mm gives

Answer
$$K_f = 1 + \frac{K_t - 1}{1 + \sqrt{a/r}} = 1 + \frac{1.65 - 1}{1 + \frac{0.313}{\sqrt{3}}} = 1.55$$

Some designers use $1/K_f$ as a Marin factor to reduce S_e. For simple loading, infinite life problems, it makes no difference whether S_e is reduced by dividing it by K_f or the nominal stress is multiplied by K_f. However, for *finite life*, since the *S-N* diagram is nonlinear, the two approaches yield differing results. There is no clear evidence pointing to which method is better. Furthermore, in Sec. 6–14, when we consider combining loads, there generally are multiple fatigue stress-concentration factors occurring at a point (e.g., K_f for bending and K_{fs} for torsion). Here, it is only practical to modify the nominal stresses. To be consistent in this text, we will exclusively use the fatigue stress-concentration factor as a multiplier of the nominal stress.

EXAMPLE 6–7 For the step-shaft of Ex. 6–6, it is determined that the fully corrected endurance limit is $S_e = 280$ MPa. Consider the shaft undergoes a fully reversing nominal stress in the fillet of $(\sigma_{rev})_{nom} = 260$ MPa. Estimate the number of cycles to failure.

Solution From Ex. 6–6, $K_f = 1.55$, and the ultimate strength is $S_{ut} = 690$ MPa $= 100$ kpsi. The maximum reversing stress is

$$(\sigma_{rev})_{max} = K_f(\sigma_{rev})_{nom} = 1.55(260) = 403 \text{ MPa}$$

From Fig. 6–18, $f = 0.845$. From Eqs. (6–14), (6–15), and (6–16)

$$a = \frac{(fS_{ut})^2}{S_e} = \frac{[0.845(690)]^2}{280} = 1214 \text{ MPa}$$

$$b = -\frac{1}{3}\log\frac{fS_{ut}}{S_e} = -\frac{1}{3}\log\left[\frac{0.845(690)}{280}\right] = -0.1062$$

Answer
$$N = \left(\frac{\sigma_{rev}}{a}\right)^{1/b} = \left(\frac{403}{1214}\right)^{1/-0.1062} = 32.3(10^3) \text{ cycles}$$

Up to this point, examples illustrated each factor in Marin's equation and stress concentrations alone. Let us consider a number of factors occurring simultaneously.

EXAMPLE 6–8

A 1015 hot-rolled steel bar has been machined to a diameter of 25 mm. It is to be placed in reversed axial loading for 70 000 cycles to failure in an operating environment of 300°C. Using ASTM minimum properties, and a reliability of 99 percent, estimate the endurance limit and fatigue strength at 70 000 cycles.

Solution

From Table A–20, $S_{ut} = 340$ MPa at 20°C. Since the rotating-beam specimen endurance limit is not known at room temperature, we determine the ultimate strength at the elevated temperature first, using Table 6–4. From Table 6–4,

$$\left(\frac{S_T}{S_{RT}}\right)_{300°} = 0.975$$

The ultimate strength at 300°C is then

$$(S_{ut})_{300°} = (S_T / S_{RT})_{300°} (S_{ut})_{20°} = 0.975(340) = 331.5 \text{ MPa}$$

The rotating-beam specimen endurance limit at 300°C is then estimated from Eq. (6–8) as

$$S'_e = 0.5(331.5) = 165.8 \text{ MPa}$$

Next, we determine the Marin factors. For the machined surface, Eq. (6–19) with Table 6–2 gives

$$k_a = aS_{ut}^b = 4.51(331.5^{-0.265}) = 0.969$$

For axial loading, from Eq. (6–21), the size factor $k_b = 1$, and from Eq. (6–26) the loading factor is $k_c = 0.85$. The temperature factor $k_d = 1$, since we accounted for the temperature in modifying the ultimate strength and consequently the endurance limit. For 99 percent reliability, from Table 6–5, $k_e = 0.814$. Finally, since no other conditions were given, the miscellaneous factor is $k_f = 1$. The endurance limit for the part is estimated by Eq. (6–18) as

Answer

$$S_e = k_a k_b k_c k_d k_e k_f S'_e$$
$$= 0.969(1)(0.85)(1)(0.814)(1)165.8 = 111 \text{ MPa}$$

For the fatigue strength at 70 000 cycles we need to construct the S-N equation. From p. 285, since $S_{ut} = 331.5 < 490$ MPa, then $f = 0.9$. From Eq. (6–14)

$$a = \frac{(f S_{ut})^2}{S_e} = \frac{[0.9(331.5)]^2}{111} = 801.929$$

and Eq. (6–15)

$$b = -\frac{1}{3} \log\left(\frac{f S_{ut}}{S_e}\right) = -\frac{1}{3} \log\left[\frac{0.9(331.5)}{111}\right] = -0.1431$$

Finally, for the fatigue strength at 70 000 cycles, Eq. (6–13) gives

Answer

$$S_f = a N^b = 891(70\ 000)^{-0.1431} = 162.51 \text{ MPa}$$

EXAMPLE 6–9

Figure 6–22*a* shows a rotating shaft simply supported in ball bearings at *A* and *D* and loaded by a nonrotating force *F* of 6.8 kN. Using ASTM "minimum" strengths, estimate the life of the part.

Solution

From Fig. 6–22*b* we learn that failure will probably occur at *B* rather than at *C* or at the point of maximum moment. Point *B* has a smaller cross section, a higher bending moment, and a higher stress-concentration factor than *C*, and the location of maximum moment has a larger size and no stress-concentration factor.

We shall solve the problem by first estimating the strength at point *B*, since the strength will be different elsewhere, and comparing this strength with the stress at the same point.

From Table A–20 we find $S_{ut} = 690$ MPa and $S_y = 580$ MPa. The endurance limit S_e' is estimated as

$$S_e' = 0.5(690) = 345 \text{ MPa}$$

From Eq. (6–19) and Table 6–2,

$$k_a = 4.51(690)^{-0.265} = 0.798$$

From Eq. (6–20),

$$k_b = (32/7.62)^{-0.107} = 0.858$$

Since $k_c = k_d = k_e = k_f = 1$,

$$S_e = 0.798(0.858)345 = 236 \text{ MPa}$$

To find the geometric stress-concentration factor K_t we enter Fig. A–15–9 with $D/d = 38/32 = 1.1875$ and $r/d = 3/32 = 0.093\ 75$ and read $K_t = 1.65$. Substituting $S_{ut} = 690/6.89 = 100$ kpsi into Eq. (6–35*a*) yields $\sqrt{a} = 0.0622\sqrt{\text{in}} = 0.313\sqrt{\text{mm}}$. Substituting this into Eq. (6–33) gives

$$K_f = 1 + \frac{K_t - 1}{1 + \sqrt{a/r}} = 1 + \frac{1.65 - 1}{1 + 0.313/\sqrt{3}} = 1.55$$

Figure 6–22

(*a*) Shaft drawing showing all dimensions in millimeters; all fillets 3-mm radius. The shaft rotates and the load is stationary; material is machined from AISI 1050 cold-drawn steel. (*b*) Bending-moment diagram.

(*a*)

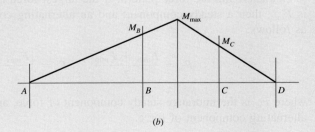

(*b*)

The next step is to estimate the bending stress at point B. The bending moment is

$$M_B = R_1 x = \frac{225F}{550}250 = \frac{225(6.8)}{550}250 = 695.5 \text{ N} \cdot \text{m}$$

Just to the left of B the section modulus is $I/c = \pi d^3/32 = \pi 32^3/32 = 3.217\,(10^3)\text{ mm}^3$. The reversing bending stress is, assuming infinite life,

$$\sigma_{\text{rev}} = K_f \frac{M_B}{I/c} = 1.55 \frac{695.5}{3.217}(10)^6 = 335.1(10^6)\text{ Pa} = 335.1 \text{ MPa}$$

This stress is greater than S_e and less than S_y. This means we have both finite life and no yielding on the first cycle.

For finite life, we will need to use Eq. (6–16). The ultimate strength, $S_{ut} = 690$ MPa $= 100$ kpsi. From Fig. 6–18, $f = 0.844$. From Eq. (6–14)

$$a = \frac{(fS_{ut})^2}{S_e} = \frac{[0.844(690)]^2}{236} = 1437 \text{ MPa}$$

and from Eq. (6–15)

$$b = -\frac{1}{3}\log\left(\frac{fS_{ut}}{S_e}\right) = -\frac{1}{3}\log\left[\frac{0.844(690)}{236}\right] = -0.1308$$

From Eq. (6–16),

Answer

$$N = \left(\frac{\sigma_{\text{rev}}}{a}\right)^{1/b} = \left(\frac{335.1}{1437}\right)^{-1/0.1308} = 68(10^3)\text{ cycles}$$

6–11 Characterizing Fluctuating Stresses

Fluctuating stresses in machinery often take the form of a sinusoidal pattern because of the nature of some rotating machinery. However, other patterns, some quite irregular, do occur. It has been found that in periodic patterns exhibiting a single maximum and a single minimum of force, the shape of the wave is not important, but the peaks on both the high side (maximum) and the low side (minimum) are important. Thus F_{max} and F_{min} in a cycle of force can be used to characterize the force pattern. It is also true that ranging above and below some baseline can be equally effective in characterizing the force pattern. If the largest force is F_{max} and the smallest force is F_{min}, then a steady component and an alternating component can be constructed as follows:

$$F_m = \frac{F_{\text{max}} + F_{\text{min}}}{2} \qquad F_a = \left|\frac{F_{\text{max}} - F_{\text{min}}}{2}\right|$$

where F_m is the midrange steady component of force, and F_a is the amplitude of the alternating component of force.

Figure 6-23

Some stress-time relations:
(a) fluctuating stress with high-frequency ripple; (b and c) nonsinusoidal fluctuating stress; (d) sinusoidal fluctuating stress; (e) repeated stress; (f) completely reversed sinusoidal stress.

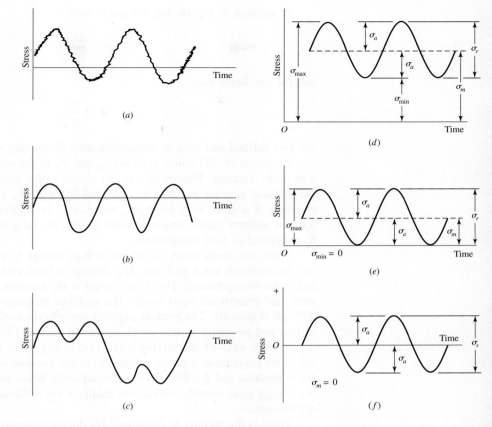

Figure 6–23 illustrates some of the various stress-time traces that occur. The components of stress, some of which are shown in Fig. 6–23d, are

σ_{min} = minimum stress σ_m = midrange component

σ_{max} = maximum stress σ_r = range of stress

σ_a = amplitude component σ_s = static or steady stress

The steady, or static, stress is *not* the same as the midrange stress; in fact, it may have any value between σ_{min} and σ_{max}. The steady stress exists because of a fixed load or preload applied to the part, and it is usually independent of the varying portion of the load. A helical compression spring, for example, is always loaded into a space shorter than the free length of the spring. The stress created by this initial compression is called the steady, or static, component of the stress. It is not the same as the midrange stress.

We shall have occasion to apply the subscripts of these components to shear stresses as well as normal stresses.

The following relations are evident from Fig. 6–23:

$$\sigma_m = \frac{\sigma_{max} + \sigma_{min}}{2}$$

$$\sigma_a = \left| \frac{\sigma_{max} - \sigma_{min}}{2} \right|$$

(6-36)

In addition to Eq. (6–36), the *stress ratio*

$$R = \frac{\sigma_{\min}}{\sigma_{\max}} \tag{6–37}$$

and the *amplitude ratio*

$$A = \frac{\sigma_a}{\sigma_m} \tag{6–38}$$

are also defined and used in connection with fluctuating stresses.

Equations (6–36) utilize symbols σ_a and σ_m as the stress components at the location under scrutiny. This means, in the absence of a notch, σ_a and σ_m are equal to the nominal stresses σ_{ao} and σ_{mo} induced by loads F_a and F_m, respectively; in the presence of a notch they are $K_f\sigma_{ao}$ and $K_f\sigma_{mo}$, respectively, as long as the material remains without plastic strain. In other words, the fatigue stress-concentration factor K_f is applied to *both* components.

When the steady stress component is high enough to induce localized notch yielding, the designer has a problem. The first-cycle local yielding produces plastic strain and strain-strengthening. This is occurring at the location where fatigue crack nucleation and growth are most likely. The material properties (S_y and S_{ut}) are new and difficult to quantify. The prudent engineer controls the concept, material and condition of use, and geometry so that no plastic strain occurs. There are discussions concerning possible ways of quantifying what is occurring under localized and general yielding in the presence of a notch, referred to as the *nominal mean stress* method, *residual stress* method, and the like.[20] The nominal mean stress method (set $\sigma_a = K_f\sigma_{ao}$ and $\sigma_m = \sigma_{mo}$) gives roughly comparable results to the residual stress method, but both are *approximations*.

There is the method of Dowling[21] for ductile materials, which, for materials with a pronounced yield point and approximated by an elastic–perfectly plastic behavior model, quantitatively expresses the steady stress component stress-concentration factor K_{fm} as

$$
\begin{aligned}
K_{fm} &= K_f & K_f|\sigma_{\max,o}| &< S_y \\
K_{fm} &= \frac{S_y - K_f\sigma_{ao}}{|\sigma_{mo}|} & K_f|\sigma_{\max,o}| &> S_y \\
K_{fm} &= 0 & K_f|\sigma_{\max,o} - \sigma_{\min,o}| &> 2S_y
\end{aligned} \tag{6–39}
$$

For the purposes of this book, for ductile materials in fatigue,

- Avoid localized plastic strain at a notch. Set $\sigma_a = K_f\sigma_{ao}$ and $\sigma_m = K_f\sigma_{mo}$.
- When plastic strain at a notch cannot be avoided, use Eqs. (6–39); or conservatively, set $\sigma_a = K_f\sigma_{ao}$ and use $K_{fm} = 1$, that is, $\sigma_m = \sigma_{mo}$.

---

[20]R. C. Juvinall, *Stress, Strain, and Strength,* McGraw-Hill, New York, 1967, articles 14.9–14.12; R. C. Juvinall and K. M. Marshek, *Fundamentals of Machine Component Design,* 4th ed., Wiley, New York, 2006, Sec. 8.11; M. E. Dowling, *Mechanical Behavior of Materials,* 3rd ed., Prentice Hall, Upper Saddle River, N.J., 2007, Secs. 10.2–10.6.

[21]Dowling, op. cit., pp. 486–487.

6–12   **Fatigue Failure Criteria for Fluctuating Stress**

Now that we have defined the various components of stress associated with a part subjected to fluctuating stress, we want to vary both the midrange stress and the stress amplitude, or alternating component, to learn something about the fatigue resistance of parts when subjected to such situations. Three methods of plotting the results of such tests are in general use and are shown in Figs. 6–24, 6–25, and 6–26.

The *modified Goodman diagram* of Fig. 6–24 has the midrange stress plotted along the abscissa and all other components of stress plotted on the ordinate, with tension in the positive direction. The endurance limit, fatigue strength, or finite-life strength, whichever applies, is plotted on the ordinate above and below the origin. The midrange-stress line is a 45° line from the origin to the tensile strength of the part. The modified Goodman diagram consists of the lines constructed to S_e (or S_f) above and below the origin. Note that the yield strength is also plotted on both axes, because yielding would be the criterion of failure if σ_{max} exceeded S_y.

Another way to display test results is shown in Fig. 6–25. Here the abscissa represents the ratio of the midrange strength S_m to the ultimate strength, with tension plotted to the right and compression to the left. The ordinate is the ratio of the alternating strength to the endurance limit. The line *BC* then represents the modified Goodman criterion of failure. Note that the existence of midrange stress in the compressive region has little effect on the endurance limit.

The very clever diagram of Fig. 6–26 is unique in that it displays four of the stress components as well as the two stress ratios. A curve representing the endurance limit for values of R beginning at $R = -1$ and ending with $R = 1$ begins at S_e on the σ_a axis and ends at S_{ut} on the σ_m axis. Constant-life curves for $N = 10^5$ and $N = 10^4$ cycles

Figure 6–24

Modified Goodman diagram showing all the strengths and the limiting values of all the stress components for a particular midrange stress.

Figure 6–25

Plot of fatigue failures for midrange stresses in both tensile and compressive regions. Normalizing the data by using the ratio of steady strength component to tensile strength S_m/S_{ut}, steady strength component to compressive strength S_m/S_{uc} and strength amplitude component to endurance limit S_a/S'_e enables a plot of experimental results for a variety of steels. *[Data source: Thomas J. Dolan, "Stress Range," Sec. 6.2 in O. J. Horger (ed.),* ASME Handbook—Metals Engineering Design, *McGraw-Hill, New York, 1953.]*

Figure 6–26

Master fatigue diagram created for AISI 4340 steel having and $S_{ut} = 1100$ and $S_y = 1025$ MPa. The stress components at A are $\sigma_{min} = 140$, $\sigma_{max} = 840$, $\sigma_m = 490$, and $\sigma_a = 350$, all in MPa. *(Source: H. J. Grover,* Fatigue of Aircraft Structures, *U.S. Government Printing Office, Washington, D.C., 1966, pp. 317, 322. See also J. A. Collins,* Failure of Materials in Mechanical Design, *Wiley, New York, 1981, p. 216.)*

have been drawn too. Any stress state, such as the one at A, can be described by the minimum and maximum components, or by the midrange and alternating components. And safety is indicated whenever the point described by the stress components lies below the constant-life line.

Figure 6–27

Fatigue diagram showing various criteria of failure. For each criterion, points on or "above" the respective line indicate failure. Some point A on the Goodman line, for example, gives the strength S_m as the limiting value of σ_m corresponding to the strength S_a, which, paired with σ_m, is the limiting value of σ_a.

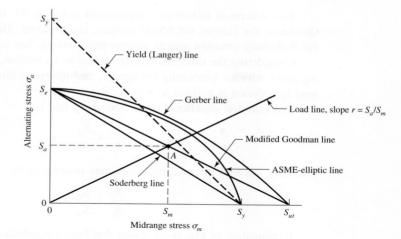

When the midrange stress is compression, failure occurs whenever $\sigma_a = S_e$ or whenever $\sigma_{\max} = S_{yc}$, as indicated by the left-hand side of Fig. 6–25. Neither a fatigue diagram nor any other failure criteria need be developed.

In Fig. 6–27, the tensile side of Fig. 6–25 has been redrawn in terms of strengths, instead of strength ratios, with the same modified Goodman criterion together with four additional criteria of failure. Such diagrams are often constructed for analysis and design purposes; they are easy to use and the results can be scaled off directly.

The early viewpoint expressed on a σ_m, σ_a diagram was that there existed a locus which divided safe from unsafe combinations of σ_m and σ_a. Ensuing proposals included the parabola of Gerber (1874), the Goodman (1890)[22] (straight) line, and the Soderberg (1930) (straight) line. As more data were generated it became clear that a fatigue criterion, rather than being a "fence," was more like a zone or band wherein the probability of failure could be estimated. We include the failure criterion of Goodman because

- It is a straight line and the algebra is linear and easy.
- It is easily graphed, every time for every problem.
- It reveals subtleties of insight into fatigue problems.
- Answers can be scaled from the diagrams as a check on the algebra.

We also caution that it is deterministic and the phenomenon is not. It is biased and we cannot quantify the bias. It is not conservative. It is a stepping-stone to understanding; it is history; and to read the work of other engineers and to have meaningful oral exchanges with them, it is necessary that you understand the Goodman approach should it arise.

Either the fatigue limit S_e or the finite-life strength S_f is plotted on the ordinate of Fig. 6–27. These values will have already been corrected using the Marin factors of Eq. (6–18). Note that the yield strength S_y is plotted on the ordinate too. This serves as a reminder that first-cycle yielding rather than fatigue might be the criterion of failure.

The midrange-stress axis of Fig. 6–27 has the yield strength S_y and the tensile strength S_{ut} plotted along it.

---

[22]It is difficult to date Goodman's work because it went through several modifications and was never published.

Five criteria of failure are diagrammed in Fig. 6–27: the Soderberg, the modified Goodman, the Gerber, the ASME-elliptic, and yielding. The diagram shows that only the Soderberg criterion guards against any yielding, but is biased low.

Considering the modified Goodman line as a criterion, point A represents a limiting point with an alternating strength S_a and midrange strength S_m. The slope of the load line shown is defined as $r = S_a/S_m$.

The criterion equation for the Soderberg line is

$$\frac{S_a}{S_e} + \frac{S_m}{S_y} = 1 \tag{6–40}$$

Similarly, we find the modified Goodman relation to be

$$\frac{S_a}{S_e} + \frac{S_m}{S_{ut}} = 1 \tag{6–41}$$

Examination of Fig. 6–25 shows that both a parabola and an ellipse have a better opportunity to pass among the midrange tension data and to permit quantification of the probability of failure. The Gerber failure criterion is written as

$$\frac{S_a}{S_e} + \left(\frac{S_m}{S_{ut}}\right)^2 = 1 \tag{6–42}$$

and the ASME-elliptic is written as

$$\left(\frac{S_a}{S_e}\right)^2 + \left(\frac{S_m}{S_y}\right)^2 = 1 \tag{6–43}$$

The *Langer* first-cycle-yielding criterion is used in connection with the fatigue curve:

$$S_a + S_m = S_y \tag{6–44}$$

The stresses $n\sigma_a$ and $n\sigma_m$ can replace S_a and S_m, where n is the design factor or factor of safety. Then, Eq. (6–40), the Soderberg line, becomes

Soderberg $$\frac{\sigma_a}{S_e} + \frac{\sigma_m}{S_y} = \frac{1}{n} \tag{6–45}$$

Equation (6–41), the modified Goodman line, becomes

mod-Goodman $$\frac{\sigma_a}{S_e} + \frac{\sigma_m}{S_{ut}} = \frac{1}{n} \tag{6–46}$$

Equation (6–42), the Gerber line, becomes

Gerber $$\frac{n\sigma_a}{S_e} + \left(\frac{n\sigma_m}{S_{ut}}\right)^2 = 1 \tag{6–47}$$

Equation (6–43), the ASME-elliptic line, becomes

ASME-elliptic $$\left(\frac{n\sigma_a}{S_e}\right)^2 + \left(\frac{n\sigma_m}{S_y}\right)^2 = 1 \tag{6–48}$$

The design equation for the Langer first-cycle-yielding is

Langer static yield $$\sigma_a + \sigma_m = \frac{S_y}{n} \tag{6–49}$$

The failure criteria are used in conjunction with a load line, $r = S_a/S_m = \sigma_a/\sigma_m$. Principal intersections are tabulated in Tables 6–6 to 6–8. Formal expressions for fatigue factor of safety are given in the lower panel of Tables 6–6 to 6–8. The first row of each table corresponds to the fatigue criterion, the second row is the static Langer criterion, and the third row corresponds to the intersection of the static and

Table 6–6

Amplitude and Steady Coordinates of Strength and Important Intersections in First Quadrant for Modified Goodman and Langer Failure Criteria

Intersecting Equations	Intersection Coordinates
$\dfrac{S_a}{S_e} + \dfrac{S_m}{S_{ut}} = 1$	$S_a = \dfrac{rS_eS_{ut}}{rS_{ut} + S_e}$
Load line $r = \dfrac{S_a}{S_m}$	$S_m = \dfrac{S_a}{r}$
$\dfrac{S_a}{S_y} + \dfrac{S_m}{S_y} = 1$	$S_a = \dfrac{rS_y}{1 + r}$
Load line $r = \dfrac{S_a}{S_m}$	$S_m = \dfrac{S_y}{1 + r}$
$\dfrac{S_a}{S_e} + \dfrac{S_m}{S_{ut}} = 1$	$S_m = \dfrac{(S_y - S_e)S_{ut}}{S_{ut} - S_e}$
$\dfrac{S_a}{S_y} + \dfrac{S_m}{S_y} = 1$	$S_a = S_y - S_m, \; r_{crit} = S_a/S_m$

Fatigue factor of safety

$$n_f = \dfrac{1}{\dfrac{\sigma_a}{S_e} + \dfrac{\sigma_m}{S_{ut}}}$$

Table 6–7

Amplitude and Steady Coordinates of Strength and Important Intersections in First Quadrant for Gerber and Langer Failure Criteria

Intersecting Equations	Intersection Coordinates
$\dfrac{S_a}{S_e} + \left(\dfrac{S_m}{S_{ut}}\right)^2 = 1$	$S_a = \dfrac{r^2S_{ut}^2}{2S_e}\left[-1 + \sqrt{1 + \left(\dfrac{2S_e}{rS_{ut}}\right)^2}\right]$
Load line $r = \dfrac{S_a}{S_m}$	$S_m = \dfrac{S_a}{r}$
$\dfrac{S_a}{S_y} + \dfrac{S_m}{S_y} = 1$	$S_a = \dfrac{rS_y}{1 + r}$
Load line $r = \dfrac{S_a}{S_m}$	$S_m = \dfrac{S_y}{1 + r}$
$\dfrac{S_a}{S_e} + \left(\dfrac{S_m}{S_{ut}}\right)^2 = 1$	$S_m = \dfrac{S_{ut}^2}{2S_e}\left[1 - \sqrt{1 + \left(\dfrac{2S_e}{S_{ut}}\right)^2\left(1 - \dfrac{S_y}{S_e}\right)}\right]$
$\dfrac{S_a}{S_y} + \dfrac{S_m}{S_y} = 1$	$S_a = S_y - S_m, \; r_{crit} = S_a/S_m$

Fatigue factor of safety

$$n_f = \dfrac{1}{2}\left(\dfrac{S_{ut}}{\sigma_m}\right)^2\dfrac{\sigma_a}{S_e}\left[-1 + \sqrt{1 + \left(\dfrac{2\sigma_mS_e}{S_{ut}\sigma_a}\right)^2}\right] \qquad \sigma_m > 0$$

Table 6–8

Amplitude and Steady Coordinates of Strength and Important Intersections in First Quadrant for ASME-Elliptic and Langer Failure Criteria

Intersecting Equations	Intersection Coordinates
$\left(\dfrac{S_a}{S_e}\right)^2 + \left(\dfrac{S_m}{S_y}\right)^2 = 1$ Load line $r = S_a/S_m$	$S_a = \sqrt{\dfrac{r^2 S_e^2 S_y^2}{S_e^2 + r^2 S_y^2}}$ $S_m = \dfrac{S_a}{r}$
$\dfrac{S_a}{S_y} + \dfrac{S_m}{S_y} = 1$ Load line $r = S_a/S_m$	$S_a = \dfrac{r S_y}{1 + r}$ $S_m = \dfrac{S_y}{1 + r}$
$\left(\dfrac{S_a}{S_e}\right)^2 + \left(\dfrac{S_m}{S_y}\right)^2 = 1$ $\dfrac{S_a}{S_y} + \dfrac{S_m}{S_y} = 1$	$S_a = 0,\ \dfrac{2 S_y S_e^2}{S_e^2 + S_y^2}$ $S_m = S_y - S_a,\ r_{\text{crit}} = S_a/S_m$

Fatigue factor of safety

$$n_f = \sqrt{\dfrac{1}{(\sigma_a/S_e)^2 + (\sigma_m/S_y)^2}}$$

fatigue criteria. The first column gives the intersecting equations and the second column the intersection coordinates.

There are two ways to proceed with a typical analysis. One method is to assume that fatigue occurs first and use one of Eqs. (6–45) to (6–48) to determine n or size, depending on the task. Most often fatigue is the governing failure mode. Then follow with a static check. If static failure governs then the analysis is repeated using Eq. (6–49).

Alternatively, one could use the tables. Determine the load line and establish which criterion the load line intersects first and use the corresponding equations in the tables.

Some examples will help solidify the ideas just discussed.

EXAMPLE 6–10

A 40-mm-diameter bar has been machined from an AISI 1050 cold-drawn bar. This part is to withstand a fluctuating tensile load varying from 0 to 70 kN. Because of the ends, and the fillet radius, a fatigue stress-concentration factor K_f is 1.85 for 10^6 or larger life. Find S_a and S_m and the factor of safety guarding against fatigue and first-cycle yielding, using (a) the Gerber fatigue line and (b) the ASME-elliptic fatigue line.

Solution

We begin with some preliminaries. From Table A–20, $S_{ut} = 690$ MPa and $S_y = 580$ MPa. Note that $F_a = F_m = 35$ kN. The Marin factors are, deterministically,

$k_a = 4.51(690)^{-0.265} = 0.798$: Eq. (6–19), Table 6–2, p. 296

$k_b = 1$ (axial loading, see k_c)

$k_c = 0.85$: Eq. (6–26), p. 290

$k_d = k_e = k_f = 1$

$S_e = 0.798(1)0.850(1)(1)(1)0.5(690) = 234$ MPa: Eqs. (6–8), (6–18), p. 290, p. 295

The nominal axial stress components σ_{ao} and σ_{mo} are

$$\sigma_{ao} = \frac{4F_a}{\pi d^2} = \frac{4(35000)}{\pi\, 0.04^2} = 27.9 \text{ MPa} \quad \sigma_{mo} = \frac{4F_m}{\pi d^2} = \frac{4(35000)}{\pi\, 0.04^2} = 27.9 \text{ MPa}$$

Applying K_f to both components σ_{ao} and σ_{mo} constitutes a prescription of no notch yielding:

$$\sigma_a = K_f \sigma_{ao} = 1.85(27.9) = 51.6 \text{ MPa} = \sigma_m$$

(*a*) Let us calculate the factors of safety first. From the bottom panel from Table 6–7 the factor of safety for fatigue is

Answer
$$n_f = \frac{1}{2}\left(\frac{690}{51.6}\right)^2\left(\frac{51.6}{234}\right)\left\{-1 + \sqrt{1 + \left[\frac{2(51.6)234}{690(51.6)}\right]^2}\right\} = 4.11$$

From Eq. (6–49) the factor of safety guarding against first-cycle yield is

Answer
$$n_y = \frac{S_y}{\sigma_a + \sigma_m} = \frac{580}{51.6 + 51.6} = 5.62$$

Thus, we see that fatigue will occur first and the factor of safety is 4.13. This can be seen in Fig. 6–28 where the load line intersects the Gerber fatigue curve first at point B. If the plots are created to true scale it would be seen that $n_f = OB/OA$.

From the first panel of Table 6–7, $r = \sigma_a/\sigma_m = 1$,

Answer
$$S_a = \frac{(1)^2\, 690^2}{2(234)}\left\{-1 + \sqrt{1 + \left[\frac{2(234)}{(1)690}\right]^2}\right\} = 211.9 \text{ MPa}$$

Figure 6–28

Principal points A, B, C, and D on the designer's diagram drawn for Gerber, Langer, and load line.

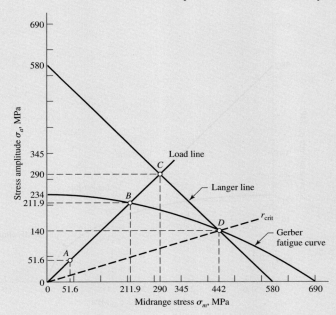

Answer
$$S_m = \frac{S_a}{r} = \frac{211.9}{1} = 211.9 \text{ MPa}$$

As a check on the previous result, $n_f = OB/OA = S_a/\sigma_a = S_m/\sigma_m = 211.9/51.6 = 4.12$ and we see total agreement.

We could have detected that fatigue failure would occur first without drawing Fig. 6–28 by calculating r_{crit}. From the third row third column panel of Table 6–7, the intersection point between fatigue and first-cycle yield is

$$S_m = \frac{690^2}{2(234)} \left[1 - \sqrt{1 + \left(\frac{2(234)}{690} \right)^2 \left(1 - \frac{580}{234} \right)} \right] = 442 \text{ MPa}$$

$$S_a = S_y - S_m = 580 - 442 = 138 \text{ MPa}$$

The critical slope is thus

$$r_{crit} = \frac{S_a}{S_m} = \frac{138}{442} = 0.312$$

which is less than the actual load line of $r = 1$. This indicates that fatigue occurs before first-cycle-yield.

(b) Repeating the same procedure for the ASME-elliptic line, for fatigue

Answer
$$n_f = \sqrt{\frac{1}{(51.6/234)^2 + (51.6/580)^2}} = 4.21$$

Again, this is less than $n_y = 5.62$ and fatigue is predicted to occur first. From the first row second column panel of Table 6–8, with $r = 1$, we obtain the coordinates S_a and S_m of point B in Fig. 6–29 as

Figure 6–29

Principal points A, B, C, and D on the designer's diagram drawn for ASME-elliptic, Langer, and load lines.

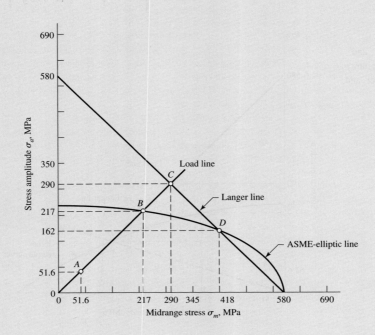

Answer
$$S_a = \sqrt{\frac{(1)^2 234^2 (580)^2}{234^2 + (1)^2 580^2}} = 217 \text{ MPa}, \quad S_m = \frac{S_a}{r} = \frac{217}{1} = 217 \text{ MPa}$$

To verify the fatigue factor of safety, $n_f = S_a/\sigma_a = 217/51.6 = 4.21$.

As before, let us calculate r_{crit}. From the third row second column panel of Table 6–8,

$$S_a = \frac{2(580) 234^2}{234^2 + 580^2} = 162 \text{ MPa}, \quad S_m = S_y - S_a = 580 - 162 = 418 \text{ MPa}$$

$$r_{crit} = \frac{S_a}{S_m} = \frac{162}{418} = 0.388$$

which again is less than $r = 1$, verifying that fatigue occurs first with $n_f = 4.21$.

The Gerber and the ASME-elliptic fatigue failure criteria are very close to each other and are used interchangeably. The ANSI/ASME Standard B106.1M–1985 uses ASME-elliptic for shafting.

EXAMPLE 6–11   A flat-leaf spring is used to retain an oscillating flat-faced follower in contact with a plate cam. The follower range of motion is 50 mm and fixed, so the alternating component of force, bending moment, and stress is fixed, too. The spring is preloaded to adjust to various cam speeds. The preload must be increased to prevent follower float or jump. For lower speeds the preload should be decreased to obtain longer life of cam and follower surfaces. The spring is a steel cantilever 0.8 m long, 50 mm wide, and 6 mm thick, as seen in Fig. 6–30a. The spring strengths are $S_{ut} = 1000$ MPa, $S_y = 880$ MPa, and $S_e = 195$ MPa fully corrected. The total cam motion is 50 mm. The designer wishes to preload the spring by deflecting it 50 mm for low speed and 125 mm for high speed.
(a) Plot the Gerber-Langer failure lines with the load line.
(b) What are the strength factors of safety corresponding to 50 mm and 125 mm preload?

Solution   We begin with preliminaries. The second area moment of the cantilever cross section is

$$I = \frac{bh^3}{12} = \frac{0.05(0.006)^3}{12} = 0.9 + 10^{-9} \text{ m}^4$$

Since, from Table A–9, beam 1, force F and deflection y in a cantilever are related by $F = 3EIy/l^3$, then stress σ and deflection y are related by

$$\sigma = \frac{Mc}{I} = \frac{0.8Fc}{I} = \frac{0.8(3EIy)}{l^3} \frac{c}{I} = \frac{2.4Ecy}{l^3} = Ky$$

where $K = \dfrac{2.4Ec}{l^3} = \dfrac{2.4(210 \times 10^9)(0.003)}{0.8^3} = 2.95 \text{ GPa/m}$

Now the minimums and maximums of y and σ can be defined by

$$y_{min} = \delta \qquad\qquad y_{max} = 0.05 + \delta$$

$$\sigma_{min} = K\delta \qquad\qquad \sigma_{max} = K(0.05 + \delta)$$

Figure 6–30

Cam follower retaining spring.
(a) Geometry; (b) designer's
fatigue diagram for Ex. 6–11.

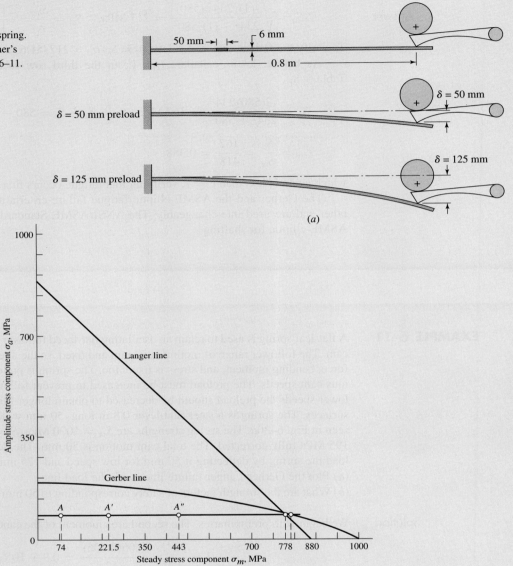

(a)

(b)

The stress components are thus

$$\sigma_a = \frac{K(0.05 + \delta) - K\delta}{2} = 0.025K$$

$$\sigma_m = \frac{K(0.05 + \delta) + K\delta}{2} = K(0.025 + \delta)$$

For $\delta = 0$, $\sigma_a = \sigma_m = 73.83 = 74$ MPa

For $\delta = 50$ mm, $\sigma_a = 74$ MPa, $\sigma_m = 2.95 \times 10^3 \, (0.025 + 0.05) = 221.5 \, MPa$

For $\delta = 125$ mm, $\sigma_a = 74$ MPa, $\sigma_m = 2.95 \times 10^3 \, (0.025 + 0.125) = 442.97 \, MPa$

(*a*) A plot of the Gerber and Langer criteria is shown in Fig. 6–30*b*. The three preload deflections of 0, 50, and 125 mm are shown as points *A*, *A'*, and *A"*. Note that since σ_a is constant at 74 MPa, the load line is horizontal and does not contain the origin. The intersection between the Gerber line and the load line is found from solving Eq. (6–42) for S_m and substituting 74 MPa for S_a:

$$S_m = S_{ut}\sqrt{1 - \frac{S_a}{S_e}} = 1000\sqrt{1 - \frac{74}{195}} = 778.3 \text{ MPa}$$

The intersection of the Langer line and the load line is found from solving Eq. (6–44) for S_m and substituting 77 MPa for S_a:

$$S_m = S_y - S_a = 880 - 74 = 806 \text{ MPa}$$

The threats from fatigue and first-cycle yielding are approximately equal.
(*b*) For $\delta = 50$ mm,

Answer
$$n_f = \frac{S_m}{\sigma_m} = \frac{778.3}{221.5} = 3.56 \quad n_y = \frac{806.2}{221.5} = 3.64$$

and for $\delta = 125$ mm,

Answer
$$n_f = \frac{778.3}{443} = 1.78 \quad n_y = \frac{806.2}{443} = 1.82$$

EXAMPLE 6–12

A steel bar undergoes cyclic loading such that $\sigma_{max} = 420$ MPa and $\sigma_{min} = -140$ MPa. For the material, $S_{ut} = 560$ MPa, $S_y = 455$ MPa, a fully corrected endurance limit of $S_e = 280$ MPa, and $f = 0.9$. Estimate the number of cycles to a fatigue failure using:
(*a*) Modified Goodman criterion.
(*b*) Gerber criterion.

Solution

From the given stresses,

$$\sigma_a = \frac{60 - (-20)}{2} = 40 \text{ kpsi} \quad \sigma_m = \frac{60 + (-20)}{2} = 20 \text{ kpsi}$$

(*a*) For the modified Goodman criterion, Eq. (6–46), the fatigue factor of safety based on infinite life is

$$n_f = \frac{1}{\dfrac{\sigma_a}{S_e} + \dfrac{\sigma_m}{S_{ut}}} = \frac{1}{\dfrac{40}{40} + \dfrac{20}{80}} = 0.8$$

This indicates a finite life is predicted. The *S-N* diagram is only applicable for completely reversed stresses. To estimate the finite life for a fluctuating stress, we will obtain an equivalent completely reversed stress that is expected to be as damaging as the fluctuating stress. A commonly used approach is to assume that since the modified Goodman line represents all stress situations with a constant life of 10^6 cycles, other constant-life lines can be generated by passing a line through (S_{ut}, 0) and a fluctuating stress point (σ_m, σ_a). The point where this line intersects the σ_a axis represents a completely reversed stress (since at this point $\sigma_m = 0$), which predicts the same life as the fluctuating stress. This completely reversed stress can be obtained by replacing S_e with σ_{rev} in Eq. (6–46) for the modified Goodman line resulting in

$$\sigma_{\text{rev}} = \frac{\sigma_a}{1 - \dfrac{\sigma_m}{S_{ut}}} = \frac{40}{1 - \dfrac{20}{80}} = 53.3 \text{ kpsi}$$

From the material properties, Eqs. (6–14) to (6–16), p. 293, give

$$a = \frac{(f S_{ut})^2}{S_e} = \frac{[0.9(80)]^2}{40} = 129.6 \text{ kpsi}$$

$$b = -\frac{1}{3} \log\left(\frac{f S_{ut}}{S_e}\right) = -\frac{1}{3} \log\left[\frac{0.9(80)}{40}\right] = -0.0851 \qquad (1)$$

$$N = \left(\frac{\sigma_{\text{rev}}}{a}\right)^{1/b} = \left(\frac{\sigma_{\text{rev}}}{129.6}\right)^{-1/0.0851}$$

Substituting σ_{rev} into Eq. (1) yields

Answer
$$N = \left(\frac{53.3}{129.6}\right)^{-1/0.0851} = 3.4(10^4) \text{ cycles}$$

(b) For Gerber, similar to part (a), from Eq. (6–47),

$$\sigma_{\text{rev}} = \frac{\sigma_a}{1 - \left(\dfrac{\sigma_m}{S_{ut}}\right)^2} = \frac{40}{1 - \left(\dfrac{20}{80}\right)^2} = 42.7 \text{ kpsi}$$

Again, from Eq. (1),

Answer
$$N = \left(\frac{42.7}{129.6}\right)^{-1/0.0851} = 4.6(10^5) \text{ cycles}$$

Comparing the answers, we see a large difference in the results. Again, the modified Goodman criterion is conservative as compared to Gerber for which the moderate difference in S_f is then magnified by a logarithmic S, N relationship.

For many *brittle* materials, the first quadrant fatigue failure criteria follows a concave upward Smith-Dolan locus represented by

$$\frac{S_a}{S_e} = \frac{1 - S_m/S_{ut}}{1 + S_m/S_{ut}} \qquad (6\text{–}50)$$

or as a design equation,

$$\frac{n\sigma_a}{S_e} = \frac{1 - n\sigma_m/S_{ut}}{1 + n\sigma_m/S_{ut}} \qquad (6\text{–}51)$$

For a radial load line of slope r, we substitute S_a/r for S_m in Eq. (6–50) and solve for S_a, obtaining the intersect

$$S_a = \frac{rS_{ut} + S_e}{2}\left[-1 + \sqrt{1 + \frac{4rS_{ut}S_e}{(rS_{ut} + S_e)^2}}\right] \qquad (6\text{–}52)$$

The fatigue diagram for a brittle material differs markedly from that of a ductile material because:

- Yielding is not involved since the material may not have a yield strength.
- Characteristically, the compressive ultimate strength exceeds the ultimate tensile strength severalfold.

- First-quadrant fatigue failure locus is concave-upward (Smith-Dolan), for example, and as flat as Goodman. Brittle materials are more sensitive to midrange stress, being lowered, but compressive midrange stresses are beneficial.
- Not enough work has been done on brittle fatigue to discover insightful generalities, so we stay in the first and a bit of the second quadrant.

The most likely domain of designer use is in the range from $-S_{ut} \leq \sigma_m \leq S_{ut}$. The locus in the first quadrant is Goodman, Smith-Dolan, or something in between. The portion of the second quadrant that is used is represented by a straight line between the points $-S_{ut}, S_{ut}$ and $0, S_e$, which has the equation

$$S_a = S_e + \left(\frac{S_e}{S_{ut}} - 1\right) S_m \qquad -S_{ut} \leq S_m \leq 0 \quad \text{(for cast iron)} \qquad (6\text{-}53)$$

Table A–24 gives properties of gray cast iron. The endurance limit stated is really $k_a k_b S'_e$ and only corrections k_c, k_d, k_e, and k_f need be made. The average k_c for axial and torsional loading is 0.9.

EXAMPLE 6–13

A grade 30 gray cast iron is subjected to a load F applied to a 25 by 10-mm cross-section link with a 6-mm-diameter hole drilled in the center as depicted in Fig 6-31a. The surfaces are machined. In the neighborhood of the hole, what is the factor of safety guarding against failure under the following conditions:
(a) The load $F = 4500$ N lbf tensile, steady.
(b) The load is 4500 N repeatedly applied.
(c) The load fluctuates between -4500 N and 1300 N without column action.
Use the Smith-Dolan fatigue locus.

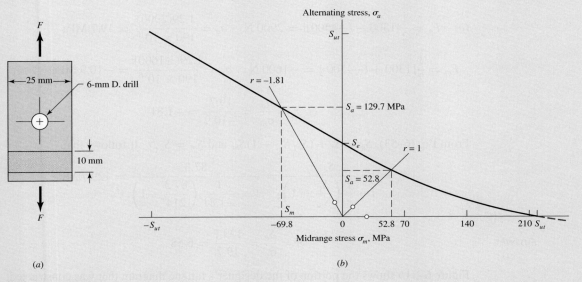

(a)

(b)

Figure 6–31

The grade 30 cast-iron part in axial fatigue with (a) its geometry displayed and (b) its designer's fatigue diagram for the circumstances of Ex. 6–13.

Solution

Some preparatory work is needed. From Table A–24, $S_{ut} = 214$ MPa, $S_{uc} = 752$ MPa, $k_a k_b S'_e = 97$ MPa. Since k_c for axial loading is 0.9, then $S_e = (k_a k_b S'_e)k_c = 97(0.9) = 87.3$ MPa. From Table A–15–1, $A = t(w - d) = 0.01(0.025 - 0.006) = 190 \times 10^{-6}$ m², $d/w = 6/25 = 0.24$, and $K_t = 2.45$. The notch sensitivity for cast iron is 0.20 (see p. 304), so

$$K_f = 1 + q(K_t - 1) = 1 + 0.20(2.45 - 1) = 1.29$$

(a) $\sigma_a = \dfrac{K_f F_a}{A} = \dfrac{1.29(0)}{A} = 0$ $\qquad$ $\sigma_m = \dfrac{K_f F_m}{A} = \dfrac{1.29(4500)}{190 \times 10^{-6}} = 30.6$ MPa

and

Answer

$$n = \frac{S_{ut}}{\sigma_m} = \frac{214}{30.6} = 6.99$$

(b) $\qquad F_a = F_m = \dfrac{F}{2} = \dfrac{4500}{2} = 2250$ N

$$\sigma_a = \sigma_m = \frac{K_f F_a}{A} = \frac{1.29(2250)}{190 \times 10^{-6}} = 15.3 \text{ MPa}$$

$$r = \frac{\sigma_a}{\sigma_m} = 1$$

From Eq. (6–52),

$$S_a = \frac{(1)31 + 12.6}{2}\left[-1 + \sqrt{1 + \frac{4(1)214(87.3)}{[(1)214 + 87.3]^2}}\right] = 52.8 \text{ MPa}$$

Answer

$$n = \frac{S_a}{\sigma_a} = \frac{52.8}{15.3} = 3.45$$

(c) $\quad F_a = \dfrac{1}{2}|1300 - (-4500)| = 2900$ N $\qquad \sigma_a = \dfrac{1.29(2900)}{190 \times 10^{-6}} = 19.7$ MPa

$F_m = \dfrac{1}{2}[1300 + (-4500)] = -1600$ N $\qquad \sigma_m = \dfrac{1.29(-1600)}{190 \times 10^{-6}} = -10.9$ MPa

$$r = \frac{\sigma_a}{\sigma_m} = \frac{19.7}{-10.9} = -1.81$$

From Eq. (6–53), $S_a = S_e + (S_e/S_{ut} - 1)S_m$ and $S_m = S_a/r$. It follows that

$$S_a = \frac{S_e}{1 - \dfrac{1}{r}\left(\dfrac{S_e}{S_{ut}} - 1\right)} = \frac{87.3}{1 - \dfrac{1}{-1.81}\left(\dfrac{87.3}{214} - 1\right)} = 129.7 \text{ MPa}$$

Answer

$$n = \frac{S_a}{\sigma_a} = \frac{129.7}{19.7} = 6.58$$

Figure 6–31b shows the portion of the designer's fatigue diagram that was constructed.

6–13 **Torsional Fatigue Strength under Fluctuating Stresses**

Extensive tests by Smith[23] provide some very interesting results on pulsating torsional fatigue. Smith's first result, based on 72 tests, shows that the existence of a torsional steady-stress component not more than the torsional yield strength has no effect on the torsional endurance limit, provided the material is *ductile, polished, notch-free,* and *cylindrical.*

Smith's second result applies to materials with stress concentration, notches, or surface imperfections. In this case, he finds that the torsional fatigue limit decreases monotonically with torsional steady stress. Since the great majority of parts will have surfaces that are less than perfect, this result indicates Gerber, ASME-elliptic, and other approximations are useful. Joerres of Associated Spring-Barnes Group, confirms Smith's results and recommends the use of the modified Goodman relation for pulsating torsion. In constructing the Goodman diagram, Joerres uses

$$S_{su} = 0.67S_{ut} \tag{6–54}$$

Also, from Chap. 5, $S_{sy} = 0.577S_{yt}$ from distortion-energy theory, and the mean load factor k_c is given by Eq. (6–26), or 0.577. This is discussed further in Chap. 10.

6–14 **Combinations of Loading Modes**

It may be helpful to think of fatigue problems as being in three categories:

- Completely reversing simple loads
- Fluctuating simple loads
- *Combinations of loading modes*

The simplest category is that of a completely reversed single stress which is handled with the *S-N* diagram, relating the alternating stress to a life. Only one type of loading is allowed here, and the midrange stress must be zero. The next category incorporates general fluctuating loads, using a criterion to relate midrange and alternating stresses (modified Goodman, Gerber, ASME-elliptic, or Soderberg). Again, only *one* type of loading is allowed at a time. The third category, which we will develop in this section, involves cases where there are combinations of different types of loading, such as combined bending, torsion, and axial.

In Sec. 6–9 we learned that a load factor k_c is used to obtain the endurance limit, and hence the result is dependent of whether the loading is axial, bending, or torsional. In this section we want to answer the question, "How do we proceed when the loading is a *mixture* of, say, axial, bending, and torsional loads?" This type of loading introduces a few complications in that there may now exist combined normal and shear stresses, each with alternating and midrange values, and several of the factors used in determining the endurance limit depend on the type of loading. There may also be multiple stress-concentration factors, one for each mode of loading. The problem of how to deal with combined stresses was encountered when developing static failure theories. The distortion energy failure theory proved to be a satisfactory method

---

[23]James O. Smith, "The Effect of Range of Stress on the Fatigue Strength of Metals," *Univ. of Ill. Eng. Exp. Sta. Bull.* 334, 1942.

of combining the multiple stresses on a stress element of a ductile material into a single equivalent von Mises stress. The same approach will be used here.

The first step is to generate *two* stress elements—one for the alternating stresses and one for the midrange stresses. Apply the appropriate fatigue stress-concentration factors to each of the stresses; i.e., apply $(K_f)_{bending}$ for the bending stresses, $(K_{fs})_{torsion}$ for the torsional stresses, and $(K_f)_{axial}$ for the axial stresses. Next, calculate an equivalent von Mises stress for each of these two stress elements, σ'_a and σ'_m. Finally, select a fatigue failure criterion (modified Goodman, Gerber, ASME-elliptic, or Soderberg) to complete the fatigue analysis. For the endurance limit, S_e, use the endurance limit modifiers, k_a, k_b, and k_c, for bending. The torsional load factor, $k_c = 0.59$ should not be applied as it is already accounted for in the von Mises stress calculation (see footnote 17 on p. 298). The load factor for the axial load can be accounted for by dividing the alternating axial stress by the axial load factor of 0.85. For example, consider the common case of a shaft with bending stresses, torsional shear stresses, and axial stresses. For this case, the von Mises stress is of the form $\sigma' = (\sigma_x^2 + 3\tau_{xy}^2)^{1/2}$. Considering that the bending, torsional, and axial stresses have alternating and midrange components, the von Mises stresses for the two stress elements can be written as

$$\sigma'_a = \left\{ \left[(K_f)_{bending}(\sigma_a)_{bending} + (K_f)_{axial}\frac{(\sigma_a)_{axial}}{0.85} \right]^2 + 3[(K_{fs})_{torsion}(\tau_a)_{torsion}]^2 \right\}^{1/2}$$

$$(6\text{--}55)$$

$$\sigma'_m = \{ [(K_f)_{bending}(\sigma_m)_{bending} + (K_f)_{axial}(\sigma_m)_{axial}]^2 + 3[(K_{fs})_{torsion}(\tau_m)_{torsion}]^2 \}^{1/2}$$

$$(6\text{--}56)$$

For first-cycle localized yielding, the maximum von Mises stress is calculated. This would be done by first adding the axial and bending alternating and midrange stresses to obtain σ_{max} and adding the alternating and midrange shear stresses to obtain τ_{max}. Then substitute σ_{max} and τ_{max} into the equation for the von Mises stress. A simpler and more conservative method is to add Eq. (6–55) and Eq. (6–56). That is, let $\sigma'_{max} \approx \sigma'_a + \sigma'_m$.

If the stress components are not in phase but have the same frequency, the maxima can be found by expressing each component in trigonometric terms, using phase angles, and then finding the sum. If two or more stress components have differing frequencies, the problem is difficult; one solution is to assume that the two (or more) components often reach an in-phase condition, so that their magnitudes are additive.

EXAMPLE 6–14	A rotating shaft is made of 42- $\times$ 4-mm AISI 1018 cold-drawn steel tubing and has a 6-mm-diameter hole drilled transversely through it. Estimate the factor of safety guarding against fatigue and static failures using the Gerber and Langer failure criteria for the following loading conditions: (*a*) The shaft is rotating and is subjected to a completely reversed torque of 120 N · m in phase with a completely reversed bending moment of 150 N · m. (*b*) The shaft is subjected to a pulsating torque fluctuating from 20 to 160 N · m and a steady bending moment of 150 N · m.
Solution	Here we follow the procedure of estimating the strengths and then the stresses, followed by relating the two.

From Table A–20 we find the minimum strengths to be $S_{ut} = 440$ MPa and $S_y = 370$ MPa. The endurance limit of the rotating-beam specimen is $0.5(440) = 220$ MPa. The surface factor, obtained from Eq. (6–19) and Table 6–2, p. 296, is

$$k_a = 4.51 S_{ut}^{-0.265} = 4.51(440)^{-0.265} = 0.899$$

From Eq. (6–20) the size factor is

$$k_b = \left(\frac{d}{7.62}\right)^{-0.107} = \left(\frac{42}{7.62}\right)^{-0.107} = 0.833$$

The remaining Marin factors are all unity, so the modified endurance strength S_e is

$$S_e = 0.899(0.833)220 = 165 \text{ MPa}$$

(*a*) Theoretical stress-concentration factors are found from Table A–16. Using $a/D = 6/42 = 0.143$ and $d/D = 34/42 = 0.810$, and using linear interpolation, we obtain $A = 0.798$ and $K_t = 2.366$ for bending; and $A = 0.89$ and $K_{ts} = 1.75$ for torsion. Thus, for bending,

$$Z_{\text{net}} = \frac{\pi A}{32D}(D^4 - d^4) = \frac{\pi(0.798)}{32(42)}[(42)^4 - (34)^4] = 3.31(10^3)\,\text{mm}^3$$

and for torsion

$$J_{\text{net}} = \frac{\pi A}{32}(D^4 - d^4) = \frac{\pi(0.89)}{32}[(42)^4 - (34)^4] = 155(10^3)\,\text{mm}^4$$

Next, using Figs. 6–20 and 6–21, pp. 303–304, with a notch radius of 3 mm we find the notch sensitivities to be 0.78 for bending and 0.96 for torsion. The two corresponding fatigue stress-concentration factors are obtained from Eq. (6–32) as

$$K_f = 1 + q(K_t - 1) = 1 + 0.78(2.366 - 1) = 2.07$$

$$K_{fs} = 1 + 0.96(1.75 - 1) = 1.72$$

The alternating bending stress is now found to be

$$\sigma_{xa} = K_f \frac{M}{Z_{\text{net}}} = 2.07 \frac{150}{3.31(10^{-6})} = 93.8(10^6)\text{Pa} = 93.8 \text{ MPa}$$

and the alternating torsional stress is

$$\tau_{xya} = K_{fs} \frac{TD}{2J_{\text{net}}} = 1.72 \frac{120(42)(10^{-3})}{2(155)(10^{-9})} = 28.0(10^6)\text{Pa} = 28.0 \text{ MPa}$$

The midrange von Mises component σ'_m is zero. The alternating component σ'_a is given by

$$\sigma'_a = (\sigma_{xa}^2 + 3\tau_{xya}^2)^{1/2} = [93.8^2 + 3(28^2)]^{1/2} = 105.6 \text{ MPa}$$

Since $S_a = S_e$, the fatigue factor of safety n_f is

Answer

$$n_f = \frac{S_a}{\sigma'_a} = \frac{165}{105.6} = 1.56$$

Figure 6–32

Designer's fatigue diagram for Ex. 6–14.

The first-cycle yield factor of safety is

Answer

$$n_y = \frac{S_y}{\sigma_a'} = \frac{370}{105.6} = 3.50$$

There is no localized yielding; the threat is from fatigue. See Fig. 6–32.

(b) This part asks us to find the factors of safety when the alternating component is due to pulsating torsion, and a steady component is due to both torsion and bending. We have $T_a = (160 - 20)/2 = 70$ N · m and $T_m = (160 + 20)/2 = 90$ N · m. The corresponding amplitude and steady-stress components are

$$\tau_{xya} = K_{fs} \frac{T_a D}{2 J_{\text{net}}} = 1.72 \frac{70(42)(10^{-3})}{2(155)(10^{-9})} = 16.3(10^6)\text{Pa} = 16.3 \text{ MPa}$$

$$\tau_{xym} = K_{fs} \frac{T_m D}{2 J_{\text{net}}} = 1.72 \frac{90(42)(10^{-3})}{2(155)(10^{-9})} = 21.0(10^6)\text{Pa} = 21.0 \text{ MPa}$$

The steady bending stress component σ_{xm} is

$$\sigma_{xm} = K_f \frac{M_m}{Z_{\text{net}}} = 2.07 \frac{150}{3.31(10^{-6})} = 93.8(10^6)\text{Pa} = 93.8 \text{ MPa}$$

The von Mises components σ_a' and σ_m' are

$$\sigma_a' = [3(16.3)^2]^{1/2} = 28.2 \text{ MPa}$$

$$\sigma_m' = [93.8^2 + 3(21)^2]^{1/2} = 100.6 \text{ MPa}$$

From Table 6–7, p. 307, the fatigue factor of safety is

Answer

$$n_f = \frac{1}{2} \left(\frac{440}{100.6} \right)^2 \frac{28.2}{165} \left\{ -1 + \sqrt{1 + \left[\frac{2(100.6)165}{440(28.2)} \right]^2} \right\} = 2.74$$

From the same table, with $r = \sigma_a'/\sigma_m' = 28.2/100.6 = 0.280$, the strengths can be shown to be $S_a = 85.5$ MPa and $S_m = 305$ MPa. See the plot in Fig. 6–32.

The first-cycle yield factor of safety n_y is

Answer

$$n_y = \frac{S_y}{\sigma_a' + \sigma_m'} = \frac{370}{28.2 + 100.6} = 2.87$$

There is no notch yielding. The likelihood of failure may first come from first-cycle yielding at the notch. See the plot in Fig. 6–32.

6–15   # Varying, Fluctuating Stresses; Cumulative Fatigue Damage

Instead of a single fully reversed stress history block composed of n cycles, suppose a machine part, at a critical location, is subjected to

- A fully reversed stress σ_1 for n_1 cycles, σ_2 for n_2 cycles, . . . , or
- A "wiggly" time line of stress exhibiting many and different peaks and valleys.

What stresses are significant, what counts as a cycle, and what is the measure of damage incurred? Consider a fully reversed cycle with stresses varying 420, 560, 280, and 420 MPa and a second fully reversed cycle −280, −420, −140, and −280 MPa as depicted in Fig. 6–33a. First, it is clear that to impose the pattern of stress in Fig. 6–33a on a part it is necessary that the time trace look like the solid lines plus the dashed lines in Fig. 6–33a. Figure 6–33b moves the snapshot to exist beginning with 560 MPa and ending with 560 MPa. Acknowledging the existence of a single stress-time trace is to discover a "hidden" cycle shown as the dashed line in Fig. 6–33b. If

Figure 6–33

Variable stress diagram prepared for assessing cumulative damage.

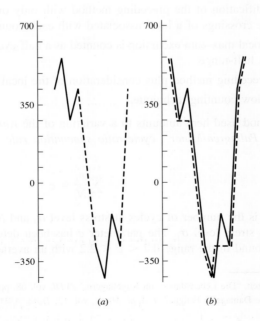

(a) (b)

there are 100 applications of the all-positive stress cycle, then 100 applications of the all-negative stress cycle, the hidden cycle is applied but once. If the all-positive stress cycle is applied alternately with the all-negative stress cycle, the hidden cycle is applied 100 times.

To ensure that the hidden cycle is not lost, begin on the snapshot with the largest (or smallest) stress and add previous history to the right side, as was done in Fig. 6–33b. Characterization of a cycle takes on a max–min–same max (or min–max–same min) form. We identify the hidden cycle first by moving along the dashed-line trace in Fig. 6–33b identifying a cycle with an 560-MPa max, a 420-MPa min, and returning to 560 MPa. Mentally deleting the used part of the trace (the dashed line) leaves a 40, 60, 40 cycle and a -40, -20, -40 cycle. Since failure loci are expressed in terms of stress amplitude component σ_a and steady component σ_m, we use Eq. (6–36) to construct the table below:

Cycle Number	σ_{max}	σ_{min}	σ_a	σ_m
1	560	-420	490	70
2	420	280	70	350
3	-140	-420	70	-210

The most damaging cycle is number 1. It could have been lost.

Methods for counting cycles include:

- Number of tensile peaks to failure.
- All maxima above the waveform mean, all minima below.
- The global maxima between crossings above the mean and the global minima between crossings below the mean.
- All positive slope crossings of levels above the mean, and all negative slope crossings of levels below the mean.
- A modification of the preceding method with only one count made between successive crossings of a level associated with each counting level.
- Each local max–min excursion is counted as a half-cycle, and the associated amplitude is half-range.
- The preceding method plus consideration of the local mean.
- Rain-flow counting technique.

The method used here amounts to a variation of the *rain-flow counting technique*.

The *Palmgren-Miner*[24] *cycle-ratio summation rule,* also called *Miner's rule,* is written

$$\sum \frac{n_i}{N_i} = c \tag{6–57}$$

where n_i is the number of cycles at stress level σ_i and N_i is the number of cycles to failure at stress level σ_i. The parameter c has been determined by experiment; it is usually found in the range $0.7 < c < 2.2$ with an average value near unity.

---

[24]A. Palmgren, "Die Lebensdauer von Kugellagern," *ZVDI,* vol. 68, pp. 339–341, 1924; M. A. Miner, "Cumulative Damage in Fatigue," *J. Appl. Mech.,* vol. 12, *Trans. ASME,* vol. 67, pp. A159–A164, 1945.

Using the deterministic formulation as a linear damage rule we write

$$D = \sum \frac{n_i}{N_i} \qquad (6\text{--}58)$$

where D is the accumulated damage. When $D = c = 1$, failure ensues.

EXAMPLE 6–15

Given a part with $S_{ut} = 1057$ MPa and at the critical location of the part, $S_e = 472.5$ MPa. For the loading of Fig. 6–33, estimate the number of repetitions of the stress-time block in Fig. 6–33 that can be made before failure

Solution

From Fig. 6–18, p. 293, for $S_{ut} = 1057$ MPa, $f = 0.795$. From Eq. (6–14),

$$a = \frac{(f S_{ut})^2}{S_e} = \frac{[0.795(1057)]^2}{472.5} = 1494.5 \text{ MPa}$$

From Eq. (6–15),

$$b = -\frac{1}{3} \log \left(\frac{f S_{ut}}{S_e} \right) = -\frac{1}{3} \log \left[\frac{0.795(1057)}{472.5} \right] = -0.0833$$

So,

$$S_f = 1494.5 N^{-0.0833} \text{ is } N = \left(\frac{S_f}{1494.5} \right)^{-1/0.0833} \qquad (1), (2)$$

We prepare to add two columns to the previous table. Using the Gerber fatigue criterion, Eq. (6–47), p. 314, with $S_e = S_f$, and $n = 1$, we can write

$$S_f = \begin{cases} \dfrac{\sigma_a}{1 - (\sigma_m/S_{ut})^2} & \sigma_m > 0 \\[2mm] S_e & \sigma_m \leq 0 \end{cases} \qquad (3)$$

where S_f is the fatigue strength associated with a completely reversed stress, σ_{rev}, equivalent to the fluctuating stresses [see Ex. 6–12, part (b)].

Cycle 1: $r = \sigma_a/\sigma_m = 490/70 = 7$, and the strength amplitude from Table 6–7, p. 315, is

$$S_a = \frac{7^2 1057^2}{2(472.5)} \left\{ -1 + \sqrt{1 + \left[\frac{2(472.5)}{7(1057)} \right]^2} \right\} = 470.4 \text{ MPa}$$

Since $\sigma_a > S_a$, that is, $490 > 472.5$, life is reduced. From Eq. (3),

$$S_f = \frac{490}{1 - (70/1057)^2} = 492.1 \text{ MPa}$$

and from Eq. (2)

$$N = \left(\frac{492.1}{1494.5} \right)^{-1/0.0833} = 619(10^3) \text{ cycles}$$

Cycle 2: $r = 70/350 = 0.2$, and the strength amplitude is

$$S_a = \frac{0.2^2 1057^2}{2(472.5)} \left\{ -1 + \sqrt{1 + \left[\frac{2(472.5)}{0.2(1057)} \right]^2} \right\} = 169.4 \text{ MPa}$$

Since $\sigma_a < S_a$, that is $70 < 169.4$, then $S_f = S_e$ and indefinite life follows. Thus, $N \to \infty$.

Cycle 3: $r = 70/-210 = -0.333$, and since $\sigma_m < 0$, $S_f = S_e$, indefinite life follows and $N \rightarrow \infty$

Cycle Number	S_f, MPa	N, cycles
1	492.1	$619(10^3)$
2	472.5	∞
3	472.5	∞

From Eq. (6–58) the damage per block is

$$D = \sum \frac{n_i}{N_i} = N\left[\frac{1}{619(10^3)} + \frac{1}{\infty} + \frac{1}{\infty}\right] = \frac{N}{619(10^3)}$$

Answer   Setting $D = 1$ yields $N = 619(10^3)$ cycles.

To further illustrate the use of the Miner rule, let us consider a steel having the properties $S_{ut} = 560$ MPa, $S'_{e,0} = 280$ MPa, and $f = 0.9$, where we have used the designation $S'_{e,0}$ instead of the more usual S'_e to indicate the endurance limit of the *virgin,* or *undamaged, material.* The log S–log N diagram for this material is shown in Fig. 6–34 by the heavy solid line. From Eqs. (6–14) and (6–15), p. 293, we find that $a = 907$ MPa and $b = -0.085\,091$. Now apply, say, a reversed stress $\sigma_1 = 420$ MPa for $n_1 = 3000$ cycles. Since $\sigma_1 > S'_{e,0}$, the endurance limit will be damaged, and we wish to find the new endurance limit $S'_{e,1}$ of the damaged material using the Miner rule. The equation of the virgin material failure line in Fig. 6–34 in the 10^3 to 10^6 cycle range is

$$S_f = aN^b = 907N^{-0.085\,091}$$

The cycles to failure at stress level $\sigma_1 = 420$ MPa are

$$N_1 = \left(\frac{\sigma_1}{907}\right)^{-1/0.085\,091} = \left(\frac{420}{907}\right)^{-1/0.085\,091} = 8520 \text{ cycles}$$

Figure 6–34

Use of the Miner rule to predict the endurance limit of a material that has been overstressed for a finite number of cycles.

Figure 6–34 shows that the material has a life $N_1 = 8520$ cycles at 420 MPa, and consequently, after the application of σ_1 for 3000 cycles, there are $N_1 - n_1 = 5520$ cycles of life remaining at σ_1. This locates the finite-life strength $S_{f,1}$ of the damaged material, as shown in Fig. 6–34. To get a second point, we ask the question: With n_1 and N_1 given, how many cycles of stress $\sigma_2 = S'_{e,0}$ can be applied before the damaged material fails? This corresponds to n_2 cycles of stress reversal, and hence, from Eq. (6–58), we have

$$\frac{n_1}{N_1} + \frac{n_2}{N_2} = 1 \qquad\qquad (a)$$

Solving for n_2 gives

$$n_2 = (N_1 - n_1)\frac{N_2}{N_1} \qquad\qquad (b)$$

Then

$$n_2 = [8.52(10^3) - 3(10^3)]\frac{10^6}{8.52(10^3)} = 0.648(10^6) \text{ cycles}$$

This corresponds to the finite-life strength $S_{f,2}$ in Fig. 6–34. A line through $S_{f,1}$ and $S_{f,2}$ is the log S–log N diagram of the damaged material according to the Miner rule. Two points, $(N_1 - n_1, \sigma_1)$ and (n_2, σ_2), determine the new equation for the line, $S_f = a'N^{b'}$. Thus, $\sigma_1 = a'(N_1 - n_1)^{b'}$, and $\sigma_2 = a'n_2^{b'}$. Dividing the two equations, taking the logarithm of the results, and solving for b' gives

$$b' = \frac{\log(\sigma_1/\sigma_2)}{\log\left(\dfrac{N_1 - n_1}{n_2}\right)}$$

Substituting n_2 from Eq. (b) and simplifying gives

$$b' = \frac{\log(\sigma_1/\sigma_2)}{\log(N_1/N_2)}$$

For the undamaged material, $N_1 = (\sigma_1/a)^{1/b}$ and $N_2 = (\sigma_2/a)^{1/b}$, then

$$b' = \frac{\log(\sigma_1/\sigma_2)}{\log[(\sigma_1/a)^{1/b}/(\sigma_2/a)^{1/b}]} = \frac{\log(\sigma_1/\sigma_2)}{(1/b)\log(\sigma_1/\sigma_2)} = b$$

This means that the damaged material line has the same slope as the virgin material line, and the two lines are parallel. The value of a' is then found from $a' = S_f/N^b$.

For the case we are illustrating, $a' = 420/[5.52(10^3)]^{-0.085\ 091} = 874.286$ MPa, and thus the new endurance limit is $S'_{e,1} = a'N_e^b = 874.286$ MPa $[(10)^6]^{-0.085\ 091} = 207$ MPa.

Though the Miner rule is quite generally used, it fails in two ways to agree with experiment. First, note that this theory states that the static strength S_{ut} is damaged, that is, decreased, because of the application of σ_1; see Fig. 6–34 at $N = 10^3$ cycles. Experiments fail to verify this prediction.

The Miner rule, as given by Eq. (6–58), does not account for the order in which the stresses are applied, and hence ignores any stresses less than $S'_{e,0}$. But it can be

---

[25]S. S. Manson, A. J. Nachtigall, C. R. Ensign, and J. C. Fresche, "Further Investigation of a Relation for Cumulative Fatigue Damage in Bending," *Trans. ASME, J. Eng. Ind.*, ser. B, vol. 87, No. 1, pp. 25–35, February 1965.

Figure 6–35

Use of the Manson method to predict the endurance limit of a material that has been overstressed for a finite number of cycles.

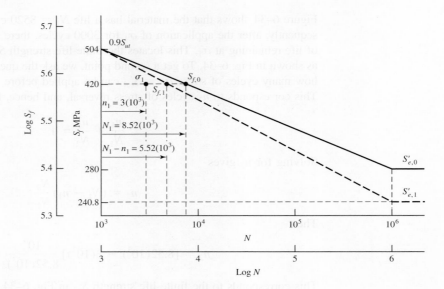

seen in Fig. 6–34 that a stress σ_3 in the range $S'_{e,1} < \sigma_3 < S'_{e,0}$ would cause damage if applied after the endurance limit had been damaged by the application of σ_1.

Manson's[25] approach overcomes both of the deficiencies noted for the Palmgren-Miner method; historically it is a much more recent approach, and it is just as easy to use. Except for a slight change, we shall use and recommend the Manson method in this book. Manson plotted the S–log N diagram instead of a log S–log N plot as is recommended here. Manson also resorted to experiment to find the point of convergence of the S–log N lines corresponding to the static strength, instead of arbitrarily selecting the intersection of $N = 10^3$ cycles with $S = 0.9S_{ut}$ as is done here. Of course, it is always better to use experiment, but our purpose in this book has been to use the simple test data to learn as much as possible about fatigue failure.

The method of Manson, as presented here, consists in having all log S–log N lines, that is, lines for both the damaged and the virgin material, converge to the same point, $0.9S_{ut}$ at 10^3 cycles. In addition, the log S–log N lines must be constructed in the same historical order in which the stresses occur.

The data from the preceding example are used for illustrative purposes. The results are shown in Fig. 6–35. Note that the strength $S_{f,1}$ corresponding to $N_1 - n_1 = 5.52(10^3)$ cycles is found in the same manner as before. Through this point and through $0.9S_{ut}$ at 10^3 cycles, draw the heavy dashed line to meet $N = 10^6$ cycles and define the endurance limit $S'_{e,1}$ of the damaged material. Again, with two points on the line, $b' = [\log (504/420)]/\log [(10^3)/5.52 (10^3)] = -0.106\ 722$, and $a' = 420/[5.52(10^3)]^{-0.106\ 722} = 1053.4$ MPa. In this case, the new endurance limit is $S'_{e,1} = a'N_e^{b'} = 1053.4\ (10^6)^{-0.106\ 722} = 240.8$ MPa, which is somewhat less than that found by the Miner method.

It is now easy to see from Fig. 6–35 that a reversed stress $\sigma = 252$ MPa, say, would not harm the endurance limit of the virgin material, no matter how many cycles it might be applied. However, if $\sigma = 252$ MPa should be applied *after* the material was damaged by $\sigma_1 = 420$ MPa, then additional damage would be done.

Both these rules involve a number of computations, which are repeated every time damage is estimated. For complicated stress-time traces, this might be every cycle. Clearly a computer program is useful to perform the tasks, including scanning the trace and identifying the cycles.

Collins said it well: "In spite of all the problems cited, the Palmgren linear damage rule is frequently used because of its simplicity and the experimental fact that other more complex damage theories do not always yield a significant improvement in failure prediction reliability."[26]

6–16 Surface Fatigue Strength

The surface fatigue mechanism is not definitively understood. The contact-affected zone, in the absence of surface shearing tractions, entertains compressive principal stresses. Rotary fatigue has its cracks grown at or near the surface in the presence of tensile stresses that are associated with crack propagation, to catastrophic failure. There are shear stresses in the zone, which are largest just below the surface. Cracks seem to grow from this stratum until small pieces of material are expelled, leaving pits on the surface. Because engineers had to design durable machinery before the surface fatigue phenomenon was understood in detail, they had taken the posture of conducting tests, observing pits on the surface, and declaring failure at an arbitrary projected area of hole, and they related this to the Hertzian contact pressure. This compressive stress did not produce the failure directly, but whatever the failure mechanism, whatever the stress type that was instrumental in the failure, the contact stress was an *index* to its magnitude.

Buckingham[27] conducted a number of tests relating the fatigue at 10^8 cycles to endurance strength (Hertzian contact pressure). While there is evidence of an endurance limit at about $3(10^7)$ cycles for cast materials, hardened steel rollers showed no endurance limit up to $4(10^8)$ cycles. Subsequent testing on hard steel shows no endurance limit. Hardened steel exhibits such high fatigue strengths that its use in resisting surface fatigue is widespread.

Our studies thus far have dealt with the failure of a machine element by yielding, by fracture, and by fatigue. The endurance limit obtained by the rotating-beam test is frequently called the *flexural endurance limit,* because it is a test of a rotating beam. In this section we shall study a property of *mating materials* called the *surface endurance shear*. The design engineer must frequently solve problems in which two machine elements mate with one another by rolling, sliding, or a combination of rolling and sliding contact. Obvious examples of such combinations are the mating teeth of a pair of gears, a cam and follower, a wheel and rail, and a chain and sprocket. A knowledge of the surface strength of materials is necessary if the designer is to create machines having a long and satisfactory life.

When two surfaces roll or roll and slide against one another with sufficient force, a pitting failure will occur after a certain number of cycles of operation. Authorities are not in complete agreement on the exact mechanism of the pitting; although the subject is quite complicated, they do agree that the Hertz stresses, the number of cycles, the surface finish, the hardness, the degree of lubrication, and the temperature all influence the strength. In Sec. 3–19 it was learned that, when two surfaces are pressed together, a maximum shear stress is developed slightly below the contacting surface. It is postulated by some authorities that a surface fatigue failure is initiated by this maximum shear stress and then is propagated rapidly to the surface. The lubricant then enters the crack that is formed and, under pressure, eventually wedges the chip loose.

---

[26]J. A. Collins, *Failure of Materials in Mechanical Design,* John Wiley & Sons, New York, 1981, p. 243.

[27]Earle Buckingham, *Analytical Mechanics of Gears,* McGraw-Hill, New York, 1949.

To determine the surface fatigue strength of mating materials, Buckingham designed a simple machine for testing a pair of contacting rolling surfaces in connection with his investigation of the wear of gear teeth. Buckingham and, later, Talbourdet gathered large numbers of data from many tests so that considerable design information is now available. To make the results useful for designers, Buckingham defined a *load-stress factor*, also called a *wear factor*, which is derived from the Hertz equations. Equations (3–73) and (3–74), p. 138, for contacting cylinders are found to be

$$b = \sqrt{\frac{2F}{\pi l} \frac{(1 - \nu_1^2)/E_1 + (1 - \nu_2^2)/E_2}{(1/d_1) + (1/d_2)}} \tag{6–59}$$

$$p_{max} = \frac{2F}{\pi bl} \tag{6–60}$$

where b = half width of rectangular contact area

F = contact force

l = length of cylinders

ν = Poisson's ratio

E = modulus of elasticity

d = cylinder diameter

It is more convenient to use the cylinder radius, so let $2r = d$. If we then designate the length of the cylinders as w (for width of gear, bearing, cam, etc.) instead of l and remove the square root sign, Eq. (6–59) becomes

$$b^2 = \frac{4F}{\pi w} \frac{(1 - \nu_1^2)/E_1 + (1 - \nu_2^2)/E_2}{1/r_1 + 1/r_2} \tag{6–61}$$

We can define a *surface endurance strength* S_C using

$$p_{max} = \frac{2F}{\pi bw} \tag{6–62}$$

as

$$S_C = \frac{2F}{\pi bw} \tag{6–63}$$

which may also be called *contact strength*, the *contact fatigue strength*, or the *Hertzian endurance strength*. The strength is the contacting pressure which, after a specified number of cycles, will cause failure of the surface. Such failures are often called *wear* because they occur over a very long time. They should not be confused with abrasive wear, however. By squaring Eq. (6–63), substituting b^2 from Eq. (6–61), and rearranging, we obtain

$$\frac{F}{w}\left(\frac{1}{r_1} + \frac{1}{r_2}\right) = \pi S_C^2 \left[\frac{1 - \nu_1^2}{E_1} + \frac{1 - \nu_2^2}{E_2}\right] = K_1 \tag{6–64}$$

The left expression consists of parameters a designer may seek to control independently. The central expression consists of material properties that come with the material and condition specification. The third expression is the parameter K_1, Buckingham's load-stress factor, determined by a test fixture with values F, w, r_1, r_2 and the number of cycles associated with the first tangible evidence of fatigue. In gear studies a similar K factor is used:

$$K_g = \frac{K_1}{4} \sin \phi \qquad (6\text{-}65)$$

where ϕ is the tooth pressure angle, and the term $[(1 - v_1^2)/E_1 + (1 - v_2^2)/E_2]$ is defined as $1/(\pi C_P^2)$, so that

$$S_C = C_P \sqrt{\frac{F}{w}\left(\frac{1}{r_1} + \frac{1}{r_2}\right)} \qquad (6\text{-}66)$$

Buckingham and others reported K_1 for 10^8 cycles and nothing else. This gives only one point on the $S_C N$ curve. For cast metals this may be sufficient, but for wrought steels, heat-treated, some idea of the slope is useful in meeting design goals of other than 10^8 cycles.

Experiments show that K_1 versus N, K_g versus N, and S_C versus N data are rectified by log-log transformation. This suggests that

$$K_1 = \alpha_1 N^{\beta_1} \qquad K_g = aN^b \qquad S_C = \alpha N^\beta$$

The three exponents are given by

$$\beta_1 = \frac{\log (K_1/K_2)}{\log (N_1/N_2)} \qquad b = \frac{\log (K_{g1}/K_{g2})}{\log (N_1/N_2)} \qquad \beta = \frac{\log (S_{C1}/S_{C2})}{\log (N_1/N_2)} \qquad (6\text{-}67)$$

Data on induction-hardened steel on steel give $(S_C)_{10^7} = 1897$ MPa and $(S_C)_{10^8} = 1673$ MPa, so β, from Eq. (6–67), is

$$\beta = \frac{\log (1897/1673)}{\log (10^7/10^8)} = -0.055$$

It may be of interest that the American Gear Manufacturers Association (AGMA) uses $\beta = -0.056$ between $10^4 < N < 10^{10}$ if the designer has no data to the contrary beyond 10^7 cycles.

A longstanding correlation in steels between S_C and H_B at 10^8 cycles is

$$(S_C)_{10^8} = \begin{cases} 0.4H_B - 10 \text{ kpsi} \\ 2.76H_B - 70 \text{ MPa} \end{cases} \qquad (6\text{-}68)$$

AGMA uses

$$_{0.99}(S_C)_{10^7} = 0.327H_B + 26 \text{ kpsi} \qquad (6\text{-}69)$$

Equation (6–66) can be used in design to find an allowable surface stress by using a design factor. Since this equation is nonlinear in its stress-load transformation, the designer must decide if loss of function denotes inability to carry the load. If so, then to find the allowable stress, one divides the load F by the design factor n_d:

$$\sigma_C = C_P \sqrt{\frac{F}{wn_d}\left(\frac{1}{r_1} + \frac{1}{r_2}\right)} = \frac{C_P}{\sqrt{n_d}}\sqrt{\frac{F}{w}\left(\frac{1}{r_1} + \frac{1}{r_2}\right)} = \frac{S_C}{\sqrt{n_d}}$$

and $n_d = (S_C/\sigma_C)^2$. If the loss of function is focused on stress, then $n_d = S_C/\sigma_C$. It is recommended that an engineer

- Decide whether loss of function is failure to carry load or stress.
- Define the design factor and factor of safety accordingly.
- Announce what he or she is using and why.
- Be prepared to defend his or her position.

In this way everyone who is party to the communication knows what a design factor (or factor of safety) of 2 means and adjusts, if necessary, the judgmental perspective.

6–17 Road Maps and Important Design Equations for the Stress-Life Method

As stated in Sec. 6–15, there are three categories of fatigue problems. The important procedures and equations for deterministic stress-life problems are presented here.

Completely Reversing Simple Loading

1 Determine S'_e either from test data or

p. 290
$$S'_e = \begin{cases} 0.5S_{ut} & S_{ut} \le 200 \text{ kpsi } (1400 \text{ MPa}) \\ 100 \text{ kpsi} & S_{ut} > 200 \text{ kpsi} \\ 700 \text{ MPa} & S_{ut} > 1400 \text{ MPa} \end{cases} \quad (6\text{–}8)$$

2 Modify S'_e to determine S_e.

p. 295
$$S_e = k_a k_b k_c k_d k_e k_f S'_e \quad (6\text{–}18)$$

$$k_a = aS_{ut}^b \quad (6\text{–}19)$$

Table 6–2

Parameters for Marin Surface Modification Factor, Eq. (6–19)

Surface Finish	Factor a S_{ut}, kpsi	S_{ut}, MPa	Exponent b
Ground	1.34	1.58	−0.085
Machined or cold-drawn	2.70	4.51	−0.265
Hot-rolled	14.4	57.7	−0.718
As-forged	39.9	272.	−0.995

Rotating shaft. For bending or torsion,

p. 296
$$k_b = \begin{cases} (d/0.3)^{-0.107} = 0.879d^{-0.107} & 0.11 \le d \le 2 \text{ in} \\ 0.91d^{-0.157} & 2 < d \le 10 \text{ in} \\ (d/7.62)^{-0.107} = 1.24d^{-0.107} & 2.79 \le d \le 51 \text{ mm} \\ 1.51d^{-0.157} & 51 < 254 \text{ mm} \end{cases} \quad (6\text{–}20)$$

For axial,

$$k_b = 1 \quad (6\text{–}21)$$

Nonrotating member. Use Table 6–3, p. 298, for d_e and substitute into Eq. (6–20) for d.

p. 298
$$k_c = \begin{cases} 1 & \text{bending} \\ 0.85 & \text{axial} \\ 0.59 & \text{torsion} \end{cases} \quad (6\text{–}26)$$

p. 299 Use Table 6–4 for k_d, or

$$k_d = 0.975 + 0.432(10^{-3})T_F - 0.115(10^{-5})T_F^2 \\ + 0.104(10^{-8})T_F^3 - 0.595(10^{-12})T_F^4 \quad (6\text{–}27)$$

pp. 300–301, k_e

Reliability, %	Transformation Variate z_a	Reliability Factor k_e
Table 6–5		
50	0	1.000
90	1.288	0.897
95	1.645	0.868
99	2.326	0.814
99.9	3.091	0.753
99.99	3.719	0.702
99.999	4.265	0.659
99.9999	4.753	0.620

Table 6–5

Reliability Factor k_e Corresponding to 8 Percent Standard Deviation of the Endurance Limit

pp. 301–302, k_f

3 Determine fatigue stress-concentration factor, K_f or K_{fs}. First, find K_t or K_{ts} from Table A–15.

p. 303 $$K_f = 1 + q(K_t - 1) \qquad \text{or} \qquad K_{fs} = 1 + q(K_{ts} - 1) \qquad (6\text{–}32)$$

Obtain q from either Fig. 6–20 or 6–21, pp. 303–304.

Alternatively,

p. 304 $$K_f = 1 + \frac{K_t - 1}{1 + \sqrt{a/r}} \qquad (6\text{–}33)$$

For $\sqrt{a}$ in units of $\sqrt{\text{in}}$, and S_{ut} in kpsi

Bending or axial: $\sqrt{a} = 0.246 - 3.08(10^{-3})S_{ut} + 1.51(10^{-5})S_{ut}^2 - 2.67(10^{-8})S_{ut}^3$

$$(6\text{–}35a)$$

Torsion: $\sqrt{a} = 0.190 - 2.51(10^{-3})S_{ut} + 1.35(10^{-5})S_{ut}^2 - 2.67(10^{-8})S_{ut}^3 \qquad (6\text{–}35b)$

4 Apply K_f or K_{fs} by *either* dividing S_e by it *or* multiplying it with the purely reversing stress, *not* both.

5 Determine fatigue life constants a and b. If $S_{ut} \geq 490$ MPa, determine f from Fig. 6–18, p. 293. If $S_{ut} < 490$ MPa, let $f = 0.9$.

p. 293 $$a = (f S_{ut})^2 / S_e \qquad (6\text{–}14)$$

$$b = -[\log (f S_{ut}/S_e)]/3 \qquad (6\text{–}15)$$

6 Determine fatigue strength S_f at N cycles, or, N cycles to failure at a reversing stress σ_{rev}

(*Note*: this only applies to purely reversing stresses where $\sigma_m = 0$).

p. 293 $$S_f = aN^b \qquad (6\text{–}13)$$

$$N = (\sigma_{\text{rev}}/a)^{1/b} \qquad (6\text{–}16)$$

Fluctuating Simple Loading

For S_e, K_f or K_{fs}, see previous subsection.

1 Calculate σ_m and σ_a. Apply K_f to both stresses.

p. 309 $$\sigma_m = (\sigma_{\max} + \sigma_{\min})/2 \qquad \sigma_a = |\sigma_{\max} - \sigma_{\min}|/2 \qquad (6\text{–}36)$$

2 Apply to a fatigue failure criterion, p. 314

$\sigma_m \geq 0$

Soderburg	$\sigma_a/S_e + \sigma_m/S_y = 1/n$	(6–45)
mod-Goodman	$\sigma_a/S_e + \sigma_m/S_{ut} = 1/n$	(6–46)
Gerber	$n\sigma_a/S_e + (n\sigma_m/S_{ut})^2 = 1$	(6–47)
ASME-elliptic	$(\sigma_a/S_e)^2 + (\sigma_m/S_y)^2 = 1/n^2$	(6–48)

$\sigma_m < 0$

p. 313 $\qquad\qquad\qquad\qquad\qquad \sigma_a = S_e/n$

Torsion. Use the same equations as apply for $\sigma_m \geq 0$, except replace σ_m and σ_a with τ_m and τ_a, use $k_c = 0.59$ for S_e, replace S_{ut} with $S_{su} = 0.67S_{ut}$ [Eq. (6–54), p. 325], and replace S_y with $S_{sy} = 0.577S_y$ [Eq. (5–21), p. 239]

3 Check for localized yielding.

p. 314 $\qquad\qquad\qquad\qquad \sigma_a + \sigma_m = S_y/n \qquad\qquad\qquad$ (6–49)

or, for torsion, $\qquad\qquad\qquad \tau_a + \tau_m = 0.577S_y/n$

4 For finite-life fatigue strength, equivalent completely reversed stress (see Ex. 6–12, pp. 321–322).

$$\text{mod-Goodman} \qquad \sigma_\text{rev} = \frac{\sigma_a}{1 - (\sigma_m/S_{ut})}$$

$$\text{Gerber} \qquad \sigma_\text{rev} = \frac{\sigma_a}{1 - (\sigma_m/S_{ut})^2}$$

If determining the finite life N with a factor of safety n, substitute σ_rev/n for σ_rev in Eq. (6–16). That is,

$$N = \left(\frac{\sigma_\text{rev}/n}{a}\right)^{1/b}$$

Combination of Loading Modes

See previous subsections for earlier definitions.

1 Calculate von Mises stresses for alternating and midrange stress states, σ_a' and σ_m'. When determining S_e, do not use K_c nor divide by K_f or K_{fs}. Apply K_f and/or K_{fs} directly to each specific alternating and midrange stress. If axial stress is present divide the alternating axial stress by $K_c = 0.85$. For the special case of combined bending, torsional shear, and axial stresses

p. 326

$$\sigma_a' = \left\{ \left[(K_f)_\text{bending}(\sigma_a)_\text{bending} + (K_f)_\text{axial}\frac{(\sigma_a)_\text{axial}}{0.85} \right]^2 + 3[(K_{fs})_\text{torsion}(\tau_a)_\text{torsion}]^2 \right\}^{1/2}$$

(6–55)

$$\sigma_m' = \{ [(K_f)_\text{bending}(\sigma_m)_\text{bending} + (K_f)_\text{axial}(\sigma_m)_\text{axial}]^2 + 3[(K_{fs})_\text{torsion}(\tau_m)_\text{torsion}]^2 \}^{1/2}$$

(6–56)

2 Apply stresses to fatigue criterion [see Eqs. (6–45) to (6–48), p. 314 in previous subsection].

3 Conservative check for localized yielding using von Mises stresses.

p. 314 $$\sigma_a' + \sigma_m' = S_y/n \tag{6–49}$$

PROBLEMS

Problems marked with an asterisk (*) are linked to problems in other chapters, as summarized in Table 1–2 of Sec. 1–17, p. 34.

6-1 A 10-mm steel drill rod was heat-treated and ground. The measured hardness was found to be 300 Brinell. Estimate the endurance strength in MPa if the rod is used in rotating bending.

6-2 Estimate S_e' in MPa for the following materials:
(a) AISI 1020 CD steel.
(b) AISI 1080 HR steel.
(c) 2024 T3 aluminum.
(d) AISI 4130 steel heat-treated to a tensile strength of 1750 MPa.

6-3 A steel rotating-beam test specimen has an ultimate strength of 840 MPa. Estimate the life of the specimen if it is tested at a completely reversed stress amplitude of 490 MPa.

6-4 A steel rotating-beam test specimen has an ultimate strength of 1600 MPa. Estimate the life of the specimen if it is tested at a completely reversed stress amplitude of 900 MPa.

6-5 A steel rotating-beam test specimen has an ultimate strength of 1610 MPa. Estimate the fatigue strength corresponding to a life of 150 kcycles of stress reversal.

6-6 Repeat Prob. 6–5 with the specimen having an ultimate strength of 1100 MPa.

6-7 A steel rotating-beam test specimen has an ultimate strength of 1050 MPa and a yield strength of 945 MPa. It is desired to test low-cycle fatigue at approximately 500 cycles. Check if this is possible without yielding by determining the necessary reversed stress amplitude.

6-8 Derive Eq. (6–17). Rearrange the equation to solve for N.

6-9 For the interval $10^3 \le N \le 10^6$ cycles, develop an expression for the axial fatigue strength $(S_f')_{ax}$ for the polished specimens of 4130 used to obtain Fig. 6–10. The ultimate strength is $S_{ut} = 875$ MPa and the endurance limit is $(S_e')_{ax} = 350$ MPa.

6-10 Estimate the endurance strength of a 32-mm-diameter rod of AISI 1040 steel having a machined finish and heat-treated to a tensile strength of 710 MPa, loaded in rotating bending.

6-11 Two steels are being considered for manufacture of as-forged connecting rods subjected to bending loads. One is AISI 4340 Cr-Mo-Ni steel capable of being heat-treated to a tensile strength of 1820 MPa. The other is a plain carbon steel AISI 1040 with an attainable S_{ut} of 791 MPa. Each rod is to have a size giving an equivalent diameter d_e of 20 mm. Determine the endurance limit for each material. Is there any advantage to using the alloy steel for this fatigue application?

6–12 A 25-mm-diameter solid round bar has a groove 2.5-mm deep with a 2.5-mm radius machined into it. The bar is made of AISI 1020 CD steel and is subjected to a purely reversing torque of 200 N · m. For the *S-N* curve of this material, let $f = 0.9$.

(*a*) Estimate the number of cycles to failure.

(*b*) If the bar is also placed in an environment with a temperature of 750°F, estimate the number of cycles to failure.

6–13 A solid square rod is cantilevered at one end. The rod is 0.6 m long and supports a completely reversing transverse load at the other end of ±2 kN. The material is AISI 1080 hot-rolled steel. If the rod must support this load for 10^4 cycles with a design factor of 1.5, what dimension should the square cross section have? Neglect any stress concentrations at the support end.

6–14 A rectangular bar is cut from an AISI 1018 cold-drawn steel flat. The bar is 60 mm wide by 10 mm in thick and has a 12-mm-dia. hole drilled through the center as depicted in Table A–15–1. The bar is concentrically loaded in push-pull fatigue by axial forces F_a, uniformly distributed across the width. Using a design factor of $n_d = 1.8$, estimate the largest force F_a that can be applied ignoring column action.

6–15 A solid round bar with diameter of 50 mm has a groove cut to a diameter of 45 mm, with a radius of 2.5 mm. The bar is not rotating. The bar is loaded with a repeated bending load that causes the bending moment at the groove to fluctuate between 0 and 2825 N · m. The bar is hot-rolled AISI 1095, but the groove has been machined. Determine the factor of safety for fatigue based on infinite life using the modified Goodman criterion, and the factor of safety for yielding.

6–16 The rotating shaft shown in the figure is machined from AISI 1020 CD steel. It is subjected to a force of $F = 6$ kN. Find the minimum factor of safety for fatigue based on infinite life. If the life is not infinite, estimate the number of cycles. Be sure to check for yielding.

Problem 6–16
Dimensions in millimeters

6–17 The shaft shown in the figure is machined from AISI 1040 CD steel. The shaft rotates at 1600 rpm and is supported in rolling bearings at A and B. The applied forces are $F_1 = 10$ kN and $F_2 = 4$ kN. Determine the minimum fatigue factor of safety based on achieving infinite life. If infinite life is not predicted, estimate the number of cycles to failure. Also check for yielding.

Problem 6–17

6–18   Solve Prob. 6–17 except with forces $F_1 = 4.8$ kN and $F_2 = 9.6$ kN.

6–19   Bearing reactions R_1 and R_2 are exerted on the shaft shown in the figure, which rotates at 1150 rev/min and supports a 45-kN bending force. Use a 1095 HR steel. Specify a diameter d using a design factor of $n_d = 1.6$ for a life of 10 hr. The surfaces are machined.

Problem 6–19

6–20   A bar of steel has the minimum properties $S_e = 276$ MPa, $S_y = 413$ MPa, and $S_{ut} = 551$ MPa. The bar is subjected to a steady torsional stress of 103 MPa and an alternating bending stress of 172 MPa. Find the factor of safety guarding against a static failure, and either the factor of safety guarding against a fatigue failure or the expected life of the part. For the fatigue analysis use:
   (*a*) Modified Goodman criterion.
   (*b*) Gerber criterion.
   (*c*) ASME-elliptic criterion.

6–21   Repeat Prob. 6–20 but with a steady torsional stress of 138 MPa and an alternating bending stress of 69 MPa.

6–22   Repeat Prob. 6–20 but with a steady torsional stress of 103 MPa, an alternating torsional stress of 69 MPa, and an alternating bending stress of 83 MPa.

6–23   Repeat Prob. 6–20 but with an alternating torsional stress of 207 MPa.

6–24   Repeat Prob. 6–20 but with an alternating torsional stress of 103 MPa and a steady bending stress of 103 MPa.

6–25   The cold-drawn AISI 1040 steel bar shown in the figure is subjected to a completely reversed axial load fluctuating between 28 kN in compression to 28 kN in tension. Estimate the fatigue factor of safety based on achieving infinite life and the yielding factor of safety. If infinite life is not predicted, estimate the number of cycles to failure.

Problem 6–25

6–26   Repeat Prob. 6–25 for a load that fluctuates from 12 kN to 28 kN. Use the Modified Goodman, Gerber, and ASME-elliptic criteria and compare their predictions.

6–27   Using the modified Goodman criterion for infinite life, repeat Prob. 6–25 for each of the following loading conditions:
   (*a*) 0 kN to 28 kN
   (*b*) 12 kN to 28 kN
   (*c*) −28 kN to 12 kN

6-28   The figure shows a formed round-wire cantilever spring subjected to a varying force. The hardness tests made on 50 springs gave a minimum hardness of 400 Brinell. It is apparent from the mounting details that there is no stress concentration. A visual inspection of the springs indicates that the surface finish corresponds closely to a hot-rolled finish. Ignore curvature effects on the bending stress. What number of applications is likely to cause failure? Solve using:

(*a*) Modified Goodman criterion.

(*b*) Gerber criterion.

Problem 6–28

6-29   The figure is a drawing of a 4- by 20-mm latching spring. A preload is obtained during assembly by shimming under the bolts to obtain an estimated initial deflection of 2 mm. The latching operation itself requires an additional deflection of exactly 4 mm. The material is ground high-carbon steel, bent then hardened and tempered to a minimum hardness of 490 Bhn. The inner radius of the bend is 4 mm. Estimate the yield strength to be 90 percent of the ultimate strength.

(*a*) Find the maximum and minimum latching forces.

(*b*) Determine the fatigue factor of safety for infinite life, using the modified Goodman criterion.

Problem 6–29
Dimensions in millimeters

6-30   The figure shows the free-body diagram of a connecting-link portion having stress concentration at three sections. The dimensions are $r = 6$ mm, $d = 20$ mm, $h = 12$ mm, $w_1 = 90$ mm, and $w_2 = 60$ mm. The forces F fluctuate between a tension of 18 kN and a compression of 72 kN. Neglect column action and find the least factor of safety if the material is cold-drawn AISI 1018 steel.

Problem 6–30

6–31   Solve Prob. 6–30 except let $w_1 = 60$ mm, $w_2 = 36$ mm, and the force fluctuates between a tension of 72 kN and a compression of 18 kN. Use the modified Goodman criteria.

6–32   For the part in Prob. 6–30, recommend a fillet radius r that will cause the fatigue factor of safety to be the same at the hole and at the fillet.

6–33   The torsional coupling in the figure is composed of a curved beam of square cross section that is welded to an input shaft and output plate. A torque is applied to the shaft and cycles from zero to T. The cross section of the beam has dimensions of 5×5 mm, and the centroidal axis of the beam describes a curve of the form $r = 20 + 10\,\theta/\pi$, where r and θ are in inches and radians, respectively ($0 \leq \theta \leq 4\pi$). The curved beam has a machined surface with yield and ultimate strength values of 420 and 770 MPa, respectively.

(a) Determine the maximum allowable value of T such that the coupling will have an infinite life with a factor of safety, $n = 3$, using the modified Goodman criterion.

(b) Repeat part (a) using the Gerber criterion.

(c) Using T found in part (b), determine the factor of safety guarding against yield.

Problem 6–33

(Dimensions in mm)

6–34   Repeat Prob. 6–33 ignoring curvature effects on the bending stress.

6–35   A part is loaded with a combination of bending, axial, and torsion such that the following stresses are created at a particular location:

Bending: Completely reversed, with a maximum stress of 60 MPa
Axial: Constant stress of 20 MPa
Torsion: Repeated load, varying from 0 MPa to 50 MPa

Assume the varying stresses are in phase with each other. The part contains a notch such that $K_{f,bending} = 1.4$, $K_{f,axial} = 1.1$, and $K_{f,torsion} = 2.0$. The material properties are $S_y = 300$ MPa and $S_u = 400$ MPa. The completely adjusted endurance limit is found to be $S_e = 200$ MPa. Find the factor of safety for fatigue based on infinite life, using the modified Goodman criterion. If the life is not infinite, estimate the number of cycles. Be sure to check for yielding.

6–36   Repeat the requirements of Prob. 6–35 with the following loading conditions:

Bending: Fluctuating stress from –40 MPa to 150 MPa
Axial: None
Torsion: Mean stress of 90 MPa, with an alternating stress of 10 percent of the mean stress

6–37* to
6–46*

For the problem specified in the table, build upon the results of the original problem to determine the minimum factor of safety for fatigue based on infinite life, using the modified Goodman criterion. The shaft rotates at a constant speed, has a constant diameter, and is made from cold-drawn AISI 1018 steel.

Problem Number	Original Problem, Page Number
6–37*	3–68, 151
6–38*	3–69, 151
6–39*	3–70, 151
6–40*	3–71, 151
6–41*	3–72, 152
6–42*	3–73, 152
6–43*	3–74, 152
6–44*	3–76, 153
6–45*	3–77, 153
6–46*	3–79, 153

6–47* to
6–50*

For the problem specified in the table, build upon the results of the original problem to determine the minimum factor of safety for fatigue based on infinite life, using the modified Goodman criterion. If the life is not infinite, estimate the number of cycles. The force F is applied as a repeated load. The material is AISI 1018 CD steel. The fillet radius at the wall is 2 mm, with theoretical stress concentrations of 1.5 for bending, 1.2 for axial, and 2.1 for torsion.

Problem Number	Original Problem, Page Number
6–47*	3–80, 153
6–48*	3–81, 154
6–49*	3–82, 154
6–50*	3–83, 154

6–51* to
6–53*

For the problem specified in the table, build upon the results of the original problem to determine the minimum factor of safety for fatigue at point A, based on infinite life, using the modified Goodman criterion. If the life is not infinite, estimate the number of cycles. The force F is applied as a repeated load. The material is AISI 1018 CD steel.

Problem Number	Original Problem, Page Number
6–51*	3–84, 154
6–52*	3–85, 155
6–53*	3–86, 155

6–54

Solve Prob. 6–17 except include a steady torque of 280 N · m in being transmitted through the shaft between the points of application of the forces.

6–55

Solve Prob. 6–18 except include a steady torque of 250 N · m being transmitted through the shaft between the points of application of the forces.

6-56   In the figure shown, shaft A, made of AISI 1010 hot-rolled steel, is welded to a fixed support and is subjected to loading by equal and opposite forces F via shaft B. A theoretical stress-concentration factor K_{ts} of 1.6 is induced in the shaft by the 3-mm weld fillet. The length of shaft A from the fixed support to the connection at shaft B is 1 m. The load F cycles from 0.5 to 2 kN.

(a) For shaft A, find the factor of safety for infinite life using the modified Goodman fatigue failure criterion.

(b) Repeat part (a) using the Gerber fatigue failure criterion.

Problem 6–56

6-57   A schematic of a clutch-testing machine is shown. The steel shaft rotates at a constant speed ω. An axial load is applied to the shaft and is cycled from zero to P. The torque T induced by the clutch face onto the shaft is given by

$$T = \frac{fP(D + d)}{4}$$

where D and d are defined in the figure and f is the coefficient of friction of the clutch face. The shaft is machined with $S_y = 800$ MPa and $S_{ut} = 1000$ MPa. The theoretical stress-concentration factors for the fillet are 3.0 and 1.8 for the axial and torsional loading, respectively.

Assume the load variation P is synchronous with shaft rotation. With $f = 0.3$, find the maximum allowable load P such that the shaft will survive a minimum of 10^6 cycles with a factor of safety of 3. Use the modified Goodman criterion. Determine the corresponding factor of safety guarding against yielding.

Problem 6–57

6–58 For the clutch of Prob. 6–57, the external load P is cycled between 20 kN and 80 kN. Assuming that the shaft is rotating synchronous with the external load cycle, estimate the number of cycles to failure. Use the modified Goodman fatigue failure criteria.

6–59 A flat leaf spring has fluctuating stress of $\sigma_{max} = 360$ MPa and $\sigma_{min} = 160$ MPa applied for $8\ (10^4)$ cycles. If the load changes to $\sigma_{max} = 320$ MPa and $\sigma_{min} = -200$ MPa, how many cycles should the spring survive, using the modified Goodman criterion? The material is AISI 1020 CD and has a fully corrected endurance strength of $S_e = 175$ MPa. Assume that $f = 0.9$.
(a) Use Miner's method.
(b) Use Manson's method.

6–60 A rotating-beam specimen with an endurance limit of 350 MPa and an ultimate strength of 700 MPa is cycled 20 percent of the time at 490 MPa, 50 percent at 385 MPa, and 30 percent at 280 MPa. Let $f = 0.9$ and estimate the number of cycles to failure.

6–61 A machine part will be cycled at ± 350 MPa for $5\ (10^3)$ cycles. Then the loading will be changed to ± 260 MPa for $5\ (10^4)$ cycles. Finally, the load will be changed to ± 225 MPa. How many cycles of operation can be expected at this stress level? For the part, $S_{ut} = 530$ MPa, $f = 0.9$, and has a fully corrected endurance strength of $S_e = 210$ MPa.
(a) Use Miner's method.
(b) Use Manson's method.

6–62 The material properties of a machine part are $S_{ut} = 595$ MPa, $f = 0.86$, and a fully corrected endurance limit of $S_e = 315$ MPa. The part is to be cycled at $\sigma_a = 245$ MPa and $\sigma_m = 210$ MPa for $12\ (10^3)$ cycles. Using the Gerber criterion, estimate the new endurance limit after cycling.
(a) Use Miner's method.
(b) Use Manson's method.

6–63 Repeat Prob. 6–62 using the Goodman criterion.

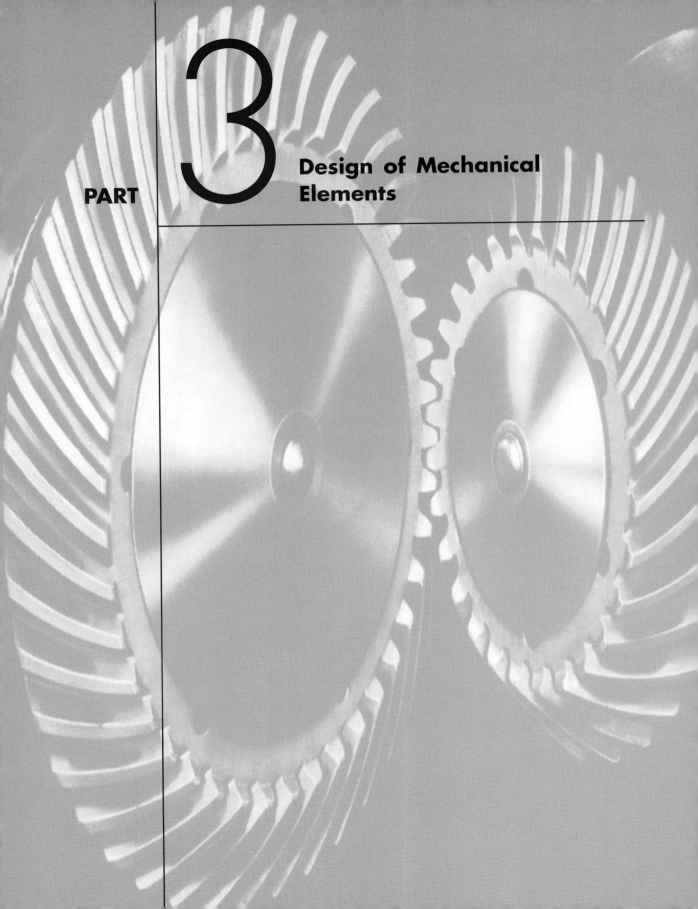

PART

3

Design of Mechanical Elements

7

Shafts and Shaft Components

7–1 Introduction

A *shaft* is a rotating member, usually of circular cross section, used to transmit power or motion. It provides the axis of rotation, or oscillation, of elements such as gears, pulleys, flywheels, cranks, sprockets, and the like and controls the geometry of their motion. An *axle* is a nonrotating member that carries no torque and is used to support rotating wheels, pulleys, and the like. The automotive axle is not a true axle; the term is a carryover from the horse-and-buggy era, when the wheels rotated on nonrotating members. A nonrotating axle can readily be designed and analyzed as a static beam, and will not warrant the special attention given in this chapter to the rotating shafts which are subject to fatigue loading.

There is really nothing unique about a shaft that requires any special treatment beyond the basic methods already developed in previous chapters. However, because of the ubiquity of the shaft in so many machine design applications, there is some advantage in giving the shaft and its design a closer inspection. A complete shaft design has much interdependence on the design of the components. The design of the machine itself will dictate that certain gears, pulleys, bearings, and other elements will have at least been partially analyzed and their size and spacing tentatively determined. Chapter 18 provides a complete case study of a power transmission, focusing on the overall design process. In this chapter, details of the shaft itself will be examined, including the following:

- Material selection
- Geometric layout
- Stress and strength
    Static strength
    Fatigue strength
- Deflection and rigidity
    Bending deflection
    Torsional deflection
    Slope at bearings and shaft-supported elements
    Shear deflection due to transverse loading of short shafts
- Vibration due to natural frequency

In deciding on an approach to shaft sizing, it is necessary to realize that a stress analysis at a specific point on a shaft can be made using only the shaft geometry in the vicinity of that point. Thus the geometry of the entire shaft is not needed. In design it is usually possible to locate the critical areas, size these to meet the strength requirements, and then size the rest of the shaft to meet the requirements of the shaft-supported elements.

The deflection and slope analyses cannot be made until the geometry of the entire shaft has been defined. Thus deflection is a function of the geometry *everywhere,* whereas the stress at a section of interest is a function of *local geometry.* For this reason, shaft design allows a consideration of stress first. Then, after tentative values for the shaft dimensions have been established, the determination of the deflections and slopes can be made.

7–2 Shaft Materials

Deflection is not affected by strength, but rather by stiffness as represented by the modulus of elasticity, which is essentially constant for all steels. For that reason, rigidity cannot be controlled by material decisions, but only by geometric decisions.

Necessary strength to resist loading stresses affects the choice of materials and their treatments. Many shafts are made from low carbon, cold-drawn or hot-rolled steel, such as AISI 1020-1050 steels.

Significant strengthening from heat treatment and high alloy content are often not warranted. Fatigue failure is reduced moderately by increase in strength, and then only to a certain level before adverse effects in endurance limit and notch sensitivity begin to counteract the benefits of higher strength. A good practice is to start with an inexpensive, low or medium carbon steel for the first time through the design calculations. If strength considerations turn out to dominate over deflection, then a higher strength material should be tried, allowing the shaft sizes to be reduced until excess deflection becomes an issue. The cost of the material and its processing must be weighed against the need for smaller shaft diameters. When warranted, typical alloy steels for heat treatment include AISI 1340-50, 3140-50, 4140, 4340, 5140, and 8650.

Shafts usually don't need to be surface hardened unless they serve as the actual journal of a bearing surface. Typical material choices for surface hardening include carburizing grades of AISI 1020, 4320, 4820, and 8620.

Cold drawn steel is usually used for diameters under about 3 inches. The nominal diameter of the bar can be left unmachined in areas that do not require fitting of components. Hot rolled steel should be machined all over. For large shafts requiring much material removal, the residual stresses may tend to cause warping. If concentricity is important, it may be necessary to rough machine, then heat treat to remove residual stresses and increase the strength, then finish machine to the final dimensions.

In approaching material selection, the amount to be produced is a salient factor. For low production, turning is the usual primary shaping process. An economic viewpoint may require removing the least material. High production may permit a volume-conservative shaping method (hot or cold forming, casting), and minimum material in the shaft can become a design goal. Cast iron may be specified if the production quantity is high, and the gears are to be integrally cast with the shaft.

Properties of the shaft locally depend on its history—cold work, cold forming, rolling of fillet features, heat treatment, including quenching medium, agitation, and tempering regimen.[1]

Stainless steel may be appropriate for some environments.

7–3 Shaft Layout

The general layout of a shaft to accommodate shaft elements, e.g., gears, bearings, and pulleys, must be specified early in the design process in order to perform a free body force analysis and to obtain shear-moment diagrams. The geometry of a shaft is generally that of a stepped cylinder. The use of shaft shoulders is an excellent means of axially locating the shaft elements and to carry any thrust loads. Figure 7–1 shows an example of a stepped shaft supporting the gear of a worm-gear speed reducer. Each shoulder in the shaft serves a specific purpose, which you should attempt to determine by observation.

---

[1]See Joseph E. Shigley, Charles R. Mischke, and Thomas H. Brown, Jr. (eds-in-chief), *Standard Handbook of Machine Design,* 3rd ed., McGraw-Hill, New York, 2004. For cold-worked property prediction see Chap. 29, and for heat-treated property prediction see Chaps. 29 and 33.

Figure 7–1

A vertical worm-gear speed reducer. *(Courtesy of the Cleveland Gear Company.)*

Figure 7–2

(*a*) Choose a shaft configuration to support and locate the two gears and two bearings. (*b*) Solution uses an integral pinion, three shaft shoulders, key and keyway, and sleeve. The housing locates the bearings on their outer rings and receives the thrust loads. (*c*) Choose fan-shaft configuration. (*d*) Solution uses sleeve bearings, a straight-through shaft, locating collars, and setscrews for collars, fan pulley, and fan itself. The fan housing supports the sleeve bearings.

The geometric configuration of a shaft to be designed is often simply a revision of existing models in which a limited number of changes must be made. If there is no existing design to use as a starter, then the determination of the shaft layout may have many solutions. This problem is illustrated by the two examples of Fig. 7–2. In Fig. 7–2*a* a geared countershaft is to be supported by two bearings. In Fig. 7–2*c* a fanshaft is to be configured. The solutions shown in Fig. 7–2*b* and 7–2*d* are not necessarily the best ones, but they do illustrate how the shaft-mounted devices are fixed and located in the axial direction, and how provision is made for torque transfer from one element to another. There are no absolute rules for specifying the general layout, but the following guidelines may be helpful.

Axial Layout of Components

The axial positioning of components is often dictated by the layout of the housing and other meshing components. In general, it is best to support load-carrying components between bearings, such as in Fig. 7–2a, rather than cantilevered outboard of the bearings, such as in Fig. 7–2c. Pulleys and sprockets often need to be mounted outboard for ease of installation of the belt or chain. The length of the cantilever should be kept short to minimize the deflection.

Only two bearings should be used in most cases. For extremely long shafts carrying several load-bearing components, it may be necessary to provide more than two bearing supports. In this case, particular care must be given to the alignment of the bearings.

Shafts should be kept short to minimize bending moments and deflections. Some axial space between components is desirable to allow for lubricant flow and to provide access space for disassembly of components with a puller. Load bearing components should be placed near the bearings, again to minimize the bending moment at the locations that will likely have stress concentrations, and to minimize the deflection at the load-carrying components.

The components must be accurately located on the shaft to line up with other mating components, and provision must be made to securely hold the components in position. The primary means of locating the components is to position them against a shoulder of the shaft. A shoulder also provides a solid support to minimize deflection and vibration of the component. Sometimes when the magnitudes of the forces are reasonably low, shoulders can be constructed with retaining rings in grooves, sleeves between components, or clamp-on collars. In cases where axial loads are very small, it may be feasible to do without the shoulders entirely, and rely on press fits, pins, or collars with setscrews to maintain an axial location. See Fig. 7–2b and 7–2d for examples of some of these means of axial location.

Supporting Axial Loads

In cases where axial loads are not trivial, it is necessary to provide a means to transfer the axial loads into the shaft, then through a bearing to the ground. This will be particularly necessary with helical or bevel gears, or tapered roller bearings, as each of these produces axial force components. Often, the same means of providing axial location, e.g., shoulders, retaining rings, and pins, will be used to also transmit the axial load into the shaft.

It is generally best to have only one bearing carry the axial load, to allow greater tolerances on shaft length dimensions, and to prevent binding if the shaft expands due to temperature changes. This is particularly important for long shafts. Figures 7–3 and 7–4 show examples of shafts with only one bearing carrying the axial load against a shoulder, while the other bearing is simply press-fit onto the shaft with no shoulder.

Providing for Torque Transmission

Most shafts serve to transmit torque from an input gear or pulley, through the shaft, to an output gear or pulley. Of course, the shaft itself must be sized to support the torsional stress and torsional deflection. It is also necessary to provide a means of transmitting the torque between the shaft and the gears. Common torque-transfer elements are:

- Keys
- Splines
- Setscrews

Figure 7–3

Tapered roller bearings used in a mowing machine spindle. This design represents good practice for the situation in which one or more torque-transfer elements must be mounted outboard. *(Source: Redrawn from material furnished by The Timken Company.)*

Figure 7–4

A bevel-gear drive in which both pinion and gear are straddle-mounted. *(Source: Redrawn from material furnished by Gleason Machine Division.)*

- Pins
- Press or shrink fits
- Tapered fits

In addition to transmitting the torque, many of these devices are designed to fail if the torque exceeds acceptable operating limits, protecting more expensive components.

Details regarding hardware components such as *keys, pins,* and *setscrews* are addressed in detail in Sec. 7–7. One of the most effective and economical means of transmitting moderate to high levels of torque is through a key that fits in a groove in the shaft and gear. Keyed components generally have a slip fit onto the shaft, so assembly and disassembly are easy. The key provides for positive angular orientation of the component, which is useful in cases where phase angle timing is important.

Splines are essentially stubby gear teeth formed on the outside of the shaft and on the inside of the hub of the load-transmitting component. Splines are generally much more expensive to manufacture than keys, and are usually not necessary for simple torque transmission. They are typically used to transfer high torques. One feature of a spline is that it can be made with a reasonably loose slip fit to allow for large axial motion between the shaft and component while still transmitting torque. This is useful for connecting two shafts where relative motion between them is common, such as in connecting a power takeoff (PTO) shaft of a tractor to an implement. SAE and ANSI publish standards for splines. Stress-concentration factors are greatest where the spline ends and blends into the shaft, but are generally quite moderate.

For cases of low torque transmission, various means of transmitting torque are available. These include pins, setscrews in hubs, tapered fits, and press fits.

Press and shrink fits for securing hubs to shafts are used both for torque transfer and for preserving axial location. The resulting stress-concentration factor is usually quite small. See Sec. 7–8 for guidelines regarding appropriate sizing and tolerancing to transmit torque with press and shrink fits. A similar method is to use a split hub with screws to clamp the hub to the shaft. This method allows for disassembly and lateral adjustments. Another similar method uses a two-part hub consisting of a split inner member that fits into a tapered hole. The assembly is then tightened to the shaft with screws, which forces the inner part into the wheel and clamps the whole assembly against the shaft.

Tapered fits between the shaft and the shaft-mounted device, such as a wheel, are often used on the overhanging end of a shaft. Screw threads at the shaft end then permit the use of a nut to lock the wheel tightly to the shaft. This approach is useful because it can be disassembled, but it does not provide good axial location of the wheel on the shaft.

At the early stages of the shaft layout, the important thing is to select an appropriate means of transmitting torque, and to determine how it affects the overall shaft layout. It is necessary to know where the shaft discontinuities, such as keyways, holes, and splines, will be in order to determine critical locations for analysis.

Assembly and Disassembly

Consideration should be given to the method of assembling the components onto the shaft, and the shaft assembly into the frame. This generally requires the largest diameter in the center of the shaft, with progressively smaller diameters towards the ends to allow components to be slid on from the ends. If a shoulder is needed on both sides of a component, one of them must be created by such means as a retaining ring or by a sleeve between two components. The gearbox itself will need means to physically position the shaft into its bearings, and the bearings into the frame. This is typically accomplished by providing access through the housing to the bearing at one end of the shaft. See Figs. 7–5 through 7–8 for examples.

Figure 7–5

Arrangement showing bearing inner rings press-fitted to shaft while outer rings float in the housing. The axial clearance should be sufficient only to allow for machinery vibrations. Note the labyrinth seal on the right.

Figure 7–6

Similar to the arrangement of Fig. 7–5 except that the outer bearing rings are preloaded.

Figure 7–7

In this arrangement the inner ring of the left-hand bearing is locked to the shaft between a nut and a shaft shoulder. The locknut and washer are AFBMA standard. The snap ring in the outer race is used to positively locate the shaft assembly in the axial direction. Note the floating right-hand bearing and the grinding runout grooves in the shaft.

Figure 7–8

This arrangement is similar to Fig. 7–7 in that the left-hand bearing positions the entire shaft assembly. In this case the inner ring is secured to the shaft using a snap ring. Note the use of a shield to prevent dirt generated from within the machine from entering the bearing.

When components are to be press-fit to the shaft, the shaft should be designed so that it is not necessary to press the component down a long length of shaft. This may require an extra change in diameter, but it will reduce manufacturing and assembly cost by only requiring the close tolerance for a short length.

Consideration should also be given to the necessity of disassembling the components from the shaft. This requires consideration of issues such as accessibility of retaining rings, space for pullers to access bearings, openings in the housing to allow pressing the shaft or bearings out, etc.

7–4 Shaft Design for Stress

Critical Locations

It is not necessary to evaluate the stresses in a shaft at every point; a few potentially critical locations will suffice. Critical locations will usually be on the outer surface, at axial locations where the bending moment is large, where the torque is present, and where stress concentrations exist. By direct comparison of various points along the shaft, a few critical locations can be identified upon which to base the design. An assessment of typical stress situations will help.

Most shafts will transmit torque through a portion of the shaft. Typically the torque comes into the shaft at one gear and leaves the shaft at another gear. A free body diagram of the shaft will allow the torque at any section to be determined. The torque is often relatively constant at steady state operation. The shear stress due to the torsion will be greatest on outer surfaces.

The bending moments on a shaft can be determined by shear and bending moment diagrams. Since most shaft problems incorporate gears or pulleys that introduce forces in two planes, the shear and bending moment diagrams will generally be needed in two planes. Resultant moments are obtained by summing moments as vectors at points of interest along the shaft. The phase angle of the moments is not important since the shaft rotates. A steady bending moment will produce a completely reversed moment on a rotating shaft, as a specific stress element will alternate from compression to tension in every revolution of the shaft. The normal stress due to bending moments will be greatest on the outer surfaces. In situations where a bearing is located at the end of the shaft, stresses near the bearing are often not critical since the bending moment is small.

Axial stresses on shafts due to the axial components transmitted through helical gears or tapered roller bearings will almost always be negligibly small compared to the bending moment stress. They are often also constant, so they contribute little to fatigue. Consequently, it is usually acceptable to neglect the axial stresses induced by the gears and bearings when bending is present in a shaft. If an axial load is applied to the shaft in some other way, it is not safe to assume it is negligible without checking magnitudes.

Shaft Stresses

Bending, torsion, and axial stresses may be present in both midrange and alternating components. For analysis, it is simple enough to combine the different types of stresses into alternating and midrange von Mises stresses, as shown in Sec. 6–14, p. 325. It is sometimes convenient to customize the equations specifically for shaft applications. Axial loads are usually comparatively very small at critical locations where bending and torsion dominate, so they will be left out of the following equations. The fluctuating stresses due to bending and torsion are given by

$$\sigma_a = K_f \frac{M_a c}{I} \qquad \sigma_m = K_f \frac{M_m c}{I} \qquad (7\text{--}1)$$

$$\tau_a = K_{fs} \frac{T_a r}{J} \qquad \tau_m = K_{fs} \frac{T_m r}{J} \qquad (7\text{--}2)$$

where M_m and M_a are the midrange and alternating bending moments, T_m and T_a are the midrange and alternating torques, and K_f and K_{fs} are the fatigue stress-concentration factors for bending and torsion, respectively.

Assuming a solid shaft with round cross section, appropriate geometry terms can be introduced for c, I, r, and J resulting in

$$\sigma_a = K_f \frac{32M_a}{\pi d^3} \qquad \sigma_m = K_f \frac{32M_m}{\pi d^3} \qquad (7\text{--}3)$$

$$\tau_a = K_{fs} \frac{16T_a}{\pi d^3} \qquad \tau_m = K_{fs} \frac{16T_m}{\pi d^3} \qquad (7\text{--}4)$$

Using the distortion energy failure theory, the von Mises stress is given by Eq. (5–15), p. 237, with $\sigma_x = \sigma$, the bending stress, $\sigma_y = 0$, and $\tau_{xy} = \tau$, the torsional shear stress. Thus, for rotating round solid shafts, neglecting axial loads, the fluctuating von Mises stresses are given by

$$\sigma_a' = (\sigma_a^2 + 3\tau_a^2)^{1/2} = \left[\left(\frac{32K_f M_a}{\pi d^3}\right)^2 + 3\left(\frac{16K_{fs}T_a}{\pi d^3}\right)^2\right]^{1/2} \tag{7-5}$$

$$\sigma_m' = (\sigma_m^2 + 3\tau_m^2)^{1/2} = \left[\left(\frac{32K_f M_m}{\pi d^3}\right)^2 + 3\left(\frac{16K_{fs}T_m}{\pi d^3}\right)^2\right]^{1/2} \tag{7-6}$$

Note that the stress-concentration factors are sometimes considered optional for the midrange components with ductile materials, because of the capacity of the ductile material to yield locally at the discontinuity.

These equivalent alternating and midrange stresses can be evaluated using an appropriate failure curve on the modified Goodman diagram (See Sec. 6–12, p. 311, and Fig. 6–27). For example, the fatigue failure criteria for the modified Goodman line as expressed previously in Eq. (6–46) is

$$\frac{1}{n} = \frac{\sigma_a'}{S_e} + \frac{\sigma_m'}{S_{ut}}$$

Substitution of σ_a' and σ_m' from Eqs. (7–5) and (7–6) results in

$$\frac{1}{n} = \frac{16}{\pi d^3}\left\{\frac{1}{S_e}[4(K_f M_a)^2 + 3(K_{fs}T_a)^2]^{1/2} + \frac{1}{S_{ut}}[4(K_f M_m)^2 + 3(K_{fs}T_m)^2]^{1/2}\right\}$$

For design purposes, it is also desirable to solve the equation for the diameter. This results in

$$d = \left(\frac{16n}{\pi}\left\{\frac{1}{S_e}[4(K_f M_a)^2 + 3(K_{fs}T_a)^2]^{1/2} + \frac{1}{S_{ut}}[4(K_f M_m)^2 + 3(K_{fs}T_m)^2]^{1/2}\right\}\right)^{1/3}$$

Similar expressions can be obtained for any of the common failure criteria by substituting the von Mises stresses from Eqs. (7–5) and (7–6) into any of the failure criteria expressed by Eqs. (6–45) through (6–48), p. 314. The resulting equations for several of the commonly used failure curves are summarized below. The names given to each set of equations identifies the significant failure theory, followed by a fatigue failure locus name. For example, DE-Gerber indicates the stresses are combined using the distortion energy (DE) theory, and the Gerber criteria is used for the fatigue failure.

DE-Goodman

$$\frac{1}{n} = \frac{16}{\pi d^3}\left\{\frac{1}{S_e}[4(K_f M_a)^2 + 3(K_{fs}T_a)^2]^{1/2} + \frac{1}{S_{ut}}[4(K_f M_m)^2 + 3(K_{fs}T_m)^2]^{1/2}\right\} \tag{7-7}$$

$$d = \left(\frac{16n}{\pi}\left\{\frac{1}{S_e}[4(K_f M_a)^2 + 3(K_{fs}T_a)^2]^{1/2} + \frac{1}{S_{ut}}[4(K_f M_m)^2 + 3(K_{fs}T_m)^2]^{1/2}\right\}\right)^{1/3} \tag{7-8}$$

DE-Gerber

$$\frac{1}{n} = \frac{8A}{\pi d^3 S_e} \left\{ 1 + \left[1 + \left(\frac{2BS_e}{AS_{ut}} \right)^2 \right]^{1/2} \right\} \tag{7-9}$$

$$d = \left(\frac{8nA}{\pi S_e} \left\{ 1 + \left[1 + \left(\frac{2BS_e}{AS_{ut}} \right)^2 \right]^{1/2} \right\} \right)^{1/3} \tag{7-10}$$

where

$$A = \sqrt{4(K_f M_a)^2 + 3(K_{fs} T_a)^2}$$
$$B = \sqrt{4(K_f M_m)^2 + 3(K_{fs} T_m)^2}$$

DE-ASME Elliptic

$$\frac{1}{n} = \frac{16}{\pi d^3} \left[4 \left(\frac{K_f M_a}{S_e} \right)^2 + 3 \left(\frac{K_{fs} T_a}{S_e} \right)^2 + 4 \left(\frac{K_f M_m}{S_y} \right)^2 + 3 \left(\frac{K_{fs} T_m}{S_y} \right)^2 \right]^{1/2} \tag{7-11}$$

$$d = \left\{ \frac{16n}{\pi} \left[4 \left(\frac{K_f M_a}{S_e} \right)^2 + 3 \left(\frac{K_{fs} T_a}{S_e} \right)^2 + 4 \left(\frac{K_f M_m}{S_y} \right)^2 + 3 \left(\frac{K_{fs} T_m}{S_y} \right)^2 \right]^{1/2} \right\}^{1/3} \tag{7-12}$$

DE-Soderberg

$$\frac{1}{n} = \frac{16}{\pi d^3} \left\{ \frac{1}{S_e} [4(K_f M_a)^2 + 3(K_{fs} T_a)^2]^{1/2} + \frac{1}{S_y} [4(K_f M_m)^2 + 3(K_{fs} T_m)^2]^{1/2} \right\} \tag{7-13}$$

$$d = \left(\frac{16n}{\pi} \left\{ \frac{1}{S_e} [4(K_f M_a)^2 + 3(K_{fs} T_a)^2]^{1/2} + \frac{1}{S_y} [4(K_f M_m)^2 + 3(K_{fs} T_m)^2]^{1/2} \right\} \right)^{1/3} \tag{7-14}$$

For a rotating shaft with constant bending and torsion, the bending stress is completely reversed and the torsion is steady. Equations (7–7) through (7–14) can be simplified by setting M_m and T_a equal to 0, which simply drops out some of the terms.

Note that in an analysis situation in which the diameter is known and the factor of safety is desired, as an alternative to using the specialized equations above, it is always still valid to calculate the alternating and mid-range stresses using Eqs. (7–5) and (7–6), and substitute them into one of the equations for the failure criteria, Eqs. (6–45) through (6–48), and solve directly for n. In a design situation, however, having the equations pre-solved for diameter is quite helpful.

It is always necessary to consider the possibility of static failure in the first load cycle. The Soderberg criteria inherently guards against yielding, as can be seen by noting that its failure curve is conservatively within the yield (Langer) line on Fig. 6–27, p. 313. The ASME Elliptic also takes yielding into account, but is not entirely conservative throughout its entire range. This is evident by noting that it crosses the yield line in Fig. 6–27. The Gerber and modified Goodman criteria do not guard against yielding, requiring a separate check for yielding. A von Mises maximum stress is calculated for this purpose.

$$\sigma'_{max} = [(\sigma_m + \sigma_a)^2 + 3(\tau_m + \tau_a)^2]^{1/2}$$

$$= \left[\left(\frac{32K_f(M_m + M_a)}{\pi d^3} \right)^2 + 3 \left(\frac{16K_{fs}(T_m + T_a)}{\pi d^3} \right)^2 \right]^{1/2}$$

(7–15)

To check for yielding, this von Mises maximum stress is compared to the yield strength, as usual.

$$n_y = \frac{S_y}{\sigma'_{max}}$$

(7–16)

For a quick, conservative check, an estimate for σ'_{max} can be obtained by simply adding σ'_a and σ'_m. $(\sigma'_a + \sigma'_m)$ will always be greater than or equal to σ'_{max}, and will therefore be conservative.

EXAMPLE 7–1

At a machined shaft shoulder the small diameter d is 28 mm, the large diameter D is 42 mm, and the fillet radius is 2.8 mm. The bending moment is 142.4 N·m and the steady torsion moment is 124.3 N·m. The heat-treated steel shaft has an ultimate strength of $S_{ut} = 735$ MPa and a yield strength of $S_y = 574$ MPa. The reliability goal is 0.99.

(a) Determine the fatigue factor of safety of the design using each of the fatigue failure criteria described in this section.

(b) Determine the yielding factor of safety.

Solution

(a) $D/d = 42/28 = 1.50$, $r/d = 2.8/28 = 0.10$, $K_t = 1.68$ (Fig. A–15–9), $K_{ts} = 1.42$ (Fig. A–15–8), $q = 0.85$ (Fig. 6–20), $q_{shear} = 0.92$ (Fig. 6–21).

From Eq. (6–32),

$$K_f = 1 + 0.85(1.68 - 1) = 1.58$$

$$K_{fs} = 1 + 0.92(1.42 - 1) = 1.39$$

Eq. (6–8):

$$S'_e = 0.5(735) = 367.5 \text{ MPa}$$

Eq. (6–19):

$$k_a = 4.51(735)^{-0.265} = 0.787$$

Eq. (6–20):

$$k_b = \left(\frac{28}{7.62} \right)^{-0.107} = 0.870$$

$$k_c = k_d = k_f = 1$$

Table 6–6: $$k_e = 0.814$$

$$S_e = 0.787(0.870)0.814(367.5) = 205 \text{ MPa}$$

For a rotating shaft, the constant bending moment will create a completely reversed bending stress.

$$M_a = 142.4 \text{ N} \cdot \text{m} \qquad T_m = 124.3 \text{ N} \cdot \text{m} \qquad M_m = T_a = 0$$

Applying Eq. (7–7) for the DE-Goodman criteria gives

$$\frac{1}{n} = \frac{16}{\pi(0.028)^3} \left\{ \frac{\left[4\,(1.58 \cdot 142.4)^2\right]^{1/2}}{205 \times 10^6} + \frac{\left[3\,(1.39 \cdot 124.3)^2\right]^{1/2}}{735 \times 10^6} \right\} = 0.615$$

Answer
$$n = 1.62 \qquad \text{DE-Goodman}$$

Similarly, applying Eqs. (7–9), (7–11), and (7–13) for the other failure criteria,

Answer
$$n = 1.87 \qquad \text{DE-Gerber}$$

Answer
$$n = 1.88 \qquad \text{DE-ASME Elliptic}$$

Answer
$$n = 1.56 \qquad \text{DE-Soderberg}$$

For comparison, consider an equivalent approach of calculating the stresses and applying the fatigue failure criteria directly. From Eqs. (7–5) and (7–6),

$$\sigma_a' = \left[\left(\frac{32 \cdot 1.58 \cdot 142.4}{\pi\,(0.028)^3} \right)^2 \right]^{1/2} = 104.4 \text{ MPa}$$

$$\sigma_m' = \left[3 \left(\frac{16 \cdot 1.39 \cdot 124.3}{\pi(0.028)^3} \right)^2 \right]^{1/2} = 69.4 \text{ MPa}$$

Taking, for example, the Goodman failure criteria, application of Eq. (6–46) gives

$$\frac{1}{n} = \frac{\sigma_a'}{S_e} + \frac{\sigma_m'}{S_{ut}} = \frac{104.4}{205} + \frac{69.4}{735} = 0.604$$

$$n = 1.62$$

which is identical with the previous result. The same process could be used for the other failure criteria.

(b) For the yielding factor of safety, determine an equivalent von Mises maximum stress using Eq. (7–15).

$$\sigma_{max}' = \left[\left(\frac{32(1.58)\,(142.4)}{\pi\,(0.028)^3} \right)^2 + 3 \left(\frac{16(1.39)\,(124.3)}{\pi\,(0.028)^3} \right)^2 \right]^{1/2} = 125.4 \text{ MPa}$$

Answer
$$n_y = \frac{S_y}{\sigma_{max}'} = \frac{574}{125.4} = 4.58$$

For comparison, a quick and very conservative check on yielding can be obtained by replacing $\sigma'_{\max}$ with $\sigma'_a + \sigma'_m$. This just saves the extra time of calculating $\sigma'_{\max}$ if σ'_a and σ'_m have already been determined. For this example,

$$n_y = \frac{S_y}{\sigma'_a + \sigma'_m} = \frac{574}{104.4 + 69.4} = 3.3$$

which is quite conservative compared with $n_y = 4.58$.

that are not known the first time through the process. Fortunately, since these elements are usually of standard proportions, it is possible to estimate the stress-concentration factors for initial design of the shaft. These stress concentrations will be fine-tuned in successive iterations, once the details are known.

Shoulders for bearing and gear support should match the catalog recommendation for the specific bearing or gear. A look through bearing catalogs shows that a typical bearing calls for the ratio of D/d to be between 1.2 and 1.5. For a first approximation, the worst case of 1.5 can be assumed. Similarly, the fillet radius at the shoulder needs to be sized to avoid interference with the fillet radius of the mating component. There is a significant variation in typical bearings in the ratio of fillet radius versus bore diameter, with r/d typically ranging from around 0.02 to 0.06. A quick look at the stress concentration charts (Figures A–15–8 and A–15–9) shows that the stress concentrations for bending and torsion increase significantly in this range. For example, with $D/d = 1.5$ for bending, $K_t = 2.7$ at $r/d = 0.02$, and reduces to $K_t = 2.1$ at $r/d = 0.05$, and further down to $K_t = 1.7$ at $r/d = 0.1$. This indicates that this is an area where some attention to detail could make a significant difference. Fortunately, in most cases the shear and bending moment diagrams show that bending moments are quite low near the bearings, since the bending moments from the ground reaction forces are small.

In cases where the shoulder at the bearing is found to be critical, the designer should plan to select a bearing with generous fillet radius, or consider providing for a larger fillet radius on the shaft by relieving it into the base of the shoulder as shown in Fig. 7–9a. This effectively creates a dead zone in the shoulder area that does not carry the bending stresses, as shown by the stress flow lines. A shoulder relief groove as shown in Fig. 7–9b can accomplish a similar purpose. Another option is to cut a large-radius relief groove into the small diameter of the shaft, as

(a)   (b)   (c)

Figure 7–9

Techniques for reducing stress concentration at a shoulder supporting a bearing with a sharp radius. (a) Large radius undercut into the shoulder. (b) Large radius relief groove into the back of the shoulder. (c) Large radius relief groove into the small diameter.

shown in Fig. 7–9c. This has the disadvantage of reducing the cross-sectional area, but is often used in cases where it is useful to provide a relief groove before the shoulder to prevent the grinding or turning operation from having to go all the way to the shoulder.

For the standard shoulder fillet, for estimating K_t values for the first iteration, an r/d ratio should be selected so K_t values can be obtained. For the worst end of the spectrum, with $r/d = 0.02$ and $D/d = 1.5$, K_t values from the stress concentration charts for shoulders indicate 2.7 for bending, 2.2 for torsion, and 3.0 for axial.

A keyway will produce a stress concentration near a critical point where the load-transmitting component is located. The stress concentration in an end-milled keyseat is a function of the ratio of the radius r at the bottom of the groove and the shaft diameter d. For early stages of the design process, it is possible to estimate the stress concentration for keyways regardless of the actual shaft dimensions by assuming a typical ratio of $r/d = 0.02$. This gives $K_t = 2.14$ for bending and $K_{ts} = 3.0$ for torsion, assuming the key is in place.

Figures A–15–16 and A–15–17 give values for stress concentrations for flat-bottomed grooves such as those used for retaining rings. By examining typical retaining ring specifications in vendor catalogs, it can be seen that the groove width is typically slightly greater than the groove depth, and the radius at the bottom of the groove is around $1/10$ of the groove width. From Figs. A–15–16 and A–15–17, stress-concentration factors for typical retaining ring dimensions are around 5 for bending and axial, and 3 for torsion. Fortunately, the small radius will often lead to a smaller notch sensitivity, reducing K_f.

Table 7–1 summarizes some typical stress-concentration factors for the first iteration in the design of a shaft. Similar estimates can be made for other features. The point is to notice that stress concentrations are essentially normalized so that they are dependent on ratios of geometry features, not on the specific dimensions. Consequently, by estimating the appropriate ratios, the first iteration values for stress concentrations can be obtained. These values can be used for initial design, then actual values inserted once diameters have been determined.

Table 7–1

First Iteration Estimates for Stress-Concentration Factors K_t and K_{ts}.

Warning: These factors are only estimates for use when actual dimensions are not yet determined. Do *not* use these once actual dimensions are available.

	Bending	Torsional	Axial
Shoulder fillet—sharp ($r/d = 0.02$)	2.7	2.2	3.0
Shoulder fillet—well rounded ($r/d = 0.1$)	1.7	1.5	1.9
End-mill keyseat ($r/d = 0.02$)	2.14	3.0	—
Sled runner keyseat	1.7	—	—
Retaining ring groove	5.0	3.0	5.0

Missing values in the table are not readily available.

EXAMPLE 7-2

This example problem is part of a larger case study. See Chap. 18 for the full context.

A double reduction gearbox design has developed to the point that the general layout and axial dimensions of the countershaft carrying two spur gears has been proposed, as shown in Fig. 7–10. The gears and bearings are located and supported by shoulders, and held in place by retaining rings. The gears transmit torque through keys. Gears have been specified as shown, allowing the tangential and radial forces transmitted through the gears to the shaft to be determined as follows.

$$W_{23}^t = 540 \text{ lbf} \qquad W_{54}^t = 2431 \text{ lbf}$$

$$W_{23}^r = 197 \text{ lbf} \qquad W_{54}^r = 885 \text{ lbf}$$

where the superscripts *t* and *r* represent tangential and radial directions, respectively; and, the subscripts 23 and 54 represent the forces exerted by gears 2 and 5 (not shown) on gears 3 and 4, respectively.

Proceed with the next phase of the design, in which a suitable material is selected, and appropriate diameters for each section of the shaft are estimated, based on providing sufficient fatigue and static stress capacity for infinite life of the shaft, with minimum safety factors of 1.5.

Figure 7-10

Shaft layout for Ex. 7–2. Dimensions in millimeters.

Solution

Perform free body diagram analysis to get reaction forces at the bearings.

$R_{Az} = 422$ N

$R_{Ay} = 1439$ N

$R_{Bz} = 8822$ N

$R_{By} = 3331$ N

From ΣM_x, find the torque in the shaft between the gears, $T = W_{23}^t(d_3/2) = 2400(0.3/2)$ $= 360$ N $\cdot$ m.

Generate shear-moment diagrams for two planes.

Combine orthogonal planes as vectors to get total moments, e.g., at J, $\sqrt{485^2 + 183^2}$ $= 518$ N $\cdot$ m.

Start with point I, where the bending moment is high, there is a stress concentration at the shoulder, and the torque is present.

At I, $M_a = 468$ N $\cdot$ m, $T_m = 360$ N $\cdot$ m, $M_m = T_a = 0$

Assume generous fillet radius for gear at I.

From Table 7–1, estimate $K_t = 1.7$, $K_{ts} = 1.5$. For quick, conservative first pass, assume $K_f = K_t$, $K_{fs} = K_{ts}$.

Choose inexpensive steel, 1020 CD, with $S_{ut} = 469$ MPa. For S_e,

Eq. (6–19)

$$k_a = a S_{ut}^b = 4.51(469)^{-0.265} = 0.883$$

Guess $k_b = 0.9$. Check later when d is known.

$$k_c = k_d = k_e = 1$$

Eq. (6–18)

$$S_e = (0.883)(0.9)(0.5)(469) = 186 \text{ MPa}$$

For first estimate of the small diameter at the shoulder at point I, use the DE-Goodman criterion of Eq. (7–8). This criterion is good for the initial design, since it is simple and conservative. With $M_m = T_a = 0$, Eq. (7–8) reduces to

$$d = \left\{ \frac{16n}{\pi} \left(\frac{2(K_f M_a)}{S_e} + \frac{[3(K_{fs} T_m)^2]^{1/2}}{S_{ut}} \right) \right\}^{1/3}$$

$$d = \left\{ \frac{16(1.5)}{\pi} \left(\frac{2(1.7)(468)}{186 \times 10^6} + \frac{\{3[(1.5)(360)]^2\}^{1/2}}{469 \times 10^6} \right) \right\}^{1/3}$$

$$d = 0.0432 \text{ m} = 43.2 \text{ mm}$$

All estimates have probably been conservative, so select the next standard size below 43.2 mm and check, $d = 42$ mm.

A typical D/d ratio for support at a shoulder is $D/d = 1.2$, thus, $D = 1.2 \times 42 = 50.4$ mm. Use $D = 50$ mm. A nominal 50-mm cold-drawn shaft diameter can be used. Check if estimates were acceptable.

$$D/d = 50/42 = 1.19$$

Assume fillet radius $r = d/10 \cong 4$ mm, $r/d = 0.1$

$$K_t = 1.6 \text{ (Fig. A–15–9)}, \quad q = 0.82 \text{ (Fig. 6–20)}$$

Eq. (6–32)

$$K_f = 1 + 0.82(1.6 - 1) = 1.49$$

$$K_{ts} = 1.35 \text{ (Fig. A–15–8)}, \quad q_s = 0.95 \text{ (Fig. 6–21)}$$

$$K_{fs} = 1 + 0.95(1.35 - 1) = 1.33$$

$$k_a = 0.883 \text{ (no change)}$$

Eq. (6–20)

$$k_b = \left(\frac{42}{7.62} \right)^{-0.107} = 0.833$$

$$S_e = (0.883)(0.833)(0.5)(469) = 172 \text{ MPa}$$

Eq. (7–5)

$$\sigma'_a = \frac{32 K_f M_a}{\pi d^3} = \frac{32(1.49)(468)}{\pi (0.042)^3} = 96 \text{ MPa}$$

Eq. (7–6)

$$\sigma'_m = \left[3 \left(\frac{16 K_{fs} T_m}{\pi d^3} \right)^2 \right]^{1/2} = \frac{\sqrt{3}(16)(1.33)(360)}{\pi (0.042)^3} = 57 \text{ MPa}$$

Using Goodman criterion

$$\frac{1}{n_f} = \frac{\sigma'_a}{S_e} + \frac{\sigma'_m}{S_{ut}} = \frac{96}{172} + \frac{57}{469} = 0.68$$

$$n_f = 1.47$$

Note that we could have used Eq. (7-7) directly.
Check yielding.

$$n_y = \frac{S_y}{\sigma'_{max}} > \frac{S_y}{\sigma'_a + \sigma'_m} = \frac{363}{96 + 57} = 2.57$$

Also check this diameter at the end of the keyway, just to the right of point I, and at the groove at point K. From moment diagram, estimate M at end of keyway to be $M = 443$ N·m

Assume the radius at the bottom of the keyway will be the standard $r/d = 0.02$, $r = 0.02$, $d = 0.02(42) = 0.84$ mm

$$K_t = 2.14 \text{ (Fig. A-15-18)}, \quad q = 0.65 \text{ (Fig. 6-20)}$$
$$K_f = 1 + 0.65(2.14 - 1) = 1.74$$
$$K_{ts} = 3.0 \text{ (Fig. A-15-19)}, \quad q_s = 0.9 \text{ (Fig. 6-21)}$$
$$K_{fs} = 1 + 0.9(3 - 1) = 2.8$$
$$\sigma'_a = \frac{32K_f M_a}{\pi d^3} = \frac{32(1.74)(443)}{\pi(0.042)^3} = 106 \text{ MPa}$$
$$\sigma'_m = \sqrt{3}(16)\frac{K_{fs} T_m}{\pi d^3} = \frac{\sqrt{3}(16)(2.8)(443)}{\pi(0.042)^3} = 148 \text{ MPa}$$
$$\frac{1}{n_f} = \frac{\sigma'_a}{S_e} + \frac{\sigma'_m}{S_{ut}} = \frac{106}{172} + \frac{148}{469} = 0.93$$
$$n_f = 1.08$$

The keyway turns out to be more critical than the shoulder. We can either increase the diameter or use a higher strength material. Unless the deflection analysis shows a need for larger diameters, let us choose to increase the strength. We started with a very low strength and can afford to increase it to avoid larger sizes. Try 1050 CD with $S_{ut} = 690$ MPa.

Recalculate factors affected by S_{ut}, i.e., $k_a \rightarrow S_e$; $q \rightarrow K_f \rightarrow \sigma'_a$

$$k_a = 4.51(690)^{-0.265} = 0.797, \quad S_e = 0.797(0.833)(0.5)(690) = 229 \text{ MPa}$$
$$q = 0.72, \quad K_f = 1 + 0.72(2.14 - 1) = 1.82$$
$$\sigma'_a = \frac{32(1.82)(443)}{\pi(0.042)^3} = 110.8 \text{ MPa}$$
$$\frac{1}{n_f} = \frac{110.8}{229} + \frac{148}{690} = 0.7$$
$$n_f = 1.43$$

Since the Goodman criterion is conservative, we will accept this as close enough to the requested 1.5.

Check at the groove at K, since K_t for flat-bottomed grooves are often very high. From the torque diagram, note that no torque is present at the groove. From the moment diagram, $M_a = 283$ N·m, $M_m = T_a = T_m = 0$. To quickly check if this location is potentially critical, just use $K_f = K_t = 5.0$ as an estimate, from Table 7-1.

$$\sigma_a = \frac{32K_f M_a}{\pi d^3} = \frac{32(5)(283)}{\pi(0.042)^3} = 194.5 \text{ MPa}$$
$$n_f = \frac{S_e}{\sigma_a} = \frac{229}{194.5} = 1.18$$

This is low. We will look up data for a specific retaining ring to obtain K_f more accurately. With a quick on-line search of a retaining ring specification using the website www.globalspec.com, appropriate groove specifications for a retaining ring for a shaft diameter of 42 mm are obtained as follows: width, $a = 1.73$ mm; depth, $t = 1.22$ mm; and corner radius at bottom of groove, $r = 0.25$ mm. From Fig. A–15–16, with $r/t = 0.25/1.22 = 0.205$, and $a/t = 1.73/1.22 = 1.42$

$$K_t = 4.3, q = 0.65 \text{ (Fig. 6–20)}$$

$$K_f = 1 + 0.65(4.3 - 1) = 3.15$$

$$\sigma_a = \frac{32K_f M_a}{\pi d^3} = \frac{32(3.15)(283)}{\pi(0.042)^3} = 122.6 \text{ MPa}$$

$$n_f = \frac{S_e}{\sigma_a} = \frac{229}{122.6} = 1.87$$

Quickly check if point M might be critical. Only bending is present, and the moment is small, but the diameter is small and the stress concentration is high for a sharp fillet required for a bearing. From the moment diagram, $M_a = 113$ N·m, and $M_m = T_m = T_a = 0$.

Estimate $K_t = 2.7$ from Table 7–1, $d = 25$ mm, and fillet radius r to fit a typical bearing.

$$r/d = 0.02, r = 0.02(25) = 0.5$$

$$q = 0.7 \text{ (Fig. 6–20)}$$

$$K_f = 1 + (0.7)(2.7 - 1) = 2.19$$

$$\sigma_a = \frac{32K_f M_a}{\pi d^3} = \frac{32(2.19)(113)}{\pi(0.025)^3} = 161 \text{ MPa}$$

$$n_f = \frac{S_e}{\sigma_a} = \frac{229}{161} = 1.42$$

This should be OK. It is close enough to recheck after the bearing is selected.

With the diameters specified for the critical locations, fill in trial values for the rest of the diameters, taking into account typical shoulder heights for bearing and gear support.

$$D_1 = D_7 = 25 \text{ mm}$$

$$D_2 = D_6 = 35 \text{ mm}$$

$$D_3 = D_5 = 42 \text{ mm}$$

$$D_4 = 50 \text{ mm}$$

The bending moments are much less on the left end of the shaft, so D_1, D_2, and D_3 could be smaller. However, unless weight is an issue, there is little advantage to requiring more material removal. Also, the extra rigidity may be needed to keep deflections small.

Table 7–2

Typical Maximum Ranges for Slopes and Transverse Deflections

Slopes	
Tapered roller	0.0005–0.0012 rad
Cylindrical roller	0.0008–0.0012 rad
Deep-groove ball	0.001–0.003 rad
Spherical ball	0.026–0.052 rad
Self-align ball	0.026–0.052 rad
Uncrowned spur gear	<0.0005 rad

Transverse Deflections	
Spur gears with $P < 10$ teeth/cm	0.25 mm
Spur gears with $11 < P < 19$	0.125 mm
Spur gears with $20 < P < 50$	0.075 mm

7–5 Deflection Considerations

Deflection analysis at even a single point of interest requires complete geometry information for the entire shaft. For this reason, it is desirable to design the dimensions at critical locations to handle the stresses, and fill in reasonable estimates for all other dimensions, before performing a deflection analysis. Deflection of the shaft, both linear and angular, should be checked at gears and bearings. Allowable deflections will depend on many factors, and bearing and gear catalogs should be used for guidance on allowable misalignment for specific bearings and gears. As a rough guideline, typical ranges for maximum slopes and transverse deflections of the shaft centerline are given in Table 7–2. The allowable transverse deflections for spur gears are dependent on the size of the teeth, as represented by the diametral pitch P, which equals the number of teeth divided by the pitch diameter.

In Sec. 4–4 several beam deflection methods are described. For shafts, where the deflections may be sought at a number of different points, integration using either singularity functions or numerical integration is practical. In a stepped shaft, the cross-sectional properties change along the shaft at each step, increasing the complexity of integration, since both M and I vary. Fortunately, only the gross geometric dimensions need to be included, as the local factors such as fillets, grooves, and keyways do not have much impact on deflection. Example 4–7 demonstrates the use of singularity functions for a stepped shaft. Many shafts will include forces in multiple planes, requiring either a three-dimensional analysis, or the use of superposition to obtain deflections in two planes which can then be summed as vectors.

A deflection analysis is straightforward, but it is lengthy and tedious to carry out manually, particularly for multiple points of interest. Consequently, practically all shaft deflection analysis will be evaluated with the assistance of software. Any general-purpose finite-element software can readily handle a shaft problem (see Chap. 19). This is practical if the designer is already familiar with using the software and with how to properly model the shaft. Special-purpose software solutions for 3-D shaft analysis are available, but somewhat expensive if only used occasionally. Software requiring very little training is readily available for planar beam analysis, and can be downloaded from the internet. Example 7–3 demonstrates how to incorporate such a program for a shaft with forces in multiple planes.

EXAMPLE 7-3

This example problem is part of a larger case study. See Chap. 18 for the full context.

In Ex. 7-2, a preliminary shaft geometry was obtained on the basis of design for stress. The resulting shaft is shown in Fig. 7-10, with proposed diameters of

$$D_1 = D_7 = 25 \text{ mm}$$

$$D_2 = D_6 = 35 \text{ mm}$$

$$D_3 = D_5 = 40 \text{ mm}$$

$$D_4 = 50 \text{ mm}$$

Check that the deflections and slopes at the gears and bearings are acceptable. If necessary, propose changes in the geometry to resolve any problems.

Solution

A simple planar beam analysis program will be used. By modeling the shaft twice, with loads in two orthogonal planes, and combining the results, the shaft deflections can readily be obtained. For both planes, the material is selected (steel with $E = 210$ GPa), the shaft lengths and diameters are entered, and the bearing locations are specified. Local details like grooves and keyways are ignored, as they will have insignificant effect on the deflections. Then the tangential gear forces are entered in the horizontal xz plane model, and the radial gear forces are entered in the vertical xy plane model. The software can calculate the bearing reaction forces, and numerically integrate to generate plots for shear, moment, slope, and deflection, as shown in Fig. 7-11.

Figure 7-11

Shear, moment, slope, and deflection plots from two planes. (*Source: Beam 2D Stress Analysis, Orand Systems, Inc.*)

Point of Interest	xz Plane	xy Plane	Total
Left bearing slope	0.02263 deg	0.01770 deg	0.02872 deg
			0.000501 rad
Right bearing slope	0.05711 deg	0.02599 deg	0.06274 deg
			0.001095 rad
Left gear slope	0.02067 deg	0.01162 deg	0.02371 deg
			0.000414 rad
Right gear slope	0.02155 deg	0.01149 deg	0.02442 deg
			0.000426 rad
Left gear deflection	0.0189 mm	0.0129 mm	0.0229 mm
Right gear deflection	0.0397 mm	0.0188 mm	0.0439 mm

Table 7–3

Slope and Deflection Values at Key Locations

The deflections and slopes at points of interest are obtained from the plots, and combined with orthogonal vector addition, that is, $\delta = \sqrt{\delta_{xz}^2 + \delta_{xy}^2}$. Results are shown in Table 7–3.

Whether these values are acceptable will depend on the specific bearings and gears selected, as well as the level of performance expected. According to the guidelines in Table 7–2, all of the bearing slopes are well below typical limits for ball bearings. The right bearing slope is within the typical range for cylindrical bearings. Since the load on the right bearing is relatively high, a cylindrical bearing might be used. This constraint should be checked against the specific bearing specifications once the bearing is selected.

The gear slopes and deflections more than satisfy the limits recommended in Table 7–2. It is recommended to proceed with the design, with an awareness that changes that reduce rigidity should warrant another deflection check.

Once deflections at various points have been determined, if any value is larger than the allowable deflection at that point, a larger shaft diameter is warranted. Since I is proportional to d^4, a new diameter can be found from

$$d_{\text{new}} = d_{\text{old}} \left| \frac{n_d y_{\text{old}}}{y_{\text{all}}} \right|^{1/4} \qquad (7\text{–}17)$$

where y_{all} is the allowable deflection at that station and n_d is the design factor. Similarly, if any slope is larger than the allowable slope θ_{all}, a new diameter can be found from

$$d_{\text{new}} = d_{\text{old}} \left| \frac{n_d (dy/dx)_{\text{old}}}{(\text{slope})_{\text{all}}} \right|^{1/4} \qquad (7\text{–}18)$$

where $(\text{slope})_{\text{all}}$ is the allowable slope. As a result of these calculations, determine the largest $d_{\text{new}}/d_{\text{old}}$ ratio, then multiply *all* diameters by this ratio. The tight constraint will be just tight, and all others will be loose. Don't be too concerned about end journal sizes, as their influence is usually negligible. The beauty of the method is that the deflections need to be completed just once and constraints can be rendered loose but for one, with diameters all identified without reworking every deflection.

EXAMPLE 7–4   For the shaft in Ex. 7–3, it was noted that the slope at the right bearing is near the limit for a cylindrical roller bearing. Determine an appropriate increase in diameters to bring this slope down to 0.0005 rad.

Solution   Applying Eq. (7–17) to the deflection at the right bearing gives

$$d_{\text{new}} = d_{\text{old}} \left| \frac{n_d \text{slope}_{\text{old}}}{\text{slope}_{\text{all}}} \right|^{1/4} = 25 \left| \frac{(1)(0.001095)}{(0.0005)} \right|^{1/4} = 30.4 \text{ mm}$$

Multiplying all diameters by the ratio

$$\frac{d_{\text{new}}}{d_{\text{old}}} = \frac{30.4}{25} = 1.216$$

gives a new set of diameters,

$$D_1 = D_7 = 30.4 \text{ mm}$$
$$D_2 = D_6 = 42.6 \text{ mm}$$
$$D_3 = D_5 = 49.4 \text{ mm}$$
$$D_4 = 60.8 \text{ mm}$$

Repeating the beam deflection analysis of Ex. 7–3 with these new diameters produces a slope at the right bearing of 0.0125 mm, with all other deflections less than their previous values.

The transverse shear V at a section of a beam in flexure imposes a shearing deflection, which is superposed on the bending deflection. Usually such shearing deflection is less than 1 percent of the transverse bending deflection, and it is seldom evaluated. However, when the shaft length-to-diameter ratio is less than 10, the shear component of transverse deflection merits attention. There are many short shafts. A tabular method is explained in detail elsewhere,[2] including examples.

For right-circular cylindrical shafts in torsion the angular deflection θ is given in Eq. (4–5). For a stepped shaft with individual cylinder length l_i and torque T_i, the angular deflection can be estimated from

$$\theta = \sum \theta_i = \sum \frac{T_i l_i}{G_i J_i} \tag{7–19}$$

or, for a constant torque throughout homogeneous material, from

$$\theta = \frac{T}{G} \sum \frac{l_i}{J_i} \tag{7–20}$$

This should be treated only as an estimate, since experimental evidence shows that the actual θ is larger than given by Eqs. (7–19) and (7–20).[3]

---

[2]C.R. Mischke, "Tabular Method for Transverse Shear Deflection," Sec. 17.3 in Joseph E. Shigley, Charles R. Mischke, and Thomas H. Brown, Jr. (eds.), *Standard Handbook of Machine Design,* 3rd ed., McGraw-Hill, New York, 2004.

[3]R. Bruce Hopkins, *Design Analysis of Shafts and Beams,* McGraw-Hill, New York, 1970, pp. 93–99.

If torsional stiffness is defined as $k_i = T_i/\theta_i$ and, since $\theta_i = T_i/k_i$ and $\theta = \sum \theta_i = \sum(T_i/k_i)$, for constant torque $\theta = T \sum(1/k_i)$, it follows that the torsional stiffness of the shaft k in terms of segment stiffnesses is

$$\frac{1}{k} = \sum \frac{1}{k_i} \tag{7–21}$$

7–6 Critical Speeds for Shafts

When a shaft is turning, eccentricity causes a centrifugal force deflection, which is resisted by the shaft's flexural rigidity EI. As long as deflections are small, no harm is done. Another potential problem, however, is called *critical speeds:* at certain speeds the shaft is unstable, with deflections increasing without upper bound. It is fortunate that although the dynamic deflection shape is unknown, using a static deflection curve gives an excellent estimate of the lowest critical speed. Such a curve meets the boundary condition of the differential equation (zero moment and deflection at both bearings) and the shaft energy is not particularly sensitive to the exact shape of the deflection curve. Designers seek first critical speeds at least twice the operating speed.

The shaft, because of its own mass, has a critical speed. The ensemble of attachments to a shaft likewise has a critical speed that is much lower than the shaft's intrinsic critical speed. Estimating these critical speeds (and harmonics) is a task of the designer. When geometry is simple, as in a shaft of uniform diameter, simply supported, the task is easy. It can be expressed[4] as

$$\omega_1 = \left(\frac{\pi}{l}\right)^2 \sqrt{\frac{EI}{m}} = \left(\frac{\pi}{l}\right)^2 \sqrt{\frac{gEI}{A\gamma}} \tag{7–22}$$

where m is the mass per unit length, A the cross-sectional area, and γ the specific weight. For an ensemble of attachments, Rayleigh's method for lumped masses gives[5]

$$\omega_1 = \sqrt{\frac{g \sum w_i y_i}{\sum w_i y_i^2}} \tag{7–23}$$

where w_i is the weight of the ith location and y_i is the deflection at the ith body location. It is possible to use Eq. (7–23) for the case of Eq. (7–22) by partitioning the shaft into segments and placing its weight force at the segment centroid as seen in Fig. 7–12.

Figure 7–12

(*a*) A uniform-diameter shaft for Eq. (7–22). (*b*) A segmented uniform-diameter shaft for Eq. (7–23).

(*a*)

(*b*)

---

[4]William T. Thomson and Marie Dillon Dahleh, *Theory of Vibration with Applications,* Prentice Hall, 5th ed., 1998, p. 273.

[5]Thomson, op. cit., p. 357.

Figure 7–13

The influence coefficient δ_{ij} is the deflection at i due to a unit load at j.

Computer assistance is often used to lessen the difficulty in finding transverse deflections of a stepped shaft. Rayleigh's equation overestimates the critical speed.

To counter the increasing complexity of detail, we adopt a useful viewpoint. Inasmuch as the shaft is an elastic body, we can use *influence coefficients*. An influence coefficient is the transverse deflection at location i on a shaft due to a unit load at location j on the shaft. From Table A–9–6 we obtain, for a simply supported beam with a single unit load as shown in Fig. 7–13,

$$\delta_{ij} = \begin{cases} \dfrac{b_j x_i}{6EIl} (l^2 - b_j^2 - x_i^2) & x_i \le a_i \\[2ex] \dfrac{a_j(l - x_i)}{6EIl} (2lx_i - a_j^2 - x_i^2) & x_i > a_i \end{cases} \tag{7–24}$$

For three loads the influence coefficients may be displayed as

i	j		
	1	**2**	**3**
1	δ_{11}	δ_{12}	δ_{13}
2	δ_{21}	δ_{22}	δ_{23}
3	δ_{31}	δ_{32}	δ_{33}

Maxwell's reciprocity theorem[6] states that there is a symmetry about the main diagonal, composed of δ_{11}, δ_{22}, and δ_{33}, of the form $\delta_{ij} = \delta_{ji}$. This relation reduces the work of finding the influence coefficients. From the influence coefficients above, one can find the deflections y_1, y_2, and y_3 of Eq. (7–23) as follows:

$$\begin{aligned} y_1 &= F_1\delta_{11} + F_2\delta_{12} + F_3\delta_{13} \\ y_2 &= F_1\delta_{21} + F_2\delta_{22} + F_3\delta_{23} \\ y_3 &= F_1\delta_{31} + F_2\delta_{32} + F_3\delta_{33} \end{aligned} \tag{7–25}$$

The forces F_i can arise from weight attached w_i or centrifugal forces $m_i\omega^2 y_i$. The equation set (7–25) written with inertial forces can be displayed as

$$\begin{aligned} y_1 &= m_1\omega^2 y_1\delta_{11} + m_2\omega^2 y_2\delta_{12} + m_3\omega^2 y_3\delta_{13} \\ y_2 &= m_1\omega^2 y_1\delta_{21} + m_2\omega^2 y_2\delta_{22} + m_3\omega^2 y_3\delta_{23} \\ y_3 &= m_1\omega^2 y_1\delta_{31} + m_2\omega^2 y_2\delta_{32} + m_3\omega^2 y_3\delta_{33} \end{aligned}$$

[6]Thomson, op. cit., p. 167.

which can be rewritten as

$$(m_1\delta_{11} - 1/\omega^2)y_1 + (m_2\delta_{12})y_2 + (m_3\delta_{13})y_3 = 0$$

$$(m_1\delta_{21})y_1 + (m_2\delta_{22} - 1/\omega^2)y_2 + (m_3\delta_{23})y_3 = 0 \qquad (a)$$

$$(m_1\delta_{31})y_1 + (m_2\delta_{32})y_2 + (m_3\delta_{33} - 1/\omega^2)y_3 = 0$$

Equation set (a) is three simultaneous equations in terms of y_1, y_2, and y_3. To avoid the trivial solution $y_1 = y_2 = y_3 = 0$, the determinant of the coefficients of y_1, y_2, and y_3 must be zero (eigenvalue problem). Thus,

$$\begin{vmatrix} (m_1\delta_{11} - 1/\omega^2) & m_2\delta_{12} & m_3\delta_{13} \\ m_1\delta_{21} & (m_2\delta_{22} - 1/\omega^2) & m_3\delta_{23} \\ m_1\delta_{31} & m_2\delta_{32} & (m_3\delta_{33} - 1/\omega^2) \end{vmatrix} = 0 \qquad (7\text{--}26)$$

which says that a deflection other than zero exists only at three distinct values of ω, the critical speeds. Expanding the determinant, we obtain

$$\left(\frac{1}{\omega^2}\right)^3 - (m_1\delta_{11} + m_2\delta_{22} + m_3\delta_{33})\left(\frac{1}{\omega^2}\right)^2 + \cdots = 0 \qquad (7\text{--}27)$$

The three roots of Eq. (7–27) can be expressed as $1/\omega_1^2$, $1/\omega_2^2$, and $1/\omega_3^2$. Thus Eq. (7–27) can be written in the form

$$\left(\frac{1}{\omega^2} - \frac{1}{\omega_1^2}\right)\left(\frac{1}{\omega^2} - \frac{1}{\omega_2^2}\right)\left(\frac{1}{\omega^2} - \frac{1}{\omega_3^2}\right) = 0$$

or

$$\left(\frac{1}{\omega^2}\right)^3 - \left(\frac{1}{\omega_1^2} + \frac{1}{\omega_2^2} + \frac{1}{\omega_3^2}\right)\left(\frac{1}{\omega^2}\right)^2 + \cdots = 0 \qquad (7\text{--}28)$$

Comparing Eqs. (7–27) and (7–28) we see that

$$\frac{1}{\omega_1^2} + \frac{1}{\omega_2^2} + \frac{1}{\omega_3^2} = m_1\delta_{11} + m_2\delta_{22} + m_3\delta_{33} \qquad (7\text{--}29)$$

If we had only a single mass m_1 alone, the critical speed would be given by $1/\omega^2 = m_1\delta_{11}$. We denote this critical speed as ω_{11} (which considers only m_1 acting alone). Likewise for m_2 or m_3 acting alone, we similarly define the terms $1/\omega_{22}^2 = m_2\delta_{22}$ or $1/\omega_{33}^2 = m_3\delta_{33}$, respectively. Thus, Eq. (7–29) can be rewritten as

$$\frac{1}{\omega_1^2} + \frac{1}{\omega_2^2} + \frac{1}{\omega_3^2} = \frac{1}{\omega_{11}^2} + \frac{1}{\omega_{22}^2} + \frac{1}{\omega_{33}^2} \qquad (7\text{--}30)$$

If we order the critical speeds such that $\omega_1 < \omega_2 < \omega_3$, then $1/\omega_1^2$ is much greater than $1/\omega_2^2$ and $1/\omega_3^2$. So the first, or fundamental, critical speed ω_1 can be approximated by

$$\frac{1}{\omega_1^2} \approx \frac{1}{\omega_{11}^2} + \frac{1}{\omega_{22}^2} + \frac{1}{\omega_{33}^2} \qquad (7\text{--}31)$$

This idea can be extended to an n-body shaft:

$$\frac{1}{\omega_1^2} \approx \sum_{i=1}^{n} \frac{1}{\omega_{ii}^2} \qquad (7\text{--}32)$$

This is called *Dunkerley's equation*. By ignoring the higher mode term(s), the first critical speed estimate is *lower* than actually is the case.

Since Eq. (7–32) has no loads appearing in the equation, it follows that if each load could be placed at some convenient location transformed into an equivalent load, then the critical speed of an array of loads could be found by summing the equivalent loads, all placed at a single convenient location. For the load at station 1, placed at the center of span, denoted with the subscript c, the equivalent load is found from

$$\omega_{11}^2 = \frac{1}{m_1 \delta_{11}} = \frac{g}{w_1 \delta_{11}} = \frac{g}{w_{1c} \delta_{cc}}$$

or

$$w_{1c} = w_1 \frac{\delta_{11}}{\delta_{cc}} \qquad (7\text{–}33)$$

EXAMPLE 7–5

Consider a simply supported steel shaft as depicted in Fig. 7–14, with 1 in diameter and a 31-in span between bearings, carrying two gears weighing 35 and 55 lbf.
(a) Find the influence coefficients.
(b) Find $\sum wy$ and $\sum wy^2$ and the first critical speed using Rayleigh's equation, Eq. (7–23).
(c) From the influence coefficients, find ω_{11} and ω_{22}.
(d) Using Dunkerley's equation, Eq. (7–32), estimate the first critical speed.
(e) Use superposition to estimate the first critical speed.
(f) Estimate the shaft's intrinsic critical speed. Suggest a modification to Dunkerley's equation to include the effect of the shaft's mass on the first critical speed of the attachments.

Solution   (a)

$$I = \frac{\pi d^4}{64} = \frac{\pi (1)^4}{64} = 0.049\,09 \text{ in}^4$$

$$6EIl = 6(30)10^6 (0.049\,09)31 = 0.2739(10^9) \text{ lbf} \cdot \text{in}^3$$

Figure 7–14

(a) A 25 mm uniform-diameter shaft for Ex. 7–5. (b) Superposing of equivalent loads at the center of the shaft for the purpose of finding the first critical speed.

From Eq. set (7–24),

$$\delta_{11} = \frac{600(175)(775^2 - 600^2 - 175^2)}{18.5 \times 10^{12}} = 0.001\ 19 \text{ mm/N}$$

$$\delta_{22} = \frac{275(500)(775^2 - 275^2 - 500^2)}{18.5 \times 10^{12}} = 0.002\ 04 \text{ mm/N}$$

$$\delta_{12} = \delta_{21} = \frac{275(175)(775^2 - 275^2 - 175^2)}{18.5 \times 10^{12}} = 0.001\ 29 \text{ mm/N}$$

Answer

	i	
i	**1**	**2**
1	0.001 19	0.001 29
2	0.001 29	0.002 04

$$y_1 = w_1\delta_{11} + w_2\delta_{12} = 175(0.001\ 19) + 275(0.001\ 29) = 0.56 \text{ mm}$$

$$y_2 = w_1\delta_{21} + w_2\delta_{22} = 175(0.001\ 29) + 275(0.002\ 04) = 0.79 \text{ mm}$$

(b)
$$\sum w_i y_i = 175(0.56) + 275(0.79) = 315.3 \text{ N·mm}$$

Answer
$$\sum w_i y_i^2 = 175(0.56)^2 + 275(0.79)^2 = 226.5 \text{ N·mm}^2$$

Answer
$$\omega = \sqrt{\frac{9810(315.3)}{226.5}} = 117 \text{ rad/s}$$

(c)

Answer
$$\frac{1}{\omega_{11}^2} = \frac{w_1}{g}\delta_{11}$$

$$\omega_{11} = \sqrt{\frac{g}{w_1\delta_{11}}} = \sqrt{\frac{9810}{175(0.001\ 19)}} = 217 \text{ rad/s}$$

Answer
$$\omega_{22} = \sqrt{\frac{g}{w_2\delta_{22}}} = \sqrt{\frac{9810}{275\ (0.002\ 04)}} = 132 \text{ rad/s}$$

(d)
$$\frac{1}{\omega_1^2} \doteq \sum \frac{1}{\omega_{ii}^2} = \frac{1}{217^2} + \frac{1}{132^2} = 7.863(10^{-5}) \tag{1}$$

Answer
$$\omega_1 \doteq \sqrt{\frac{1}{7.863(10^{-5})}} = 113 \text{ rad/s}$$

which is less than part *b*, as expected.
(e) From Eq. (7–24)

$$\delta_{cc} = \frac{b_{cc}x_{cc}(l^2 - b_{cc}^2 - x_{cc}^2)}{6EIl} = \frac{387.5(387.5)(775^2 - 387.5^2 - 387.5^2)}{18.5 \times 10^{12}}$$

$$= 0.002\ 44 \text{ mm/N}$$

From Eq. (7–33),

$$w_{1c} = w_1 \frac{\delta_{11}}{\delta_{cc}} = 175 \frac{0.001\ 19}{0.002\ 44} = 85.3\ \text{N}$$

$$w_{2c} = w_2 \frac{\delta_{22}}{\delta_{cc}} = 275 \frac{0.002\ 04}{0.002\ 44} = 229.9\ \text{N}$$

Answer

$$\omega = \sqrt{\frac{g}{\delta_{cc} \sum w_{ic}}} = \sqrt{\frac{9810}{0.002\ 44(85.3 + 229.9)}} = 112.9\ \text{rad/s}$$

which, except for rounding, agrees with part *d*, as expected.

(*f*) For the shaft, $E = 207\ 000\ \text{N/mm}^2$, $\gamma = 76.6 \times 10^{-6}\ \text{N/mm}^3$, and A $= \pi.\,(25^2)/4$ $= 491\ \text{mm}^2$. Considering the shaft alone, the critical speed, from Eq. (7–22), is

Answer

$$\omega_s = \left(\frac{\pi}{l}\right)^2 \sqrt{\frac{gEI}{A\gamma}} = \left(\frac{\pi}{775}\right)^2 \sqrt{\frac{9810(207\ 000)19\ 175}{491(76.6 \times 10^{-6})}}$$

$$= 529\ \text{rad/s}$$

We can simply add $1/\omega_s^2$ to the right side of Dunkerley's equation, Eq. (1), to include the shaft's contribution,

Answer

$$\frac{1}{\omega_1^2} \doteq \frac{1}{529^2} + 7.863(10^{-5}) = 8.22(10^{-5})$$

$$\omega_1 \doteq 110\ \text{rad/s}$$

which is slightly less than part *d*, as expected.

The shaft's first critical speed ω_s is just one more single effect to add to Dunkerley's equation. Since it does not fit into the summation, it is usually written up front.

Answer

$$\frac{1}{\omega_1^2} \doteq \frac{1}{\omega_s^2} + \sum_{i=1}^{n} \frac{1}{\omega_{ii}^2} \tag{7–34}$$

Common shafts are complicated by the stepped-cylinder geometry, which makes the influence-coefficient determination part of a numerical solution.

7–7 Miscellaneous Shaft Components

Setscrews

Unlike bolts and cap screws, which depend on tension to develop a clamping force, the setscrew depends on compression to develop the clamping force. The resistance to axial motion of the collar or hub relative to the shaft is called *holding power*. This holding power, which is really a force resistance, is due to frictional resistance of the contacting portions of the collar and shaft as well as any slight penetration of the setscrew into the shaft.

Figure 7–15 shows the point types available with socket setscrews. These are also manufactured with screwdriver slots and with square heads.

Table 7–4 lists values of the seating torque and the corresponding holding power for inch-series setscrews. The values listed apply to both axial holding power, for

Figure 7–15

Socket setscrews: (*a*) flat point; (*b*) cup point; (*c*) oval point; (*d*) cone point; (*e*) half-dog point.

Table 7–4

Typical Holding Power (Force) for Socket Setscrews*

Source: Unbrako Division, SPS Technologies, Jenkintown, Pa.

Size, mm	Seating Torque, N·m	Holding Power, N
#0	0.11	222
#1	0.2	289
#2	0.2	378
#3	0.5	534
#4	0.5	712
#5	1.1	890
#6	1.1	1112
#8	2.2	1713
#10	4.0	2403
6	9.8	4450
8	18.6	6675
10	32.8	8900
11	48.6	11 125
12	70.0	13 350
14	70.0	15 575
16	149.7	17 800
20	271.2	22 250
22	587.6	26 700
25	813.6	31 150

*Based on alloy-steel screw against steel shaft, class 3A coarse or fine threads in class 2B holes, and cup-point socket setscrews.

resisting thrust, and the tangential holding power, for resisting torsion. Typical factors of safety are 1.5 to 2.0 for static loads and 4 to 8 for various dynamic loads.

Setscrews should have a length of about half of the shaft diameter. Note that this practice also provides a rough rule for the radial thickness of a hub or collar.

Keys and Pins

Keys and pins are used on shafts to secure rotating elements, such as gears, pulleys, or other wheels. Keys are used to enable the transmission of torque from the shaft to the shaft-supported element. Pins are used for axial positioning and for the transfer of torque or thrust or both.

Figure 7–16 shows a variety of keys and pins. Pins are useful when the principal loading is shear and when both torsion and thrust are present. Taper pins are sized according to the diameter at the large end. Some of the most useful sizes of these are listed in Table 7–5. The diameter at the small end is

$$d = D - 0.0208L \qquad (7\text{–}35)$$

where d = diameter at small end, mm
  D = diameter at large end, mm
  L = length, mm

Figure 7–16

(a) Square key; (b) round key; (c and d) round pins; (e) taper pin; (f) split tubular spring pin. The pins in parts (e) and (f) are shown longer than necessary, to illustrate the chamfer on the ends, but their lengths should be kept smaller than the hub diameters to prevent injuries due to projections on rotating parts.

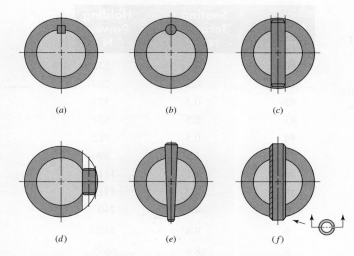

(a)   (b)   (c)

(d)   (e)   (f)

Table 7–5

Dimensions at Large End of Some Standard Taper Pins—Millimeter Series

Size	Commercial		Precision	
	Maximum	**Minimum**	**Maximum**	**Minimum**
4/0	2.802	2.751	2.794	2.769
2/0	3.614	3.564	3.607	3.581
0	3.995	3.945	3.988	3.962
2	4.935	4.884	4.928	4.902
4	6.383	6.332	6.375	6.350
6	8.694	8.644	8.687	8.661
8	12.530	12.479	12.522	12.497

Table 7–6

Millimeter Dimensions for Some Standard Square- and Rectangular-Key Applications

Source: Joseph E. Shigley, "Unthreaded Fasteners," Chap. 24 in Joseph E. Shigley, Charles R. Mischke, and Thomas H. Brown, Jr. (eds.), *Standard Handbook of Machine Design,* 3rd ed., McGraw-Hill, New York, 2004.

Shaft Diameter		Key Size		
Over	To (Incl.)	w	h	Keyway Depth
8	11	2	2	1
11	14	3	2	1
		3	3	1.5
14	22	5	3	1.5
		5	5	2
22	30	6	5	2
		6	6	3
30	36	8	6	3
		8	8	5
36	44	10	6	3
		10	10	5
44	58	12	10	5
		12	12	6
58	70	16	12	5.5
		16	16	8
70	80	20	12	6
		20	20	10

For less important applications, a dowel pin or a drive pin can be used. A large variety of these are listed in manufacturers' catalogs.[7]

The square key, shown in Fig. 7–16*a*, is also available in rectangular sizes. Standard sizes of these, together with the range of applicable shaft diameters, are listed in Table 7–6. The shaft diameter determines standard sizes for width, height, and key depth. The designer chooses an appropriate key length to carry the torsional load. Failure of the key can be by direct shear, or by bearing stress. Example 7–6 demonstrates the process to size the length of a key. The maximum length of a key is limited by the hub length of the attached element, and should generally not exceed about 1.5 times the shaft diameter to avoid excessive twisting with the angular deflection of the shaft. Multiple keys may be used as necessary to carry greater loads, typically oriented at 90° from one another. Excessive safety factors should be avoided in key design, since it is desirable in an overload situation for the key to fail, rather than more costly components.

Stock key material is typically made from low carbon cold-rolled steel, and is manufactured such that its dimensions never exceed the nominal dimension. This allows standard cutter sizes to be used for the keyseats. A setscrew is sometimes used along with a key to hold the hub axially, and to minimize rotational backlash when the shaft rotates in both directions.

---

[7]See also Joseph E. Shigley, "Unthreaded Fasteners," Chap. 24. In Joseph E. Shigley, Charles R. Mischke, and Thomas H. Brown, Jr. (eds.), *Standard Handbook of Machine Design,* 3rd ed., McGraw-Hill, New York, 2004.

Figure 7–17

(a) Gib-head key;
(b) Woodruff key.

Taper 3 in 300 mm

(a)

(b)

The gib-head key, in Fig. 7–17a, is tapered so that, when firmly driven, it acts to prevent relative axial motion. This also gives the advantage that the hub position can be adjusted for the best axial location. The head makes removal possible without access to the other end, but the projection may be hazardous.

The Woodruff key, shown in Fig. 7–17b, is of general usefulness, especially when a wheel is to be positioned against a shaft shoulder, since the keyslot need not be machined into the shoulder stress concentration region. The use of the Woodruff key also yields better concentricity after assembly of the wheel and shaft. This is especially important at high speeds, as, for example, with a turbine wheel and shaft. Woodruff keys are particularly useful in smaller shafts where their deeper penetration helps prevent key rolling. Dimensions for some standard Woodruff key sizes can be found in Table 7–7, and Table 7–8 gives the shaft diameters for which the different keyseat widths are suitable.

Pilkey[8] gives values for stress concentrations in an end-milled keyseat, as a function of the ratio of the radius r at the bottom of the groove and the shaft diameter d. For fillets cut by standard milling-machine cutters, with a ratio of $r/d = 0.02$, Peterson's charts give $K_t = 2.14$ for bending and $K_{ts} = 2.62$ for torsion without the key in place, or $K_{ts} = 3.0$ for torsion with the key in place. The stress concentration at the end of the keyseat can be reduced somewhat by using a sled-runner keyseat, eliminating the abrupt end to the keyseat, as shown in Fig. 7–17. It does, however, still have the sharp radius in the bottom of the groove on the sides. The sled-runner keyseat can only be used when definite longitudinal key positioning is not necessary. It is also not as suitable near a shoulder. Keeping the end of a keyseat at least a distance of $d/10$ from

---

[8]W. D. Pilkey, *Peterson's Stress-Concentration Factors,* 2nd ed., John Wiley & Sons, New York, 1997, pp. 408–409.

Table 7–7

Dimensions of
Woodruff Keys—
Millimeter Series

| Key Size | | Height | Offset | Keyseat Depth | |
w	D	b	e	Shaft	Hub
1.5	6	3	0.4	1.8	0.95
1.5	10	4	0.4	3.4	0.95
2	10	4	0.4	3.0	1.3
2	12	5	1	3.8	1.3
2	16	6	1.5	5.0	1.3
3	12	5	1	3.4	1.7
3	16	6	1.5	4.6	1.7
3	20	8	1.5	6.2	1.7
5	16	6	1.5	4.2	2.1
5	20	8	1.5	5.8	2.1
5	22	10	1.5	7.4	2.1
5	20	8	1.5	5.4	2.5
5	22	10	1.5	7.0	2.5
5	25	11	1.5	8.6	2.5
6	22	10	1.5	6.2	3.3
6	25	11	1.5	7.8	3.3
6	30	14	2	10.6	3.3
8	25	11	1.5	7.0	4.1
8	30	14	2	9.8	4.1
8	38	16	3	12.2	4.1
10	30	14	2	9.0	4.9
10	38	16	3	11.4	4.9

Table 7–8

Sizes of Woodruff Keys
Suitable for Various
Shaft Diameters

| Keyseat Width, mm | Shaft Diameter, mm | |
	From	To (inclusive)
1.5	8	12
2	10	22
3	10	38
5	12	40
5	14	50
6	18	58
8	20	60
10	25	66

Figure 7-18

Typical uses for retaining rings: (*a*) External ring and (*b*) its application; (*c*) internal ring and (*d*) its application.

Retaining ring

Retaining ring

(*a*)   (*b*)   (*c*)   (*d*)

the start of the shoulder fillet will prevent the two stress concentrations from combining with each other.[9]

Retaining Rings

A retaining ring is frequently used instead of a shaft shoulder or a sleeve to axially position a component on a shaft or in a housing bore. As shown in Fig. 7–18, a groove is cut in the shaft or bore to receive the spring retainer. For sizes, dimensions, and axial load ratings, the manufacturers' catalogs should be consulted.

Appendix Tables A–15–16 and A–15–17 give values for stress-concentration factors for flat-bottomed grooves in shafts, suitable for retaining rings. For the rings to seat nicely in the bottom of the groove, and support axial loads against the sides of the groove, the radius in the bottom of the groove must be reasonably sharp, typically about one-tenth of the groove width. This causes comparatively high values for stress-concentration factors, around 5 for bending and axial, and 3 for torsion. Care should be taken in using retaining rings, particularly in locations with high bending stresses.

EXAMPLE 7–6

A UNS G10350 steel shaft, heat-treated to a minimum yield strength of 525 MPa, has a diameter of 36 mm. The shaft rotates at 600 rev/min and transmits 30 kW through a gear. Select an appropriate key for the gear.

Solution

A 10-mm square key is selected, UNS G10200 cold-drawn steel being used. The design will be based on a yield strength of 455 MPa. A factor of safety of 2.80 will be employed in the absence of exact information about the nature of the load.

The torque is obtained from the horsepower equation

$$\text{Angular speed } w = 600(2)\pi/60 = 62.8 \text{ rad/s}$$
$$T = 30\,000/62.8 = 478 \text{ N·m}$$

From Fig. 7–19, the force F at the surface of the shaft is

$$F = \frac{T}{r} = \frac{478}{0.018} = 26\,556 \text{ N}$$

By the distortion-energy theory, the shear strength is

$$S_{sy} = 0.577S_y = (0.577)(455) = 262.5 \text{ MPa}$$

| Figure 7–19

---

[9]Ibid, p. 381.

Failure by shear across the area *ab* will create a stress of $\tau = F/tl$. Substituting the strength divided by the factor of safety for τ gives

$$\frac{S_{sy}}{n} = \frac{F}{tl} \quad \text{or} \quad \frac{262.5 \times 10^6}{2.80} = \frac{26556}{0.01l}$$

or $l = 0.0283$ mm. To resist crushing, the area of one-half the face of the key is used:

$$\frac{S_y}{n} = \frac{F}{tl/2} \quad \text{or} \quad \frac{455 \times 10^6}{2.80} = \frac{26556}{0.01l/2}$$

and $l = 0.0327$ mm. The hub length of a gear is usually greater than the shaft diameter, for stability. If the key, in this example, is made equal in length to the hub, it would therefore have ample strength, since it would probably be 36 mm or longer.

7–8   Limits and Fits

The designer is free to adopt any geometry of fit for shafts and holes that will ensure the intended function. There is sufficient accumulated experience with commonly recurring situations to make standards useful. There are two standards for limits and fits in the United States, one based on inch units and the other based on metric units.[10] These differ in nomenclature, definitions, and organization. No point would be served by separately studying each of the two systems. The metric version is the newer of the two and is well organized, and so here we present only the metric version but include a set of inch conversions to enable the same system to be used with either system of units.

In using the standard, *capital letters always refer to the hole; lowercase letters are used for the shaft.*

The definitions illustrated in Fig. 7–20 are explained as follows:

- *Basic size* is the size to which limits or deviations are assigned and is the same for both members of the fit.
- *Deviation* is the algebraic difference between a size and the corresponding basic size.
- *Upper deviation* is the algebraic difference between the maximum limit and the corresponding basic size.
- *Lower deviation* is the algebraic difference between the minimum limit and the corresponding basic size.
- *Fundamental deviation* is either the upper or the lower deviation, depending on which is closer to the basic size.
- *Tolerance* is the difference between the maximum and minimum size limits of a part.
- *International tolerance grade* numbers (IT) designate groups of tolerances such that the tolerances for a particular IT number have the same relative level of accuracy but vary depending on the basic size.
- *Hole basis* represents a system of fits corresponding to a basic hole size. The fundamental deviation is H.

---

[10]*Preferred Limits and Fits for Cylindrical Parts,* ANSI B4.1-1967. *Preferred Metric Limits and Fits,* ANSI B4.2-1978.

Figure 7–20

Definitions applied to a cylindrical fit.

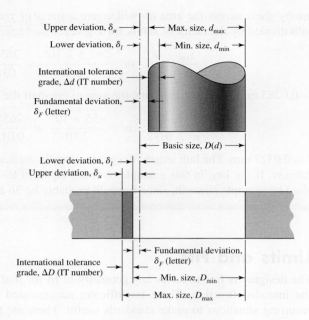

- *Shaft basis* represents a system of fits corresponding to a basic shaft size. The fundamental deviation is h. The shaft-basis system is not included here.

The magnitude of the tolerance zone is the variation in part size and is the same for both the internal and the external dimensions. The tolerance zones are specified in international tolerance grade numbers, called IT numbers. The smaller grade numbers specify a smaller tolerance zone. These range from IT0 to IT16, but only grades IT6 to IT11 are needed for the preferred fits. These are listed in Tables A–11 to A–14 for basic sizes up to 16 in or 400 mm.

The standard uses *tolerance position letters,* with capital letters for internal dimensions (holes) and lowercase letters for external dimensions (shafts). As shown in Fig. 7–20, the fundamental deviation locates the tolerance zone relative to the basic size.

Table 7–9 shows how the letters are combined with the tolerance grades to establish a preferred fit. The ISO symbol for the hole for a sliding fit with a basic size of 32 mm is 32H7. Inch units are not a part of the standard. However, the designation $(1\frac{3}{8}$ in$)$ H7 includes the same information and is recommended for use here. In both cases, the capital letter H establishes the fundamental deviation and the number 7 defines a tolerance grade of IT7.

For the sliding fit, the corresponding shaft dimensions are defined by the symbol 32g6 $[(1\frac{3}{8}$ in$)$g6$]$.

The fundamental deviations for shafts are given in Tables A–11 and A–13. For letter codes c, d, f, g, and h,

Upper deviation = fundamental deviation
Lower deviation = upper deviation − tolerance grade

For letter codes k, n, p, s, and u, the deviations for shafts are

Lower deviation = fundamental deviation
Upper deviation = lower deviation + tolerance grade

Table 7-9	**Type of Fit**	**Description**	**Symbol**
Descriptions of Preferred Fits Using the Basic Hole System	Clearance	*Loose running fit:* for wide commercial tolerances or allowances on external members	H11/c11
Source: Preferred Metric Limits and Fits, ANSI B4.2-1978. See also BS 4500.		*Free running fit:* not for use where accuracy is essential, but good for large temperature variations, high running speeds, or heavy journal pressures	H9/d9
		Close running fit: for running on accurate machines and for accurate location at moderate speeds and journal pressures	H8/f7
		Sliding fit: where parts are not intended to run freely, but must move and turn freely and locate accurately	H7/g6
		Locational clearance fit: provides snug fit for location of stationary parts, but can be freely assembled and disassembled	H7/h6
	Transition	*Locational transition fit:* for accurate location, a compromise between clearance and interference	H7/k6
		Locational transition fit: for more accurate location where greater interference is permissible	H7/n6
	Interference	*Locational interference fit:* for parts requiring rigidity and alignment with prime accuracy of location but without special bore pressure requirements	H7/p6
		Medium drive fit: for ordinary steel parts or shrink fits on light sections, the tightest fit usable with cast iron	H7/s6
		Force fit: suitable for parts that can be highly stressed or for shrink fits where the heavy pressing forces required are impractical	H7/u6

The lower deviation H (for holes) is zero. For these, the upper deviation equals the tolerance grade.

As shown in Fig. 7–20, we use the following notation:

$$D = \text{basic size of hole}$$
$$d = \text{basic size of shaft}$$
$$\delta_u = \text{upper deviation}$$
$$\delta_l = \text{lower deviation}$$
$$\delta_F = \text{fundamental deviation}$$
$$\Delta D = \text{tolerance grade for hole}$$
$$\Delta d = \text{tolerance grade for shaft}$$

Note that these quantities are all deterministic. Thus, for the hole,

$$D_{\max} = D + \Delta D \qquad D_{\min} = D \tag{7–36}$$

For shafts with clearance fits c, d, f, g, and h,

$$d_{\max} = d + \delta_F \qquad d_{\min} = d + \delta_F - \Delta d \tag{7–37}$$

For shafts with interference fits k, n, p, s, and u,

$$d_{\min} = d + \delta_F \qquad d_{\max} = d + \delta_F + \Delta d \tag{7–38}$$

EXAMPLE 7–7

A journal bearing and bushing need to be described. The nominal size is 25 mm. What dimensions are needed for a 25-mm basic size with a close running fit if this is a lightly loaded journal and bushing assembly?

Solution

Choose the basic size as 25 mm. From Table 7–9, for 25 mm, the fit is H8/f7. From Table A–11, the tolerance grade are $\Delta D = 0.033$ mm and $\Delta d = 0.021$ mm.

Hole:

Answer
$$D_{max} = D + (\Delta D)_{hole} = 25 + 0.033 = 25.033 \text{ mm}$$

Answer
$$D_{min} = D = 25 \text{ mm}$$

Shaft: From Table A–12: Fundamental deviation $= -0.02$ mm

Answer
$$d_{max} = d + \delta_F = 25.0000 + (-0.02) = 24.98 \text{ mm}$$

Answer
$$d_{min} = d + \delta_F - \Delta d = 25.0000 + (-0.02) - 0.021 = 24.959 \text{ mm}$$

Alternatively,

Answer
$$d_{min} = d_{max} - \Delta d = 24.98 - 0.021 = 24.595 \text{ mm}$$

EXAMPLE 7–8

Find the hole and shaft limits for a medium drive fit using a basic hole size of 50 mm.

Solution

The symbol for the fit, from Table 7–9, is H7/s6. For the hole, we use Table A–11 and find the IT7 grade to be $\Delta D = 0.025$ mm. Thus, from Eq. (7–36),

Answer
$$D_{max} = D + \Delta D = 50 + 0.025 = 50.025 \text{ mm}$$

Answer
$$D_{min} = D = 50 \text{ mm}$$

The IT6 tolerance for the shaft is $\Delta d = 0.016$ mm. Also, from Table A–12, the fundamental deviation is $\delta_F = 0.043$ mm. Using Eq. (7–38), we get for the shaft that

Answer
$$d_{min} = d + \delta_F = 50 + 0.043 = 50.043 \text{ mm}$$

Answer
$$d_{max} = d + \delta_F + \Delta d = 50 + 0.043 + 0.016 = 50.059 \text{ mm}$$

Stress and Torque Capacity in Interference Fits

Interference fits between a shaft and its components can sometimes be used effectively to minimize the need for shoulders and keyways. The stresses due to an interference fit can be obtained by treating the shaft as a cylinder with a uniform external pressure, and the hub as a hollow cylinder with a uniform internal pressure. Stress equations for these situations were developed in Sec. 3–16, and will be converted here from radius terms into diameter terms to match the terminology of this section.

The pressure p generated at the interface of the interference fit, from Eq. (3–56) converted into terms of diameters, is given by

$$p = \frac{\delta}{\dfrac{d}{E_o}\left(\dfrac{d_o^2 + d^2}{d_o^2 - d^2} + \nu_o\right) + \dfrac{d}{E_i}\left(\dfrac{d^2 + d_i^2}{d^2 - d_i^2} - \nu_i\right)} \tag{7–39}$$

or, in the case where both members are of the same material,

$$p = \frac{E\delta}{2d^3}\left[\frac{(d_o^2 - d^2)(d^2 - d_i^2)}{d_o^2 - d_i^2}\right] \tag{7–40}$$

where d is the nominal shaft diameter, d_i is the inside diameter (if any) of the shaft, d_o is the outside diameter of the hub, E is Young's modulus, and ν is Poisson's ratio, with subscripts o and i for the outer member (hub) and inner member (shaft), respectively. The term δ is the *diametral* interference between the shaft and hub, that is, the difference between the shaft outside diameter and the hub inside diameter.

$$\delta = d_{\text{shaft}} - d_{\text{hub}} \tag{7–41}$$

Since there will be tolerances on both diameters, the maximum and minimum pressures can be found by applying the maximum and minimum interferences. Adopting the notation from Fig. 7–20, we write

$$\delta_{\min} = d_{\min} - D_{\max} \tag{7–42}$$

$$\delta_{\max} = d_{\max} - D_{\min} \tag{7–43}$$

where the diameter terms are defined in Eqs. (7–36) and (7–38). The maximum interference should be used in Eq. (7–39) or (7–40) to determine the maximum pressure to check for excessive stress.

From Eqs. (3–58) and (3–59), with radii converted to diameters, the tangential stresses at the interface of the shaft and hub are

$$\sigma_{t,\,\text{shaft}} = -p\frac{d^2 + d_i^2}{d^2 - d_i^2} \tag{7–44}$$

$$\sigma_{t,\,\text{hub}} = p\frac{d_o^2 + d^2}{d_o^2 - d^2} \tag{7–45}$$

The radial stresses at the interface are simply

$$\sigma_{r,\,\text{shaft}} = -p \tag{7–46}$$

$$\sigma_{r,\,\text{hub}} = -p \tag{7–47}$$

The tangential and radial stresses are orthogonal, and should be combined using a failure theory to compare with the yield strength. If either the shaft or hub yields during assembly, the full pressure will not be achieved, diminishing the torque that can be transmitted. The interaction of the stresses due to the interference fit with the other stresses in the shaft due to shaft loading is not trivial. Finite-element analysis of the interface would be appropriate when warranted. A stress element on the surface of a rotating shaft will experience a completely reversed bending stress in the longitudinal direction, as well as the steady compressive stresses in the tangential and radial directions. This is a three-dimensional stress element. Shear stress due to torsion in the shaft may also be present. Since the stresses due to the press fit are compressive, the fatigue situation is usually actually improved. For this reason, it may be acceptable to simplify the shaft analysis by ignoring the steady compressive stresses due to the press fit. There is, however, a stress concentration effect in the shaft bending stress near the ends of the hub, due to the sudden change from compressed to uncompressed material. The design of the hub geometry, and, therefore, its uniformity and rigidity, can have a significant effect on the specific value of the stress-concentration factor, making it difficult to report generalized values. For first estimates, values are typically not greater than 2.

The amount of torque that can be transmitted through an interference fit can be estimated with a simple friction analysis at the interface. The friction force is the product of the coefficient of friction f and the normal force acting at the interface. The normal force can be represented by the product of the pressure p and the surface area A of interface. Therefore, the friction force F_f is

$$F_f = fN = f(pA) = f[p2\pi(d/2)l] = \pi f pld \qquad (7\text{--}48)$$

where l is the length of the hub. This friction force is acting with a moment arm of $d/2$ to provide the torque capacity of the joint, so

$$T = F_f d/2 = \pi f pld(d/2)$$
$$T = (\pi/2)f pld^2 \qquad (7\text{--}49)$$

The minimum interference, from Eq. (7–42), should be used to determine the minimum pressure to check for the maximum amount of torque that the joint should be designed to transmit without slipping.

PROBLEMS

Problems marked with an asterisk (*) are linked to problems in other chapters, as summarized in Table 1–2 of Sec. 1–17, p. 34.

7-1 A shaft is loaded in bending and torsion such that $M_a = 70$ N · m, $T_a = 45$ N · m, $M_m = 55$ N · m, and $T_m = 35$ N · m. For the shaft, $S_u = 700$ MPa and $S_y = 560$ MPa, and a fully corrected endurance limit of $S_e = 210$ MPa is assumed. Let $K_f = 2.2$ and $K_{fs} = 1.8$. With a design factor of 2.0 determine the minimum acceptable diameter of the shaft using the

(a) DE-Gerber criterion.

(b) DE-ASME Elliptic criterion.

(c) DE-Soderberg criterion.

(d) DE-Goodman criterion.

Discuss and compare the results.

7-2 The section of shaft shown in the figure is to be designed to approximate relative sizes of $d = 0.75D$ and $r = D/20$ with diameter d conforming to that of standard rolling-bearing bore sizes. The shaft is to be made of SAE 2340 steel, heat-treated to obtain minimum strengths in the shoulder area of 1226 MPa ultimate tensile strength and 1130 MPa yield strength with a Brinell hardness not less than 370. At the shoulder the shaft is subjected to a completely reversed bending moment of 70 N · m, accompanied by a steady torsion of 45 N · m. Use a design factor of 2.5 and size the shaft for an infinite life using the DE-ASME Elliptic criterion.

Problem 7–2

Section of a shaft containing a grinding-relief groove. Unless otherwise specified, the diameter at the root of the groove $d_r = d - 2r$, and though the section of diameter d is ground, the root of the groove is still a machined surface.

7–3 The rotating solid steel shaft is simply supported by bearings at points B and C and is driven by a gear (not shown) which meshes with the spur gear at D, which has a 150-mm pitch diameter. The force F from the drive gear acts at a pressure angle of 20°. The shaft transmits a torque to point A of $T_A = 340$ N · m. The shaft is machined from steel with $S_y = 420$ MPa and $S_{ut} = 560$ MPa. Using a factor of safety of 2.5, determine the minimum allowable diameter of the 250-mm section of the shaft based on (*a*) a static yield analysis using the distortion energy theory and (*b*) a fatigue-failure analysis. Assume sharp fillet radii at the bearing shoulders for estimating stress-concentration factors.

Problem 7–3

7–4 A geared industrial roll shown in the figure is driven at 300 rev/min by a force F acting on a 75-mm-diameter pitch circle as shown. The roll exerts a normal force of 5200 N/m of roll length on the material being pulled through. The material passes under the roll. The coefficient of friction is 0.40. Develop the moment and shear diagrams for the shaft modeling the roll force as (*a*) a concentrated force at the center of the roll, and (*b*) a uniformly distributed force along the roll. These diagrams will appear on two orthogonal planes.

Problem 7–4

Material moves under the roll.
Dimensions in inches.

7–5 Design a shaft for the situation of the industrial roll of Prob. 7–4 with a design factor of 2 and a reliability goal of 0.999 against fatigue failure. Plan for a ball bearing on the left and a cylindrical roller on the right. For deformation use a factor of safety of 2.

7–6 The figure shows a proposed design for the industrial roll shaft of Prob. 7–4. Hydrodynamic film bearings are to be used. All surfaces are machined except the journals, which are ground and polished. The material is 1035 HR steel. Perform a design assessment. Is the design satisfactory?

Problem 7–6
Bearing shoulder fillets 0.72 mm, others 1.6 mm. Sled-runner keyway is 90 mm long. Dimensions in mm

7–7* to 7–16* For the problem specified in the table, build upon the results of the original problem to obtain a preliminary design of the shaft by performing the following tasks:

(*a*) Sketch a general shaft layout, including means to locate the components and to transmit the torque. Estimates for the component widths are acceptable at this point.

(*b*) Specify a suitable material for the shaft.

(*c*) Determine critical diameters of the shaft based on infinite fatigue life with a design factor of 1.5. Check for yielding.

(*d*) Make any other dimensional decisions necessary to specify all diameters and axial dimensions. Sketch the shaft to scale, showing all proposed dimensions.

(*e*) Check the deflections at the gears, and the slopes at the gears and the bearings for satisfaction of the recommended limits in Table 7–2. Assume the deflections for any pulleys are not likely to be critical. If any of the deflections exceed the recommended limits, make appropriate changes to bring them all within the limits.

Problem Number	Original Problem, Page Number
7–7*	3–68, 151
7–8*	3–69, 151
7–9*	3–70, 151
7–10*	3–71, 151
7–11*	3–72, 152
7–12*	3–73, 152
7–13*	3–74, 152
7–14*	3–76, 153
7–15*	3–77, 153
7–16*	3–79, 153

7–17 In the double-reduction gear train shown, shaft *a* is driven by a motor attached by a flexible coupling attached to the overhang. The motor provides a torque of 10 N · m at a speed of 1200 rpm. The gears have 20° pressure angles, with diameters shown in the figure. Use AISI 1020 cold-drawn steel. Design one of the shafts (as specified by the instructor) with a design factor of 1.5 by performing the following tasks.

(*a*) Sketch a general shaft layout, including means to locate the gears and bearings, and to transmit the torque.

(*b*) Perform a force analysis to find the bearing reaction forces, and generate shear and bending moment diagrams.

(*c*) Determine potential critical locations for stress design.

(*d*) Determine critical diameters of the shaft based on fatigue and static stresses at the critical locations.

(*e*) Make any other dimensional decisions necessary to specify all diameters and axial dimensions. Sketch the shaft to scale, showing all proposed dimensions.

(*f*) Check the deflection at the gear, and the slopes at the gear and the bearings to satisfy the recommended limits in Table 7–2.

(*g*) If any of the deflections exceed the recommended limits, make appropriate changes to bring them all within the limits.

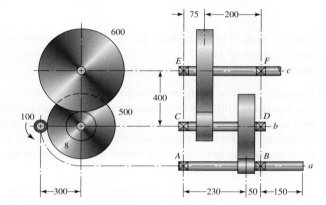

Problem 7–17
Dimensions in mm.

7–18 In the figure is a proposed shaft design to be used for the input shaft *a* in Prob. 7–17. A ball bearing is planned for the left bearing, and a cylindrical roller bearing for the right.

(*a*) Determine the minimum fatigue factor of safety by evaluating at any critical locations. Use the DE-ASME Elliptic fatigue criterion.

(*b*) Check the design for adequacy with respect to deformation, according to the recommendations in Table 7–2.

Problem 7–18

Shoulder fillets at bearing seat 0.75-mm radius, others 3-mm radius, except right-hand bearing seat transition, 6 mm. The material is 1030 HR. Keyways 10 mm wide by 5 mm deep. Dimensions in mm.

7–19* The shaft shown in the figure is proposed for the application defined in Prob. 3–72, p. 152. The material is AISI 1018 cold-drawn steel. The gears seat against the shoulders, and have hubs with setscrews to lock them in place. The effective centers of the gears for force transmission are shown. The keyseats are cut with standard endmills. The bearings are press-fit against the shoulders. Determine the minimum fatigue factor of safety using the DE-Gerber fatigue criterion.

Problem 7–19*
1.6 mm. Dimensions in mm.

7–20* Continue Prob. 7–19 by checking that the deflections satisfy the suggested minimums for bearings and gears in Table 7–2. If any of the deflections exceed the recommended limits, make appropriate changes to bring them all within the limits.

7–21* The shaft shown in the figure is proposed for the application defined in Prob. 3–73, p. 152. The material is AISI 1018 cold-drawn steel. The gears seat against the shoulders, and have hubs with setscrews to lock them in place. The effective centers of the gears for force transmission are shown. The keyseats are cut with standard endmills. The bearings are press-fit against the shoulders. Determine the minimum fatigue factor of safety using the DE-Gerber failure criterion.

Problem 7–21*
All fillets 2 mm. Dimensions in mm.

7–22* Continue Prob. 7–21 by checking that the deflections satisfy the suggested minimums for bearings and gears in Table 7–2. If any of the deflections exceed the recommended limits, make appropriate changes to bring them all within the limits.

7–23 The shaft shown in the figure is driven by a gear at the right keyway, drives a fan at the left keyway, and is supported by two deep-groove ball bearings. The shaft is made from AISI 1020

Problem 7–23
Dimensions in millimeters.

cold-drawn steel. At steady-state speed, the gear transmits a radial load of 1.1 kN and a tangential load of 3 kN at a pitch diameter of 200 mm.

(a) Determine fatigue factors of safety at any potentially critical locations using the DE-Gerber failure criterion.

(b) Check that deflections satisfy the suggested minimums for bearings and gears.

7–24   An AISI 1020 cold-drawn steel shaft with the geometry shown in the figure carries a transverse load of 7 kN and a torque of 107 N · m. Examine the shaft for strength and deflection. If the largest allowable slope at the bearings is 0.001 rad and at the gear mesh is 0.0005 rad, what is the factor of safety guarding against damaging distortion? Using the DE-ASME Elliptic criterion, what is the factor of safety guarding against fatigue failure? If the shaft turns out to be unsatisfactory, what would you recommend to correct the problem?

Problem 7–24
Dimensions in millimeters.

All fillets 2 mm

7–25   A shaft is to be designed to support the spur pinion and helical gear shown in the figure on two bearings spaced 700 mm center-to-center. Bearing *A* is a cylindrical roller and is to take only radial load; bearing *B* is to take the thrust load of 900 N produced by the helical gear and its share of the radial load. The bearing at *B* can be a ball bearing. The radial loads of both gears are in the same plane, and are 2.7 kN for the pinion and 900 N for the gear. The shaft speed is 1200 rev/min. Design the shaft. Make a sketch to scale of the shaft showing all fillet sizes, keyways, shoulders, and diameters. Specify the material and its heat treatment.

Problem 7–25
Dimensions in millimeters.

7–26   A heat-treated steel shaft is to be designed to support the spur gear and the overhanging worm shown in the figure. A bearing at *A* takes pure radial load. The bearing at *B* takes the worm-thrust load for either direction of rotation. The dimensions and the loading are shown in the figure; note that the radial loads are in the same plane. Make a complete design of the shaft, including a sketch of the shaft showing all dimensions. Identify the material and its heat treatment (if necessary). Provide an assessment of your final design. The shaft speed is 310 rev/min.

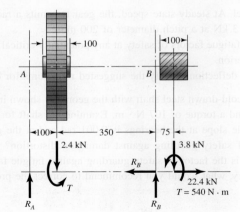

Problem 7–26
Dimensions in millimeters.

7–27 A bevel-gear shaft mounted on two 40-mm 02-series ball bearings is driven at 1720 rev/min by a motor connected through a flexible coupling. The figure shows the shaft, the gear, and the bearings. The shaft has been giving trouble—in fact, two of them have already failed—and the down time on the machine is so expensive that you have decided to redesign the shaft yourself rather than order replacements. A hardness check of the two shafts in the vicinity of the fracture of the two shafts showed an average of 198 Bhn for one and 204 Bhn of the other. As closely as you can estimate the two shafts failed at a life measure between 600 000 and 1 200 000 cycles of operation. The surfaces of the shaft were machined, but not ground. The fillet sizes were not measured, but they correspond with the recommendations for the ball bearings used. You know that the load is a pulsating or shock-type load, but you have no idea of the magnitude, because the shaft drives an indexing mechanism, and the forces are inertial. The keyways are 10 mm wide by 5 mm deep. The straight-toothed bevel pinion drives a 48-tooth bevel gear. Specify a new shaft in sufficient detail to ensure a long and trouble-free life.

Problem 7–27
Dimensions in millimeters

7–28 A 25-mm-diameter uniform steel shaft is 600 mm long between bearings.
(a) Find the lowest critical speed of the shaft.
(b) If the goal is to double the critical speed, find the new diameter.
(c) A half-size model of the original shaft has what critical speed?

7–29 Demonstrate how rapidly Rayleigh's method converges for the uniform-diameter solid shaft of Prob. 7–28, by partitioning the shaft into first one, then two, and finally three elements.

7–30 Compare Eq. (7–27) for the angular frequency of a two-disk shaft with Eq. (7–28), and note that the constants in the two equations are equal.
(a) Develop an expression for the *second* critical speed.
(b) Estimate the second critical speed of the shaft addressed in Ex. 7–5, parts *a* and *b*.

7–31 For a uniform-diameter shaft, does hollowing the shaft increase or decrease the critical speed? Determine the ratio of the critical speeds for a solid shaft of diameter d to a hollow shaft of inner diameter $d/2$ and outer diameter d.

7–32 The steel shaft shown in the figure carries a 100-N gear on the left and a 175-N gear on the right. Estimate the first critical speed due to the loads, the shaft's critical speed without the loads, and the critical speed of the combination.

Problem 7–32
Dimensions in mm.

7–33 A transverse drilled and reamed hole can be used in a solid shaft to hold a pin that locates and holds a mechanical element, such as the hub of a gear, in axial position, and allows for the transmission of torque. Since a small-diameter hole introduces high stress concentration, and a larger diameter hole erodes the area resisting bending and torsion, investigate the existence of a pin diameter with minimum adverse affect on the shaft. Specifically, determine the pin diameter, as a percentage of the shaft diameter, that minimizes the peak stress in the shaft. (*Hint:* Use Table A–16.)

7–34* The shaft shown in Prob. 7–19 is proposed for the application defined in Prob. 3–72, p. 152. Specify a square key for gear *B*, using a factor of safety of 1.1.

7–35* The shaft shown in Prob. 7–21 is proposed for the application defined in Prob. 3–73, p. 152. Specify a square key for gear *B*, using a factor of safety of 1.1.

7–36 A guide pin is required to align the assembly of a two-part fixture. The nominal size of the pin is 15 mm. Make the dimensional decisions for a 15-mm basic size locational clearance fit.

7–37 An interference fit of a cast-iron hub of a gear on a steel shaft is required. Make the dimensional decisions for a 45-mm basic size medium drive fit.

7–38 A pin is required for forming a linkage pivot. Find the dimensions required for a 45-mm basic size pin and clevis with a sliding fit.

7–39 A journal bearing and bushing need to be described. The nominal size is 32 mm. What dimensions are needed for a 32 mm basic size with a close running fit if this is a lightly loaded journal and bushing assembly?

7–40 A ball bearing has been selected with the bore size specified in the catalog as 35.000 mm to 35.020 mm. Specify appropriate minimum and maximum shaft diameters to provide a locational interference fit.

7–41 A shaft diameter is carefully measured to be 36.05 mm. A bearing is selected with a catalog specification of the bore diameter range from 36 mm to 36.025 mm. Determine if this is an acceptable selection if a locational interference fit is desired.

7–42 A gear and shaft with nominal diameter of 35 mm are to be assembled with a *medium drive fit,* as specified in Table 7–9. The gear has a hub, with an outside diameter of 60 mm, and an overall length of 50 mm. The shaft is made from AISI 1020 CD steel, and the gear is made from steel that has been through hardened to provide $S_u = 700$ MPa and $S_y = 600$ MPa.

(a) Specify dimensions with tolerances for the shaft and gear bore to achieve the desired fit.

(b) Determine the minimum and maximum pressures that could be experienced at the interface with the specified tolerances.

(c) Determine the worst-case static factors of safety guarding against yielding at assembly for the shaft and the gear based on the distortion energy failure theory.

(d) Determine the maximum torque that the joint should be expected to transmit without slipping, i.e., when the interference pressure is at a minimum for the specified tolerances.

8

Screws, Fasteners, and the Design of Nonpermanent Joints

Chapter Outline

The helical-thread screw was undoubtably an extremely important mechanical invention. It is the basis of power screws, which change angular motion to linear motion to transmit power or to develop large forces (presses, jacks, etc.), and threaded fasteners, an important element in nonpermanent joints.

This book presupposes a knowledge of the elementary methods of fastening. Typical methods of fastening or joining parts use such devices as bolts, nuts, cap screws, setscrews, rivets, spring retainers, locking devices, pins, keys, welds, and adhesives. Studies in engineering graphics and in metal processes often include instruction on various joining methods, and the curiosity of any person interested in mechanical engineering naturally results in the acquisition of a good background knowledge of fastening methods. Contrary to first impressions, the subject is one of the most interesting in the entire field of mechanical design.

One of the key targets of current design for manufacture is to reduce the number of fasteners. However, there will always be a need for fasteners to facilitate disassembly for whatever purposes. For example, jumbo jets such as Boeing's 747 require as many as 2.5 million fasteners, some of which cost several dollars apiece. To keep costs down, aircraft manufacturers, and their subcontractors, constantly review new fastener designs, installation techniques, and tooling.

The number of innovations in the fastener field over any period you might care to mention has been tremendous. An overwhelming variety of fasteners are available for the designer's selection. Serious designers generally keep specific notebooks on fasteners alone. Methods of joining parts are extremely important in the engineering of a quality design, and it is necessary to have a thorough understanding of the performance of fasteners and joints under all conditions of use and design.

8–1 Thread Standards and Definitions

The terminology of screw threads, illustrated in Fig. 8–1, is explained as follows:

The *pitch* is the distance between adjacent thread forms measured parallel to the thread axis. The pitch in U.S. units is the reciprocal of the number of thread forms per inch N.

The *major diameter d* is the largest diameter of a screw thread.

The *minor* (or root) *diameter d_r* is the smallest diameter of a screw thread.

The pitch diameter d_p is a theoretical diameter between the major and minor diameters.

The *lead l*, not shown, is the distance the nut moves parallel to the screw axis when the nut is given one turn. For a single thread, as in Fig. 8–1, the lead is the same as the pitch.

A *multiple-threaded* product is one having two or more threads cut beside each other (imagine two or more strings wound side by side around a pencil). Standardized products such as screws, bolts, and nuts all have single threads; a *double-threaded* screw has a lead equal to twice the pitch, a *triple-threaded* screw has a lead equal to 3 times the pitch, and so on.

All threads are made according to the *right-hand rule* unless otherwise noted. That is, if the bolt is turned clockwise, the bolt advances toward the nut.

The *American National (Unified)* thread standard has been approved in this country and in Great Britain for use on all standard threaded products. The thread angle is 60° and the crests of the thread may be either flat or rounded.

Figure 8–2 shows the thread geometry of the metric M and MJ profiles. The M profile replaces the inch class and is the basic ISO 68 profile with 60° symmetric

Figure 8–1

Terminology of screw threads. Sharp vee threads shown for clarity; the crests and roots are actually flattened or rounded during the forming operation.

Figure 8–2

Basic profile for metric M and MJ threads.
d = major diameter
d_r = minor diameter
d_p = pitch diameter
p = pitch
$H = \frac{\sqrt{3}}{2} p$

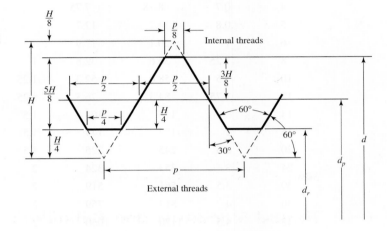

threads. The MJ profile has a rounded fillet at the root of the external thread and a larger minor diameter of both the internal and external threads. This profile is especially useful where high fatigue strength is required.

Tables 8–1 and 8–2 will be useful in specifying and designing threaded parts. Note that the thread size is specified by giving the pitch p for metric sizes and by giving the number of threads per inch N for the Unified sizes. The screw sizes in Table 8–2 with diameter under $\frac{1}{4}$ in are numbered or gauge sizes. The second column in Table 8–2 shows that a No. 8 screw has a nominal major diameter of 0.1640 in.

A great many tensile tests of threaded rods have shown that an unthreaded rod having a diameter equal to the mean of the pitch diameter and minor diameter will have the same tensile strength as the threaded rod. The area of this unthreaded rod is called the tensile-stress area A_t of the threaded rod; values of A_t are listed in both tables.

Two major Unified thread series are in common use: UN and UNR. The difference between these is simply that a root radius must be used in the UNR series. Because of reduced thread stress-concentration factors, UNR series threads have improved fatigue strengths. Unified threads are specified by stating the nominal major diameter, the number of threads per inch, and the thread series, for example, $\frac{5}{8}$ in-18 UNRF or 0.625 in-18 UNRF.

Metric threads are specified by writing the diameter and pitch in millimeters, in that order. Thus, M12 × 1.75 is a thread having a nominal major diameter of 12 mm and a pitch of 1.75 mm. Note that the letter M, which precedes the diameter, is the clue to the metric designation.

Table 8–1

Diameters and Areas of Coarse-Pitch and Fine-Pitch Metric Threads.*

Nominal Major Diameter d mm	Coarse-Pitch Series			Fine-Pitch Series		
	Pitch p mm	Tensile-Stress Area A_t mm^2	Minor-Diameter Area A_r mm^2	Pitch p mm	Tensile-Stress Area A_t mm^2	Minor-Diameter Area A_r mm^2
1.6	0.35	1.27	1.07			
2	0.40	2.07	1.79			
2.5	0.45	3.39	2.98			
3	0.5	5.03	4.47			
3.5	0.6	6.78	6.00			
4	0.7	8.78	7.75			
5	0.8	14.2	12.7			
6	1	20.1	17.9			
8	1.25	36.6	32.8	1	39.2	36.0
10	1.5	58.0	52.3	1.25	61.2	56.3
12	1.75	84.3	76.3	1.25	92.1	86.0
14	2	115	104	1.5	125	116
16	2	157	144	1.5	167	157
20	2.5	245	225	1.5	272	259
24	3	353	324	2	384	365
30	3.5	561	519	2	621	596
36	4	817	759	2	915	884
42	4.5	1120	1050	2	1260	1230
48	5	1470	1380	2	1670	1630
56	5.5	2030	1910	2	2300	2250
64	6	2680	2520	2	3030	2980
72	6	3460	3280	2	3860	3800
80	6	4340	4140	1.5	4850	4800
90	6	5590	5360	2	6100	6020
100	6	6990	6740	2	7560	7470
110				2	9180	9080

*The equations and data used to develop this table have been obtained from ANSI B1.1-1974 and B18.3.1-1978. The minor diameter was found from the equation $d_r = d - 1.226\ 869p$, and the pitch diameter from $d_p = d - 0.649\ 519p$. The mean of the pitch diameter and the minor diameter was used to compute the tensile-stress area.

Square and Acme threads, whose profiles are shown in Fig. 8–3a and b, respectively, are used on screws when power is to be transmitted. Table 8–3 lists the preferred pitches for inch-series Acme threads. However, other pitches can be and often are used, since the need for a standard for such threads is not great.

Modifications are frequently made to both Acme and square threads. For instance, the squar e thread is sometimes modified by cutting the space between the teeth so as to have an included thread angle of 10 to 15°. This is not difficult, since these threads are usually cut with a single-point tool anyhow; the modification retains most of the high efficiency inherent in square threads and makes the cutting simpler. Acme

Table 8–2

Diameters and Area of Unified Screw Threads UNC and UNF*

Size Designation	Nominal Major Diameter in	Coarse Series—UNC			Fine Series—UNF		
		Threads per Inch N	Tensile-Stress Area A_t in²	Minor-Diameter Area A_r in²	Threads per Inch N	Tensile-Stress Area A_t in²	Minor-Diameter Area A_r in²
0	0.0600				80	0.001 80	0.001 51
1	0.0730	64	0.002 63	0.002 18	72	0.002 78	0.002 37
2	0.0860	56	0.003 70	0.003 10	64	0.003 94	0.003 39
3	0.0990	48	0.004 87	0.004 06	56	0.005 23	0.004 51
4	0.1120	40	0.006 04	0.004 96	48	0.006 61	0.005 66
5	0.1250	40	0.007 96	0.006 72	44	0.008 80	0.007 16
6	0.1380	32	0.009 09	0.007 45	40	0.010 15	0.008 74
8	0.1640	32	0.014 0	0.011 96	36	0.014 74	0.012 85
10	0.1900	24	0.017 5	0.014 50	32	0.020 0	0.017 5
12	0.2160	24	0.024 2	0.020 6	28	0.025 8	0.022 6
$\frac{1}{4}$	0.2500	20	0.031 8	0.026 9	28	0.036 4	0.032 6
$\frac{5}{16}$	0.3125	18	0.052 4	0.045 4	24	0.058 0	0.052 4
$\frac{3}{8}$	0.3750	16	0.077 5	0.067 8	24	0.087 8	0.080 9
$\frac{7}{16}$	0.4375	14	0.106 3	0.093 3	20	0.118 7	0.109 0
$\frac{1}{2}$	0.5000	13	0.141 9	0.125 7	20	0.159 9	0.148 6
$\frac{9}{16}$	0.5625	12	0.182	0.162	18	0.203	0.189
$\frac{5}{8}$	0.6250	11	0.226	0.202	18	0.256	0.240
$\frac{3}{4}$	0.7500	10	0.334	0.302	16	0.373	0.351
$\frac{7}{8}$	0.8750	9	0.462	0.419	14	0.509	0.480
1	1.0000	8	0.606	0.551	12	0.663	0.625
$1\frac{1}{4}$	1.2500	7	0.969	0.890	12	1.073	1.024
$1\frac{1}{2}$	1.5000	6	1.405	1.294	12	1.581	1.521

*This table was compiled from ANSI B1.1-1974. The minor diameter was found from the equation $d_r = d - 1.299\ 038p$, and the pitch diameter from $d_p = d - 0.649\ 519p$. The mean of the pitch diameter and the minor diameter was used to compute the tensile-stress area.

Figure 8–3

(a) Square thread;
(b) Acme thread.

Table 8–3

Preferred Pitches for Acme Threads

d, in	$\frac{1}{4}$	$\frac{5}{16}$	$\frac{3}{8}$	$\frac{1}{2}$	$\frac{5}{8}$	$\frac{3}{4}$	$\frac{7}{8}$	1	$1\frac{1}{4}$	$1\frac{1}{2}$	$1\frac{3}{4}$	2	$2\frac{1}{2}$	3
p, in	$\frac{1}{16}$	$\frac{1}{14}$	$\frac{1}{12}$	$\frac{1}{10}$	$\frac{1}{8}$	$\frac{1}{6}$	$\frac{1}{6}$	$\frac{1}{5}$	$\frac{1}{5}$	$\frac{1}{4}$	$\frac{1}{4}$	$\frac{1}{4}$	$\frac{1}{3}$	$\frac{1}{2}$

threads are sometimes modified to a stub form by making the teeth shorter. This results in a larger minor diameter and a somewhat stronger screw.

8–2 ## The Mechanics of Power Screws

A power screw is a device used in machinery to change angular motion into linear motion, and, usually, to transmit power. Familiar applications include the lead screws of lathes, and the screws for vises, presses, and jacks.

An application of power screws to a power-driven jack is shown in Fig. 8–4. You should be able to identify the worm, the worm gear, the screw, and the nut. Is the worm gear supported by one bearing or two?

Figure 8–4

The Joyce worm-gear screw jack. (*Courtesy Joyce-Dayton Corp., Dayton, Ohio.*)

Figure 8–5

Portion of a power screw.

Figure 8–6

Force diagrams: (a) lifting the load; (b) lowering the load.

In Fig. 8–5 a square-threaded power screw with single thread having a mean diameter d_m, a pitch p, a lead angle λ, and a helix angle ψ is loaded by the axial compressive force F. We wish to find an expression for the torque required to raise this load, and another expression for the torque required to lower the load.

First, imagine that a single thread of the screw is unrolled or developed (Fig. 8–6) for exactly a single turn. Then one edge of the thread will form the hypotenuse of a right triangle whose base is the circumference of the mean-thread-diameter circle and whose height is the lead. The angle λ, in Figs. 8–5 and 8–6, is the lead angle of the thread. We represent the summation of all the axial forces acting upon the normal thread area by F. To raise the load, a force P_R acts to the right (Fig. 8–6a), and to lower the load, P_L acts to the left (Fig. 8–6b). The friction force is the product of the coefficient of friction f with the normal force N, and acts to oppose the motion. The system is in equilibrium under the action of these forces, and hence, for raising the load, we have

$$\sum F_x = P_R - N \sin \lambda - fN \cos \lambda = 0$$

$$\sum F_y = -F - fN \sin \lambda + N \cos \lambda = 0$$

(a)

In a similar manner, for lowering the load, we have

$$\sum F_x = -P_L - N \sin \lambda + fN \cos \lambda = 0$$

$$\sum F_y = -F + fN \sin \lambda + N \cos \lambda = 0$$

(b)

Since we are not interested in the normal force N, we eliminate it from each of these sets of equations and solve the result for P. For raising the load, this gives

$$P_R = \frac{F(\sin \lambda + f \cos \lambda)}{\cos \lambda - f \sin \lambda} \tag{c}$$

and for lowering the load,

$$P_L = \frac{F(f \cos \lambda - \sin \lambda)}{\cos \lambda + f \sin \lambda} \tag{d}$$

Next, divide the numerator and the denominator of these equations by $\cos \lambda$ and use the relation $\tan \lambda = l/\pi d_m$ (Fig. 8–6). We then have, respectively,

$$P_R = \frac{F[(l/\pi d_m) + f]}{1 - (fl/\pi d_m)} \tag{e}$$

$$P_L = \frac{F[f - (l/\pi d_m)]}{1 + (fl/\pi d_m)} \tag{f}$$

Finally, noting that the torque is the product of the force P and the mean radius $d_m/2$, for raising the load we can write

$$T_R = \frac{F d_m}{2} \left(\frac{l + \pi f d_m}{\pi d_m - f l} \right) \tag{8–1}$$

where T_R is the torque required for two purposes: to overcome thread friction and to raise the load.

The torque required to lower the load, from Eq. (f), is found to be

$$T_L = \frac{F d_m}{2} \left(\frac{\pi f d_m - l}{\pi d_m + f l} \right) \tag{8–2}$$

This is the torque required to overcome a part of the friction in lowering the load. It may turn out, in specific instances where the lead is large or the friction is low, that the load will lower itself by causing the screw to spin without any external effort. In such cases, the torque T_L from Eq. (8–2) will be negative or zero. When a positive torque is obtained from this equation, the screw is said to be *self-locking*. Thus the condition for self-locking is

$$\pi f d_m > l$$

Now divide both sides of this inequality by πd_m. Recognizing that $l/\pi d_m = \tan \lambda$, we get

$$f > \tan \lambda \tag{8–3}$$

This relation states that self-locking is obtained whenever the coefficient of thread friction is equal to or greater than the tangent of the thread lead angle.

An expression for efficiency is also useful in the evaluation of power screws. If we let $f = 0$ in Eq. (8–1), we obtain

$$T_0 = \frac{F l}{2\pi} \tag{g}$$

which, since thread friction has been eliminated, is the torque required only to raise the load. The thread efficiency is thus defined as

$$e = \frac{T_0}{T_R} = \frac{Fl}{2\pi T_R}$$ (8–4)

The preceding equations have been developed for square threads where the normal thread loads are parallel to the axis of the screw. In the case of Acme or other threads, the normal thread load is inclined to the axis because of the thread angle 2α and the lead angle λ. Since lead angles are small, this inclination can be neglected and only the effect of the thread angle (Fig. 8–7a) considered. The effect of the angle α is to increase the frictional force by the wedging action of the threads. Therefore the frictional terms in Eq. (8–1) must be divided by cos α. For raising the load, or for tightening a screw or bolt, this yields

$$T_R = \frac{Fd_m}{2} \left(\frac{l + \pi f d_m \sec \alpha}{\pi d_m - fl \sec \alpha} \right)$$ (8–5)

In using Eq. (8–5), remember that it is an approximation because the effect of the lead angle has been neglected.

For power screws, the Acme thread is not as efficient as the square thread, because of the additional friction due to the wedging action, but it is often preferred because it is easier to machine and permits the use of a split nut, which can be adjusted to take up for wear.

Usually a third component of torque must be applied in power-screw applications. When the screw is loaded axially, a thrust or collar bearing must be employed between the rotating and stationary members in order to carry the axial component. Figure 8–7b shows a typical thrust collar in which the load is assumed to be concentrated at the mean collar diameter d_c. If f_c is the coefficient of collar friction, the torque required is

$$T_c = \frac{F f_c d_c}{2}$$ (8–6)

For large collars, the torque should probably be computed in a manner similar to that employed for disk clutches (see Sec. 16–5).

Nominal body stresses in power screws can be related to thread parameters as follows. The maximum nominal shear stress τ in torsion of the screw body can be expressed as

$$\tau = \frac{16T}{\pi d_r^3}$$ (8–7)

Figure 8–7

(a) Normal thread force is increased because of angle α; (b) thrust collar has frictional diameter d_c.

(a)

(b)

The axial stress σ in the body of the screw due to load F is

$$\sigma = \frac{F}{A} = \frac{4F}{\pi d_r^2} \tag{8-8}$$

in the absence of column action. For a short column the J. B. Johnson buckling formula is given by Eq. (4–43), which is

$$\left(\frac{F}{A}\right)_{crit} = S_y - \left(\frac{S_y}{2\pi}\frac{l}{k}\right)^2\frac{1}{CE} \tag{8-9}$$

Nominal thread stresses in power screws can be related to thread parameters as follows. The bearing stress in Fig. 8–8, σ_B, is

$$\sigma_B = -\frac{F}{\pi d_m n_t p/2} = -\frac{2F}{\pi d_m n_t p} \tag{8-10}$$

where n_t is the number of engaged threads. The bending stress at the root of the thread σ_b is found from

$$Z = \frac{I}{c} = \frac{(\pi d_r n_t)(p/2)^2}{6} = \frac{\pi}{24}d_r n_t p^2 \qquad M = \frac{Fp}{4}$$

so

$$\sigma_b = \frac{M}{Z} = \frac{Fp}{4}\frac{24}{\pi d_r n_t p^2} = \frac{6F}{\pi d_r n_t p} \tag{8-11}$$

The transverse shear stress τ at the center of the root of the thread due to load F is

$$\tau = \frac{3V}{2A} = \frac{3}{2}\frac{F}{\pi d_r n_t p/2} = \frac{3F}{\pi d_r n_t p} \tag{8-12}$$

and at the top of the root it is zero. The von Mises stress σ' at the top of the root "plane" is found by first identifying the orthogonal normal stresses and the shear

Figure 8–8

Geometry of square thread useful in finding bending and transverse shear stresses at the thread root.

stresses. From the coordinate system of Fig. 8–8, we note

$$\sigma_x = \frac{6F}{\pi d_r n_t p} \qquad \tau_{xy} = 0$$

$$\sigma_y = -\frac{4F}{\pi d_r^2} \qquad \tau_{yz} = \frac{16T}{\pi d_r^3}$$

$$\sigma_z = 0 \qquad \tau_{zx} = 0$$

then use Eq. (5–14) of Sec. 5–5.

The screw-thread form is complicated from an analysis viewpoint. Remember the origin of the tensile-stress area A_t, which comes from experiment. A power screw lifting a load is in compression and its thread pitch is *shortened* by elastic deformation. Its engaging nut is in tension and its thread pitch is *lengthened*. The engaged threads cannot share the load equally. Some experiments show that the first engaged thread carries 0.38 of the load, the second 0.25, the third 0.18, and the seventh is free of load. In estimating thread stresses by the equations above, substituting $0.38F$ for F and setting n_t to 1 will give the largest level of stresses in the thread-nut combination.

EXAMPLE 8–1

A square-thread power screw has a major diameter of 32 mm and a pitch of 4 mm with double threads, and it is to be used in an application similar to that in Fig. 8–4. The given data include $f = f_c = 0.08$, $d_c = 40$ mm, and $F = 6.4$ kN per screw.
(a) Find the thread depth, thread width, pitch diameter, minor diameter, and lead.
(b) Find the torque required to raise and lower the load.
(c) Find the efficiency during lifting the load.
(d) Find the body stresses, torsional and compressive.
(e) Find the bearing stress.
(f) Find the thread bending stress at the root of the thread.
(g) Determine the von Mises stress at the root of the thread.
(h) Determine the maximum shear stress at the root of the thread.

Solution

(a) From Fig. 8–3a the thread depth and width are the same and equal to half the pitch, or 2 mm. Also

Answer

$$d_m = d - p/2 = 32 - 4/2 = 30 \text{ mm}$$

$$d_r = d - p = 32 - 4 = 28 \text{ mm}$$

$$l = np = 2(4) = 8 \text{ mm}$$

(b) Using Eqs. (8–1) and (8–6), the torque required to turn the screw against the load is

$$T_R = \frac{Fd_m}{2}\left(\frac{l + \pi f d_m}{\pi d_m - fl}\right) + \frac{Ff_c d_c}{2}$$

$$= \frac{6.4(30)}{2}\left[\frac{8 + \pi(0.08)(30)}{\pi(30) - 0.08(8)}\right] + \frac{6.4(0.08)40}{2}$$

Answer

$$= 15.94 + 10.24 = 26.18 \text{ N} \cdot \text{m}$$

Using Eqs. (8–2) and (8–6), we find the load-lowering torque is

$$T_L = \frac{Fd_m}{2}\left(\frac{\pi f d_m - l}{\pi d_m + fl}\right) + \frac{Ff_c d_c}{2}$$

$$= \frac{6.4(30)}{2}\left[\frac{\pi(0.08)30 - 8}{\pi(30) + 0.08(8)}\right] + \frac{6.4(0.08)(40)}{2}$$

Answer
$$= -0.466 + 10.24 = 9.77 \text{ N} \cdot \text{m}$$

The minus sign in the first term indicates that the screw alone is not self-locking and would rotate under the action of the load except for the fact that the collar friction is present and must be overcome, too. Thus the torque required to rotate the screw "with" the load is less than is necessary to overcome collar friction alone.

(c) The overall efficiency in raising the load is

Answer
$$e = \frac{Fl}{2\pi T_R} = \frac{6.4(8)}{2\pi(26.18)} = 0.311$$

(d) The body shear stress τ due to torsional moment T_R at the outside of the screw body is

Answer
$$\tau = \frac{16T_R}{\pi d_r^3} = \frac{16(26.18)(10^3)}{\pi(28^3)} = 6.07 \text{ MPa}$$

The axial nominal normal stress σ is

Answer
$$\sigma = -\frac{4F}{\pi d_r^2} = -\frac{4(6.4)10^3}{\pi(28^2)} = -10.39 \text{ MPa}$$

(e) The bearing stress σ_B is, with one thread carrying 0.38F,

Answer
$$\sigma_B = -\frac{2(0.38F)}{\pi d_m(1)p} = -\frac{2(0.38)(6.4)10^3}{\pi(30)(1)(4)} = -12.9 \text{ MPa}$$

(f) The thread-root bending stress σ_b with one thread carrying 0.38F is

Answer
$$\sigma_b = \frac{6(0.38F)}{\pi d_r(1)p} = \frac{6(0.38)(6.4)10^3}{\pi(28)(1)4} = 41.5 \text{ MPa}$$

(g) The transverse shear at the extreme of the root cross section due to bending is zero. However, there is a circumferential shear stress at the extreme of the root cross section of the thread as shown in part (d) of 6.07 MPa. The three-dimensional stresses, after Fig. 8–8, noting the y coordinate is into the page, are

$$\sigma_x = 41.5 \text{ MPa} \qquad \tau_{xy} = 0$$

$$\sigma_y = -10.39 \text{ MPa} \qquad \tau_{yz} = 6.07 \text{ MPa}$$

$$\sigma_z = 0 \qquad \tau_{zx} = 0$$

For the von Mises stress, Eq. (5–14) of Sec. 5–5 can be written as

Answer
$$\sigma' = \frac{1}{\sqrt{2}}\{(41.5 - 0)^2 + [0 - (-10.39)]^2 + (-10.39 - 41.5)^2 + 6(6.07)^2\}^{1/2}$$

$$= 48.7 \text{ MPa}$$

Alternatively, you can determine the principal stresses and then use Eq. (5–12) to find the von Mises stress. This would prove helpful in evaluating τ_{max} as well. The principal stresses can be found from Eq. (3–15); however, sketch the stress element and note that there are no shear stresses on the x face. This means that σ_x is a principal stress. The remaining stresses can be transformed by using the plane stress equation, Eq. (3–13). Thus, the remaining principal stresses are

$$\frac{-10.39}{2} \pm \sqrt{\left(\frac{-10.39}{2}\right)^2 + 6.07^2} = 2.79, -13.18 \text{ MPa}$$

Ordering the principal stresses gives $\sigma_1, \sigma_2, \sigma_3 = 41.5, 2.79, -13.18$ MPa. Substituting these into Eq. (5–12) yields

Answer
$$\sigma' = \left\{\frac{[41.5 - 2.79]^2 + [2.79 - (-13.18)]^2 + [-13.18 - 41.5]^2}{2}\right\}^{1/2}$$
$$= 48.7 \text{ MPa}$$

(h) The maximum shear stress is given by Eq. (3–16), where $\tau_{max} = \tau_{1/3}$, giving

Answer
$$\tau_{max} = \frac{\sigma_1 - \sigma_3}{2} = \frac{41.5 - (-13.18)}{2} = 27.3 \text{ MPa}$$

Table 8–4	**Screw Material**	**Nut Material**	**Safe p_b, MPa**	**Notes**
Screw Bearing Pressure p_b	Steel	Bronze	17.2–24.1	Low speed
	Steel	Bronze	11.0–17.2	≤50 mm/s
Source: H. A. Rothbart and T. H. Brown, Jr., *Mechanical Design Handbook,* 2nd ed., McGraw-Hill, New York, 2006.		Cast iron	6.9–17.2	≤40 mm/s
	Steel	Bronze	5.5–9.7	100–200 mm/s
		Cast iron	4.1–6.9	100–200 mm/s
	Steel	Bronze	1.0–1.7	≥250 mm/s

Ham and Ryan[1] showed that the coefficient of friction in screw threads is independent of axial load, practically independent of speed, decreases with heavier lubricants, shows little variation with combinations of materials, and is best for steel on bronze. Sliding coefficients of friction in power screws are about 0.10–0.15.

Table 8–4 shows safe bearing pressures on threads, to protect the moving surfaces from abnormal wear. Table 8–5 shows the coefficients of sliding friction for common material pairs. Table 8–6 shows coefficients of starting and running friction for common material pairs.

---

[1]Ham and Ryan, *An Experimental Investigation of the Friction of Screw-threads,* Bulletin 247, University of Illinois Experiment Station, Champaign-Urbana, Ill., June 7, 1932.

Table 8–5

Coefficients of Friction f for Threaded Pairs

Source: H. A. Rothbart and T. H. Brown, Jr., *Mechanical Design Handbook,* 2nd ed., McGraw-Hill, New York, 2006.

Screw Material	Nut Material			
	Steel	**Bronze**	**Brass**	**Cast Iron**
Steel, dry	0.15–0.25	0.15–0.23	0.15–0.19	0.15–0.25
Steel, machine oil	0.11–0.17	0.10–0.16	0.10–0.15	0.11–0.17
Bronze	0.08–0.12	0.04–0.06	—	0.06–0.09

Table 8–6

Thrust-Collar Friction Coefficients

Source: H. A. Rothbart and T. H. Brown, Jr., *Mechanical Design Handbook,* 2nd ed., McGraw-Hill, New York, 2006.

Combination	Running	Starting
Soft steel on cast iron	0.12	0.17
Hard steel on cast iron	0.09	0.15
Soft steel on bronze	0.08	0.10
Hard steel on bronze	0.06	0.08

8–3 Threaded Fasteners

Figure 8–9 is a drawing of a standard hexagon-head bolt. Points of stress concentration are at the fillet, at the start of the threads (runout), and at the thread-root fillet in the plane of the nut when it is present. See Table A–29 for dimensions. The diameter of the washer face is the same as the width across the flats of the hexagon. The thread length of inch-series bolts, where d is the nominal diameter, is

$$L_T = \begin{cases} 2d + \frac{1}{4} \text{ in} & L \le 6 \text{ in} \\ 2d + \frac{1}{2} \text{ in} & L > 6 \text{ in} \end{cases} \tag{8–13}$$

and for metric bolts is

$$L_T = \begin{cases} 2d + 6 & L \le 125 \quad d \le 48 \\ 2d + 12 & 125 < L \le 200 \\ 2d + 25 & L > 200 \end{cases} \tag{8–14}$$

where the dimensions are in millimeters. The ideal bolt length is one in which only one or two threads project from the nut after it is tightened. Bolt holes may have burrs or sharp edges after drilling. These could bite into the fillet and increase stress concentration. Therefore, washers must always be used under the bolt head to prevent this. They should be of hardened steel and loaded onto the bolt so that the rounded edge of the stamped hole faces the washer face of the bolt. Sometimes it is necessary to use washers under the nut too.

The purpose of a bolt is to clamp two or more parts together. The clamping load stretches or elongates the bolt; the load is obtained by twisting the nut until the bolt has elongated almost to the elastic limit. If the nut does not loosen, this bolt tension remains as the preload or clamping force. When tightening, the mechanic should, if

Figure 8–9

Hexagon-head bolt; note the washer face, the fillet under the head, the start of threads, and the chamfer on both ends. Bolt lengths are always measured from below the head.

Figure 8–10

Typical cap-screw heads: (*a*) fillister head; (*b*) flat head; (*c*) hexagonal socket head. Cap screws are also manufactured with hexagonal heads similar to the one shown in Fig. 8–9, as well as a variety of other head styles. This illustration uses one of the conventional methods of representing threads.

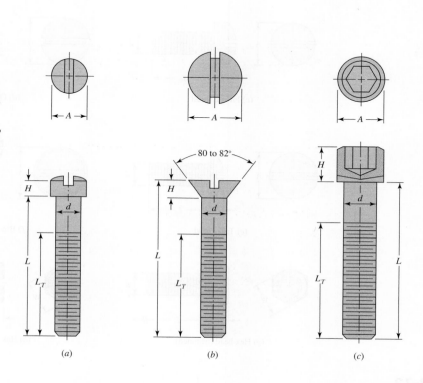

possible, hold the bolt head stationary and twist the nut; in this way the bolt shank will not feel the thread-friction torque.

The head of a hexagon-head cap screw is slightly thinner than that of a hexagon-head bolt. Dimensions of hexagon-head cap screws are listed in Table A–30. Hexagon-head cap screws are used in the same applications as bolts and also in applications in which one of the clamped members is threaded. Three other common cap-screw head styles are shown in Fig. 8–10.

A variety of machine-screw head styles are shown in Fig. 8–11. Inch-series machine screws are generally available in sizes from No. 0 to about $\frac{3}{8}$ in.

Several styles of hexagonal nuts are illustrated in Fig. 8–12; their dimensions are given in Table A–31. The material of the nut must be selected carefully to match that of the bolt. During tightening, the first thread of the nut tends to take the entire load; but yielding occurs, with some strengthening due to the cold work that takes place, and the load is eventually divided over about three nut threads. For this reason you should never reuse nuts; in fact, it can be dangerous to do so.

Figure 8–11

Types of heads used on machine screws.

(a) Round head

(b) Flat head

(c) Fillister head

(d) Oval head

(e) Truss head

(f) Binding head

(g) Hex head (trimmed)

(h) Hex head (upset)

Figure 8–12

Hexagonal nuts: (a) end view, general; (b) washer-faced regular nut; (c) regular nut chamfered on both sides; (d) jam nut with washer face; (e) jam nut chamfered on both sides.

(a)　　　(b)　　　(c)　　　(d)　　　(e)

8–4   Joints—Fastener Stiffness

When a connection is desired that can be disassembled without destructive methods and that is strong enough to resist external tensile loads, moment loads, and shear loads, or a combination of these, then the simple bolted joint using hardened-steel washers is a good solution. Such a joint can also be dangerous unless it is properly designed and assembled by a *trained* mechanic.

Figure 8–13

A bolted connection loaded in tension by the forces P. Note the use of two washers. Note how the threads extend into the body of the connection. This is usual and is desired. l is the grip of the connection.

Figure 8–14

Section of cylindrical pressure vessel. Hexagon-head cap screws are used to fasten the cylinder head to the body. Note the use of an O-ring seal. l is the effective grip of the connection (see Table 8–7).

A section through a tension-loaded bolted joint is illustrated in Fig. 8–13. Notice the clearance space provided by the bolt holes. Notice, too, how the bolt threads extend into the body of the connection.

As noted previously, the purpose of the bolt is to clamp the two, or more, parts together. Twisting the nut stretches the bolt to produce the clamping force. This clamping force is called the *pretension* or *bolt preload*. It exists in the connection after the nut has been properly tightened no matter whether the external tensile load P is exerted or not.

Of course, since the members are being clamped together, the clamping force that produces tension in the bolt induces compression in the members.

Figure 8–14 shows another tension-loaded connection. This joint uses cap screws threaded into one of the members. An alternative approach to this problem (of not using a nut) would be to use studs. A stud is a rod threaded on both ends. The stud is screwed into the lower member first; then the top member is positioned and fastened down with hardened washers and nuts. The studs are regarded as permanent, and so the joint can be disassembled merely by removing the nut and washer. Thus the threaded part of the lower member is not damaged by reusing the threads.

The *spring rate* is a limit as expressed in Eq. (4–1). For an elastic member such as a bolt, as we learned in Eq. (4–2), it is the ratio between the force applied to the member and the deflection produced by that force. We can use Eq. (4–4) and the results of Prob. 4–1 to find the stiffness constant of a fastener in any bolted connection.

The *grip l* of a connection is the total thickness of the clamped material. In Fig. 8–13 the grip is the sum of the thicknesses of both members and both washers. In Fig. 8–14 the effective grip is given in Table 8–7.

The stiffness of the portion of a bolt or screw within the clamped zone will generally consist of two parts, that of the unthreaded shank portion and that of the

Table 8–7

Suggested Procedure for Finding Fastener Stiffness

(a) *(b)*

Given fastener diameter d and pitch p in mm or number of threads per inch

Washer thickness: t from Table A–32 or A–33

Nut thickness [Fig. (*a*) only]: H from Table A–31

Grip length:

For Fig. (*a*): l = thickness of all material squeezed between face of bolt and face of nut

For Fig. (*b*): $l = \begin{cases} h + t_2/2, & t_2 < d \\ h + d/2, & t_2 \geq d \end{cases}$

Fastener length (round up using Table A–17*):

For Fig. (*a*): $L > l + H$

For Fig. (*b*): $L > h + 1.5d$

Threaded length L_T: Inch series:

$$L_T = \begin{cases} 2d + \frac{1}{4}\text{ in}, & L \leq 6 \text{ in} \\ 2d + \frac{1}{2}\text{ in}, & L > 6 \text{ in} \end{cases}$$

Metric series:

$$L_T = \begin{cases} 2d + 6 \text{ mm}, & L \leq 125 \text{ mm}, d \leq 48 \text{ mm} \\ 2d + 12 \text{ mm}, & 125 < L \leq 200 \text{ mm} \\ 2d + 25 \text{ mm}, & L > 200 \text{ mm} \end{cases}$$

Length of unthreaded portion in grip: $l_d = L - L_T$

Length of threaded portion in grip: $l_t = l - l_d$

Area of unthreaded portion: $A_d = \pi d^2/4$

Area of threaded portion: A_t from Table 8–1 or 8–2

Fastener stiffness: $k_b = \dfrac{A_d A_t E}{A_d l_t + A_t l_d}$

---

*Bolts and cap screws may not be available in all the preferred lengths listed in Table A–17. Large fasteners may not be available in fractional inches or in millimeter lengths ending in a nonzero digit. Check with your bolt supplier for availability.

threaded portion. Thus the stiffness constant of the bolt is equivalent to the stiffnesses of two springs in series. Using the results of Prob. 4–1, we find

$$\frac{1}{k} = \frac{1}{k_1} + \frac{1}{k_2} \quad \text{or} \quad k = \frac{k_1 k_2}{k_1 + k_2} \tag{8-15}$$

for two springs in series. From Eq. (4–4), the spring rates of the threaded and unthreaded portions of the bolt in the clamped zone are, respectively,

$$k_t = \frac{A_t E}{l_t} \qquad k_d = \frac{A_d E}{l_d} \tag{8-16}$$

where A_t = tensile-stress area (Tables 8–1, 8–2)

l_t = length of threaded portion of grip

A_d = major-diameter area of fastener

l_d = length of unthreaded portion in grip

Substituting these stiffnesses in Eq. (8–15) gives

$$k_b = \frac{A_d A_t E}{A_d l_t + A_t l_d} \tag{8-17}$$

where k_b is the estimated effective stiffness of the bolt or cap screw in the clamped zone. For short fasteners, the one in Fig. 8–14, for example, the unthreaded area is small and so the first of the expressions in Eq. (8–16) can be used to find k_b. For long fasteners, the threaded area is relatively small, and so the second expression in Eq. (8–16) can be used. Table 8–7 is useful.

8–5 Joints—Member Stiffness

In the previous section, we determined the stiffness of the fastener in the clamped zone. In this section, we wish to study the stiffnesses of the members in the clamped zone. Both of these stiffnesses must be known in order to learn what happens when the assembled connection is subjected to an external tensile loading.

There may be more than two members included in the grip of the fastener. All together these act like compressive springs in series, and hence the total spring rate of the members is

$$\frac{1}{k_m} = \frac{1}{k_1} + \frac{1}{k_2} + \frac{1}{k_3} + \cdots + \frac{1}{k_i} \tag{8-18}$$

If one of the members is a soft gasket, its stiffness relative to the other members is usually so small that for all practical purposes the others can be neglected and only the gasket stiffness used.

If there is no gasket, the stiffness of the members is rather difficult to obtain, except by experimentation, because the compression region spreads out between the bolt head and the nut and hence the area is not uniform. There are, however, some cases in which this area can be determined.

Ito has used ultrasonic techniques to determine the pressure distribution at the member interface.[2] The results show that the pressure stays high out to about 1.5 bolt

---

[2] Y. Ito, J. Toyoda, and S. Nagata, "Interface Pressure Distribution in a Bolt-Flange Assembly," ASME paper no. 77-WA/DE-11, 1977.

Figure 8–15

Compression of a member with the equivalent elastic properties represented by a frustum of a hollow cone. Here, l represents the grip length.

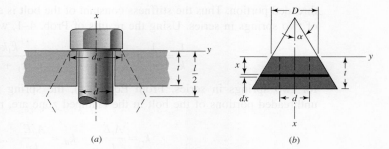

(a) (b)

radii. The pressure, however, falls off farther away from the bolt. Thus Ito suggests the use of Rotscher's pressure-cone method for stiffness calculations with a variable cone angle. This method is quite complicated, and so here we choose to use a simpler approach using a fixed cone angle.

Figure 8–15 illustrates the general cone geometry using a half-apex angle α. An angle $\alpha = 45°$ has been used, but Little reports that this overestimates the clamping stiffness.[3] When loading is restricted to a washer-face annulus (hardened steel, cast iron, or aluminum), the proper apex angle is smaller. Osgood reports a range of $25°$ $\le \alpha \le 33°$ for most combinations.[4] In this book we shall use $\alpha = 30°$ except in cases in which the material is insufficient to allow the frusta to exist.

Referring now to Fig. 8–15b, the contraction of an element of the cone of thickness dx subjected to a compressive force P is, from Eq. (4–3),

$$d\delta = \frac{P\, dx}{EA} \qquad (a)$$

The area of the element is

$$A = \pi(r_o^2 - r_i^2) = \pi\left[\left(x \tan \alpha + \frac{D}{2}\right)^2 - \left(\frac{d}{2}\right)^2\right]$$

$$= \pi\left(x \tan \alpha + \frac{D+d}{2}\right)\left(x \tan \alpha + \frac{D-d}{2}\right) \qquad (b)$$

Substituting this in Eq. (a) and integrating gives a total contraction of

$$\delta = \frac{P}{\pi E}\int_0^t \frac{dx}{[x \tan \alpha + (D+d)/2][x \tan \alpha + (D-d)/2]} \qquad (c)$$

Using a table of integrals, we find the result to be

$$\delta = \frac{P}{\pi E d \tan \alpha}\ln \frac{(2t \tan \alpha + D - d)(D + d)}{(2t \tan \alpha + D + d)(D - d)} \qquad (d)$$

Thus the spring rate or stiffness of this frustum is

$$k = \frac{P}{\delta} = \frac{\pi E d \tan \alpha}{\ln \dfrac{(2t \tan \alpha + D - d)(D + d)}{(2t \tan \alpha + D + d)(D - d)}} \qquad (8\text{–}19)$$

---

[3]R. E. Little, "Bolted Joints: How Much Give?" *Machine Design,* Nov. 9, 1967.

[4]C. C. Osgood, "Saving Weight on Bolted Joints," *Machine Design,* Oct. 25, 1979.

With $\alpha = 30°$, this becomes

$$k = \frac{0.5774\pi Ed}{\ln\dfrac{(1.155t + D - d)(D + d)}{(1.155t + D + d)(D - d)}} \tag{8–20}$$

Equation (8–20), or (8–19), must be solved separately for each frustum in the joint. Then individual stiffnesses are assembled to obtain k_m using Eq. (8–18).

If the members of the joint have the same Young's modulus E with symmetrical frusta back to back, then they act as two identical springs in series. From Eq. (8–18) we learn that $k_m = k/2$. Using the grip as $l = 2t$ and d_w as the diameter of the washer face, from Eq. (8–19) we find the spring rate of the members to be

$$k_m = \frac{\pi Ed \tan\alpha}{2\ln\dfrac{(l\tan\alpha + d_w - d)(d_w + d)}{(l\tan\alpha + d_w + d)(d_w - d)}} \tag{8–21}$$

The diameter of the washer face is about 50 percent greater than the fastener diameter for standard hexagon-head bolts and cap screws. Thus we can simplify Eq. (8–21) by letting $d_w = 1.5d$. If we also use $\alpha = 30°$, then Eq. (8–21) can be written as

$$k_m = \frac{0.5774\pi Ed}{2\ln\left(5\dfrac{0.5774l + 0.5d}{0.5774l + 2.5d}\right)} \tag{8–22}$$

It is easy to program the numbered equations in this section, and you should do so. The time spent in programming will save many hours of formula plugging.

To see how good Eq. (8–21) is, solve it for k_m/Ed:

$$\frac{k_m}{Ed} = \frac{\pi\tan\alpha}{2\ln\left[\dfrac{(l\tan\alpha + d_w - d)(d_w + d)}{(l\tan\alpha + d_w + d)(d_w - d)}\right]}$$

Earlier in the section use of $\alpha = 30°$ was recommended for hardened steel, cast iron, or aluminum members. Wileman, Choudury, and Green conducted a finite element study of this problem.[5] The results, which are depicted in Fig. 8–16, agree with the $\alpha = 30°$ recommendation, coinciding exactly at the aspect ratio $d/l = 0.4$. Additionally, they offered an exponential curve-fit of the form

$$\frac{k_m}{Ed} = A\exp(Bd/l) \tag{8–23}$$

with constants A and B defined in Table 8–8. Equation (8–23) offers a simple calculation for member stiffness k_m. However, it is very important to note that the *entire joint* must be made up of the *same material*. For departure from these conditions, Eq. (8–20) remains the basis for approaching the problem.

---

[5]J. Wileman, M. Choudury, and I. Green, "Computation of Member Stiffness in Bolted Connections," *Trans. ASME, J. Mech. Design*, vol. 113, December 1991, pp. 432–437.

Figure 8–16

The dimensionless plot of stiffness versus aspect ratio of the members of a bolted joint, showing the relative accuracy of methods of Rotscher, Mischke, and Motosh, compared to a finite-element analysis (FEA) conducted by Wileman, Choudury, and Green.

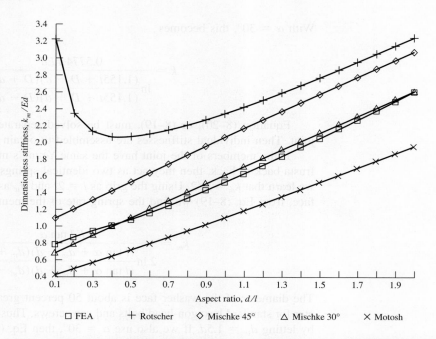

Aspect ratio, d/l

☐ FEA   ✛ Rotscher   ◇ Mischke 45°   △ Mischke 30°   ✕ Motosh

Table 8–8

Stiffness Parameters of Various Member Materials[†]

[†]*Source:* J. Wileman, M. Choudury, and I. Green, "Computation of Member Stiffness in Bolted Connections," *Trans. ASME, J. Mech. Design,* vol. 113, December 1991, pp. 432–437.

Material Used	Poisson Ratio	Elastic GPa	Modulus Mpsi	A	B
Steel	0.291	207	30.0	0.787 15	0.628 73
Aluminum	0.334	71	10.3	0.796 70	0.638 16
Copper	0.326	119	17.3	0.795 68	0.635 53
Gray cast iron	0.211	100	14.5	0.778 71	0.616 16
General expression				0.789 52	0.629 14

EXAMPLE 8–2

As shown in Fig. 8–17a, two plates are clamped by washer-faced $\frac{1}{2}$ in-20 UNF $\times$ 1$\frac{1}{2}$ in SAE grade 5 bolts each with a standard $\frac{1}{2}$ N steel plain washer.

(a) Determine the member spring rate k_m if the top plate is steel and the bottom plate is gray cast iron.

(b) Using the method of conical frusta, determine the member spring rate k_m if both plates are steel.

(c) Using Eq. (8–23), determine the member spring rate k_m if both plates are steel. Compare the results with part (b).

(d) Determine the bolt spring rate k_b.

Solution

From Table A–32, the thickness of a standard $\frac{1}{2}$ N plain washer is 0.095 in.

(a) As shown in Fig. 8–17b, the frusta extend halfway into the joint the distance

$$\frac{1}{2}(0.5 + 0.75 + 0.095) = 0.6725 \text{ in}$$

Figure 8–17

Dimensions in inches.

(a)   (b)

The distance between the joint line and the dotted frusta line is $0.6725 - 0.5 - 0.095 = 0.0775$ in. Thus, the top frusta consist of the steel washer, steel plate, and 0.0775 in of the cast iron. Since the washer and top plate are both steel with $E = 30(10^6)$ psi, they can be considered a single frustum of 0.595 in thick. The outer diameter of the frustum of the steel member at the joint interface is $0.75 + 2(0.595) \tan 30° = 1.437$ in. The outer diameter at the midpoint of the entire joint is $0.75 + 2(0.6725) \tan 30° = 1.527$ in. Using Eq. (8–20), the spring rate of the steel is

$$k_1 = \frac{0.5774\pi(30)(10^6)0.5}{\ln\left\{\dfrac{[1.155(0.595) + 0.75 - 0.5](0.75 + 0.5)}{[1.155(0.595) + 0.75 + 0.5](0.75 - 0.5)}\right\}} = 30.80(10^6) \text{ lbf/in}$$

From Tables 8–8 or A–5, for gray cast iron, $E = 14.5$ Mpsi. Thus for the upper cast-iron frustum

$$k_2 = \frac{0.5774\pi(14.5)(10^6)0.5}{\ln\left\{\dfrac{[1.155(0.0775) + 1.437 - 0.5](1.437 + 0.5)}{[1.155(0.0775) + 1.437 + 0.5](1.437 - 0.5)}\right\}} = 285.5(10^6) \text{ lbf/in}$$

For the lower cast-iron frustum

$$k_3 = \frac{0.5774\pi(14.5)(10^6)0.5}{\ln\left\{\dfrac{[1.155(0.6725) + 0.75 - 0.5](0.75 + 0.5)}{[1.155(0.6725) + 0.75 + 0.5](0.75 - 0.5)}\right\}} = 14.15(10^6) \text{ lbf/in}$$

The three frusta are in series, so from Eq. (8–18)

$$\frac{1}{k_m} = \frac{1}{30.80(10^6)} + \frac{1}{285.5(10^6)} + \frac{1}{14.15(10^6)}$$

Answer   This results in $k_m = 9.378 \ (10^6)$ lbf/in.
          (b) If the entire joint is steel, Eq. (8–22) with $l = 2(0.6725) = 1.345$ in gives

Answer $$k_m = \frac{0.5774\pi(30.0)(10^6)0.5}{2\ln\left\{5\left[\dfrac{0.5774(1.345) + 0.5(0.5)}{0.5774(1.345) + 2.5(0.5)}\right]\right\}} = 14.64(10^6) \text{ lbf/in.}$$

(c) From Table 8–8, $A = 0.787\ 15$, $B = 0.628\ 73$. Equation (8–23) gives

Answer $$k_m = 30(10^6)(0.5)(0.787\ 15)\exp[0.628\ 73(0.5)/1.345] = 14.92(10^6) \text{ lbf/in}$$

For this case, the difference between the results for Eqs. (8–22) and (8–23) is less than 2 percent.

(d) Following the procedure of Table 8–7, the threaded length of a 0.5-in bolt is $L_T = 2(0.5) + 0.25 = 1.25$ in. The length of the unthreaded portion is $l_d = 1.5 - 1.25 = 0.25$ in. The length of the unthreaded portion in grip is $l_t = 1.345 - 0.25 = 1.095$ in. The major diameter area is $A_d = (\pi/4)(0.5^2) = 0.196\ 3$ in^2. From Table 8–2, the tensile-stress area is $A_t = 0.159\ 9$ in^2. From Eq. (8–17)

Answer $$k_b = \frac{0.196\ 3(0.159\ 9)30(10^6)}{0.196\ 3(1.095) + 0.159\ 9(0.25)} = 3.69(10^6) \text{ lbf/in}$$

8–6   Bolt Strength

In the specification standards for bolts, the strength is specified by stating SAE or ASTM minimum quantities, the *minimum proof strength*, or *minimum proof load*, and the *minimum tensile strength*. The *proof load* is the maximum load (force) that a bolt can withstand without acquiring a permanent set. The *proof strength* is the quotient of the proof load and the tensile-stress area. The proof strength thus corresponds roughly to the proportional limit and corresponds to 0.0001-in permanent set in the fastener (first measurable deviation from elastic behavior). Tables 8–9, 8–10, and 8–11 provide *minimum* strength specifications for steel bolts. The values of the mean proof strength, the mean tensile strength, and the corresponding standard deviations are not part of the specification codes, so it is the designer's responsibility to obtain these values, perhaps by laboratory testing, if designing to a reliability specification.

The SAE specifications are found in Table 8–9. The bolt grades are numbered according to the tensile strengths, with decimals used for variations at the same strength level. Bolts and screws are available in all grades listed. Studs are available in grades 1, 2, 4, 5, 8, and 8.1. Grade 8.1 is not listed.

ASTM specifications are listed in Table 8–10. ASTM threads are shorter because ASTM deals mostly with structures; structural connections are generally loaded in shear, and the decreased thread length provides more shank area.

Specifications for metric fasteners are given in Table 8–11.

It is worth noting that all specification-grade bolts made in this country bear a manufacturer's mark or logo, in addition to the grade marking, on the bolt head. Such marks confirm that the bolt meets or exceeds specifications. If such marks are missing, the bolt may be imported; for imported bolts there is no obligation to meet specifications.

Table 8–9

SAE Specifications for Steel Bolts

SAE Grade No.	Size Range Inclusive, in	Minimum Proof Strength,* kpsi	Minimum Tensile Strength,* kpsi	Minimum Yield Strength,* kpsi	Material	Head Marking
1	$\frac{1}{4}$–$1\frac{1}{2}$	33	60	36	Low or medium carbon	
2	$\frac{1}{4}$–$\frac{3}{4}$	55	74	57	Low or medium carbon	
	$\frac{7}{8}$–$1\frac{1}{2}$	33	60	36		
4	$\frac{1}{4}$–$1\frac{1}{2}$	65	115	100	Medium carbon, cold-drawn	
5	$\frac{1}{4}$–1	85	120	92	Medium carbon, Q&T	
	$1\frac{1}{8}$–$1\frac{1}{2}$	74	105	81		
5.2	$\frac{1}{4}$–1	85	120	92	Low-carbon martensite, Q&T	
7	$\frac{1}{4}$–$1\frac{1}{2}$	105	133	115	Medium-carbon alloy, Q&T	
8	$\frac{1}{4}$–$1\frac{1}{2}$	120	150	130	Medium-carbon alloy, Q&T	
8.2	$\frac{1}{4}$–1	120	150	130	Low-carbon martensite, Q&T	

*Minimum strengths are strengths exceeded by 99 percent of fasteners.

Bolts in fatigue axial loading fail at the fillet under the head, at the thread runout, and at the first thread engaged in the nut. If the bolt has a standard shoulder under the head, it has a value of K_f from 2.1 to 2.3, *and* this shoulder fillet is protected from scratching or scoring by a washer. If the thread runout has a 15° or less half-cone angle, the stress is higher at the first engaged thread in the nut. Bolts are sized by examining the loading at the plane of the washer face of the nut. This is the weakest part of the bolt *if and only if* the conditions above are satisfied (washer protection of the shoulder fillet and thread runout $\leq 15°$). Inattention to this requirement has led to a record of 15 percent fastener fatigue failure under the head, 20 percent at thread runout, and 65 percent where the designer is focusing attention. It does little good to concentrate on the plane of the nut washer face if it is not the weakest location.

Table 8–10

ASTM Specifications for Steel Bolts

ASTM Designation No.	Size Range, Inclusive, in	Minimum Proof Strength,* kpsi	Minimum Tensile Strength,* kpsi	Minimum Yield Strength,* kpsi	Material	Head Marking
A307	$\frac{1}{4}-1\frac{1}{2}$	33	60	36	Low carbon	
A325, type 1	$\frac{1}{2}-1$	85	120	92	Medium carbon, Q&T	A325
	$1\frac{1}{8}-1\frac{1}{2}$	74	105	81		
A325, type 2	$\frac{1}{2}-1$	85	120	92	Low-carbon, martensite, Q&T	A325
	$1\frac{1}{8}-1\frac{1}{2}$	74	105	81		
A325, type 3	$\frac{1}{2}-1$	85	120	92	Weathering steel, Q&T	A325
	$1\frac{1}{8}-1\frac{1}{2}$	74	105	81		
A354, grade BC	$\frac{1}{4}-2\frac{1}{2}$	105	125	109	Alloy steel, Q&T	BC
	$2\frac{3}{4}-4$	95	115	99		
A354, grade BD	$\frac{1}{4}-4$	120	150	130	Alloy steel, Q&T	
A449	$\frac{1}{4}-1$	85	120	92	Medium-carbon, Q&T	
	$1\frac{1}{8}-1\frac{1}{2}$	74	105	81		
	$1\frac{3}{4}-3$	55	90	58		
A490, type 1	$\frac{1}{2}-1\frac{1}{2}$	120	150	130	Alloy steel, Q&T	A490
A490, type 3	$\frac{1}{2}-1\frac{1}{2}$	120	150	130	Weathering steel, Q&T	A490

*Minimum strengths are strengths exceeded by 99 percent of fasteners.

Table 8–11

Metric Mechanical-Property Classes for Steel Bolts, Screws, and Studs

Property Class	Size Range, Inclusive	Minimum Proof Strength,* MPa	Minimum Tensile Strength,* MPa	Minimum Yield Strength,* MPa	Material	Head Marking
4.6	M5–M36	225	400	240	Low or medium carbon	4.6
4.8	M1.6–M16	310	420	340	Low or medium carbon	4.8
5.8	M5–M24	380	520	420	Low or medium carbon	5.8
8.8	M16–M36	600	830	660	Medium carbon, Q&T	8.8
9.8	M1.6–M16	650	900	720	Medium carbon, Q&T	9.8
10.9	M5–M36	830	1040	940	Low-carbon martensite, Q&T	10.9
12.9	M1.6–M36	970	1220	1100	Alloy, Q&T	12.9

*Minimum strengths are strengths exceeded by 99 percent of fasteners.

Nuts are graded so that they can be mated with their corresponding grade of bolt. The purpose of the nut is to have its threads deflect to distribute the load of the bolt more evenly to the nut. The nut's properties are controlled in order to accomplish this. The grade of the nut should be the grade of the bolt.

8–7 Tension Joints—The External Load

Let us now consider what happens when an external tensile load P, as in Fig. 8–13, is applied to a bolted connection. It is to be assumed, of course, that the clamping force, which we will call the *preload* F_i, has been correctly applied by tightening the nut *before* P is applied. The nomenclature used is:

F_i = preload

P_{total} = Total external tensile load applied to the joint

$$P = \text{external tensile load per bolt}$$

$$P_b = \text{portion of } P \text{ taken by bolt}$$

$$P_m = \text{portion of } P \text{ taken by members}$$

$$F_b = P_b + F_i = \text{resultant bolt load}$$

$$F_m = P_m - F_i = \text{resultant load on members}$$

$$C = \text{fraction of external load } P \text{ carried by bolt}$$

$$1 - C = \text{fraction of external load } P \text{ carried by members}$$

$$N = \text{Number of bolts in the joint}$$

If N bolts equally share the total external load, then

$$P = P_{\text{total}}/N \tag{a}$$

The load P is tension, and it causes the connection to stretch, or elongate, through some distance δ. We can relate this elongation to the stiffnesses by recalling that k is the force divided by the deflection. Thus

$$\delta = \frac{P_b}{k_b} \quad \text{and} \quad \delta = \frac{P_m}{k_m} \tag{b}$$

or

$$P_m = P_b \frac{k_m}{k_b} \tag{c}$$

Since $P = P_b + P_m$, we have

$$P_b = \frac{k_b P}{k_b + k_m} = CP \tag{d}$$

and

$$P_m = P - P_b = (1 - C)P \tag{e}$$

where

$$C = \frac{k_b}{k_b + k_m} \tag{f}$$

is called the *stiffness constant of the joint*. The resultant bolt load is

$$F_b = P_b + F_i = CP + F_i \qquad F_m < 0 \tag{8-24}$$

and the resultant load on the connected members is

$$F_m = P_m - F_i = (1 - C)P - F_i \qquad F_m < 0 \tag{8-25}$$

Of course, these results are valid only as long as some clamping load remains in the members; this is indicated by the qualifier in the equations.

Table 8–12 is included to provide some information on the relative values of the stiffnesses encountered. The grip contains only two members, both of steel, and no

Table 8–12

Computation of Bolt and Member Stiffnesses. Steel members clamped using a $\frac{1}{2}$ in-13 NC steel bolt.

$$C = \frac{k_b}{k_b + k_m}$$

Bolt Grip, in	Stiffnesses, M lbf/in			
	k_b	k_m	C	$1 - C$
2	2.57	12.69	0.168	0.832
3	1.79	11.33	0.136	0.864
4	1.37	10.63	0.114	0.886

washers. The ratios C and $1 - C$ are the coefficients of P in Eqs. (8–24) and (8–25), respectively. They describe the proportion of the external load taken by the bolt and by the members, respectively. In all cases, the members take over 80 percent of the external load. Think how important this is when fatigue loading is present. Note also that making the grip longer causes the members to take an even greater percentage of the external load.

8–8 Relating Bolt Torque to Bolt Tension

Having learned that a high preload is very desirable in important bolted connections, we must next consider means of ensuring that the preload is actually developed when the parts are assembled.

If the overall length of the bolt can actually be measured with a micrometer when it is assembled, the bolt elongation due to the preload F_i can be computed using the formula $\delta = F_i l/(AE)$. Then the nut is simply tightened until the bolt elongates through the distance δ. This ensures that the desired preload has been attained.

The elongation of a screw cannot usually be measured, because the threaded end is often in a blind hole. It is also impractical in many cases to measure bolt elongation. In such cases the wrench torque required to develop the specified preload must be estimated. Then torque wrenching, pneumatic-impact wrenching, or the turn-of-the-nut method may be used.

The torque wrench has a built-in dial that indicates the proper torque.

With impact wrenching, the air pressure is adjusted so that the wrench stalls when the proper torque is obtained, or in some wrenches, the air automatically shuts off at the desired torque.

The turn-of-the-nut method requires that we first define the meaning of snug-tight. The *snug-tight* condition is the tightness attained by a few impacts of an impact wrench, or the full effort of a person using an ordinary wrench. When the snug-tight condition is attained, all additional turning develops useful tension in the bolt. The turn-of-the-nut method requires that you compute the fractional number of turns necessary to develop the required preload from the snug-tight condition. For example, for heavy hexagonal structural bolts, the turn-of-the-nut specification states that the nut should be turned a minimum of 180° from the snug-tight condition under optimum conditions. Note that this is also about the correct rotation for the wheel nuts of a passenger car. Problems 8–15 to 8–17 illustrate the method further.

Although the coefficients of friction may vary widely, we can obtain a good estimate of the torque required to produce a given preload by combining Eqs. (8–5) and (8–6):

$$T = \frac{F_i d_m}{2}\left(\frac{l + \pi f d_m \sec \alpha}{\pi d_m - f l \sec \alpha}\right) + \frac{F_i f_c d_c}{2} \tag{a}$$

Table 8–13

Distribution of Preload
F_i for 20 Tests of
Unlubricated Bolts
Torqued to 90 N · m

23.6,	27.6,	28.0,	29.4,	30.3,	30.7,	32.9,	33.8,	33.8,	33.8,
34.7,	35.6,	35.6,	37.4,	37.8,	37.8,	39.2,	40.0,	40.5,	42.7

Mean value $\bar{F}_i = 34.3$ kN. Standard deviation, $\hat{\sigma} = 4.91$ kN.

where d_m is the average of the major and minor diameters. Since $\tan \lambda = l/\pi d_m$, we divide the numerator and denominator of the first term by πd_m and get

$$T = \frac{F_i d_m}{2} \left(\frac{\tan \lambda + f \sec \alpha}{1 - f \tan \lambda \sec \alpha} \right) + \frac{F_i f_c d_c}{2} \tag{b}$$

The diameter of the washer face of a hexagonal nut is the same as the width across flats and equal to $1\frac{1}{2}$ times the nominal size. Therefore the mean collar diameter is $d_c = (d + 1.5d)/2 = 1.25d$. Equation (b) can now be arranged to give

$$T = \left[\left(\frac{d_m}{2d} \right) \left(\frac{\tan \lambda + f \sec \alpha}{1 - f \tan \lambda \sec \alpha} \right) + 0.625 f_c \right] F_i d \tag{c}$$

We now define a *torque coefficient* K as the term in brackets, and so

$$K = \left(\frac{d_m}{2d} \right) \left(\frac{\tan \lambda + f \sec \alpha}{1 - f \tan \lambda \sec \alpha} \right) + 0.625 f_c \tag{8–26}$$

Equation (c) can now be written

$$T = KF_i d \tag{8–27}$$

The coefficient of friction depends upon the surface smoothness, accuracy, and degree of lubrication. On the average, both f and f_c are about 0.15. The interesting fact about Eq. (8–26) is that $K \approx 0.20$ for $f = f_c = 0.15$ no matter what size bolts are employed and no matter whether the threads are coarse or fine.

Blake and Kurtz have published results of numerous tests of the torquing of bolts.[6] By subjecting their data to a statistical analysis, we can learn something about the distribution of the torque coefficients and the resulting preload. Blake and Kurtz determined the preload in quantities of unlubricated and lubricated bolts of size $\frac{1}{2}$ in-20 UNF when torqued to 800 lbf · in. This corresponds roughly to an M12 × 1.25 bolt torqued to 90 N · m. The statistical analyses of these two groups of bolts, converted to SI units, are displayed in Tables 8–13 and 8–14.

We first note that both groups have about the same mean preload, 34 kN. The unlubricated bolts have a standard deviation of 4.9 kN. The lubricated bolts have a standard deviation of 3 kN.

The means obtained from the two samples are nearly identical, approximately 34 kN; using Eq. (8–27), we find, for both samples, $K = 0.208$.

Bowman Distribution, a large manufacturer of fasteners, recommends the values shown in Table 8–15. In this book we shall use these values and use $K = 0.2$ when the bolt condition is not stated.

---

[6]J. C. Blake and H. J. Kurtz, "The Uncertainties of Measuring Fastener Preload," *Machine Design*, vol. 37, Sept. 30, 1965, pp. 128–131.

Table 8–14

Distribution of Preload F_i for 10 Tests of Lubricated Bolts Torqued to 90 N · m

30.3,	32.5,	32.5,	32.9,	32.9,	33.8,	34.3,	34.7,	37.4,	40.5

Mean value, $\bar{F}_i = 34.18$ kN. Standard deviation, $\hat{\sigma} = 2.88$ kN.

Table 8–15

Torque Factors K for Use with Eq. (8–27)

Bolt Condition	K
Nonplated, black finish	0.30
Zinc-plated	0.20
Lubricated	0.18
Cadmium-plated	0.16
With Bowman Anti-Seize	0.12
With Bowman-Grip nuts	0.09

EXAMPLE 8–3

A $\frac{3}{4}$ in-16 UNF $\times$ $2\frac{1}{2}$ in SAE grade 5 bolt is subjected to a load P of 6 kip in a tension joint. The initial bolt tension is $F_i = 25$ kip. The bolt and joint stiffnesses are $k_b = 6.50$ and $k_m = 13.8$ Mlbf/in, respectively.

(a) Determine the preload and service load stresses in the bolt. Compare these to the SAE minimum proof strength of the bolt.

(b) Specify the torque necessary to develop the preload, using Eq. (8–27).

(c) Specify the torque necessary to develop the preload, using Eq. (8–26) with $f = f_c = 0.15$.

Solution

From Table 8–2, $A_t = 0.373$ in^2.

(a) The preload stress is

Answer

$$\sigma_i = \frac{F_i}{A_t} = \frac{25}{0.373} = 67.02 \text{ kpsi}$$

The stiffness constant is

$$C = \frac{k_b}{k_b + k_m} = \frac{6.5}{6.5 + 13.8} = 0.320$$

From Eq. (8–24), the stress under the service load is

Answer

$$\sigma_b = \frac{F_b}{A_t} = \frac{CP + F_i}{A_t} = C\frac{P}{A_t} + \sigma_i$$

$$= 0.320\frac{6}{0.373} + 67.02 = 72.17 \text{ kpsi}$$

From Table 8–9, the SAE minimum proof strength of the bolt is $S_p = 85$ kpsi. The preload and service load stresses are respectively 21 and 15 percent less than the proof strength.

(b) From Eq. (8–27), the torque necessary to achieve the preload is

Answer
$$T = KF_i d = 0.2(25)(10^3)(0.75) = 3750 \text{ lbf} \cdot \text{in}$$

(c) The minor diameter can be determined from the minor area in Table 8–2. Thus $d_r = \sqrt{4A_r/\pi} = \sqrt{4(0.351)/\pi} = 0.6685$ in. Thus, the mean diameter is $d_m = (0.75 + 0.6685)/2 = 0.7093$ in. The lead angle is

$$\lambda = \tan^{-1}\frac{l}{\pi d_m} = \tan^{-1}\frac{1}{\pi d_m N} = \tan^{-1}\frac{1}{\pi(0.7093)(16)} = 1.6066°$$

For $\alpha = 30°$, Eq. (8–26) gives

$$T = \left\{ \left[\frac{0.7093}{2(0.75)}\right]\left[\frac{\tan 1.6066° + 0.15(\sec 30°)}{1 - 0.15(\tan 1.6066°)(\sec 30°)}\right] + 0.625(0.15) \right\} 25(10^3)(0.75)$$

$$= 3551 \text{ lbf} \cdot \text{in}$$

which is 5.3 percent less than the value found in part (b).

8–9 Statically Loaded Tension Joint with Preload

Equations (8–24) and (8–25) represent the forces in a bolted joint with preload. The tensile stress in the bolt can be found as in Ex. 8–3 as

$$\sigma_b = \frac{F_b}{A_t} = \frac{CP + F_i}{A_t} \tag{a}$$

Thus, the yielding factor of safety guarding against the static stress exceeding the proof strength is

$$n_p = \frac{S_p}{\sigma_b} = \frac{S_p}{(CP + F_i)/A_t} \tag{b}$$

or

$$n_p = \frac{S_p A_t}{CP + F_i} \tag{8–28}$$

Since it is common to load a bolt close to the proof strength, the yielding factor of safety is often not much greater than unity. Another indicator of yielding that is sometimes used is a *load factor,* which is applied only to the load P as a guard against overloading. Applying such a load factor, n_L, to the load P in Eq. (a), and equating it to the proof strength gives

$$\frac{Cn_L P + F_i}{A_t} = S_p \tag{c}$$

Solving for the load factor gives

$$n_L = \frac{S_p A_t - F_i}{CP} \tag{8–29}$$

It is also essential for a safe joint that the external load be smaller than that needed to cause the joint to separate. If separation does occur, then the entire external load will be imposed on the bolt. Let P_0 be the value of the external load that would cause joint separation. At separation, $F_m = 0$ in Eq. (8–25), and so

$$(1 - C)P_0 - F_i = 0 \qquad (d)$$

Let the factor of safety against joint separation be

$$n_0 = \frac{P_0}{P} \qquad (e)$$

Substituting $P_0 = n_0P$ in Eq. (d), we find

$$n_0 = \frac{F_i}{P(1 - C)} \qquad (8\text{–}30)$$

as a load factor guarding against joint separation.

Figure 8–18 is the stress-strain diagram of a good-quality bolt material. Notice that there is no clearly defined yield point and that the diagram progresses smoothly up to fracture, which corresponds to the tensile strength. This means that no matter how much preload is given the bolt, it will retain its load-carrying capacity. This is what keeps the bolt tight and determines the joint strength. The pretension is the "muscle" of the joint, and its magnitude is determined by the bolt strength. If the full bolt strength is not used in developing the pretension, then money is wasted and the joint is weaker.

Good-quality bolts can be preloaded into the plastic range to develop more strength. Some of the bolt torque used in tightening produces torsion, which increases the principal tensile stress. However, this torsion is held only by the friction of the bolt head and nut; in time it relaxes and lowers the bolt tension slightly. Thus, as a rule, a bolt will either fracture during tightening, or not at all.

Above all, do not rely too much on wrench torque; it is not a good indicator of preload. Actual bolt elongation should be used whenever possible—especially with fatigue loading. In fact, if high reliability is a requirement of the design, then preload should always be determined by bolt elongation.

Russell, Burdsall & Ward Inc. (RB&W) recommendations for preload are 60 kpsi for SAE grade 5 bolts for nonpermanent connections, and that A325 bolts (equivalent to SAE grade 5) used in structural applications be tightened to proof load or beyond

Figure 8–18

Typical stress-strain diagram for bolt materials showing proof strength S_p, yield strength S_y, and ultimate tensile strength S_{ut}.

(85 kpsi up to a diameter of 1 in).[7] Bowman recommends a preload of 75 percent of proof load,[8] which is about the same as the RB&W recommendations for reused bolts. In view of these guidelines, it is recommended for both static and fatigue loading that the following be used for preload:

$$F_i = \begin{cases} 0.75F_p & \text{for nonpermanent connections, reused fasteners} \\ 0.90F_p & \text{for permanent connections} \end{cases} \qquad (8\text{--}31)$$

where F_p is the proof load, obtained from the equation

$$F_p = A_t S_p \qquad (8\text{--}32)$$

Here S_p is the proof strength obtained from Tables 8–9 to 8–11. For other materials, an approximate value is $S_p = 0.85S_y$. Be very careful not to use a soft material in a threaded fastener. For high-strength steel bolts used as structural steel connectors, if advanced tightening methods are used, tighten to yield.

You can see that the RB&W recommendations on preload are in line with what we have encountered in this chapter. The purposes of development were to give the reader the perspective to appreciate Eqs. (8–31) and a methodology with which to handle cases more specifically than the recommendations.

EXAMPLE 8–4

Figure 8–19 is a cross section of a grade 25 cast-iron pressure vessel. A total of N bolts are to be used to resist a separating force of 36 kip.

(a) Determine k_b, k_m, and C.

(b) Find the number of bolts required for a load factor of 2 where the bolts may be reused when the joint is taken apart.

(c) With the number of bolts obtained in part (b), determine the realized load factor for overload, the yielding factor of safety, and the load factor for joint separation.

Solution

(a) The grip is $l = 1.50$ in. From Table A–31, the nut thickness is $\frac{35}{64}$ in. Adding two threads beyond the nut of $\frac{2}{11}$ in gives a bolt length of

$$L = \frac{35}{64} + 1.50 + \frac{2}{11} = 2.229 \text{ in}$$

From Table A–17 the next fraction size bolt is $L = 2\frac{1}{4}$ in. From Eq. (8–13), the thread length is $L_T = 2(0.625) + 0.25 = 1.50$ in. Thus, the length of the unthreaded portion

| Figure 8–19

$\frac{5}{8}$ in-11 UNC $\times$ 2$\frac{1}{4}$ in grade 5 finished hex head bolt

No. 25 CI

$\frac{3}{4}$ in

$\frac{3}{4}$ in

[7]Russell, Burdsall & Ward Inc., *Helpful Hints for Fastener Design and Application*, Mentor, Ohio, 1965, p. 42.

[8]Bowman Distribution–Barnes Group, *Fastener Facts*, Cleveland, 1985, p. 90.

in the grip is $l_d = 2.25 - 1.50 = 0.75$ in. The threaded length in the grip is $l_t = l - l_d = 0.75$ in. From Table 8–2, $A_t = 0.226$ in². The major-diameter area is $A_d = \pi(0.625)^2/4 = 0.3068$ in². The bolt stiffness is then

Answer
$$k_b = \frac{A_d A_t E}{A_d l_t + A_t l_d} = \frac{0.3068(0.226)(30)}{0.3068(0.75) + 0.226(0.75)}$$

$$= 5.21 \text{ Mlbf/in}$$

From Table A–24, for no. 25 cast iron we will use $E = 14$ Mpsi. The stiffness of the members, from Eq. (8–22), is

Answer
$$k_m = \frac{0.5774\pi E d}{2 \ln\left(5\dfrac{0.5774l + 0.5d}{0.5774l + 2.5d}\right)} = \frac{0.5774\pi(14)(0.625)}{2 \ln\left[5\dfrac{0.5774(1.5) + 0.5(0.625)}{0.5774(1.5) + 2.5(0.625)}\right]}$$

$$= 8.95 \text{ Mlbf/in}$$

If you are using Eq. (8–23), from Table 8–8, $A = 0.778\ 71$ and $B = 0.616\ 16$, and

$$k_m = EdA \exp(Bd/l)$$

$$= 14(0.625)(0.778\ 71) \exp[0.616\ 16(0.625)/1.5]$$

$$= 8.81 \text{ Mlbf/in}$$

which is only 1.6 percent lower than the previous result.

From the first calculation for k_m, the stiffness constant C is

Answer
$$C = \frac{k_b}{k_b + k_m} = \frac{5.21}{5.21 + 8.95} = 0.368$$

(*b*) From Table 8–9, $S_p = 85$ kpsi. Then, using Eqs. (8–31) and (8–32), we find the recommended preload to be

$$F_i = 0.75 A_t S_p = 0.75(0.226)(85) = 14.4 \text{ kip}$$

For N bolts, Eq. (8–29) can be written

$$n_L = \frac{S_p A_t - F_i}{C(P_{\text{total}}/N)} \tag{1}$$

or

$$N = \frac{C n_L P_{\text{total}}}{S_p A_t - F_i} = \frac{0.368(2)(36)}{85(0.226) - 14.4} = 5.52$$

Answer Six bolts should be used to provide the specified load factor.

(*c*) With six bolts, the load factor actually realized is

Answer
$$n_L = \frac{85(0.226) - 14.4}{0.368(36/6)} = 2.18$$

From Eq. (8–28), the yielding factor of safety is

Answer
$$n_p = \frac{S_p A_t}{C(P_{\text{total}}/N) + F_i} = \frac{85(0.226)}{0.368(36/6) + 14.4} = 1.16$$

From Eq. (8–30), the load factor guarding against joint separation is

Answer
$$n_0 = \frac{F_i}{(P_{\text{total}}/N)(1 - C)} = \frac{14.4}{(36/6)(1 - 0.368)} = 3.80$$

8–10 Gasketed Joints

If a full gasket of area A_g is present in the joint, the gasket pressure p is found by dividing the force in the member by the gasket area per bolt. Thus, for N bolts,

$$p = -\frac{F_m}{A_g/N} \qquad (a)$$

With a load factor n, Eq. (8–25) can be written as

$$F_m = (1 - C)nP - F_i \qquad (b)$$

Substituting this into Eq. (a) gives the gasket pressure as

$$p = [F_i - nP(1 - C)]\frac{N}{A_g} \qquad (8\text{–}33)$$

In full-gasketed joints uniformity of pressure on the gasket is important. To maintain adequate uniformity of pressure adjacent bolts should not be placed more than six nominal diameters apart on the bolt circle. To maintain wrench clearance, bolts should be placed at least three diameters apart. A rough rule for bolt spacing around a bolt circle is

$$3 \le \frac{\pi D_b}{Nd} \le 6 \qquad (8\text{–}34)$$

where D_b is the diameter of the bolt circle and N is the number of bolts.

8–11 Fatigue Loading of Tension Joints

Tension-loaded bolted joints subjected to fatigue action can be analyzed directly by the methods of Chap. 6. Table 8–16 lists average fatigue stress-concentration factors for the fillet under the bolt head and also at the beginning of the threads on the bolt shank. These are already corrected for notch sensitivity. Designers should be aware that situations may arise in which it would be advisable to investigate these factors more closely, since they are only average values. Peterson observes that the distribution of typical bolt failures is about 15 percent under the head, 20 percent at the end of the thread, and 65 percent in the thread at the nut face.[9]

Use of rolled threads is the predominant method of thread-forming in screw fasteners. In thread-rolling, the amount of cold work and strain-strengthening is unknown to the designer; therefore, fully corrected (including K_f) axial endurance strength is reported in Table 8–17. Since K_f is included as an endurance strength reducer in Table 8–17, it should not be applied as a stress increaser when using values from this table. For cut threads, the methods of Chap. 6 are useful. Anticipate that the endurance strengths will be considerably lower.

Table 8–16

Fatigue Stress-Concentration Factors K_f for Threaded Elements

SAE Grade	Metric Grade	Rolled Threads	Cut Threads	Fillet
0 to 2	3.6 to 5.8	2.2	2.8	2.1
4 to 8	6.6 to 10.9	3.0	3.8	2.3

[9]W. D. Pilkey and D. F. Pilkey, *Peterson's Stress-Concentration Factors,* 3rd ed., John Wiley & Sons, New York, 2008, p. 411.

Table 8–17

Fully Corrected Endurance Strengths for Bolts and Screws with Rolled Threads*

Grade or Class	Size Range	Endurance Strength
SAE 5	$\frac{1}{4}$–1 in	18.6 kpsi
	$1\frac{1}{8}$–$1\frac{1}{2}$ in	16.3 kpsi
SAE 7	$\frac{1}{4}$–$1\frac{1}{2}$ in	20.6 kpsi
SAE 8	$\frac{1}{4}$–$1\frac{1}{2}$ in	23.2 kpsi
ISO 8.8	M16–M36	129 MPa
ISO 9.8	M1.6–M16	140 MPa
ISO 10.9	M5–M36	162 MPa
ISO 12.9	M1.6–M36	190 MPa

*Repeatedly applied, axial loading, fully corrected, including K_f as a strength reducer.

For a general case with a constant preload, and an external load on a per bolt basis fluctuating between $P_{\min}$ and $P_{\max}$, a bolt will experience fluctuating forces such that

$$F_{b\min} = CP_{\min} + F_i \qquad (a)$$

$$F_{b\max} = CP_{\max} + F_i \qquad (b)$$

The alternating stress experienced by a bolt is

$$\sigma_a = \frac{(F_{b\max} - F_{b\min})/2}{A_t} = \frac{(CP_{\max} + F_i) - (CP_{\min} + F_i)}{2A_t}$$

$$\sigma_a = \frac{C(P_{\max} - P_{\min})}{2A_t} \qquad (8\text{–}35)$$

The midrange stress experienced by a bolt is

$$\sigma_m = \frac{(F_{b\max} + F_{b\min})/2}{A_t} = \frac{(CP_{\max} + F_i) + (CP_{\min} + F_i)}{2A_t}$$

$$\sigma_m = \frac{C(P_{\max} + P_{\min})}{2A_t} + \frac{F_i}{A_t} \qquad (8\text{–}36)$$

A load line typically experienced by a bolt is shown in Fig. 8–20, where the stress starts from the preload stress and increases with a constant slope of $\sigma_a/(\sigma_m - \sigma_i)$. The Goodman failure line is also shown in Fig. 8–20. The fatigue factor of safety can be found by intersecting the load line and the Goodman line to find the intersection point (S_m, S_a). The load line is given by

Load line:
$$S_a = \frac{\sigma_a}{\sigma_m - \sigma_i}(S_m - \sigma_i) \qquad (a)$$

The Goodman line, rearranging Eq. (6–40), p. 314, is

Goodman line:
$$S_a = S_e - \frac{S_e}{S_{ut}}S_m \qquad (b)$$

Equating Eqs. (a) and (b), solving for S_m, then substituting S_m back into Eq. (b) yields

$$S_a = \frac{S_e\sigma_a(S_{ut} - \sigma_i)}{S_{ut}\sigma_a + S_e(\sigma_m - \sigma_i)} \qquad (c)$$

Figure 8–20

Designer's fatigue diagram showing a Goodman failure line and a commonly used load line for a constant preload and a fluctuating load.

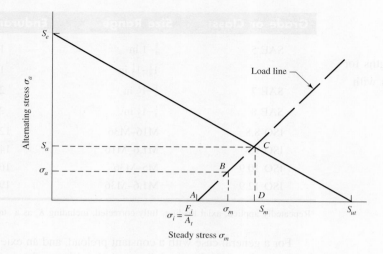

The fatigue factor of safety is given by

$$n_f = \frac{S_a}{\sigma_a} \tag{8-37}$$

Substituting Eq. (c) into Eq. (8–37) gives

$$n_f = \frac{S_e(S_{ut} - \sigma_i)}{S_{ut}\sigma_a + S_e(\sigma_m - \sigma_i)} \tag{8-38}$$

The same approach can be used for the other failure curves, though the algebra is a bit more tedious to put in equation form such as Eq. (8–38). An easier approach would be to solve in stages numerically, first S_m, then S_a, and finally n_f.

Often, the type of fatigue loading encountered in the analysis of bolted joints is one in which the externally applied load fluctuates between zero and some maximum force P. This would be the situation in a pressure cylinder, for example, where a pressure either exists or does not exist. For such cases, Eqs. (8–35) and (8–36) can be simplified by setting $P_{max} = P$ and $P_{min} = 0$, resulting in

$$\sigma_a = \frac{CP}{2A_t} \tag{8-39}$$

$$\sigma_m = \frac{CP}{2A_t} + \frac{F_i}{A_t} \tag{8-40}$$

Note that Eq. (8–40) can be viewed as the sum of the alternating stress and the preload stress. If the preload is considered to be constant, the load line relationship between the alternating and midrange stresses can be treated as

$$\sigma_m = \sigma_a + \sigma_i \tag{8-41}$$

This load line has a slope of unity, and is a special case of the load line shown in Fig. 8–20. With the simplifications in the algebra, we can now proceed as before to obtain the fatigue factor of safety using each of the typical failure criteria, duplicated here from Eqs. (6–41), (6–42), and (6–43).

Goodman:

$$\frac{S_a}{S_e} + \frac{S_m}{S_{ut}} = 1 \tag{8-42}$$

Gerber:

$$\frac{S_a}{S_e} + \left(\frac{S_m}{S_{ut}}\right)^2 = 1 \tag{8-43}$$

ASME-elliptic:

$$\left(\frac{S_a}{S_e}\right)^2 + \left(\frac{S_m}{S_p}\right)^2 = 1 \tag{8-44}$$

Now if we intersect Eq. (8–41) and each of Eqs. (8–42) to (8–44) to solve for S_a, and apply Eq. (8–37), we obtain fatigue factors of safety for each failure criteria in a repeated loading situation.

Goodman:

$$n_f = \frac{S_e(S_{ut} - \sigma_i)}{\sigma_a(S_{ut} + S_e)} \tag{8-45}$$

Gerber:

$$n_f = \frac{1}{2\sigma_a S_e}[S_{ut}\sqrt{S_{ut}^2 + 4S_e(S_e + \sigma_i)} - S_{ut}^2 - 2\sigma_i S_e] \tag{8-46}$$

ASME-elliptic:

$$n_f = \frac{S_e}{\sigma_a(S_p^2 + S_e^2)}(S_p\sqrt{S_p^2 + S_e^2 - \sigma_i^2} - \sigma_i S_e) \tag{8-47}$$

Note that Eqs. (8–45) to (8–47) are only applicable for repeated loads. If K_f is being applied to the stresses, rather than to S_e, be sure to apply it to both σ_a and σ_m. Otherwise, the slope of the load line will not remain 1 to 1.

If desired, σ_a from Eq. (8–39) and $\sigma_i = F_i/A_t$ can be directly substituted into any of Eqs. (8–45) to (8–47). If we do so for the Goodman criteria in Eq. (8–45), we obtain

$$n_f = \frac{2S_e(S_{ut}A_t - F_i)}{CP(S_{ut} + S_e)} \tag{8-48}$$

when preload F_i is present. With no preload, $C = 1$, $F_i = 0$, and Eq. (8–48) becomes

$$n_{f0} = \frac{2S_e S_{ut}A_t}{P(S_{ut} + S_e)} \tag{8-49}$$

Preload is beneficial for resisting fatigue when n_f/n_{f0} is greater than unity. For Goodman, Eqs. (8–48) and (8–49) with $n_f/n_{f0} \geq 1$ puts an upper bound on the preload F_i of

$$F_i \leq (1 - C)S_{ut}A_t \tag{8-50}$$

If this cannot be achieved, and n_f is unsatisfactory, use the Gerber or ASME-elliptic criterion to obtain a less conservative assessment. If the design is still not satisfactory, additional bolts and/or a different size bolt may be called for.

Bolts loosen, as they are friction devices, and cyclic loading and vibration as well as other effects allow the fasteners to lose tension with time. How does one fight loosening? Within strength limitations, the higher the preload the better. A rule of thumb is that preloads of 60 percent of proof load rarely loosen. If more is better, how much more? Well, not enough to create reused fasteners as a future threat. Alternatively, fastener-locking schemes can be employed.

After solving for the fatigue factor of safety, you should also check the possibility of yielding, using the proof strength

$$n_p = \frac{S_p}{\sigma_m + \sigma_a} \qquad (8\text{–}51)$$

which is equivalent to Eq. (8–28).

EXAMPLE 8–5

Figure 8–21 shows a connection using cap screws. The joint is subjected to a fluctuating force whose maximum value is 5 kip per screw. The required data are: cap screw, 5/8 in-11 UNC, SAE 5; hardened-steel washer, $t_w = \frac{1}{16}$ in thick; steel cover plate, $t_1 = \frac{5}{8}$ in, $E_s = 30$ Mpsi; and cast-iron base, $t_2 = \frac{5}{8}$ in, $E_{ci} = 16$ Mpsi.
(a) Find k_b, k_m, and C using the assumptions given in the caption of Fig. 8–21.
(b) Find all factors of safety and explain what they mean.

Solution

(a) For the symbols of Figs. 8–15 and 8–21, $h = t_1 + t_w = 0.6875$ in, $l = h + d/2 = 1$ in, and $D_2 = 1.5d = 0.9375$ in. The joint is composed of three frusta; the upper two frusta are steel and the lower one is cast iron.

For the upper frustum: $t = l/2 = 0.5$ in, $D = 0.9375$ in, and $E = 30$ Mpsi. Using these values in Eq. (8–20) gives $k_1 = 46.46$ Mlbf/in.

For the middle frustum: $t = h - l/2 = 0.1875$ in and $D = 0.9375 + 2(l - h)$ tan $30° = 1.298$ in. With these and $E_s = 30$ Mpsi, Eq. (8–20) gives $k_2 = 197.43$ Mlbf/in.

The lower frustum has $D = 0.9375$ in, $t = l - h = 0.3125$ in, and $E_{ci} = 16$ Mpsi. The same equation yields $k_3 = 32.39$ Mlbf/in.

Substituting these three stiffnesses into Eq. (8–18) gives $k_m = 17.40$ Mlbf/in. The cap screw is short and threaded all the way. Using $l = 1$ in for the grip and $A_t = 0.226$ in^2 from Table 8–2, we find the stiffness to be $k_b = A_t E/l = 6.78$ Mlbf/in. Thus the joint constant is

Answer

$$C = \frac{k_b}{k_b + k_m} = \frac{6.78}{6.78 + 17.40} = 0.280$$

Figure 8–21

Pressure-cone frustum member model for a cap screw. For this model the significant sizes are

$$l = \begin{cases} h + t_2/2 & t_2 < d \\ h + d/2 & t_2 \geq d \end{cases}$$

$D_1 = d_w + l \tan \alpha = 1.5d + 0.577l$
$D_2 = d_w = 1.5d$
where l = effective grip. The solutions are for $\alpha = 30°$ and $d_w = 1.5d$.

(*b*) Equation (8–30) gives the preload as

$$F_i = 0.75F_p = 0.75A_tS_p = 0.75(0.226)(85) = 14.4 \text{ kip}$$

where from Table 8–9, $S_p = 85$ kpsi for an SAE grade 5 cap screw. Using Eq. (8–28), we obtain the load factor as the yielding factor of safety is

Answer

$$n_p = \frac{S_pA_t}{CP + F_i} = \frac{85(0.226)}{0.280(5) + 14.4} = 1.22$$

This is the traditional factor of safety, which compares the maximum bolt stress to the proof strength.
　　Using Eq. (8–29),

Answer

$$n_L = \frac{S_pA_t - F_i}{CP} = \frac{85(0.226) - 14.4}{0.280(5)} = 3.44$$

This factor is an indication of the overload on P that can be applied without exceeding the proof strength.
　　Next, using Eq. (8–30), we have

Answer

$$n_0 = \frac{F_i}{P(1 - C)} = \frac{14.4}{5(1 - 0.280)} = 4.00$$

If the force P gets too large, the joint will separate and the bolt will take the entire load. This factor guards against that event.
　　For the remaining factors, refer to Fig. 8–22. This diagram contains the modified Goodman line, the Gerber line, the proof-strength line, and the load line. The intersection

Figure 8–22

Designer's fatigue diagram for preloaded bolts, drawn to scale, showing the modified Goodman line, the Gerber line, and the Langer proof-strength line, with an exploded view of the area of interest. The strengths used are $S_p = 85$ kpsi, $S_e = 18.6$ kpsi, and $S_{ut} = 120$ kpsi. The coordinates are A, $\sigma_i = 63.72$ kpsi; B, $\sigma_a = 3.10$ kpsi, $\sigma_m = 66.82$ kpsi; C, $S_a = 7.55$ kpsi, $S_m = 71.29$ kpsi; D, $S_a = 10.64$ kpsi, $S_m = 74.36$ kpsi; E, $S_a = 11.32$ kpsi, $S_m = 75.04$ kpsi.

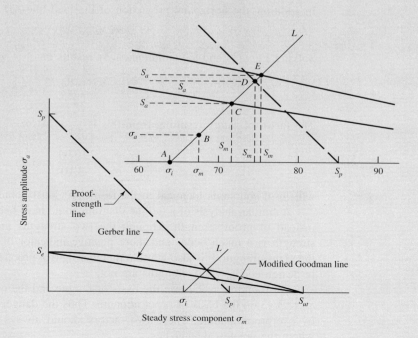

of the load line L with the respective failure lines at points C, D, and E defines a set of strengths S_a and S_m at each intersection. Point B represents the stress state σ_a, σ_m. Point A is the preload stress σ_i. Therefore the load line begins at A and makes an angle having a unit slope. This angle is 45° only when both stress axes have the same scale.

The factors of safety are found by dividing the distances AC, AD, and AE by the distance AB. Note that this is the same as dividing S_a for each theory by σ_a.

The quantities shown in the caption of Fig. 8–22 are obtained as follows:

Point A

$$\sigma_i = \frac{F_i}{A_t} = \frac{14.4}{0.226} = 63.72 \text{ kpsi}$$

Point B

$$\sigma_a = \frac{CP}{2A_t} = \frac{0.280(5)}{2(0.226)} = 3.10 \text{ kpsi}$$

$$\sigma_m = \sigma_a + \sigma_i = 3.10 + 63.72 = 66.82 \text{ kpsi}$$

Point C

This is the modified Goodman criteria. From Table 8–17, we find $S_e = 18.6$ kpsi. Then, using Eq. (8–45), the factor of safety is found to be

Answer
$$n_f = \frac{S_e(S_{ut} - \sigma_i)}{\sigma_a(S_{ut} + S_e)} = \frac{18.6(120 - 63.72)}{3.10(120 + 18.6)} = 2.44$$

Point D

This is on the proof-strength line where

$$S_m + S_a = S_p \tag{1}$$

In addition, the horizontal projection of the load line AD is

$$S_m = \sigma_i + S_a \tag{2}$$

Solving Eqs. (1) and (2) simultaneously results in

$$S_a = \frac{S_p - \sigma_i}{2} = \frac{85 - 63.72}{2} = 10.64 \text{ kpsi}$$

The factor of safety resulting from this is

Answer
$$n_p = \frac{S_a}{\sigma_a} = \frac{10.64}{3.10} = 3.43$$

which, of course, is identical to the result previously obtained by using Eq. (8–29).

A similar analysis of a fatigue diagram could have been done using yield strength instead of proof strength. Though the two strengths are somewhat related, proof strength is a much better and more positive indicator of a fully loaded bolt than is the yield strength. It is also worth remembering that proof-strength values are specified in design codes; yield strengths are not.

We found $n_f = 2.44$ on the basis of fatigue and the modified Goodman line, and $n_p = 3.43$ on the basis of proof strength. Thus the danger of failure is by fatigue, not by overproof loading. These two factors should always be compared to determine where the greatest danger lies.

Point E

For the Gerber criterion, from Eq. (8–46), the safety factor is

Answer

$$n_f = \frac{1}{2\sigma_a S_e}[S_{ut}\sqrt{S_{ut}^2 + 4S_e(S_e + \sigma_i)} - S_{ut}^2 - 2\sigma_i S_e]$$

$$= \frac{1}{2(3.10)(18.6)}[120\sqrt{120^2 + 4(18.6)(18.6 + 63.72)} - 120^2 - 2(63.72)(18.6)]$$

$$= 3.65$$

which is greater than $n_p = 3.43$ and contradicts the conclusion earlier that the danger of failure is fatigue. Figure 8–22 clearly shows the conflict where point D lies between points C and E. Again, the conservative nature of the Goodman criterion explains the discrepancy and the designer must form his or her own conclusion.

8–12   Bolted and Riveted Joints Loaded in Shear[10]

Riveted and bolted joints loaded in shear are treated exactly alike in design and analysis.

Figure 8–23a shows a riveted connection loaded in shear. Let us now study the various means by which this connection might fail.

Figure 8–23b shows a failure by bending of the rivet or of the riveted members. The bending moment is approximately $M = Ft/2$, where F is the shearing force and t is the grip of the rivet, that is, the total thickness of the connected parts. The bending stress in the members or in the rivet is, neglecting stress concentration,

$$\sigma = \frac{M}{I/c} \tag{8-52}$$

where I/c is the section modulus for the weakest member or for the rivet or rivets, depending upon which stress is to be found. The calculation of the bending stress in this manner is an assumption, because we do not know exactly how the load is distributed to the rivet or the relative deformations of the rivet and the members. Although this equation can be used to determine the bending stress, it is seldom used in design; instead its effect is compensated for by an increase in the factor of safety.

In Fig. 8–23c failure of the rivet by pure shear is shown; the stress in the rivet is

$$\tau = \frac{F}{A} \tag{8-53}$$

where A is the cross-sectional area of all the rivets in the group. It may be noted that it is standard practice in structural design to use the nominal diameter of the rivet rather than the diameter of the hole, even though a hot-driven rivet expands and nearly fills up the hole.

---

[10]The design of bolted and riveted connections for boilers, bridges, buildings, and other structures in which danger to human life is involved is strictly governed by various construction codes. When designing these structures, the engineer should refer to the *American Institute of Steel Construction Handbook*, the American Railway Engineering Association specifications, or the Boiler Construction Code of the American Society of Mechanical Engineers.

Figure 8–23

Modes of failure in shear loading of a bolted or riveted connection: (*a*) shear loading; (*b*) bending of rivet; (*c*) shear of rivet; (*d*) tensile failure of members; (*e*) bearing of rivet on members or bearing of members on rivet; (*f*) shear tear-out; (*g*) tensile tear-out.

(*a*)   (*b*)   (*c*)   (*d*)

(*e*)   (*f*)   (*g*)

Rupture of one of the connected members or plates by pure tension is illustrated in Fig. 8–23*d*. The tensile stress is

$$\sigma = \frac{F}{A} \tag{8–54}$$

where A is the net area of the plate, that is, the area reduced by an amount equal to the area of all the rivet holes. For brittle materials and static loads and for either ductile or brittle materials loaded in fatigue, the stress-concentration effects must be included. It is true that the use of a bolt with an initial preload and, sometimes, a rivet will place the area around the hole in compression and thus tend to nullify the effects of stress concentration, but unless definite steps are taken to ensure that the preload does not relax, it is on the conservative side to design as if the full stress-concentration effect were present. The stress-concentration effects are not considered in structural design, because the loads are static and the materials ductile.

In calculating the area for Eq. (8–54), the designer should, of course, use the combination of rivet or bolt holes that gives the smallest area.

Figure 8–23*e* illustrates a failure by crushing of the rivet or plate. Calculation of this stress, which is usually called a *bearing stress,* is complicated by the distribution of the load on the cylindrical surface of the rivet. The exact values of the forces acting upon the rivet are unknown, and so it is customary to assume that the components of these forces are uniformly distributed over the projected contact area of the rivet. This gives for the stress

$$\sigma = -\frac{F}{A} \tag{8–55}$$

where the projected area for a single rivet is $A = td$. Here, t is the thickness of the thinnest plate and d is the rivet or bolt diameter.

Edge shearing, or tearing, of the margin is shown in Fig. 8–23*f* and *g*, respectively. In structural practice this failure is avoided by spacing the rivets at least $1\frac{1}{2}$ diameters away from the edge. Bolted connections usually are spaced an even greater distance than this for satisfactory appearance, and hence this type of failure may usually be neglected.

In a rivet joint, the rivets all share the load in shear, bearing in the rivet, bearing in the member, and shear in the rivet. Other failures are participated in by only some of the joint. In a bolted joint, shear is taken by clamping friction, and bearing does not exist. When bolt preload is lost, one bolt begins to carry the shear and bearing until yielding slowly brings other fasteners in to share the shear and bearing. Finally, all participate, and this is the basis of most bolted-joint analysis if loss of bolt preload is complete. The usual analysis involves

- Bearing in the bolt (all bolts participate)
- Bearing in members (all holes participate)
- Shear of bolt (all bolts participate eventually)
- Distinguishing between thread and shank shear
- Edge shearing and tearing of member (edge bolts participate)
- Tensile yielding of member across bolt holes
- Checking member capacity

EXAMPLE 8–6

The bolted connection shown in Figure 8–24 uses SAE grade 5 bolts. The members are hot-rolled AISI 1018 steel. A tensile shear load $F = 4000$ lbf is applied to the connection. Find the factor of safety for all possible modes of failure.

Solution

Members: $S_y = 32$ kpsi

Bolts: $S_y = 92$ kpsi, $S_{sy} = (0.577)92 = 53.08$ kpsi

Shear of bolts

$$A_s = 2\left[\frac{\pi(0.375)^2}{4}\right] = 0.221 \text{ in}^2$$

$$\tau = \frac{F_s}{A_s} = \frac{4}{0.221} = 18.1 \text{ kpsi}$$

Answer

$$n = \frac{S_{sy}}{\tau} = \frac{53.08}{18.1} = 2.93$$

| **Figure 8–24**

$\frac{5}{8}$ in — 1$\frac{1}{8}$ in — $\frac{5}{8}$ in

$\frac{5}{8}$ in

$\frac{5}{8}$ in

$\frac{3}{8}$ in-16 UNC

$\frac{1}{4}$ in

$\frac{1}{4}$ in

Bearing on bolts

$$A_b = 2(0.25)(0.375) = 0.188 \text{ in}^2$$

$$\sigma_b = \frac{-4}{0.188} = -21.3 \text{ kpsi}$$

Answer

$$n = \frac{S_y}{|\sigma_b|} = \frac{92}{|-21.3|} = 4.32$$

Bearing on members

Answer

$$n = \frac{S_{yc}}{|\sigma_b|} = \frac{32}{|-21.3|} = 1.50$$

Tension of members

$$A_t = (2.375 - 0.75)(1/4) = 0.406 \text{ in}^2$$

$$\sigma_t = \frac{4}{0.406} = 9.85 \text{ kpsi}$$

Answer

$$n = \frac{S_y}{A_t} = \frac{32}{9.85} = 3.25$$

Shear Joints with Eccentric Loading

In the previous example, the load distributed equally to the bolts since the load acted along a line of symmetry of the fasteners. The analysis of a shear joint undergoing eccentric loading requires locating the center of relative motion between the two members. In Fig. 8–25 let A_1 to A_5 be the respective cross-sectional areas of a group of five pins, or hot-driven rivets, or tight-fitting shoulder bolts. Under this assumption the rotational pivot point lies at the centroid of the cross-sectional area pattern of the pins, rivets, or bolts. Using statics, we learn that the centroid G is located by the coordinates $\bar{x}$ and $\bar{y}$, where x_i and y_i are the distances to the ith area center:

$$\bar{x} = \frac{A_1 x_1 + A_2 x_2 + A_3 x_3 + A_4 x_4 + A_5 x_5}{A_1 + A_2 + A_3 + A_4 + A_5} = \frac{\sum_1^n A_i x_i}{\sum_1^n A_i}$$

$$\bar{y} = \frac{A_1 y_1 + A_2 y_2 + A_3 y_3 + A_4 y_4 + A_5 y_5}{A_1 + A_2 + A_3 + A_4 + A_5} = \frac{\sum_1^n A_i y_i}{\sum_1^n A_i}$$

(8–56)

In many instances the centroid can be located by symmetry.

An example of eccentric loading of fasteners is shown in Fig. 8–26. This is a portion of a machine frame containing a beam subjected to the action of a bending load. In this case, the beam is fastened to vertical members at the ends with specially prepared load-sharing bolts. You will recognize the schematic representation in Fig. 8–26b as a statically indeterminate beam with both ends fixed and with moment and shear reactions at each end.

Figure 8–25

Centroid of pins, rivets, or bolts.

For convenience, the centers of the bolts at the left end of the beam are drawn to a larger scale in Fig. 8–26c. Point O represents the centroid of the group, and it is assumed in this example that all the bolts are of the same diameter. Note that the forces shown in Fig. 8–26c are the *resultant* forces acting on the pins with a net force and moment equal and opposite to the *reaction* loads V_1 and M_1 acting at O. The total load taken by each bolt will be calculated in three steps. In the first step the shear V_1 is divided equally among the bolts so that each bolt takes $F' = V_1/n$, where n refers to the number of bolts in the group and the force F' is called the *direct load*, or *primary shear*.

It is noted that an equal distribution of the direct load to the bolts assumes an absolutely rigid member. The arrangement of the bolts or the shape and size of the members sometimes justifies the use of another assumption as to the division of the load. The direct loads F_n' are shown as vectors on the loading diagram (Fig. 8–26c).

The *moment load*, or *secondary shear*, is the additional load on each bolt due to the moment M_1. If r_A, r_B, r_C, etc., are the radial distances from the centroid to the center of each bolt, the moment and moment loads are related as follows:

$$M_1 = F_A'' r_A + F_B'' r_B + F_C'' r_C + \cdots \qquad (a)$$

Figure 8–26

(*a*) Beam bolted at both ends with distributed load;
(*b*) free-body diagram of beam;
(*c*) enlarged view of bolt group centered at O showing primary and secondary resultant shear forces.

(*a*)

(*b*)

(*c*)

where the F'' are the moment loads. The force taken by each bolt depends upon its radial distance from the centroid; that is, the bolt farthest from the centroid takes the greatest load, while the nearest bolt takes the smallest. We can therefore write

$$\frac{F''_A}{r_A} = \frac{F''_B}{r_B} = \frac{F''_C}{r_C} \qquad (b)$$

where again, the diameters of the bolts are assumed equal. If not, then one replaces F'' in Eq. (b) with the shear stresses $\tau'' = 4F''/\pi d^2$ for each bolt. Solving Eqs. (a) and (b) simultaneously, we obtain

$$F''_n = \frac{M_1 r_n}{r_A^2 + r_B^2 + r_C^2 + \cdots} \qquad (8\text{--}57)$$

where the subscript n refers to the particular bolt whose load is to be found. Each moment load is a force vector perpendicular to the radial line from the centroid to the bolt center.

In the third step the direct and moment loads are added vectorially to obtain the resultant load on each bolt. Since all the bolts or rivets are usually the same size, only that bolt having the maximum load need be considered. When the maximum load is found, the strength may be determined by using the various methods already described.

EXAMPLE 8–7	Shown in Fig. 8–27 is a 15- by 200-mm rectangular steel bar cantilevered to a 250-mm steel channel using four tightly fitted bolts located at A, B, C, and D. Assume the bolt threads do not extend into the joint.

For the $F = 16$ kN load shown find
(a) The resultant load on each bolt
(b) The maximum shear stress in each bolt
(c) The maximum bearing stress
(d) The critical bending stress in the bar

Solution	(a) Point O, the centroid of the bolt group in Fig. 8–27, is found by symmetry. If a free-body diagram of the beam were constructed, the shear reaction V would pass through O and the moment reactions M would be about O. These reactions are

$$V = 16 \text{ kN} \qquad M = 16(300 + 50 + 75) = 6800 \text{ N} \cdot \text{m}$$

In Fig. 8–28, the bolt group has been drawn to a larger scale and the reactions and resultants are shown. The distance from the centroid to the center of each bolt is

$$r = \sqrt{(60)^2 + (75)^2} = 96.0 \text{ mm}$$

Figure 8–27

Dimensions in millimeters.

Figure 8–28

The resultants are found as follows. The primary shear load per bolt is

$$F' = \frac{V}{n} = \frac{16}{4} = 4 \text{ kN}$$

Since the r_n are equal, the secondary shear forces are equal, and Eq. (8–57) becomes

$$F'' = \frac{Mr}{4r^2} = \frac{M}{4r} = \frac{6800}{4(96.0)} = 17.7 \text{ kN}$$

The primary and secondary shear forces are plotted to scale in Fig. 8–28 and the resultants obtained by using the parallelogram rule. The magnitudes are found by measurement (or analysis) to be

Answer
$$F_A = F_B = 21.0 \text{ kN}$$

Answer
$$F_C = F_D = 14.8 \text{ kN}$$

(b) Bolts A and B are critical because they carry the largest shear load. The problem stated to assume that the bolt threads are not to extend into the joint. This would require special bolts. If standard nuts and bolts were used, the bolts would need to be 46 mm long with a thread length of $L_T = 38$ mm. Thus the unthreaded portion of the bolt is $46 - 38 = 8$ mm long. This is less than the 15 mm for the plate in Fig. 8–28, and the bolts would tend to shear along the minor diameter at a stress of $\tau = F/A_s = 21.0(10)^3/144 = 146$ MPa. Using bolts not extending into the joint, or shoulder bolts, is preferred. For this example, the body area of each bolt is $A = \pi(16^2)/4 = 201.1$ mm², resulting in a shear stress of

Answer
$$\tau = \frac{F}{A} = \frac{21.0(10)^3}{201.1} = 104 \text{ MPa}$$

(c) The channel is thinner than the bar, and so the largest bearing stress is due to the pressing of the bolt against the channel web. The bearing area is $A_b = td = 10(16) = 160$ mm². Thus the bearing stress is

Answer
$$\sigma = -\frac{F}{A_b} = -\frac{21.0(10)^3}{160} = -131 \text{ MPa}$$

(d) The critical bending stress in the bar is assumed to occur in a section parallel to the y axis and through bolts A and B. At this section the bending moment is

$$M = 16(300 + 50) = 5600 \text{ N} \cdot \text{m}$$

The second moment of area through this section is obtained as follows:

$$I = I_\text{bar} - 2(I_\text{holes} + \bar{d}^2 A)$$

$$= \frac{15(200)^3}{12} - 2\left[\frac{15(16)^3}{12} + (60)^2(15)(16)\right] = 8.26(10)^6 \text{ mm}^4$$

Then

Answer
$$\sigma = \frac{Mc}{I} = \frac{5600(100)}{8.26(10)^6}(10)^3 = 67.8 \text{ MPa}$$

PROBLEMS

8-1   A power screw is 25 mm in diameter and has a thread pitch of 5 mm.

(a) Find the thread depth, the thread width, the mean and root diameters, and the lead, provided square threads are used.

(b) Repeat part (a) for Acme threads.

8-2   Using the information in the footnote of Table 8–1, show that the tensile-stress area is

$$A_t = \frac{\pi}{4}(d - 0.938\ 194p)^2$$

8-3   Show that for zero collar friction the efficiency of a square-thread screw is given by the equation

$$e = \tan \lambda \frac{1 - f \tan \lambda}{\tan \lambda + f}$$

Plot a curve of the efficiency for lead angles up to 45°. Use $f = 0.08$.

8-4   A single-threaded power screw is 25 mm in diameter with a pitch of 5 mm. A vertical load on the screw reaches a maximum of 5 kN. The coefficients of friction are 0.06 for the collar and 0.09 for the threads. The frictional diameter of the collar is 45 mm. Find the overall efficiency and the torque to "raise" and "lower" the load.

8-5   The machine shown in the figure can be used for a tension test but not for a compression test. Why? Can both screws have the same hand?

Problem 8–5

8-6   The press shown for Prob. 8–5 has a rated load of 5000 lbf. The twin screws have Acme threads, a diameter of 2 in, and a pitch of $\frac{1}{4}$ in. Coefficients of friction are 0.05 for the threads and 0.08 for the collar bearings. Collar diameters are 3.5 in. The gears have an efficiency of 95 percent and a speed ratio of 60:1. A slip clutch, on the motor shaft, prevents overloading. The full-load motor speed is 1720 rev/min.

(a) When the motor is turned on, how fast will the press head move?

(b) What should be the horsepower rating of the motor?

8-7   For the screw clamp shown, a force is applied at the end of the handle $3\frac{1}{2}$ in from the screw centerline. The $\frac{3}{8}$ in diameter handle is made of cold-drawn AISI 1006 steel. The screw is $\frac{3}{4}$ in-10 UNC and is 8 in long, overall. The maximum possible length of screw in the clamping region is 6 in.

(a) What screw torque will cause the handle to bend permanently?

(b) What clamping force will the answer to part (a) cause if the collar friction is neglected and if the thread friction is 0.15?

(c) What clamping force will cause the screw to buckle?

(d) Are there any other stresses or possible failures to be checked?

Problem 8–7

8-8   The C clamp shown in the figure for Prob. 8–7 uses a $\frac{3}{4}$ in-6 Acme thread. The frictional coefficients are 0.15 for the threads and for the collar. The collar, which in this case is the anvil striker's swivel joint, has a friction diameter of 1 in. Calculations are to be based on a maximum force of 8 lbf applied to the handle at a radius of $3\frac{1}{2}$ in from the screw centerline. Find the clamping force.

8-9   Find the power required to drive a 40-mm power screw having double square threads with a pitch of 6 mm. The nut is to move at a velocity of 48 mm/s and move a load of $F = 10$ kN. The frictional coefficients are 0.10 for the threads and 0.15 for the collar. The frictional diameter of the collar is 60 mm.

8-10   A single square-thread power screw has an input power of 3 kW at a speed of 1 rev/s. The screw has a diameter of 40 mm and a pitch of 8 mm. The frictional coefficients are 0.14 for the threads and 0.09 for the collar, with a collar friction radius of 50 mm. Find the axial resisting load F and the combined efficiency of the screw and collar.

8-11   An M14 × 2 hex-head bolt with a nut is used to clamp together two 15-mm steel plates.

(a) Determine a suitable length for the bolt, rounded up to the nearest 5 mm.

(b) Determine the bolt stiffness.

(c) Determine the stiffness of the members.

8-12   Repeat Prob. 8–11 with the addition of one 14R metric plain washer under the nut.

8-13   Repeat Prob. 8–11 with one of the plates having a threaded hole to eliminate the nut.

8–14   A 2-in steel plate and a 1-in cast-iron plate are compressed with one bolt and nut. The bolt is $\frac{1}{2}$ in-13 UNC.
(a) Determine a suitable length for the bolt, rounded up to the nearest $\frac{1}{4}$ in.
(b) Determine the bolt stiffness.
(c) Determine the stiffness of the members.

8–15   Repeat Prob. 8–14 with the addition of one $\frac{1}{2}$ N American Standard plain washer under the head of the bolt, and another identical washer under the nut.

8–16   Repeat Prob. 8–14 with the cast-iron plate having a threaded hole to eliminate the nut.

8–17   Two identical aluminum plates are each 2 in thick, and are compressed with one bolt and nut. Washers are used under the head of the bolt and under the nut.
Washer properties: steel; ID = 0.531 in; OD = 1.062 in; thickness = 0.095 in
Nut properties: steel; height = $\frac{7}{16}$ in
Bolt properties: $\frac{1}{2}$ in-13 UNC grade 8
Plate properties: aluminum; E = 10.3 Mpsi; S_u = 47 kpsi; S_y = 25 kpsi
(a) Determine a suitable length for the bolt, rounded up to the nearest $\frac{1}{4}$ in.
(b) Determine the bolt stiffness.
(c) Determine the stiffness of the members.

8–18   Repeat Prob. 8–17 with no washer under the head of the bolt, and two washers stacked under the nut.

8–19   A 30-mm thick AISI 1020 steel plate is sandwiched between two 10-mm thick 2024-T3 aluminum plates and compressed with a bolt and nut with no washers. The bolt is M10 × 1.5, property class 5.8.
(a) Determine a suitable length for the bolt, rounded up to the nearest 5 mm.
(b) Determine the bolt stiffness.
(c) Determine the stiffness of the members.

8–20   Repeat Prob. 8–19 with the bottom aluminum plate replaced by one that is 20 mm thick.

8–21   Repeat Prob. 8–19 with the bottom aluminum plate having a threaded hole to eliminate the nut.

8–22   Two 20-mm steel plates are to be clamped together with a bolt and nut. Specify a coarse thread metric bolt to provide a joint constant C of approximately 0.2.

8–23   A 2-in steel plate and a 1-in cast-iron plate are to be compressed with one bolt and nut. Specify a UNC bolt to provide a joint constant C of approximately 0.2.

8–24   An aluminum bracket with a $\frac{1}{2}$-in thick flange is to be clamped to a steel column with a $\frac{3}{4}$-in wall thickness. A cap screw passes through a hole in the bracket flange, and threads into a tapped hole through the column wall. Specify a UNC cap screw to provide a joint constant C of approximately 0.25.

8–25   An M14 × 2 hex-head bolt with a nut is used to clamp together two 20-mm steel plates. Compare the results of finding the overall member stiffness by use of Eqs. (8–20), (8–22), and (8–23).

8–26   A $\frac{3}{4}$ in-16 UNF series SAE grade 5 bolt has a $\frac{3}{4}$-in ID steel tube 10 in long, clamped between washer faces of bolt and nut by turning the nut snug and adding one-third of a turn. The tube OD is the washer-face diameter d_w = 1.5d = 1.5(0.75) = 1.125 in = OD.
(a) Determine the bolt stiffness, the tube stiffness, and the joint constant C.
(b) When the one-third turn-of-nut is applied, what is the initial tension F_i in the bolt?

$\frac{3}{4}$ in-16 UNF grade 5

1.125 in

Problem 8–26

10 in

8–27 From your experience with Prob. 8–26, generalize your solution to develop a turn-of-nut equation

$$N_t = \frac{\theta}{360°} = \left(\frac{k_b + k_m}{k_b k_m}\right) F_i N$$

where N_t = turn of the nut, in rotations, from snug tight

θ = turn of the nut in degrees

N = number of thread/in ($1/p$ where p is pitch)

F_i = initial preload

k_b, k_m = spring rates of the bolt and members, respectively

Use this equation to find the relation between torque-wrench setting T and turn-of-nut N_t. ("Snug tight" means the joint has been tightened to perhaps half the intended preload to flatten asperities on the washer faces and the members. Then the nut is loosened and retightened finger tight, and the nut is rotated the number of degrees indicated by the equation. Properly done, the result is competitive with torque wrenching.)

8–28 RB&W[11] recommends turn-of-nut from snug fit to preload as follows: 1/3 turn for bolt grips of 1–4 diameters, 1/2 turn for bolt grips 4–8 diameters, and 2/3 turn for grips of 8–12 diameters. These recommendations are for structural steel fabrication (permanent joints), producing preloads of 100 percent of proof strength and beyond. Machinery fabricators with fatigue loadings and possible joint disassembly have much smaller turns-of-nut. The RB&W recommendation enters the nonlinear plastic deformation zone.

For Ex. 8–4, use Eq. (8–27) with $K = 0.2$ to estimate the torque necessary to establish the desired preload. Then, using the results from Prob. 8–27, determine the turn of the nut in degrees. How does this compare with the RB&W recommendations?

8–29 For a bolted assembly with six bolts, the stiffness of each bolt is $k_b = 3$ Mlbf/in and the stiffness of the members is $k_m = 12$ Mlbf/in per bolt. An external load of 80 kips is applied to the entire joint. Assume the load is equally distributed to all the bolts. It has been determined to use $\frac{1}{2}$ in-13 UNC grade 8 bolts with rolled threads. Assume the bolts are preloaded to 75 percent of the proof load.

(a) Determine the yielding factor of safety.

(b) Determine the overload factor of safety.

(c) Determine the factor of safety based on joint separation.

8–30   For the bolted assembly of Prob. 8–29, it is desired to find the range of torque that a mechanic could apply to initially preload the bolts without expecting failure once the joint is loaded. Assume a torque coefficient of $K = 0.2$.

(*a*) Determine the maximum bolt preload that can be applied without exceeding the proof strength of the bolts.

(*b*) Determine the minimum bolt preload that can be applied while avoiding joint separation.

(*c*) Determine the value of torque in units of lbf · ft that should be specified for preloading the bolts if it is desired to preload to the midpoint of the values found in parts (*a*) and (*b*).

8–31   For a bolted assembly with eight bolts, the stiffness of each bolt is $k_b = 1.0$ MN/mm and the stiffness of the members is $k_m = 2.6$ MN/mm per bolt. The joint is subject to occasional disassembly for maintenance and should be preloaded accordingly. Assume the external load is equally distributed to all the bolts. It has been determined to use M6 × 1 class 5.8 bolts with rolled threads.

(*a*) Determine the maximum external load P_{max} that can be applied to the entire joint without exceeding the proof strength of the bolts.

(*b*) Determine the maximum external load P_{max} that can be applied to the entire joint without causing the members to come out of compression.

8–32   For a bolted assembly, the stiffness of each bolt is $k_b = 4$ Mlbf/in and the stiffness of the members is $k_m = 12$ Mlbf/in per bolt. The joint is subject to occasional disassembly for maintenance and should be preloaded accordingly. A fluctuating external load is applied to the entire joint with $P_{max} = 80$ kips and $P_{min} = 20$ kips. Assume the load is equally distributed to all the bolts. It has been determined to use $\frac{1}{2}$ in-13 UNC grade 8 bolts with rolled threads.

(*a*) Determine the minimum number of bolts necessary to avoid yielding of the bolts.

(*b*) Determine the minimum number of bolts necessary to avoid joint separation.

**8–33 to
8–36**   The figure illustrates the non-permanent connection of a steel cylinder head to a grade 30 cast-iron pressure vessel using N bolts. A confined gasket seal has an effective sealing diameter D. The cylinder stores gas at a maximum pressure p_g. For the specifications given in the table for the specific problem assigned, select a suitable bolt length from the preferred sizes in Table A–17, then determine the yielding factor of safety n_p, the load factor n_L, and the joint separation factor n_0.

Problems 8–33 to 8–36

Problem Number	8–33	8–34	8–35	8–36
A	20 mm	$\frac{1}{2}$ in	20 mm	$\frac{3}{8}$ in
B	20 mm	$\frac{5}{8}$ in	25 mm	$\frac{1}{2}$ in
C	100 mm	3.5 in	0.8 m	3.25 in
D	150 mm	4.25 in	0.9 m	3.5 in
E	200 mm	6 in	1.0 m	5.5 in
F	300 mm	8 in	1.1 m	7 in
N	10	10	36	8
p_g	6 MPa	1500 psi	550 kPa	1200 psi
Bolt grade	ISO 9.8	SAE 5	ISO 10.9	SAE 8
Bolt spec.	M12 × 1.75	$\frac{1}{2}$ in -13	M10 × 1.5	$\frac{7}{16}$ in -14

**8–37 to
8–40**

Repeat the requirements for the problem specified in the table if the bolts and nuts are replaced with cap screws that are threaded into tapped holes in the cast-iron cylinder.

Problem Number	Originating Problem Number
8–37	8–33
8–38	8–34
8–39	8–35
8–40	8–36

**8–41 to
8–44**

For the pressure vessel defined in the problem specified in the table, redesign the bolt specifications to satisfy all of the following requirements.

- Use coarse-thread bolts selecting a class from Table 8–11 for Probs. 8–41 and 8–43, or a grade from Table 8–9 for Probs. 8–42 and 8–44.

- To ensure adequate gasket sealing around the bolt circle, use enough bolts to provide a maximum center-to-center distance between bolts of four bolt diameters.

- Obtain a joint stiffness constant C between 0.2 and 0.3 to ensure most of the pressure load is carried by the members.

- The bolts may be reused, so the yielding factor of safety should be at least 1.1.

- The overload factor and the joint separation factor should allow for the pressure to exceed the expected pressure by 15 percent.

Problem Number	Originating Problem Number
8–41	8–33
8–42	8–34
8–43	8–35
8–44	8–36

8-45 Bolts distributed about a bolt circle are often called upon to resist an external bending moment as shown in the figure. The external moment is 12 kip · in and the bolt circle has a diameter of 8 in. The neutral axis for bending is a diameter of the bolt circle. What needs to be determined is the most severe external load seen by a bolt in the assembly.

(a) View the effect of the bolts as placing a line load around the bolt circle whose intensity F_b', in pounds per inch, varies linearly with the distance from the neutral axis according to the relation $F_b' = F_{b,\text{max}}' R \sin\theta$. The load on any particular bolt can be viewed as the effect of the line load over the arc associated with the bolt. For example, there are 12 bolts shown in the figure. Thus each bolt load is assumed to be distributed on a 30° arc of the bolt circle. Under these conditions, what is the largest bolt load?

(b) View the largest load as the intensity $F_{b,\text{max}}'$ multiplied by the arc length associated with each bolt and find the largest bolt load.

(c) Express the load on any bolt as $F = F_{\text{max}} \sin\theta$, sum the moments due to all the bolts, and estimate the largest bolt load. Compare the results of these three approaches to decide how to attack such problems in the future.

Problem 8–45

Bolted connection subjected to bending.

8-46 The figure shows a cast-iron bearing block that is to be bolted to a steel ceiling joist and is to support a gravity load of 18 kN. Bolts used are M24 ISO 8.8 with coarse threads and with 4.6-mm-thick steel washers under the bolt head and nut. The joist flanges are 20 mm in thickness, and the dimension A, shown in the figure, is 20 mm. The modulus of elasticity of the bearing block is 135 GPa.

Problem 8–46

(a) Find the wrench torque required if the fasteners are lubricated during assembly and the joint is to be permanent.

(b) Determine the factors of safety guarding against yielding, overload, and joint separation.

8–47 The upside-down steel A frame shown in the figure is to be bolted to steel beams on the ceiling of a machine room using ISO grade 8.8 bolts. This frame is to support the 40-kN vertical load as illustrated. The total bolt grip is 48 mm, which includes the thickness of the steel beam, the A-frame feet, and the steel washers used. The bolts are size M20 × 2.5.

(a) What tightening torque should be used if the connection is permanent and the fasteners are lubricated?

(b) Determine the factors of safety guarding against yielding, overload, and joint separation.

Drill 2 holes for
M20 × 2.5 bolts

Problem 8–47

$W = 40$ kN

8–48 For the bolted assembly in Prob. 8–29, assume the external load is a repeated load. Determine the fatigue factor of safety for the bolts using the following failure criteria:

(a) Goodman.

(b) Gerber.

(c) ASME-elliptic.

8–49 For a bolted assembly with eight bolts, the stiffness of each bolt is $k_b = 1.0$ MN/mm and the stiffness of the members is $k_m = 2.6$ MN/mm per bolt. The bolts are preloaded to 75 percent of proof strength. Assume the external load is equally distributed to all the bolts. The bolts are M6 × 1 class 5.8 with rolled threads. A fluctuating external load is applied to the entire joint with $P_{max} = 60$ kN and $P_{min} = 20$ kN.

(a) Determine the yielding factor of safety.

(b) Determine the overload factor of safety.

(c) Determine the factor of safety based on joint separation.

(d) Determine the fatigue factor of safety using the Goodman criterion.

8–50 For the bolted assembly in Prob. 8–32, assume 10 bolts are used. Determine the fatigue factor of safety using the Goodman criterion.

8–51 to
8–54 For the pressure cylinder defined in the problem specified in the table, the gas pressure is cycled between zero and p_g. Determine the fatigue factor of safety for the bolts using the following failure criteria:

(a) Goodman.

(b) Gerber.

(c) ASME-elliptic.

Problem Number	Originating Problem Number
8–51	8–33
8–52	8–34
8–53	8–35
8–54	8–36

8–55 to 8–58 For the pressure cylinder defined in the problem specified in the table, the gas pressure is cycled between p_g and $p_g/2$. Determine the fatigue factor of safety for the bolts using the Goodman criterion.

Problem Number	Originating Problem Number
8–55	8–33
8–56	8–34
8–57	8–35
8–58	8–36

8–59 A 1-in-diameter hot-rolled AISI 1144 steel rod is hot-formed into an eyebolt similar to that shown in the figure for Prob. 3–122, with an inner 3-in-diameter eye. The threads are 1 in-12 UNF and are die-cut.

(a) For a repeatedly applied load collinear with the thread axis, using the Gerber criterion, is fatigue failure more likely in the thread or in the eye?

(b) What can be done to strengthen the bolt at the weaker location?

(c) If the factor of safety guarding against a fatigue failure is $n_f = 2$, what repeatedly applied load can be applied to the eye?

8–60 The section of the sealed joint shown in the figure is loaded by a force cycling between 4 and 6 kips. The members have $E = 16$ Mpsi. All bolts have been carefully preloaded to $F_i = 25$ kip each.

Problem 8–60

$\frac{3}{4}$ in-16 UNF $\times$ $2\frac{1}{2}$ in
SAE grade 5

$1\frac{1}{2}$ in

No. 40 CI

(a) Determine the yielding factor of safety.

(b) Determine the overload factor of safety.

(c) Determine the factor of safety based on joint separation.

(d) Determine the fatigue factor of safety using the Goodman criterion.

8–61 Suppose the welded steel bracket shown in the figure is bolted underneath a structural-steel ceiling beam to support a fluctuating vertical load imposed on it by a pin and yoke. The bolts are $\frac{1}{2}$-in coarse-thread SAE grade 8, tightened to recommended preload for nonpermanent assembly. The stiffnesses have already been computed and are $k_b = 4$ Mlbf/in and $k_m = 16$ Mlbf/in.

Problem 8–61

(a) Assuming that the bolts, rather than the welds, govern the strength of this design, determine the safe repeated load that can be imposed on this assembly using the Goodman criterion with the load line in Fig. 8–20 and a fatigue design factor of 2.

(b) Compute the static load factors based on the load found in part (a).

8–62 Using the Gerber fatigue criterion and a fatigue-design factor of 2, determine the external repeated load P that a $1\frac{1}{4}$-in SAE grade 5 coarse-thread bolt can take compared with that for a fine-thread bolt. The joint constants are $C = 0.30$ for coarse- and 0.32 for fine-thread bolts. Assume the bolts are preloaded to 75 percent of the proof load.

8–63 An M30 × 3.5 ISO 8.8 bolt is used in a joint at recommended preload, and the joint is subject to a repeated tensile fatigue load of $P = 65$ kN per bolt. The joint constant is $C = 0.28$. Find the static load factors and the factor of safety guarding against a fatigue failure based on the Gerber fatigue criterion.

8–64 The figure shows a fluid-pressure linear actuator (hydraulic cylinder) in which $D = 4$ in, $t = \frac{3}{8}$ in, $L = 12$ in, and $w = \frac{3}{4}$ in. Both brackets as well as the cylinder are of steel. The actuator has been designed for a working pressure of 2000 psi. Six $\frac{3}{8}$-in SAE grade 5 coarse-thread bolts are used, tightened to 75 percent of proof load. Assume the bolts are unthreaded within the grip.

(a) Find the stiffnesses of the bolts and members, assuming that the entire cylinder is compressed uniformly and that the end brackets are perfectly rigid.

(b) Using the Gerber fatigue criterion, find the factor of safety guarding against a fatigue failure.

(c) What pressure would be required to cause total joint separation?

Problem 8–64

8–65   Using the Goodman fatigue criterion, repeat Prob. 8–64 with the working pressure cycling between 1200 psi and 2000 psi.

8–66   The figure shows a bolted lap joint that uses SAE grade 5 bolts. The members are made of cold-drawn AISI 1020 steel. Assume the bolt threads do not extend into the joint. Find the safe tensile shear load F that can be applied to this connection to provide a minimum factor of safety of 2 for the following failure modes: shear of bolts, bearing on bolts, bearing on members, and tension of members.

Problem 8–66

8–67   The bolted connection shown in the figure uses SAE grade 8 bolts. The members are hot-rolled AISI 1040 steel. A tensile shear load $F = 5000$ lbf is applied to the connection. Assume the bolt threads do not extend into the joint. Find the factor of safety for all possible modes of failure.

Problem 8–67

8–68   A bolted lap joint using ISO class 5.8 bolts and members made of cold-drawn SAE 1040 steel is shown in the figure. Assume the bolt threads do not extend into the joint. Find the tensile shear load F that can be applied to this connection to provide a minimum factor of safety of 2.5 for the following failure modes: shear of bolts, bearing on bolts, bearing on members, and tension of members.

Problem 8–68
Dimensions in millimeters.

8–69
The bolted connection shown in the figure is subjected to a tensile shear load of 90 kN. The bolts are ISO class 5.8 and the material is cold-drawn AISI 1015 steel. Assume the bolt threads do not extend into the joint. Find the factor of safety of the connection for all possible modes of failure.

Problem 8–69
Dimensions in millimeters.

8–70
The figure shows a connection that employs three SAE grade 4 bolts. The tensile shear load on the joint is 5000 lbf. The members are cold-drawn bars of AISI 1020 steel. Assume the bolt threads do not extend into the joint. Find the factor of safety for each possible mode of failure.

Problem 8–70

8–71
A beam is made up by bolting together two cold-drawn bars of AISI 1018 steel as a lap joint, as shown in the figure. The bolts used are ISO 5.8. Assume the bolt threads do not extend into the joint. Ignoring any twisting, determine the factor of safety of the connection.

Problem 8–71
Dimensions in millimeters.

8–72
Standard design practice, as exhibited by the solutions to Probs. 8–66 to 8–70, is to assume that the bolts, or rivets, share the shear equally. For many situations, such an assumption may lead to an unsafe design. Consider the yoke bracket of Prob. 8–61, for example. Suppose this bracket is bolted to a wide-flange *column* with the centerline through the two bolts in the vertical direction. A vertical load through the yoke-pin hole at distance *B* from the column flange would place a shear load on the bolts as well as a tensile load. The tensile load comes about because the bracket tends to pry itself about the bottom corner, much like a claw

hammer, exerting a large tensile load on the upper bolt. In addition, it is almost certain that both the spacing of the bolt holes and their diameters will be slightly different on the column flange from what they are on the yoke bracket. Thus, unless yielding occurs, only one of the bolts will take the shear load. The designer has no way of knowing which bolt this will be.

In this problem the bracket is 8 in long, $A = \frac{1}{2}$ in, $B = 3$ in, $C = 6$ in, and the column flange is $\frac{1}{2}$ in thick. The bolts are $\frac{1}{2}$ in-13 UNC $\times$ $1\frac{1}{2}$ in SAE grade 4. The nuts are tightened to 75 percent of proof load. The vertical yoke-pin load is 2500 lbf. If the upper bolt takes all the shear load as well as the tensile load, determine a static factor of safety for the bolt, based on the von Mises stress exceeding the proof strength.

8–73   The bearing of Prob. 8–46 is bolted to a vertical surface and supports a horizontal shaft. The bolts used have coarse threads and are M20 ISO 5.8. The joint constant is $C = 0.25$, and the dimensions are $A = 20$ mm, $B = 50$ mm, and $C = 160$ mm. The bearing base is 240 mm long. The bearing load is 14 kN. The bolts are tightened to 75 percent of proof load. Determine a static factor of safety for the bolt, based on the von Mises stress exceeding the proof strength. Use worst-case loading, as discussed in Prob. 8–72.

8–74   A split-ring clamp-type shaft collar such as is described in Prob. 5–67 must resist an axial load of 1000 lbf. Using a design factor of $n_d = 3$ and a coefficient of friction of 0.12, specify an SAE Grade 5 cap screw using fine threads. What wrench torque should be used if a lubricated screw is used?

8–75   A vertical channel 152 $\times$ 76 (see Table A–7) has a cantilever beam bolted to it as shown. The channel is hot-rolled AISI 1006 steel. The bar is of hot-rolled AISI 1015 steel. The shoulder bolts are M10 $\times$ 1.5 ISO 5.8. Assume the bolt threads do not extend into the joint. For a design factor of 2.0, find the safe force F that can be applied to the cantilever.

Problem 8–75
Dimensions in millimeters.

8–76   The cantilever bracket is bolted to a column with three M12 $\times$ 1.75 ISO 5.8 bolts. The bracket is made from AISI 1020 hot-rolled steel. Assume the bolt threads do not extend into the joint. Find the factors of safety for the following failure modes: shear of bolts, bearing of bolts, bearing of bracket, and bending of bracket.

Problem 8–76
Dimensions in millimeters.

8–77 A $\frac{3}{8}$- × 2-in AISI 1018 cold-drawn steel bar is cantilevered to support a static load of 250 lbf as illustrated. The bar is secured to the support using two $\frac{3}{8}$ in-16 UNC SAE grade 4 bolts. Assume the bolt threads do not extend into the joint. Find the factor of safety for the following modes of failure: shear of bolt, bearing on bolt, bearing on member, and strength of member.

Problem 8–77

8–78 The figure shows a welded fitting which has been tentatively designed to be bolted to a channel so as to transfer the 2000-lbf load into the channel. The channel and the two fitting plates are of hot-rolled stock having a minimum S_y of 42 kpsi. The fitting is to be bolted using six SAE grade 4 shoulder bolts. Assume the bolt threads do not extend into the joint. Check the strength of the design by computing the factor of safety for all possible modes of failure.

Problem 8–78

8–79 A cantilever is to be attached to the flat side of a 6-in, 13.0-lbf/in channel used as a column. The cantilever is to carry a load as shown in the figure. To a designer the choice of a bolt array is usually an a priori decision. Such decisions are made from a background of knowledge of the effectiveness of various patterns.

Problem 8-79

(*a*) If two fasteners are used, should the array be arranged vertically, horizontally, or diagonally? How would you decide?

(*b*) If three fasteners are used, should a linear or triangular array be used? For a triangular array, what should be the orientation of the triangle? How would you decide?

8-80 Using your experience with Prob. 8-79, specify an optimal bolt pattern for two bolts for the bracket in Prob. 8-79 and size the bolts.

8-81 Using your experience with Prob. 8-79, specify an optimal bolt pattern for three bolts for the bracket in Prob. 8-79 and size the bolts.

9

Welding, Bonding, and the Design of Permanent Joints

Form can more readily pursue function with the help of joining processes such as welding, brazing, soldering, cementing, and gluing—processes that are used extensively in manufacturing today. Whenever parts have to be assembled or fabricated, there is usually good cause for considering one of these processes in preliminary design work. Particularly when sections to be joined are thin, one of these methods may lead to significant savings. The elimination of individual fasteners, with their holes and assembly costs, is an important factor. Also, some of the methods allow rapid machine assembly, furthering their attractiveness.

Riveted permanent joints were common as the means of fastening rolled steel shapes to one another to form a permanent joint. The childhood fascination of seeing a cherry-red hot rivet thrown with tongs across a building skeleton to be unerringly caught by a person with a conical bucket, to be hammered pneumatically into its final shape, is all but gone. Two developments relegated riveting to lesser prominence. The first was the development of high-strength steel bolts whose preload could be controlled. The second was the improvement of welding, competing both in cost and in latitude of possible form.

9–1 Welding Symbols

A weldment is fabricated by welding together a collection of metal shapes, cut to particular configurations. During welding, the several parts are held securely together, often by clamping or jigging. The welds must be precisely specified on working drawings, and this is done by using the welding symbol, shown in Fig. 9–1, as standardized by the American Welding Society (AWS). The arrow of this symbol points to the joint to be welded. The body of the symbol contains as many of the following elements as are deemed necessary:

- Reference line
- Arrow

Figure 9–1

The AWS standard welding symbol showing the location of the symbol elements.

Finish symbol
Contour symbol
Root opening; depth of filling for plug and slot welds
Size; size or strength for resistance welds
Reference line

Groove angle; included angle of countersink for plug welds
Length of weld
Pitch (center-to-center spacing) of welds
Arrow connecting reference line to arrow side of joint, to grooved member, or both

Specification; process; or other reference
Tail (may be omitted when reference is not used)
Basic weld symbol or detail reference

Field weld symbol
Weld all around symbol
Number of spot or projection welds

(sides) R (Other side)
(Both) (Arrow side)
(N)
F / A
T S L – P

- Basic weld symbols as in Fig. 9–2
- Dimensions and other data
- Supplementary symbols
- Finish symbols
- Tail
- Specification or process

The *arrow side* of a joint is the line, side, area, or near member to which the arrow points. The side opposite the arrow side is the *other side*.

Figures 9–3 to 9–6 illustrate the types of welds used most frequently by designers. For general machine elements most welds are fillet welds, though butt welds are used a great deal in designing pressure vessels. Of course, the parts to be joined must be arranged so that there is sufficient clearance for the welding operation. If unusual joints are required because of insufficient clearance or because of the section shape, the design may be a poor one and the designer should begin again and endeavor to synthesize another solution.

Since heat is used in the welding operation, there are metallurgical changes in the parent metal in the vicinity of the weld. Also, residual stresses may be introduced because of clamping or holding or, sometimes, because of the order of welding. Usually these residual stresses are not severe enough to cause concern; in some cases

Figure 9–2

Arc- and gas-weld symbols.

			Type of weld				
Bead	Fillet	Plug or slot	Groove				
			Square	V	Bevel	U	J
⌒	◣	▽	‖	∨	⋁	⋃	⋃

Figure 9–3

Fillet welds. (*a*) The number indicates the leg size; the arrow should point only to one weld when both sides are the same. (*b*) The symbol indicates that the welds are intermittent and staggered 60 mm along on 200-mm centers.

(*a*)   (*b*)

Figure 9–4

The circle on the weld symbol indicates that the welding is to go all around.

Figure 9–5

Butt or groove welds:
(a) square butt-welded on
both sides; (b) single V with
60° bevel and root opening
of 2 mm; (c) double V;
(d) single bevel.

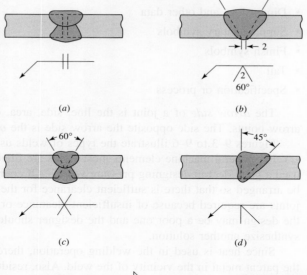

Figure 9–6

Special groove welds:
(a) T joint for thick plates;
(b) U and J welds for thick
plates; (c) corner weld (may
also have a bead weld on
inside for greater strength but
should not be used for heavy
loads); (d) edge weld for
sheet metal and light loads.

a light heat treatment after welding has been found helpful in relieving them. When the parts to be welded are thick, preheating will also be of benefit. If the reliability of the component is to be quite high, a testing program should be established to learn what changes or additions to the operations are necessary to ensure the best quality.

9–2 Butt and Fillet Welds

Figure 9–7a shows a single V-groove weld loaded by the tensile force F. For either tension or compression loading, the average normal stress is

$$\sigma = \frac{F}{hl} \tag{9–1}$$

where h is the weld throat and l is the length of the weld, as shown in the figure. Note that the value of h does not include the reinforcement. The reinforcement can be desirable, but it varies somewhat and does produce stress concentration at point A in the figure. If fatigue loads exist, it is good practice to grind or machine off the reinforcement.

Figure 9–7

A typical butt joint.

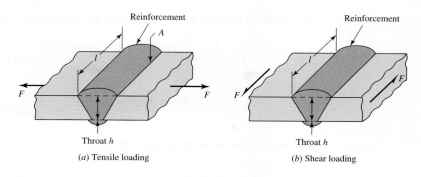

(a) Tensile loading (b) Shear loading

Figure 9–8

A transverse fillet weld.

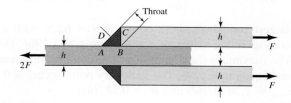

Figure 9–9

Free body from Fig. 9–8.

The average stress in a butt weld due to shear loading (Fig. 9–7b) is

$$\tau = \frac{F}{hl}$$
(9–2)

Figure 9–8 illustrates a typical transverse fillet weld. In Fig. 9–9 a portion of the welded joint has been isolated from Fig. 9–8 as a free body. At angle θ the forces on each weldment consist of a normal force F_n and a shear force F_s. Summing forces in the x and y directions gives

$$F_s = F \sin \theta$$
(a)

$$F_n = F \cos \theta$$
(b)

Using the law of sines for the triangle in Fig. 9–9 yields

$$\frac{t}{\sin 45°} = \frac{h}{\sin(180° - 45° - \theta)} = \frac{h}{\sin(135° - \theta)} = \frac{\sqrt{2}h}{\cos \theta + \sin \theta}$$

Solving for the throat thickness t gives

$$t = \frac{h}{\cos \theta + \sin \theta}$$
(c)

The nominal stresses at the angle θ in the weldment, τ and σ, are

$$\tau = \frac{F_s}{A} = \frac{F \sin\theta(\cos\theta + \sin\theta)}{hl} = \frac{F}{hl}(\sin\theta\cos\theta + \sin^2\theta) \qquad (d)$$

$$\sigma = \frac{F_n}{A} = \frac{F \cos\theta(\cos\theta + \sin\theta)}{hl} = \frac{F}{hl}(\cos^2\theta + \sin\theta\cos\theta) \qquad (e)$$

The von Mises stress σ' at angle θ is

$$\sigma' = (\sigma^2 + 3\tau^2)^{1/2} = \frac{F}{hl}[(\cos^2\theta + \sin\theta\cos\theta)^2 + 3(\sin^2\theta + \sin\theta\cos\theta)^2]^{1/2} \qquad (f)$$

The largest von Mises stress occurs at $\theta = 62.5°$ with a value of $\sigma' = 2.16F/(hl)$. The corresponding values of τ and σ are $\tau = 1.196F/(hl)$ and $\sigma = 0.623F/(hl)$.

The maximum shear stress can be found by differentiating Eq. (d) with respect to θ and equating to zero. The stationary point occurs at $\theta = 67.5°$ with a corresponding $\tau_{max} = 1.207F/(hl)$ and $\sigma = 0.5F/(hl)$.

There are some experimental and analytical results that are helpful in evaluating Eqs. (d) through (f) and the consequences. A model of the transverse fillet weld of Fig. 9–8 is easily constructed for photoelastic purposes and has the advantage of a balanced loading condition. Norris constructed such a model and reported the stress distribution along the sides AB and BC of the weld.[1] An approximate graph of the results he obtained is shown as Fig. 9–10a. Note that stress concentrations exist at A and B on the horizontal leg and at B on the vertical leg. Norris states that he could not determine the stresses at A and B with any certainty.

Salakian presents data for the stress distribution across the throat of a fillet weld (Fig. 9–10b).[2] This graph is of particular interest because we have just learned that it is the throat stresses that are used in design. Again, the figure shows a stress concentration at point B. Note that Fig. 9–10a applies either to the weld metal or to the parent metal, and that Fig. 9–10b applies only to the weld metal.

Equations (a) through (f) and their consequences seem familiar, and we can become comfortable with them. The net result of photoelastic and finite element analysis of transverse fillet weld geometry is more like that shown in Fig. 9–10 than those given by mechanics of materials or elasticity methods. The most important concept here is that we have *no analytical approach that predicts the existing stresses*. The geometry of the fillet is crude by machinery standards, and even if it were ideal, the macrogeometry is too abrupt and complex for our methods. There are also subtle bending stresses due to eccentricities. Still, in the absence of robust analysis, weldments must be specified and the resulting joints must be safe. The approach has been to use a simple *and conservative* model, verified by testing as conservative. The approach has been to

• Consider the external loading to be carried by shear forces on the throat area of the weld. By ignoring the normal stress on the throat, the shearing stresses are inflated sufficiently to render the model conservative.

---

[1] C. H. Norris, "Photoelastic Investigation of Stress Distribution in Transverse Fillet Welds," *Welding J.,* vol. 24, 1945, p. 557s.

[2] A. G. Salakian and G. E. Claussen, "Stress Distribution in Fillet Welds: A Review of the Literature," *Welding J.,* vol. 16, May 1937, pp. 1–24.

Figure 9–10

Stress distribution in fillet welds: (*a*) stress distribution on the legs as reported by Norris; (*b*) distribution of principal stresses and maximum shear stress as reported by Salakian.

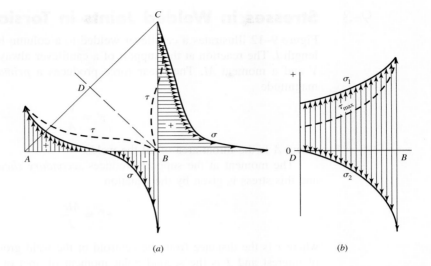

(*a*)   (*b*)

Figure 9–11

Parallel fillet welds.

- Use distortion energy for significant stresses.
- Circumscribe typical cases by code.

For this model, the basis for weld analysis or design employs

$$\tau = \frac{F}{0.707hl} = \frac{1.414F}{hl} \qquad (9\text{--}3)$$

which assumes the entire force F is accounted for by a shear stress in the minimum throat area. Note that this inflates the maximum estimated shear stress by a factor of $1.414/1.207 = 1.17$. Further, consider the parallel fillet welds shown in Fig. 9–11 where, as in Fig. 9–8, each weld transmits a force F. However, in the case of Fig. 9–11, the maximum shear stress *is* at the minimum throat area and corresponds to Eq. (9–3).

Under circumstances of combined loading we

- Examine primary shear stresses due to external forces.
- Examine secondary shear stresses due to torsional and bending moments.
- Estimate the strength(s) of the parent metal(s).
- Estimate the strength of deposited weld metal.
- Estimate permissible load(s) for parent metal(s).
- Estimate permissible load for deposited weld metal.

9–3 Stresses in Welded Joints in Torsion

Figure 9–12 illustrates a cantilever welded to a column by two fillet welds each of length *l*. The reaction at the support of a cantilever always consists of a shear force *V* and a moment *M*. The shear force produces a *primary shear* in the welds of magnitude

$$\tau' = \frac{V}{A} \tag{9-4}$$

where *A* is the throat area of all the welds.

The moment at the support produces *secondary shear* or *torsion* of the welds, and this stress is given by the equation

$$\tau'' = \frac{Mr}{J} \tag{9-5}$$

where *r* is the distance from the centroid of the weld group to the point in the weld of interest and *J* is the second polar moment of area of the weld group about the centroid of the group. When the sizes of the welds are known, these equations can be solved and the results combined to obtain the maximum shear stress. Note that *r* is usually the farthest distance from the centroid of the weld group.

Figure 9–13 shows two welds in a group. The rectangles represent the throat areas of the welds. Weld 1 has a throat thickness $t_1 = 0.707h_1$, and weld 2 has a throat thickness $t_2 = 0.707h_2$. Note that h_1 and h_2 are the respective weld sizes. The throat area of both welds together is

$$A = A_1 + A_2 = t_1 d + t_2 b \tag{a}$$

This is the area that is to be used in Eq. (9–4).

The *x* axis in Fig. 9–13 passes through the centroid G_1 of weld 1. The second moment of area about this axis is

$$I_x = \frac{t_1 d^3}{12}$$

Similarly, the second moment of area about an axis through G_1 parallel to the *y* axis is

$$I_y = \frac{dt_1^3}{12}$$

Figure 9–12

This is a *moment connection;* such a connection produces *torsion* in the welds. The shear stresses shown are resultant stresses.

| Figure 9–13

Thus the second polar moment of area of weld 1 about its own centroid is

$$J_{G1} = I_x + I_y = \frac{t_1 d^3}{12} + \frac{dt_1^3}{12} \tag{b}$$

In a similar manner, the second polar moment of area of weld 2 about its centroid is

$$J_{G2} = \frac{bt_2^3}{12} + \frac{t_2 b^3}{12} \tag{c}$$

The centroid G of the weld group is located at

$$\bar{x} = \frac{A_1 x_1 + A_2 x_2}{A} \qquad \bar{y} = \frac{A_1 y_1 + A_2 y_2}{A}$$

Using Fig. 9–13 again, we see that the distances r_1 and r_2 from G_1 and G_2 to G, respectively, are

$$r_1 = [(\bar{x} - x_1)^2 + \bar{y}^2]^{1/2} \qquad r_2 = [(y_2 - \bar{y})^2 + (x_2 - \bar{x})^2]^{1/2}$$

Now, using the parallel-axis theorem, we find the second polar moment of area of the weld group to be

$$J = (J_{G1} + A_1 r_1^2) + (J_{G2} + A_2 r_2^2) \tag{d}$$

This is the quantity to be used in Eq. (9–5). The distance r must be measured from G and the moment M computed about G.

The reverse procedure is that in which the allowable shear stress is given and we wish to find the weld size. The usual procedure is to estimate a probable weld size and then to use iteration.

Observe in Eqs. (b) and (c) the quantities t_1^3 and t_2^3, respectively, which are the cubes of the weld thicknesses. These quantities are small and can be neglected. This leaves the terms $t_1 d^3/12$ and $t_2 b^3/12$, which make J_{G1} and J_{G2} linear in the weld width. Setting the weld thicknesses t_1 and t_2 to unity leads to the idea of treating each fillet weld as a line. The resulting second moment of area is then a *unit second polar moment of area*. The advantage of treating the weld size as a line is that the value of J_u is the same regardless of the weld size. Since the throat width of a fillet weld is $0.707h$, the relationship between J and the unit value is

$$J = 0.707 h J_u \tag{9–6}$$

in which J_u is found by conventional methods for an area having unit width. The transfer formula for J_u must be employed when the welds occur in groups, as in Fig. 9–12. Table 9–1 lists the throat areas and the unit second polar moments of area for the most common fillet welds encountered. The example that follows is typical of the calculations normally made.

Table 9–1

Torsional Properties of Fillet Welds*

Weld	Throat Area	Location of G	Unit Second Polar Moment of Area
1.	$A = 0.707hd$	$\bar{x} = 0$ $\bar{y} = d/2$	$J_u = d^3/12$
2.	$A = 1.414hd$	$\bar{x} = b/2$ $\bar{y} = d/2$	$J_u = \dfrac{d(3b^2 + d^2)}{6}$
3.	$A = 0.707h(b + d)$	$\bar{x} = \dfrac{b^2}{2(b + d)}$ $\bar{y} = \dfrac{d^2}{2(b + d)}$	$J_u = \dfrac{(b + d)^4 - 6b^2d^2}{12(b + d)}$
4.	$A = 0.707h(2b + d)$	$\bar{x} = \dfrac{b^2}{2b + d}$ $\bar{y} = d/2$	$J_u = \dfrac{8b^3 + 6bd^2 + d^3}{12} - \dfrac{b^4}{2b + d}$
5.	$A = 1.414h(b + d)$	$\bar{x} = b/2$ $\bar{y} = d/2$	$J_u = \dfrac{(b + d)^3}{6}$
6.	$A = 1.414\pi hr$		$J_u = 2\pi r^3$

*G is the centroid of weld group; h is weld size; plane of torque couple is in the plane of the paper; all welds are of unit width.

EXAMPLE 9-1

A 50-kN load is transferred from a welded fitting into a 200-mm steel channel as illustrated in Fig. 9–14. Estimate the maximum stress in the weld.

Solution[3]

(*a*) Label the ends and corners of each weld by letter. See Fig. 9–15. Sometimes it is desirable to label each weld of a set by number.

(*b*) Estimate the primary shear stress τ'. As shown in Fig. 9–14, each plate is welded to the channel by means of three 6-mm fillet welds. Figure 9–15 shows that we have divided the load in half and are considering only a single plate. From case 4 of Table 9–1 we find the throat area as

$$A = 0.707(6)[2(56) + 190] = 1280 \text{ mm}^2$$

Then the primary shear stress is

$$\tau' = \frac{V}{A} = \frac{25(10)^3}{1280} = 19.5 \text{ MPa}$$

(*c*) Draw the τ' stress, to scale, at each lettered corner or end. See Fig. 9–16.

(*d*) Locate the centroid of the weld pattern. Using case 4 of Table 9–1, we find

$$\bar{x} = \frac{(56)^2}{2(56) + 190} = 10.4 \text{ mm}$$

This is shown as point O on Figs. 9–15 and 9–16.

(*e*) Find the distances r_i (see Fig. 9–16):

$$r_A = r_B = [(190/2)^2 + (56 - 10.4)^2]^{1/2} = 105 \text{ mm}$$
$$r_C = r_D = [(190/2)^2 + (10.4)^2]^{1/2} = 95.6 \text{ mm}$$

These distances can also be scaled from the drawing.

Figure 9–14

Dimensions in millimeters.

Figure 9–15

Diagram showing the weld geometry on a single plate; all dimensions in millimeters. Note that V and M represent the reaction loads applied by the welds *to the plate*.

---

[3]We are indebted to Professor George Piotrowski of the University of Florida for the detailed steps, presented here, of his method of weld analysis. R.G.B, J.K.N.

Figure 9–16

Free-body diagram of one of
the side plates.

(*f*) Find *J*. Using case 4 of Table 9–1 again, with Eq. (9–6), we get

$$J = 0.707(6) \left[\frac{8(56)^3 + 6(56)(190)^2 + (190)^3}{12} - \frac{(56)^4}{2(56) + 190} \right]$$

$$= 7.07(10)^6 \text{ mm}^4$$

(*g*) Find *M*:

$$M = Fl = 25(100 + 10.4) = 2760 \text{ N} \cdot \text{m}$$

(*h*) Estimate the secondary shear stresses τ'' at each lettered end or corner:

$$\tau_A'' = \tau_B'' = \frac{Mr}{J} = \frac{2760(10)^3(105)}{7.07(10)^6} = 41.0 \text{ MPa}$$

$$\tau_C'' = \tau_D'' = \frac{2760(10)^3(95.6)}{7.07(10)^6} = 37.3 \text{ MPa}$$

(*i*) Draw the τ'' stress at each corner and end. See Fig. 9–16. Note that this is a free-body
diagram of one of the side plates, and therefore the τ' and τ'' stresses represent what the
channel is doing to the plate (through the welds) to hold the plate in equilibrium.
(*j*) At each point labeled, combine the two stress components as vectors (since they
apply to the same area). At point *A*, the angle that τ_A'' makes with the vertical, α, is
also the angle r_A makes with the horizontal, which is $\alpha = \tan^{-1}(45.6/95) = 25.64°$.
This angle also applies to point *B*. Thus

$$\tau_A = \tau_B = \sqrt{(19.5 - 41.0 \sin 25.64°)^2 + (41.0 \cos 25.64°)^2} = 37.0 \text{ MPa}$$

Similarly, for *C* and *D*, $\beta = \tan^{-1}(10.4/95) = 6.25°$. Thus

$$\tau_C = \tau_D = \sqrt{(19.5 + 37.3 \sin 6.25°)^2 + (37.3 \cos 6.25°)^2} = 43.9 \text{ MPa}$$

(*k*) Identify the most highly stressed point:

Answer

$$\tau_{\max} = \tau_C = \tau_D = 43.9 \text{ MPa}$$

9-4   Stresses in Welded Joints in Bending

Figure 9–17a shows a cantilever welded to a support by fillet welds at top and bottom. A free-body diagram of the beam would show a shear-force reaction V and a moment reaction M. The shear force produces a primary shear in the welds of magnitude

$$\tau' = \frac{V}{A} \tag{a}$$

where A is the total throat area.

The moment M induces a horizontal shear stress component in the welds. Treating the two welds of Fig. 9–17b as lines we find the unit second moment of area to be

$$I_u = \frac{bd^2}{2} \tag{b}$$

The second moment of area I, based on weld throat area, is

$$I = 0.707hI_u = 0.707h\frac{bd^2}{2} \tag{c}$$

The nominal throat shear stress is now found to be

$$\tau'' = \frac{Mc}{I} = \frac{Md/2}{0.707hbd^2/2} = \frac{1.414M}{bdh} \tag{d}$$

The model gives the coefficient of 1.414, in contrast to the predictions of Sec. 9–2 of 1.197 from distortion energy, or 1.207 from maximum shear. The conservatism of the model's 1.414 is not that it is simply larger than either 1.196 or 1.207, but the tests carried out to validate the model show that it is large enough.

The second moment of area in Eq. (d) is based on the distance d between the two welds. If this moment is found by treating the two welds as having rectangular footprints, the distance between the weld throat centroids is approximately $(d + h)$. This would produce a slightly larger second moment of area, and result in a smaller level of stress. This method of treating welds as a line does not interfere with the conservatism of the model. It also makes Table 9–2 possible with all the conveniences that ensue.

The vertical (primary) shear of Eq. (a) and the horizontal (secondary) shear of Eq. (d) are then combined as vectors to give

$$\tau = (\tau'^2 + \tau''^2)^{1/2} \tag{e}$$

Figure 9–17

A rectangular cross-section cantilever welded to a support at the top and bottom edges.

(a)   (b) Weld pattern

Table 9–2

Bending Properties of Fillet Welds*

Weld	Throat Area	Location of G	Unit Second Moment of Area
1.	$A = 0.707hd$	$\bar{x} = 0$ $\bar{y} = d/2$	$I_u = \dfrac{d^3}{12}$
2.	$A = 1.414hd$	$\bar{x} = b/2$ $\bar{y} = d/2$	$I_u = \dfrac{d^3}{6}$
3.	$A = 1.414hb$	$\bar{x} = b/2$ $\bar{y} = d/2$	$I_u = \dfrac{bd^2}{2}$
4.	$A = 0.707h(2b + d)$	$\bar{x} = \dfrac{b^2}{2b + d}$ $\bar{y} = d/2$	$I_u = \dfrac{d^2}{12}(6b + d)$
5.	$A = 0.707h(b + 2d)$	$\bar{x} = b/2$ $\bar{y} = \dfrac{d^2}{b + 2d}$	$I_u = \dfrac{2d^3}{3} - 2d^2\bar{y} + (b + 2d)\bar{y}^2$
6.	$A = 1.414h(b + d)$	$\bar{x} = b/2$ $\bar{y} = d/2$	$I_u = \dfrac{d^2}{6}(3b + d)$
7.	$A = 0.707h(b + 2d)$	$\bar{x} = b/2$ $\bar{y} = \dfrac{d^2}{b + 2d}$	$I_u = \dfrac{2d^3}{3} - 2d^2\bar{y} + (b + 2d)\bar{y}^2$

Table 9–2

(Continued)

Weld	Throat Area	Location of G	Unit Second Moment of Area
8.	$A = 1.414h(b + d)$	$\bar{x} = b/2$ $\bar{y} = d/2$	$I_u = \dfrac{d^2}{6}(3b + d)$
9.	$A = 1.414\pi hr$		$I_u = \pi r^3$

*I_u, unit second moment of area, is taken about a horizontal axis through G, the centroid of the weld group, h is weld size; the plane of the bending couple is normal to the plane of the paper and parallel to the y-axis; all welds are of the same size.

9–5 The Strength of Welded Joints

The matching of the electrode properties with those of the parent metal is usually not so important as speed, operator appeal, and the appearance of the completed joint. The properties of electrodes vary considerably, but Table 9–3 lists the minimum properties for some electrode classes.

It is preferable, in designing welded components, to select a steel that will result in a fast, economical weld even though this may require a sacrifice of other qualities such as machinability. Under the proper conditions, all steels can be welded, but best results will be obtained if steels having a UNS specification between G10140 and G10230 are chosen. All these steels have a tensile strength in the hot-rolled condition in the range of 410 to 480 MPa.

The designer can choose factors of safety or permissible working stresses with more confidence if he or she is aware of the values of those used by others. One of the best standards to use is the American Institute of Steel Construction (AISC) code for building construction.[4] The permissible stresses are now based on the yield strength of the material instead of the ultimate strength, and the code permits the use of a variety of ASTM structural steels having yield strengths varying from 230 to 340 MPa. Provided the loading is the same, the code permits the same stress in the weld metal as in the parent metal. For these ASTM steels, $S_y = 0.5S_u$. Table 9–4 lists the formulas specified by the code for calculating these permissible stresses for various loading conditions. The factors of safety implied by this code are easily calculated. For tension, $n = 1/0.60 = 1.67$. For shear, $n = 0.577/0.40 = 1.44$, using the distortion-energy theory as the criterion of failure.

It is important to observe that the electrode material is often the strongest material present. If a bar of AISI 1010 steel is welded to one of 1018 steel, the weld metal is actually a mixture of the electrode material and the 1010 and 1018 steels.

---

[4]For a copy, either write the AISC, 400 N. Michigan Ave., Chicago, IL 60611, or contact on the Internet at www.aisc.org.

Table 9–3

Minimum Weld-Metal Properties

AWS Electrode Number*	Tensile Strength kpsi (MPa)	Yield Strength, kpsi (MPa)	Percent Elongation
E60xx	62 (427)	50 (345)	17–25
E70xx	70 (482)	57 (393)	22
E80xx	80 (551)	67 (462)	19
E90xx	90 (620)	77 (531)	14–17
E100xx	100 (689)	87 (600)	13–16
E120xx	120 (827)	107 (737)	14

*The American Welding Society (AWS) specification code numbering system for electrodes. This system uses an E prefixed to a four- or five-digit numbering system in which the first two or three digits designate the approximate tensile strength. The last digit includes variables in the welding technique, such as current supply. The next-to-last digit indicates the welding position, as, for example, flat, or vertical, or overhead. The complete set of specifications may be obtained from the AWS upon request.

Table 9–4

Stresses Permitted by the AISC Code for Weld Metal

Type of Loading	Type of Weld	Permissible Stress	n*
Tension	Butt	$0.60S_y$	1.67
Bearing	Butt	$0.90S_y$	1.11
Bending	Butt	$0.60–0.66S_y$	1.52–1.67
Simple compression	Butt	$0.60S_y$	1.67
Shear	Butt or fillet	$0.30S_{ut}^{†}$	

*The factor of safety n has been computed by using the distortion-energy theory.

†Shear stress on base metal should not exceed $0.40S_y$ of base metal.

Furthermore, a welded cold-drawn bar has its cold-drawn properties replaced with the hot-rolled properties in the vicinity of the weld. Finally, remembering that the weld metal is usually the strongest, do check the stresses in the parent metals.

The AISC code, as well as the AWS code, for bridges includes permissible stresses when fatigue loading is present. The designer will have no difficulty in using these codes, but their empirical nature tends to obscure the fact that they have been established by means of the same knowledge of fatigue failure already discussed in Chap. 6. Of course, for structures covered by these codes, the actual stresses *cannot* exceed the permissible stresses; otherwise the designer is legally liable. But in general, codes tend to conceal the actual margin of safety involved.

The fatigue stress-concentration factors listed in Table 9–5 are suggested for use. These factors should be used for the parent metal as well as for the weld metal. Table 9–6 gives steady-load information and minimum fillet sizes.

Table 9–5

Fatigue Stress-Concentration Factors, K_{fs}

Type of Weld	K_{fs}
Reinforced butt weld	1.2
Toe of transverse fillet weld	1.5
End of parallel fillet weld	2.7
T-butt joint with sharp corners	2.0

Table 9-6

Allowable Steady Loads and Minimum Fillet Weld Sizes

Schedule A: Allowable Load for Various Sizes of Fillet Welds

	Strength Level of Weld Metal (EXX)						
	60*	70*	80	90*	100	110*	120
τ =	124	145	165	186	207	228	248
Allowable shear stress on throat, MPa of fillet weld or partial penetration groove weld							
†f =	87.67h	102.52h	116.66h	131.5h	146.35h	161.2h	175.34h
Allowable Unit Force on Fillet Weld, N/mm							
Leg Size h, mm	Allowable Unit Force for Various Sizes of Fillet Welds N/mm						
25	2192	2563	2916	3288	3659	4030	4383
22	1929	2255	2566	2893	3220	3546	3857
20	1753	2050	2333	2630	2927	3224	3506
16	1403	1640	1866	2104	2342	2579	2805
12	1052	1230	1400	1578	1756	1934	2104
11	964	1127	1283	1447	1610	1773	1927
10	877	1025	1167	1315	1463	1612	1753
8	701	820	933	1052	1171	1290	1403
6	526	615	700	789	878	967	1052
5	438	513	583	658	732	806	877
3	263	308	350	395	439	484	526
2	175	205	233	263	293	322	351

Schedule B: Minimum Fillet Weld Size, h

Material Thickness of Thicker Part Joined, mm		Weld Size, mm
*To 6 incl.		3
Over 6	To 12	5
Over 12	To 20	6
†Over 20	To 38	8
Over 38	To 58	10
Over 58	To 150	12
Over 150		16

Not to exceed the thickness of the thinner part.
*Minimum size for bridge application does not go below 5 mm.
†For minimum fillet weld size, schedule does not go above 8 mm fillet weld for every 20 mm material.

*Fillet welds actually tested by the joint AISC-AWS Task Committee.
†f = 0.707h τall.

Source: From Omer W. Blodgett (ed.), Stress Allowables Affect Weldment Design, D412, The James F. Lincoln Arc Welding Foundation, Cleveland, May 1991, p. 3. Reprinted by permission of Lincoln Electric Company.

9–6 Static Loading

Some examples of statically loaded joints are useful in comparing and contrasting the conventional method of analysis and the welding code methodology.

EXAMPLE 9–2

A 12-mm by 50-mm rectangular-cross-section 1015 bar carries a static load of 68 kN. It is welded to a gusset plate with a 10-mm fillet weld 50 mm long on both sides with an E70XX electrode as depicted in Fig. 9–18. Use the welding code method.

(a) Is the weld metal strength satisfactory?

(b) Is the attachment strength satisfactory?

Solution

(a) From Table 9–6, allowable force per unit length for a 10-mm E70 electrode metal is 0.98×10^6 N/m of weldment; thus

$$F = 980l = 980(100) = 98 \text{ kN}$$

Since $98 > 68$ kN, weld metal strength is satisfactory.

(b) Check shear in attachment adjacent to the welds. From Table 9–4 and Table A–20, from which $S_y = 190$ MPa, the allowable attachment shear stress is

$$\tau_{\text{all}} = 0.4\,S_y = 0.4(190) = 76 \text{ MPa}$$

The shear stress τ on the base metal adjacent to the weld is

$$\tau = \frac{F}{2hl} = \frac{68\,000}{2(0.01)(0.05)} = 68 \text{ MPa}$$

Since $\tau_{\text{all}} \geq \tau$, the attachment is satisfactory near the weld beads. The tensile stress in the shank of the attachment σ is

$$\sigma = \frac{F}{tl} = \frac{68\,000}{(0.012)(0.05)} = 113 \text{ MPa}$$

The allowable tensile stress σ_{all}, from Table 9–4, is $0.6S_y$ and, with welding code safety level preserved,

$$\sigma_{\text{all}} = 0.6S = 0.6(190) = 114 \text{ MPa}$$

Since $\sigma_{\text{all}} \geq \sigma$, the shank tensile stress is satisfactory

---

| Figure 9–18

EXAMPLE 9–3   A specially rolled A36 structural steel section for the attachment has a cross section as shown in the figure and has yield and ultimate tensile strengths of 248 and 400 MPa, respectively. It is statically loaded through the attachment centroid by a load of F = 107 kN. Unsymmetrical weld tracks can compensate for eccentricity such that there is no moment to be resisted by the welds. Specify the weld track lengths l_1 and l_2 for an 8-mm fillet weld using an E70XX electrode. This is part of a design problem in which the design variables include weld lengths and the fillet leg size.

Solution   The y coordinate of the section centroid of the attachment is

$$\bar{y} = \frac{\sum y_i A_i}{\sum A_i} = \frac{25(20)50 + 75(10)50}{10(50) + 20(50)} = 41.7 \text{ mm}$$

Summing moments about point B to zero gives

$$\sum M_B = 0 = -F_1 b + F\bar{y} = -F_1(0.1) + 107(0.0417)$$

from which

$$F_1 = 44.6 \text{ kN}$$

It follows that

$$F_2 = 107 - 44.6 = 62.4 \text{ kN}$$

The weld throat areas have to be in the ratio 62.4/44.6 = 1.4, that is, $l_2 = 1.4l_1$. The weld length design variables are coupled by this relation, so l_1 is the weld length design variable. The other design variable is the fillet weld leg size h, which has been decided by the problem statement. From Table 9–4, the allowable shear stress on the throat τ_{all} is

$$\tau_{\text{all}} = 0.3(483) = 144.9 \text{ MPa}$$

The shear stress τ on the 45° throat is

$$\tau = \frac{F}{(0.707)h(l_1 + l_2)} = \frac{F}{(0.707)h(l_1 + 1.4l_1)}$$

$$= \frac{F}{(0.707)h(2.4l_1)} = \tau_{\text{all}} = 144.9 \text{ MPa}$$

from which the weld length l_1 is

$$l_1 = \frac{107\,000}{144.9 \times 10^6(0.707)0.008(2.4)} = 54.3 \text{ mm}$$

and

$$l_2 = 1.4l_1 = 1.4(54.3) = 76 \text{ mm}$$

 Figure 9–19

These are the weld-bead lengths required by weld metal strength. The attachment shear stress allowable in the base metal, from Table 9–4, is

$$\tau_{\text{all}} = 0.4S_y = 0.4(248) = 99.2 \text{ MPa}$$

The shear stress τ in the base metal adjacent to the weld is

$$\tau = \frac{F}{h(l_1 + l_2)} = \frac{F}{h(l_1 + 1.4l_1)} = \frac{F}{h(2.4l_1)} = \tau_{\text{all}} = 99.2 \text{ MPa}$$

from which

$$l_1 = \frac{F}{14.4h(2.4)} = \frac{107\,000}{99.2 \times 10^6(0.008)} = 56.2 \text{ mm}$$

$$l_2 = 1.4l_1 = 1.4(56.2) = 78.7 \text{ mm}$$

These are the weld-bead lengths required by base metal (attachment) strength. The base metal controls the weld lengths. For the allowable tensile stress σ_{all} in the shank of the attachment, the AISC allowable for tension members is $0.6S_y$; therefore,

$$\sigma_{\text{all}} = 0.6S_y = 0.6(248) = 148.8 \text{ MPa}$$

The nominal tensile stress σ is *uniform* across the attachment cross section because of the load application at the centroid. The stress σ is

$$\sigma = \frac{F}{A} = \frac{107\,000}{(0.02)(0.05) + 0.05(0.01)} = 71.3 \text{ MPa}$$

Since $\sigma_{\text{all}} \geq \sigma$, the shank section is satisfactory. With l_1 set to a nominal 58 mm, l_2 should be $1.4(58) = 81.2$ mm.

Decision

Set $l_1 = 58$ mm, $l_2 = 82$ mm. The small magnitude of the departure from $l_2/l_1 = 1.4$ is not serious. The joint is essentially moment-free.

EXAMPLE 9–4

Perform an adequacy assessment of the statically loaded welded cantilever carrying 2.2 kN depicted in Fig. 9–20. The cantilever is made of AISI 1018 HR steel and welded with a 10-mm fillet weld as shown in the figure. An E6010 electrode was used, and the design factor was 3.0.

(a) Use the conventional method for the weld metal.

(b) Use the conventional method for the attachment (cantilever) metal.

(c) Use a welding code for the weld metal.

Solution   (*a*) From Table 9–3, $S_y = 345$ MPa, $S_{ut} = 427$ MPa. From Table 9–2, second pattern, $b = 10$ mm, $d = 50$ mm, so

$$A = 1.414hd = 1.414(10)50 = 707 \text{ mm}^2$$

$$I_u = d^3/6 = 50^3/6 = 20\ 833 \text{ mm}^3$$

$$I = 0.707hI_u = 0.707(10)20\ 833 = 147\ 289 \text{ mm}^4$$

| Figure 9–20

Primary shear:

$$\tau' = \frac{F}{A} = \frac{2200}{707} = 3.1 \text{ MPa}$$

Secondary shear:

$$\tau'' = \frac{Mr}{I} = \frac{2200(150)25}{147\ 289} = 56 \text{ MPa}$$

The shear magnitude τ is the Pythagorean combination

$$\tau = (\tau'^2 + \tau''^2)^{1/2} = (3.1^2 + 56^2)^{1/2} = 56.1 \text{ MPa}$$

The factor of safety based on a minimum strength and the distortion-energy criterion is

Answer
$$n = \frac{S_{sy}}{\tau} = \frac{0.577(345)}{56.1} = 3.55$$

Since $n \geq n_d$, that is, $3.39 \geq 3.0$, the weld metal has satisfactory strength.

(*b*) From Table A–20, minimum strengths are $S_{ut} = 400$ MPa and $S_y = 220$ MPa. Then

$$\sigma = \frac{M}{I/c} = \frac{M}{bd^2/6} = \frac{2200(150)}{10(50^2)/6} = 79.2 \text{ MPa}$$

Answer
$$n = \frac{S_y}{\sigma} = \frac{220}{79.2} = 2.78$$

Since $n < n_d$, that is, $2.78 < 3.0$, the joint is unsatisfactory as to the attachment strength. (*c*) From part (*a*), $\tau = 56.1$ MPa. For an E6010 electrode Table 9–6 gives the allowable shear stress τ_{all} as 124 MPa. Since $\tau < \tau_{\text{all}}$, the weld is satisfactory. Since the code already has a design factor of $0.577(345)/124 = 1.6$ included at the equality, the corresponding factor of safety to part (*a*) is

Answer
$$n = 1.6\frac{124}{56.1} = 3.54$$

which is consistent.

9-7 Fatigue Loading

The conventional methods will be provided here. In fatigue, the Gerber criterion is best; however, you will find that the modified Goodman criterion is in common use. For the surface factor of Eq. 6–19, an as-forged surface should always be assumed for weldments unless a superior finish is specified and obtained.

Some examples of fatigue loading of welded joints follow.

EXAMPLE 9–5

The 1018 steel strap of Fig. 9–21 has a 4500-N, completely reversed load applied. Determine the factor of safety of the weldment for infinite life.

Solution

From Table A–20 for the 1018 attachment metal the strengths are $S_{ut} = 400$ MPa and $S_y = 220$ MPa. For the E6010 electrode, $S_{ut} = 430$ MPa and $S_y = 340$ MPa. The fatigue stress-concentration factor, from Table 9–5, is $K_{fs} = 2.7$. From Table 6–2, p. 280, $k_a = 272(400)^{-0.995} = 0.7$. The shear area is:

$$A = 2(0.707)10(50) = 707 \text{ mm}^2$$

For a uniform shear stress on the throat, $k_b = 1$.
From Eq. (6–26), p. 290, for torsion (shear),

$$k_c = 0.59 \qquad k_d = k_e = k_f = 1$$

From Eqs. (6–8), p. 282, and (6–18), p. 287,

$$S_{se} = 0.70(1)0.59(1)(1)(1)0.5(400) = 82.6 \text{ MPa}$$

$$K_{fs} = 2.7 \qquad F_a = 4500 \text{ N} \qquad F_m = 0$$

Only primary shear is present:

$$\tau_a' = \frac{K_{fs} F_a}{A} = \frac{2.7(4500)}{707} = 17.2 \text{ MPa} \qquad \tau_m' = 0 \text{ MPa}$$

| Figure 9–21

In the absence of a midrange component, the fatigue factor of safety n_f is given by

Answer

$$n_f = \frac{S_{se}}{\tau_a'} = \frac{82.8}{17.2} = 4.81$$

EXAMPLE 9–6   The 1018 steel strap of Fig. 9–22 has a repeatedly applied load of 9000 N ($F_e = F_m = 4500$ N). Determine the fatigue factor of safety fatigue strength of the weldment.

Solution   From Table 6–2, p. 296, $k_a = 272(400)^{-0.995} = 0.7$.

$$A = 2(0.707)10(50) = 707 \text{ mm}^2$$

For uniform shear stress on the throat $k_b = 1$.
From Eq. (6–26), p. 298, $k_c = 0.59$. From Eqs. (6–8), p. 290, and (6–18), p. 295,

$$S_{se} = 0.7(1)0.59(1)(1)(1)0.5(400) = 82.6 \text{ MPa}$$

From Table 9–5, $K_{fs} = 2$. Only primary shear is present:

$$\tau_a' = \tau_m' = \frac{K_{fs} F_a}{A} = \frac{2(4500)}{707} = 12.7 \text{ MPa}$$

From Eq. (6–53), p. 323, $S_{su} \doteq 0.67 S_{ut}$. This, together with the Gerber fatigue failure criterion for shear stresses from Table 6–7, p. 315, gives

$$n_f = \frac{1}{2}\left(\frac{0.67 S_{ut}}{\tau_m}\right)^2 \frac{\tau_a}{S_{se}} \left[-1 + \sqrt{1 + \left(\frac{2\tau_m S_{se}}{0.67 S_{ut} \tau_a}\right)^2} \right]$$

Answer
$$n_f = \frac{1}{2}\left[\frac{0.67(400)}{12.7}\right]^2 \frac{12.7}{82.6} \left\{ -1 + \sqrt{1 + \left[\frac{2(12.7)82.6}{0.67(400)12.7}\right]^2} \right\} = 5.98$$

| **Figure 9–22**

Figure 9–23

(*a*) Spot welding; (*b*) seam welding.

(*a*)　　　　　　　(*b*)

9–8   Resistance Welding

The heating and consequent welding that occur when an electric current is passed through several parts that are pressed together is called *resistance welding*. *Spot welding* and *seam welding* are forms of resistance welding most often used. The advantages of resistance welding over other forms are the speed, the accurate regulation of time and heat, the uniformity of the weld, and the mechanical properties that result. In addition the process is easy to automate, and filler metal and fluxes are not needed.

The spot- and seam-welding processes are illustrated schematically in Fig. 9–23. Seam welding is actually a series of overlapping spot welds since the current is applied in pulses as the work moves between the rotating electrodes.

Failure of a resistance weld occurs either by shearing of the weld or by tearing of the metal around the weld. Because of the possibility of tearing, it is good practice to avoid loading a resistance-welded joint in tension. Thus, for the most part, design so that the spot or seam is loaded in pure shear. The shear stress is then simply the load divided by the area of the spot. Because the thinner sheet of the pair being welded may tear, the strength of spot welds is often specified by stating the load per spot based on the thickness of the thinnest sheet. Such strengths are best obtained by experiment.

Somewhat larger factors of safety should be used when parts are fastened by spot welding rather than by bolts or rivets, to account for the metallurgical changes in the materials due to the welding.

9–9   Adhesive Bonding[5]

The use of polymeric adhesives to join components for structural, semistructural, and nonstructural applications has expanded greatly in recent years as a result of the unique advantages adhesives may offer for certain assembly processes and the development of new adhesives with improved robustness and environmental acceptability. The increasing complexity of modern assembled structures and the diverse types of materials used have led to many joining applications that would not be possible with more conventional joining techniques. Adhesives are also being used either in conjunction

---

[5]For a more extensive discussion of this topic, see J. E. Shigley and C. R. Mischke, *Mechanical Engineering Design*, 6th ed., McGraw-Hill, New York, 2001, Sec. 9–11. This section was prepared with the assistance of Professor David A. Dillard, Professor of Engineering Science and Mechanics and Director of the Center for Adhesive and Sealant Science, Virginia Polytechnic Institute and State University, Blacksburg, Virginia, and with the encouragement and technical support of the Bonding Systems Division of 3M, Saint Paul, Minnesota.

Figure 9–24

Diagram of an automobile body showing at least 15 locations adhesives and sealants could be used or are being used. Particular note should be made of the windshield (8), which is considered a load-bearing structure in modern automobiles and is adhesively bonded. Also attention should be paid to hem flange bonding (1), in which adhesives are used to bond and seal. Adhesives are used to bond friction surfaces in brakes and clutches (10). Antiflutter adhesive bonding (2) helps control deformation of hood and trunk lids under wind shear. Thread-sealing adhesives are used in engine applications (12). (From A. V. Pocius, *Adhesion and Adhesives Technology,* 2nd edition, Hanser Publishers, Munich, 2002. Reprinted by permission.)

with or to replace mechanical fasteners and welds. Reduced weight, sealing capabilities, and reduced part count and assembly time, as well as improved fatigue and corrosion resistance, all combine to provide the designer with opportunities for customized assembly. The worldwide size of the adhesive and sealant industry is approximately 40 billion Euro dollars, and the United States market is about 12 billion US dollars.[6] Figure 9–24 illustrates the numerous places where adhesives are used on a modern automobile. Indeed, the fabrication of many modern vehicles, devices, and structures is dependent on adhesives.

In well-designed joints and with proper processing procedures, use of adhesives can result in significant reductions in weight. Eliminating mechanical fasteners eliminates the weight of the fasteners, and also may permit the use of thinner-gauge materials because stress concentrations associated with the holes are eliminated. The capability of polymeric adhesives to dissipate energy can significantly reduce noise, vibration, and harshness (NVH), crucial in modern automobile performance. Adhesives can be used to assemble heat-sensitive materials or components that might be damaged by drilling holes for mechanical fasteners. They can be used to join dissimilar materials or thin-gauge stock that cannot be joined through other means.

Types of Adhesive

There are numerous adhesive types for various applications. They may be classified in a variety of ways depending on their chemistry (e.g., epoxies, polyurethanes, polyimides), their form (e.g., paste, liquid, film, pellets, tape), their type (e.g., hot melt, reactive hot melt, thermosetting, pressure sensitive, contact), or their load-carrying capability (structural, semistructural, or nonstructural).

Structural adhesives are relatively strong adhesives that are normally used well below their glass transition temperature; common examples include epoxies and certain acrylics. Such adhesives can carry significant stresses, and they lend themselves to structural

---

[6]From E. M. Petrie, *Handbook of Adhesives and Sealants,* 2nd ed., McGraw-Hill, New York, 2007.

applications. For many engineering applications, semistructural applications (where failure would be less critical) and nonstructural applications (of headliners, etc., for aesthetic purposes) are also of significant interest to the design engineer, providing cost-effective means required for assembly of finished products. These include *contact adhesives,* where a solution or emulsion containing an elastomeric adhesive is coated onto both adherends, the solvent is allowed to evaporate, and then the two adherends are brought into contact. Examples include rubber cement and adhesives used to bond laminates to countertops. *Pressure-sensitive adhesives* are very low modulus elastomers that deform easily under small pressures, permitting them to wet surfaces. When the substrate and adhesive are brought into intimate contact, van der Waals forces are sufficient to maintain the contact and provide relatively durable bonds. Pressure-sensitive adhesives are normally purchased as tapes or labels for nonstructural applications, although there are also double-sided foam tapes that can be used in semistructural applications. As the name implies, *hot melts* become liquid when heated, wetting the surfaces and then cooling into a solid polymer. These materials are increasingly applied in a wide array of engineering applications by more sophisticated versions of the glue guns in popular use. *Anaerobic adhesives* cure within narrow spaces deprived of oxygen; such materials have been widely used in mechanical engineering applications to lock bolts or bearings in place. Cure in other adhesives may be induced by exposure to ultraviolet light or electron beams, or it may be catalyzed by certain materials that are ubiquitous on many surfaces, such as water.

Table 9–7 presents important strength properties of commonly used adhesives.

Table 9–7

Mechanical Performance of Various Types of Adhesives

Adhesive Chemistry or Type	Room Temperature Lap-Shear Strength, MPa (psi)		Peel Strength per Unit Width, kN/m (lbf/in)	
Pressure-sensitive	0.01–0.07	(2–10)	0.18–0.88	(1–5)
Starch-based	0.07–0.7	(10–100)	0.18–0.88	(1–5)
Cellosics	0.35–3.5	(50–500)	0.18–1.8	(1–10)
Rubber-based	0.35–3.5	(50–500)	1.8–7	(10–40)
Formulated hot melt	0.35–4.8	(50–700)	0.88–3.5	(5–20)
Synthetically designed hot melt	0.7–6.9	(100–1000)	0.88–3.5	(5–20)
PVAc emulsion (white glue)	1.4–6.9	(200–1000)	0.88–1.8	(5–10)
Cyanoacrylate	6.9–13.8	(1000–2000)	0.18–3.5	(1–20)
Protein-based	6.9–13.8	(1000–2000)	0.18–1.8	(1–10)
Anaerobic acrylic	6.9–13.8	(1000–2000)	0.18–1.8	(1–10)
Urethane	6.9–17.2	(1000–2500)	1.8–8.8	(10–50)
Rubber-modified acrylic	13.8–24.1	(2000–3500)	1.8–8.8	(10–50)
Modified phenolic	13.8–27.6	(2000–4000)	3.6–7	(20–40)
Unmodified epoxy	10.3–27.6	(1500–4000)	0.35–1.8	(2–10)
Bis-maleimide	13.8–27.6	(2000–4000)	0.18–3.5	(1–20)
Polyimide	13.8–27.6	(2000–4000)	0.18–0.88	(1–5)
Rubber-modified epoxy	20.7–41.4	(3000–6000)	4.4–14	(25–80)

Source: From A. V. Pocius, *Adhesion and Adhesives Technology,* 2nd ed., Hanser Gardner Publishers, Ohio, 2002. Reprinted by permission.

Figure 9–25

Common types of lap joints
used in mechanical design:
(*a*) single lap; (*b*) double lap;
(*c*) scarf; (*d*) bevel; (*e*) step;
(*f*) butt strap; (*g*) double butt
strap; (*h*) tubular lap. *(Adapted
from R. D. Adams, J. Comyn,
and W. C. Wake,* Structural
Adhesive Joints in Engineering,
*2nd ed., Chapman and Hall,
New York, 1997.)*

Stress Distributions

Good design practice normally requires that adhesive joints be constructed in such
a manner that the adhesive carries the load in shear rather than tension. Bonds
are typically much stronger when loaded in shear rather than in tension across the
bond plate. Lap-shear joints represent an important family of joints, both for test
specimens to evaluate adhesive properties and for actual incorporation into practical designs. Generic types of lap joints that commonly arise are illustrated in
Fig. 9–25.

The simplest analysis of lap joints suggests the applied load is uniformly distributed over the bond area. Lap joint test results, such as those obtained following
the ASTM D1002 for single-lap joints, report the "apparent shear strength" as the
breaking load divided by the bond area. Although this simple analysis can be adequate for stiff adherends bonded with a soft adhesive over a relatively short bond
length, significant peaks in shear stress occur except for the most flexible adhesives.
In an effort to point out the problems associated with such practice, ASTM D4896
outlines some of the concerns associated with taking this simplistic view of stresses
within lap joints.

In 1938, O. Volkersen presented an analysis of the lap joint, known as the *shear-
lag model*. It provides valuable insights into the shear-stress distributions in a host
of lap joints. Bending induced in the single-lap joint due to eccentricity significantly
complicates the analysis, so here we will consider a symmetric double-lap joint to

Figure 9–26

Double-lap joint.

illustrate the principles. The shear-stress distribution for the double lap joint of Fig. 9–26 is given by

$$\tau(x) = \frac{P\omega}{4b \, \sinh(\omega l/2)} \cos h \, (\omega x) + \left[\frac{P\omega}{4b \, \cos h \, (\omega l/2)} \left(\frac{2E_o t_o - E_i t_i}{2E_o t_o + E_i t_i} \right) \right.$$

$$\left. + \frac{(\alpha_i - \alpha_o)\Delta T\omega}{[1/(E_o t_o) + 2/(E_i t_i)] \cosh(\omega l/2)} \right] \sinh(\omega x)$$

(9–7)

where

$$\omega = \sqrt{\frac{G}{h}\left(\frac{1}{E_o t_o} + \frac{2}{E_i t_i}\right)}$$

and E_o, t_o, α_o, and E_i, t_i, α_i, are the modulus, thickness, coefficient of thermal expansion for the outer and inner adherend, respectively; G, h, b, and l are the shear modulus, thickness, width, and length of the adhesive, respectively; and ΔT is a change in temperature of the joint. If the adhesive is cured at an elevated temperature such that the stress-free temperature of the joint differs from the service temperature, the mismatch in thermal expansion of the outer and inner adherends induces a thermal shear across the adhesive.

EXAMPLE 9–7

The double-lap joint depicted in Fig. 9–26 consists of aluminum outer adherends and an inner steel adherend. The assembly is cured at 250°F and is stress-free at 200°F. The completed bond is subjected to an axial load of 9000 N at a service temperature of 70°F. The width b is 25 mm, the length of the bond l is 25 mm. Additional information is tabulated below:

	G, GPa	E, GPa	α, mm/(mm · °F)	Thickness, mm
Adhesive	1.4		$55(10^{-6})$	0.5
Outer adherend		69	$13.3(10^{-6})$	3.8
Inner adherend		207	$6.0(10^{-6})$	2.5

Sketch a plot of the shear stress as a function of the length of the bond due to (a) thermal stress, (b) load-induced stress, and (c) the sum of stresses in a and b; and (d) find where the largest shear stress is maximum.

Solution   In Eq. (9–7) the parameter ω is given by

$$\omega = \sqrt{\frac{G}{h}\left(\frac{1}{E_o t_o} + \frac{2}{E_i t_i}\right)}$$

$$= \sqrt{\frac{1400}{0.5}\left[\frac{1}{69\,000(3.8)} + \frac{2}{207\,000(2.5)}\right]} = 0.147 \text{ mm}^{-1}$$

(a) For the thermal component, $\alpha_i - \alpha_o = 6(10^{-6}) - 13.3(10^{-6}) = -7.3(10^{-6})$ in/(in · °F), $\Delta T = 70 - 200 = -130$°F,

$$\tau_{th}(x) = \frac{(\alpha_i - \alpha_o)\Delta T \omega \sinh(\omega x)}{(1/E_o t_o + 2/E_i t_i)\cosh(\omega l/2)}$$

$$\tau_{th}(x) = \frac{-7.3(10^{-6})(-130)0.147 \sinh(0.147x)}{\left[\dfrac{1}{69\,000(3.8)} + \dfrac{2}{207\,000(2.5)}\right]\cosh\left[\dfrac{0.147(25)}{2}\right]}$$

$$= 5.642 \sinh(0.147x)$$

The thermal stress is plotted in Fig. (9–27) and tabulated at $x = -12.7$, 0, and 12.7 in the table below.

(b) The bond is "balanced" ($E_o t_o = E_i t_i/2$), so the load-induced stress is given by

$$\tau_P(x) = \frac{P\omega \cosh(\omega x)}{4b \sinh(\omega l/2)} = \frac{9000(0.147)\cosh(0.147x)}{4(25)\sinh[0.147(25)/2]} = 4.322 \cosh(0.147) \quad (1)$$

The load-induced stress is plotted in Fig. (9–27) and tabulated at $x = -12.7$, 0, and 12.7 in the table below.

(c) Total stress table (in MPa):

	$\tau(-12.7)$	$\tau(0)$	$\tau(12.7)$
Thermal only	°17	0	17
Load-induced only	13.3	4.2	13.3
Combined	°3.8	4.2	30.3

(d) The maximum shear stress predicted by the shear-lag model will always occur at the ends. See the plot in Fig. 9–27. Since the residual stresses are always present, significant shear stresses may already exist prior to application of the load. The large stresses present for the combined-load case could result in local yielding of a ductile adhesive or failure of a more brittle one. The significance of the thermal stresses serves as a caution against joining dissimilar adherends when large temperature changes are involved. Note also that the average shear stress due to the load is

Figure 9–27

Plot for Ex. 9–7.

$\tau_{avg} = P/(2bl) = 7.2$ MPa. Equation (1) produced a maximum of 13.2 MPa, almost double the average.

Although design considerations for single-lap joints are beyond the scope of this chapter, one should note that the load eccentricity is an important aspect in the stress state of single-lap joints. Adherend bending can result in shear stresses that may be as much as double those given for the double-lap configuration (for a given total bond area). In addition, peel stresses can be quite large and often account for joint failure. Finally, plastic bending of the adherends can lead to high strains, which less ductile adhesives cannot withstand, leading to bond failure as well. Bending stresses in the adherends at the end of the overlap can be four times greater than the average stress within the adherend; thus, they must be considered in the design. Figure 9–28 shows the shear and peel stresses present in a typical single-lap joint that corresponds to the ASTM D1002 test specimen. Note that the shear stresses are significantly larger than predicted by the Volkersen analysis, a result of the increased adhesive strains associated with adherend bending.

Joint Design

Some basic guidelines that should be used in adhesive joint design include:

- Design to place bondline in shear, not peel. Beware of peel stresses focused at bond terminations. When necessary, reduce peel stresses through tapering the adherend ends, increasing bond area where peel stresses occur, or utilizing rivets at bond terminations where peel stresses can initiate failures.

- Where possible, use adhesives with adequate ductility. The ability of an adhesive to yield reduces the stress concentrations associated with the ends of joints and increases the toughness to resist debond propagation.

- Recognize environmental limitations of adhesives and surface preparation methods. Exposure to water, solvents, and other diluents can significantly degrade adhesive performance in some situations, through displacing the adhesive from the surface or degrading the polymer. Certain adhesives may be susceptible to environmental stress cracking in the presence of certain solvents. Exposure to ultraviolet light can also degrade adhesives.

Figure 9–28

Stresses within a single-lap joint. (*a*) Lap-joint tensile forces have a line of action that is not initially parallel to the adherend sides. (*b*) As the load increases the adherends and bond bend. (*c*) In the locality of the end of an adherend peel and shear stresses appear, and the peel stresses often induce joint failure. (*d*) The seminal Goland and Reissner stress predictions (*J. Appl. Mech.*, vol. 77, 1944) are shown. (*Note that the predicted shear-stress maximum is higher than that predicted by the Volkersen shear-lag model because of adherend bending.*)

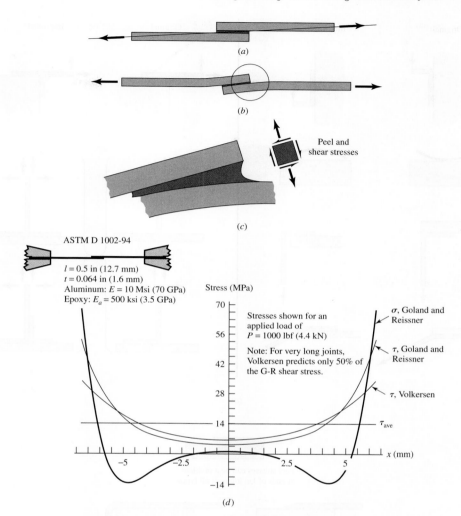

(*a*)

(*b*)

Peel and shear stresses

(*c*)

ASTM D 1002-94

$l = 0.5$ in (12.7 mm)
$t = 0.064$ in (1.6 mm)
Aluminum: $E = 10$ Msi (70 GPa)
Epoxy: $E_a = 500$ ksi (3.5 GPa)

Stress (MPa)

Stresses shown for an applied load of
$P = 1000$ lbf (4.4 kN)

Note: For very long joints, Volkersen predicts only 50% of the G-R shear stress.

σ, Goland and Reissner

τ, Goland and Reissner

τ, Volkersen

τ_{ave}

x (mm)

(*d*)

• Design in a way that permits or facilitates inspections of bonds where possible. A missing rivet or bolt is often easy to detect, but debonds or unsatisfactory adhesive bonds are not readily apparent.

• Allow for sufficient bond area so that the joint can tolerate some debonding before going critical. This increases the likelihood that debonds can be detected. Having some regions of the overall bond at relatively low stress levels can significantly improve durability and reliability.

• Where possible, bond to multiple surfaces to offer support to loads in any direction. Bonding an attachment to a single surface can place peel stresses on the bond, whereas bonding to several adjacent planes tends to permit arbitrary loads to be carried predominantly in shear.

• Adhesives can be used in conjunction with spot welding. The process is known as *weld bonding*. The spot welds serve to fixture the bond until it is cured.

Figure 9–29 presents examples of improvements in adhesive bonding.

Original   Improved   Original   Improved

(a)

Peel stresses can be a problem
at ends of lap joints of all types

Tapered to reduce peel   Mechanically reduce peel

Rivet, spot weld, or
bolt to reduce peel   Larger bond area to reduce peel

(b)

Figure 9–29

Design practices that improve adhesive bonding. (*a*) Gray load vectors are to be avoided as resulting strength is poor. (*b*) Means to reduce peel stresses in lap-type joints.

References

Good references are available for analyzing and designing adhesive bonds, including the following:

R. D. Adams, J. Comyn, and W. C. Wake, *Structural Adhesive Joints in Engineering,* 2nd ed., Chapman and Hall, New York, 1997.

G. P. Anderson, S. J. Bennett, and K. L. DeVries, *Analysis and Testing of Adhesive Bonds,* Academic Press, New York, 1977.

H. F. Brinson (ed.), *Engineered Materials Handbook, vol. 3: Adhesives and Sealants,* ASM International, Metals Park, Ohio, 1990.

A. J. Kinloch, *Adhesion and Adhesives: Science and Technology,* Chapman and Hall, New York, 1987.

A. J. Kinloch (ed.), *Durability of Structural Adhesives,* Applied Science Publishers, New York, 1983.

R. W. Messler, Jr., *Joining of Materials and Structures,* Elsevier Butterworth-Heinemann, Mass., 2004.

E. M. Petrie, *Handbook of Adhesives and Sealants,* 2nd ed., McGraw-Hill, New York, 2007.

A. V. Pocius, *Adhesion and Adhesives Technology: An Introduction,* 2nd ed., Hanser Gardner, Ohio, 1997.

The Internet is also a good source of information. For example, try this website: www.3m.com/adhesives.

PROBLEMS

9–1 to 9–4

The figure shows a horizontal steel bar of thickness h loaded in steady tension and welded to a vertical support. Find the load F that will cause an allowable shear stress, τ_{allow}, in the throats of the welds.

Problem Number	b	d	h	τ_{allow}
9–1	50 mm	50 mm	5 mm	140 MPa
9–2	50 mm	50 mm	8 mm	140 MPa
9–3	50 mm	30 mm	5 mm	140 MPa
9–4	100 mm	50 mm	8 mm	140 MPa

Problems 9–1 to 9–4

9–5 to 9–8

For the weldments of Probs. 9–1 to 9–4, the electrodes are specified in the table. For the electrode metal indicated, what is the allowable load on the weldment?

Problem Number	Reference Problem	Electrode
9–5	9–1	E7010
9–6	9–2	E6010
9–7	9–3	E7010
9–8	9–4	E6010

9–9 to 9–12 The materials for the members being joined in Probs. 9–1 to 9–4 are specified below. What load on the weldment is allowable because member metal is incorporated in the welds?

Problem Number	Reference Problem	Bar	Vertical Support
9–9	9–1	1018 CD	1018 HR
9–10	9–2	1020 CD	1020 CD
9–11	9–3	1035 HR	1035 CD
9–12	9–4	1035 HR	1020 CD

9–13 to 9–16 A steel bar of thickness h is welded to a vertical support as shown in the figure. What is the shear stress in the throat of the welds due to the force F?

Problem Number	b	d	h	F
9–13	50 mm	50 mm	5 mm	100 kN
9–14	50 mm	50 mm	8 mm	140 kN
9–15	50 mm	30 mm	5 mm	100 kN
9–16	100 mm	50 mm	8 mm	140 kN

Problems 9–13 to 9–16

9–17 to 9–20 A steel bar of thickness h, to be used as a beam, is welded to a vertical support by two fillet welds as shown in the figure.

(a) Find the safe bending force F if the allowable shear stress in the welds is τ_{allow}.

(b) In part a, you found a simple expression for F in terms of the allowable shear stress. Find the allowable load if the electrode is E7010, the bar is hot-rolled 1020, and the support is hot-rolled 1015.

Problem Number	b	c	d	h	τ_{allow}
9–17	50 mm	150 mm	50 mm	5 mm	140 MPa
9–18	50 mm	150 mm	50 mm	8 mm	140 MPa
9–19	50 mm	150 mm	30 mm	5 mm	140 MPa
9–20	100 mm	150 mm	50 mm	8 mm	140 MPa

Problems 9–17 to 9–20

9–21 to 9–24

The figure shows a weldment just like that for Probs. 9–17 to 9–20 except there are four welds instead of two. Find the safe bending force F if the allowable shear stress in the welds is τ_{allow}.

Problem Number	b	c	d	h	τ_{allow}
9–21	50 mm	150 mm	50 mm	5 mm	140 MPa
9–22	50 mm	150 mm	50 mm	8 mm	140 MPa
9–23	50 mm	150 mm	30 mm	5 mm	140 MPa
9–24	100 mm	150 mm	50 mm	8 mm	140 MPa

Problems 9–21 to 9–24

9–25 to 9–28

The weldment shown in the figure is subjected to an alternating force F. The hot-rolled steel bar has a thickness h and is of AISI 1010 steel. The vertical support is likewise AISI 1010 HR steel. The electrode is given in the table below. Estimate the fatigue load F the bar will carry if three fillet welds are used.

Problem Number	b	d	h	Electrode
9–25	50 mm	50 mm	5 mm	E6010
9–26	50 mm	50 mm	8 mm	E6010
9–27	50 mm	30 mm	5 mm	E7010
9–28	100 mm	50 mm	8 mm	E7010

Problems 9–25 to 9–28

9-29   The permissible shear stress for the weldment illustrated is 140 MPa. Estimate the load, F, that will cause this stress in the weldment throat.

Problem 9-29

9-30 to 9-31   A steel bar of thickness h is subjected to a bending force F. The vertical support is stepped such that the horizontal welds are b_1 and b_2 long. Determine F if the maximum allowable shear stress is τ_{allow}.

Problem Number	b_1	b_2	c	d	h	τ_{allow}
9-30	50 mm	100 mm	150 mm	100 mm	8 mm	140 MPa
9-31	30 mm	50 mm	150 mm	50 mm	5 mm	140 MPa

Problems 9-30 to 9-31

9-32   In the design of weldments in torsion it is helpful to have a hierarchical perception of the relative efficiency of common patterns. For example, the weld-bead patterns shown in Table 9–1 can be ranked for desirability. Assume the space available is an $a \times a$ square. Use a formal figure of merit that is directly proportional to J and inversely proportional to the volume of weld metal laid down:

$$\text{fom} = \frac{J}{\text{vol}} = \frac{0.707hJ_u}{(h^2/2)l} = 1.414\frac{J_u}{hl}$$

A tactical figure of merit could omit the constant, that is, $\text{fom}' = J_u/(hl)$. Rank the six patterns of Table 9–1 from most to least efficient.

9-33   The space available for a weld-bead pattern subject to bending is $a \times a$. Place the patterns of Table 9–2 in hierarchical order of efficiency of weld metal placement to resist bending. A formal figure of merit can be directly proportion to I and inversely proportional to the volume of weld metal laid down:

$$\text{fom} = \frac{I}{\text{vol}} = \frac{0.707hI_u}{(h^2/2)l} = 1.414\frac{I_u}{hl}$$

The tactical figure of merit can omit the constant 1.414, that is, fom' $= I_u/(hl)$. Omit the patterns intended for T beams and I beams. Rank the remaining seven.

9–34 The attachment shown in the figure is made of 1018 HR steel 12 mm thick. The static force is 100 kN. The member is 75 mm wide. Specify the weldment (give the pattern, electrode number, type of weld, length of weld, and leg size).

Problem 9–34
Dimensions in millimeters.

9–35 The attachment shown carries a static bending load of 12 kN. The attachment length, l_1, is 225 mm. Specify the weldment (give the pattern, electrode number, type of weld, length of weld, and leg size).

Problem 9–35
Dimensions in millimeters.

9–36 The attachment in Prob. 9–35 has not had its length determined. The static force is 12 kN. Specify the weldment (give the pattern, electrode number, type of weld, length of bead, and leg size). Specify the attachment length.

9–37 A vertical column of A36 structural steel is 250 mm wide. An attachment has been designed to the point shown in the figure. The static load of 80 kN is applied, and the clearance a of 156 mm has to be equaled or exceeded. The attachment is also 1018 hot-rolled steel, to be made from 12-mm plate with weld-on bosses when all dimensions are known. Specify the weldment (give the pattern, electrode number, type of weld, length of weld bead, and leg size). Specify also the length l_1 for the attachment.

Problem 9–37

9–38 Write a computer program to assist with a task such as that of Prob. 9–37 with a rectangular weld-bead pattern for a torsional shear joint. In doing so solicit the force F, the clearance a, and the largest allowable shear stress. Then, as part of an iterative loop, solicit the dimensions b and d of the rectangle. These can be your design variables. Output all the parameters after the leg size has been determined by computation. In effect this will be your adequacy assessment when you stop iterating. Include the figure of merit $J_u/(hl)$ in the output. The fom and the leg size h with available width will give you a useful insight into the nature of this class of welds. Use your program to verify your solutions to Prob. 9–37.

9–39 Fillet welds in joints resisting bending are interesting in that they can be simpler than those resisting torsion. From Prob. 9–33 you learned that your objective is to place weld metal as far away from the weld-bead centroid as you can, but distributed in an orientation parallel to the x axis. Furthermore, placement on the top and bottom of the built-in end of a cantilever with rectangular cross section results in parallel weld beads, each element of which is in the ideal position. The object of this problem is to study the full weld bead and the interrupted weld-bead pattern. Consider the case of Fig. 9–17, p. 479, with $F = 40$ kN, the beam length is 250 mm, $b = 200$ mm, and $d = 200$ mm. For the second case, for the interrupted weld consider a centered gap of $b_1 = 50$ mm existing in the top and bottom welds. Study the two cases with $\tau_{all} = 88$ MPa. What do you notice about τ, σ, and τ_{max}? Compare the fom'.

9–40 For a rectangular weld-bead track resisting bending, develop the necessary equations to treat cases of vertical welds, horizontal welds, and weld-all-around patterns with depth d and width b and allowing central gaps in parallel beads of length b_1 and d_1. Do this by superposition of parallel tracks, vertical tracks subtracting out the gaps. Then put the two together for a rectangular weld bead with central gaps of length b_1 and d_1. Show that the results are

$$A = 1.414(b - b_1 + d - d_1)h$$

$$I_u = \frac{(b - b_1)d^2}{2} + \frac{d^3 - d_1^3}{6}$$

$$I = 0.707hI_u$$

$$l = 2(b - b_1) + 2(d - d_1)$$

$$\text{fom} = \frac{I_u}{hl}$$

9–41 Write a computer program based on the Prob. 9–40 protocol. Solicit the largest allowable shear stress, the force F, and the clearance a, as well as the dimensions b and d. Begin an iterative loop by soliciting b_1 and d_1. Either or both of these can be your design variables. Program to find the leg size corresponding to a shear-stress level at the maximum allowable at a corner. Output all your parameters including the figure of merit. Use the program to check any previous problems to which it is applicable. Play with it in a "what if" mode and learn from the trends in your parameters.

9–42 When comparing two different weldment patterns it is useful to observe the resistance to bending or torsion and the volume of weld metal deposited. The *measure of effectiveness,* defined as second moment of area divided by weld-metal volume, is useful. If a 150-mm by 200-mm section of a cantilever carries a static 40-kN bending load 250 mm from the weldment plane, with an allowable shear stress of 88 MPa realized, compare horizontal weldments with vertical weldments by determining the measure of effectiveness for each weld pattern. The horizontal beads are to be 150 mm long and the vertical beads, 200 mm long.

9–43 to 9–45   A 50-mm dia. steel bar is subjected to the loading indicated. Locate and estimate the maximum shear stress in the weld throat.

Problem Number	F	T
9–43	0	2 kN · m
9–44	8 kN	0
9–45	8 kN	2 kN · m

Problems 9–43 to 9–45

9–46   For Prob. 9–45, determine the weld size if the maximum allowable shear stress is 120 MPa.

9–47   Find the maximum shear stress in the throat of the weld metal in the figure.

Problem 9–47
Dimensions in millimeters.

9–48   The figure shows a welded steel bracket loaded by a static force F. Estimate the factor of safety if the allowable shear stress in the weld throat is 120 MPa.

Problem 9–48

9–49 The figure shows a formed sheet-steel bracket. Instead of securing it to the support with machine screws, welding all around the bracket support flange has been proposed. If the combined shear stress in the weld metal is limited to 6 MPa, estimate the total load W the bracket will support. The dimensions of the top flange are the same as the mounting flange.

Problem 9–49

Structural support is A26 structural steel, bracket is 1020 press cold-formed steel. The weld electrode is E6010.

9–50 Without bracing, a machinist can exert only about 400 N on a wrench or tool handle. The lever shown in the figure has $t = 12$ mm and $w = 50$ mm. We wish to specify the fillet-weld size to secure the lever to the tubular part at A. Both parts are of steel, and the shear stress in the weld throat should not exceed 20 MPa. Find a safe weld size.

Problem 9–50

9–51 Estimate the safe static load F for the weldment shown in the figure if an E6010 electrode is used and the design factor is to be 2. The steel members are 1015 hot-rolled steel. Use conventional analysis.

Problem 9–51

Dimensions in millimeters.

9–52   Brackets, such as the one shown, are used in mooring small watercraft. Failure of such brackets is usually caused by bearing pressure of the mooring clip against the side of the hole. Our purpose here is to get an idea of the static and dynamic margins of safety involved. We use a bracket 6 mm thick made of hot-rolled 1018 steel. We then assume wave action on the boat will create force F no greater than 5 kN.

(a) Identify the moment M that produces a shear stress on the throat resisting bending action with a "tension" at A and "compression" at C.

(b) Find the force component F_y that produces a shear stress at the throat resisting a "tension" throughout the weld.

(c) Find the force component F_x that produces an in-line shear throughout the weld.

(d) Find A, I_u, and I using Table 9–2, in part.

(e) Find the shear stress τ_1 at A due to F_y and M, the shear stress τ_2 due to F_x, and combine to find τ.

(f) Find the factor of safety guarding against shear yielding in the weldment.

(g) Find the factor of safety guarding against a static failure in the parent metal at the weld.

(h) Find the factor of safety guarding against a fatigue failure in the weld metal using a Gerber failure criterion.

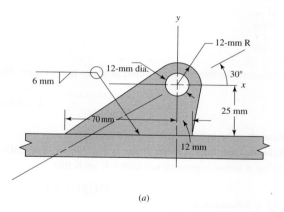

(a)

Problem 9–52

Small watercraft mooring bracket.

9–53   For the sake of perspective it is always useful to look at the matter of scale. Double all dimensions in Prob. 9–18 and find the allowable load. By what factor has it increased? First make a guess, then carry out the computation. Would you expect the same ratio if the load had been variable?

9–54   Hardware stores often sell plastic hooks that can be mounted on walls with pressure-sensitive adhesive foam tape. Two designs are shown in (a) and (b) of the figure. Indicate which one you would buy and why.

Problem 9–54

9–55   For a balanced double-lap joint cured at room temperature, Volkersen's equation simplifies to

$$\tau(x) = \frac{P\omega \cosh(\omega x)}{4b \sinh(\omega l/2)} = A_1 \cosh(\omega x)$$

(a) Show that the average stress $\bar{\tau}$ is $P/(2bl)$.
(b) Show that the largest shear stress is $P\omega/[4b \tanh(\omega l/2)]$.
(c) Define a stress-augmentation factor K such that

$$\tau(l/2) = K\bar{\tau}$$

and it follows that

$$K = \frac{P\omega}{4b \tanh(\omega l/2)} \frac{2bl}{P} = \frac{\omega l/2}{\tanh(\omega l/2)} = \frac{\omega l}{2} \frac{\exp(\omega l/2) + \exp(-\omega l/2)}{\exp(\omega l/2) - \exp(-\omega l/2)}$$

9–56   Program the shear-lag solution for the shear-stress state into your computer using Eq. (9–7). Determine the maximum shear stress for each of the following scenarios:

Part	E_a, GPa	t_o, mm	t_i, mm	E_o, GPa	E_i, GPa	h, mm
a	1.4	3	6	207	207	0.13
b	1.4	3	6	207	207	0.4
c	1.4	3	3	207	207	0.13
d	1.4	3	6	207	69	0.13

Provide plots of the actual stress distributions predicted by this analysis. You may omit thermal stresses from the calculations, assuming that the service temperature is similar to the stress-free temperature. If the allowable shear stress is 550 MPa and the load to be carried is 1.2 kN, estimate the respective factors of safety for each geometry. Let $l = 30$ mm and $b = 25$ mm.

10

Mechanical Springs

When a designer wants rigidity, negligible deflection is an acceptable approximation as long as it does not compromise function. Flexibility is sometimes needed and is often provided by metal bodies with cleverly controlled geometry. These bodies can exhibit flexibility to the degree the designer seeks. Such flexibility can be linear or nonlinear in relating deflection to load. These devices allow controlled application of force or torque; the storing and release of energy can be another purpose. Flexibility allows temporary distortion for access and the immediate restoration of function. Because of machinery's value to designers, springs have been intensively studied; moreover, they are mass-produced (and therefore low cost), and ingenious configurations have been found for a variety of desired applications. In this chapter we will discuss the more frequently used types of springs, their necessary parametric relationships, and their design.

In general, springs may be classified as wire springs, flat springs, or special-shaped springs, and there are variations within these divisions. Wire springs include helical springs of round or square wire, made to resist and deflect under tensile, compressive, or torsional loads. Flat springs include cantilever and elliptical types, wound motor- or clock-type power springs, and flat spring washers, usually called Belleville springs.

10–1 Stresses in Helical Springs

Figure 10–1a shows a round-wire helical compression spring loaded by the axial force F. We designate D as the *mean coil diameter* and d as the *wire diameter*. Isolate a section in the spring, as shown in Fig. 10–1b. For equilibrium, the isolated section contains a direct shear force F and a torsional moment $T = FD/2$.

The maximum shear stress in the wire may be computed by superposition of the direct shear stress given by Eq. (3–23), p. 103, with $V = F$ and the torsional shear stress given by Eq. (3–37), p. 115. The result is

$$\tau_{max} = \frac{Tr}{J} + \frac{F}{A} \qquad (a)$$

at the *inside* fiber of the spring. Substitution of $\tau_{max} = \tau$, $T = FD/2$, $r = d/2$, $J = \pi d^4/32$, and $A = \pi d^2/4$ gives

$$\tau = \frac{8FD}{\pi d^3} + \frac{4F}{\pi d^2} \qquad (b)$$

Figure 10–1

(*a*) Axially loaded helical spring; (*b*) free-body diagram showing that the wire is subjected to a direct shear and a torsional shear.

(*a*)

$T = FD/2$

(*b*)

Now we define the *spring index*

$$C = \frac{D}{d} \tag{10-1}$$

which is a measure of coil curvature. The preferred value of C ranges from 4 to 12.[1] With this relation, Eq. (*b*) can be rearranged to give

$$\tau = K_s \frac{8FD}{\pi d^3} \tag{10-2}$$

where K_s is a *shear stress-correction factor* and is defined by the equation

$$K_s = \frac{2C + 1}{2C} \tag{10-3}$$

The use of square or rectangular wire is not recommended for springs unless space limitations make it necessary. Springs of special wire shapes are not made in large quantities, unlike those of round wire; they have not had the benefit of refining development and hence may not be as strong as springs made from round wire. When space is severely limited, the use of nested round-wire springs should always be considered. They may have an economical advantage over the special-section springs, as well as a strength advantage.

10-2   The Curvature Effect

Equation (10-2) is based on the wire being straight. However, the curvature of the wire causes a localized increase in stress on the inner surface of the coil, which can be accounted for with a curvature factor. This factor can be applied in the same way as a stress concentration factor. For static loading, the curvature factor is normally neglected because any localized yielding leads to localized strain strengthening. For fatigue applications, the curvature factor should be included.

Unfortunately, it is necessary to find the curvature factor in a roundabout way. The reason for this is that the published equations also include the effect of the direct shear stress. Suppose K_s in Eq. (10-2) is replaced by another K factor, which corrects for both curvature and direct shear. Then this factor is given by either of the equations

$$K_W = \frac{4C - 1}{4C - 4} + \frac{0.615}{C} \tag{10-4}$$

$$K_B = \frac{4C + 2}{4C - 3} \tag{10-5}$$

The first of these is called the *Wahl factor,* and the second, the *Bergsträsser factor.*[2] Since the results of these two equations differ by the order of 1 percent, Eq. (10-5) is preferred. The curvature correction factor can now be obtained by canceling out

---

[1]*Design Handbook: Engineering Guide to Spring Design,* Associated Spring-Barnes Group Inc., Bristol, CT, 1987.

[2]Cyril Samónov, "Some Aspects of Design of Helical Compression Springs," *Int. Symp. Design and Synthesis,* Tokyo, 1984.

the effect of the direct shear. Thus, using Eq. (10–5) with Eq. (10–3), the curvature correction factor is found to be

$$K_c = \frac{K_B}{K_s} = \frac{2C(4C + 2)}{(4C - 3)(2C + 1)} \tag{10–6}$$

Now, K_s, K_B or K_W, and K_c are simply stress-correction factors applied multiplicatively to Tr/J at the critical location to estimate a particular stress. There is *no* stress-concentration factor. In this book we will use

$$\tau = K_B \frac{8FD}{\pi d^3} \tag{10–7}$$

to predict the largest shear stress.

10–3 Deflection of Helical Springs

The deflection-force relations are quite easily obtained by using Castigliano's theorem. The total strain energy for a helical spring is composed of a torsional component and a shear component. From Eqs. (4–18) and (4–20), p. 176, the strain energy is

$$U = \frac{T^2 l}{2GJ} + \frac{F^2 l}{2AG} \tag{a}$$

Substituting $T = FD/2$, $l = \pi DN$, $J = \pi d^4/32$, and $A = \pi d^2/4$ results in

$$U = \frac{4F^2 D^3 N}{d^4 G} + \frac{2F^2 DN}{d^2 G} \tag{b}$$

where $N = N_a$ = number of active coils. Then using Castigliano's theorem, Eq. (4–26), p. 179, to find total deflection y gives

$$y = \frac{\partial U}{\partial F} = \frac{8FD^3 N}{d^4 G} + \frac{4FDN}{d^2 G} \tag{c}$$

Since $C = D/d$, Eq. (c) can be rearranged to yield

$$y = \frac{8FD^3 N}{d^4 G} \left(1 + \frac{1}{2C^2}\right) \approx \frac{8FD^3 N}{d^4 G} \tag{10–8}$$

The spring rate, also called the *scale* of the spring, is $k = F/y$, and so

$$k \approx \frac{d^4 G}{8D^3 N} \tag{10–9}$$

10–4 Compression Springs

The four types of ends generally used for compression springs are illustrated in Fig. 10–2. The terminal end of each spring is only shown on the right-end of the spring. A spring with *plain ends* has a noninterrupted helicoid; the ends are the same as if a long spring had been cut into sections. A spring with plain ends that are *squared* or *closed* is obtained by deforming the ends to a zero-degree helix angle. Springs should always be both squared and ground for important applications, because a better transfer of the load is obtained.

Table 10–1 shows how the type of end used affects the number of coils and the spring length.[3] Note that the digits 0, 1, 2, and 3 appearing in Table 10–1 are often

---

[3]For a thorough discussion and development of these relations, see Cyril Samónov, "Computer-Aided Design of Helical Compression Springs," ASME paper No. 80-DET-69, 1980.

Figure 10–2

Types of ends for compression springs: (a) both ends plain; (b) both ends squared; (c) both ends squared and ground; (d) both ends plain and ground.

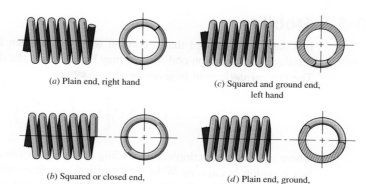

(a) Plain end, right hand

(c) Squared and ground end, left hand

(b) Squared or closed end, right hand

(d) Plain end, ground, left hand

Table 10–1

Formulas for the Dimensional Characteristics of Compression-Springs. (N_a = Number of Active Coils)

Source: From *Design Handbook,* 1987, p. 32. Courtesy of Associated Spring.

		Type of Spring Ends		
Term	**Plain**	**Plain and Ground**	**Squared or Closed**	**Squared and Ground**
End coils, N_e	0	1	2	2
Total coils, N_t	N_a	$N_a + 1$	$N_a + 2$	$N_a + 2$
Free length, L_0	$pN_a + d$	$p(N_a + 1)$	$pN_a + 3d$	$pN_a + 2d$
Solid length, L_s	$d(N_t + 1)$	dN_t	$d(N_t + 1)$	dN_t
Pitch, p	$(L_0 - d)/N_a$	$L_0/(N_a + 1)$	$(L_0 - 3d)/N_a$	$(L_0 - 2d)/N_a$

used without question. *Some of these need closer scrutiny as they may not be integers.* This depends on how a springmaker forms the ends. Forys[4] pointed out that squared and ground ends give a solid length L_s of

$$L_s = (N_t - a)d$$

where a varies, with an average of 0.75, so the entry dN_t in Table 10–1 may be overstated. The way to check these variations is to take springs from a particular springmaker, close them solid, and measure the solid height. Another way is to look at the spring and count the wire diameters in the solid stack.

Set removal or *presetting* is a process used in the manufacture of compression springs to induce useful residual stresses. It is done by making the spring longer than needed and then compressing it to its solid height. This operation *sets* the spring to the required final free length and, since the torsional yield strength has been exceeded, induces residual stresses opposite in direction to those induced in service. Springs to be preset should be designed so that 10 to 30 percent of the initial free length is removed during the operation. If the stress at the solid height is greater than 1.3 times the torsional yield strength, distortion may occur. If this stress is much less than 1.1 times, it is difficult to control the resulting free length.

Set removal increases the strength of the spring and so is especially useful when the spring is used for energy-storage purposes. However, set removal should not be used when springs are subject to fatigue.

---

[4]Edward L. Forys, "Accurate Spring Heights," *Machine Design,* vol. 56, no. 2, January 26, 1984.

10–5   Stability

In Chap. 4 we learned that a column will buckle when the load becomes too large. Similarly, compression coil springs may buckle when the deflection becomes too large. The critical deflection is given by the equation

$$y_{cr} = L_0 C_1' \left[1 - \left(1 - \frac{C_2'}{\lambda_{eff}^2} \right)^{1/2} \right] \tag{10-10}$$

where y_{cr} is the deflection corresponding to the onset of instability. Samónov[5] states that this equation is cited by Wahl[6] and verified experimentally by Haringx.[7] The quantity λ_{eff} in Eq. (10–10) is the *effective slenderness ratio* and is given by the equation

$$\lambda_{eff} = \frac{\alpha L_0}{D} \tag{10-11}$$

C_1' and C_2' are dimensionless elastic constants defined by the equations

$$C_1' = \frac{E}{2(E - G)}$$

$$C_2' = \frac{2\pi^2 (E - G)}{2G + E}$$

Equation (10–11) contains the *end-condition constant* α. This depends upon how the ends of the spring are supported. Table 10–2 gives values of α for usual end conditions. Note how closely these resemble the end conditions for columns.

Absolute stability occurs when, in Eq. (10–10), the term C_2'/λ_{eff}^2 is greater than unity. This means that the condition for absolute stability is that

$$L_0 < \frac{\pi D}{\alpha} \left[\frac{2(E - G)}{2G + E} \right]^{1/2} \tag{10-12}$$

Table 10–2

End-Condition Constants α for Helical Compression Springs*

End Condition	Constant α
Spring supported between flat parallel surfaces (fixed ends)	0.5
One end supported by flat surface perpendicular to spring axis (fixed); other end pivoted (hinged)	0.707
Both ends pivoted (hinged)	1
One end clamped; other end free	2

*Ends supported by flat surfaces must be squared and ground.

[5]Cyril Samónov "Computer-Aided Design," op. cit.

[6]A. M. Wahl, *Mechanical Springs,* 2d ed., McGraw-Hill, New York, 1963.

[7]J. A. Haringx, "On Highly Compressible Helical Springs and Rubber Rods and Their Application for Vibration-Free Mountings," I and II, *Philips Res. Rep.,* vol. 3, December 1948, pp. 401–449, and vol. 4, February 1949, pp. 49–80.

For steels, this turns out to be

$$L_0 < 2.63\frac{D}{\alpha}$$ (10–13)

For squared and ground ends supported between flat parallel surfaces, $\alpha = 0.5$ and $L_0 < 5.26D$.

10–6 Spring Materials

Springs are manufactured either by hot- or cold-working processes, depending upon the size of the material, the spring index, and the properties desired. In general, pre-hardened wire should not be used if $D/d < 4$ or if d > 6 mm. Winding of the spring induces residual stresses through bending, but these are normal to the direction of the torsional working stresses in a coil spring. Quite frequently in spring manufacture, they are relieved, after winding, by a mild thermal treatment.

A great variety of spring materials are available to the designer, including plain carbon steels, alloy steels, and corrosion-resisting steels, as well as nonferrous materials such as phosphor bronze, spring brass, beryllium copper, and various nickel alloys. Descriptions of the most commonly used steels will be found in Table 10–3. The UNS steels listed in Appendix A should be used in designing hot-worked, heavy-coil springs, as well as flat springs, leaf springs, and torsion bars.

Spring materials may be compared by an examination of their tensile strengths; these vary so much with wire size that they cannot be specified until the wire size is known. The material and its processing also, of course, have an effect on tensile strength. It turns out that the graph of tensile strength versus wire diameter is almost a straight line for some materials when plotted on log-log paper. Writing the equation of this line as

$$S_{ut} = \frac{A}{d^m}$$ (10–14)

furnishes a good means of estimating minimum tensile strengths when the intercept A and the slope m of the line are known. Values of these constants have been worked out from recent data and are given for strengths in units of kpsi and MPa in Table 10–4. In Eq. (10–14) when d is measured in millimeters, then A is in MPa · mm^m and when d is measured in inches, then A is in kpsi · in^m.

Although the torsional yield strength is needed to design the spring and to analyze the performance, spring materials customarily are tested only for tensile strength—perhaps because it is such an easy and economical test to make. A very rough estimate of the torsional yield strength can be obtained by assuming that the tensile yield strength is between 60 and 90 percent of the tensile strength. Then the distortion-energy theory can be employed to obtain the torsional yield strength ($S_{sy} = 0.577S_y$). This approach results in the range

$$0.35S_{ut} \le S_{sy} \le 0.52S_{ut}$$ (10–15)

for steels.

For wires listed in Table 10–5, the maximum allowable shear stress in a spring can be seen in column 3. Music wire and hard-drawn steel spring wire have a low end of range $S_{sy} = 0.45S_{ut}$. Valve spring wire, Cr-Va, Cr-Si, and other (not shown) hardened and tempered carbon and low-alloy steel wires as a group have $S_{sy} \ge 0.50S_{ut}$. Many nonferrous materials (not shown) as a group have $S_{sy} \ge 0.35S_{ut}$. In view of this,

Table 10–3

High-Carbon and Alloy Spring Steels

Source: From Harold C. R. Carlson, "Selection and Application of Spring Materials," *Mechanical Engineering,* vol. 78, 1956, pp. 331–334.

Name of Material	Similar Specifications	Description
Music wire, 0.80–0.95C	UNS G10850 AISI 1085 ASTM A228-51	This is the best, toughest, and most widely used of all spring materials for small springs. It has the highest tensile strength and can withstand higher stresses under repeated loading than any other spring material. Available in diameters 0.12 to 3 mm (0.005 to 0.125 in). Do not use above 120°C (250°F) or at subzero temperatures.
Oil-tempered wire, 0.60–0.70C	UNS G10650 AISI 1065 ASTM 229-41	This general-purpose spring steel is used for many types of coil springs where the cost of music wire is prohibitive and in sizes larger than available in music wire. Not for shock or impact loading. Available in diameters 3 to 12 mm (0.125 to 0.5000 in), but larger and smaller sizes may be obtained. Not for use above 180°C (350°F) or at subzero temperatures.
Hard-drawn wire, 0.60–0.70C	UNS G10660 AISI 1066 ASTM A227-47	This is the cheapest general-purpose spring steel and should be used only where life, accuracy, and deflection are not too important. Available in diameters 0.8 to 12 mm (0.031 to 0.500 in). Not for use above 120°C (250°F) or at subzero temperatures.
Chrome-vanadium	UNS G61500 AISI 6150 ASTM 231-41	This is the most popular alloy spring steel for conditions involving higher stresses than can be used with the high-carbon steels and for use where fatigue resistance and long endurance are needed. Also good for shock and impact loads. Widely used for aircraft-engine valve springs and for temperatures to 220°C (425°F). Available in annealed or pretempered sizes 0.8 to 12 mm (0.031 to 0.500 in) in diameter.
Chrome-silicon	UNS G92540 AISI 9254	This alloy is an excellent material for highly stressed springs that require long life and are subjected to shock loading. Rockwell hardnesses of C50 to C53 are quite common, and the material may be used up to 250°C (475°F). Available from 0.8 to 12 mm (0.031 to 0.500 in) in diameter.

Table 10–4

Constants A and m of $S_{ut} = A/d^m$ for Estimating Minimum Tensile Strength of Common Spring Wires

Source: From *Design Handbook,* 1987, p. 19. Courtesy of Associated Spring.

Material	ASTM No.	Exponent m	Diameter, in	A, kpsi · in^m	Diameter, mm	A, MPa · mm^m	Relative Cost of Wire
Music wire*	A228	0.145	0.004–0.256	201	0.10–6.5	2211	2.6
OQ&T wire[†]	A229	0.187	0.020–0.500	147	0.5–12.7	1855	1.3
Hard-drawn wire[‡]	A227	0.190	0.028–0.500	140	0.7–12.7	1783	1.0
Chrome-vanadium wire[§]	A232	0.168	0.032–0.437	169	0.8–11.1	2005	3.1
Chrome-silicon wire[‖]	A401	0.108	0.063–0.375	202	1.6–9.5	1974	4.0
302 Stainless wire[#]	A313	0.146	0.013–0.10	169	0.3–2.5	1867	7.6–11
		0.263	0.10–0.20	128	2.5–5	2065	
		0.478	0.20–0.40	90	5–10	2911	
Phosphor-bronze wire**	B159	0	0.004–0.022	145	0.1–0.6	1000	8.0
		0.028	0.022–0.075	121	0.6–2	913	
		0.064	0.075–0.30	110	2–7.5	932	

*Surface is smooth, free of defects, and has a bright, lustrous finish.

[†]Has a slight heat-treating scale which must be removed before plating.

[‡]Surface is smooth and bright with no visible marks.

[§]Aircraft-quality tempered wire, can also be obtained annealed.

[‖]Tempered to Rockwell C49, but may be obtained untempered.

[#]Type 302 stainless steel.

**Temper CA510.

Joerres[8] uses the maximum allowable torsional stress for static application shown in Table 10–6. For specific materials for which you have torsional yield information use this table as a guide. Joerres provides set-removal information in Table 10–6, that $S_{sy} \geq 0.65S_{ut}$ increases strength through cold work, but at the cost of an additional operation by the springmaker. Sometimes the additional operation can be done by the manufacturer during assembly. Some correlations with carbon steel springs show that the tensile yield strength of spring wire in torsion can be estimated from $0.75S_{ut}$. The corresponding estimate of the yield strength in shear based on distortion energy theory is $S_{sy} = 0.577(0.75)S_{ut} = 0.433S_{ut} \approx 0.45S_{ut}$. Samónov discusses the problem of allowable stress and shows that

$$S_{sy} = \tau_{all} = 0.56S_{ut} \qquad (10\text{–}16)$$

for high-tensile spring steels, which is close to the value given by Joerres for hardened alloy steels. He points out that this value of allowable stress is specified by Draft Standard 2089 of the German Federal Republic when Eq. (10–2) is used without stress-correction factor.

---

[8]Robert E. Joerres, "Springs," Chap. 6 in Joseph E. Shigley, Charles R. Mischke, and Thomas H. Brown, Jr. (eds.), *Standard Handbook of Machine Design,* 3rd ed., McGraw-Hill, New York, 2004.

Table 10–5

Mechanical Properties of Some Spring Wires

Material	Elastic Limit, Percent of S_{ut} Tension	Torsion	Diameter d, in	mm	E Mpsi	GPa	G Mpsi	GPa
Music wire A228	65–75	45–60	<0.032	<8	29.5	203.4	12.0	82.7
			0.033–0.063	0.8–1.6	29.0	200	11.85	81.7
			0.064–0.125	1.61–3	28.5	196.5	11.75	81.0
			>0.125	>3	28.0	193	11.6	80.0
HD spring A227	60–70	45–55	<0.032	<8	28.8	198.6	11.7	80.7
			0.033–0.063	0.8–1.6	28.7	197.9	11.6	80.0
			0.064–0.125	1.61–3	28.6	197.2	11.5	79.3
			>0.125	>3	28.5	196.5	11.4	78.6
Oil tempered A239	85–90	45–50			28.5	196.5	11.2	77.2
Valve spring A230	85–90	50–60			29.5	203.4	11.2	77.2
Chrome-vanadium A231	88–93	65–75			29.5	203.4	11.2	77.2
A232	88–93				29.5	203.4	11.2	77.2
Chrome-silicon A401	85–93	65–75			29.5	203.4	11.2	77.2
Stainless steel								
A313*	65–75	45–55			28	193	10	69.0
17-7PH	75–80	55–60			29.5	208.4	11	75.8
414	65–70	42–55			29	200	11.2	77.2
420	65–75	45–55			29	200	11.2	77.2
431	72–76	50–55			30	206	11.5	79.3
Phosphor-bronze B159	75–80	45–50			15	103.4	6	41.4
Beryllium-copper B197	70	50			17	117.2	6.5	44.8
	75	50–55			19	131	7.3	50.3
Inconel alloy X-750	65–70	40–45			31	213.7	11.2	77.2

*Also includes 302, 304, and 316.

Note: See Table 10–6 for allowable torsional stress design values.

Table 10–6

Maximum Allowable Torsional Stresses for Helical Compression Springs in Static Applications

Source: Robert E. Joerres, "Springs," Chap. 6 in Joseph E. Shigley, Charles R. Mischke, and Thomas H. Brown, Jr. (eds.), *Standard Handbook of Machine Design,* 3rd ed., McGraw-Hill, New York, 2004.

Material	Maximum Percent of Tensile Strength Before Set Removed (includes K_W or K_B)	After Set Removed (includes K_s)
Music wire and cold-drawn carbon steel	45	60–70
Hardened and tempered carbon and low-alloy steel	50	65–75
Austenitic stainless steels	35	55–65
Nonferrous alloys	35	55–65

EXAMPLE 10-1   A helical compression spring is made of no. 16 music wire. The outside diameter of the spring is 11 mm. The ends are squared and there are $12\frac{1}{2}$ total turns.
(a) Estimate the torsional yield strength of the wire.
(b) Estimate the static load corresponding to the yield strength.
(c) Estimate the scale of the spring.
(d) Estimate the deflection that would be caused by the load in part (b).
(e) Estimate the solid length of the spring.
(f) What length should the spring be to ensure that when it is compressed solid and then released, there will be no permanent change in the free length?
(g) Given the length found in part (f), is buckling a possibility?
(h) What is the pitch of the body coil?

Solution   (a) From Table A–28, the wire diameter is $d = 0.94$ mm. From Table 10–4, we find $A = 2211$ MPa·mm^m and $m = 0.145$. Therefore, from Eq. (10–14)

$$S_{ut} = \frac{A}{d^m} = \frac{2211}{0.94^{0.145}} = 2231 \text{ MPa}$$

Then, from Table 10–6,

Answer
$$S_{sy} = 0.45 S_{ut} = 0.45(2231) = 1004 \text{ MPa}$$

(b) The mean spring coil diameter is $D = 11 - 0.94 = 10.06$ mm, and so the spring index is $C = 10.06/0.94 = 10.7$. Then, from Eq. (10–6),

$$K_B = \frac{4C + 2}{4C - 3} = \frac{4(10.7) + 2}{4(10.7) - 3} = 1.126$$

Now rearrange Eq. (10–7) replacing K_s and τ with K_B and S_{ys}, respectively, and solve for F:

Answer
$$F = \frac{\pi d^3 S_{sy}}{8 K_B D} = \frac{\pi (0.94^3) 1004}{8(1.126) 10.06} = 31 \text{ N}$$

(c) From Table 10–1, $N_a = 12.5 - 2 = 10.5$ turns. In Table 10–5, $G = 81\,700$ MPa, and the scale of the spring is found to be, from Eq. (10–9),

$$k = \frac{d^4 G}{8 D^3 N_a} = \frac{0.94^4 (81\,700)}{8(10.06^3) 10.5} = 0.9 \text{ N/mm}$$

Answer

(d)
$$y = \frac{F}{k} = \frac{31}{0.9} = 34.4 \text{ mm}$$

Answer

(e) From Table 10–1,

$$L_s = (N_t + 1)d = (12.5 + 1)0.94 = 12.7 \text{ mm}$$

Answer

(f)
$$L_0 = y + L_s = 34.4 + 12.7 = 47.1 \text{ mm}$$

Answer   (g) To avoid buckling, Eq. (10–13) and Table 10–2 give

$$L_0 < 2.63 \frac{D}{\alpha} = 2.63 \frac{10.06}{0.5} = 52.9 \text{ mm}$$

Answer

Mathematically, a free length of 47.1 mm is less than 52.9 mm, and buckling is unlikely. However, the forming of the ends will control how close α is to 0.5. This has to be investigated and an inside rod or exterior tube or hole may be needed.

(*h*) Finally, from Table 10–1, the pitch of the body coil is

$$p = \frac{L_0 - 3d}{N_a} = \frac{47.1 - 3(0.94)}{10.5} = 4.4 \text{ mm}$$

10–7 Helical Compression Spring Design for Static Service

The preferred range of the spring index is $4 \leq C \leq 12$, with the lower indexes being more difficult to form (because of the danger of surface cracking) and springs with higher indexes tending to tangle often enough to require individual packing. This can be the first item of the design assessment. The recommended range of active turns is $3 \leq N_a \leq 15$. To maintain linearity when a spring is about to close, it is necessary to avoid the gradual touching of coils (due to nonperfect pitch). A helical coil spring force-deflection characteristic is ideally linear. Practically, it is nearly so, but not at each end of the force-deflection curve. The spring force is not reproducible for very small deflections, and near closure, nonlinear behavior begins as the number of active turns diminishes as coils begin to touch. The designer confines the spring's operating point to the central 75 percent of the curve between no load, $F = 0$, and closure, $F = F_s$. Thus, the maximum operating force should be limited to $F_{\max} \leq \frac{7}{8}F_s$. Defining the fractional overrun to closure as ξ, where

$$F_s = (1 + \xi)F_{\max} \tag{10–17}$$

it follows that

$$F_s = (1 + \xi)F_{\max} = (1 + \xi)\left(\frac{7}{8}\right)F_s$$

From the outer equality $\xi = 1/7 = 0.143 \approx 0.15$. Thus, it is recommended that $\xi \geq 0.15$.

In addition to the relationships and material properties for springs, we now have some recommended design conditions to follow, namely:

$$4 \leq C \leq 12 \tag{10–18}$$

$$3 \leq N_a \leq 15 \tag{10–19}$$

$$\xi \geq 0.15 \tag{10–20}$$

$$n_s \geq 1.2 \tag{10–21}$$

where n_s is the factor of safety at closure (solid height).

When considering designing a spring for high volume production, the figure of merit can be the cost of the wire from which the spring is wound. The fom would be proportional to the relative material cost, weight density, and volume:

$$\text{fom} = -(\text{relative material cost})\frac{\gamma\pi^2 d^2 N_t D}{4} \tag{10–22}$$

For comparisons between steels, the specific weight γ can be omitted.

Spring design is an open-ended process. There are many decisions to be made, and many possible solution paths as well as solutions. In the past, charts, nomographs, and "spring design slide rules" were used by many to simplify the spring design problem. Today, the computer enables the designer to create programs in many different formats—direct programming, spreadsheet, MATLAB, etc. Commercial programs are also available.[9] There are almost as many ways to create a spring-design program as there are programmers. Here, we will suggest one possible design approach.

Design Strategy

Make the a priori decisions, with hard-drawn steel wire the first choice (relative material cost is 1.0). Choose a wire size d. With all decisions made, generate a column of parameters: d, D, C, OD or ID, N_a, L_s, L_0, $(L_0)_{cr}$, n_s, and fom. By incrementing wire sizes available, we can scan the table of parameters and apply the design recommendations by inspection. After wire sizes are eliminated, choose the spring design with the highest figure of merit. This will give the optimal design despite the presence of a discrete design variable d and aggregation of equality and inequality constraints. The column vector of information can be generated by using the flowchart displayed in Fig. 10–3. It is general enough to accommodate to the situations of as-wound and

Figure 10–3

Helical coil compression spring design flowchart for static loading.

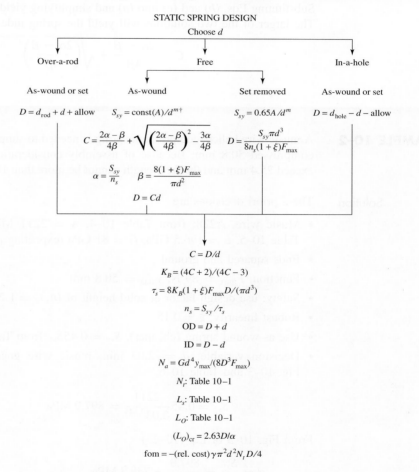

STATIC SPRING DESIGN
Choose d

Over-a-rod	Free	In-a-hole
As-wound or set	As-wound	Set removed	As-wound or set

$D = d_{rod} + d + \text{allow}$ $S_{sy} = \text{const}(A)/d^{m\dagger}$ $S_{sy} = 0.65A/d^m$ $D = d_{hole} - d - \text{allow}$

$$C = \frac{2\alpha - \beta}{4\beta} + \sqrt{\left(\frac{2\alpha - \beta}{4\beta}\right)^2 - \frac{3\alpha}{4\beta}} \qquad D = \frac{S_{sy}\pi d^3}{8n_s(1 + \xi)F_{max}}$$

$$\alpha = \frac{S_{sy}}{n_s} \qquad \beta = \frac{8(1 + \xi)F_{max}}{\pi d^2}$$

$$D = Cd$$

$$C = D/d$$

$$K_B = (4C + 2)/(4C - 3)$$

$$\tau_s = 8K_B(1 + \xi)F_{max}D/(\pi d^3)$$

$$n_s = S_{sy}/\tau_s$$

$$\text{OD} = D + d$$

$$\text{ID} = D - d$$

$$N_a = Gd^4 y_{max}/(8D^3 F_{max})$$

N_t: Table 10–1

L_s: Table 10–1

L_0: Table 10–1

$$(L_0)_{cr} = 2.63D/\alpha$$

$$\text{fom} = -(\text{rel. cost})\gamma\pi^2 d^2 N_t D/4$$

Print or display: d, D, C, OD, ID, N_a, N_t, L_s, L_0, $(L_0)_{cr}$, n_s, fom
Build a table, conduct design assessment by inspection
Eliminate infeasible designs by showing active constraints
Choose among satisfactory designs using the figure of merit

[†]const is found from Table 10–6.

---

[9]For example, see *Advanced Spring Design*, a program developed jointly between the Spring Manufacturers Institute (SMI), www.smihq.org, and Universal Technical Systems, Inc. (UTS), www.uts.com.

set-removed springs, operating over a rod, or in a hole free of rod or hole. In as-wound springs the controlling equation must be solved for the spring index as follows. From Eq. (10–3) with $\tau = S_{sy}/n_s$, $C = D/d$, K_B from Eq. (10–6), and Eq. (10–17),

$$\frac{S_{sy}}{n_s} = K_B \frac{8F_s D}{\pi d^3} = \frac{4C + 2}{4C - 3}\left[\frac{8(1 + \xi)F_{max}C}{\pi d^2}\right] \qquad (a)$$

Let

$$\alpha = \frac{S_{sy}}{n_s} \qquad (b)$$

$$\beta = \frac{8(1 + \xi)F_{max}}{\pi d^2} \qquad (c)$$

Substituting Eqs. (b) and (c) into (a) and simplifying yields a quadratic equation in C. The larger of the two solutions will yield the spring index

$$C = \frac{2\alpha - \beta}{4\beta} + \sqrt{\left(\frac{2\alpha - \beta}{4\beta}\right)^2 - \frac{3\alpha}{4\beta}} \qquad (10–23)$$

EXAMPLE 10–2

A music wire helical compression spring is needed to support an 89-N load after being compressed 50.8 mm. Because of assembly considerations the solid height cannot exceed 25.4 mm and the free length cannot be more than 101.6 mm. Design the spring.

Solution

The a priori decisions are

- Music wire, A228; from Table 10–4, $A = 2211$ MPa·mm^m; $m = 0.145$; from Table 10–5, $E = 196.5$ GPa, $G = 81$ GPa (expecting $d > 1.61$ mm)
- Ends squared and ground
- Function: $F_{max} = 89$ N, $y_{max} = 50.8$ mm
- Safety: use design factor at solid height of $(n_s)_d = 1.2$
- Robust linearity: $\xi = 0.15$
- Use as-wound spring (cheaper), $S_{sy} = 0.45 S_{ut}$ from Table 10–6
- Decision variable: $d = 2.03$ mm, music wire gage #30, Table A–26. From Fig. 10–3 and Table 10–6,

$$S_{sy} = 0.45\frac{2211}{2.03^{0.145}} = 897.9 \text{ MPa}$$

From Fig. 10–3 or Eq. (10–23)

$$\alpha = \frac{S_{sy}}{n_s} = \frac{897.9}{1.2} = 748.3 \text{ MPa}$$

$$\beta = \frac{8(1 + \xi)F_{max}}{\pi d^2} = \frac{8(1 + 0.15)89}{\pi(2.03^2)} = 63.2 \text{ MPa}$$

$$C = \frac{2(748.3) - 63.2}{4(63.2)} + \sqrt{\left[\frac{2(748.3) - 63.2}{4(63.2)}\right]^2 - \frac{3(748.3)}{4(63.2)}} = 10.5$$

Continuing with Fig. 10–3:

$$D = Cd = 10.5(2.03) = 21.3 \text{ mm}$$

$$K_B = \frac{4(10.5) + 2}{4(10.5) - 3} = 1.128$$

$$\tau_s = 1.128 \frac{8(1 + 0.15)89(21.3)}{\pi(2.03)^3} = 748 \text{ MPa}$$

$$n_s = \frac{897.9}{748} = 1.2$$

$$OD = 21.3 + 2.03 = 23.3 \text{ mm}$$

$$N_a = \frac{2.03^4(81\,000)50.8}{8(21.3)^3 89} = 10.16 \text{ turns}$$

$$N_t = 10.16 + 2 = 12.16 \text{ total turns}$$

$$L_s = 2.03(12.16) = 24.3 \text{ mm}$$

$$L_0 = 24.3 + (1 + 0.15)50.8 = 82.7 \text{ mm}$$

$$(L)_{cr} = 2.63(21.3/0.5) = 112 \text{ mm}$$

$$\text{fom} = -2.6 \frac{\pi^2(2.03)^2 12.16(21.3)}{4(25.4)^3} = -0.417$$

Repeat the above for other wire diameters and form a table (easily accomplished with a spreadsheet program):

d:	1.6	1.7	1.8	1.9	2.03	2.1	2.3	2.4
D	9.9	12.2	14.6	17.5	21.3	25.2	31.1	36.0
C	6.2	7.2	8.1	9.2	10.5	12.0	13.5	15.0
OD	11.5	13.8	16.4	19.3	23.3	27.8	32.8	38.4
N_a	39.1	26.9	19.3	14.2	10.2	7.3	5.4	4.1
L_s	65.2	48.8	38.1	30.7	24.3	19.9	16.8	14.6
L_0	123.8	107.3	96.6	89.1	82.7	78.3	75.2	73
$(L_0)_{cr}$	52.1	63.7	76.9	91.5	112	160.5	161.0	189.9
n_s	1.2	1.2	1.2	1.2	1.2	1.2	1.2	1.2
fom	−0.41	−0.40	−0.40	−0.404	−0.42	−0.44	−0.47	−0.51

Now examine the table and perform the adequacy assessment. The constraint $3 \le N_a \le 15$ rules out wire diameters less than 1.9 mm. The spring index constraint $4 \le C \le 12$ rules out diameters larger than 2.1 mm. The $L_s \le 1$ constraint rules out diameters less than 0.203 mm. The $L_0 \le 4$ constraint rules out diameters less than 1.8 mm. The buckling criterion rules out free lengths longer than $(L_0)_{cr}$, which rules out diameters less than 1.9 mm. The factor of safety n_s is exactly 1.20 because the mathematics forced it. Had the spring been in a hole or over a rod, the helix diameter would be chosen without reference to $(n_s)_d$. The result is that there are only two springs in the feasible domain, one with a wire diameter of 0.203 mm and the other with a wire diameter of 2.1 mm. The figure of merit decides and the decision is the design with 0.203 mm wire diameter.

Having designed a spring, will we have it made to our specifications? Not necessarily. There are vendors who stock literally thousands of music wire compression springs. By browsing their catalogs, we will usually find several that are close. Maximum deflection and maximum load are listed in the display of characteristics. Check to see if this allows soliding without damage. Often it does not. Spring rates may only be close. At the very least this situation allows a small number of springs to be ordered "off the shelf" for testing. The decision often hinges on the economics of special order versus the acceptability of a close match.

Spring design is not a closed-form approach and requires iteration. Example 10–2 provided an iterative approach to spring design for static service by first selecting the wire diameter. The diameter selection can be rather arbitrary. In the next example, we will first select a value for the spring index C, which is within the recommended range.

EXAMPLE 10–3

Design a compression spring with plain ends using hard-drawn wire. The deflection is to be 56 mm when the force is 90 N and to close solid when the force is 24 lbf. Upon closure, use a design factor of 1.2 guarding against yielding. Select the smallest gauge W&M (Washburn & Moen) wire.

The extension spring shown in Figure 10.4 has full-twisted loop ends. The material is AISI 1065 OQ&T wire. The spring has 84 coils and is close-wound with a preload of 85 N.

(a) Find the closed length of the spring.
(b) Find the torsional stress in the spring corresponding to the preload.
(c) Estimate the spring rate.
(d) What load would cause permanent deformation?
(e) What is the spring deflection corresponding to the load found in part d?

Solution

Given: $N_b = 84$ coils, $F_i = 85$ N, OQ&T steel, OD $= 38$ mm, $d = 4$ mm, $D = 38 - 4 = 34$ mm

(a) Eq. (10–39):

$$L_0 = 2(D - d) + (N_b + 1)d$$

Answer

$$= 2(34 - 4) + (84 + 1)(4) = 400 \text{ mm}$$

or

$$2d + L_0 = 2(4) + 400 = 408 \text{ mm overall.}$$

(b)

$$C = \frac{D}{d} = \frac{34}{4} = 8.5$$

$$K_B = \frac{4(8.5) + 2}{4(8.5) - 3} = 1.161$$

Answer
$$\tau_i = K_B \left[\frac{8F_i D}{\pi d^3}\right] = 1.161 \left[\frac{8(85)(34)}{\pi(4)^3}\right] = 133.5 \text{ MPa}$$

(c) From Table 10–5 use: $G = 79.3$ GPa and $E = 196.5$ GPa

$$N_a = N_b + \frac{G}{E} = 84 + \frac{79.3}{196.5} = 84.4 \text{ turns}$$

Answer
$$k = \frac{d^4 G}{8D^3 N_a} = \frac{(4)^4(79\ 300)}{8(34)^3(84.4)} = 0.765 \text{ N/mm}$$

(d) Table 10–4:

$$A = 1855 \text{ MPa} \cdot \text{mm}^m, \quad m = 0.187$$

$$S_{ut} = \frac{1855}{(4)^{0.187}} = 1431.4 \text{ MPa}$$

$$S_y = 0.75(1431.4) = 1073.6 \text{ MPa}$$

$$S_{sy} = 0.75(1431.4) = 715.7 \text{ MPa}$$

Body

$$F = \frac{\pi d^3 S_{sy}}{\pi K_B D}$$

$$= \frac{\pi(4)^3(715.7)}{8(1.161)(34)} = 455.7 \text{ N}$$

Torsional stress on hook point B

$$C_2 = \frac{2r_2}{d} = \frac{2(6 + 4/2)}{4} = 4$$

$$(K)_B = \frac{4C_2 - 1}{4C_2 - 4} = \frac{4(4) - 1}{4(4) - 4} = 1.25$$

$$F = \frac{\pi(4)^3(715.7)}{8(1.25)(34)} = 423.2 \text{ N}$$

Normal stress on hook point A

$$C_1 = \frac{2r_1}{d} = \frac{34}{4} = 8.5$$

$$(K)_A = \frac{4C_1^2 - C_1 - 1}{4C_1(C_1 - 1)} = \frac{4(8.5)^2 - 8.5 - 1}{4(8.5)(8.5 - 1)} = 1.096$$

$$S_{yt} = \sigma = F\left[\frac{16(K)_A D}{\pi d^3} + \frac{4}{\pi d^2}\right]$$

| Figure 10–4

6 mm R.   38 mm   4 mm   12-mm R.

$$F = \frac{1073.6}{[16(1.096)(34)]/[\pi(4)^3] + \{4/[\pi(4)^2]\}} = 352.6 \text{ N}$$

Answer

$$= \min(455.7, 423.2, 352.6) = 352.6 \text{ N}$$

(*e*) Eq. (10–48):

Answer

$$y = \frac{F - F_i}{k} = \frac{352.6 - 85}{0.765} = 349.8 \text{ mm}$$

10–8   Critical Frequency of Helical Springs

If a wave is created by a disturbance at one end of a swimming pool, this wave will travel down the length of the pool, be reflected back at the far end, and continue in this back-and-forth motion until it is finally damped out. The same effect occurs in helical springs, and it is called *spring surge*. If one end of a compression spring is held against a flat surface and the other end is disturbed, a compression wave is created that travels back and forth from one end to the other exactly like the swimming-pool wave.

Spring manufacturers have taken slow-motion movies of automotive valve-spring surge. These pictures show a very violent surging, with the spring actually jumping out of contact with the end plates. Figure 10–5 is a photograph of a failure caused by such surging.

When helical springs are used in applications requiring a rapid reciprocating motion, the designer must be certain that the physical dimensions of the spring are not such as to create a natural vibratory frequency close to the frequency of the applied force; otherwise, resonance may occur, resulting in damaging stresses, since the internal damping of spring materials is quite low.

The governing equation for the translational vibration of a spring placed between two flat and parallel plates is the wave equation

$$\frac{\partial^2 u}{\partial x^2} = \frac{W}{kgl^2} \frac{\partial^2 u}{\partial t^2} \tag{10–24}$$

where k = spring rate

g = acceleration due to gravity

l = length of spring between plates

W = weight of spring

x = coordinate along length of spring

u = motion of any particle at distance x

The solution to this equation is harmonic and depends on the given physical properties as well as the end conditions of the spring. The harmonic, *natural*, frequencies for a spring placed between two flat and parallel plates, in radians per second, are

$$\omega = m\pi \sqrt{\frac{kg}{W}} \qquad m = 1, 2, 3, \ldots$$

Figure 10–5

Valve-spring failure in an overrevved engine. Fracture is along the 45° line of maximum principal stress associated with pure torsional loading. *(Personal photograph of Larry D. Mitchell, coauthor of* Mechanical Engineering Design, *4th ed., McGraw-Hill, New York, 1983.)*

where the fundamental frequency is found for $m = 1$, the second harmonic for $m = 2$, and so on. We are usually interested in the frequency in cycles per second; since $\omega = 2\pi f$, we have, for the fundamental frequency in hertz,

$$f = \frac{1}{2}\sqrt{\frac{kg}{W}} \tag{10–25}$$

assuming the spring ends are always in contact with the plates.

Wolford and Smith show that the frequency is

$$f = \frac{1}{4}\sqrt{\frac{kg}{W}} \tag{10–26}$$

where the spring has one end against a flat plate and the other end free.[10] They also point out that Eq. (10–25) applies when one end is against a flat plate and the other end is driven with a sine-wave motion.

The weight of the active part of a helical spring is

$$W = AL\gamma = \frac{\pi d^2}{4}(\pi D N_a)(\gamma) = \frac{\pi^2 d^2 D N_a \gamma}{4} \tag{10–27}$$

where γ is the specific weight.

The fundamental critical frequency should be greater than 15 to 20 times the frequency of the force or motion of the spring in order to avoid resonance with the

---

[10]J. C. Wolford and G. M. Smith, "Surge of Helical Springs," *Mech. Eng. News,* vol. 13, no. 1, February 1976, pp. 4–9.

harmonics. If the frequency is not high enough, the spring should be redesigned to increase k or decrease W.

10–9 **Fatigue Loading of Helical Compression Springs**

Springs are almost always subject to fatigue loading. In many instances the number of cycles of required life may be small, say, several thousand for a padlock spring or a toggle-switch spring. But the valve spring of an automotive engine must sustain millions of cycles of operation without failure; so it must be designed for infinite life.

To improve the fatigue strength of dynamically loaded springs, shot peening can be used. It can increase the torsional fatigue strength by 20 percent or more. Shot size is about 0.4 mm, so spring coil wire diameter and pitch must allow for complete coverage of the spring surface.

The best data on the torsional endurance limits of spring steels are those reported by Zimmerli.[11] He discovered the surprising fact that size, material, and tensile strength have no effect on the endurance limits (infinite life only) of spring steels in sizes under 10 mm. We have already observed that endurance limits tend to level out at high tensile strengths (Fig. 6–17), p. 291, but the reason for this is not clear. Zimmerli suggests that it may be because the original surfaces are alike or because plastic flow during testing makes them the same. Unpeened springs were tested from a minimum torsional stress of 138 MPa to a maximum of 620 MPa and peened springs in the range 138 MPa to 930 MPa. The corresponding endurance strength components for infinite life were found to be

Unpeened:

$$S_{sa} = 241 \text{ MPa} \qquad S_{sm} = 379 \text{ MPa} \qquad (10\text{–}28)$$

Peened:

$$S_{sa} = 398 \text{ MPa} \qquad S_{sm} = 534 \text{ MPa} \qquad (10\text{–}29)$$

For example, given an unpeened spring with $S_{su} = 1480$ MPa, the Gerber ordinate intercept for shear, from Eq. (6–42), p. 314, is

$$S_{se} = \frac{S_{sa}}{1 - \left(\dfrac{S_{sm}}{S_{su}}\right)^2} = \frac{241}{1 - \left(\dfrac{379}{1480}\right)^2} = 257.9 \text{ MPa}$$

For the Goodman failure criterion, the intercept would be 331.1 MPa. Each possible wire size would change these numbers, since S_{su} would change.

An extended study[12] of available literature regarding torsional fatigue found that for polished, notch-free, cylindrical specimens subjected to torsional shear stress, the maximum alternating stress that may be imposed without causing failure is *constant* and independent of the mean stress in the cycle provided that the maximum stress range does not equal or exceed the torsional yield strength of the metal. With notches and abrupt section changes this consistency is not found. Springs are free of notches and surfaces are often very smooth. This failure criterion is known as the *Sines failure criterion* in torsional fatigue.

---

[11]F. P. Zimmerli, "Human Failures in Spring Applications," *The Mainspring*, no. 17, Associated Spring Corporation, Bristol, Conn., August–September 1957.

[12]Oscar J. Horger (ed.), *Metals Engineering: Design Handbook*, McGraw-Hill, New York, 1953, p. 84.

In constructing certain failure criteria on the designers' torsional fatigue diagram, the torsional modulus of rupture S_{su} is needed. We shall continue to employ Eq. (6–54), p. 325, which is

$$S_{su} = 0.67S_{ut} \tag{10–30}$$

In the case of shafts and many other machine members, fatigue loading in the form of completely reversed stresses is quite ordinary. Helical springs, on the other hand, are never used as both compression and extension springs. In fact, they are usually assembled with a preload so that the working load is additional. Thus the stress-time diagram of Fig. 6–23d, p. 309, expresses the usual condition for helical springs. The worst condition, then, would occur when there is no preload, that is, when $\tau_{min} = 0$.

Now, we define

$$F_a = \frac{F_{max} - F_{min}}{2} \tag{10–31a}$$

$$F_m = \frac{F_{max} + F_{min}}{2} \tag{10–31b}$$

where the subscripts have the same meaning as those of Fig. 6–23d when applied to the axial spring force F. Then the shear stress amplitude is

$$\tau_a = K_B \frac{8F_a D}{\pi d^3} \tag{10–32}$$

where K_B is the Bergsträsser factor, obtained from Eq. (10–5), and corrects for both direct shear and the curvature effect. As noted in Sec. 10–2, the Wahl factor K_W can be used instead, if desired.

The midrange shear stress is given by the equation

$$\tau_m = K_B \frac{8F_m D}{\pi d^3} \tag{10–33}$$

EXAMPLE 10–4

An as-wound helical compression spring, made of music wire, has a wire size of 2.3 mm, an outside coil diameter of 14 mm, a free length of 98 mm, 21 active coils, and both ends squared and ground. The spring is unpeened. This spring is to be assembled with a preload of 25 N and will operate with a maximum load of 156 N during use.
(a) Estimate the factor of safety guarding against fatigue failure using a torsional Gerber fatigue failure criterion with Zimmerli data.
(b) Repeat part (a) using the Sines torsional fatigue criterion (steady stress component has no effect), with Zimmerli data.
(c) Repeat using a torsional Goodman failure criterion with Zimmerli data.
(d) Estimate the critical frequency of the spring.

Solution

The mean coil diameter is $D = 14 - 2.3 = 11.7$ mm. The spring index is $C = D/d = 11.7/2.3 = 5.09$. Then

$$K_B = \frac{4C+2}{4C-3} = \frac{4(5.09)+2}{4(5.09)-3} = 1.288$$

From Eqs. (10–31),

$$F_a = \frac{156 - 25}{2} = 65.5 \text{ N} \qquad F_m = \frac{156 + 25}{2} = 90.5 \text{ N}$$

The alternating shear-stress component is found from Eq. (10–32) to be

$$\tau_a = K_B \frac{8F_a D}{\pi d^3} = (1.288) \frac{8(65.5)11.7}{\pi(2.3)^3} = 206.6 \text{ MPa}$$

Equation (10–33) gives the midrange shear-stress component

$$\tau_m = K_B \frac{8F_m D}{\pi d^3} = (1.288) \frac{8(90.5)11.7}{\pi(2.3)^3} = 285.4 \text{ MPa}$$

From Table 10–4 we find $A = 2211$ MPa $\cdot$ mm^m and $m = 0.145$. The ultimate tensile strength is estimated from Eq. (10–14) as

$$S_{ut} = \frac{A}{d^m} = \frac{2211}{2.3^{0.145}} = 1959 \text{ MPa}$$

Also the shearing ultimate strength is estimated from

$$S_{su} = 0.67S_{ut} = 0.67(1959) = 1312 \text{ MPa}$$

The load-line slope $r = \tau_a/\tau_m = 206.6/285.4 = 0.72$.

(a) The Gerber ordinate intercept for the Zimmerli data, Eq. (10–28), is

$$S_{se} = \frac{S_{sa}}{1 - (S_{sm}/S_{su})^2} = \frac{241}{1 - (379/1312)^2} = 263 \text{ MPa}$$

The amplitude component of strength S_{sa}, from Table 6–7, p. 315, is

$$S_{sa} = \frac{r^2 S_{su}^2}{2S_{se}} \left[-1 + \sqrt{1 + \left(\frac{2S_{se}}{r S_{su}}\right)^2} \right]$$

$$= \frac{0.72^2 1312^2}{2(263)} \left\{ -1 + \sqrt{1 + \left[\frac{2(263)}{0.72(1312)}\right]^2} \right\} = 245.3 \text{ MPa}$$

and the fatigue factor of safety n_f is given by

Answer

$$n_f = \frac{S_{sa}}{\tau_a} = \frac{245.3}{206.6} = 1.19$$

(b) The Sines failure criterion ignores S_{sm} so that, for the Zimmerli data with $S_{sa} = 241$ MPa,

Answer

$$n_f = \frac{S_{sa}}{\tau_a} = \frac{241}{206.6} = 1.17$$

(c) The ordinate intercept S_{se} for the Goodman failure criterion with the Zimmerli data is

$$S_{se} = \frac{S_{sa}}{1 - (S_{sm}/S_{su})} = \frac{241}{1 - (379/1312)} = 338.9 \text{ MPa}$$

The amplitude component of the strength S_{sa} for the Goodman criterion, from Table 6–6, p. 315, is

$$S_{sa} = \frac{r S_{se} S_{su}}{r S_{su} + S_{se}} = \frac{0.72(338.9)1312}{0.72(1312) + 338.9} = 249.4 \text{ MPa}$$

The fatigue factor of safety is given by

Answer

$$n_f = \frac{S_{sa}}{\tau_a} = \frac{249.4}{206.6} = 1.21$$

(*d*) Using Eq. (10–9) and Table 10–5, we estimate the spring rate as

$$k = \frac{d^4 G}{8 D^3 N_a} = \frac{2.3^4(81\,000)}{8(11.7)^3 21} = 8.4 \text{ N/mm}$$

From Eq. (10–27) we estimate the spring weight as

$$W = \frac{\pi^2(2.3^2)11.7(21)82 \times 10^{-6}}{4} = 0.26 \text{ N}$$

and from Eq. (10–25) the frequency of the fundamental wave is

Answer

$$f_n = \frac{1}{2}\left[\frac{8400(9.81)}{0.26}\right]^{1/2} = 281 \text{ Hz}$$

If the operating or exciting frequency is more than $281/20 = 14.1$ Hz, the spring may have to be redesigned.

We used three approaches to estimate the fatigue factor of safety in Ex. 10–4. The results, in order of smallest to largest, were 1.18 (Sines), 1.21 (Gerber), and 1.23 (Goodman). Although the results were very close to one another, using the Zimmerli data as we have, the Sines criterion will always be the most conservative and the Goodman the least. If we perform a fatigue analysis using strength properties as was done in Chap. 6, different results would be obtained, but here the Goodman criterion would be more conservative than the Gerber criterion. Be prepared to see designers or design software using any one of these techniques. This is why we cover them. Which criterion is correct? Remember, we are performing *estimates* and only testing will reveal the truth—*statistically*.

10–10 Helical Compression Spring Design for Fatigue Loading

Let us begin with the statement of a problem. In order to compare a static spring to a dynamic spring, we shall design the spring in Ex. 10–2 for dynamic service.

EXAMPLE 10–5

A music wire helical compression spring with infinite life is needed to resist a dynamic load that varies from 20 to 80 N at 5 Hz while the end deflection varies from 12 to 50 mm. Because of assembly considerations, the solid height cannot exceed 25 mm and the free length cannot be more than 100 mm. The springmaker has the following wire sizes in stock: 1.7, 1.8, 2.0, 2.15, 2.3, 2.4, 2.6, and 2.8 mm.

Solution

The a priori decisions are:

- Material and condition: for music wire, $A = 2211$ MPa·mm^m, $m = 0.145$, (81 GPa; relative cost is 2.6
- Surface treatment: unpeened
- End treatment: squared and ground
- Robust linearity: $\xi = 0.15$
- Set: use in as-wound condition
- Fatigue-safe: $n_f = 1.5$ using the Sines-Zimmerli fatigue-failure criterion
- Function: $F_{min} = 20$ N, $F_{max} = 80$ N, $y_{min} = 12$ mm, $y_{max} = 50$ mm, sp operates free (no rod or hole)
- Decision variable: wire size d

The figure of merit will be the volume of wire to wind the spring, Eq. (10–22). design strategy will be to set wire size d, build a table, inspect the table, and ch the satisfactory spring with the highest figure of merit.

Solution

Set $d = 2.8$ mm. Then

$$F_a = \frac{80 - 20}{2} = 30 \text{ N} \qquad F_m = \frac{80 + 20}{2} = 50 \text{ N}$$

$$k = \frac{F_{max}}{y_{max}} = \frac{80}{50} = 1.6 \text{ N/mm}$$

$$S_{ut} = \frac{2211}{2.8^{0.145}} = 1904 \text{ MPa}$$

$$S_{su} = 0.67(1904) = 1276 \text{ MPa}$$

$$S_{sy} = 0.45(1904) = 857 \text{ MPa}$$

From Eq. (10–28), with the Sines criterion, $S_{se} = S_{sa} = 241$ MPa. Equation (10 can be used to determine C with S_{se}, n_f, and F_a in place of S_{sy}, n_s, and $(1 + \xi)$ respectively. Thus,

$$\alpha = \frac{S_{se}}{n_f} = \frac{241}{1.5} = 161 \text{ MPa}$$

$$\beta = \frac{8F_a}{\pi d^2} = \frac{8(30)}{\pi(2.8^2)} = 9.7 \text{ MPa}$$

$$C = \frac{2(161) - 9.7}{4(9.7)} + \sqrt{\left[\frac{2(161) - 9.7}{4(9.7)}\right]^2 - \frac{3(161)}{4(9.7)}} = 15.3$$

$$D = Cd = 15.3(2.8) = 42.8 \text{ mm}$$

$$F_s = (1 + \xi)F_{max} = (1 + 0.15)80 = 92\text{N}$$

$$N_a = \frac{d^4 G}{8D^3 k} = \frac{2.8^4(8\ 1000)}{8(42.8)^3 1.6} = 4.96 \text{ turns}$$

$$N_t = N_a + 2 = 4.96 + 2 = 6.96 \text{ turns}$$

$$L_s = dN_t = 2.8\ (6.96) = 19.5 \text{ mm}$$

$$L_0 = L_s + \frac{F_s}{k} = 19.5 + 92/1.6 = 77 \text{ mm}$$

$$\text{ID} = 42.8 - 2.8 = 40 \text{ mm}$$

$$\text{OD} = 42.8 + 2.8 = 45.6 \text{ mm}$$

$$y_s = L_0 - L_s = 77 - 19.5 = 57.5 \text{ mm}$$

$$(L_0)_{cr} < \frac{2.63D}{\alpha} = 2.63\frac{(42.8)}{0.5} = 225 \text{ mm}$$

$$K_B = \frac{4(15.3) + 2}{4(15.3) - 3} = 1.09$$

$$W = \frac{\pi^2 d^2 DN_a \gamma}{4} = \frac{\pi^2 2.8^2(42.8)4.96(82 \times 10^{-6})}{4} = 0.34 \text{ N}$$

$$f_n = 0.5\sqrt{\frac{9.81k}{W}} = 0.5\sqrt{\frac{9.81(1600)}{0.34}} = 107 \text{ Hz}$$

$$\tau_a = K_B \frac{8F_a D}{\pi d^3} = 1.09 \frac{8(30)42.8}{\pi 2.8^3} = 162 \text{ MPa}$$

$$\tau_m = \tau_a \frac{F_m}{F_a} = 162(50)/30 = 270 \text{ MPa}$$

$$\tau_s = \tau_a \frac{F_s}{F_a} = 162(92)/30 = 497 \text{ MPa}$$

$$n_f = \frac{S_{sa}}{\tau_a} = 241/162 = 1.5$$

$$n_s = \frac{S_{sy}}{\tau_s} = 857/497 = 1.7$$

$$\text{fom} = -(\text{relative material cost})\pi^2 d^2 N_t D/4$$

$$= -2.6\pi^2(2.8)^2 6.96(42.8)/[4(25)^3] = -1.01$$

Inspection of the results shows that all conditions are satisfied except for $4 \leq C \leq 12$. Repeat the process using the other available wire sizes and develop the following table:

d:	1.7	1.8	2.0	2.15	2.3	2.4	2.6	2.8
D	9.5	10.6	15.4	18.7	22.5	26.4	34.8	42.8
ID	6.3	7.2	11.9	15.0	18.6	22.6	32.1	40
OD	9.9	10.9	16.1	19.4	23.2	27.5	37.4	45.6
C	5.6	5.9	7.7	8.7	9.8	11.0	13.4	15.3
N_a	105.2	84.7	37.0	25.21	17.61	12.73	7.13	4.96
L_s	194.3	161.5	81.5	59.9	45.8	36.1	24.3	19.5
L_0	270.3	234.1	145.6	121.7	106.0	95.3	82.3	77
$(L_0)_{cr}$	42.6	47.6	80.8	90.7	110.1	131.8	182.8	225
n_f	1.5	1.5	1.5	1.5	1.5	1.5	1.5	1.5
n_s	1.82	1.81	1.78	1.77	1.75	1.74	1.71	1.7
f_n	86.7	88.9	96.0	98.8	101.0	102.8	105.6	107
fom	−1.17	−1.12	−0.98	−0.95	−0.93	−0.93	−0.96	−1.01

The problem-specific inequality constraints are

$$L_s \leq 25 \text{ mm}$$

$$L_0 \leq 100 \text{ mm}$$

$$f_n \geq 5(20) = 100 \text{ Hz}$$

The general constraints are

$$3 \leq N_a \leq 15$$

$$4 \leq C \leq 12$$

$$(L_0)_{cr} > L_0$$

We see that none of the diameters satisfy the given constraints. The 2.6-mm-diameter wire is the closest to satisfying all requirements. The value of $C = 13.4$ is not a serious deviation and can be tolerated. However, the tight constraint on L_s needs to be addressed. If the assembly conditions can be relaxed to accept a solid height of 24.3 mm, we have a solution. If not, the only other possibility is to use the 2.8-mm diameter and accept a value $C = 15.3$, individually package the springs, and possibly reconsider supporting the spring in service.

10–11 Extension Springs

Extension springs differ from compression springs in that they carry tensile loading, they require some means of transferring the load from the support to the body of the spring, and the spring body is wound with an initial tension. The load transfer can be done with a threaded plug or a swivel hook; both of these add to the cost of the finished product, and so one of the methods shown in Fig. 10–6 is usually employed.

Stresses in the body of the extension spring are handled the same as compression springs. In designing a spring with a hook end, bending and torsion in the hook must

Figure 10–6

Types of ends used on extension springs. *(Courtesy of Associated Spring.)*

(a) Machine half loop–open

(b) Raised hook

(c) Short twisted loop

(d) Full twisted loop

Figure 10–7

Ends for extension springs. (*a*) Usual design; stress at A is due to combined axial force and bending moment. (*b*) Side view of part *a*; stress is mostly torsion at B. (*c*) Improved design; stress at A is due to combined axial force and bending moment. (*d*) Side view of part *c*; stress at B is mostly torsion.

(a)

(b)

(c)

(d)

Note: Radius r_1 is in the plane of the end coil for curved beam bending stress. Radius r_2 is at a right angle to the end coil for torsional shear stress.

be included in the analysis. In Fig. 10–7*a* and *b* a commonly used method of designing the end is shown. The maximum tensile stress at A, due to bending and axial loading, is given by

$$\sigma_A = F\left[(K)_A \frac{16D}{\pi d^3} + \frac{4}{\pi d^2}\right] \tag{10–34}$$

where $(K)_A$ is a bending stress-correction factor for curvature, given by

$$(K)_A = \frac{4C_1^2 - C_1 - 1}{4C_1(C_1 - 1)} \qquad C_1 = \frac{2r_1}{d} \qquad (10\text{–}35)$$

The maximum torsional stress at point B is given by

$$\tau_B = (K)_B \frac{8FD}{\pi d^3} \qquad (10\text{–}36)$$

where the stress-correction factor for curvature, $(K)_B$, is

$$(K)_B = \frac{4C_2 - 1}{4C_2 - 4} \qquad C_2 = \frac{2r_2}{d} \qquad (10\text{–}37)$$

Figure 10–7c and d show an improved design due to a reduced coil diameter.

When extension springs are made with coils in contact with one another, they are said to be *close-wound*. Spring manufacturers prefer some initial tension in close-wound springs in order to hold the free length more accurately. The corresponding load-deflection curve is shown in Fig. 10–8a, where y is the extension beyond the

Figure 10-8

(a) Geometry of the force F and extension y curve of an extension spring; (b) geometry of the extension spring; and (c) torsional stresses due to initial tension as a function of spring index C in helical extension springs.

free length L_0 and F_i is the initial tension in the spring that must be exceeded before the spring deflects. The load-deflection relation is then

$$F = F_i + ky \qquad (10\text{–}38)$$

where k is the spring rate. The free length L_0 of a spring measured inside the end loops or hooks as shown in Fig. 10–7b can be expressed as

$$L_0 = 2(D - d) + (N_b + 1)d = (2C - 1 + N_b)d \qquad (10\text{–}39)$$

where D is the mean coil diameter, N_b is the number of body coils, and C is the spring index. With ordinary twisted end loops as shown in Fig. 10–8b, to account for the deflection of the loops in determining the spring rate k, the equivalent number of active helical turns N_a for use in Eq. (10–9) is

$$N_a = N_b + \frac{G}{E} \qquad (10\text{–}40)$$

where G and E are the shear and tensile moduli of elasticity, respectively (see Prob. 10–38).

The initial tension in an extension spring is created in the winding process by twisting the wire as it is wound onto the mandrel. When the spring is completed and removed from the mandrel, the initial tension is locked in because the spring cannot get any shorter. The amount of initial tension that a springmaker can routinely incorporate is as shown in Fig. 10–8c. The preferred range can be expressed in terms of the *uncorrected torsional* stress τ_i as

$$\tau_i = \frac{231}{\exp(0.105C)} \pm 6.9\left(4 - \frac{C - 3}{6.5}\right) \text{ MPa} \qquad (10\text{–}41)$$

where C is the spring index.

Guidelines for the maximum allowable corrected stresses for static applications of extension springs are given in Table 10–7.

| | **Percent of Tensile Strength** | | |
| | **In Torsion** | | **In Bending** |
Materials	**Body**	**End**	**End**
Patented, cold-drawn or hardened and tempered carbon and low-alloy steels	45–50	40	75
Austenitic stainless steel and nonferrous alloys	35	30	55

Table 10-7

Maximum Allowable Stresses (K_W or K_B corrected) for Helical Extension Springs in Static Applications

Source: From *Design Handbook,* 1987, p. 52. Courtesy of Associated Spring.

This information is based on the following conditions: set not removed and low temperature heat treatment applied. For springs that require high initial tension, use the same percent of tensile strength as for end.

EXAMPLE 10-6 A hard-drawn steel wire extension spring has a wire diameter of 0.9 mm, an outside coil diameter of 6.3 mm, hook radii of $r_1 = 2.7$ mm and $r_2 = 2.3$ mm, and an initial tension of 5 N. The number of body turns is 12.17. From the given information:

(a) Determine the physical parameters of the spring.
(b) Check the initial preload stress conditions.
(c) Find the factors of safety under a static 23 N load.

Solution (a)
$$D = \text{OD} - d = 6.3 - 0.9 = 5.4 \text{ mm}$$

$$C = \frac{D}{d} = \frac{5.4}{0.9} = 6.0$$

$$K_B = \frac{4C + 2}{4C - 3} = 1.24$$

Eq. (10–40) and Table 10–5:

$$N_a = N_b + G/E = 12.17 + 79/198 = 12.57 \text{ turns}$$

Eq. (10–9):
$$k = \frac{d^4 G}{8 D^3 N_a} = \frac{0.9^4 (79\,000)}{8(5.4^3)12.57} = 3.27 \text{ N/mm}$$

Eq. (10–39): $L_0 = (2C - 1 + N_b)d = [2(6.0) - 1 + 12.17] \, 0.9 = 20.9 \text{ mm}$

The deflection under the service load is

$$y_{\max} = \frac{F_{\max} - F_i}{k} = \frac{23 - 5}{3.27} = 5.5 \text{ mm}$$

where the spring length becomes $L = L_0 + y = 20.9 + 5.5 = 26.4$ mm

(b) The uncorrected initial stress is given by Eq. (10–2) without the correction factor. That is,

$$(\tau_i)_{\text{uncorr}} = \frac{8 F_i D}{\pi d^3} = \frac{8(5)(5.4)}{\pi(0.9^3)} = 94.3 \text{ MPa}$$

The preferred range is given by Eq. (10–41) and for this case is

$$(\tau_i)_{\text{pref}} = \frac{231}{\exp(0.105C)} \pm 6.9 \left(4 - \frac{C - 3}{6.5} \right)$$

$$= \frac{231}{\exp[0.105(6.0)]} \pm 6.9 \left(4 - \frac{6.0 - 3}{6.5} \right)$$

$$= 123 \pm 24.4 = 147.4, \, 98.6 \text{ MPa}$$

Answer Thus, the initial tension of 94.3 MPa is in the preferred range.

(c) For hard-drawn wire, Table 10–4 gives $m = 0.190$ and $A = 1783$ MPa · mm^m. From Eq. (10–14)

$$S_{ut} = \frac{A}{d^m} = \frac{1783}{0.9^{0.190}} = 1819 \text{ MPa}$$

For torsional shear in the main body of the spring, from Table 10–7,

$$S_{sy} = 0.45 S_{ut} = 0.45(1819) = 818.6 \text{ MPa}$$

The shear stress under the service load is

$$\tau_{max} = \frac{8K_B F_{max} D}{\pi d^3} = \frac{8(1.24)23(5.4)}{\pi(0.9^3)} = 538 \text{ MPa}$$

Thus, the factor of safety is

Answer

$$n = \frac{S_{sy}}{\tau_{max}} = \frac{818.6}{538} = 1.52$$

For the end-hook bending at A,

$$C_1 = 2r_1/d = 2(2.7)/0.9 = 6$$

From Eq. (10–35)

$$(K)_A = \frac{4C_1^2 - C_1 - 1}{4C_1(C_1 - 1)} = \frac{4(6^2) - 6 - 1}{4(6)(6 - 1)} = 1.14$$

From Eq. (10–34)

$$\sigma_A = F_{max}\left[(K)_A \frac{16D}{\pi d^3} + \frac{4}{\pi d^2}\right]$$

$$= 23\left[1.14\frac{16(5.4)}{\pi(0.9^3)} + \frac{4}{\pi(0.9^2)}\right] = 1025.3 \text{ MPa}$$

The yield strength, from Table 10–7, is given by

$$S_y = 0.75S_{ut} = 0.75(1819) = 1364.3 \text{ MPa}$$

The factor of safety for end-hook bending at A is then

Answer

$$n_A = \frac{S_y}{\sigma_A} = \frac{1364.3}{1025.3} = 1.33$$

For the end-hook in torsion at B, from Eq. (10–37)

$$C_2 = 2r_2/d = 2(2.3)/0.9 = 5.1$$

$$(K)_B = \frac{4C_2 - 1}{4C_2 - 4} = \frac{4(5.1) - 1}{4(5.1) - 4} = 1.18$$

and the corresponding stress, given by Eq. (10–36), is

$$\tau_B = (K)_B\frac{8F_{max} D}{\pi d^3} = 1.18\frac{8(23)5.4}{\pi(0.9^3)} = 511.9 \text{ MPa}$$

Using Table 10–7 for yield strength, the factor of safety for end-hook torsion at B is

Answer

$$n_B = \frac{(S_{sy})_B}{\tau_B} = \frac{0.4(1819)}{511.9} = 1.42$$

Yield due to bending of the end hook will occur first.

Next, let us consider a fatigue problem.

EXAMPLE 10–7

The helical coil extension spring of Ex. 10–6 is subjected to a dynamic loading from 6.5 to 20 N. Estimate the factors of safety using the Gerber failure criterion for (a) coil fatigue, (b) coil yielding, (c) end-hook bending fatigue at point A of Fig. 10–7a, and (d) end-hook torsional fatigue at point B of Fig. 10–7b.

Solution

A number of quantities are the same as in Ex. 10–6: $d = 0.9$ mm, $S_{ut} = 1819$ MPa, $D = 5.4$ mm, $r_1 = 2.7$ mm, $C = 6$, $K_B = 1.24$, $(K)_A = 1.14$, $(K)_B = 1.18$, $N_b = 12.17$ turns, $L_0 = 20.9$ mm, $k = 3.27$ N/mm, $F_i = 5$ N, and $(\tau_i)_{uncorr} = 94.3$ MPa. Then

$$F_a = (F_{max} - F_{min})/2 = (20 - 6.5)/2 = 6.75 \text{ N}$$

$$F_m = (F_{max} + F_{min})/2 = (20 + 6.5)/2 = 13.25 \text{ N}$$

The strengths from Ex. 10–6 include $S_{ut} = 1819$ MPa, $S_y = 1364.3$ MPa, and $S_{sy} = 818.6$ MPa. The ultimate shear strength is estimated from Eq. (10–30) as

$$S_{su} = 0.67 S_{ut} = 0.67(1819) = 1218.7 \text{ MPa}$$

(a) Body-coil fatigue:

$$\tau_a = \frac{8 K_B F_a D}{\pi d^3} = \frac{8(1.24)6.75(5.4)}{\pi(0.9^3)} = 157.9 \text{ MPa}$$

$$\tau_m = \frac{F_m}{F_a} \tau_a = \frac{13.25}{6.75} 157.9 = 310 \text{ MPa}$$

Using the Zimmerli data of Eq. (10–28) gives

$$S_{se} = \frac{S_{sa}}{1 - \left(\dfrac{S_{sm}}{S_{su}}\right)^2} = \frac{241}{1 - \left(\dfrac{379}{1218.7}\right)^2} = 266.8 \text{ MPa}$$

From Table 6–7, p. 307, the Gerber fatigue criterion for shear is

Answer

$$(n_f)_{body} = \frac{1}{2}\left(\frac{S_{su}}{\tau_m}\right)^2 \frac{\tau_a}{S_{se}}\left[-1 + \sqrt{1 + \left(2\frac{\tau_m}{S_{su}}\frac{S_{se}}{\tau_a}\right)^2}\right]$$

$$= \frac{1}{2}\left(\frac{1218.7}{310}\right)^2 \frac{157.9}{266.8}\left[-1 + \sqrt{1 + \left(2\frac{310}{1218.7}\frac{266.8}{157.9}\right)^2}\right] = 1.46$$

(b) The load-line for the coil body begins at $S_{sm} = \tau_i$ and has a slope $r = \tau_a/(\tau_m - \tau_i)$. It can be shown that the intersection with the yield line is given by $(S_{sa})_y = [r/(r + 1)](S_{sy} - \tau_i)$. Consequently, $\tau_i = (F_i/F_a)\tau_a = (5/6.75)157.9 = 117$ MPa, $r = 157.9/(310 - 117) = 0.82$, and

$$(S_{sa})_y = \frac{0.82}{0.82 + 1}(818.6 - 117) = 316 \text{ MPa}$$

Thus,

Answer

$$(n_y)_{body} = \frac{(S_{sa})_y}{\tau_a} = \frac{316}{157.9} = 2.0$$

(c) End-hook bending fatigue: using Eqs. (10–34) and (10–35) gives

$$\sigma_a = F_a \left[(K)_A \frac{16D}{\pi d^3} + \frac{4}{\pi d^2} \right]$$

$$= 6.75 \left[1.14 \frac{16(5.4)}{\pi(0.9^3)} + \frac{4}{\pi(0.9^2)} \right] = 301 \text{ MPa}$$

$$\sigma_m = \frac{F_m}{F_a} \sigma_a = \frac{13.25}{6.75} 301 = 590.9 \text{ MPa}$$

To estimate the tensile endurance limit using the distortion-energy theory,

$$S_e = S_{se}/0.577 = 266.8/0.577 = 462.4 \text{ MPa}$$

Using the Gerber criterion for tension gives

Answer

$$(n_f)_A = \frac{1}{2} \left(\frac{S_{ut}}{\sigma_m} \right)^2 \frac{\sigma_a}{S_e} \left[-1 + \sqrt{1 + \left(2 \frac{\sigma_m}{S_{ut}} \frac{S_e}{\sigma_a} \right)^2} \right]$$

$$= \frac{1}{2} \left(\frac{1819}{590.9} \right)^2 \frac{301}{462.4} \left[-1 + \sqrt{1 + \left(2 \frac{590.9}{1819} \frac{462.4}{301} \right)^2} \right] = 1.27$$

(d) End-hook torsional fatigue: from Eq. (10–36)

$$(\tau_a)_B = (K)_B \frac{8 F_a D}{\pi d^3} = 1.18 \frac{8(6.75)(5.4)}{\pi(0.9^3)} = 150.2 \text{ MPa}$$

$$(\tau_m)_B = \frac{F_m}{F_a} (\tau_a)_B = \frac{13.25}{6.75} 150.2 = 294.8 \text{ MPa}$$

Then, again using the Gerber criterion, we obtain

Answer

$$(n_f)_B = \frac{1}{2} \left(\frac{S_{su}}{\tau_m} \right)^2 \frac{\tau_a}{S_{se}} \left[-1 + \sqrt{1 + \left(2 \frac{\tau_m}{S_{su}} \frac{S_{se}}{\tau_a} \right)^2} \right]$$

$$= \frac{1}{2} \left(\frac{1218.7}{294.8} \right)^2 \frac{150.2}{266.8} \left[-1 + \sqrt{1 + \left(2 \frac{294.8}{1218.7} \frac{266.8}{150.2} \right)^2} \right] = 1.53$$

The analyses in Exs. 10–6 and 10–7 show how extension springs differ from compression springs. The end hooks are usually the weakest part, with bending usually controlling. We should also appreciate that a fatigue failure separates the extension spring under load. Flying fragments, lost load, and machine shutdown are threats to personal safety as well as machine function. For these reasons higher design factors are used in extension-spring design than in the design of compression springs.

In Ex. 10–7 we estimated the endurance limit for the hook in bending using the Zimmerli data, which are based on torsion in compression springs and the distortion theory. An alternative method is to use Table 10–8, which is based on a stress-ratio of $R = \tau_{\min}/\tau_{\max} = 0$. For this case, $\tau_a = \tau_m = \tau_{\max}/2$. Label the strength

Table 10–8

Maximum Allowable Stresses for ASTM A228 and Type 302 Stainless Steel Helical Extension Springs in Cyclic Applications

Source: From Design Handbook, 1987, p. 52. Courtesy of Associated Spring.

| Number of Cycles | Percent of Tensile Strength | | |
| | In Torsion | | In Bending |
	Body	End	End
10^5	36	34	51
10^6	33	30	47
10^7	30	28	45

This information is based on the following conditions: not shot-peened, no surging and ambient environment with a low temperature heat treatment applied. Stress ratio = 0.

values of Table 10–8 as S_r for bending or S_{sr} for torsion. Then for torsion, for example, $S_{sa} = S_{sm} = S_{sr}/2$ and the Gerber ordinate intercept, given by Eq. (6–42) for shear, is

$$S_{se} = \frac{S_{sa}}{1 - (S_{sm}/S_{su})^2} = \frac{S_{sr}/2}{1 - \left(\dfrac{S_{sr}/2}{S_{su}}\right)^2} \qquad (10\text{–}42)$$

So in Ex. 10–7 an estimate for the bending endurance limit from Table 10–8 would be

$$S_r = 0.45S_{ut} = 0.45(1819) = 818.6 \text{ MPa}$$

and from Eq. (10–42)

$$S_e = \frac{S_r/2}{1 - [S_r/(2S_{ut})]^2} = \frac{818.6/2}{1 - \left(\dfrac{818.6/2}{1819}\right)^2} = 431 \text{ MPa}$$

Using this in place of 462.4 MPa in Ex. 10–7 results in $(n_f)_A = 1.03$, a reduction of 5 percent.

10–12 Helical Coil Torsion Springs

When a helical coil spring is subjected to end torsion, it is called a *torsion spring*. It is usually close-wound, as is a helical coil extension spring, but with negligible initial tension. There are single-bodied and double-bodied types as depicted in Fig. 10–9. As shown in the figure, torsion springs have ends configured to apply torsion to the coil body in a convenient manner, with short hook, hinged straight offset, straight torsion, and special ends. The ends ultimately connect a force at a distance from the coil axis to apply a torque. The most frequently encountered (and least expensive) end is the straight torsion end. If intercoil friction is to be avoided completely, the spring can be wound with a pitch that just separates the body coils. Helical coil torsion springs are usually used with a rod or arbor for reactive support when ends cannot be built in, to maintain alignment, and to provide buckling resistance if necessary.

The wire in a torsion spring is in bending, in contrast to the torsion encountered in helical coil compression and extension springs. The springs are designed to wind tighter in service. As the applied torque increases, the inside diameter of the coil decreases. Care must be taken so that the coils do not interfere with the pin, rod, or arbor. The bending mode in the coil might seem to invite square- or rectangular-cross-section wire, but cost, range of materials, and availability discourage its use.

Figure 10–9

Torsion springs. *(Courtesy of Associated Spring.)*

Special ends

Short hook ends

Hinge ends

Double torsion

Straight offset

Straight torsion

Table 10–9

End Position Tolerances for Helical Coil Torsion Springs (for D/d Ratios up to and Including 16)

Source: From *Design Handbook,* 1987, p. 52. Courtesy of Associated Spring.

Total Coils	Tolerance: ± Degrees*
Up to 3	8
Over 3–10	10
Over 10–20	15
Over 20–30	20
Over 30	25

*Closer tolerances available on request.

Torsion springs are familiar in clothespins, window shades, and animal traps, where they may be seen around the house, and out of sight in counterbalance mechanisms, ratchets, and a variety of other machine components. There are many stock springs that can be purchased off-the-shelf from a vendor. This selection can add economy of scale to small projects, avoiding the cost of custom design and small-run manufacture.

Describing the End Location

In specifying a torsion spring, the ends must be located relative to each other. Commercial tolerances on these relative positions are listed in Table 10–9. The simplest scheme for expressing the initial unloaded location of one end with respect to the other is in terms of an angle β defining the partial turn present in the coil body as $N_p = \beta/360°$, as shown in Fig. 10–10. For analysis purposes the nomenclature of Fig. 10–10 can be used. Communication with a springmaker is often in terms of the back-angle α.

The number of body turns N_b is the number of turns in the free spring body by count. The body-turn count is related to the initial position angle β by

$$N_b = \text{integer} + \frac{\beta}{360°} = \text{integer} + N_p$$

Figure 10–10

The free-end location angle is β. The rotational coordinate θ is proportional to the product Fl. Its back angle is α. For all positions of the moving end $\theta + \alpha = \Sigma = $ constant.

where N_p is the number of partial turns. The above equation means that N_b takes on noninteger, discrete values such as 5.3, 6.3, 7.3, . . . , with successive differences of 1 as possibilities in designing a specific spring. This consideration will be discussed later.

Bending Stress

A torsion spring has bending induced in the coils, rather than torsion. This means that residual stresses built in during winding are in the same direction but of opposite sign to the working stresses that occur during use. The strain-strengthening locks in residual stresses opposing working stresses *provided* the load is always applied in the winding sense. Torsion springs can operate at bending stresses exceeding the yield strength of the wire from which it was wound.

The bending stress can be obtained from curved-beam theory expressed in the form

$$\sigma = K\frac{Mc}{I}$$

where K is a stress-correction factor. The value of K depends on the shape of the wire cross section and whether the stress sought is at the inner or outer fiber. Wahl analytically determined the values of K to be, for round wire,

$$K_i = \frac{4C^2 - C - 1}{4C(C - 1)} \qquad K_o = \frac{4C^2 + C - 1}{4C(C + 1)} \tag{10–43}$$

where C is the spring index and the subscripts i and o refer to the inner and outer fibers, respectively. In view of the fact that K_o is always less than unity, we shall use K_i to estimate the stresses. When the bending moment is $M = Fr$ and the section modulus $I/c = d^3/32$, we express the bending equation as

$$\sigma = K_i\frac{32Fr}{\pi d^3} \tag{10–44}$$

which gives the bending stress for a round-wire torsion spring.

Deflection and Spring Rate

For torsion springs, angular deflection can be expressed in radians or revolutions (turns). If a term contains revolution units the term will be expressed with a prime sign. The spring rate k' is expressed in units of torque/revolution (lbf · in/rev or N · mm/rev) and moment is proportional to angle θ' expressed in turns rather than radians. The spring rate, if linear, can be expressed as

$$k' = \frac{M_1}{\theta_1'} = \frac{M_2}{\theta_2'} = \frac{M_2 - M_1}{\theta_2' - \theta_1'} \tag{10–45}$$

where the moment M can be expressed as Fl or Fr.

The angle subtended by the end deflection of a cantilever, when viewed from the built-in ends, is y/l rad. From Table A–9–1,

$$\theta_e = \frac{y}{l} = \frac{Fl^2}{3EI} = \frac{Fl^2}{3E(\pi d^4/64)} = \frac{64Ml}{3\pi d^4 E} \tag{10–46}$$

For a straight torsion end spring, end corrections such as Eq. (10–46) must be added to the body-coil deflection. The strain energy in bending is, from Eq. (4–23),

$$U = \int \frac{M^2\, dx}{2EI}$$

For a torsion spring, $M = Fl = Fr$, and integration must be accomplished over the length of the body-coil wire. The force F will deflect through a distance $r\theta$ where θ is the angular deflection of the coil body, in radians. Applying Castigliano's theorem gives

$$r\theta = \frac{\partial U}{\partial F} = \int_0^{\pi DN_b} \frac{\partial}{\partial F}\left(\frac{F^2 r^2\, dx}{2EI}\right) = \int_0^{\pi DN_b} \frac{Fr^2\, dx}{EI}$$

Substituting $I = \pi d^4/64$ for round wire and solving for θ gives

$$\theta = \frac{64FrDN_b}{d^4 E} = \frac{64MDN_b}{d^4 E}$$

The total angular deflection in radians is obtained by adding Eq. (10–46) for each end of lengths l_1, l_2:

$$\theta_t = \frac{64MDN_b}{d^4 E} + \frac{64Ml_1}{3\pi d^4 E} + \frac{64Ml_2}{3\pi d^4 E} = \frac{64MD}{d^4 E}\left(N_b + \frac{l_1 + l_2}{3\pi D}\right) \tag{10–47}$$

The equivalent number of active turns N_a is expressed as

$$N_a = N_b + \frac{l_1 + l_2}{3\pi D} \tag{10–48}$$

The spring rate k in torque per radian is

$$k = \frac{Fr}{\theta_t} = \frac{M}{\theta_t} = \frac{d^4 E}{64DN_a} \tag{10–49}$$

The spring rate may also be expressed as torque per turn. The expression for this is obtained by multiplying Eq. (10–49) by 2π rad/turn. Thus spring rate k' (units torque/turn) is

$$k' = \frac{2\pi d^4 E}{64DN_a} = \frac{d^4 E}{10.2DN_a} \tag{10–50}$$

Tests show that the effect of friction between the coils and arbor is such that the constant 10.2 should be increased to 10.8. The equation above becomes

$$k' = \frac{d^4 E}{10.8DN_a} \tag{10–51}$$

(units torque per turn). Equation (10–51) gives better results. Also Eq. (10–47) becomes

$$\theta_t' = \frac{10.8MD}{d^4 E}\left(N_b + \frac{l_1 + l_2}{3\pi D}\right) \tag{10–52}$$

Torsion springs are frequently used over a round bar or pin. When the load is applied to a torsion spring, the spring winds up, causing a decrease in the inside diameter of the coil body. It is necessary to ensure that the inside diameter of the coil never becomes equal to or less than the diameter of the pin, in which case loss of spring function would ensue. The helix diameter of the coil D' becomes

$$D' = \frac{N_b D}{N_b + \theta'_c} \tag{10-53}$$

where θ'_c is the angular deflection of the body of the coil in number of turns, given by

$$\theta'_c = \frac{10.8 M D N_b}{d^4 E} \tag{10-54}$$

The new inside diameter $D'_i = D' - d$ makes the diametral clearance Δ between the body coil and the pin of diameter D_p equal to

$$\Delta = D' - d - D_p = \frac{N_b D}{N_b + \theta'_c} - d - D_p \tag{10-55}$$

Equation (10–55) solved for N_b is

$$N_b = \frac{\theta'_c (\Delta + d + D_p)}{D - \Delta - d - D_p} \tag{10-56}$$

which gives the number of body turns corresponding to a specified diametral clearance of the arbor. This angle may not be in agreement with the necessary partial-turn remainder. Thus the diametral clearance may be exceeded but not equaled.

Static Strength

First column entries in Table 10–6 can be divided by 0.577 (from distortion-energy theory) to give

$$S_y = \begin{cases} 0.78 S_{ut} & \text{Music wire and cold-drawn carbon steels} \\ 0.87 S_{ut} & \text{OQ\&T carbon and low-alloy steels} \\ 0.61 S_{ut} & \text{Austenitic stainless steel and nonferrous alloys} \end{cases} \tag{10-57}$$

Fatigue Strength

Since the spring wire is in bending, the Sines equation is not applicable. The Sines model is in the presence of pure torsion. Since Zimmerli's results were for compression springs (wire in pure torsion), we will use the repeated bending stress ($R = 0$) values provided by Associated Spring in Table 10–10. As in Eq. (10–40) we will use the Gerber fatigue-failure criterion incorporating the Associated Spring $R = 0$ fatigue strength S_r:

$$S_e = \frac{S_r/2}{1 - \left(\dfrac{S_r/2}{S_{ut}}\right)^2} \tag{10-58}$$

The value of S_r (and S_e) has been corrected for size, surface condition, and type of loading, but not for temperature or miscellaneous effects. The Gerber fatigue criterion is now defined. The strength-amplitude component is given by Table 6–7, p. 315, as

$$S_a = \frac{r^2 S_{ut}^2}{2 S_e} \left[-1 + \sqrt{1 + \left(\frac{2 S_e}{r S_{ut}}\right)^2} \right] \tag{10-59}$$

Table 10–10

Maximum Recommended Bending Stresses (K_B Corrected) for Helical Torsion Springs in Cyclic Applications as Percent of S_{ut}

Source: Courtesy of Associated Spring.

Fatigue Life, Cycles	ASTM A228 and Type 302 Stainless Steel		ASTM A230 and A232	
	Not Shot-Peened	Shot-Peened*	Not Shot-Peened	Shot-Peened*
10^5	53	62	55	64
10^6	50	60	53	62

This information is based on the following conditions: no surging, springs are in the "as-stress-relieved" condition.

*Not always possible.

where the slope of the load line is $r = M_a/M_m$. The load line is radial through the origin of the designer's fatigue diagram. The factor of safety guarding against fatigue failure is

$$n_f = \frac{S_a}{\sigma_a} \tag{10–60}$$

Alternatively, we can find n_f directly by using Table 6–7, p. 315:

$$n_f = \frac{1}{2}\frac{\sigma_a}{S_e}\left(\frac{S_{ut}}{\sigma_m}\right)^2\left[-1 + \sqrt{1 + \left(2\frac{\sigma_m}{S_{ut}}\frac{S_e}{\sigma_a}\right)^2}\right] \tag{10–61}$$

EXAMPLE 10–8

A stock spring is shown in Fig. 10–11. It is made from 1.8-mm-diameter music wire and has $4\frac{1}{4}$ body turns with straight torsion ends. It works over a pin of 10 mm diameter. The coil outside diameter is 15 mm.

(a) Find the maximum operating torque and corresponding rotation for static loading.

(b) Estimate the inside coil diameter and pin diametral clearance when the spring is subjected to the torque in part (a).

Figure 10–11

Angles α, β, and θ are measured between the straight-end centerline translated to the coil axis. Coil OD is 15 mm.

(c) Estimate the fatigue factor of safety n_f if the applied moment varies between $M_{min} = 0.1$ to $M_{max} = 0.5$ N·m.

Solution

(a) For music wire, from Table 10–4 we find that $A = 2211$ MPa · mm^m and $m = 0.145$. Therefore,

$$S_{ut} = \frac{A}{d^m} = \frac{2210}{(1.8)^{0.145}} = 2029 \text{ MPa}$$

Using Eq. (10–57) gives

$$S_y = 0.78 S_{ut} = 0.78(2029) = 1582 \text{ MPa}$$

The mean coil diameter is $D = 15 - 1.8 = 13.2$ mm. The spring index $C = D/d = 13.2/1.8 = 7.33$. The bending stress correction factor K_i from Eq. (10–43), is

$$K_i = \frac{4(7.33)^2 - 7.33 - 1}{4(7.33)(7.33 - 1)} = 1.113$$

Now rearrange Eq. (10–44), substitute S_y for σ, and solve for the maximum torque Fr to obtain

$$M_{max} = (Fr)_{max} = \frac{\pi d^3 S_y}{32 K_i} = \frac{\pi (1.8)^3 1583}{32(1.113)} = 814 \text{ N·mm}$$

Note that no factor of safety has been used. Next, from Eq. (10–54) and Table 10–5, the number of turns of the coil body θ'_c is

$$\theta'_c = \frac{10.8 M D N_b}{d^4 E} = \frac{10.8(814)13.2(4.25)}{1.8^4(196\,000)} = 0.24 \text{ turn}$$

Answer

$$(\theta'_c)_{\deg} = 0.24(360°) = 86.4°$$

The active number of turns N_a, from Eq. (10–48), is

$$N_a = N_b + \frac{l_1 + l_2}{3\pi D} = 4.25 + \frac{25 + 25}{3\pi(13.2)} = 4.65 \text{ turns}$$

The spring rate of the complete spring, from Eq. (10–51), is

$$k' = \frac{1.8^4(196\,000)}{10.8(13.2)4.65} = 3104 \text{ N·mm}$$

The number of turns of the complete spring θ' is

$$\theta' = \frac{M}{k'} = \frac{814}{3104} = 0.26 \text{ turn}$$

Answer

$$(\theta'_s)_{\deg} = 0.26(360°) = 93.6°$$

(b) With no load, the mean coil diameter of the spring is 13.2 mm. From Eq. (10–53),

$$D' = \frac{N_b D}{N_b + \theta'_c} = \frac{4.25(13.2)}{4.25 + 0.24} = 12.5 \text{ mm}$$

The diametral clearance between the inside of the spring coil and the pin at load is

Answer
$$\Delta = D' - d - D_p = 12.5 - 1.8 - 10 = 0.7 \text{ mm}$$

(c) Fatigue:

$$M_a = (M_{\max} - M_{\min})/2 = (500 - 100)/2 = 200 \text{ N·mm}$$

$$M_m = (M_{\max} + M_{\min})/2 = (500 + 100)/2 = 300 \text{ N·mm}$$

$$r = \frac{M_a}{M_m} = \frac{2}{3}$$

$$\sigma_a = K_i \frac{32 M_a}{\pi d^3} = 1.113 \frac{32(200)}{\pi 1.8^3} = 388.8 \text{ MPa}$$

$$\sigma_m = \frac{M_m}{M_a} \sigma_a = \frac{3}{2}(388.8) = 583.2 \text{ MPa}$$

From Table 10–10, $S_r = 0.50 S_{ut} = 0.50(2029) = 1014.5$. Then

$$S_e = \frac{1014.5/2}{1 - \left(\dfrac{1014.5/2}{2029}\right)^2} = 541 \text{ MPa}$$

The amplitude component of the strength S_a, from Eq. (10–59), is

$$S_a = \frac{(2/3)^2 2029^2}{2(541)} \left[-1 + \sqrt{1 + \left(\frac{2}{2/3} \frac{541}{2029}\right)^2} \right] = 474.4 \text{ MPa}$$

The fatigue factor of safety is

Answer
$$n_f = \frac{S_a}{\sigma_a} = \frac{474.4}{388.8} = 1.22$$

10–13 Belleville Springs

The inset of Fig. 10–12 shows the cross-section of a coned-disk spring, commonly called a *Belleville spring*. Although the mathematical treatment is beyond the scope of this book, you should at least become familiar with the remarkable characteristics of these springs.

Aside from the obvious advantage that a Belleville spring occupies only a small space, variation in the h/t ratio will produce a wide variety of load-deflection curve shapes, as illustrated in Fig. 10–12. For example, using an h/t ratio of 2.83 or larger gives an S curve that might be useful for snap-acting mechanisms. A reduction of the ratio to a value between 1.41 and 2.1 causes the central portion of the curve to become horizontal, which means that the load is constant over a considerable deflection range.

A higher load for a given deflection may be obtained by nesting, that is, by stacking the springs in parallel. On the other hand, stacking in series provides a larger deflection for the same load, but in this case there is danger of instability.

Figure 10–12

Load-deflection curves for Belleville springs. *(Courtesy of Associated Spring.)*

10–14   **Miscellaneous Springs**

The extension spring shown in Fig. 10–13 is made of slightly curved strip steel, not flat, so that the force required to uncoil it remains constant; thus it is called a *constant-force spring*. This is equivalent to a zero spring rate. Such springs can also be manu-factured having either a positive or a negative spring rate.

A *volute spring,* shown in Fig. 10–14*a*, is a wide, thin strip, or "flat," of material wound on the flat so that the coils fit inside one another. Since the coils do not stack, the solid height of the spring is the width of the strip. A variable-spring scale, in a compression volute spring, is obtained by permitting the coils to contact the support. Thus, as the deflection increases, the number of active coils decreases. The volute spring has another important advantage that cannot be obtained with round-wire springs: if the coils are wound so as to contact or slide on one another during action, the sliding friction will serve to damp out vibrations or other unwanted transient disturbances.

Figure 10–13

Constant-force spring. *(Courtesy of Vulcan Spring & Mfg. Co. Telford, PA. www.vulcanspring.com.)*

Figure 10–14

(a) A volute spring; (b) a flat triangular spring.

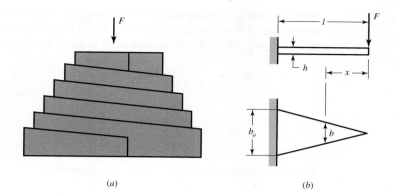

(a)　　　　　　　　　　　　　(b)

A *conical spring,* as the name implies, is a coil spring wound in the shape of a cone (see Prob. 10–28). Most conical springs are compression springs and are wound with round wire. But a volute spring is a conical spring too. Probably the principal advantage of this type of spring is that it can be wound so that the solid height is only a single wire diameter.

Flat stock is used for a great variety of springs, such as clock springs, power springs, torsion springs, cantilever springs, and hair springs; frequently it is specially shaped to create certain spring actions for fuse clips, relay springs, spring washers, snap rings, and retainers.

In designing many springs of flat stock or strip material, it is often economical and of value to proportion the material to obtain a constant stress throughout the spring material. A uniform-section cantilever spring has a stress

$$\sigma = \frac{M}{I/c} = \frac{Fx}{I/c} \qquad (a)$$

which is proportional to the distance x if I/c is a constant. But there is no reason why I/c need be a constant. For example, one might design such a spring as that shown in Fig. 10–14b, in which the thickness h is constant but the width b is permitted to vary. Since, for a rectangular section, $I/c = bh^2/6$, we have, from Eq. (a),

$$\frac{bh^2}{6} = \frac{Fx}{\sigma}$$

or

$$b = \frac{6Fx}{h^2\sigma} \qquad (b)$$

Since b is linearly related to x, the width b_o at the base of the spring is

$$b_o = \frac{6Fl}{h^2\sigma} \qquad (10\text{–}62)$$

Good approximations for deflections can be found easily by using Castigliano's theorem. To demonstrate this, assume that deflection of the triangular flat spring is primarily due to bending and we can neglect the transverse shear force.[13] The bending moment as a function of x is $M = -Fx$ and the beam width at x can be expressed as $b = b_o x/l$. Thus, the deflection of F is given by Eq. (4–31), p. 180, as

$$
y = \int_0^l \frac{M(\partial M/\partial F)}{EI} \, dx = \frac{1}{E} \int_0^l \frac{-Fx(-x)}{\frac{1}{12}(b_o x/l)h^3} \, dx
$$

$$
= \frac{12Fl}{b_o h^3 E} \int_0^l x \, dx = \frac{6Fl^3}{b_o h^3 E}
$$

(10–63)

Thus the spring constant, $k = F/y$, is estimated as

$$
k = \frac{b_o h^3 E}{6l^3}
$$

(10–64)

The methods of stress and deflection analysis illustrated in previous sections of this chapter have served to illustrate that springs may be analyzed and designed by using the fundamentals discussed in the earlier chapters of this book. This is also true for most of the miscellaneous springs mentioned in this section, and you should now experience no difficulty in reading and understanding the literature of such springs.

10–15 Summary

In this chapter we have considered helical coil springs in considerable detail in order to show the importance of viewpoint in approaching engineering problems, their analysis, and design. For compression springs undergoing static and fatigue loads, the complete design process was presented. This was not done for extension and torsion springs, as the process is the same, although the governing conditions are not. The governing conditions, however, were provided and extension to the design process from what was provided for the compression spring should be straightforward. Problems are provided at the end of the chapter, and it is hoped that the reader will develop additional, similar, problems to tackle.

As spring problems become more computationally involved, programmable calculators and computers must be used. Spreadsheet programming is very popular for repetitive calculations. As mentioned earlier, commercial programs are available. With these programs, backsolving can be performed; that is, when the final objective criteria are entered, the program determines the input values.

PROBLEMS

10–1   Within the range of recommended values of the spring index, C, determine the maximum and minimum percentage difference between the Bergsträsser factor, K_B, and the Wahl factor, K_W.

---

[13]Note that, because of shear, the width of the beam cannot be zero at $x = 0$. So, there is already some simplification in the design model. All of this can be accounted for in a more sophisticated model.

10–2 It is instructive to examine the question of the units of the parameter A of Eq. (10–14). Show that for U.S. customary units the units for A_{uscu} are kpsi $\cdot$ in^m and for SI units are MPa $\cdot$ mm^m for A_{SI}. which make the dimensions of both A_{uscu} and A_{SI} different for every material to which Eq. (10–14) applies. Also show that the conversion from A_{uscu} to A_{SI} is given by

$$A_{\text{SI}} = 6.895(25.40)^m A_{\text{uscu}}$$

10–3 A helical compression spring is wound using 2.5-mm-diameter music wire. The spring has an outside diameter of 31 mm with plain ground ends, and 14 total coils.
(a) Estimate the spring rate.
(b) What force is needed to compress this spring to closure?
(c) What should the free length be to ensure that when the spring is compressed solid the torsional stress does not exceed the yield strength?
(d) Is there a possibility that the spring might buckle in service?

10–4 The spring in Prob. 10–3 is to be used with a static load of 130 N. Perform a design assessment represented by Eqs. (10–13) and (10–18) through (10–21) if the spring is closed to solid height.

10–5 A helical compression spring is made with oil-tempered wire with wire diameter of 5 mm, mean coil diameter of 50 mm, a total of 12 coils, a free length of 125 mm, with squared ends.
(a) Find the solid length.
(b) Find the force necessary to deflect the spring to its solid length.
(c) Find the factor of safety guarding against yielding when the spring is compressed to its solid length.

10–6 A helical compression spring is to be made of oil-tempered wire of 4-mm diameter with a spring index of $C = 10$. The spring is to operate inside a hole, so buckling is not a problem and the ends can be left plain. The free length of the spring should be 80 mm. A force of 50 N should deflect the spring 15 mm.
(a) Determine the spring rate.
(b) Determine the minimum hole diameter for the spring to operate in.
(c) Determine the total number of coils needed.
(d) Determine the solid length.
(e) Determine a static factor of safety based on the yielding of the spring if it is compressed to its solid length.

10–7 A helical compression spring is made of hard-drawn spring steel wire 2 mm in diameter and has an outside diameter of 22 mm. The ends are plain and ground, and there are $8\frac{1}{2}$ total coils.
(a) The spring is wound to a free length, which is the largest possible with a solid-safe property. Find this free length.
(b) What is the pitch of this spring?
(c) What force is needed to compress the spring to its solid length?
(d) Estimate the spring rate.
(e) Will the spring buckle in service?

10–8 The spring of Prob. 10–7 is to be used with a static load of 75 N. Perform a design assessment represented by Eqs. (10–13) and (10–18) through (10–21) if the spring closed to solid height.

10–9 to 10–19 Listed in the tables are six springs. Investigate these squared-and-ground-ended helical compression springs to see if they are solid-safe. If not, what is the largest free length to which they can be wound using $n_s = 1.2$?

Problem Number	d, mm	OD, mm	L₀, mm	Nₜ	Material
10–9	0.15	0.9	16	40	A228 music wire
10–10	0.3	3.0	20.6	15.1	B159 phosphor-bronze
10–11	1.0	6.0	19.1	10.4	A313 stainless steel
10–12	3.5	50	74.6	5.25	A227 hard-drawn steel
10–13	3.7	25	95.3	13.0	A229 OQ&T steel
10–14	4.9	76	228.6	8.0	A232 chrome-vanadium

	d, mm	OD, mm	L₀, mm	Nₜ	Material
10–15	0.25	0.95	12.1	38	A313 stainless steel
10–16	1.2	6.5	15.7	10.2	A228 music wire
10–17	3.5	50.6	75.5	5.5	A229 OQ&T spring steel
10–18	3.8	31.4	71.4	12.8	B159 phosphor-bronze
10–19	4.5	69.2	215.6	8.2	A232 chrome-vanadium

10–20 Consider the steel spring in the illustration.

(a) Find the pitch, solid height, and number of active turns.

(b) Find the spring rate. Assume the material is A227 HD steel.

(c) Find the force F_s required to close the spring solid.

(d) Find the shear stress in the spring due to the force F_s.

Problem 10–20

— 120 mm —
50 mm
3.4 mm

10–21 A static service music wire helical compression spring is needed to support a 90-N load after being compressed 50 mm. The solid height of the spring cannot exceed 38 mm. The free length must not exceed 100 mm. The static factor of safety must equal or exceed 1.2. For robust linearity use a fractional overrun to closure ξ of 0.15. There are two springs to be designed. Start with a wire diameter of 1.9 mm.

(a) The spring must operate over a 20-mm rod. A 1.2-mm diametral clearance allowance should be adequate to avoid interference between the rod and the spring due to out-of-round coils. Design the spring.

(b) The spring must operate in a 25-mm-diameter hole. A 1.2-mm diametral clearance allowance should be adequate to avoid interference between the spring and the hole due to swelling of the spring diameter as the spring is compressed and out-of-round coils. Design the spring.

10–22 Solve Prob. 10–21 by iterating with an initial value of $C = 10$. If you have already solved Prob. 10–21, compare the steps and the results.

10-23 A holding fixture for a workpiece 37.5 mm thick at the clamp locations is being designed. The detail of one of the clamps is shown in the figure. A spring is required to drive the clamp upward when removing the workpiece with a starting force of 45 N. The clamp screw has an M10 × 1.25 thread. Allow a diametral clearance of 1.25 mm between it and the uncompressed spring. It is further specified that the free length of the spring should be $L_0 \leq 48$ mm, the solid height $L_s \leq 31.5$ mm, and the safety factor when closed solid should be $n_s \geq 1.2$. Starting with $d = 2$ mm, design a suitable helical coil compression spring for this fixture. For A227 HD steel, wire diameters are available in 0.2-mm increments between 0.2 to 3.2 mm.

Problem 10-23
Clamping fixture.

10-24 Solve Prob. 10-23 by iterating with an initial value of $C = 8$. If you have already solved Prob. 10-23, compare the steps and the results.

10-25 Your instructor will provide you with a stock spring supplier's catalog, or pages reproduced from it. Accomplish the task of Prob. 10-23 by selecting an available stock spring. (This is design by *selection*.)

10-26 A compression spring is needed to fit over a 12-mm diameter rod. To allow for some clearance, the inside diameter of the spring is to be 15 mm. To ensure a reasonable coil, use a spring index of 10. The spring is to be used in a machine by compressing it from a free length of 125 mm in through a stroke of 75 mm to its solid length. The spring should have squared and ground ends, unpeened, and is to be made from cold-drawn wire.
(a) Determine a suitable wire diameter.
(b) Determine a suitable total number of coils.
(c) Determine the spring constant.
(d) Determine the static factor of safety when compressed to solid length.
(e) Determine the fatigue factor of safety when repeatedly cycled from free length to solid length. Use the Gerber-Zimmerli fatigue-failure criterion.

10-27 A compression spring is needed to fit within a 25 mm diameter hole. To allow for some clearance, the outside diameter of the spring is to be no larger than 22-mm. To ensure a reasonable coil, use a spring index of 8. The spring is to be used in a machine by compressing it from a free length of 75 mm to a solid length of 25 mm. The spring should have squared ends, unpeened, to be made from music wire.

(a) Determine a suitable wire diameter.

(b) Determine a suitable total number of coils.

(c) Determine the spring constant.

(d) Determine the static factor of safety when compressed to solid length.

(e) Determine the fatigue factor of safety when repeatedly cycled from free length to solid length. Use the Gerber-Zimmerli fatigue-failure criterion.

10–28   A helical compression spring is to be cycled between 600 N and 1.2 kN with a 25-mm stroke. The number of cycles is low, so fatigue is not an issue. The coil must fit in a 52-mm diameter hole with a 2-mm clearance all the way around the spring. Use unpeened oil tempered wire with squared and ground ends.

(a) Determine a suitable wire diameter, using a spring index of $C = 7$.

(b) Determine a suitable mean coil diameter.

(c) Determine the necessary spring constant.

(d) Determine a suitable total number of coils.

(e) Determine the necessary free length so that if the spring were compressed to its solid length, there would be no yielding.

10–29   The figure shows a conical compression helical coil spring where R_1 and R_2 are the initial and final coil radii, respectively, d is the diameter of the wire, and N_a is the total number of active coils. The wire cross section primarily transmits a torsional moment, which changes with the coil radius. Let the coil radius be given by

$$R = R_1 + \frac{R_2 - R_1}{2\pi N_a}\theta$$

where θ is in radians. Use Castigliano's method to estimate the spring rate as

$$k = \frac{d^4 G}{16 N_a (R_2 + R_1)(R_2^2 + R_1^2)}$$

Problem 10–29

10–30 A helical coil compression spring is needed for food service machinery. The load varies from a minimum of 20 N to a maximum of 90 N. The spring rate k is to be 1660 N/m. The outside diameter of the spring cannot exceed 62 mm. The springmaker has available suitable dies for drawing 2-, 2.3-, 2.6-, and 3-mm-diameter wire. Using a fatigue design factor n_f of 1.5, and the Gerber-Zimmerli fatigue-failure criterion, design a suitable spring.

10–31 Solve Prob. 10–30 using the Goodman-Zimmerli fatigue-failure criterion.

10–32 Solve Prob. 10–30 using the Sines-Zimmerli fatigue-failure criterion.

10–33 Design the spring of Ex. 10–5 using the Gerber-Zimmerli fatigue-failure criterion.

10–34 Solve Prob. 10–33 using the Goodman-Zimmerli fatigue-failure criterion.

10–35 A hard-drawn spring steel extension spring is to be designed to carry a static load of 90 N with an extension of 12 mm using a design factor of $n_y = 1.5$ in bending. Use full-coil end hooks with the fullest bend radius of $r = D/2$ and $r_2 = 2d$. The free length must be less than 75 mm, and the body turns must be fewer than 30. (Note: Integer and half-integer body turns allow end hooks to be placed in the same plane. However, this adds extra cost and is done only when necessary.)

10–36 Design a compression spring with plain ends using hard-drawn wire. The deflection is to be 56 mm when the force is 90 N and to close solid when the force is 100 N. Upon closure, use a design factor of 1.2 guarding against yielding. Select the smallest gauge W&M (Washburn & Moen) wire.

10–37 Design an infinite-life helical coil extension spring with full end loops and generous loop-bend radii for a minimum load of 40 N and a maximum load of 80 N, with an accompanying stretch of 6 mm. The spring is for food-service equipment and must be stainless steel. The outside diameter of the coil cannot exceed 25 mm, and the free length cannot exceed 62 mm. Using a fatigue design factor of $n_f = 2$, complete the design. Use the Gerber criterion with Table 10–8.

10–38 Prove Eq. (10–40). *Hint:* Using Castigliano's theorem, determine the deflection due to bending of an end hook alone as if the hook were fixed at the end connecting it to the body of the spring. Consider the wire diameter d small as compared to the mean radius of the hook, $R = D/2$. Add the deflections of the end hooks to the deflection of the main body to determine the final spring constant, then equate it to Eq. (10–9).

10–39 The figure shows a finger exerciser used by law-enforcement officers and athletes to strengthen their grip. It is formed by winding A227 hard-drawn steel wire around a mandrel to obtain $2\frac{1}{2}$ turns when the grip is in the closed position. After winding, the wire is cut to leave the two legs as handles. The plastic handles are then molded on, the grip is squeezed together, and a wire clip is placed around the legs to obtain initial "tension" and to space the handles for the best initial gripping position. The clip is formed like a figure 8 to prevent it from coming off. When the grip is in the closed position, the stress in the spring should not exceed the permissible stress.
(a) Determine the configuration of the spring before the grip is assembled.
(b) Find the force necessary to close the grip.

Problem 9–39
Dimensions in millimeters.

10–40 The rat trap shown in the figure uses two opposite-image torsion springs. The wire has a diameter of 2 mm, and the outside diameter of the spring in the position shown is 12 mm. Each spring has 11 turns. Use of a fish scale revealed a force of about 36 N is needed to set the trap.

(a) Find the probable configuration of the spring prior to assembly.

(b) Find the maximum stress in the spring when the trap is set.

Problem 10–40

10–41 Wire form springs can be made in a variety of shapes. The clip shown operates by applying a force F. The wire diameter is d, the length of the straight section is l, and Young's modulus is E. Consider the effects of bending only, with $d \ll R$.

(a) Use Castigliano's theorem to determine the spring rate k.

(b) Determine the spring rate if the clip is made from 2-mm diameter A227 hard-drawn steel wire with $R = 6$ mm and $l = 25$ mm.

(c) For part (b), estimate the value of the load F, which will cause the wire to yield.

Problem 10–41

10–42   For the wire form shown, the wire diameter is d, the length of the straight section is l, and Young's modulus is E. Consider the effects of bending only, with $d \ll R$.

(a) Use Castigliano's method to determine the spring rate k.

(b) Determine the spring rate if the form is made from 0.7-mm diameter A313 stainless wire with 15 mm in and 12 mm.

(c) For part (b), estimate the value of the load F, which will cause the wire to yield.

Problem 10–42

10–43   Figure 10–14b shows a spring of constant thickness and constant stress. A constant stress spring can be designed where the width b is constant as shown.

(a) Determine how h varies as a function of x.

(b) Given Young's modulus E, determine the spring rate k in terms of E, l, b, and h_o. Verify the units of k.

Problem 10–43

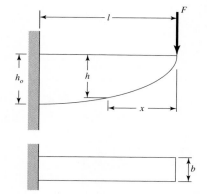

10–44   Using the experience gained with Prob. 10–30, write a computer program that would help in the design of helical coil compression springs.

10–45   Using the experience gained with Prob. 10–37, write a computer program that would help in the design of a helical coil extension spring.

10-42 For the wire form shown, the wire diameter is d, the length of the straight section is A, and Young's modulus is E. Consider the effects of bending only, with $d \ll R$.

(a) Use Castigliano's method to determine the spring rate k.

(b) Determine the spring rate if the form is made from 0.2-mm diameter A313 stainless wire with 15 mm in and 12 mm.

(c) For part (b), estimate the value of the load F, which will cause the wire to yield.

Problem 10-42

10-43 Figure 10-11b shows a spring of constant thickness and constant stress. A constant-stress spring can be designed where the width b is constant as shown.

(a) Determine how b varies as a function of x.

(b) Given Young's modulus E, determine the spring rate k in terms of E, b, o, and δ_{max}. Verify the units of k.

Problem 10-43

10-44 Using the experience gained with Prob. 10-30, write a computer program that would help in the design of helical coil compression springs.

10-45 Using the experience gained with Prob. 10-37, write a computer program that would help in the design of a helical coil extension spring.

11

Rolling-Contact Bearings

Chapter Outline

The terms *rolling-contact bearing, antifriction bearing,* and *rolling bearing* are all used to describe that class of bearing in which the main load is transferred through elements in rolling contact rather than in sliding contact. In a rolling bearing the starting friction is about twice the running friction, but still it is negligible in comparison with the starting friction of a sleeve bearing. Load, speed, and the operating viscosity of the lubricant do affect the frictional characteristics of a rolling bearing. It is probably a mistake to describe a rolling bearing as "antifriction," but the term is used generally throughout the industry.

From the mechanical designer's standpoint, the study of antifriction bearings differs in several respects when compared with the study of other topics because the bearings they specify have already been designed. The specialist in antifriction-bearing design is confronted with the problem of designing a group of elements that compose a rolling bearing: these elements must be designed to fit into a space whose dimensions are specified; they must be designed to receive a load having certain characteristics; and finally, these elements must be designed to have a satisfactory life when operated under the specified conditions. Bearing specialists must therefore consider such matters as fatigue loading, friction, heat, corrosion resistance, kinematic problems, material properties, lubrication, machining tolerances, assembly, use, and cost. From a consideration of all these factors, bearing specialists arrive at a compromise that, in their judgment, is a good solution to the problem as stated.

We begin with an overview of bearing types; then we note that bearing life cannot be described in deterministic form. We introduce the invariant, the statistical distribution of bearing life, which is described by the Weibull distribution. There are some useful deterministic equations addressing load versus life at constant reliability, and the catalog rating as rating life is introduced. The load-life-reliability relationship combines statistical and deterministic relationships, which gives the designer a way to move from the desired load and life to the catalog rating in *one* equation.

Ball bearings also resist thrust, and a unit of thrust does different damage per revolution than a unit of radial load, so we must find the equivalent pure radial load that does the same damage as the existing radial and thrust loads. Next, variable loading, stepwise and continuous, is approached, and the equivalent pure radial load doing the same damage is quantified. Oscillatory loading is mentioned.

With this preparation we have the tools to consider the selection of ball and cylindrical roller bearings. The question of misalignment is quantitatively approached.

Tapered roller bearings have some complications, and our experience so far contributes to understanding them.

Having the tools to find the proper catalog ratings, we make decisions (selections), we perform a design assessment, and the bearing reliability is quantified. Lubrication and mounting conclude our introduction. Vendors' manuals should be consulted for specific details relating to bearings of their manufacture.

11-1 Bearing Types

Bearings are manufactured to take pure radial loads, pure thrust loads, or a combination of the two kinds of loads. The nomenclature of a ball bearing is illustrated in Fig. 11–1, which also shows the four essential parts of a bearing. These are the outer ring, the inner ring, the balls or rolling elements, and the separator. In low-priced bearings, the separator is sometimes omitted, but it has the important function of separating the elements so that rubbing contact will not occur.

Figure 11–1

Nomenclature of a ball bearing. (*General Motors Corp. Used with permission, GM Media Archives.*)

Figure 11–2

Various types of ball bearings.

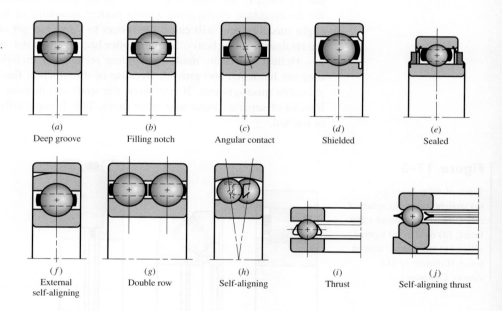

(a)	(b)	(c)	(d)	(e)
Deep groove	Filling notch	Angular contact	Shielded	Sealed

(f)	(g)	(h)	(i)	(j)
External self-aligning	Double row	Self-aligning	Thrust	Self-aligning thrust

In this section we include a selection from the many types of standardized bearings that are manufactured. Most bearing manufacturers provide engineering manuals and brochures containing lavish descriptions of the various types available. In the small space available here, only a meager outline of some of the most common types can be given. So you should include a survey of bearing manufacturers' literature in your studies of this section.

Some of the various types of standardized bearings that are manufactured are shown in Fig. 11–2. The single-row deep-groove bearing will take radial load as well as some thrust load. The balls are inserted into the grooves by moving the inner ring to an eccentric position. The balls are separated after loading, and the separator is then inserted. The use of a filling notch (Fig. 11–2b) in the inner and outer rings enables a greater number of balls to be inserted, thus increasing the load capacity. The thrust capacity is decreased, however, because of the bumping of the balls against the edge of the notch when thrust loads are present. The angular-contact bearing (Fig. 11–2c) provides a greater thrust capacity.

All these bearings may be obtained with shields on one or both sides. The shields are not a complete closure but do offer a measure of protection against dirt. A variety of bearings are manufactured with seals on one or both sides. When the seals are on both sides, the bearings are lubricated at the factory. Although a sealed bearing is supposed to be lubricated for life, a method of relubrication is sometimes provided.

Single-row bearings will withstand a small amount of shaft misalignment of deflection, but where this is severe, self-aligning bearings may be used. Double-row bearings are made in a variety of types and sizes to carry heavier radial and thrust loads. Sometimes two single-row bearings are used together for the same reason, although a double-row bearing will generally require fewer parts and occupy less space. The one-way ball thrust bearings (Fig. 11–2i) are made in many types and sizes.

Some of the large variety of standard roller bearings available are illustrated in Fig. 11–3. Straight roller bearings (Fig. 11–3a) will carry a greater radial load than ball bearings of the same size because of the greater contact area. However, they have the disadvantage of requiring almost perfect geometry of the raceways and rollers. A slight misalignment will cause the rollers to skew and get out of line. For this reason, the retainer must be heavy. Straight roller bearings will not, of course, take thrust loads.

Helical rollers are made by winding rectangular material into rollers, after which they are hardened and ground. Because of the inherent flexibility, they will take considerable misalignment. If necessary, the shaft and housing can be used for raceways instead of separate inner and outer races. This is especially important if radial space is limited.

Figure 11–3

Types of roller bearings: (a) straight roller; (b) spherical roller, thrust; (c) tapered roller, thrust; (d) needle; (e) tapered roller; (f) steep-angle tapered roller. (*Courtesy of The Timken Company.*)

(a)   (b)   (c)

(d)   (e)   (f)

The spherical-roller thrust bearing (Fig. 11–3b) is useful where heavy loads and misalignment occur. The spherical elements have the advantage of increasing their contact area as the load is increased.

Needle bearings (Fig. 11–3d) are very useful where radial space is limited. They have a high load capacity when separators are used, but may be obtained without separators. They are furnished both with and without races.

Tapered roller bearings (Fig. 11–3e, f) combine the advantages of ball and straight roller bearings, since they can take either radial or thrust loads or any combination of the two, and in addition, they have the high load-carrying capacity of straight roller bearings. The tapered roller bearing is designed so that all elements in the roller surface and the raceways intersect at a common point on the bearing axis.

The bearings described here represent only a small portion of the many available for selection. Many special-purpose bearings are manufactured, and bearings are also made for particular classes of machinery. Typical of these are:

- Instrument bearings, which are high-precision and are available in stainless steel and high-temperature materials
- Nonprecision bearings, usually made with no separator and sometimes having split or stamped sheet-metal races
- Ball bushings, which permit either rotation or sliding motion or both
- Bearings with flexible rollers

11–2 Bearing Life

When the ball or roller of rolling-contact bearings rolls, contact stresses occur on the inner ring, the rolling element, and on the outer ring. Because the curvature of the contacting elements in the axial direction is different from that in the radial direction, the equations for these stresses are more involved than in the Hertz equations presented in Chap. 3. If a bearing is clean and properly lubricated, is mounted and sealed against the entrance of dust and dirt, is maintained in this condition, and is operated at reasonable temperatures, then metal fatigue will be the only cause of failure. Inasmuch as metal fatigue implies many millions of stress applications successfully endured, we need a quantitative life measure. Common life measures are

- Number of revolutions of the inner ring (outer ring stationary) until the first tangible evidence of fatigue
- Number of hours of use at a standard angular speed until the first tangible evidence of fatigue

The commonly used term is *bearing life,* which is applied to either of the measures just mentioned. It is important to realize, as in all fatigue, life as defined above is a stochastic variable and, as such, has both a distribution and associated statistical parameters. The life measure of an individual bearing is defined as the total number of revolutions (or hours at a constant speed) of bearing operation until the failure criterion is developed. Under ideal conditions, the fatigue failure consists of spalling of the load-carrying surfaces. The American Bearing Manufacturers Association (ABMA) standard states that the failure criterion is the first evidence of fatigue. The fatigue criterion used by the Timken Company laboratories is the spalling or pitting of an area of 0.01 in^2. Timken also observes that the useful life of the bearing may extend considerably beyond this point. This is an operational definition of fatigue failure in rolling bearings.

The *rating life* is a term sanctioned by the ABMA and used by most manufacturers. The rating life of a group of nominally identical ball or roller bearings is defined as the number of revolutions (or hours at a constant speed) that 90 percent of a group of bearings will achieve or exceed before the failure criterion develops. The terms *minimum life*, L_{10} *life*, and B_{10} *life* are also used as synonyms for rating life. The rating life is the 10th percentile location of the bearing group's revolutions-to-failure distribution.

Median life is the 50th percentile life of a group of bearings. The term *average life* has been used as a synonym for median life, contributing to confusion. When many groups of bearings are tested, the median life is between 4 and 5 times the L_{10} life.

Each bearing manufacturer will choose a specific rating life for which load ratings of its bearings are reported. The most commonly used rating life is 10^6 revolutions. The Timken Company is a well-known exception, rating its bearings at 3 000 hours at 500 rev/min, which is $90(10^6)$ revolutions. These levels of rating life are actually quite low for today's bearings, but since rating life is an arbitrary reference point, the traditional values have generally been maintained.

11–3 Bearing Load Life at Rated Reliability

When nominally identical groups are tested to the life-failure criterion at different loads, the data are plotted on a graph as depicted in Fig. 11–4 using a log-log transformation. To establish a single point, load F_1 and the rating life of group one $(L_{10})_1$ are the coordinates that are logarithmically transformed. The reliability associated with this point, and all other points, is 0.90. Thus we gain a glimpse of the load-life function at 0.90 reliability. Using a regression equation of the form

$$FL^{1/a} = \text{constant} \qquad (11\text{–}1)$$

the result of many tests for various kinds of bearings result in

- $a = 3$ for ball bearings
- $a = 10/3$ for roller bearings (cylindrical and tapered roller)

A *catalog load rating* is defined as the radial load that causes 10 percent of a group of bearings to fail at the bearing manufacturer's rating life. We shall denote the catalog load rating as C_{10}. The catalog load rating is often referred to as a *Basic Dynamic Load Rating*, or sometimes just Basic Load Rating, if the manufacturer's rating life is 10^6 revolutions. The radial load that would be necessary to cause failure at such a low life would be unrealistically high. Consequently, the Basic Load Rating should be viewed as a reference value, and not as an actual load to be achieved by a bearing.

Figure 11–4

Typical bearing load-life log-log curve.

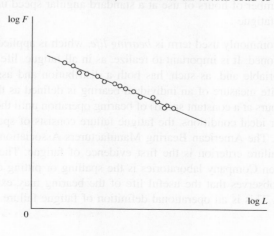

In selecting a bearing for a given application, it is necessary to relate the desired load and life requirements to the published catalog load rating corresponding to the catalog rating life. From Eq. (11–1) we can write

$$F_1 L_1^{1/a} = F_2 L_2^{1/a}$$ (11–2)

where the subscripts 1 and 2 can refer to any set of load and life conditions. Letting F_1 and L_1 correlate with the catalog load rating and rating life, and F_2 and L_2 correlate with desired load and life for the application, we can express Eq. (11–2) as

$$F_R L_R^{1/a} = F_D L_D^{1/a}$$ (a)

where the units of L_R and L_D are revolutions, and the subscripts R and D stand for Rated and Desired.

It is sometimes convenient to express the life in hours at a given speed. Accordingly, any life L in revolutions can be expressed as

$$L = 60 \, \mathscr{L} n$$ (b)

where $\mathscr{L}$ is in hours, n is in rev/min, and 60 min/h is the appropriate conversion factor. Incorporating Eq. (b) into Eq. (a),

$$F_R (\mathscr{L}_R n_R 60)^{1/a} = F_D (\mathscr{L}_D n_D 60)^{1/a}$$ (c)

catalog rating, lbf or kN   desired speed, rev/min

rating life in hours   desired life, hours

rating speed, rev/min   desired radial load, lbf or kN

Solving Eq. (c) for F_R, and noting that it is simply an alternate notation for the catalog load rating C_{10}, we obtain an expression for a catalog load rating as a function of the desired load, desired life, and catalog rating life.

$$C_{10} = F_R = F_D \left(\frac{L_D}{L_R} \right)^{1/a} = F_D \left(\frac{\mathscr{L}_D n_D 60}{\mathscr{L}_R n_R 60} \right)^{1/a}$$ (11–3)

It is sometimes convenient to define $x_D = L_D / L_R$ as a dimensionless *multiple of rating life.*

EXAMPLE 11–1

Consider SKF, which rates its bearings for 1 million revolutions. If you desire a life of 5000 h at 1725 rev/min with a load of 2 kN with a reliability of 90 percent, for which catalog rating would you search in an SKF catalog?

Solution

The rating life is $L_{10} = L_R = \mathscr{L}_R n_R 60 = 10^6$ revolutions. From Eq. (11–3),

Answer

$$C_{10} = F_D \left(\frac{\mathscr{L}_D n_D 60}{\mathscr{L}_R n_R 60} \right)^{1/a} = 400 \left[\frac{5000(1725)60}{10^6} \right]^{1/3} = 16.1 \text{ kN}$$

11–4   Reliability versus Life—The Weibull Distribution

At constant load, the life measure distribution of rolling-contact bearings is right skewed. Because of its robust ability to adjust to varying amounts of skewness, the *three-parameter Weibull distribution* is used exclusively for expressing the reliability of rolling-contact bearings. Unlike the development of the normal distribution in Sec. 1–12, we will begin with the definition of the reliability, R, for a Weibull distribution of the life measure, x, as

$$R = \exp\left[-\left(\frac{x - x_0}{\theta - x_0}\right)^b\right] \tag{11–4}$$

where the three parameters are[1]

x_0 = guaranteed, or "minimum," value of x

θ = characteristic parameter. For rolling-contact bearings, this corresponds to the 63.2121 percentile value of x

b = shape parameter that controls the skewness. For rolling-contact bearings, $b \approx 1.5$

The life measure is expressed in dimensionless form as $x = L/L_{10}$.

From Eq. (1–8), $R = 1 - p$, where p is the probability of a value of x occurring between $-\vartheta$ and x, and is the integral of the probability distribution, $f(x)$, between those limits. Accordingly, $f(x) = -dR/dx$. Thus, from the derivative of Eq. (11–4), the Weibull probability density function, $f(x)$, is given by

$$f(x) = \begin{cases} \dfrac{b}{\theta - x_0}\left(\dfrac{x - x_0}{\theta - x_0}\right)^{b-1} \exp\left[-\left(\dfrac{x - x_0}{\theta - x_0}\right)^b\right] & x \geq x_0 \geq 0 \\ 0 & x < x_0 \end{cases} \tag{11–5}$$

The mean and standard deviation of $f(x)$ are

$$\mu_x = x_0 + (\theta - x_0)\Gamma(1 + 1/b) \tag{11–6}$$

$$\hat{\sigma}_x = (\theta - x_0)\sqrt{\Gamma(1 + 2/b) - \Gamma^2(1 + 1/b)} \tag{11–7}$$

where Γ is the *gamma function,* and is found tabulated in Table A–34.

Given a specific required reliability, solving Eq. (11–4) for x yields

$$x = x_0 + (\theta - x_0)\left(\ln\frac{1}{R}\right)^{1/b} \tag{11–8}$$

EXAMPLE 11–2   Construct the distributional properties of a 02–30 millimeter deep-groove ball bearing if the Weibull parameters are $x_0 = 0.020$, $\theta = 4.459$, and $b = 1.483$.

Find the mean, median, 10th percentile life, standard deviation, and coefficient of variation.

---

[1]To estimate the Weibull parameters from data, see J. E. Shigley and C. R. Mischke, *Mechanical Engineering Design*, 5th ed., McGraw-Hill, New York, 1989, Sec. 4–12, Ex. 4–10.

| Solution | From Eq. (11–6) and interpolating Table A–34, the mean dimensionless life is |

| Answer | $$\mu_x = x_0 + (\theta - x_0)\Gamma(1 + 1/b)$$ |

$$= 0.020 + (4.459 - 0.020)\Gamma(1 + 1/1.483)$$

$$= 0.020 + 4.439\Gamma(1.67431) = 0.020 + 4.439(0.9040) = 4.033$$

This says that the average bearing life is $4.033\ L_{10}$.

The median dimensionless life corresponds to $R = 0.50$, or L_{50}, and from Eq. (11–8) is

| Answer |

$$x_{0.50} = x_0 + (\theta - x_0)\left(\ln\frac{1}{0.50}\right)^{1/b}$$

$$= 0.020 + (4.459 - 0.020)\left(\ln\frac{1}{0.50}\right)^{1/1.483} = 3.487$$

or, $L = 3.487\ L_{10}$.

The 10th percentile value of the dimensionless life x is

| Answer |

$$x_{0.10} = 0.020 + (4.459 - 0.020)\left(\ln\frac{1}{0.90}\right)^{1/1.483} \approx 1 \qquad \text{(as it should be)}$$

The standard deviation of the dimensionless life, given by Eq. (11–7), is

| Answer |

$$\hat{\sigma}_x = (\theta - x_0)\sqrt{\Gamma(1 + 2/b) - \Gamma^2(1 + 1/b)}$$

$$= (4.459 - 0.020)\sqrt{\Gamma(1 + 2/1.483) - \Gamma^2(1 + 1/1.483)}$$

$$= 4.439\sqrt{\Gamma(2.349) - \Gamma^2(1.674)} = 4.439\sqrt{1.2023 - 0.9040^2}$$

$$= 2.755$$

The coefficient of variation of the dimensionless life is

| Answer |

$$C_x = \frac{\hat{\sigma}_x}{\mu_x} = \frac{2.755}{4.033} = 0.683$$

11–5 Relating Load, Life, and Reliability

This is the designer's problem. The desired load is not the manufacturer's test load or catalog entry. The desired speed is different from the vendor's test speed, and the reliability expectation is typically much higher than the 0.90 accompanying the catalog entry. Figure 11–5 shows the situation. The catalog information is plotted as point A, whose coordinates are (the logs of) C_{10} and $x_{10} = L_{10}/L_{10} = 1$, a point on the 0.90 reliability contour. The design point is at D, with the coordinates (the logs of) F_D and x_D, a point that is on the $R = R_D$ reliability contour. The designer must move from point D to point A via point B as follows. Along a constant reliability contour (BD), Eq. (11–2) applies:

$$F_B x_B^{1/a} = F_D x_D^{1/a}$$

from which

$$F_B = F_D\left(\frac{x_D}{x_B}\right)^{1/a} \qquad (a)$$

Figure 11–5

Constant reliability contours. Point A represents the catalog rating C_{10} at $x = L/L_{10} = 1$. Point B is on the target reliability design line R_D, with a load of C_{10}. Point D is a point on the desired reliability contour exhibiting the design life $x_D = L_D/L_{10}$ at the design load F_D.

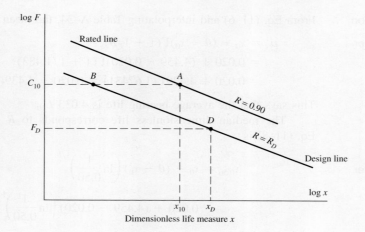

Along a constant load line (AB), Eq. (11–4) applies:

$$R_D = \exp\left[-\left(\frac{x_B - x_0}{\theta - x_0}\right)^b\right]$$

Solving for x_B gives

$$x_B = x_0 + (\theta - x_0)\left(\ln\frac{1}{R_D}\right)^{1/b}$$

Now substitute this in Eq. (a) to obtain

$$F_B = F_D\left(\frac{x_D}{x_B}\right)^{1/a} = F_D\left[\frac{x_D}{x_0 + (\theta - x_0)[\ln(1/R_D)]^{1/b}}\right]^{1/a}$$

Noting that $F_B = C_{10}$, and including an application factor a_f with the design load,

$$C_{10} = a_f F_D\left[\frac{x_D}{x_0 + (\theta - x_0)[\ln(1/R_D)]^{1/b}}\right]^{1/a} \tag{11–9}$$

The application factor serves as a factor of safety to increase the design load to take into account overload, dynamic loading, and uncertainty. Typical load application factors for certain types of applications will be discussed shortly.

Eq. (11–9) can be simplified slightly for calculator entry by noting that

$$\ln\frac{1}{R_D} = \ln\frac{1}{1 - p_f} = \ln(1 + p_f + \cdots) \approx p_f = 1 - R_D$$

where p_f is the probability for failure. Equation (11–9) can be written as

$$C_{10} \approx a_f F_D\left[\frac{x_D}{x_0 + (\theta - x_0)(1 - R_D)^{1/b}}\right]^{1/a} \qquad R \geq 0.90 \tag{11–10}$$

Either Eq. (11–9) or Eq. (11–10) may be used to convert from a design situation with a desired load, life, and reliability to a catalog load rating based on a rating life at 90 percent reliability. Note that when $R_D = 0.90$, the denominator is equal to one, and the equation reduces to Eq. (11–3). The Weibull parameters are usually provided in the manufacturer's catalog. Typical values are given on p. 601 at the beginning of the end-of-chapter problems.

EXAMPLE 11–3   The design load on a ball bearing is 1840 N and an application factor of 1.2 is appropriate. The speed of the shaft is to be 400 rev/min, the life to be 30 kh with a reliability of 0.99. What is the C_{10} catalog entry to be sought (or exceeded) when searching for a deep-groove bearing in a manufacturer's catalog on the basis of 10^6 revolutions for rating life? The Weibull parameters are $x_0 = 0.02$, $(\theta - x_0) = 4.439$, and $b = 1.483$.

Solution

$$x_D = \frac{L_D}{L_R} = \frac{60 \mathcal{L}_D n_D}{L_{10}} = \frac{60(30\,000)400}{10^6} = 720$$

Thus, the design life is 720 times the L_{10} life. For a ball bearing, $a = 3$. Then, from Eq. (11–10),

Answer

$$C_{10} = (1.2)(1.84)\left[\frac{720}{0.02 + 4.439(1 - 0.99)^{1/1.483}}\right]^{1/3} = 32.8 \text{ k}N$$

Shafts generally have two bearings. Often these bearings are different. If the bearing reliability of the shaft with its pair of bearings is to be R, then R is related to the individual bearing reliabilities R_A and R_B, using Eq. (1–9), as

$$R = R_A R_B$$

First, we observe that if the product $R_A R_B$ equals R, then, in general, R_A and R_B are both greater than R. Since the failure of either or both of the bearings results in the shutdown of the shaft, then A or B or both can create a failure. Second, in sizing bearings one can begin by making R_A and R_B equal to the square root of the reliability goal, $\sqrt{R}$. In Ex. 11–3, if the bearing was one of a pair, the reliability goal would be $\sqrt{0.99}$, or 0.995. The bearings selected are discrete in their reliability property in your problem, so the selection procedure "rounds up," and the overall reliability exceeds the goal R. Third, it may be possible, if $R_A > \sqrt{R}$, to round down on B yet have the product $R_A R_B$ still exceed the goal R.

11–6   Combined Radial and Thrust Loading

A ball bearing is capable of resisting radial loading and a thrust loading. Furthermore, these can be combined. Consider F_a and F_r to be the axial thrust and radial loads, respectively, and F_e to be the *equivalent radial load* that does the same damage as the combined radial and thrust loads together. A rotation factor V is defined such that $V = 1$ when the inner ring rotates and $V = 1.2$ when the outer ring rotates. Two dimensionless groups can now be formed: $F_e/(VF_r)$ and $F_a/(VF_r)$. When these two dimensionless groups are plotted as in Fig. 11–6, the data fall in a gentle curve that is well approximated by two straight-line segments. The abscissa e is defined by the intersection of the two lines. The equations for the two lines shown in Fig. 11–6 are

$$\frac{F_e}{VF_r} = 1 \qquad \text{when} \quad \frac{F_a}{VF_r} \le e \qquad (11\text{–}11a)$$

$$\frac{F_e}{VF_r} = X + Y\frac{F_a}{VF_r} \qquad \text{when} \quad \frac{F_a}{VF_r} > e \qquad (11\text{–}11b)$$

Figure 11–6

The relationship of dimensionless group $F_e/(VF_r)$ and $F_a/(VF_r)$ and the straight-line segments representing the data.

where, as shown, X is the ordinate intercept and Y is the slope of the line for $F_a/(VF_r) > e$. It is common to express Eqs. (11–11a) and (11–11b) as a single equation,

$$F_e = X_i V F_r + Y_i F_a \qquad (11\text{–}12)$$

where $i = 1$ when $F_a/(VF_r) \leq e$ and $i = 2$ when $F_a/(VF_r) > e$. The X and Y factors depend upon the geometry and construction of the specific bearing. Table 11–1 lists representative values of X_1, Y_1, X_2, and Y_2 as a function of e, which in turn is a function of F_a/C_0, where C_0 is the basic static load rating. The *basic static load rating* is the load that will produce a total permanent deformation in the raceway and rolling

Table 11–1

Equivalent Radial Load Factors for Ball Bearings

F_a/C_0	e	$F_a/(VF_r) \leq e$		$F_a/(VF_r) > e$	
		X_1	Y_1	X_2	Y_2
0.014*	0.19	1.00	0	0.56	2.30
0.021	0.21	1.00	0	0.56	2.15
0.028	0.22	1.00	0	0.56	1.99
0.042	0.24	1.00	0	0.56	1.85
0.056	0.26	1.00	0	0.56	1.71
0.070	0.27	1.00	0	0.56	1.63
0.084	0.28	1.00	0	0.56	1.55
0.110	0.30	1.00	0	0.56	1.45
0.17	0.34	1.00	0	0.56	1.31
0.28	0.38	1.00	0	0.56	1.15
0.42	0.42	1.00	0	0.56	1.04
0.56	0.44	1.00	0	0.56	1.00

*Use 0.014 if $F_a/C_0 < 0.014$.

element at any contact point of 0.0001 times the diameter of the rolling element. The basic static load rating is typically tabulated, along with the basic dynamic load rating C_{10}, in bearing manufacturers' publications. See Table 11–2, for example.

In these equations, the rotation factor V is intended to correct for the rotating-ring conditions. The factor of 1.2 for outer-ring rotation is simply an acknowledgment that the fatigue life is reduced under these conditions. Self-aligning bearings are an exception: they have $V = 1$ for rotation of either ring.

Since straight or cylindrical roller bearings will take no axial load, or very little, the Y factor is always zero.

The ABMA has established standard boundary dimensions for bearings, which define the bearing bore, the outside diameter (OD), the width, and the fillet sizes on the shaft and housing shoulders. The basic plan covers all ball and straight roller bearings in the metric sizes. The plan is quite flexible in that, for a given bore, there is an assortment of widths and outside diameters. Furthermore, the outside diameters selected are such that, for a particular outside diameter, one can usually find a variety of bearings having different bores and widths.

Table 11–2

Dimensions and Load Ratings for Single-Row 02-Series Deep-Groove and Angular-Contact Ball Bearings

Bore, mm	OD, mm	Width, mm	Fillet Radius, mm	Shoulder Diameter, mm		Load Ratings, kN			
				d_S	d_H	Deep Groove		Angular Contact	
						C_{10}	C_0	C_{10}	C_0
10	30	9	0.6	12.5	27	5.07	2.24	4.94	2.12
12	32	10	0.6	14.5	28	6.89	3.10	7.02	3.05
15	35	11	0.6	17.5	31	7.80	3.55	8.06	3.65
17	40	12	0.6	19.5	34	9.56	4.50	9.95	4.75
20	47	14	1.0	25	41	12.7	6.20	13.3	6.55
25	52	15	1.0	30	47	14.0	6.95	14.8	7.65
30	62	16	1.0	35	55	19.5	10.0	20.3	11.0
35	72	17	1.0	41	65	25.5	13.7	27.0	15.0
40	80	18	1.0	46	72	30.7	16.6	31.9	18.6
45	85	19	1.0	52	77	33.2	18.6	35.8	21.2
50	90	20	1.0	56	82	35.1	19.6	37.7	22.8
55	100	21	1.5	63	90	43.6	25.0	46.2	28.5
60	110	22	1.5	70	99	47.5	28.0	55.9	35.5
65	120	23	1.5	74	109	55.9	34.0	63.7	41.5
70	125	24	1.5	79	114	61.8	37.5	68.9	45.5
75	130	25	1.5	86	119	66.3	40.5	71.5	49.0
80	140	26	2.0	93	127	70.2	45.0	80.6	55.0
85	150	28	2.0	99	136	83.2	53.0	90.4	63.0
90	160	30	2.0	104	146	95.6	62.0	106	73.5
95	170	32	2.0	110	156	108	69.5	121	85.0

Figure 11–7

The basic ABMA plan for boundary dimensions. These apply to ball bearings, straight roller bearings, and spherical roller bearings, but not to inch-series ball bearings or tapered roller bearings. The contour of the corner is not specified. It may be rounded or chamfered, but it must be small enough to clear the fillet radius specified in the standards.

This basic ABMA plan is illustrated in Fig. 11–7. The bearings are identified by a two-digit number called the *dimension-series code.* The first number in the code is from the *width series,* 0, 1, 2, 3, 4, 5, and 6. The second number is from the *diameter series* (outside), 8, 9, 0, 1, 2, 3, and 4. Figure 11–7 shows the variety of bearings that may be obtained with a particular bore. Since the dimension-series code does not reveal the dimensions directly, it is necessary to resort to tabulations. The 02 series is used here as an example of what is available. See Table 11–2.

The housing and shaft shoulder diameters listed in the tables should be used whenever possible to secure adequate support for the bearing and to resist the maximum thrust loads (Fig. 11–8). Table 11–3 lists the dimensions and load ratings of some straight roller bearings.

To assist the designer in the selection of bearings, most of the manufacturers' handbooks contain data on bearing life for many classes of machinery, as well as information on load-application factors. Such information has been accumulated the hard way, that is, by experience, and the beginner designer should utilize this information until he or she gains enough experience to know when deviations are possible. Table 11–4 contains recommendations on bearing life for some classes of machinery. The load-application factors in Table 11–5 serve the same purpose as factors of safety; use them to increase the equivalent load before selecting a bearing.

Figure 11–8

Shaft and housing shoulder diameters d_S and d_H should be adequate to ensure good bearing support.

Table 11–3

Dimensions and Basic Load Ratings for Cylindrical Roller Bearings

| | 02-Series | | | | 03-Series | | | |
Bore, mm	OD, mm	Width, mm	Load Rating, kN C_{10}	C_0	OD, mm	Width, mm	Load Rating, kN C_{10}	C_0
25	52	15	16.8	8.8	62	17	28.6	15.0
30	62	16	22.4	12.0	72	19	36.9	20.0
35	72	17	31.9	17.6	80	21	44.6	27.1
40	80	18	41.8	24.0	90	23	56.1	32.5
45	85	19	44.0	25.5	100	25	72.1	45.4
50	90	20	45.7	27.5	110	27	88.0	52.0
55	100	21	56.1	34.0	120	29	102	67.2
60	110	22	64.4	43.1	130	31	123	76.5
65	120	23	76.5	51.2	140	33	138	85.0
70	125	24	79.2	51.2	150	35	151	102
75	130	25	93.1	63.2	160	37	183	125
80	140	26	106	69.4	170	39	190	125
85	150	28	119	78.3	180	41	212	149
90	160	30	142	100	190	43	242	160
95	170	32	165	112	200	45	264	189
100	180	34	183	125	215	47	303	220
110	200	38	229	167	240	50	391	304
120	215	40	260	183	260	55	457	340
130	230	40	270	193	280	58	539	408
140	250	42	319	240	300	62	682	454
150	270	45	446	260	320	65	781	502

Table 11–4

Bearing-Life Recommendations for Various Classes of Machinery

Type of Application	Life, kh
Instruments and apparatus for infrequent use	Up to 0.5
Aircraft engines	0.5–2
Machines for short or intermittent operation where service interruption is of minor importance	4–8
Machines for intermittent service where reliable operation is of great importance	8–14
Machines for 8-h service that are not always fully utilized	14–20
Machines for 8-h service that are fully utilized	20–30
Machines for continuous 24-h service	50–60
Machines for continuous 24-h service where reliability is of extreme importance	100–200

Table 11–5

Load-Application Factors

Type of Application	Load Factor
Precision gearing	1.0–1.1
Commercial gearing	1.1–1.3
Applications with poor bearing seals	1.2
Machinery with no impact	1.0–1.2
Machinery with light impact	1.2–1.5
Machinery with moderate impact	1.5–3.0

EXAMPLE 11–4

An SKF 6210 angular-contact ball bearing has an axial load F_a of 1780 N and a radial load F_r of 2225 N applied with the outer ring stationary. The basic static load rating C_0 is 19 800 N and the basic load rating C_{10} is 35 150 N. Estimate the L_{10} life at a speed of 720 rev/min.

Solution

$V = 1$ and $F_a/C_0 = 1780/19\ 800 = 0.090$. Interpolate for e in Table 11–1:

F_a/C_0	e	
0.084	0.28	
0.090	e	from which $e = 0.285$
0.110	0.30	

$F_a/(VF_r) = 1780/[(1)2225] = 0.8 > 0.285$. Thus, interpolate for Y_2:

F_a/C_0	Y_2	
0.084	1.55	
0.090	Y_2	from which $Y_2 = 1.527$
0.110	1.45	

From Eq. (11–9),

$$F_e = X_2 V F_r + Y_2 F_a = 0.56(1)2225 + 1.527(1780) = 3964 \text{ kN}$$

With $L_D = L_{10}$ and $F_D = F_e$, solving Eq. (11–3) for L_{10} gives

Answer

$$\mathscr{L}_{10} = \frac{60\mathscr{L}_R n_R}{60 n_D}\left(\frac{C_{10}}{F_e}\right)^a = \frac{10^6}{60(720)}\left(\frac{35\ 150}{3964}\right)^3 = 161\ 395 \text{ h}$$

We now know how to combine a steady radial load and a steady thrust load into an equivalent steady radial load F_e that inflicts the same damage per revolution as the radial-thrust combination.

11-7 Variable Loading

Bearing loads are frequently variable and occur in some identifiable patterns:

- Piecewise constant loading in a cyclic pattern
- Continuously variable loading in a repeatable cyclic pattern
- Random variation

Equation (11–1) can be written as

$$F^a L = \text{constant} = K \qquad\qquad (a)$$

Note that F may already be an equivalent steady radial load for a radial–thrust load combination. Figure 11–9 is a plot of F^a as ordinate and L as abscissa for Eq. (a). If a load level of F_1 is selected and run to the failure criterion, then the area under the F_1-L_1 trace is numerically equal to K. The same is true for a load level F_2; that is, the area under the F_2-L_2 trace is numerically equal to K. The linear damage theory says that in the case of load level F_1, the area from $L = 0$ to $L = L_A$ does damage measured by $F_1^a L_A = D$.

Consider the piecewise continuous cycle depicted in Fig. 11–10. The loads F_{ei} are equivalent steady radial loads for combined radial–thrust loads. The damage done by loads F_{e1}, F_{e2}, and F_{e3} is

$$D = F_{e1}^a l_1 + F_{e2}^a l_2 + F_{e3}^a l_3 \qquad\qquad (b)$$

Figure 11–9

Plot of F^a as ordinate and L as abscissa for $F^a L = $ constant. The linear damage hypothesis says that in the case of load F_1, the area under the curve from $L = 0$ to $L = L_A$ is a measure of the damage $D = F_1^a L_A$. The complete damage to failure is measured by $C_{10}^a L_B$.

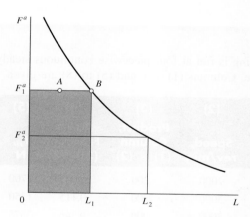

Figure 11–10

A three-part piecewise-continuous periodic loading cycle involving loads F_{e1}, F_{e2}, and F_{e3}. F_{eq} is the equivalent steady load inflicting the same damage when run for $l_1 + l_2 + l_3$ revolutions, doing the same damage D per period.

where l_i is the number of revolutions at life L_i. The equivalent steady load F_{eq} when run for $l_1 + l_2 + l_3$ revolutions does the same damage D. Thus

$$D = F_{eq}^a(l_1 + l_2 + l_3) \qquad (c)$$

Equating Eqs. (b) and (c), and solving for F_{eq}, we get

$$F_{eq} = \left[\frac{F_{e1}^a l_1 + F_{e2}^a l_2 + F_{e3}^a l_3}{l_1 + l_2 + l_3}\right]^{1/a} = \left[\sum f_i F_{ei}^a\right]^{1/a} \qquad (11\text{-}13)$$

where f_i is the fraction of revolution run up under load F_{ei}. Since l_i can be expressed as $n_i t_i$, where n_i is the rotational speed at load F_{ei} and t_i is the duration of that speed, then it follows that

$$F_{eq} = \left[\frac{\sum n_i t_i F_{ei}^a}{\sum n_i t_i}\right]^{1/a} \qquad (11\text{-}14)$$

The character of the individual loads can change, so an application factor (a_f) can be prefixed to each F_{ei} as $(a_{fi}F_{ei})^a$; then Eq. (11–13) can be written

$$F_{eq} = \left[\sum f_i(a_{fi}F_{ei})^a\right]^{1/a} \qquad L_{eq} = \frac{K}{F_{eq}^a} \qquad (11\text{-}15)$$

EXAMPLE 11-5

A ball bearing is run at four piecewise continuous steady loads as shown in the following table. Columns (1), (2), and (5) to (8) are given.

(1) Time Fraction	(2) Speed, rev/min	(3) Product, Column (1) × (2)	(4) Turns Fraction, (3)/∑(3)	(5) F_{ri}, N	(6) F_{ai}, N	(7) F_{ei}, N	(8) a_{fi}	(9) $a_{fi}F_{ei}$, N
0.1	2000	200	0.077	2700	1350	3573	1.10	3930
0.1	3000	300	0.115	1350	1350	2817	1.25	3521
0.3	3000	900	0.346	3375	1350	3951	1.10	4346
0.5	2400	1200	0.462	1688	1350	3006	1.25	3758
		2600	1.000					

Columns 1 and 2 are multiplied to obtain column 3. The column 3 entry is divided by the sum of column 3, 2600, to give column 4. Columns 5, 6, and 7 are the radial, axial, and equivalent loads respectively. Column 8 is the appropriate application factor. Column 9 is the product of columns 7 and 8.

Solution

From Eq. (11–13), with $a = 3$, the equivalent radial load F_e is

Answer

$$F_e = [0.077(3930)^3 + 0.115(3521)^3 + 0.346(4346)^3 + 0.462(3758)^3]^{1/3} = 3971 \text{ N}$$

Sometimes the question after several levels of loading is: How much life is left if the next level of stress is held until failure? Failure occurs under the linear damage hypothesis when the damage D equals the constant $K = F^a L$. Taking the first form of Eq. (11–13), we write

$$F_{eq}^a L_{eq} = F_{e1}^a l_1 + F_{e2}^a l_2 + F_{e3}^a l_3$$

and note that

$$K = F_{e1}^a L_1 = F_{e2}^a L_2 = F_{e3}^a L_3$$

and K also equals

$$K = F_{e1}^a l_1 + F_{e2}^a l_2 + F_{e3}^a l_3 = \frac{K}{L_1} l_1 + \frac{K}{L_2} l_2 + \frac{K}{L_3} l_3 = K \sum \frac{l_i}{L_i}$$

From the outer parts of the preceding equation we obtain

$$\sum \frac{l_i}{L_i} = 1 \qquad (11\text{–}16)$$

This equation was advanced by Palmgren in 1924, and again by Miner in 1945. See Eq. (6–58), p. 331.

The second kind of load variation mentioned is continuous, periodic variation, depicted by Fig. 11–11. The differential damage done by F^a during rotation through the angle $d\theta$ is

$$dD = F^a \, d\theta$$

An example of this would be a cam whose bearings rotate with the cam through the angle $d\theta$. The total damage during a complete cam rotation is given by

$$D = \int dD = \int_0^\phi F^a \, d\theta = F_{eq}^a \phi$$

from which, solving for the equivalent load, we obtain

$$F_{eq} = \left[\frac{1}{\phi} \int_0^\phi F^a \, d\theta \right]^{1/a} \qquad L_{eq} = \frac{K}{F_{eq}^a} \qquad (11\text{–}17)$$

The value of ϕ is often 2π, although other values occur. Numerical integration is often useful to carry out the indicated integration, particularly when a is not an integer and trigonometric functions are involved. We have now learned how to find the steady equivalent load that does the same damage as a continuously varying cyclic load.

Figure 11-11

A continuous load variation of a cyclic nature whose period is ϕ.

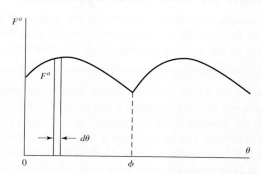

EXAMPLE 11–6

The operation of a particular rotary pump involves a power demand of $P = \bar{P} + A' \sin \theta$ where $\bar{P}$ is the average power. The bearings feel the same variation as $F = \bar{F} + A \sin \theta$. Develop an application factor a_f for this application of ball bearings.

Solution

From Eq. (11–17), with $a = 3$,

$$F_{eq} = \left(\frac{1}{2\pi} \int_0^{2\pi} F^a \, d\theta\right)^{1/a} = \left(\frac{1}{2\pi} \int_0^{2\pi} (\bar{F} + A \sin \theta)^3 \, d\theta\right)^{1/3}$$

$$= \left[\frac{1}{2\pi}\left(\int_0^{2\pi} \bar{F}^3 \, d\theta + 3\bar{F}^2 A \int_0^{2\pi} \sin \theta \, d\theta + 3\bar{F}A^2 \int_0^{2\pi} \sin^2 \theta \, d\theta \right.\right.$$

$$\left.\left. + A^3 \int_0^{2\pi} \sin^3 \theta \, d\theta\right)\right]^{1/3}$$

$$F_{eq} = \left[\frac{1}{2\pi}(2\pi\bar{F}^3 + 0 + 3\pi\bar{F}A^2 + 0)\right]^{1/3} = \bar{F}\left[1 + \frac{3}{2}\left(\frac{A}{\bar{F}}\right)^2\right]^{1/3}$$

In terms of $\bar{F}$, the application factor is

Answer

$$a_f = \left[1 + \frac{3}{2}\left(\frac{A}{\bar{F}}\right)^2\right]^{1/3}$$

We can present the result in tabular form:

$A/\bar{F}$	a_f
0	1
0.2	1.02
0.4	1.07
0.6	1.15
0.8	1.25
1.0	1.36

11–8 Selection of Ball and Cylindrical Roller Bearings

We have enough information concerning the loading of rolling-contact ball and roller bearings to develop the steady equivalent radial load that will do as much damage to the bearing as the existing loading. Now let's put it to work.

EXAMPLE 11–7

Shown in Figure 11–12 is a gear-driven squeeze roll that mates with an idler roll. The roll is designed to exert a normal force of 5.25 N/mm of roll length and a pull of 4.2 N/mm on the material being processed. The roll speed is 300 rev/min, and a design life of 30 kh is desired. Use an application factor of 1.2, and select a pair of angular-contact 02-series ball bearings from Table 11–2 to be mounted at 0 and A. Use the same size bearings at both locations and a combined reliability of at least 0.92.

Figure 11–12

Dimensions in mm

Solution   Assume concentrated forces as shown.

$$P_z = 200(4.2) = 840 \text{ N}$$

$$P_y = 200(5.25) = 1050 \text{ N}$$

$$T = 840(50) = 42\,000 \text{ N} \cdot \text{mm}$$

$$\sum T^x = -42\,000 + 38F \cos 20° = 0$$

$$F = \frac{42\,000}{38(0.940)} = 1176.2 \text{ N}$$

$$\sum M_O^z = 145P_y + 290R_A^y - 360F \sin 20° = 0;$$

thus

$$145(1050) + 290R_A^y - 360(1176.2)(0.342) = 0$$

$$R_A^y = -25.6 \text{ N}$$

$$\sum M_O^y = -145P_z - 290R_A^z - 360F \cos 20° = 0;$$

thus

$$-145(840) - 290R_A^z - 360(1176.2)(0.940) = 0$$

$$R_A^z = -1792 \text{ N};$$

$$R_A = [(-1792)^2 + (-25.6)^2]^{1/2} = 1792 \text{ N}$$

$$\sum F^z = R_O^z + P_z + R_A^z + F\cos 20° = 0$$

$$R_O^z + 840 - 1792 + 1176.2(0.940) = 0$$

$$R_O^z = -153.3 \text{ N}$$

$$\sum F^y = R_O^y + P_y + R_A^y - F\sin 20° = 0$$

$$R_O^y + 1050 - 25.6 - 1176.2(0.342) = 0$$

$$R_O^y = -622 \text{ N}$$

$$R_O = [(-153.3)^2 + (-622)^2]^{1/2} = 640.6 \text{ N}$$

So the reaction at A governs.

Reliability Goal: $\sqrt{0.92} = 0.96$

$$F_D = 1.2(1792) = 2150.4 \text{ N}$$

$$x_D = 30\,000(300)(60/10^6) = 540$$

$$C_{10} = 2150.4 \left\{ \frac{540}{0.02 + 4.439[\ln(1/0.96)]^{1/1.483}} \right\}^{1/3}$$

$$= 21.59 \text{ kN}$$

A 02–35 bearing will do.

Answer

Decision: Specify an angular-contact 02–35 mm ball bearing for the locations at A and O. Check combined reliability.

11–9 Selection of Tapered Roller Bearings

Tapered roller bearings have a number of features that make them complicated. As we address the differences between tapered roller and ball and cylindrical roller bearings, note that the underlying fundamentals are the same, but that there are differences in detail. Moreover, bearing and cup combinations are not necessarily priced in proportion to capacity. Any catalog displays a mix of high-production, low-production, and successful special-order designs. Bearing suppliers have computer programs that will take your problem descriptions, give intermediate design assessment information, and list a number of satisfactory cup-and-cone combinations in order of decreasing cost. Company sales offices provide access to comprehensive engineering services to help designers select and apply their bearings. At a large original equipment manufacturer's plant, there may be a resident bearing company representative.

Bearing suppliers provide a wealth of engineering information and detail in their catalogs and engineering guides, both in print and online. It is strongly recommended that the designer become familiar with the specifics of the supplier. It will usually utilize a similar approach as presented here, but may include various modifying factors for such things as temperature and lubrication. Many of the suppliers will provide online software tools to aid in bearing selection. The engineer will always benefit from a general understanding of the theory utilized in such software tools. Our goal here is to introduce the vocabulary, show congruence to fundamentals that were learned earlier, offer examples, and develop confidence. Finally, problems should reinforce the learning experience.

The four components of a tapered roller bearing assembly are the

- Cone (inner ring)
- Cup (outer ring)
- Tapered rollers
- Cage (spacer-retainer)

The assembled bearing consists of two separable parts: (1) the cone assembly: the cone, the rollers, and the cage; and (2) the cup. Bearings can be made as single-row, two-row, four-row, and thrust-bearing assemblies. Additionally, auxiliary components such as spacers and closures can be used. Figure 11–13 shows the nomenclature of a tapered roller bearing, and the point G through which radial and axial components of load act.

A tapered roller bearing can carry both radial and thrust (axial) loads, or any combination of the two. However, even when an external thrust load is not present, the radial load will induce a thrust reaction within the bearing because of the taper. To avoid the separation of the races and the rollers, this thrust must be resisted by an equal and opposite force. One way of generating this force is to always use at least two tapered roller bearings on a shaft. Two bearings can be mounted with the cone backs facing each other, in a configuration called *direct mounting,* or with the cone fronts facing each other, in what is called *indirect mounting.*

Figure 11–14 shows a pair of tapered roller bearings mounted directly (*b*) and indirectly (*a*) with the bearing reaction locations A_0 and B_0 shown for the shaft. For the shaft as a beam, the span is a_e, the effective spread. It is through points A_0 and B_0 that the radial loads act perpendicular to the shaft axis, and the thrust loads act along the shaft axis. The geometric spread a_g for the direct mounting is greater than for the indirect mounting. With indirect mounting the bearings are closer together compared to the direct mounting; however, the system stability is the same (a_e is the same in both cases). Thus direct and indirect mounting involve space and compactness needed or desired, but with the same system stability.

In addition to the usual ratings and geometry information, catalog data for tapered roller bearings will include the location of the effective force center. Two sample pages from a Timken catalog are shown in Fig. 11–15.

Figure 11–13

Nomenclature of a tapered roller bearing. Point G is the location of the effective load center; use this point to estimate the radial bearing load. *(Courtesy of The Timken Company.)*

Figure 11–14

Comparison of mounting stability between indirect and direct mountings. *(Courtesy of The Timken Company.)*

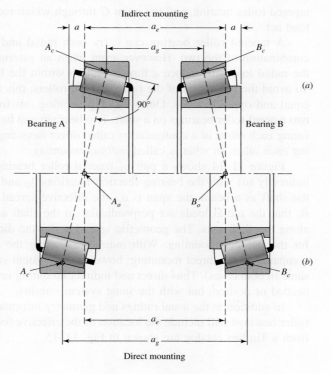

A radial load on a tapered roller bearing will induce a thrust reaction. The *load zone* includes about half the rollers and subtends an angle of approximately 180°. Using the symbol F_i for the induced thrust load from a radial load with a 180° load zone, Timken provides the equation

$$F_i = \frac{0.47F_r}{K} \tag{11-18}$$

where the K factor is geometry-specific, and is the ratio of the radial load rating to the thrust load rating. The K factor can be first approximated with 1.5 for a radial bearing and 0.75 for a steep angle bearing in the preliminary selection process. After a possible bearing is identified, the exact value of K for each bearing can be found in the bearing catalog.

A shaft supported by a pair of direct-mounted tapered roller bearings is shown in Fig. 11–16. Force vectors are shown as applied to the shaft. F_{rA} and F_{rB} are the radial loads carried by the bearings, applied at the effective force centers G_A and G_B. The induced loads F_{iA} and F_{iB} due to the effect of the radial loads on the tapered bearings are also shown. Additionally, there may be an externally applied thrust load F_{ae} on the shaft from some other source, such as the axial load on a helical gear. Since the bearings experience both radial and thrust loads, it is necessary to determine equivalent radial loads. Following the form of Eq. (11–12), where $F_e = XVF_r + YF_a$, Timken recommends using $X = 0.4$ and $V = 1$ for all cases, and using the K factor for the specific bearing for Y. This gives an equation of the form

$$F_e = 0.4F_r + KF_a \tag{a}$$

The axial load F_a is the net axial load carried by the bearing due to the combination of the induced axial load from the other bearing and the external axial load. However, only one of the bearings will carry the net axial load, and which one it is depends on the direction the bearings are mounted, the relative magnitudes of the induced loads, the direction of the external load, and whether the shaft or the housing is the moving part. Timken handles it with a table containing each of the configurations and a sign convention on the external loads. It further requires the application to be oriented horizontally with left and right bearings that must match the left and right sign conventions. Here, we will present a method that gives equivalent results, but that is perhaps more conducive to visualizing and understanding the logic behind it.

First, determine visually which bearing is being "squeezed" by the external thrust load, and label it as bearing A. Label the other bearing as bearing B. For example, in Fig. 11–16, the external thrust F_{ae} causes the shaft to push to the left against the cone of the left bearing, squeezing it against the rollers and the cup. On the other hand, it tends to pull apart the cup from the right bearing. The left bearing is therefore labeled as bearing A. If the direction of F_{ae} were reversed, then the right bearing would be labeled as bearing A. This approach to labeling the bearing being squeezed by the external thrust is applied similarly regardless of whether the bearings are mounted directly or indirectly, regardless of whether the shaft or the housing carries the external thrust, and regardless of the orientation of the assembly. To clarify by

SINGLE-ROW STRAIGHT BORE

bore	outside diameter	width	rating at 500 rpm for 3000 hours L_{10} one-row radial	thrust	factor	eff. load center	part numbers cone	cup	cone max shaft fillet radius	width	backing shoulder diameters		cup max housing fillet radius	width	backing shoulder diameters	
d	**D**	**T**	**N lbf**	**N lbf**	**K**	**a②**	cone	cup	**R①**	**B**	**d_b**	**d_a**	**r①**	**C**	**D_b**	**D_a**
25.000	52.000	16.250	8190	5260	1.56	−3.6	◆30205	◆30205	1.0	15.000	30.5	29.0	1.0	13.000	46.0	48.5
0.9843	2.0472	0.6398	1840	1180		−0.14			0.04	0.5906	1.20	1.14	0.04	0.5118	1.81	1.91
25.000	52.000	19.250	9520	9510	1.00	−3.0	◆32205-B	◆32205-B	1.0	18.000	34.0	31.0	1.0	15.000	43.5	49.5
0.9843	2.0472	0.7579	2140	2140		−0.12			0.04	0.7087	1.34	1.22	0.04	0.5906	1.71	1.95
25.000	52.000	22.000	13200	7960	1.66	−7.6	◆33205	◆33205	1.0	22.000	34.0	30.5	1.0	18.000	44.5	49.0
0.9843	2.0472	0.8661	2980	1790		−0.30			0.04	0.8661	1.34	1.20	0.04	0.7087	1.75	1.93
25.000	62.000	18.250	13000	6680	1.95	−5.1	◆30305	◆30305	1.5	17.000	32.5	30.0	1.5	15.000	55.0	57.0
0.9843	2.4409	0.7185	2930	1500		−0.20			0.06	0.6693	1.28	1.18	0.06	0.5906	2.17	2.24
25.000	62.000	25.250	17400	8930	1.95	−9.7	◆32305	◆32305	1.5	24.000	35.0	31.5	1.5	20.000	54.0	57.0
0.9843	2.4409	0.9941	3910	2010		−0.38			0.06	0.9449	1.38	1.24	0.06	0.7874	2.13	2.24
25.159	50.005	13.495	6990	4810	1.45	−2.8	07096	07196	1.5	14.260	31.5	29.5	1.0	9.525	44.5	47.0
0.9905	1.9687	0.5313	1570	1080		−0.11			0.06	0.5614	1.24	1.16	0.04	0.3750	1.75	1.85
25.400	50.005	13.495	6990	4810	1.45	−2.8	07100	07196	1.0	14.260	30.5	29.5	1.0	9.525	44.5	47.0
1.0000	1.9687	0.5313	1570	1080		−0.11			0.04	0.5614	1.20	1.16	0.04	0.3750	1.75	1.85
25.400	50.005	13.495	6990	4810	1.45	−2.8	07100-S	07196	1.5	14.260	31.5	29.5	1.0	9.525	44.5	47.0
1.0000	1.9687	0.5313	1570	1080		−0.11			0.06	0.5614	1.24	1.16	0.04	0.3750	1.75	1.85
25.400	50.292	14.224	7210	4620	1.56	−3.3	L44642	L44610	3.5	14.732	36.0	29.5	1.3	10.668	44.5	47.0
1.0000	1.9800	0.5600	1620	1040		−0.13			0.14	0.5800	1.42	1.16	0.05	0.4200	1.75	1.85
25.400	50.292	14.224	7210	4620	1.56	−3.3	L44643	L44610	1.3	14.732	31.5	29.5	1.3	10.668	44.5	47.0
1.0000	1.9800	0.5600	1620	1040		−0.13			0.05	0.5800	1.24	1.16	0.05	0.4200	1.75	1.85
25.400	51.994	15.011	6990	4810	1.45	−2.8	07100	07204	1.0	14.260	30.5	29.5	1.3	12.700	45.0	48.0
1.0000	2.0470	0.5910	1570	1080		−0.11			0.04	0.5614	1.20	1.16	0.05	0.5000	1.77	1.89
25.400	56.896	19.368	10900	5740	1.90	−6.9	1780	1729	0.8	19.837	30.5	30.0	1.3	15.875	49.0	51.0
1.0000	2.2400	0.7625	2450	1290		−0.27			0.03	0.7810	1.20	1.18	0.05	0.6250	1.93	2.01
25.400	57.150	19.431	11700	10900	1.07	−3.0	M84548	M84510	1.5	19.431	36.0	33.0	1.5	14.732	48.5	54.0
1.0000	2.2500	0.7650	2620	2450		−0.12			0.06	0.7650	1.42	1.30	0.06	0.5800	1.91	2.13
25.400	58.738	19.050	11600	6560	1.77	−5.8	1986	1932	1.3	19.355	32.5	30.5	1.3	15.080	52.0	54.0
1.0000	2.3125	0.7500	2610	1470		−0.23			0.05	0.7620	1.28	1.20	0.05	0.5937	2.05	2.13
25.400	59.530	23.368	13900	13000	1.07	−5.1	M84249	M84210	0.8	23.114	36.0	32.5	1.5	18.288	49.5	56.0
1.0000	2.3437	0.9200	3140	2930		−0.20			0.03	0.9100	1.42	1.27	0.06	0.7200	1.95	2.20
25.400	60.325	19.842	11000	6550	1.69	−5.1	15578	15523	1.3	17.462	32.5	30.5	1.5	15.875	51.0	54.0
1.0000	2.3750	0.7812	2480	1470		−0.20			0.05	0.6875	1.28	1.20	0.06	0.6250	2.01	2.13
25.400	61.912	19.050	12100	7280	1.67	−5.8	15101	15243	0.8	20.638	32.5	31.5	2.0	14.288	54.0	58.0
1.0000	2.4375	0.7500	2730	1640		−0.23			0.03	0.8125	1.28	1.24	0.08	0.5625	2.13	2.28
25.400	62.000	19.050	12100	7280	1.67	−5.8	15100	15245	3.5	20.638	38.0	31.5	1.3	14.288	55.0	58.0
1.0000	2.4409	0.7500	2730	1640		−0.23			0.14	0.8125	1.50	1.24	0.05	0.5625	2.17	2.28
25.400	62.000	19.050	12100	7280	1.67	−5.8	15101	15245	0.8	20.638	32.5	31.5	1.3	14.288	55.0	58.0
1.0000	2.4409	0.7500	2730	1640		−0.23			0.03	0.8125	1.28	1.24	0.05	0.5625	2.17	2.28

Figure 11–15 (Continued on next page)

Catalog entry of single-row straight-bore Timken roller bearings, in part. (*Courtesy of The Timken Company.*)

SINGLE-ROW STRAIGHT BORE

| | | | rating at 500 rpm for 3000 hours L$_{10}$ | | factor | eff. load center | part numbers | | cone | | | | cup | | | |
| bore | outside diameter | width | one-row radial | thrust | | | cone | cup | max shaft fillet radius | width | backing shoulder diameters | | max housing fillet radius | width | backing shoulder diameters | |
d	D	T	N lbf	N lbf	K	a②			R①	B	d$_b$	d$_a$	r①	C	D$_b$	D$_a$
25.400 1.0000	62.000 2.4409	19.050 0.7500	12100 2730	7280 1640	1.67	−5.8 −0.23	15102	15245	1.5 0.06	20.638 0.8125	34.0 1.34	31.5 1.24	1.3 0.05	14.288 0.5625	55.0 2.17	58.0 2.28
25.400 1.0000	62.000 2.4409	20.638 0.8125	12100 2730	7280 1640	1.67	−5.8 −0.23	15101	15244	0.8 0.03	20.638 0.8125	32.5 1.28	31.5 1.24	1.3 0.05	15.875 0.6250	55.0 2.17	58.0 2.28
25.400 1.0000	63.500 2.5000	20.638 0.8125	12100 2730	7280 1640	1.67	−5.8 −0.23	15101	15250	0.8 0.03	20.638 0.8125	32.5 1.28	31.5 1.24	1.3 0.05	15.875 0.6250	56.0 2.20	59.0 2.32
25.400 1.0000	63.500 2.5000	20.638 0.8125	12100 2730	7280 1640	1.67	−5.8 −0.23	15101	15250X	0.8 0.03	20.638 0.8125	32.5 1.28	31.5 1.24	1.5 0.06	15.875 0.6250	55.0 2.17	59.0 2.32
25.400 1.0000	64.292 2.5312	21.433 0.8438	14500 3250	13500 3040	1.07	−3.3 −0.13	M86643	M86610	1.5 0.06	21.433 0.8438	38.0 1.50	36.5 1.44	1.5 0.06	16.670 0.6563	54.0 2.13	61.0 2.40
25.400 1.0000	65.088 2.5625	22.225 0.8750	13100 2950	16400 3690	0.80	−2.3 −0.09	23100	23256	1.5 0.06	21.463 0.8450	39.0 1.54	34.5 1.36	1.5 0.06	15.875 0.6250	53.0 2.09	63.0 2.48
25.400 1.0000	66.421 2.6150	23.812 0.9375	18400 4140	8000 1800	2.30	−9.4 −0.37	2687	2631	1.3 0.05	25.433 1.0013	33.5 1.32	31.5 1.24	1.3 0.05	19.050 0.7500	58.0 2.28	60.0 2.36
25.400 1.0000	68.262 2.6875	22.225 0.8750	15300 3440	10900 2450	1.40	−5.1 −0.20	02473	02420	0.8 0.03	22.225 0.8750	34.5 1.36	33.5 1.32	1.5 0.06	17.462 0.6875	59.0 2.32	63.0 2.48
25.400 1.0000	72.233 2.8438	25.400 1.0000	18400 4140	17200 3870	1.07	−4.6 −0.18	HM88630	HM88610	0.8 0.03	25.400 1.0000	39.5 1.56	39.5 1.56	2.3 0.09	19.842 0.7812	60.0 2.36	69.0 2.72
25.400 1.0000	72.626 2.8593	30.162 1.1875	22700 5110	13000 2910	1.76	−10.2 −0.40	3189	3120	0.8 0.03	29.997 1.1810	35.5 1.40	35.0 1.38	3.3 0.13	23.812 0.9375	61.0 2.40	67.0 2.64
26.157 1.0298	62.000 2.4409	19.050 0.7500	12100 2730	7280 1640	1.67	−5.8 −0.23	15103	15245	0.8 0.03	20.638 0.8125	33.0 1.30	32.5 1.28	1.3 0.05	14.288 0.5625	55.0 2.17	58.0 2.28
26.162 1.0300	63.100 2.4843	23.812 0.9375	18400 4140	8000 1800	2.30	−9.4 −0.37	2682	2630	1.5 0.06	25.433 1.0013	34.5 1.36	32.0 1.26	0.8 0.03	19.050 0.7500	57.0 2.24	59.0 2.32
26.162 1.0300	66.421 2.6150	23.812 0.9375	18400 4140	8000 1800	2.30	−9.4 −0.37	2682	2631	1.5 0.06	25.433 1.0013	34.5 1.36	32.0 1.26	1.3 0.05	19.050 0.7500	58.0 2.28	60.0 2.36
26.975 1.0620	58.738 2.3125	19.050 0.7500	11600 2610	6560 1470	1.77	−5.8 −0.23	1987	1932	0.8 0.03	19.355 0.7620	32.5 1.28	31.5 1.24	1.3 0.05	15.080 0.5937	52.0 2.05	54.0 2.13
† 26.988 † 1.0625	50.292 1.9800	14.224 0.5600	7210 1620	4620 1040	1.56	−3.3 −0.13	L44649	L44610	3.5 0.14	14.732 0.5800	37.5 1.48	31.0 1.22	1.3 0.05	10.668 0.4200	44.5 1.75	47.0 1.85
† 26.988 † 1.0625	60.325 2.3750	19.842 0.7812	11000 2480	6550 1470	1.69	−5.1 −0.20	15580	15523	3.5 0.14	17.462 0.6875	38.5 1.52	32.0 1.26	1.5 0.06	15.875 0.6250	51.0 2.01	54.0 2.13
† 26.988 † 1.0625	62.000 2.4409	19.050 0.7500	12100 2730	7280 1640	1.67	−5.8 −0.23	15106	15245	0.8 0.03	20.638 0.8125	33.5 1.32	33.0 1.30	1.3 0.05	14.288 0.5625	55.0 2.17	58.0 2.28
† 26.988 † 1.0625	66.421 2.6150	23.812 0.9375	18400 4140	8000 1800	2.30	−9.4 −0.37	2688	2631	1.5 0.06	25.433 1.0013	35.0 1.38	33.0 1.30	1.3 0.05	19.050 0.7500	58.0 2.28	60.0 2.36
28.575 1.1250	56.896 2.2400	19.845 0.7813	11600 2610	6560 1470	1.77	−5.8 −0.23	1985	1930	0.8 0.03	19.355 0.7620	34.0 1.34	33.5 1.32	0.8 0.03	15.875 0.6250	51.0 2.01	54.0 2.11
28.575 1.1250	57.150 2.2500	17.462 0.6875	11000 2480	6550 1470	1.69	−5.1 −0.20	15590	15520	3.5 0.14	17.462 0.6875	39.0 1.56	33.5 1.32	1.5 0.06	13.495 0.5313	51.0 2.01	53.0 2.09
28.575 1.1250	58.738 2.3125	19.050 0.7500	11600 2610	6560 1470	1.77	−5.8 −0.23	1985	1932	0.8 0.03	19.355 0.7620	34.0 1.34	33.5 1.32	1.3 0.05	15.080 0.5937	52.0 2.05	54.0 2.13
28.575 1.1250	58.738 2.3125	19.050 0.7500	11600 2610	6560 1470	1.77	−5.8 −0.23	1988	1932	3.5 0.14	19.355 0.7620	39.5 1.56	33.5 1.32	1.3 0.05	15.080 0.5937	52.0 2.05	54.0 2.13
28.575 1.1250	60.325 2.3750	19.842 0.7812	11000 2480	6550 1470	1.69	−5.1 −0.20	15590	15523	3.5 0.14	17.462 0.6875	39.5 1.56	33.5 1.32	1.5 0.06	15.875 0.6250	51.0 2.01	54.0 2.13
28.575 1.1250	60.325 2.3750	19.845 0.7813	11600 2610	6560 1470	1.77	−5.8 −0.23	1985	1931	0.5 0.03	19.355 0.7620	34.0 1.34	33.5 1.32	1.3 0.05	15.875 0.6250	52.0 2.05	55.0 2.17

① These maximum fillet radii will be cleared by the bearing corners.
② Minus value indicates center is inside cone backface.
† For standard class **ONLY**, the maximum metric size is a whole mm value.

* For "J" part tolerances—see metric tolerances, page 73, and fitting practice, page 65.

◆ ISO cone and cup combinations are designated with a common part number and should be purchased as an assembly.
For ISO bearing tolerances—see metric tolerances, page 73, and fitting practice, page 65.

Figure 11–15

(*Continued*)

Figure 11–16

Direct-mounted tapered roller bearings, showing radial, induced thrust, and external thrust loads.

example, consider the vertical shaft and cylinder in Fig. 11–17 with direct-mounted bearings. In Fig. 11–17a, an external load is applied in the upward direction to a rotating shaft, compressing the top bearing, which should be labeled as bearing A. On the other hand, in Fig. 11–17b, an upward external load is applied to a rotating outer cylinder with a stationary shaft. In this case, the lower bearing is being squeezed and should be labeled as bearing A. If there is no external thrust, then either bearing can arbitrarily be labeled as bearing A.

Second, determine which bearing actually carries the net axial load. Generally, it would be expected that bearing A would carry the axial load, since the external thrust F_{ae} is directed toward A, along with the induced thrust F_{iB} from bearing B. However, if the induced thrust F_{iA} from bearing A happens to be larger than the combination of the external thrust and the thrust induced by bearing B, then bearing B will carry the net thrust load. We will use Eq. (a) for the bearing carrying the thrust load. Timken recommends leaving the other bearing at its original radial load, rather than reducing it due to the negative net thrust load. The results are presented in equation form below, where the induced thrusts are defined by Eq. (11–18).

$$\text{If} \qquad F_{iA} \le (F_{iB} + F_{ae}) \qquad \begin{cases} F_{eA} = 0.4F_{rA} + K_A(F_{iB} + F_{ae}) & \text{(11–19a)} \\ F_{eB} = F_{rB} & \text{(11–19b)} \end{cases}$$

$$\text{If} \qquad F_{iA} > (F_{iB} + F_{ae}) \qquad \begin{cases} F_{eB} = 0.4F_{rB} + K_B(F_{iA} - F_{ae}) & \text{(11–20a)} \\ F_{eA} = F_{rA} & \text{(11–20b)} \end{cases}$$

In any case, if the equivalent radial load is ever less than the original radial load, then the original radial load should be used.

Once the equivalent radial loads are determined, they should be used to find the catalog rating load using any of Eqs. (11–3), (11–9), or (11–10) as before. Timken uses a Weibell model with $x_0 = 0$, $\theta = 4.48$, and $b = 3/2$. Note that since K_A and K_B are dependent on the specific bearing chosen, it may be necessary to iterate the process.

Figure 11-17

Examples of determining which bearing carries the external thrust load. In each case, the compressed bearing is labeled as bearing A.
(a) External thrust applied to rotating shaft; (b) External thrust applied to rotating cylinder.

Bearing A

F_{ae}

Bearing B

Bearing B

F_{ae}

Bearing A

(a)

(b)

EXAMPLE 11-8

The shaft depicted in Fig. 11–18a carries a helical gear with a tangential force of 3980 N, a radial force of 1770 N, and a thrust force of 1690 N at the pitch cylinder with directions shown. The pitch diameter of the gear is 200 mm. The shaft runs at a speed of 800 rev/min, and the span (effective spread) between the direct-mount bearings is 150 mm. The design life is to be 5000 h and an application factor of 1 is appropriate. If the reliability of the bearing set is to be 0.99, select suitable single-row tapered-roller Timken bearings.

Solution

The reactions in the xz plane from Fig. 11–18b are

$$R_{zA} = \frac{3980(50)}{150} = 1327 \text{ N}$$

$$R_{zB} = \frac{3980(100)}{150} = 2653 \text{ N}$$

The reactions in the xy plane from Fig. 11–18c are

$$R_{yA} = \frac{1770(50)}{150} + \frac{169\,000}{150} = 1716.7 = 1717 \text{ N}$$

$$R_{yB} = \frac{1770(100)}{150} - \frac{169\,000}{150} = 53.3 \text{ N}$$

Figure 11-18

Essential geometry of helical gear and shaft. Length dimensions in mm, loads in N, couple in N · mm. (*a*) Sketch (not to scale) showing thrust, radial, and tangential forces. (*b*) Forces in *xz* plane. (*c*) Forces in *xy* plane.

The radial loads F_{rA} and F_{rB} are the vector additions of R_{yA} and R_{zA}, and R_{yB} and R_{zB}, respectively:

$$F_{rA} = (R_{zA}^2 + R_{yA}^2)^{1/2} = (1327^2 + 1717^2)^{1/2} = 2170 \text{ N}$$

$$F_{rB} = (R_{zB}^2 + R_{yB}^2)^{1/2} = (2653^2 + 53.3^2)^{1/2} = 2654 \text{ N}$$

Trial 1: With direct mounting of the bearings and application of the external thrust to the shaft, the squeezed bearing is bearing *A* as labeled in Fig. 11–18*a*. Using *K* of 1.5 as the initial guess for each bearing, the induced loads from the bearings are

$$F_{iA} = \frac{0.47 F_{rA}}{K_A} = \frac{0.47(2170)}{1.5} = 680 \text{ N}$$

$$F_{iB} = \frac{0.47 F_{rB}}{K_B} = \frac{0.47(2654)}{1.5} = 832 \text{ N}$$

Since F_{iA} is clearly less than $F_{iB} + F_{ae}$, bearing *A* carries the net thrust load, and Eq. (11–19) is applicable. Therefore, the dynamic equivalent loads are

$$F_{eA} = 0.4 F_{rA} + K_A(F_{iB} + F_{ae}) = 0.4(2170) + 1.5(832 + 1690) = 4651 \text{ N}$$

$$F_{eB} = F_{rB} = 2654 \text{ N}$$

The multiple of rating life is

$$x_D = \frac{L_D}{L_R} = \frac{\mathscr{L}_D n_D 60}{L_R} = \frac{(5000)(800)(60)}{90(10^6)} = 2.67$$

Estimate R_D as $\sqrt{0.99} = 0.995$ for each bearing. For bearing A, from Eq. (11–10) the catalog entry C_{10} should equal or exceed

$$C_{10} = (1)(4651)\left[\frac{2.67}{(4.48)(1-0.995)^{2/3}}\right]^{3/10} = 11\,486\text{ N}$$

From Fig. 11–15, tentatively select type TS 15100 cone and 15245 cup, which will work: $K_A = 1.67$, $C_{10} = 12\,100$ N.

For bearing B, from Eq. (11–10), the catalog entry C_{10} should equal or exceed

$$C_{10} = (1)2654\left[\frac{2.67}{(4.48)(1-0.995)^{2/3}}\right]^{3/10} = 6554\text{ N}$$

Tentatively select the bearing identical to bearing A, which will work: $K_B = 1.67$, $C_{10} = 12\,100$ N.

Trial 2: Repeat the process with $K_A = K_B = 1.67$ from tentative bearing selection.

$$F_{iA} = \frac{0.47F_{rA}}{K_A} = \frac{0.47(2170)}{1.67} = 611\text{ N}$$

$$F_{iB} = \frac{0.47F_{rB}}{K_B} = \frac{0.47(2654)}{1.67} = 747\text{ N}$$

Since F_{iA} is still less than $F_{iB} + F_{ae}$, Eq. (11–19) is still applicable.

$$F_{eA} = 0.4F_{rA} + K_A(F_{iB} + F_{ae}) = 0.4(2170) + 1.67(747 + 1690) = 4938\text{ N}$$

$$F_{eB} = F_{rB} = 2654\text{ N}$$

For bearing A, from Eq. (11–10) the corrected catalog entry C_{10} should equal or exceed

$$C_{10} = (1)(4938)\left[\frac{2.67}{(4.48)(1-0.995)^{2/3}}\right]^{3/10} = 12\,195\text{ N}$$

Although this catalog entry exceeds slightly the tentative selection for bearing A, we will keep it since the reliability of bearing B exceeds 0.995. In the next section we will quantitatively show that the combined reliability of bearing A and B will exceed the reliability goal of 0.99.

For bearing B, $F_{eB} = F_{rB} = 2654$ N. From Eq. (11–10),

$$C_{10} = (1)2654\left[\frac{2.67}{(4.48)(1-0.995)^{2/3}}\right]^{3/10} = 6554\text{ N}$$

Select cone and cup 15100 and 15245, respectively, for both bearing A and B. Note from Fig. 11–14 the effective load center is located at $a = -5.8$ mm, that is, 5.8 mm into the cup from the back. Thus the shoulder-to-shoulder dimension should be $150 - 2(5.8) = 138.4$ mm. Note that in each iteration of Eq. (11–10) to find the catalog load rating, the bracketed portion of the equation is identical and need not be re-entered on a calculator each time.

11–10   Design Assessment for Selected Rolling-Contact Bearings

In textbooks, machine elements typically are treated singly. This can lead the reader to the presumption that a design assessment involves only that element, in this case a rolling-contact bearing. The immediately adjacent elements (the shaft journal and the housing bore) have immediate influence on the performance. Other elements, further removed (gears producing the bearing load), also have influence. Just as some say, "If you pull on something in the environment, you find that it is attached to everything else." This should be intuitively obvious to those involved with machinery. How, then, can one check shaft attributes that aren't mentioned in a problem statement? Possibly, because the bearing hasn't been designed yet (in fine detail). All this points out the necessary iterative nature of designing, say, a speed reducer. If power, speed, and reduction are stipulated, then gear sets can be roughed in, their sizes, geometry, and location estimated, shaft forces and moments identified, bearings tentatively selected, seals identified; the bulk is beginning to make itself evident, the housing and lubricating scheme as well as the cooling considerations become clearer, shaft overhangs and coupling accommodations appear. It is time to iterate, now addressing each element again, knowing much more about all of the others. When you have completed the necessary iterations, you will know what you need for the design assessment for the bearings. In the meantime you do as much of the design assessment as you can, avoiding bad selections, even if tentative. Always keep in mind that you eventually have to do it all in order to pronounce your completed design satisfactory.

An outline of a design assessment for a rolling contact bearing includes, at a minimum,

- Bearing reliability for the load imposed and life expected
- Shouldering on shaft and housing satisfactory
- Journal finish, diameter and tolerance compatible
- Housing finish, diameter and tolerance compatible
- Lubricant type according to manufacturer's recommendations; lubricant paths and volume supplied to keep operating temperature satisfactory
- Preloads, if required, are supplied

Since we are focusing on rolling-contact bearings, we can address bearing reliability quantitatively, as well as shouldering. Other quantitative treatment will have to wait until the materials for shaft and housing, surface quality, and diameters and tolerances are known.

Bearing Reliability

Equation (11–9) can be solved for the reliability R_D in terms of C_{10}, the basic load rating of the selected bearing:

$$R = \exp\left(-\left\{\frac{x_D\left(\dfrac{a_f F_D}{C_{10}}\right)^a - x_0}{\theta - x_0}\right\}^b\right) \tag{11–21}$$

Equation (11–10) can likewise be solved for R_D:

$$R \approx 1 - \left\{ \frac{x_D \left(\dfrac{a_f F_D}{C_{10}} \right)^a - x_0}{\theta - x_0} \right\}^b \qquad R \ge 0.90 \qquad (11\text{–}22)$$

EXAMPLE 11–9

In Ex. 11–3, the minimum required load rating for 99 percent reliability, at $x_D = L_D/L_{10} = 540$, is $C_{10} = 29.7$ kN. From Table 11–2 a 02-40 mm deep-groove ball bearing would satisfy the requirement. If the bore in the application had to be 70 mm or larger (selecting a 02-70 mm deep-groove ball bearing), what is the resulting reliability?

Solution

From Table 11–2, for a 02-70 mm deep-groove ball bearing, $C_{10} = 61.8$ kN. Using Eq. (11–19), recalling from Ex. 11–3 that $a_f = 1.2$, $F_D = 1840$ N, $x_0 = 0.02$, $(\theta - x_0) = 4.439$, and $b = 1.483$, we can write

Answer

$$R \approx 1 - \left\{ \frac{540 \left[\dfrac{1.2(1840)}{61\,800} \right]^3 - 0.02}{4.439} \right\}^{1.483} = 0.999\,962$$

which, as expected, is much higher than 0.99 from Ex. 11–3.

In tapered roller bearings, or other bearings for a two-parameter Weibull distribution, Eq. (11–21) becomes, for $x_0 = 0$, $\theta = 4.48$, $b = \frac{3}{2}$,

$$R = \exp \left\{ - \left[\frac{x_D}{\theta [C_{10}/(a_f F_D)]^a} \right]^b \right\}$$

$$= \exp \left\{ - \left[\frac{x_D}{4.48 [C_{10}/(a_f F_D)]^{10/3}} \right]^{3/2} \right\} \qquad (11\text{–}23)$$

and Eq. (11–22) becomes

$$R \approx 1 - \left\{ \frac{x_D}{\theta [C_{10}/(a_f F_D)]^a} \right\}^b = 1 - \left\{ \frac{x_D}{4.48 [C_{10}/(a_f F_D)]^{10/3}} \right\}^{3/2} \qquad (11\text{–}24)$$

EXAMPLE 11–10

In Ex. 11–8 bearings A and B (cone 15100 and cup 15245) have $C_{10} = 12\,100$ N. What is the reliability of the pair of bearings A and B?

Solution

The desired life x_D was $5000(800)60/[90(10^6)] = 2.67$ rating lives. Using Eq. (11–24) for bearing A, where from Ex. 11–8, $F_D = F_{eA} = 4938$ N, and $a_f = 1$, gives

$$R_A \approx 1 - \left\{ \frac{2.67}{4.48[12\ 100/(1 \times 4938)]^{10/3}} \right\}^{3/2} = 0.994\ 791$$

which is less than 0.995, as expected. Using Eq. (11–24) for bearing B with $F_D = F_{eB} = 2654$ N gives

$$R_B \approx 1 - \left\{ \frac{2.67}{4.48[12\ 100/(1 \times 2654)]^{10/3}} \right\}^{3/2} = 0.999\ 766$$

Answer

The reliability of the bearing pair is

$$R = R_A R_B = 0.994\ 791(0.999\ 766) = 0.994\ 558$$

which is greater than the overall reliability goal of 0.99. When two bearings are made identical for simplicity, or reducing the number of spares, or other stipulation, and the loading is not the same, both can be made smaller and still meet a reliability goal. If the loading is disparate, then the more heavily loaded bearing can be chosen for a reliability goal just slightly larger than the overall goal.

An additional example is useful to show what happens in cases of pure thrust loading.

EXAMPLE 11–11

Consider a constrained housing as depicted in Fig. 11–19 with two direct-mount tapered roller bearings resisting an external thrust F_{ae} of 8000 N. The shaft speed is 950 rev/min, the desired life is 10 000 h, the expected shaft diameter is approximately 1 in. The reliability goal is 0.95. The application factor is appropriately $a_f = 1$.
(a) Choose a suitable tapered roller bearing for A.
(b) Choose a suitable tapered roller bearing for B.
(c) Find the reliabilities R_A, R_B, and R.

Solution

(a) By inspection, note that the left bearing carries the axial load and is properly labeled as bearing A. The bearing reactions at A are

$$F_{rA} = F_{rB} = 0$$

$$F_{aA} = F_{ae} = 8000 \text{ N}$$

Since bearing B is unloaded, we will start with $R = R_A = 0.95$.
With no radial loads, there are no induced thrust loads. Eq. (11–19) is applicable.

$$F_{eA} = 0.4F_{rA} + K_A(F_{iB} + F_{ae}) = K_A F_{ae}$$

If we set $K_A = 1$, we can find C_{10} in the thrust column and avoid iteration:

$$F_{eA} = (1)8000 = 8000 \text{ N}$$

$$F_{eB} = F_{rB} = 0$$

The multiple of rating life is

$$x_D = \frac{L_D}{L_R} = \frac{\mathscr{L}_D n_D 60}{L_R} = \frac{(10\,000)(950)(60)}{90(10^6)} = 6.333$$

Then, from Eq. (11–10), for bearing A

$$C_{10} = a_f F_{eA} \left[\frac{x_D}{4.48(1 - R_D)^{2/3}} \right]^{3/10}$$

$$= (1)8000 \left[\frac{6.33}{4.48(1 - 0.95)^{2/3}} \right]^{3/10} = 16\,159 \text{ N}$$

Answer

Figure 11–15 presents one possibility in the 1-in bore (25.4-mm) size: cone, HM88630, cup HM88610 with a thrust rating $(C_{10})_a = 17\,200$ N.

Figure 11–19

The constrained housing of Ex. 11–11.

Bearing A Bearing B

$F_{ae} = 8000$ N

Answer

(b) Bearing B experiences no load, and the cheapest bearing of this bore size will do, including a ball or roller bearing.

(c) The actual reliability of bearing A, from Eq. (11–24), is

Answer

$$R_A \approx 1 - \left\{ \frac{x_D}{4.48[C_{10}/(a_f F_D)]^{10/3}} \right\}^{3/2}$$

$$\approx 1 - \left\{ \frac{6.333}{4.48[17\,200/(1 \times 8000)]^{10/3}} \right\}^{3/2} = 0.963$$

which is greater than 0.95, as one would expect. For bearing B,

Answer

$$F_D = F_{eB} = 0$$

$$R_B \approx 1 - \left[\frac{6.333}{0.85(17\,200/0)^{10/3}} \right]^{3/2} = 1 - 0 = 1$$

as one would expect. The combined reliability of bearings A and B as a pair is

Answer

$$R = R_A R_B = 0.963(1) = 0.963$$

which is greater than the reliability goal of 0.95, as one would expect.

Matters of Fit

Table 11–2 (and Fig. 11–8), which shows the rating of single-row, 02-series, deep-groove and angular-contact ball bearings, includes shoulder diameters recommended for the shaft seat of the inner ring and the shoulder diameter of the outer ring, denoted d_S and d_H, respectively. The shaft shoulder can be greater than d_S but not enough to obstruct the annulus. It is important to maintain concentricity and perpendicularity with the shaft centerline, and to that end the shoulder diameter should equal or exceed d_S. The housing shoulder diameter d_H is to be equal to or less than d_H to maintain concentricity and perpendicularity with the housing bore axis. Neither the shaft shoulder nor the housing shoulder features should allow interference with the free movement of lubricant through the bearing annulus.

In a tapered roller bearing (Fig. 11–15), the cup housing shoulder diameter should be equal to or less than D_b. The shaft shoulder for the cone should be equal to or greater than d_b. Additionally, free lubricant flow is not to be impeded by obstructing any of the annulus. In splash lubrication, common in speed reducers, the lubricant is thrown to the housing cover (ceiling) and is directed in its draining by ribs to a bearing. In direct mounting, a tapered roller bearing pumps oil from outboard to inboard. An oil passageway to the outboard side of the bearing needs to be provided. The oil returns to the sump as a consequence of bearing pump action. With an indirect mount, the oil is directed to the inboard annulus, the bearing pumping it to the outboard side. An oil passage from the outboard side to the sump has to be provided.

11–11   Lubrication

The contacting surfaces in rolling bearings have a relative motion that is both rolling and sliding, and so it is difficult to understand exactly what happens. If the relative velocity of the sliding surfaces is high enough, then the lubricant action is hydrodynamic (see Chap. 12). *Elastohydrodynamic lubrication* (EHD) is the phenomenon that occurs when a lubricant is introduced between surfaces that are in pure rolling contact. The contact of gear teeth and that found in rolling bearings and in cam-and-follower surfaces are typical examples. When a lubricant is trapped between two surfaces in rolling contact, a tremendous increase in the pressure within the lubricant film occurs. But viscosity is exponentially related to pressure, and so a very large increase in viscosity occurs in the lubricant that is trapped between the surfaces. Leibensperger[2] observes that the change in viscosity in and out of contact pressure is equivalent to the difference between cold asphalt and light sewing machine oil.

The purposes of an antifriction-bearing lubricant may be summarized as follows:

1   To provide a film of lubricant between the sliding and rolling surfaces

2   To help distribute and dissipate heat

3   To prevent corrosion of the bearing surfaces

4   To protect the parts from the entrance of foreign matter

Either oil or grease may be employed as a lubricant. The following rules may help in deciding between them.

Use Grease When	Use Oil When
1. The temperature is not over 200°F.	1. Speeds are high.
2. The speed is low.	2. Temperatures are high.
3. Unusual protection is required from the entrance of foreign matter.	3. Oiltight seals are readily employed.
4. Simple bearing enclosures are desired.	4. Bearing type is not suitable for grease lubrication.
5. Operation for long periods without attention is desired.	5. The bearing is lubricated from a central supply which is also used for other machine parts.

11–12 Mounting and Enclosure

There are so many methods of mounting antifriction bearings that each new design is a real challenge to the ingenuity of the designer. The housing bore and shaft outside diameter must be held to very close limits, which of course is expensive. There are usually one or more counterboring operations, several facing operations and drilling, tapping, and threading operations, all of which must be performed on the shaft, housing, or cover plate. Each of these operations contributes to the cost of production, so that the designer, in ferreting out a trouble-free and low-cost mounting, is faced with a difficult and important problem. The various bearing manufacturers' handbooks give many mounting details in almost every design area. In a text of this nature, however, it is possible to give only the barest details.

The most frequently encountered mounting problem is that which requires one bearing at each end of a shaft. Such a design might use one ball bearing at each end, one tapered roller bearing at each end, or a ball bearing at one end and a straight roller bearing at the other. One of the bearings usually has the added function of positioning or axially locating the shaft. Figure 11–20 shows a very common solution to this problem. The inner rings are backed up against the shaft shoulders and are held in position by round nuts threaded onto the shaft. The outer ring of the left-hand bearing is backed up against a housing shoulder and is held in position by a device that is not shown. The outer ring of the right-hand bearing floats in the housing.

There are many variations possible on the method shown in Fig. 11–20. For example, the function of the shaft shoulder may be performed by retaining rings, by the hub of a gear or pulley, or by spacing tubes or rings. The round nuts may be replaced by retaining rings or by washers locked in position by screws, cotters, or taper pins. The housing shoulder may be replaced by a retaining ring; the outer ring of the bearing may be grooved for a retaining ring, or a flanged outer ring may be used. The force against the outer ring of the left-hand bearing is usually applied by the cover plate, but if no thrust is present, the ring may be held in place by retaining rings.

---

[2]R. L. Leibensperger, "When Selecting a Bearing," *Machine Design,* vol. 47, no. 8, April 3, 1975, pp. 142–147.

Figure 11–20

A common bearing mounting.

Figure 11–21

An alternative bearing mounting to that in Fig. 11–20.

Figure 11–21 shows an alternative method of mounting in which the inner races are backed up against the shaft shoulders as before but no retaining devices are required. With this method the outer races are completely retained. This eliminates the grooves or threads, which cause stress concentration on the overhanging end, but it requires accurate dimensions in an axial direction or the employment of adjusting means. This method has the disadvantage that if the distance between the bearings is great, the temperature rise during operation may expand the shaft enough to destroy the bearings.

It is frequently necessary to use two or more bearings at one end of a shaft. For example, two bearings could be used to obtain additional rigidity or increased load capacity or to cantilever a shaft. Several two-bearing mountings are shown in Fig. 11–22. These may be used with tapered roller bearings, as shown, or with ball bearings. In either case it should be noted that the effect of the mounting is to preload the bearings in an axial direction.

Figure 11–23 shows another two-bearing mounting. Note the use of washers against the cone backs.

When maximum stiffness and resistance to shaft misalignment is desired, pairs of angular-contact ball bearings (Fig. 11–2) are often used in an arrangement called *duplexing*. Bearings manufactured for duplex mounting have their rings ground with an offset, so that when a pair of bearings is tightly clamped together, a preload is automatically established. As shown in Fig. 11–24, three mounting arrangements are used. The face-to-face mounting, called DF, will take heavy radial loads and thrust loads from either direction. The DB mounting (back to back) has the greatest aligning stiffness and is also good for heavy radial loads and thrust loads from either direction. The tandem arrangement, called the DT mounting, is used where the thrust is always

Figure 11–22

Two-bearing mountings. *(Courtesy of The Timken Company.)*

(a)   (b)

Figure 11–23

Mounting for a washing-machine spindle. *(Courtesy of The Timken Company.)*

Figure 11–24

Arrangements of angular ball bearings. (*a*) DF mounting; (*b*) DB mounting; (*c*) DT mounting. *(Courtesy of The Timken Company.)*

(a)   (b)   (c)

in the same direction; since the two bearings have their thrust functions in the same direction, a preload, if required, must be obtained in some other manner.

Bearings are usually mounted with the rotating ring a press fit, whether it be the inner or outer ring. The stationary ring is then mounted with a push fit. This permits the stationary ring to creep in its mounting slightly, bringing new portions of the ring into the load-bearing zone to equalize wear.

Preloading

The object of preloading is to remove the internal clearance usually found in bearings, to increase the fatigue life, and to decrease the shaft slope at the bearing. Figure 11–25 shows a typical bearing in which the clearance is exaggerated for clarity.

Preloading of straight roller bearings may be obtained by:

1 Mounting the bearing on a tapered shaft or sleeve to expand the inner ring

2 Using an interference fit for the outer ring

3 Purchasing a bearing with the outer ring preshrunk over the rollers

Figure 11–25

Clearance in an off-the-shelf bearing, exaggerated for clarity.

Ball bearings are usually preloaded by the axial load built in during assembly. However, the bearings of Fig. 11–24a and b are preloaded in assembly because of the differences in widths of the inner and outer rings.

It is always good practice to follow manufacturers' recommendations in determining preload, since too much will lead to early failure.

Alignment

The permissible misalignment in bearings depends on the type of bearing and the geometric and material properties of the specific bearing. Manufacturers' catalogs should be referenced for detailed specifications on a given bearing. In general, cylindrical and tapered roller bearings require alignments that are closer than deep-groove ball bearings. Spherical ball bearings and self-aligning bearings are the most forgiving. Table 7–2, p. 371, gives typical maximum ranges for each type of bearing. The life of the bearing decreases significantly when the misalignment exceeds the allowable limits.

Additional protection against misalignment is obtained by providing the full shoulders (see Fig. 11–8) recommended by the manufacturer. Also, if there is any misalignment at all, it is good practice to provide a safety factor of around 2 to account for possible increases during assembly.

Enclosures

To exclude dirt and foreign matter and to retain the lubricant, the bearing mountings must include a seal. The three principal methods of sealings are the felt seal, the commercial seal, and the labyrinth seal (Fig. 11–26).

Felt seals may be used with grease lubrication when the speeds are low. The rubbing surfaces should have a high polish. Felt seals should be protected from dirt by placing them in machined grooves or by using metal stampings as shields.

The *commercial seal* is an assembly consisting of the rubbing element and, generally, a spring backing, which are retained in a sheet-metal jacket. These seals are usually made by press fitting them into a counterbored hole in the bearing cover. Since they obtain the sealing action by rubbing, they should not be used for high speeds.

The *labyrinth seal* is especially effective for high-speed installations and may be used with either oil or grease. It is sometimes used with flingers. At least three grooves should be used, and they may be cut on either the bore or the outside diameter. The clearance may vary from 0.25 to 1.0 mm, depending upon the speed and temperature.

Figure 11–26

Typical sealing methods. *(General Motors Corp. Used with permission, GM Media Archives.)*

(a) Felt seal

(b) Commercial seal

(c) Labyrinth seal

PROBLEMS

Problems marked with an asterisk (*) are linked to problems in other chapters, as summarized in Table 1–2 of Sec. 1–17, p. 34.

Since each bearing manufacturer makes individual decisions with respect to materials, treatments, and manufacturing processes, manufacturers' experiences with bearing life distribution differ. In solving the following problems, we will use the experience of two manufacturers, tabulated in Table 11–6.

Table 11–6

Typical Weibull Parameters for Two Manufacturers

Manufacturer	Rating Life, Revolutions	Weibull Parameters Rating Lives		
		x_0	θ	b
1	$90(10^6)$	0	4.48	1.5
2	$1(10^6)$	0.02	4.459	1.483

Tables 11–2 and 11–3 are based on manufacturer 2.

11–1 A certain application requires a ball bearing with the inner ring rotating, with a design life of 25 kh at a speed of 350 rev/min. The radial load is 2.5 kN and an application factor of 1.2 is appropriate. The reliability goal is 0.90. Find the multiple of rating life required, x_D, and the catalog rating C_{10} with which to enter a bearing table. Choose a 02-series deep-groove ball bearing from Table 11–2, and estimate the reliability in use.

11–2 An angular-contact, inner ring rotating, 02-series ball bearing is required for an application in which the life requirement is 50 kh at 480 rev/min. The design radial load is 2745 N. The application factor is 1.4. The reliability goal is 0.90. Find the multiple of rating life x_D required and the catalog rating C_{10} with which to enter Table 11–2. Choose a bearing and estimate the existing reliability in service.

11–3 The other bearing on the shaft of Prob. 11–2 is to be a 03-series cylindrical roller bearing with inner ring rotating. For a 7342-N radial load, find the catalog rating C_{10} with which to enter Table 11–3. The reliability goal is 0.90. Choose a bearing and estimate its reliability in use.

11–4 Problems 11–2 and 11–3 raise the question of the reliability of the bearing pair on the shaft. Since the combined reliabilities R is $R_1 R_2$, what is the reliability of the two bearings (probability that either or both will not fail) as a result of your decisions in Probs. 11–2 and 11–3? What does this mean in setting reliability goals for each of the bearings of the pair on the shaft?

11–5 Combine Probs. 11–2 and 11–3 for an overall reliability of $R = 0.90$. Reconsider your selections, and meet this overall reliability goal.

11–6 A straight (cylindrical) roller bearing is subjected to a radial load of 20 kN. The life is to be 8000 h at a speed of 950 rev/min and exhibit a reliability of 0.95. What basic load rating should be used in selecting the bearing from a catalog of manufacturer 2 in Table 11–6?

11–7 Two ball bearings from different manufacturers are being considered for a certain application. Bearing A has a catalog rating of 2.0 kN based on a catalog rating system of 3 000 hours at 500 rev/min. Bearing B has a catalog rating of 7.0 kN based on a catalog that rates at 10^6 cycles. For a given application, determine which bearing can carry the larger load.

11–8 to 11–13 For the bearing application specifications given in the table for the assigned problem, determine the Basic Load Rating for a ball bearing with which to enter a bearing catalog of manufacturer 2 in Table 11–6. Assume an application factor of one.

Problem Number	Radial Load	Design Life	Desired Reliability
11–8	2 kN	10^9 rev	90%
11–9	8 kN	5 kh, 900 rev/min	90%
11–10	4 kN	8 kh, 500 rev/min	90%
11–11	3 kN	5 yrs, 40 h/week, 400 rev/min	95%
11–12	9 kN	10^8 rev	99%
11–13	8 kN	5 kh, 900 rev/min	96%

11–14* to 11–17* For the problem specified in the table, build upon the results of the original problem to obtain a Basic Load Rating for a ball bearing at C with a 95 percent reliability, assuming distribution data from manufacturer 2 in Table 11–6. The shaft rotates at 1200 rev/min, and the desired bearing life is 15 kh. Use an application factor of 1.2.

Problem Number	Original Problem, Page Number
11–14*	3–68, 151
11–15*	3–69, 151
11–16*	3–70, 151
11–17*	3–71, 151

11–18* For the shaft application defined in Prob. 3–77, p. 153, the input shaft EG is driven at a constant speed of 191 rev/min. Obtain a Basic Load Rating for a ball bearing at A for a life of 12 kh with a 95 percent reliability, assuming distribution data from manufacturer 2 in Table 11–6.

11–19* For the shaft application defined in Prob. 3–79, p. 153, the input shaft EG is driven at a constant speed of 280 rev/min. Obtain a Basic Load Rating for a cylindrical roller bearing at A for a life of 14 kh with a 98 percent reliability, assuming distribution data from manufacturer 2 in Table 11–6.

11–20 An 02-series single-row deep-groove ball bearing with a 65-mm bore (see Tables 11–1 and 11–2 for specifications) is loaded with a 3-kN axial load and a 7-kN radial load. The outer ring rotates at 500 rev/min.

(*a*) Determine the equivalent radial load that will be experienced by this particular bearing.

(*b*) Determine whether this bearing should be expected to carry this load with a 95 percent reliability for 10 kh.

11–21 An 02-series single-row deep-groove ball bearing with a 30-mm bore (see Tables 11–1 and 11–2 for specifications) is loaded with a 2-kN axial load and a 5-kN radial load. The inner ring rotates at 400 rev/min.

(*a*) Determine the equivalent radial load that will be experienced by this particular bearing.

(*b*) Determine the predicted life (in revolutions) that this bearing could be expected to give in this application with a 99 percent reliability.

11–22 to 11–26

An 02-series single-row deep-groove ball bearing is to be selected from Table 11–2 for the application conditions specified in the table. Assume Table 11–1 is applicable if needed. Specify the smallest bore size from Table 11–2 that can satisfy these conditions.

Problem Number	Radial Load	Axial Load	Design Life	Ring Rotating	Desired Reliability
11–22	8 kN	0 kN	10^9 rev	Inner	90%
11–23	8 kN	2 kN	10 kh, 400 rev/min	Inner	99%
11–24	8 kN	3 kN	10^8 rev	Outer	90%
11–25	10 kN	5 kN	12 kh, 300 rev/min	Inner	95%
11–26	9 kN	3 kN	10^8 rev	Outer	99%

11–27*

The shaft shown in the figure is proposed as a preliminary design for the application defined in Prob. 3–72, p. 152. The effective centers of the gears for force transmission are shown. The dimensions for the bearing surfaces (indicated with cross markings) have been estimated. The shaft rotates at 1200 rev/min, and the desired bearing life is 15 kh with a 95 percent reliability in each bearing, assuming distribution data from manufacturer 2 in Table 11–6. Use an application factor of 1.2.

(a) Obtain a Basic Load Rating for a ball bearing at the right end.

(b) Use an online bearing catalog to find a specific bearing that satisfies the needed Basic Load Rating and the geometry requirements. If necessary, indicate appropriate adjustments to the dimensions of the bearing surface.

Problem 11–27*
All fillets 2 mm. Dimensions in millimeters.

11–28*

Repeat the requirements of Prob. 11–27 for the bearing at the left end of the shaft.

11–29*

The shaft shown in the figure is proposed as a preliminary design for the application defined in Prob. 3–73, p. 152. The effective centers of the gears for force transmission are shown. The dimensions for the bearing surfaces (indicated with cross markings) have been estimated. The shaft rotates at 900 rev/min, and the desired bearing life is 12 kh with a 98 percent reliability in each bearing, assuming distribution data from manufacturer 2 in Table 11–6. Use an application factor of 1.2.

(a) Obtain a Basic Load Rating for a ball bearing at the right end.

(b) Use an online bearing catalog to find a specific bearing that satisfies the needed Basic Load Rating and the geometry requirements. If necessary, indicate appropriate adjustments to the dimensions of the bearing surface.

Problem 11–29*

All fillets 2 mm. Dimensions in millimeters.

11–30* Repeat the requirements of Prob. 11–29 for the bearing at the left end of the shaft.

11–31 The second shaft on a parallel-shaft 18.7 kW foundry crane speed reducer contains a helical gear with a pitch diameter of 205 mm. Helical gears transmit components of force in the tangential, radial, and axial directions (see Chap. 13). The components of the gear force transmitted to the second shaft are shown in Fig. 11–12, at point A. The bearing reactions at C and D, assuming simple-supports, are also shown. A ball bearing is to be selected for location C to accept the thrust, and a cylindrical roller bearing is to be utilized at location D. The life goal of the speed reducer is 10 kh, with a reliability factor for the ensemble of all four bearings (both shafts) to equal or exceed 0.96 for the Weibull parameters of Ex. 11–3. The application factor is to be 1.2.

(*a*) Select the roller bearing for location D.

(*b*) Select the ball bearing (angular contact) for location C, assuming the inner ring rotates.

Figure 11–12

Forces in pounds applied to the second shaft of the helical gear speed reducer of Ex. 11–7.

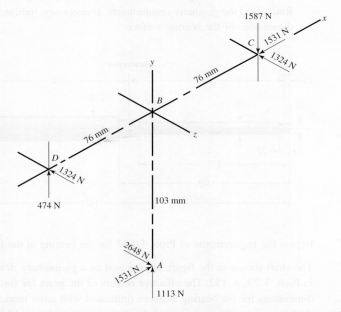

11–32 The figure shown is a geared countershaft with an overhanging pinion at C. Select an angular-contact ball bearing from Table 11–2 for mounting at O and a straight roller bearing from Table 11–3 for mounting at B. The force on gear A is $F_A = 2670$ N, and the shaft is to run at a speed of 480 rev/min. Solution of the statics problem gives force of bearings against the shaft at O as $\mathbf{R}_O = -1722\mathbf{j} + 2078$ N, and at B as $\mathbf{R}_B = 1406\mathbf{j} - 7187$ N. Specify the bearings required, using an application factor of 1.4, a desired life of 50 kh, and a combined reliability goal of 0.9, assuming distribution date from manufacturer 2 in Table 11-6.

Problem 11–32
Dimensions in mm.

11–33 The figure is a schematic drawing of a countershaft that supports two V-belt pulleys. The countershaft runs at 1500 rev/min and the bearings are to have a life of 60 kh at a combined reliability of 0.98, assuming distribution data from manufacturer 2 in Table 11–6. The belt tension on the loose side of pulley A is 15 percent of the tension on the tight side. Select deep-groove bearings from Table 11–2 for use at O and E, using an application factor of unity.

Problem 11–33
Dimensions in millimeters.

11–34 A gear-reduction unit uses the countershaft depicted in the figure. Find the two bearing reactions. The bearings are to be angular-contact ball bearings, having a desired life of 50 kh when used at 300 rev/min. Use 1.2 for the application factor and a reliability goal for the bearing pair of 0.96, assuming distribution data from manufacturer 2 in Table 11–6. Select the bearings from Table 11–2.

Problem 11–34
Dimensions in mm.

11–35   The worm shaft shown in part *a* of the figure transmits 1007 W at 600 rev/min. A static force analysis gave the results shown in part *b* of the figure. Bearing *A* is to be an angular-contact ball bearing selected from Table 11–2, mounted to take the 2470-N thrust load. The bearing at *B* is to take only the radial load, so an 02-series cylindrical roller bearing from Table 11–3 will be employed. Use an application factor of 1.2, a desired life of 30 kh, and a combined reliability goal of 0.99, assuming distribution data from manufacturer 2 in Table 11–6. Specify each bearing.

Problem 11–35
(*a*) Worm and worm gear;
(*b*) force analysis of worm shaft,
forces in pounds.

11–36   In bearings tested at 2000 rev/min with a steady radial load of 18 kN, a set of bearings showed an L_{10} life of 115 h and an L_{80} life of 600 h. The basic load rating of this bearing is 39.6 kN. Estimate the Weibull shape factor *b* and the characteristic life θ for a two-parameter model. This manufacturer rates ball bearings at 1 million revolutions.

11–37   A 16-tooth pinion drives the double-reduction spur-gear train in the figure. All gears have 25° pressure angles. The pinion rotates ccw at 1200 rev/min and transmits power to the gear train. The shaft has not yet been designed, but the free bodies have been generated. The shaft speeds are 1200 rev/min, 240 rev/min, and 80 rev/min. A bearing study is commencing with a 10-kh life and a gearbox bearing ensemble reliability of 0.99, assuming distribution data from manufacturer 2 in Table 11–6. An application factor of 1.2 is appropriate. For each shaft, specify a matched pair of 02-series cylindrical roller bearings from Table 11–3.

Problem 11–37
(*a*) Drive detail; (*b*) force analysis on shafts. Forces in Newtons; linear dimensions in millimeters.

11–38   Estimate the remaining life in revolutions of an 02-30 mm angular-contact ball bearing already subjected to 200 000 revolutions with a radial load of 18 kN, if it is now to be subjected to a change in load to 30 kN.

11–39   The same 02-30 angular-contact ball bearing as in Prob. 11–38 is to be subjected to a two-step loading cycle of 4 min with a loading of 18 kN, and one of 6 min with a loading of 30 kN. This cycle is to be repeated until failure. Estimate the total life in revolutions, hours, and loading cycles.

11–40   A countershaft is supported by two tapered roller bearings using an indirect mounting. The radial bearing loads are 2240 N for the left-hand bearing and 4380 N for the right-hand bearing. An axial load of 800 N is carried by the left bearing. The shaft rotates at 400 rev/min and is to have a desired life of 40 kh. Use an application factor of 1.4 and a combined reliability goal of 0.90, assuming distribution data from manufacturer 1 in Table 11–6. Using an initial K = 1.5, find the required radial rating for each bearing. Select the bearings from Fig. 11–15.

11–41*   For the shaft application defined in Prob. 3–74, p. 152, perform a preliminary specification for tapered roller bearings at C and D. A bearing life of 10^8 revolutions is desired with a 90 percent combined reliability for the bearing set, assuming distribution data from manufacturer 1 in Table 11–6. Should the bearings be oriented with direct mounting or indirect mounting for the axial thrust to be carried by the bearing at C? Assuming bearings are available with $K = 1.5$, find the required radial rating for each bearing. For this preliminary design, assume an application factor of one.

11–42* For the shaft application defined in Prob. 3–76, p. 153, perform a preliminary specification for tapered roller bearings at A and B. A bearing life of 500 million revolutions is desired with a 90 percent combined reliability for the bearing set, assuming distribution data from manufacturer 1 in Table 11–6. Should the bearings be oriented with direct mounting or indirect mounting for the axial thrust to be carried by the bearing at A? Assuming bearings are available with $K = 1.5$, find the required radial rating for each bearing. For this preliminary design, assume an application factor of one.

11–43 An outer hub rotates around a stationary shaft, supported by two tapered roller bearings as shown in Fig. 11–23. The device is to operate at 250 rev/min, 8 hours per day, 5 days per week, for 5 years, before bearing replacement is necessary. A reliability of 90 percent on each bearing is acceptable. A free body analysis determines the radial force carried by the upper bearing to be 12 kN and the radial force at the lower bearing to be 25 kN. In addition, the outer hub applies a downward force of 5 kN. Assuming bearings are available from manufacturer 1 in Table 11–6 with $K = 1.5$, find the required radial rating for each bearing. Assume an application factor of 1.2.

11–44 The gear-reduction unit shown has a gear that is press fit onto a cylindrical sleeve that rotates around a stationary shaft. The helical gear transmits an axial thrust load T of 1 kN as shown in the figure. Tangential and radial loads (not shown) are also transmitted through the gear, producing radial ground reaction forces at the bearings of 3.5 kN for bearing A and 2.5 kN for bearing B. The desired life for each bearing is 90 kh at a speed of 150 rev/min with a 90 percent reliability. The first iteration of the shaft design indicates approximate diameters of 28 mm at A and 25 mm at B. Assuming distribution data from manufacturer 1 in Table 11–6, select suitable tapered roller bearings from Fig. 11–15.

Problem 11–44
(Courtesy of The Timken Company.)

12

Lubrication and Journal Bearings

The object of lubrication is to reduce friction, wear, and heating of machine parts that move relative to each other. A lubricant is any substance that, when inserted between the moving surfaces, accomplishes these purposes. In a sleeve bearing, a shaft, or *journal,* rotates or oscillates within a sleeve, or *bushing,* and the relative motion is sliding. In an antifriction bearing, the main relative motion is rolling. A follower may either roll or slide on the cam. Gear teeth mate with each other by a combination of rolling and sliding. Pistons slide within their cylinders. All these applications require lubrication to reduce friction, wear, and heating.

The field of application for journal bearings is immense. The crankshaft and connecting-rod bearings of an automotive engine must operate for thousands of miles at high temperatures and under varying load conditions. The journal bearings used in the steam turbines of power-generating stations are said to have reliabilities approaching 100 percent. At the other extreme there are thousands of applications in which the loads are light and the service relatively unimportant; a simple, easily installed bearing is required, using little or no lubrication. In such cases an antifriction bearing might be a poor answer because of the cost, the elaborate enclosures, the close tolerances, the radial space required, the high speeds, or the increased inertial effects. Instead, a nylon bearing requiring no lubrication, a powder-metallurgy bearing with the lubrication "built in," or a bronze bearing with ring oiling, wick feeding, or solid-lubricant film or grease lubrication might be a very satisfactory solution. Recent metallurgy developments in bearing materials, combined with increased knowledge of the lubrication process, now make it possible to design journal bearings with satisfactory lives and very good reliabilities.

Much of the material we have studied thus far in this book has been based on fundamental engineering studies, such as statics, dynamics, the mechanics of solids, metal processing, mathematics, and metallurgy. In the study of lubrication and journal bearings, additional fundamental studies, such as chemistry, fluid mechanics, thermodynamics, and heat transfer, must be utilized in developing the material. While we shall not utilize all of them in the material to be included here, you can now begin to appreciate better how the study of mechanical engineering design is really an integration of most of your previous studies and a directing of this total background toward the resolution of a single objective.

12–1 Types of Lubrication

Five distinct forms of lubrication may be identified:

1. Hydrodynamic
2. Hydrostatic
3. Elastohydrodynamic
4. Boundary
5. Solid film

Hydrodynamic lubrication means that the load-carrying surfaces of the bearing are separated by a relatively thick film of lubricant, to prevent metal-to-metal contact, and that the stability thus obtained can be explained by the laws of fluid mechanics. Hydrodynamic lubrication does not depend upon the introduction of the lubricant under pressure, though that may occur; but it does require the existence of an adequate supply at all times. The film pressure is created by the moving surface itself pulling the lubricant into a wedge-shaped zone at a velocity sufficiently high to create the pressure necessary to separate the surfaces against the load on the bearing. Hydrodynamic lubrication is also called *full-film,* or *fluid, lubrication.*

Hydrostatic lubrication is obtained by introducing the lubricant, which is sometimes air or water, into the load-bearing area at a pressure high enough to separate the surfaces with a relatively thick film of lubricant. So, unlike hydrodynamic lubrication, this kind of lubrication does not require motion of one surface relative to another. We shall not deal with hydrostatic lubrication in this book, but the subject should be considered in designing bearings where the velocities are small or zero and where the frictional resistance is to be an absolute minimum.

Elastohydrodynamic lubrication is the phenomenon that occurs when a lubricant is introduced between surfaces that are in rolling contact, such as mating gears or rolling bearings. The mathematical explanation requires the Hertzian theory of contact stress and fluid mechanics.

Insufficient surface area, a drop in the velocity of the moving surface, a lessening in the quantity of lubricant delivered to a bearing, an increase in the bearing load, or an increase in lubricant temperature resulting in a decrease in viscosity—any one of these—may prevent the buildup of a film thick enough for full-film lubrication. When this happens, the highest asperities may be separated by lubricant films only several molecular dimensions in thickness. This is called *boundary lubrication*. The change from hydrodynamic to boundary lubrication is not at all a sudden or abrupt one. It is probable that a mixed hydrodynamic- and boundary-type lubrication occurs first, and as the surfaces move closer together, the boundary-type lubrication becomes predominant. The viscosity of the lubricant is not of as much importance with boundary lubrication as is the chemical composition.

When bearings must be operated at extreme temperatures, a *solid-film lubricant* such as graphite or molybdenum disulfide must be used because the ordinary mineral oils are not satisfactory. Much research is currently being carried out in an effort to find composite bearing materials with low wear rates as well as small frictional coefficients.

12-2 Viscosity

In Fig. 12–1 let plate A be moving with a velocity U on a film of lubricant of thickness h. We imagine the film as composed of a series of horizontal layers and the force F causing these layers to deform or slide on one another just like a deck of cards. The layers in contact with the moving plate are assumed to have a velocity U; those in contact with the stationary surface are assumed to have a zero velocity. Intermediate layers have velocities that depend upon their distances y from the stationary surface. Newton's viscous effect states that the shear stress in the fluid is proportional to the rate of change of velocity with respect to y. Thus

$$\tau = \frac{F}{A} = \mu \frac{du}{dy} \qquad (12\text{–}1)$$

| Figure 12–1

where μ is the constant of proportionality and defines *absolute viscosity,* also called *dynamic viscosity.* The derivative du/dy is the rate of change of velocity with distance and may be called the rate of shear, or the velocity gradient. The viscosity μ is thus a measure of the internal frictional resistance of the fluid. For most lubricating fluids, the rate of shear is constant, and $du/dy = U/h$. Thus, from Eq. (12–1),

$$\tau = \frac{F}{A} = \mu \frac{U}{h} \tag{12–2}$$

Fluids exhibiting this characteristic are said to be *Newtonian fluids.* The unit of viscosity in the ips system is seen to be the pound-force-second per square inch; this is the same as stress or pressure multiplied by time. The ips unit is called the *reyn,* in honor of Sir Osborne Reynolds.

The absolute viscosity is measured by the pascal-second (Pa · s) in SI; this is the same as a Newton-second per square meter. The conversion from ips units to SI is the same as for stress. For example, multiply the absolute viscosity in reyns by 6890 to convert to units of Pa · s.

The American Society of Mechanical Engineers (ASME) has published a list of cgs units that are not to be used in ASME documents.[1] This list results from a recommendation by the International Committee of Weights and Measures (CIPM) that the use of cgs units with special names be discouraged. Included in this list is a unit of force called the *dyne* (dyn), a unit of dynamic viscosity called the *poise* (P), and a unit of kinematic viscosity called the *stoke* (St). All of these units have been, and still are, used extensively in lubrication studies.

The poise is the cgs unit of dynamic or absolute viscosity, and its unit is the dyne-second per square centimeter (dyn · s/cm^2). It has been customary to use the centipoise (cP) in analysis, because its value is more convenient. When the viscosity is expressed in centipoises, it is designated by Z. The conversion from cgs units to SI and ips units is as follows:

$$\mu(\text{Pa} \cdot \text{s}) = (10)^{-3} Z \, (\text{cP})$$

$$\mu(\text{reyn}) = \frac{Z \, (\text{cP})}{6.89(10)^6}$$

$$\mu(\text{mPa} \cdot \text{s}) = 6.89 \, \mu' \, (\mu\text{reyn})$$

In using ips units, the microreyn (μreyn) is often more convenient. The symbol μ' will be used to designate viscosity in μreyn such that $\mu = \mu'/(10^6)$.

The ASTM standard method for determining viscosity uses an instrument called the Saybolt Universal Viscosimeter. The method consists of measuring the time in seconds for 60 mL of lubricant at a specified temperature to run through a tube 17.6 mm in diameter and 12.25 mm long. The result is called the *kinematic viscosity,* and in the past the unit of the square centimeter per second has been used. One square centimeter per second is defined as a *stoke.* By the use of the *Hagen-Poiseuille law,* the kinematic viscosity based upon seconds Saybolt, also called *Saybolt Universal viscosity* (SUV) in seconds, is

$$Z_k = \left(0.22t - \frac{180}{t} \right) \tag{12–3}$$

where Z_k is in centistokes (cSt) and t is the number of seconds Saybolt.

---

[1]*ASME Orientation and Guide for Use of Metric Units,* 2nd ed., American Society of Mechanical Engineers, 1972, p. 13.

Figure 12–2

A comparison of the viscosities of various fluids.

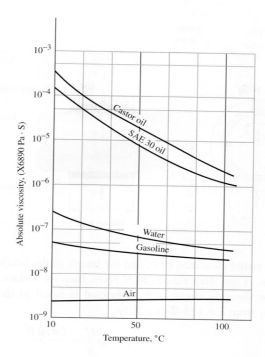

In SI, the kinematic viscosity ν has the unit of the square meter per second (m²/s), and the conversion is

$$\nu(\text{m}^2/\text{s}) = 10^{-6} Z_k \,(\text{cSt})$$

Thus, Eq. (12–3) becomes

$$\nu = \left(0.22t - \frac{180}{t}\right)(10^{-6})$$ (12–4)

To convert to dynamic viscosity, we multiply ν by the density in SI units. Designating the density as ρ with the unit of the kilogram per cubic meter, we have

$$\mu = \rho\left(0.22t - \frac{180}{t}\right)(10^{-6})$$ (12–5)

where μ is in pascal-seconds.

Figure 12–2 shows the absolute viscosity in the ips system of a number of fluids often used for lubrication purposes and their variation with temperature.

12–3 Petroff's Equation

The phenomenon of bearing friction was first explained by Petroff on the assumption that the shaft is concentric with its bushing. Though we shall seldom make use of Petroff's method of analysis in the material to follow, it is important because it defines groups of dimensionless parameters and because the coefficient of friction predicted by this law turns out to be quite good even when the shaft is not concentric.

Let us now consider a vertical shaft rotating in a guide bearing. It is assumed that the bearing carries a very small load, that the clearance space is completely filled with oil, and that leakage is negligible (Fig. 12–3). We denote the radius of the shaft

Figure 12–3

Petroff's lightly loaded journal bearing consisting of a shaft journal and a bushing with an axial-groove internal lubricant reservoir. The linear velocity gradient is shown in the end view. The clearance c is several thousandths of an inch and is grossly exaggerated for presentation purposes.

by r, the radial clearance by c, and the length of the bearing by l, all dimensions being in inches. If the shaft rotates at N rev/s, then its surface velocity is $U = 2\pi rN$ in/s. Since the shearing stress in the lubricant is equal to the velocity gradient times the viscosity, from Eq. (12–2) we have

$$\tau = \mu\frac{U}{h} = \frac{2\pi r\mu N}{c} \tag{a}$$

where the radial clearance c has been substituted for the distance h. The force required to shear the film is the stress times the area. The torque is the force times the lever arm r. Thus

$$T = (\tau A)(r) = \left(\frac{2\pi r\mu N}{c}\right)(2\pi rl)(r) = \frac{4\pi^2 r^3 l\mu N}{c} \tag{b}$$

If we now designate a small force on the bearing by W, in pounds-force, then the pressure P, in Newtons N per square meter of projected area, is $P = W/2rl$. The frictional force is fW, where f is the coefficient of friction, and so the frictional torque is

$$T = fWr = (f)(2rlP)(r) = 2r^2 flP \tag{c}$$

Substituting the value of the torque from Eq. (c) in Eq. (b) and solving for the coefficient of friction, we find

$$f = 2\pi^2 \frac{\mu N}{P}\frac{r}{c} \tag{12–6}$$

Equation (12–6) is called *Petroff's equation* and was first published in 1883. The two quantities $\mu N/P$ and r/c are very important parameters in lubrication. Substitution of the appropriate dimensions in each parameter will show that they are dimensionless.

The *bearing characteristic number,* or the *Sommerfeld number,* is defined by the equation

$$S = \left(\frac{r}{c}\right)^2 \frac{\mu N}{P} \tag{12–7}$$

The Sommerfeld number is very important in lubrication analysis because it contains many of the parameters that are specified by the designer. Note that it is also dimensionless. The quantity r/c is called the *radial clearance ratio*. If we multiply both sides of Eq. (12–6) by this ratio, we obtain the interesting relation

$$f\frac{r}{c} = 2\pi^2 \frac{\mu N}{P}\left(\frac{r}{c}\right)^2 = 2\pi^2 S \qquad (12\text{–}8)$$

12–4   Stable Lubrication

The difference between boundary and hydrodynamic lubrication can be explained by reference to Fig. 12–4. This plot of the change in the coefficient of friction versus the bearing characteristic $\mu N/P$ was obtained by the McKee brothers in an actual test of friction.[2] The plot is important because it defines stability of lubrication and helps us to understand hydrodynamic and boundary, or thin-film, lubrication.

Recall Petroff's bearing model in the form of Eq. (12–6) predicts that f is proportional to $\mu N/P$, that is, a straight line from the origin in the first quadrant. On the coordinates of Fig. 12–4 the locus to the right of point C is an example. Petroff's model presumes thick-film lubrication, that is, no metal-to-metal contact, the surfaces being completely separated by a lubricant film.

The McKee abscissa was ZN/P (centipoise × rev/min/psi) and the value of abscissa B in Fig. 12–4 was 30. The corresponding $\mu N/P$ (reyn × rev/s/psi) is $0.33(10^{-6})$. Designers keep $\mu N/P \geq 1.7(10^{-6})$, which corresponds to $ZN/P \geq 150$. A design constraint to keep thick-film lubrication is to be sure that

$$\frac{\mu N}{P} \geq 1.7(10^{-6}) \qquad (a)$$

Suppose we are operating to the right of line BA and something happens, say, an increase in lubricant temperature. This results in a lower viscosity and hence a smaller value of $\mu N/P$. The coefficient of friction decreases, not as much heat is generated in shearing the lubricant, and consequently the lubricant temperature drops. Thus the region to the right of line BA defines *stable lubrication* because variations are self-correcting.

To the left of line BA, a decrease in viscosity would increase the friction. A temperature rise would ensue, and the viscosity would be reduced still more. The result would be compounded. Thus the region to the left of line BA represents *unstable lubrication*.

It is also helpful to see that a small viscosity, and hence a small $\mu N/P$, means that the lubricant film is very thin and that there will be a greater possibility of some metal-to-metal contact, and hence of more friction. Thus, point C represents what is probably the beginning of metal-to-metal contact as $\mu N/P$ becomes smaller.

Figure 12–4

The variation of the coefficient of friction f with $\mu N/P$.

[2] S. A. McKee and T. R. McKee, "Journal Bearing Friction in the Region of Thin Film Lubrication," *SAE J.*, vol. 31, 1932, pp. (T)371–377.

12–5 Thick-Film Lubrication

Let us now examine the formation of a lubricant film in a journal bearing. Figure 12–5a shows a journal that is just beginning to rotate in a clockwise direction. Under starting conditions, the bearing will be dry, or at least partly dry, and hence the journal will climb or roll up the right side of the bearing as shown in Fig. 12–5a.

Now suppose a lubricant is introduced into the top of the bearing as shown in Fig. 12–5b. The action of the rotating journal is to pump the lubricant around the bearing in a clockwise direction. The lubricant is pumped into a wedge-shaped space and forces the journal over to the other side. A *minimum film thickness* h_0 occurs, not at the bottom of the journal, but displaced clockwise from the bottom as in Fig. 12–5b. This is explained by the fact that a film pressure in the converging half of the film reaches a maximum somewhere to the left of the bearing center.

Figure 12–5 shows how to decide whether the journal, under hydrodynamic lubrication, is eccentrically located on the right or on the left side of the bearing. Visualize the journal beginning to rotate. Find the side of the bearing upon which the journal tends to roll. Then, if the lubrication is hydrodynamic, mentally place the journal on the opposite side.

The nomenclature of a journal bearing is shown in Fig. 12–6. The dimension c is the *radial clearance* and is the difference in the radii of the bushing and journal. In Fig. 12–6 the center of the journal is at O and the center of the bearing at O'. The

Figure 12–5

Formation of a film.

(a) Dry

(b) Lubricated

Figure 12–6

Nomenclature of a partial journal bearing.

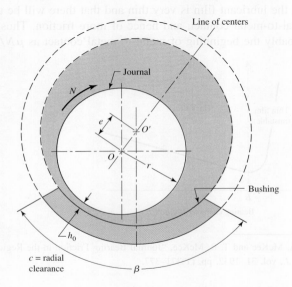

distance between these centers is the *eccentricity* and is denoted by *e*. The *minimum film thickness* is designated by h_0, and it occurs at the line of centers. The film thickness at any other point is designated by *h*. We also define an *eccentricity ratio ε* as

$$\epsilon = \frac{e}{c}$$

The bearing shown in the figure is known as a *partial bearing*. If the radius of the bushing is the same as the radius of the journal, it is known as a *fitted bearing*. If the bushing encloses the journal, as indicated by the dashed lines, it becomes a *full bearing*. The angle *β* describes the angular length of a partial bearing. For example, a 120° partial bearing has the angle *β* equal to 120°.

12–6 Hydrodynamic Theory

The present theory of hydrodynamic lubrication originated in the laboratory of Beauchamp Tower in the early 1880s in England. Tower had been employed to study the friction in railroad journal bearings and learn the best methods of lubricating them. It was an accident or error, during the course of this investigation, that prompted Tower to look at the problem in more detail and that resulted in a discovery that eventually led to the development of the theory.

Figure 12–7 is a schematic drawing of the journal bearing that Tower investigated. It is a partial bearing, having a diameter of 101.6 mm, a length of 152.4 mm, and a bearing arc of 157°, and having bath-type lubrication, as shown. The coefficients of friction obtained by Tower in his investigations on this bearing were quite low, which is now not surprising. After testing this bearing, Tower later drilled a 12.7-mm-diameter lubricator hole through the top. But when the apparatus was set in motion, oil flowed out of this hole. In an effort to prevent this, a cork stopper was used, but this popped out, and so it was necessary to drive a wooden plug into the hole. When the wooden plug was pushed out too, Tower, at this point, undoubtedly realized that he was on the verge of discovery. A pressure gauge connected to the hole indicated a pressure of more than twice the unit bearing load. Finally, he investigated the bearing film pressures in detail throughout the bearing width and length and reported a distribution similar to that of Fig. 12–8.[3]

The results obtained by Tower had such regularity that Osborne Reynolds concluded that there must be a definite equation relating the friction, the pressure, and

Figure 12–7

Schematic representation of the partial bearing used by Tower.

[3]Beauchamp Tower, "First Report on Friction Experiments," *Proc. Inst. Mech. Eng.,* November 1883, pp. 632–666; "Second Report," ibid., 1885, pp. 58–70; "Third Report," ibid., 1888, pp. 173–205; "Fourth Report," ibid., 1891, pp. 111–140.

Figure 12–8

Approximate pressure-distribution curves obtained by Tower.

the velocity. The present mathematical theory of lubrication is based upon Reynolds' work following the experiment by Tower.[4] The original differential equation, developed by Reynolds, was used by him to explain Tower's results. The solution is a challenging problem that has interested many investigators ever since then, and it is still the starting point for lubrication studies.

Reynolds pictured the lubricant as adhering to both surfaces and being pulled by the moving surface into a narrowing, wedge-shaped space so as to create a fluid or film pressure of sufficient intensity to support the bearing load. One of the important simplifying assumptions resulted from Reynolds' realization that the fluid films were so thin in comparison with the bearing radius that the curvature could be neglected. This enabled him to replace the curved partial bearing with a flat bearing, called a *plane slider bearing*. Other assumptions made were:

1 The lubricant obeys Newton's viscous effect, Eq. (12–1).

2 The forces due to the inertia of the lubricant are neglected.

3 The lubricant is assumed to be incompressible.

4 The viscosity is assumed to be constant throughout the film.

5 The pressure does not vary in the axial direction.

Figure 12–9a shows a journal rotating in the clockwise direction supported by a film of lubricant of variable thickness h on a partial bearing, which is fixed. We specify that the journal has a constant surface velocity U. Using Reynolds' assumption that curvature can be neglected, we fix a right-handed xyz reference system to the stationary bearing. We now make the following additional assumptions:

6 The bushing and journal extend infinitely in the z direction; this means there can be no lubricant flow in the z direction.

7 The film pressure is constant in the y direction. Thus the pressure depends only on the coordinate x.

8 The velocity of any particle of lubricant in the film depends only on the coordinates x and y.

We now select an element of lubricant in the film (Fig. 12–9a) of dimensions dx, dy, and dz, and compute the forces that act on the sides of this element. As shown in

---

[4]Osborne Reynolds, "Theory of Lubrication, Part I," *Phil. Trans. Roy. Soc. London,* 1886.

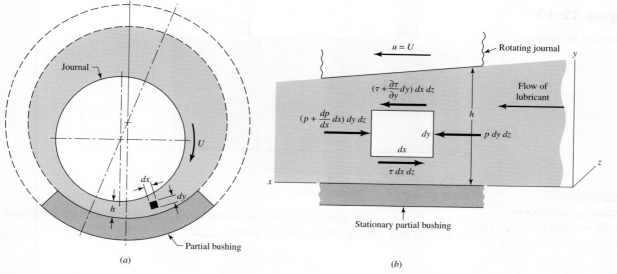

| Figure 12–9

Fig. 12–9b, normal forces, due to the pressure, act upon the right and left sides of the element, and shear forces, due to the viscosity and to the velocity, act upon the top and bottom sides. Summing the forces in the x direction gives

$$\sum F_x = p \, dy \, dz - \left(p + \frac{dp}{dx}dx\right) dy \, dz - \tau \, dx \, dz + \left(\tau + \frac{\partial \tau}{\partial y}dy\right) dx \, dz = 0 \quad (a)$$

This reduces to

$$\frac{dp}{dx} = \frac{\partial \tau}{\partial y} \quad (b)$$

From Eq. (12–1), we have

$$\tau = \mu \frac{\partial u}{\partial y} \quad (c)$$

where the partial derivative is used because the velocity u depends upon both x and y. Substituting Eq. (c) in Eq. (b), we obtain

$$\frac{dp}{dx} = \mu \frac{\partial^2 u}{\partial y^2} \quad (d)$$

Holding x constant, we now integrate this expression twice with respect to y. This gives

$$\frac{\partial u}{\partial y} = \frac{1}{\mu}\frac{dp}{dx}y + C_1$$

$$u = \frac{1}{2\mu}\frac{dp}{dx}y^2 + C_1 y + C_2 \quad (e)$$

Note that the act of holding x constant means that C_1 and C_2 can be functions of x. We now assume that there is no slip between the lubricant and the boundary surfaces. This gives two sets of boundary conditions for evaluating C_1 and C_2:

At $y = 0, \ u = 0$

At $y = h, \ u = U$ $\quad (f)$

Figure 12–10

Velocity of the lubricant.

Notice, in the second condition, that h is a function of x. Substituting these conditions in Eq. (e) and solving for C_1 and C_2 gives

$$C_1 = \frac{U}{h} - \frac{h}{2\mu} \frac{dp}{dx} \qquad C_2 = 0$$

or

$$u = \frac{1}{2\mu} \frac{dp}{dx} (y^2 - hy) + \frac{U}{h} y \tag{12–9}$$

This equation gives the velocity distribution of the lubricant in the film as a function of the coordinate y and the pressure gradient dp/dx. The equation shows that the velocity distribution across the film (from $y = 0$ to $y = h$) is obtained by superposing a parabolic distribution onto a linear distribution. Figure 12–10 shows the superposition of these distributions to obtain the velocity for particular values of x and dp/dx. In general, the parabolic term may be additive or subtractive to the linear term, depending upon the sign of the pressure gradient. When the pressure is maximum, $dp/dx = 0$ and the velocity is

$$u = \frac{U}{h} y \tag{g}$$

which is a linear relation.

We next define Q as the volume of lubricant flowing in the x direction per unit time. By using a width of unity in the z direction, the volume may be obtained by the expression

$$Q = \int_0^h u \, dy \tag{h}$$

Substituting the value of u from Eq. (12–9) and integrating gives

$$Q = \frac{Uh}{2} - \frac{h^3}{12\mu} \frac{dp}{dx} \tag{i}$$

The next step uses the assumption of an incompressible lubricant and states that the flow is the same for any cross section. Thus

$$\frac{dQ}{dx} = 0$$

From Eq. (*i*),

$$\frac{dQ}{dx} = \frac{U}{2}\frac{dh}{dx} - \frac{d}{dx}\left(\frac{h^3}{12\mu}\frac{dp}{dx}\right) = 0$$

or

$$\frac{d}{dx}\left(\frac{h^3}{\mu}\frac{dp}{dx}\right) = 6U\frac{dh}{dx} \qquad (12\text{--}10)$$

which is the classical Reynolds equation for one-dimensional flow. It neglects side leakage, that is, flow in the z direction. A similar development is used when side leakage is not neglected. The resulting equation is

$$\frac{\partial}{\partial x}\left(\frac{h^3}{\mu}\frac{\partial p}{\partial x}\right) + \frac{\partial}{\partial z}\left(\frac{h^3}{\mu}\frac{\partial p}{\partial z}\right) = 6U\frac{\partial h}{\partial x} \qquad (12\text{--}11)$$

There is no general analytical solution to Eq. (12–11); approximate solutions have been obtained by using electrical analogies, mathematical summations, relaxation methods, and numerical and graphical methods. One of the important solutions is due to Sommerfeld[5] and may be expressed in the form

$$\frac{r}{c}f = \phi\left[\left(\frac{r}{c}\right)^2\frac{\mu N}{P}\right] \qquad (12\text{--}12)$$

where ϕ indicates a functional relationship. Sommerfeld found the functions for half-bearings and full bearings by using the assumption of no side leakage.

12-7 Design Considerations

We may distinguish between two groups of variables in the design of sliding bearings. In the first group are those whose values either are given or are under the control of the designer. These are:

1 The viscosity μ
2 The load per unit of projected bearing area, P
3 The speed N
4 The bearing dimensions r, c, β, and l

Of these four variables, the designer usually has no control over the speed, because it is specified by the overall design of the machine. Sometimes the viscosity is specified in advance, as, for example, when the oil is stored in a sump and is used for lubricating and cooling a variety of bearings. The remaining variables, and sometimes the viscosity, may be controlled by the designer and are therefore the *decisions* the designer makes. In other words, when these four decisions are made, the design is complete.

In the second group are the dependent variables. The designer cannot control these except indirectly by changing one or more of the first group. These are:

1 The coefficient of friction f
2 The temperature rise ΔT
3 The volume flow rate of oil Q
4 The minimum film thickness h_0

[5]A. Sommerfeld, "Zur Hydrodynamischen Theorie der Schmiermittel-Reibung" ("On the Hydrodynamic Theory of Lubrication"), *Z. Math. Physik*, vol. 50, 1904, pp. 97–155.

$$N_b = 0, N_f = 0$$
$$N = |N_j + 0 - 2(0)| = N_j$$
(a)

$$N_b = 0, N_f = N_j$$
$$N = |N_j + 0 - 2N_j| = N_j$$
(b)

$$N_b = 0, N_f = \frac{N_j}{2}$$
$$N = |N_j + 0 - 2N_j/2| = 0$$
(c)

$$N_b = N_j, N_f = 0$$
$$N = |N_j + N_j - 2(0)| = 2N_j$$
(d)

Figure 12–11

How the significant speed varies. (*a*) Common bearing case. (*b*) Load vector moves at the same speed as the journal. (*c*) Load vector moves at half journal speed, no load can be carried. (*d*) Journal and bushing move at same speed, load vector stationary, capacity halved.

This group of variables tells us how well the bearing is performing, and hence we may regard them as *performance factors*. Certain limitations on their values must be imposed by the designer to ensure satisfactory performance. These limitations are specified by the characteristics of the bearing materials and of the lubricant. The fundamental problem in bearing design, therefore, is to define satisfactory limits for the second group of variables and then to decide upon values for the first group such that these limitations are not exceeded.

Significant Angular Speed

In the next section we will examine several important charts relating key variables to the Sommerfeld number. To this point we have assumed that only the journal rotates and it is the journal rotational speed that is used in the Sommerfeld number. It has been discovered that the angular speed N that is significant to hydrodynamic film bearing performance is[6]

$$N = |N_j + N_b - 2N_f| \qquad (12\text{–}13)$$

where N_j = journal angular speed, rev/s

N_b = bearing angular speed, rev/s

N_f = load vector angular speed, rev/s

When determining the Sommerfeld number for a general bearing, use Eq. (12–13) when entering N. Figure 12–11 shows several situations for determining N.

Trumpler's Design Criteria for Journal Bearings

Because the bearing assembly creates the lubricant pressure to carry a load, it reacts to loading by changing its eccentricity, which reduces the minimum film thickness h_0 until the load is carried. What is the limit of smallness of h_0? Close examination reveals that the moving adjacent surfaces of the journal and bushing are not smooth but consist of a series of asperities that pass one another, separated by a lubricant film. In starting

---

[6]Paul Robert Trumpler, *Design of Film Bearings,* Macmillan, New York, 1966, pp. 103–119.

a bearing under load from rest there is metal-to-metal contact and surface asperities are broken off, free to move and circulate with the oil. Unless a filter is provided, this debris accumulates. Such particles have to be free to tumble at the section containing the minimum film thickness without snagging in a togglelike configuration, creating additional damage and debris. Trumpler, an accomplished bearing designer, provides a throat of at least 200 μ in to pass particles from ground surfaces.[7] He also provides for the influence of size (tolerances tend to increase with size) by stipulating

$$h_0 \geq 0.00508 + 0.000\,04d \text{ mm} \tag{a}$$

where d is the journal diameter in millimeters.

A lubricant is a mixture of hydrocarbons that reacts to increasing temperature by vaporizing the lighter components, leaving behind the heavier. This process (bearings have lots of time) slowly increases the viscosity of the remaining lubricant, which increases heat generation rate and elevates lubricant temperatures. This sets the stage for future failure. For light oils, Trumpler limits the maximum film temperature T_{max} to

$$T_{max} \leq 121\,°C \tag{b}$$

Some oils can operate at slightly higher temperatures. Always check with the lubricant manufacturer.

A journal bearing often consists of a ground steel journal working against a softer, usually nonferrous, bushing. In starting under load there is metal-to-metal contact, abrasion, and the generation of wear particles, which, over time, can change the geometry of the bushing. The starting load divided by the projected area is limited to

$$\frac{W_{st}}{lD} \leq 2068 \text{ kPa} \tag{c}$$

If the load on a journal bearing is suddenly increased, the increase in film temperature in the annulus is immediate. Since ground vibration due to passing trucks, trains, and earth tremors is often present, Trumpler used a design factor of 2 or more on the running load, but not on the starting load of Eq. (c):

$$n_d \geq 2 \tag{d}$$

Many of Trumpler's designs are operating today, long after his consulting career was over; clearly they constitute good advice to the beginning designer.

12-8 The Relations of the Variables

Before proceeding to the problem of design, it is necessary to establish the relationships between the variables. Albert A. Raimondi and John Boyd, of Westinghouse Research Laboratories, used an iteration technique to solve Reynolds' equation on the digital computer.[8] This is the first time such extensive data have been available for use by designers, and consequently we shall employ them in this book.[9]

---

[7]Op. cit., pp. 192–194.

[8]A. A. Raimondi and John Boyd, "A Solution for the Finite Journal Bearing and Its Application to Analysis and Design, Parts I, II, and III," *Trans. ASLE*, vol. 1, no. 1, in *Lubrication Science and Technology,* Pergamon, New York, 1958, pp. 159–209.

[9]See also the earlier companion paper, John Boyd and Albert A. Raimondi, "Applying Bearing Theory to the Analysis and Design of Journal Bearings, Parts I and II," *J. Appl. Mechanics*, vol. 73, 1951, pp. 298–316.

The Raimondi and Boyd papers were published in three parts and contain 45 detailed charts and 6 tables of numerical information. In all three parts, charts are used to define the variables for length-diameter (l/d) ratios of 1:4, 1:2, and 1 and for beta angles of 60 to 360°. Under certain conditions the solution to the Reynolds equation gives negative pressures in the diverging portion of the oil film. Since a lubricant cannot usually support a tensile stress, Part III of the Raimondi-Boyd papers assumes that the oil film is ruptured when the film pressure becomes zero. Part III also contains data for the infinitely long bearing; since it has no ends, this means that there is no side leakage. The charts appearing in this book are from Part III of the papers, and are for full journal bearings ($\beta = 360°$) only. Space does not permit the inclusion of charts for partial bearings. This means that you must refer to the charts in the original papers when beta angles of less than 360° are desired. The notation is very nearly the same as in this book, so no problems should arise.

Viscosity Charts (Figs. 12–12 to 12–14)

One of the most important assumptions made in the Raimondi-Boyd analysis is that *viscosity of the lubricant is constant as it passes through the bearing.* But since work is done on the lubricant during this flow, the temperature of the oil is higher when it leaves the loading zone than it was on entry. And the viscosity charts clearly indicate that the viscosity drops off significantly with a rise in temperature. Since the analysis is based on a constant viscosity, our problem now is to determine the value of viscosity to be used in the analysis.

Some of the lubricant that enters the bearing emerges as a side flow, which carries away some of the heat. The balance of the lubricant flows through the load-bearing zone and carries away the balance of the heat generated. In determining the viscosity to be used we shall employ a temperature that is the average of the inlet and outlet temperatures, or

$$T_{av} = T_1 + \frac{\Delta T}{2} \tag{12–14}$$

where T_1 is the inlet temperature and ΔT is the temperature rise of the lubricant from inlet to outlet. Of course, the viscosity used in the analysis must correspond to T_{av}.

Viscosity varies considerably with temperature in a nonlinear fashion. The ordinates in Figs. 12–12 to 12–14 are not logarithmic, as the decades are of differing vertical length. These graphs represent the temperature versus viscosity functions for common grades of lubricating oils in both customary engineering and SI units. We have the temperature versus viscosity function only in graphical form, unless curve fits are developed. See Table 12–1.

One of the objectives of lubrication analysis is to determine the oil outlet temperature when the oil and its inlet temperature are specified. This is a trial-and-error type of problem. In an analysis, the temperature rise will first be estimated. This allows for the viscosity to be determined from the chart. With the value of the viscosity, the analysis is performed where the temperature rise is then computed. With this, a new estimate of the temperature rise is established. This process is continued until the estimated and computed temperatures agree.

To illustrate, suppose we have decided to use SAE 30 oil in an application in which the oil inlet temperature is $T_1 = 180°F$. We begin by estimating that the

Figure 12–12

Viscosity–temperature chart
in U.S. customary units.
(Raimondi and Boyd.)

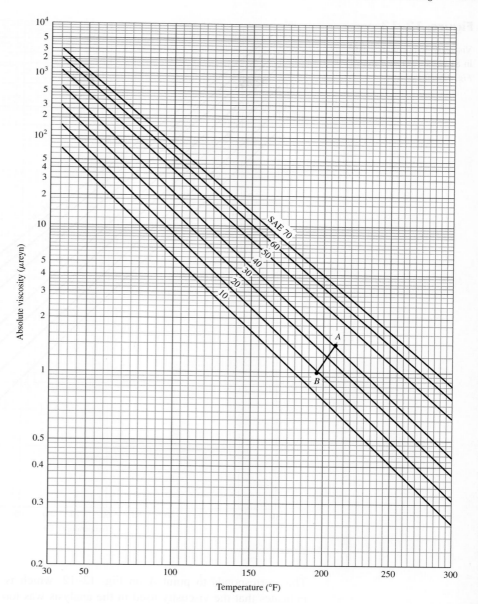

temperature rise will be $\Delta T = 30°F$. Then, from Eq. (12–14),

$$T_{av} = T_1 + \frac{\Delta T}{2} = 180 + \frac{30}{2} = 195°F$$

From Fig. 12–12 we follow the SAE 30 line and find that $\mu = 1.40$ μreyn at 195°F. So we use this viscosity (in an analysis to be explained in detail later) and find that the temperature rise is actually $\Delta T = 54°F$. Thus Eq. (12–14) gives

$$T_{av} = 180 + \frac{54}{2} = 207°F$$

Figure 12–13

Viscosity–temperature chart in SI units. *(Adapted from Fig. 12–12.)*

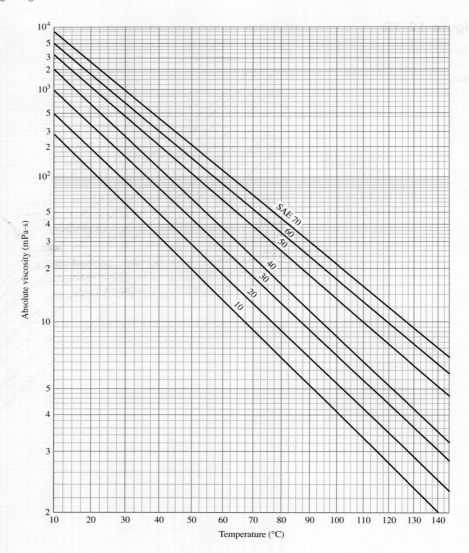

This corresponds to point *A* on Fig. 12–12, which is above the SAE 30 line and indicates that the viscosity used in the analysis was too high.

For a second guess, try $\mu = 1.00$ μreyn. Again we run through an analysis and this time find that $\Delta T = 30°$F. This gives an average temperature of

$$T_{\text{av}} = 180 + \frac{30}{2} = 195°\text{F}$$

and locates point *B* on Fig. 12–12.

If points *A* and *B* are fairly close to each other and on opposite sides of the SAE 30 line, a straight line can be drawn between them with the intersection locating the correct values of viscosity and average temperature to be used in the analysis. For this illustration, we see from the viscosity chart that they are $T_{\text{av}} = 203°$F and $\mu = 1.20$ μreyn.

Figure 12–14

Chart for multiviscosity lubricants. This chart was derived from known viscosities at two points, 100 and 210°F, and the results are believed to be correct for other temperatures.

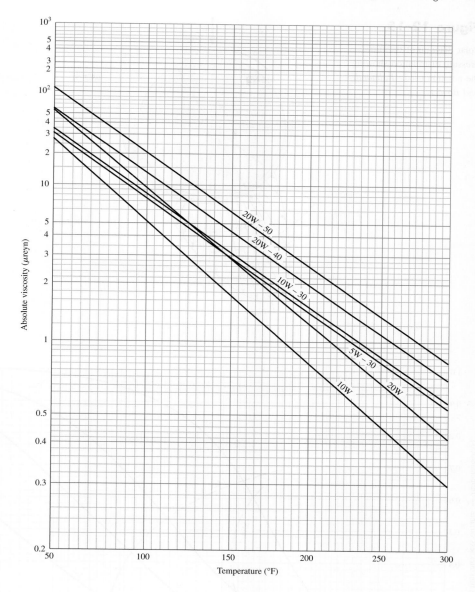

Table 12–1

Curve Fits* to Approximate the Viscosity versus Temperature Functions for SAE Grades 10 to 60

Source: A. S. Seireg and S. Dandage, "Empirical Design Procedure for the Thermodynamic Behavior of Journal Bearings," *J. Lubrication Technology,* vol. 104, April 1982, pp. 135–148.

Oil Grade, SAE	Viscosity μ_0, reyn	Constant b, °F
10	$0.0158(10^{-6})$	1157.5
20	$0.0136(10^{-6})$	1271.6
30	$0.0141(10^{-6})$	1360.0
40	$0.0121(10^{-6})$	1474.4
50	$0.0170(10^{-6})$	1509.6
60	$0.0187(10^{-6})$	1564.0

$^*\mu = \mu_0 \exp\left[b/(T + 95)\right]$, T in °F.

Figure 12–15

Polar diagram of the film–pressure distribution showing the notation used. *(Raimondi and Boyd.)*

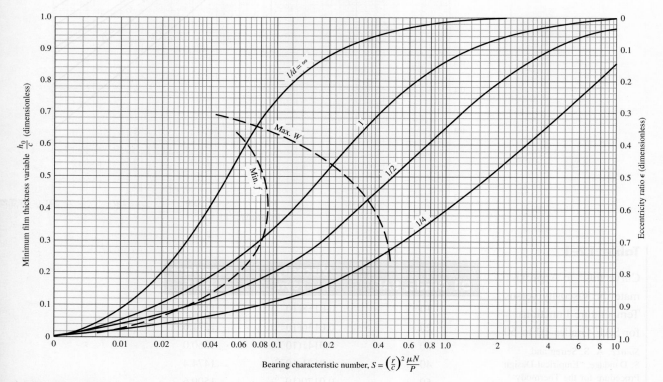

Figure 12–16

Chart for minimum film thickness variable and eccentricity ratio. The left boundary of the zone defines the optimal h_0 for minimum friction; the right boundary is optimum h_0 for load. *(Raimondi and Boyd.)*

Figure 12–17

Chart for determining the position of the minimum film thickness h_0. (*Raimondi and Boyd.*)

The remaining charts from Raimondi and Boyd relate several variables to the Sommerfeld number. These variables are

Minimum film thickness (Figs. 12–16 and 12–17)
Coefficient of friction (Fig. 12–18)
Lubricant flow (Figs. 12–19 and 12–20)
Film pressure (Figs. 12–21 and 12–22)

Figure 12–15 shows the notation used for the variables. We will describe the use of these curves in a series of four examples using the same set of given parameters.

Minimum Film Thickness

In Fig. 12–16, *the minimum film thickness* variable h_0/c and *eccentricity ratio* $\epsilon = e/c$ are plotted against the Sommerfeld number S with contours for various values of l/d. The corresponding angular position of the minimum film thickness is found in Fig. 12–17.

EXAMPLE 12–1

Determine h_0 and e using the following given parameters: $\mu = 0.027\ 56$ Pa · s, $N = 30$ rev/s, $W = 2210$ N (bearing load), $r = 19$ mm, $c = 0.038$ mm, and $l = 38$ mm.

Solution

The nominal bearing pressure (in projected area of the journal) is

$$P = \frac{W}{2rl} = \frac{2210}{2(0.019)0.038} = 1.531 \text{ MPa}$$

The Sommerfeld number is, from Eq. (12–7), where $N = N_j = 30$ rev/s,

$$S = \left(\frac{r}{c}\right)^2 \left(\frac{\mu N}{P}\right) = \left(\frac{19}{0.038}\right)^2 \left[\frac{0.027\ 56(30)}{1.531(10^6)}\right] = 0.135$$

Also, $l/d = 38/[2(19)] = 1$. Entering Fig. 12–16 with $S = 0.135$ and $l/d = 1$ gives $h_0/c = 0.42$ and $\epsilon = 0.58$. The quantity h_0/c is called the *minimum film thickness variable*. Since $c = 0.038$ mm, the minimum film thickness h_0 is

$$h_0 = 0.42(0.038) = 0.016 \text{ mm}$$

We can find the angular location ϕ of the minimum film thickness from the chart of Fig. 12–17. Entering with $S = 0.135$ and $l/d = 1$ gives $\phi = 53°$.

The eccentricity ratio is $\epsilon = e/c = 0.58$. This means the eccentricity e is

$$e = 0.58(0.038) = 0.022 \text{ mm}$$

Note that if the journal is centered in the bushing, $e = 0$ and $h_0 = c$, corresponding to a very light (zero) load. Since $e = 0$, $\epsilon = 0$. As the load is increased the journal displaces downward; the limiting position is reached when $h_0 = 0$ and $e = c$, that is, when the journal touches the bushing. For this condition the eccentricity ratio is unity. Since $h_0 = c - e$, dividing both sides by c, we have

$$\frac{h_0}{c} = 1 - \epsilon$$

Design optima are sometimes *maximum load*, which is a load-carrying characteristic of the bearing, and sometimes *minimum parasitic power loss* or *minimum coefficient of friction*. Dashed lines appear on Fig. 12–16 for maximum load and minimum coefficient of friction, so you can easily favor one of maximum load or minimum coefficient of friction, but not both. The zone between the two dashed-line contours might be considered a desirable location for a design point.

Coefficient of Friction

The friction chart, Fig. 12–18, has the *friction variable* $(r/c)f$ plotted against Sommerfeld number S with contours for various values of the l/d ratio.

EXAMPLE 12–2

Using the parameters given in Ex. 12–1, determine the coefficient of friction, the torque to overcome friction, and the power loss to friction.

Solution

We enter Fig. 12–18 with $S = 0.135$ and $l/d = 1$ and find $(r/c)f = 3.50$. The coefficient of friction f is

$$f = 3.50 \, c/r = 3.50(0.038/19) = 0.0070$$

The friction torque on the journal is

$$T = f Wr = 0.007(2210)0.019 = 0.2939 \text{ N} \cdot \text{m}$$

The power loss is

$$(hp)_{\text{loss}} = \frac{TN}{1050} = \frac{2.62(30)}{1050} = 0.075 \text{ hp}$$

$$TN (2\pi) = 0.2939(30)(2\pi) = 55.4 \text{ W}$$

Coefficient-of-friction variable $\frac{r}{c}f$ (dimensionless)

$l/d = 1/4$

$1/2$

1

∞

Bearing characteristic number, $S = \left(\frac{r}{c}\right)^2 \frac{\mu N}{P}$

Figure 12–18

Chart for coefficient-of-friction variable; note that Petroff's equation is the asymptote. *(Raimondi and Boyd.)*

Lubricant Flow

Figures 12–19 and 12–20 are used to determine the lubricant flow and side flow.

EXAMPLE 12–3

Continuing with the parameters of Ex. 12–1, determine the total volumetric flow rate Q and the side flow rate Q_s.

Solution

To estimate the lubricant flow, enter Fig. 12–19 with $S = 0.135$ and $l/d = 1$ to obtain $Q/(rcNl) = 4.28$. The total volumetric flow rate is

$$Q = 4.58rcNl = 4.28(19)0.038(30)38 = 3523 \text{ mm}^3/\text{s}$$

From Fig. 12–20 we find the *flow ratio* $Q_s/Q = 0.655$ and Q_s is

$$Q_s = 0.655Q = 0.655(3523) = 2308 \text{ mm}^3/\text{s}$$

Figure 12–19

Chart for flow variable.
Note: Not for pressure-fed
bearings. *(Raimondi and Boyd.)*

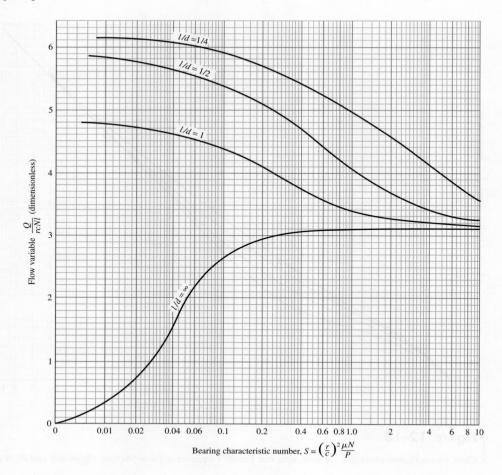

Figure 12–20

Chart for determining the ratio
of side flow to total flow.
(Raimondi and Boyd.)

Figure 12–21

Chart for determining the maximum film pressure. *Note:* Not for pressure-fed bearings. *(Raimondi and Boyd.)*

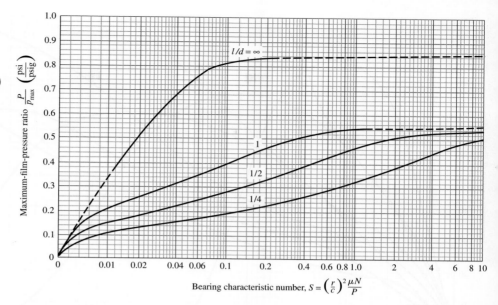

The side leakage Q_s is from the lower part of the bearing, where the internal pressure is above atmospheric pressure. The leakage forms a fillet at the journal-bushing external junction, and it is carried by journal motion to the top of the bushing, where the internal pressure is below atmospheric pressure and the gap is much larger, to be "sucked in" and returned to the lubricant sump. That portion of side leakage that leaks away from the bearing has to be made up by adding oil to the bearing sump periodically by maintenance personnel.

Film Pressure

The maximum pressure developed in the film can be estimated by finding the pressure ratio P/p_{max} from the chart in Fig. 12–21. The locations where the terminating and maximum pressures occur, as defined in Fig 12–15, are determined from Fig. 12–22.

EXAMPLE 12–4

Using the parameters given in Ex. 12–1, determine the maximum film pressure and the locations of the maximum and terminating pressures.

Solution

Entering Fig. 12–21 with $S = 0.135$ and $l/d = 1$, we find $P/p_{max} = 0.42$. The maximum pressure p_{max} is therefore

$$p_{max} = \frac{P}{0.42} = \frac{1.531}{0.42} = 3.645 \text{ MPa}$$

With $S = 0.135$ and $l/d = 1$, from Fig. 12–22, $\theta_{p_{max}} = 18.5°$ and the terminating position θ_{p_0} is 75°.

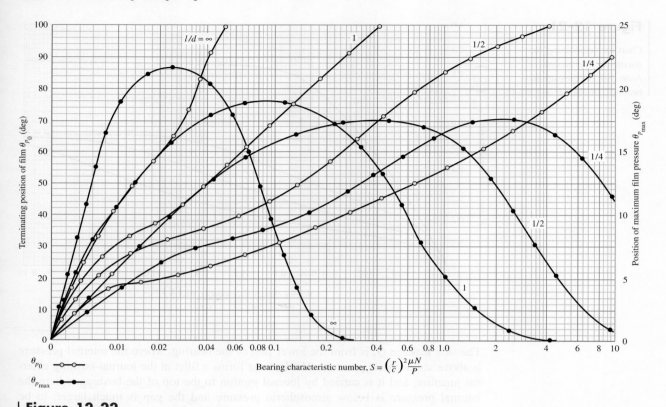

Figure 12–22

Chart for finding the terminating position of the lubricant film and the position of maximum film pressure. *(Raimondi and Boyd.)*

Examples 12–1 to 12–4 demonstrate how the Raimondi and Boyd charts are used. It should be clear that we do not have journal-bearing parametric relations as equations, but in the form of charts. Moreover, the examples were simple because the steady-state equivalent viscosity was given. We will now show how the average film temperature (and the corresponding viscosity) is found from energy considerations.

Lubricant Temperature Rise

The temperature of the lubricant rises until the rate at which work is done by the journal on the film through fluid shear is the same as the rate at which heat is transferred to the greater surroundings. The specific arrangement of the bearing plumbing affects the quantitative relationships. See Fig. 12–23. A lubricant sump (internal or external to the bearing housing) supplies lubricant at sump temperature T_s to the bearing annulus at temperature $T_s = T_1$. The lubricant passes once around the bushing and is delivered at a higher lubricant temperature $T_1 + \Delta T$ to the sump. Some of the lubricant leaks out of the bearing at a mixing-cup temperature of $T_1 + \Delta T/2$ and is returned to the sump. The sump may be a keyway-like groove in the bearing cap or a larger chamber up to half the bearing circumference. It can occupy "all" of the bearing cap of a split bearing. In such a bearing the side leakage occurs from the lower portion and is sucked back in, into the ruptured film arc. The sump could be well removed from the journal-bushing interface.

Figure 12–23

Schematic of a journal bearing with an external sump with cooling; lubricant makes one pass before returning to the sump.

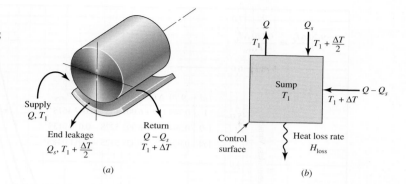

(a)   (b)

Let

$$Q = \text{volumetric oil-flow rate into the bearing, m}^3\text{/s}$$

$$Q_s = \text{volumetric side-flow leakage rate out of the bearing and to the sump, m}^3\text{/s}$$

$$Q - Q_s = \text{volumetric oil-flow discharge from annulus to sump, m}^3\text{/s}$$

$$T_1 = \text{oil inlet temperature (equal to sump temperature } T_s), \,°\text{C}$$

$$\Delta T = \text{temperature rise in oil between inlet and outlet, }°\text{C}$$

$$\rho = \text{lubricant density, kg/m}^3$$

$$C_p = \text{specific heat capacity of lubricant, kJ/(kg} \cdot °\text{C)}$$

$$J = \text{Joulean heat equivalent, N} \cdot \text{m/J}$$

$$H = \text{heat rate, J/s or W}$$

Using the sump as a control region, we can write an enthalpy balance. Using T_1 as the datum temperature gives

$$H_{\text{loss}} = \rho C_p Q_s \Delta T/2 + \rho C_p (Q - Q_s)\Delta T = \rho C_p Q \Delta T \left(1 - 0.5\frac{Q_s}{Q}\right) \qquad (a)$$

The thermal energy loss at steady state H_{loss} is equal to the rate the journal does work on the film is $H_{\text{loss}} = \dot{W} = 2\pi T N/J$. The torque $T = f W r$, the load in terms of pressure is $W = 2 Prl$, and multiplying numerator and denominator by the clearance c gives

$$H_{\text{loss}} = \frac{4\pi \, PrlNc}{J} \frac{rf}{c} \qquad (b)$$

Equating Eqs. (a) and (b) and rearranging results in

$$\frac{J\rho C_p \, \Delta T}{4\pi \, P} = \frac{rf/c}{(1 - 0.5 Q_s/Q)[Q/(rcNl)]} \qquad (c)$$

For common petroleum lubricants $\rho = 862$ kg/m^3, $C_p = 0.42$ 1.758 kJ/(kg $\cdot$ °C), and $J = 995(10)^6$ m $\cdot$ N/J; therefore the left term of Eq. (c) is

$$\frac{J\rho C_p \, \Delta T}{4\pi \, P} = \frac{995(10)^6 \, 862(1.758)\,\Delta T_C}{4\pi \, P_{\text{MPa}}} = 0.12 \,\Delta T_C/P_{\text{MPa}}$$

thus

$$\frac{0.12\,\Delta T_C}{P_{\text{MPa}}} = \frac{rf/c}{\left(1 - \frac{1}{2}Q_s/Q\right)[Q/(rcN_j l)]} \qquad (12\text{--}15)$$

Figure 12–24

Figures 12–18, 12–19, and 12–20 combined to reduce iterative table look-up. *(Source: Chart based on work of Raimondi and Boyd boundary condition (2), i.e., no negative lubricant pressure developed. Chart is for full journal bearing using single lubricant pass, side flow emerges with temperature rise $\Delta T/2$, thru flow emerges with temperature rise ΔT, and entire flow is supplied at datum sump temperature.)*

where ΔT_F is the temperature rise in °C and P_{MPa} is the bearing pressure in MPa. The right side of Eq. (12–15) can be evaluated from Figs. 12–18, 12–19, and 12–20 for various Sommerfeld numbers and l/d ratios to give Fig. 12–24. It is easy to show that the left side of Eq. (12–15) can be expressed as $0.120\Delta T_C/P_{\text{MPa}}$ where ΔT_C is expressed in °C and the pressure P_{MPa} is expressed in MPa. The ordinate in Fig. 12–24 is either $9.70\,\Delta T_F/P_{\text{psi}}$ or $0.120\,\Delta T_C/P_{\text{MPa}}$, which is not surprising since both are dimensionless in proper units and *identical in magnitude*. Since solutions to bearing problems involve iteration and reading many graphs can introduce errors, Fig. 12–24 reduces three graphs to one, a step in the proper direction.

Interpolation

For l/d ratios other than the ones given in the charts, Raimondi and Boyd have provided the following interpolation equation

$$y = \frac{1}{(l/d)^3}\left[-\frac{1}{8}\left(1-\frac{l}{d}\right)\left(1-2\frac{l}{d}\right)\left(1-4\frac{l}{d}\right)y_\infty + \frac{1}{3}\left(1-2\frac{l}{d}\right)\left(1-4\frac{l}{d}\right)y_1\right.$$

$$\left.-\frac{1}{4}\left(1-\frac{l}{d}\right)\left(1-4\frac{l}{d}\right)y_{1/2} + \frac{1}{24}\left(1-\frac{l}{d}\right)\left(1-2\frac{l}{d}\right)y_{1/4}\right] \qquad (12\text{–}16)$$

where y is the desired variable within the interval $\infty > l/d > \frac{1}{4}$ and y_∞, y_1, $y_{1/2}$, and $y_{1/4}$ are the variables corresponding to l/d ratios of ∞, 1, $\frac{1}{2}$, and $\frac{1}{4}$, respectively.

12–9 Steady-State Conditions in Self-Contained Bearings

The case in which the lubricant carries away all of the enthalpy increase from the journal-bushing pair has already been discussed. Bearings in which the warm lubricant stays within the bearing housing will now be addressed. These bearings are called *self-contained* bearings because the lubricant sump is within the bearing housing and the lubricant is cooled within the housing. These bearings are described as *pillow-block* or *pedestal* bearings. They find use on fans, blowers, pumps, and motors, for example. Integral to design considerations for these bearings is dissipating heat from the bearing housing to the surroundings at the same rate that enthalpy is being generated within the fluid film.

In a self-contained bearing the sump can be positioned as a keywaylike cavity in the bushing, the ends of the cavity not penetrating the end planes of the bushing. Film oil exits the annulus at about one-half of the relative peripheral speeds of the journal and bushing and slowly tumbles the sump lubricant, mixing with the sump contents. Since the film in the top "half" of the cap has cavitated, it contributes essentially nothing to the support of the load, but it does contribute friction. Bearing caps are in use in which the "keyway" sump is expanded peripherally to encompass the top half of the bearing. This reduces friction for the same load, but the included angle β of the bearing has been reduced to 180°. Charts for this case were included in the Raimondi and Boyd paper.

The heat given up by the bearing housing may be estimated from the equation

$$H_{\text{loss}} = \hbar_{\text{CR}} A (T_b - T_\infty) \tag{12–17}$$

where H_{loss} = heat dissipated, J/s or W$\hbar_{\text{CR}}$

$\hbar_{\text{CR}}$ = combined overall coefficient of radiation and convection heat transfer, W/(m² · °C)

A = surface area of bearing housing, m²

T_b = surface temperature of the housing, °C

T_∞ = ambient temperature, °C

The overall coefficient $\hbar_{\text{CR}}$ depends on the material, surface coating, geometry, even the roughness, the temperature difference between the housing and surrounding objects, and air velocity. After Karelitz,[10] and others, in ordinary industrial environments, the overall coefficient   can be treated as a constant. Some representative values are

$$\hbar_{\text{CR}} = \begin{cases} 11.4 \text{ W}/(\text{m}^2 \cdot {}^\circ\text{C}) & \text{for still air} \\ 15.3 \text{ W}/(\text{m}^2 \cdot {}^\circ\text{C}) & \text{for shaft-stirred air} \\ 33.5 \text{ W}/(\text{m}^2 \cdot {}^\circ\text{C}) & \text{for air moving at 24.5 m/s} \end{cases} \tag{12–18}$$

An expression similar to Eq. (12–17) can be written for the temperature difference $T_f - T_b$ between the lubricant film and the housing surface. This is possible because the bushing and housing are metal and very nearly isothermal. If one defines $\overline{T}_f$ as

---

[10]G. B. Karelitz, "Heat Dissipation in Self-Contained Bearings," *Trans. ASME*, Vol. 64, 1942, p. 463; D. C. Lemmon and E. R. Booser, "Bearing Oil-Ring Performance," *Trans. ASME*, J. Bas. Engin., Vol. 88, 1960, p. 327.

| Table 12–2

Lubrication System	Conditions	Range of α
Oil ring	Moving air	1–2
	Still air	$\frac{1}{2}$–1
Oil bath	Moving air	$\frac{1}{2}$–1
	Still air	$\frac{1}{5}$–$\frac{2}{5}$

the *average* film temperature (halfway between the lubricant inlet temperature T_s and the outlet temperature $T_s + \Delta T$), then the following proportionality has been observed between $\overline{T}_f - T_b$ and the difference between the housing surface temperature and the ambient temperature, $T_b - T_\infty$:

$$\overline{T}_f - T_b = \alpha(T_b - T_\infty) \qquad (a)$$

where $\overline{T}_f$ is the average film temperature and α is a constant depending on the lubrication scheme and the bearing housing geometry. Equation (a) may be used to estimate the bearing housing temperature. Table 12–2 provides some guidance concerning suitable values of α. The work of Karelitz allows the broadening of the application of the charts of Raimondi and Boyd, to be applied to a variety of bearings beyond the natural circulation pillow-block bearing.

Solving Eq. (a) for T_b and substituting into Eq. (12–17) gives the bearing heat loss rate to the surroundings as

$$H_{\text{loss}} = \frac{\hbar_{\text{CR}}A}{1 + \alpha}(\overline{T}_f - T_\infty) \qquad (12\text{–}19a)$$

and rewriting Eq. (a) gives

$$T_b = \frac{\overline{T}_f + \alpha T_\infty}{1 + \alpha} \qquad (12\text{–}19b)$$

In beginning a steady-state analysis the average film temperature is unknown, hence the viscosity of the lubricant in a self-contained bearing is unknown. Finding the equilibrium temperatures is an iterative process wherein a trial average film temperature (and the corresponding viscosity) is used to compare the heat generation rate and the heat loss rate. An adjustment is made to bring these two heat rates into agreement. This can be done on paper with a tabular array to help adjust $\overline{T}_f$ to achieve equality between heat generation and loss rates. A root-finding algorithm can be used. Even a simple one can be programmed for a digital computer.

Because of the shearing action there is a uniformly distributed energy release in the lubricant that heats the lubricant as it works its way around the bearing. The temperature is uniform in the radial direction but increases from the sump temperature T_s by an amount ΔT during the lubricant pass. The exiting lubricant mixes with the sump contents, being cooled to sump temperature. The lubricant in the sump is cooled because the bushing and housing metal are at a nearly uniform lower temperature because of heat losses by convection and radiation to the surroundings at ambient temperature T_∞. In the usual configurations of such bearings, the bushing and housing metal temperature is approximately midway between the average film temperature $\overline{T}_f = T_s + \Delta T/2$ and the ambient temperature T_∞. The heat generation rate H_{gen}, at steady state, is equal to the work rate from the frictional torque T. Then from Eq. (b), Sec. 12–3, the torque is $T = 4\pi^2 r^3 l\mu/c$, resulting in

$$H_{gen} = \frac{4\pi^2 r^3 l\mu N}{c}(2\pi N) = \frac{248\mu N^2 l r^3}{c} \qquad (b)$$

Equating this to Eq. (12–19a) and solving for T_f gives

$$\bar{T}_f = T_\infty + 248(1 + \alpha)\frac{\mu N^2 l r^3}{\hbar_{CR}Ac} \qquad (12\text{–}20)$$

EXAMPLE 12–5

Consider a pillow-block bearing with a keyway sump, whose journal rotates at 900 rev/min in shaft-stirred air at 21°C with $\alpha = 1$. The lateral area of the bearing is 25 800 mm². The lubricant is SAE grade 20 oil. The gravity radial load is 450 N and the l/d ratio is unity. The bearing has a journal diameter of 50 mm + 0.000/−0.002 25 mm, a bushing bore of 50.05 + 0.008/−0.000 mm. For a minimum clearance assembly estimate the steady-state temperatures as well as the minimum film thickness and coefficient of friction.

Solution

The minimum radial clearance, c_{min}, is

$$c_{min} = \frac{50.05 - 50}{2} = 0.025 \text{ mm}$$

$$P = \frac{W}{ld} = \frac{450}{(50)50} = 0.18 \text{ MPa}$$

$$S = \left(\frac{r}{c}\right)^2 \frac{\mu N}{P} = \left(\frac{25}{0.025}\right)^2 \frac{\mu 15}{0.18(10^6)} = 83.3 \mu$$

The friction loss, F_f, is found as follows:

$$F_f = fWr2\pi N = WNc2\pi\left(\frac{fr}{c}\right) = 450(15)2\pi\left(\frac{fr}{c}\right) = 42\,411\left(\frac{fr}{c}\right) \text{ J/s}$$

The heat generation rate H_{gen}, is

$$H_{gen} = F_f = 42\,411(fr/c) = J/s \text{ or W}$$

From Eq. (12–19a) with $\hbar_{CR} = 15.3$ W/(m² °C), the rate of heat loss to the environment H_{loss} is

$$H_{loss} = \frac{\hbar_{CR}A}{\alpha + 1}(\bar{T}_f - 21) = \frac{15.3(0.0258)}{(1 + 1)}(\bar{T}_f - 21) = 1974(\bar{T}_f - 21) J/s \text{ or W}$$

Build a table as follows for trial values of $\bar{T}_f$ of 87.8 and 90.5°C:

Trial $\bar{T}_f$	μ'	S	fr/c	H_{gen}	H_{loss}
87.8	0.008 73	0.73	3.4	145 049	131 863
90.5	0.007 82	0.65	3.06	129 880	137 193

The temperature at which $H_{gen} = H_{loss} = 135.4$ kJ/s is 89.7°C. Rounding $\overline{T}_f$ to 90°C we find $\mu = 0.0081$ Pa · s and $S = 83.3(0.0081) = 0.67$. From Fig. 12–24, $\dfrac{0.12\Delta T_c}{PMPa} = 4.25$ and thus

$$\Delta T_C = 4.25P/0.12 = 4.25(0.18)/0.12 = 6.4°C$$

$$T_1 = T_s = \overline{T}_f - \Delta T/2 = 90 - 6.4/2 = 86.8°C$$

$$T_{max} = T_1 + \Delta T_F = 86.8 + 6.4 = 93.2°C$$

From Eq. (12–19b)

$$T_b = \frac{T_f + \alpha T_\infty}{1 + \alpha} = \frac{90 + (1)21}{1 + 1} = 55.5°C$$

with $S = 0.67$, the minimum film thickness from Fig. 12–16 is

$$h_0 = \frac{h_0}{c}c = 0.79(0.025) = 0.0198 \text{ mm}$$

The coefficient of friction from Fig. 12–18 is

$$f = \frac{fr}{c}\frac{c}{r} = 12.8\frac{0.025}{25} = 0.012\,8$$

The parasitic friction torque T is

$$T = fWr = 0.012\,8(450)(0.025) = 0.144 \text{ N} \cdot \text{m}$$

12–10   Clearance

In designing a journal bearing for thick-film lubrication, the engineer must select the grade of oil to be used, together with suitable values for P, N, r, c, and l. A poor selection of these or inadequate control of them during manufacture or in use may result in a film that is too thin, so that the oil flow is insufficient, causing the bearing to overheat and, eventually, fail. Furthermore, the radial clearance c is difficult to hold accurate during manufacture, and it may increase because of wear. What is the effect of an entire range of radial clearances, expected in manufacture, and what will happen to the bearing performance if c increases because of wear? Most of these questions can be answered and the design optimized by plotting curves of the performance as functions of the quantities over which the designer has control.

Figure 12–25 shows the results obtained when the performance of a particular bearing is calculated for a whole range of radial clearances and is plotted with clearance as the independent variable. The bearing used for this graph is the one of Examples 12–1 to 12–4 with SAE 20 oil at an inlet temperature of 100°F. The graph shows that if the clearance is too tight, the temperature will be too high and the minimum film thickness too low. High temperatures may cause the bearing to fail by fatigue. If the oil film is too thin, dirt particles may be unable to pass without scoring or may embed themselves in the bearing. In either event, there will be excessive wear and friction, resulting in high temperatures and possible seizing.

To investigate the problem in more detail, Table 12–3 was prepared using the two types of preferred running fits that seem to be most useful for journal-bearing design

Figure 12–25

A plot of some performance characteristics of the bearing of Exs. 12–1 to 12–4 for radial clearances of 0.0127 to 0.0762 mm. The bearing outlet temperature is designated T_2. New bearings should be designed for the shaded zone, because wear will move the operating point to the right.

Table 12–3

Maximum, Minimum, and Average Clearances for 38 mm-Diameter Journal Bearings Based on Type of Fit

Type of Fit	Symbol	Clearance c, mm		
		Maximum	**Average**	**Minimum**
Close-running	H8/f7	0.044 45	0.028 58	0.012 70
Free-running	H9/d9	0.100 33	0.069 85	0.039 37

Table 12–4

Performance of 38 mm-Diameter Journal Bearing with Various Clearances. (SAE 20 Lubricant, $T_1 = 100°F$, $N = 30$ r/s, $W = 500$ lbf, $L = 1.5$ in)

c, mm	T_2, °C	h_0, mm	f	Q, mm³/s	H, J/s
0.012 70	108	0.009 65	0.011 3	999	0.090 3
0.028 58	61	0.016 51	0.009 0	2507	0.071 4
0.039 37	56	0.019 56	0.008 7	3571	0.069 3
0.044 45	53	0.019 30	0.008 4	4129	0.067 2
0.069 85	48	0.018 54	0.007 9	6864	0.063
0.100 33	45	0.017 53	0.007 7	10108	0.062

(see Table 7–9), p. 389. The results shown in Table 12–3 were obtained by using Eqs. (7–36) and (7–37) of Sec. 7–8. Notice that there is a slight overlap, but the range of clearances for the free-running fit is about twice that of the close-running fit.

The six clearances of Table 12–3 were used in a computer program to obtain the numerical results shown in Table 12–4. These conform to the results of Fig. 12–25, too. Both the table and the figure show that a tight clearance results in a high temperature. Figure 12–26 can be used to estimate an upper temperature limit when the characteristics of the application are known.

It would seem that a large clearance will permit the dirt particles to pass through and also will permit a large flow of oil, as indicated in Table 12–4. This lowers the temperature and increases the life of the bearing. However, if the clearance becomes

Figure 12-26

Temperature limits for mineral oils. The lower limit is for oils containing antioxidants and applies when oxygen supply is unlimited. The upper limit applies when insignificant oxygen is present. The life in the shaded zone depends on the amount of oxygen and catalysts present.

(Source: M. J. Neale (ed.), Tribology Handbook, Section B1, Newnes-Butterworth, London, 1975.)

too large, the bearing becomes noisy and the minimum film thickness begins to decrease again.

In between these two limitations there exists a rather large range of clearances that will result in satisfactory bearing performance.

When both the production tolerance and the future wear on the bearing are considered, it is seen, from Fig. 12–25, that the best compromise is a clearance range slightly to the left of the top of the minimum-film-thickness curve. In this way, future wear will move the operating point to the right and increase the film thickness and decrease the operating temperature.

12–11 Pressure-Fed Bearings

The load-carrying capacity of self-contained natural-circulating journal bearings is quite restricted. The factor limiting better performance is the heat-dissipation capability of the bearing. A first thought of a way to increase heat dissipation is to cool the sump with an external fluid such as water. The high-temperature problem is in the film where the heat is generated but cooling is not possible in the film until later. This does not protect against exceeding the maximum allowable temperature of the lubricant. A second alternative is to reduce the *temperature rise* in the film by dramatically increasing the rate of lubricant flow. The lubricant itself is reducing the temperature rise. A water-cooled sump may still be in the picture. To increase lubricant flow, an external pump must be used with lubricant supplied at pressures of tens of pounds per square inch gage. Because the lubricant is supplied to the bearing under pressure, such bearings are called *pressure-fed bearings*.

To force a greater flow through the bearing and thus obtain an increased cooling effect, a common practice is to use a circumferential groove at the center of the bearing, with an oil-supply hole located opposite the load-bearing zone. Such a bearing is shown in Fig. 12–27. The effect of the groove is to create two half-bearings, each having a smaller l/d ratio than the original. The groove divides the pressure-distribution curve into two lobes and reduces the minimum film thickness, but it has wide acceptance among lubrication engineers because such bearings carry more load without overheating.

To set up a method of solution for oil flow, we shall assume a groove ample enough that the pressure drop in the groove itself is small. Initially we will neglect eccentricity and then apply a correction factor for this condition. The oil flow, then, is the amount that flows out of the two halves of the bearing in the direction of the concentric shaft. If we neglect the rotation of the shaft, the flow of the lubricant is

Figure 12–27

Centrally located full annular groove. *(Courtesy of the Cleveland Graphite Bronze Company, Division of Clevite Corporation.)*

Figure 12–28

Flow of lubricant from a pressure-fed bearing having a central annular groove.

caused by the supply pressure p_s, shown in Fig. 12–28. Laminar flow is assumed, with the pressure varying linearly from $p = p_s$ at $x = 0$, to $p = 0$ at $x = l'$. Consider the static equilibrium of an element of thickness dx, height $2y$, and unit depth. Note particularly that the origin of the reference system has been chosen at the midpoint of the clearance space and symmetry about the x axis is implied with the shear stresses τ being equal on the top and bottom surfaces. The equilibrium equation in the x direction is

$$-2y(p + dp) + 2yp + 2\tau\, dx = 0 \qquad (a)$$

Expanding and canceling terms, we find that

$$\tau = y\frac{dp}{dx} \qquad (b)$$

Newton's equation for viscous flow [Eq. (12–1)] is

$$\tau = \mu\frac{du}{dy} \qquad (c)$$

Now eliminating τ from Eqs. (b) and (c) gives

$$\frac{du}{dy} = \frac{1}{\mu}\frac{dp}{dx}y \qquad (d)$$

Treating dp/dx as a constant and integrating with respect to y gives

$$u = \frac{1}{2\mu}\frac{dp}{dx}y^2 + C_1 \qquad (e)$$

At the boundaries, where $y = \pm c/2$, the velocity u is zero. Using one of these conditions in Eq. (e) gives

$$0 = \frac{1}{2\mu} \frac{dp}{dx} \left(\frac{c}{2}\right)^2 + C_1$$

or

$$C_1 = -\frac{c^2}{8\mu} \frac{dp}{dx}$$

Substituting C_1 in Eq. (e) yields

$$u = \frac{1}{8\mu} \frac{dp}{dx} (4y^2 - c^2) \qquad (f)$$

Assuming the pressure varies linearly from p_s to 0 at $x = 0$ to l', respectively, the pressure can be written as

$$p = p_s - \frac{p_s}{l'} x \qquad (g)$$

and therefore the pressure gradient is given by

$$\frac{dp}{dx} = -\frac{p_s}{l'} \qquad (h)$$

We can now substitute Eq. (h) in Eq. (f) and the relationship between the oil velocity and the coordinate y is

$$u = \frac{p_s}{8\mu l'} (c^2 - 4y^2) \qquad (12\text{–}21)$$

Figure 12–29 shows a graph of this relation fitted into the clearance space c so that you can see how the velocity of the lubricant varies from the journal surface to the bearing surface. The distribution is parabolic, as shown, with the maximum velocity occurring at the center, where $y = 0$. The magnitude is, from Eq. (12–21),

$$u_{\max} = \frac{p_s c^2}{8\mu l'} \qquad (i)$$

To consider eccentricity, as shown in Fig. 12–30, the film thickness is $h = c - e \cos \theta$. Substituting h for c in Eq. (i), with the average ordinate of a parabola being two-thirds the maximum, the average velocity at any angular position θ is

$$u_{\mathrm{av}} = \frac{2}{3} \frac{p_s h^2}{8\mu l'} = \frac{p_s}{12\mu l'} (c - e \cos \theta)^2 \qquad (j)$$

We still have a little further to go in this analysis, so please be patient. Now that we have an expression for the lubricant velocity, we can compute the amount of lubricant

Figure 12–29

Parabolic distribution of the lubricant velocity.

| Figure 12-30

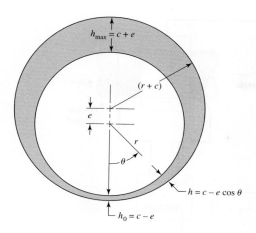

that flows out both ends. The elemental side flow at any position θ (Fig. 12–30) is

$$dQ_s = 2u_{av}\,dA = 2u_{av}(rh\,d\theta) \tag{k}$$

where dA is the elemental area. Substituting u_{av} from Eq. (*j*) and (*h*) from Fig. 12–30 gives

$$dQ_s = \frac{p_s r}{6\mu l'}(c - e\cos\theta)^3\,d\theta \tag{l}$$

Integrating around the bearing gives the total side flow as

$$Q_s = \int dQ_s = \frac{p_s r}{6\mu l'}\int_0^{2\pi}(c - e\cos\theta)^3\,d\theta = \frac{p_s r}{6\mu l'}(2\pi c^3 + 3\pi\,ce^2)$$

Rearranging, with $\epsilon = e/c$, gives

$$Q_s = \frac{\pi p_s rc^3}{3\mu l'}(1 + 1.5\epsilon^2) \tag{12-22}$$

In analyzing the performance of pressure-fed bearings, the bearing length should be taken as l', as defined in Fig. 12–28. The characteristic pressure in each of the two bearings that constitute the pressure-fed bearing assembly P is given by

$$P = \frac{W/2}{2rl'} = \frac{W}{4rl'} \tag{12-23}$$

The charts for flow variable and flow ratio (Figs. 12–19 and 12–20) do not apply to pressure-fed bearings. Also, the maximum film pressure given by Fig. 12–21 must be increased by the oil supply pressure p_s to obtain the total film pressure.

Since the oil flow has been increased by forced feed, Eq. (12–14) will give a temperature rise that is too high because the side flow carries away all the heat generated. The plumbing in a pressure-fed bearing is depicted schematically in Fig. 12–31. The oil leaves the sump at the externally maintained temperature T_s at the volumetric rate Q_s. The heat gain of the fluid passing through the bearing is

$$H_{gain} = 2\rho C_p(Q_s/2)\Delta T = \rho C_p Q_s \Delta T \tag{m}$$

At steady state, the rate at which the journal does frictional work on the fluid film is

$$H_f = \frac{2\pi TN}{J} = \frac{2\pi f\,WrN}{J} = \frac{2\pi\,WNc}{J}\frac{fr}{c} \tag{n}$$

Figure 12–31

Pressure-fed centrally located full annular-groove journal bearing with external, coiled lubricant sump.

Equating the heat gain to the frictional work and solving for ΔT gives

$$\Delta T = \frac{2\pi WNc}{J\rho C_p Q_s}\frac{fr}{c}\qquad(o)$$

Substituting Eq. (12–22) for Q_s in the equation for ΔT gives

$$\Delta T = \frac{2\pi}{J\rho C_p}WNc\frac{fr}{c}\frac{3\mu l'}{(1+1.5\epsilon^2)\pi p_s rc^3}$$

The Sommerfeld number may be expressed as

$$S = \left(\frac{r}{c}\right)^2\frac{\mu N}{P} = \left(\frac{r}{c}\right)^2\frac{4rl'\mu N}{W}$$

Solving for $\mu Nl'$ in the Sommerfeld expression; substituting in the ΔT expression; and using $J = 9336$ lbf · in/Btu, $\rho = 0.0311$ lbm/in³, and $C_p = 0.42$ Btu/(lbm · °F), we find

$$\Delta T_F = \frac{3(fr/c)SW^2}{2J\rho C_p p_s r^4}\frac{1}{(1+1.5\epsilon^2)} = \frac{0.0123(fr/c)SW^2}{(1+1.5\epsilon^2)p_s r^4}\qquad(12\text{–}24)$$

where ΔT_F is ΔT in °F. The corresponding equation in SI units uses the bearing load W in kN, lubricant supply pressure p_s in kPa, and the journal radius r in mm:

$$\Delta T_C = \frac{978(10^6)}{1+1.5\epsilon^2}\frac{(fr/c)SW^2}{p_s r^4}\qquad(12\text{–}25)$$

An analysis example of a pressure-fed bearing will be useful.

EXAMPLE 12–6   A circumferential-groove pressure-fed bearing is lubricated with SAE grade 20 oil supplied at a gauge pressure of 207 kPa. The journal diameter d_j is 44.45 mm, with a unilateral tolerance of -0.05 mm. The central circumferential bushing has a diameter d_b of 44.53 mm, with a unilateral tolerance of $+0.1$ mm. The l'/d ratio of the two "half-bearings" that constitute the complete pressure-fed bearing is $1/2$. The journal angular speed is 3000 rev/min, or 50 rev/s, and the radial steady load is 4 kN. The external sump is maintained at 49°C as long as the necessary heat transfer does not exceed 800 Btu/h.

(a) Find the steady-state average film temperature.

(b) Compare h_0, T_{max}, and P_{st} with the Trumpler criteria.

(c) Estimate the volumetric side flow Q_s, the heat loss rate H_{loss}, and the parasitic friction torque.

Solution   (a)

$$r = \frac{d_j}{2} = \frac{44.45}{2} = 22.23 \text{ mm}$$

$$c_{min} = \frac{(d_b)_{min} - (d_j)_{max}}{2} = \frac{44.53 - 44.45}{2} = 0.04 \text{ mm}$$

Since $l'/d = 1/2$, $l' = d/2 = r = 22.23$ mm. Then the pressure due to the load is

$$P = \frac{W}{4rl'} = \frac{4}{4(0.022\ 23)^2} = 2024 \text{ kPa}$$

The Sommerfeld number S can be expressed as

$$S = \left(\frac{r}{c}\right)^2 \frac{\mu N}{P} = \left(\frac{22.23}{0.04}\right)^2 \frac{\mu'}{2.024} \frac{50}{(10^6)} = 7.63\mu \qquad (1)$$

We will use a tabulation method to find the average film temperature. The first trial average film temperature $\overline{T}_f$ will be 77°C. Using the Seireg curve fit of Table 12–1, we obtain

$$\mu' = 6.89(10^{-3})0.0136 \exp[1271.6/(1.8(77) + 127)] = 0.011\ 24 \text{ Pa·s}$$

From Eq. (1)

$$7.63\mu = 7.63(0.011\ 24) = 0.0858$$

From Fig. (12–18), $fr/c = 3.3$, and from Fig. (12–16), $\epsilon = 0.80$. From Eq. (12–25),

$$\Delta T_C = \frac{978(10^6)3.3(0.0858)4^2}{[1 + 1.5(0.80)^2]207(22.23)^4} = 44.7°C$$

$$T_{av} = T_s + \frac{\Delta T}{2} = 49 + \frac{44.7}{2} = 71.4°C$$

We form a table, adding a second line with $\overline{T}_f = 71.4°C$:

Trial $\overline{T}_f$	μ	S	fr/c	ϵ	ΔT_F	T_{av}
77	0.011 24	0.0858	3.3	0.8	44.7	71.4
75.8	0.011 53	0.088	3.39	0.792	53.6	75.8

If the iteration had not closed, one could plot trial $\overline{T}_f$ against resulting T_{av} and draw a straight line between them, the intersection with a $\overline{T}_f = T_{av}$ line defining the new trial $\overline{T}_f$.

Answer The result of this tabulation is $\bar{T}_f = 75.8$, $\Delta T_F = 53.6°C$, and $T_{max} = 49 + 53.6 = 102.6°C$

(b) Since $h_0 = (1 - \epsilon)c$,

$$h_0 = (1 - 0.792)0.04 = 0.0083 \text{ mm}$$

The required four Trumpler criteria, from "Significant Angular Speed" in Sec. 12–7 are

$$h_0 \geq 0.005\ 08 + 0.000\ 04(44.45) = 0.006\ 86 \text{ mm} \qquad \text{(OK)}$$

Answer $$T_{max} = T_s + \Delta T = 49 + 53.6 = 102.6°C \qquad \text{(OK)}$$

$$P_{st} = \frac{W_{st}}{4rl'} = \frac{4}{4(0.022\ 23)^2} = 2024 \text{ kPa} \qquad \text{(OK)}$$

The factor of safety on the load is approximately unity. (Not OK.)

(c) From Eq. (12–22),

Answer $$Q_s = \frac{\pi(207\ 000)22.23(0.04)^3}{3(0.011\ 53)22.23}[1 + 1.5(0.792)^2] = 2335 \text{ mm}^3/\text{s}$$

$$H_{loss} = \rho C_p Q_s \Delta T = 861(1758)(2335 \times 10^{-9})53.6 = 1.89 \text{ J/s}$$

The parasitic friction torque T is

Answer $$T = fWr = \frac{fr}{c}Wr = 3.39(4)(0.04/2) = 0.27 \text{ N} \cdot \text{m}$$

12–12 Loads and Materials

Some help in choosing unit loads and bearing materials is afforded by Tables 12–5 and 12–6. Since the diameter and length of a bearing depend upon the unit load, these tables will help the designer to establish a starting point in the design.

Table 12–5

Range of Unit Loads in Current Use for Sleeve Bearings

Application	Unit Load	
	psi	MPa
Diesel engines:		
Main bearings	900–1700	6–12
Crankpin	1150–2300	8–15
Wristpin	2000–2300	14–15
Electric motors	120–250	0.8–1.5
Steam turbines	120–250	0.8–1.5
Gear reducers	120–250	0.8–1.5
Automotive engines:		
Main bearings	600–750	4–5
Crankpin	1700–2300	10–15
Air compressors:		
Main bearings	140–280	1–2
Crankpin	280–500	2–4
Centrifugal pumps	100–180	0.6–1.2

The length-diameter ratio l/d of a bearing depends upon whether it is expected to run under thin-film-lubrication conditions. A long bearing (large l/d ratio) reduces the coefficient of friction and the side flow of oil and therefore is desirable where thin-film or boundary-value lubrication is present. On the other hand, where forced-feed or positive lubrication is present, the l/d ratio should be relatively small. The short bearing length results in a greater flow of oil out of the ends, thus keeping the bearing cooler. Current practice is to use an l/d ratio of about unity, in general, and then to increase this ratio if thin-film lubrication is likely to occur and to decrease it for thick-film lubrication or high temperatures. If shaft deflection is likely to be severe, a short bearing should be used to prevent metal-to-metal contact at the ends of the bearings.

You should always consider the use of a partial bearing if high temperatures are a problem, because relieving the non-load-bearing area of a bearing can very substantially reduce the heat generated.

The two conflicting requirements of a good bearing material are that it must have a satisfactory compressive and fatigue strength to resist the externally applied loads and that it must be soft and have a low melting point and a low modulus of elasticity. The second set of requirements is necessary to permit the material to wear or break in, since the material can then conform to slight irregularities and absorb and release foreign particles. The resistance to wear and the coefficient of friction are also important because all bearings must operate, at least for part of the time, with thin-film or boundary lubrication.

Additional considerations in the selection of a good bearing material are its ability to resist corrosion and, of course, the cost of producing the bearing. Some of the commonly used materials are listed in Table 12–6, together with their composition and characteristics.

Bearing life can be increased very substantially by depositing a layer of babbitt, or other white metal, in thicknesses from 0.025 to 0.35 mm over steel backup material. In fact, a copper-lead layer on steel to provide strength, combined with a babbitt overlay to enhance surface conformability and corrosion resistance, makes an excellent bearing.

Small bushings and thrust collars are often expected to run with thin-film or boundary lubrication. When this is the case, improvements over a solid bearing material can

Table 12–6

Some Characteristics of Bearing Alloys

Alloy Name	Thickness, mm	SAE Number	Clearance Ratio r/c	Load Capacity	Corrosion Resistance
Tin-base babbitt	0.56	12	600–1000	1.0	Excellent
Lead-base babbitt	0.56	15	600–1000	1.2	Very good
Tin-base babbitt	0.1	12	600–1000	1.5	Excellent
Lead-base babbitt	0.1	15	600–1000	1.5	Very good
Leaded bronze	Solid	792	500–1000	3.3	Very good
Copper-lead	0.56	480	500–1000	1.9	Good
Aluminum alloy	Solid		400–500	3.0	Excellent
Silver plus overlay	0.33	17P	600–1000	4.1	Excellent
Cadmium (1.5% Ni)	0.56	18	400–500	1.3	Good
Trimetal 88*				4.1	Excellent
Trimetal 77†				4.1	Very good

*This is a 0.203-mm layer of copper-lead on a steel back plus 0.025 mm of tin-base babbitt.

†This is a 0.33-mm layer of copper-lead on a steel back plus 0.025 mm of lead-base babbitt.

be made to add significantly to the life. A powder-metallurgy bushing is porous and permits the oil to penetrate into the bushing material. Sometimes such a bushing may be enclosed by oil-soaked material to provide additional storage space. Bearings are frequently ball-indented to provide small basins for the storage of lubricant while the journal is at rest. This supplies some lubrication during starting. Another method of reducing friction is to indent the bearing wall and to fill the indentations with graphite.

With all these tentative decisions made, a lubricant can be selected and the hydro-dynamic analysis made as already presented. The values of the various performance parameters, if plotted as in Fig. 12–25, for example, will then indicate whether a satisfactory design has been achieved or additional iterations are necessary.

12–13 Bearing Types

A bearing may be as simple as a hole machined into a cast-iron machine member. It may still be simple yet require detailed design procedures, as, for example, the two-piece grooved pressure-fed connecting-rod bearing in an automotive engine. Or it may be as elaborate as the large water-cooled, ring-oiled bearings with built-in reservoirs used on heavy machinery.

Figure 12–32 shows two types of bushings. The solid bushing is made by casting, by drawing and machining, or by using a powder-metallurgy process. The lined bush-ing is usually a split type. In one method of manufacture the molten lining material is cast continuously on thin strip steel. The babbitted strip is then processed through presses, shavers, and broaches, resulting in a lined bushing. Any type of grooving may be cut into the bushings. Bushings are assembled as a press fit and finished by boring, reaming, or burnishing.

Flanged and straight two-piece bearings are shown in Fig. 12–33. These are avail-able in many sizes in both thick- and thin-wall types, with or without lining material. A locking lug positions the bearing and effectively prevents axial or rotational move-ment of the bearing in the housing.

Some typical groove patterns are shown in Fig. 12–34. In general, the lubricant may be brought in from the end of the bushing, through the shaft, or through the bushing. The flow may be intermittent or continuous. The preferred practice is to bring

Figure 12–32

Sleeve bushings.

(a) Solid bushing   (b) Lined bushing

Figure 12–33

Two-piece bushings.

(a) Flanged   (b) Straight

Figure 12–34

Developed views of typical groove patterns. *(Courtesy of the Cleveland Graphite Bronze Company, Division of Clevite Corporation.)*

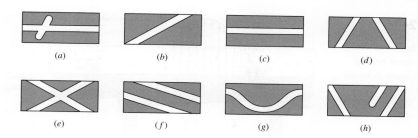

(a)   (b)   (c)   (d)

(e)   (f)   (g)   (h)

the oil in at the center of the bushing so that it will flow out both ends, thus increasing the flow and cooling action.

12–14 Thrust Bearings

This chapter is devoted to the study of the mechanics of lubrication and its application to the design and analysis of journal bearings. The design and analysis of thrust bearings is an important application of lubrication theory, too. A detailed study of thrust bearings is not included here, because it would not contribute anything significantly different and because of space limitations. Having studied this chapter, you should experience no difficulty in reading the literature on thrust bearings and applying that knowledge to actual design situations.[11]

Figure 12–35 shows a fixed-pad thrust bearing consisting essentially of a runner sliding over a fixed pad. The lubricant is brought into the radial grooves and pumped into the wedge-shaped space by the motion of the runner. Full-film, or hydrodynamic, lubrication is obtained if the speed of the runner is continuous and sufficiently high, if the lubricant has the correct viscosity, and if it is supplied in sufficient quantity. Figure 12–36 provides a picture of the pressure distribution under conditions of full-film lubrication.

We should note that bearings are frequently made with a flange, as shown in Fig. 12–37. The flange positions the bearing in the housing and also takes a thrust load. Even when it is grooved, however, and has adequate lubrication, such an arrangement is not theoretically a hydrodynamically lubricated thrust bearing. The reason for this is that the clearance space is not wedge-shaped but has a uniform thickness. Similar reasoning would apply to various designs of thrust washers.

Figure 12–35

Fixed-pad thrust bearing. *(Courtesy of Westinghouse Electric Corporation.)*

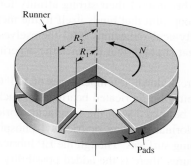

Runner

R_2

R_1

N

Pads

---

[11]Harry C. Rippel, *Cast Bronze Thrust Bearing Design Manual,* International Copper Research Association, Inc., 825 Third Ave., New York, NY 10022, 1967. CBBI, 14600 Detroit Ave., Cleveland, OH, 44107, 1967.

Figure 12–36

Pressure distribution of lubricant in a thrust bearing. *(Courtesy of Copper Research Corporation.)*

Figure 12–37

Flanged sleeve bearing takes both radial and thrust loads.

12–15 Boundary-Lubricated Bearings

When two surfaces slide relative to each other with only a partial lubricant film between them, *boundary lubrication* is said to exist. Boundary- or thin-film lubrication occurs in hydrodynamically lubricated bearings when they are starting or stopping, when the load increases, when the supply of lubricant decreases, or whenever other operating changes happen to occur. There are, of course, a very large number of cases in design in which boundary-lubricated bearings must be used because of the type of application or the competitive situation.

The coefficient of friction for boundary-lubricated surfaces may be greatly decreased by the use of animal or vegetable oils mixed with the mineral oil or grease. Fatty acids, such as stearic acid, palmitic acid, or oleic acid, or several of these, which occur in animal and vegetable fats, are called *oiliness agents*. These acids appear to reduce friction, either because of their strong affinity for certain metallic surfaces or because they form a soap film that binds itself to the metallic surfaces by a chemical reaction. Thus the fatty-acid molecules bind themselves to the journal and bearing surfaces with such great strength that the metallic asperities of the rubbing metals do not weld or shear.

Fatty acids will break down at temperatures of 121°C or more, causing increased friction and wear in thin-film-lubricated bearings. In such cases the *extreme-pressure,* or EP, lubricants may be mixed with the fatty-acid lubricant. These are composed of chemicals such as chlorinated esters or tricresyl phosphate, which form an organic film between the rubbing surfaces. Though the EP lubricants make it possible to operate at higher temperatures, there is the added possibility of excessive chemical corrosion of the sliding surfaces.

When a bearing operates partly under hydrodynamic conditions and partly under dry or thin-film conditions, a *mixed-film lubrication* exists. If the lubricant is supplied by hand oiling, by drop or mechanical feed, or by wick feed, for example, the bearing is operating under mixed-film conditions. In addition to occurring with a scarcity of lubricant, mixed-film conditions may be present when

- The viscosity is too low.
- The bearing speed is too low.
- The bearing is overloaded.
- The clearance is too tight.
- Journal and bearing are not properly aligned.

Relative motion between surfaces in contact in the presence of a lubricant is called *boundary lubrication*. This condition is present in hydrodynamic film bearings during starting, stopping, overloading, or lubricant deficiency. Some bearings are boundary lubricated (or dry) at all times. To signal this an adjective is placed before the word "bearing." Commonly applied adjectives (to name a few) are thin-film, boundary friction, Oilite, Oiles, and bushed-pin. The applications include situations in which thick film will not develop and there are low journal speed, oscillating journal, padded slides, light loads, and lifetime lubrication. The characteristics include considerable friction, the ability to tolerate expected wear without losing function, and light loading. Such bearings are limited by lubricant temperature, speed, pressure, galling, and cumulative wear. Table 12–7 gives some properties of a range of bushing materials.

Linear Sliding Wear

Consider the sliding block depicted in Fig. 12–38, moving along a plate with contact pressure P' acting over area A, in the presence of a coefficient of sliding friction f_s. The linear measure of wear w is expressed in inches or millimeters. The work done by force $f_s PA$ during displacement S is $f_s PAS$ or $f_s PAVt$, where V is the sliding velocity and t is time. The material volume removed due to wear is wA and is proportional to the work done, that is, $wA \propto f_s PAVt$, or

$$wA = KPAVt$$

	Table 12–7				
Some Materials for Boundary-Lubricated Bearings and Their Operating Limits	**Material**	**Maximum Load, MPa**	**Maximum Temperature, °C**	**Maximum Speed, m/s**	**Maximum PV Value***
	Cast bronze	31.0	163	7.6	1.76
	Porous bronze	31.0	66	7.6	1.76
	Porous iron	55.2	66	4.1	1.76
	Phenolics	41.4	93	12.7	0.53
	Nylon	7.0	93	5.1	0.11
	Teflon	3.5	260	0.5	0.035
	Reinforced Teflon	17.2	260	5.1	0.35
	Teflon fabric	413.7	260	0.25	0.88
	Delrin	7.0	82	5.1	0.105
	Carbon-graphite	4.1	399	12.7	0.53
	Rubber	0.3	66	20.3	
	Wood	13.8	66	10.2	0.53

*P = load, MPa; V = speed, m/s.

Figure 12–38

Sliding block subjected to wear.

Table 12–8

Wear Factors in U.S. Customary Units*

Source: Oiles America Corp., Plymouth, MI 48170.

Bushing Material	Wear Factor K	Limiting PV
Oiles 800	$6037(10^{-20})$	0.63
Oiles 500	$1207(10^{-20})$	1.64
Polyactal copolymer	$100\,615(10^{-20})$	0.18
Polyactal homopolymer	$127\,730(10^{-20})$	0.11
66 nylon	$402\,460(10^{-20})$	0.07
66 nylon +15% PTFE	$26\,160(10^{-20})$	0.25
+15% PTFE +30% glass	$32\,200(10^{-20})$	0.35
+2.5% MoS$_2$	$402\,460(10^{-20})$	0.07
6 nylon	$402\,460(10^{-20})$	0.07
Polycarbonate +15% PTFE	$150\,920(10^{-20})$	0.25
Sintered bronze	$205\,250(10^{-20})$	0.3
Phenol +25% glass fiber	$16\,100(10^{-20})$	0.4

*dim$[K]$ = m^3 · s/(N · m · s), dim $[PV]$ = MPa · m/s.

Table 12–9

Coefficients of Friction

Source: Oiles America Corp., Plymouth, MI 48170.

Type	Bearing	f_s
Placetic	Oiles 80	0.05
Composite	Drymet ST	0.03
	Toughmet	0.05
Met	Cermet M	0.05
	Oiles 2000	0.03
	Oiles 300	0.03
	Oiles 500SP	0.03

where K is the proportionality factor, which includes f_s, and is determined from laboratory testing. The linear wear is then expressed as

$$w = KPVt \qquad (12\text{–}26)$$

In US customary units, P is expressed in psi, V in fpm (i.e., ft/min), and t in hours. This makes the units of K in^3 · min/(lbf · ft · h). SI units commonly used for K are cm^3 · min/(kgf · m · h), where 1 kgf = 9.806 N. Tables 12–8 and 12–9 give some wear factors and coefficients of friction from one manufacturer.

Table 12-10

Motion-Related Factor f_1

Mode of Motion	Characteristic Pressure P, MPa		Velocity V, mm/s	f_1*
Rotary	5 or less		16.8 or less	1.0
			16.8–168.8	1.0–1.3
			168.8–508	1.3–1.8
	5–25		16.8 or less	1.5
			16.8–168.8	1.5–2.0
			168.8–508	2.0–2.7
Oscillatory	5 or less	>30°	16.8 or less	1.3
			16.8–508	1.3–2.4
		<30°	16.8 or less	2.0
			16.8–508	2.0–3.6
	5–25	>30°	16.8 or less	2.0
			16.8–508	2.0–3.2
		<30°	16.8 or less	3.0
			16.8–508	3.0–4.8
Reciprocating	5 or less		168.8 or less	1.5
			168.8–508	1.5–3.8
	5–25		168.8 or less	2.0
			168.8–508	2.0–7.5

*Values of f_1 based on results over an extended period of time on automotive manufacturing machinery.

Table 12-11

Environmental Factor f_2

Source: Oiles America Corp., Plymouth, MI 48170.

Ambient Temperature, °C	Foreign Matter	f_2
60 or lower	No	1.0
60 or lower	Yes	3.0–6.0
60–99	No	3.0–6.0
60–99	Yes	6.0–12.0

It is useful to include a modifying factor f_1 depending on motion type, load, and speed and an environment factor f_2 to account for temperature and cleanliness conditions (see Tables 12–10 and 12–11). These factors account for departures from the laboratory conditions under which K was measured. Equation (12–26) can now be written as

$$w = f_1 f_2 KPVt \tag{12-27}$$

Wear, then, is proportional to PV, material property K, operating conditions f_1 and f_2, and time t.

Bushing Wear

Consider a pin of diameter D, rotating at speed N, in a bushing of length L, and supporting a stationary radial load F. The nominal pressure P is given by

$$P = \frac{F}{DL} \tag{12-28}$$

Figure 12–39

Pressure distribution on a boundary-lubricated bushing.

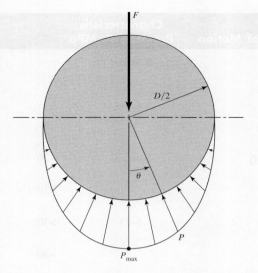

and if N is in rev/min and D is in inches, velocity in ft/min is given by

$$V = \frac{\pi DN}{12} \tag{12-29}$$

Thus PV, in psi $\cdot$ ft/min, is

$$PV = \frac{F}{DL}\frac{\pi DN}{12} = \frac{\pi}{12}\frac{FN}{L} \tag{12-30}$$

Note the independence of PV from the journal diameter D.

A time-wear equation similar to Eq. (12–27) can be written. However, before doing so, it is important to note that Eq. (12–28) provides the nominal value of P. Figure 12–39 provides a more accurate representation of the pressure distribution, which can be written as

$$p = P_{\max}\cos\theta \qquad -\frac{\pi}{2} \le \theta \le \frac{\pi}{2}$$

The vertical component of $p\,dA$ is $p\,dA\cos\theta = [pL(D/2)\,d\theta]\cos\theta = P_{\max}(DL/2)\cos^2\theta\,d\theta$. Integrating this from $\theta = -\pi/2$ to $\pi/2$ yields F. Thus,

$$\int_{-\pi/2}^{\pi/2} P_{\max}\left(\frac{DL}{2}\right)\cos^2\theta\,d\theta = \frac{\pi}{4}P_{\max}DL = F$$

or

$$P_{\max} = \frac{4}{\pi}\frac{F}{DL} \tag{12-31}$$

Substituting V from Eq. (12–29) and $P_{\max}$ for P from Eq. (12–31) into Eq. (12–27) gives

$$w = f_1 f_2 K \frac{4}{\pi}\frac{F}{DL}\frac{\pi DNt}{12} = \frac{f_1 f_2 KFNt}{3L} \tag{12-32}$$

In designing a bushing, because of various trade-offs it is recommended that the length/diameter ratio be in the range

$$0.5 \le L/D \le 2 \tag{12-33}$$

EXAMPLE 12–7   An Oiles SP 500 alloy brass bushing is 25 mm long with a 25-mm dia bore and operates in
a clean environment at 21°C. The allowable wear without loss of function is 0.125 mm. The
radial load is 2250 N. The shaft speed is 200 rev/min. Estimate the number of revolutions
for radial wear to be 0.125 mm. See Fig. 12–40 and Table 12–12.

Solution   From Table 12–8, $K = 1207(10^{-20})$ m$^3 \cdot$ s/(N $\cdot$ m $\cdot$ s). $P = 2250/[(25)(25)] = 3.6$
MPa, $V = \pi DN = \pi(25)(200)/60 = 262$ mm/s

Tables 12–10 and 12–11:

$$f_1 = 1.8, \quad f_2 = 1$$

Table 12–12:

$$PV_{max} = 1640 \text{ MPA} \cdot \text{mm/s}, \quad P_{max} = 24.5 \text{ MPa}, \quad V_{max} = 510 \text{ mm/s}$$

$$P_{max} = \frac{4}{\pi} \frac{F}{DL} = \frac{4(2250)}{\pi(25)(25)} = 4.58 \text{ MPa} < 24.5 \text{ MPa} \qquad O.K.$$

$$P = \frac{F}{DL} = 3.6 \text{ MPa} \quad V = 262 \text{ mm/s}$$

$$PV = 3.6(262) = 943.2 \text{ MPa} \cdot \text{mm/s} < 1640 \text{ MPa} \cdot \text{mm/s} \qquad O.K.$$

Figure 12–40

Journal/bushing for Ex. 12–7.

Table 12–12

Oiles 500 SP (SPBN ·
SPWN) Service Range
and Properties

Source: Oiles America Corp.,
Plymouth, MI 48170.

Service Range	Units	Allowable
Characteristic pressure P_{max}	MPa	<24.5
Velocity V_{max}	m/s	<0.51
PV product	MPa. m/s	<1.64
Temperature T	°C	<149

Properties	Test Method, Units	Value
Tensile strength	(ASTM E8) psi	>758
Elongation	(ASTM E8) %	>12
Compressive strength	(ASTM E9) psi	67.4
Brinell hardness	(ASTM E10) HB	>210
Coefficient of thermal expansion	(10^{-5}) °C	>1.6
Specific gravity		8.2

Solving Eq. (12–32) for t

$$t = \frac{\pi D L w}{4 f_1 f_2 K V F} = \frac{\pi (0.025)(0.025)\, 0.125}{4(1.8)(1)(1207)\,(10^{-14})(262)(2250)} = 4\,790\,891 \text{ s} = 1331 \text{ h}$$

Answer   Cycles $= Nt = 200(79\,848) = 16(10^6)$ rev

Temperature Rise

At steady state, the rate at which work is done against bearing friction equals the rate at which heat is transferred from the bearing housing to the surroundings by convection and radiation. The rate of heat generation in J/s is given by $f_s FV/J$, or

$$H_{\text{gen}} = \frac{f_s F(\pi D)(60N)}{12J} = \frac{5\pi f_s FDN}{J} \qquad (12\text{–}34)$$

where N is journal speed in rev/min and $J = 1$ m $\cdot$ N/J. The rate at which heat is transferred to the surroundings, in J/s, is

$$H_{\text{loss}} = \hbar_{\text{CR}} A \Delta T = \hbar_{\text{CR}} A (T_b - T_\infty) = \frac{\hbar_{\text{CR}} A}{2}(T_f - T_\infty) \qquad (12\text{–}35)$$

where A = housing surface area, m^2

$\hbar_{\text{CR}}$ = overall combined coefficient of heat transfer, W/(m$^2 \cdot$ °C)

T_b = housing metal temperature, °C

T_f = lubricant temperature, °C

The empirical observation that T_b is about midway between T_f and T_∞ has been incorporated in Eq. (12–35). Equating Eqs. (12–34) and (12–35) gives

$$T_f = T_\infty + \frac{10\pi f_s FDN}{J \hbar_{\text{CR}} A} \qquad (12\text{–}36)$$

Although this equation seems to indicate the temperature rise $T_f - T_\infty$ is independent of length L, the housing surface area generally is a function of L. The housing surface area can be initially estimated, and as tuning of the design proceeds, improved results will converge. If the bushing is to be housed in a pillow block, the surface area can be roughly estimated from

$$A \approx \frac{2\pi D L}{144} \qquad (12\text{–}37)$$

Substituting Eq. (12–37) into Eq. (12–36) gives

$$T_f \approx T_\infty + \frac{10\pi f_s FDN}{J \hbar_{\text{CR}}(2\pi DL/144)} = T_\infty + \frac{720 f_s FN}{J \hbar_{\text{CR}} L} \qquad (12\text{–}38)$$

EXAMPLE 12–8

Choose an Oiles 500 bushing to give a maximum wear of 0.025 mm for 800 h of use with a 5 rev/s journal and 220 N radial load. Use $\hbar_{CR} = 13.3$ W/(m$^2 \cdot$ °C), $T_{max} = 149$°C, $f_s = 0.03$, and a design factor $n_d = 2$. Table 12–13 lists the available bushing sizes from the manufacturer.

Solution

With a design factor n_d, substitute $n_d F$ for F. First estimate the bushing length using Eq. (12-32) with $f_1 = f_2 = 1$, and $K = 1207(10^{-20})$ from Table 12–8:

$$L = \frac{4 f_1 f_2 K n_d F N t}{w} = \frac{4(1)1(1207)10^{-20}(2)220(5)800(3600)}{0.025(10^{-3})} = 0.0122 \text{ m} \quad (1)$$
$$= 12.2 \text{ mm}$$

Table 12–13

Available Bushing Sizes (in millimeters) of One Manufacturer*

ID	OD	12	16	20	22	25	30	38	44	50	62	75	90	100	125
12	20	•	•	•	•	•									
16	22		•	•		•		•							
20	28		•	•		•		•							
22	30			•		•	•								
25	35			•			•	•	•	•					
25	38			•		•		•		•					
30	40					•	•	•		•					
38	50					•	•	•	•	•					
44	58						•	•	•	•	•	•	•	•	
50	60							•		•	•	•	•		
58	70							•		•	•	•	•		
62	75							•		•		•			
70	84							•		•	•	•			
75	90										•	•	•	•	
90	103										•		•		•
100	120										•		•		•
115	135											•		•	•
125	150											•		•	•

*In a display such as this a manufacturer is likely to show catalog numbers where the • appears.

From Eq. (12–38) with $f_s = 0.03$ from Table 12–9, $\hbar_{CR} = 13.3$ W/(m² · °C), and $n_d F$ for F,

$$L \doteq \frac{f_s n_d F N}{J \hbar_{CR}(T_f - T_\infty)} = \frac{0.03(2)220(5)}{13.3(149 - 21)} = 0.0388 \text{ m} = 38.8 \text{ mm}$$

The two results bracket L such that 12.2 mm $\leq L \leq$ 38.8 mm. As a start let $L = 25$ mm. From Table 12–13, we select $D = 25$ mm from the midrange of available bushings.

Trial 1: $D = L = 25$ mm.

Eq. (12–31): $P_{\max} = \dfrac{4}{\pi} \dfrac{n_d F}{DL} = \dfrac{4}{\pi} \dfrac{2(220)}{(0.025)(0.025)} = 0.9$ MPa < 24.5 MPa   (OK)

$$P = \frac{n_d F}{DL} = \frac{2(220)}{25(25)} = 0.7 \text{ MPa}$$

Eq. (12–29): $V = \pi DN = \pi(0.025)5 = 0.39$ m/s < 0.51 m/s   (OK)

$$PV = 0.7(0.39) = 0.27 \text{ MPa} \cdot \text{m/s} < 1.64 \text{ MPa} \cdot \text{m/s}$$ (OK)

V	f₁	
0.17	1.3	
0.39	f_1	$=> f_1 = 1.64$
0.51	1.8	

Our second estimate is $L \geq 12.2(1.64) = 20$ mm. From Table 12–13, there are many 20-mm bushings to select from. The smallest diameter in Table 12–13 is $D = 12$ mm. This gives an L/D ratio of 1.5, which is acceptable according to Eq. (12–33).

Trial 2: $D = 12.5$ mm, $L = 20$ mm.

$$V = \pi DN = \pi 0.0125(5) = 0.2 \text{ m/s} < 0.51 \text{ m/s}$$ (OK)

$$P_{\max} = \frac{4}{\pi} \frac{n_d F}{DL} = \frac{4}{\pi} \frac{2(220)}{12.5(20)} = 2.2 \text{ MPa} < 24.5 \text{ MPa}$$ (OK)

$$P = \frac{n_d F}{DL} = \frac{2(220)}{12.5(20)} = 1.76 \text{ MPa}$$

$$PV = 1.7(0.2) = 0.34 \text{ MPa} \cdot \text{m/s} < 1.64 \text{ MPa} \cdot \text{m/s}$$ (OK)

Answer   Select any of the bushings from the trials, where the optimum, from trial 2, is $D = 12$ mm and L = 20 mm. Other factors may enter in the overall design that make the other bushings more appropriate.

PROBLEMS

12–1   A full journal bearing has a journal diameter of 25 mm, with a unilateral tolerance of -0.03 mm. The bushing bore has a diameter of 25.03 mm and a unilateral tolerance of 0.04 mm. The l/d ratio is 1/2. The load is 1.2 kN and the journal runs at 1100 rev/min. If the average viscosity is 55 mPa · s, find the minimum film thickness, the power loss, and the side flow for the minimum clearance assembly.

12–2   A full journal bearing has a journal diameter of 32 mm, with a unilateral tolerance of −0.012 mm. The bushing bore has a diameter of 32.05 mm and a unilateral tolerance of 0.032 mm. The bearing is 64 mm long. The journal load is 1.75 kN and it runs at a speed of 900 rev/min. Using an average viscosity of 55 mPa · s find the minimum film thickness, the maximum film pressure, and the total oil-flow rate for the minimum clearance assembly.

12–3   A journal bearing has a journal diameter of 75 mm, with a unilateral tolerance of −0.025 mm. The bushing bore has a diameter of 75.125 mm and a unilateral tolerance of 0.1 mm. The bushing is 37.5 mm long. The journal speed is 600 rev/min and the load is 3450 N. For both SAE 10 and SAE 40, lubricants, find the minimum film thickness and the maximum film pressure for an operating temperature of 62°C for the minimum clearance assembly.

12–4   A journal bearing has a journal diameter of 75 mm with a unilateral tolerance of −0.075 mm. The bushing bore has a diameter of 75.15 mm and a unilateral tolerance of 0.1 mm. The bushing is 75 mm long and supports a 2585-N load. The journal speed is 750 rev/min. Find the minimum oil film thickness and the maximum film pressure for both SAE 20W and SAE 20W-40 lubricants, for the tightest assembly if the operating film temperature is 150°F.

12–5   A full journal bearing has a journal with a diameter of 50 mm and a unilateral tolerance of −0.03 mm. The bushing has a bore with a diameter of 50.06 mm and a unilateral tolerance of 0.05 mm. The bushing is 25 mm long and supports a load of 2590 N at a speed of 800 rev/min. Find the minimum film thickness, the power loss, and the total lubricant flow if the average film temperature is 54°C and SAE 20 lubricant is used. The tightest assembly is to be analyzed.

12–6   A full journal bearing has a shaft journal diameter of 25 mm with a unilateral tolerance of −0.01 mm. The bushing bore has a diameter of 25.04 mm with a unilateral tolerance of 0.03 mm. The l/d ratio is unity. The bushing load is 1.25 kN, and the journal rotates at 1200 rev/min. Analyze the minimum clearance assembly if the average viscosity is 50 mPa · s to find the minimum oil film thickness, the power loss, and the percentage of side flow.

12–7   A full journal bearing has a shaft journal with a diameter of 30 mm and a unilateral tolerance of −0.015 mm. The bushing bore has a diameter of 30.05 mm with a unilateral tolerance of 0.035 mm. The bushing bore is 50 mm in length. The bearing load is 2.75 kN and the journal rotates at 1120 rev/min. Analyze the minimum clearance assembly and find the minimum film thickness, the coefficient of friction, and the total oil flow if the average viscosity is 60 mPa · s.

12–8   A journal bearing has a shaft diameter of 75.00 mm with a unilateral tolerance of −0.02 mm. The bushing bore has a diameter of 75.10 mm with a unilateral tolerance of 0.06 mm. The bushing is 36 mm long and supports a load of 2 kN. The journal speed is 720 rev/min. For the minimum clearance assembly find the minimum film thickness, the heat loss rate, and the maximum lubricant pressure for SAE 20 and SAE 40 lubricants operating at an average film temperature of 60°C.

12–9   A full journal bearing is 28 mm long. The shaft journal has a diameter of 56 mm with a unilateral tolerance of −0.012 mm. The bushing bore has a diameter of 56.05 mm with a unilateral tolerance of 0.012 mm. The load is 2.4 kN and the journal speed is 900 rev/min. For the minimum clearance assembly find the minimum oil-film thickness, the power loss, and the side flow if the operating temperature is 65°C and SAE 40 lubricating oil is used.

12–10   A 32- × 32-mm sleeve bearing supports a load of 3160 N and has a journal speed of 3600 rev/min. An SAE 10 oil is used having an average temperature of 71°C. Using Fig. 12–16, estimate the radial clearance for minimum coefficient of friction f and for maximum load-carrying capacity W. The difference between these two clearances is called the clearance range. Is the resulting range attainable in manufacture?

12-11 A full journal bearing has a shaft diameter of 80 mm with a unilateral tolerance of −0.01 mm. The l/d ratio is unity. The bushing has a bore diameter of 80.08 mm with a unilateral tolerance of 0.03 mm. The SAE 40 oil supply is in an axial-groove sump with a steady-state temperature of 60°C. The radial load is 3 kN. Estimate the average film temperature, the minimum film thickness, the heat loss rate, and the lubricant side-flow rate for the minimum clearance assembly, if the journal speed is 8 rev/s.

12-12 A 64- × 64-mm sleeve bearing uses grade 20 lubricant. The axial-groove sump has a steady-state temperature of 43°C. The shaft journal has a diameter of 63.5 mm with a unilateral tolerance of −0.025 mm. The bushing bore has a diameter of 63.6 mm with a unilateral tolerance of 0.025 mm. The journal speed is 1120 rev/min and the radial load is 5.34 kN. Estimate

(a) The magnitude and location of the minimum oil-film thickness.
(b) The eccentricity.
(c) The coefficient of friction.
(d) The power loss rate.
(e) Both the total and side oil-flow rates.
(f) The maximum oil-film pressure and its angular location.
(g) The terminating position of the oil film.
(h) The average temperature of the side flow.
(i) The oil temperature at the terminating position of the oil film.

12-13 A set of sleeve bearings has a specification of shaft journal diameter of 31.25 mm with a unilateral tolerance of −0.025 mm. The bushing bore has a diameter of 31.3 mm with a unilateral tolerance of 0.075 mm. The bushing is 31.25 mm in long. The radial load is 1113 N and the shaft rotational speed is 1750 rev/min. The lubricant is SAE 10 oil and the axial-groove sump temperature at steady state T_s is 49°C. For the c_{min}, c_{median}, and c_{max} assemblies analyze the bearings and observe the changes in S, ϵ, f, Q, Q_s, ΔT, T_{max}, $\overline{T}_f$, and h_p.

12-14 An interpolation equation was given by Raimondi and Boyd, and it is displayed as Eq. (12–16). This equation is a good candidate for a computer program. Write such a program for interactive use. Once ready for service it can save time and reduce errors. Another version of this program can be used with a subprogram that contains curve fits to Raimondi and Boyd charts for computer use.

12-15 A natural-circulation pillow-block bearing has a journal diameter d of 62.5 mm with a unilateral tolerance of −0.025 mm. The bushing bore diameter B is 62.6 mm with a unilateral tolerance of 0.1 mm. The shaft runs at an angular speed of 1120 rev/min; the bearing uses SAE grade 20 oil and carries a steady load of 1350 N in shaft-stirred air at 21°C. The lateral area of the pillow-block housing is 38 700 mm². Perform a design assessment using minimum radial clearance for a load of 2700 N and 1350 N. Use Trumpler's criteria and assume that both l/d and α are unity.

12-16 An eight-cylinder diesel engine has a front main bearing with a journal diameter of 87.5 mm and a unilateral tolerance of −0.025 mm. The bushing bore diameter is 87.625 mm with a unilateral tolerance of +0.125 mm. The bushing length is 50 mm. The pressure-fed bearing has a central annular groove 6.25 mm wide. The SAE 30 oil comes from a sump at 49°C using a supply pressure of 345 kPa. The sump's heat-dissipation capacity is 1.46 kW per bearing. For a minimum radial clearance, a speed of 2000 rev/min, and a radial load of 20.7 kN, find the average film temperature and apply Trumpler's criteria in your design assessment.

12-17 A pressure-fed bearing has a journal diameter of 50.00 mm with a unilateral tolerance of −0.05 mm. The bushing bore diameter is 50.084 mm with a unilateral tolerance of 0.10 mm. The length of the bushing is 55 mm. Its central annular groove is 5 mm wide and is fed by

SAE 30 oil is 55°C at 200 kPa supply gauge pressure. The journal speed is 2880 rev/min carrying a load of 10 kN. The sump can dissipate 300 watts per bearing if necessary. For minimum radial clearances, perform a design assessment using Trumpler's criteria.

12–18   Design a central annular-groove pressure-fed bearing with an l'/d ratio of 0.5, using SAE grade 20 oil, the lubricant supplied at 207 kPa. The exterior oil cooler can maintain the sump temperature at 49°C for heat dissipation rates up to 438 W. The load to be carried is 3860 N at 3000 rev/min. The groove width is 6.25 mm. Use nominal journal diameter d as one design variable and c as the other. Use Trumpler's criteria for your adequacy assessment.

12–19   Repeat design problem Prob. 12–18 using the nominal bushing bore B as one decision variable and the radial clearance c as the other. Again, Trumpler's criteria to be used.

12–20   Table 12–1 gives the Seireg and Dandage curve fit approximation for the absolute viscosity in customary U.S. engineering units. Show that in SI units of mPa · s and a temperature of C degrees Celsius, the viscosity can be expressed as

$$\mu = 6.89(10^6)\mu_0 \exp[(b/(1.8C + 127))]$$

where μ_0 and b are from Table 12–1. If the viscosity μ_0' is expressed in μreyn, then

$$\mu = 6.89\mu_0' \exp[(b/(1.8C + 127))]$$

What is the viscosity of a grade 50 oil at 70°C? Compare your results with Fig. 12–13.

12–21   For Prob. 12–18 a satisfactory design is

$$d = 50^{+0}_{-0.025}\text{ mm} \qquad b = 50.125^{+0.075}_{-0}\text{ mm}$$

Double the size of the bearing dimensions and quadruple the load to 16 kN.
(a) Analyze the scaled-up bearing for median assembly.
(b) Compare the results of a similar analysis for the 50-mm bearing, median assembly.

12–22   An Oiles SP 500 alloy brass bushing is 25 mm long with a 25-mm bore and operates in a clean environment at 21°C. The allowable wear without loss of function is 0.125 mm. The radial load is 3150 N. The peripheral velocity is 0.17 m/s. Estimate the number of revolitions for radial wear to be 0.125 mm. See the figure and Table 12−12 from the manufacturer.

12–23   Choose an Oiles SP 500 alloy brass bushing to give a maximum wear of 0.05 mm in for 1000 h of use with a 400 rev/min journal and 450 N radial load. Use $\hbar_{CR} = 15.3$ W/(m² · °C), $T_{max} = 149$°C, $f_s = 0.03$, and a design factor $n_d = 2$. Table 12–13 lists the bushing sizes available from the manufacturer.

Problem 12-22

2250 N

25mm

25 mm

13 Gears—General

Chapter Outline

This chapter addresses gear geometry, the kinematic relations, and the forces transmitted by the four principal types of gears: spur, helical, bevel, and worm gears. The forces transmitted between meshing gears supply torsional moments to shafts for motion and power transmission and create forces and moments that affect the shaft and its bearings. The next two chapters will address stress, strength, safety, and reliability of the four types of gears.

13–1 Types of Gears

Spur gears, illustrated in Fig. 13–1, have teeth parallel to the axis of rotation and are used to transmit motion from one shaft to another, parallel, shaft. Of all types, the spur gear is the simplest and, for this reason, will be used to develop the primary kinematic relationships of the tooth form.

Helical gears, shown in Fig. 13–2, have teeth inclined to the axis of rotation. Helical gears can be used for the same applications as spur gears and, when so used, are not as noisy, because of the more gradual engagement of the teeth during meshing. The inclined tooth also develops thrust loads and bending couples, which are not present with spur gearing. Sometimes helical gears are used to transmit motion between nonparallel shafts.

Bevel gears, shown in Fig. 13–3, have teeth formed on conical surfaces and are used mostly for transmitting motion between intersecting shafts. The figure actually illustrates *straight-tooth bevel gears. Spiral bevel gears* are cut so the tooth is no longer straight, but forms a circular arc. *Hypoid gears* are quite similar to spiral bevel gears except that the shafts are offset and nonintersecting.

Figure 13–1

Spur gears are used to transmit rotary motion between parallel shafts.

Figure 13–2

Helical gears are used to transmit motion between parallel or nonparallel shafts.

Figure 13–3

Bevel gears are used to transmit rotary motion between intersecting shafts.

Figure 13–4

Worm gearsets are used to transmit rotary motion between nonparallel and nonintersecting shafts.

Worms and *worm gears,* shown in Fig. 13–4, represent the fourth basic gear type. As shown, the worm resembles a screw. The direction of rotation of the worm gear, also called the worm wheel, depends upon the direction of rotation of the worm and upon whether the worm teeth are cut right-hand or left-hand. Worm gearsets are also made so that the teeth of one or both wrap partly around the other. Such sets are called *single-enveloping* and *double-enveloping* worm gearsets. Worm gearsets are mostly used when the speed ratios of the two shafts are quite high, say, 3 or more.

13–2 Nomenclature

The terminology of spur-gear teeth is illustrated in Fig. 13–5. The *pitch circle* is a theoretical circle upon which all calculations are usually based; its diameter is the *pitch diameter.* The pitch circles of a pair of mating gears are tangent to each other. A *pinion* is the smaller of two mating gears. The larger is often called the *gear.*

The *circular pitch p* is the distance, measured on the pitch circle, from a point on one tooth to a corresponding point on an adjacent tooth. Thus the circular pitch is equal to the sum of the *tooth thickness* and the *width of space.*

Figure 13-5

Nomenclature of spur-gear
teeth.

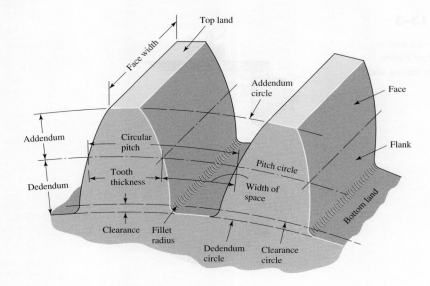

The *module m* is the ratio of the pitch diameter to the number of teeth. The customary unit of length used is the millimeter. The module is the index of tooth size in SI.

The *diametral pitch P* is the ratio of the number of teeth on the gear to the pitch
diameter. Thus, it is the reciprocal of the module. Since diametral pitch is used only
with U.S. units, it is expressed as teeth per inch.

The *addendum a* is the radial distance between the *top land* and the pitch circle.
The *dedendum b* is the radial distance from the *bottom land* to the pitch circle. The
whole depth h_t is the sum of the addendum and the dedendum.

The *clearance circle* is a circle that is tangent to the addendum circle of the mating gear. The *clearance c* is the amount by which the dedendum in a given gear exceeds
the addendum of its mating gear. The *backlash* is the amount by which the width of
a tooth space exceeds the thickness of the engaging tooth measured on the pitch circles.

You should prove for yourself the validity of the following useful relations:

$$P = \frac{N}{d} \tag{13-1}$$

$$m = \frac{d}{N} \tag{13-2}$$

$$p = \frac{\pi d}{N} = \pi m \tag{13-3}$$

$$pP = \pi \tag{13-4}$$

where P = diametral pitch, teeth per inch

N = number of teeth

d = pitch diameter, in or mm

m = module, mm

p = circular pitch, in or mm

13–3 Conjugate Action

The following discussion assumes the teeth to be perfectly formed, perfectly smooth, and absolutely rigid. Such an assumption is, of course, unrealistic, because the application of forces will cause deflections.

Mating gear teeth acting against each other to produce rotary motion are similar to cams. When the tooth profiles, or cams, are designed to produce a constant angular-velocity ratio during meshing, these are said to have *conjugate action*. In theory, at least, it is possible arbitrarily to select any profile for one tooth and then to find a profile for the meshing tooth that will give conjugate action. One of these solutions is the *involute profile*, which, with few exceptions, is in universal use for gear teeth and is the only one with which we will be concerned.

When one curved surface pushes against another (Fig. 13–6), the point of contact occurs where the two surfaces are tangent to each other (point *c*), and the forces at any instant are directed along the common normal *ab* to the two curves. The line *ab*, representing the direction of action of the forces, is called the *line of action*. The line of action will intersect the line of centers *O-O* at some point *P*. The angular-velocity ratio between the two arms is inversely proportional to their radii to the point *P*. Circles drawn through point *P* from each center are called *pitch circles,* and the radius of each circle is called the *pitch radius*. Point *P* is called the *pitch point*.

Figure 13–6 is useful in making another observation. A pair of gears is really a pair of cams that act through a small arc and, before running off the involute contour, are replaced by another identical pair of cams. The cams can run in either direction and are configured to transmit a constant angular-velocity ratio. If involute curves are used, the gears tolerate changes in center-to-center distance with *no* variation in constant angular-velocity ratio. Furthermore, the rack profiles are straight-flanked, making primary tooling simpler.

To transmit motion at a constant angular-velocity ratio, the pitch point must remain fixed; that is, all the lines of action for every instantaneous point of contact must pass through the same point *P*. In the case of the involute profile, it will be shown that all points of contact occur on the same straight line *ab*, that all normals to the tooth profiles at the point of contact coincide with the line *ab*, and, thus, that these profiles transmit uniform rotary motion.

Figure 13–6

Cam *A* and follower *B* in contact. When the contacting surfaces are involute profiles, the ensuing conjugate action produces a constant angular-velocity ratio.

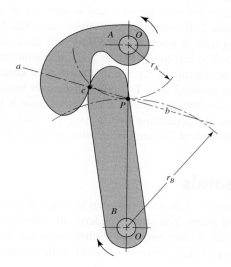

Figure 13–7

(a) Generation of an involute;
(b) involute action.

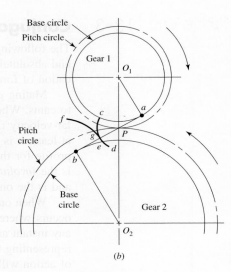

(a)

(b)

13–4   Involute Properties

An involute curve may be generated as shown in Fig. 13–7a. A partial flange B is attached to the cylinder A, around which is wrapped a cord *def*, which is held tight. Point b on the cord represents the tracing point, and as the cord is wrapped and unwrapped about the cylinder, point b will trace out the involute curve *ac*. The radius of the curvature of the involute varies continuously, being zero at point a and a maximum at point c. At point b the radius is equal to the distance *be*, since point b is instantaneously rotating about point e. Thus the generating line *de* is normal to the involute at all points of intersection and, at the same time, is always tangent to the cylinder A. The circle on which the involute is generated is called the *base circle*.

Let us now examine the involute profile to see how it satisfies the requirement for the transmission of uniform motion. In Fig. 13–7b, two gear blanks with fixed centers at O_1 and O_2 are shown having base circles whose respective radii are O_1a and O_2b. We now imagine that a cord is wound clockwise around the base circle of gear 1, pulled tight between points a and b, and wound counterclockwise around the base circle of gear 2. If, now, the base circles are rotated in different directions so as to keep the cord tight, a point g on the cord will trace out the involutes *cd* on gear 1 and *ef* on gear 2. The involutes are thus generated simultaneously by the tracing point. The tracing point, therefore, represents the point of contact, while the portion of the cord *ab* is the generating line. The point of contact moves along the generating line; the generating line does not change position, because it is always tangent to the base circles; and since the generating line is always normal to the involutes at the point of contact, the requirement for uniform motion is satisfied.

13–5   Fundamentals

Among other things, it is necessary that you actually be able to draw the teeth on a pair of meshing gears. You should understand, however, that you are not doing this for manufacturing or shop purposes. Rather, we make drawings of gear teeth to obtain an understanding of the problems involved in the meshing of the mating teeth.

Figure 13–8

Construction of an involute curve.

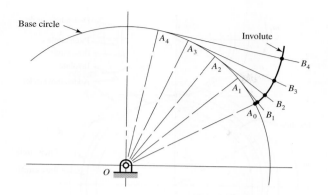

First, it is necessary to learn how to construct an involute curve. As shown in Fig. 13–8, divide the base circle into a number of equal parts, and construct radial lines OA_0, OA_1, OA_2, etc. Beginning at A_1, construct perpendiculars A_1B_1, A_2B_2, A_3B_3, etc. Then along A_1B_1 lay off the distance A_1A_0, along A_2B_2 lay off twice the distance A_1A_0, etc., producing points through which the involute curve can be constructed.

To investigate the fundamentals of tooth action, let us proceed step by step through the process of constructing the teeth on a pair of gears.

When two gears are in mesh, their pitch circles roll on one another without slipping. Designate the pitch radii as r_1 and r_2 and the angular velocities as ω_1 and ω_2, respectively. Then the pitch-line velocity is

$$V = |r_1\omega_1| = |r_2\omega_2|$$

Thus the relation between the radii on the angular velocities is

$$\left|\frac{\omega_1}{\omega_2}\right| = \frac{r_2}{r_1} \tag{13–5}$$

Suppose now we wish to design a speed reducer such that the input speed is 1800 rev/min and the output speed is 1200 rev/min. This is a ratio of 3:2; the gear pitch diameters would be in the same ratio, for example, a 4-in pinion driving a 6-in gear. The various dimensions found in gearing are always based on the pitch circles.

Suppose we specify that an 18-tooth pinion is to mesh with a 30-tooth gear and that the diametral pitch of the gearset is to be 2 teeth per inch. Then, from Eq. (13–1), the pitch diameters of the pinion and gear are, respectively,

$$d_1 = \frac{N_1}{P} = \frac{18}{2} = 9 \text{ in} \qquad d_2 = \frac{N_2}{P} = \frac{30}{2} = 15 \text{ in}$$

The first step in drawing teeth on a pair of mating gears is shown in Fig. 13–9. The center distance is the sum of the pitch radii, in this case 12 in. So locate the pinion and gear centers O_1 and O_2, 12 in apart. Then construct the pitch circles of radii r_1 and r_2. These are tangent at P, the *pitch point*. Next draw line ab, the common tangent, through the pitch point. We now designate gear 1 as the driver, and since it is rotating counterclockwise, we draw a line cd through point P at an angle ϕ to the common tangent ab. The line cd has three names, all of which are in general use. It is called the *pressure line*, the *generating line*, and the *line of action*. It represents the direction in which the resultant force acts between the gears. The angle ϕ is called the *pressure angle*, and it usually has values of 20 or 25°, though $14\frac{1}{2}°$ was once used.

Figure 13–9

Circles of a gear layout.

Figure 13–10

Base circle radius can be related to the pressure angle ϕ and the pitch circle radius by $r_b = r \cos \phi$.

Next, on each gear draw a circle tangent to the pressure line. These circles are the *base circles*. Since they are tangent to the pressure line, the pressure angle determines their size. As shown in Fig. 13–10, the radius of the base circle is

$$r_b = r \cos \phi \qquad (13\text{--}6)$$

where r is the pitch radius.

Now generate an involute on each base circle as previously described and as shown in Fig. 13–9. This involute is to be used for one side of a gear tooth. It is not necessary to draw another curve in the reverse direction for the other side of the tooth, because we are going to use a template which can be turned over to obtain the other side.

The addendum and dedendum distances for standard interchangeable teeth are, as we shall learn later, $1/P$ and $1.25/P$, respectively. Therefore, for the pair of gears we are constructing,

$$a = \frac{1}{P} = \frac{1}{2} = 0.500 \text{ in} \qquad b = \frac{1.25}{P} = \frac{1.25}{2} = 0.625 \text{ in}$$

Using these distances, draw the addendum and dedendum circles on the pinion and on the gear as shown in Fig. 13–9.

Next, using heavy drawing paper, or preferably, a sheet of 0.015- to 0.020-in clear plastic, cut a template for each involute, being careful to locate the gear centers properly with respect to each involute. Figure 13–11 is a reproduction of the template used to create some of the illustrations for this book. Note that only one side of the tooth profile is formed on the template. To get the other side, turn the template over. For some problems you might wish to construct a template for the entire tooth.

Figure 13–11

A template for drawing gear teeth.

Figure 13–12

Tooth action.

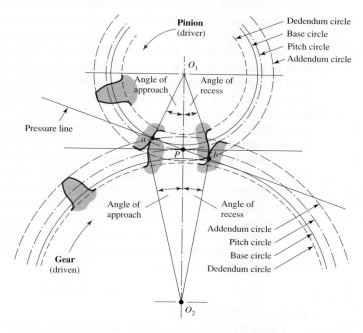

To draw a tooth, we must know the tooth thickness. From Eq. (13–4), the circular pitch is

$$p = \frac{\pi}{P} = \frac{\pi}{2} = 1.57 \text{ in}$$

Therefore, the tooth thickness is

$$t = \frac{p}{2} = \frac{1.57}{2} = 0.785 \text{ in}$$

measured on the pitch circle. Using this distance for the tooth thickness as well as the tooth space, draw as many teeth as desired, using the template, after the points have been marked on the pitch circle. In Fig. 13–12 only one tooth has been drawn on each gear. You may run into trouble in drawing these teeth if one of the base circles happens to be larger than the dedendum circle. The reason for this is that the involute begins at the base circle and is undefined below this circle. So, in drawing gear teeth, we usually draw a radial line for the profile below the base circle. The actual shape, however, will depend upon the kind of machine tool used to form the teeth in manufacture, that is, how the profile is generated.

The portion of the tooth between the clearance circle and the dedendum circle includes the fillet. In this instance the clearance is

$$c = b - a = 0.625 - 0.500 = 0.125 \text{ in}$$

The construction is finished when these fillets have been drawn.

Referring again to Fig. 13–12, the pinion with center at O_1 is the driver and turns counterclockwise. The pressure, or generating, line is the same as the cord used in Fig. 13–7a to generate the involute, and contact occurs along this line. The initial contact will take place when the flank of the driver comes into contact with the tip of the driven tooth. This occurs at point a in Fig. 13–12, where the addendum circle of the driven gear crosses the pressure line. If we now construct tooth profiles through point a and draw radial lines from the intersections of these profiles with the pitch circles to the gear centers, we obtain the *angle of approach* for each gear.

As the teeth go into mesh, the point of contact will slide up the side of the driving tooth so that the tip of the driver will be in contact just before contact ends. The final point of contact will therefore be where the addendum circle of the driver crosses the pressure line. This is point b in Fig. 13–12. By drawing another set of tooth profiles through b, we obtain the *angle of recess* for each gear in a manner similar to that of finding the angles of approach. The sum of the angle of approach and the angle of recess for either gear is called the *angle of action*. The line ab is called *the line of action*.

We may imagine a *rack* as a spur gear having an infinitely large pitch diameter. Therefore, the rack has an infinite number of teeth and a base circle which is an infinite distance from the pitch point. The sides of involute teeth on a rack are straight lines making an angle to the line of centers equal to the pressure angle. Figure 13–13 shows an involute rack in mesh with a pinion. Corresponding sides on involute teeth are parallel curves; the *base pitch* is the constant and fundamental distance between them along a common normal as shown in Fig. 13–13. The base pitch is related to the circular pitch by the equation

$$p_b = p_c \cos \phi \qquad (13\text{–}7)$$

where p_b is the base pitch.

Figure 13–14 shows a pinion in mesh with an *internal,* or *ring, gear.* Note that both of the gears now have their centers of rotation on the same side of the pitch point. Thus the positions of the addendum and dedendum circles with respect to the pitch circle are reversed; the addendum circle of the internal gear lies *inside* the pitch circle. Note, too, from Fig. 13–14, that the base circle of the internal gear lies inside the pitch circle near the addendum circle.

Another interesting observation concerns the fact that the operating diameters of the pitch circles of a pair of meshing gears need not be the same as the respective design pitch diameters of the gears, though this is the way they have been constructed in Fig. 13–12. If we increase the center distance, we create two new operating pitch circles having larger diameters because they must be tangent to each other at the pitch

Figure 13–13

Involute-toothed pinion and rack.

Base pitch

p_b

ϕ

p_c

Circular pitch

Figure 13–14

Internal gear and pinion.

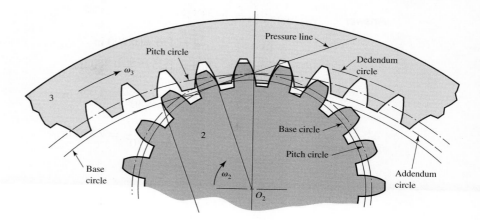

point. Thus the pitch circles of gears really do not come into existence until a pair of gears are brought into mesh.

Changing the center distance has no effect on the base circles, because these were used to generate the tooth profiles. Thus the base circle is basic to a gear. Increasing the center distance increases the pressure angle and decreases the length of the line of action, but the teeth are still conjugate, the requirement for uniform motion transmission is still satisfied, and the angular-velocity ratio has not changed.

EXAMPLE 13–1   A parallel helical gearset uses a 17-tooth pinion driving a 34-tooth gear. The pinion has a right-hand helix angle of 30°, a normal pressure angle of 20°, and a normal diametral pitch of 5 teeth/in. Find:
(a) The normal, transverse, and axial circular pitches
(b) The normal base circular pitch
(c) The transverse diametral pitch and the transverse pressure angle
(d) The addendum, dedendum, and pitch diameter of each gear

Solution

(a) $p_n = \pi/5 = 0.6283$ in

$p_t = p_n/\cos\psi = 0.6283/\cos 30° = 0.7255$ in

$p_x = p_t/\tan\psi = 0.7255/\tan 30° = 1.25$ in

Answer   (b) Eq. (13–7): $p_{nb} = p_n \cos\phi_n = 0.6283 \cos 20° = 0.590$ in

Answer   (c) $P_t = P_n \cos\psi = 5 \cos 30° = 4.33$ teeth/in

$\phi_t = \tan^{-1}(\tan\phi_n/\cos\psi) = \tan^{-1}(\tan 20°/\cos 30°) = 22.8°$

Answer   (d) Table 13–4:

$$a = 1/5 = 0.200 \text{ in}$$

$$b = 1.25/5 = 0.250 \text{ in}$$

$$d_P = \frac{17}{5 \cos 30°} = 3.926 \text{ in}$$

$$d_G = \frac{34}{5 \cos 30°} = 7.852 \text{ in}$$

13–6 Contact Ratio

The zone of action of meshing gear teeth is shown in Fig. 13–15. We recall that tooth contact begins and ends at the intersections of the two addendum circles with the pressure line. In Fig. 13–15 initial contact occurs at a and final contact at b. Tooth profiles drawn through these points intersect the pitch circle at A and B, respectively. As shown, the distance AP is called the *arc of approach* q_a, and the distance PB, the *arc of recess* q_r. The sum of these is the *arc of action* q_t.

Now, consider a situation in which the arc of action is exactly equal to the circular pitch, that is, $q_t = p$. This means that one tooth and its space will occupy the entire arc AB. In other words, when a tooth is just beginning contact at a, the previous tooth is simultaneously ending its contact at b. Therefore, during the tooth action from a to b, there will be exactly one pair of teeth in contact.

Next, consider a situation in which the arc of action is greater than the circular pitch, but not very much greater, say, $q_t \approx 1.2p$. This means that when one pair of teeth is just entering contact at a, another pair, already in contact, will not yet have reached b. Thus,

Figure 13–15

Definition of contact ratio.

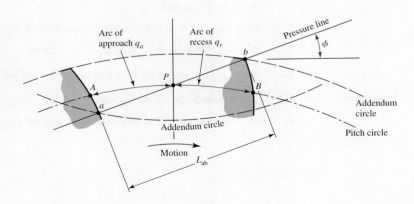

for a short period of time, there will be two teeth in contact, one in the vicinity of A and another near B. As the meshing proceeds, the pair near B must cease contact, leaving only a single pair of contacting teeth, until the procedure repeats itself.

Because of the nature of this tooth action, either one or two pairs of teeth in contact, it is convenient to define the term *contact ratio* m_c as

$$m_c = \frac{q_t}{p} \qquad (13\text{–}8)$$

a number that indicates the average number of pairs of teeth in contact. Note that this ratio is also equal to the length of the path of contact divided by the base pitch. Gears should not generally be designed having contact ratios less than about 1.20, because inaccuracies in mounting might reduce the contact ratio even more, increasing the possibility of impact between the teeth as well as an increase in the noise level.

An easier way to obtain the contact ratio is to measure the line of action ab instead of the arc distance AB. Since ab in Fig. 13–15 is tangent to the base circle when extended, the base pitch p_b must be used to calculate m_c instead of the circular pitch as in Eq. (13–8). If the length of the line of action is L_{ab}, the contact ratio is

$$m_c = \frac{L_{ab}}{p \cos \phi} \qquad (13\text{–}9)$$

in which Eq. (13–7) was used for the base pitch.

13–7 Interference

The contact of portions of tooth profiles that are not conjugate is called *interference*. Consider Fig. 13–16. Illustrated are two 16-tooth gears that have been cut to the now obsolete $14\frac{1}{2}^{\circ}$ pressure angle. The driver, gear 2, turns clockwise. The initial and final points of contact are designated A and B, respectively, and are located on the pressure line. Now notice that the points of tangency of the pressure line with the base circles C and D are located *inside* of points A and B. Interference is present.

The interference is explained as follows. Contact begins when the tip of the driven tooth contacts the flank of the driving tooth. In this case the flank of the driving tooth first makes contact with the driven tooth at point A, and this occurs *before* the involute portion of the driving tooth comes within range. In other words, contact is occurring below the base circle of gear 2 on the *noninvolute* portion of the flank. The actual effect is that the involute tip or face of the driven gear tends to dig out the noninvolute flank of the driver.

In this example the same effect occurs again as the teeth leave contact. Contact should end at point D or before. Since it does not end until point B, the effect is for the tip of the driving tooth to dig out, or interfere with, the flank of the driven tooth.

When gear teeth are produced by a generation process, interference is automatically eliminated because the cutting tool removes the interfering portion of the flank. This effect is called *undercutting;* if undercutting is at all pronounced, the undercut tooth is considerably weakened. Thus the effect of eliminating interference by a generation process is merely to substitute another problem for the original one.

---

[1]Robert Lipp, "Avoiding Tooth Interference in Gears," *Machine Design,* Vol. 54, No. 1, 1982, pp. 122–124.

Figure 13–16

Interference in the action
of gear teeth.

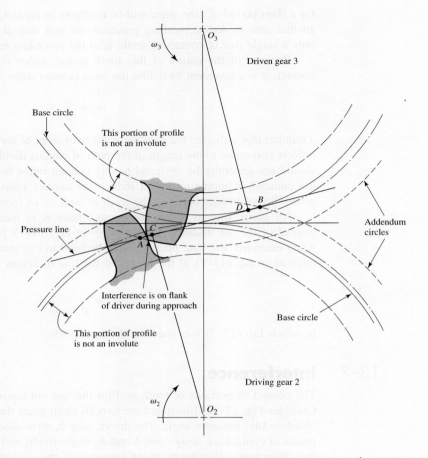

The smallest number of teeth on a spur pinion and gear,[1] one-to-one gear ratio, which can exist without interference is N_P. This number of teeth for spur gears is given by

$$N_P = \frac{2k}{3 \sin^2 \phi} (1 + \sqrt{1 + 3 \sin^2 \phi})$$ (13–10)

where $k = 1$ for full-depth teeth, 0.8 for stub teeth and ϕ = pressure angle.

For a 20° pressure angle, with $k = 1$,

$$N_P = \frac{2(1)}{3 \sin^2 20°} (1 + \sqrt{1 + 3 \sin^2 20°}) = 12.3 = 13 \text{ teeth}$$

Thus 13 teeth on pinion and gear are interference-free. Realize that 12.3 teeth is possible in meshing arcs, but for fully rotating gears, 13 teeth represents the least number. For a $14\frac{1}{2}°$ pressure angle, $N_P = 23$ teeth, so one can appreciate why few $14\frac{1}{2}°$-tooth systems are used, as the higher pressure angles can produce a smaller pinion with accompanying smaller center-to-center distances.

If the mating gear has more teeth than the pinion, that is, $m_G = N_G/N_P = m$ is more than one, then the smallest number of teeth on the pinion without interference is given by

$$N_P = \frac{2k}{(1 + 2m) \sin^2 \phi} (m + \sqrt{m^2 + (1 + 2m) \sin^2 \phi})$$ (13–11)

For example, if $m = 4$, $\phi = 20°$,

$$N_P = \frac{2(1)}{[1 + 2(4)] \sin^2 20°} [4 + \sqrt{4^2 + [1 + 2(4)] \sin^2 20°}] = 15.4 = 16 \text{ teeth}$$

Thus a 16-tooth pinion will mesh with a 64-tooth gear without interference.

The largest gear with a specified pinion that is interference-free is

$$N_G = \frac{N_P^2 \sin^2 \phi - 4k^2}{4k - 2N_P \sin^2 \phi} \tag{13–12}$$

For example, for a 13-tooth pinion with a pressure angle ϕ of 20°,

$$N_G = \frac{13^2 \sin^2 20° - 4(1)^2}{4(1) - 2(13) \sin^2 20°} = 16.45 = 16 \text{ teeth}$$

For a 13-tooth spur pinion, the maximum number of gear teeth possible without interference is 16.

The smallest spur pinion that will operate with a rack without interference is

$$N_P = \frac{2(k)}{\sin^2 \phi} \tag{13–13}$$

For a 20° pressure angle full-depth tooth the smallest number of pinion teeth to mesh with a rack is

$$N_P = \frac{2(1)}{\sin^2 20°} = 17.1 = 18 \text{ teeth}$$

Since gear-shaping tools amount to contact with a rack, and the gear-hobbing process is similar, the minimum number of teeth to prevent interference to prevent undercutting by the hobbing process is equal to the value of N_P when N_G is infinite.

The importance of the problem of teeth that have been weakened by undercutting cannot be overemphasized. Of course, interference can be eliminated by using more teeth on the pinion. However, if the pinion is to transmit a given amount of power, more teeth can be used only by increasing the pitch diameter.

Interference can also be reduced by using a larger pressure angle. This results in a smaller base circle, so that more of the tooth profile becomes involute. The demand for smaller pinions with fewer teeth thus favors the use of a 25° pressure angle even though the frictional forces and bearing loads are increased and the contact ratio decreased.

13–8   The Forming of Gear Teeth

There are a large number of ways of forming the teeth of gears, such as *sand casting, shell molding, investment casting, permanent-mold casting, die casting,* and *centrifugal casting.* Teeth can also be formed by using the *powder-metallurgy process;* or, by using *extrusion,* a single bar of aluminum may be formed and then sliced into gears. Gears that carry large loads in comparison with their size are usually made of steel and are cut with either *form cutters* or *generating cutters.* In form cutting, the tooth space takes the exact form of the cutter. In generating, a tool having a shape different from the tooth profile is moved relative to the gear blank to obtain the proper tooth shape. One of the newest and most promising of the methods of forming teeth is called *cold forming,* or *cold rolling,* in which dies are rolled against steel blanks to form the teeth. The mechanical properties of the metal are greatly improved by the rolling process, and a high-quality generated profile is obtained at the same time.

Gear teeth may be machined by milling, shaping, or hobbing. They may be finished by shaving, burnishing, grinding, or lapping.

Gears made of thermoplastics such as nylon, polycarbonate, acetal are quite popular and are easily manufactured by *injection molding*. These gears are of low to moderate precision, low in cost for high production quantities, and capable of light loads, and can run without lubrication.

Milling

Gear teeth may be cut with a form milling cutter shaped to conform to the tooth space. With this method it is theoretically necessary to use a different cutter for each gear, because a gear having 25 teeth, for example, will have a different-shaped tooth space from one having, say, 24 teeth. Actually, the change in space is not too great, and it has been found that eight cutters may be used to cut with reasonable accuracy any gear in the range of 12 teeth to a rack. A separate set of cutters is, of course, required for each pitch.

Shaping

Teeth may be generated with either a pinion cutter or a rack cutter. The pinion cutter (Fig. 13–17) reciprocates along the vertical axis and is slowly fed into the gear blank to the required depth. When the pitch circles are tangent, both the cutter and the blank rotate slightly after each cutting stroke. Since each tooth of the cutter is a cutting tool, the teeth are all cut after the blank has completed one rotation. The sides of an involute rack tooth are straight. For this reason, a rack-generating tool provides an accurate method of cutting gear teeth. This is also a shaping operation and is illustrated by the drawing of Fig. 13–18. In operation, the cutter reciprocates and is first fed into the gear blank until the pitch circles are tangent. Then, after each cutting stroke, the gear blank

Figure 13–17

Generating a spur gear with a pinion cutter. (*Courtesy of Boston Gear Works, Inc.*)

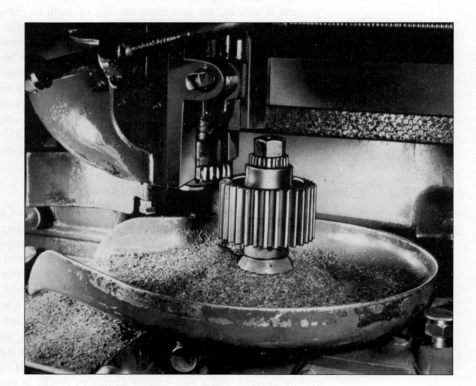

Figure 13–18

Shaping teeth with a rack. (This is a drawing-board figure that J. E. Shigley executed over 35 years ago in response to a question from a student at the University of Michigan.)

Gear blank rotates in this direction

Rack cutter reciprocates in a direction perpendicular to this page

Figure 13–19

Hobbing a worm gear. *(Courtesy of Boston Gear Works, Inc.)*

and cutter roll slightly on their pitch circles. When the blank and cutter have rolled a distance equal to the circular pitch, the cutter is returned to the starting point, and the process is continued until all the teeth have been cut.

Hobbing

The hobbing process is illustrated in Fig. 13–19. The hob is simply a cutting tool that is shaped like a worm. The teeth have straight sides, as in a rack, but the hob axis must be turned through the lead angle in order to cut spur-gear teeth. For this reason, the teeth generated by a hob have a slightly different shape from those generated by a rack cutter. Both the hob and the blank must be rotated at the proper angular-velocity ratio. The hob is then fed slowly across the face of the blank until all the teeth have been cut.

Finishing

Gears that run at high speeds and transmit large forces may be subjected to additional dynamic forces if there are errors in tooth profiles. Errors may be diminished somewhat by finishing the tooth profiles. The teeth may be finished, after cutting, by either shaving or burnishing. Several shaving machines are available that cut off a minute amount of metal, bringing the accuracy of the tooth profile within the limits of 250 μin.

Burnishing, like shaving, is used with gears that have been cut but not heat-treated. In burnishing, hardened gears with slightly oversize teeth are run in mesh with the gear until the surfaces become smooth.

Grinding and lapping are used for hardened gear teeth after heat treatment. The grinding operation employs the generating principle and produces very accurate teeth. In lapping, the teeth of the gear and lap slide axially so that the whole surface of the teeth is abraded equally.

13-9 Straight Bevel Gears

When gears are used to transmit motion between intersecting shafts, some form of bevel gear is required. A bevel gearset is shown in Fig. 13–20. Although bevel gears are usually made for a shaft angle of 90°, they may be produced for almost any angle. The teeth may be cast, milled, or generated. Only the generated teeth may be classed as accurate.

The terminology of bevel gears is illustrated in Fig. 13–20. The pitch of bevel gears is measured at the large end of the tooth, and both the circular pitch and the pitch diameter are calculated in the same manner as for spur gears. It should be noted that the clearance is uniform. The pitch angles are defined by the pitch cones meeting at the apex, as shown in the figure. They are related to the tooth numbers as follows:

$$\tan \gamma = \frac{N_P}{N_G} \qquad \tan \Gamma = \frac{N_G}{N_P} \tag{13-14}$$

Figure 13-20

Terminology of bevel gears.

where the subscripts P and G refer to the pinion and gear, respectively, and where γ and Γ are, respectively, the pitch angles of the pinion and gear.

Figure 13–20 shows that the shape of the teeth, when projected on the back cone, is the same as in a spur gear having a radius equal to the back-cone distance r_b. This is called Tredgold's approximation. The number of teeth in this imaginary gear is

$$N' = \frac{2\pi r_b}{p} \tag{13–15}$$

where N' is the *virtual number of teeth* and p is the circular pitch measured at the large end of the teeth. Standard straight-tooth bevel gears are cut by using a 20° pressure angle, unequal addenda and dedenda, and full-depth teeth. This increases the contact ratio, avoids undercuts, and increases the strength of the pinion.

13–10   Parallel Helical Gears

Helical gears, used to transmit motion between parallel shafts, are shown in Fig. 13–2. The helix angle is the same on each gear, but one gear must have a right-hand helix and the other a left-hand helix. The shape of the tooth is an involute helicoid and is illustrated in Fig. 13–21. If a piece of paper cut in the shape of a parallelogram is wrapped around a cylinder, the angular edge of the paper becomes a helix. If we unwind this paper, each point on the angular edge generates an involute curve. This surface obtained when every point on the edge generates an involute is called an *involute helicoid*.

The initial contact of spur-gear teeth is a line extending all the way across the face of the tooth. The initial contact of helical-gear teeth is a point that extends into a line as the teeth come into more engagement. In spur gears the line of contact is parallel to the axis of rotation; in helical gears the line is diagonal across the face of the tooth. It is this gradual engagement of the teeth and the smooth transfer of load from one tooth to another that gives helical gears the ability to transmit heavy loads at high speeds. Because of the nature of contact between helical gears, the contact ratio is of only minor importance, and it is the contact area, which is proportional to the face width of the gear, that becomes significant.

Helical gears subject the shaft bearings to both radial and thrust loads. When the thrust loads become high or are objectionable for other reasons, it may be desirable to use double helical gears. A double helical gear (herringbone) is equivalent to two helical gears of opposite hand, mounted side by side on the same shaft. They develop opposite thrust reactions and thus cancel out the thrust load.

When two or more single helical gears are mounted on the same shaft, the hand of the gears should be selected to produce the minimum thrust load.

Figure 13–21
An involute helicoid.

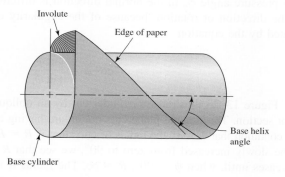

Figure 13–22

Nomenclature of helical gears.

Figure 13–22 represents a portion of the top view of a helical rack. Lines *ab* and *cd* are the centerlines of two adjacent helical teeth taken on the same pitch plane. The angle ψ is the *helix angle*. The distance *ac* is the *transverse circular pitch* p_t in the plane of rotation (usually called the *circular pitch*). The distance *ae* is the *normal circular pitch* p_n and is related to the transverse circular pitch as follows:

$$p_n = p_t \cos \psi \tag{13–16}$$

The distance *ad* is called the *axial pitch* p_x and is related by the expression

$$p_x = \frac{p_t}{\tan \psi} \tag{13–17}$$

Since $p_n P_n = \pi$, the *normal diametral pitch* is

$$P_n = \frac{P_t}{\cos \psi} \tag{13–18}$$

The pressure angle ϕ_n in the normal direction is different from the pressure angle ϕ_t in the direction of rotation, because of the angularity of the teeth. These angles are related by the equation

$$\cos \psi = \frac{\tan \phi_n}{\tan \phi_t} \tag{13–19}$$

Figure 13–23 illustrates a cylinder cut by an oblique plane *ab* at an angle ψ to a right section. The oblique plane cuts out an arc having a radius of curvature of R. For the condition that $\psi = 0$, the radius of curvature is $R = D/2$. If we imagine the angle ψ to be slowly increased from zero to 90°, we see that R begins at a value of $D/2$ and increases until, when $\psi = 90°$, $R = \infty$. The radius R is the apparent pitch radius of

Figure 13–23

A cylinder cut by an oblique plane.

(a)

(b)

a helical-gear tooth when viewed in the direction of the tooth elements. A gear of the same pitch and with the radius R will have a greater number of teeth, because of the increased radius. In helical-gear terminology this is called the *virtual number of teeth.* It can be shown by analytic geometry that the virtual number of teeth is related to the actual number by the equation

$$N' = \frac{N}{\cos^3 \psi}$$

(13–20)

where N' is the virtual number of teeth and N is the actual number of teeth. It is necessary to know the virtual number of teeth in design for strength and also, sometimes, in cutting helical teeth. This apparently larger radius of curvature means that few teeth may be used on helical gears, because there will be less undercutting.

EXAMPLE 13–2

A stock helical gear has a normal pressure angle of 22°, a helix angle of 32°, and a transverse diametral pitch of 3.0 mm, and has 24 teeth. Find:
(a) The pitch diameter
(b) The transverse, the normal, and the axial pitches
(c) The normal diametral pitch
(d) The transverse pressure angle

Solution

Answer   (a) $d = Nm_t = 24(3) = 72$ mm

Answer   (b) $p_t = \pi m_t = \pi(3) = 9.42478$ mm

Answer $$p_n = p_t \cos \psi = 9.42478 = 9.42478 \, cos \, (32) = 7.99267 \text{ mm}$$

Answer $$p_x = \frac{p_t}{\tan \psi} = \frac{9.42478}{\tan 32°} = 15.08280 \text{ mm}$$

Answer (c) $$m_n = m_t \cos \psi = 3 \cos 32° = 2.54414 \text{ mm}$$

Answer (d) $$\phi_t = \tan^{-1}\left(\frac{\tan \phi_n}{\cos \psi}\right) = \tan^{-1}\left(\frac{\tan 22°}{\cos 32°}\right) = 25.47402°$$

Just like teeth on spur gears, helical-gear teeth can interfere. Equation (13–19) can be solved for the pressure angle ϕ_t in the tangential (rotation) direction to give

$$\phi_t = \tan^{-1}\left(\frac{\tan \phi_n}{\cos \psi}\right)$$

The smallest tooth number N_P of a helical-spur pinion that will run without interference[2] with a gear with the same number of teeth is

$$N_P = \frac{2k \cos \psi}{3 \sin^2 \phi_t}(1 + \sqrt{1 + 3 \sin^2 \phi_t}) \tag{13–21}$$

For example, if the normal pressure angle ϕ_n is 20°, the helix angle ψ is 30°, then ϕ_t is

$$\phi_t = \tan^{-1}\left(\frac{\tan 20°}{\cos 30°}\right) = 22.80°$$

$$N_P = \frac{2(1) \cos 30°}{3 \sin^2 22.80°}(1 + \sqrt{1 + 3 \sin^2 22.80°}) = 8.48 = 9 \text{ teeth}$$

For a given gear ratio $m_G = N_G/N_P = m$, the smallest pinion tooth count is

$$N_P = \frac{2k \cos \psi}{(1 + 2m) \sin^2 \phi_t}[m + \sqrt{m^2 + (1 + 2m) \sin^2 \phi_t}] \tag{13–22}$$

The largest gear with a specified pinion is given by

$$N_G = \frac{N_P^2 \sin^2 \phi_t - 4k^2 \cos^2 \psi}{4k \cos \psi - 2N_P \sin^2 \phi_t} \tag{13–23}$$

For example, for a nine-tooth pinion with a pressure angle ϕ_n of 20°, a helix angle ψ of 30°, and recalling that the tangential pressure angle ϕ_t is 22.80°,

$$N_G = \frac{9^2 \sin^2 22.80° - 4(1)^2 \cos^2 30°}{4(1) \cos 30° - 2(9) \sin^2 22.80°} = 12.02 = 12$$

The smallest pinion that can be run with a rack is

$$N_P = \frac{2k \cos \psi}{\sin^2 \phi_t} \tag{13–24}$$

---

[2]Op. cit., Robert Lipp, *Machine Design*, pp. 122–124.

For a normal pressure angle ϕ_n of 20° and a helix angle ψ of 30°, and $\phi_t = 22.80°$,

$$N_P = \frac{2(1)\cos 30°}{\sin^2 22.80°} = 11.5 = 12 \text{ teeth}$$

For helical-gear teeth the number of teeth in mesh across the width of the gear will be greater than unity and a term called *face-contact ratio* is used to describe it. This increase of contact ratio, and the gradual sliding engagement of each tooth, results in quieter gears.

13–11   Worm Gears

The nomenclature of a worm gearset is shown in Fig. 13–24. The worm and worm gear of a set have the same hand of helix as for crossed helical gears, but the helix angles are usually quite different. The helix angle on the worm is generally quite large, and that on the gear very small. Because of this, it is usual to specify the lead angle λ on the worm and helix angle ψ_G on the gear; the two angles are equal for a 90° shaft angle. The worm lead angle is the complement of the worm helix angle, as shown in Fig. 13–24.

In specifying the pitch of worm gearsets, it is customary to state the *axial pitch* p_x of the worm and the *transverse circular pitch* p_t, often simply called the circular pitch, of the mating gear. These are equal if the shaft angle is 90°. The pitch diameter of the gear is the diameter measured on a plane containing the worm axis, as shown in Fig. 13–24; it is the same as for spur gears and is

$$d_G = \frac{N_G p_t}{\pi} \tag{13-25}$$

Figure 13–24

Nomenclature of a single-enveloping worm gearset.

Since it is not related to the number of teeth, the worm may have any pitch diameter; this diameter should, however, be the same as the pitch diameter of the hob used to cut the worm-gear teeth. Generally, the pitch diameter of the worm should be selected to fall into the range

$$\frac{C^{0.875}}{3.0} \le d_W \le \frac{C^{0.875}}{1.7} \tag{13–26}$$

where C is the center distance. These proportions appear to result in optimum horsepower capacity of the gearset.

The *lead L* and the *lead angle* λ of the worm have the following relations:

$$L = p_x N_W \tag{13–27}$$

$$\tan \lambda = \frac{L}{\pi d_W} \tag{13–28}$$

13–12 Tooth Systems[3]

A *tooth system* is a standard that specifies the relationships involving the addendum, dedendum, working depth, tooth thickness, and pressure angle. The standards were originally planned to attain interchangeability of gears of all tooth numbers, but of the same pressure angle and pitch.

Table 13–1 contains the standards most used for spur gears. A $14\frac{1}{2}^{\circ}$ pressure angle was once used for these but is now obsolete; the resulting gears had to be comparatively larger to avoid interference problems.

Table 13–2 is particularly useful in selecting the pitch or module of a gear. Cutters are generally available for the sizes shown in this table.

Table 13–3 lists the standard tooth proportions for straight bevel gears. These sizes apply to the large end of the teeth. The nomenclature is defined in Fig. 13–20.

Standard tooth proportions for helical gears are listed in Table 13–4. Tooth proportions are based on the normal pressure angle; these angles are standardized the

Table 13–1

Standard and Commonly Used Tooth Systems for Spur Gears

Tooth System	Pressure Angle ϕ, deg	Addendum a	Dedendum b
Full depth	20	$1/P$ or m	$1.25/P$ or $1.25m$
			$1.35/P$ or $1.35m$
	$22\frac{1}{2}$	$1/P$ or m	$1.25/P$ or $1.25m$
			$1.35/P$ or $1.35m$
	25	$1/P$ or m	$1.25/P$ or $1.25m$
			$1.35/P$ or $1.35m$
Stub	20	$0.8/P$ or $0.8m$	$1/P$ or m

---

[3]Standardized by the American Gear Manufacturers Association (AGMA). Write AGMA for a complete list of standards, because changes are made from time to time. The address is: 1001 N. Fairfax Street, Suite 500, Alexandria, VA 22314-1587; or, www.agma.org.

Table 13–2

Tooth Sizes in General Uses

Diametral Pitch P(teeth/in)	
Coarse	2, $2\frac{1}{4}$, $2\frac{1}{2}$, 3, 4, 6, 8, 10, 12, 16
Fine	20, 24, 32, 40, 48, 64, 80, 96, 120, 150, 200

Module m(mm/tooth)	
Preferred	1, 1.25, 1.5, 2, 2.5, 3, 4, 5, 6, 8, 10, 12, 16, 20, 25, 32, 40, 50
Next Choice	1.125, 1.375, 1.75, 2.25, 2.75, 3.5, 4.5, 5.5, 7, 9, 11, 14, 18, 22, 28, 36, 45

Table 13–3

Tooth Proportions for 20° Straight Bevel-Gear Teeth

Item	Formula
Working depth	$h_k = 2.0/P \ [= 2.0\ m]$
Clearance	$c = (0.188/P) + 0.002$ in $[= 0.188\ m + 0.05\ \text{mm}]$
Addendum of gear	$a_G = \dfrac{0.54}{P} + \dfrac{0.460}{P(m_{90})^2} \left[= 0.54\ m + \dfrac{0.46\ m}{(m_{90})^2}\right]$
Gear ratio	$m_G = N_G/N_P$
Equivalent 90° ratio	$m_{90} = m_G$ when $\Gamma = 90°$
	$m_{90} = \sqrt{m_G \dfrac{\cos \gamma}{\cos \Gamma}}$ when $\Gamma \neq 90°$
Face width	$F = 0.3A_0$ or $F = \dfrac{10}{P}$, whichever is smaller $\left[F = \dfrac{A_0}{3}\ \text{or}\ F = 10\ m\right]$

Minimum number of teeth				
Pinion	16	15	14	13
Gear	16	17	20	30

Table 13–4

Standard Tooth Proportions for Helical Gears

Quantity*	Formula	Quantity*	Formula
Addendum	$\dfrac{1.00}{P_n}$	External gears:	
Dedendum	$\dfrac{1.25}{P_n}$	Standard center distance	$\dfrac{D + d}{2}$
Pinion pitch diameter	$\dfrac{N_P}{P_n \cos \psi}$	Gear outside diameter	$D + 2a$
Gear pitch diameter	$\dfrac{N_G}{P_n \cos \psi}$	Pinion outside diameter	$d + 2a$
Normal arc tooth thickness†	$\dfrac{\pi}{P_n} - \dfrac{B_n}{2}$	Gear root diameter	$D - 2b$
Pinion base diameter	$d \cos \phi_t$	Pinion root diameter	$d - 2b$
		Internal gears:	
Gear base diameter	$D \cos \phi_t$	Center distance	$\dfrac{D - d}{2}$
Base helix angle	$\tan^{-1}(\tan \psi \cos \phi_t)$	Inside diameter	$D - 2a$
		Root diameter	$D + 2b$

*All dimensions are in inches, and angles are in degrees.

†B_n is the normal backlash.

Table 13–5

Recommended Pressure Angles and Tooth Depths for Worm Gearing

Lead Angle λ, deg	Pressure Angle ϕ_n, deg	Addendum a	Dedendum b_G
0–15	$14\frac{1}{2}$	$0.3683p_x$	$0.3683p_x$
15–30	20	$0.3683p_x$	$0.3683p_x$
30–35	25	$0.2865p_x$	$0.3314p_x$
35–40	25	$0.2546p_x$	$0.2947p_x$
40–45	30	$0.2228p_x$	$0.2578p_x$

Figure 13–25

A graphical depiction of the face width of the worm of a worm gearset.

same as for spur gears. Though there will be exceptions, the face width of helical gears should be at least 2 times the axial pitch to obtain good helical-gear action.

Tooth forms for worm gearing have not been highly standardized, perhaps because there has been less need for it. The pressure angles used depend upon the lead angles and must be large enough to avoid undercutting of the worm-gear tooth on the side at which contact ends. A satisfactory tooth depth, which remains in about the right proportion to the lead angle, may be obtained by making the depth a proportion of the axial circular pitch. Table 13–5 summarizes what may be regarded as good practice for pressure angle and tooth depth.

The *face width* F_G of the worm gear should be made equal to the length of a tangent to the worm pitch circle between its points of intersection with the addendum circle, as shown in Fig. 13–25.

13–13 Gear Trains

Consider a pinion 2 driving a gear 3. The speed of the driven gear is

$$n_3 = \left| \frac{N_2}{N_3} n_2 \right| = \left| \frac{d_2}{d_3} n_2 \right| \qquad (13\text{–}29)$$

where n = revolutions or rev/min

N = number of teeth

d = pitch diameter

Equation (13–29) applies to any gearset regardless of whether the gears are spur, helical, bevel, or worm. The absolute-value signs are used to permit complete freedom in choosing positive and negative directions. In the case of spur and parallel helical gears, the directions in the viewing plane ordinarily correspond to the right-hand rule—positive for counterclockwise rotation and negative for clockwise rotation.

Rotational directions are somewhat more difficult to deduce for worm and crossed helical gearsets. Figure 13–26 will be of help in these situations.

The gear train shown in Fig. 13–27 is made up of five gears. Considering gear 2 to be the primary driving gear, the speed of gear 6 is

$$n_6 = -\frac{N_2}{N_3}\frac{N_3}{N_4}\frac{N_5}{N_6} n_2 \qquad (a)$$

Hence we notice that gear 3 is an idler, that its tooth numbers cancel in Eq. (a), and hence that it affects only the direction of rotation of gear 6. We notice, furthermore,

Figure 13–26

Thrust, rotation, and hand relations for crossed helical gears. Note that each pair of drawings refers to a single gearset. These relations also apply to worm gearsets. *(Reproduced by permission, Boston Gear Division, Colfax Corp.)*

(a)

Right hand

(b)

(c)

Left hand

(d)

Figure 13–27

A gear train.

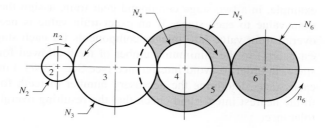

that gears 2, 3, and 5 are drivers, while 3, 4, and 6 are driven members. We define the *train value e* as

$$e = \frac{\text{product of driving tooth numbers}}{\text{product of driven tooth numbers}} \qquad (13\text{–}30)$$

Note that pitch diameters can be used in Eq. (13–30) as well. When Eq. (13–30) is used for spur gears, e is positive if the last gear rotates in the same sense as the first, and negative if the last rotates in the opposite sense.

Now we can write

$$n_L = e\, n_F \qquad (13\text{–}31)$$

where n_L is the speed of the last gear in the train and n_F is the speed of the first.

As a rough guideline, a train value of up to 10 to 1 can be obtained with one pair of gears. Greater ratios can be obtained in less space and with fewer dynamic problems by compounding additional pairs of gears. A two-stage compound gear train, such as shown in Fig. 13–28, can obtain a train value of up to 100 to 1.

The design of gear trains to accomplish a specific train value is straightforward. Since numbers of teeth on gears must be integers, it is better to determine

Figure 13–28

A two-stage compound gear train.

them first, and then obtain pitch diameters second. Determine the number of stages necessary to obtain the overall ratio, then divide the overall ratio into portions to be accomplished in each stage. To minimize package size, keep the portions as evenly divided between the stages as possible. In cases where the overall train value need only be approximated, each stage can be identical. For example, in a two-stage compound gear train, assign the square root of the overall train value to each stage. If an exact train value is needed, attempt to factor the overall train value into integer components for each stage. Then assign the smallest gear(s) to the minimum number of teeth allowed for the specific ratio of each stage, in order to avoid interference (see Sec. 13–7). Finally, applying the ratio for each stage, determine the necessary number of teeth for the mating gears. Round to the nearest integer and check that the resulting overall ratio is within acceptable tolerance.

EXAMPLE 13–3

A gearbox is needed to provide a 30:1 (± 1 percent) increase in speed, while minimizing the overall gearbox size. Specify appropriate tooth numbers.

Solution

Since the ratio is greater than 10:1, but less than 100:1, a two-stage compound gear train, such as in Figure 13–28, is needed. The portion to be accomplished in each stage is $\sqrt{30} = 5.4772$. For this ratio, assuming a typical $20°$ pressure angle, the minimum number of teeth to avoid interference is 16, according to Eq. (13–11). The number of teeth necessary for the mating gears is

Answer

$$16\sqrt{30} = 87.64 \approx 88$$

From Eq. (13–30), the overall train value is

$$e = (88/16)(88/16) = 30.25$$

This is within the 1 percent tolerance. If a closer tolerance is desired, then increase the pinion size to the next integer and try again.

EXAMPLE 13–4   A gearbox is needed to provide an *exact* 30:1 increase in speed, while minimizing the overall gearbox size. Specify appropriate teeth numbers.

Solution   The previous example demonstrated the difficulty with finding integer numbers of teeth to provide an exact ratio. In order to obtain integers, factor the overall ratio into two integer stages.

$$e = 30 = (6)(5)$$

$$N_2/N_3 = 6 \quad \text{and} \quad N_4/N_5 = 5$$

With two equations and four unknown numbers of teeth, two free choices are available. Choose N_3 and N_5 to be as small as possible without interference. Assuming a 20° pressure angle, Eq. (13–11) gives the minimum as 16. Then

$$N_2 = 6N_3 = 6(16) = 96$$

$$N_4 = 5N_5 = 5(16) = 80$$

The overall train value is then exact.

$$e = (96/16)(80/16) = (6)(5) = 30$$

It is sometimes desirable for the input shaft and the output shaft of a two-stage compound gear train to be in-line, as shown in Fig. 13–29. This configuration is called a *compound reverted gear train*. This requires the distances between the shafts to be the same for both stages of the train, which adds to the complexity of the design task. The distance constraint is

$$d_2/2 + d_3/2 = d_4/2 + d_5/2$$

Figure 13–29

A compound reverted gear train.

The diametral pitch relates the diameters and the numbers of teeth, $P = N/d$. Replacing all the diameters gives

$$N_2/(2P) + N_3/(2P) = N_4/(2P) + N_5/(2P)$$

Assuming a constant diametral pitch in both stages, we have the geometry condition stated in terms of numbers of teeth:

$$N_2 + N_3 = N_4 + N_5$$

This condition must be exactly satisfied, in addition to the previous ratio equations, to provide for the in-line condition on the input and output shafts.

EXAMPLE 13–5

A gearbox is needed to provide an exact 30:1 increase in speed, while minimizing the overall gearbox size. The input and output shafts should be in-line. Specify appropriate teeth numbers.

Solution

The governing equations are

$$N_2/N_3 = 6$$
$$N_4/N_5 = 5$$
$$N_2 + N_3 = N_4 + N_5$$

With three equations and four unknown numbers of teeth, only one free choice is available. Of the two smaller gears, N_3 and N_5, the free choice should be used to minimize N_3 since a greater gear ratio is to be achieved in this stage. To avoid interference, the minimum for N_3 is 16.

Applying the governing equations yields

$$N_2 = 6N_3 = 6(16) = 96$$
$$N_2 + N_3 = 96 + 16 = 112 = N_4 + N_5$$

Substituting $N_4 = 5N_5$ gives

$$112 = 5N_5 + N_5 = 6N_5$$
$$N_5 = 112/6 = 18.67$$

If the train value need only be approximated, then this can be rounded to the nearest integer. But for an exact solution, it is necessary to choose the initial free choice for N_3 such that solution of the rest of the teeth numbers results exactly in integers. This can be done by trial and error, letting $N_3 = 17$, then 18, etc., until it works. Or, the problem can be normalized to quickly determine the minimum free choice. Beginning again, let the free choice be $N_3 = 1$. Applying the governing equations gives

$$N_2 = 6N_3 = 6(1) = 6$$
$$N_2 + N_3 = 6 + 1 = 7 = N_4 + N_5$$

Substituting $N_4 = 5N_5$, we find

$$7 = 5N_5 + N_5 = 6N_5$$

$$N_5 = 7/6$$

This fraction could be eliminated if it were multiplied by a multiple of 6. The free choice for the smallest gear N_3 should be selected as a multiple of 6 that is greater than the minimum allowed to avoid interference. This would indicate that $N_3 = 18$. Repeating the application of the governing equations for the final time yields

$$N_2 = 6N_3 = 6(18) = 108$$

$$N_2 + N_3 = 108 + 18 = 126 = N_4 + N_5$$

$$126 = 5N_5 + N_5 = 6N_5$$

$$N_5 = 126/6 = 21$$

$$N_4 = 5N_5 = 5(21) = 105$$

Thus,

Answer

$$N_2 = 108$$

$$N_3 = 18$$

$$N_4 = 105$$

$$N_5 = 21$$

Checking, we calculate $e = (108/18)(105/21) = (6)(5) = 30$.
And checking the geometry constraint for the in-line requirement, we calculate

$$N_2 + N_3 = N_4 + N_5$$

$$108 + 18 = 105 + 21$$

$$126 = 126$$

Unusual effects can be obtained in a gear train by permitting some of the gear axes to rotate about others. Such trains are called *planetary*, or *epicyclic*, *gear trains*. Planetary trains always include a *sun gear,* a *planet carrier* or *arm,* and one or more *planet gears,* as shown in Fig. 13–30. Planetary gear trains are unusual mechanisms because they have two degrees of freedom; that is, for constrained motion, a planetary train must have two inputs. For example, in Fig. 13–30 these two inputs could be the motion of any two of the elements of the train. We might, say, specify that the sun gear rotates at 100 rev/min clockwise and that the ring gear rotates at 50 rev/min counterclockwise; these are the inputs. The output would be the motion of the arm. In most planetary trains one of the elements is attached to the frame and has no motion. Figure 13–31 shows a planetary train composed of a sun gear 2, an arm or

Figure 13–30

A planetary gear train.

Figure 13–31

A gear train on the arm of a planetary gear train.

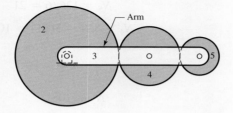

carrier 3, and planet gears 4 and 5. The angular velocity of gear 2 relative to the arm in rev/min is

$$n_{23} = n_2 - n_3 \qquad (b)$$

Also, the velocity of gear 5 relative to the arm is

$$n_{53} = n_5 - n_3 \qquad (c)$$

Dividing Eq. (c) by Eq. (b) gives

$$\frac{n_{53}}{n_{23}} = \frac{n_5 - n_3}{n_2 - n_3} \qquad (d)$$

Equation (d) expresses the ratio of gear 5 to that of gear 2, and both velocities are taken relative to the arm. Now this ratio is the same and is proportional to the tooth numbers, whether the arm is rotating or not. It is the train value. Therefore, we may write

$$e = \frac{n_5 - n_3}{n_2 - n_3} \qquad (e)$$

This equation can be used to solve for the output motion of any planetary train. It is more conveniently written in the form

$$e = \frac{n_L - n_A}{n_F - n_A} \qquad (13\text{–}32)$$

where n_F = rev/min of first gear in planetary train

n_L = rev/min of last gear in planetary train

n_A = rev/min of arm

EXAMPLE 13–6

In Fig. 13–30 the sun gear is the input, and it is driven clockwise at 100 rev/min. The ring gear is held stationary by being fastened to the frame. Find the rev/min and direction of rotation of the arm and gear 4.

Solution

Let $n_F = n_2 = -100$ rev/min, and $n_L = n_5 = 0$. For e, unlock gear 5 and fix the arm. Then, planet gear 4 and ring gear 5 rotate in the *same* direction, *opposite* of sun gear 2. Thus, e is negative and

$$e = -\left(\frac{N_2}{N_4}\right)\left(\frac{N_4}{N_5}\right) = -\left(\frac{20}{30}\right)\left(\frac{30}{80}\right) = -0.25$$

Substituting this value in Eq. (13–32) gives

$$-0.25 = \frac{0 - n_A}{(-100) - n_A}$$

or

Answer

$$n_A = -20 \text{ rev/min} = 20 \text{ rev/min clockwise}$$

To obtain the speed of gear 4, we follow the procedure outlined by Eqs. (b), (c), and (d). Thus

$$n_{43} = n_4 - n_3 \qquad n_{23} = n_2 - n_3$$

and so

$$\frac{n_{43}}{n_{23}} = \frac{n_4 - n_3}{n_2 - n_3} \qquad (1)$$

But

$$\frac{n_{43}}{n_{23}} = -\frac{20}{30} = -\frac{2}{3} \qquad (2)$$

Substituting the known values in Eq. (1) gives

$$-\frac{2}{3} = \frac{n_4 - (-20)}{(-100) - (-20)}$$

Solving gives

Answer

$$n_4 = 33\frac{1}{3} \text{ rev/min} = 33\frac{1}{3} \text{ rev/min counter-clockwise}$$

13–14   **Force Analysis—Spur Gearing**

Before beginning the force analysis of gear trains, let us agree on the notation to be used. Beginning with the numeral 1 for the frame of the machine, we shall designate the input gear as gear 2, and then number the gears successively 3, 4, etc., until we arrive at the last gear in the train. Next, there may be several shafts involved, and usually one or two gears are mounted on each shaft as well as other elements. We shall designate the shafts, using lowercase letters of the alphabet, a, b, c, etc.

With this notation we can now speak of the force exerted by gear 2 against gear 3 as F_{23}. The force of gear 2 against shaft a is F_{2a}. We can also write F_{a2} to mean the force of shaft a against gear 2. Unfortunately, it is also necessary to use superscripts to indicate directions. The coordinate directions will usually be indicated by the x, y, and z coordinates, and the radial and tangential directions by superscripts r and t. With this notation, F_{43}^{t} is the tangential component of the force of gear 4 acting against gear 3.

Figure 13–32a shows a pinion mounted on shaft a rotating clockwise at n_2 rev/min and driving a gear on shaft b at n_3 rev/min. The reactions between the mating teeth occur along the pressure line. In Fig. 13–32b the pinion has been separated from the gear and the shaft, and their effects have been replaced by forces. F_{a2} and T_{a2} are the force and torque, respectively, exerted by shaft a against pinion 2. F_{32} is the force exerted by gear 3 against the pinion. Using a similar approach, we obtain the free-body diagram of the gear shown in Fig. 13–32c.

In Fig. 13–33, the free-body diagram of the pinion has been redrawn and the forces have been resolved into tangential and radial components. We now define

$$W_t = F_{32}^{t} \qquad (a)$$

as the *transmitted load*. This tangential load is really the useful component, because the radial component F_{32}^{r} serves no useful purpose. It does not transmit power. The applied torque and the transmitted load are seen to be related by the equation

$$T = \frac{d}{2}W_t \qquad (b)$$

where we have used $T = T_{a2}$ and $d = d_2$ to obtain a general relation.

The power H transmitted through a rotating gear can be obtained from the standard relationship of the product of torque T and angular velocity ω.

$$H = T\omega = (W_t d/2)\omega \qquad (13\text{–}33)$$

Figure 13–32

Free-body diagrams of the forces and moments acting upon two gears of a simple gear train.

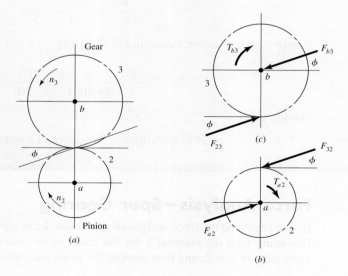

Figure 13–33

Resolution of gear forces.

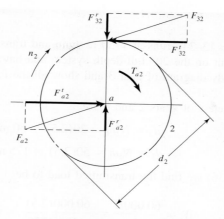

While any unit can be used in this equation, the units of the resulting power will obviously be dependent on the units of the other parameters. It will often be desirable to work with the power in either horsepower or kilowatts, and appropriate conversion factors should be used.

Since meshed gears are reasonably efficient, with losses of less than 2 percent, the power is generally treated as constant through the mesh. Consequently, with a pair of meshed gears, Eq. (13–33) will give the same power regardless of which gear is used for d and ω.

Gear data is often tabulated using *pitch-line velocity*, which is the linear velocity of a point on the gear at the radius of the pitch circle; thus $V = (d/2)\omega$. Converting this to customary units gives

$$V = \pi dn/12 \tag{13–34}$$

where V = pitch-line velocity, ft/min

   d = gear diameter, in

   n = gear speed, rev/min

Many gear design problems will specify the power and speed, so it is convenient to solve Eq. (13–33) for W_t. With the pitch-line velocity and appropriate conversion factors incorporated, Eq. (13–33) can be rearranged and expressed in U.S. customary units as

$$W_t = 33\,000\frac{H}{V} \tag{13–35}$$

where W_t = transmitted load, lbf

   H = power, hp

   V = pitch-line velocity, ft/min

The corresponding equation in SI units is

$$W_t = \frac{60\,000H}{\pi dn} \tag{13–36}$$

where W_t = transmitted load, kN

   H = power, kW

   d = gear diameter, mm

   n = speed, rev/min

EXAMPLE 13–7   Pinion 2 in Fig. 13–34a runs at 1750 rev/min and transmits 2.5 kW to idler gear 3. The teeth are cut on the 20° full-depth system and have a module of $m = 2.5$ mm. Draw a free-body diagram of gear 3 and show all the forces that act upon it.

Solution   The pitch diameters of gears 2 and 3 are

$$d_2 = N_2 m = 20(2.5) = 50 \text{ mm}$$

$$d_3 = N_3 m = 50(2.5) = 125 \text{ mm}$$

From Eq. (13–36) we find the transmitted load to be

$$W_t = \frac{60\,000H}{\pi d_2 n} = \frac{60\,000(2.5)}{\pi(50)(1750)} = 0.546 \text{ kN}$$

Thus, the tangential force of gear 2 on gear 3 is $F_{23}^t = 0.546$ kN, as shown in Fig. 13–34b. Therefore

$$F_{23}^r = F_{23}^t \tan 20° = (0.546) \tan 20° = 0.199 \text{ kN}$$

and so

$$F_{23} = \frac{F_{23}^t}{\cos 20°} = \frac{0.546}{\cos 20°} = 0.581 \text{ kN}$$

Since gear 3 is an idler, it transmits no power (torque) to its shaft, and so the tangential reaction of gear 4 on gear 3 is also equal to W_t. Therefore

$$F_{43}^t = 0.546 \text{ kN} \qquad F_{43}^r = 0.199 \text{ kN} \qquad F_{43} = 0.581 \text{ kN}$$

and the directions are shown in Fig. 13–34b.
The shaft reactions in the x and y directions are

$$F_{b3}^x = -(F_{23}^t + F_{43}^r) = -(-0.546 + 0.199) = 0.347 \text{ kN}$$

$$F_{b3}^y = -(F_{23}^r + F_{43}^t) = -(0.199 - 0.546) = 0.347 \text{ kN}$$

Figure 13–34

A gear train containing an idler gear. (a) The gear train. (b) Free-body of the idler gear.

(a)   (b)

The resultant shaft reaction is

$$F_{b3} = \sqrt{(0.347)^2 + (0.347)^2} = 0.491 \text{ kN}$$

These are shown on the figure.

13–15   Force Analysis—Bevel Gearing

In determining shaft and bearing loads for bevel-gear applications, the usual practice is to use the tangential or transmitted load that would occur if all the forces were concentrated at the midpoint of the tooth. While the actual resultant occurs somewhere between the midpoint and the large end of the tooth, there is only a small error in making this assumption. For the transmitted load, this gives

$$W_t = \frac{T}{r_{av}} \tag{13–37}$$

where T is the torque and r_{av} is the pitch radius at the midpoint of the tooth for the gear under consideration.

The forces acting at the center of the tooth are shown in Fig. 13–35. The resultant force W has three components: a tangential force W_t, a radial force W_r, and an axial force W_a. From the trigonometry of the figure,

$$W_r = W_t \tan \phi \cos \gamma$$
$$W_a = W_t \tan \phi \sin \gamma \tag{13–38}$$

Figure 13–35

Bevel-gear tooth forces.

The three forces W_t, W_r, and W_a are at right angles to each other and can be used to determine the bearing loads by using the methods of statics.

EXAMPLE 13–8

The bevel pinion in Fig. 13–36a rotates at 600 rev/min in the direction shown and transmits 3.75 kW to the gear. The mounting distances, the location of all bearings, and the average pitch radii of the pinion and gear are shown in the figure. For simplicity, the teeth have been replaced by pitch cones. Bearings A and C should take the thrust loads. Find the bearing forces on the gearshaft.

Figure 13–36

(a) Bevel gearset of Ex. 13–8.
(b) Free-body diagram of shaft CD. Dimensions in millimeters.

Solution   The pitch angles are

$$\gamma = \tan^{-1}\left(\frac{75}{225}\right) = 18.4° \qquad \Gamma = \tan^{-1}\left(\frac{225}{75}\right) = 71.6°$$

The pitch-line velocity corresponding to the average pitch radius is

$$V = 2\pi r_p n = \frac{2\pi(32)(600)}{60} = 2011 \text{ mm/s}$$

Therefore the transmitted load is

$$W_t = \frac{H}{V} = \frac{3750}{2.001} = 1865 \text{ N}$$

and from Eq. (13–38), with Γ replacing γ, we have

$$W_r = W_t \tan \phi \cos \Gamma = 1865 \tan 20° \cos 71.6° = 214 \text{ N}$$

$$W_a = W_t \tan \phi \cos \Gamma = 1865 \tan 20° \sin 71.6° = 644 \text{ N}$$

where W_t acts in the positive x direction, W_a in the $-y$ direction, as illustrated in the isometric sketch of Fig. 13–36b.

In preparing to take a sum of the moments about bearing D, define the position vector from D to G as

$$\mathbf{R}_G = 90\mathbf{i} - (60 + 32)\mathbf{j} = 90\mathbf{i} - 92\mathbf{j}$$

We shall also require a vector from D to C:

$$\mathbf{R}_C = -(60 + 90)\mathbf{j} = -150\mathbf{j}$$

Then, summing moments about D gives

$$\mathbf{R}_G \times \mathbf{W} + \mathbf{R}_C \times \mathbf{F}_C + \mathbf{T} = 0 \qquad (1)$$

When we place the details in Eq. (1), we get

$$(90\mathbf{i} - 92\mathbf{j}) \times (-214\mathbf{i} - 644\mathbf{j} + 1865\mathbf{k})$$
$$+ (-150\mathbf{j}) \times (F_C^x\mathbf{i} + F_C^y\mathbf{j} + F_C^z\mathbf{k}) + T\mathbf{j} = 0 \qquad (2)$$

After the two cross products are taken, the equation becomes

$$(-171\,580\mathbf{i} - 167\,850\mathbf{j} - 77\,712\mathbf{k}) + (-150F_C^z\mathbf{i} + 150F_C^x\mathbf{k}) + T\mathbf{j} = 0$$

from which

$$T = 168\mathbf{j} \text{ N} \cdot \text{m} \qquad F_C^x = 518 \text{ N} \qquad F_C^z = -1144 \text{ N} \qquad (3)$$

Now sum the forces to zero. Thus

$$\mathbf{F}_D + \mathbf{F}_C + \mathbf{W} = 0 \qquad (4)$$

When the details are inserted, Eq. (4) becomes

$$(F_D^x\mathbf{i} + F_D^z\mathbf{k}) + (518\mathbf{i} + F_C^y\mathbf{j} - 1144\mathbf{k}) + (-214\mathbf{i} - 644\mathbf{j} + 1865\mathbf{k}) = 0 \qquad (5)$$

First we see that $F_C^y = 644$ N, and so

Answer
$$F_C = 518i + 644j - 1144k \text{ N}$$

Then, from Eq. (5),

Answer
$$F_D = -303i - 721k \text{ N}$$

These are all shown in Fig. 13–36b in the proper directions. The analysis for the pinion shaft is quite similar.

13–16   Force Analysis—Helical Gearing

Figure 13–37 is a three-dimensional view of the forces acting against a helical-gear tooth. The point of application of the forces is in the pitch plane and in the center of the gear face. From the geometry of the figure, the three components of the total (normal) tooth force W are

$$W_r = W \sin \phi_n$$
$$W_t = W \cos \phi_n \cos \psi \qquad (13\text{–}39)$$
$$W_a = W \cos \phi_n \sin \psi$$

where W = total force
W_r = radial component
W_t = tangential component, also called the transmitted load
W_a = axial component, also called the thrust load

Figure 13–37

Tooth forces acting on a right-hand helical gear.

Usually W_t is given and the other forces are desired. In this case, it is not difficult to discover that

$$W_r = W_t \tan \phi_t$$

$$W_a = W_t \tan \psi$$

$$W = \frac{W_t}{\cos \phi_n \cos \psi}$$

(13–40)

EXAMPLE 13–9   In Fig. 13–38 a 750-W an electric motor runs at 1800 rev/min in the clockwise direction, as viewed from the positive x axis. Keyed to the motor shaft is an 18-tooth helical pinion having a normal pressure angle of 20°, a helix angle of 30°, and a normal module of 3.0 mm. The hand of the helix is shown in the figure. Make a three-dimensional sketch of the motor shaft and pinion, and show the forces acting on the pinion and the bearing reactions at A and B. The thrust should be taken out at A.

Solution   From Eq. (13–19) we find

$$\phi_t = \tan^{-1} \frac{\tan \phi_n}{\cos \psi} = \tan^{-1} \frac{\tan 20°}{\cos 30°} = 22.8°$$

Also, $m_t = m_n/\cos \psi = 3/\cos 30° = 3.46$ mm. Therefore the pitch diameter of the pinion is $d_p = 18(3.46) = 62.3$ mm. The pitch-line velocity is

$$V = \pi dn = \frac{\pi(62.3)(1800)}{12} = 5871.6 \text{ mm/s} = 5.87 \text{ m/s}$$

The transmitted load is

$$W_t = \frac{H}{V} = \frac{750}{5.87} = 128 \text{ N}$$

From Eq. (13–40) we find

$$W_r = W_t \tan \phi_t = (128) \tan 22.8° = 54 \text{ N}$$

$$W_a = W_t \tan \psi = (128) \tan 30° = 74 \text{ N}$$

$$W = \frac{W_t}{\cos \phi_n \cos \psi} = \frac{128}{\cos 20° \cos 30°} = 157 \text{ N}$$

Figure 13–38

The motor and gear train of Ex. 13–9.

Figure 13–39

Free-body diagram of motor shaft of Ex. 13–9.

These three forces, W_r = in the $-y$ direction, W_a = in the $-x$ direction, and W_t in the $+z$ direction, are shown acting at point C in Fig. 13–39. We assume bearing reactions at A and B as shown. Then $F_A^x = W_a = 74$ N. Taking moments about the z axis,

$$-(54)(325) + (74)\left(\frac{62.3}{2}\right) + 250F_B^y = 0$$

or $F_B^y = 61$ N. Summing forces in the y direction then gives $F_A^y = 7$ N. Taking moments about the y axis, next

$$250F_B^z - (128)(325) = 0$$

or $F_B^z = 166$ N. Summing forces in the z direction and solving gives $F_A^z = 38$ N. Also, the torque is $T = W_t d_p/2 = 128(62.3/2) = 3982$ N · mm.

For comparison, solve the problem again using vectors. The force at C is

$$W = -74i - 54j + 128k \text{ N}$$

Position vectors to B and C from origin A are

$$R_B = 250i \qquad R_C = 325i + 31.15j$$

Taking moments about A, we have

$$R_B \times F_B + T + R_C \times W = 0$$

Using the directions assumed in Fig. 13–39 and substituting values gives

$$250i \times (F_B^y j - F_B^z k) - Ti + (325i + 31.15j) \times (-74i - 54j + 128k) = 0$$

When the cross products are evaluated we get

$$(250F_B^y k + 250F_B^z j) - Ti + (3987i - 41\,600j - 15\,245k) = 0$$

obtaining $T = 4$ kN · mm, $F_B^y = 61$ N, and $F_B^z = 166$ N.
Next,

$$F_A = -F_B - W, \text{ and so } F_A = 74i - 7j + 38k \text{ N}.$$

13–17 Force Analysis—Worm Gearing

If friction is neglected, then the only force exerted by the gear will be the force W, shown in Fig. 13–40, having the three orthogonal components W^x, W^y, and W^z. From the geometry of the figure, we see that

$$W^x = W \cos \phi_n \sin \lambda$$
$$W^y = W \sin \phi_n \tag{13-41}$$
$$W^z = W \cos \phi_n \cos \lambda$$

We now use the subscripts W and G to indicate forces acting against the worm and gear, respectively. We note that W^y is the separating, or radial, force for both the worm and the gear. The tangential force on the worm is W^x and is W^z on the gear, assuming a 90° shaft angle. The axial force on the worm is W^z, and on the gear, W^x. Since the gear forces are opposite to the worm forces, we can summarize these relations by writing

$$W_{Wt} = -W_{Ga} = W^x$$
$$W_{Wr} = -W_{Gr} = W^y \tag{13-42}$$
$$W_{Wa} = -W_{Gt} = W^z$$

It is helpful in using Eq. (13–41) and also Eq. (13–42) to observe that *the gear axis is parallel to the x direction and the worm axis is parallel to the z direction* and that we are employing a right-handed coordinate system.

In our study of spur-gear teeth we learned that the motion of one tooth relative to the mating tooth is primarily a rolling motion; in fact, when contact occurs at the pitch point, the motion is pure rolling. In contrast, the relative motion between worm and worm-gear teeth is pure sliding, and so we must expect that friction plays an important role in the performance of worm gearing. By introducing a coefficient of friction f, we can develop another set of relations similar to those of Eq. (13–41). In Fig. 13–40 we see that the force W acting normal to the worm-tooth profile produces a frictional force $W_f = fW$, having a component $fW \cos \lambda$ in the negative x direction and another component $fW \sin \lambda$ in the positive z direction. Equation (13–41) therefore becomes

$$W^x = W(\cos \phi_n \sin \lambda + f \cos \lambda)$$
$$W^y = W \sin \phi_n \tag{13-43}$$
$$W^z = W(\cos \phi_n \cos \lambda - f \sin \lambda)$$

Figure 13–40

Drawing of the pitch cylinder of a worm, showing the forces exerted upon it by the worm gear.

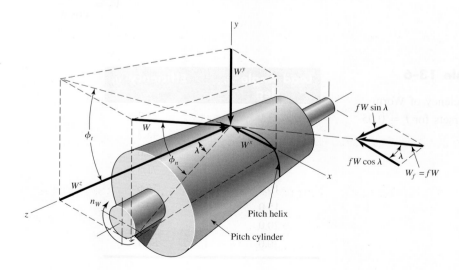

Equation (13–42), of course, still applies.

Inserting $-W_{Gt}$ from Eq. (13–42) for W^z in Eq. (13–43) and multiplying both sides by f, we find the frictional force W_f to be

$$W_f = fW = \frac{fW_{Gt}}{f \sin \lambda - \cos \phi_n \cos \lambda} \tag{13–44}$$

A useful relation between the two tangential forces, W_{Wt} and W_{Gt}, can be obtained by equating the first and third parts of Eqs. (13–42) and (13–43) and eliminating W. The result is

$$W_{Wt} = W_{Gt} \frac{\cos \phi_n \sin \lambda + f \cos \lambda}{f \sin \lambda - \cos \phi_n \cos \lambda} \tag{13–45}$$

Efficiency η can be defined by using the equation

$$\eta = \frac{W_{Wt} \text{ (without friction)}}{W_{Wt} \text{ (with friction)}} \tag{a}$$

Substitute Eq. (13–45) with $f = 0$ in the numerator of Eq. (*a*) and the same equation in the denominator. After some rearranging, you will find the efficiency to be

$$\eta = \frac{\cos \phi_n - f \tan \lambda}{\cos \phi_n + f \cot \lambda} \tag{13–46}$$

Selecting a typical value of the coefficient of friction, say $f = 0.05$, and the pressure angles shown in Table 13–5, we can use Eq. (13–46) to get some useful design information. Solving this equation for lead angles from 1 to 30° gives the interesting results shown in Table 13–6.

Many experiments have shown that the coefficient of friction is dependent on the relative or sliding velocity. In Fig. 13–41, V_G is the pitch-line velocity of the gear and V_W the pitch-line velocity of the worm. Vectorially, $\mathbf{V}_W = \mathbf{V}_G + \mathbf{V}_S$; consequently, the sliding velocity is

$$V_S = \frac{V_W}{\cos \lambda} \tag{13–47}$$

Table 13–6	Lead Angle λ, deg	Efficiency η, %
Efficiency of Worm Gearsets for $f = 0.05$	1.0	25.2
	2.5	45.7
	5.0	62.6
	7.5	71.3
	10.0	76.6
	15.0	82.7
	20.0	85.6
	30.0	88.7

Figure 13–41

Velocity components in worm gearing.

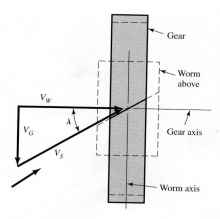

Figure 13–42

Representative values of the coefficient of friction for worm gearing. These values are based on good lubrication. Use curve *B* for high-quality materials, such as a case-hardened steel worm mating with a phosphor-bronze gear. Use curve *A* when more friction is expected, as with a cast-iron worm mating with a cast-iron worm gear.

Published values of the coefficient of friction vary as much as 20 percent, undoubtedly because of the differences in surface finish, materials, and lubrication. The values on the chart of Fig. 13–42 are representative and indicate the general trend.

EXAMPLE 13–10

A 2-tooth right-hand worm transmits 1 hp at 1200 rev/min to a 30-tooth worm gear. The gear has a transverse diametral pitch of 6 teeth/in and a face width of 1 in. The worm has a pitch diameter of 2 in and a face width of $2\frac{1}{2}$ in. The normal pressure angle is $14\frac{1}{2}°$. The materials and quality of the gearing to be used are such that curve *B* of Fig. 13–42 should be used to obtain the coefficient of friction.

(*a*) Find the axial pitch, the center distance, the lead, and the lead angle.

(*b*) Figure 13–43 is a drawing of the worm gear oriented with respect to the coordinate system described earlier in this section; the gear is supported by bearings *A* and *B*. Find the forces exerted by the bearings against the worm-gear shaft, and the output torque.

Figure 13-43

The pitch cylinders of
the worm gear train of
Ex. 13–10.

Worm pitch cylinder

1200 rev/min

Gear pitch cylinder

A

B

$1\frac{1}{2}$ in

$2\frac{1}{2}$ in

Solution   (*a*) The axial pitch is the same as the transverse circular pitch of the gear, which is

Answer
$$p_x = p_t = \frac{\pi}{P} = \frac{\pi}{6} = 0.5236 \text{ in}$$

The pitch diameter of the gear is $d_G = N_G/P = 30/6 = 5$ in. Therefore, the center distance is

Answer
$$C = \frac{d_W + d_G}{2} = \frac{2 + 5}{2} = 3.5 \text{ in}$$

From Eq. (13–27), the lead is

$$L = p_x N_W = (0.5236)(2) = 1.0472 \text{ in}$$

Answer   Also using Eq. (13–28), we find

Answer
$$\lambda = \tan^{-1}\frac{L}{\pi d_W} = \tan^{-1}\frac{1.0472}{\pi(2)} = 9.46°$$

(*b*) Using the right-hand rule for the rotation of the worm, you will see that your thumb points in the positive z direction. Now use the bolt-and-nut analogy (the worm is right-handed, as is the screw thread of a bolt), and turn the bolt clockwise with the right hand while preventing nut rotation with the left. The nut will move axially along the bolt toward your right hand. Therefore the surface of the gear (Fig. 13–43) in contact with the worm will move in the negative z direction. Thus, viewing the gear in the negative x direction, the gear rotates clockwise about the x axis
    The pitch-line velocity of the worm is

$$V_W = \frac{\pi d_W n_W}{12} = \frac{\pi(2)(1200)}{12} = 628 \text{ ft/min}$$

The speed of the gear is $n_G = (\frac{2}{30})(1200) = 80$ rev/min. Therefore the pitch-line velocity of the gear is

$$V_G = \frac{\pi d_G n_G}{12} = \frac{\pi(5)(80)}{12} = 105 \text{ ft/min}$$

Then, from Eq. (13–47), the sliding velocity V_S is found to be

$$V_S = \frac{V_W}{\cos \lambda} = \frac{628}{\cos 9.46°} = 637 \text{ ft/min}$$

Getting to the forces now, we begin with the horsepower formula

$$W_{Wt} = \frac{33\,000H}{V_W} = \frac{(33\,000)(1)}{628} = 52.5 \text{ lbf}$$

This force acts in the negative x direction, the same as in Fig. 13–40. Using Fig. 13–42, we find $f = 0.03$. Then, the first equation of Eq. (13–43) gives

$$W = \frac{W^x}{\cos \phi_n \sin \lambda + f \cos \lambda}$$

$$= \frac{52.5}{\cos 14.5° \sin 9.46° + 0.03 \cos 9.46°} = 278 \text{ lbf}$$

Also, from Eq. (13–43),

$$W^y = W \sin \phi_n = 278 \sin 14.5° = 69.6 \text{ lbf}$$
$$W^z = W(\cos \phi_n \cos \lambda - f \sin \lambda)$$
$$= 278(\cos 14.5° \cos 9.46° - 0.03 \sin 9.46°) = 264 \text{ lbf}$$

We now identify the components acting on the gear as

$$W_{Ga} = -W^x = 52.5 \text{ lbf}$$
$$W_{Gr} = -W^y = -69.6 \text{ lbf}$$
$$W_{Gt} = -W^z = -264 \text{ lbf}$$

A free-body diagram showing the forces and torsion acting on the gearshaft is shown in Fig. 13–44.

We shall make B a thrust bearing in order to place the gearshaft in compression. Thus, summing forces in the x direction gives

Answer
$$F_B^x = -52.5 \text{ lbf}$$

Taking moments about the z axis, we have

Answer
$$-(52.5)(2.5) - (69.6)(1.5) + 4F_B^y = 0 \qquad F_B^y = 58.9 \text{ lbf}$$

Taking moments about the y axis,

Answer
$$(264)(1.5) - 4F_B^z = 0 \qquad F_B^z = 99 \text{ lbf}$$

Figure 13–44

Free-body diagram for
Ex. 13–10. Forces are given
in lbf.

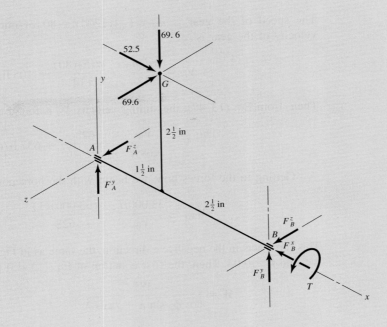

Summing forces in the y direction,

Answer
$$-69.6 + 58.9 + F_A^y = 0 \qquad F_A^y = 10.7 \text{ lbf}$$

Similarly, summing forces in the z direction,

Answer
$$-264 + 99 + F_A^z = 0 \qquad F_A^z = 165 \text{ lbf}$$

We still have one more equation to write. Summing moments about x,

Answer
$$-(264)(2.5) + T = 0 \qquad T = 660 \text{ lbf} \cdot \text{in}$$

It is because of the frictional loss that this output torque is less than the product of the gear ratio and the input torque.

PROBLEMS

Problems marked with an asterisk (*) are linked to problems in other chapters, as summarized in Table 1–2 of Sec. 1–17, p. 34.

13–1   A 17-tooth spur pinion has a diametral pitch of 8 teeth/in, runs at 1120 rev/min, and drives a gear at a speed of 544 rev/min. Find the number of teeth on the gear and the theoretical center-to-center distance.

13–2   A 15-tooth spur pinion has a module of 3 mm and runs at a speed of 1600 rev/min. The driven gear has 60 teeth. Find the speed of the driven gear, the circular pitch, and the theoretical center-to-center distance.

13-3   A spur gearset has a module of 6 mm and a velocity ratio of 4. The pinion has 16 teeth. Find the number of teeth on the driven gear, the pitch diameters, and the theoretical center-to-center distance.

13-4   A 21-tooth spur pinion mates with a 28-tooth gear. The diametral pitch is 3 teeth/in and the pressure angle is 20°. Make a drawing of the gears showing one tooth on each gear. Find and tabulate the following results: the addendum, dedendum, clearance, circular pitch, tooth thickness, and base-circle diameters; the lengths of the arc of approach, recess, and action; and the base pitch and contact ratio.

13-5   A 20° straight-tooth bevel pinion having 14 teeth and a diametral pitch of 6 teeth/in drives a 32-tooth gear. The two shafts are at right angles and in the same plane. Find:
(a) The cone distance
(b) The pitch angles
(c) The pitch diameters
(d) The face width

13-6   A gearset consists of a 16-tooth pinion driving a 40-tooth gear. The diametral pitch is 2, and the addendum and dedendum are 1/P and 1.25/P, respectively. The gears are cut using a pressure angle of 20°.
(a) Compute the circular pitch, the center distance, and the radii of the base circles.
(b) In mounting these gears, the center distance was incorrectly made $\frac{1}{4}$ in larger. Compute the new values of the pressure angle and the pitch-circle diameters.

13-7   A parallel helical gearset consists of a 19-tooth pinion driving a 57-tooth gear. The pinion has a left-hand helix angle of 30°, a normal pressure angle of 20°, and a normal module of 2.5 mm. Find:
(a) The normal, transverse, and axial circular pitches
(b) The transverse diametral pitch and the transverse pressure angle
(c) The addendum, dedendum, and pitch diameter of each gear

13-8   To avoid the problem of interference in a pair of spur gears using a 20° pressure angle, specify the minimum number of teeth allowed on the pinion for each of the following gear ratios.
(a) 2 to 1
(b) 3 to 1
(c) 4 to 1
(d) 5 to 1

13-9   Repeat Prob. 13–8 with a 25° pressure angle.

13-10   For a spur gearset with $\phi = 20°$, while avoiding interference, find:
(a) The smallest pinion tooth count that will run with itself
(b) The smallest pinion tooth count at a ratio $m_G = 2.5$, and the largest gear tooth count possible with this pinion
(c) The smallest pinion that will run with a rack

13-11   Repeat problem 13–10 for a helical gearset with $\phi_n = 20°$ and $\psi = 30°$.

13-12   The decision has been made to use $\phi_n = 20°$, $m_t = 3$ mm, and $\psi = 30°$ for a 2:1 reduction. Choose the smallest acceptable full-depth pinion and gear tooth count to avoid interference.

13-13   Repeat Problem 13–12 with $\psi = 45°$.

13-14   By employing a pressure angle larger than standard, it is possible to use fewer pinion teeth, and hence obtain smaller gears without undercutting during machining. If the gears are full-depth spur gears, what is the smallest possible pressure angle ϕ that can be obtained without undercutting for a 9-tooth pinion to mesh with a rack?

13–15 A parallel-shaft gearset consists of an 18-tooth helical pinion driving a 32-tooth gear. The pinion has a left-hand helix angle of 25°, a normal pressure angle of 20°, and a normal module of 3 mm. Find:

(a) The normal, transverse, and axial circular pitches
(b) The transverse module and the transverse pressure angle
(c) The pitch diameters of the two gears

13–16 The double-reduction helical gearset shown in the figure is driven through shaft a at a speed of 900 rev/min. Gears 2 and 3 have a normal diametral pitch of 3 mm, a 30° helix angle, and a normal pressure angle of 20°. The second pair of gears in the train, gears 4 and 5, have a normal diametral pitch of 2 mm, a 25° helix angle, and a normal pressure angle of 20°. The tooth numbers are: $N_2 = 14$, $N_3 = 54$, $N_4 = 16$, $N_5 = 36$. Find:

(a) The directions of the thrust force exerted by each gear upon its shaft
(b) The speed and direction of shaft c
(c) The center distance between shafts

Problem 13–16
Dimensions in mm.

13–17 Shaft a in the figure rotates at 600 rev/min in the direction shown. Find the speed and direction of rotation of shaft d.

Problem 13–17

13–18 The mechanism train shown consists of an assortment of gears and pulleys to drive gear 9. Pulley 2 rotates at 1200 rev/min in the direction shown. Determine the speed and direction of rotation of gear 9.

Problem 13–18

13–19 The figure shows a gear train consisting of a pair of helical gears and a pair of miter gears. The helical gears have a $17\frac{1}{2}°$ normal pressure angle and a helix angle as shown. Find:
   (a) The speed of shaft c
   (b) The distance between shafts a and b
   (c) The pitch diameter of the miter gears

Problem 13–19
Dimensions in mm.

13–20 A compound reverted gear train is to be designed as a speed increaser to provide a total increase of speed of exactly 45 to 1. With a 20° pressure angle, specify appropriate numbers of teeth to minimize the gearbox size while avoiding the interference problem in the teeth. Assume all gears will have the same diametral pitch.

13–21 Repeat Prob. 13–20 with a 25° pressure angle.

13–22 Repeat Prob. 13–20 for a gear ratio of exactly 30 to 1.

13–23 Repeat Prob. 13–20 for a gear ratio of *approximately* 45 to 1.

13–24 A gearbox is to be designed with a compound reverted gear train that transmits 25 horsepower with an input speed of 2500 rev/min. The output should deliver the power at a rotational speed in the range of 280 to 300 rev/min. Spur gears with 20° pressure angle are to be used. Determine suitable numbers of teeth for each gear, to minimize the gearbox size while providing an output speed within the specified range. Be sure to avoid an interference problem in the teeth.

13–25 The tooth numbers for the automotive differential shown in the figure are $N_2 = 16$, $N_3 = 48$, $N_4 = 14$, $N_5 = N_6 = 20$. The drive shaft turns at 900 rev/min.

(a) What are the wheel speeds if the car is traveling in a straight line on a good road surface?

(b) Suppose the right wheel is jacked up and the left wheel resting on a good road surface. What is the speed of the right wheel?

(c) Suppose, with a rear-wheel drive vehicle, the auto is parked with the right wheel resting on a wet icy surface. Does the answer to part (b) give you any hint as to what would happen if you started the car and attempted to drive on?

Problem 13–25

13–26 The figure illustrates an all-wheel drive concept using three differentials, one for the front axle, another for the rear, and the third connected to the drive shaft.

(a) Explain why this concept may allow greater acceleration.

(b) Suppose either the center or the rear differential, or both, can be locked for certain road conditions. Would either or both of these actions provide greater traction? Why?

Problem 13–26

The Audi "Quattro concept,"
showing the three differentials that
provide permanent all-wheel drive.
*(Reprinted by permission of
Audi of America, Inc.)*

13–27 In the reverted planetary train illustrated, find the speed and direction of rotation of the arm if gear 2 is unable to rotate and gear 6 is driven at 12 rev/min in the clockwise direction as viewed from the bottom of the figure.

Problem 13–27

13–28 In the gear train of Prob. 13–27, let gear 6 be driven at 85 rev/min counterclockwise (as viewed from the bottom of the figure) while gear 2 is held stationary. What is the speed and direction of rotation of the arm?

13–29 Tooth numbers for the gear train shown in the figure are $N_2 = 12$, $N_3 = 16$, and $N_4 = 12$. How many teeth must internal gear 5 have? Suppose gear 5 is fixed. What is the speed of the arm if shaft a rotates at 320 rev/min counterclockwise as viewed from the left side of the figure?

Problem 13–29

13–30 The tooth numbers for the gear train illustrated are $N_2 = 20$, $N_3 = 16$, $N_4 = 30$, $N_6 = 36$, and $N_7 = 46$. Gear 7 is fixed. If shaft a is turned through 10 revolutions, how many turns will shaft b make?

Problem 13–30

13–31   Shaft a in the figure has a power input of 75 kW at a speed of 1000 rev/min in the counterclockwise direction. The gears have a module of 5 mm and a 20° pressure angle. Gear 3 is an idler.
(a) Find the force F_{3b} that gear 3 exerts against shaft b.
(b) Find the torque T_{4c} that gear 4 exerts on shaft c.

Problem 13–31

13–32   The 24T 6-pitch 20° pinion 2 shown in the figure rotates clockwise at 1000 rev/min and is
driven at a power of 25 hp. Gears 4, 5, and 6 have 24, 36, and 144 teeth, respectively. What
torque can arm 3 deliver to its output shaft? Draw free-body diagrams of the arm and of each
gear and show all forces that act upon them.

Problem 13–32

13–33   The gears shown in the figure have a module of 12 mm and a 20° pressure angle. The pinion
rotates at 1800 rev/min clockwise and transmits 150 kW through the idler pair to gear 5 on
shaft c. What forces do gears 3 and 4 transmit to the idler shaft?

Gears—General | **719**

Problem 13–33

13–34 The figure shows a pair of shaft-mounted spur gears having a diametral pitch of 5 teeth/in with an 18-tooth 20° pinion driving a 45-tooth gear. The power input is 32 maximum at 1800 rev/min. Find the direction and magnitude of the forces acting on bearings *A*, *B*, *C*, and *D*.

Problem 13–34

13–35 The figure shows the electric-motor frame dimensions for a 30-hp 900 rev/min motor. The frame is bolted to its support using four $\frac{3}{4}$-in bolts spaced $11\frac{1}{4}$ in apart in the view shown and 14 in apart when viewed from the end of the motor. A 4 diametral pitch 20° spur pinion having 20 teeth and a face width of 2 in is keyed to and flush with the end of the motor shaft. This pinion drives another gear whose axis is in the same *xz* plane and directly behind the motor shaft. Determine the maximum shear and tensile forces on the mounting bolts based on 200 percent overload torque. Does the direction of rotation matter?

Problem 13–35
NEMA No. 364 frame; dimensions in inches. The *z* axis is directed out of the paper.

13-36 Continue Prob. 13–24 by finding the following information, assuming a diametral pitch of 6 teeth/in.

(a) Determine pitch diameters for each of the gears.

(b) Determine the pitch line velocities (in ft/min) for each set of gears.

(c) Determine the magnitudes of the tangential, radial, and total forces transmitted between each set of gears.

(d) Determine the input torque.

(e) Determine the output torque, neglecting frictional losses.

13-37 A speed-reducer gearbox containing a compound reverted gear train transmits 35 horsepower with an input speed of 1200 rev/min. Spur gears with 20° pressure angle are used, with 16 teeth on each of the small gears and 48 teeth on each of the larger gears. A diametral pitch of 10 teeth/in is proposed.

(a) Determine the speeds of the intermediate and output shafts.

(b) Determine the pitch line velocities (in ft/min) for each set of gears.

(c) Determine the magnitudes of the tangential, radial, and total forces transmitted between each set of gears.

(d) Determine the input torque.

(e) Determine the output torque, neglecting frictional losses.

13-38* For the countershaft in Prob. 3–72, p. 152, assume the gear ratio from gear B to its mating gear is 2 to 1.

(a) Determine the minimum number of teeth that can be used on gear B without an interference problem in the teeth.

(b) Using the number of teeth from part (a), what diametral pitch is required to also achieve the given 200-mm pitch diameter?

(c) Suppose the 20° pressure angle gears are exchanged for gears with 25° pressure angle, while maintaining the same pitch diameters and diametral pitch. Determine the new forces F_A and F_B if the same power is to be transmitted.

13-39* For the countershaft in Prob. 3–73, p. 152, assume the gear ratio from gear B to its mating gear is 5 to 1.

(a) Determine the minimum number of teeth that can be used on gear B without an interference problem in the teeth.

(b) Using the number of teeth from part (a), what module is required to also achieve the given 300-mm pitch diameter?

(c) Suppose the 20° pressure angle for gear A is exchanged for a gear with 25° pressure angle, while maintaining the same pitch diameters and module. Determine the new forces F_A and F_B if the same power is to be transmitted.

13-40* For the gear and sprocket assembly analyzed in Prob. 3–77, p. 153, information for the gear sizes and the forces transmitted through the gears was provided in the problem statement. In this problem, we will perform the preceding design steps necessary to acquire the information for the analysis. A motor providing 2 kW is to operate at 191 rev/min. A gear unit is needed to reduce the motor speed by half to drive a chain sprocket.

(a) Specify appropriate numbers of teeth on gears F and C to minimize the size while avoiding the interference problem in the teeth.

(b) Assuming an initial guess of 125-mm pitch diameter for gear F, what is the module that should be used for the stress analysis of the gear teeth?

(c) Calculate the input torque applied to shaft *EFG*.

(d) Calculate the magnitudes of the radial, tangential, and total forces transmitted between gears *F* and *C*.

13–41* For the gear and sprocket assembly analyzed in Prob. 3–79, p. 153, information for the gear sizes and the forces transmitted through the gears was provided in the problem statement. In this problem, we will perform the preceding design steps necessary to acquire the information for the analysis. A motor providing 0.75 kW is to operate at 70 rev/min. A gear unit is needed to double the motor speed to drive a chain sprocket.

(a) Specify appropriate numbers of teeth on gears *F* and *C* to minimize the size while avoiding the interference problem in the teeth.

(b) Assuming an initial guess of 250-mm pitch diameter for gear *F*, what is the diametral pitch that should be used for the stress analysis of the gear teeth?

(c) Calculate the input torque applied to shaft *EFG*.

(d) Calculate the magnitudes of the radial, tangential, and total forces transmitted between gears *F* and *C*.

13–42* For the bevel gearset in Probs. 3–74 and 3–76, pp. 152 and 153 respectively, shaft *AB* is rotating at 600 rev/min and transmits 7.5 kW. The gears have a 20° pressure angle.

(a) Determine the bevel angle γ for the gear on shaft *AB*.

(b) Determine the pitch-line velocity.

(c) Determine the tangential, radial, and axial forces acting on the pinion. Were the forces given in Prob. 3–74 correct?

13–43 The figure shows a 16*T* 20° straight bevel pinion driving a 32*T* gear, and the location of the bearing centerlines. Pinion shaft *a* receives 2.5 hp at 240 rev/min. Determine the bearing reactions at *A* and *B* if *A* is to take both radial and thrust loads.

Problem 13–43
Dimensions in inches.

13–44 The figure shows a 10 diametral pitch 18-tooth 20° straight bevel pinion driving a 30-tooth gear. The transmitted load is 25 lbf. Find the bearing reactions at *C* and *D* on the output shaft if *D* is to take both radial and thrust loads.

Problem 13–44
Dimensions in inches.

13–45   The gears shown in the figure have a normal diametral pitch of 3 mm, a normal pressure angle of 20°, and a 30° helix angle. The transmitted load is 3.5 kN. The pinion rotates counterclockwise about the *y* axis, as viewed from the positive *y* axis. Find the force exerted by each gear on its shaft.

Problem 13–45

13–46   The gears shown in the figure have a normal diametral pitch of 3 mm, a normal pressure angle of 20°, and a 30° helix angle. The transmitted load is 3.5 kN. Gear 2 rotates clockwise about the *y* axis, as viewed from the positive *y* axis. Gear 3 is an idler. Find the forces exerted by gears 2 and 3 on their shafts.

Problem 13–46

13–47   A gear train is composed of four helical gears with the three shaft axes in a single plane, as shown in the figure. The gears have a normal pressure angle of 20° and a 30° helix angle. Gear 2 is the driver, and is rotating counterclockwise as viewed from the top. Shaft *b* is an idler and the transmitted load from gear 2 to gear 3 is 500 lbf. The gears on shaft *b* both have a normal diametral pitch of 7 teeth/in and have 54 and 14 teeth, respectively. Find the forces exerted by gears 3 and 4 on shaft *b*.

Problem 13–47

13–48 In the figure for Prob. 13–34, pinion 2 is to be a right-hand helical gear having a helix angle of 30°, a normal pressure angle of 20°, 16 teeth, and a normal diametral pitch of 6 teeth/in. A motor delivers 25-hp to shaft *a* at a speed of 1720 rev/min clockwise about the *x* axis. Gear 3 has 42 teeth. Find the reaction exerted by bearings *C* and *D* on shaft *b*. One of these bearings is to take both radial and thrust loads. This bearing should be selected so as to place the shaft in compression.

13–49 Gear 2, in the figure, has 16 teeth, a 20° transverse pressure angle, a 15° helix angle, and a module of 4 mm. Gear 2 drives the idler on shaft *b*, which has 36 teeth. The driven gear on shaft *c* has 28 teeth. If the driver rotates at 1600 rev/min and transmits 6 kW, find the radial and thrust load on each shaft.

Problem 13–49

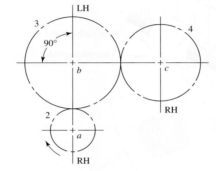

13–50 The figure shows a double-reduction helical gearset. Pinion 2 is the driver, and it receives a torque of 1200 lbf · in from its shaft in the direction shown. Pinion 2 has a normal diametral pitch of 8 teeth/in, 14 teeth, and a normal pressure angle of 20° and is cut right-handed with a helix angle of 30°. The mating gear 3 on shaft *b* has 36 teeth. Gear 4, which is the driver for the second pair of gears in the train, has a normal diametral pitch of 5 teeth/in, 15 teeth, and a normal pressure angle of 20° and is cut left-handed with a helix angle of 15°. Mating gear 5 has 45 teeth. Find the magnitude and direction of the force exerted by the bearings *C* and *D* on shaft *b* if bearing *C* can take only a radial load while bearing *D* is mounted to take both radial and thrust loads.

Problem 13–50
Dimensions in inches.

13–51 A right-hand single-tooth hardened-steel (hardness not specified) worm has a catalog rating of 2000 W at 600 rev/min when meshed with a 48-tooth cast-iron gear. The axial pitch of the worm is 25 mm, the normal pressure angle is $14\frac{1}{2}°$, the pitch diameter of the worm is 100 mm, and the face widths of the worm and gear are, respectively, 100 mm and 50 mm. Bearings are centered at locations A and B on the worm shaft. Determine which should be the thrust bearing (so that the axial load in the shaft is in compression), and find the magnitudes and directions of the forces exerted by both bearings.

Problem 13–51
Dimensions in millimeters.

13–52 The hub diameter and projection for the gear of Prob. 13–51 are 100 and 37.5 mm, respectively. The face width of the gear is 50 mm. Locate bearings C and D on opposite sides, spacing C 10 mm from the gear on the hidden face (see figure) and D 10 mm from the hub face. Choose one as the thrust bearing, so that the axial load in the shaft is in compression. Find the output torque and the magnitudes and directions of the forces exerted by the bearings on the gearshaft.

13–53 A 2-tooth left-hand worm transmits $\frac{3}{4}$ hp at 600 rev/min to a 36-tooth gear having a transverse diametral pitch of 8 teeth/in. The worm has a normal pressure angle of 20°, a pitch diameter of $1\frac{1}{2}$ in, and a face width of $1\frac{1}{2}$ in. Use a coefficient of friction of 0.05 and find the force exerted by the gear on the worm and the torque input. For the same geometry as shown for Prob. 13–51, the worm velocity is clockwise as viewed from the positive z axis.

13–54 Write a computer program that will analyze a spur gear or helical-mesh gear, accepting ϕ_n, ψ, P_t, N_P, and N_G; compute m_G, d_P, d_G, p_t, p_n, p_x, and ϕ_t; and give advice as to the smallest tooth count that will allow a pinion to run with itself without interference, run with its gear, and run with a rack. Also have it give the largest tooth count possible with the intended pinion.

14

Spur and Helical Gears

This chapter is devoted primarily to analysis and design of spur and helical gears to resist bending failure of the teeth as well as pitting failure of tooth surfaces. Failure by bending will occur when the significant tooth stress equals or exceeds either the yield strength or the bending endurance strength. A surface failure occurs when the significant contact stress equals or exceeds the surface endurance strength. The first two sections present a little of the history of the analyses from which current methodology developed.

The American Gear Manufacturers Association[1] (AGMA) has for many years been the responsible authority for the dissemination of knowledge pertaining to the design and analysis of gearing. The methods this organization presents are in general use in the United States when strength and wear are primary considerations. In view of this fact it is important that the AGMA approach to the subject be presented here.

The general AGMA approach requires a great many charts and graphs—too many for a single chapter in this book. We have omitted many of these here by choosing a single pressure angle and by using only full-depth teeth. This simplification reduces the complexity but does not prevent the development of a basic understanding of the approach. Furthermore, the simplification makes possible a better development of the fundamentals and hence should constitute an ideal introduction to the use of the general AGMA method.[2] Sections 14–1 and 14–2 are elementary and serve as an examination of the foundations of the AGMA method. Table 14–1 is largely AGMA nomenclature.

14–1 The Lewis Bending Equation

Wilfred Lewis introduced an equation for estimating the bending stress in gear teeth in which the tooth form entered into the formulation. The equation, announced in 1892, still remains the basis for most gear design today.

To derive the basic Lewis equation, refer to Fig. 14–1a, which shows a rectangular cantilever beam of cross-sectional dimensions F and t, having a length l and a load W^t, uniformly distributed across the face width F. The section modulus I/c is $Ft^2/6$, and therefore the bending stress is

$$\sigma = \frac{M}{I/c} = \frac{6W^t l}{Ft^2} \tag{a}$$

Gear designers denote the components of gear-tooth forces as W_t, W_r, W_a or W^t, W^r, W^a interchangeably. The latter notation leaves room for post-subscripts essential to free-body diagrams. For instance, for gears 2 and 3 in mesh, W^t_{23} is the transmitted

[2]The standards ANSI/AGMA 2001-D04 (revised AGMA 2001-C95) and ANSI/AGMA 2101-D04 (metric edition of ANSI/AGMA 2001-D04), *Fundamental Rating Factors and Calculation Methods for Involute Spur and Helical Gear Teeth,* are used in this chapter. The use of American National Standards is completely voluntary; their existence does not in any respect preclude people, whether they have approved the standards or not, from manufacturing, marketing, purchasing, or using products, processes, or procedures not conforming to the standards.

The American National Standards Institute does not develop standards and will in no circumstances give an interpretation of any American National Standard. Requests for interpretation of these standards should be addressed to the American Gear Manufacturers Association. [Tables or other self-supporting sections may be quoted or extracted in their entirety. Credit line should read: "Extracted from ANSI/AGMA Standard 2001-D04 or 2101-D04 *Fundamental Rating Factors and Calculation Methods for Involute Spur and Helical Gear Teeth*" with the permission of the publisher, American Gear Manufacturers Association, 1001 N. Fairfax Street, Suite 500, Alexandria, Virginia 22314-1587.] The foregoing is adapted in part from the ANSI foreword to these standards.

Table 14–1

Symbols, Their Names, and Locations

Symbol*	Name	Where Found
C_e	Mesh alignment correction factor	Eq. (14–35)
$C_f (Z_R)$	Surface condition factor	Eq. (14–16)
$C_H (Z_W)$	Hardness-ratio factor	Eq. (14–18)
C_{ma}	Mesh alignment factor	Eq. (14–34)
C_{mc}	Load correction factor	Eq. (14–31)
C_{mf}	Face load-distribution factor	Eq. (14–30)
$C_p (Z_E)$	Elastic coefficient	Eq. (14–13)
C_{pf}	Pinion proportion factor	Eq. (14–32)
C_{pm}	Pinion proportion modifier	Eq. (14–33)
d	Pitch diameter	Ex. (14–1)
d_P	Pitch diameter, pinion	Eq. (14–22)
d_G	Pitch diameter, gear	Eq. (14–22)
$F (b)$	Net face width of narrowest member	Eq. (14–15)
f_P	Pinion surface finish	Fig. 14–13
H	Power	Fig. 14–17
H_B	Brinell hardness	Ex. 14–3
H_{BG}	Brinell hardness of gear	Sec. 14–12
H_{BP}	Brinell hardness of pinion	Sec. 14–12
hp	Horsepower	Ex. 14–1
h_t	Gear-tooth whole depth	Sec. 14–16
$I (Z_I)$	Geometry factor of pitting resistance	Eq. (14–16)
$J (Y_J)$	Geometry factor for bending strength	Eq. (14–15)
K_B	Rim-thickness factor	Eq. (14–40)
K_f	Fatigue stress-concentration factor	Eq. (14–9)
$K_m (K_H)$	Load-distribution factor	Eq. (14–30)
K_o	Overload factor	Eq. (14–15)
$K_R (Y_Z)$	Reliability factor	Eq. (14–17)
K_s	Size factor	Sec. 14–10
$K_T (Y_\theta)$	Temperature factor	Eq. (14–17)
K_v	Dynamic factor	Eq. (14–27)
m	Module	Eq. (14–15)
m_B	Backup ratio	Eq. (14–39)
m_F	Face-contact ratio	Eq. (14–19)
m_G	Gear ratio (never less than 1)	Eq. (14–22)
m_N	Load-sharing ratio	Eq. (14–21)
m_t	Transverse module	Eq. (14–15)
N	Number of stress cycles	Fig. 14–14
N_G	Number of teeth on gear	Eq. (14–22)
N_P	Number of teeth on pinion	Eq. (14–22)
n	Speed, in rev/min	Eq. (13–34)

(Continued)

Symbol*	Name	Where Found
n_P	Pinion speed, in rev/min	Ex. 14–4
P	Diametral pitch	Eq. (14–2)
P_d	Transverse diametral pitch	Eq. (14–15)
p_N	Normal base pitch	Eq. (14–24)
p_n	Normal circular pitch	Eq. (14–24)
p_x	Axial pitch	Eq. (14–19)
Q_v	Quality number	Eq. (14–29)
R	Reliability	Eq. (14–38)
R_a	Root-mean-squared roughness	Fig. 14–13
r_f	Tooth fillet radius	Fig. 14–1
r_G	Pitch-circle radius, gear	In standard
r_P	Pitch-circle radius, pinion	In standard
r_{bP}	Pinion base-circle radius	Eq. (14–25)
r_{bG}	Gear base-circle radius	Eq. (14–25)
S_C	Buckingham surface endurance strength	Ex. 14–3
S_c	AGMA surface endurance strength	Eq. (14–18)
S_t	AGMA bending strength	Eq. (14–17)
S	Bearing span	Fig. 14–10
S_I	Pinion offset from center span	Fig. 14–10
S_F	Safety factor—bending	Eq. (14–41)
S_H	Safety factor—pitting	Eq. (14–42)
W^t or W_t	Transmitted load	Fig. 14–1
Y_N	Stress-cycle factor for bending strength	Fig. 14–14
Z_N	Stress-cycle factor for pitting resistance	Fig. 14–15
β	Exponent	Eq. (14–44)
σ	Bending stress, AGMA	Eq. (14–15)
σ_C	Contact stress from Hertzian relationships	Eq. (14–14)
σ_c	Contact stress from AGMA relationships	Eq. (14–16)
σ_{all}	Allowable bending stress, AGMA	Eq. (14–17)
$\sigma_{c,\text{all}}$	Allowable contact stress, AGMA	Eq. (14–18)
ϕ	Pressure angle	Eq. (14–12)
ϕ_n	Normal pressure angle	Eq. (14–24)
ϕ_t	Transverse pressure angle	Eq. (14–23)
ψ	Helix angle	Ex. 14–5

Table 14–1

Symbols, Their Names, and Locations (*Continued*)

*Where applicable, the alternate symbol for the metric standard is shown in parenthesis.

force of body 2 on body 3, and W^t_{32} is the transmitted force of body 3 on body 2. When working with double- or triple-reduction speed reducers, this notation is compact and essential to clear thinking. Since gear-force components rarely take exponents, this causes no complication. Pythagorean combinations, if necessary, can be treated with parentheses or avoided by expressing the relations trigonometrically.

| Figure 14–1

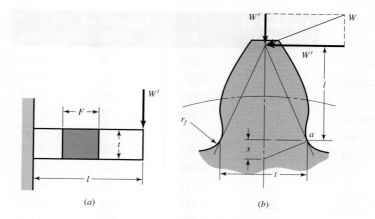

(a)   (b)

Referring now to Fig. 14–1b, we assume that the maximum stress in a gear tooth occurs at point a. By similar triangles, you can write

$$\frac{t/2}{x} = \frac{l}{t/2} \quad \text{or} \quad x = \frac{t^2}{4l} \quad \text{or} \quad l = \frac{t^2}{4x} \tag{b}$$

By rearranging Eq. (a),

$$\sigma = \frac{6W^t l}{Ft^2} = \frac{W^t}{F}\frac{1}{t^2/6l} = \frac{W^t}{F}\frac{1}{t^2/4l}\frac{4}{6} \tag{c}$$

If we now substitute the value of l from Eq. (b) in Eq. (c) and multiply the numerator and denominator by the circular pitch p, we find

$$\sigma = \frac{W^t p}{F(\frac{2}{3})\,xp} \tag{d}$$

Letting $y = 2x/(3p)$, we have

$$\sigma = \frac{W^t}{Fpy} \tag{14–1}$$

This completes the development of the original Lewis equation. The factor y is called the *Lewis form factor,* and it may be obtained by a graphical layout of the gear tooth or by digital computation.

In using this equation, most engineers prefer to employ the diametral pitch in determining the stresses. This is done by substituting $p = \pi/P$ and $y = \pi Y$ in Eq. (14–1). This gives

$$\sigma = \frac{W^t P}{FY} \tag{14–2}$$

where

$$Y = \frac{2xP}{3} \tag{14–3}$$

The use of this equation for Y means that only the bending of the tooth is considered and that the compression due to the radial component of the force is neglected. Values of Y obtained from this equation are tabulated in Table 14–2.

Table 14–2

Values of the Lewis Form Factor Y (These Values Are for a Normal Pressure Angle of 20°, Full-Depth Teeth, and a Diametral Pitch of Unity in the Plane of Rotation)

Number of Teeth	Y	Number of Teeth	Y
12	0.245	28	0.353
13	0.261	30	0.359
14	0.277	34	0.371
15	0.290	38	0.384
16	0.296	43	0.397
17	0.303	50	0.409
18	0.309	60	0.422
19	0.314	75	0.435
20	0.322	100	0.447
21	0.328	150	0.460
22	0.331	300	0.472
24	0.337	400	0.480
26	0.346	Rack	0.485

The use of Eq. (14–3) also implies that the teeth do not share the load and that the greatest force is exerted at the tip of the tooth. But we have already learned that the contact ratio should be somewhat greater than unity, say about 1.5, to achieve a quality gearset. If, in fact, the gears are cut with sufficient accuracy, the tip-load condition is not the worst, because another pair of teeth will be in contact when this condition occurs. Examination of run-in teeth will show that the heaviest loads occur near the middle of the tooth. Therefore the maximum stress probably occurs while a single pair of teeth is carrying the full load, at a point where another pair of teeth is just on the verge of coming into contact.

Dynamic Effects

When a pair of gears is driven at moderate or high speed and noise is generated, it is certain that dynamic effects are present. One of the earliest efforts to account for an increase in the load due to velocity employed a number of gears of the same size, material, and strength. Several of these gears were tested to destruction by meshing and loading them at zero velocity. The remaining gears were tested to destruction at various pitch-line velocities. For example, if a pair of gears failed at 500 lbf tangential load at zero velocity and at 250 lbf at velocity V_1, then a *velocity factor,* designated K_v, of 2 was specified for the gears at velocity V_1. Then another, identical, pair of gears running at a pitch-line velocity V_1 could be assumed to have a load equal to twice the tangential or transmitted load.

Note that the definition of dynamic factor K_v has been altered. AGMA standards ANSI/AGMA 2001-D04 and 2101-D04 contain this caution:

> Dynamic factor K_v has been redefined as the reciprocal of that used in previous AGMA standards. It is now greater than 1.0. In earlier AGMA standards it was less than 1.0.

Care must be taken in referring to work done prior to this change in the standards.

In the nineteenth century, Carl G. Barth first expressed the velocity factor, and in terms of the current AGMA standards, they are represented as

$$K_v = \frac{600 + V}{600} \qquad \text{(cast iron, cast profile)} \qquad (14\text{–}4a)$$

$$K_v = \frac{1200 + V}{1200} \qquad \text{(cut or milled profile)} \qquad (14\text{–}4b)$$

where V is the pitch-line velocity in feet per minute. It is also quite probable, because of the date that the tests were made, that the tests were conducted on teeth having a cycloidal profile instead of an involute profile. Cycloidal teeth were in general use in the nineteenth century because they were easier to cast than involute teeth. Equation (14–4a) is called the *Barth equation*. The Barth equation is often modified into Eq. (14–4b), for cut or milled teeth. Later, AGMA added

$$K_v = \frac{50 + \sqrt{V}}{50} \qquad \text{(hobbed or shaped profile)} \qquad (14\text{–}5a)$$

$$K_v = \sqrt{\frac{78 + \sqrt{V}}{78}} \qquad \text{(shaved or ground profile)} \qquad (14\text{–}5b)$$

In SI units, Eqs. (14–4a) through (14–5b) become

$$K_v = \frac{3.05 + V}{3.05} \qquad \text{(cast iron, cast profile)} \qquad (14\text{–}6a)$$

$$K_v = \frac{6.1 + V}{6.1} \qquad \text{(cut or milled profile)} \qquad (14\text{–}6b)$$

$$K_v = \frac{3.56 + \sqrt{V}}{3.56} \qquad \text{(hobbed or shaped profile)} \qquad (14\text{–}6c)$$

$$K_v = \sqrt{\frac{5.56 + \sqrt{V}}{5.56}} \qquad \text{(shaved or ground profile)} \qquad (14\text{–}6d)$$

where V is in meters per second (m/s).

Introducing the velocity factor into Eq. (14–2) gives

$$\sigma = \frac{K_v W^t P}{F Y} \qquad (14\text{–}7)$$

The metric version of this equation is

$$\sigma = \frac{K_v W^t}{F m Y} \qquad (14\text{–}8)$$

where the face width F and the module m are both in millimeters (mm). Expressing the tangential component of load W^t in newtons (N) then results in stress units of megapascals (MPa).

As a general rule, spur gears should have a face width F from 3 to 5 times the circular pitch p.

Equations (14–7) and (14–8) are important because they form the basis for the AGMA approach to the bending strength of gear teeth. They are in general use for estimating the capacity of gear drives when life and reliability are not important considerations. The equations can be useful in obtaining a preliminary estimate of gear sizes needed for various applications.

EXAMPLE 14–1

A stock spur gear is available having a diametral pitch of 4 mm, a 44-mm face, 18 teeth, and a pressure angle of 20° with full-depth teeth. The material is AISI 1020 steel in as-rolled condition. Use a design factor of $n_d = 3.5$ to rate the horsepower output of the gear corresponding to a speed of 25 rev/s and moderate applications.

Solution

The term *moderate applications* seems to imply that the gear can be rated by using the yield strength as a criterion of failure. From Table A–20, we find $S_{ut} = 379$ MPa and $S_y = 206$ MPa. A design factor of 3.5 means that the allowable bending stress is $206/3.5 = 58.86$ MPa. The pitch diameter is $Nm = 18(4) = 72$ mm, so the pitch-line velocity is

$$V = \pi dn = \pi(0.072)25 = 5.56487 \text{ m/s}$$

The velocity factor from Eq. (14–4b) is found to be

$$K_v = \frac{6.1 + V}{6.1} = \frac{6.1 + 5.65487}{6.1} = 1.92703$$

Table 14–2 gives the form factor as $Y = 0.296$ for 16 teeth. We now arrange and substitute in Eq. (14–7) as follows:

$$W^t = \frac{FY\sigma_{all}}{K_v P} = \frac{0.004(0.004)0.039(58.9)10^6}{1.92703} = 1434.54233 \text{ N}$$

The power that can be transmitted is

Answer

$$hp = W^t V = 1434.54(5.65) = 8112.15 \text{ watt}$$

It is important to emphasize that this is a rough estimate, and that this approach must not be used for important applications. The example is intended to help you understand some of the fundamentals that will be involved in the AGMA approach.

EXAMPLE 14–2

Estimate the horsepower rating of the gear in the previous example based on obtaining an infinite life in bending.

Solution

The rotating-beam endurance limit is estimated from Eq. (6–8), p. 290,

$$S'_e = 0.5S_{ut} = 0.5(379 \text{ MPa}) = 189.5 \text{ MPa}$$

To obtain the surface finish Marin factor k_a we refer to Table 6–3, p. 298, for machined surface, finding $a = 4.51$ and $b = -0.265$. Then Eq. (6–19), p. 295, gives the surface finish Marin factor k_a as

$$k_a = aS_{ut}^b = 4.51(189.5)^{-0.265} = 0.935$$

The next step is to estimate the size factor k_b. From Table 13–1, p. 688, the sum of the addendum and dedendum is

$$l = \frac{1}{P} + \frac{1.25}{P} = \frac{1}{0.25} + \frac{1.25}{0.25} = 9 \text{ mm}$$

The tooth thickness t in Fig. 14–1b is given in Sec. 14–1 [Eq. (b)] as $t = (4lx)^{1/2}$ when $x = 3Y/(2P)$ from Eq. (14–3). Therefore, since from Ex. 14–1 $Y = 0.296$ and $P = 0.25$,

$$x = \frac{3Y}{2P} = \frac{3(0.296)}{2(0.25)} = 1.776 \text{ mm}$$

then

$$t = (4lx)^{1/2} = [4(9)1.776]^{1/2} = 7.996$$

We have recognized the tooth as a cantilever beam of rectangular cross section, so the equivalent rotating-beam diameter must be obtained from Eq. (6–25), p. 297:

$$d_e = 0.808(hb)^{1/2} = 0.808(Ft)^{1/2} = 0.808[44(7.996)]^{1/2} = 15.156 \text{ mm}$$

Then, Eq. (6–20), p. 296, gives k_b as

$$k_b = \left(\frac{d_e}{0.30}\right)^{-0.107} = \left(\frac{15.156}{7.62}\right)^{-0.107} = 0.929$$

The load factor k_c from Eq. (6–26), p. 298, is unity. With no information given concerning temperature and reliability we will set $k_d = k_e = 1$.

In general, a gear tooth is subjected only to one-way bending. Exceptions include idler gears and gears used in reversing mechanisms. We will account for one-way bending by establishing a miscellaneous-effects Marin factor k_f.

For one-way bending the steady and alternating stress components are $\sigma_a = \sigma_m = \sigma/2$ where σ is the largest repeatedly applied bending stress as given in Eq. (14–7). If a material exhibited a Goodman failure locus,

$$\frac{S_a}{S_e'} + \frac{S_m}{S_{ut}} = 1$$

Since S_a and S_m are equal for one-way bending, we substitute S_a for S_m and solve the preceding equation for S_a, giving

$$S_a = \frac{S_e' S_{ut}}{S_e' + S_{ut}}$$

Now replace S_a with $\sigma/2$, and in the denominator replace S_e' with $0.5S_{ut}$ to obtain

$$\sigma = \frac{2S_e' S_{ut}}{0.5S_{ut} + S_{ut}} = \frac{2S_e'}{0.5 + 1} = 1.33S_e'$$

Now $k_f = \sigma/S_e' = 1.33S_e'/S_e' = 1.33$. However, a Gerber fatigue locus gives mean values of

$$\frac{S_a}{S_e'} + \left(\frac{S_m}{S_{ut}}\right)^2 = 1$$

Setting $S_a = S_m$ and solving the quadratic in S_a gives

$$S_a = \frac{S_{ut}^2}{2S_e'}\left(-1 + \sqrt{1 + \frac{4S_e'^2}{S_{ut}^2}}\right)$$

Setting $S_a = \sigma/2$, $S_{ut} = S_e'/0.5$ gives

$$\sigma = \frac{S_e'}{0.5^2}[-1 + \sqrt{1 + 4(0.5)^2}] = 1.66S_e'$$

and $k_f = \sigma/S_e' = 1.66$. Since a Gerber locus runs in and among fatigue data and Goodman does not, we will use $k_f = 1.66$. The Marin equation for the fully corrected endurance strength is

$$S_e = k_a k_b k_c k_d k_e k_f S_e'$$

$$= 0.935(0.929)(1)(1)(1)1.66(189.5) = 273.3 \text{ MPa}$$

For stress, we will first determine the fatigue stress-concentration factor K_f. For a 20° full-depth tooth the radius of the root fillet is denoted r_f, where

$$r_f = \frac{0.300}{P} = \frac{0.300}{0.25} = 1.2 \text{ mm}$$

From Fig. A–15–6

$$\frac{r}{d} = \frac{r_f}{t} = \frac{1.2}{7.996} = 0.15$$

Since $D/d = \infty$, we approximate with $D/d = 3$, giving $K_t = 1.68$. From Fig. 6–20, p. 303, $q = 0.62$. From Eq. (6–32), p. 303,

$$K_f = 1 + (0.62)(1.68 - 1) = 1.42$$

For a design factor of $n_d = 3.5$, as used in Ex. 14–1, applied to the load or strength, the maximum bending stress is

$$\sigma_{\max} = K_f \sigma_{\text{all}} = \frac{S_e}{n_d}$$

$$\sigma_{\text{all}} = \frac{S_e}{K_f n_d} = \frac{273 3}{1.42(3.5)} = 54.92 \text{ MPa}$$

The transmitted load W^t is

$$W^t = \frac{FY\sigma_{\text{all}}}{K_v P} = \frac{44(0.296)54.92}{1.927(0.25)} = 1485 \text{ N}$$

and the power is, with $V = 628$ ft/min from Ex. 14–1,

$$p = \frac{W^t V}{33\,000} = \frac{1485(5.6549)}{33\,000} = 8396.4 \text{ Watt}$$

Again, it should be emphasized that these results should be accepted *only* as preliminary estimates to alert you to the nature of bending in gear teeth.

In Ex. 14–2 our resources (Fig. A–15–6) did not directly address stress concentration in gear teeth. A photoelastic investigation by Dolan and Broghamer reported in 1942 constitutes a primary source of information on stress concentration.[3] Mitchiner and Mabie[4] interpret the results in term of fatigue stress-concentration factor K_f as

$$K_f = H + \left(\frac{t}{r}\right)^L \left(\frac{t}{l}\right)^M \tag{14–9}$$

where $\quad H = 0.34 - 0.458\ 366\ 2\phi$

$\qquad\quad L = 0.316 - 0.458\ 366\ 2\phi$

$\qquad\quad M = 0.290 + 0.458\ 366\ 2\phi$

$$r = \frac{(b - r_f)^2}{(d/2) + b - r_f}$$

In these equations l and t are from the layout in Fig. 14–1, ϕ is the pressure angle, r_f is the fillet radius, b is the dedendum, and d is the pitch diameter. It is left as an exercise for the reader to compare K_f from Eq. (14–9) with the results of using the approximation of Fig. A–15–6 in Ex. 14–2.

14–2   Surface Durability

In this section we are interested in the failure of the surfaces of gear teeth, which is generally called *wear*. *Pitting*, as explained in Sec. 6–16, is a surface fatigue failure due to many repetitions of high contact stresses. Other surface failures are *scoring*, which is a lubrication failure, and *abrasion*, which is wear due to the presence of foreign material.

To obtain an expression for the surface-contact stress, we shall employ the Hertz theory. In Eq. (3–74), p. 138, it was shown that the contact stress between two cylinders may be computed from the equation

$$p_{\max} = \frac{2F}{\pi b l} \tag{a}$$

where $\quad p_{\max}$ = largest surface pressure

$\qquad\quad F$ = force pressing the two cylinders together

$\qquad\quad l$ = length of cylinders

and half-width b is obtained from Eq. (3–73), p. 138, given by

$$b = \left[\frac{2F}{\pi l} \frac{(1 - \nu_1^2)/E_1 + (1 - \nu_2^2)/E_2}{1/d_1 + 1/d_2}\right]^{1/2} \tag{14–10}$$

where ν_1, ν_2, E_1, and E_2 are the elastic constants and d_1 and d_2 are the diameters, respectively, of the two contacting cylinders.

To adapt these relations to the notation used in gearing, we replace F by $W^t/\cos\phi$, d by $2r$, and l by the face width F. With these changes, we can substitute the value

---

[3]T. J. Dolan and E. I. Broghamer, *A Photoelastic Study of the Stresses in Gear Tooth Fillets*, Bulletin 335, Univ. Ill. Exp. Sta., March 1942, See also W. D. Pilkey and D. F. Pilkey, *Peterson's Stress-Concentration Factors*, 3rd ed., John Wiley & Sons, Hoboken, NJ, 2008, pp. 407–409, 434–437.

[4]R. G. Mitchiner and H. H. Mabie, "Determination of the Lewis Form Factor and the AGMA Geometry Factor J of External Spur Gear Teeth," *J. Mech. Des.*, Vol. 104, No. 1, Jan. 1982, pp. 148–158.

of b as given by Eq. (14–10) in Eq. (a). Replacing p_{max} by σ_C, the *surface compressive stress (Hertzian stress)* is found from the equation

$$\sigma_C^2 = \frac{W^t}{\pi F \cos \phi} \frac{1/r_1 + 1/r_2}{(1 - \nu_1^2)/E_1 + (1 - \nu_2^2)/E_2} \tag{14–11}$$

where r_1 and r_2 are the instantaneous values of the radii of curvature on the pinion- and gear-tooth profiles, respectively, at the point of contact. By accounting for load sharing in the value of W^t used, Eq. (14–11) can be solved for the Hertzian stress for any or all points from the beginning to the end of tooth contact. Of course, pure rolling exists only at the pitch point. Elsewhere the motion is a mixture of rolling and sliding. Equation (14–11) does not account for any sliding action in the evaluation of stress. We note that AGMA uses μ for Poisson's ratio instead of ν as is used here.

We have already noted that the first evidence of wear occurs near the pitch line. The radii of curvature of the tooth profiles at the pitch point are

$$r_1 = \frac{d_P \sin \phi}{2} \qquad r_2 = \frac{d_G \sin \phi}{2} \tag{14–12}$$

where ϕ is the pressure angle and d_P and d_G are the pitch diameters of the pinion and gear, respectively.

Note, in Eq. (14–11), that the denominator of the second group of terms contains four elastic constants, two for the pinion and two for the gear. As a simple means of combining and tabulating the results for various combinations of pinion and gear materials, AGMA defines an *elastic coefficient* C_p by the equation

$$C_p = \left[\frac{1}{\pi \left(\frac{1 - \nu_P^2}{E_P} + \frac{1 - \nu_G^2}{E_G} \right)} \right]^{1/2} \tag{14–13}$$

With this simplification, and the addition of a velocity factor K_v, Eq. (14–11) can be written as

$$\sigma_C = -C_p \left[\frac{K_v W^t}{F \cos \phi} \left(\frac{1}{r_1} + \frac{1}{r_2} \right) \right]^{1/2} \tag{14–14}$$

where the sign is negative because σ_C is a compressive stress.

EXAMPLE 14–3

The pinion of Examples 14–1 and 14–2 is to be mated with a 50-tooth gear manufactured of ASTM No. 50 cast iron. Using the tangential load of 1700 N, estimate the factor of safety of the drive based on the possibility of a surface fatigue failure.

Solution

From Table A–5 we find the elastic constants to be $E_P = 207$ GPa, $\nu_P = 0.292$, $E_G = 100$ GPa, $\nu_G = 0.211$. We substitute these in Eq. (14–13) to get the elastic coefficient as

$$C_p = \left\{ \frac{1}{\pi \left[\frac{1 - (0.292)^2}{207(10^9)} + \frac{1 - (0.211)^2}{100(10^9)} \right]} \right\}^{1/2} = 150\,927.3$$

From Example 14–1, the pinion pitch diameter is $d_P = 48$ mm. The value for the gear is $d_G = 50(3) = 150$ mm. Then Eq. (14–12) is used to obtain the radii of curvature at the pitch points. Thus

$$r_1 = \frac{48 \sin 20°}{2} = 8.2 \text{ mm} \qquad r_2 = \frac{150 \sin 20°}{2} = 25.7 \text{ mm}$$

The face width is given as $F = 38$ mm. Use $K_v = 1.5$ from Example 14–1. Substituting all these values in Eq. (14–14) with $\phi = 20°$ gives the contact stress as

$$\sigma_C = -150\,927.3 \left[\frac{1.5(1700)}{0.038 \cos 20°} \left(\frac{1}{0.0082} + \frac{1}{0.0257} \right) \right]^{1/2} = -511.5 \text{ MPa}$$

The surface endurance strength of cast iron can be estimated from

$$S_C = 2.206 H_B \text{ MPa}$$

for 10^8 cycles. Table A–22 gives $H_B = 262$ for ASTM No. 50 cast iron. Therefore $S_C = 2.206(262) = 578$ MPa. Contact stress is not linear with respect to the transmitted load [see Eq. (14–14)]. If the factor of safety is defined as the loss-of-function load divided by the imposed load, then the ratio of loads is the ratio of stresses squared. In other words,

$$n = \frac{\text{loss-of-function load}}{\text{imposed load}} = \frac{S_C^2}{\sigma_C^2} = \left(\frac{578}{511.5} \right)^2 = 1.28$$

One is free to define the factor of safety as S_C/σ_C. Awkwardness comes when one compares the factor of safety in bending fatigue with the factor of safety in surface fatigue for a particular gear. Suppose the factor of safety of this gear in bending fatigue is 1.20 and the factor of safety in surface fatigue is 1.28 as above. The threat, since 1.28 is greater than 1.20, is in bending fatigue since both numbers are based on load ratios. If the factor of safety in surface fatigue is based on $S_C/\sigma_C = \sqrt{1.28} = 1.13$, then 1.20 is greater than 1.13, but the threat is not from surface fatigue. The surface fatigue factor of safety can be defined either way. One way has the burden of requiring a squared number before numbers that instinctively seem comparable can be compared.

In addition to the dynamic factor K_v already introduced, there are transmitted load excursions, nonuniform distribution of the transmitted load over the tooth contact, and the influence of rim thickness on bending stress. Tabulated strength values can be means, ASTM minimums, or of unknown heritage. In surface fatigue there are no endurance limits. Endurance strengths have to be qualified as to corresponding cycle count, and the slope of the S-N curve needs to be known. In bending fatigue there is a definite change in slope of the S-N curve near 10^6 cycles, but some evidence indicates that an endurance limit does not exist. Gearing experience leads to cycle counts of 10^{11} or more. Evidence of diminishing endurance strengths in bending have been included in AGMA methodology.

14–3   **AGMA Stress Equations**

Two fundamental stress equations are used in the AGMA methodology, one for bending stress and another for pitting resistance (contact stress). In AGMA terminology, these are called *stress numbers,* as contrasted with actual applied stresses, and are designated by a lowercase letter s instead of the Greek lower case σ we have used in this book (and shall continue to use). The fundamental equations are

$$
\sigma = \begin{cases} W^t K_o K_v K_s \dfrac{P_d}{F} \dfrac{K_m K_B}{J} & \text{(U.S. customary units)} \\[2em] W^t K_o K_v K_s \dfrac{1}{b m_t} \dfrac{K_H K_B}{Y_J} & \text{(SI units)} \end{cases}
\tag{14–15}
$$

where for U.S. customary units (SI units),

  W^t is the tangential transmitted load, lbf (N)
  K_o is the overload factor
  K_v is the dynamic factor
  K_s is the size factor
  P_d is the transverse diametral pitch
  F (b) is the face width of the narrower member, in (mm)
  K_m (K_H) is the load-distribution factor
  K_B is the rim-thickness factor
  J (Y_J) is the geometry factor for bending strength (which includes root fillet stress-concentration factor K_f)
  (m_t) is the transverse metric module

Before you try to digest the meaning of all these terms in Eq. (14–15), view them as advice concerning items the designer should consider *whether he or she follows the voluntary standard or not*. These items include issues such as

- Transmitted load magnitude
- Overload
- Dynamic augmentation of transmitted load
- Size
- Geometry: pitch and face width
- Distribution of load across the teeth
- Rim support of the tooth
- Lewis form factor and root fillet stress concentration

The fundamental equation for pitting resistance (contact stress) is

$$
\sigma_c = \begin{cases} C_P \sqrt{W^t K_o K_v K_s \dfrac{K_m}{d_P F} \dfrac{C_f}{I}} & \text{(U.S. customary units)} \\[2em] Z_E \sqrt{W^t K_o K_v K_s \dfrac{K_H}{d_{w1} b} \dfrac{Z_R}{Z_I}} & \text{(SI units)} \end{cases}
\tag{14–16}
$$

where W^t, K_o, K_v, K_s, K_m, F, and b are the same terms as defined for Eq. (14–15). For U.S. customary units (SI units), the additional terms are

  C_p (Z_E) is an elastic coefficient, $\sqrt{\text{lbf/in}^2}$ ($\sqrt{\text{N/mm}^2}$)
  C_f (Z_R) is the surface condition factor
  d_P (d_{w1}) is the pitch diameter of the *pinion*, in (mm)
  I (Z_I) is the geometry factor for pitting resistance

The evaluation of all these factors is explained in the sections that follow. The development of Eq. (14–16) is clarified in the second part of Sec. 14–5.

14-4 AGMA Strength Equations

Instead of using the term *strength*, AGMA uses data termed *allowable stress numbers* and designates these by the symbols s_{at} and s_{ac}. It will be less confusing here if we continue the practice in this book of using the uppercase letter S to designate strength and the lowercase Greek letters σ and τ for stress. To make it perfectly clear we shall use the term *gear strength* as a replacement for the phrase *allowable stress numbers* as used by AGMA.

Following this convention, values for *gear bending strength*, designated here as S_t, are to be found in Figs. 14–2, 14–3, and 14–4, and in Tables 14–3 and 14–4. Since gear strengths are not identified with other strengths such as S_{ut}, S_e, or S_y as used elsewhere in this book, their use should be restricted to gear problems.

In this approach the strengths are modified by various factors that produce limiting values of the bending stress and the contact stress.

Figure 14-2

Allowable bending stress number for through-hardened steels, S_t. The SI equations are: $S_t = 0.533H_B + 88.3$ MPa, grade 1, and $S_t = 0.703H_B + 113$ MPa, grade 2. (*Source: ANSI/AGMA 2001-D04 and 2101-D04.*)

Figure 14-3

Allowable bending stress number for nitrided through-hardened steel gears (i.e., AISI 4140, 4340), S_t. The SI equations are: $S_t = 0.568H_B + 83.8$ MPa, grade 1, and $S_t = 0.749H_B + 110$ MPa, grade 2. (*Source: ANSI/AGMA 2001-D04 and 2101-D04.*)

Figure 14–4

Allowable bending stress numbers for nitriding steel gears, S_t. The SI equations are:
$S_t = 0.594H_B + 87.76$ MPa
Nitralloy grade 1
$S_t = 0.784H_B + 114.81$ MPa
Nitralloy grade 2
$S_t = 0.7255H_B + 63.89$ MPa
2.5% chrome, grade 1
$S_t = 0.7255H_B + 153.63$ MPa
2.5% chrome, grade 2
$S_t = 0.7255H_B + 201.91$ MPa
2.5% chrome, grade 3
(Source: ANSI/AGMA 2001-D04, 2101-D04.)

In figure:
Metallurgical and quality control procedures required
Grade 3 – 2.5% Chrome $S_t = 105.2H_B + 29\,280$ psi
Grade 2 – 2.5% Chrome $S_t = 105.2H_B + 22\,280$ psi
Grade 2 – Nitralloy $S_t = 113.8H_B + 16\,650$ psi
Grade 1 – 2.5% Chrome $S_t = 105.2H_B + 9280$ psi
Grade 1 – Nitralloy $S_t = 86.2H_B + 12\,730$ psi

Allowable bending stress numbers, S_t, kpsi (vertical axis, 30 to 70)
Core hardness, H_B (horizontal axis, 250 to 350)

Table 14–3

Repeatedly Applied Bending Strength S_t at 10^7 Cycles and 0.99 Reliability for Steel Gears

Source: ANSI/AGMA 2001-D04.

Material Designation	Heat Treatment	Minimum Surface Hardness[1]	Allowable Bending Stress Number S_t,[2] psi (MPa)		
			Grade 1	Grade 2	Grade 3
Steel[3]	Through-hardened	See Fig. 14–2	See Fig. 14–2	See Fig. 14–2	—
	Flame[4] or induction hardened[4] with type A pattern[5]	See Table 8*	45 000 (310)	55 000 (380)	—
	Flame[4] or induction hardened[4] with type B pattern[5]	See Table 8*	22 000 (151)	22 000 (151)	—
	Carburized and hardened	See Table 9*	55 000 (380)	65 000 or (448 or 70 000[6] 482)	75 000 (517)
	Nitrided[4,7] (through-hardened steels)	83.5 HR15N	See Fig. 14–3	See Fig. 14–3	—
Nitralloy 135M, Nitralloy N, and 2.5% chrome (no aluminum)	Nitrided[4,7]	87.5 HR15N	See Fig. 14–4	See Fig. 14–4	See Fig. 14–4

Notes: See ANSI/AGMA 2001-D04 for references cited in notes 1–7.

[1]Hardness to be equivalent to that at the root diameter in the center of the tooth space and face width.

[2]See tables 7 through 10 for major metallurgical factors for each stress grade of steel gears.

[3]The steel selected must be compatible with the heat treatment process selected and hardness required.

[4]The allowable stress numbers indicated may be used with the case depths prescribed in 16.1.

[5]See figure 12 for type A and type B hardness patterns.

[6]If bainite and microcracks are limited to grade 3 levels, 70 000 psi may be used.

[7]The overload capacity of nitrided gears is low. Since the shape of the effective S-N curve is flat, the sensitivity to shock should be investigated before proceeding with the design. [7]

*Tables 8 and 9 of ANSI/AGMA 2001-D04 are comprehensive tabulations of the major metallurgical factors affecting S_t and S_c of flame-hardened and induction-hardened (Table 8) and carburized and hardened (Table 9) steel gears.

Table 14–4

Repeatedly Applied Bending Strength S_t for Iron and Bronze Gears at 10^7 Cycles and 0.99 Reliability
Source: ANSI/AGMA 2001-D04.

Material	Material Designation[1]	Heat Treatment	Typical Minimum Surface Hardness[2]	Allowable Bending Stress Number, S_t,[3] psi (MPa)
ASTM A48 gray cast iron	Class 20	As cast	—	5000 (35)
	Class 30	As cast	174 HB	8500 (58)
	Class 40	As cast	201 HB	13 000 (90)
ASTM A536 ductile (nodular) Iron	Grade 60–40–18	Annealed	140 HB	22 000–33 000 (151–227)
	Grade 80–55–06	Quenched and tempered	179 HB	22 000–33 000 (151–227)
	Grade 100–70–03	Quenched and tempered	229 HB	27 000–40 000 (186–275)
	Grade 120–90–02	Quenched and tempered	269 HB	31 000–44 000 (213–275)
Bronze		Sand cast	Minimum tensile strength 40 000 psi	5700 (39)
	ASTM B–148 Alloy 954	Heat treated	Minimum tensile strength 90 000 psi	23 600 (163)

Notes:

[1]See ANSI/AGMA 2004-B89, *Gear Materials and Heat Treatment Manual.*

[2]Measured hardness to be equivalent to that which would be measured at the root diameter in the center of the tooth space and face width.

[3]The lower values should be used for general design purposes. The upper values may be used when:

High quality material is used.

Section size and design allow maximum response to heat treatment.

Proper quality control is effected by adequate inspection.

Operating experience justifies their use.

The equation for the allowable bending stress is

$$\sigma_{all} = \begin{cases} \dfrac{S_t}{S_F}\dfrac{Y_N}{K_T K_R} & \text{(U.S. customary units)} \\[2ex] \dfrac{S_t}{S_F}\dfrac{Y_N}{Y_\theta Y_Z} & \text{(SI units)} \end{cases} \tag{14–17}$$

where for U.S. customary units (SI units),

S_t is the allowable bending stress, lbf/in^2 (N/mm^2)
Y_N is the stress-cycle factor for bending stress
K_T (Y_θ) are the temperature factors
K_R (Y_Z) are the reliability factors
S_F is the AGMA factor of safety, a stress ratio

The equation for the allowable contact stress $\sigma_{c,\text{all}}$ is

$$\sigma_{c,\text{all}} = \begin{cases} \dfrac{S_c}{S_H} \dfrac{Z_N C_H}{K_T K_R} & \text{(U.S. customary units)} \\[3mm] \dfrac{S_c}{S_H} \dfrac{Z_N Z_W}{Y_\theta Y_Z} & \text{(SI units)} \end{cases} \qquad (14\text{–}18)$$

where the upper equation is in U.S. customary units and the lower equation is in SI units, Also,

S_c is the allowable contact stress, lbf/in^2 (N/mm^2)
Z_N is the stress-cycle factor
C_H (Z_W) are the hardness ratio factors for pitting resistance
K_T (Y_θ) are the temperature factors
K_R (Y_Z) are the reliability factors
S_H is the AGMA factor of safety, a stress ratio

The values for the allowable contact stress, designated here as S_c, are to be found in Fig. 14–5 and Tables 14–5, 14–6, and 14–7.

AGMA allowable stress numbers (strengths) for bending and contact stress are for

- Unidirectional loading
- 10 million stress cycles
- 99 percent reliability

Figure 14–5

Contact-fatigue strength S_c at 10^7 cycles and 0.99 reliability for through-hardened steel gears. The SI equations are:
$S_c = 2.22 H_B + 200$ MPa, grade 1, and
$S_c = 2.41 H_B + 237$ MPa, grade 2. (*Source: ANSI/AGMA 2001-D04 and 2101-D04.*)

Table 14–5

Nominal Temperature Used in Nitriding and Hardnesses Obtained

Source: Darle W. Dudley, *Handbook of Practical Gear Design,* rev. ed., McGraw-Hill, New York, 1984.

Steel	Temperature Before Nitriding, °F	Nitriding, °F	Hardness, Rockwell C Scale Case	Core
Nitralloy 135*	1150	975	62–65	30–35
Nitralloy 135M	1150	975	62–65	32–36
Nitralloy N	1000	975	62–65	40–44
AISI 4340	1100	975	48–53	27–35
AISI 4140	1100	975	49–54	27–35
31 Cr Mo V 9	1100	975	58–62	27–33

*Nitralloy is a trademark of the Nitralloy Corp., New York.

Table 14–6

Repeatedly Applied Contact Strength S_c at 10^7 Cycles and 0.99 Reliability for Steel Gears

Source: ANSI/AGMA 2001-D04.

Material Designation	Heat Treatment	Minimum Surface Hardness[1]	Allowable Contact Stress Number,[2] S_c, psi (σ_{HP}, MPa) Grade 1	Grade 2	Grade 3
Steel[3]	Through hardened[4]	See Fig. 14–5	See Fig. 14–5	See Fig. 14–5	—
	Flame[5] or induction hardened[5]	50 HRC	170 000 (1172)	190 000 (1310)	—
		54 HRC	175 000 (1206)	195 000 (1344)	—
	Carburized and hardened[5]	See Table 9*	180 000 (1240)	225 000 (1551)	275 000 (1896)
	Nitrided[5] (through hardened steels)	83.5 HR15N	150 000 (1035)	163 000 (1123)	175 000 (1206)
		84.5 HR15N	155 000 (1068)	168 000 (1158)	180 000 (1240)
2.5% chrome (no aluminum)	Nitrided[5]	87.5 HR15N	155 000 (1068)	172 000 (1186)	189 000 (1303)
Nitralloy 135M	Nitrided[5]	90.0 HR15N	170 000 (1172)	183 000 (1261)	195 000 (1344)
Nitralloy N	Nitrided[5]	90.0 HR15N	172 000 (1186)	188 000 (1296)	205 000 (1413)
2.5% chrome (no aluminum)	Nitrided[5]	90.0 HR15N	176 000 (1213)	196 000 (1351)	216 000 (1490)

Notes: See ANSI/AGMA 2001-D04 for references cited in notes 1–5.

[1]Hardness to be equivalent to that at the start of active profile in the center of the face width.

[2]See Tables 7 through 10 for major metallurgical factors for each stress grade of steel gears.

[3]The steel selected must be compatible with the heat treatment process selected and hardness required.

[4]These materials must be annealed or normalized as a minimum.

[5]The allowable stress numbers indicated may be used with the case depths prescribed in 16.1.

*Table 9 of ANSI/AGMA 2001-D04 is a comprehensive tabulation of the major metallurgical factors affecting S_t and S_c of carburized and hardened steel gears.

The factors in this section, too, will be evaluated in subsequent sections.

When two-way (reversed) loading occurs, as with idler gears, AGMA recommends using 70 percent of S_t values. This is equivalent to $1/0.70 = 1.43$ as a value of k_e in Ex. 14–2. The recommendation falls between the value of $k_e = 1.33$ for a Goodman failure locus and $k_e = 1.66$ for a Gerber failure locus.

14–5   Geometry Factors I and J (Z_I and Y_J)

We have seen how the factor Y is used in the Lewis equation to introduce the effect of tooth form into the stress equation. The AGMA factors[5] I and J are intended to accomplish the same purpose in a more involved manner.

The determination of I and J depends upon the *face-contact ratio* m_F. This is defined as

$$m_F = \frac{F}{p_x} \tag{14–19}$$

where p_x is the axial pitch and F is the face width. For spur gears, $m_F = 0$.

---

[5]A useful reference is AGMA 908-B89, *Geometry Factors for Determining Pitting Resistance and Bending Strength of Spur, Helical and Herringbone Gear Teeth.*

Table 14-7

Repeatedly Applied Contact Strength S_c 10^7 Cycles and 0.99 Reliability for Iron and Bronze Gears

Source: ANSI/AGMA 2001-D04.

Material	Material Designation[1]	Heat Treatment	Typical Minimum Surface Hardness[2]	Allowable Contact Stress Number,[3] S_c, psi (σ_{HP}, MPa)
ASTM A48 gray	Class 20	As cast	—	50 000–60 000 (344–415)
cast iron	Class 30	As cast	174 HB	65 000–75 000 (448–517)
	Class 40	As cast	201 HB	75 000–85 000 (517–586)
ASTM A536 ductile	Grade 60–40–18	Annealed	140 HB	77 000–92 000 (530–634)
(nodular) iron	Grade 80–55–06	Quenched and tempered	179 HB	77 000–92 000 (530–634)
	Grade 100–70–03	Quenched and tempered	229 HB	92 000–112 000 (634–772)
	Grade 120–90–02	Quenched and tempered	269 HB	103 000–126 000 (710–868)
Bronze	—	Sand cast	Minimum tensile strength 40 000 psi	30 000 (206)
	ASTM B-148 Alloy 954	Heat treated	Minimum tensile strength 90 000 psi	65 000 (448)

Notes:

[1]See ANSI/AGMA 2004-B89, *Gear Materials and Heat Treatment Manual.*

[2]Hardness to be equivalent to that at the start of active profile in the center of the face width.

[3]The lower values should be used for general design purposes. The upper values may be used when:
  High-quality material is used.
  Section size and design allow maximum response to heat treatment.
  Proper quality control is effected by adequate inspection.
  Operating experience justifies their use.

Low-contact-ratio (LCR) helical gears having a small helix angle or a thin face width, or both, have face-contact ratios less than unity ($m_F \leq 1$), and will not be considered here. Such gears have a noise level not too different from that for spur gears. Consequently we shall consider here only spur gears with $m_F = 0$ and conventional helical gears with $m_F > 1$.

Bending-Strength Geometry Factor J (Y_J)

The AGMA factor J employs a modified value of the Lewis form factor, also denoted by Y; a *fatigue stress-concentration factor* K_f; and a tooth *load-sharing ratio* m_N. The resulting equation for J for spur and helical gears is

$$J = \frac{Y}{K_f m_N} \qquad (14-20)$$

It is important to note that the form factor Y in Eq. (14–20) is *not* the Lewis factor at all. The value of Y here is obtained from calculations within AGMA 908-B89, and is often based on the highest point of single-tooth contact.

The factor K_f in Eq. (14–20) is called a *stress-correction factor* by AGMA. It is based on a formula deduced from a photoelastic investigation of stress concentration in gear teeth over 50 years ago.

The load-sharing ratio m_N is equal to the face width divided by the minimum total length of the lines of contact. This factor depends on the transverse contact ratio m_p, the face-contact ratio m_F, the effects of any profile modifications, and the tooth deflection. For spur gears, $m_N = 1.0$. For helical gears having a face-contact ratio $m_F > 2.0$, a conservative approximation is given by the equation

$$m_N = \frac{p_N}{0.95Z} \tag{14–21}$$

where p_N is the normal base pitch and Z is the length of the line of action in the transverse plane (distance L_{ab} in Fig. 13–15, p. 676).

Use Fig. 14–6 to obtain the geometry factor J for spur gears having a 20° pressure angle and full-depth teeth. Use Figs. 14–7 and 14–8 for helical gears having a 20° normal pressure angle and face-contact ratios of $m_F = 2$ or greater. For other gears, consult the AGMA standard.

Figure 14–6

Spur-gear geometry factors J. *Source:* The graph is from AGMA 218.01, which is consistent with tabular data from the current AGMA 908-B89. The graph is convenient for design purposes.

(a)

$$m_N = \frac{p_N}{0.95Z}$$

Value for Z is for an element of indicated
numbers of teeth and a 75-tooth mate

Normal tooth thickness of pinion and gear
tooth each reduced 0.024 in to provide 0.048 in
total backlash for one normal diametral pitch

(b)

Figure 14–7

Helical-gear geometry factors J'. *Source:* The graph is from AGMA 218.01, which is consistent with tabular data from
the current AGMA 908-B89. The graph is convenient for design purposes.

Surface-Strength Geometry Factor I (Z_I)

The factor I is also called the *pitting-resistance geometry factor* by AGMA. We will
develop an expression for I by noting that the sum of the reciprocals of Eq. (14–14),
from Eq. (14–12), can be expressed as

$$\frac{1}{r_1} + \frac{1}{r_2} = \frac{2}{\sin \phi_t}\left(\frac{1}{d_P} + \frac{1}{d_G}\right) \tag{a}$$

where we have replaced ϕ by ϕ_t, the transverse pressure angle, so that the relation will
apply to helical gears too. Now define *speed ratio* m_G as

$$m_G = \frac{N_G}{N_P} = \frac{d_G}{d_P} \tag{14–22}$$

Figure 14–8

J'-factor multipliers for use with Fig. 14–7 to find J. *Source:* The graph is from AGMA 218.01, which is consistent with tabular data from the current AGMA 908-B89. The graph is convenient for design purposes.

The modifying factor can be applied to the J factor when other than 75 teeth are used in the mating element

Equation (*a*) can now be written

$$\frac{1}{r_1} + \frac{1}{r_2} = \frac{2}{d_P \sin \phi_t} \frac{m_G + 1}{m_G} \tag{b}$$

Now substitute Eq. (*b*) for the sum of the reciprocals in Eq. (14–14). The result is found to be

$$\sigma_c = |\sigma_C| = C_p \left[\frac{K_v W^t}{d_P F} \frac{1}{\dfrac{\cos \phi_t \sin \phi_t}{2} \dfrac{m_G}{m_G + 1}} \right]^{1/2} \tag{c}$$

The geometry factor I for external spur and helical gears is the denominator of the second term in the brackets in Eq. (*c*). By adding the load-sharing ratio m_N, we obtain a factor valid for both spur and helical gears. The equation is then written as

$$I = \begin{cases} \dfrac{\cos \phi_t \sin \phi_t}{2m_N} \dfrac{m_G}{m_G + 1} & \text{external gears} \\[3mm] \dfrac{\cos \phi_t \sin \phi_t}{2m_N} \dfrac{m_G}{m_G - 1} & \text{internal gears} \end{cases} \tag{14–23}$$

where $m_N = 1$ for spur gears. In solving Eq. (14–21) for m_N, note that

$$p_N = p_n \cos \phi_n \tag{14–24}$$

where p_n is the normal circular pitch. The quantity Z, for use in Eq. (14–21), can be obtained from the equation

$$Z = [(r_P + a)^2 - r_{bP}^2]^{1/2} + [(r_G + a)^2 - r_{bG}^2]^{1/2} - (r_P + r_G) \sin \phi_t \tag{14–25}$$

where r_P and r_G are the pitch radii and r_{bP} and r_{bG} the base-circle radii of the pinion and gear, respectively.[6] Recall from Eq. (13–6), the radius of the base circle is

$$r_b = r \cos \phi_t \tag{14–26}$$

---

[6]For a development, see Joseph E. Shigley and John J. Uicker Jr., *Theory of Machines and Mechanisms*, McGraw-Hill, New York, 1980, p. 262.

Certain precautions must be taken in using Eq. (14–25). The tooth profiles are not conjugate below the base circle, and consequently, if either one or the other of the first two terms in brackets is larger than the third term, then it should be replaced by the third term. In addition, the effective outside radius is sometimes less than $r + a$, owing to removal of burrs or rounding of the tips of the teeth. When this is the case, always use the effective outside radius instead of $r + a$.

14–6   The Elastic Coefficient C_p (Z_E)

Values of C_p may be computed directly from Eq. (14–13) or obtained from Table 14–8.

14–7   Dynamic Factor K_v

As noted earlier, dynamic factors are used to account for inaccuracies in the manufacture and meshing of gear teeth in action. *Transmission error* is defined as the departure from uniform angular velocity of the gear pair. Some of the effects that produce transmission error are:

- Inaccuracies produced in the generation of the tooth profile; these include errors in tooth spacing, profile lead, and runout
- Vibration of the tooth during meshing due to the tooth stiffness
- Magnitude of the pitch-line velocity
- Dynamic unbalance of the rotating members
- Wear and permanent deformation of contacting portions of the teeth
- Gearshaft misalignment and the linear and angular deflection of the shaft
- Tooth friction

In an attempt to account for these effects, AGMA has defined a set of *quality numbers*, Q_v.[7] These numbers define the tolerances for gears of various sizes manufactured to a specified accuracy. Quality numbers 3 to 7 will include most commercial-quality gears. Quality numbers 8 to 12 are of precision quality. The following equations for the dynamic factor are based on these Q_v numbers:

$$K_v = \begin{cases} \left(\dfrac{A + \sqrt{V}}{A}\right)^B & V \text{ in ft/min} \\[3mm] \left(\dfrac{A + \sqrt{200V}}{A}\right)^B & V \text{ in m/s} \end{cases} \tag{14–27}$$

where

$$A = 50 + 56(1 - B)$$
$$B = 0.25(12 - Q_v)^{2/3} \tag{14–28}$$

---

[7]AGMA 2000-A88. ANSI/AGMA 2001-D04, adopted in 2004, replaced the quality number Q_v with the transmission accuracy level number A_v and incorporated ANSI/AGMA 2015-1-A01. A_v ranges from 6 to 12, with lower numbers representing greater accuracy. The Q_v approach was maintained as an alternate approach, and resulting K_v values are comparable.

Table 14-8

Elastic Coefficient C_p (Z_E), $\sqrt{\text{psi}}$ ($\sqrt{\text{MPa}}$)   Source: AGMA 218.01

Pinion Material	Pinion Modulus of Elasticity E_P psi (MPa)*	Gear Material and Modulus of Elasticity E_G, lbf/in² (MPa)*					
		Steel 30×10^6 (2×10^5)	Malleable Iron 25×10^6 (1.7×10^5)	Nodular Iron 24×10^6 (1.7×10^5)	Cast Iron 22×10^6 (1.5×10^5)	Aluminum Bronze 17.5×10^6 (1.2×10^5)	Tin Bronze 16×10^6 (1.1×10^5)
Steel	30×10^6 (2×10^5)	2300 (191)	2180 (181)	2160 (179)	2100 (174)	1950 (162)	1900 (158)
Malleable iron	25×10^6 (1.7×10^5)	2180 (181)	2090 (174)	2070 (172)	2020 (168)	1900 (158)	1850 (154)
Nodular iron	24×10^6 (1.7×10^5)	2160 (179)	2070 (172)	2050 (170)	2000 (166)	1880 (156)	1830 (152)
Cast iron	22×10^6 (1.5×10^5)	2100 (174)	2020 (168)	2000 (166)	1960 (163)	1850 (154)	1800 (149)
Aluminum bronze	17.5×10^6 (1.2×10^5)	1950 (162)	1900 (158)	1880 (156)	1850 (154)	1750 (145)	1700 (141)
Tin bronze	16×10^6 (1.1×10^5)	1900 (158)	1850 (154)	1830 (152)	1800 (149)	1700 (141)	1650 (137)

Poisson's ratio = 0.30.

*When more exact values for modulus of elasticity are obtained from roller contact tests, they may be used.

Figure 14–9

Dynamic factor K_v. The equations to these curves are given by Eq. (14–27) and the end points by Eq. (14–29). *(ANSI/AGMA 2001-D04, Annex A)*

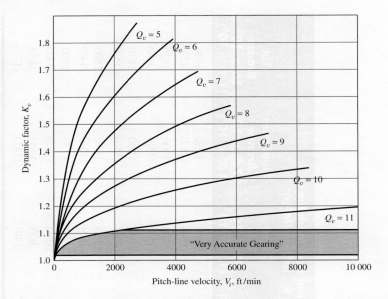

Figure 14–9 graphically represents Eq. (14–27). The maximum recommended pitch-line velocity for a given quality number is represented by the end point of each Q_v curve, and is given by

$$(V_t)_{\max} = \begin{cases} [A + (Q_v - 3)]^2 & \text{ft/min} \\[2mm] \dfrac{[A + (Q_v - 3)]^2}{200} & \text{m/s} \end{cases} \qquad (14\text{–}29)$$

14–8 Overload Factor K_o

The overload factor K_o is intended to make allowance for all externally applied loads in excess of the nominal tangential load W^t in a particular application (see Figs. 14–17 and 14–18 for tables). Examples include variations in torque from the mean value due to firing of cylinders in an internal combustion engine or reaction to torque variations in a piston pump drive. There are other similar factors such as application factor or service factor. These factors are established after considerable field experience in a particular application.[8]

14–9 Surface Condition Factor C_f (Z_R)

The surface condition factor C_f or Z_R is used only in the pitting resistance equation, Eq. (14–16). It depends on

- Surface finish as affected by, but not limited to, cutting, shaving, lapping, grinding, shotpeening
- Residual stress
- Plastic effects (work hardening)

Standard surface conditions for gear teeth have not yet been established. When a detrimental surface finish effect is known to exist, AGMA specifies a value of C_f greater than unity.

---

[8]An extensive list of service factors appears in Howard B. Schwerdlin, "Couplings," Chap. 16 in Joseph E. Shigley, Charles R. Mischke, and Thomas H. Brown, Jr. (eds.), *Standard Handbook of Machine Design*, 3rd ed., McGraw-Hill, New York, 2004.

14-10   Size Factor K_s

The size factor reflects nonuniformity of material properties due to size. It depends upon

- Tooth size
- Diameter of part
- Ratio of tooth size to diameter of part
- Face width
- Area of stress pattern
- Ratio of case depth to tooth size
- Hardenability and heat treatment

Standard size factors for gear teeth have not yet been established for cases where there is a detrimental size effect. In such cases AGMA recommends a size factor greater than unity. If there is no detrimental size effect, use unity.

AGMA has identified and provided a symbol for size factor. Also, AGMA suggests $K_s = 1$, which makes K_s a placeholder in Eqs. (14–15) and (14–16) until more information is gathered. Following the standard in this manner is a failure to apply all of your knowledge. From Table 13–1, p. 688, $l = a + b = 2.25/P$. The tooth thickness t in Fig. 14–6 is given in Sec. 14–1, Eq. (b), as $t = \sqrt{4lx}$ where $x = 3Y/(2P)$ from Eq. (14–3). From Eq. (6–25), p. 297, the equivalent diameter d_e of a rectangular section in bending is $d_e = 0.808\sqrt{Ft}$. From Eq. (6–20), p. 296, $k_b = (d_e/0.3)^{-0.107}$. Noting that K_s is the reciprocal of k_b, we find the result of all the algebraic substitution is

$$K_s = \frac{1}{k_b} = 1.192\left(\frac{F\sqrt{Y}}{P}\right)^{0.0535} \qquad (a)$$

K_s can be viewed as Lewis's geometry incorporated into the Marin size factor in fatigue. You may set $K_s = 1$, or you may elect to use the preceding Eq. (a). This is a point to discuss with your instructor. We will use Eq. (a) to remind you that you have a choice. If K_s in Eq. (a) is less than 1, use $K_s = 1$.

14-11   Load-Distribution Factor K_m (K_H)

The load-distribution factor modified the stress equations to reflect nonuniform distribution of load across the line of contact. The ideal is to locate the gear "midspan" between two bearings at the zero slope place when the load is applied. However, this is not always possible. The following procedure is applicable to

- Net face width to pinion pitch diameter ratio $F/d_P \le 2$
- Gear elements mounted between the bearings
- Face widths up to 40 in
- Contact, when loaded, across the full width of the narrowest member

The load-distribution factor under these conditions is currently given by the *face load distribution factor*, C_{mf}, where

$$K_m = C_{mf} = 1 + C_{mc}(C_{pf}C_{pm} + C_{ma}C_e) \qquad (14\text{–}30)$$

where

$$C_{mc} = \begin{cases} 1 & \text{for uncrowned teeth} \\ 0.8 & \text{for crowned teeth} \end{cases} \qquad (14\text{--}31)$$

$$C_{pf} = \begin{cases} \dfrac{F}{10d} - 0.025 & F \leq 1 \text{ in} \\[2mm] \dfrac{F}{10d} - 0.0375 + 0.0125F & 1 < F \leq 17 \text{ in} \\[2mm] \dfrac{F}{10d} - 0.1109 + 0.0207F - 0.000\,228F^2 & 17 < F \leq 40 \text{ in} \\[2mm] \dfrac{b}{10d} - 0.025 & b \leq 25 \text{ mm} \\[2mm] \dfrac{b}{10d} - 0.0375 + 4.92(10^{-4})b & 25 < b \leq 425 \text{ mm} \\[2mm] \dfrac{b}{10d} - 0.1109 + 8.15(10^{-4})b - 3.53(10^{-7})b^2 & 425 < b \leq 1000 \text{ mm} \end{cases} \qquad (14\text{--}32)$$

Note that for values of $F/(10d_P) < 0.05$, $F/(10d_P) = 0.05$ is used.

$$C_{pm} = \begin{cases} 1 & \text{for straddle-mounted pinion with } S_1/S < 0.175 \\ 1.1 & \text{for straddle-mounted pinion with } S_1/S \geq 0.175 \end{cases} \qquad (14\text{--}33)$$

$$C_{ma} = A + BF + CF^2 \qquad \text{(see Table 14–9 for values of } A, B, \text{ and } C) \qquad (14\text{--}34)$$

$$C_e = \begin{cases} 0.8 & \text{for gearing adjusted at assembly, or compatibility} \\ & \text{is improved by lapping, or both} \\ 1 & \text{for all other conditions} \end{cases} \qquad (14\text{--}35)$$

See Fig. 14–10 for definitions of S and S_1 for use with Eq. (14–33), and see Fig. 14–11 for graph of C_{ma}.

Table 14–9

Empirical Constants A, B, and C for Eq. (14–34), Face Width F in Inches*

Source: ANSI/AGMA 2001-D04.

Condition	A	B	C
Open gearing	0.247	0.0167	$-0.765(10^{-4})$
Commercial, enclosed units	0.127	0.0158	$-0.930(10^{-4})$
Precision, enclosed units	0.0675	0.0128	$-0.926(10^{-4})$
Extraprecision enclosed gear units	0.00360	0.0102	$-0.822(10^{-4})$

*See ANSI/AGMA 2101-D04, pp. 20–22, for SI formulation.

Figure 14–10

Definition of distances S and S_1 used in evaluating C_{pm}, Eq. (14–33). (ANSI/AGMA 2001-D04.)

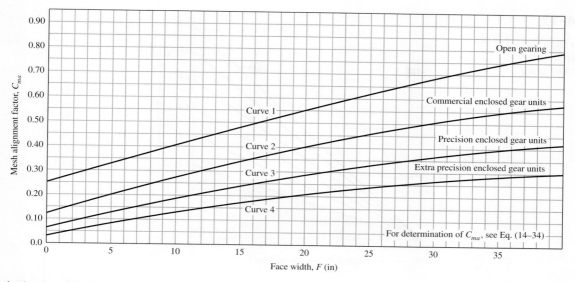

Figure 14-11

Mesh alignment factor C_{ma}. Curve-fit equations in Table 14–9. *(ANSI/AGMA 2001-D04.)*

14-12   Hardness-Ratio Factor C_H (Z_W)

The pinion generally has a smaller number of teeth than the gear and consequently is subjected to more cycles of contact stress. If both the pinion and the gear are through-hardened, then a uniform surface strength can be obtained by making the pinion harder than the gear. A similar effect can be obtained when a surface-hardened pinion is mated with a through-hardened gear. The hardness-ratio factor C_H is used *only for the gear.* Its purpose is to adjust the surface strengths for this effect. For the pinion, $C_H = 1$. For the gear, C_H is obtained from the equation

$$C_H = 1.0 + A'(m_G - 1.0) \tag{14-36}$$

where

$$A' = 8.98(10^{-3})\left(\frac{H_{BP}}{H_{BG}}\right) - 8.29(10^{-3}) \qquad 1.2 \le \frac{H_{BP}}{H_{BG}} \le 1.7$$

The terms H_{BP} and H_{BG} are the Brinell hardness (10-mm ball at 3000-kg load) of the pinion and gear, respectively. The term m_G is the speed ratio and is given by Eq. (14–22). See Fig. 14–12 for a graph of Eq. (14–36). For

$$\frac{H_{BP}}{H_{BG}} < 1.2, \qquad A' = 0$$

$$\frac{H_{BP}}{H_{BG}} > 1.7, \qquad A' = 0.006\,98$$

When surface-hardened pinions with hardnesses of 48 Rockwell C scale (Rockwell C48) or harder are run with through-hardened gears (180–400 Brinell), a work hardening occurs. The C_H factor is a function of pinion surface finish f_P and the mating gear hardness. Figure 14–13 displays the relationships:

$$C_H = 1 + B'(450 - H_{BG}) \tag{14-37}$$

Figure 14–12

Hardness-ratio factor C_H (through-hardened steel). *(ANSI/AGMA 2001-D04.)*

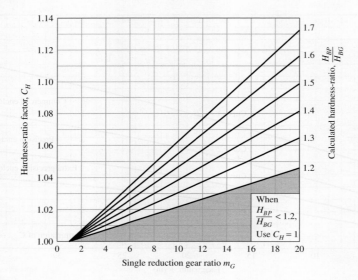

Figure 14–13

Hardness-ratio factor C_H (surface-hardened steel pinion). *(ANSI/AGMA 2001-D04.)*

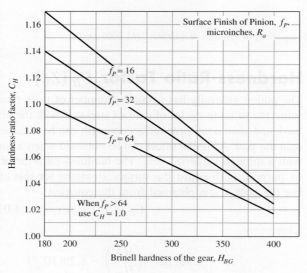

where $B' = 0.000\ 75\ \exp[-0.0112 f_P]$ and f_P is the surface finish of the pinion expressed as root-mean-square roughness R_a in μ in.

14–13   Stress-Cycle Factors Y_N and Z_N

The AGMA strengths as given in Figs. 14–2 through 14–4, in Tables 14–3 and 14–4 for bending fatigue, and in Fig. 14–5 and Tables 14–5 and 14–6 for contact-stress fatigue are based on 10^7 load cycles applied. The purpose of the stress-cycle factors Y_N and Z_N is to modify the gear strength for lives other than 10^7 cycles. Values for these factors are given in Figs. 14–14 and 14–15. Note that for 10^7 cycles $Y_N = Z_N = 1$ on each graph. Note also that the equations for Y_N and Z_N change on either side of 10^7 cycles. For life goals slightly higher than 10^7 cycles, the mating gear may be experiencing fewer than 10^7 cycles and the equations for $(Y_N)_P$ and $(Y_N)_G$ can be different. The same comment applies to $(Z_N)_P$ and $(Z_N)_G$.

Figure 14–14

Repeatedly applied bending strength stress-cycle factor Y_N. *(ANSI/AGMA 2001-D04.)*

Figure 14–15

Pitting resistance stress-cycle factor Z_N. *(ANSI/AGMA 2001-D04.)*

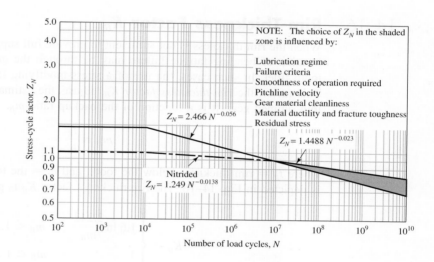

14–14   Reliability Factor K_R (Y_Z)

The reliability factor accounts for the effect of the statistical distributions of material fatigue failures. Load variation is not addressed here. The gear strengths S_t and S_c are based on a reliability of 99 percent. Table 14–10 is based on data developed by the U.S. Navy for bending and contact-stress fatigue failures.

The functional relationship between K_R and reliability is highly nonlinear. When interpolation is required, linear interpolation is too crude. A log transformation to each quantity produces a linear string. A least-squares regression fit is

$$K_R = \begin{cases} 0.658 - 0.0759 \ln(1 - R) & 0.5 < R < 0.99 \\ 0.50 - 0.109 \ln(1 - R) & 0.99 \leq R \leq 0.9999 \end{cases} \qquad (14\text{–}38)$$

For cardinal values of R, take K_R from the table. Otherwise use the logarithmic interpolation afforded by Eqs. (14–38).

Table 14–10

Reliability Factors K_R (Y_Z)

Source: ANSI/AGMA 2001-D04.

Reliability	K_R (Y_Z)
0.9999	1.50
0.999	1.25
0.99	1.00
0.90	0.85
0.50	0.70

14–15   Temperature Factor K_T (Y_θ)

For oil or gear-blank temperatures up to 250°F (120°C), use $K_T = Y_\theta = 1.0$. For higher temperatures, the factor should be greater than unity. Heat exchangers may be used to ensure that operating temperatures are considerably below this value, as is desirable for the lubricant.

14–16   Rim-Thickness Factor K_B

When the rim thickness is not sufficient to provide full support for the tooth root, the location of bending fatigue failure may be through the gear rim rather than at the tooth fillet. In such cases, the use of a stress-modifying factor K_B is recommended. This factor, the *rim-thickness factor K_B*, adjusts the estimated bending stress for the thin-rimmed gear. It is a function of the backup ratio m_B,

$$m_B = \frac{t_R}{h_t} \tag{14–39}$$

where t_R = rim thickness below the tooth, and h_t = the tooth height. The geometry is depicted in Fig. 14–16. The rim-thickness factor K_B is given by

$$K_B = \begin{cases} 1.6 \ln \dfrac{2.242}{m_B} & m_B < 1.2 \\ 1 & m_B \geq 1.2 \end{cases} \tag{14–40}$$

Figure 14–16

Rim-thickness factor K_B.
(ANSI/AGMA 2001-D04.)

For $m_B < 1.2$
$K_B = 1.6 \ln \left(\dfrac{2.242}{m_B} \right)$

For $m_B \geq 1.2$
$K_B = 1.0$

$m_B = \dfrac{t_R}{h_t}$

Figure 14–16 also gives the value of K_B graphically. The rim-thickness factor K_B is applied in addition to the 0.70 reverse-loading factor when applicable.

14–17 Safety Factors S_F and S_H

The ANSI/AGMA standards 2001-D04 and 2101-D04 contain a safety factor S_F guarding against bending fatigue failure and safety factor S_H guarding against pitting failure.

The definition of S_F, from Eq. (14–17), for U.S. customary units, is

$$S_F = \frac{S_t Y_N/(K_T K_R)}{\sigma} = \frac{\text{fully corrected bending strength}}{\text{bending stress}} \qquad (14\text{--}41)$$

where σ is estimated from Eq. (14–15), for U.S. customary units. It is a strength-over-stress definition in a case where the stress is linear with the transmitted load.

The definition of S_H, from Eq. (14–18), is

$$S_H = \frac{S_c Z_N C_H/(K_T K_R)}{\sigma_c} = \frac{\text{fully corrected contact strength}}{\text{contact stress}} \qquad (14\text{--}42)$$

when σ_c is estimated from Eq. (14–16). This, too, is a strength-over-stress definition but in a case where the stress is *not* linear with the transmitted load W^t.

While the definition of S_H does not interfere with its intended function, a caution is required when comparing S_F with S_H in an analysis in order to ascertain the nature and severity of the threat to loss of function. To render S_H linear with the transmitted load, W^t it could have been defined as

$$S_H = \left(\frac{\text{fully corrected contact strength}}{\text{contact stress imposed}}\right)^2 \qquad (14\text{--}43)$$

with the exponent 2 for linear or helical contact, or an exponent of 3 for crowned teeth (spherical contact). With the definition, Eq. (14–42), compare S_F with S_H^2 (or S_H^3 for crowned teeth) when trying to identify the threat to loss of function with confidence.

The role of the overload factor K_o is to include predictable excursions of load beyond W^t based on experience. A safety factor is intended to account for unquantifiable elements in addition to K_o. When designing a gear mesh, the quantity S_F becomes a design factor $(S_F)_d$ within the meanings used in this book. The quantity S_F evaluated as part of a design assessment is a factor of safety. This applies equally well to the quantity S_H.

14–18 Analysis

Description of the procedure based on the AGMA standard is highly detailed. The best review is a "road map" for bending fatigue and contact-stress fatigue. Figure 14–17 identifies the bending stress equation, the endurance strength in bending equation, and the factor of safety S_F. Figure 14–18 displays the contact-stress equation, the contact fatigue endurance strength equation, and the factor of safety S_H. The equations in these figures are in terms of U.S. customary units. Similar roadmaps can readily be generated in terms of SI units.

The following example of a gear mesh analysis is intended to make all the details presented concerning the AGMA method more familiar.

SPUR GEAR BENDING
Based on ANSI/AGMA 2001-D04 (U.S. customary units)

$$d_P = \frac{N_P}{P_d}$$

$$V = \frac{\pi d n}{12}$$

$$W^t = \frac{33\,000\,H}{V}$$

Gear
bending
stress
equation
Eq. (14–15)

$$\sigma = W^t K_o K_v K_s \frac{P_d}{F} \frac{K_m K_B}{J}$$

1 [or Eq. (a), Sec. 14–10]; p. 751

Eq. (14–30); p. 751

Eq. (14–40); p. 756

Fig. 14–6; p. 745

Eq. (14–27); p. 748

Table below

Gear
bending
endurance
strength
equation
Eq. (14–17)

$$\sigma_{all} = \frac{S_t}{S_F} \frac{Y_N}{K_T K_R}$$

$_{0.99}(S_t)_{10^7}$ Tables 14–3, 14–4; pp. 740, 741

Fig. 14–14; p. 755

Table 14–10, Eq. (14–38); pp. 756, 755

1 if $T < 250°F$

Bending
factor of
safety
Eq. (14–41)

$$S_F = \frac{S_t Y_N/(K_T K_R)}{\sigma}$$

Remember to compare S_F with S_H^2 when deciding whether bending
or wear is the threat to function. For crowned gears compare S_F with S_H^3.

Table of Overload Factors, K_o

Power source	Driven Machine		
	Uniform	Moderate shock	Heavy shock
Uniform	1.00	1.25	1.75
Light shock	1.25	1.50	2.00
Medium shock	1.50	1.75	2.25

Figure 14–17

Roadmap of gear bending equations based on AGMA standards. *(ANSI/AGMA 2001-D04.)*

SPUR GEAR WEAR
Based on ANSI/AGMA 2001-D04 (U.S. customary units)

$$d_P = \frac{N_P}{P_d}$$

$$V = \frac{\pi d n}{12}$$

$$W^t = \frac{33\,000\,\text{H}}{V}$$

1 [or Eq. (a), Sec. 14–10]; p. 751

Eq. (14–30); p. 751

Gear
contact
stress
equation
Eq. (14–16)

$$\sigma_c = C_p \left(W^t K_o K_v K_s \frac{K_m}{d_P F} \frac{C_f}{I} \right)^{1/2}$$

1

Eq. (14–23); p. 747

Eq. (14–27); p. 748

Table below

Eq. (14–13), Table 14–8; pp. 736, 749

$0.99(S_c)_{10}7$ Tables 14–6, 14–7; pp. 743, 744

Fig. 14–15; p. 755

Gear
contact
endurance
strength
Eq. (14–18)

$$\sigma_{c,\text{all}} = \frac{S_c\, Z_N\, C_H}{S_H\, K_T\, K_R}$$

Section 14–12, gear only; pp. 753, 754

Table 14–10, Eq. (14–38); pp. 756, 755

1 if $T < 250°F$

Wear
factor of
safety
Eq. (14–42)

Gear only

$$S_H = \frac{S_c\, Z_N\, C_H / (K_T\, K_R)}{\sigma_c}$$

Remember to compare S_F with S_H^2 when deciding whether bending
or wear is the threat to function. For crowned gears compare S_F with S_H^3.

Table of Overload Factors, K_o

Power source	Driven Machine		
	Uniform	Moderate shock	Heavy shock
Uniform	1.00	1.25	1.75
Light shock	1.25	1.50	2.00
Medium shock	1.50	1.75	2.25

Figure 14–18

Roadmap of gear wear equations based on AGMA standards. *(ANSI/AGMA 2001-D04.)*

EXAMPLE 14–4 A 17-tooth 20° pressure angle spur pinion rotates at 1800 rev/min and transmits 2982.8 Watt to a 52-tooth disk gear. The 0.4 teeth/mm, the 38.1 mm, and the quality standard is No. 6. The gears are straddle-mounted with bearings immediately adjacent. The pinion is a grade 1 steel with a hardness of 240 Brinell tooth surface and through-hardened core. The gear is steel, through-hardened also, grade 1 material, with a Brinell hardness of 200, tooth surface and core. Poisson's ratio is 0.30, $J_P = 0.30$, $J_G = 0.40$, and Young's modulus is 206.8 GPa. The loading is smooth because of motor and load. Assume a pinion life of 10^8 cycles and a reliability of 0.90, and use $Y_N = 1.3558N^{-0.0178}$, $Z_N = 1.4488N^{-0.023}$. The tooth profile is uncrowned. This is a commercial enclosed gear unit.

(a) Find the factor of safety of the gears in bending.
(b) Find the factor of safety of the gears in wear.
(c) By examining the factors of safety, identify the threat to each gear and to the mesh.

Solution   There will be many terms to obtain so use Figs. 14–17 and 14–18 as guides to what is needed.

$$d_P = N_P/P_d = 17/0.4 = 42.5 \text{ m} \quad d_G = 52/0.4 = 130 \text{ mm}$$

$$V = \frac{\pi d_P n_P}{60} = \frac{\pi(42.5)1800}{60} = 4005.53 \text{ mm/s}$$

$$W^t = \frac{33\,000\,H}{V} = \frac{33\,000(2982.8)}{0.40055} = 744.67 \text{ N}$$

Assuming uniform loading, $K_o = 1$. To evaluate K_v, from Eq. (14–28) with a quality number $Q_v = 6$,

$$B = 0.25(12 - 6)^{2/3} = 0.8255$$

$$A = 50 + 56(1 - 0.8255) = 59.77$$

Then from Eq. (14–27) the dynamic factor is

$$K_v = \left(\frac{59.77 + \sqrt{200(0.4)}}{59.77}\right)^{0.8255} = 1.37$$

To determine the size factor, K_s, the Lewis form factor is needed. From Table 14–2, with $N_P = 17$ teeth, $Y_P = 0.303$. Interpolation for the gear with $N_G = 52$ teeth yields $Y_G = 0.412$. Thus from Eq. (a) of Sec. 14–10, with $F = 38.1$ mm,

$$(K_s)_P = 1.192\left(\frac{38.1\sqrt{0.303}}{0.4}\right)^{0.0535} = 1.4708$$

$$(K_s)_G = 1.192\left(\frac{38.1\sqrt{0.412}}{0.4}\right)^{0.0535} = 1.4829$$

The load distribution factor K_m is determined from Eq. (14–30), where five terms are needed. They are, where 38.1 mm when needed:

> Uncrowned, Eq. (14–30): $C_{mc} = 1$,
> Eq. (14–32): $C_{pf} = 38.1/[10(42.5)] - 0.0375 + 4.92(10e\text{-}4)\,(38.1) = 0.071$
> Bearings immediately adjacent, Eq. (14–33): $C_{pm} = 1$
> Commercial enclosed gear units (Fig. 14–11): $C_{ma} = 0.15$
> Eq. (14–35): $C_e = 1$

Thus,

$$K_m = 1 + C_{mc}(C_{pf}C_{pm} + C_{ma}C_e) = 1 + (1)[0.071(1) + 0.15(1)] = 1.22$$

Assuming constant thickness gears, the rim-thickness factor $K_B = 1$. The speed ratio is $m_G = N_G/N_P = 52/17 = 3.059$. The load cycle factors given in the problem statement, with $N(\text{pinion}) = 10^8$ cycles and $N(\text{gear}) = 10^8/m_G = 10^8/3.059$ cycles, are

$$(Y_N)_P = 1.3558(10^8)^{-0.0178} = 0.977$$

$$(Y_N)_G = 1.3558(10^8/3.059)^{-0.0178} = 0.996$$

From Table 14.10, with a reliability of 0.9, $K_R = 0.85$. From Fig. 14–18, the temperature and surface condition factors are $K_T = 1$ and $C_f = 1$. From Eq. (14–23), with $m_N = 1$ for spur gears,

$$I = \frac{\cos 20° \sin 20°}{2} \frac{3.059}{3.059 + 1} = 0.121$$

From Table 14–8, $C_p = 191\sqrt{\text{MPa}}$.

Next, we need the terms for the gear endurance strength equations. From Table 14–3, for grade 1 steel with $H_{BP} = 240$ and $H_{BG} = 200$, we use Fig. 14–2, which gives

$$(S_t)_P = 0.533(240) + 88.3 = 216.22 \text{ MPa}$$

$$(S_t)_G = 0.533(200) + 88.3 = 194.9 \text{ MPa}$$

Similarly, from Table 14–6, we use Fig. 14–5, which gives

$$(S_c)_P = 2.22(240) + 200 = 732.8 \text{ MPa}$$

$$(S_c)_G = 2.22(200) + 200 = 644 \text{ MPa}$$

From Fig. 14–15,

$$(Z_N)_P = 1.4488(10^8)^{-0.023} = 0.948$$

$$(Z_N)_G = 1.4488(10^8/3.059)^{-0.023} = 0.973$$

For the hardness ratio factor C_H, the hardness ratio is $H_{BP}/H_{BG} = 240/200 = 1.2$. Then, from Sec. 14–12,

$$A' = 8.98(10^{-3})(H_{BP}/H_{BG}) - 8.29(10^{-3})$$

$$= 8.98(10^{-3})(1.2) - 8.29(10^{-3}) = 0.002\,49$$

Thus, from Eq. (14–36),

$$C_H = 1 + 0.002\,49(3.059 - 1) = 1.005$$

(*a*) **Pinion tooth bending.** Substituting the appropriate terms for the pinion into Eq. (14–15) gives

$$\sigma)_P = \left(W^t K_o K_v K_s \frac{P_d}{F} \frac{K_m K_B}{J}\right)_P = 744.67(1)1.377(1.4708)\frac{0.4}{38.1}\frac{1.22(1)}{0.30}$$

$$= 64.44 \text{ MPa}$$

Substituting the appropriate terms for the pinion into Eq. (14–41) gives

Answer

$$S_F)_P = \left(\frac{S_t Y_N/(K_T K_R)}{\sigma}\right)_P = \frac{216.22(0.977)/[1(0.85)]}{64.44} = 3.856$$

Gear tooth bending. Substituting the appropriate terms for the gear into Eq. (14–15) gives

$$(\sigma)_G = 744.67(1)1.377(1.4829)\frac{0.4}{38.1}\frac{1.22(1)}{0.40} = 48.73 \text{ MPa}$$

Substituting the appropriate terms for the gear into Eq. (14–41) gives

Answer

$$S_F)_G = \frac{194.9(0.996)/[1(0.85)]}{48.73} = 4.688$$

(*b*) **Pinion tooth wear.** Substituting the appropriate terms for the pinion into Eq. (14–16) gives

$$(\sigma_c)_P = C_P \left(W^t K_o K_v K_s \frac{K_m}{d_P F} \frac{C_f}{I}\right)_P^{1/2}$$

$$= 191 \left[744.67(1)1.377(1.4708)\frac{1.22}{42.5(38.1)}\frac{1}{0.121}\right]^{1/2} = 585.3 \text{ MPa}$$

Substituting the appropriate terms for the pinion into Eq. (14–42) gives

Answer

$$(S_H)_P = \left[\frac{S_c Z_N/(K_T K_R)}{\sigma_c}\right]_P = \frac{732.8(0.948)/[1(0.85)]}{585.3} = 1.397$$

Gear tooth wear. The only term in Eq. (14–16) that changes for the gear is K_s. Thus,

$$(\sigma_c)_G = \left[\frac{(K_s)_G}{(K_s)_P}\right]^{1/2} (\sigma_c)_P = \left(\frac{1.4829}{1.4708}\right)^{1/2} 585.3 = 587.71 \text{ MPa}$$

Substituting the appropriate terms for the gear into Eq. (14–42) with $C_H = 1.005$ gives

Answer

$$(S_H)_G = \frac{644(0.973)1.005/[1(0.85)]}{587.71} = 1.261$$

(*c*) For the pinion, we compare $(S_F)_P$ with $(S_H)_P^2$, or 3.856 with $1.3972 = 1\ 9520$, the threat in the pinion is from wear. For the gear, we compare $(S_F)_G$ with $(S_H)_G^2$, or 4.688 with $1.261^2 = 1.59$, so the threat in the gear is also from wear.

There are perspectives to be gained from Ex. 14–4. First, the pinion is overly strong in bending compared to wear. The performance in wear can be improved by surface-hardening techniques, such as flame or induction hardening, nitriding, or carburizing and case hardening, as well as shot peening. This in turn permits the gearset to be made smaller. Second, in bending, the gear is stronger than the pinion, indicating that both the gear core hardness and tooth size could be reduced; that is, we may increase P and reduce the diameters of the gears, or perhaps allow a cheaper material. Third, in wear, surface strength equations have the ratio $(Z_N)/K_R$. The values of $(Z_N)_P$ and $(Z_N)_G$ are affected by gear ratio m_G. The designer can control strength by specifying surface hardness. This point will be elaborated later.

Having followed a spur-gear analysis in detail in Ex. 14–4, it is timely to analyze a helical gearset under similar circumstances to observe similarities and differences.

EXAMPLE 14–5

A 17-tooth 20° normal pitch-angle helical pinion with a right-hand helix angle of 30° rotates at 1800 rev/min when transmitting 2982.8 Watt to a 52-tooth helical gear. The normal diametral pitch 0.4 teeth/mm, the face width is 38.1 mm, and the set has a quality number of 6. The gears are straddle-mounted with bearings immediately adjacent. The pinion and gear are made from a through-hardened steel with surface and core hardnesses of 240 Brinell on the pinion and surface and core hardnesses of 200 Brinell on the gear. The transmission is smooth, connecting an electric motor and a centrifugal pump. Assume a pinion life of 10^8 cycles and a reliability of 0.9 and use the upper curves in Figs. 14–14 and 14–15.

(a) Find the factors of safety of the gears in bending.
(b) Find the factors of safety of the gears in wear.
(c) By examining the factors of safety identify the threat to each gear and to the mesh.

Solution

All of the parameters in this example are the same as in Ex. 14–4 with the exception that we are using helical gears. Thus, several terms will be the same as Ex. 14–4. The reader should verify that the following terms remain unchanged: $K_o = 1$, $Y_P = 0.303$, $Y_G = 0.412$, $m_G = 3.059$, $(K_s)_P = 1.4708$, $(K_s)_G = 1.4829$, $(Y_N)_P = 0.977$, $(Y_N)_G = 0.996$, $K_R = 0.85$, $K_T = 1$, $C_f = 1$, $C_p = 191\sqrt{\text{MPa}}$, $(S_t)_P = 216.22$ MPa, $(S_t)_G = 194.9$ MPa, $(S_c)_P = 732.8$ MPa, $(S_c)_G = 644$ MPa, $(Z_N)_P = 0.948$, $(Z_N)_G = 0.973$, and $C_H = 1.005$.

For helical gears, the transverse diametral pitch, given by Eq. (13–18), p. 684, is

$$P_t = P_n \cos \psi = 0.4 \cos 30° = 0.3464 \text{ teeth/mm}$$

Thus, the pitch diameters are $d_P = N_P/P_t = 17/0.3464 = 49.075$ mm and $d_G = 52/0.3464 = 150.11$ mm. The pitch-line velocity and transmitted force are

$$V = \frac{\pi d_P n_P}{60} = \frac{\pi(49.075)1800}{60} = 4625.19 \text{ mm/s}$$

$$W^t = \frac{33\,000H}{V} = \frac{33\,000(2982.8)}{4625.19} = 644.9 \text{ N}$$

As in Ex. 14–4, for the dynamic factor, $B = 0.8255$ and $A = 59.77$. Thus, Eq. (14–27) gives

$$K_v = \left(\frac{59.77 + \sqrt{200(4.625)}}{59.77}\right)^{0.8255} = 1.40$$

The geometry factor I for helical gears requires a little work. First, the transverse pressure angle is given by Eq. (13–19) p. 684,

$$\phi_t = \tan^{-1}\left(\frac{\tan \phi_n}{\cos \psi}\right) = \tan^{-1}\left(\frac{\tan 20°}{\cos 30°}\right) = 22.80°$$

The radii of the pinion and gear are $r_P = 49.075/2 = 24.537$ in and $r_G = 150.111/2 = 22.62$ in, respectively. The addendum is $a = 1/P_n = 1/0.4 = 2.5$, and the base-circle radii of the pinion and gear are given by Eq. (13–6), p. 672, with $\phi = \phi_t$:

$$(r_b)_P = r_P \cos \phi_t = 24.537 \cos 22.80° = 22.62 \text{ mm}$$

$$(r_b)_G = 3.002 \cos 22.80° = 69.191$$

From Eq. (14–25), the surface strength geometry factor

$$Z = \sqrt{(24.537 + 2.5)^2 - 22.62^2} + \sqrt{(75.056 + 2.5)^2 - 69.191^2}$$
$$- (24.537 + 75.056) \sin 22.80°$$
$$= 14.8105 + 35.0354 - 38.5938 = 11.2520 \text{ in}$$

Since the first two terms are less than 38.5938, the equation for Z stands. From Eq. (14–24) the normal circular pitch p_N is

$$p_N = p_n \cos \phi_n = \frac{\pi}{P_n} \cos 20° = \frac{\pi}{0.4} \cos 20° = 7.3803 \text{ mm}$$

From Eq. (14–21), the load sharing ratio

$$m_N = \frac{p_N}{0.95Z} = \frac{7.3803}{0.95(11.2520)} = 0.6904$$

Substituting in Eq. (14–23), the geometry factor I is

$$I = \frac{\sin 22.80° \cos 22.80°}{2(0.6904)} \frac{3.06}{3.06 + 1} = 0.195$$

From Fig. 14–7, geometry factors $J'_P = 0.45$ and $J'_G = 0.54$. Also from Fig. 14–8 the J-factor multipliers are 0.94 and 0.98, correcting J'_P and J'_G to

$$J_P = 0.45(0.94) = 0.423$$

$$J_G = 0.54(0.98) = 0.529$$

The load-distribution factor K_m is estimated from Eq. (14–32):

$$C_{pf} = \frac{38.1}{10(49.075)} - 0.0375 + 4.92e\text{-}4(38.1) = 0.0569$$

with $C_{mc} = 1$, $C_{pm} = 1$, $C_{ma} = 0.15$ from Fig. 14–11, and $C_e = 1$. Therefore, from Eq. (14–30),

$$K_m = 1 + (1)[0.0569(1) + 0.15(1)] = 1.207$$

(a) **Pinion tooth bending.** Substituting the appropriate terms into Eq. (14–15) using P_t gives

$$(\sigma)_P = \left(W^t K_o K_v K_s \frac{P_t}{F} \frac{K_m K_B}{J} \right)_P = 644.904(1)1.404(1.043) \frac{0.3464}{38.1} \frac{1.208(1)}{0.423}$$
$$= 34.5545 \text{ MPa}$$

Substituting the appropriate terms for the pinion into Eq. (14–41) gives

Answer
$$(S_F)_P = \left(\frac{S_t Y_N / (K_T K_R)}{\sigma}\right)_P = \frac{216.22(0.977)/[1(0.85)]}{3445} = 7.19$$

Gear tooth bending. Substituting the appropriate terms for the gear into Eq. (14–15) gives

$$(\sigma)_G = 644.9035(1)1.404(1.052)\frac{0.3464}{38.1}\frac{1.208(1)}{0.529} = 27.85 \text{ MPa}$$

Substituting the appropriate terms for the gear into Eq. (14–41) gives

Answer
$$(S_F)_G = \frac{194.9(0.996)/[1(0.85)]}{27.85} = 8.2$$

(b) **Pinion tooth wear.** Substituting the appropriate terms for the pinion into Eq. (14–16) gives

$$(\sigma_c)_P = C_p \left(W^t K_o K_v K_s \frac{K_m}{d_P F}\frac{C_f}{I}\right)^{1/2}_P$$

$$= 191\left[644.9035(1)1.404(1.043)\frac{1.208}{49.075(38.1)}\frac{1}{0.195}\right]^{1/2} = 403 \text{ MPa}$$

Substituting the appropriate terms for the pinion into Eq. (14–42) gives

Answer
$$(S_H)_P = \left(\frac{S_c Z_N / (K_T K_R)}{\sigma_c}\right)_P = \frac{732.8(0.948)/[1(0.85)]}{403} = 2.03$$

Gear tooth wear. The only term in Eq. (14–16) that changes for the gear is K_s. Thus,

$$(\sigma_c)_G = \left[\frac{(K_s)_G}{(K_s)_P}\right]^{1/2}(\sigma_c)_P = \left(\frac{1.052}{1.043}\right)^{1/2}403 = 404 \text{ MPa}$$

Substituting the appropriate terms for the gear into Eq. (14–42) with $C_H = 1.005$ gives

Answer
$$(S_H)_G = \frac{644(0.973)1.005/[1(0.85)]}{404} = 1.8325$$

(c) For the pinion we compare S_F with S_H^2, or 7.19 with $2.03^2 = 4.12$, so the threat in the pinion is from wear. For the gear we compare S_F with S_H^2, or 8.20 with $1.83^2 = 3.358$, so the threat is also from wear in the gear. For the meshing gearset wear controls.

It is worthwhile to compare Ex. 14–4 with Ex. 14–5. The spur and helical gearsets were placed in nearly identical circumstances. The helical gear teeth are of greater length because of the helix and identical face widths. The pitch diameters of the helical gears are larger. The J factors and the I factor are larger, thereby reducing stresses. The result is larger factors of safety. In the design phase the gearsets in Ex. 14–4 and Ex. 14–5 can be made smaller with control of materials and relative hardnesses.

Now that examples have given the AGMA parameters substance, it is time to examine some desirable (and necessary) relationships between material properties of spur gears in mesh. In bending, the AGMA equations are displayed side by side:

$$\sigma_P = \left(W^t K_o K_v K_s \frac{P_d}{F} \frac{K_m K_B}{J}\right)_P \qquad \sigma_G = \left(W^t K_o K_v K_s \frac{P_d}{F} \frac{K_m K_B}{J}\right)_G$$

$$(S_F)_P = \left(\frac{S_t Y_N / (K_T K_R)}{\sigma}\right)_P \qquad (S_F)_G = \left(\frac{S_t Y_N / (K_T K_R)}{\sigma}\right)_G$$

Equating the factors of safety, substituting for stress and strength, canceling identical terms (K_s virtually equal or exactly equal), and solving for $(S_t)_G$ gives

$$(S_t)_G = (S_t)_P \frac{(Y_N)_P}{(Y_N)_G} \frac{J_P}{J_G} \tag{a}$$

The stress-cycle factor Y_N comes from Fig. 14–14, where for a particular hardness, $Y_N = \alpha N^\beta$. For the pinion, $(Y_N)_P = \alpha N_P^\beta$, and for the gear, $(Y_N)_G = \alpha(N_P/m_G)^\beta$. Substituting these into Eq. (a) and simplifying gives

$$(S_t)_G = (S_t)_P m_G^\beta \frac{J_P}{J_G} \tag{14–44}$$

Normally, $m_G > 1$ and $J_G > J_P$, so Eq. (14–44) shows that the gear can be less strong (lower Brinell hardness) than the pinion for the same safety factor.

EXAMPLE 14–6

In a set of spur gears, a 250-Brinell 14-tooth 16-pitch 20° full-depth pinion meshes with a 60-tooth gear. Both gear and pinion are of grade 1 through-hardened steel. Using $\beta = -0.023$, what hardness can the gear have for the same factor of safety?

Solution

For through-hardened grade 1 steel the pinion strength $(S_t)_P$ is given in Fig. 14–2:

$$(S_t)_P = 0.533(250) + 88.3 = 221.55 \text{ MPa}$$

From Fig. 14–6 the form factors are $J_P = 0.32$ and $J_G = 0.41$. Equation (14–44) gives

$$(S_t)_G = 221.55 \left(\frac{60}{14}\right)^{-0.023} \frac{0.32}{0.41} = 167.23 \text{ MPa}$$

Use the equation in Fig. 14–2 again.

Answer

$$(H_B)_G = \frac{167.23 - 88.3}{77.3} = 148.26 \text{ Brinell}$$

The AGMA contact-stress equations also are displayed side by side:

$$(\sigma_c)_P = C_p \left(W^t K_o K_v K_s \frac{K_m}{d_P F} \frac{C_f}{I}\right)_P^{1/2} \qquad (\sigma_c)_G = C_p \left(W^t K_o K_v K_s \frac{K_m}{d_P F} \frac{C_f}{I}\right)_G^{1/2}$$

$$(S_H)_P = \left(\frac{S_c Z_N / (K_T K_R)}{\sigma_c}\right)_P \qquad (S_H)_G = \left(\frac{S_c Z_N C_H / (K_T K_R)}{\sigma_c}\right)_G$$

Equating the factors of safety, substituting the stress relations, and canceling identical terms including K_s gives, after solving for $(S_c)_G$,

$$(S_c)_G = (S_c)_P \frac{(Z_N)_P}{(Z_N)_G} \left(\frac{1}{C_H}\right)_G = (S_C)_P m_G^\beta \left(\frac{1}{C_H}\right)_G$$

where, as in the development of Eq. (14–44), $(Z_N)_P/(Z_N)_G = m_G^\beta$ and the value of β for wear comes from Fig. 14–15. Since C_H is so close to unity, it is usually neglected; therefore

$$(S_c)_G = (S_c)_P m_G^\beta \tag{14–45}$$

EXAMPLE 14–7　　For $\beta = -0.056$ for a through-hardened steel, grade 1, continue Ex. 14–6 for wear.

Solution　　From Fig. 14–5,

$$(S_c)_P = 2.22(300) + 200 = 866 \text{ MPa}$$

From Eq. (14–45),

$$(S_c)_G = (S_c)_P \left(\frac{64}{18}\right)^{-0.056} = 866 \left(\frac{64}{18}\right)^{-0.056} = 807 \text{ MPa}$$

Answer

$$(H_B)_G = \frac{807 - 200}{2.22} = 273 \text{ Brinell}$$

which is slightly less than the pinion hardness of 300 Brinell.

Equations (14–44) and (14–45) apply as well to helical gears.

14–19　Design of a Gear Mesh

A useful decision set for spur and helical gears includes

- Function: load, speed, reliability, life, K_o
- Unquantifiable risk: design factor n_d ⎫
- Tooth system: ϕ, ψ, addendum, dedendum, root fillet radius ⎬ a priori decisions
- Gear ratio m_G, N_P, N_G
- Quality number Q_v ⎭

- Diametral pitch P_d ⎫
- Face width F ⎪
- Pinion material, core hardness, case hardness ⎬ design decisions
- Gear material, core hardness, case hardness ⎭

The first item to notice is the dimensionality of the decision set. There are four design decision categories, eight different decisions if you count them separately. This is a larger number than we have encountered before. It is important to use a design strategy that is convenient in either longhand execution or computer implementation. The design decisions have been placed in order of importance (impact on the amount of work to be redone in iterations). The steps, after the a priori decisions have been made are

- Choose a diametral pitch.
- Examine implications on face width, pitch diameters, and material properties. If not satisfactory, return to pitch decision for change.
- Choose a pinion material and examine core and case hardness requirements. If not satisfactory, return to pitch decision and iterate until no decisions are changed.
- Choose a gear material and examine core and case hardness requirements. If not satisfactory, return to pitch decision and iterate until no decisions are changed.

With these plan steps in mind, we can consider them in more detail.

First select a trial diametral pitch.

Pinion bending:

- Select a median face width for this pitch, $4\pi/P$
- Find the range of necessary ultimate strengths
- Choose a material and a core hardness
- Find face width to meet factor of safety in bending
- Choose face width
- Check factor of safety in bending

Gear bending:

- Find necessary companion core hardness
- Choose a material and core hardness
- Check factor of safety in bending

Pinion wear:

- Find necessary S_c and attendant case hardness
- Choose a case hardness
- Check factor of safety in wear

Gear wear:

- Find companion case hardness
- Choose a case hardness
- Check factor of safety in wear

Completing this set of steps will yield a satisfactory design. Additional designs with diametral pitches adjacent to the first satisfactory design will produce several among which to choose. A figure of merit is necessary in order to choose the best. Unfortunately, a figure of merit in gear design is complex in an academic environment because material and processing costs vary. The possibility of using a process depends on the manufacturing facility if gears are made in house.

After examining Ex. 14–4 and Ex. 14–5 and seeing the wide range of factors of safety, one might entertain the notion of setting all factors of safety equal.[9] In

---

[9]In designing gears it makes sense to define the factor of safety in wear as $(S)_H^2$ for uncrowned teeth, so that there is no mix-up. ANSI, in the preface to ANSI/AGMA 2001-D04 and 2101-D04, states "the use is completely voluntary . . . does not preclude anyone from using . . . procedures . . . not conforming to the standards."

steel gears, wear is usually controlling and $(S_H)_P$ and $(S_H)_G$ can be brought close to equality. The use of softer cores can bring down $(S_F)_P$ and $(S_F)_G$, but there is value in keeping them higher. A tooth broken by bending fatigue not only can destroy the gear set, but can bend shafts, damage bearings, and produce inertial stresses up and downstream in the power train, causing damage elsewhere if the gear box locks.

EXAMPLE 14–8

Design a 4:1 spur-gear reduction for a 74.57 kW, three-phase squirrel-cage induction motor running at 1120 rev/min. The load is smooth, providing a reliability of 0.95 at 10^9 revolutions of the pinion. Gearing space is meager. Use Nitralloy 135M, grade 1 material to keep the gear size small. The gears are heat-treated first then nitrided.

Solution

Make the a priori decisions:

- Function: 74.57 kW, 1120 rev/min, $R = 0.95$, $N = 10^9$ cycles, $K_o = 1$
- Design factor for unquantifiable exingencies: $n_d = 2$
- Tooth system: $\phi_n = 20°$
- Tooth count: $N_P = 18$ teeth, $N_G = 72$ teeth (no interference, Sec. 13–7, p. 677)
- Quality number: $Q_v = 6$, use grade 1 material
- Assume $m_B \geq 1.2$ in Eq. (14–40), $K_B = 1$

Pitch: Select a trial diametral pitch of $P_d = 0.1575$ teeth/mm. Thus, $d_P = 18/0.1575 = 114.3$ mm and $d_G = 72/0.1575 = 457.2$ mm. From Table 14–2, $Y_P = 0.309$, $Y_G = 0.4324$ (interpolated). From Fig. 14–6, $J_P = 0.32$, $J_G = 0.415$.

$$V = \frac{\pi d_P n_P}{60} = \frac{\pi (114.3)1120}{60} = 6702.90 \text{ mm/s}$$

$$W^t = \frac{33\,000\,H}{V} = \frac{33\,000\,(74570)}{6702.9 \times 1e\text{-}3} = 11125.03 \text{ N}$$

From Eqs. (14–28) and (14–27),

$$B = 0.25(12 - Q_v)^{2/3} = 0.25(12 - 6)^{2/3} = 0.8255$$

$$A = 50 + 56(1 - 0.8255) = 59.77$$

$$K_v = \left(\frac{59.77 + \sqrt{1319}}{59.77}\right)^{0.8255} = 1.480$$

From Eq. (14–38), $K_R = 0.658 - 0.0759\ln(1 - 0.95) = 0.885$. From Fig. 14–14,

$$(Y_N)_P = 1.3558(10^9)^{-0.0178} = 0.938$$

$$(Y_N)_G = 1.3558(10^9/4)^{-0.0178} = 0.961$$

From Fig. 14–15,

$$(Z_N)_P = 1.4488(10^9)^{-0.023} = 0.900$$

$$(Z_N)_G = 1.4488(10^9/4)^{-0.023} = 0.929$$

From the recommendation after Eq. (14–8), $3p \leq F \leq 5p$. Try $F = 4p = 4\pi/P = 4\pi/0.1575 = 79.8$ mm. From Eq. (a), Sec. 14–10,

$$K_s = 1.192 \left(\frac{F\sqrt{Y}}{P}\right)^{0.0535} = 1.192 \left(\frac{79.8\sqrt{0.309}}{0.1576}\right)^{0.0535} = 1.612$$

From Eqs. (14–31), (14–33) and (14–35), $C_{mc} = C_{pm} = C_e = 1$. From Fig. 14–11, $C_{ma} = 0.175$ for commercial enclosed gear units. From Eq. (14–32), $F/(10d_P) = 79.8/[10(114.30)] = 0.0698$. Thus,

$$C_{pf} = 0.0698 - 0.0375 + 4.92 \times 1e\text{-}4\ (79.8) = 0.0717$$

From Eq. (14–30),

$$K_m = 1 + (1)[0.0717(1) + 0.175(1)] = 1.247$$

From Table 14–8, for steel gears, $C_p = 191\sqrt{\text{MPa}}$. From Eq. (14–23), with $m_G = 4$ and $m_N = 1$,

$$I = \frac{\cos 20° \sin 20°}{2} \frac{4}{4+1} = 0.1286$$

Pinion tooth bending. With the above estimates of K_s and K_m from the trial diametral pitch, we check to see if the mesh width F is controlled by bending or wear considerations. Equating Eqs. (14–15) and (14–17), substituting $n_d W^t$ for W^t, and solving for the face width $(F)_{\text{bend}}$ necessary to resist bending fatigue, we obtain

$$(F)_{\text{bend}} = n_d W^t K_o K_v K_s P_d \frac{K_m K_B}{J_P} \frac{K_T K_R}{S_t Y_N} \tag{1}$$

Equating Eqs. (14–16) and (14–18), substituting $n_d W^t$ for W^t, and solving for the face width $(F)_{\text{wear}}$ necessary to resist wear fatigue, we obtain

$$(F)_{\text{wear}} = \left(\frac{C_p K_T K_R}{S_c Z_N}\right)^2 n_d W^t K_o K_v K_s \frac{K_m C_f}{d_P I} \tag{2}$$

From Table 14–5 the hardness range of Nitralloy 135M is Rockwell C32–36 (302–335 Brinell). Choosing a midrange hardness as attainable, using 320 Brinell. From Fig. 14–4,

$$S_t = 0.594(320) + 87.76 = 277.84 \text{ MPa}$$

Inserting the numerical value of S_t in Eq. (1) to estimate the face width gives

$$(F)_{\text{bend}} = 2(11125.03)(1)1.48(1.14)0.1575 \frac{1.247(1)(1)0.885}{0.32(277.84)0.938} = 110.95 \text{ mm}$$

From Table 14–6 for Nitralloy 135M, $S_c = 1172$ MPa. Inserting this in Eq. (2), we find

$$(F)_{\text{wear}} = \left(\frac{191(1)(0.885)}{1172\,(0.900)}\right)^2 2(11125.03)1(1.48)1.14 \frac{1.247(1)}{114.3(0.1286)} = 116.15 \text{ mm}$$

Decision   Make face width 120 mm. Correct K_s and K_m:

$$K_s = 1.192 \left(\frac{120\sqrt{0.309}}{0.1575} \right)^{0.0535} = 1.6475$$

$$\frac{F}{10d_P} = \frac{120}{10(114.3)} = 0.1050$$

$$C_{pf} = 0.1050 - .0375 + 4.92 \times 1e\text{-}4 \, (120) = 0.1268$$

$$K_m = 1 + (1)[0.1268(1) + .175(1)] = 1.3018$$

The bending stress induced by W^t in bending, from Eq. (14–15), is

$$(\sigma)_P = 11125.03(1)1.48(1.147) \; \frac{0.1575}{120} \; \frac{1.259(1)}{0.32} = 145.16 \text{ MPa}$$

The AGMA factor of safety in bending of the pinion, from Eq. (14–41), is

$$(S_F)_P = \frac{277.84(0.938)/1(0.885)]}{145.16} = 2.03$$

Decision   **Gear tooth bending.** Use cast gear blank because of the 18-in pitch diameter. Use the same material, heat treatment, and nitriding. The load-induced bending stress is in the ratio of J_P/J_G. Then

$$(\sigma)_G = 145.16 \; \frac{0.32}{0.415} = 112 \text{ MPa}$$

The factor of safety of the gear in bending is

$$(S_F)_G = \frac{277.84(0.961)/1(0.885)]}{112} = 2.69$$

Pinion tooth wear. The contact stress, given by Eq. (14–16), is

$$(\sigma_c)_P = 191 \left[11125.03(1)1.48(1.147) \; \frac{1.259}{114.3(120)} \; \frac{1}{0.129} \right]^{1/2} = 828 \text{ MPa}$$

The factor of safety from Eq. (14–42), is

$$(S_H)_P = \frac{1172(0.900)/[1(0.885)]}{828} = 1.437$$

By our definition of factor of safety, pinion bending is $(S_F)_P = 2.03$, and wear is $(S_H)_P^2 = (1.437)^2 = 2.07$.

Gear tooth wear. The hardness of the gear and pinion are the same. Thus, from Fig. 14–12, $C_H = 1$, the contact stress on the gear is the same as the pinion, $(\sigma_c)_G = 828$ MPa. The wear strength is also the same, $S_c = 1172$ MPa. The factor of safety of the gear in wear is

$$(S_H)_G = \frac{1172 \, (0.929)/[1(0.885)]}{828} = 1.48$$

So, for the gear in bending, $(S_F)_G = 2.69$, and wear $(S_H)_G^2 = (1.48)^2 = 2.20$.

Rim. Keep $m_B \geq 1.2$. The whole depth is h_t = addendum + dedendum = $1/P_d$ + $1.25/P_d = 2.25/P_d = 2.25/0.1575 = 14.2875$. The rim thickness t_R is

$$t_R \geq m_B h_t = 1.2(14.2875) = 17.1450 \text{ mm}$$

In the design of the gear blank, be sure the rim thickness exceeds 17.1450 mm; if it does not, review and modify this mesh design.

This design example showed a satisfactory design for a four-pitch spur-gear mesh. Material could be changed, as could pitch. There are a number of other satisfactory designs, thus a figure of merit is needed to identify the best.

One can appreciate that gear design was one of the early applications of the digital computer to mechanical engineering. A design program should be interactive, presenting results of calculations, pausing for a decision by the designer, and showing the consequences of the decision, with a loop back to change a decision for the better. The program can be structured in totem-pole fashion, with the most influential decision at the top, then tumbling down, decision after decision, ending with the ability to change the current decision or to begin again. Such a program would make a fine class project. Troubleshooting the coding will reinforce your knowledge, adding flexibility as well as bells and whistles in subsequent terms.

Standard gears may not be the most economical design that meets the functional requirements, because no application is standard in all respects.[10] Methods of designing custom gears are well understood and frequently used in mobile equipment to provide good weight-to-performance index. The required calculations including optimizations are within the capability of a personal computer.

PROBLEMS

Problems marked with an asterisk (*) are linked to problems in other chapters, as summarized in Table 1–2 of Sec. 1–17, p. 34.

Because the results will vary depending on the method used, the problems are presented by section.

14–1   A steel spur pinion has a pitch of 3 mm, 22 full-depth teeth, and a 20° pressure angle. The pinion runs at a speed of 1200 rev/min and transmits 11 kW to a 60-tooth gear. If the face width is 50 mm, estimate the bending stress.

14–2   A steel spur pinion has a diametral pitch of 10 teeth/in, 18 teeth cut full-depth with a 20° pressure angle, and a face width of 1 in. This pinion is expected to transmit 2 hp at a speed of 600 rev/min. Determine the bending stress.

14–3   A steel spur pinion has a module of 1.25 mm, 18 teeth cut on the 20° full-depth system, and a face width of 12 mm. At a speed of 1800 rev/min, this pinion is expected to carry a steady load of 0.5 kW. Determine the bending stress.

---

[10]See H. W. Van Gerpen, C. K. Reece, and J. K. Jensen, *Computer Aided Design of Custom Gears*, Van Gerpen–Reece Engineering, Cedar Falls, Iowa, 1996.

14–4   A steel spur pinion has 16 teeth cut on the 20° full-depth system with a module of 8 mm and a face width of 90 mm. The pinion rotates at 150 rev/min and transmits 6 kW to the mating steel gear. What is the bending stress?

14–5   A steel spur pinion has a module of 1 mm and 16 teeth cut on the 20° full-depth system and is to carry 0.15 kW at 400 rev/min. Determine a suitable face width based on an allowable bending stress of 150 MPa.

14–6   A 20° full-depth steel spur pinion has 20 teeth and a module of 2 mm and is to transmit 0.5 kW at a speed of 200 rev/min. Find an appropriate face width if the bending stress is not to exceed 75 MPa.

14–7   A 20° full-depth steel spur pinion has a module of 5 mm and 24 teeth and transmits 4.5 kW at a speed of 50 rev/min. Find an appropriate face width if the allowable bending stress is 140 MPa.

14–8   A steel spur pinion is to transmit 20 hp at a speed of 400 rev/min. The pinion is cut on the 20° full-depth system and has a diametral pitch of 4 teeth/in and 16 teeth. Find a suitable face width based on an allowable stress of 12 kpsi.

14–9   A 20° full-depth steel spur pinion with 18 teeth is to transmit 2.5 hp at a speed of 600 rev/min. Determine appropriate values for the face width and diametral pitch based on an allowable bending stress of 10 kpsi.

14–10   A 20° full-depth steel spur pinion is to transmit 1.5 kW hp at a speed of 900 rev/min. If the pinion has 18 teeth, determine suitable values for the module and face width. The bending stress should not exceed 75 MPa.

14–11   A speed reducer has 20° full-depth teeth and consists of a 20-tooth steel spur pinion driving a 60-tooth cast-iron gear. The horsepower transmitted is 11 kW at a pinion speed of 1200 rev/min. For a module of 4 mm and a face width of 50 mm, find the contact stress.

14–12   A gear drive consists of a 16-tooth 20° steel spur pinion and a 48-tooth cast-iron gear having a pitch of 12 teeth/in. For a power input of 1.5 hp at a pinion speed of 700 rev/min, select a face width based on an allowable contact stress of 100 kpsi.

14–13   A gearset has a module of 5 mm, a 20° pressure angle, and a 24-tooth cast-iron spur pinion driving a 48-tooth cast-iron gear. The pinion is to rotate at 50 rev/min. What horsepower input can be used with this gearset if the contact stress is limited to 690 MPa and $F = 60$ mm?

14–14   A 20° 20-tooth cast-iron spur pinion having a module of 4 mm drives a 32-tooth cast-iron gear. Find the contact stress if the pinion speed is 1000 rev/min, the face width is 50 mm, and 10 kW of power is transmitted.

14–15   A steel spur pinion and gear have a diametral pitch of 12 teeth/in, milled teeth, 17 and 30 teeth, respectively, a 20° pressure angle, a face width of $\frac{7}{8}$ in, and a pinion speed of 525 rev/min. The tooth properties are $S_{ut} = 76$ kpsi, $S_y = 42$ kpsi and the Brinell hardness is 149. Use the Gerber criteria to compensate for one-way bending. For a design factor of 2.25, what is the power rating of the gearset?

14–16   A milled-teeth steel pinion and gear pair have $S_{ut} = 113$ kpsi, $S_y = 86$ kpsi and a hardness at the involute surface of 262 Brinell. The diametral pitch is 3 teeth/in, the face width is 2.5 in, and the pinion speed is 870 rev/min. The tooth counts are 20 and 100. Use the Gerber criteria to compensate for one-way bending. For a design factor of 1.5, rate the gearset for power considering both bending and wear.

14–17 A 20° full-depth steel spur pinion rotates at 1145 rev/min. It has a module of 6 mm, a face width of 75 mm, and 16 milled teeth. The ultimate tensile strength at the involute is 900 MPa exhibiting a Brinell hardness of 260. The gear is steel with 30 teeth and has identical material strengths. Use the Gerber criteria to compensate for one-way bending. For a design factor of 1.3 find the power rating of the gearset based on the pinion and the gear resisting bending and wear fatigue.

14–18 A steel spur pinion has a pitch of 6 teeth/in, 17 full-depth milled teeth, and a pressure angle of 20°. The pinion has an ultimate tensile strength at the involute surface of 116 kpsi, a Brinell hardness of 232, and a yield strength of 90 kpsi. Its shaft speed is 1120 rev/min, its face width is 2 in, and its mating gear has 51 teeth. Use a design factor of 2.

    (*a*) Pinion bending fatigue imposes what power limitation? Use the Gerber criteria to compensate for one-way bending.

    (*b*) Pinion surface fatigue imposes what power limitation? The gear has identical strengths to the pinion with regard to material properties.

    (*c*) Determine power limitations due to gear bending and wear.

    (*d*) Specify the power rating for the gearset.

14–19 A commercial enclosed gear drive consists of a 20° spur pinion having 16 teeth driving a 48-tooth gear. The pinion speed is 300 rev/min, the face width 50 mm, and the diametral pitch 4 mm. The gears are grade 1 steel, through-hardened at 200 Brinell, made to No. 6 quality standards, uncrowned, and are to be accurately and rigidly mounted. Assume a pinion life of 10^8 cycles and a reliability of 0.90. Determine the AGMA bending and contact stresses and the corresponding factors of safety if 4 kW is to be transmitted.

14–20 A 20° spur pinion with 20 teeth and a module of 2.5 mm transmits 120 W to a 36-tooth gear. The pinion speed is 100 rev/min, and the gears are grade 1, 18-mm face width, through-hardened steel at 200 Brinell, uncrowned, manufactured to a No. 6 quality standard, and considered to be of open gearing quality installation. Find the AGMA bending and contact stresses and the corresponding factors of safety for a pinion life of 10^8 cycles and a reliability of 0.95.

14–21 Repeat Prob. 14–19 using helical gears each with a 20° normal pitch angle and a helix angle of 30° and a normal diametral pitch of 6 teeth/in.

14–22 A spur gearset has 17 teeth on the pinion and 51 teeth on the gear. The pressure angle is 20° and the overload factor $K_o = 1$. The diametral pitch is 6 teeth/in and the face width is 2 in. The pinion speed is 1120 rev/min and its cycle life is to be 10^8 revolutions at a reliability $R = 0.99$. The quality number is 5. The material is a through-hardened steel, grade 1, with Brinell hardnesses of 232 core and case of both gears. For a design factor of 2, rate the gearset for these conditions using the AGMA method.

14–23 In Sec. 14–10, Eq. (*a*) is given for K_s based on the procedure in Ex. 14–2. Derive this equation.

14–24 A speed-reducer has 20° full-depth teeth, and the single-reduction spur-gear gearset has 22 and 60 teeth. The diametral pitch is 4 teeth/in and the face width is $3\frac{1}{4}$ in. The pinion shaft speed is 1145 rev/min. The life goal of 5-year 24-hour-per-day service is about $3(10^9)$ pinion revolutions. The absolute value of the pitch variation is such that the quality number is 6. The materials are 4340 through-hardened grade 1 steels, heat-treated to 250 Brinell, core and case, both gears. The load is moderate shock and the power is smooth. For a reliability of 0.99, rate the speed reducer for power.

14–25 The speed reducer of Prob. 14–24 is to be used for an application requiring 40 hp at 1145 rev/min. For the gear and the pinion, estimate the AGMA factors of safety for bending and wear, that is, $(S_F)_P$, $(S_F)_G$, $(S_H)_P$, and $(S_H)_G$. By examining the factors of safety, identify the threat to each gear and to the mesh.

14–26 The gearset of Prob. 14–24 needs improvement of wear capacity. Toward this end the gears are nitrided so that the grade 1 materials have hardnesses as follows: The pinion core is 250 and the pinion case hardness is 390 Brinell, and the gear core hardness is 250 core and 390 case. Estimate the power rating for the new gearset.

14–27 The gearset of Prob. 14–24 has had its gear specification changed to 9310 for carburizing and surface hardening with the result that the pinion Brinell hardnesses are 285 core and 580–600 case, and the gear hardnesses are 285 core and 580–600 case. Estimate the power rating for the new gearset.

14–28 The gearset of Prob. 14–27 is going to be upgraded in material to a quality of grade 2 (9310) steel. Estimate the power rating for the new gearset.

14–29 Matters of scale always improve insight and perspective. Reduce the physical size of the gearset in Prob. 14–24 by one-half and note the result on the estimates of transmitted load W^t and power.

14–30 AGMA procedures with cast-iron gear pairs differ from those with steels because life predictions are difficult; consequently $(Y_N)_P$, $(Y_N)_G$, $(Z_N)_P$, and $(Z_N)_G$ are set to unity. The consequence of this is that the fatigue strengths of the pinion and gear materials are the same. The reliability is 0.99 and the life is 10^7 revolution of the pinion ($K_R = 1$). For longer lives the reducer is derated in power. For the pinion and gear set of Prob. 14–24, use grade 40 cast iron for both gears ($H_B = 201$ Brinell). Rate the reducer for power with S_F and S_H equal to unity.

14–31 Spur-gear teeth have rolling and slipping contact (often about 8 percent slip). Spur gears tested to wear failure are reported at 10^8 cycles as Buckingham's surface fatigue load-stress factor K. This factor is related to Hertzian contact strength S_C by

$$S_C = \sqrt{\frac{1.4K}{(1/E_1 + 1/E_2)\sin\phi}}$$

where ϕ is the normal pressure angle. Cast iron grade 20 gears with $\phi = 14\frac{1}{2}°$ and 20° pressure angle exhibit a minimum K of 0.56 and 0.77 MPa, respectively. How does this compare with $S_C = 2.2H_B$ MPa?

14–32 You've probably noticed that although the AGMA method is based on two equations, the details of assembling all the factors is computationally intensive. To reduce error and omissions, a computer program would be useful. Write a program to perform a power rating of an existing gearset, then use Prob. 14–24, 14–26, 14–27, 14–28, and 14–29 to test your program by comparing the results to your longhand solutions.

14–33 In Ex. 14–5 use nitrided grade 1 steel (4140) which produces Brinell hardnesses of 250 core and 500 at the surface (case). Use the upper fatigue curves on Figs. 14–14 and 14–15. Estimate the power capacity of the mesh with factors of safety of $S_F = S_H = 1$.

14–34 In Ex. 14–5 use carburized and case-hardened gears of grade 1. Carburizing and case-hardening can produce a 550 Brinell case. The core hardnesses are 200 Brinell. Estimate the power capacity of the mesh with factors of safety of $S_F = S_H = 1$, using the lower fatigue curves in Figs. 14–14 and 14–15.

14–35 In Ex. 14–5, use carburized and case-hardened gears of grade 2 steel. The core hardnesses are 200, and surface hardnesses are 600 Brinell. Use the lower fatigue curves of Figs. 14–14 and 14–15. Estimate the power capacity of the mesh using $S_F = S_H = 1$. Compare the power capacity with the results of Prob. 14–34.

14–36* The countershaft in Prob. 3–72, p. 152, is part of a speed reducing compound gear train using 20° spur gears. A gear on the input shaft drives gear A. Gear B drives a gear on the output shaft. The input shaft runs at 2400 rev/min. Each gear reduces the speed (and thus increases the torque) by a 2 to 1 ratio. All gears are to be of the same material. Since gear B is the smallest gear, transmitting the largest load, it will likely be critical, so a preliminary analysis is to be performed on it. Use a module of 4 mm, a face-width of 4 times the circular pitch, a Grade 2 steel through-hardened to a Brinell hardness of 300, and a desired life of 11 kW with a 95 percent reliability. Determine factors of safety for bending and wear.

14–37* The countershaft in Prob. 3–73, p. 152, is part of a speed reducing compound gear train using 20° spur gears. A gear on the input shaft drives gear A with a 2 to 1 speed reduction. Gear B drives a gear on the output shaft with a 5 to 1 speed reduction. The input shaft runs at 1800 rev/min. All gears are to be of the same material. Since gear B is the smallest gear, transmitting the largest load, it will likely be critical, so a preliminary analysis is to be performed on it. Use a module of 18.75 mm/tooth, a face-width of 4 times the circular pitch, a Grade 2 steel through-hardened to a Brinell hardness of 300, and a desired life of 12 kh with a 98 percent reliability. Determine factors of safety for bending and wear.

14–38* Build on the results of Prob. 13–40, p. 720, to find factors of safety for bending and wear for gear F. Both gears are made from Grade 2 carburized and hardened steel. Use a face-width of 4 times the circular pitch. The desired life is 12 kh with a 95 percent reliability.

14–39* Build on the results of Prob. 13–41, p. 721, to find factors of safety for bending and wear for gear C. Both gears are made from Grade 2 carburized and hardened steel. Use a face-width of 4 times the circular pitch. The desired life is 14 kh with a 98 percent reliability.

Chapter Outline

The American Gear Manufacturers Association (AGMA) has established standards for the analysis and design of the various kinds of bevel and worm gears. Chapter 14 was an introduction to the AGMA methods for spur and helical gears and contains many of the definitions of terms used in this chapter. AGMA has established similar methods for other types of gearing, which all follow the same general approach.

15-1 Bevel Gearing—General

Bevel gears may be classified as follows:

- Straight bevel gears
- Spiral bevel gears
- Zerol bevel gears
- Hypoid gears
- Spiroid gears

A straight bevel gear was illustrated in Fig. 13–35, p. 701. These gears are usually used for pitch-line velocities up to 1000 ft/min (5 m/s) when the noise level is not an important consideration. They are available in many stock sizes and are less expensive to produce than other bevel gears, especially in small quantities.

A *spiral bevel* gear is shown in Fig. 15–1; the definition of the *spiral angle* is illustrated in Fig. 15–2. These gears are recommended for higher speeds and where the noise level is an important consideration. Spiral bevel gears are the bevel counterpart of the helical gear; it can be seen in Fig. 15–1 that the pitch surfaces and the nature of contact are the same as for straight bevel gears except for the differences brought about by the spiral-shaped teeth.

The *Zerol bevel gear* is a patented gear having curved teeth but with a zero spiral angle. The axial thrust loads permissible for Zerol bevel gears are not as large as those for the spiral bevel gear, and so they are often used instead of straight bevel gears. The Zerol bevel gear is generated by the same tool used for regular spiral bevel gears. For design purposes, use the same procedure as for straight bevel gears and then simply substitute a Zerol bevel gear.

Figure 15-1

Spiral bevel gears. *(Courtesy of Gleason Works, Rochester, N.Y.)*

Figure 15–2

Cutting spiral-gear teeth on the basic crown rack.

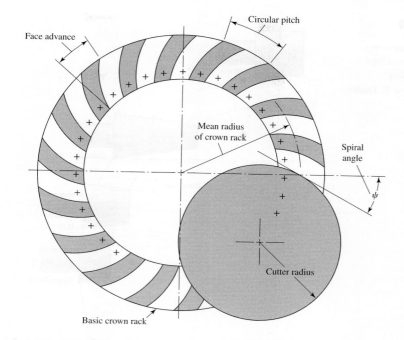

Face advance

Circular pitch

Mean radius of crown rack

Spiral angle

ψ

Cutter radius

Basic crown rack

Figure 15–3

Hypoid gears. *(Courtesy of Gleason Works, Rochester, N.Y.)*

It is frequently desirable, as in the case of automotive differential applications, to have gearing similar to bevel gears but with the shafts offset. Such gears are called *hypoid gears,* because their pitch surfaces are hyperboloids of revolution. The tooth action between such gears is a combination of rolling and sliding along a straight line and has much in common with that of worm gears. Figure 15–3 shows a pair of hypoid gears in mesh.

Figure 15–4 is included to assist in the classification of spiral bevel gearing. It is seen that the hypoid gear has a relatively small shaft offset. For larger offsets, the pinion begins to resemble a tapered worm and the set is then called *spiroid gearing.*

Figure 15–4

Comparison of intersecting-
and offset-shaft bevel-type
gearings. (*From* Gear
Handbook *by Darle W. Dudley,
1962, pp. 2–24.*)

15–2 **Bevel-Gear Stresses and Strengths**

In a typical bevel-gear mounting, Fig. 13–36, p. 702, for example, one of the gears is often mounted outboard of the bearings. This means that the shaft deflections can be more pronounced and can have a greater effect on the nature of the tooth contact. Another difficulty that occurs in predicting the stress in bevel-gear teeth is the fact that the teeth are tapered. Thus, to achieve perfect line contact passing through the cone center, the teeth should bend more at the large end than at the small end. To obtain this condition requires that the load be proportionately greater at the large end. Because of this varying load across the face of the tooth, it is desirable to have a fairly short face width.

Because of the complexity of bevel, spiral bevel, Zerol bevel, hypoid, and spiroid gears, as well as the limitations of space, only a portion of the applicable standards that refer to straight-bevel gears is presented here.[1] Table 15–1 gives the symbols used in ANSI/AGMA 2003-B97.

Fundamental Contact Stress Equation

$$s_c = \sigma_c = C_p \left(\frac{W^t}{F d_P I} K_o K_v K_m C_s C_{xc} \right)^{1/2} \quad \text{(U.S. customary units)}$$

(15–1)

$$\sigma_H = Z_E \left(\frac{1000 W^t}{b d Z_1} K_A K_\omega K_{H\beta} Z_x Z_{xc} \right)^{1/2} \quad \text{(SI units)}$$

The first term in each equation is the AGMA symbol, whereas σ_c, our normal notation, is directly equivalent.

---

[1]Figures 15–5 to 15–13 and Tables 15–1 to 15–7 have been extracted from ANSI/AGMA 2003-B97, *Rating the Pitting Resistance and Bending Strength of Generated Straight Bevel, Zerol Bevel and Spiral Bevel Gear Teeth* with the permission of the publisher, the American Gear Manufacturers Association, 1001 N. Fairfax Street, Suite 500, Alexandria, VA, 22314-1587.

Table 15–1

Symbols Used in Bevel Gear Rating Equations, ANSI/AGMA 2003-B97 Standard   *Source: ANSI/AGMA 2003-B97.*

AGMA Symbol	ISO Symbol	Description	Units
A_m	R_m	Mean cone distance	in (mm)
A_0	R_e	Outer cone distance	in (mm)
C_H	Z_W	Hardness ratio factor for pitting resistance	
C_i	Z_i	Inertia factor for pitting resistance	
C_L	Z_{NT}	Stress cycle factor for pitting resistance	
C_p	Z_E	Elastic coefficient	$[\text{lbf/in}^2]^{0.5}$ $([\text{N/mm}^2]^{0.5})$
C_R	Z_Z	Reliability factor for pitting	
C_{SF}		Service factor for pitting resistance	
C_S	Z_x	Size factor for pitting resistance	
C_{xc}	Z_{xc}	Crowning factor for pitting resistance	
D, d	d_{e2}, d_{e1}	Outer pitch diameters of gear and pinion, respectively	in (mm)
E_G, E_P	E_2, E_1	Young's modulus of elasticity for materials of gear and pinion, respectively	lbf/in^2 (N/mm^2)
e	e	Base of natural (Napierian) logarithms	
F	b	Net face width	in (mm)
F_{eG}, F_{eP}	b'_2, b'_1	Effective face widths of gear and pinion, respectively	in (mm)
f_P	R_{a1}	Pinion surface roughness	μin (μm)
H_{BG}	H_{B2}	Minimum Brinell hardness number for gear material	HB
H_{BP}	H_{B1}	Minimum Brinell hardness number for pinion material	HB
h_c	$E_{ht\,min}$	Minimum total case depth at tooth middepth	in (mm)
h_e	h'_c	Minimum effective case depth	in (mm)
$h_{e\,lim}$	$h'_{c\,lim}$	Suggested maximum effective case depth limit at tooth middepth	in (mm)
I	Z_I	Geometry factor for pitting resistance	
J	Y_J	Geometry factor for bending strength	
J_G, J_P	Y_{J2}, Y_{J1}	Geometry factor for bending strength for gear and pinion, respectively	
K_F	Y_F	Stress correction and concentration factor	
K_i	Y_i	Inertia factor for bending strength	
K_L	Y_{NT}	Stress cycle factor for bending strength	
K_m	$K_{H\beta}$	Load distribution factor	
K_o	K_A	Overload factor	
K_R	Y_z	Reliability factor for bending strength	
K_S	Y_X	Size factor for bending strength	
K_{SF}		Service factor for bending strength	
K_T	K_θ	Temperature factor	
K_v	K_v	Dynamic factor	
K_x	Y_β	Lengthwise curvature factor for bending strength	
	m_{et}	Outer transverse module	
	m_{mt}	Mean transverse module	(mm)
	m_{mn}	Mean normal module	(mm)
m_{NI}	ε_{NI}	Load sharing ratio, pitting	(mm)
m_{NJ}	ε_{NJ}	Load sharing ratio, bending	
N	z_2	Number of gear teeth	
N_L	n_L	Number of load cycles	
n	z_1	Number of pinion teeth	
n_P	n_1	Pinion speed	rev/min

(Continued)

Table 15–1

Symbols Used in Bevel Gear Rating Equations, ANSI/AGMA 2003-B97 Standard (*Continued*)

AGMA Symbol	ISO Symbol	Description	Units
P	P	Design power through gear pair	hp (kW)
P_a	P_a	Allowable transmitted power	hp (kW)
P_{ac}	P_{az}	Allowable transmitted power for pitting resistance	hp (kW)
P_{acu}	P_{azu}	Allowable transmitted power for pitting resistance at unity service factor	hp (kW)
P_{at}	P_{ay}	Allowable transmitted power for bending strength	hp (kW)
P_{atu}	P_{ayu}	Allowable transmitted power for bending strength at unity service factor	hp (kW)
P_d		Outer transverse diametral pitch	teeth/in
P_m		Mean transverse diametral pitch	teeth/in
P_{mn}		Mean normal diametral pitch	teeth/in
Q_v	Q_v	Transmission accuracy number	
q	q	Exponent used in formula for lengthwise curvature factor	
R, r	r_{mpt2}, r_{mpt1}	Mean transverse pitch radii for gear and pinion, respectively	in (mm)
R_t, r_t	r_{myo2}, r_{myo1}	Mean transverse radii to point of load application for gear and pinion, respectively	in (mm)
r_c	r_{c0}	Cutter radius used for producing Zerol bevel and spiral bevel gears	in (mm)
s	g_c	Length of the instantaneous line of contact between mating tooth surfaces	in (mm)
s_{ac}	$\sigma_{H\,\lim}$	Allowable contact stress number	lbf/in^2 (N/mm^2)
s_{at}	$\sigma_{F\,\lim}$	Bending stress number (allowable)	lbf/in^2 (N/mm^2)
s_c	σ_H	Calculated contact stress number	lbf/in^2 (N/mm^2)
s_F	s_F	Bending safety factor	
s_H	s_H	Contact safety factor	
s_t	σ_F	Calculated bending stress number	lbf/in^2 (N/mm^2)
s_{wc}	σ_{HP}	Permissible contact stress number	lbf/in^2 (N/mm^2)
s_{wt}	σ_{FP}	Permissible bending stress number	lbf/in^2 (N/mm^2)
T_P	T_1	Operating pinion torque	lbf in (Nm)
T_T	θ_T	Operating gear blank temperature	°F(°C)
t_0	s_{ai}	Normal tooth top land thickness at narrowest point	in (mm)
U_c	U_c	Core hardness coefficient for nitrided gear	lbf/in^2 (N/mm^2)
U_H	U_H	Hardening process factor for steel	lbf/in^2 (N/mm^2)
v_t	v_{et}	Pitch-line velocity at outer pitch circle	ft/min (m/s)
Y_{KG}, Y_{KP}	Y_{K2}, Y_{K1}	Tooth form factors including stress-concentration factor for gear and pinion, respectively	
μ_G, μ_p	ν_2, ν_1	Poisson's ratio for materials of gear and pinion, respectively	
ρ_0	ρy_o	Relative radius of profile curvature at point of maximum contact stress between mating tooth surfaces	in (mm)
ϕ	α_n	Normal pressure angle at pitch surface	
ϕ_t	α_{wt}	Transverse pressure angle at pitch point	
ψ	β_m	Mean spiral angle at pitch surface	
ψ_b	β_{mb}	Mean base spiral angle	

Permissible Contact Stress Number (Strength) Equation

$$s_{wc} = (\sigma_c)_{\text{all}} = \frac{s_{ac}C_L C_H}{S_H K_T C_R} \quad \text{(U.S. customary units)}$$

$$\sigma_{HP} = \frac{\sigma_{H\lim}Z_{NT}Z_W}{S_H K_\theta Z_Z} \quad \text{(SI units)}$$

(15–2)

Bending Stress

$$s_t = \frac{W^t}{F}P_d K_o K_v \frac{K_s K_m}{K_x J} \quad \text{(U.S. customary units)}$$

$$\sigma_F = \frac{1000W^t}{b}\frac{K_A K_v}{m_{et}}\frac{Y_x K_{H\beta}}{Y_\beta Y_J} \quad \text{(SI units)}$$

(15–3)

Permissible Bending Stress Equation

$$s_{wt} = \frac{s_{at}K_L}{S_F K_T K_R} \quad \text{(U.S. customary units)}$$

$$\sigma_{FP} = \frac{\sigma_{F\lim}Y_{NT}}{S_F K_\theta Y_z} \quad \text{(SI units)}$$

(15–4)

15–3   AGMA Equation Factors

Overload Factor K_o (K_A)

The overload factor makes allowance for any externally applied loads in excess of the nominal transmitted load. Table 15–2, from Appendix A of 2003-B97, is included for your guidance.

Safety Factors S_H and S_F

The factors of safety S_H and S_F as defined in 2003-B97 are adjustments to strength, not load, and consequently cannot be used as is to assess (by comparison) whether the threat is from wear fatigue or bending fatigue. Since W^t is the same for the pinion and gear, the comparison of $\sqrt{S_H}$ to S_F allows direct comparison.

Dynamic Factor K_v

In 2003-C87 AGMA changed the definition of K_v to its reciprocal but used the same symbol. Other standards have yet to follow this move. The dynamic factor K_v makes

Table 15–2

Overload Factors K_o (K_A)

Source: ANSI/AGMA 2003-B97.

Character of Prime Mover	Character of Load on Driven Machine			
	Uniform	Light Shock	Medium Shock	Heavy Shock
Uniform	1.00	1.25	1.50	1.75 or higher
Light shock	1.10	1.35	1.60	1.85 or higher
Medium shock	1.25	1.50	1.75	2.00 or higher
Heavy shock	1.50	1.75	2.00	2.25 or higher

Note: This table is for speed-decreasing drives. For speed-increasing drives, add $0.01(N/n)^2$ or $0.01(z_2/z_1)^2$ to the above factors.

Figure 15–5

Dynamic factor K_v.
(Source: ANSI/AGMA 2003-B97.)

allowance for the effect of gear-tooth quality related to speed and load, and the increase in stress that follows. AGMA uses a *transmission accuracy number* Q_v to describe the precision with which tooth profiles are spaced along the pitch circle. Figure 15–5 shows graphically how pitch-line velocity and transmission accuracy number are related to the dynamic factor K_v. Curve fits are

$$K_v = \left(\frac{A + \sqrt{v_t}}{A}\right)^B \qquad \text{(U.S. customary units)}$$

$$K_v = \left(\frac{A + \sqrt{200v_{et}}}{A}\right)^B \qquad \text{(SI units)}$$

(15–5)

where

$$A = 50 + 56(1 - B)$$
$$B = 0.25(12 - Q_v)^{2/3}$$

(15–6)

and $v_t(v_{et})$ is the pitch-line velocity at outside pitch diameter, expressed in ft/min (m/s):

$$v_t = \pi d_P n_P / 12 \qquad \text{(U.S. customary units)}$$

$$v_{et} = 5.236(10^{-5})d_1 n_1 \qquad \text{(SI units)}$$

(15–7)

The maximum recommended pitch-line velocity is associated with the abscissa of the terminal points of the curve in Fig. 15–5:

$$v_{t\,max} = [A + (Q_v - 3)]^2 \qquad \text{(U.S. customary units)}$$

$$v_{et\,max} = \frac{[A + (Q_v - 3)]^2}{200} \qquad \text{(SI units)}$$

(15–8)

where $v_{t\,max}$ and $v_{et\,max}$ are in ft/min and m/s, respectively.

Size Factor for Pitting Resistance C_s (Z_x)

$$C_s = \begin{cases} 0.5 & F < 0.5 \text{ in} \\ 0.125F + 0.4375 & 0.5 \leq F \leq 4.5 \text{ in} \\ 1 & F > 4.5 \text{ in} \end{cases} \quad \text{(U.S. customary units)}$$

$$Z_x = \begin{cases} 0.5 & b < 12.7 \text{ mm} \\ 0.004\,92b + 0.4375 & 12.7 \leq b \leq 114.3 \text{ mm} \\ 1 & b > 114.3 \text{ mm} \end{cases} \quad \text{(SI units)}$$

(15–9)

Size Factor for Bending K_s (Y_x)

$$K_S = \begin{cases} 0.4867 + 0.2132/P_d & 0.5 \leq P_d \leq 16 \text{ teeth/in} \\ 0.5 & P_d > 16 \text{ teeth/in} \end{cases} \quad \text{(U.S. customary units)}$$

$$Y_x = \begin{cases} 0.5 & m_{et} < 1.6 \text{ mm} \\ 0.4867 + 0.008\,339m_{et} & 1.6 \leq m_{et} \leq 50 \text{ mm} \end{cases} \quad \text{(SI units)}$$

(15–10)

Load-Distribution Factor K_m ($K_{H\beta}$)

$$K_m = K_{mb} + 0.0036F^2 \quad \text{(U.S. customary units)}$$
$$K_{H\beta} = K_{mb} + 5.6(10^{-6})b^2 \quad \text{(SI units)}$$

(15–11)

where

$$K_{mb} = \begin{cases} 1.00 & \text{both members straddle-mounted} \\ 1.10 & \text{one member straddle-mounted} \\ 1.25 & \text{neither member straddle-mounted} \end{cases}$$

Crowning Factor for Pitting C_{xc} (Z_{xc})

The teeth of most bevel gears are crowned in the lengthwise direction during manufacture to accommodate the deflection of the mountings.

$$C_{xc} = Z_{xc} = \begin{cases} 1.5 & \text{properly crowned teeth} \\ 2.0 & \text{or larger uncrowned teeth} \end{cases}$$

(15–12)

Lengthwise Curvature Factor for Bending Strength K_x (Y_β)

For straight-bevel gears,

$$K_x = Y_\beta = 1$$

(15–13)

Pitting Resistance Geometry Factor I (Z_I)

Figure 15–6 shows the geometry factor I (Z_I) for straight-bevel gears with a 20° pressure angle and 90° shaft angle. Enter the figure ordinate with the number of pinion teeth, move to the number of gear-teeth contour, and read from the abscissa.

Bending Strength Geometry Factor J (Y_J)

Figure 15–7 shows the geometry factor J for straight-bevel gears with a 20° pressure angle and 90° shaft angle.

Figure 15–6

Contact geometry factor I (Z_I) for coniflex straight-bevel gears with a 20° normal pressure angle and a 90° shaft angle. *(Source: ANSI/AGMA 2003-B97.)*

Figure 15–7

Bending factor J (Y_J) for coniflex straight-bevel gears with a 20° normal pressure angle and 90° shaft angle. *(Source: ANSI/AGMA 2003-B97.)*

Stress-Cycle Factor for Pitting Resistance C_L (Z_{NT})

$$C_L = \begin{cases} 2 & 10^3 \leq N_L < 10^4 \\ 3.4822N_L^{-0.0602} & 10^4 \leq N_L \leq 10^{10} \end{cases}$$

$$Z_{NT} = \begin{cases} 2 & 10^3 \leq n_L < 10^4 \\ 3.4822n_L^{-0.0602} & 10^4 \leq n_L \leq 10^{10} \end{cases}$$

(15–14)

See Fig. 15–8 for a graphical presentation of Eqs. (15–14).

Stress-Cycle Factor for Bending Strength K_L (Y_{NT})

$$K_L = \begin{cases} 2.7 & 10^2 \leq N_L < 10^3 \\ 6.1514N_L^{-0.1192} & 10^3 \leq N_L < 3(10^6) \\ 1.683N_L^{-0.0323} & 3(10^6) \leq N_L \leq 10^{10} \quad \text{critical} \\ 1.3558N_L^{-0.0178} & 3(10^6) \leq N_L \leq 10^{10} \quad \text{general} \end{cases}$$

$$Y_{NT} = \begin{cases} 2.7 & 10^2 \leq n_L < 10^3 \\ 6.1514n_L^{-0.1192} & 10^3 \leq n_L < 3(10^6) \\ 1.683n_L^{-0.0323} & 3(10^6) \leq n_L \leq 10^{10} \quad \text{critical} \\ 1.3558n_L^{-0.0178} & 3(10^6) \leq n_L \leq 10^{10} \quad \text{general} \end{cases}$$

(15–15)

See Fig. 15–9 for a plot of Eqs. (15–15).

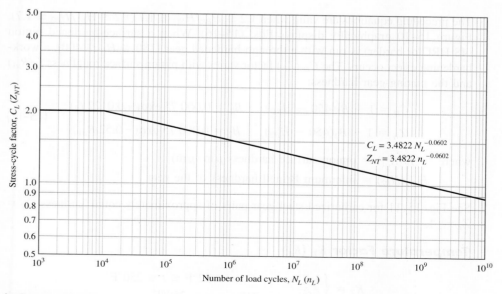

Figure 15–8

Contact stress-cycle factor for pitting resistance C_L (Z_{NT}) for carburized case-hardened steel bevel gears. *(Source: ANSI/AGMA 2003-B97.)*

Figure 15-9

Stress-cycle factor for bending strength K_L (Y_{NT}) for carburized case-hardened steel bevel gears. *(Source: ANSI/AGMA 2003-B97.)*

Hardness-Ratio Factor C_H (Z_W)

$$C_H = 1 + B_1(N/n - 1) \qquad B_1 = 0.008\,98(H_{BP}/H_{BG}) - 0.008\,29$$

$$Z_W = 1 + B_1(z_2/z_1 - 1) \qquad B_1 = 0.008\,98(H_{B1}/H_{B2}) - 0.008\,29 \qquad (15\text{-}16)$$

The preceding equations are valid when $1.2 \leq H_{BP}/H_{BG} \leq 1.7$ $(1.2 \leq H_{B1}/H_{B2} \leq 1.7)$. Figure 15-10 graphically displays Eqs. (15-16). When a surface-hardened pinion (48 HRC or harder) is run with a through-hardened gear $(180 \leq H_B \leq 400)$, a work-hardening effect occurs. The C_H (Z_W) factor varies with pinion surface roughness $f_P(R_{a1})$ and the mating-gear hardness:

$$C_H = 1 + B_2(450 - H_{BG}) \qquad B_2 = 0.000\,75 \exp(-0.0122\,f_P)$$

$$Z_W = 1 + B_2(450 - H_{B2}) \qquad B_2 = 0.000\,75 \exp(-0.52\,R_{a1}) \qquad (15\text{-}17)$$

where $f_P(R_{a1}) = $ pinion surface hardness μin (μm)

$H_{BG}(H_{B2}) = $ minimum Brinell hardness of the gear

See Fig. 15-11 for carburized steel gear pairs of approximately equal hardness $C_H = Z_W = 1$.

Temperature Factor K_T (K_θ)

$$K_T = \begin{cases} 1 & 32°\text{F} \leq t \leq 250°\text{F} \\ (460 + t)/710 & t > 250°\text{F} \end{cases}$$

$$K_\theta = \begin{cases} 1 & 0°\text{C} \leq \theta \leq 120°\text{C} \\ (273 + \theta)/393 & \theta > 120°\text{C} \end{cases} \qquad (15\text{-}18)$$

Figure 15–10

Hardness-ratio factor C_H (Z_W) for through-hardened pinion and gear.
(*Source: ANSI/AGMA 2003-B97.*)

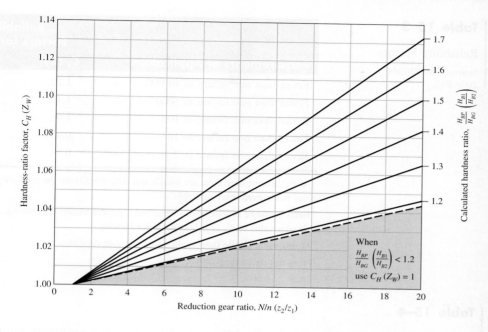

Figure 15–11

Hardness-ratio factor C_H (Z_W) for surface-hardened pinions.
(*Source: ANSI/AGMA 2003-B97.*)

Reliability Factors C_R (Z_Z) and K_R (Y_Z)

Table 15–3 displays the reliability factors. Note that $C_R = \sqrt{K_R}$ and $Z_Z = \sqrt{Y_Z}$. Logarithmic interpolation equations are

$$Y_Z = K_R = \begin{cases} 0.50 - 0.25 \log(1 - R) & 0.99 \le R \le 0.999 & \text{(15–19)} \\ 0.70 - 0.15 \log(1 - R) & 0.90 \le R < 0.99 & \text{(15–20)} \end{cases}$$

The reliability of the stress (fatigue) numbers allowable in Tables 15–4, 15–5, 15–6, and 15–7 is 0.99.

Table 15–3

Reliability Factors

Source: ANSI/AGMA 2003-B97.

Requirements of Application	Reliability Factors for Steel*	
	C_R (Z_Z)	K_R (Y_Z)[†]
Fewer than one failure in 10 000	1.22	1.50
Fewer than one failure in 1000	1.12	1.25
Fewer than one failure in 100	1.00	1.00
Fewer than one failure in 10	0.92	0.85[‡]
Fewer than one failure in 2	0.84	0.70[§]

*At the present time there are insufficient data concerning the reliability of bevel gears made from other materials.

[†]Tooth breakage is sometimes considered a greater hazard than pitting. In such cases a greater value of K_R (Y_Z) is selected for bending.

[‡]At this value plastic flow might occur rather than pitting.

[§]From test data extrapolation.

Table 15–4

Allowable Contact Stress Number for Steel Gears, s_{ac} $(\sigma_{H\,\lim})$ *Source: ANSI/AGMA 2003-B97.*

Material Designation	Heat Treatment	Minimum Surface* Hardness	Allowable Contact Stress Number, s_{ac} $(\sigma_{H\,\lim})$ lbf/in^2 (N/mm^2)		
			Grade 1[†]	Grade 2[†]	Grade 3[†]
Steel	Through-hardened[‡]	Fig. 15–12	Fig. 15–12	Fig. 15–12	Fig. 15–12
	Flame or induction hardened[§]	50 HRC	175 000 (1210)	190 000 (1310)	
	Carburized and case hardened[§]	2003-B97 Table 8	200 000 (1380)	225 000 (1550)	250 000 (1720)
AISI 4140	Nitrided[§]	84.5 HR15N		145 000 (1000)	
Nitralloy 135M	Nitrided[§]	90.0 HR15N		160 000 (1100)	

*Hardness to be equivalent to that at the tooth middepth in the center of the face width.

[†]See ANSI/AGMA 2003-B97, Tables 8 through 11, for metallurgical factors for each stress grade of steel gears.

[‡]These materials must be annealed or normalized as a minumum.

[§]The allowable stress numbers indicated may be used with the case depths prescribed in 21.1, ANSI/AGMA 2003-B97.

Elastic Coefficient for Pitting Resistance C_p (Z_E)

$$C_p = \sqrt{\frac{1}{\pi[(1 - \nu_P^2)/E_P + (1 - \nu_G^2)/E_G]}}$$

$$Z_E = \sqrt{\frac{1}{\pi[(1 - \nu_1^2)/E_1 + (1 - \nu_2^2)/E_2]}}$$

$$(15–21)$$

Table 15–5

Allowable Contact Stress Number for Iron Gears, s_{ac} ($\sigma_{H\,lim}$)   *Source: ANSI/AGMA 2003-B97.*

Material	Material Designation ASTM	Material Designation ISO	Heat Treatment	Typical Minimum Surface Hardness	Allowable Contact Stress Number, s_{ac} ($\sigma_{H\,lim}$) lbf/in² (N/mm²)
Cast iron	ASTM A48	ISO/DR 185			
	Class 30	Grade 200	As cast	175 HB	50 000 (345)
	Class 40	Grade 300	As cast	200 HB	65 000 (450)
Ductile (nodular) iron	ASTM A536	ISO/DIS 1083			
	Grade 80-55-06	Grade 600-370-03	Quenched	180 HB	94 000 (650)
	Grade 120-90-02	Grade 800-480-02	and tempered	300 HB	135 000 (930)

Table 15–6

Allowable Bending Stress Numbers for Steel Gears, s_{at} ($\sigma_{F\,lim}$)   *Source: ANSI/AGMA 2003-B97.*

Material Designation	Heat Treatment	Minimum Surface Hardness	Bending Stress Number (Allowable), s_{at} ($\sigma_{F\,lim}$) lbf/in² (N/mm²) Grade 1*	Grade 2*	Grade 3*
Steel	Through-hardened	Fig. 15–13	Fig. 15–13	Fig. 15–13	
	Flame or induction hardened				
	Unhardened roots	50 HRC	15 000 (85)	13 500 (95)	
	Hardened roots		22 500 (154)		
	Carburized and case hardened[†]	2003-B97 Table 8	30 000 (205)	35 000 (240)	40 000 (275)
AISI 4140	Nitrided[†,‡]	84.5 HR15N		22 000 (150)	
Nitralloy 135M	Nitrided[†,‡]	90.0 HR15N		24 000 (165)	

*See ANSI/AGMA 2003-B97, Tables 8–11, for metallurgical factors for each stress grade of steel gears.

[†]The allowable stress numbers indicated may be used with the case depths prescribed in 21.1, ANSI/AGMA 2003-B97.

[‡]The overload capacity of nitrided gears is low. Since the shape of the effective S-N curve is flat, the sensitivity to shock should be investigated before proceeding with the design.

where $\qquad C_p$ = elastic coefficient, 2290 $\sqrt{\text{psi}}$ for steel

$\qquad\qquad Z_E$ = elastic coefficient, 190 $\sqrt{\text{N/mm}^2}$ for steel

$\qquad E_P$ and E_G = Young's moduli for pinion and gear respectively, psi

$\qquad E_1$ and E_2 = Young's moduli for pinion and gear respectively, N/mm²

Allowable Contact Stress

Tables 15–4 and 15–5 provide values of $s_{ac}(\sigma_H)$ for steel gears and for iron gears, respectively. Figure 15–12 graphically displays allowable stress for grade 1 and 2 materials.

Table 15–7

Allowable Bending Stress Number for Iron Gears, s_{at} ($\sigma_{F\,lim}$) *Source: ANSI/AGMA 2003-B97.*

Material	Material Designation ASTM	ISO	Heat Treatment	Typical Minimum Surface Hardness	Bending Stress Number (Allowable), s_{at} ($\sigma_{F\,lim}$) lbf/in² (N/mm²)
Cast iron	ASTM A48	ISO/DR 185			
	Class 30	Grade 200	As cast	175 HB	4500 (30)
	Class 40	Grade 300	As cast	200 HB	6500 (45)
Ductile (nodular) iron	ASTM A536 Grade 80-55-06 Grade 120-90-02	ISO/DIS 1083 Grade 600-370-03 Grade 800-480-02	Quenched and tempered	180 HB 300 HB	10 000 (70) 13 500 (95)

Figure 15–12

Allowable contact stress number for through-hardened steel gears, s_{ac}($\sigma_{H\,lim}$). *(Source: ANSI/AGMA 2003-B97.)*

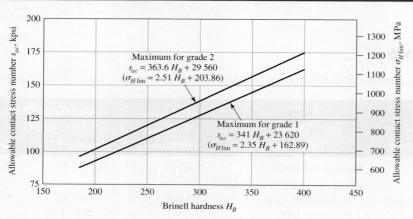

The equations are

$$s_{ac} = 341H_B + 23\,620 \text{ psi} \qquad \text{grade 1}$$
$$\sigma_{H\,lim} = 2.35H_B + 162.89 \text{ MPa} \qquad \text{grade 1}$$
$$s_{ac} = 363.6H_B + 29\,560 \text{ psi} \qquad \text{grade 2}$$
$$\sigma_{H\,lim} = 2.51H_B + 203.86 \text{ MPa} \qquad \text{grade 2}$$

(15–22)

Allowable Bending Stress Numbers

Tables 15–6 and 15–7 provide s_{at} ($\sigma_{F\,lim}$) for steel gears and for iron gears, respectively. Figure 15–13 shows graphically allowable bending stress s_{at} ($\sigma_{H\,lim}$) for through-hardened steels. The equations are

$$s_{at} = 44H_B + 2100 \text{ psi} \qquad \text{grade 1}$$
$$\sigma_{F\,lim} = 0.30H_B + 14.48 \text{ MPa} \qquad \text{grade 1}$$
$$s_{at} = 48H_B + 5980 \text{ psi} \qquad \text{grade 2}$$
$$\sigma_{H\,lim} = 0.33H_B + 41.24 \text{ MPa} \qquad \text{grade 2}$$

(15–23)

Reversed Loading

AGMA recommends use of 70 percent of allowable strength in cases where tooth load is completely reversed, as in idler gears and reversing mechanisms.

Summary

Figure 15–14 is a "road map" for straight-bevel gear wear relations using 2003-B97. Figure 15–15 is a similar guide for straight-bevel gear bending using 2003-B97.

Figure 15–13

Allowable bending stress number for through-hardened steel gears, $s_{at}(\sigma_{F\,\text{lim}})$. *(Source: ANSI/AGMA 2003-B97.)*

Figure 15–14

"Road map" summary of principal straight-bevel gear wear equations and their parameters.

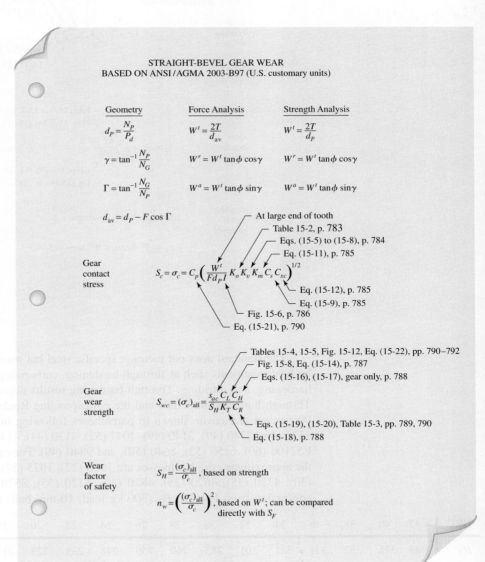

STRAIGHT-BEVEL GEAR WEAR
BASED ON ANSI/AGMA 2003-B97 (U.S. customary units)

Geometry	Force Analysis	Strength Analysis
$d_P = \dfrac{N_P}{P_d}$	$W^t = \dfrac{2T}{d_{\text{av}}}$	$W^t = \dfrac{2T}{d_P}$
$\gamma = \tan^{-1}\dfrac{N_P}{N_G}$	$W^r = W^t \tan\phi \, \cos\gamma$	$W^r = W^t \tan\phi \, \cos\gamma$
$\Gamma = \tan^{-1}\dfrac{N_G}{N_P}$	$W^a = W^t \tan\phi \, \sin\gamma$	$W^a = W^t \tan\phi \, \sin\gamma$
$d_{\text{av}} = d_P - F\cos\Gamma$		

Gear contact stress
$$S_c = \sigma_c = C_p\left(\frac{W^t}{Fd_P I} K_o K_v K_m C_s C_{xc}\right)^{1/2}$$

At large end of tooth
Table 15-2, p. 783
Eqs. (15-5) to (15-8), p. 784
Eq. (15-11), p. 785
Eq. (15-12), p. 785
Eq. (15-9), p. 785
Fig. 15-6, p. 786
Eq. (15-21), p. 790

Gear wear strength
$$S_{wc} = (\sigma_c)_{\text{all}} = \frac{s_{ac}\, C_L\, C_H}{S_H\, K_T\, C_R}$$

Tables 15-4, 15-5, Fig. 15-12, Eq. (15-22), pp. 790–792
Fig. 15-8, Eq. (15-14), p. 787
Eqs. (15-16), (15-17), gear only, p. 788
Eqs. (15-19), (15-20), Table 15-3, pp. 789, 790
Eq. (15-18), p. 788

Wear factor of safety
$$S_H = \frac{(\sigma_c)_{\text{all}}}{\sigma_c}, \text{ based on strength}$$

$$n_w = \left(\frac{(\sigma_c)_{\text{all}}}{\sigma_c}\right)^2, \text{ based on } W^t; \text{ can be compared directly with } S_F$$

Figure 15–15

"Road map" summary of principal straight-bevel gear bending equations and their parameters.

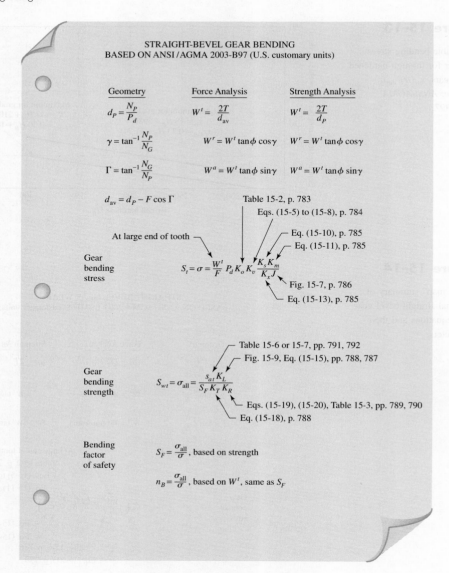

The standard does not mention specific steel but mentions the hardness attainable by heat treatments such as through-hardening, carburizing and case-hardening, flame-hardening, and nitriding. Through-hardening results depend on size (diametral pitch). Through-hardened materials and the corresponding Rockwell C-scale hardness at the 90 percent martensite shown in parentheses following include 1045 (50), 1060 (54), 1335 (46), 2340 (49), 3140 (49), 4047 (52), 4130 (44), 4140 (49), 4340 (49), 5145 (51), E52100 (60), 6150 (53), 8640 (50), and 9840 (49). For carburized case-hard materials the approximate core hardnesses are 1015 (22), 1025 (37), 1118 (33), 1320 (35), 2317 (30), 4320 (35), 4620 (35), 4820 (35), 6120 (35), 8620 (35), and E9310 (30). The conversion from HRC to H_B (300-kg load, 10-mm ball) is

HRC	42	40	38	36	34	32	30	28	26	24	22	20	18	16	14	12	10
H_B	388	375	352	331	321	301	285	269	259	248	235	223	217	207	199	192	187

Most bevel-gear sets are made from carburized case-hardened steel, and the factors incorporated in 2003-B97 largely address these high-performance gears. For through-hardened gears, 2003-B97 is silent on K_L and C_L, and Figs. 15–8 and 15–9 should prudently be considered as approximate.

15-4 Straight-Bevel Gear Analysis

EXAMPLE 15-1

A pair of identical straight-tooth miter gears listed in a catalog has a module of 5 at the large end, 25 teeth, a 27.5-mm face width, and a 20° normal pressure angle; the gears are grade 1 steel through-hardened with a core and case hardness of 180 Brinell. The gears are uncrowned and intended for general industrial use. They have a quality number of $Q_v = 7$. It is likely that the application intended will require outboard mounting of the gears. Use a safety factor of 1, a 10^7 cycle life, and a 0.99 reliability.

(a) For a speed of 600 rev/min find the power rating of this gearset based on AGMA bending strength.

(b) For the same conditions as in part (a) find the power rating of this gearset based on AGMA wear strength.

(c) For a reliability of 0.995, a gear life of 10^9 revolutions, and a safety factor of $S_F = S_H = 1.5$, find the power rating for this gearset using AGMA strengths.

Solution

From Figs. 15–14 and 15–15,

$$d_P = n_P m_{et} 25(5) = 125 \text{ mm}$$

$$v_{et} = \pi d_P n_P/60 = \pi(0.125)600/60 = 3.93 \text{ m/s}$$

Overload factor: uniform-uniform loading, Table 15–2, $K_A = 1.00$.
Safety factor: $S_F = 1$, $S_H = 1$.
Dynamic factor K_v: from Eq. (15–6),

$$B = 0.25(12 - 7)^{2/3} = 0.731$$

$$A = 50 + 56(1 - 0.731) = 65.06$$

$$K_v = \left(\frac{65.06 + \sqrt{200(3.93)}}{65.06}\right)^{0.731} = 1.299$$

From Eq. (15–8),

$$v_{et, \text{max}} = [65.06 + (7 - 3)]^2/200 = 23.8 \text{ m/s}$$

$v_{et} < v_{et \text{ max}}$, that is, $3.93 < 23.8$ m/s, therefore K_v is valid. From Eq. (15–10),

$$Y_x = 0.4867 + 0.008\,339(5) = 0.528$$

From Eq. (15–11),

$$K_{mb} = 1.25 \quad \text{and} \quad K_{HB} = 1.25 + 5.6(10^{-6})(27.5)^2 = 1.254$$

From Eq. (15–13), $Y_B = 1$. From Fig. 15–6, $Z_I = 0.065$; from Fig. 15–7, $Y_J = 0.216$, $Y_{JG} = 0.216$. $J_G = 0.216$. From Eq. (15–15),

$$Y_{NT} = 1.683(10^7)^{-0.0323} = 0.999\,96 \doteq 1$$

From Eq. (15–14),

$$Z_{NT} = 3.4822(10^7)^{-0.0602} = 1.32$$

Since $H_{B1}/H_{B2} = 1$, then from Fig. 15–10, $Z_W = 1$. From Eqs. (15–13) and (15–18), $Y_B = 1$ and $K_\theta = 1$, respectively. From Eq. (15–20),

$$Y_Z = 0.70 - 0.15\log(1 - 0.99) = 1, \qquad Z_Z = \sqrt{Y_Z} = \sqrt{1} = 1$$

(*a*) *Bending:* From Eq. (15–23),

$$\sigma_{Flim} = 0.3(180) + 14.48 = 68.48 \text{ MPa}$$

From Eq. (15–3),

$$\sigma_F = \frac{1000W^t}{b}\frac{K_A K_V}{m_{et}}\frac{Y_X K_{HB}}{Y_B Y_J} = \frac{W^t}{27.5(5)}(1)1.299\frac{0.528(1.254)}{(1)0.216}$$

$$= 0.029 W^t$$

From Eq. (15–4),

$$\sigma_{FP} = \frac{\sigma_{Flim} Y_{NT}}{S_F K_\theta Y_Z} = \frac{64.48(1)}{(1)(1)(1)} = 64.48 \text{ MPa}$$

Equating σ_F and σ_{FP},

$$0.029 W^t = 64.48 \qquad W^t = 2223 \text{ N}$$

Answer

$$H = W^t v_{et} = 2223(3.93) = 8736 \text{ W}$$

(*b*) *Wear:* From Fig. 15–12,

$$\sigma_{H lim} = 2.35(180) + 162.89 = 585.9 \text{ MPa}$$

From Eq. (15–2),

$$\sigma_{HP} = \frac{(\sigma_{H lim})_P Z_{NT} Z_W}{S_H K_\theta Z_Z} = \frac{585.9(1.32)(1)}{(1)(1)(1)} = 773.4 \text{ MPa}$$

Now $Z_E = 190 \sqrt{\text{N/mm}^2}$ from definitions following Eq. (15–21). From Eq. (15–9),

$$Z_X = 0.004\,92(27.5) + 0.4375 = 0.573$$

From Eq. (15–12), $Z_{xc} = 2$. Substituting in Eq. (15–1) gives

$$\sigma_H = Z_E \left(\frac{1000W^t}{bdz_1} K_A K_v K_{HB} Z_x Z_{xc}\right)^{1/2}$$

$$= 190\left[\frac{W^t}{27.5(125)0.065}(1)1.299(1.254)0.573(2)\right]^{1/2} = 17.37\sqrt{W^t}$$

Equating σ_H and σ_{HP} gives

$$17.37\sqrt{W^t} = 773.4, \qquad W^t = 1982 \text{ N}$$

$$H = 1982(3.93) = 7789 \text{ W}$$

Rated power for the gearset is

Answer

$$H = \min(8736, 7789) = 7789 \text{ W}$$

(c) Life goal 10^9 cycles, $R = 0.995$, $S_F = S_H = 1.5$, and from Eq. (15–15),

$$Y_{NT} = 1.683(10^9)^{-0.0323} = 0.8618$$

From Eq. (15–19),

$$Y_z = 0.50 - 0.25 \log(1 - 0.995) = 1.075, \quad Z_z = \sqrt{Y_z} = \sqrt{1.075} = 1.037$$

From Eq. (15–14),

$$Z_{NT} = 3.4822(10^9)^{-0.0602} = 1$$

Bending: From Eq. (15–23) and part (a), $\sigma_{Flim} = 64.48$ MPa. From Eq. (15–3),

$$\sigma_F = \frac{1000W^t}{27.5(5)}(1)1.299\frac{0.528(1.254)}{(1)0.216} = 0.029\,W^t$$

From Eq. (15–4),

$$\sigma_{FP} = \frac{\sigma_{Flim}\,Y_{NT}}{S_F K_\theta\,Y_Z} = \frac{64.48(0.8618)}{1.5(1)1.075} = 34.5 \text{ MPa}$$

Equating σ_F to σ_{FP} gives

$$0.029\,W^t = 34.5 \qquad W^t = 1190 \text{ N}$$

$$H = 1190(3.73) = 4438.7 \text{ W}$$

Wear: From Eq. (15–22), and part (b), $\sigma_{H\,lim} = 585.9$ MPa. Substituting into Eq. (15–2) gives

$$\sigma_{HP} = \frac{(\sigma_{H\,lim})_P Z_{NT} Z_W}{S_H K_\theta Z_Z} = \frac{585.9(1)(1)}{1.5(1)1.037} = 376.7 \text{ MPa}$$

Substituting into Eq. (15–1) gives, from part (b), $\sigma_H = 17.37\sqrt{W^t}$. Equating σ_H to $(\sigma_H)_P$ gives

$$376.7 = 17.37\sqrt{W^t} \qquad W^t = 470 \text{ N}$$

The wear power is

$$H = 470(3.73) = 1753 \text{ W}$$

Answer   The mesh rated power is $H = \min(4438.7, 1753) = 1753$ W.

15–5 Design of a Straight-Bevel Gear Mesh

A useful decision set for straight-bevel gear design is

- Function: power, speed, m_G, R
- Design factor: n_d ⎫
- Tooth system ⎬ A priori decisions
- Tooth count: N_P, N_G ⎭

- Pitch and face width: P_d, F ⎫
- Quality number: Q_v ⎪
- Gear material, core and case hardness ⎬ Design decisions
- Pinion material, core and case hardness ⎭

In bevel gears the quality number is linked to the wear strength. The J factor for the gear can be smaller than for the pinion. Bending strength is not linear with face width, because added material is placed at the small end of the teeth. Consequently, face width is roughly prescribed as

$$F = \min(0.3A_0, 10/P_d) \tag{15–24}$$

where A_0 is the cone distance (see Fig. 13–20, p. 682), given by

$$A_0 = \frac{d_P}{2 \sin \gamma} = \frac{d_G}{2 \sin \Gamma} \tag{15–25}$$

EXAMPLE 15–2

Design a straight-bevel gear mesh for shaft centerlines that intersect perpendicularly, to deliver 7.00 hp at 1000 rev/min with a gear ratio of 4:1, temperature of 300°F, normal pressure angle of 20°, using a design factor of 3. The load is uniform-uniform. Although the minimum number of teeth on the pinion is 13, which will mesh with 31 or more teeth without interference, use a pinion of 22 teeth. The material is to be AGMA grade 1 and the teeth are to be crowned. The reliability goal is 0.995 with a pinion life of 10^9 revolutions.

Solution

First we list the a priori decisions and their immediate consequences.

Function: 7.00 hp at 1000 rev/min, gear ratio $n_G = 4$, 300°F environment, neither gear straddle-mounted, $K_{mb} = 1.25$ [Eq. (15–11)], $R = 0.995$ at 10^9 revolutions of the pinion,

Eq. (15–14): $(C_L)_G = 3.4822(10^9/3)^{-0.0602} = 1.068$

$(C_L)_P = 3.4822(10^9)^{-0.0602} = 1$

Eq. (15–15): $(K_L)_G = 1.683(10^9/3)^{-0.0323} = 0.8929$

$(K_L)_P = 1.683(10^9)^{-0.0323} = 0.8618$

Eq. (15–19): $K_R = 0.50 - 0.25 \log(1 - 0.995) = 1.075$

$C_R = \sqrt{K_R} = \sqrt{1.075} = 1.037$

Eq. (15–18): $K_T = C_T = (460 + 300)/710 = 1.070$

Design factor: $n_d = 3$, $S_F = 3$, $S_H = \sqrt{3} = 1.732$.

Tooth system: crowned, straight-bevel gears, normal pressure angle 20°,

Eq. (15–13): $$K_x = 1$$

Eq. (15–12): $$C_{xc} = 1.5.$$

With $N_P = 22$ teeth, $N_G = (4)22 = 88$ teeth and from Fig. 15–14,

$$\gamma = \tan^{-1}(N_P/N_G) = \tan^{-1}(22/88) = 14.04° \qquad \Gamma = \tan^{-1}(88/22) = 75.96°$$

From Figs. 15–6 and 15–7, $I = 0.0825$, $J_P = 0.248$ and $J_G = 0.202$. Note that $J_P > J_G$.

Decision 1: Trial diametral pitch, $P_d = 8$ teeth/in.

Eq. (15–10): $K_s = 0.4867 + 0.2132/8 = 0.5134$

$$d_P = N_P/P_d = 22/8 = 2.75 \text{ in}$$

$$d_G = 2.75(4) = 11 \text{ in}$$

$$v_t = \pi d_P n_P/12 = \pi(2.75)1000/12 = 719.95 \text{ ft/min}$$

$$W^t = 33\,000\,\text{hp}/v_t = 33\,000(7)/719.95 = 320.86 \text{ lbf}$$

Eq. (15–25): $A_0 = d_P/(2 \sin\gamma) = 2.75/(2 \sin 14.04) = 5.67 \text{ in}$

Eq. (15–24):

$$F = \min(0.3A_0, 10/P_d) = \min[0.3(5.67), 10/8] = \min(1.70, 1.25) = 1.25 \text{ in}$$

Decision 2: Let $F = 1.25$ in. Then,

Eq. (15–9): $$C_s = 0.125(1.25) + 0.4375 = 0.5937$$

Eq. (15–11): $$K_m = 1.25 + 0.0036(1.25)^2 = 1.256$$

Decision 3: Let the transmission accuracy number be 6. Then, from Eq. (15–6),

$$B = 0.25(12 - 6)^{2/3} = 0.8255$$

$$A = 50 + 56(1 - 0.8255) = 59.77$$

Eq. (15–5): $$K_v = \left(\frac{59.77 + \sqrt{719.95}}{59.77}\right)^{0.8255} = 1.358$$

Decision 4: Pinion and gear material and treatment. Carburize and case-harden grade ASTM 1320 to

Core 21 HRC (H_B is 229 Brinell)
Case 55-64 HRC (H_B is 515 Brinell)

From Table 15–4, $s_{ac} = 200\,000$ psi and from Table 15–6, $s_{at} = 30\,000$ psi.

Gear bending: From Eq. (15–3), the bending stress is

$$(s_t)_G = \frac{W^t}{F} P_d K_o K_v \frac{K_s K_m}{K_x J_G} = \frac{320.86}{1.25} 8(1)1.358 \frac{0.5134(1.256)}{(1)0.202}$$

$$= 8899.16 \text{ psi}$$

The bending strength, from Eq. (15–4), is given by

$$(s_{wt})_G = \left(\frac{s_{at} K_L}{S_F K_T K_R} \right)_G = \frac{30\ 000(0.8929)}{3(1.070)1.075} = 7757.6 \text{ psi}$$

The strength exceeds the stress by a factor of $7757.6/8899.16 = 0.87$ giving an actual factor of safety of $(S_F)_G = 3(0.87) = 2.62$.

Pinion bending: The bending stress can be found from

$$(s_t)_P = (s_t)_G \frac{J_G}{J_P} = 8899.16 \frac{0.202}{0.248} = 7248.51$$

The bending strength, again from Eq. (15–4), is given by

$$(s_{wt})_P = \left(\frac{s_{at} K_L}{S_F K_T K_R} \right)_P = \frac{30\ 000\ (0.8618)}{3(1.070)1.075} = 7487.15 \text{ psi}$$

The strength exceeds the stress by a factor of $7487.15/7248.51 = 1.03$, giving an actual factor of safety of $(S_F)_P = 3(1.03) = 2.90$.

Gear wear: The load-induced contact stress for the pinion and gear, from Eq. (15–1), is

$$s_c = C_p \left(\frac{W^t}{F d_P I} K_o K_v K_m C_s C_{xc} \right)^{1/2}$$

$$= 2290 \left[\frac{320.86}{1.25(2.75)0.0825} (1)1.358(1.256)0.5937(1.5) \right]^{1/2}$$

$$= 94926.61 \text{ psi}$$

From Eq. (15–2) the contact strength of the gear is

$$(s_{wc})_G = \left(\frac{s_{ac} C_L C_H}{S_H K_T C_R} \right)_G = \frac{200\ 000(1.068)(1)}{\sqrt{3}(1.070)1.037} = 111156.35 \text{ psi}$$

The strength exceeds the stress by a factor of $111156/94926.61 = 1.17$, giving an actual factor of safety of $(S_H)_G^2 = 1.17^2(3) = 4.11$.

Pinion wear: From Eq. (15–2) the contact strength of the pinion is

$$(s_{wc})_P = \left(\frac{s_{ac} C_L C_H}{S_H K_T C_R} \right)_P = \frac{200\ 000(1)(1)}{\sqrt{3}(1.070)1.037} = 104042.695 \text{ psi}$$

The strength exceeds the stress by a factor of $111156.35/104042.70 = 1.068$ giving an actual factor of safety of $(S_H)_P^2 = 1.068^2(3) = 3.42$.

The actual factors of safety are 2.24, 2.66, 3.21, and 2.28. Making a direct comparison of the factors, we note that the threat from gear bending and pinion wear are practically equal. We also note that three of the ratios are comparable. Our goal would be to make changes in the design decisions that drive the factors closer to 2. The next step would be to adjust the design variables. It is obvious that an iterative process is involved. We need a figure of merit to order the designs. A computer program clearly is desirable.

15–6   **Worm Gearing—AGMA Equation**

Sections 13–11 and 13–17 introduced wormgearing and its force analysis and efficiency. Here, we will present a condensed version of the AGMA recommendations for cylindrical (single-enveloping) wormgearing[2]. For brevity, the equations will be shown for U.S. customary units only. Similar equations for SI units are available in the AGMA standards.

Since they are essentially nonenveloping worm gears, the *crossed helical* gears, shown in Fig. 15–16, can be considered with other worm gearing. Crossed helical gears, and worm gears too, usually have a 90° shaft angle, though this need not be so. The relation between the shaft and helix angles is

$$\sum = \psi_P \pm \psi_G \qquad (15\text{–}26)$$

where $\sum$ is the shaft angle. The plus sign is used when both helix angles are of the same hand, and the minus sign when they are of opposite hand. The subscript P in Eq. (15–26) refers to the pinion (worm); the subscript W is used for this same purpose. The subscript G refers to the gear, also called *gear wheel, worm wheel,* or simply the *wheel.* Table 15–8 gives cylindrical worm dimensions common to worm and gear.

Figure 15–16

View of the pitch cylinders of a pair of crossed helical gears.

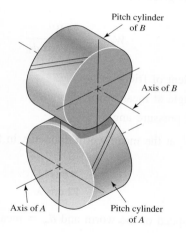

Pitch cylinder of B

Axis of B

Axis of A

Pitch cylinder of A

Table 15–8

Cylindrical Worm Dimensions Common to Both Worm and Gear*

Quantity	Symbol	ϕ_n 14.5° $N_W \leq 2$	ϕ_n 20° $N_W \leq 2$	ϕ_n 25° $N_W > 2$
Addendum	a	$0.3183p_x$	$0.3183p_x$	$0.286p_x$
Dedendum	b	$0.3683p_x$	$0.3683p_x$	$0.349p_x$
Whole depth	h_t	$0.6866p_x$	$0.6866p_x$	$0.635p_x$

*The table entries are for a tangential diametral pitch of the gear of $P_t = 1$.

[2]ANSI/AGMA 6034-B92, February 1992, *Practice for Enclosed Cylindrical Wormgear Speed-Reducers and Gear Motors;* and ANSI/AGMA 6022-C93, Dec. 1993, *Design Manual for Cylindrical Wormgearing.* Note: Equations (15–32) to (15–38) are contained in Annex C of 6034-B92 for informational purposes only. To comply with ANSI/AGMA 6034-B92, use the tabulations of these rating factors provided in the standard.

Good proportions indicate the worm pitch diameter d falls in the range

$$\frac{C^{0.875}}{3} \leq d \leq \frac{C^{0.875}}{1.6} \tag{15–27}$$

where C is the center-to-center distance. AGMA relates the allowable tangential force on the worm-gear tooth $(W^t)_{\text{all}}$ to other parameters by

$$(W^t)_{\text{all}} = C_s D_m^{0.8} F_e C_m C_v \tag{15–28}$$

where C_s = materials factor

D_m = mean gear diameter, mm

F_e = effective face width of the gear (actual face width, but not to exceed $0.67d_m$, the mean worm diameter), mm

C_m = ratio correction factor

C_v = velocity factor

The friction force W_f is given by

$$W_f = \frac{f W^t}{\cos \lambda \cos \phi_n} \tag{15–29}$$

where f = coefficient of friction

λ = lead angle at mean worm diameter

ϕ_n = normal pressure angle

The sliding velocity V_s at the mean worm diameter, in feet per minute, is

$$V_s = \frac{\pi n_W d_m}{12 \cos \lambda} \tag{15–30}$$

where n_W = rotative speed of the worm and d_m = mean worm diameter. The torque at the worm gear is

$$T_G = \frac{W^t D_m}{2} \tag{15–31}$$

where D_m is the mean gear diameter.

The parameters in Eq. (15–28) are, quantitatively,

$$C_s = 720 + 10.37 C^3 \qquad C \leq 3 \text{ in} \tag{15–32}$$

For sand-cast gears,

$$C_s = \begin{cases} 1000 & C > 3 \quad D_m \leq 2.5 \text{ in} \\ 1190 - 477 \log D_m & C > 3 \quad D_m > 2.5 \text{ in} \end{cases} \tag{15–33}$$

For chilled-cast gears,

$$C_s = \begin{cases} 1000 & C > 3 \quad D_m \leq 8 \text{ in} \\ 1412 - 456 \log D_m & C > 3 \quad D_m > 8 \text{ in} \end{cases} \tag{15–34}$$

For centrifugally cast gears,

$$C_s = \begin{cases} 1000 & C > 3 \quad D_m \leq 25 \text{ in} \\ 1251 - 180 \log D_m & C > 3 \quad D_m > 25 \text{ in} \end{cases} \tag{15-35}$$

The ratio correction factor C_m for gear ratio m_G is given by

$$C_m = \begin{cases} 0.02\sqrt{-m_G^2 + 40m_G - 76} + 0.46 & 3 < m_G \leq 20 \\ 0.0107\sqrt{-m_G^2 + 56m_G + 5145} & 20 < m_G \leq 76 \\ 1.1483 - 0.006\,58m_G & m_G > 76 \end{cases} \tag{15-36}$$

The velocity factor C_v is given by

$$C_v = \begin{cases} 0.659 \exp(-0.0011V_s) & V_s < 700 \text{ ft/min} \\ 13.31\,V_s^{-0.571} & 700 \leq V_s < 3000 \text{ ft/min} \\ 65.52\,V_s^{-0.774} & V_s > 3000 \text{ ft/min} \end{cases} \tag{15-37}$$

AGMA reports the coefficient of friction f as

$$f = \begin{cases} 0.15 & V_s = 0 \\ 0.124 \exp(-0.074V_s^{0.645}) & 0 < V_s \leq 10 \text{ ft/min} \\ 0.103 \exp(-0.110V_s^{0.450}) + 0.012 & V_s > 10 \text{ ft/min} \end{cases} \tag{15-38}$$

Now we examine some worm-gear mesh geometry. The addendum a and dedendum b are

$$a = \frac{p_x}{\pi} = 0.3183p_x \tag{15-39}$$

$$b = \frac{1.157p_x}{\pi} = 0.3683p_x \tag{15-40}$$

The full depth h_t is

$$h_t = \begin{cases} \dfrac{2.157p_x}{\pi} = 0.6866p_x & p_x \geq 0.16 \text{ in} \\ \dfrac{2.200p_x}{\pi} + 0.002 = 0.7003p_x + 0.002 & p_x < 0.16 \text{ in} \end{cases} \tag{15-41}$$

The worm outside diameter d_o is

$$d_o = d + 2a \tag{15-42}$$

The worm root diameter d_r is

$$d_r = d - 2b \tag{15-43}$$

The worm-gear throat diameter D_t is

$$D_t = D + 2a \tag{15-44}$$

where D is the worm-gear pitch diameter. The worm-gear root diameter D_r is

$$D_r = D - 2b \tag{15–45}$$

The clearance c is

$$c = b - a \tag{15–46}$$

The worm face width (maximum) $(F_W)_{max}$ is

$$(F_W)_{max} = 2\sqrt{\left(\frac{D_t}{2}\right)^2 - \left(\frac{D}{2} - a\right)^2} = 2\sqrt{2Da} \tag{15–47}$$

which was simplified using Eq. (15–44). The worm-gear face width F_G is

$$F_G = \begin{cases} 2d_m/3 & p_x > 0.16 \text{ in} \\ 1.125\sqrt{(d_o + 2c)^2 - (d_o - 4a)^2} & p_x \le 0.16 \text{ in} \end{cases} \tag{15–48}$$

The heat loss rate H_{loss} from the worm-gear case in ft · lbf/min is

$$H_{loss} = 33\,000(1 - e)H_{in} \tag{15–49}$$

where e is efficiency, given by Eq. (13–46), p. 708, and H_{in} is the input horsepower from the worm. The overall coefficient $\hbar_{CR}$ for combined convective and radiative heat transfer from the worm-gear case in ft · lbf/(min · in^2 · °F) is

$$\hbar_{CR} = \begin{cases} \dfrac{n_W}{6494} + 0.13 & \text{no fan on worm shaft} \\[2mm] \dfrac{n_W}{3939} + 0.13 & \text{fan on worm shaft} \end{cases} \tag{15–50}$$

The temperature of the oil sump t_s is given by

$$t_s = t_a + \frac{H_{loss}}{\hbar_{CR}A} = \frac{33\,000(1 - e)(H)_{in}}{\hbar_{CR}A} + t_a \tag{15–51}$$

where A is the case lateral area in in^2, and t_a is the ambient temperature in °F. Bypassing Eqs. (15–49), (15–50), and (15–51) one can apply the AGMA recommendation for minimum lateral area A_{min} in in^2 using

$$A_{min} = 43.20C^{1.7} \tag{15–52}$$

Because worm teeth are inherently much stronger than worm-gear teeth, they are not considered. The teeth in worm gears are short and thick on the edges of the face; midplane they are thinner as well as curved. Buckingham[3] adapted the Lewis equation for this case:

$$\sigma_a = \frac{W_G^t}{p_n F_e y} \tag{15–53}$$

where $p_n = p_x \cos \lambda$ and y is the Lewis form factor related to circular pitch. For $\phi_n = 14.5°$, $y = 0.100$; $\phi_n = 20°$, $y = 0.125$; $\phi_n = 25°$, $y = 0.150$; $\phi_n = 30°$, $y = 0.175$.

---

[3]Earle Buckingham, *Analytical Mechanics of Gears,* McGraw-Hill, New York, 1949, p. 495.

15-7 **Worm-Gear Analysis**

Compared to other gearing systems worm-gear meshes have a much lower mechanical efficiency. Cooling, for the benefit of the lubricant, becomes a design constraint sometimes resulting in what appears to be an oversize gear case in light of its contents. If the heat can be dissipated by natural cooling, or simply with a fan on the wormshaft, simplicity persists. Water coils within the gear case or lubricant outpumping to an external cooler is the next level of complexity. For this reason, gear-case area is a design decision.

To reduce cooling load, use multiple-thread worms. Also keep the worm pitch diameter as small as possible.

Multiple-thread worms can remove the self-locking feature of many worm-gear drives. When the worm drives the gearset, the mechanical efficiency e_W is given by

$$e_W = \frac{\cos \phi_n - f \tan \lambda}{\cos \phi_n + f \cot \lambda} \tag{15-54}$$

With the gear driving the gearset, the mechanical efficiency e_G is given by

$$e_G = \frac{\cos \phi_n - f \cot \lambda}{\cos \phi_n + f \tan \lambda} \tag{15-55}$$

To ensure that the worm gear will drive the worm,

$$f_{\text{stat}} < \cos \phi_n \tan \lambda \tag{15-56}$$

where values of f_{stat} can be found in ANSI/AGMA 6034-B92. To prevent the worm gear from driving the worm, refer to clause 9 of 6034-B92 for a discussion of self-locking in the static condition.

It is important to have a way to relate the tangential component of the gear force W_G^t to the tangential component of the worm force W_W^t, which includes the role of friction and the angularities of ϕ_n and λ. Refer to Eq. (13–45), p. 708, solved for W_W^t:

$$W_W^t = W_G^t \frac{\cos \phi_n \sin \lambda + f \cos \lambda}{\cos \phi_n \cos \lambda - f \sin \lambda} \tag{15-57}$$

In the absence of friction

$$W_W^t = W_G^t \tan \lambda$$

The mechanical efficiency of most gearing is very high, which allows power in and power out to be used almost interchangeably. Worm gearsets have such poor efficiencies that we work with, and speak of, output power. The magnitude of the gear transmitted force W_G^t can be related to the output horsepower H_0, the application factor K_a, the efficiency e, and design factor n_d by

$$W_G^t = \frac{33\ 000 n_d H_0 K_a}{V_G e} \tag{15-58}$$

We use Eq. (15–57) to obtain the corresponding worm force W_W^t. It follows that the worm and gear transmitted powers in hp are

$$H_W = \frac{W_W^t V_W}{33\ 000} = \frac{\pi d_W n_W W_W^t}{12(33\ 000)} \tag{15-59}$$

$$H_G = \frac{W_G^t V_G}{33\ 000} = \frac{\pi d_G n_G W_G^t}{12(33\ 000)} \tag{15-60}$$

Table 15–9

Largest Lead Angle
Associated with a
Normal Pressure Angle
ϕ_n for Worm Gearing

ϕ_n	Maximum Lead Angle λ_{max}
14.5°	16°
20°	25°
25°	35°
30°	45°

From Eq. (13–44), p. 708,

$$W_f = \frac{f W_G^t}{f \sin \lambda - \cos \phi_n \cos \lambda} \tag{15–61}$$

The sliding velocity of the worm at the pitch cylinder V_s is

$$V_s = \frac{\pi d n_W}{12 \cos \lambda} \tag{15–62}$$

and the friction power H_f is given by

$$H_f = \frac{|W_f| V_s}{33\,000} \text{ hp} \tag{15–63}$$

Table 15–9 gives the largest lead angle λ_{max} associated with normal pressure angle ϕ_n.

EXAMPLE 15–3

A single-threaded steel worm rotates at 1725 rev/min, meshing with a 56-tooth worm gear transmitting 1 hp to the output shaft. The pitch diameter of the worm is 1.50 in. The tangential diametral pitch of the gear is 8 teeth per inch and the normal pressure angle is 20°. The ambient temperature is 70°F, the application factor is 1.25, the design factor is 1, the gear face is 0.5 in, the lateral case area is 850 in², and the gear is sand-cast bronze.
(a) Determine and evaluate the geometric properties of the gears.
(b) Determine the transmitted gear forces and the mesh efficiency.
(c) Is the mesh sufficient to handle the loading?
(d) Estimate the lubricant sump temperature.

Solution

$N_W = 1$, $N_G = 56$, $P_t = 8$ teeth/in, $d = 1.5$ in, $H_O = 1$ hp, $\phi_n = 20°$, $ta = 70°F$, $K_a = 1.25$, $n_d = 1$, $F_e = 2$ in, $A = 850$ in².

(a) $\qquad m_G = N_G/N_W = 56$, $\qquad D = N_G/P_t = 56/8 = 7.0$ in

$\qquad p_x = \pi/8 = 0.3927$ in, $\qquad C = 1.5 + 7 = 8.5$ in

Answer

Eq. (15–39): $\qquad a = p_x/\pi = 0.3927/\pi = 0.125$ in

Eq. (15–40): $\qquad b = 0.3683 p_x = 0.1446$ in

Eq. (15–41): $\qquad h_t = 0.6866 p_x = 0.2696$ in

Eq. (15–42): $\qquad d_0 = 1.5 + 2(0.125) = 1.75$ in

Eq. (15–43): $d_r = 1.5 - 2(0.1446) = 1.2108$ in

Eq. (15–44): $D_t = 7 + 2(0.125) = 7.25$ in

Eq. (15–45): $D_r = 7 - 2(0.1446) = 6.711$ in

Eq. (15–46): $c = 0.1446 - 0.125 = 0.0196$ in

Eq. (15–47): $$(F_W)_{\max} = 2\sqrt{2(7)0.125} = 2.646 \text{ in}$$

$$V_W = \pi(1.5)(1725/12) = 677.4 \text{ ft/min}$$

$$V_G = \frac{\pi(7)(1725/56)}{12} = 56.45 \text{ ft/min}$$

Eq. (13–28): $L = p_x N_W = 0.3927$ in, $\lambda = \tan^{-1}\left(\dfrac{0.3927}{\pi(1.5)}\right) = 4.764°$

$$P_n = \frac{P_t}{\cos \lambda} = \frac{8}{\cos 4.764°} = 8.028$$

$$p_n = \frac{\pi}{P_n} = 0.3913 \text{ in}$$

(b) Eq. (15–62): $$V_s = \frac{\pi(1.5)(1725)}{12 \cos 4.764°} = 679.8 \text{ ft/min}$$

Eq. (15–38): $f = 0.103 \exp\left[-0.110(679.8)^{0.450}\right] + 0.012 = 0.0250$

Eq. (15–54): The efficiency is

Answer $$e = \frac{\cos \phi_n - f \tan \lambda}{\cos \phi_n + f \cot \lambda} = \frac{\cos 20° - 0.0250 \tan 4.764°}{\cos 20° + 0.0250 \cot 4.764°} = 0.7563$$

Answer Eq. (15–58): $$W_G^t = \frac{33\,000\, n_d H_o K_a}{V_G e} = \frac{33\,000(1)(1)(1.25)}{56.45(0.7563)} = 966 \text{ lbf}$$

Eq. (15–57): $W_W^t = W_G^t \left(\dfrac{\cos \phi_n \sin \lambda + f \cos \lambda}{\cos \phi_n \cos \lambda - f \sin \lambda}\right)$

Answer $$= 966 \left(\frac{\cos 20° \sin 4.764° + 0.025 \cos 4.764°}{\cos 20° \cos 4.764° - 0.025 \sin 4.764°}\right) = 106.4 \text{ lbf}$$

(c) Eq. (15–33): $C_s = 1190 - 477 \log 7.0 = 787$

Eq. (15–36): $C_m = 0.0107 \sqrt{-56^2 + 56(56) + 5145} = 0.767$

Eq. (15–37): $C_v = 0.659 \exp[-0.0011(679.8)] = 0.312$

Eq. (15–38): $(W^t)_{\text{all}} = 787(7)^{0.8}(2)(0.767)(0.312) = 1787 \text{ lbf}$

Since $W_G^t < (W^t)_{\text{all}}$, the mesh will survive at least 25 000 h.

Eq. (15–61): $$W_f = \frac{0.025(966)}{0.025 \sin 4.764° - \cos 20° \cos 4.764°} = -29.5 \text{ lbf}$$

Eq. (15–63): $H_f = \dfrac{29.5(679.8)}{33\,000} = 0.608$ hp

$$H_W = \dfrac{106.4(677.4)}{33\,000} = 2.18 \text{ hp}$$

$$H_G = \dfrac{966(56.45)}{33\,000} = 1.65 \text{ hp}$$

Answer   The mesh is sufficient,

$$P_n = P_t/\cos\lambda = 8/\cos 4.764° = 8.028$$

$$p_n = \pi/8.028 = 0.3913 \text{ in}$$

$$\sigma_G = \dfrac{966}{0.3913(0.5)(0.125)} = 39\,500 \text{ psi}$$

The stress is high. At the rated horsepower,

$$\sigma_G = \dfrac{1}{1.65} 39\,500 = 23\,940 \text{ psi} \quad \text{acceptable}$$

(d) Eq. (15–52): $A_{min} = 43.2(8.5)^{1.7} = 1642 \text{ in}^2 < 1700 \text{ in}^2$

Eq. (15–49): $H_{loss} = 33\,000(1 - 0.7563)(2.18) = 17\,530 \text{ ft} \cdot \text{lbf/min}$

Assuming a fan exists on the worm shaft,

Eq. (15–50): $\hbar_{CR} = \dfrac{1725}{3939} + 0.13 = 0.568 \text{ ft} \cdot \text{lbf/(min} \cdot \text{in}^2 \cdot °\text{F)}$

Answer   Eq. (15–51): $t_s = 70 + \dfrac{17\,530}{0.568(1700)} = 88.2°\text{F}$

15–8   Designing a Worm-Gear Mesh

A usable decision set for a worm-gear mesh includes

- Function: power, speed, m_G, K_a
- Design factor: n_d
- Tooth system
- Materials and processes
- Number of threads on the worm: N_W
- Axial pitch of worm: p_x
- Pitch diameter of the worm: d_W
- Face width of gear: F_G
- Lateral area of case: A

} A priori decisions

} Design decisions

Reliability information for worm gearing is not well developed at this time. The use of Eq. (15–28) together with the factors C_s, C_m, and C_v, with an alloy steel case-hardened worm together with customary nonferrous worm-wheel materials, will result in lives in excess of 25 000 h. The worm-gear materials in the experience base are principally bronzes:

- Tin- and nickel-bronzes (chilled-casting produces hardest surfaces)
- Lead-bronze (high-speed applications)
- Aluminum- and silicon-bronze (heavy load, slow-speed application)

The factor C_s for bronze in the spectrum sand-cast, chilled-cast, and centrifugally cast increases in the same order.

Standardization of tooth systems is not as far along as it is in other types of gearing. For the designer this represents freedom of action, but acquisition of tooling for tooth-forming is more of a problem for in-house manufacturing. When using a subcontractor the designer must be aware of what the supplier is capable of providing with on-hand tooling.

Axial pitches for the worm are usually integers, and quotients of integers are common. Typical pitches are $\frac{1}{4}$, $\frac{5}{16}$, $\frac{3}{8}$, $\frac{1}{2}$, $\frac{3}{4}$, 1, $\frac{5}{4}$, $\frac{6}{4}$, $\frac{7}{4}$, and 2, but others are possible. Table 15–8 shows dimensions common to both worm gear and cylindrical worm for proportions often used. Teeth frequently are stubbed when lead angles are 30° or larger.

Worm-gear design is constrained by available tooling, space restrictions, shaft center-to-center distances, gear ratios needed, and the designer's experience. ANSI/AGMA 6022-C93, *Design Manual for Cylindrical Wormgearing* offers the following guidance. Normal pressure angles are chosen from 14.5°, 17.5°, 20°, 22.5°, 25°, 27.5°, and 30°. The recommended minimum number of gear teeth is given in Table 15–10. The normal range of the number of threads on the worm is 1 through 10. Mean worm pitch diameter is usually chosen in the range given by Eq. (15–27).

A design decision is the axial pitch of the worm. Since acceptable proportions are couched in terms of the center-to-center distance, which is not yet known, one chooses a trial axial pitch p_x. Having N_W and a trial worm diameter d,

$$N_G = m_G N_W \qquad P_t = \frac{\pi}{p_x} \qquad D = \frac{N_G}{P_t}$$

Table 15–10

Minimum Number of Gear Teeth for Normal Pressure Angle ϕ_n

ϕ_n	$(N_G)_{min}$
14.5	40
17.5	27
20	21
22.5	17
25	14
27.5	12
30	10

Then

$$(d)_{lo} = C^{0.875}/3 \qquad (d)_{hi} = C^{0.875}/1.6$$

Examine $(d)_{lo} \leq d \leq (d)_{hi}$, and refine the selection of mean worm-pitch diameter to d_1 if necessary. Recompute the center-to-center distance as $C = (d_1 + D)/2$. There is even an opportunity to make C a round number. Choose C and set

$$d_2 = 2C - D$$

Equations (15–39) through (15–48) apply to one usual set of proportions.

EXAMPLE 15–4

Design a 15-hp 13:1 worm-gear speed-reducer mesh for a lumber mill planer feed drive for 3- to 10-h daily use. A 1200-rev/min squirrel-cage induction motor drives the planer feed ($K_a = 1.25$), and the ambient temperature is 70°F.

Solution

Function: $H_0 = 15$ hp, $m_G = 13$, $n_W = 1200$ rev/min.
Design factor: $n_d = 1.2$.
Materials and processes: case-hardened alloy steel worm, sand-cast bronze gear.
Worm threads: double, $N_W = 2$, $N_G = m_G N_W = 13(2) = 26$ gear teeth acceptable for $\phi_n = 20°$, according to Table 15–10.
Decision 1: Choose an axial pitch of worm $p_x = 1.5$ in. Then,

$$P_t = \pi/p_x = \pi/1.5 = 2.0944$$

$$D = N_G/P_t = 26/2.0944 = 12.41 \text{ in}$$

Eq. (15–39): $a = 0.3183p_x = 0.3183(1.5) = 0.4775$ in (addendum)

Eq. (15–40): $b = 0.3683(1.5) = 0.5525$ in (dedendum)

Eq. (15–41): $h_t = 0.6866(1.5) = 1.030$ in

Decision 2: Choose a mean worm diameter $d = 2.000$ in. Then

$$C = (d + D)/2 = (2.000 + 12.41)/2 = 7.207 \text{ in}$$

$$(d)_{lo} = 7.207^{0.875}/3 = 1.877 \text{ in}$$

$$(d)_{hi} = 7.207^{0.875}/1.6 = 3.52 \text{ in}$$

The range, given by Eq. (15–27), is $1.877 \leq d \leq 3.52$ in, which is satisfactory. Try $d = 2.500$ in. Recompute C:

$$C = (2.5 + 12.414)/2 = 7.457 \text{ in}$$

The range is now $1.715 \leq d \leq 3.216$ in, which is still satisfactory. Decision: $d = 2.500$ in. Then

Eq. (13–27): $\qquad\qquad L = p_x N_W = 1.5(2) = 3.000$ in

Eq. (13–28):

$\lambda = \tan^{-1}[L/(\pi d)] = \tan^{-1}[3/(\pi 2.5)] = 20.905°$ (from Table 15–9 lead angle OK)

Eq. (15–62): $\qquad V_s = \dfrac{\pi d n_W}{12 \cos \lambda} = \dfrac{\pi (2.5)1200}{12 \cos 20.905°} = 840.74$ ft/min

$\qquad\qquad\qquad V_W = \dfrac{\pi d n_W}{12} = \dfrac{\pi (2.5)1200}{12} = 785.40$ ft/min

$\qquad\qquad\qquad V_G = \dfrac{\pi D n_G}{12} = \dfrac{\pi (12.414)1200/13}{12} = 300$ ft/min

Eq. (15–33): $\quad C_s = 1190 - 477 \log 12.414 = 668.20$

Eq. (15–36): $\quad C_m = 0.02\sqrt{1313 + 40(13) - 76} + 0.46 = 0.792$

Eq. (15–37): $\quad C_v = 13.31(840.74)^{-0.571} = 0.285$

Eq. (15–38): $\qquad f = 0.103 \exp[-0.11(840.74)^{0.45}] + 0.012 = 0.02256^5$

Eq. (15–54): $\quad e_W = \dfrac{\cos 20° - 0.0191 \tan 20.905°}{\cos 20° + 0.0191 \cot 20.905°} = 0.942$

(If the worm gear drives, $e_G = 0.939$.) To ensure nominal 10-hp output, with adjustments for K_a, n_d, and e,

Eq. (15–57): $\quad W_W^t = 1222 \dfrac{\cos 20° \sin 20.905° + 0.0191 \cos 20.905°}{\cos 20° \cos 20.905° - 0.0191 \sin 20.905°} = 495.4$ lbf

Eq. (15–58): $\quad W_G^t = \dfrac{33\,000(1.2)15(1.25)}{300(0.942)} = 2627.39$ lbf

Eq. (15–59): $\quad H_W = \dfrac{\pi (2.5)1200(495.4)}{12(33\,000)} = 11.79$ hp

Eq. (15–60): $\quad H_G = \dfrac{\pi (12.414)1200/13(1222)}{12(33\,000)} = 23.885$ hp

Eq. (15–61): $\quad W_f = \dfrac{0.023(2627.39)}{0.023 \sin 20.905° - \cos 20° \cos 20.905°} = -68.15$ lbf

Eq. (15–63): $\quad H_f = \dfrac{|-68.15|840.74}{33\,000} = 1.736$ hp

With $C_s = 702.8$, $C_m = 0.772$, and $C_v = 0.232$,

$$(F_e)_{\text{req}} = \frac{W_G^t}{C_s D^{0.8} C_m C_v} = \frac{2627.39}{702.8(12.41)^{0.8}\, 0.772\, (0.232)} = 2.35 \text{ in}$$

Decision 3: The available range of $(F_e)_G$ is $1.479 \leq (F_e)_G \leq 2d/3$ or $1.667 \leq (F_e)_G \leq 2.35$in. Set $(F_e)_G = 2.01$ in.

Eq. (15–28): $\quad W_{\text{all}}^t = 668.20(12.414)^{0.8}\, 2.01\, (0.772)\, 0.232 = 2268.46$ lbf

This is lower than 2627.38854 lbf.

Decision 4:

Eq. (15–50): $\hbar_{CR} = \dfrac{n_W}{6494} + 0.13 = \dfrac{1200}{6494} + 0.13 = 0.315 \text{ ft} \cdot \text{lbf}/(\text{min} \cdot \text{in}^2 \cdot {}^\circ\text{F})$

Eq. (15–49): $H_{loss} = 33\,000(1 - e)H_W = 33\,000(1 - 0.93224)11.92 = 26647.69 \text{ ft} \cdot \text{lbf/min}$

The AGMA area, from Eq. (15–52), is $A_{min} = 43.2C^{1.7} = 43.2(7.207)^{1.7} = 1314.76 \text{ in}^2$. A rough estimate of the lateral area for 6-in clearances:

Vertical:	$d + D + 6 = 2.5 + 12.414 + 6 = 20.914 \text{ in}$
Width:	$D + 6 = 12.414 + 6 = 18.414 \text{ in}$
Thickness:	$d + 6 = 2.5 + 6 = 8.5 \text{ in}$
Area:	$20.914(18.412) + 2(8.5)20.914 + 18.412(8.5) \doteq 1282.29 \text{ in}^2$

Expect an area of 1300 in². Choose: Air-cooled, no fan on worm, with an ambient temperature of 70°F.

$$t_s = t_a + \frac{H_{loss}}{\hbar_{CR}A} = 70 + \frac{26647.69}{0.315(1300)} = 70 + 66.02 = 144.5{}^\circ\text{F}$$

Lubricant is safe with some margin for smaller area.

Eq. (13–18): $\qquad P_n = \dfrac{P_t}{\cos\lambda} = \dfrac{2.094}{\cos 20.905{}^\circ} = 2.242$

$$p_n = \frac{\pi}{P_n} = \frac{\pi}{2.242} = 1.401 \text{ in}$$

Gear bending stress, for reference, is

Eq. (15–53): $\qquad \sigma = \dfrac{W_G^t}{p_n F_e y} = \dfrac{2654.90}{1.401(2.01)\,0.125} = 7545.10 \text{psi}$

The risk is from wear, which is addressed by the AGMA method that provides $(W_G^t)_{all}$.

15–9 Buckingham Wear Load

A precursor to the AGMA method was the method of Buckingham, which identified an allowable wear load in worm gearing. Buckingham showed that the allowable gear-tooth loading for wear can be estimated from

$$(W_G^t)_{all} = K_w d_G F_e \qquad (15\text{–}64)$$

where $\quad K_w =$ worm-gear load factor

$\quad d_G =$ gear-pitch diameter

$\quad F_e =$ worm-gear effective face width

[5]Note: From ANSI/AGMA 6034-B92, the rating factors are $C_s = 703$, $C_m = 0.773$, $C_v = 0.2345$, and $f = 0.01995$.

Table 15-11

Wear Factor K_w for Worm Gearing

Source: Earle Buckingham, *Design of Worm and Spiral Gears,* Industrial Press, New York, 1981.

Material		Thread Angle ϕ_n			
Worm	**Gear**	**$14\frac{1}{2}°$**	**20°**	**25°**	**30°**
Hardened steel*	Chilled bronze	90	125	150	180
Hardened steel*	Bronze	60	80	100	120
Steel, 250 BHN (min.)	Bronze	36	50	60	72
High-test cast iron	Bronze	80	115	140	165
Gray iron†	Aluminum	10	12	15	18
High-test cast iron	Gray iron	90	125	150	180
High-test cast iron	Cast steel	22	31	37	45
High-test cast iron	High-test cast iron	135	185	225	270
Steel 250 BHN (min.)	Laminated phenolic	47	64	80	95
Gray iron	Laminated phenolic	70	96	120	140

*Over 500 BHN surface.
†For steel worms, multiply given values by 0.6.

Table 15–11 gives values for K_w for worm gearsets as a function of the material pairing and the normal pressure angle.

EXAMPLE 15-5

Estimate the allowable gear wear load $(W_G^t)_{\text{all}}$ for the gearset of Ex. 15–4 using Buckingham's wear equation.

Solution

From Table 15–11 for a hardened steel worm and a bronze bear, K_w is given as 80 for $\phi_n = 20°$. Equation (15–64) gives

$$(W_G^t)_{\text{all}} = 80(10.504)1.5 = 1260 \text{ lbf}$$

which is larger than the 1239 lbf of the AGMA method. The method of Buckingham does not have refinements of the AGMA method. [Is $(W_G^t)_{\text{all}}$ linear with gear diameter?]

For material combinations not addressed by AGMA, Buckingham's method allows quantitative treatment.

PROBLEMS

15-1

An uncrowned straight-bevel pinion has 20 teeth, a module of 4 mm, and a transmission accuracy number of 6. Both the pinion and gear are made of through-hardened steel with a Brinell hardness of 300. The driven gear has 60 teeth. The gearset has a life goal of 10^9 revolutions of the pinion with a reliability of 0.999. The shaft angle is 90°, and the pinion speed is 900 rev/min. The face width is 32 mm, and the normal pressure angle is 20°. The pinion is mounted outboard of its bearings, and the gear is straight-mounted. Based on the AGMA bending strength, what is the power rating of the gearset? Use $K_0 = 1$ and $S_F = S_H = 1$.

15–2 For the gearset and conditions of Prob. 15–1, find the power rating based on the AGMA surface durability.

15–3 An uncrowned straight-bevel pinion has 30 teeth, a diametral pitch of 6, and a transmission accuracy number of 6. The driven gear has 60 teeth. Both are made of No. 30 cast iron. The shaft angle is 90°. The face width is 1.25 in, the pinion speed is 900 rev/min, and the normal pressure angle is 20°. The pinion is mounted outboard of its bearings and the bearings of the gear straddle it. What is the power rating based on AGMA bending strength? Note: For cast iron gearsets reliability information has not yet been developed. We say that if the life is greater than 10^7 revolutions, then set $K_L = 1$, $C_L = 1$, $C_R = 1$, $K_R = 1$, and apply a factor of safety. Use $S_F = 2$ and $S_H = \sqrt{2}$.

15–4 For the gearset and conditions of Prob. 15–3, find the power rating based on AGMA surface durability.

15–5 An uncrowned straight-bevel pinion has 22 teeth, a module of 4 mm, and a transmission accuracy number of 5. The pinion and the gear are made of through-hardened steel, both having core and case hardnesses of 180 Brinell. The pinion drives the 24-tooth bevel gear. The shaft angle is 90°, the pinion speed is 1800 rev/min, the face width is 25 mm, and the normal pressure angle is 20°. Both gears have an outboard mounting. Find the power rating based on AGMA pitting resistance if the life goal is 10^9 revolutions of the pinion at 0.999 reliability.

15–6 For the gearset and conditions of Prob. 15–5, find the power rating for AGMA bending strength.

15–7 In straight-bevel gearing, there are some analogs to Eqs. (14–44) and (14–45) pp. 766 and 767, respectively. If we have a pinion core with a hardness of $(H_B)_{11}$ and we try equal power ratings, the transmitted load W^t can be made equal in all four cases. It is possible to find these relations:

	Core	Case
Pinion	$(H_B)_{11}$	$(H_B)_{12}$
Gear	$(H_B)_{21}$	$(H_B)_{22}$

(a) For carburized case-hardened gear steel with core AGMA bending strength $(s_{at})_G$ and pinion core strength $(s_{at})_P$, show that the relationship is

$$(s_{at})_G = (s_{at})_P \frac{J_p}{J_G} m_G^{-0.0323}$$

This allows $(H_B)_{21}$ to be related to $(H_B)_{11}$.

(b) Show that the AGMA contact strength of the gear case $(s_{ac})_G$ can be related to the AGMA core bending strength of the pinion core $(s_{at})_P$ by

$$(s_{ac})_G = \frac{C_p}{(C_L)_G C_H} \sqrt{\frac{S_H^2}{S_F} \frac{(s_{at})_P (K_L)_P K_x J_P K_T C_s C_{xc}}{N_P I K_s}}$$

If factors of safety are applied to the transmitted load W_t, then $S_H = \sqrt{S_F}$ and S_H^2/S_F is unity. The result allows $(H_B)_{22}$ to be related to $(H_B)_{11}$.

(c) Show that the AGMA contact strength of the gear $(s_{ac})_G$ is related to the contact strength of the pinion $(s_{ac})_P$ by

$$(s_{ac})_P = (s_{ac})_G m_G^{0.0602} C_H$$

15–8 Refer to your solution to Probs. 15–1 and 15–2. If the pinion core hardness is 300 Brinell, use the relations from Prob. 15–7 to determine the required hardness of the gear core and the case hardnesses of both gears to ensure equal power ratings.

15–9 Repeat Probs. 15–1 and 15–2 with the hardness protocol

	Core	Case
Pinion	300	373
Gear	339	345

which can be established by the relations in Prob. 15–7, and see if the result matches transmitted loads W^t in all four cases.

15–10 A catalog of stock bevel gears lists a power rating of 5.2 hp at 1200 rev/min pinion speed for a straight-bevel gearset consisting of a 20-tooth pinion driving a 40-tooth gear. This gear pair has a 20° normal pressure angle, a face width of 0.71 in, a diametral pitch of 10 teeth/in, and is through-hardened to 300 BHN. Assume the gears are for general industrial use, are generated to a transmission accuracy number of 5, and are uncrowned. Also assume the gears are rated for a life of 3×10^6 revolutions with a 99 percent reliability. Given these data, what do you think about the stated catalog power rating?

15–11 Apply the relations of Prob. 15–7 to Ex. 15–1 and find the Brinell case hardness of the gears for equal allowable load W^t in bending and wear. Check your work by reworking Ex. 15–1 to see if you are correct. How would you go about the heat treatment of the gears?

15–12 Your experience with Ex. 15–1 and problems based on it will enable you to write an interactive computer program for power rating of through-hardened steel gears. Test your understanding of bevel-gear analysis by noting the ease with which the coding develops. The hardness protocol developed in Prob. 15–7 can be incorporated at the end of your code, first to display it, then as an option to loop back and see the consequences of it.

15–13 Use your experience with Prob. 15–11 and Ex. 15–2 to design an interactive computer-aided design program for straight-steel bevel gears, implementing the ANSI/AGMA 2003-B97 standard. It will be helpful to follow the decision set in Sec. 15–5, allowing the return to earlier decisions for revision as the consequences of earlier decisions develop.

15–14 A single-thread steel worm rotates at 180.64 rad/s, meshing with a 24-tooth worm gear transmitting 0.7457 kW to the output shaft. The worm pitch diameter is 3 in and the tangential diametral pitch of the gear is 4 teeth/in. The normal pressure angle is 14.5°. The ambient temperature is 294.26 K. The application factor is 1.25 and the design factor is 1; gear face width is 12.7 mm, lateral case area 5483.86 cm², and the gear is chill-cast bronze.

(*a*) Find the gear geometry.
(*b*) Find the transmitted gear forces and the mesh efficiency.
(*c*) Is the mesh sufficient to handle the loading?
(*d*) Estimate the lubricant sump temperature.

15–15 to 15–22 As in Ex. 15–4, design a cylindrical worm-gear mesh to connect a squirrel-cage induction motor to a liquid agitator. The motor speed is 1125 rev/min, and the velocity ratio is to be 10:1. The output power requirement is 25 hp. The shaft axes are 90° to each other. An overload factor K_o (see Table 15–2) makes allowance for external dynamic excursions of load from the nominal or average load W^t. For this service $K_o = 1.25$ is appropriate. Additionally, a design factor n_d of 1.1 is to be included to address other unquantifiable risks. For Probs. 15–15 to 15–17 use the AGMA method for $(W_G^t)_{all}$ whereas for Probs. 15–18 to 15–22, use the Buckingham method. See Table 15–12.

Table 15–12

Table Supporting
Problems 15-15 to 15-22

Problem No.	Method	Materials	
		Worm	**Gear**
15–15	AGMA	Steel, HRC 58	Sand-cast bronze
15–16	AGMA	Steel, HRC 58	Chilled-cast bronze
15–17	AGMA	Steel, HRC 58	Centrifugal-cast bronze
15–18	Buckingham	Steel, 500 Bhn	Chilled-cast bronze
15–19	Buckingham	Steel, 500 Bhn	Cast bronze
15–20	Buckingham	Steel, 250 Bhn	Cast bronze
15–21	Buckingham	High-test cast iron	Cast bronze
15–22	Buckingham	High-test cast iron	High-test cast iron

16

Clutches, Brakes, Couplings, and Flywheels

Chapter Outline

This chapter is concerned with a group of elements usually associated with rotation that have in common the function of storing and/or transferring rotational energy. Because of this similarity of function, clutches, brakes, couplings, and flywheels are treated together in this chapter.

A simplified dynamic representation of a friction clutch or brake is shown in Fig. 16–1a. Two inertias, I_1 and I_2, traveling at the respective angular velocities ω_1 and ω_2, one of which may be zero in the case of brakes, are to be brought to the same speed by engaging the clutch or brake. Slippage occurs because the two elements are running at different speeds and energy is dissipated during actuation, resulting in a temperature rise. In analyzing the performance of these devices we shall be interested in:

1 The actuating force
2 The torque transmitted
3 The energy loss
4 The temperature rise

The torque transmitted is related to the actuating force, the coefficient of friction, and the geometry of the clutch or brake. This is a problem in statics, which will have to be studied separately for each geometric configuration. However, temperature rise is related to energy loss and can be studied without regard to the type of brake or clutch, because the geometry of interest is that of the heat-dissipating surfaces.

The various types of devices to be studied may be classified as follows:

1 Rim types with internal expanding shoes
2 Rim types with external contracting shoes
3 Band types
4 Disk or axial types
5 Cone types
6 Miscellaneous types

A flywheel is an inertial energy-storage device. It absorbs mechanical energy by increasing its angular velocity and delivers energy by decreasing its velocity. Figure 16–1b is a mathematical representation of a flywheel. An input torque T_i, corresponding to a coordinate θ_i, will cause the flywheel speed to increase. And a load or output torque T_o, with coordinate θ_o, will absorb energy from the flywheel and cause it to slow down. We shall be interested in designing flywheels so as to obtain a specified amount of speed regulation.

Figure 16–1

(a) Dynamic representation of a clutch or brake; (b) mathematical representation of a flywheel.

16–1   Static Analysis of Clutches and Brakes

Many types of clutches and brakes can be analyzed by following a general procedure. The procedure entails the following tasks:

- Estimate, model, or measure the pressure distribution on the friction surfaces.
- Find a relationship between the largest pressure and the pressure at any point.
- Use the conditions of static equilibrium to find the braking force or torque and the support reactions.

Let us apply these tasks to the doorstop depicted in Fig. 16–2a. The stop is hinged at pin A. A normal pressure distribution $p(u)$ is shown under the friction pad as a function of position u, taken from the right edge of the pad. A similar distribution of shearing frictional traction is on the surface, of intensity $fp(u)$, in the direction of the motion of the floor relative to the pad, where f is the coefficient of friction. The width of the pad into the page is w_2. The net force in the y direction and moment about C from the pressure are respectively,

$$N = w_2 \int_0^{w_1} p(u)\, du = p_{av} w_1 w_2 \tag{a}$$

$$w_2 \int_0^{w_1} p(u)u\, du = \bar{u} w_2 \int_0^{w_1} p(u)\, du = p_{av} w_1 w_2 \bar{u} \tag{b}$$

We sum the forces in the x-direction to obtain

$$\sum F_x = R_x \mp w_2 \int_0^{w_1} fp(u)\, du = 0$$

where $-$ or $+$ is for rightward or leftward relative motion of the floor, respectively. Assuming f constant, solving for R_x gives

$$R_x = \pm w_2 \int_0^{w_+} fp(u)\, du = \pm f w_1 w_2 p_{av} \tag{c}$$

Summing the forces in the y direction gives

$$\sum F_y = -F + w_2 \int_0^{w_1} p(u)\, du + R_y = 0$$

from which

$$R_y = F - w_2 \int_0^{w1} p(u)\, du = F - p_{av} w_1 w_2 \tag{d}$$

for either direction. Summing moments about the pin located at A we have

$$\sum M_A = Fb - w_2 \int_0^{w1} p(u)(c + u)\, du \mp afw_2 \int_0^{w_1} p(u)\, du = 0$$

A brake shoe is *self-energizing* if its moment sense helps set the brake, *self-deenergizing* if the moment resists setting the brake. Continuing,

$$F = \frac{w_2}{b}\left[\int_0^{w_1} p(u)(c + u)\, du \pm af \int_0^{w_1} p(u)\, du \right] \tag{e}$$

Figure 16–2

A common doorstop.
(a) Free body of the doorstop.
(b) Trapezoidal pressure distribution on the foot pad based on linear deformation of pad. (c) Free-body diagram for leftward movement of the floor, uniform pressure, Ex. 16–1. (d) Free-body diagram for rightward movement of the floor, uniform pressure, Ex. 16–1. (e) Free-body diagram for leftward movement of the floor, trapezoidal pressure, Ex. 16–1. For (c), (d), and (e), units of forces and dimensions are lbf and in, respectively.

Can F be equal to or less than zero? Only during rightward motion of the floor when the expression in brackets in Eq. (*e*) is equal to or less than zero. We set the brackets to zero or less:

$$\int_0^{w_1} p(u)(c + u)\, du - af \int_0^{w_1} p(u)\, du \leq 0$$

from which

$$f_{cr} \geq \frac{1}{a} \frac{\displaystyle\int_0^{w_1} p(u)(c + u)\, du}{\displaystyle\int_0^{w_1} p(u)\, du} = \frac{1}{a} \frac{c \displaystyle\int_0^{w_1} p(u)\, du + \int_0^{w_1} p(u)u\, du}{\displaystyle\int_0^{w_1} p(u)\, du} \tag{f}$$

$$f_{cr} \geq \frac{c + \bar{u}}{a}$$

where $\bar{u}$ is the distance of the center of pressure from the right edge of the pad. The conclusion that a *self-acting* or *self-locking* phenomenon is present is independent of our knowledge of the normal pressure distribution $p(u)$. Our ability to *find* the critical value of the coefficient of friction f_{cr} is dependent on our knowledge of $p(u)$, from which we derive $\bar{u}$.

EXAMPLE 16–1

The doorstop depicted in Fig. 16–2a has the following dimensions: $a = 100$ mm, $b = 50$ mm, $c = 40$ mm, $w_1 = 25$ mm, $w_2 = 18$ mm, where w_2 is the depth of the pad into the plane of the paper.

(*a*) For a leftward relative movement of the floor, an actuating force F of 45 N, a coefficient of friction of 0.4, use a uniform pressure distribution p_{av}, find R_x, R_y, p_{av}, and the largest pressure p_a.

(*b*) Repeat part *a* for rightward relative movement of the floor.

(*c*) Model the normal pressure to be the "crush" of the pad, much as if it were composed of many small helical coil springs. Find R_x, R_y, p_{av}, and p_a for leftward relative movement of the floor and other conditions as in part *a*.

(*d*) For rightward relative movement of the floor, is the doorstop a self-acting brake?

Solution   (*a*)

Eq. (*c*): $R_x = fp_{av}w_1w_2 = 0.4(25)18\, p_{av} = 180\, p_{av}$

Eq. (*d*): $R_y = F - p_{av}w_1w_2 = 45 - p_{av}(25)(18) = 45 - 450\, p_{av}$

Eq. (*e*): $F = \frac{w_2}{b}\left[\int_0^{25} p_{av}(c + u)\, du + af \int_0^{25} p_{av}\, du\right]$

$$= \frac{w_2}{b}\left(p_{av}c \int_0^{25} du + p_{av} \int_0^{25} u\, du + afp_{av} \int_0^{25} du\right)$$

$$= \frac{w_2 p_{av}}{b}\left(25c + \frac{25^2}{2} + 25af\right)$$

$$= 832.5\, p_{av}$$

Solving for p_{av} gives

$$p_{av} = \frac{F}{922.5} = \frac{45}{832.5} = 0.054 \text{ MPa}$$

We evaluate R_x and R_y as

Answer

$$R_x = 180(0.054) = 9.7 \text{ N}$$

Answer

$$R_y = 45 - 450(0.05) = 20.7 \text{ N}$$

The normal force N on the pad is $F - R_y = 45 - 22.5 = 24.3$ N, upward. The line of action is through the center of pressure, which is at the center of the pad..The friction force is $fN = 0.4(24.3) = 9.7$ N directed to the left. A check of the moments about A gives

$$\sum M_A = Fb - fNa - N(w_1/2 + c)$$

$$= 45(50) - 0.4(24.3)100 - 24.3(25/2 + 40) = 0$$

Answer   The maximum pressure $p_a = p_{av} = 0.054$ MPa.

(b)

Eq. (c): $R_x = -f p_{av} w_1 w_2 = -0.4(25)(18) p_{av} = -180 p_{av}$

Eq. (d): $R_y = F - p_{av} w_1 w_2 = 45 - p_{av}(25)(18) = 45 - 450 p_{av}$

Eq. (e): $F = \frac{w_2}{b} \left[\int_0^{25} p_{av}(c + u) \, du - af \int_0^{25} p_{av} \, du \right]$

$$= \frac{w_2}{b} \left(p_{av} c \int_0^{25} du + p_{av} \int_0^{25} u \, du - af p_{av} \int_0^{25} du \right)$$

$$= 112.5 p_{av}$$

from which

$$p_{av} = \frac{F}{112.5} = \frac{45}{112.5} = 0.4 \text{ MPa}$$

which makes

Answer

$$R_x = -180(0.4) = 72 \text{ N}$$

Answer

$$R_y = 45 - 450(0.4) = -135 \text{ N}$$

The normal force N on the pad is $45 + 135 = 180$ N upward. The friction shearing force is $fN = 0.4(180) = 72$ N to the right. We now check the moments about A:

$$M_A = fNa + Fb - N(c + 0.5) = 72(100) + 45(50) - 180(40 + 12.5) = 0$$

Note the change in average pressure from 0.05 MPa in part a to 0.4 MPa. Also note how directions of forces have changed. The maximum pressure p_a is the same as p_{av}, which has changed from 0.05 MPa to 0.4 MPa.

(c) We will model the deformation of the pad as follows. If the doorstop rotates $\Delta\phi$ counterclockwise, the right and left edges of the pad will deform down y_1 and y_2, respectively (Fig. 16–2b). From similar triangles, $y_1/(r_1 \, \Delta\phi) = c/r_1$ and $y_2/(r_2 \, \Delta\phi) = (c + w_1)/r_2$. Thus, $y_1 = c \, \Delta\phi$ and $y_2 = (c + w_1) \, \Delta\phi$. This means that y is directly proportional to the horizontal distance from the pivot point A; that is, $y = C_1 v$, where

C_1 is a constant (see Fig. 16–2b). Assuming the pressure is directly proportional to deformation, then $p(v) = C_2 v$, where C_2 is a constant. In terms of u, the pressure is $p(u) = C_2(c + u) = C_2(40 + u)$.

Eq. (e):

$$F = \frac{w_2}{b}\left[\int_0^{w_1} p(u)c\,du + \int_0^{w_1} p(u)u\,du + af\int_0^{w_1} p(u)\,du\right]$$

$$= \frac{18}{50}\left[\int_0^{25} C_2(40 + u)40du + \int_0^{25} C_2(40 + u)u\,du + af\int_0^{25} C_2(40 + u)du\right]$$

$$= 0.36C_2[1600(25) + 40(25)^2 + (25)^3/2 + 100(0.4)(40(25) - (25)^2/2)]$$

$$= 33\,592.5C_2$$

Since $F = 45$ N, then $C_2 = 45/33\,592.5 = 0.001\,34$ MPa/mm, and $p(u) = 0.001\,34$ $(40 + u)$. The average pressure is given by

Answer $$p_{av} = \frac{1}{w_1}\int_0^{w_1} p(u)\,du = \frac{1}{25}\int_0^{25} 0.001\,34(40 + u)du = 0.001\,34(40 + 12.5) = 0.07 \text{ MPa}$$

The maximum pressure occurs at $u = 25$ mm, and is

Answer $$p_a = 0.001\,34(40 + 25) = 0.087 \text{ MPa}$$

Equations (c) and (d) of Sec. 16–1 are still valid. Thus,

Answer $$R_x = 180\,p_{av} = 180(0.07) = 12.6 \text{ N}$$

$$R_y = 45 - 450\,p_{av} = 45 - 450(0.07) = 13.5 \text{ N}$$

The average pressure is $p_{av} = 0.07$ MPa and the maximum pressure is $p_a = 0.087$ MPa, which is approximately 24 percent higher than the average pressure. The presumption that the pressure was uniform in part a (because the pad was small, or because the arithmetic would be easier?) underestimated the peak pressure. Modeling the pad as a one-dimensional springset is better, but the pad is really a three-dimensional continuum. A theory of elasticity approach or a finite element modeling may be overkill, given uncertainties inherent in this problem, but it still represents better modeling.

(d) To evaluate $\bar{u}$ we need to evaluate two integrations

$$\int_0^c p(u)u\,du = \int_0^{25} 0.001\,34(40 + u)udu = 27.2 \text{ N}$$

$$\int_0^c p(u)\,du = \int_0^{25} 0.001\,34(40 + u)du = 1.76 \text{ N}$$

Thus $\bar{u} = 27.2/1.76 = 15.5$ mm. Then, from Eq. (f) of Sec. 16–1, the critical coefficient of friction is

Answer $$f_{cr} \geq \frac{c + \bar{u}}{a} = \frac{40 + 15.5}{100} = 0.56$$

The doorstop friction pad does not have a high enough coefficient of friction to make the doorstop a self-acting brake. The configuration must change and/or the pad material specification must be changed to sustain the function of a doorstop.

16–2 Internal Expanding Rim Clutches and Brakes

The internal-shoe rim clutch shown in Fig. 16–3 consists essentially of three elements: the mating frictional surface, the means of transmitting the torque to and from the surfaces, and the actuating mechanism. Depending upon the operating mechanism, such clutches are further classified as *expanding-ring, centrifugal, magnetic, hydraulic,* and *pneumatic.*

The expanding-ring clutch is often used in textile machinery, excavators, and machine tools where the clutch may be located within the driving pulley. Expanding-ring clutches benefit from centrifugal effects; transmit high torque, even at low speeds; and require both positive engagement and ample release force.

The centrifugal clutch is used mostly for automatic operation. If no spring is used, the torque transmitted is proportional to the square of the speed. This is particularly useful for electric-motor drives where, during starting, the driven machine comes up to speed without shock. Springs can also be used to prevent engagement until a certain motor speed is reached, but some shock may occur.

Magnetic clutches are particularly useful for automatic and remote-control systems. Such clutches are also useful in drives subject to complex load cycles (see Sec. 11–7).

Hydraulic and pneumatic clutches are also useful in drives having complex loading cycles and in automatic machinery, or in robots. Here the fluid flow can be controlled remotely using solenoid valves. These clutches are also available as disk, cone, and multiple-plate clutches.

In braking systems, the *internal-shoe* or *drum* brake is used mostly for automotive applications.

To analyze an internal-shoe device, refer to Fig. 16–4, which shows a shoe pivoted at point A, with the actuating force acting at the other end of the shoe. Since the shoe is long, we cannot make the assumption that the distribution of normal forces is uniform. The mechanical arrangement permits no pressure to be applied at the heel, and we will therefore assume the pressure at this point to be zero.

It is the usual practice to omit the friction material for a short distance away from the heel (point A). This eliminates interference, and the material would contribute little to the performance anyway, as will be shown. In some designs the hinge pin is made movable to provide additional heel pressure. This gives the effect of a floating

Figure 16–3

An internal expanding centrifugal-acting rim clutch. (*Courtesy of the Hilliard Corporation.*)

Figure 16-4

Internal friction shoe geometry.

Figure 16-5

The geometry associated with an arbitrary point on the shoe.

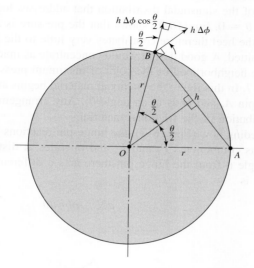

shoe. (Floating shoes will not be treated in this book, although their design follows the same general principles.)

Let us consider the pressure p acting upon an element of area of the frictional material located at an angle θ from the hinge pin (Fig. 16–4). We designate the maximum pressure p_a located at an angle θ_a from the hinge pin. To find the pressure distribution on the periphery of the internal shoe, consider point B on the shoe (Fig. 16–5). As in Ex. 16–1, if the shoe deforms by an infinitesimal rotation $\Delta\phi$ about the pivot point A, deformation perpendicular to AB is $h\,\Delta\phi$. From the isosceles triangle AOB, $h = 2r\sin(\theta/2)$, so

$$h\,\Delta\phi = 2r\,\Delta\phi\sin(\theta/2)$$

The deformation perpendicular to the rim is $h\,\Delta\phi\cos(\theta/2)$, which is

$$h\,\Delta\phi\cos(\theta/2) = 2r\,\Delta\phi\sin(\theta/2)\cos(\theta/2) = r\,\Delta\phi\sin\theta$$

Thus, the deformation, and consequently the pressure, is proportional to $\sin\theta$. In terms of the pressure at B and where the pressure is a maximum, this means

$$\frac{p}{\sin\theta} = \frac{p_a}{\sin\theta_a} \tag{a}$$

Figure 16-6

Defining the angle θ_a at which the maximum pressure p_a occurs when (a) shoe exists in zone $\theta_1 \le \theta_2 \le \pi/2$ and (b) shoe exists in zone $\theta_1 \le \pi/2 \le \theta_2$.

Rearranging gives

$$p = \frac{p_a}{\sin \theta_a} \sin \theta \qquad (16\text{-}1)$$

This pressure distribution has interesting and useful characteristics:

- The pressure distribution is sinusoidal with respect to the angle θ.
- If the shoe is short, as shown in Fig. 16–6a, the largest pressure *on the shoe* is p_a occurring at the end of the shoe, θ_2.
- If the shoe is long, as shown in Fig. 16–6b, the largest pressure on the shoe is p_a occurring at $\theta_a = 90°$.

Since limitations on friction materials are expressed in terms of the largest allowable pressure on the lining, the designer wants to think in terms of p_a and not about the amplitude of the sinusoidal distribution that addresses locations off the shoe.

When $\theta = 0$, Eq. (16–1) shows that the pressure is zero. The frictional material located at the heel therefore contributes very little to the braking action and might as well be omitted. A good design would concentrate as much frictional material as possible in the neighborhood of the point of maximum pressure. Such a design is shown in Fig. 16–7. In this figure the frictional material begins at an angle θ_1, measured from the hinge pin A, and ends at an angle θ_2. Any arrangement such as this will give a good distribution of the frictional material.

Proceeding now (Fig. 16–7), the hinge-pin reactions are R_x and R_y. The actuating force F has components F_x and F_y and operates at distance c from the hinge pin. At any angle θ from the hinge pin there acts a differential normal force dN whose magnitude is

$$dN = pbr \, d\theta \qquad (b)$$

Figure 16-7

Forces on the shoe.

where b is the face width (perpendicular to the page) of the friction material. Substituting the value of the pressure from Eq. (16–1), the normal force is

$$dN = \frac{p_a br \sin \theta \, d\theta}{\sin \theta_a} \qquad (c)$$

The normal force dN has horizontal and vertical components $dN \cos \theta$ and $dN \sin \theta$, as shown in the figure. The frictional force fdN has horizontal and vertical components whose magnitudes are $fdN \sin \theta$ and $fdN \cos \theta$, respectively. By applying the conditions of static equilibrium, we may find the actuating force F, the torque T, and the pin reactions R_x and R_y.

We shall find the actuating force F, using the condition that the summation of the moments about the hinge pin is zero. The frictional forces have a moment arm about the pin of $r - a \cos \theta$. The moment M_f of these frictional forces is

$$M_f = \int fdN(r - a \cos \theta) = \frac{fp_a br}{\sin \theta_a} \int_{\theta_1}^{\theta_2} \sin \theta (r - a \cos \theta) \, d\theta \qquad (16\text{--}2)$$

which is obtained by substituting the value of dN from Eq. (c). It is convenient to integrate Eq. (16–2) for each problem, and we shall therefore retain it in this form. The moment arm of the normal force dN about the pin is $a \sin \theta$. Designating the moment of the normal forces by M_N and summing these about the hinge pin give

$$M_N = \int dN(a \sin \theta) = \frac{p_a bra}{\sin \theta_a} \int_{\theta_1}^{\theta_2} \sin^2 \theta \, d\theta \qquad (16\text{--}3)$$

The actuating force F must balance these moments. Thus

$$F = \frac{M_N - M_f}{c} \qquad (16\text{--}4)$$

We see here that a condition for zero actuating force exists. In other words, if we make $M_N = M_f$, self-locking is obtained, and no actuating force is required. This furnishes us with a method for obtaining the dimensions for some self-energizing action. Thus the dimension a in Fig. 16–7 must be such that

$$M_N > M_f \qquad (16\text{--}5)$$

The torque T applied to the drum by the brake shoe is the sum of the frictional forces fdN times the radius of the drum:

$$T = \int frdN = \frac{fp_a br^2}{\sin \theta_a} \int_{\theta_1}^{\theta_2} \sin \theta \, d\theta$$

$$= \frac{fp_a br^2 (\cos \theta_1 - \cos \theta_2)}{\sin \theta_a} \qquad (16\text{--}6)$$

The hinge-pin reactions are found by taking a summation of the horizontal and vertical forces. Thus, for R_x, we have

$$R_x = \int dN \cos \theta - \int fdN \sin \theta - F_x$$

$$= \frac{p_a br}{\sin \theta_a} \left(\int_{\theta_1}^{\theta_2} \sin \theta \cos \theta \, d\theta - f \int_{\theta_1}^{\theta_2} \sin^2 \theta \, d\theta \right) - F_x \qquad (d)$$

The vertical reaction is found in the same way:

$$R_y = \int dN \sin\theta + \int f \, dN \cos\theta - F_y$$

$$= \frac{p_a br}{\sin\theta_a}\left(\int_{\theta_1}^{\theta_2} \sin^2\theta \, d\theta + f\int_{\theta_1}^{\theta_2} \sin\theta\cos\theta \, d\theta\right) - F_y \tag{e}$$

The direction of the frictional forces is reversed if the rotation is reversed. Thus, for counterclockwise rotation the actuating force is

$$F = \frac{M_N + M_f}{c} \tag{16-7}$$

and since both moments have the same sense, the self-energizing effect is lost. Also, for counterclockwise rotation the signs of the frictional terms in the equations for the pin reactions change, and Eqs. (d) and (e) become

$$R_x = \frac{p_a br}{\sin\theta_a}\left(\int_{\theta_1}^{\theta_2} \sin\theta\cos\theta \, d\theta + f\int_{\theta_1}^{\theta_2} \sin^2\theta \, d\theta\right) - F_x \tag{f}$$

$$R_y = \frac{p_a br}{\sin\theta_a}\left(\int_{\theta_1}^{\theta_2} \sin^2\theta \, d\theta - f\int_{\theta_1}^{\theta_2} \sin\theta\cos\theta \, d\theta\right) - F_y \tag{g}$$

Equations (d), (e), (f), and (g) can be simplified to ease computations. Thus, let

$$A = \int_{\theta_1}^{\theta_2} \sin\theta\cos\theta \, d\theta = \left(\frac{1}{2}\sin^2\theta\right)_{\theta_1}^{\theta_2}$$

$$B = \int_{\theta_1}^{\theta_2} \sin^2\theta \, d\theta = \left(\frac{\theta}{2} - \frac{1}{4}\sin 2\theta\right)_{\theta_1}^{\theta_2} \tag{16-8}$$

Then, for clockwise rotation as shown in Fig. 16–7, the hinge-pin reactions are

$$R_x = \frac{p_a br}{\sin\theta_a}(A - fB) - F_x$$

$$R_y = \frac{p_a br}{\sin\theta_a}(B + fA) - F_y \tag{16-9}$$

For counterclockwise rotation, Eqs. (f) and (g) become

$$R_x = \frac{p_a br}{\sin\theta_a}(A + fB) - F_x$$

$$R_y = \frac{p_a br}{\sin\theta_a}(B - fA) - F_y \tag{16-10}$$

In using these equations, the reference system always has its origin at the center of the drum. The positive x axis is taken through the hinge pin. The positive y axis is always in the direction of the shoe, even if this should result in a left-handed system. The following assumptions are implied by the preceding analysis:

1 The pressure at any point on the shoe is assumed to be proportional to the distance from the hinge pin, being zero at the heel. This should be considered from the standpoint that pressures specified by manufacturers are averages rather than maxima.

2 The effect of centrifugal force has been neglected. In the case of brakes, the shoes are not rotating, and no centrifugal force exists. In clutch design, the effect of this force must be considered in writing the equations of static equilibrium.

3 The shoe is assumed to be rigid. Since this cannot be true, some deflection will occur, depending upon the load, pressure, and stiffness of the shoe. The resulting pressure distribution may be different from that which has been assumed.

4 The entire analysis has been based upon a coefficient of friction that does not vary with pressure. Actually, the coefficient may vary with a number of conditions, including temperature, wear, and environment.

EXAMPLE 16-2 The brake shown in Fig. 16–8 is 300 mm in diameter and is actuated by a mechanism that exerts the same force F on each shoe. The shoes are identical and have a face width of 32 mm. The lining is a molded asbestos having a coefficient of friction of 0.32 and a pressure limitation of 1000 kPa. Estimate the maximum
(a) Actuating force F.
(b) Braking capacity.
(c) Hinge-pin reactions.

Solution (a) The right-hand shoe is self-energizing, and so the force F is found on the basis that the maximum pressure will occur on this shoe. Here $\theta_1 = 0°$, $\theta_2 = 126°$, $\theta_a = 90°$, and $\sin \theta_a = 1$. Also,

$$a = \sqrt{(112)^2 + (50)^2} = 122.7 \text{ mm}$$

Integrating Eq. (16–2) from 0 to θ_2 yields

$$M_f = \frac{f p_a b r}{\sin \theta_a} \left[\left(-r \cos \theta \right)_0^{\theta_2} - a \left(\frac{1}{2} \sin^2 \theta \right)_0^{\theta_2} \right]$$

$$= \frac{f p_a b r}{\sin \theta_a} \left(r - r \cos \theta_2 - \frac{a}{2} \sin^2 \theta_2 \right)$$

Figure 16-8

Brake with internal expanding shoes; dimensions in millimeters.

Changing all lengths to meters, we have

$$M_f = (0.32)[1000(10)^3](0.032)(0.150)$$

$$\times \left[0.150 - 0.150 \cos 126° - \left(\frac{0.1227}{2}\right) \sin^2 126° \right]$$

$$= 304 \text{ N} \cdot \text{m}$$

The moment of the normal forces is obtained from Eq. (16–3). Integrating from 0 to θ_2 gives

$$M_N = \frac{p_a b r a}{\sin \theta_a} \left(\frac{\theta}{2} - \frac{1}{4} \sin 2\theta \right)_0^{\theta_2}$$

$$= \frac{p_a b r a}{\sin \theta_a} \left(\frac{\theta_2}{2} - \frac{1}{4} \sin 2\theta_2 \right)$$

$$= [1000(10)^3](0.032)(0.150)(0.1227) \left\{ \frac{\pi}{2} \frac{126}{180} - \frac{1}{4} \sin[(2)(126°)] \right\}$$

$$= 788 \text{ N} \cdot \text{m}$$

From Eq. (16–4), the actuating force is

Answer

$$F = \frac{M_N - M_f}{c} = \frac{788 - 304}{100 + 112} = 2.28 \text{ kN}$$

(b) From Eq. (16–6), the torque applied by the right-hand shoe is

$$T_R = \frac{f p_a b r^2 (\cos \theta_1 - \cos \theta_2)}{\sin \theta_a}$$

$$= \frac{0.32[1000(10)^3](0.032)(0.150)^2(\cos 0° - \cos 126°)}{\sin 90°} = 366 \text{ N} \cdot \text{m}$$

The torque contributed by the left-hand shoe cannot be obtained until we learn its maximum operating pressure. Equations (16–2) and (16–3) indicate that the frictional and normal moments are proportional to this pressure. Thus, for the left-hand shoe,

$$M_N = \frac{788 p_a}{1000} \qquad M_f = \frac{304 p_a}{1000}$$

Then, from Eq. (16–7),

$$F = \frac{M_N + M_f}{c}$$

or

$$2.28 = \frac{(788/1000)p_a + (304/1000)p_a}{100 + 112}$$

Solving gives $p_a = 443$ kPa. Then, from Eq. (16–6), the torque on the left-hand shoe is

$$T_L = \frac{f p_a b r^2 (\cos \theta_1 - \cos \theta_2)}{\sin \theta_a}$$

Since $\sin \theta_a = \sin 90° = 1$, we have

$$T_L = 0.32[443(10)^3](0.032)(0.150)^2(\cos 0° - \cos 126°) = 162 \text{ N} \cdot \text{m}$$

The braking capacity is the total torque:

Answer
$$T = T_R + T_L = 366 + 162 = 528 \text{ N} \cdot \text{m}$$

(c) In order to find the hinge-pin reactions, we note that $\sin \theta_a = 1$ and $\theta_1 = 0$. Then Eq. (16–8) gives

$$A = \frac{1}{2} \sin^2 \theta_2 = \frac{1}{2} \sin^2 126° = 0.3273$$

$$B = \frac{\theta_2}{2} - \frac{1}{4} \sin 2\theta_2 = \frac{\pi(126)}{2(180)} - \frac{1}{4} \sin[(2)(126°)] = 1.3373$$

Also, let

$$D = \frac{p_a b r}{\sin \theta_a} = \frac{1000(0.032)(0.150)}{1} = 4.8 \text{ kN}$$

where $p_a = 1000$ kPa for the right-hand shoe. Then, using Eq. (16–9), we have

$$R_x = D(A - fB) - F_x = 4.8[0.3273 - 0.32(1.3373)] - 2.28 \sin 24°$$
$$= -1.410 \text{ kN}$$

$$R_y = D(B + fA) - F_y = 4.8[1.3373 + 0.32(0.3273)] - 2.28 \cos 24°$$
$$= 4.839 \text{ kN}$$

The resultant on this hinge pin is

Answer
$$R = \sqrt{(-1.410)^2 + (4.839)^2} = 5.04 \text{ kN}$$

The reactions at the hinge pin of the left-hand shoe are found using Eqs. (16–10) for a pressure of 443 kPa. They are found to be $R_x = 0.678$ kN and $R_y = 0.538$ kN. The resultant is

Answer
$$R = \sqrt{(0.678)^2 + (0.538)^2} = 0.866 \text{ kN}$$

The reactions for both hinge pins, together with their directions, are shown in Fig. 16–9.

| Figure 16–9

This example dramatically shows the benefit to be gained by arranging the shoes to be self-energizing. If the left-hand shoe were turned over so as to place the hinge pin at the top, it could apply the same torque as the right-hand shoe. This would make the capacity of the brake $(2)(366) = 732$ N $\cdot$ m instead of the present 528 N $\cdot$ m, a 30 percent improvement. In addition, some of the friction material at the heel could be eliminated without seriously affecting the capacity, because of the low pressure in this area. This change might actually improve the overall design because the additional rim exposure would improve the heat-dissipation capacity.

16–3 External Contracting Rim Clutches and Brakes

The patented clutch-brake of Fig. 16–10 has external contracting friction elements, but the actuating mechanism is pneumatic. Here we shall study only pivoted external shoe brakes and clutches, though the methods presented can easily be adapted to the clutch-brake of Fig. 16–10.

Operating mechanisms can be classified as:

1 Solenoids

2 Levers, linkages, or toggle devices

3 Linkages with spring loading

4 Hydraulic and pneumatic devices

The static analysis required for these devices has already been covered in Sec. 3–1. The methods there apply to any mechanism system, including all those used in brakes and clutches. It is not necessary to repeat the material in Chap. 3 that applies directly to such mechanisms. Omitting the operating mechanisms from consideration allows us to concentrate on brake and clutch performance without the extraneous influences introduced by the need to analyze the statics of the control mechanisms.

The notation for external contracting shoes is shown in Fig. 16–11. The moments of the frictional and normal forces about the hinge pin are the same as for the internal

Figure 16–10

An external contracting clutch-brake that is engaged by expanding the flexible tube with compressed air. *(Courtesy of Twin Disc Clutch Company.)*

Figure 16–11

Notation of external contracting shoes.

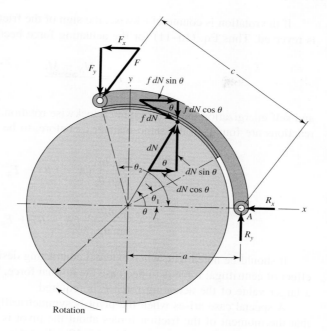

expanding shoes. Equations (16–2) and (16–3) apply and are repeated here for convenience:

$$M_f = \frac{f p_a b r}{\sin \theta_a} \int_{\theta_1}^{\theta_2} \sin \theta (r - a \cos \theta) \, d\theta \tag{16–2}$$

$$M_N = \frac{p_a b r a}{\sin \theta_a} \int_{\theta_1}^{\theta_2} \sin^2 \theta \, d\theta \tag{16–3}$$

Both these equations give positive values for clockwise moments (Fig. 16–11) when used for external contracting shoes. The actuating force must be large enough to balance both moments:

$$F = \frac{M_N + M_f}{c} \tag{16–11}$$

The horizontal and vertical reactions at the hinge pin are found in the same manner as for internal expanding shoes. They are

$$R_x = \int dN \cos \theta + \int f \, dN \sin \theta - F_x \tag{a}$$

$$R_y = \int f \, dN \cos \theta - \int dN \sin \theta + F_y \tag{b}$$

By using Eq. (16–8) and Eq. (c) from Sec. 16–2, we have

$$R_x = \frac{p_a b r}{\sin \theta_a}(A + fB) - F_x$$

$$R_y = \frac{p_a b r}{\sin \theta_a}(fA - B) + F_y \tag{16–12}$$

If the rotation is counterclockwise, the sign of the frictional term in each equation is reversed. Thus Eq. (16–11) for the actuating force becomes

$$F = \frac{M_N - M_f}{c} \tag{16–13}$$

and self-energization exists for counterclockwise rotation. The horizontal and vertical reactions are found, in the same manner as before, to be

$$R_x = \frac{p_a b r}{\sin \theta_a}(A - fB) - F_x$$

$$R_y = \frac{p_a b r}{\sin \theta_a}(-fA - B) + F_y \tag{16–14}$$

It should be noted that, when external contracting designs are used as clutches, the effect of centrifugal force is to decrease the normal force. Thus, as the speed increases, a larger value of the actuating force F is required.

A special case arises when the pivot is symmetrically located and also placed so that the moment of the friction forces about the pivot is zero. The geometry of such a brake will be similar to that of Fig. 16–12a. To get a pressure-distribution relation, we note that lining wear is such as to retain the cylindrical shape, much as a milling machine cutter feeding in the x direction would do to the shoe held in a vise. See Fig. 16–12b. This means the abscissa component of wear is w_0 for all positions θ. If wear in the radial direction is expressed as $w(\theta)$, then

$$w(\theta) = w_0 \cos \theta$$

Using Eq. (12–26), p. 654, to express radial wear $w(\theta)$ as

$$w(\theta) = KPVt$$

Figure 16–12

(a) Brake with symmetrical pivoted shoe; (b) wear of brake lining.

(b)

(a)

where K is a material constant, P is pressure, V is rim velocity, and t is time. Then, denoting P as $p(\theta)$ above and solving for $p(\theta)$ gives

$$p(\theta) = \frac{w(\theta)}{KVt} = \frac{w_0 \cos \theta}{KVt}$$

Since all elemental surface areas of the friction material see the same rubbing speed for the same duration, $w_0/(KVt)$ is a constant and

$$p(\theta) = (\text{constant}) \cos \theta = p_a \cos \theta \qquad (c)$$

where p_a is the maximum value of $p(\theta)$.

Proceeding to the force analysis, we observe from Fig. 16–12a that

$$dN = pbr \, d\theta \qquad (d)$$

or

$$dN = p_a br \cos \theta \, d\theta \qquad (e)$$

The distance a to the pivot is chosen by finding where the moment of the frictional forces M_f is zero. First, this ensures that reaction R_y is at the correct location to establish symmetrical wear. Second, a cosinusoidal pressure distribution is sustained, preserving our predictive ability. Symmetry means $\theta_1 = \theta_2$, so

$$M_f = 2 \int_0^{\theta_2} (f \, dN)(a \cos \theta - r) = 0$$

Substituting Eq. (e) gives

$$2 f p_a br \int_0^{\theta_2} (a \cos^2 \theta - r \cos \theta) \, d\theta = 0$$

from which

$$a = \frac{4r \sin \theta_2}{2\theta_2 + \sin 2\theta_2} \qquad (16\text{--}15)$$

The distance a depends on the pressure distribution. Mislocating the pivot makes M_f zero about a different location, so the brake lining adjusts its local contact pressure, through wear, to compensate. The result is unsymmetrical wear, retiring the shoe lining, hence the shoe, sooner.

With the pivot located according to Eq. (16–15), the moment about the pin is zero, and the horizontal and vertical reactions are

$$R_x = 2 \int_0^{\theta_2} dN \cos \theta = 2 \int_0^{\theta_2} (p_a br \cos \theta \, d\theta) \cos \theta = \frac{p_a br}{2}(2\theta_2 + \sin 2\theta_2) \qquad (16\text{--}16)$$

where, because of symmetry,

$$\int f \, dN \sin \theta = 0$$

Also,

$$R_y = 2 \int_0^{\theta_2} f \, dN \cos \theta = 2 \int_0^{\theta_2} f(p_a br \cos \theta \, \delta\theta) \cos \theta = \frac{p_a brf}{2}(2\theta_2 + \sin 2\theta_2) \qquad (16\text{--}17)$$

where

$$\int dN \sin \theta = 0$$

also because of symmetry. Note, too, that $R_x = N$ and $R_y = fN$, as might be expected for the particular choice of the dimension a. Also, it can be shown that the torque is

$$T = afN \qquad (16\text{-}18)$$

16-4 Band-Type Clutches and Brakes

Flexible clutch and brake bands are used in power excavators and in hoisting and other machinery. The analysis follows the notation of Fig. 16–13.

Because of friction and the rotation of the drum, the actuating force P_2 is less than the pin reaction P_1. Any element of the band, of angular length $d\theta$, will be in equilibrium under the action of the forces shown in the figure. Summing these forces in the vertical direction, we have

$$(P + dP) \sin \frac{d\theta}{2} + P \sin \frac{d\theta}{2} - dN = 0 \qquad (a)$$

$$dN = P\,d\theta \qquad (b)$$

since for small angles $\sin d\theta/2 = d\theta/2$. Summing the forces in the horizontal direction gives

$$(P + dP) \cos \frac{d\theta}{2} - P \cos \frac{d\theta}{2} - f\,dN = 0 \qquad (c)$$

$$dP - f\,dN = 0 \qquad (d)$$

since for small angles, $\cos(d\theta/2) \approx 1$. Substituting the value of dN from Eq. (b) in (d) and integrating give

$$\int_{P_2}^{P_1} \frac{dP}{P} = f \int_0^\phi d\theta \qquad \text{or} \qquad \ln \frac{P_1}{P_2} = f\phi$$

Figure 16–13

Forces on a brake band.

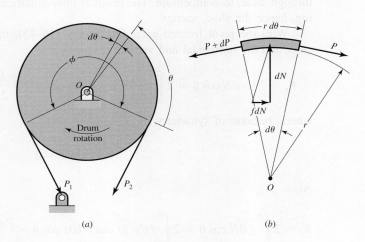

(a)

(b)

and thus

$$\frac{P_1}{P_2} = e^{f\phi} \qquad (16\text{--}19)$$

The torque may be obtained from the equation

$$T = (P_1 - P_2)\frac{D}{2} \qquad (16\text{--}20)$$

The normal force dN acting on an element of area of width b and length $r\,d\theta$ is

$$dN = pbr\,d\theta \qquad (e)$$

where p is the pressure. Substitution of the value of dN from Eq. (b) gives
$$P\,d\theta = pbr\,d\theta$$

Therefore

$$p = \frac{P}{br} = \frac{2P}{bD} \qquad (16\text{--}21)$$

The pressure is therefore proportional to the tension in the band. The maximum pressure p_a will occur at the toe and has the value

$$p_a = \frac{2P_1}{bD} \qquad (16\text{--}22)$$

16–5   Frictional-Contact Axial Clutches

An axial clutch is one in which the mating frictional members are moved in a direction parallel to the shaft. One of the earliest of these is the cone clutch, which is simple in construction and quite powerful. However, except for relatively simple installations, it has been largely displaced by the disk clutch employing one or more disks as the operating members. Advantages of the disk clutch include the freedom from centrifugal effects, the large frictional area that can be installed in a small space, the more effective heat-dissipation surfaces, and the favorable pressure distribution. Figure 16–14 shows a

Figure 16–14

Cross-sectional view of a single-plate clutch; A, driver; B, driven plate (keyed to driven shaft); C, actuator.

Figure 16–15

An oil-actuated multiple-disk clutch-brake for operation in an oil bath or spray. It is especially useful for rapid cycling. (*Courtesy of Twin Disc Clutch Company.*)

Figure 16–16

Disk friction member.

single-plate disk clutch; a multiple-disk clutch-brake is shown in Fig. 16–15. Let us now determine the capacity of such a clutch or brake in terms of the material and geometry.

Figure 16–16 shows a friction disk having an outside diameter D and an inside diameter d. We are interested in obtaining the axial force F necessary to produce a certain torque T and pressure p. Two methods of solving the problem, depending upon the construction of the clutch, are in general use. If the disks are rigid, then the greatest amount of wear will at first occur in the outer areas, since the work of friction is greater in those areas. After a certain amount of wear has taken place, the pressure distribution will change so as to permit the wear to be uniform. This is the basis of the first method of solution.

Another method of construction employs springs to obtain a uniform pressure over the area. It is this assumption of uniform pressure that is used in the second method of solution.

Uniform Wear

After initial wear has taken place and the disks have worn down to a point where uniform wear is established, the axial wear can be expressed by Eq. (12–27), p. 655, as

$$w = f_1 f_2 KPV t$$

where the pressure P and velocity V can vary in the wear area. For uniform wear, w is constant, therefore PV is constant. Setting $p = P$, and $V = r\omega$, where ω is the angular velocity of the rotating member, we find that in the wear area, pr is constant. The maximum pressure p_a occurs where r is minimum, $r = d/2$, and thus

$$pr = p_a \frac{d}{2} \qquad (a)$$

We can take an expression from Eq. (a), which is the condition for having the same amount of work done at radius r as is done at radius $d/2$. Referring to Fig. 16–16, we have an element of area of radius r and thickness dr. The area of this element is $2\pi r\, dr$, so that the normal force acting upon this element is $dF = 2\pi pr\, dr$. We can find the total normal force by letting r vary from $d/2$ to $D/2$ and integrating. Thus, with Eq. (a),

$$F = \int_{d/2}^{D/2} 2\pi pr\, dr = \pi p_a d \int_{d/2}^{D/2} dr = \frac{\pi p_a d}{2}(D - d) \qquad (16\text{–}23)$$

The torque is found by integrating the product of the frictional force and the radius:

$$T = \int_{d/2}^{D/2} 2\pi f pr^2\, dr = \pi f p_a d \int_{d/2}^{D/2} r\, dr = \frac{\pi f p_a d}{8}(D^2 - d^2) \qquad (16\text{–}24)$$

By substituting the value of F from Eq. (16–23) we may obtain a more convenient expression for the torque. Thus

$$T = \frac{Ff}{4}(D + d) \qquad (16\text{–}25)$$

In use, Eq. (16–23) gives the actuating force for the selected maximum pressure p_a. This equation holds for any number of friction pairs or surfaces. Equation (16–25), however, gives the torque capacity for only a single friction surface.

Uniform Pressure

When uniform pressure p_a can be assumed over the area of the disk, the actuating force F is simply the product of the pressure and the area. This gives

$$F = \frac{\pi p_a}{4}(D^2 - d^2) \qquad (16\text{–}26)$$

As before, the torque is found by integrating the product of the frictional force and the radius:

$$T = 2\pi f p_a \int_{d/2}^{D/2} r^2\, dr = \frac{\pi f p_a}{12}(D^3 - d^3) \qquad (16\text{–}27)$$

From Eq. (16–26) we can rewrite Eq. (16–27) as

$$T = \frac{Ff}{3}\frac{D^3 - d^3}{D^2 - d^2} \qquad (16\text{–}28)$$

It should be noted for both equations that the torque is for a single pair of mating surfaces. This value must therefore be multiplied by the number of pairs of surfaces in contact.

Figure 16–17

Dimensionless plot of Eqs. (b) and (c).

Let us express Eq. (16–25) for torque during uniform wear as

$$\frac{T}{fFD} = \frac{1 + d/D}{4} \qquad (b)$$

and Eq. (16–28) for torque during uniform pressure (new clutch) as

$$\frac{T}{fFD} = \frac{1}{3}\frac{1 - (d/D)^3}{1 - (d/D)^2} \qquad (c)$$

and plot these in Fig. 16–17. What we see is a dimensionless presentation of Eqs. (b) and (c) which reduces the number of variables from five (T, f, F, D, and d) to three (T/FD, f, and d/D) which are dimensionless. This is the method of Buckingham. The dimensionless groups (called pi terms) are

$$\pi_1 = \frac{T}{FD} \qquad \pi_2 = f \qquad \pi_3 = \frac{d}{D}$$

This allows a five-dimensional space to be reduced to a three-dimensional space. Further, because of the "multiplicative" relation between f and T in Eqs. (b) and (c), it is possible to plot π_1/π_2 versus π_3 in a two-dimensional space (the plane of a sheet of paper) to view all cases over the domain of existence of Eqs. (b) and (c) and to compare, without risk of oversight! By examining Fig. 16–17 we can conclude that a new clutch, Eq. (b), always transmits more torque than an old clutch, Eq. (c). Furthermore, since clutches of this type are typically proportioned to make the diameter ratio d/D fall in the range $0.6 \le d/D \le 1$, the largest discrepancy between Eq. (b) and Eq. (c) will be

$$\frac{T}{fFD} = \frac{1 + 0.6}{4} = 0.400 \qquad \text{(old clutch, uniform wear)}$$

$$\frac{T}{fFD} = \frac{1}{3}\frac{1 - 0.6^3}{1 - 0.6^2} = 0.4083 \qquad \text{(new clutch, uniform pressure)}$$

so the proportional error is $(0.4083 - 0.400)/0.400 = 0.021$, or about 2 percent. Given the uncertainties in the actual coefficient of friction and the certainty that new clutches get old, there is little reason to use anything but Eqs. (16–23), (16–24), and (16–25).

16–6   **Disk Brakes**

As indicated in Fig. 16–16, there is no fundamental difference between a disk clutch and a disk brake. The analysis of the preceding section applies to disk brakes too.

We have seen that rim or drum brakes can be designed for self-energization. While this feature is important in reducing the braking effort required, it also has a disadvantage. When drum brakes are used as vehicle brakes, only a slight change in the coefficient of friction will cause a large change in the pedal force required for braking. A not unusual 30 percent reduction in the coefficient of friction due to a temperature change or moisture, for example, can result in a 50 percent change in the pedal force required to obtain the same braking torque obtainable prior to the change. The disk brake has no self-energization, and hence is not so susceptible to changes in the coefficient of friction.

Another type of disk brake is the *floating caliper brake,* shown in Fig. 16–18. The caliper supports a single floating piston actuated by hydraulic pressure. The action is much like that of a screw clamp, with the piston replacing the function of the screw. The floating action also compensates for wear and ensures a fairly constant pressure over the area of the friction pads. The seal and boot of Fig. 16–18 are designed to obtain clearance by backing off from the piston when the piston is released.

Caliper brakes (named for the nature of the actuating linkage) and disk brakes (named for the shape of the unlined surface) press friction material against the face(s)

Figure 16–18

An automotive disk brake. (*Courtesy DaimlerChrysler Corporation.*)

Figure 16–19

Geometry of contact area of an annular-pad segment of a caliper brake.

of a rotating disk. Depicted in Fig. 16–19 is the geometry of an annular-pad brake contact area. The governing axial wear equation is Eq. (12–27), p. 655,

$$w = f_1 f_2 KPVt$$

The coordinate $\bar{r}$ locates the line of action of force F that intersects the y axis. Of interest also is the effective radius r_e, which is the radius of an equivalent shoe of infinitesimal radial thickness. If p is the local contact pressure, the actuating force F and the friction torque T are given by

$$F = \int_{\theta_1}^{\theta_2} \int_{r_i}^{r_o} pr \, dr \, d\theta = (\theta_2 - \theta_1) \int_{r_i}^{r_o} pr \, dr \qquad (16\text{–}29)$$

$$T = \int_{\theta_1}^{\theta_2} \int_{r_i}^{r_o} fpr^2 \, dr \, d\theta = (\theta_2 - \theta_1)f \int_{r_i}^{r_o} pr^2 \, dr \qquad (16\text{–}30)$$

The equivalent radius r_e can be found from $f\,Fr_e = T$, or

$$r_e = \frac{T}{fF} = \frac{\displaystyle\int_{r_i}^{r_o} pr^2 \, dr}{\displaystyle\int_{r_i}^{r_o} pr \, dr} \qquad (16\text{–}31)$$

The locating coordinate $\bar{r}$ of the activating force is found by taking moments about the x axis:

$$M_x = F\bar{r} = \int_{\theta_1}^{\theta_2} \int_{r_i}^{r_o} pr(r \sin \theta)dr \, d\theta = (\cos \theta_1 - \cos \theta_2) \int_{r_i}^{r_o} pr^2 \, dr$$

$$(16\text{–}32)$$

$$\bar{r} = \frac{M_x}{F} = \frac{(\cos \theta_1 - \cos \theta_2)}{\theta_2 - \theta_1} r_e$$

Uniform Wear

It is clear from Eq. (12–27) that for the axial wear to be the same everywhere, the product PV must be a constant. From Eq. (a), Sec. 16–5, the pressure p can be expressed in terms of the largest allowable pressure p_a (which occurs at the inner radius r_i) as

$p = p_a r_i/r$. Equation (16–29) becomes

$$F = (\theta_2 - \theta_1)p_a r_i(r_o - r_i) \qquad (16\text{–}33)$$

Equation (16–30) becomes

$$T = (\theta_2 - \theta_1)fp_a r_i \int_{r_i}^{r_o} r\, dr = \frac{1}{2}(\theta_2 - \theta_1)fp_a r_i(r_o^2 - r_i^2) \qquad (16\text{–}34)$$

Equation (16–31) becomes

$$r_e = \frac{p_a r_i \displaystyle\int_{r_i}^{r_o} r\, dr}{p_a r_i \displaystyle\int_{r_i}^{r_o} dr} = \frac{r_o^2 - r_i^2}{2}\frac{1}{r_o - r_i} = \frac{r_o + r_i}{2} \qquad (16\text{–}35)$$

Equation (16–32) becomes

$$\bar{r} = \frac{\cos\theta_1 - \cos\theta_2}{\theta_2 - \theta_1}\frac{r_o + r_i}{2} \qquad (16\text{–}36)$$

Uniform Pressure

In this situation, approximated by a new brake, $p = p_a$. Equation (16–29) becomes

$$F = (\theta_2 - \theta_1)p_a \int_{r_i}^{r_o} r\, dr = \frac{1}{2}(\theta_2 - \theta_1)p_a(r_o^2 - r_i^2) \qquad (16\text{–}37)$$

Equation (16–30) becomes

$$T = (\theta_2 - \theta_1)fp_a \int_{r_i}^{r_o} r^2\, dr = \frac{1}{3}(\theta_2 - \theta_1)fp_a(r_o^3 - r_i^3) \qquad (16\text{–}38)$$

Equation (16–31) becomes

$$r_e = \frac{p_a \displaystyle\int_{r_i}^{r_o} r^2\, dr}{p_a \displaystyle\int_{r_i}^{r_o} r\, dr} = \frac{r_o^3 - r_i^3}{3}\frac{2}{r_o^2 - r_i^2} = \frac{2}{3}\frac{r_o^3 - r_i^3}{r_o^2 - r_i^2} \qquad (16\text{–}39)$$

Equation (16–32) becomes

$$\bar{r} = \frac{\cos\theta_1 - \cos\theta_2}{\theta_2 - \theta_1}\frac{2}{3}\frac{r_o^3 - r_i^3}{r_o^2 - r_i^2} = \frac{2}{3}\frac{r_o^3 - r_i^3}{r_o^2 - r_i^2}\frac{\cos\theta_1 - \cos\theta_2}{\theta_2 - \theta_1} \qquad (16\text{–}40)$$

EXAMPLE 16–3   Two annular pads, $r_i = 90$ mm, $r_o = 130$ mm, subtend an angle of 108°, have a coefficient of friction of 0.42, and are actuated by a pair of hydraulic cylinders 38 mm in diameter. The torque requirement is 1300 N · m. For uniform wear
(a) Find the largest normal pressure p_a.
(b) Estimate the actuating force F.
(c) Find the equivalent radius r_e and force location $\bar{r}$.
(d) Estimate the required hydraulic pressure.

Solution

(a) From Eq. (16–34), with $T = 1300/2 = 650$ N · m for each pad,

Answer

$$p_a = \frac{2T}{(\theta_2 - \theta_1) f r_i \left(r_o^2 - r_i^2\right)}$$

$$= \frac{2(650\,000)}{(144° - 36°)(\pi/180)0.42(90)(130^2 - 90^2)} = 2.07 \text{ MPa}$$

(b) From Eq. (16–33),

Answer

$$F = (\theta_2 - \theta_1) p_a r_i (r_o - r_i) = (144° - 36°)(\pi/180)2.07(90)(130 - 40)$$

$$= 14069.26 \text{ N}$$

(c) From Eq. (16–35),

Answer

$$r_e = \frac{r_o + r_i}{2} = \frac{130 - 90}{2} = 110.00 \text{ mm}$$

From Eq. (16–36),

Answer

$$\bar{r} = \frac{\cos \theta_1 - \cos \theta_2}{\theta_2 - \theta_1} \frac{r_o + r_i}{2} = \frac{\cos 36° - \cos 144°}{(144° - 36°)(\pi/180)} \frac{130 + 90}{2}$$

$$= 94.42 \text{ mm}$$

(d) Each cylinder supplies the actuating force, 16 681 N.

Answer

$$p_{\text{hydraulic}} = \frac{F}{A_P} = \frac{14069.26}{\pi(38^2/4)} = 12.41 \text{ MPa}$$

Circular (Button or Puck) Pad Caliper Brake

Figure 16–20 displays the pad geometry. Numerical integration is necessary to analyze this brake since the boundaries are difficult to handle in closed form. Table 16–1 gives the parameters for this brake as determined by Fazekas. The effective radius is given by

$$r_e = \delta e \qquad (16–41)$$

The actuating force is given by

$$F = \pi R^2 p_{\text{av}} \qquad (16–42)$$

Figure 16–20

Geometry of circular pad of a caliper brake.

and the torque is given by

$$T = f F r_e \qquad (16–43)$$

Table 16–1

Parameters for a Circular-Pad Caliper Brake

Source: G. A. Fazekas, "On Circular Spot Brakes," *Trans. ASME, J. Engineering for Industry,* vol. 94, Series B, No. 3, August 1972, pp. 859–863.

$\dfrac{R}{e}$	$\delta = \dfrac{r_e}{e}$	$\dfrac{p_{max}}{p_{av}}$
0.0	1.000	1.000
0.1	0.983	1.093
0.2	0.969	1.212
0.3	0.957	1.367
0.4	0.947	1.578
0.5	0.938	1.875

EXAMPLE 16–4

A button-pad disk brake uses dry sintered metal pads. The pad radius is 10 mm, and its center is 50 mm from the axis of rotation of the 88-mm-diameter disk. Using half of the largest allowable pressure, $p_{max} = 2.8$ MPa, find the actuating force and the brake torque. The coefficient of friction is 0.37.

Solution

Since the pad radius $R = 10$ mm and eccentricity $e = 50$ mm,

$$\frac{R}{e} = \frac{10}{50} = 0.20$$

From Table 16–1, by interpolation, $\delta = 0.969$ and $p_{max}/p_{av} = 1.212$. It follows that the effective radius e is found from Eq. (16–41):

$$r_e = \delta e = 0.969(50) = 48.45 \text{ mm}$$

and the average pressure is

$$p_{av} = \frac{p_{max}/2}{1.290} = \frac{2.8/2}{1.290} = 1.16 \text{ MPa}$$

The actuating force F is found from Eq. (16–42) to be

Answer

$$F = \pi R^2 p_{av} = \pi(10)^2 1.16 = 362.89 \text{ N} \qquad \text{(one side)}$$

The brake torque T is

Answer

$$T = fFr_e = 0.37\,(362.89)\,48.45 = 6505.35 \text{ N} \cdot \text{mm} \qquad \text{(one side)}$$

16–7   Cone Clutches and Brakes

The drawing of a *cone clutch* in Fig. 16–21 shows that it consists of a *cup* keyed or splined to one of the shafts, a *cone* that must slide axially on splines or keys on the mating shaft, and a helical *spring* to hold the clutch in engagement. The clutch is disengaged by means of a fork that fits into the shifting groove on the friction cone. The *cone angle* α and the diameter and face width of the cone are the important geometric design parameters. If the cone angle is too small, say, less than about 8°, then

Figure 16–21

Cross section of a cone clutch.

Figure 16–22

Contact area of a cone clutch.

the force required to disengage the clutch may be quite large. The wedging effect lessens rapidly when larger cone angles are used. Depending upon the characteristics of the friction materials, a good compromise can usually be found using cone angles between 10 and 15°.

To find a relation between the operating force F and the torque transmitted, designate the dimensions of the friction cone as shown in Figure 16–22. As in the case of the axial clutch, we can obtain one set of relations for a uniform-wear and another set for a uniform-pressure assumption.

Uniform Wear

The pressure relation is the same as for the axial clutch:

$$p = p_a \frac{d}{2r} \tag{a}$$

Next, referring to Fig. 16–22, we see that we have an element of area dA of radius r and width $dr/\sin \alpha$. Thus $dA = (2\pi r dr)/\sin \alpha$. As shown in Fig. 16–22, the operating force will be the integral of the axial component of the differential force $p\,dA$. Thus

$$F = \int p\,dA \sin \alpha = \int_{d/2}^{D/2} \left(p_a \frac{d}{2r}\right)\left(\frac{2\pi r\,dr}{\sin \alpha}\right)(\sin \alpha)$$

$$= \pi p_a d \int_{d/2}^{D/2} dr = \frac{\pi p_a d}{2}(D - d) \qquad (16\text{–}44)$$

which is the same result as in Eq. (16–23).

The differential friction force is $f p\,dA$, and the torque is the integral of the product of this force with the radius. Thus

$$T = \int rfp\,dA = \int_{d/2}^{D/2} (rf)\left(p_a \frac{d}{2r}\right)\left(\frac{2\pi r\,dr}{\sin \alpha}\right)$$

$$= \frac{\pi f p_a d}{\sin \alpha} \int_{d/2}^{D/2} r\,dr = \frac{\pi f p_a d}{8 \sin \alpha}(D^2 - d^2) \qquad (16\text{–}45)$$

Note that Eq. (16–24) is a special case of Eq. (16–45), with $\alpha = 90°$. Using Eq. (16–44), we find that the torque can also be written

$$T = \frac{Ff}{4 \sin \alpha}(D + d) \qquad (16\text{–}46)$$

Uniform Pressure

Using $p = p_a$, the actuating force is found to be

$$F = \int p_a\,dA \sin \alpha = \int_{d/2}^{D/2} (p_a)\left(\frac{2\pi r\,dr}{\sin \alpha}\right)(\sin \alpha) = \frac{\pi p_a}{4}(D^2 - d^2) \quad (16\text{–}47)$$

The torque is

$$T = \int rfp_a\,dA = \int_{d/2}^{D/2} (rfp_a)\left(\frac{2\pi r\,dr}{\sin \alpha}\right) = \frac{\pi f p_a}{12 \sin \alpha}(D^3 - d^3) \qquad (16\text{–}48)$$

Using Eq. (16–47) in Eq. (16–48) gives

$$T = \frac{Ff}{3 \sin \alpha}\frac{D^3 - d^3}{D^2 - d^2} \qquad (16\text{–}49)$$

As in the case of the axial clutch, we can write Eq. (16–46) dimensionlessly as

$$\frac{T \sin \alpha}{f Fd} = \frac{1 + d/D}{4} \qquad (b)$$

and write Eq. (16–49) as

$$\frac{T \sin \alpha}{f Fd} = \frac{1}{3}\frac{1 - (d/D)^3}{1 - (d/D)^2} \qquad (c)$$

This time there are six (T, α, f, F, D, and d) parameters and four pi terms:

$$\pi_1 = \frac{T}{FD} \qquad \pi_2 = f \qquad \pi_3 = \sin \alpha \qquad \pi_4 = \frac{d}{D}$$

As in Fig. 16–17, we plot $T \sin \alpha/(fFD)$ as ordinate and d/D as abscissa. The plots and conclusions are the same. There is little reason for using equations other than Eqs. (16–44), (16–45), and (16–46).

16-8 Energy Considerations

When the rotating members of a machine are caused to stop by means of a brake, the kinetic energy of rotation must be absorbed by the brake. This energy appears in the brake in the form of heat. In the same way, when the members of a machine that are initially at rest are brought up to speed, slipping must occur in the clutch until the driven members have the same speed as the driver. Kinetic energy is absorbed during slippage of either a clutch or a brake, and this energy appears as heat.

We have seen how the torque capacity of a clutch or brake depends upon the coefficient of friction of the material and upon a safe normal pressure. However, the character of the load may be such that, if this torque value is permitted, the clutch or brake may be destroyed by its own generated heat. The capacity of a clutch is therefore limited by two factors, the characteristics of the material and the ability of the clutch to dissipate heat. In this section we shall consider the amount of heat generated by a clutching or braking operation. If the heat is generated faster than it is dissipated, we have a temperature-rise problem; that is the subject of the next section.

To get a clear picture of what happens during a simple clutching or braking operation, refer to Fig. 16–1a, which is a mathematical model of a two-inertia system connected by a clutch. As shown, inertias I_1 and I_2 have initial angular velocities of ω_1 and ω_2, respectively. During the clutch operation both angular velocities change and eventually become equal. We assume that the two shafts are rigid and that the clutch torque is constant.

Writing the equation of motion for inertia 1 gives

$$I_1\ddot{\theta}_1 = -T \qquad (a)$$

where $\ddot{\theta}_1$ is the angular acceleration of I_1 and T is the clutch torque. A similar equation for I_2 is

$$I_2\ddot{\theta}_2 = T \qquad (b)$$

We can determine the instantaneous angular velocities $\dot{\theta}_1$ and $\dot{\theta}_2$ of I_1 and I_2 after any period of time t has elapsed by integrating Eqs. (a) and (b). The results are

$$\dot{\theta}_1 = -\frac{T}{I_1}t + \omega_1 \qquad (c)$$

$$\dot{\theta}_2 = \frac{T}{I_2}t + \omega_2 \qquad (d)$$

where $\dot{\theta}_1 = \omega_1$ and $\dot{\theta}_2 = \omega_2$ at $t = 0$. The difference in the velocities, sometimes called the relative velocity, is

$$\dot{\theta} = \dot{\theta}_1 - \dot{\theta}_2 = -\frac{T}{I_1}t + \omega_1 - \left(\frac{T}{I_2}t + \omega_2\right)$$

$$= \omega_1 - \omega_2 - T\left(\frac{I_1 + I_2}{I_1 I_2}\right)t \qquad (16\text{–}50)$$

The clutching operation is completed at the instant in which the two angular velocities $\dot{\theta}_1$ and $\dot{\theta}_2$ become equal. Let the time required for the entire operation be t_1. Then $\dot{\theta} = 0$ when $\dot{\theta}_1 = \dot{\theta}_2$, and so Eq. (16–50) gives the time as

$$t_1 = \frac{I_1 I_2(\omega_1 - \omega_2)}{T(I_1 + I_2)} \qquad (16\text{–}51)$$

This equation shows that the time required for the engagement operation is directly proportional to the velocity difference and inversely proportional to the torque.

We have assumed the clutch torque to be constant. Therefore, using Eq. (16–50), we find the rate of energy-dissipation during the clutching operation to be

$$u = T\dot{\theta} = T\left[\omega_1 - \omega_2 - T\left(\frac{I_1 + I_2}{I_1 I_2}\right)t\right] \tag{e}$$

This equation shows that the energy-dissipation rate is greatest at the start, when $t = 0$.

The total energy dissipated during the clutching operation or braking cycle is obtained by integrating Eq. (e) from $t = 0$ to $t = t_1$. The result is found to be

$$
\begin{aligned}
E &= \int_0^{t_1} u\, dt = T \int_0^{t_1}\left[\omega_1 - \omega_2 - T\left(\frac{I_1 + I_2}{I_1 I_2}\right)t\right] dt \\
&= \frac{I_1 I_2 (\omega_1 - \omega_2)^2}{2(I_1 + I_2)}
\end{aligned}
\tag{16–52}
$$

where Eq. (16–51) was employed. Note that the energy dissipated is proportional to the velocity difference squared and is independent of the clutch torque.

Note that E in Eq. (16–52) is the energy lost or dissipated and is the energy that is absorbed by the clutch or brake. If the inertias are expressed in U.S. customary units (lbf $\cdot$ in $\cdot$ s^2), then the energy absorbed by the clutch assembly is in in $\cdot$ lbf. Using these units, the heat generated in Btu is

$$H = \frac{E}{9336} \tag{16–53}$$

In SI, the inertias are expressed in kilogram-meter^2 units, and the energy dissipated is expressed in joules.

16–9   Temperature Rise

The temperature rise of the clutch or brake assembly can be approximated by the classic expression

$$\Delta T = \frac{H}{C_p W} \tag{16–54}$$

where ΔT = temperature rise, °F

C_p = specific heat capacity, Btu/(lbm $\cdot$ °F); use 0.12 for steel or cast iron

W = mass of clutch or brake parts, lbm

A similar equation can be written for SI units. It is

$$\Delta T = \frac{E}{C_p m} \tag{16–55}$$

where ΔT = temperature rise, °C

C_p = specific heat capacity; use 500 J/kg $\cdot$ °C for steel or cast iron

m = mass of clutch or brake parts, kg

The temperature-rise equations above can be used to explain what happens when a clutch or brake is operated. However, there are so many variables involved that it would be most unlikely that such an analysis would even approximate experimental results. For this reason such analyses are most useful, for repetitive cycling, in pinpointing those design parameters that have the greatest effect on performance.

If an object is at initial temperature T_1 in an environment of temperature T_∞, then Newton's cooling model is expressed as

$$\frac{T - T_\infty}{T_1 - T_\infty} = \exp\left(-\frac{\hbar_{CR}A}{WC_p}t\right) \qquad (16\text{-}56)$$

where T = temperature at time t, °C
  T_1 = initial temperature, °C
  T_∞ = environmental temperature, °C
  $\hbar_{CR}$ = overall coefficient of heat transfer W/(m² · °C)
  A = lateral surface area, m²
  W = mass of the object, kg
  C_p = specific heat capacity of the object, J/(kg · °C)

Figure 16–23 shows an application of Eq. (16–56). The curve ABC is the exponential decline of temperature given by Eq. (16–56). At time t_B a second application of the brake occurs. The temperature quickly rises to temperature T_2, and a new cooling curve is started. For repetitive brake applications, subsequent temperature peaks $T_3, T_4, \ldots$, may be higher than the previous peaks if insufficient cooling has occurred between applications. If this is a production situation with brake applications every t_1 seconds, then a steady state develops in which all the peaks T_{max} and all the valleys T_{min} are repetitive.

The heat-dissipation capacity of disk brakes has to be planned to avoid reaching disk and pad temperatures that are detrimental to the parts. When a disk brake has a rhythm such as discussed above, then the rate of heat transfer is described by another Newtonian equation:

$$H_{loss} = \hbar_{CR}A(T - T_\infty) = (h_r + f_v h_c)A(T - T_\infty) \qquad (16\text{-}57)$$

Figure 16–23

The effect of clutching or braking operations on temperature. T_∞ is the ambient temperature. Note that the temperature rise ΔT may be different for each operation.

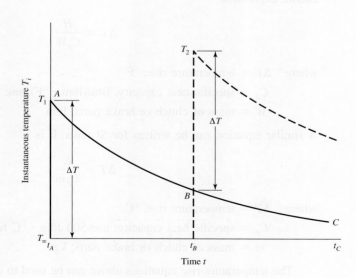

where H_{loss} = rate of energy loss, J/s or W

$ħ_{\text{CR}}$ = overall coefficient of heat transfer, W/(m² · °C)

h_r = radiation component of $ħ_{\text{CR}}$, W/(m² · °C), Fig. 16–24a

h_c = convective component of $ħ_{\text{CR}}$, W/(m² · °C), Fig. 16–24a

f_v = ventilation factor, Fig. 16–24b

T = disk temperature, °C

T_∞ = ambient temperature, °C

The energy E absorbed by the brake stopping a rotary inertia I in terms of original and final angular velocities ω_o and ω_f is the change in kinetic energy, $I(\omega_o^2 - \omega_f^2)/2$.

Figure 16–24

(a) Heat-transfer coefficient in still air. (b) Ventilation factors. (*Courtesy of Tolo-o-matic.*)

(a)

(b)

Expressing this in Btu,

$$E = \frac{1}{2} \frac{I}{9336} (\omega_o^2 - \omega_f^2)$$ (16–58)

The temperature rise ΔT due to a single stop is

$$\Delta T = \frac{E}{WC}$$ (16–59)

$T_{\max}$ has to be high enough to transfer E Btu in t_1 seconds. For steady state, rearrange Eq. (16–56) as

$$\frac{T_{\min} - T_\infty}{T_{\max} - T_\infty} = \exp(-\beta t_1)$$

where $\beta = \hbar_{CR} A/(W C_p)$. Cross-multiply, multiply the equation by -1, add $T_{\max}$ to both sides, set $T_{\max} - T_{\min} = \Delta T$, and rearrange, obtaining

$$T_{\max} = T_\infty + \frac{\Delta T}{1 - \exp(-\beta t_1)}$$ (16–60)

EXAMPLE 16–5

A caliper brake is used 24 times per hour to arrest a machine shaft from a speed of 250 rev/min to rest. The ventilation of the brake provides a mean air speed of 8 m/s. The equivalent rotary inertia of the machine as seen from the brake shaft is 32 kg · m · s. The disk is steel with a density $\gamma = 7800$ kg/m^3, a specific heat capacity of 0.45 kJ/(kg · °C), a diameter of 150 mm, a thickness of 6 mm. The pads are dry sintered metal. The lateral area of the brake surface is 0.032 m^2. Find $T_{\max}$ and $T_{\min}$ for the steady-state operation.

Solution

$$t_1 = 60^2/24 = 150 \text{ s}$$

Assuming a temperature rise of $T_{\max} - T_\infty = 100$°C, from Fig. 16–24a,

$$h_r = 8.8 \text{W}/(\text{m}^2 \cdot \text{°C})$$

$$h_c = 5.9 \text{W}/(\text{m}^2 \cdot \text{°C})$$

Fig.16–24b: $$f_v = 4.8$$

$$\hbar_{CR} = h_r + f_v h_c = 8.8 + 4.8(5.9) = 37.1 \text{W}/(\text{m}^2 \cdot \text{°C})$$

The mass of the disk is

$$W = \frac{\pi \gamma D^2 h}{4} = \frac{\pi (10)^3 7.8(0.15)^2 0.006}{4} = 0.83 \text{ kg}$$

Eq. (16–58): $$E = \frac{1}{2} I(\omega_o^2 - \omega_f^2) = \frac{32}{2} \left(\frac{2\pi}{60} 250\right)^2 = 11 \text{ kJ}$$

$$\beta = \frac{\hbar_{CR} A}{W C_p} = \frac{37.1(0.032)}{0.83(0.45)10^3} = 3.179(10^{-3}) \text{ s}^{-1}$$

Eq. (16–59):
$$\Delta T = \frac{E}{WC_p} = \frac{11}{0.83(0.45)} = 29.5°C$$

Answer   Eq. (16–60):
$$T_{max} = 21 + \frac{29.5}{1 - \exp[-3.179(10^{-3})150]} = 98.8°C$$

Answer
$$T_{min} = 98.8 - 29.5 = 69.3°C$$

The predicted temperature rise here is $T_{max} - T_\infty = 77.8°C$. Iterating with revised values of h_r and h_c from Fig. 16–24a, we can make the solution converge to $T_{max} = 104°C$ and $T_{min} = 77°C$.

Table 16–3 for dry sintered metal pads gives a continuous operating maximum temperature of 300–350°C. There is no danger of overheating.

16–10   Friction Materials

A brake or friction clutch should have the following lining material characteristics to a degree that is dependent on the severity of service:

- High and reproducible coefficient of friction
- Imperviousness to environmental conditions, such as moisture
- The ability to withstand high temperatures, together with good thermal conductivity and diffusivity, as well as high specific heat capacity
- Good resiliency
- High resistance to wear, scoring, and galling
- Compatible with the environment
- Flexibility

Table 16–2 gives area of friction surface required for several braking powers. Table 16–3 gives important characteristics of some friction materials for brakes and clutches.

Table 16–2

Area of Friction Material Required for a Given Average Braking Power   *Sources:* M. J. Neale, *The Tribology Handbook,* Butterworth, London, 1973; *Friction Materials for Engineers,* Ferodo Ltd., Chapel-en-le-frith, England, 1968.

		Ratio of Area to Average Braking Power, $(10^{-6})m^2/(joules/s)$		
Duty Cycle	Typical Applications	Band and Drum Brakes	Plate Disk Brakes	Caliper Disk Brakes
Infrequent	Emergency brakes	52	171	17.1
Intermittent	Elevators, cranes, and winches	171	434	43
Heavy-duty	Excavators, presses	342–422	832	86

Table 16-3

Characteristics of Friction Materials for Brakes and Clutches *Sources:* Ferodo Ltd., Chapel-en-le-frith, England; Scan-pac, Mequon, Wisc.; Raybestos, New York, N.Y. and Stratford, Conn.; Gatke Corp., Chicago, Ill.; General Metals Powder Co., Akron, Ohio; D. A. B. Industries, Troy, Mich.; Friction Products Co., Medina, Ohio.

Material	Friction Coefficient f	Maximum Pressure p_{max}, MPa	Maximum Temperature Instantaneous, °C	Maximum Temperature Continuous, °C	Maximum Velocity V_{max}, m/s	Applications
Cermet	0.32	1.0	815	400		Brakes and clutches
Sintered metal (dry)	0.29–0.33	2.1–2.8	500–550	300–350	18	Clutches and caliper disk brakes
Sintered metal (wet)	0.06–0.08	3.4	500	300	18	Clutches
Rigid molded asbestos (dry)	0.35–0.41	0.7	350–400	180	18	Drum brakes and clutches
Rigid molded asbestos (wet)	0.06	2.1	350	180	18	Industrial clutches
Rigid molded asbestos pads	0.31–0.49	5.2	500–750	230–350	24	Disk brakes
Rigid molded nonasbestos	0.33–0.63	0.7–1.0		260–400	24–38	Clutches and brakes
Semirigid molded asbestos	0.37–0.41	0.7	350	150	18	Clutches and brakes
Flexible molded asbestos	0.39–0.45	0.7	350–400	150–180	18	Clutches and brakes
Wound asbestos yarn and wire	0.38	0.7	350	150	18	Vehicle clutches
Woven asbestos yarn and wire	0.38	0.7	260	130	18	Industrial clutches and brakes
Woven cotton	0.47	0.7	110	75	18	Industrial clutches and brakes
Resilient paper (wet)	0.09–0.15	2.8	150		$PV < 18$ MPa · m/s	Clutches and transmission bands

The manufacture of friction materials is a highly specialized process, and it is advisable to consult manufacturers' catalogs and handbooks, as well as manufacturers directly, in selecting friction materials for specific applications. Selection involves a consideration of the many characteristics as well as the standard sizes available.

The *woven-cotton lining* is produced as a fabric belt that is impregnated with resins and polymerized. It is used mostly in heavy machinery and is usually supplied in rolls up to 15 mm in length. Thicknesses available range from 3 to 25 mm, in widths up to about 300 mm.

A *woven-asbestos lining* is made in a similar manner to the cotton lining and may also contain metal particles. It is not quite as flexible as the cotton lining and comes in a smaller range of sizes. Along with the cotton lining, the asbestos lining was widely used as a brake material in heavy machinery.

Molded-asbestos linings contain asbestos fiber and friction modifiers; a thermoset polymer is used, with heat, to form a rigid or semirigid molding. The principal use was in drum brakes.

Molded-asbestos pads are similar to molded linings but have no flexibility; they were used for both clutches and brakes.

Sintered-metal pads are made of a mixture of copper and/or iron particles with friction modifiers, molded under high pressure and then heated to a high temperature to fuse the material. These pads are used in both brakes and clutches for heavy-duty applications.

Cermet pads are similar to the sintered-metal pads and have a substantial ceramic content.

Table 16–4 lists properties of typical brake linings. The linings may consist of a mixture of fibers to provide strength and ability to withstand high temperatures, various friction particles to obtain a degree of wear resistance as well as a higher coefficient of friction, and bonding materials.

Table 16–5 includes a wider variety of clutch friction materials, together with some of their properties. Some of these materials may be run wet by allowing them to dip in oil or to be sprayed by oil. This reduces the coefficient of friction somewhat but carries away more heat and permits higher pressures to be used.

Table 16–4

Some Properties of Brake Linings

	Woven Lining	Molded Lining	Rigid Block
Compressive strength, kpsi	10–15	10–18	10–15
Compressive strength, MPa	70–100	70–125	70–100
Tensile strength, kpsi	2.5–3	4–5	3–4
Tensile strength, MPa	17–21	27–35	21–27
Max. temperature, °F	400–500	500	750
Max. temperature, °C	200–260	260	400
Max. speed, ft/min	7500	5000	7500
Max. speed, m/s	38	25	38
Max. pressure, psi	50–100	100	150
Max. pressure, kPa	340–690	690	1000
Frictional coefficient, mean	0.45	0.47	0.40–45

Table 16–5

Friction Materials for Clutches

Material	Friction Coefficient		Max. Temperature		Max. Pressure	
	Wet	Dry	°F	°C	psi	kPa
Cast iron on cast iron	0.05	0.15–0.20	600	320	150–250	1000–1750
Powdered metal* on cast iron	0.05–0.1	0.1–0.4	1000	540	150	1000
Powdered metal* on hard steel	0.05–0.1	0.1–0.3	1000	540	300	2100
Wood on steel or cast iron	0.16	0.2–0.35	300	150	60–90	400–620
Leather on steel or cast iron	0.12	0.3–0.5	200	100	10–40	70–280
Cork on steel or cast iron	0.15–0.25	0.3–0.5	200	100	8–14	50–100
Felt on steel or cast iron	0.18	0.22	280	140	5–10	35–70
Woven asbestos* on steel or cast iron	0.1–0.2	0.3–0.6	350–500	175–260	50–100	350–700
Molded asbestos* on steel or cast iron	0.08–0.12	0.2–0.5	500	260	50–150	350–1000
Impregnated asbestos* on steel or cast iron	0.12	0.32	500–750	260–400	150	1000
Carbon graphite on steel	0.05–0.1	0.25	700–1000	370–540	300	2100

*The friction coefficient can be maintained with ±5 percent for specific materials in this group.

16–11 Miscellaneous Clutches and Couplings

The square-jaw clutch shown in Fig. 16–25a is one form of positive-contact clutch. These clutches have the following characteristics:

1 They do not slip.

2 No heat is generated.

3 They cannot be engaged at high speeds.

4 Sometimes they cannot be engaged when both shafts are at rest.

5 Engagement at any speed is accompanied by shock.

The greatest differences among the various types of positive clutches are concerned with the design of the jaws. To provide a longer period of time for shift action during engagement, the jaws may be ratchet-shaped, spiral-shaped, or gear-tooth-shaped. Sometimes a great many teeth or jaws are used, and they may be cut either circumferentially, so that they engage by cylindrical mating, or on the faces of the mating elements.

Although positive clutches are not used to the extent of the frictional-contact types, they do have important applications where synchronous operation is required, as, for example, in power presses or rolling-mill screw-downs.

Devices such as linear drives or motor-operated screwdrivers must run to a definite limit and then come to a stop. An overload-release type of clutch is required for these applications. Figure 16–25b is a schematic drawing illustrating the principle of

Figure 16–25

(a) Square-jaw clutch;
(b) overload release clutch using a detent.

←— Shift lever

(a)   (b)

Figure 16–26

Shaft couplings. (a) Plain.
(b) Light-duty toothed coupling. (c) BOST-FLEX® through-bore design having elastomer insert to transmit torque by compression; insert permits 1° misalignment.
(d) Three-jaw coupling available with bronze, rubber, or polyurethane insert to minimize vibration. (*Reproduced by permission, Boston Gear Division, Colfax Corp.*)

(a)   (b)

(c)   (d)

operation of such a clutch. These clutches are usually spring-loaded so as to release at a predetermined torque. The clicking sound which is heard when the overload point is reached is considered to be a desirable signal.

Both fatigue and shock loads must be considered in obtaining the stresses and deflections of the various portions of positive clutches. In addition, wear must generally be considered. The application of the fundamentals discussed in Parts 1 and 2 of this book is usually sufficient for the complete design of these devices.

An overrunning clutch or coupling permits the driven member of a machine to "freewheel" or "overrun" because the driver is stopped or because another source of power increases the speed of the driven mechanism. The construction uses rollers or balls mounted between an outer sleeve and an inner member having cam flats machined around the periphery. Driving action is obtained by wedging the rollers between the sleeve and the cam flats. This clutch is therefore equivalent to a pawl and ratchet with an infinite number of teeth.

There are many varieties of overrunning clutches available, and they are built in capacities up to hundreds of horsepower. Since no slippage is involved, the only power loss is that due to bearing friction and windage.

The shaft couplings shown in Fig. 16–26 are representative of the selection available in catalogs.

16-12 Flywheels

The equation of motion for the flywheel represented in Fig. 16–1b is

$$\sum M = T_i(\theta_i, \dot{\theta}_i) - T_o(\theta_o, \dot{\theta}_o) - I\ddot{\theta} = 0$$

or

$$I\ddot{\theta} = T_i(\theta_i, \omega_i) - T_o(\theta_o, \omega_o) \qquad (a)$$

where T_i is considered positive and T_o negative, and where $\dot{\theta}$ and $\ddot{\theta}$ are the first and second time derivatives of θ, respectively. Note that both T_i and T_o may depend for their values on the angular displacements θ_i and θ_o as well as their angular velocities ω_i and ω_o. In many cases the torque characteristic depends upon only one of these. Thus, the torque delivered by an induction motor depends upon the speed of the motor. In fact, motor manufacturers publish charts detailing the torque-speed characteristics of their various motors.

When the input and output torque functions are given, Eq. (a) can be solved for the motion of the flywheel using well-known techniques for solving linear and non-linear differential equations. We can dispense with this here by assuming a rigid shaft, giving $\theta_i = \theta = \theta_o$ and $\omega_i = \omega = \omega_o$. Thus, Eq. ($a$) becomes

$$I\ddot{\theta} = T_i(\theta, \omega) - T_o(\theta, \omega) \qquad (b)$$

When the two torque functions are known and the starting values of the displacement θ and velocity ω are given, Eq. (b) can be solved for θ, ω, and $\ddot{\theta}$ as functions of time. However, we are not really interested in the instantaneous values of these terms at all. Primarily we want to know the overall performance of the flywheel. What should its moment of inertia be? How do we match the power source to the load? And what are the resulting performance characteristics of the system that we have selected?

To gain insight into the problem, a hypothetical situation is diagrammed in Fig. 16–27. An input power source subjects a flywheel to a constant torque T_i while the shaft rotates from θ_1 to θ_2. This is a positive torque and is plotted upward. Equation (b) indicates that a positive acceleration $\ddot{\theta}$ will be the result, and so the shaft velocity increases from ω_1 to ω_2. As shown, the shaft now rotates from θ_2 to θ_3 with zero torque and hence, from Eq. (b), with zero acceleration. Therefore $\omega_3 = \omega_2$. From θ_3 to θ_4 a load, or output torque, of constant magnitude is applied, causing the shaft to slow down from ω_3 to ω_4. Note that the output torque is plotted in the negative direction in accordance with Eq. (b).

The work input to the flywheel is the area of the rectangle between θ_1 and θ_2, or

$$U_i = T_i(\theta_2 - \theta_1) \qquad (c)$$

| Figure 16–27

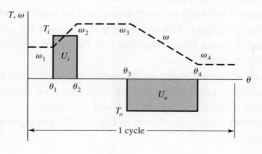

The work output of the flywheel is the area of the rectangle from θ_3 to θ_4, or

$$U_o = T_o(\theta_4 - \theta_3) \tag{d}$$

If U_o is greater than U_i, the load uses more energy than has been delivered to the flywheel and so ω_4 will be less than ω_1. If $U_o = U_i$, ω_4 will be equal to ω_1 because the gains and losses are equal; we are assuming no friction losses. And finally, ω_4 will be greater than ω_1 if $U_i > U_o$.

We can also write these relations in terms of kinetic energy. At $\theta = \theta_1$ the flywheel has a velocity of ω_1 rad/s, and so its kinetic energy is

$$E_1 = \frac{1}{2}I\omega_1^2 \tag{e}$$

At $\theta = \theta_2$ the velocity is ω_2, and so

$$E_2 = \frac{1}{2}I\omega_2^2 \tag{f}$$

Thus the change in kinetic energy is

$$E_2 - E_1 = \frac{1}{2}I(\omega_2^2 - \omega_1^2) \tag{16-61}$$

Many of the torque displacement functions encountered in practical engineering situations are so complicated that they must be integrated by numerical methods. Figure 16–28, for example, is a typical plot of the engine torque for one cycle of motion of a single-cylinder internal combustion engine. Since a part of the torque curve is negative, the flywheel must return part of the energy back to the engine. Integrating this curve from $\theta = 0$ to 4π and dividing the result by 4π yields the mean torque T_m available to drive a load during the cycle.

It is convenient to define a *coefficient of speed fluctuation* as

$$C_s = \frac{\omega_2 - \omega_1}{\omega} \tag{16-62}$$

Figure 16–28

Relation between torque and crank angle for a one-cylinder, four-stroke–cycle internal combustion engine.

Crank torque T

T_m

180°   360°   540°   720°

Crank angle θ

where ω is the nominal angular velocity, given by

$$\omega = \frac{\omega_2 + \omega_1}{2} \qquad (16\text{-}63)$$

Equation (16–61) can be factored to give

$$E_2 - E_1 = \frac{I}{2}(\omega_2 - \omega_1)(\omega_2 + \omega_1)$$

Since $\omega_2 - \omega_1 = C_s\omega$ and $\omega_2 + \omega_1 = 2\omega$, we have

$$E_2 - E_1 = C_s I \omega^2 \qquad (16\text{-}64)$$

Equation (16–64) can be used to obtain an appropriate flywheel inertia corresponding to the energy change $E_2 - E_1$.

EXAMPLE 16–6

Table 16–6 lists values of the torque used to plot Fig. 16–28. The nominal speed of the engine is to be 250 rad/s.

(a) Integrate the torque-displacement function for one cycle and find the energy that can be delivered to a load during the cycle.

(b) Determine the mean torque T_m (see Fig. 16–28).

(c) The greatest energy fluctuation is approximately between $\theta = 15°$ and $\theta = 150°$ on the torque diagram; see Fig. 16–28 and note that $T_o = -T_m$. Using a coefficient of speed fluctuation $C_s = 0.1$, find a suitable value for the flywheel inertia.

(d) Find ω_2 and ω_1.

Solution

(a) Using $n = 48$ intervals of $\Delta\theta = 4\pi/48$, numerical integration of the data of Table 16–6 yields $E = 388$ J. This is the energy that can be delivered to the load.

Table 16–6

Plotting Data for Fig. 16–28

θ, deg	T, N·m	θ, deg	T, N·m	θ, deg	T, N·m	θ, deg	T, N·m
0	0	195	−12	375	−9	555	−12
15	316	210	−23	390	−14	570	−23
30	236	225	−29	405	−10	585	−33
45	275	240	−36	420	0.9	600	−40
60	244	255	−35	435	14	615	−42
75	208	270	−27	450	27	630	−41
90	180	285	−14	465	35	645	−35
105	137	300	−0.9	480	36	660	−30
120	120	315	10	495	29	675	−31
135	91	330	14	510	23	690	−62
150	60	345	9	525	12	705	−86
165	21	360	0	540	0	720	0
180	0						

Answer   (b)
$$T_m = \frac{388}{4\pi} = 30.9 \text{ N} \cdot \text{m}$$

(c) The largest positive loop on the torque-displacement diagram occurs between $\theta = 0°$ and $\theta = 180°$. We select this loop as yielding the largest speed change. Subtracting 30.9 N · m from the values in Table 16–6 for this loop gives, respectively, −30.9, 29.3, 21.1, 25.0, 21.9, 18.2, 15.3, 10.9, 9.2, 6.2, 3.1, −1.0, and −30.9 N · m. Numerically integrating $T - T_m$ with respect to θ yields $E_2 - E_1 = 408$ J. We now solve Eq. (16–64) for I. This gives

Answer
$$I = \frac{E_2 - E_1}{C_s \omega^2} = \frac{408}{0.1(250)^2} = 0.065 \text{ kg} \cdot \text{s}^2 \text{ m}$$

(d) Equations (16–62) and (16–63) can be solved simultaneously for ω_2 and ω_1. Substituting appropriate values in these two equations yields

Answer
$$\omega_2 = \frac{\omega}{2}(2 + C_s) = \frac{250}{2}(2 + 0.1) = 262.5 \text{ rad/s}$$

Answer
$$\omega_1 = 2\omega - \omega_2 = 2(250) - 262.5 = 237.5 \text{ rad/s}$$

These two speeds occur at $\theta = 180°$ and $\theta = 0°$, respectively.

Punch-press torque demand often takes the form of a severe impulse and the running friction of the drive train. The motor overcomes the minor task of overcoming friction while attending to the major task of restoring the flywheel's angular speed. The situation can be idealized as shown in Fig. 16–29. Neglecting the running friction, Euler's equation can be written as

$$T(\theta_1 - 0) = \frac{1}{2}I(\omega_1^2 - \omega_2^2) = E_2 - E_1$$

where the only significant inertia is that of the flywheel. Punch presses can have the motor and flywheel on one shaft, then, through a gear reduction, drive a slider-crank

Figure 16–29

(a) Punch-press torque demand during punching. (b) Squirrel-cage electric motor torque-speed characteristic.

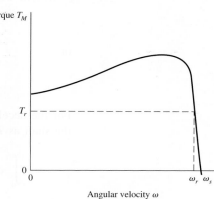

mechanism that carries the punching tool. The motor can be connected to the punch continuously, creating a punching rhythm, or it can be connected on command through a clutch that allows one punch and a disconnect. The motor and flywheel must be sized for the most demanding service, which is steady punching. The work done is given by

$$W = \int_{\theta_1}^{\theta_2} [T(\theta) - T] d\theta = \frac{1}{2} I(\omega_{max}^2 - \omega_{min}^2)$$

This equation can be arranged to include the coefficient of speed fluctuation C_s as follows:

$$W = \frac{1}{2} I(\omega_{max}^2 - \omega_{min}^2) = \frac{I}{2} (\omega_{max} - \omega_{min})(\omega_{max} + \omega_{min})$$

$$= \frac{I}{2} (C_s \overline{\omega})(2\omega_0) = IC_s \overline{\omega} \omega_0$$

When the speed fluctuation is low, $\omega_0 \approx \overline{\omega}$, and

$$I = \frac{W}{C_s \overline{\omega}^2}$$

An induction motor has a linear torque characteristic $T = a\omega + b$ in the range of operation. The constants a and b can be found from the nameplate speed ω_r and the synchronous speed ω_s:

$$a = \frac{T_r - T_s}{\omega_r - \omega_s} = \frac{T_r}{\omega_r - \omega_s} = -\frac{T_r}{\omega_s - \omega_r}$$

$$b = \frac{T_r \omega_s - T_s \omega_r}{\omega_s - \omega_r} = \frac{T_r \omega_s}{\omega_s - \omega_r}$$

(16–65)

For example, a 3-kW three-phase squirrel-cage ac motor rated at 1125 rev/min has a torque of $3000/\|1125(2)\pi/60 = 25.5$ N $\cdot$ m. The rated angular velocity is $\omega_r = 2\pi n_r/60 = 2\pi(1125)/60 = 117.81$ rad/s, and the synchronous angular velocity $\omega_s = 2\pi(1200)/60 = 125.66$ rad/s. Thus $a = -3.25$ N $\cdot$ m $\cdot$ s/rad, and $b = 408$ N $\cdot$ m, and we can express $T(\omega)$ as $a\omega + b$. During the interval from t_1 to t_2 the motor accelerates the flywheel according to $I\ddot{\theta} = T_M$ (i.e., $Td\omega/dt = T_M$). Separating the equation $T_M = Id\omega/dt$ we have

$$\int_{t_1}^{t_2} dt = \int_{\omega_r}^{\omega_2} \frac{I d\omega}{T_M} = I \int_{\omega_r}^{\omega_2} \frac{d\omega}{a\omega + b} = \frac{I}{a} \ln \frac{a\omega_2 + b}{a\omega_r + b} = \frac{I}{a} \ln \frac{T_2}{T_r}$$

or

$$t_2 - t_1 = \frac{I}{a} \ln \frac{T_2}{T_r}$$

(16–66)

For the deceleration interval when the motor and flywheel feel the punch torque on the shaft as T_L, $(T_M - T_L) = Id\omega/dt$, or

$$\int_0^{t_1} dt = I \int_{\omega_2}^{\omega_r} \frac{d\omega}{T_M - T_L} = I \int_{\omega_2}^{\omega_r} \frac{d\omega}{a\omega + b - T_L} = \frac{I}{a} \ln \frac{a\omega_r + b - T_L}{a\omega_2 + b - T_L}$$

or

$$t_1 = \frac{I}{a} \ln \frac{T_r - T_L}{T_2 - T_L}$$

(16–67)

We can divide Eq. (16–66) by Eq. (16–67) to obtain

$$\frac{T_2}{T_r} = \left(\frac{T_L - T_r}{T_L - T_2}\right)^{(t_2 - t_1)/t_1} \tag{16–68}$$

Equation (16–68) can be solved for T_2 numerically. Having T_2 the flywheel inertia is, from Eq. (16–66),

$$I = \frac{a(t_2 - t_1)}{\ln(T_2/T_r)} \tag{16–69}$$

It is important that a be in units of N · m · s/rad so that I has proper units. The constant a should not be in lbf · in per rev/min or N · m per rev/s.

PROBLEMS

16-1 The figure shows an internal rim-type brake having an inside rim diameter of 300 mm and a dimension $R = 125$ mm. The shoes have a face width of 40 mm and are both actuated by a force of 2.2 kN. The drum rotates clockwise. The mean coefficient of friction is 0.28.

(a) Find the maximum pressure and indicate the shoe on which it occurs.

(b) Estimate the braking torque effected by each shoe, and find the total braking torque.

(c) Estimate the resulting hinge-pin reactions.

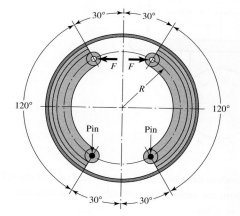

Problem 16–1

16-2 For the brake in Prob. 16–1, consider the pin and actuator locations to be the same. However, instead of 120°, let the friction surface of the brake shoes be 90° and centrally located. Find the maximum pressure and the total braking torque.

16-3 In the figure for Prob. 16–1, the inside rim diameter is 280 mm and the dimension R is 90 mm. The shoes have a face width of 30 mm. Find the braking torque and the maximum pressure for each shoe if the actuating force is 1 kN, the drum rotation is counterclockwise, and $f = 0.30$.

16-4 The figure shows a 400-mm-diameter brake drum with four internally expanding shoes. Each of the hinge pins A and B supports a pair of shoes. The actuating mechanism is to be arranged to produce the same force F on each shoe. The face width of the shoes is 75 mm. The material used permits a coefficient of friction of 0.24 and a maximum pressure of 1000 kPa.

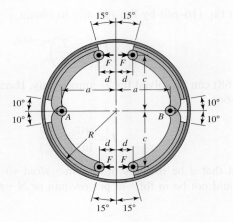

Problem 16–4

The dimensions in millimeters are
$a = 150$, $c = 165$, $R = 200$,
and $d = 50$.

(a) Determine the maximum actuating force.
(b) Estimate the brake capacity.
(c) Noting that rotation may be in either direction, estimate the hinge-pin reactions.

16–5 The block-type hand brake shown in the figure has a face width of 30 mm and a mean coefficient of friction of 0.25. For an estimated actuating force of 400 N, find the maximum pressure on the shoe and find the braking torque.

Problem 16–5

Dimensions in inches.

16–6 Suppose the standard deviation of the coefficient of friction in Prob. 16–5 is $\hat{\sigma}_f = 0.025$, where the deviation from the mean is due entirely to environmental conditions. Find the brake torques corresponding to $\pm 3\hat{\sigma}_f$.

16–7 The brake shown in the figure has a coefficient of friction of 0.30, a face width of 50 mm, and a limiting shoe lining pressure of 1 MPa. Find the limiting actuating force F and the torque capacity.

16–8 Refer to the symmetrical pivoted external brake shoe of Fig. 16–12 and Eq. (16–15). Suppose the pressure distribution was uniform, that is, the pressure p is independent of θ. What would the pivot distance a' be? If $\theta_1 = \theta_2 = 60°$, compare a with a'.

16–9 The shoes on the brake depicted in the figure subtend a 90° arc on the drum of this external pivoted-shoe brake. The actuation force P is applied to the lever. The rotation direction of the drum is counterclockwise, and the coefficient of friction is 0.30.
(a) What should the dimension e be, in order to eliminate frictional moments on each shoe?
(b) Draw the free-body diagrams of the handle lever and both shoe levers, with forces expressed in terms of the actuation force P.
(c) Does the direction of rotation of the drum affect the braking torque?

Problem 16–7
Dimensions in millimeters.

Problem 16–9
Dimensions in 15 millimeters.

16–10   Problem 16–9 is preliminary to analyzing the brake. A rigid molded non-asbestos lining is used dry in the brake of Prob. 16–9 on a cast iron drum. The shoes are 190 mm wide and subtend a 90° arc. Conservatively estimate the maximum allowable actuation force and the braking torque.

16–11   The maximum band interface pressure on the brake shown in the figure is 620 kPa. Use a 350 mm-diameter drum, a band width of 25 mm, a coefficient of friction of 0.30, and an angle-of-wrap of 270°. Find the maximum band tensions and the torque capacity.

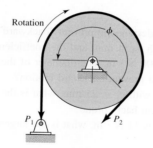

Problem 16–11

16–12 The drum for the band brake in Prob. 16–11 is 300 mm in diameter. The band selected has a mean coefficient of friction of 0.28 and a width of 80 mm. It can safely support a tension of 7.6 kN. If the angle of wrap is 270°, find the maximum lining pressure and the corresponding torque capacity.

16–13 The brake shown in the figure has a coefficient of friction of 0.30 and is to operate using a maximum force F of 400 N. If the band width is 50 mm, find the maximum band tensions and the braking torque.

Problem 16–13
Dimensions in millimeters.

16–14 The figure depicts a band brake whose drum rotates counterclockwise at 200 rev/min. The drum diameter is 400 mm and the band lining 75 mm wide. The coefficient of friction is 0.20. The maximum lining interface pressure is 480 kPa.

(a) Find the maximum brake torque, necessary force P, and steady-state power.

(b) Complete the free-body diagram of the drum. Find the bearing radial load that a pair of straddle-mounted bearings would have to carry.

(c) What is the lining pressure p at both ends of the contact arc?

Problem 16–14

16–15 The figure shows a band brake designed to prevent "backward" rotation of the shaft. The angle of wrap is 270°, the band width is 54 mm, and the coefficient of friction is 0.20. The torque to be resisted by the brake is 200 N · m. The diameter of the pulley is 210 mm.

(a) What dimension c_1 will just prevent backward motion?

(b) If the rocker was designed with $c_1 = 25$ mm, what is the maximum pressure between the band and drum at 200 N · m back torque?

(c) If the back-torque demand is 11 N · m, what is the largest pressure between the band and drum?

Problem 16–15

56 mm

P_2 c_1

P_1

Rocker detail

16–16   A plate clutch has a single pair of mating friction surfaces 250-mm OD by 175-mm ID. The mean value of the coefficient of friction is 0.30, and the actuating force is 4 kN.

   (a) Find the maximum pressure and the torque capacity using the uniform-wear model.

   (b) Find the maximum pressure and the torque capacity using the uniform-pressure model.

16–17   A hydraulically operated multidisk plate clutch has an effective disk outer diameter of 165 mm and an inner diameter of 100 mm. The coefficient of friction is 0.24, and the limiting pressure is 830 kPa. There are six planes of sliding present.

   (a) Using the uniform wear model, estimate the limiting axial force F and the torque T.

   (b) Let the inner diameter of the friction pairs d be a variable. Complete the following table:

d, in	50	75	100	125	150
T, N·m					

   (c) What does the table show?

16–18   Look again at Prob. 16–17.

   (a) Show how the optimal diameter d^* is related to the outside diameter D.

   (b) What is the optimal inner diameter?

   (c) What does the tabulation show about maxima?

   (d) Common proportions for such plate clutches lie in the range $0.45 \leq d/D \leq 0.80$. Is the result in part a useful?

16–19   A cone clutch has $D = 330$ mm, $d = 306$ mm, a cone length of 60 mm, and a coefficient of friction of 0.26. A torque of 200 N . m is to be transmitted. For this requirement, estimate the actuating force and maximum pressure by both models.

16–20   Show that for the caliper brake the $T/(fFD)$ versus d/D plots are the same as Eqs. (b) and (c) of Sec. 16–5.

16–21   A two-jaw clutch has the dimensions shown in the figure and is made of ductile steel. The clutch has been designed to transmit 2 kW at 500 rev/min. Find the bearing and shear stresses in the key and the jaws.

Problem 16–21
Dimensions in millimeters.

16–22 A brake has a normal braking torque of 320 N . m and heat-dissipating cast-iron surfaces whose mass is 18 kg. Suppose a load is brought to rest in 8.3 s from an initial angular speed of 1800 rev/min using the normal braking torque; estimate the temperature rise of the heat-dissipating surfaces.

16–23 A cast-iron flywheel has a rim whose OD is 1.5 m and whose ID is 1.4 m. The flywheel weight is to be such that an energy fluctuation of 6.75 J will cause the angular speed to vary no more than 240 to 260 rev/min. Estimate the coefficient of speed fluctuation. If the weight of the spokes is neglected, what should be the width of the rim?

16–24 A single-geared blanking press has a stroke of 200 mm and a rated capacity of 320 kN. A cam-driven ram is assumed to be capable of delivering the full press load at constant force during the last 15 percent of a constant-velocity stroke. The camshaft has an average speed of 90 rev/min and is geared to the flywheel shaft at a 6:1 ratio. The total work done is to include an allowance of 16 percent for friction.
(*a*) Estimate the maximum energy fluctuation.
(*b*) Find the rim weight for an effective diameter of 1.2 m and a coefficient of speed fluctuation of 0.10.

16–25 Using the data of Table 16–6, find the mean output torque and flywheel inertia required for a three-cylinder in-line engine corresponding to a nominal speed of 2400 rev/min. Use $C_s = 0.30$.

16–26 When a motor armature inertia, a pinion inertia, and a motor torque reside on a motor shaft, and a gear inertia, a load inertia, and a load torque exist on a second shaft, it is useful to reflect all the torques and inertias to one shaft, say, the armature shaft. We need some rules to make such reflection easy. Consider the pinion and gear as disks of pitch radius.

- A torque on a second shaft is reflected to the motor shaft as the load torque divided by the negative of the stepdown ratio.
- An inertia on a second shaft is reflected to the motor shaft as its inertia divided by the stepdown ratio squared.
- The inertia of a disk gear on a second shaft in mesh with a disk pinion on the motor shaft is reflected to the pinion shaft as the *pinion* inertia multiplied by the stepdown ratio squared.

(*a*) Verify the three rules.
(*b*) Using the rules, reduce the two-shaft system in the figure to a motor-shaft shish kebab equivalent. Correctly done, the dynamic response of the shish kebab and the real system are identical.
(*c*) For a stepdown ratio of $n = 10$ compare the shish kebab inertias.

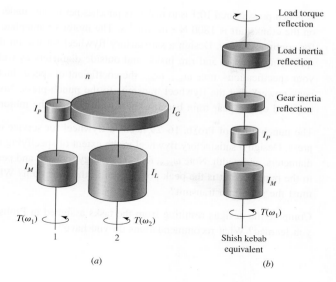

Problem 16–26
Dimensions in millimeters.

16–27 Apply the rules of Prob. 16–26 to the three-shaft system shown in the figure to create a motor shaft shish kebab.

(a) Show that the equivalent inertia I_e is given by

$$I_e = I_M + I_P + n^2 I_P + \frac{I_P}{n^2} + \frac{m^2 I_P}{n^2} + \frac{I_L}{m^2 n^2}$$

(b) If the overall gear reduction R is a constant nm, show that the equivalent inertia becomes

$$I_e = I_M + I_P + n^2 I_P + \frac{I_P}{n^2} + \frac{R^2 I_P}{n^4} + \frac{I_L}{R^2}$$

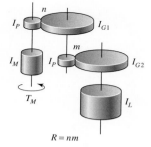

Problem 16–27

(c) If the problem is to minimize the gear-train inertia, find the ratios n and m for the values of $I_P = 1$, $I_M = 10$, $I_L = 100$, and $R = 10$.

16–28 For the conditions of Prob. 16–27, make a plot of the equivalent inertia I_e as ordinate and the stepdown ratio n as abscissa in the range $1 \le n \le 10$. How does the minimum inertia compare to the single-step inertia?

16–29 A punch-press geared 10:1 is to make six punches per minute under circumstances where the torque on the crankshaft is 1800 N · m for $\frac{1}{2}$ s. The motor's nameplate reads 2.2 kW at 1125 rev/min for continuous duty. Design a satisfactory flywheel for use on the motor shaft to the extent of specifying material and rim inside and outside diameters as well as its width. As you prepare your specifications, note ω_{max}, ω_{min}, the coefficient of speed fluctuation C_s, energy transfer, and peak powerΣ that the flywheel transmits to the punch-press. Note power and shock conditions imposed on the gear train because the flywheel is on the motor shaft.

16–30 The punch-press of ProΣb. 16–29 needs a flywheel for service on the crankshaft of the punch-press. Design a satisfactory flywheel to the extent of specifying material, rim inside and outside diameters, and width. Note ω_{max}, ω_{min}, C_s, energy transfer, and peak power the flywheel transmits to the punch. What is the peak power seen in the gear train? What power and shock conditions must the gear-train transmit?

16–31 Compare the designs resulting from the tasks assigned in Probs. 16–29 and 16–30. What have you learned? What recommendations do you have?

17

Flexible Mechanical Elements

Chapter Outline

Belts, ropes, chains, and other similar elastic or flexible machine elements are used in conveying systems and in the transmission of power over comparatively long distances. It often happens that these elements can be used as a replacement for gears, shafts, bearings, and other relatively rigid power-transmission devices. In many cases their use simplifies the design of a machine and substantially reduces the cost.

In addition, since these elements are elastic and usually quite long, they play an important part in absorbing shock loads and in damping out and isolating the effects of vibration. This is an important advantage as far as machine life is concerned.

Most flexible elements do not have an infinite life. When they are used, it is important to establish an inspection schedule to guard against wear, aging, and loss of elasticity. The elements should be replaced at the first sign of deterioration.

17–1   Belts

The four principal types of belts are shown, with some of their characteristics, in Table 17–1. *Crowned pulleys* are used for flat belts, and *grooved pulleys,* or *sheaves,* for round and V belts. Timing belts require *toothed wheels,* or *sprockets.* In all cases, the pulley axes must be separated by a certain minimum distance, depending upon the belt type and size, to operate properly. Other characteristics of belts are:

- They may be used for long center distances.
- Except for timing belts, there is some slip and creep, and so the angular-velocity ratio between the driving and driven shafts is neither constant nor exactly equal to the ratio of the pulley diameters.
- In some cases an idler or tension pulley can be used to avoid adjustments in center distance that are ordinarily necessitated by age or the installation of new belts.

Figure 17–1 illustrates the geometry of open and closed flat-belt drives. For a flat belt with this drive the belt tension is such that the sag or droop is visible in Fig. 17–2a, when the belt is running. Although the top is preferred for the loose side of the belt, for other belt types either the top or the bottom may be used, because their installed tension is usually greater.

Two types of reversing drives are shown in Fig. 17–2. Notice that both sides of the belt contact the driving and driven pulleys in Figs. 17–2b and 17–2c, and so these drives cannot be used with V belts or timing belts.

Table 17–1

Characteristics of Some Common Belt Types (Figures are cross sections except for the timing belt, which is a side view).

Belt Type	Figure	Joint	Size Range	Center Distance
Flat		Yes	$t = \begin{cases} 0.03 \text{ to } 0.20 \text{ in} \\ 0.75 \text{ to } 5 \text{ mm} \end{cases}$	No upper limit
Round		Yes	$d = \frac{1}{8}$ to $\frac{3}{4}$ in	No upper limit
V		None	$b = \begin{cases} 0.31 \text{ to } 0.91 \text{ in} \\ 8 \text{ to } 19 \text{ mm} \end{cases}$	Limited
Timing		None	$p = 2$ mm and up	Limited

Figure 17–1

Flat-belt geometry. (*a*) Open
belt. (*b*) Crossed belt.

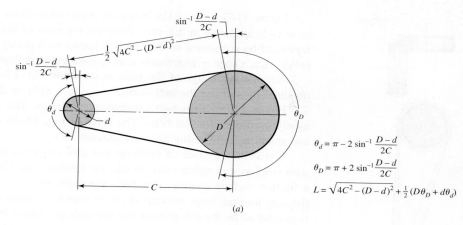

$$\theta_d = \pi - 2\sin^{-1}\frac{D-d}{2C}$$

$$\theta_D = \pi + 2\sin^{-1}\frac{D-d}{2C}$$

$$L = \sqrt{4C^2 - (D-d)^2} + \tfrac{1}{2}(D\theta_D + d\theta_d)$$

(*a*)

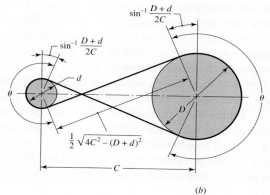

$$\theta = \pi + 2\sin^{-1}\frac{D+d}{2C}$$

$$L = \sqrt{4C^2 - (D+d)^2} + \tfrac{1}{2}(D+d)\theta$$

(*b*)

Figure 17–2

Nonreversing and reversing
belt drives. (*a*) Nonreversing
open belt. (*b*) Reversing
crossed belt. Crossed belts
must be separated to prevent
rubbing if high-friction
materials are used.
(*c*) Reversing open-belt drive.

Midpoint

Figure 17–3

Quarter-twist belt drive; an
idler guide pulley must be
used if motion is to be in both
directions.

Figure 17–3 shows a flat-belt drive with out-of-plane pulleys. The shafts need not be at right angles as in this case. Note the top view of the drive in Fig. 17–3. The pulleys must be positioned so that the belt leaves each pulley in the midplane of the other pulley face. Other arrangements may require guide pulleys to achieve this condition.

Another advantage of flat belts is shown in Fig. 17–4, where clutching action is obtained by shifting the belt from a loose to a tight or driven pulley.

Figure 17–5 shows two variable-speed drives. The drive in Fig. 17–5a is commonly used only for flat belts. The drive of Fig. 17–5b can also be used for V belts and round belts by using grooved sheaves.

Flat belts are made of urethane and also of rubber-impregnated fabric reinforced with steel wire or nylon cords to take the tension load. One or both surfaces may have a friction surface coating. Flat belts are quiet, they are efficient at high speeds, and they can transmit large amounts of power over long center distances. Usually, flat belting is purchased by the roll and cut and the ends are joined by using special kits furnished by the manufacturer. Two or more flat belts running side by side, instead of a single wide belt, are often used to form a conveying system.

A V belt is made of fabric and cord, usually cotton, rayon, or nylon, and impregnated with rubber. In contrast with flat belts, V belts are used with similar sheaves and at shorter center distances. V belts are slightly less efficient than flat belts, but a number of them can be used on a single sheave, thus making a multiple drive. V belts are made only in certain lengths and have no joints.

Timing belts are made of rubberized fabric and steel wire and have teeth that fit into grooves cut on the periphery of the sprockets. The timing belt does not stretch or slip and consequently transmits power at a constant angular-velocity ratio. The fact that the belt is toothed provides several advantages over ordinary belting. One of these is that no initial tension is necessary, so that fixed-center drives may be used. Another

Loose pulley   Driven

Shift fork

Driver

Figure 17–4

This drive eliminates the need
for a clutch. Flat belt can be
shifted left or right by use of
a fork.

Fork

(a)

(b)

Figure 17–5

Variable-speed belt drives.

is the elimination of the restriction on speeds; the teeth make it possible to run at nearly any speed, slow or fast. Disadvantages are the first cost of the belt, the necessity of grooving the sprockets, and the attendant dynamic fluctuations caused at the belt-tooth meshing frequency.

17-2 Flat- and Round-Belt Drives

Modern flat-belt drives consist of a strong elastic core surrounded by an elastomer. These drives have distinct advantages over gear drives or V-belt drives. A flat-belt drive has an efficiency of about 98 percent, which is about the same as for a gear drive. On the other hand, the efficiency of a V-belt drive ranges from about 70 to 96 percent.[1] Flat-belt drives produce very little noise and absorb more torsional vibration from the system than either V-belt or gear drives.

When an open-belt drive (Fig. 17–1a) is used, the contact angles are found to be

$$\theta_d = \pi - 2 \sin^{-1} \frac{D - d}{2C}$$

$$\theta_D = \pi + 2 \sin^{-1} \frac{D - d}{2C}$$

(17–1)

where D = diameter of large pulley

d = diameter of small pulley

C = center distance

θ = angle of contact

The length of the belt is found by summing the two arc lengths with twice the distance between the beginning and end of contact. The result is

$$L = [4C^2 - (D - d)^2]^{1/2} + \frac{1}{2}(D\theta_D + d\theta_d)$$

(17–2)

A similar set of equations can be derived for the crossed belt of Fig. 17–2b. For this belt, the angle of wrap is the same for both pulleys and is

$$\theta = \pi + 2 \sin^{-1} \frac{D + d}{2C}$$

(17–3)

The belt length for crossed belts is found to be

$$L = [4C^2 - (D + d)^2]^{1/2} + \frac{1}{2}(D + d)\theta$$

(17–4)

Firbank[2] explains flat-belt-drive theory in the following way. A change in belt tension due to friction forces between the belt and pulley will cause the belt to elongate or contract and move relative to the surface of the pulley. This motion is caused by *elastic creep* and is associated with sliding friction as opposed to static friction. The action at the driving pulley, through that portion of the angle of contact that is actually transmitting power, is such that the belt moves more slowly than the surface speed of the pulley because of the elastic creep. The angle of contact is made up of

[1] A. W. Wallin, "Efficiency of Synchronous Belts and V-Belts," *Proc. Nat. Conf. Power Transmission,* vol. 5, Illinois Institute of Technology, Chicago, Nov. 7–9, 1978, pp. 265–271.

[2] T. C. Firbank, *Mechanics of the Flat Belt Drive,* ASME paper no. 72-PTG-21.

the *effective arc,* through which power is transmitted, and the *idle arc.* For the driving pulley the belt first contacts the pulley with a *tight-side tension* F_1 and a velocity V_1, which is the same as the surface velocity of the pulley. The belt then passes through the idle arc with no change in F_1 or V_1. Then creep or sliding contact begins, and the belt tension changes in accordance with the friction forces. At the end of the effective arc the belt leaves the pulley with a *loose-side tension* F_2 and a reduced speed V_2.

Firbank has used this theory to express the mechanics of flat-belt drives in mathematical form and has verified the results by experiment. His observations include the finding that substantially more power is transmitted by static friction than sliding friction. He also found that the coefficient of friction for a belt having a nylon core and leather surface was typically 0.7, but that it could be raised to 0.9 by employing special surface finishes.

Our model will assume that the friction force on the belt is proportional to the normal pressure along the arc of contact. We seek first a relationship between the tight side tension and slack side tension, similar to that of band brakes but incorporating the consequences of movement, that is, centrifugal tension in the belt. In Fig. 17–6 we see a free body of a small segment of the belt. The differential force dS is due to centrifugal force, dN is the normal force between the belt and pulley, and $f\,dN$ is the shearing traction due to friction at the point of slip. The belt width is b and the thickness is t. The belt mass per unit length is m. The centrifugal force dS can be expressed as

Figure 17–6

Free body of an infinitesimal element of a flat belt in contact with a pulley.

$$dS = (mr\,d\theta)r\omega^2 = mr^2\omega^2\,d\theta = mV^2\,d\theta = F_c\,d\theta \tag{a}$$

where V is the belt speed. Summing forces radially gives

$$\sum F_r = -(F + dF)\frac{d\theta}{2} - F\frac{d\theta}{2} + dN + dS = 0$$

Ignoring the higher-order term, we have

$$dN = F\,d\theta - dS \tag{b}$$

Summing forces tangentially gives

$$\sum F_t = -f\,dN - F + (F + dF) = 0$$

from which, incorporating Eqs. (*a*) and (*b*), we obtain

$$dF = f\,dN = fF\,d\theta - f\,dS = fF\,d\theta - fmr^2\omega^2\,d\theta$$

or

$$\frac{dF}{d\theta} - fF = -fmr^2\omega^2 \tag{c}$$

The solution to this nonhomogeneous first-order linear differential equation is

$$F = A\exp(f\theta) + mr^2\omega^2 \tag{d}$$

where A is an arbitrary constant. Assuming θ starts at the loose side, the boundary condition that F at $\theta = 0$ equals F_2 gives $A = F_2 - mr^2\omega^2$. The solution is

$$F = (F_2 - mr^2\omega^2)\exp(f\theta) + mr^2\omega^2 \tag{17-5}$$

At the end of the angle of wrap ϕ, the tight side,

$$F|_{\theta=\phi} = F_1 = (F_2 - mr^2\omega^2)\exp(f\phi) + mr^2\omega^2 \tag{17-6}$$

Now we can write

$$\frac{F_1 - mr^2\omega^2}{F_2 - mr^2\omega^2} = \frac{F_1 - F_c}{F_2 - F_c} = \exp(f\phi) \tag{17-7}$$

where, from Eq. (a), $F_c = mr^2\omega^2$. It is also useful that Eq. (17-7) can be written as

$$F_1 - F_2 = (F_1 - F_c)\frac{\exp(f\phi) - 1}{\exp(f\phi)} \tag{17-8}$$

Now F_c is found as follows: with n being the rotational speed, in rev/min, of the pulley of diameter d, in inches, the belt speed is

$$V = \pi \, dn/12 \qquad \text{ft/min}$$

The weight w of a foot of belt is given in terms of the weight density γ in N/m^3 as $w = \gamma bt$ N/m where b and t are in inches. F_c is written as

$$F_c = \frac{w}{g}\left(\frac{V}{60}\right)^2 = \frac{w}{32.17}\left(\frac{V}{60}\right)^2 \tag{e}$$

Figure 17-7 shows a free body diagram of a pulley and part of the belt. The tight side tension F_1 and the loose side tension F_2 have the following additive components:

$$F_1 = F_i + F_c + \Delta F/2 = F_i + F_c + T/d \tag{f}$$

$$F_2 = F_i + F_c - \Delta F/2 = F_i + F_c - T/d \tag{g}$$

where
$$F_i = \text{initial tension}$$
$$F_c = \text{hoop tension due to centrifugal force}$$
$$\Delta F/2 = \text{tension due to the transmitted torque } T$$
$$d = \text{diameter of the pulley}$$

The difference between F_1 and F_2 is related to the pulley torque. Subtracting Eq. (g) from Eq. (f) gives

$$F_1 - F_2 = \frac{2T}{d} \tag{h}$$

Adding Eqs. (f) and (g) gives

$$F_1 + F_2 = 2F_i + 2F_c$$

Figure 17-7

Forces and torques on a pulley.

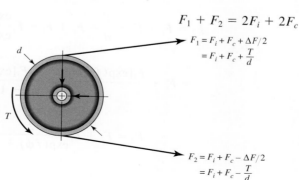

from which

$$F_i = \frac{F_1 + F_2}{2} - F_c \tag{i}$$

Dividing Eq. (i) by Eq. (h), manipulating, and using Eq. (17–7) gives

$$\frac{F_i}{T/d} = \frac{(F_1 + F_2)/2 - F_c}{(F_1 - F_2)/2} = \frac{F_1 + F_2 - 2F_c}{F_1 - F_2} = \frac{(F_1 - F_c) + (F_2 - F_c)}{(F_1 - F_c) - (F_2 - F_c)}$$

$$= \frac{(F_1 - F_c)/(F_2 - F_c) + 1}{(F_1 - F_c)/(F_2 - F_c) - 1} = \frac{\exp(f\phi) + 1}{\exp(f\phi) - 1}$$

from which

$$F_i = \frac{T}{d}\frac{\exp(f\phi) + 1}{\exp(f\phi) - 1} \tag{17–9}$$

Equation (17–9) give us a fundamental insight into flat belting. If F_i equals zero, then T equals zero: no initial tension, no torque transmitted. The torque is in proportion to the initial tension. This means that if there is to be a satisfactory flat-belt drive, the initial tension must be (1) provided, (2) sustained, (3) in the proper amount, and (4) maintained by routine inspection.

From Eq. (f), incorporating Eq. (17–9) gives

$$F_1 = F_i + F_c + \frac{T}{d} = F_c + F_i + F_i\frac{\exp(f\phi) - 1}{\exp(f\phi) + 1}$$

$$= F_c + \frac{F_i[\exp(f\phi) + 1] + F_i[\exp(f\phi) - 1]}{\exp(f\phi) + 1}$$

$$F_1 = F_c + F_i\frac{2\exp(f\phi)}{\exp(f\phi) + 1} \tag{17–10}$$

From Eq. (g), incorporating Eq. (17–9) gives

$$F_2 = F_i + F_c - \frac{T}{d} = F_c + F_i - F_i\frac{\exp(f\phi) - 1}{\exp(f\phi) + 1}$$

$$= F_c + \frac{F_i[\exp(f\phi) + 1] - F_i[\exp(f\phi) - 1]}{\exp(f\phi) + 1}$$

$$F_2 = F_c + F_i\frac{2}{\exp(f\phi) + 1} \tag{17–11}$$

Equation (17–7) is called the *belting equation,* but Eqs. (17–9), (17–10), and (17–11) reveal how belting works. We plot Eqs. (17–10) and (17–11) as shown in Fig. 17–8 against F_i as abscissa. The initial tension needs to be sufficient so that the difference between the F_1 and F_2 curve is $2T/d$. With no torque transmitted, the least possible belt tension is $F_1 = F_2 = F_c$.

The transmitted horsepower is given by

$$H = \frac{(F_1 - F_2)V}{33\,000} \tag{j}$$

where the forces are in lbf and V is in ft/min. Manufacturers provide specifications for their belts that include allowable tension F_a (or stress σ_{all}), the tension being expressed in units of force per unit width. Belt life is usually several years. The severity of flexing at the pulley and its effect on life is reflected in a pulley correction factor C_p. Speed in excess of 3 m/s and its effect on life is reflected in a velocity correction factor C_v. For polyamide and urethane belts use $C_v = 1$. For leather belts see Fig. 17–9. A service factor K_s is used for excursions of load from nominal, applied

Figure 17–8

Plot of initial tension F_i against belt tension F_1 or F_2, showing the intercept F_c, the equations of the curves, and where $2T/d$ is to be found.

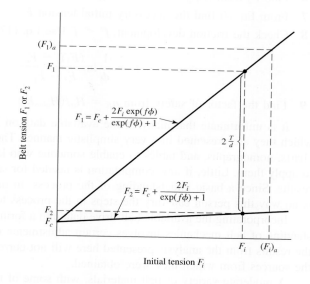

Figure 17–9

Velocity correction factor C_v for leather belts for various thicknesses. *(Data source: Machinery's Handbook, 20th ed., Industrial Press, New York, 1976, p. 1047.)*

to the nominal power as $H_d = H_{\text{nom}}K_s n_d$, where n_d is the design factor for exigencies. These effects are incorporated as follows:

$$(F_1)_a = bF_a C_p C_v \qquad (17\text{-}12)$$

where $(F_1)_a$ = allowable largest tension, N

$\qquad b$ = belt width, mm

$\qquad F_a$ = manufacturer's allowed tension, N/mm

$\qquad C_p$ = pulley correction factor (Table 17–4)

$\qquad C_v$ = velocity correction factor

The steps in analyzing a flat-belt drive can include (see Ex. 17–1)

1   Find $\exp(f\phi)$ from belt-drive geometry and friction

2   From belt geometry and speed find F_c

3   From $T = 63\ 025H_{\text{nom}}K_s n_d/n$ find necessary torque

4   From torque T find the necessary $(F_1)_a - F_2 = 2T/d$

5   From Tables 17–2 and 17–4, and Eq. (17–12) determine $(F_1)_a$.

6   Find F_2 from $(F_1)_a - [(F_1)_a - F_2]$

7   From Eq. (i) find the necessary initial tension F_i

8   Check the friction development, $f' < f$. Use Eq. (17–7) solved for f':

$$f' = \frac{1}{\phi}\ln\frac{(F_1)_a - F_c}{F_2 - F_c}$$

9   Find the factor of safety from $n_{fs} = H_a/(H_{\text{nom}}K_s)$

It is unfortunate that many of the available data on belting are from sources in which they are presented in a very simplistic manner. These sources use a variety of charts, nomographs, and tables to enable someone who knows nothing about belting to apply them. Little, if any, computation is needed for such a person to obtain valid results. Since a basic understanding of the process, in many cases, is lacking, there is no way this person can vary the steps in the process to obtain a better design.

Incorporating the available belt-drive data into a form that provides a good understanding of belt mechanics involves certain adjustments in the data. Because of this, the results from the analysis presented here will not correspond exactly with those of the sources from which they were obtained.

A moderate variety of belt materials, with some of their properties, are listed in Table 17–2. These are sufficient for solving a large variety of design and analysis problems. The design equation to be used is Eq. (j).

The values given in Table 17–2 for the allowable belt tension are based on a belt speed of 3 m/s. For higher speeds, use Fig. 17–9 to obtain C_v values for leather belts. For polyamide and urethane belts, use $C_v = 1.0$.

The service factors K_s for V-belt drives, given in Table 17–15 in Sec. 17–3, are also recommended here for flat- and round-belt drives.

Minimum pulley sizes for the various belts are listed in Tables 17–2 and 17–3. The pulley correction factor accounts for the amount of bending or flexing of the belt and how this affects the life of the belt. For this reason it is dependent on the size and material of the belt used. See Table 17–4. Use $C_p = 1.0$ for urethane belts.

Flat-belt pulleys should be crowned to keep belts from running off the pulleys. If only one pulley is crowned, it should be the larger one. Both pulleys must be crowned whenever the pulley axes are not in a horizontal position. Use Table 17–5 for the crown height.

Table 17-2

Properties of Some Flat- and Round-Belt Materials. (Diameter = d, thickness = t, width = w)

Material	Specification	Size, mm	Minimum Pulley Diameter, mm	Allowable Tension per Unit Width at 3 m/s, (10^3) N/m	Specific Weight, kN/m^3	Coefficient of Friction
Leather	1 ply	$t = 4.5$	75	5	9.5–12.2	0.4
		$t = 5$	90	6	9.5–12.2	0.4
	2 ply	$t = 7$	115	7	9.5–12.2	0.4
		$t = 8$	150	9	9.5–12.2	0.4
		$t = 9$	230	10	9.5–12.2	0.4
Polyamide[b]	F–0[c]	$t = 0.8$	15	1.8	9.5	0.5
	F–1[c]	$t = 1.3$	25	6	9.5	0.5
	F–2[c]	$t = 1.8$	60	10	13.8	0.5
	A–2[c]	$t = 2.8$	60	10	10.0	0.8
	A–3[c]	$t = 3.3$	110	18	11.4	0.8
	A–4[c]	$t = 5.0$	240	30	10.6	0.8
	A–5[c]	$t = 6.4$	340	48	10.6	0.8
Urethane[d]	$w = 12.7$	$t = 1.6$	See	1.0^e	10.3–12.2	0.7
	$w = 19$	$t = 2.0$	Table	1.7^e	10.3–12.2	0.7
	$w = 32$	$t = 2.3$	17–3	3.3^e	10.3–12.2	0.7
	Round	$d = 6$	See	1.4^e	10.3–12.2	0.7
		$d = 10$	Table	3.3^e	10.3–12.2	0.7
		$d = 12$	17–3	5.8^e	10.3–12.2	0.7
		$d = 20$		13^e	10.3–12.2	0.7

[a]Add 2 in to pulley size for belts 8 in wide or more.

[b]Source: *Habasit Engineering Manual*, Habasit Belting, Inc., Chamblee (Atlanta), Ga.

[c]Friction cover of acrylonitrile-butadiene rubber on both sides.

[d]Source: Eagle Belting Co., Des Plaines, Ill.

[e]At 6% elongation; 12% is maximum allowable value.

Table 17-3

Minimum Pulley Sizes for Flat and Round Urethane Belts. (Listed are the Pulley Diameters in mm)

Source: Eagle Belting Co., Des Plaines, Ill.

Belt Style	Belt Size, mm	Ratio of Pulley Speed to Belt Length, rev/(m · s)		
		Up to 14	14 to 27	28 — 55
Flat	12.7 × 1.6	9.7	11.2	12.7
	19 × 2.0	12.7	16	19
	32 × 2.3	12.7	16	19
Round	6	38.1	44.5	50.8
	10	57.1	66.5	76.2
	12	76.2	88.9	101.6
	20	127	152	177.8

Table 17–4

Pulley Correction Factor C_P for Flat Belts*

Material	Small-Pulley Diameter, mm					
	40 – 100	115 – 200	220 – 310	355 – 405	460 – 800	Over 800
Leather	0.5	0.6	0.7	0.8	0.9	1.0
Polyamide, F–0	0.95	1.0	1.0	1.0	1.0	1.0
F–1	0.70	0.92	0.95	1.0	1.0	1.0
F–2	0.73	0.86	0.96	1.0	1.0	1.0
A–2	0.73	0.86	0.96	1.0	1.0	1.0
A–3	—	0.70	0.87	0.94	0.96	1.0
A–4	—	—	0.71	0.80	0.85	0.92
A–5	—	—	—	0.72	0.77	0.91

*Average values of C_P for the given ranges were approximated from curves in the *Habasit Engineering Manual,* Habasit Belting, Inc., Chamblee (Atlanta), Ga.

Table 17–5

Crown Height and ISO Pulley Diameters for Flat Belts*

ISOCrown Pulley Diameter, mm	ISO Height, mm	Pulley Diameter, mm	Crown Height, in	
			w ≤ 250 mm	w > 250 mm
40, 50, 62	0.3	315, 355	0.75	0.75
70, 80	0.3	315, 355	1.0	1.0
90, 100, 115	0.3	570, 635, 710	1.3	1.3
125, 142	0.4	800, 900	1.3	1.5
160, 180	0.5	1015	1.3	1.5
200, 230	0.6	1140, 1270, 1420	1.5	2.0
250, 285	0.75	1600, 1800, 2030	1.8	2.5

*Crown should be rounded, not angled; maximum roughness is R_a = AA 1500 μmm.

EXAMPLE 17–1

A polyamide A-3 flat belt 150 mm wide is used to transmit 11 kW under light shock conditions where $K_s = 1.25$, and a factor of safety equal to or greater than 1.1 is appropriate. The pulley rotational axes are parallel and in the horizontal plane. The shafts are 2.4 m apart. The 150-mm driving pulley rotates at 1750 rev/min in such a way that the loose side is on top. The driven pulley is 450 mm in diameter. See Fig. 17–10. The factor of safety is for unquantifiable exigencies.

(a) Estimate the centrifugal tension F_c and the torque T.
(b) Estimate the allowable F_1, F_2, F_i and allowable power H_a.
(c) Estimate the factor of safety. Is it satisfactory?

Figure 17–10

The flat-belt drive of Ex. 17–1. (Drawing is not to scale)

Solution   (a) Eq. (17–1): $\phi = \theta_d = \pi - 2\sin^{-1}\left[\dfrac{450-150}{2(2400)}\right] = 3.0165$ rad

$$\exp(f\phi) = \exp[0.8(3.0165)] = 11.17$$

$$V = \pi(0.15)1750/60 = 13.7 \text{ m/s}$$

Table 17–2: $w = \gamma bt = 11\,000(0.15)0.0033 = 5.4$ N/m

Answer   Eq. (e): $F_c = \dfrac{w}{g}V^2 = \dfrac{5.4}{9.81}(13.7)^2 = 103.3$ N

$$T = \frac{H_{\text{nom}}K_s n_d}{2\pi n} = \frac{1.25(1.1)11000}{2\pi 1750/60}$$

Answer $= 82.5$ N · m

(b) The necessary $(F_1)_a - F_2$ to transmit the torque T, from Eq. (h), is

$$(F_1)_a - F_2 = \frac{2T}{d} = \frac{2(82)}{0.15} = 1093.3 \text{ N}$$

From Table 17–2 $F_a = 18$ kN/m. For polyamide belts $C_v = 1$, and from Table 17–4 $C_p = 0.70$. From Eq. (17–12) the allowable largest belt tension $(F_1)_a$ is

Answer $$(F_1)_a = bF_a C_p C_v = 0.15(18000)0.70(1) = 1890 \text{ N}$$

then

Answer $$F_2 = (F_1)_a - [(F_1)_a - F_2] = 1890 - 1093 = 796.7 \text{ N}$$

and from Eq. (i)

$$F_i = \frac{(F_1)_a + F_2}{2} - F_c = \frac{1890 + 796.7}{2} - 103 = 1240.4 \text{ N}$$

Answer   The combination $(F_1)_a$, F_2, and F_i will transmit the design power of $11(1.25)(1.1) = 15.125$ kW and protect the belt. We check the friction development by solving Eq. (17–7) for f':

$$f' = \frac{1}{\phi}\ln\frac{(F_1)_a - F_c}{F_2 - F_c} = \frac{1}{3.0165}\ln\frac{1890 - 103}{797 - 103} = 0.314$$

From Table 17–2, $f = 0.8$. Since $f' < f$, that is, $0.314 < 0.80$, there is no danger of slipping.

(c)

Answer $$n_{fs} = \frac{H}{H_{\text{nom}}K_s} = \frac{15.125}{11(1.25)} = 1.1 \qquad \text{(as expected)}$$

Answer   The belt is satisfactory and the maximum allowable belt tension exists. If the initial tension is maintained, the capacity is the design power of 15.125 kW.

Initial tension is the key to the functioning of the flat belt as intended. There are ways of controlling initial tension. One way is to place the motor and drive pulley on a pivoted mounting plate so that the weight of the motor, pulley, and mounting plate and a share of the belt weight induces the correct initial tension and maintains it. A second way is the use of a spring-loaded idler pulley, adjusted to the same task. Both of these methods accommodate to temporary or permanent belt stretch. See Fig. 17–11.

Because flat belts were used for long center-to-center distances, the weight of the belt itself can provide the initial tension. The static belt deflects to an approximate catenary curve, and the dip from a straight belt can be measured against a stretched music wire. This provides a way of measuring and adjusting the dip. From catenary theory the dip is related to the initial tension by

$$d = \frac{L^2 w}{8F_i} \qquad (17\text{–}13)$$

where $d = dip$, m

$C = $ center-to-center distance, m

$w = $ weight per foot of the belt, N/m

$F_i = $ initial tension, N

In Ex. 17–1 the *dip* corresponding to a 1240-N initial tension is

$$d = \frac{(2.4^2)5.4}{8(1240)} = 0.0031 \text{ m} = 3.1 \text{ mm}$$

Figure 17–11

Belt-tensioning schemes.
(*a*) Weighted idler pulley.
(*b*) Pivoted motor mount.
(*c*) Catenary-induced tension.

A decision set for a flat belt can be

- Function: power, speed, durability, reduction, service factor, C
- Design factor: n_d
- Initial tension maintenance
- Belt material
- Drive geometry, d, D
- Belt thickness: t
- Belt width: b

Depending on the problem, some or all of the last four could be design variables. Belt cross-sectional area is really the design decision, but available belt thicknesses and widths are discrete choices. Available dimensions are found in suppliers' catalogs.

EXAMPLE 17–2

Design a flat-belt drive to connect horizontal shafts on 4.8 m centers. The velocity ratio is to be 2.25:1. The angular speed of the small driving pulley is 860 rev/min, and the nominal power transmission is to be 44 760 W under very light shock.

Solution

- Function: $H_{nom} = 44\ 760$ W, 860 rev/min, 2.25:1 ratio, $K_s = 1.15$, $C = 4.8$ m
- Design factor: $n_d = 1.05$
- Initial tension maintenance: catenary
- Belt material: polyamide
- Drive geometry, d, D
- Belt thickness: t
- Belt width: b

The last four could be design variables. Let's make a few more a priori decisions.

Decision $d = 400$ mm, $D = 2.25d = 900$ mm.

Decision   Use polyamide A-3 belt; therefore $t = 3.3$ mm and $C_v = 1$.
Now there is one design decision remaining to be made, the belt width b.

Table 17–2: $\gamma = 11.4$ kN/m^3 $f = 0.8$ $F_a = 18$ kN/m at 600 rev/min

Table 17–4: $C_p = 0.94$

Eq. (17–12): $F_{1a} = b(18\ 000)0.94(1) = 16\ 920b$ N   (1)

$$H_d = H_{nom}K_s n_d = 44\ 760(1.15)1.05 = 54\ 048 \text{ W}$$

$$T = \frac{H_d}{2\pi n} = \frac{54\ 048}{2\pi\ 860/60} = 600 \text{ N} \cdot \text{m}$$

Estimate $\exp(f\phi)$ for full friction development:

Eq. (17–1): $$\phi = \theta_d = \pi - 2\sin^{-1}\frac{900 - 400}{2(4800)} = 3.037 \text{ rad}$$

$$\exp(f\phi) = \exp[0.80(3.037)] = 11.35$$

Estimate centrifugal tension F_c in terms of belt width b:

$$w = \gamma bt = (11\ 400)b(0.0033) = 37.6b \text{ N/m}$$

$$V = \pi dn = \pi(0.4)860/60 = 18 \text{ m/s}$$

Eq. (e): $$F_c = \frac{w}{g}V^2 = \frac{(37.6)b(18)^2}{9.81} = 1241.8b \text{ N} \qquad (2)$$

For design conditions, that is, at H_d power level, using Eq. (h) gives

$$(F_1)_a - F_2 = 2T/d = 2(600)/0.4 = 3000 \text{ N} \qquad (3)$$

$$F_2 = (F_1)_a - [(F_1)_a - F_2] = 16\ 920b - 3000 \text{ N} \qquad (4)$$

Using Eq. (i) gives

$$F_i = \frac{(F_1)_a + F_2}{2} - F_c = \frac{16\ 920b + 16\ 920b - 3000}{2} - 1241.8b = 15\ 678.2b - 1500 \text{ N} \qquad (5)$$

Place friction development at its highest level, using Eq. (17–7):

$$f\phi = \ln\frac{(F_1)_a - F_c}{F_2 - F_c} = \ln\frac{16\ 920b - 1241.8b}{16\ 920b - 3000 - 1241.8b} = \ln\frac{15\ 678.2b}{15\ 678.2b - 3000}$$

Solving the preceding equation for belt width b at which friction is fully developed gives

$$b = \frac{3000}{15\ 678.2}\frac{\exp(f\phi)}{\exp(f\phi) - 1} = \frac{3000}{15\ 678.2}\frac{11.38}{11.38 - 1} = 0.210 \text{ m} = 210 \text{ mm}$$

A belt width greater than 210 mm will develop friction less than $f = 0.80$. The manufacturer's data indicate that the next available larger width is 250 mm.

Decision   Use 250 mm-wide belt.
It follows that for a 250-mm-wide belt

Eq. (2): $$F_c = 1241.8(0.25) = 310 \text{ N}$$

Eq. (1): $$(F_1)_a = 16\ 920(0.25) = 4230 \text{ N}$$

Eq. (4): $$F_2 = 4230 - 3000 = 1230 \text{ N}$$

Eq. (5): $$F_i = 15\ 678.2(0.25) - 1500 = 2420 \text{ N}$$

The transmitted power, from Eq. (3), is

$$H_t = [(F_1)_a - F_2]V = 3000(18) = 54\ 000 \text{ W}$$

and the level of friction development f', from Eq. (17–7) is

$$f' = \frac{1}{\phi}\ln\frac{(F_1)_a - F_c}{F_2 - F_c} = \frac{1}{3.037}\ln\frac{4230 - 310}{1230 - 310} = 0.477$$

which is less than $f = 0.8$, and thus is satisfactory. Had a 225-mm belt width been available, the analysis would show $(F_1)_a = 3807$ N, $F_2 = 807$ N, $F_i = 2028$ N, and $f' = 0.63$. With a figure of merit available reflecting cost, thicker belts (A-4 or A-5) could be examined to ascertain which of the satisfactory alternatives is best. From Eq. (17–13) the catenary dip is

$$dip = \frac{L^2w}{8F_i} = \frac{4.8^2(37.6)0.25}{8(2420)} = 0.011 \text{ m} = 11 \text{ mm}$$

Figure 17–12

Flat-belt tensions.

(a)

(b)

Figure 17–12 illustrates the variation of flexible flat-belt tensions at some cardinal points during a belt pass.

Flat Metal Belts

Thin flat metal belts with their attendant strength and geometric stability could not be fabricated until laser welding and thin rolling technology made possible belts as thin as 0.05 mm and as narrow as 0.65 mm. The introduction of perforations allows no-slip applications. Thin metal belts exhibit

- High strength-to-weight ratio
- Dimensional stability
- Accurate timing
- Usefulness to temperatures up to 370°C
- Good electrical and thermal conduction properties

In addition, stainless steel alloys offer "inert," nonabsorbent belts suitable to hostile (corrosive) environments, and can be made sterile for food and pharmaceutical applications.

Thin metal belts can be classified as friction drives, timing or positioning drives, or tape drives. Among friction drives are plain, metal-coated, and perforated belts. Crowned pulleys are used to compensate for tracking errors.

Figure 17–13 shows a thin flat metal belt with the tight tension F_1 and the slack side tension F_2 revealed. The relationship between F_1 and F_2 and the driving torque T is the same as in Eq. (h). Equations (17–9), (17–10), and (17–11) also apply, with the hoop tension due to centrifugal force typically neglected for the very thin metal belts. The largest allowable tension, as in Eq. (17–12), is posed in terms of stress in metal belts. A bending stress is created by making the belt conform to the pulley, and its tensile magnitude σ_b is given by

$$\sigma_b = \frac{Et}{(1 - \nu^2)D} = \frac{E}{(1 - \nu^2)(D/t)} \qquad (17\text{–}14)$$

Figure 17–13

Metal-belt tensions and torques.

(a)

(b)

where E = Young's modulus

t = belt thickness

ν = Poisson's ratio

D = pulley diameter

The tensile stresses $(\sigma)_1$ and $(\sigma)_2$ imposed by the belt tensions F_1 and F_2 are

$$(\sigma)_1 = F_1/(bt) \quad \text{and} \quad (\sigma)_2 = F_2/(bt)$$

The largest tensile stress is $(\sigma_b)_1 + F_1/(bt)$ and the smallest is $(\sigma_b)_2 + F_2/(bt)$. During a belt pass both levels of stress appear.

Although the belts are of simple geometry, the method of Marin is not used because the condition of the butt weldment (to form the loop) is not accurately known, and the testing of weld coupons is difficult. The belts are run to failure on two equal-sized pulleys. Information concerning fatigue life, as shown in Table 17–6, is obtainable. Tables 17–7 and 17–8 give additional information.

Table 17–6 shows metal belt life expectancies for a stainless steel belt. From Eq. (17–14) with $E = 190$ GPa and $\nu = 0.29$, the bending stresses corresponding to the four entries of the table are 337, 527, 633 and 1054 MPa. Using a natural log transformation on stress and passes shows that the regression line ($r = -0.96$) is

$$\sigma = 14\ 169\ 982 N_p^{-0.407} = 14.17(10^6) N_p^{-0.407} \tag{17–15}$$

where N_p is the number of belt passes.

Table 17–6

Belt Life for Stainless Steel Friction Drives*

$\dfrac{D}{t}$	Belt Passes
625	$\geq 10^6$
400	$0.500 \cdot 10^6$
333	$0.165 \cdot 10^6$
200	$0.085 \cdot 10^6$

*Data courtesy of Belt Technologies, Agawam, Mass.

Table 17–7

Minimum Pulley
Diameter*

Belt Thickness, mm	Minimum Pulley Diameter, mm
0.05	30
0.08	45
0.13	75
0.20	125
0.25	150
0.38	255
0.50	315
1.00	635

*Data courtesy of Belt Technologies, Agawam, Mass.

Table 17–8

Typical Material
Properties, Metal Belts*

Alloy	Yield Strength, MPa	Young's Modulus, GPa	Poisson's Ratio
301 or 302 stainless steel	1206	193	0.285
BeCu	1170	117	0.220
1075 or 1095 carbon steel	1585	207	0.287
Titanium	1034	103	—
Inconel	1103	207	0.284

*Data courtesy of Belt Technologies, Agawam, Mass.

The selection of a metal flat belt can consist of the following steps:

1 Find $\exp(f\phi)$ from geometry and friction

2 Find endurance strength

$$S_f = 14.17(10^6)N_p^{-0.407} \qquad 301, 302 \text{ stainless}$$

$$S_f = S_y/3 \qquad\qquad\qquad \text{others}$$

3 Allowable tension

$$(F_1)_a = \left[S_f - \frac{Et}{(1-\nu^2)D} \right] tb = ab$$

4 $\Delta F = 2T/D$

5 $F_2 = (F_1)_a - \Delta F = ab - \Delta F$

6 $F_i = \dfrac{(F_1)_a + F_2}{2} = \dfrac{ab + ab - \Delta F}{2} = ab - \dfrac{\Delta F}{2}$

7 $b_{\min} = \dfrac{\Delta F}{a} \dfrac{\exp(f\phi)}{\exp(f\phi) - 1}$

8 Choose $b > b_{\min}$, $(F_1)_a = ab$, $F_2 = ab - \Delta F$, $F_i = ab - \Delta F/2$, $T = \Delta FD/2$

9 Check frictional development f':

$$f' = \frac{1}{\phi} \ln \frac{(F_1)_a}{F_2} \qquad f' < f$$

EXAMPLE 17–3

A friction-drive stainless steel metal belt runs over two 100-mm metal pulleys ($f = 0.35$). The belt thickness is to be 0.08 mm. For a life exceeding 10^6 belt passes with smooth torque ($K_s = 1$), (*a*) select the belt if the torque is to be 3.5 N · m, and (*b*) find the initial tension F_i.

Solution

(*a*) From step 1, $\phi = \theta_d = \pi$, therefore $\exp(0.35\pi) = 3.00$. From step 2,

$$(S_f)_{10^6} = 97\,702(10^6)^{-0.407} = 353 \text{ MPa}$$

From steps 3, 4, 5, and 6,

$$F_{1a} = \left[353(10^6) - \frac{193(10^9)0.08(10^{-3})}{(1 - 0.285^2)0.1} \right] 0.08(10^{-3})b = 14\,796b \text{ N} \qquad (1)$$

$$\Delta F = 2T/D = 2(3.5)/0.1 = 70 \text{ N·m}$$

$$F_2 = F_{1a} - \Delta F = 14\,796b - 70 \text{ N} \qquad (2)$$

$$F_i = \frac{F_{1a} + F_2}{2} = \frac{14\,796b + 70}{2} \text{ N} \qquad (3)$$

From step 7,

$$b_{\min} = \frac{\Delta F}{a} \frac{\exp(f\phi)}{\exp(f\phi) - 1} = \frac{70}{14\,796} \frac{3.00}{3.00 - 1} = 0.0071 \text{ m} = 7.1 \text{ mm}$$

Decision

Select an available 19-mm-wide belt 0.08 mm thick.

Eq. (1): $\qquad\qquad F_{1a} = 14\,796(0.019) = 281 \text{ N}$

Eq. (2): $\qquad\qquad F_2 = 281 - 70 = 211 \text{ N}$

Eq. (3): $\qquad\qquad F_i = (281 + 211)/2 = 246 \text{ N}$

$$f' = \frac{1}{\phi} \ln \frac{F_1}{F_2} = \frac{1}{\pi} \ln \frac{281}{211} = 0.0912$$

Note $f' < f$, that is, $0.0882 < 0.35$.

17–3 V Belts

The cross-sectional dimensions of V belts have been standardized by manufacturers, with each section designated by a letter of the alphabet for sizes in inch dimensions. Metric sizes are designated in numbers. Though these have not been included here, the procedure for analyzing and designing them is the same as presented here. Dimensions, minimum sheave diameters, and the horsepower range for each of the lettered sections are listed in Table 17–9.

Table 17–9

Standard V-Belt Sections

Belt Section	Width a, mm	Thickness b, mm	Minimum Sheave Diameter, mm	kW Range, One or More Belts
A	12	8.5	75	0.2–7.5
B	16	11	135	0.7–18.5
C	22	13	230	11–75
D	30	19	325	37–186
E	38	25	540	75 and up

Table 17–10

Inside Circumferences of Standard V Belts

Section	Circumference, mm
A	650, 775, 825, 875, 950, 1050, 1150, 1200, 1275, 1325, 1375, 1425, 1500, 1550, 1600, 1650, 1700, 1775, 1875, 1950, 2000, 2125, 2250, 2400, 2625, 2800, 3000, 3200
B	875, 950, 1050, 1150, 1200, 1275, 1325, 1375, 1425, 1500, 1550, 1600, 1650, 1700, 1775, 1875, 1950, 2000, 2125, 2250, 2400, 2625, 2800, 3000, 3200, 3275, 3400, 3450, 3950, 4325, 4500, 4875, 5250, 6000, 6750, 7500
C	1275, 1500, 1700, 1875, 2025, 2125, 2250, 2400, 2625, 2800, 3000, 3200, 3400, 3600, 3950, 4050, 4350, 4500, 4875, 5250, 6000, 6750, 7500, 8250, 9000, 9750, 10 500
D	3000, 3200, 3600, 3950, 4050, 4350, 4500, 4875, 5250, 6000, 6750, 7500, 8250, 9000, 9750, 10 500, 12 000, 13 500, 15 000, 16 500
E	4500, 4875, 5250, 6000, 6750, 7500, 8250, 9000, 9750, 10 500, 12 000, 13 500, 15 000, 16 500

Table 17–11

Length Conversion Dimensions (Add the listed quantity to the inside circumference to obtain the pitch length in mm).

Belt section	A	B	C	D	E
Quantity to be added	32	45	72	82	112

To specify a V belt, give the belt-section letter, followed by the inside circumference in millimeters (standard circumferences are listed in Table 17–10). For example, B75 is a B-section belt having an inside circumference of 1875 mm.

Calculations involving the belt length are usually based on the pitch length. For any given belt section, the pitch length is obtained by adding a quantity to the inside circumference (Tables 17–10 and 17–11). For example, a B75 belt has a pitch length of 1920 mm. Similarly, calculations of the velocity ratios are made using the pitch diameters of the sheaves, and for this reason the stated diameters are usually understood to be the pitch diameters even though they are not always so specified.

The groove angle of a sheave is made somewhat smaller than the belt-section angle. This causes the belt to wedge itself into the groove, thus increasing friction. The exact value of this angle depends on the belt section, the sheave diameter, and the angle of contact. If it is made too much smaller than the belt, the force required to pull the belt out of the groove as the belt leaves the pulley will be excessive. Optimum values are given in the commercial literature.

The minimum sheave diameters have been listed in Table 17–9. For best results, a V belt should be run quite fast: 20 m/s is a good speed. Trouble may be encountered if the belt runs much faster than 25 m/s or much slower than 5 m/s.

The *pitch length* L_p and the center-to-center distance C are

$$L_p = 2C + \pi(D + d)/2 + (D - d)^2/(4C) \qquad (17\text{–}16a)$$

$$C = 0.25\left\{\left[L_p - \frac{\pi}{2}(D + d)\right] + \sqrt{\left[L_p - \frac{\pi}{2}(D + d)\right]^2 - 2(D - d)^2}\right\} \qquad (17\text{–}16b)$$

where D = pitch diameter of the large sheave and d = pitch diameter of the small sheave.

In the case of flat belts, there is virtually no limit to the center-to-center distance. Long center-to-center distances are not recommended for V belts because the excessive vibration of the slack side will shorten the belt life materially. In general, the center-to-center distance should be no greater than 3 times the sum of the sheave diameters and no less than the diameter of the larger sheave. Link-type V belts have less vibration, because of better balance, and hence may be used with longer center-to-center distances.

The basis for power ratings of V belts depends somewhat on the manufacturer; it is not often mentioned quantitatively in vendors' literature but is available from vendors. The basis may be a number of hours, 24 000 h, for example, or a life of 10^8 or 10^9 belt passes. Since the number of belts must be an integer, an undersized belt set that is augmented by one belt can be substantially oversized. Table 17–12 gives power ratings of standard V belts.

The rating, whether in terms of hours or belt passes, is for a belt running on equal-diameter sheaves (180° of wrap), of moderate length, and transmitting a steady load. Deviations from these laboratory test conditions are acknowledged by multiplicative adjustments. If the tabulated power of a belt for a C-section belt is 7.06 kW for a 300-mm-diameter sheave at a peripheral speed of 15 m/s (Table 17–12), then, when the belt is used under other conditions, the tabulated value H_{tab} is adjusted as follows:

$$H_a = K_1 K_2 H_{\text{tab}} \qquad (17\text{–}17)$$

where H_a = allowable power, per belt

K_1 = angle-of-wrap (ϕ) correction factor, Table 17–13

K_2 = belt length correction factor, Table 17–14

The allowable power can be near to H_{tab}, depending upon circumstances.

In a V belt the effective coefficient of friction f' is $f/\sin(\phi/2)$, which amounts to an augmentation by a factor of about 3 due to the grooves. The effective coefficient of friction f' is sometimes tabulated against *sheave* groove angles of 30°, 34°, and 38°, the tabulated values being 0.50, 0.45, and 0.40, respectively, revealing a belt material-on-metal coefficient of friction of 0.13 for each case. The Gates Rubber Company declares its effective coefficient of friction to be 0.5123 for grooves. Thus

$$\frac{F_1 - F_c}{F_2 - F_c} = \exp(0.5123\phi) \qquad (17\text{–}18)$$

The design power is given by

$$H_d = H_{\text{nom}}K_s n_d \qquad (17\text{–}19)$$

where H_{nom} is the nominal power, K_s is the service factor given in Table 17–15, and n_d is the design factor. The number of belts, N_b, is usually the next higher integer to H_d/H_a.

Table 17–12

Horsepower (kW) Ratings of Standard V Belts

Belt Section	Sheave Pitch Diameter, mm	Belt Speed, m/s				
		5	**10**	**15**	**20**	**25**
A	65	0.35	0.46	0.40	0.11	
	75	0.49	0.75	0.84	0.69	0.28
	85	0.60	0.98	1.17	1.64	0.84
	95	0.69	1.16	1.43	1.49	1.28
	105	0.77	1.30	1.64	1.78	1.63
	115	0.83	1.41	1.82	2.01	1.93
	125 and up	0.87	1.51	1.97	2.21	2.16
B	105	0.80	1.18	1.25	0.94	0.16
	115	0.95	1.48	1.71	1.55	0.92
	125	1.07	1.74	2.09	2.06	1.57
	135	1.19	1.95	2.42	2.49	2.10
	145	1.28	2.14	2.69	2.87	2.57
	155	1.36	2.31	2.94	3.19	2.98
	165	1.43	2.45	3.16	3.48	3.34
	175 and up	1.50	2.58	3.35	3.74	3.66
C	150	1.37	1.98	2.03	1.40	
	175	1.85	2.94	3.46	3.31	2.33
	200	2.21	3.66	4.54	4.74	4.12
	225	2.49	4.21	5.38	5.86	5.51
	250	2.72	4.66	6.05	7.16	6.63
	275	2.89	5.03	6.59	7.46	7.53
	300 and up	3.05	5.33	7.06	8.13	8.28
D	250	3.09	4.57	4.89	3.80	1.01
	275	3.73	5.84	6.80	6.34	4.19
	300	4.26	6.91	8.36	8.50	6.85
	325	4.71	7.83	9.70	10.30	9.10
	350	5.09	8.58	10.89	11.79	11.04
	375	5.42	9.25	11.86	13.13	12.68
	400	5.71	9.85	12.76	14.32	14.17
	425 and up	5.98	10.37	13.50	15.37	15.44
E	400	6.48	10.44	13.06	13.50	11.41
	450	7.40	12.46	15.82	17.16	16.04
	500	8.13	13.95	18.05	20.07	19.69
	550	8.73	15.14	19.84	22.53	22.75
	600	9.25	16.11	21.34	24.54	25.22
	650	9.70	17.01	22.60	26.19	27.38
	700 and up	10.00	17.68	23.72	27.68	29.17

That is,

$$N_b \geq \frac{H_d}{H_a} \qquad N_b = 1, 2, 3, \ldots \tag{17–20}$$

Designers work on a per-belt basis.

The flat-belt tensions shown in Fig. 17–12 ignored the tension induced by bending the belt about the pulleys. This is more pronounced with V belts, as shown in Fig. 17–14.

The centrifugal tension F_c is given by

$$F_c = K_c \left(\frac{V}{1000}\right)^2 \tag{17–21}$$

where K_c is from Table 17–16.

Table 17–13

Angle of Contact Correction Factor K_1 for VV* and V-Flat Drives

$\dfrac{D-d}{C}$	θ, deg	K_1	
		VV	**V Flat**
0.00	180	1.00	0.75
0.10	174.3	0.99	0.76
0.20	166.5	0.97	0.78
0.30	162.7	0.96	0.79
0.40	156.9	0.94	0.80
0.50	151.0	0.93	0.81
0.60	145.1	0.91	0.83
0.70	139.0	0.89	0.84
0.80	132.8	0.87	0.85
0.90	126.5	0.85	0.85
1.00	120.0	0.82	0.82
1.10	113.3	0.80	0.80
1.20	106.3	0.77	0.77
1.30	98.9	0.73	0.73
1.40	91.1	0.70	0.70
1.50	82.8	0.65	0.65

*A curve fit for the VV column in terms of θ is
$K_1 = 0.143\,543 + 0.007\,468\,\theta - 0.000\,015\,052\,\theta^2$
in the range $90° \leq \theta \leq 180°$.

Table 17–14

Belt-Length Correction Factor K_2^*

Length Factor	Nominal Belt Length, m				
	A Belts	**B Belts**	**C Belts**	**D Belts**	**E Belts**
0.85	Up to 0.88	Up to 1.15	Up to 1.88	Up to 3.2	
0.90	0.95–1.15	1.2–1.5	2.03–2.4	3.6–4.05	Up to 4.88
0.95	1.2–1.38	1.55–1.88	2.63–3.0	4.33–5.25	5.25–6.0
1.00	1.5–1.88	1.95–2.43	3.2–3.95	6.0	6.75–7.5
1.05	1.95–2.25	2.63–3.0	4.05–4.88	6.75–8.25	8.25–9.75
1.10	2.4–2.8	3.2–3.6	5.25–6.0	9.0–10.5	10.5–12.0
1.15	3.0 and up	3.95–4.5	6.75–7.5	12.0	13.5–15.0
1.20		4.88 and up	8.25 and up	13.5 and up	16.5

*Multiply the rated power per belt by this factor to obtain the corrected power.

Table 17–15

Suggested Service Factors K_S for V-Belt Drives

Driven Machinery	Source of Power	
	Normal Torque Characteristic	**High or Nonuniform Torque**
Uniform	1.0 to 1.2	1.1 to 1.3
Light shock	1.1 to 1.3	1.2 to 1.4
Medium shock	1.2 to 1.4	1.4 to 1.6
Heavy shock	1.3 to 1.5	1.5 to 1.8

Figure 17–14

V-belt tensions.

(a)

(b)

Table 17–16

Some V-Belt Parameters*

Belt Section	K_b	K_c
A	220	0.561
B	576	0.965
C	1 600	1.716
D	5 680	3.498
E	10 850	5.041
3V	230	0.425
5V	1098	1.217
8V	4830	3.288

*Data courtesy of Gates Rubber Co., Denver, Colo.

The power that is transmitted per belt is based on $\Delta F = F_1 - F_2$, where

$$\Delta F = \frac{H_d/N_b}{\pi n d} \tag{17–22}$$

then from Eq. (17–8) the largest tension F_1 is given by

$$F_1 = F_c + \frac{\Delta F \exp(f\phi)}{\exp(f\phi) - 1} \tag{17–23}$$

From the definition of ΔF, the least tension F_2 is

$$F_2 = F_1 - \Delta F \tag{17–24}$$

From Eq. (j) in Sec. 17–2

$$F_i = \frac{F_1 + F_2}{2} - F_c \tag{17–25}$$

The factor of safety is

$$n_{fs} = \frac{H_a N_b}{H_{\text{nom}} K_s} \tag{17-26}$$

Durability (life) correlations are complicated by the fact that the bending induces flexural stresses in the belt; the corresponding belt tension that induces the same maximum tensile stress is $(F_b)_1$ at the driving sheave and $(F_b)_2$ at the driven pulley. These equivalent tensions are added to F_1 as

$$T_1 = F_1 + (F_b)_1 = F_1 + \frac{K_b}{d}$$

$$T_2 = F_1 + (F_b)_2 = F_1 + \frac{K_b}{D}$$

where K_b is given in Table 17–16. The equation for the tension versus pass trade-off used by the Gates Rubber Company is of the form

$$T^b N_P = K^b$$

where N_P is the number of passes and b is approximately 11. See Table 17–17. The Miner rule is used to sum damage incurred by the two tension peaks:

$$\frac{1}{N_P} = \left(\frac{K}{T_1}\right)^{-b} + \left(\frac{K}{T_2}\right)^{-b}$$

or

$$N_P = \left[\left(\frac{K}{T_1}\right)^{-b} + \left(\frac{K}{T_2}\right)^{-b}\right]^{-1} \tag{17-27}$$

The lifetime t in hours is given by

$$t = \frac{N_P L_p}{3600} \tag{17-28}$$

Table 17–17	**Belt Section**	**10^8 to 10^9 Force Peaks**		**10^9 to 10^{10} Force Peaks**		**Minimum Sheave Diameter, mm**
		K	**b**	**K**	**b**	
Durability Parameters for Some V-Belt Sections	A	2999	11.089			75
Source: M. E. Spotts, *Design*	B	5309	10.926			125
of Machine Elements, 6th ed.	C	9069	11.173			215
Prentice Hall, Englewood	D	18 726	11.105			325
Cliffs, N.J., 1985.	E	26 791	11.100			540
	3V	3240	12.464	4726	10.153	66
	5V	7360	12.593	10 653	10.283	177
	8V	16 189	12.629	23 376	10.319	312

The constants K and b have their ranges of validity. If $N_P > 10^9$, report that $N_P = 10^9$ and $t > N_P L_p/(3600V)$ without placing confidence in numerical values beyond the validity interval. See the statement about N_P and t near the conclusion of Ex. 17–4.

The analysis of a V-belt drive can consist of the following steps:

- Find V, L_p, C, ϕ, and $\exp(0.5123\phi)$
- Find H_d, H_a, and N_b from H_d/H_a and round up
- Find F_c, ΔF, F_1, F_2, and F_i, and n_{fs}
- Find belt life in number of passes, or hours, if possible

EXAMPLE 17–4

A 7.46-kW split-phase motor running at 1750 rev/min is used to drive a rotary pump, which operates 24 hours per day. An engineer has specified a 188-mm small sheave, a 280-mm large sheave, and three B2800 belts. The service factor of 1.2 was augmented by 0.1 because of the continuous-duty requirement. Analyze the drive and estimate the belt life in passes and hours.

Solution

The peripheral speed V of the belt is

$$V = \pi \, dn = \pi(0.188)1750/60 = 17 \text{ m/s}$$

Table 17–11: $L_p = L + L_c = 2800 + 45 = 2845$ mm

Eq. (17–16b):
$$C = 0.25 \left\{ \left[2845 - \frac{\pi}{2}(280 + 188) \right] \right.$$

$$\left. + \sqrt{\left[2845 - \frac{\pi}{2}(280 + 188) \right]^2 - 2(280 - 188)^2} \right\}$$

$$= 1054 \text{ mm}$$

Eq. (17–1): $\phi = \theta_d = \pi - 2\sin^{-1}(280 - 188)/[2(1054)] = 3.054$ rad

$$\exp[0.5123(3.054)] = 4.781$$

Interpolating in Table 17–12 for $V = 17$ m/s gives $H_{tab} = 3.5$ kW. The wrap angle in degrees is $3.054(180)/\pi = 175°$. From Table 17–13, $K_1 = 0.99$. From Table 17–14, $K_2 = 1.05$. Thus, from Eq. (17–17),

$$H_a = K_1 K_2 H_{tab} = 0.99(1.05)3.5 = 3.64 \text{ kW}$$

Eq. (17–19): $H_d = H_{nom} K_s n_d = 7.46(1.2 + 0.1)(1) = 9.7$ kW

Eq. (17–20): $N_b \geq H_d/H_a = 9.7/3.64 = 2.67 \rightarrow 3$

From Table 17–16, $K_c = 0.965$. Thus, from Eq. (17–21),

$$F_c = 0.965(17/2.4)^2 = 48.4 \text{ N}$$

Eq.(17–22): $$\Delta F = \frac{9700/3}{\pi(1750/60)0.188} = 188 \text{ N}$$

Eq. (17–23): $$F_1 = 48.4 + \frac{188(4.781)}{4.781 - 1} = 286 \text{ N}$$

Eq. (17–24): $$F_2 = F_1 - \Delta F = 286 - 188 = 98 \text{ N}$$

Eq. (17–25): $$F_i = \frac{286 + 98}{2} - 48.4 = 144 \text{ N}$$

Eq. (17–26): $$n_{fs} = \frac{H_a N_b}{H_{\text{nom}} K_s} = \frac{3.64(3)}{7.46(1.3)} = 1.13$$

Life: From Table 17–16, $K_b = 576$.

$$F_{b1} = \frac{K_b}{d} = \frac{65}{0.188} = 346 \text{ N}$$

$$F_{b2} = 65/0.28 = 232 \text{ N}$$

$$T_1 = F_1 + F_{b1} = 286 + 346 = 632 \text{ N}$$

$$T_2 = F_1 + F_{b2} = 286 + 232 = 518 \text{ N}$$

From Table 17–17, $K = 5309$ and $b = 10.926$.

Eq. (17–27): $$N_P = \left[\left(\frac{5309}{632} \right)^{-10.926} + \left(\frac{5309}{518} \right)^{-10.926} \right]^{-1} = 11(10^9) \text{ passes}$$

Answer Since N_P is out of the validity range of Eq. (17–27), life is reported as greater than 10^9 passes. Then

Answer Eq. (17–28): $$t > \frac{10^9(2.845)}{3600(17)} = 46\,487 \text{ h}$$

17–4 Timing Belts

A timing belt is made of a rubberized fabric coated with a nylon fabric, and has steel wire within to take the tension load. It has teeth that fit into grooves cut on the periphery of the pulleys (Fig. 17–15). A timing belt does not stretch appreciably or slip and consequently transmits power at a constant angular-velocity ratio. No initial

Figure 17–15

Timing-belt drive showing portions of the pulley and belt. Note that the pitch diameter of the pulley is greater than the diametral distance across the top lands of the teeth.

Table 17–18

Standard Pitches
of Timing Belts

Service	Designation	Pitch p, mm
Extra light	XL	5
Light	L	10
Heavy	H	12
Extra heavy	XH	22
Double extra heavy	XXH	30

tension is needed. Such belts can operate over a very wide range of speeds, have efficiencies in the range of 97 to 99 percent, require no lubrication, and are quieter than chain drives. There is no chordal-speed variation, as in chain drives (see Sec. 17–5), and so they are an attractive solution for precision-drive requirements.

The steel wire, the tension member of a timing belt, is located at the belt pitch line (Fig. 17–15). Thus the pitch length is the same regardless of the thickness of the backing.

The five standard inch-series pitches available are listed in Table 17–18 with their letter designations. Standard pitch lengths are available in sizes from 150 to 4500 mm. Pulleys come in sizes from 15 mm pitch diameter up to 900 mm and with groove numbers from 10 to 120.

The design and selection process for timing belts is so similar to that for V belts that the process will not be presented here. As in the case of other belt drives, the manufacturers will provide an ample supply of information and details on sizes and strengths.

17–5   Roller Chain

Basic features of chain drives include a constant ratio, since no slippage or creep is involved; long life; and the ability to drive a number of shafts from a single source of power.

Roller chains have been standardized as to sizes by the ANSI. Figure 17–16 shows the nomenclature. The pitch is the linear distance between the centers of the rollers. The width is the space between the inner link plates. These chains are manufactured in single, double, triple, and quadruple strands. The dimensions of standard sizes are listed in Table 17–19.

Figure 17–16

Portion of a double-strand roller chain.

Table 17–19

Dimensions of American
Standard Roller
Chains—Single Strand

Source: Compiled from ANSI
B29.1-1975.

ANSI Chain Number	Pitch, in (mm)	Width, in (mm)	Minimum Tensile Strength, lbf (N)	Average Weight, lbf/ft (N/m)	Roller Diameter, in (mm)	Multiple-Strand Spacing, in (mm)
25	0.250 (6.35)	0.125 (3.18)	780 (3 470)	0.09 (1.31)	0.130 (3.30)	0.252 (6.40)
35	0.375 (9.52)	0.188 (4.76)	1 760 (7 830)	0.21 (3.06)	0.200 (5.08)	0.399 (10.13)
41	0.500 (12.70)	0.25 (6.35)	1 500 (6 670)	0.25 (3.65)	0.306 (7.77)	— —
40	0.500 (12.70)	0.312 (7.94)	3 130 (13 920)	0.42 (6.13)	0.312 (7.92)	0.566 (14.38)
50	0.625 (15.88)	0.375 (9.52)	4 880 (21 700)	0.69 (10.1)	0.400 (10.16)	0.713 (18.11)
60	0.750 (19.05)	0.500 (12.7)	7 030 (31 300)	1.00 (14.6)	0.469 (11.91)	0.897 (22.78)
80	1.000 (25.40)	0.625 (15.88)	12 500 (55 600)	1.71 (25.0)	0.625 (15.87)	1.153 (29.29)
100	1.250 (31.75)	0.750 (19.05)	19 500 (86 700)	2.58 (37.7)	0.750 (19.05)	1.409 (35.76)
120	1.500 (38.10)	1.000 (25.40)	28 000 (124 500)	3.87 (56.5)	0.875 (22.22)	1.789 (45.44)
140	1.750 (44.45)	1.000 (25.40)	38 000 (169 000)	4.95 (72.2)	1.000 (25.40)	1.924 (48.87)
160	2.000 (50.80)	1.250 (31.75)	50 000 (222 000)	6.61 (96.5)	1.125 (28.57)	2.305 (58.55)
180	2.250 (57.15)	1.406 (35.71)	63 000 (280 000)	9.06 (132.2)	1.406 (35.71)	2.592 (65.84)
200	2.500 (63.50)	1.500 (38.10)	78 000 (347 000)	10.96 (159.9)	1.562 (39.67)	2.817 (71.55)
240	3.00 (76.70)	1.875 (47.63)	112 000 (498 000)	16.4 (239)	1.875 (47.62)	3.458 (87.83)

Figure 17–17 shows a sprocket driving a chain and rotating in a counterclockwise direction. Denoting the chain pitch by p, the pitch angle by γ, and the pitch diameter of the sprocket by D, from the trigonometry of the figure we see

$$\sin\frac{\gamma}{2} = \frac{p/2}{D/2} \quad \text{or} \quad D = \frac{p}{\sin(\gamma/2)} \tag{a}$$

Since $\gamma = 360°/N$, where N is the number of sprocket teeth, Eq. (a) can be written

$$D = \frac{p}{\sin(180°/N)} \tag{17–29}$$

The angle $\gamma/2$, through which the link swings as it enters contact, is called the *angle of articulation*. It can be seen that the magnitude of this angle is a function of the number of teeth. Rotation of the link through this angle causes impact between

Figure 17-17

Engagement of a chain and sprocket.

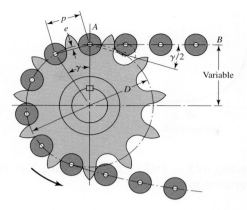

the rollers and the sprocket teeth and also wear in the chain joint. Since the life of a properly selected drive is a function of the wear and the surface fatigue strength of the rollers, it is important to reduce the angle of articulation as much as possible.

The number of sprocket teeth also affects the velocity ratio during the rotation through the pitch angle γ. At the position shown in Fig. 17–17, the chain AB is tangent to the pitch circle of the sprocket. However, when the sprocket has turned an angle of $\gamma/2$, the chain line AB moves closer to the center of rotation of the sprocket. This means that the chain line AB is moving up and down, and that the lever arm varies with rotation through the pitch angle, all resulting in an uneven chain exit velocity. You can think of the sprocket as a polygon in which the exit velocity of the chain depends upon whether the exit is from a corner, or from a flat of the polygon. Of course, the same effect occurs when the chain first enters into engagement with the sprocket.

The chain velocity V is defined as the number of feet coming off the sprocket per unit time. Thus the chain velocity in feet per minute is

$$V = \frac{Npn}{12} \tag{17-30}$$

where N = number of sprocket teeth

p = chain pitch, in

n = sprocket speed, rev/min

The maximum exit velocity of the chain is

$$v_{max} = \frac{\pi Dn}{12} = \frac{\pi np}{12\sin(\gamma/2)} \tag{b}$$

where Eq. (a) has been substituted for the pitch diameter D. The minimum exit velocity occurs at a diameter d, smaller than D. Using the geometry of Fig. 17–17, we find

$$d = D\cos\frac{\gamma}{2} \tag{c}$$

Thus the minimum exit velocity is

$$v_{min} = \frac{\pi dn}{12} = \frac{\pi np}{12}\frac{\cos(\gamma/2)}{\sin(\gamma/2)} \tag{d}$$

| Figure 17–18

Now substituting $\gamma/2 = 180°/N$ and employing Eqs. (17–30), (b), and (d), we find the speed variation to be

$$\frac{\Delta V}{V} = \frac{v_{max} - v_{min}}{V} = \frac{\pi}{N}\left[\frac{1}{\sin(180°/N)} - \frac{1}{\tan(180°/N)}\right] \qquad (17\text{–}31)$$

This is called the *chordal speed variation* and is plotted in Fig. 17–18. When chain drives are used to synchronize precision components or processes, due consideration must be given to these variations. For example, if a chain drive synchronized the cutting of photographic film with the forward drive of the film, the lengths of the cut sheets of film might vary too much because of this chordal speed variation. Such variations can also cause vibrations within the system.

Although a large number of teeth is considered desirable for the driving sprocket, in the usual case it is advantageous to obtain as small a sprocket as possible, and this requires one with a small number of teeth. For smooth operation at moderate and high speeds it is considered good practice to use a driving sprocket with at least 17 teeth; 19 or 21 will, of course, give a better life expectancy with less chain noise. Where space limitations are severe or for very slow speeds, smaller tooth numbers may be used by sacrificing the life expectancy of the chain.

Driven sprockets are not made in standard sizes over 120 teeth, because the pitch elongation will eventually cause the chain to "ride" high long before the chain is worn out. The most successful drives have velocity ratios up to 6:1, but higher ratios may be used at the sacrifice of chain life.

Roller chains seldom fail because they lack tensile strength; they more often fail because they have been subjected to a great many hours of service. Actual failure may be due either to wear of the rollers on the pins or to fatigue of the surfaces of the rollers. Roller-chain manufacturers have compiled tables that give the horsepower capacity corresponding to a life expectancy of 15 kh for various sprocket speeds. These capacities are tabulated in Table 17–20 for 17-tooth sprockets. Table 17–21 displays available tooth counts on sprockets of one supplier. Table 17–22 lists the tooth correction factors for other than 17 teeth. Table 17–23 shows the multiple-strand factors K_2.

The capacities of chains are based on the following:

- 15 000 h at full load
- Single strand
- ANSI proportions
- Service factor of unity
- 100 pitches in length
- Recommended lubrication

- Elongation maximum of 3 percent
- Horizontal shafts
- Two 17-tooth sprockets

The fatigue strength of link plates governs capacity at lower speeds. The American Chain Association (ACA) publication *Chains for Power Transmission and Materials Handling* (1982) gives, for single-strand chain, the nominal power H_1, link-plate limited, as

$$H_1 = 0.003 N_1^{1.08} n_1^{0.9} p^{(3-0.07p)} \qquad \text{kW} \tag{17-32}$$

and the nominal power H_2, roller-limited, as

$$H_2 = \frac{746 K_r N_1^{1.5} p^{0.8}}{n_1^{1.5}} \qquad \text{kW} \tag{17-33}$$

where N_1 = number of teeth in the smaller sprocket

n_1 = sprocket speed, rev/min

p = pitch of the chain, mm

K_r = 29 for chain numbers 25, 35; 3.4 for chain 41; and 17 for chains 40–240

Table 17–20

Rated Horsepower Capacity of Single-Strand Single-Pitch Roller Chain for a 17-Tooth Sprocket

Source: Compiled from ANSI B29.1-1975 information only section, and from B29.9-1958.

Sprocket Speed, rev/min	ANSI Chain Number					
	25	35	40	41	50	60
50	0.05	0.16	0.37	0.20	0.72	1.24
100	0.09	0.29	0.69	0.38	1.34	2.31
150	0.13*	0.41*	0.99*	0.55*	1.92*	3.32
200	0.16*	0.54*	1.29	0.71	2.50	4.30
300	0.23	0.78	1.85	1.02	3.61	6.20
400	0.30*	1.01*	2.40	1.32	4.67	8.03
500	0.37	1.24	2.93	1.61	5.71	9.81
600	0.44*	1.46*	3.45*	1.90*	6.72*	11.6
700	0.50	1.68	3.97	2.18	7.73	13.3
800	0.56*	1.89*	4.48*	2.46*	8.71*	15.0
900	0.62	2.10	4.98	2.74	9.69	16.7
1000	0.68*	2.31*	5.48	3.01	10.7	18.3
1200	0.81	2.73	6.45	3.29	12.6	21.6
1400	0.93*	3.13*	7.41	2.61	14.4	18.1
1600	1.05*	3.53*	8.36	2.14	12.8	14.8
1800	1.16	3.93	8.96	1.79	10.7	12.4
2000	1.27*	4.32*	7.72*	1.52*	9.23*	10.6
2500	1.56	5.28	5.51*	1.10*	6.58*	7.57
3000	1.84	5.64	4.17	0.83	4.98	5.76
Type A		**Type B**			**Type C**	

*Estimated from ANSI tables by linear interpolation.

Note: Type A—manual or drip lubrication; type B—bath or disk lubrication; type C—oil-stream lubrication.

(*Continued*)

Table 17–20

Rated Horsepower Capacity of Single-Strand Single-Pitch Roller Chain for a 17-Tooth Sprocket

(*Continued*)

Sprocket Speed, rev/min		ANSI Chain Number							
		80	100	120	140	160	180	200	240
50	Type A	2.88	5.52	9.33	14.4	20.9	28.9	38.4	61.8
100		5.38	10.3	17.4	26.9	39.1	54.0	71.6	115
150		7.75	14.8	25.1	38.8	56.3	77.7	103	166
200		10.0	19.2	32.5	50.3	72.9	101	134	215
300		14.5	27.7	46.8	72.4	105	145	193	310
400		18.7	35.9	60.6	93.8	136	188	249	359
500	Type B	22.9	43.9	74.1	115	166	204	222	0
600		27.0	51.7	87.3	127	141	155	169	
700		31.0	59.4	89.0	101	112	123	0	
800		35.0	63.0	72.8	82.4	91.7	101		
900		39.9	52.8	61.0	69.1	76.8	84.4		
1000		37.7	45.0	52.1	59.0	65.6	72.1		
1200		28.7	34.3	39.6	44.9	49.9	0		
1400		22.7	27.2	31.5	35.6	0			
1600		18.6	22.3	25.8	0				
1800		15.6	18.7	21.6					
2000		13.3	15.9	0					
2500		9.56	0.40						
3000		7.25	0						

Type C	Type C′

Note: Type A—manual or drip lubrication; type B—bath or disk lubrication; type C—oil-stream lubrication; type C′—type C, but this is a galling region; submit design to manufacturer for evaluation.

Table 17–21

Single-Strand Sprocket Tooth Counts Available from One Supplier*

No.	Available Sprocket Tooth Counts
25	8-30, 32, 34, 35, 36, 40, 42, 45, 48, 54, 60, 64, 65, 70, 72, 76, 80, 84, 90, 95, 96, 102, 112, 120
35	4-45, 48, 52, 54, 60, 64, 65, 68, 70, 72, 76, 80, 84, 90, 95, 96, 102, 112, 120
41	6-60, 64, 65, 68, 70, 72, 76, 80, 84, 90, 95, 96, 102, 112, 120
40	8-60, 64, 65, 68, 70, 72, 76, 80, 84, 90, 95, 96, 102, 112, 120
50	8-60, 64, 65, 68, 70, 72, 76, 80, 84, 90, 95, 96, 102, 112, 120
60	8-60, 62, 63, 64, 65, 66, 67, 68, 70, 72, 76, 80, 84, 90, 95, 96, 102, 112, 120
80	8-60, 64, 65, 68, 70, 72, 76, 78, 80, 84, 90, 95, 96, 102, 112, 120
100	8-60, 64, 65, 67, 68, 70, 72, 74, 76, 80, 84, 90, 95, 96, 102, 112, 120
120	9-45, 46, 48, 50, 52, 54, 55, 57, 60, 64, 65, 67, 68, 70, 72, 76, 80, 84, 90, 96, 102, 112, 120
140	9-28, 30, 31, 32, 33, 34, 35, 36, 37, 39, 40, 42, 43, 45, 48, 54, 60, 64, 65, 68, 70, 72, 76, 80, 84, 96
160	8-30, 32–36, 38, 40, 45, 46, 50, 52, 53, 54, 56, 57, 60, 62, 63, 64, 65, 66, 68, 70, 72, 73, 80, 84, 96
180	13-25, 28, 35, 39, 40, 45, 54, 60
200	9-30, 32, 33, 35, 36, 39, 40, 42, 44, 45, 48, 50, 51, 54, 56, 58, 59, 60, 63, 64, 65, 68, 70, 72
240	9-30, 32, 35, 36, 40, 44, 45, 48, 52, 54, 60

*Morse Chain Company, Ithaca, NY, Type B hub sprockets.

Table 17–22

Tooth Correction Factors, K_1

Number of Teeth on Driving Sprocket	K_1 Pre-extreme Horsepower	K_1 Post-extreme Horsepower
11	0.62	0.52
12	0.69	0.59
13	0.75	0.67
14	0.81	0.75
15	0.87	0.83
16	0.94	0.91
17	1.00	1.00
18	1.06	1.09
19	1.13	1.18
20	1.19	1.28
N	$(N_1/17)^{1.08}$	$(N_1/17)^{1.5}$

Table 17–23

Multiple-Strand Factors, K_2

Number of Strands	K_2
1	1.0
2	1.7
3	2.5
4	3.3
5	3.9
6	4.6
8	6.0

The constant 0.003 becomes 0.00165 for no. 41 lightweight chain. The nominal horsepower in Table 17–20 is $H_{nom} = \min(H_1, H_2)$. For example, for $N_1 = 17$, $n_1 = 1000$ rev/min, no. 40 chain with $p = 12.5$ mm, from Eq. (17–32),

$$H_1 = 0.003(17)^{1.08} \, 1000^{0.9} 12.5/25.4^{[3-0.07(12.5/25.4)]} = 3.92 \text{ kW}$$

From Eq. (17–33),

$$H_2 = \frac{746(17)17^{1.5}(12.5/25.4^{0.8})}{1000^{1.5}} = 15.94 \text{ kW}$$

The tabulated value in Table 17–20 is $H_{tab} = \min(3.92, 15.94) = 3.92$ kW.

It is preferable to have an odd number of teeth on the driving sprocket (17, 19, . . .) and an even number of pitches in the chain to avoid a special link. The approximate length of the chain L in pitches is

$$\frac{L}{p} \doteq \frac{2C}{p} + \frac{N_1 + N_2}{2} + \frac{(N_2 - N_1)^2}{4\pi^2 C/p} \tag{17–34}$$

The center-to-center distance C is given by

$$C = \frac{p}{4}\left[-A + \sqrt{A^2 - 8\left(\frac{N_2 - N_1}{2\pi}\right)^2}\right] \tag{17–35}$$

where

$$A = \frac{N_1 + N_2}{2} - \frac{L}{p} \qquad (17\text{-}36)$$

The allowable power H_a is given by

$$H_a = K_1 K_2 H_{\text{tab}} \qquad (17\text{-}37)$$

where K_1 = correction factor for tooth number other than 17 (Table 17–22)

K_2 = strand correction (Table 17–23)

The horsepower that must be transmitted H_d is given by

$$H_d = H_{\text{nom}} K_s n_d \qquad (17\text{-}38)$$

where K_s is a service factor to account for non-uniform loads, and n_d is a design factor.

Equation (17–32) is the basis of the pre-extreme power entries (vertical entries) of Table 17–20, and the chain power is limited by link-plate fatigue. Equation (17–33) is the basis for the post-extreme power entries of these tables, and the chain power performance is limited by impact fatigue. The entries are for chains of 100 pitch length and 17-tooth sprocket. For a deviation from this

$$H_2 = 1000 \left[K_r \left(\frac{N_1}{n_1} \right)^{1.5} p^{0.8} \left(\frac{L_p}{100} \right)^{0.4} \left(\frac{15\,000}{h} \right)^{0.4} \right] \qquad (17\text{-}39)$$

where L_p is the chain length in pitches and h is the chain life in hours. Viewed from a deviation viewpoint, Eq. (17–39) can be written as a trade-off equation in the following form:

$$\frac{H_2^{2.5} h}{N_1^{3.75} L_p} = \text{constant} \qquad (17\text{-}40)$$

If tooth-correction factor K_1 is used, then omit the term $N_1^{3.75}$.

In Eq. (17–40) one would expect the h/L_p term because doubling the hours can require doubling the chain length, other conditions constant, for the same number of cycles. Our experience with contact stresses leads us to expect a load (tension) life relation of the form $F^a L = \text{constant}$. In the more complex circumstance of roller-bushing impact, the Diamond Chain Company has identified $a = 2.5$.

The maximum speed (rev/min) for a chain drive is limited by galling between the pin and the bushing. Tests suggest

$$n_1 \le 1000 \left[\frac{82.5}{7.95^p (1.0278)^{N_1} (1.323)^{F/1000}} \right]^{1/(1.59 \log p + 1.873)} \quad \text{rev/min}$$

where F is the chain tension in lbf.

EXAMPLE 17–5   Select drive components for a 2:1 reduction, 68 kW input at 300 rev/min, moderate shock, an abnormally long 18-hour day, poor lubrication, cold temperatures, dirty surroundings, short drive $C/p = 25$.

Solution   *Function:* $H_{\text{nom}} = 68$ kW, $n_1 = 300$ rev/min, $C/p = 25$, $K_s = 1.3$
Design factor: $n_d = 1.5$

Sprocket teeth: $N_1 = 17$ teeth, $N_2 = 34$ teeth, $K_1 = 1$, $K_2 = 1, 1.7, 2.5, 3.3$
Chain number of strands:

$$H_{\text{tab}} = \frac{n_d K_s H_{\text{nom}}}{K_1 K_2} = \frac{1.5(1.3)68}{(1)K_2} = \frac{132.6}{K_2}$$

Form a table:

Number of Strands	132.6/K2 (Table 17–23)	Chain Number (Table 17–20)	Lubrication Type
1	$132.6/1 = 132.6$	200	C'
2	$132.6/1.7 = 78$	160	C
3	$132/2.5 = 53.04$	140	B
4	$132/3.3 = 40.18$	140	B

Decision   3 strands of number 140 chain (H_{tab} is 54 kW).
Number of pitches in the chain:

$$\frac{L}{p} = \frac{2C}{p} + \frac{N_1 + N_2}{2} + \frac{(N_2 - N_1)^2}{4\pi^2 C/p}$$

$$= 2(25) + \frac{17 + 34}{2} + \frac{(34 - 17)^2}{4\pi^2(25)} = 75.79 \text{ pitches}$$

Decision   Use 76 pitches. Then $L/p = 76$.
Identify the center-to-center distance: From Eqs. (17–35) and (17–36),

$$A = \frac{N_1 + N_2}{2} - \frac{L}{p} = \frac{17 + 34}{2} - 76 = -50.5$$

$$C = \frac{p}{4}\left[-A + \sqrt{A^2 - 8\left(\frac{N_2 - N_1}{2\pi}\right)^2}\right]$$

$$= \frac{p}{4}\left[50.5 + \sqrt{50.5^2 - 8\left(\frac{34 - 17}{2\pi}\right)^2}\right] = 25.104p$$

For a 140 chain, $p = 44.45$ mm. Thus,

$$C = 25.104p = 25.104(44.45) = 1115.9 \text{ mm}$$

Lubrication: Type B
Comment: This is operating on the pre-extreme portion of the power, so durability estimates other than 15 000 h are not available. Given the poor operating conditions, life will be much shorter.

Lubrication of roller chains is essential in order to obtain a long and trouble-free life. Either a drip feed or a shallow bath in the lubricant is satisfactory. A medium or light mineral oil, without additives, should be used. Except for unusual conditions, heavy oils and greases are not recommended, because they are too viscous to enter the small clearances in the chain parts.

17-6 Wire Rope

Wire rope is made with two types of winding, as shown in Fig. 17–19. The *regular lay*, which is the accepted standard, has the wire twisted in one direction to form the strands, and the strands twisted in the opposite direction to form the rope. In the completed rope the visible wires are approximately parallel to the axis of the rope. Regular-lay ropes do not kink or untwist and are easy to handle.

Lang-lay ropes have the wires in the strand and the strands in the rope twisted in the same direction, and hence the outer wires run diagonally across the axis of the rope. Lang-lay ropes are more resistant to abrasive wear and failure due to fatigue than are regular-lay ropes, but they are more likely to kink and untwist.

Standard ropes are made with a hemp core, which supports and lubricates the strands. When the rope is subjected to heat, either a steel center or a wire-strand center must be used.

Wire rope is designated as, for example, a 28-mm 6 × 7 haulage rope. The first figure is the diameter of the rope (Fig. 17–19c). The second and third figures are the number of strands and the number of wires in each strand, respectively. Table 17–24 lists some of the various ropes that are available, together with their characteristics and properties. The area of the metal in standard hoisting and haulage rope is about $A_m = 0.38d^2$.

When a wire rope passes around a sheave, there is a certain amount of readjustment of the elements. Each of the wires and strands must slide on several others, and presumably some individual bending takes place. It is probable that in this complex action there exists some stress concentration. The stress in one of the wires of a rope passing around a sheave may be calculated as follows. From solid mechanics, we have

$$M = \frac{EI}{\rho} \quad \text{and} \quad M = \frac{\sigma I}{c} \tag{a}$$

where the quantities have their usual meaning. Eliminating M and solving for the stress gives

$$\sigma = \frac{Ec}{\rho} \tag{b}$$

For the radius of curvature ρ, we can substitute the sheave radius $D/2$. Also, $c = d_w/2$, where d_w is the wire diameter. These substitutions give

$$\sigma = E_r \frac{d_w}{D} \tag{c}$$

where E_r is the *modulus of elasticity of the rope*, not the wire. To understand this equation, observe that the individual wire makes a corkscrew figure in space and if you pull on it to determine E it will stretch or give more than its native E would

Figure 17–19

Types of wire rope; both lays are available in either right or left hand.

(a) Regular lay

(b) Lang lay

(c) Section of 6 × 7 rope

Table 17-24

Wire-Rope Data   *Source: Compiled from American Steel and Wire Company Handbook.*

Rope	Weight per meter, (10^{-3})N	Minimum Sheave Diameter, mm	Standard Sizes d, mm	Material	Size of Outer Wires	Modulus of Elasticity,* GPa	Strength,† MPa
6 × 7 haulage	$33.92d^2$	$42d$	6 – 38	Monitor steel	$d/9$	96	690
				Plow steel	$d/9$	96	608
				Mild plow steel	$d/9$	96	524
6 × 19 standard hoisting	$36.18d^2$	$26d$ – $34d$	6 – 70	Monitor steel	$d/13$ – $d/16$	83	730
				Plow steel	$d/13$ – $d/16$	83	640
				Mild plow steel	$d/13$ – $d/16$	83	550
6 × 37 special flexible	$35.05d^2$	$18d$	6 – 90	Monitor steel	$d/22$	76	690
				Plow steel	$d/22$	76	608
8 × 19 extra flexible	$32.79d^2$	$21d$ – $26d$	6 – 38	Monitor steel	$d/15$ – $d/19$	69	634
				Plow steel	$d/15$ – $d/19$	69	550
7 × 7 aircraft	$38.45d^2$	—	1.6 – 10	Corrosion-resistant steel	—	—	850
				Carbon steel	—	—	850
7 × 9 aircraft	$39.58d^2$	—	3 – 36	Corrosion-resistant steel	—	—	930
				Carbon steel	—	—	986
19-wire aircraft	$48.62d^2$	—	0.8 – 8	Corrosion-resistant steel	—	—	1137
				Carbon steel	—	—	1137

*The modulus of elasticity is only approximate; it is affected by the loads on the rope and, in general, increases with the life of the rope.

†The strength is based on the nominal area of the rope. The figures given are only approximate and are based on 25-mm rope sizes and 6-mm aircraft-cable sizes.

suggest. Therefore E is still the modulus of elasticity of the *wire*, but in its peculiar configuration as part of the rope, its modulus is smaller. For this reason we say that E_r in Eq. (c) is the modulus of elasticity of the rope, not the wire, recognizing that one can quibble over the name used.

Equation (c) gives the tensile stress σ in the outer wires. The sheave diameter is represented by D. This equation reveals the importance of using a large-diameter sheave. The suggested minimum sheave diameters in Table 17–24 are based on a D/d_w ratio of 400. If possible, the sheaves should be designed for a larger ratio. For elevators and mine hoists, D/d_w is usually taken from 800 to 1000. If the ratio is less than 200, heavy loads will often cause a permanent set in the rope.

A wire rope tension giving the same tensile stress as the sheave bending is called the *equivalent bending load* F_b, given by

$$F_b = \sigma A_m = \frac{E_r d_w A_m}{D} \tag{17-41}$$

A wire rope may fail because the static load exceeds the ultimate strength of the rope. Failure of this nature is generally not the fault of the designer, but rather that of the operator in permitting the rope to be subjected to loads for which it was not designed.

The first consideration in selecting a wire rope is to determine the static load. This load is composed of the following items:

- The known or dead weight
- Additional loads caused by sudden stops or starts
- Shock loads
- Sheave-bearing friction

When these loads are summed, the total can be compared with the ultimate strength of the rope to find a factor of safety. However, the ultimate strength used in this determination must be reduced by the strength loss that occurs when the rope passes over a curved surface such as a stationary sheave or a pin; see Fig. 17–20.

For an average operation, use a factor of safety of 5. Factors of safety up to 8 or 9 are used if there is danger to human life and for very critical situations. Table 17–25

Figure 17–20

Percent strength loss due to different D/d ratios; derived from standard test data for 6×19 and 6×17 class ropes. *(Materials provided by the Wire Rope Technical Board (WRTB), Wire Rope Users Manual Third Edition, Second printing. Reprinted by permission.)*

Table 17–25

Minimum Factors of Safety for Wire Rope*

Source: Compiled from a variety of sources, including ANSI A17.1-1978.

Track cables	3.2	Passenger elevators, m/s:		
Guys	3.5	0.25		7.60
Mine shafts, m:		1.52		9.20
Up to 152.5	8.0	4.06		11.25
305–610	7.0	6.10		11.80
610–915	6.0	7.62		11.90
Over 915	5.0	Freight elevators, m/s:		
Hoisting	5.0	0.25		6.65
Haulage	6.0	1.52		8.20
		4.06		10.00
Cranes and derricks	6.0	6.10		10.50
Electric hoists	7.0	7.62		10.55
Hand elevators	5.0	Powered dumbwaiters, m/s:		
Private elevators	7.5	0.25		4.8
Hand dumbwaiter	4.5	1.52		6.6
Grain elevators	7.5	4.06		8.0

*Use of these factors does not preclude a fatigue failure.

lists minimum factors of safety for a variety of design situations. Here, the factor of safety is defined as

$$n = \frac{F_u}{F_t}$$

where F_u is the ultimate wire load and F_t is the largest working tension.

Once you have made a tentative selection of a rope based upon static strength, the next consideration is to ensure that the wear life of the rope and the sheave or sheaves meets certain requirements. When a loaded rope is bent over a sheave, the rope stretches like a spring, rubs against the sheave, and causes wear of both the rope and the sheave. The amount of wear that occurs depends upon the pressure of the rope in the sheave groove. This pressure is called the *bearing pressure;* a good estimate of its magnitude is given by

$$P = \frac{2F}{dD} \tag{17–42}$$

where F = tensile force on rope
d = rope diameter
D = sheave diameter

The allowable pressures given in Table 17–26 are to be used only as a rough guide; they may not prevent a fatigue failure or severe wear. They are presented here because they represent past practice and furnish a starting point in design.

A fatigue diagram not unlike an *S-N* diagram can be obtained for wire rope. Such a diagram is shown in Fig. 17–21. Here the ordinate is the pressure-strength ratio p/S_u, and S_u is the ultimate tensile strength of the *wire*. The abscissa is the number of bends that occur in the total life of the rope. The curve implies that a wire rope has a fatigue limit; but this is not true at all. A wire rope that is used over sheaves will eventually

Table 17–26

Maximum Allowable Bearing Pressures of Ropes on Sheaves (in MPa)

Source: Wire Rope Users Manual, AISI, 1979.

		Sheave Material			
Rope	Wood[a]	Cast Iron[b]	Cast Steel[c]	Chilled Cast Irons[d]	Manganese Steel[e]
Regular lay:					
6 × 7	1.0	2.1	3.8	4.5	10.1
6 × 19	1.7	3.3	6.2	7.6	16.6
6 × 37	2.1	4.0	7.4	9.1	20.7
8 × 19	2.4	4.7	8.7	10.7	24.1
Lang lay:					
6 × 7	1.1	2.4	4.1	4.9	11.4
6 × 19	1.9	3.8	6.9	8.3	19.0
6 × 37	2.3	4.6	8.1	10.0	22.8

[a]On end grain of beech, hickory, or gum.

[b]For H_B(min.) = 125.

[c]30–40 carbon; H_B(min.) = 160.

[d]Use only with uniform surface hardness.

[e]For high speeds with balanced sheaves having ground surfaces.

Figure 17–21

Experimentally determined relation between the fatigue life of wire rope and the sheave pressure.

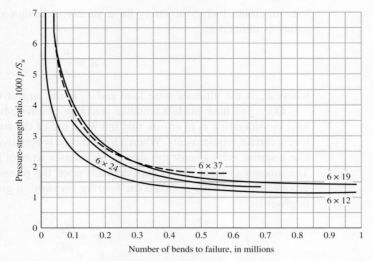

fail in fatigue or in wear. However, the graph does show that the rope will have a long life if the ratio p/S_u is less than 0.001. Substitution of this ratio in Eq. (17–42) gives

$$S_u = \frac{2000F}{dD} \tag{17–43}$$

where S_u is the ultimate strength of the *wire,* not the rope, and the units of S_u are related to the units of F. This interesting equation contains the wire strength, the load, the rope diameter, and the sheave diameter—all four variables in a single equation! Dividing both sides of Eq. (17–42) by the ultimate strength of the wires S_u and solving for F gives

$$F_f = \frac{(p/S_u)S_u dD}{2} \tag{17–44}$$

where F_f is interpreted as the allowable fatigue tension as the wire is flexed a number of times corresponding to p/S_u selected from Fig. 17–21 for a particular rope and life expectancy. The factor of safety can be defined in fatigue as

$$n_f = \frac{F_f - F_b}{F_t} \tag{17–45}$$

where F_f is the rope tension strength under flexing and F_t is the tension at the place where the rope is flexing. Unfortunately, the designer often has vendor information that tabulates ultimate rope tension and gives no ultimate-strength S_u information concerning the wires from which the rope is made. Some guidance in strength of individual wires is

Improved plow steel (monitor) $1655 < S_u < 1930$ MPa
Plow steel $1448 < S_u < 1655$ MPa
Mild plow steel $1241 < S_u < 1448$ MPa

In wire-rope usage, the factor of safety is defined for static loading as

$$n_s = \frac{F_u - F_b}{F_t} \tag{17–46}$$

where F_b is the rope tension that would induce the same outer-wire stress as that given by Eq. (*c*). Be careful when comparing recommended static factors of safety to Eq. (17–46), as n_s is sometimes defined as F_u/F_t. The factor of safety in fatigue loading can be defined as in Eq. (17–45), or by using a static analysis and compensating with a large factor of safety applicable to static loading, as in Table 17–25. When using factors of safety expressed in codes, standards, corporate design manuals, or wire-rope manufacturers' recommendations or from the literature, be sure to ascertain upon which basis the factor of safety is to be evaluated, and proceed accordingly.

If the rope is made of plow steel, the wires are probably hard-drawn AISI 1070 or 1080 carbon steel. Referring to Table 10–3, p. 516, we see that this lies somewhere between hard-drawn spring wire and music wire. But the constants m and A needed to solve Eq. (10–14), p. 515, for S_u are lacking.

Practicing engineers who desire to solve Eq. (17–43) should determine the wire strength S_u for the rope under consideration by unraveling enough wire to test for the Brinell hardness. Then S_u can be found using Eq. (2–21), p. 52. Fatigue failure in wire rope is not sudden, as in solid bodies, but progressive, and shows as the breaking of an outside wire. This means that the beginning of fatigue can be detected by periodic routine inspection.

Figure 17–22 is another graph showing the gain in life to be obtained by using large D/d ratios. In view of the fact that the life of wire rope used over sheaves is only finite, it is extremely important that the designer specify and insist that periodic inspection, lubrication, and maintenance procedures be carried out during the life of the rope. Table 17–27 gives useful properties of some wire ropes.

For a mine-hoist problem we present here the working equations, some repeated from the preceding presentation. The wire rope tension F_t due to load and acceleration/deceleration is

$$F_t = \left(\frac{W}{m} + wl\right)\left(1 + \frac{a}{g}\right) \tag{17–47}$$

Figure 17–22

Service-life curve based on bending and tensile stresses only. This curve shows that the life corresponding to $D/d = 48$ is twice that of $D/d = 33$. *(Materials provided by the Wire Rope Technical Board (WRTB),* Wire Rope Users Manual Third Edition, *Second printing. Reprinted by permission.)*

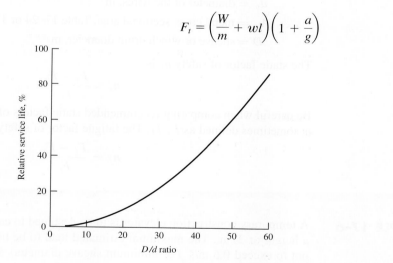

Table 17–27

Some Useful Properties of 6 × 7, 6 × 19, and 6 × 37 Wire Ropes

Wire Rope	Weight per meter w, (10^{-3}) N/m	Weight per meter Including Core w, (10^{-3}) N/m	Minimum Sheave Diameter D, mm	Better Sheave Diameter D, mm	Diameter of Wires d_w, mm	Area of Metal A_m, mm²	Rope Young's Modulus E_r, GPa
6 × 7	$33.92d^2$		$42d$	$72d$	$0.111d$	$0.38d^2$	13×10^6
6 × 19	$36.18d^2$	$39.8d^2$	$30d$	$45d$	$0.067d$	$0.40d^2$	12×10^6
6 × 37	$35.05d^2$	$38.67d^2$	$18d$	$27d$	$0.048d$	$0.40d^2$	12×10^6

where W = weight at the end of the rope (cage and load), N

m = number of wire ropes supporting the load

w = weight/foot of the wire rope, N/m

l = suspended length of rope, m

a = maximum acceleration/deceleration experienced, m/s^2

g = acceleration of gravity, m/s^2

The fatigue tensile strength in pounds for a specified life F_f is

$$F_f = \frac{(p/S_u)S_u Dd}{2} \qquad (17\text{--}47)$$

where (p/S_u) = specified life, from Fig. 17–21

S_u = ultimate tensile strength of the wires, Pa

D = sheave or winch drum diameter, m

d = nominal wire rope size, m

The *equivalent bending load* F_b is

$$F_b = \frac{E_r d_w A_m}{D} \qquad (17\text{--}48)$$

where E_r = Young's modulus for the wire rope, Table 17–24 or 17–27, Pa

d_w = diameter of the wires, m

A_m = metal cross-sectional area, Table 17–24 or 17–28, m^2

D = sheave or winch drum diameter, m

The static factor of safety n_s is

$$n_s = \frac{F_u - F_b}{F_t} \qquad (17\text{--}49)$$

Be careful when comparing recommended static factors of safety to Eq. (17–49), as n_s is sometimes defined as F_u/F_t. The fatigue factor of safety n_f is

$$n_f = \frac{F_f - F_b}{F_t} \qquad (17\text{--}50)$$

EXAMPLE 17–6

A temporary construction elevator is to be designed to carry workers and materials to a height of 27 m. The maximum estimated load to be hoisted is 22 kN at a velocity not to exceed 0.6 m/s. For minimum sheave diameters and acceleration of 1.2 m/s^2, specify the number of ropes required if the 25-mm plow-steel 6 × 19 hoisting strand is used.

Solution

Since this is a design task, a decision set is useful.

A priori decisions:

- Function: load, height, acceleration, velocity, life goal
- Design Factor: n_d
- Material: IPS, PS, MPS or other
- Rope: Lay, number of strands, number of wires per strand

Decision variables:

- Nominal wire size: d
- Number of load-supporting wires: m

From experience with Prob. 17–29, a 25-mm diameter rope is not likely to have much of a life, so approach the problem with the d and m decisions open.

Function: 22 kN load, 27 m lift, acceleration = 1.2 m/s^2, velocity = 0.6 m/s, life goal = 10^5 cycles

Design Factor: $n_d = 2$

Material: IPS

Rope: Regular lay, 25-mm plow-steel 6 × 19 hoisting

Design variables

Choose 750 mm D_{min}. Table 17–27: $w = 0.0362d^2$ N/m

$$wl = 0.0362d^2 (27) = 0.253d^2 \text{ N, ea.}$$

Eq. (17–46):

$$F_t = \left(\frac{W}{m} + wl\right)\left(1 + \frac{a}{g}\right) = \left(\frac{22\,000}{m} + 0.253d^2\right)\left(1 + \frac{1.2}{9.81}\right)$$

$$= \frac{24\,691}{m} + 0.284d^2 \text{ N}, \quad \text{each wire}$$

Eq. (17–47):

$$F_f = \frac{(p/S_u)S_u Dd}{2}$$

From Fig. 17–21 for 10^5 cycles $p/S_u = 0.004$; $S_u = 1655$ MPa, based on metal area.

$$F_f = \frac{0.004(1655)(750)d}{2} = 2482d \text{ N} \quad \text{each wire}$$

Eq. (17–48) and Table 17–27:

$$F_b = \frac{E_w d_w A_m}{D} = \frac{83\,000(0.067d)(0.4d^2)}{750} = 2.97d^3 \text{ N}, \quad \text{each wire}$$

Eq. (17–45):

$$n_f = \frac{F_f - F_b}{F_t} = \frac{2482d - 2.97d^3}{(24\,691/m) + 0.284d^2}$$

We could use a computer program to build a table similar to that of Ex. 17–6. Alternatively, we could recognize that $0.284d^2$ is small compared to $24\,691/m$, and therefore eliminate the $0.284d^2$ term.

$$n_f \doteq \frac{2482d - 2.97d^3}{24\,691/m} = \frac{m}{24\,691}(2482d - 2.97d^3)$$

Maximize n_f,

$$\frac{\partial n_f}{\partial d} = 0 = \frac{m}{24\,691}[2482 - 3(2.97)d^2]$$

From which

$$d* = \sqrt{\frac{2482}{8.91}} = 16.7 \text{ mm}$$

Back-substituting

$$n_f = \frac{m}{24\,691}[2482(16.7) - 2.97(16.7)^3] =$$

Thus $n_f = 1.12, 2.24, 3.35, 4.47$ for $m = 1, 2, 3, 4$ respectively. If we choose $d = 1.25$ mm, then $m = 2$.

$$n_f = \frac{2482(12.5) - 2.97(12.5)^3}{(24\,691/2) + 0.284(12.5)^2} = 2.036$$

This slightly less than $n_d = 2$

Decisions #1: d = 12.5 mm

Answer   *Decisions #2: m = 2* ropes supporting load. Rope should be inspected weekly for any signs of fatigue (broken outer wires).

Comment: Table 17–25 gives n for freight elevators in terms of velocity.

$$F_u = (S_u)_{\text{nom}} A_{\text{nom}} = 730\left(\frac{\pi d^2}{4}\right) = 573.3d^2 \text{ N}, \quad \text{each wire}$$

$$n = \frac{F_u}{F_t} = \frac{573.3(12.5)^2}{(24\,691/2) + 2.97(12.5)^2} = 7.0$$

By comparison, interpolation for 0.6 m/s gives 7.08-close. The category of construction hoists is not addressed in Table 17–25. We should investigate this before proceeding further.

17–7   Flexible Shafts

One of the greatest limitations of the solid shaft is that it cannot transmit motion or power around corners. It is therefore necessary to resort to belts, chains, or gears, together with bearings and the supporting framework associated with them. The flexible shaft may often be an economical solution to the problem of transmitting motion around corners. In addition to the elimination of costly parts, its use may reduce noise considerably.

There are two main types of flexible shafts: the power-drive shaft for the transmission of power in a single direction, and the remote-control or manual-control shaft for the transmission of motion in either direction.

The construction of a flexible shaft is shown in Fig. 17–24. The cable is made by winding several layers of wire around a central core. For the power-drive shaft, rotation should be in a direction such that the outer layer is wound up. Remote-control cables have a different lay of the wires forming the cable, with more wires in each layer, so that the torsional deflection is approximately the same for either direction of rotation.

Flexible shafts are rated by specifying the torque corresponding to various radii of curvature of the casing. A 380-mm radius of curvature, for example, will give from 2 to 5 times more torque capacity than a 175-mm radius. When flexible shafts are used

Figure 17–24

Flexible shaft: (*a*) construction details; (*b*) a variety of configurations. *(Courtesy of S. S. White Technologies, Inc.)*

Mandrel   First Layer   Last Layer
(4 Wires)   (7 Wires)

(*a*)

(*b*)

in a drive in which gears are also used, the gears should be placed so that the flexible shaft runs at as high a speed as possible. This permits the transmission of the maximum amount of horsepower.

PROBLEMS

17–1   Example 17–2 resulted in selection of a 10-in-wide A-3 polyamide flat belt. Show that the value of F_1 restoring f to 0.80 is

$$F_1 = \frac{(\Delta F + F_c) \exp f\phi - F_c}{\exp f\phi - 1}$$

and compare the initial tensions.

17–2   A 150-mm-wide polyamide F-1 flat belt is used to connect a 50-mm-diameter pulley to drive a larger pulley with an angular velocity ratio of 0.5. The center-to-center distance is 2.7 m. Theangular speed of the small pulley is 1750 rev/min as it delivers 1.5 kW. The service is such that a service factor K_s of 1.25 is appropriate.

(*a*) Find F_c, F_i, F_{1a}, and F_2.
(*b*) Find H_a, n_{fs}, and belt length.
(*c*) Find the dip.

17–3   Perspective and insight can be gained by doubling all geometric dimensions and observing the effect on problem parameters. Take the drive of Prob. 17–2, double the dimensions, and compare.

17–4 A flat-belt drive is to consist of two 1.2-m-diameter cast-iron pulleys spaced 4.8 m apart. Select a belt type to transmit 45 kW at a pulley speed of 380 rev/min. Use a service factor of 1.1 and a design factor of 1.0.

17–5 In solving problems and examining examples, you probably have noticed some recurring forms:

$$w = 12\gamma bt = (12\gamma t)b = a_1 b$$

$$(F_1)_a = F_a b C_p C_v = (F_a C_p C_v)b = a_0 b$$

$$F_c = \frac{wV^2}{g} = \frac{a_1 b}{32.174}\left(\frac{V}{60}\right)^2 = a_2 b$$

$$(F_1)_a - F_2 = 2T/d = 33\,000 H_d/V = 33\,000 H_{\text{nom}} K_s n_d/V$$

$$F_2 = (F_1)_a - [(F_1)_a - F_2] = a_0 b - 2T/d$$

$$f\phi = \ln\frac{(F_1)_a - F_c}{F_2 - F_c} = \ln\frac{(a_0 - a_2)b}{(a_0 - a_2)b - 2T/d}$$

Show that

$$b = \frac{1}{a_0 - a_2}\frac{33\,000 H_d}{V}\frac{\exp(f\phi)}{\exp(f\phi) - 1}$$

17–6 Return to Ex. 17–1 and complete the following.
(a) What is the minimum initial tension, F_i, that would put the drive as built at the point of slip?
(b) With the tension from part a, find the belt width b that exhibits $n_{fs} = n_d = 1.1$.
(c) For the belt width from part b find the corresponding $(F_1)_a$, F_c, F_i, F_2, power, and n_{fs}, assuming operation at the tension limit.
(d) What have you learned?

17–7 Take the drive of Example 17–1 and double the belt width. Compare F_c, F_i, $(F_1)_a$, F_2, H_a, n_{fs}, and dip, assuming operation at the tension limit.

17–8 Belted pulleys place loads on shafts, inducing bending and loads on bearings. Examine Fig. 17–7 and develop an expression for the load the belt places on the pulley, and then apply it to Ex. 17–2.

17–9 The line shaft illustrated in the figure is used to transmit power from an electric motor by means of flat-belt drives to various machines. Pulley A is driven by a vertical belt from the motor pulley. A belt from pulley B drives a machine tool at an angle of 70° from the vertical and at a center-to-center distance of 2.7 m. Another belt from pulley C drives a grinder at a center-to-center distance of 3.4 m. Pulley C has a double width to permit belt shifting as shown in Fig. 17–4. The belt from pulley D drives a dust-extractor fan whose axis is located horizontally .4 m from the axis of the lineshaft. Additional data are

Machine	Speed, rev/min	Power, kW	Lineshaft Pulley	Diameter, mm
Machine tool	400	9.3	B	400
Grinder	300	3.4	C	350
Dust extractor	500	6.0	D	450

Problem 17–9
(Courtesy of Dr. Ahmed F. Abdel
Azim, Zagazig University, Cairo.)

From *D*

From *C*

From *B*

70°   60°

3 m

Motor pulley:
Dia. = 300 mm
Speed = 900 rev/min

8 ft

The power requirements, listed above, account for the overall efficiencies of the equipment. The two line-shaft bearings are mounted on hangers suspended from two overhead wide-flange beams. Select the belt types and sizes for each of the four drives. Make provision for replacing belts from time to time because of wear or permanent stretch.

17–10 Two shafts 6 m apart, with axes in the same horizontal plane, are to be connected with a flat belt in which the driving pulley, powered by a six-pole squirrel-cage induction motor with a 75-kW rating at 1140 rev/min, drives the second shaft at half its angular speed. The driven shaft drives light-shock machinery loads. Select a flat belt.

17–11 The mechanical efficiency of a flat-belt drive is approximately 98 percent. Because of its high value, the efficiency is often neglected. If a designer should choose to include it, where would he or she insert it in the flat-belt protocol?

17–12 In metal belts, the centrifugal tension F_c is ignored as negligible. Convince yourself that this is a reasonable problem simplification.

17–13 A designer has to select a metal-belt drive to transmit a power of H_{nom} under circumstances where a service factor of K_s and a design factor of n_d are appropriate. The design goal becomes $H_d = H_{nom}K_s n_d$. Use Eq. (17–8) with negligible centrifugal force to show that the minimum belt width is given by

$$b_{min} = \frac{1}{a}\left(\frac{33\,000H_d}{V}\right)\frac{\exp f\theta}{\exp f\theta - 1}$$

where a is the constant from $(F_1)_a = ab$.

17–14 Design a friction metal flat-belt drive to connect a 1-kW, four-pole squirrel-cage motor turning at 1750 rev/min to a shaft 380 mm away, running at half speed. The circumstances are such that a service factor of 1.2 and a design factor of 1.05 are appropriate. The life goal is 10^6 belt passes, $f = 0.35$, and the environmental considerations require a stainless steel belt.

17–15 A beryllium-copper metal flat belt with $S_f = 390$ MPa is to transmit 3.7 kW at 1125 rev/min with a life goal of 10^6 belt passes between two shafts 20 in apart whose centerlines are in a horizontal plane. The coefficient of friction between belt and pulley is 0.32. The conditions are such that a service factor of 1.25 and a design factor of 1.1 are appropriate. The driven shaft rotates at one-third the motor-pulley speed. Specify your belt, pulley sizes, and initial tension at installation.

17–16   For the conditions of Prob. 17–15 use a 1095 plain carbon-steel heat-treated belt. Conditions at the driving pulley hub require a pulley outside diameter of 75 mm or more. Specify your belt, pulley sizes, and initial tension at installation.

17–17   A single V belt is to be selected to deliver engine power to the wheel-drive transmission of a ridingtractor. A 3.7285 kW single-cylinder engine is used. At most, 60 percent of this power is transmitted tothe belt. The driving sheave has a diameter of 157.48 mm, the driven, 304.8 mm. The belt selected shouldbe as close to a 2336.8 mm pitch length as possible. The engine speed is governor-controlled to a maxi-mum of 324.63 rad/s. Select a satisfactory belt and assess the factor of safety and the belt life inpasses.

17–18   Two B2125 V belts are used in a drive composed of a 135-mm driving sheave, rotating at 1200 rev/min, and a 400-mm driven sheave. Find the power capacity of the drive based on a service factor of 1.25, and find the center-to-center distance.

17–19   A 45-kW four-cylinder internal combustion engine is used to drive a medium-shock brick-making machine under a schedule of two shifts per day. The drive consists of two 650-mm sheaves spaced about 3.6 m apart, with a sheave speed of 400 rev/min. Select a Gates Rubber V-belt arrangement. Find the factor of safety, and estimate the life in passes and hours.

17–20   A reciprocating air compressor has a 1.5-m-diameter flywheel 350 mm wide, and it operates at 170 rev/min. An eight-pole squirrel-cage induction motor has nameplate data 37 kW at 875 rev/min.

(a) Design V-belt drive

(b) Can cutting the V-belt grooves in the flywheel be avoided by using a V-flat drive?

17–21   The geometric implications of a V-flat drive are interesting.

(a) If the earth's equator was an inextensible string, snug to the spherical earth, and you spliced 1.8 m of string into the equatorial cord and arranged it to be concentric to the equator, how far off the ground is the string?

(b) Using the solution to part a, formulate the modifications to the expressions for m_G, θ_d and θ_D, L_p, and C.

(c) As a result of this exercise, how would you revise your solution to part b of Prob. 17–20?

17–22   A 1.5-kW electric motor running at 1720 rev/min is to drive a blower at a speed of 240 rev/min. Select a V-belt drive for this application and specify standard V belts, sheave sizes, and the resulting center-to-center distance. The motor size limits the center distance to at least 550 mm.

17–23   The standard roller-chain number indicates the chain pitch in inches, construction proportions, series, and number of strands as follows:

This convention makes the pitch directly readable from the chain number. In Ex. 17–5 ascertain the pitch from the selected chain number and confirm from Table 17–19.

17–24   Equate Eqs. (17–32) and (17–33) to find the rotating speed n_1 at which the power equates and marks the division between the premaximum and the postmaximum power domains.
(a) Show that

$$n_1 = \left[\frac{0.25(10^6) K_r N_1^{0.42}}{p^{(2.2-0.07p)}} \right]^{1/2.4}$$

(b) Find the speed n_1 for a no. 60 chain, $p = 19$ mm, $N_1 = 17$, $K_r = 17$, and confirm from Table 17–20.
(c) At which speeds is Eq. (17–40) applicable?

17–25   A double-strand no. 60 roller chain is used to transmit power between a 13-tooth driving sprocket rotating at 300 rev/min and a 52-tooth driven sprocket.
(a) What is the allowable horsepower of this drive?
(b) Estimate the center-to-center distance if the chain length is 82 pitches.
(c) Estimate the torque and bending force on the driving shaft by the chain if the actual horsepower transmitted is 30 percent less than the corrected (allowable) power.

17–26   A four-strand no. 40 roller chain transmits power from a 21-tooth driving sprocket to an 84-tooth driven sprocket. The angular speed of the driving sprocket is 2000 rev/min.
(a) Estimate the chain length if the center-to-center distance has to be about 508 mm.
(b) Neglecting chain length effects, estimate the tabulated horsepower entry H'_{tab} for a 20 000-h life goal.
(c) Estimate the allowable horsepower for a 20 000-h life.
(d) Estimate the tension in the chain at the allowable power.

17–27   A 700 rev/min 18.65-kW squirrel-cage induction motor is to drive a two-cylinder reciprocating pump, out-of-doors under a shed. A service factor K_s of 1.5 and a design factor of 1.1 are appropriate. The pump speed is 140 rev/min. Select a suitable chain and sprocket sizes.

17–28   A centrifugal pump is driven by a 37.3-kW synchronous motor at a speed of 1800 rev/min. The pump is to operate at 900 rev/min. Despite the speed, the load is smooth ($K_s = 1.2$). For a design factor of 1.1 specify a chain and sprockets that will realize a 50 000-h life goal. Let the sprockets be 19T and 38T.

17–29   A mine hoist uses a 50-mm 6 × 19 monitor-steel wire rope. The rope is used to haul loads of 36 kN from the shaft 145 m deep. The drum has a diameter of 1.8 m, the sheaves are of good-quality cast steel, and the smallest is 1 m in diameter.
(a) Using a maximum hoisting speed of 6 m/s and a maximum acceleration of 0.6 m/s², estimate the stresses in the rope.
(b) Estimate the various factors of safety.

17–30   A temporary construction elevator is to be designed to carry workers and materials to a height of 27.43 m. The maximum estimated load to be hoisted is 22.24 kN at a velocity not to exceed 0.61 m/s. For minimum sheave diameters and acceleration of 1.22 m/s², specify the number of ropes required if the 12.7 mm plow-steel 15.24 × 48.26 cm hoisting strand is used.

17–31   A 0.61 km mine hoist operates with a 1.83 mm drum using 15.24 × 48.26 cm monitor-steel wire rope. The cage and load weigh 35.59 kN, and the cage is subjected to an acceleration of 0.61 m/s² when starting.
(a) For a single-strand hoist how does the factor of safety $n_f = F_f/F_t$, neglecting bending, vary with the choice of rope diameter?
(b) For four supporting strands of wire rope attached to the cage, how does the factor of safety vary with the choice of rope diameter?

17–32 Generalize the results of Prob. 17–31 by representing the factor of safety n as

$$n_f = \frac{ad}{(b/m) + cd^2}$$

where m is the number of ropes supporting the cage, and a, b, and c are constants. Show that the optimal diameter is $d^* = [b/(mc)]^{1/2}$ and the corresponding maximum attainable factor of safety is $n_f^* = a[m/(bc)]^{1/2}/2$.

17–33 From your results in Prob. 17–32, show that to meet a fatigue factor of safety n_1 the optimal solution is

$$m = \frac{4bcn_1}{a^2} \text{ ropes}$$

having a diameter of

$$d = \frac{a}{2cn_1}$$

Solve Prob. 17–31 if a factor of safety of 2 is required. Show what to do in order to accommodate to the necessary discreteness in the rope diameter d and the number of ropes m.

17–34 For Prob. 17–29 estimate the elongation of the rope if a 31.14 kN loaded mine cart is placed on the cage which weighs 4.45 kN. The results of Prob. 4–7 may be useful.

Computer Programs

In approaching the ensuing computer problems, the following suggestions may be helpful:

- Decide whether an analysis program or a design program would be more useful. In problems as simple as these, you will find the programs similar. For maximum instructional benefit, try the design problem.

- Creating a design program without a figure of merit precludes ranking alternative designs but does not hinder the attainment of satisfactory designs. Your instructor can provide the class design library with commercial catalogs, which not only have price information but define available sizes.

- Quantitative understanding and logic of interrelations are required for programming. Difficulty in programming is a signal to you and your instructor to increase your understanding. The following programs can be accomplished in 100 to 500 lines of code.

- Make programs interactive and user-friendly.

- Let the computer do what it can do best; the user should do what a human can do best.

- Assume the user has a copy of the text and can respond to prompts for information.

- If interpolating in a table is in order, solicit table entries in the neighborhood, and let the computer crunch the numbers.

- In decision steps, allow the user to make the necessary decision, even if it is undesirable. This allows learning of consequences and the use of the program for analysis.

- Display a lot of information in the summary. Show the decision set used up-front for user perspective.

- When a summary is complete, adequacy assessment can be accomplished with ease, so consider adding this feature.

17–35 Your experience with Probs. 17–1 through 17–11 has placed you in a position to write an interactive computer program to design/select flat-belt drive components. A possible decision set is

A Priori Decisions

- Function: H_{nom}, rev/min, velocity ratio, approximate C
- Design factor: n_d
- Initial tension maintenance: catenary
- Belt material: t, d_{min}, allowable tension, density, f
- Drive geometry: d, D
- Belt thickness: t (in material decision)

Design Decisions

- Belt width: b

17–36 Problems 17–12 through 17–16 have given you some experience with flat metal friction belts, indicating that a computer program could be helpful in the design/selection process. A possible decision set is

A Priori Decisions

- Function: H_{nom}, rev/min, velocity ratio approximate C
- Design factor: n_d
- Belt material: S_y, E, ν, d_{min}
- Drive geometry: d, D
- Belt thickness: t

Design Decisions

- Belt width: b
- Length of belt (often stan dard loop periphery)

17–37 Problems 17–17 through 17–22 have given you enough experience with V belts to convince you that a computer program would be helpful in the design/selection of V-belt drive components. Write such a program.

17–38 Experience with Probs. 17–23 through 17–28 can suggest an interactive computer program to help in the design/selection process of roller-chain elements. A possible decision set is

A Priori Decisions

- Function: power, speed, space, K_s, life goal
- Design factor: n_d
- Sprocket tooth counts: N_1, N_2, K_1, K_2

Design Decisions

- Chain number
- Strand count
- Lubrication system
- Chain length in pitches

(center-to-center distance for reference)

18

Power Transmission Case Study

925

Transmission of power from a source, such as an engine or motor, through a machine to an output actuation is one of the most common machine tasks. An efficient means of transmitting power is through rotary motion of a shaft that is supported by bearings. Gears, belt pulleys, or chain sprockets may be incorporated to provide for torque and speed changes between shafts. Most shafts are cylindrical (solid or hollow), and include stepped diameters with shoulders to accommodate the positioning and support of bearings, gears, etc.

The design of a system to transmit power requires attention to the design and selection of individual components (gears, bearings, shaft, etc.). However, as is often the case in design, these components are not independent. For example, in order to design the shaft for stress and deflection, it is necessary to know the applied forces. If the forces are transmitted through gears, it is necessary to know the gear specifications in order to determine the forces that will be transmitted to the shaft. But stock gears come with certain bore sizes, requiring knowledge of the necessary shaft diameter. It is no surprise that the design process is interdependent and iterative, but where should a designer start?

The nature of machine design textbooks is to focus on each component separately. This chapter will focus on an overview of a power transmission system design, demonstrating how to incorporate the details of each component into an overall design process. A typical two-stage gear reduction such as shown in Fig. 18–1 will be assumed for this discussion. The design sequence is similar for variations of this particular transmission system.

The following outline will help clarify a logical design sequence. Discussion of how each part of the outline affects the overall design process will be given in sequence in this chapter. Details on the specifics for designing and selecting major components are covered in separate chapters, particularly Chap. 7 on shaft design, Chap. 11 on bearing selection, and Chaps. 13 and 14 on gear specification. A complete case study is presented as a specific vehicle to demonstrate the process.

Figure 18–1

A compound reverted gear train.

CASE STUDY PART 1
PROBLEM SPECIFICATION

Section 1–17, p. 24, presents the background for this case study involving a speed reducer. A two-stage, compound reverted gear train such as shown in Fig. 18–1 will be designed. In this chapter, the design of the intermediate shaft and its components is presented, taking into account the other shafts as necessary.

A subset of the pertinent design specifications that will be needed for this part of the design are given here.

Power to be delivered: 15 kW
Input speed: 1750 rpm
Output speed: 82–88 rev/min
Usually low shock levels, occasional moderate shock
Input and output shafts extend 100 mm outside gearbox
Maximum gearbox size: 350-mm × 350-mm base, 550-mm height
Output shaft and input shaft in-line
Gear and bearing life > 12 000 hours; infinite shaft life

18–1 Design Sequence for Power Transmission

There is not a precise sequence of steps for any design process. By nature, design is an iterative process in which it is necessary to make some tentative choices, and to build a skeleton of a design, and to determine which parts of the design are critical. However, much time can be saved by understanding the dependencies between the parts of the problem, allowing the designer to know what parts will be affected by any given change. In this section, only an outline is presented, with a short explanation of each step. Further details will be discussed in the following sections.

- *Power and torque requirements.* Power considerations should be addressed first, as this will determine the overall sizing needs for the entire system. Any necessary speed or torque ratio from input to output must be determined before addressing gear/pulley sizing.

- *Gear specification.* Necessary gear ratios and torque transmission issues can now be addressed with selection of appropriate gears. Note that a full force analysis of the shafts is not yet needed, as only the transmitted loads are required to specify the gears.

- *Shaft layout.* The general layout of the shaft, including axial location of gears and bearings must now be specified. Decisions on how to transmit the torque from the gears to the shaft need to be made (keys, splines, etc.), as well as how to hold gears and bearings in place (retaining rings, press fits, nuts, etc.). However, it is not necessary at this point to size these elements, since their standard sizes allow estimation of stress-concentration factors.

- *Force analysis.* Once the gear/pulley diameters are known, and the axial locations of the gears and bearings are known, the free-body, shear force, and bending moment diagrams for the shafts can be produced. Forces at the bearings can be determined.

- *Shaft material selection.* Since fatigue design depends so heavily on the material choice, it is usually easier to make a reasonable material selection first, then check for satisfactory results.

- *Shaft design for stress (fatigue and static).* At this point, a stress design of the shaft should look very similar to a typical design problem from the shaft chapter (Chap. 7). Shear force and bending moment diagrams are known, critical locations can be predicted, approximate stress concentrations can be used, and estimates for shaft diameters can be determined.

- *Shaft design for deflection.* Since deflection analysis is dependent on the entire shaft geometry, it is saved until this point. With all shaft geometry now estimated, the critical deflections at the bearing and gear locations can be checked by analysis.

- *Bearing selection.* Specific bearings from a catalog may now be chosen to match the estimated shaft diameters. The diameters can be adjusted slightly as necessary to match the catalog specifications.

- *Key and retaining ring selection.* With shaft diameters settling in to stable values, appropriate keys and retaining rings can be specified in standard sizes. This should make little change in the overall design if reasonable stress-concentration factors were assumed in previous steps.

- *Final analysis.* Once everything has been specified, iterated, and adjusted as necessary for any specific part of the task, a complete analysis from start to finish will provide a final check and specific safety factors for the actual system.

18–2 Power and Torque Requirements

Power transmission systems will typically be specified by a power capacity, for example, a 40-horsepower gearbox. This rating specifies the combination of torque and speed that the unit can endure. Remember that, in the ideal case, *power in* equals *power out*, so that we can refer to the power being the same throughout the system. In reality, there are small losses due to factors like friction in the bearings and gears. In many transmission systems, the losses in the rolling bearings will be negligible. Gears have a reasonably high efficiency, with about 1 to 2 percent power loss in a pair of meshed gears. Thus, in the double-reduction gearbox in Fig. 18–1, with two pairs of meshed gears the output power is likely to be about 2 to 4 percent less than the input power. Since this is a small loss, it is common to speak of simply the power of the system, rather than input power and output power. Flat belts and timing belts have efficiencies typically in the mid to upper 90 percent range. V belts and worm gears have efficiencies that may dip much lower, requiring a distinction between the necessary input power to obtain a desired output power.

Torque, on the other hand, is typically not constant throughout a transmission system. Remember that power equals the product of torque and speed. Since *power in = power out*, we know that for a gear train

$$H = T_i \omega_i = T_o \omega_o \qquad (18-1)$$

With a constant power, a gear ratio to decrease the angular velocity will simultaneously increase torque. The gear ratio, or train value, for the gear train is

$$e = \omega_o / \omega_i = T_i / T_o \qquad (18-2)$$

A typical power transmission design problem will specify the desired power capacity, along with either the input and output angular velocities, or the input and output torques. There will usually be a tolerance specified for the output values. After the specific gears are specified, the actual output values can be determined.

18–3 Gear Specification

With the gear train value known, the next step is to determine appropriate gears. As a rough guideline, a train value of up to 10 to 1 can be obtained with one pair of gears. Greater ratios can be obtained by compounding additional pairs of gears (See Sec. 13–13, p. 690). The compound reverted gear train in Fig. 18–1 can obtain a train value of up to 100 to 1.

Since numbers of teeth on gears must be integers, it is best to design with teeth numbers rather than diameters. See Ex. 13–3, 13–4, and 13–5, pp. 692–694, for details on designing appropriate numbers of teeth to satisfy the gear train value and any necessary geometry condition, such as in-line condition of input and output shaft. Care should be taken at this point to find the best combination of teeth numbers to minimize the overall package size. If the train value only needs to be approximate, use this flexibility to try different options of tooth numbers to minimize the package size. A difference of one tooth on the smallest gear can result in a significant increase in size of the overall package.

If designing for large production quantities, gears can be purchased in large enough quantities that it is not necessary to worry about preferred sizes. For small lot production, consideration should be given to the tradeoffs between smaller gearbox size and extra cost for odd gear sizes that are difficult to purchase off the shelf. If stock gears are to be used, their availability in prescribed numbers of teeth with anticipated diametral pitch should be checked at this time. If necessary, iterate the design for numbers of teeth that are available.

CASE STUDY PART 2
SPEED, TORQUE, AND GEAR RATIOS

Continue the case study by determining appropriate tooth counts to reduce the input speed of $\omega_i = 1750$ rev/min to an output speed within the range

$$82 \text{ rev/min} < \omega_o < 88 \text{ rev/min}$$

Once final tooth counts are specified, determine values of
   (a) Speeds for the intermediate and output shafts
   (b) Torques for the input, intermediate and output shafts, to transmit 16 kW.

Solution

Use the notation for gear numbers from Fig. 18–1. Choose mean value for initial design, $\omega_5 = 85$ rev/min.

$$e = \frac{\omega_5}{\omega_2} = \frac{85}{1750} = \frac{1}{20.59} \qquad \text{Eq. (18–2)}$$

For a compound reverted geartrain,

$$e = \frac{1}{20.59} = \frac{N_2}{N_3} \frac{N_4}{N_5} \qquad \text{Eq. (13–30), p. 699}$$

For smallest package size, let both stages be the same reduction. Also, by making the two stages identical, the in-line condition on the input and output shaft will automatically be satisfied.

$$\frac{N_2}{N_3} = \frac{N_4}{N_5} = \sqrt{\frac{1}{20.59}} = \frac{1}{4.54}$$

For this ratio, the minimum number of teeth from Eq. (13-11), p. 686, is 16.

$$N_2 = N_4 = 16 \text{ teeth}$$

$$N_3 = 4.54(N_2) = 72.64$$

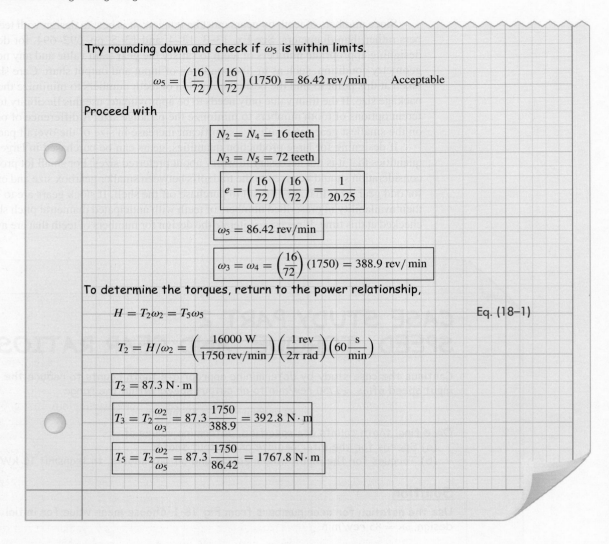

Try rounding down and check if ω_5 is within limits.

$$\omega_5 = \left(\frac{16}{72}\right)\left(\frac{16}{72}\right)(1750) = 86.42 \text{ rev/min} \qquad \text{Acceptable}$$

Proceed with

$$N_2 = N_4 = 16 \text{ teeth}$$

$$N_3 = N_5 = 72 \text{ teeth}$$

$$e = \left(\frac{16}{72}\right)\left(\frac{16}{72}\right) = \frac{1}{20.25}$$

$$\omega_5 = 86.42 \text{ rev/min}$$

$$\omega_3 = \omega_4 = \left(\frac{16}{72}\right)(1750) = 388.9 \text{ rev/min}$$

To determine the torques, return to the power relationship,

$$H = T_2\omega_2 = T_5\omega_5 \qquad\qquad \text{Eq. (18–1)}$$

$$T_2 = H/\omega_2 = \left(\frac{16000 \text{ W}}{1750 \text{ rev/min}}\right)\left(\frac{1 \text{ rev}}{2\pi \text{ rad}}\right)\left(60\frac{\text{s}}{\text{min}}\right)$$

$$T_2 = 87.3 \text{ N} \cdot \text{m}$$

$$T_3 = T_2\frac{\omega_2}{\omega_3} = 87.3\frac{1750}{388.9} = 392.8 \text{ N} \cdot \text{m}$$

$$T_5 = T_2\frac{\omega_2}{\omega_5} = 87.3\frac{1750}{86.42} = 1767.8 \text{ N} \cdot \text{m}$$

If a maximum size for the gearbox has been specified in the problem specification, a minimum diametral pitch (maximum tooth size) can be estimated at this point by writing an expression for the gearbox size in terms of gear diameters, and converting to numbers of teeth through the diametral pitch. For example, from Fig. 18–1, the overall height of the gearbox is

$$Y = d_3 + d_2/2 + d_5/2 + 2/P + \text{clearances} + \text{wall thicknesses}$$

where the $2/P$ term accounts for the addendum height of the teeth on gears 3 and 5 that extend beyond the pitch diameters. Substituting $d_i = N_i/P$ gives

$$Y = N_3/P + N_2/(2P) + N_5/(2P) + 2/P + \text{clearances} + \text{wall thicknesses}$$

Solving this for P, we find

$$P = (N_3 + N_2/2 + N_5/2 + 2)/(Y - \text{clearances} - \text{wall thicknesses}) \quad \text{(18–3)}$$

This is the minimum value that can be used for diametral pitch, and therefore the maximum tooth size, to stay within the overall gearbox constraint. It should be rounded *up* to the next standard diametral pitch, which reduces the maximum tooth size.

The AGMA approach, as described in Chap. 14, for both bending and contact stress should be applied next to determine suitable gear parameters. The primary design parameters to be specified by the designer include material, diametral pitch, and face width. A recommended procedure is to start with an estimated diametral pitch. This allows determination of gear diameters ($d = N/P$), pitch-line velocities [Eq. (13–34), p. 699], and transmitted loads [Eq. (13–35) or (13–36), p. 699]. Typical spur gears are available with face widths from 3 to 5 times the circular pitch p. Using an average of 4, a first estimate can be made for face width $F = 4p = 4\pi/P$. Alternatively, the designer can simply perform a quick search of on-line gear catalogs to find available face widths for the diametral pitch and number of teeth.

Next, the AGMA equations in Chap. 14 can be used to determine appropriate material choices to provide desired safety factors. It is generally most efficient to attempt to analyze the most critical gear first, as it will determine the limiting values of diametral pitch and material strength. Usually, the critical gear will be the smaller gear, on the high-torque (low-speed) end of the gearbox.

If the required material strengths are too high, such that they are either too expensive or not available, iteration with a smaller diametral pitch (larger tooth) will help. Of course, this will increase the overall gearbox size. Often the excessive stress will be in one of the small gears. Rather than increase the tooth size for all gears, it is sometimes better to reconsider the design of tooth counts, shifting more of the gear ratio to the pair of gears with less stress, and less ratio to the pair of gears with the excessive stress. This will allow the offending gear to have more teeth and therefore larger diameter, decreasing its stress.

If contact stress turns out to be more limiting than bending stress, consider gear materials that have been heat treated or case hardened to increase the surface strength. Adjustments can be made to the diametral pitch if necessary to achieve a good balance of size, material, and cost. If the stresses are all much lower than the material strengths, a larger diametral pitch is in order, which will reduce the size of the gears and the gearbox.

Everything up to this point should be iterated until acceptable results are obtained, as this portion of the design process can usually be accomplished independently from the next stages of the process. The designer should be satisfied with the gear selection before proceeding to the shaft. Selection of specific gears from catalogs at this point will be helpful in later stages, particularly in knowing overall width, bore size, recommended shoulder support, and maximum fillet radius.

CASE STUDY PART 3
GEAR SPECIFICATION

Continue the case study by specifying appropriate gears, including pitch diameter, diametral pitch, face width, and material. Achieve safety factors of at least 1.2 for wear and bending.

Solution

Estimate the minimum diametral pitch for overall gearbox height = 22 in.

From Eq. (18-3) and Fig. 18-1,

$$P_{min} = \frac{\left(N_3 + \dfrac{N_2}{2} + \dfrac{N_5}{2} + 2\right)}{(Y - \text{clearances} - \text{wall thickness})}$$

Allow 1.5 in for clearances and wall thicknesses:

$$P_{min} = \frac{\left(72 + \dfrac{16}{2} + \dfrac{72}{2} + 2\right)}{(22 - 1.5)} = 5.76 \text{ teeth/in}$$

Start with $P = 6$ teeth/in

$$d_2 = d_4 = N_2/P = 16/6 = 2.67 \text{ in}$$
$$d_3 = d_5 = 72/6 = 12.0 \text{ in}$$

Shaft speeds were previously determined to be

$$\omega_2 = 1750 \text{ rev/min} \qquad \omega_3 = \omega_4 = 388.9 \text{ rev/min} \qquad \omega_5 = 86.4 \text{ rev/min}$$

Get pitch-line velocities and transmitted loads for later use.

$$V_{23} = \frac{\pi d_2 \omega_2}{12} = \frac{\pi (2.67)(1750)}{12} = 1223 \text{ ft/min} \qquad \text{Eq. (13–34), p. 707}$$

$$V_{45} = \frac{\pi d_5 \omega_5}{12} = 271.5 \text{ ft/min}$$

$$W_{23}^t = 33\,000 \frac{H}{V_{23}} = 33\,000\left(\frac{20}{1223}\right) = 540.0 \text{ lbf} \qquad \text{Eq. (13–35), p. 707}$$

$$W_{45}^t = 33\,000 \frac{H}{V_{45}} = 2431 \text{ lbf}$$

Start with gear 4, since it is the smallest gear, transmitting the largest load. It will likely be critical. Start with wear by contact stress, since it is often the limiting factor.

Gear 4 Wear

$$I = \frac{\cos 20° \sin 20°}{2(1)}\left(\frac{4.5}{4.5 + 1}\right) = 0.1315 \qquad \text{Eq. (14–23), p. 755}$$

For K_v, assume $Q_v = 7$. $B = 0.731$, $A = 65.1$ Eq. (14–29), p. 756

$$K_v = \left(\frac{65.1 + \sqrt{271.5}}{65.1}\right)^{0.731} = 1.18 \qquad \text{Eq. (14–27), p. 756}$$

Face width F is typically from 3 to 5 times circular pitch. Try

$$F = 4\left(\frac{\pi}{P}\right) = 4\left(\frac{\pi}{6}\right) = 2.09 \text{ in.}$$

Since gear specifications are readily available on the Internet, we might as well check for commonly available face widths. On www.globalspec.com, entering $P = 6$ teeth/in and $d = 2.67$ in, stock spur gears from several sources have face widths of 1.5 in or 2.0 in. These are also available for the meshing gear 5 with $d = 12$ in.
   Choose $F = 2.0$ in.

For K_m, $C_{pf} = 0.0624$ Eq. (14–32), p. 760

$$C_{mc} = 1 \text{ uncrowned teeth} \qquad \text{Eq. (14–31), p. 760}$$

$$C_{pm} = 1 \text{ straddle-mounted} \qquad \text{Eq. (14–33), p. 760}$$

$$C_{ma} = 0.15 \text{ commercial enclosed unit} \qquad \text{Eq. (14–34), p. 760}$$

$$C_e = 1 \qquad \text{Eq. (14–35), p. 760}$$

$$K_m = 1.21 \qquad \text{Eq. (14–30), p. 759}$$

$$C_p = 2300 \qquad \text{Table 14–8, p. 757}$$

$$K_o = K_s = C_f = 1$$

$$\sigma_c = 2300\sqrt{\frac{2431(1.18)(1.21)}{2.67(2)(0.1315)}} = \underline{161\,704\,\text{psi}} \qquad \text{Eq. (14–16), p. 746}$$

Get factors for $\sigma_{c.\text{all}}$. For life factor Z_N, get number of cycles for specified life of 12 000 h.

$$L_4 = (12\,000\,\text{h})\left(60\frac{\text{min}}{\text{h}}\right)\left(389\frac{\text{rev}}{\text{min}}\right) = 2.8\times10^8\,\text{rev}$$

$$Z_N = 0.9 \qquad \text{Fig. 14–15, p. 763}$$

$$K_R = K_T = C_H = 1$$

For a design factor of 1.2,

$$\sigma_{c.\text{all}} = S_c Z_N / S_H = \sigma_c \qquad \text{Eq. (14–18), p. 750}$$

$$S_c = \frac{S_H \sigma_c}{Z_N} = \frac{1.2(161\,700)}{0.9} = \underline{215\,605\,\text{psi}}$$

From Table 14-6, p. 751, this strength is achievable with Grade 2 carburized and hardened with $S_c = 225\,000$ psi. To find the achieved factor of safety, $n_c = \sigma_{c,\text{all}}/\sigma_c$ with $S_H = 1$. The factor of safety for wear of gear 4 is

$$n_c = \frac{\sigma_{c,\text{all}}}{\sigma_c} = \frac{S_c Z_N}{\sigma_c} = \frac{225\,000(0.9)}{161\,700} = \underline{1.25}$$

Gear 4 Bending

$$J = 0.27 \qquad \text{Fig. 14–6, p. 753}$$

$$K_B = 1$$

Everything else is the same as before.

$$\sigma = W_t K_v \frac{P_d}{F}\frac{K_m}{J} = (2431)(1.18)\left(\frac{6}{2}\right)\left(\frac{1.21}{0.27}\right) \qquad \text{Eq. (14–15), p. 746}$$

$$\sigma = \underline{38\,566\,\text{psi}}$$

$$Y_N = 0.9 \qquad \text{Fig. 14–14, p. 763}$$

Using Grade 2 carburized and hardened, same as chosen for wear, find $S_t = 65\,000$ psi (Table 14–3, p. 748).

$$\sigma_{\text{all}} = S_t Y_N = 58\,500 \text{ psi}$$

The factor of safety for bending of gear 4 is

$$n = \frac{\sigma_{\text{all}}}{\sigma} = \frac{58\,500}{38\,570} = 1.52$$

Gear 5 Bending and Wear

Everything is the same as for gear 4, except $J, Y_N,$ and Z_N.

$$J = 0.41 \qquad\qquad\qquad\qquad\qquad \text{Fig. 14–6, p. 753}$$

$$L_5 = (12\,000\text{h})(60 \text{ min/h})(86.4 \text{ rev/min}) = 6.2 \times 10^7 \text{ rev}$$

$$Y_N = 0.97 \qquad\qquad\qquad\qquad\qquad \text{Fig. 14–14, p. 763}$$

$$Z_N = 1.0 \qquad\qquad\qquad\qquad\qquad \text{Fig. 14–15, p. 763}$$

$$\sigma_c = 2300\sqrt{\frac{2431(1.18)(1.21)}{2.67(2)(0.1315)}} = 161\,704 \text{ psi}$$

$$\sigma = (2431)(1.18)\left(\frac{6}{2}\right)\left(\frac{1.21}{0.41}\right) = 25\,397 \text{ psi}$$

Choose Grade 2 carburized and hardened, the same as gear 4

$$n_c = \frac{\sigma_{c.\text{all}}}{\sigma_c} = \frac{225\,000}{161\,704} = 1.39$$

$$n = \frac{\sigma_{\text{all}}}{\sigma} = \frac{65\,000(0.97)}{25\,397} = 2.48$$

Gear 2 Wear

Gears 2 and 3 are evaluated similarly. Only selected results are shown.

$$K_v = 1.37$$

Try $F = 1.5$ in, since the loading is less on gears 2 and 3.

$$K_m = 1.19$$

All other factors are the same as those for gear 4.

$$\sigma_c = 2300\sqrt{\frac{(539.7)(1.37)(1.19)}{2.67(1.5)(0.1315)}} = 94\,010 \text{ psi}$$

$$L_2 = (12\,000 \text{ h})(60 \text{ min/h})(1750 \text{ rev/min}) = 1.26 \times 10^9 \text{ rev} \qquad Z_N = 0.8$$

Try grade 1 flame-hardened, $S_c = 170\,000$ psi

$$n_c = \frac{\sigma_{c.\text{all}}}{\sigma_c} = \frac{170\,000(0.8)}{94\,000} = 1.45$$

Gear 2 Bending

$$J = 0.27 \qquad Y_N = 0.88$$

$$\sigma = 539.7(1.37)\frac{(6)(1.19)}{(1.5)(0.27)} = 13\,035 \text{ psi}$$

$$n = \frac{\sigma_{all}}{\sigma} = \frac{45\,000(0.88)}{13\,040} = 3.04$$

Gear 3 Wear and Bending

$$J = 0.41 \qquad Y_N = 0.9 \qquad Z_N = 0.9$$

$$\sigma_c = 2300\sqrt{\frac{(539.7)(1.37)(1.19)}{2.67(1.5)(0.1315)}} = 94\,000 \text{ psi}$$

$$\sigma = 539.7(1.37)\frac{(6)(1.19)}{1.5(0.41)} = 8584 \text{ psi}$$

Try Grade 1 steel, through-hardened to 300 H_B. From Fig. 14-2, p. 747, $S_t = 36\,000$ psi and from Fig. 14-5, p. 750, $S_c = 126\,000$ psi.

$$n_c = \frac{126\,000(0.9)}{94\,000} = 1.21$$

$$n = \frac{\sigma_{all}}{\sigma} = \frac{36\,000(0.9)}{8584} = 3.77$$

In summary, the resulting gear specifications are:

All gears, $P = 6$ teeth/in

Gear 2, Grade 1 flame-hardened, $S_c = 170\,000$ psi and $S_t = 45\,000$ psi
$d_2 = 2.67$ in, face width $= 1.5$ in

Gear 3, Grade 1 through-hardened to 300 H_B, $S_c = 126\,000$ psi and $S_t = 36\,000$ psi
$d_3 = 12.0$ in, face width $= 1.5$ in

Gear 4, Grade 2 carburized and hardened, $S_c = 225\,000$ psi and $S_t = 65\,000$ psi
$d_4 = 2.67$ in, face width $= 2.0$ in

Gear 5, Grade 2 carburized and hardened, $S_c = 225\,000$ psi and $S_t = 65\,000$ psi
$d_5 = 12.0$ in, face width $= 2.0$ in

18–4 Shaft Layout

The general layout of the shafts, including axial location of gears and bearings, must now be specified in order to perform a free-body force analysis and to obtain shear force and bending moment diagrams. If there is no existing design to use as a starter, then the determination of the shaft layout may have many solutions. Section 7–3, p. 353, discusses the issues involved in shaft layout. In this section the focus will be on how the decisions relate to the overall process.

A free-body force analysis can be performed without knowing shaft diameters, but can not be performed without knowing axial distances between gears and bearings. It is extremely important to keep axial distances small. Even small forces can create large bending moments if the moment arms are large. Also, recall that beam deflection equations typically include length terms raised to the third power.

It is worth examining the entirety of the gearbox at this time, to determine what factors drive the length of the shaft and the placement of the components. A rough sketch, such as shown in Fig. 18–2, is sufficient for this purpose.

CASE STUDY PART 4
SHAFT LAYOUT

Continue the case study by preparing a sketch of the gearbox sufficient to determine the axial dimensions. In particular, estimate the overall length, and the distance between the gears of the intermediate shaft, in order to fit with the mounting requirements of the other shafts.

Solution

Fig. 18–2 shows the rough sketch. It includes all three shafts, with consideration of how the bearings are to mount in the case. The gear widths are known at this point. Bearing widths are guessed, allowing a little more space for larger bearings on the intermediate shaft where bending moments will be greater. Small changes in bearing widths will have minimal effect on the force analysis, since the location of the ground reaction force will change very little. The 100-mm distance between the two gears on the countershaft is dictated by the requirements of the input and output shafts, including the space for the case to mount the bearings. Small allotments are given for the retaining rings, and for space behind the bearings. Adding it all up gives the intermediate shaft length as 290 mm.

Figure 18–2

Sketch for shaft layout. Dimensions are in millimeters.

Wider face widths on gears require more shaft length. Originally, gears with hubs were considered for this design to allow the use of set screws instead of high-stress-concentration retaining rings. However, the extra hub lengths added several inches to the shaft lengths and the gearbox housing.

Several points are worth noting in the layout in Fig. 18–2. The gears and bearings are positioned against shoulders, with retaining rings to hold them in position. While it is desirable to place gears near the bearings, a little extra space is provided between them to accommodate any housing that extends behind the bearing, and to allow for a bearing puller to have space to access the back of the bearing. The extra change in diameter between the bearings and the gears allows the shoulder height for the bearing and the bore size for the gear to be different. This diameter can have loose tolerances and a large fillet radius.

Each bearing is restrained axially on its shaft, but only one bearing on each shaft is axially fixed in the housing, allowing for slight axial thermal expansion of the shafts.

18–5 Force Analysis

Once the gear diameters are known, and the axial locations of the components are set, the free-body diagrams and shear force and bending moment diagrams for the shafts can be produced. With the known transmitted loads, determine the radial and axial loads transmitted through the gears (see Secs. 13–14 through 13–17, pp. 697–706). From summation of forces and moments on each shaft, ground reaction forces at the bearings can be determined. For shafts with gears and pulleys, the forces and moments will usually have components in two planes along the shaft. For rotating shafts, usually only the resultant magnitude is needed, so force components at bearings are summed as vectors. Shear force and bending moment diagrams are usually obtained in two planes, then summed as vectors at any point of interest. A torque diagram should also be generated to clearly visualize the transfer of torque from an input component, through the shaft, and to an output component.

See the beginning of Ex. 7–2, p. 366, for the force analysis portion of the case study for the intermediate shaft. The bending moment is largest at gear 4. This is predictable, since gear 4 is smaller, and must transmit the same torque that entered the shaft through the much larger gear 3.

While the force analysis is not difficult to perform manually, if beam software is to be used for the deflection analysis, it will necessarily calculate reaction forces, along with shear force and bending moment diagrams in the process of calculating deflections. The designer can enter guessed values for diameters into the software at this point, just to get the force information, and later enter actual diameters to the same model to determine deflections.

18–6 Shaft Material Selection

A trial material for the shaft can be selected at any point before the stress design of the shaft, and can be modified as necessary during the stress design process. Section 7–2, p. 352, provides details for decisions regarding material selection. For the case study, an inexpensive steel, 1020 CD, is initially selected. After the stress analysis, a slightly higher strength 1050 CD is chosen to reduce the critical stresses without further increasing the shaft diameters.

18–7 Shaft Design for Stress

The critical shaft diameters are to be determined by stress analysis at critical locations. Section 7–4, p. 358, provides a detailed examination of the issues involved in shaft design for stress.

CASE STUDY PART 5
DESIGN FOR STRESS

Proceed with the next phase of the case study design, in which appropriate diameters for each section of the shaft are estimated, based on providing sufficient fatigue and static stress capacity for infinite life of the shaft, with minimum safety factor of 1.5.

Solution

The solution to this phase of the design is presented in Ex. 7–2, p. 366.

Since the bending moment is highest at gear 4, potentially critical stress points are at its shoulder, keyway, and retaining ring groove. It turns out that the keyway is the critical location. It seems that shoulders often get the most attention. This example demonstrates the danger of neglecting other stress concentration sources, such as keyways.

The material choice was changed in the course of this phase, choosing to pay for a higher strength to limit the shaft diameter to 2 in. If the shaft were to get much bigger, the small gear would not be able to provide an adequate bore size. If it becomes necessary to increase the shaft diameter any more, the gearing specification will need to be redesigned.

18–8 Shaft Design for Deflection

Section 7–5, p. 371, provides a detailed discussion of deflection considerations for shafts. Typically, a deflection problem in a shaft will not cause catastrophic failure of the shaft, but will lead to excess noise and vibration, and premature failure of the gears or bearings.

CASE STUDY PART 6
DEFLECTION CHECK

Proceed with the next phase of the case study by checking that deflections and slopes at the gears and bearings on the intermediate shaft are within acceptable ranges.

Solution

The solution to this phase of the design is presented in Ex. 7–3, p. 372.

It turns out that in this problem all the deflections are within recommended limits for bearings and gears. This is not always the case, and it would be a poor choice to neglect the deflection analysis. In a first iteration of this case study, with longer shafts due to using gears with hubs, the deflections were more critical than the stresses.

18–9 Bearing Selection

Bearing selection is straightforward now that the bearing reaction forces and the approximate bore diameters are known. See Chap. 11 for general details on bearing selection. Rolling-contact bearings are available with a wide range of load capacities and dimensions, so it is usually not a problem to find a suitable bearing that is close to the estimated bore diameter and width.

CASE STUDY PART 7
BEARING SELECTION

Continue the case study by selecting appropriate bearings for the intermediate shaft, with a reliability of 99 percent. The problem specifies a design life of 12 000 h. The intermediate shaft speed is 389 rev/min. The estimated bore size is 25 mm, and the estimated bearing width is 25 mm.

Solution

From the free-body diagram (see Ex. 7-2, p. 366),

$$R_{Az} = 422 \text{ N} \qquad R_{Ay} = 1439 \text{ N} \qquad R_A = 1500 \text{ N}$$

$$R_{Bg} = 8822 \text{ N} \qquad R_{By} = 3331 \text{ N} \qquad R_B = 9430 \text{ N}$$

At the shaft speed of 389 rev/min, the design life of 12 000 h correlates to a bearing life of $L_D = (12\,000\,\text{h})(60\,\text{min/h})(389\,\text{rev/min}) = 2.8 \times 10^8$ rev.

Start with bearing B since it has the higher loads and will likely raise any lurking problems. From Eq. (11-7), p. 578, assuming a ball bearing with $a = 3$ and $L = 2.8 \times 10^6$ rev,

$$F_{RB} = 9430 \left[\frac{2.8 \times 10^8 / 10^6}{0.02 + 4.439(1 - 0.99)^{1/1.483}} \right]^{1/3} = 102.4 \text{ kN}$$

A check on the Internet for available bearings (www.globalspec.com is one good starting place) shows that this load is relatively high for a ball bearing with bore size in the neighborhood of 25 mm. Try a cylindrical roller bearing. Recalculating F_{RB} with the exponent $a = 3/10$ for roller bearings, we obtain

$$F_{RB} = 80.7 \text{ kN}$$

Cylindrical roller bearings are available from several sources in this range. A specific one is chosen from SKF, a common supplier of bearings, with the following specifications:

Cylindrical roller bearing at right end of shaft
$C = 83$ kN, ID $= 30$ mm, OD $= 72$ mm, $W = 27$ mm
Shoulder diameter $= 37$ mm to 39 mm, and maximum fillet radius $= 1.1$ mm

For bearing A, again assuming a ball bearing,

$$F_{RA} = 1500\left[\frac{2.8 \times 10^8/10^6}{0.02 + 4.439(1-0.99)^{1/1.483}}\right]^{1/3} = 16.3 \text{ kN}$$

A specific ball bearing is chosen from the SKF Internet catalog.

Deep-groove ball bearing at left end of shaft

$C = 20.3$ kN, ID = 25 mm, OD = 62 mm, $W = 19$ mm

Shoulder diameter = 32 mm to 35 mm, and maximum fillet radius = 2 mm

At this point, the actual bearing dimensions can be checked against the initial assumptions. For bearing B the bore diameter of 30 mm is slightly larger than the original 25 mm. There is no reason for this to be a problem as long as there is room for the shoulder diameter. The original estimate for shoulder support diameters was 35 mm. As long as this diameter is less than 42 mm, the next step of the shaft, there should not be any problem. In the case study, the recommended shoulder support diameters are within the acceptable range. The original estimates for stress concentration at the bearing shoulder assumed a fillet radius such that $r/d = 0.02$. The actual bearings selected have ratios of 0.036 and 0.080. This allows the fillet radii to be increased from the original design, decreasing the stress-concentration factors.

The bearing widths are close to the original estimates. Slight adjustments should be made to the shaft dimensions to match the bearings. No redesign should be necessary.

18–10 Key and Retaining Ring Selection

The sizing and selection of keys is discussed in Sec. 7–7, p. 380, with an example in Ex. 7–6, p. 386. The cross-sectional size of the key will be dictated to correlate with the shaft size (see Tables 7–6 and 7–8, pp. 383, 385), and must certainly match an integral keyway in the gear bore. The design decision includes the length of the key, and if necessary an upgrade in material choice.

The key could fail by shearing across the key, or by crushing due to bearing stress. For a square key, it turns out that checking only the crushing failure is adequate, since the shearing failure will be less critical according to the distortion energy failure theory, and equal according to the maximum shear stress failure theory. Check Ex. 7–6 to investigate why.

CASE STUDY PART 8
KEY DESIGN

Continue the case study by specifying appropriate keys for the two gears on the intermediate shaft to provide a factor of safety of 2. The gears are to be custom bored and keyed to the required specifications. Previously obtained information includes the following:

Transmitted torque: $T = 350$ N·m

Bore diameters: $d_3 = d_4 = 42$ mm

Gear hub lengths: $l_3 = 38$ mm, $l_4 = 50$ mm

Solution

From Table 7-6, p. 383, for a shaft diameter of 42 mm, choose a square key with side dimension $t = 10$ mm. Choose 1020 CD material, with $S_y = 390$ MPa. The force on the key at the surface of the shaft is

$$F = \frac{T}{r} = \frac{350}{0.042/2} = 16.67 \text{ kN}$$

Checking for failure by crushing, we find the area of one-half the face of the key is used.

$$n = \frac{S_y}{\sigma} = \frac{S_y}{F/(tl/2)}$$

Solving for l gives

$$l = \frac{2Fn}{tS_y} = \frac{2(16670)(2)}{(0.01)(390 \times 10^6)} = 17 \text{ mm}$$

Since both gears have the same bore diameter and transmit the same torque, the same key specification can be used for both.

Retaining ring selection is simply a matter of checking catalog specifications. The retaining rings are listed for nominal shaft diameter, and are available with different axial load capacities. Once selected, the designer should make note of the depth of the groove, the width of the groove, and the fillet radius in the bottom of the groove. The catalog specification for the retaining ring also includes an edge margin, which is the minimum distance to the next smaller diameter change. This is to ensure support for the axial load carried by the ring. It is important to check stress-concentration factors with actual dimensions, as these factors can be rather large. In the case study, a specific retaining ring was already chosen during the stress analysis (see Ex. 7-2, p. 366) at the potentially critical location of gear 4. The other locations for retaining rings were not at points of high stress, so it is not necessary to worry about the stress concentration due to the retaining rings in these locations. Specific retaining rings should be selected at this time to complete the dimensional specifications of the shaft.

For the case study, retaining rings specifications are entered into globalspec, and specific rings are selected from Truarc Co., with the following specifications:

	Both Gears	**Left Bearing**	**Right Bearing**
Nominal Shaft diameter	42 mm	25 mm	30 mm
Groove diameter	38 ± 0.125 mm	24 ± 0.1mm	28 ± 0.1 mm
Groove width	$1.7^{+0.1}_{-0.0}$ mm	$1.2^{+0.1}_{-0.0}$ mm	$1.4^{+0.1}_{-0.0}$ mm
Nominal groove depth	1.2 mm	0.8 mm	0.9 mm
Max groove fillet radius	0.25 mm	0.25 mm	0.25 mm
Minimum edge margin	3.6 mm	2.6 mm	2.6 mm
Allowable axial thrust	52.7 kN	26.7 kN	31.2 kN

These are within the estimates used for the initial shaft layout, and should not require any redesign. The final shaft should be updated with these dimensions.

| Figure 18-3

18–11 **Final Analysis**

At this point in the design, everything seems to check out. Final details include determining dimensions and tolerances for appropriate fits with the gears and bearings. See Sec. 7–8, p. 387, for details on obtaining specific fits. Any small changes from the nominal diameters already specified will have negligible effect on the stress and deflection analysis. However, for manufacturing and assembly purposes, the designer should not overlook the tolerance specification. Improper fits can lead to failure of the design. Lack of attention to tolerance specification can make the part unfunctional or overly expensive to manufacture. Further insight on tolerance specification is given in Sec. 1–14, p. 27. The final drawing for the intermediate shaft is shown in Fig. 18–3. This drawing shows the important dimensions and dimensional tolerances in a form that is generally considered satisfactory for small production quantities where direct attention is given to manufacturing methods. A more robust method of part specification that also addresses the allowed variations from perfect form (e.g., straightness or concentricity) is known as Geometric Dimensioning and Tolerancing and is introduced in Chap. 20.

For documentation purposes, and for a check on the design work, the design process should conclude with a complete analysis of the final design. Remember that analysis is much more straightforward than design, so the investment of time for the final analysis will be relatively small.

PROBLEMS

18–1 For the case study problem, design the input shaft, including complete specification of the gear, bearings, key, retaining rings, and shaft.

18–2 For the case study problem, design the output shaft, including complete specification of the gear, bearings, key, retaining rings, and shaft.

18–3 For the case study problem, use helical gears and design the intermediate shaft. Compare your results with the spur gear design presented in this chapter.

18–4 Perform a final analysis for the resulting design of the intermediate shaft of the case study problem presented in this chapter. Produce a final drawing with dimensions and tolerances for the shaft. Does the final design satisfy all the requirements? Identify the critical aspects of the design with the lowest factor of safety.

18–5 For the case study problem, change the power requirement to 40 horsepower. Design the intermediate shaft, including complete specification of the gears, bearings, keys, retaining rings, and shaft.

PART 4 **Special Topics**

19

Finite-Element Analysis

Chapter Outline

Mechanical components in the form of simple bars, beams, etc., can be analyzed quite easily by basic methods of mechanics that provide closed-form solutions. Actual components, however, are rarely so simple, and the designer is forced to less effective approximations of closed-form solutions, experimentation, or numerical methods. There are a great many numerical techniques used in engineering applications for which the digital computer is very useful. In mechanical design, where computer-aided design (CAD) software is heavily employed, the analysis method that integrates well with CAD is *finite-element analysis* (FEA). The mathematical theory and applications of the method are vast. There are also a number of commercial FEA software packages that are available, such as ANSYS, NASTRAN, ALGOR, etc.

The purpose of this chapter is only to expose the reader to some of the fundamental aspects of FEA, and therefore the coverage is extremely introductory in nature. For further detail, the reader is urged to consult the many references cited at the end of this chapter. Figure 19–1 shows a finite-element model of a connecting rod that was developed to study the effects of dynamic elastohydrodynamic lubrication on bearing and structural performance.[1]

There are a multitude of FEA applications such as static and dynamic, linear and nonlinear, stress and deflection analysis; free and forced vibrations; heat transfer (which can be combined with stress and deflection analysis to provide thermally induced stresses and deflections); elastic instability (buckling); acoustics; electrostatics

Figure 19–1

Model of a connecting rod using ANSYS finite-element software. (*a*) Meshed model; (*b*) stress contours. *Courtesy of S. Boedo (see footnote 1).*

(a)

(b)

---

[1]S. Boedo, "Elastohydrodynamic Lubrication of Conformal Bearing Systems," *Proceedings of 2002 ANSYS Users Conference,* Pittsburgh, PA, April 22–24, 2002.

and magnetics (which can be combined with heat transfer); fluid dynamics; piping analysis; and multiphysics. For the purposes of this chapter, we will limit ourselves to basic mechanics analyses.

An actual mechanical component is a continuous elastic structure (continuum). FEA divides (discretizes) the structure into small but finite, well-defined, elastic substructures (elements). By using polynomial functions, together with matrix operations, the continuous elastic behavior of each element is developed in terms of the element's material and geometric properties. Loads can be applied within the element (gravity, dynamic, thermal, etc.), on the surface of the element, or at the *nodes* of the element. The element's nodes are the fundamental governing entities of the element, as it is the node where the element connects to other elements, where elastic properties of the element are eventually established, where boundary conditions are assigned, and where forces (contact or body) are ultimately applied. A node possesses *degrees of freedom* (dof's). Degrees of freedom are the independent translational and rotational motions that can exist at a node. At most, a node can possess three translational and three rotational degrees of freedom. Once each element within a structure is defined *locally* in matrix form, the elements are then *globally* assembled (attached) through their common nodes (dof's) into an overall system matrix. Applied loads and boundary conditions are then specified and through matrix operations the values of all unknown displacement degrees of freedom are determined. Once this is done, it is a simple matter to use these displacements to determine strains and stresses through the constitutive equations of elasticity.

19–1   The Finite-Element Method

The modern development of the finite-element method began in the 1940s in the field of structural mechanics with the work of Hrennikoff,[2] McHenry,[3] and Newmark,[4] who used a lattice of line elements (rods and beams) for the solution of stresses in continuous solids. In 1943, from a 1941 lecture, Courant[5] suggested piecewise polynomial interpolation over triangular subregions as a method to model torsional problems. With the advent of digital computers in the 1950s it became practical for engineers to write and solve the stiffness equations in matrix form.[6,7,8] A classic paper by Turner, Clough, Martin, and Topp published in 1956 presented the matrix stiffness equations for the

---

[2]A. Hrennikoff, "Solution of Problems in Elasticity by the Frame Work Method," *Journal of Applied Mechanics,* Vol. 8, No. 4, pp. 169–175, December 1941.

[3]D. McHenry, "A Lattice Analogy for the Solution of Plane Stress Problems," *Journal of Institution of Civil Engineers,* Vol. 21, pp. 59–82, December 1943.

[4]N. M. Newmark, "Numerical Methods of Analysis in Bars, Plates, and Elastic Bodies," *Numerical Methods in Analysis in Engineering* (ed. L. E. Grinter), Macmillan, 1949.

[5]R. Courant, "Variational Methods for the Solution of Problems of Equilibrium and Vibrations," *Bulletin of the American Mathematical Society,* Vol. 49, pp. 1–23, 1943.

[6]S. Levy, "Structural Analysis and Influence Coefficients for Delta Wings," *Journal of Aeronautical Sciences,* Vol. 20, No. 7, pp. 449–454, July 1953.

[7]J. H. Argyris, "Energy Theorems and Structural Analysis," *Aircraft Engineering,* October, November, December 1954 and February, March, April, May 1955.

[8]J. H. Argyris and S. Kelsey, *Energy Theorems and Structural Analysis,* Butterworths, London, 1960 (reprinted from *Aircraft Engineering,* 1954–55).

truss, beam, and other elements.[9] The expression *finite element* is first attributed to Clough.[10] Since these early beginnings, a great deal of effort has been expended in the development of the finite element method in the areas of element formulations and computer implementation of the entire solution process. The major advances in computer technology include the rapidly expanding computer hardware capabilities, efficient and accurate matrix solver routines, and computer graphics for ease in the visual preprocessing stages of model building, including automatic adaptive mesh generation, and in the postprocessing stages of reviewing the solution results. A great abundance of literature has been presented on the subject, including many textbooks. A partial list of some textbooks, introductory and more comprehensive, is given at the end of this chapter.

Since the finite-element method is a numerical technique that discretizes the domain of a continuous structure, errors are inevitable. These errors are:

1   **Computational errors.** These are due to round-off errors from the computer floating-point calculations and the formulations of the numerical integration schemes that are employed. Most commercial finite-element codes concentrate on reducing these errors, and consequently the analyst generally is concerned with discretization factors.

2   **Discretization errors.** The geometry and the displacement distribution of a true structure continuously vary. Using a finite number of elements to model the structure introduces errors in matching geometry and the displacement distribution due to the inherent mathematical limitations of the elements.

For an example of discretization errors, consider the constant thickness, thin plate structure shown in Fig. 19–2a. Figure 19–2b shows a finite-element model of the

(a) (b)

Figure 19–2

Structural problem. (a) Idealized model; (b) finite-element model.

---

[9]M. J. Turner, R. W. Clough, H. C. Martin, and L. J. Topp, "Stiffness and Deflection Analysis of Complex Structures," *Journal of Aeronautical Sciences,* Vol. 23, No. 9, pp. 805–824, September 1956.

[10]R. W. Clough, "The Finite Element Method in Plane Stress Analysis," *Proceedings of the Second Conference on Electronic Computation,* American Society of Civil Engineers, Pittsburgh, PA, pp. 345–378, September 1960.

structure where three-node, plane stress, simplex triangular elements are employed. This element type has a flaw that creates two basic problems. The element has straight sides that remain straight after deformation. The strains throughout the plane stress triangular element are constant. The first problem, a geometric one, is the modeling of curved edges. Note that the surface of the model with a large curvature appears poorly modeled, whereas the surface of the hole seems to be reasonably modeled. The second problem, which is much more severe, is that the strains in various regions of the actual structure are changing rapidly, and the constant strain element will provide only an approximation of the average strain at the center of the element. So, in a nutshell, the results predicted by this model will be extremely poor. The results can be improved by significantly increasing the number of elements (increased mesh density). Alternatively, using a better element, such as an eight-node quadrilateral, which is more suited to the application, will provide the improved results. Because of higher-order interpolation functions, the eight-node quadrilateral element can model curved edges and provides a higher-order function for the strain distribution.

In Fig. 19–2b, the triangular elements are shaded and the nodes of the elements are represented by the black dots. Forces and constraints can be placed only at the nodes. The nodes of the simplex triangular plane stress elements have only two degrees of freedom, corresponding to translation in the plane. Thus, the solid black, simple support triangles on the left edge represent the fixed support of the model. Also, the distributed load can be applied only to three nodes as shown. The modeled load has to be statically consistent with the actual load.

19–2 Element Geometries

Many geometric shapes of elements are used in finite-element analysis for specific applications. The various elements used in a general-purpose commercial FEA software code constitute what is referred to as the *element library* of the code. Elements can be placed in the following categories: *line elements, surface elements, solid elements,* and *special-purpose elements.* Table 19–1 provides some, but not all, of the

Table 19–1

Sample Finite-Element Library

Element Type	None	Shape	Number of Nodes	Applications
Line	Truss		2	Pin-ended bar in tension or compression
	Beam		2	Bending
	Frame		2	Axial, torsional, and bending. With or without load stiffening.

(Continued)

Figure 19–1 (*Continued*)

Element Type	None	Shape	Number of Nodes	Applications
Surface	4-node quadri- lateral		4	Plane stress or strain, axisymmetry, shear panel, thin flat plate in bending
	8-node quadri- lateral		8	Plane stress or strain, thin plate or shell in bending
	3-node triangular		3	Plane stress or strain, axisymmetry, shear panel, thin flat plate in bending. Prefer quad where possible. Used for transitions of quads.
	6-node triangular		6	Plane stress or strain, axisymmetry, thin plate or shell in bending. Prefer quad where possible. Used for transitions of quads.
Solid[†]	8-node hexagonal (brick)		8	Solid, thick plate
	6-node pentagonal (wedge)		6	Solid, thick plate. Used for transitions.
	4-node tetrahedron (tet)		4	Solid, thick plate. Used for transitions.
Special purpose	Gap		2	Free displacement for prescribed compressive gap
	Hook		2	Free displacement for prescribed extension gap
	Rigid		Variable	Rigid constraints between nodes

[†]These elments are also available with midside nodes.

types of elements available for finite-element analysis for structural problems. Not all elements support all degrees of freedom. For example, the 3-D truss element supports only three translational degrees of freedom at each node. Connecting elements with differing dof's generally requires some manual modification. For example, consider connecting a truss element to a frame element. The frame element supports all six dof's at each node. A truss member, when connected to it, can rotate freely at the connection.

19–3 The Finite-Element Solution Process

We will describe the finite-element solution process on a very simple one-dimensional problem, using the linear truss element. A truss element is a bar loaded in tension or compression and is of constant cross-sectional area A, length l, and elastic modulus E. The basic truss element has two nodes, and for a one-dimensional problem, each node will have only one degree of freedom. A truss element can be modeled as a simple linear spring with a spring rate, given by Eq. (4–4), p. 163, as

$$k = \frac{AE}{l} \tag{19–1}$$

Consider a spring element (e) of spring rate k_e, with nodes i and j, as shown in Fig. 19–3. Nodes and elements will be numbered. So, to avoid confusion as to what a number corresponds to, elements will be numbered within parentheses. Assuming all forces f and displacements u directed toward the right as positive, the forces at each node can be written as

$$f_{i,e} = k_e(u_i - u_j) = k_e u_i - k_e u_j$$
$$f_{j,e} = k_e(u_j - u_i) = -k_e u_i + k_e u_j \tag{19–2}$$

The two equations can be written in matrix form as

$$\left\{ \begin{matrix} f_{i,e} \\ f_{i,e} \end{matrix} \right\} = \begin{bmatrix} k_e & -k_e \\ -k_e & k_e \end{bmatrix} \left\{ \begin{matrix} u_i \\ u_j \end{matrix} \right\} \tag{19–3}$$

Next, consider a two-spring system as shown in Fig. 19–4a. Here we have numbered the nodes and elements. We have also labeled the forces at each node. However, these forces are the total external forces at each node, F_1, F_2, and F_3. If we draw separate free-body diagrams we will expose the internal forces as shown in Fig. 19–4b.

Figure 19–3

A simple spring element.

Figure 19–4

A two-element spring system. (*a*) System model, (*b*) separate free-body diagrams.

Using Eq. (19–3) for each spring gives

Element 1
$$\left\{ \begin{array}{c} f_{1,1} \\ f_{2,1} \end{array} \right\} = \left[\begin{array}{cc} k_1 & -k_1 \\ -k_1 & k_1 \end{array} \right] \left\{ \begin{array}{c} u_1 \\ u_2 \end{array} \right\} \qquad (19\text{–}4a)$$

Element 2
$$\left\{ \begin{array}{c} f_{2,2} \\ f_{3,2} \end{array} \right\} = \left[\begin{array}{cc} k_2 & -k_2 \\ -k_2 & k_2 \end{array} \right] \left\{ \begin{array}{c} u_2 \\ u_3 \end{array} \right\} \qquad (19\text{–}4b)$$

The total force at each node is the external force, $F_1 = f_{1,1}$, $F_2 = f_{2,1} \times f_{2,2}$, and $F_3 = f_{3,2}$. Combining the two matrices in terms of the external forces gives

$$\left\{ \begin{array}{c} f_{1,1} \\ f_{2,1} + f_{2,2} \\ f_3 \end{array} \right\} = \left\{ \begin{array}{c} F_1 \\ F_2 \\ F_3 \end{array} \right\} = \left[\begin{array}{ccc} k_1 & -k_1 & 0 \\ -k_1 & (k_1 + k_2) & -k_2 \\ 0 & -k_2 & k_2 \end{array} \right] \left\{ \begin{array}{c} u_1 \\ u_2 \\ u_3 \end{array} \right\} \qquad (19\text{–}5)$$

If we know the displacement of a node, then the force at the node will be unknown. For example, in Fig. 19–4a, the displacement of node 1 at the wall is zero, so F_1 is the unknown reaction force (note, up to this point, we have not applied a static solution of the system). If we do not know the displacement of a node, then we know the force. For example, in Fig. 19–4a, the displacements at nodes 2 and 3 are unknown, and the forces F_2 and F_3 are to be specified. To see how the remainder of the solution process can be implemented, let us consider the following example.

EXAMPLE 19–1 Consider the aluminum step-shaft shown in Fig. 19–5a. The areas of sections AB and BC are 64.5 mm² and 96.8 mm², respectively. The lengths of sections AB and BC are 250 mm and 300 mm, respectively. A force $F = 4500$ N is applied to B. Initially, a gap of $\varepsilon = 0.05$ mm exists between end C and the right rigid wall. Determine the wall reactions, the internal forces in the members, and the deflection

Figure 19–5

(a) Step shaft; (b) spring model.

of point B. Let $E = 69$ GPa and assume that end C hits the wall. Check the validity of the assumption.

Solution

The step-shaft is modeled by the two-spring system of Fig. 19–5b where

$$k_1 = \left(\frac{AE}{l}\right)_{AB} = \frac{64.5(69\ 000)}{250} = 17\ 802\ \text{N/mm}$$

$$k_2 = \left(\frac{AE}{l}\right)_{BC} = \frac{96.8(69\ 000)}{300} = 22\ 264\ \text{N/mm}$$

With $u_1 = 0$, $F_2 = 4900$ N and the assumption that $u_3 = \epsilon = 0.05$ mm, Eq. (19-5) becomes

$$\begin{Bmatrix} F_1 \\ 4900 \\ F_3 \end{Bmatrix} = 10^3 \begin{bmatrix} 17.802 & -17.802 & 0 \\ -17.802 & 40.066 & -22.264 \\ 0 & -22.264 & 22.264 \end{bmatrix} \begin{Bmatrix} 0 \\ u_2 \\ 0.05 \end{Bmatrix} \quad (1)$$

For large problems, there is a systematic method of solving equations like Eq. (1), called *partitioning* or *the elimination approach*.[11] However, for this simple problem, the solution is quite simple. From the second equation of the matrix equation

$$4900 = 10^3[-17.802(0) + 40.066u_2 - 22.264(0.05)]$$

or,

Answer

$$u_B = u_2 = \frac{4900/10^3 + 22.264(0.05)}{40.066} = 0.15\ \text{mm}$$

Since $u_B > \epsilon$, it is verified that point C hits the wall.

The reactions at the walls are F_1 and F_3. From the first and third equations of matrix Eq. (1),

Answer

$$F_1 = 10^3[-17.802u_2] = 10^3[-17.802(0.15)] = -2670\ \text{N}$$

---

[11]See T. R. Chandrupatla and A. D. Belegundu, *Introduction to Finite Elements in Engineering*, 4th ed., Prentice Hall, Upper Saddle River, NJ, 2012, pp. 71–75.

and

Answer

$$F_3 = 10^3[-22.264u_2 + 22.264(0.05)]$$

$$= 10^3[-22.264(0.15) + 22.264(0.05)] = -2226 \text{ N}$$

Since F_3 is negative, this also verifies that C hits the wall. Note that $F_1 + F_3 = -2670 - 2226 = -4896$ N, balancing the applied force (with no statics equations necessary).

For internal forces, it is necessary to return to the individual (local) equations. From Eq. (19–4a),

$$\left\{ \begin{array}{c} f_{1,1} \\ f_{2,1} \end{array} \right\} = \left[\begin{array}{cc} k_1 & -k_1 \\ -k_1 & k_1 \end{array} \right] \left\{ \begin{array}{c} u_1 \\ u_2 \end{array} \right\} = 10^3 \left[\begin{array}{cc} 17.802 & -17.802 \\ -17.802 & 17.802 \end{array} \right] \left\{ \begin{array}{c} 0 \\ 0.15 \end{array} \right\} = \left\{ \begin{array}{c} -2670 \\ 2670 \end{array} \right\} \text{N}$$

Answer

Since $f_{1,1}$ is directed to the left and $f_{2,1}$ is directed to the right, the element is in tension, with a force of 2670 N. If the stress is desired, it is simply $\sigma_{AB} = f_{2,1}/A_{AB} = 2670/64.5 = 41.4$ MPa.

For element *BC*, from Eq. (19.4b),

$$\left\{ \begin{array}{c} f_{2,2} \\ f_{3,2} \end{array} \right\} = \left[\begin{array}{cc} k_2 & -k_2 \\ -k_2 & k_2 \end{array} \right] \left\{ \begin{array}{c} u_2 \\ u_3 \end{array} \right\} = 10^5 \left[\begin{array}{cc} 22.264 & -22.264 \\ -22.264 & 22.264 \end{array} \right] \left\{ \begin{array}{c} 0.15 \\ 0.05 \end{array} \right\} = \left\{ \begin{array}{c} 2226 \\ -2226 \end{array} \right\} \text{N}$$

Answer

Since $f_{2,2}$ is directed to the right and $f_{3,2}$ is directed to the left, the element is in compression, with a force of 2226 N. If the stress is desired, it is simply $\sigma_{BC} = -f_{2,2}/A_{BC} = -2226/96.8 = -23$ MPa.

19–4 Mesh Generation

The network of elements and nodes that discretize a region is referred to as a *mesh*. The *mesh density* increases as more elements are placed within a given region. *Mesh refinement* is when the mesh is modified from one analysis of a model to the next analysis to yield improved results. Results generally improve when the mesh density is increased in areas of high stress gradients and/or when geometric transition zones are meshed smoothly. Generally, but not always, the FEA results converge toward the exact results as the mesh is continuously refined. To assess improvement, in regions where high stress gradients appear, the structure can be remeshed with a higher mesh density at this location. If there is a minimal change in the maximum stress value, it is reasonable to presume that the solution has converged. There are three basic ways to generate an element mesh—manually, semiautomatically, or fully automatically.

1 Manual mesh generation. This is how the element mesh was created in the early days of the finite-element method. This is a very labor intensive method of creating the mesh, and except for some quick modifications of a model is it rarely done. *Note:* Care must be exercised in editing an input text file. With some FEA software, other files such as the preprocessor binary graphics file may not change. Consequently, the files may no longer be compatible with each other.

2 **Semiautomatic mesh generation.** Over the years, computer algorithms have been developed that enable the modeler to automatically mesh regions of the structure that he or she has divided up, using well-defined boundaries. Since the modeler has to define these regions, the technique is deemed *semiautomatic*. The development of the many computer algorithms for mesh generation emanates from the field of computer graphics. If the reader desires more information on this subject, a review of the literature available from this field is recommended.

3 **Fully automatic mesh generation.** Many software vendors have concentrated their efforts on developing fully automatic mesh generation, and in some instances, with automatic *self-adaptive* mesh refinement. The obvious goal is to significantly reduce the modeler's preprocessing time and effort to arrive at a final well-constructed FEA mesh. Once the complete boundary of the structure is defined, without subdivisions as in semiautomatic mesh generation and with a minimum of user intervention, various schemes are available to discretize the region with *one element type*. For plane elastic problems the boundary is defined by a series of internal and external geometric lines and the element type to be automeshed would be the plane elastic element. For thin-walled structures, the geometry would be defined by three-dimensional surface representations and the automeshed element type would be the three-dimensional plate element. For solid structures, the boundary could be constructed by using *constructive solid geometry* (*CSG*) or *boundary representation* (*B-rep*) techniques. The finite-element types for automeshing would be the brick and/or tetrahedron element(s).

Automatic self-adaptive mesh refinement programs estimate the error of the FEA solution. On the basis of the error, the mesh is automatically revised and reanalyzed. The process is repeated until some convergence or termination criterion is satisfied.

Returning to the thin-plate model of Fig. 19–2, the boundaries of the structure are constructed as shown in Fig. 19–6a. The boundaries were then automeshed as shown in Fig. 19–6b, where 294 elements and 344 nodes were generated. Note the uniformity of the element generation at the boundaries. The finite-element solver then generated the deflections and von Mises stresses shown in Fig. 19–6c. The maximum von Mises stress at the location shown is 4110.4 psi. The model was then automeshed with an increased mesh density as shown in Fig. 19–6d, where the model has 1008 elements and 1096 nodes. The results are shown in Fig. 19–6e where the maximum von Mises stress is found to be 4184.9 psi, which is only 1.8 percent higher. In all likelihood, the solution has nearly converged. *Note:* The stress contours of Figs. 19–6c and e are better visualized in color.

When stress concentrations are present, it is necessary to have a very fine mesh at the stress-concentration region in order to get realistic results. What is important is that the mesh density needs to be increased only in the region around the stress concentration and that the transition mesh from the rest of the structure to the stress-concentration region be gradual. An abrupt mesh transition, in itself, will have the same effect as a stress concentration. Stress concentration will be discussed further in Sec. 19–7, Modeling Techniques.

Figure 19–6

Automatic meshing the thin-plate model of Fig. 19–2. (*a*) Model boundaries; (*b*) automesh with 294 elements and 344 nodes; (*c*) deflected (exaggerated scale) with stress contours; (*d*) automesh with 1008 elements and 1096 nodes, (*e*) deflected (exaggerated scale) with stress contours.

19–5   **Load Application**

There are two basic forms of specifying loads on a structure—nodal and element loading. However, element loads are eventually applied to the nodes by using equivalent nodal loads. One aspect of load application is related to Saint-Venant's principle. If one is not concerned about the stresses near points of load application, it is not necessary to attempt to distribute the loading very precisely. The net force and/or moment can be applied to a single node, provided the element supports the dof associated with the force and/or moment at the node. However, the analyst should not be surprised, or concerned, when reviewing the results and the stresses in the vicinity of the load application point are found to be very large. Concentrated moments can be applied to the nodes of beam and most plate elements. However, concentrated moments cannot be applied to truss, two-dimensional plane elastic, axisymmetric, or brick elements. They do not support rotational degrees of freedom. A pure moment can be applied to these elements only by using forces in the form of a couple. From the mechanics of statics, a couple can be generated by using two or more forces acting in a plane where the net force from the forces is zero. The net moment from the forces is a vector perpendicular to the plane of the forces and is the summation of the moments from the forces taken about any common point.

Element loads include static loads due to gravity (weight), thermal effects, surface loads such as uniform and hydrostatic pressure, and dynamic loads due to constant acceleration and steady-state rotation (centrifugal acceleration). As stated earlier, element loads are converted by the software to equivalent nodal loads and in the end are treated as concentrated loads applied to nodes.

For gravity loading, the gravity constant in appropriate units and the direction of gravity must be supplied by the modeler. If the model length and force units are inches and lbf, $g = 386.1$ ips^2. If the model length and force units are meters and Newtons, $g = 9.81$ m/s^2. The gravity direction is normally toward the center of the earth.

For thermal loading, the thermal expansion coefficient α must be given for each material, as well as the initial temperature of the structure, and the final nodal temperatures. Most software packages have the capability of first performing a finite-element heat transfer analysis on the structure to determine the final nodal temperatures. The temperature results are written to a file, which can be transferred to the static stress analysis. Here the heat transfer model should have the same nodes and element type the static stress analysis model has.

Surface loading can generally be applied to most elements. For example, uniform or linear transverse line loads (force/length) can be specified on beams. Uniform and linear pressure can normally be applied on the edges of two-dimensional plane and axisymmetric elements. Lateral pressure can be applied on plate elements, and pressure can be applied on the surface of solid brick elements. Each software package has its unique manner in which to specify these surface loads, usually in a combination of text and graphic modes.

19-6 Boundary Conditions

The simulation of boundary conditions and other forms of constraint is probably the single most difficult part of the accurate modeling of a structure for a finite-element analysis. In specifying constraints, it is relatively easy to make mistakes of omission or misrepresentation. It may be necessary for the analyst to test different approaches to model esoteric constraints such as bolted joints, welds, etc., which are not as simple as the idealized pinned or fixed joints. Testing should be confined to simple problems and not to a large, complex structure. Sometimes, when the exact nature of a boundary condition is uncertain, only limits of behavior may be possible. For example, we have modeled shafts with bearings as being simply supported. It is more likely that the support is something between simply supported and fixed, and we could analyze both constraints to establish the limits. However, by assuming simply supported, the results of the solution are conservative for stress and deflections. That is, the solution would predict stresses and deflections larger than the actual.

For another example, consider beam 16 in Table A–9. The horizontal beam is uniformly loaded and is fixed at both ends. Although not explicitly stated, tables such as these assume that the beams are not restrained in the horizontal direction. That is, it is assumed that the beam can slide horizontally in the supports. If the ends were completely or partially restrained, a beam-column solution would be necessary.[12] With a finite-element analysis, a special element, a beam with stiffening, could be used.

Multipoint constraint equations are quite often used to model boundary conditions or rigid connections between elastic members. When used in the latter form, the equations are acting as elements and are thus referred to as *rigid elements*. Rigid elements can rotate or translate only rigidly.

Boundary elements are used to force specific nonzero displacements on a structure. Boundary elements can also be useful in modeling boundary conditions that are askew from the global coordinate system.

19–7 Modeling Techniques

With today's CAD packages and automatic mesh generators, it is an easy task to create a solid model and mesh the volume with finite elements. With today's computing speeds and with gobs of computer memory, it is very easy to create a model with extremely large numbers of elements and nodes. The finite-element modeling techniques of the past now seem passé and unnecessary. However, much unnecessary time can be spent on a very complex model when a much simpler model will do. The complex model may not even provide an accurate solution, whereas a simpler one will. What is important is what solution the analyst is looking for: deflections, stresses, or both?

For example, consider the steel step-shaft of Ex. 4–7, p. 173, repeated here as Fig. 19–7a. Let the fillets at the steps have a radius of 0.5 mm. If only deflections and slopes were sought at the steps, a highly meshed solid model would not yield much more than the simple five-element beam model, shown in Fig. 19–7b, would. The fillets at the steps, which could not be modeled easily with beam elements, would not contribute much to a difference in results between the two models. Nodes are necessary wherever boundary conditions, applied forces, and changes in cross section and/or material occur. The displacement results for the FEA model are shown in Fig. 19–7c.

The FE model of Fig. 19–7b is not capable of providing the stress at the fillet of the step at D. Here, a full-blown solid model would have to be developed and meshed, using solid elements with a high mesh density at the fillet as shown in Fig. 19–8a. Here, the steps at the bearing supports are not modeled, as we are concerned only with the stress concentration at $x = 215$ mm. The brick and tetrahedron elements do not support

---

[12]See R. B. Budynas, *Advanced Strength and Applied Stress Analysis,* 2nd ed., McGraw-Hill, New York, 1999, pp. 471–482.

(a) Dimensions in mm

(b)

Displacements/rotations (degrees) of nodes

Node No.	x Translation	y Translation	z Translation	θ_x Rotation (deg)	θ_y Rotation (deg)	θ_z Rotation (deg)
1	0.0000 E + 00	0.0000 E + 00	0.0000 E + 00	0.0000 E + 00	0.0000 E + 00	−9.7930 E − 02
2	0.0000 E + 00	−8.4951 E − 04	0.0000 E + 00	0.0000 E + 00	0.0000 E + 00	−9.6179 E − 02
3	0.0000 E + 00	−9.3649 E − 03	0.0000 E + 00	0.0000 E + 00	0.0000 E + 00	−7.9874 E − 03
4	0.0000 E + 00	−9.3870 E − 03	0.0000 E + 00	0.0000 E + 00	0.0000 E + 00	2.8492 E − 03
5	0.0000 E + 00	−6.0507 E − 04	0.0000 E + 00	0.0000 E + 00	0.0000 E + 00	6.8558 E − 02
6	0.0000 E + 00	0.0000 E + 00	0.0000 E + 00	0.0000 E + 00	0.0000 E + 00	6.9725 E − 02

(c)

Figure 19–7

(a) Steel step shaft of Ex. 4–7; (b) finite-element model using five beam elements; (c) displacement results for FEA model.

rotational degrees of freedom. To model the simply supported boundary condition at the left end, nodes along the z axis were constrained from translating in the x and y directions. Nodes along the y axis were constrained from translating in the z direction. Nodes on the right end on an axis parallel with the z axis through the center of the shaft were constrained from translating in the y direction, and nodes on an axis parallel with the y axis through the center of the shaft were constrained from translating in the z direction. This ensures no rigid-body translation or rotation and no overconstraint at the ends. The maximum tensile stress at the fillet at the beam bottom is found to be $\sigma_{\max} = 164.8$ MPa. Performing an analytical check at the step yields $D/d = 44/38 = 1.158$, and $r/d = 0.5/38 = 0.0133$. Figure A–15–9 is not very accurate for these values. Resorting

Figure 19–8

(*a*) Solid model of the step-shaft of Ex. 4–7 using 56384 brick and tetrahedron elements; (*b*) view of stress contours at step, rotated 180° about *x* axis, showing maximum tension.

(*a*)

σ_{max} = 164.8 MPa

(*b*)

to another source,[13] the stress-concentration factor is found to be $K_t = 3.00$. The reaction at the right support is $R_F = (200/500)2670 = 1068$ N. The bending moment at the start of the fillet is $M = 1068(288) = 307\ 584$ N · mm. The analytical prediction of the maximum stress is thus

$$\sigma_{max} = K_t \left(\frac{32M}{\pi d^3} \right) = 3.00 \left[\frac{32(307\ 584)}{\pi(38^3)} \right] = 171.3 \text{ MPa}$$

The finite-element model is 4.5 percent lower. If more elements were used in the fillet region, the results would undoubtedly be closer. However, the results are within engineering acceptability.

If we want to check deflections, we should compare the results with the three-element beam model, not the five-element model. This is because we did not model the bearing steps in the solid model. The vertical deflection, at $x = 215$ mm, for the

---

[13]See, W. D. Pilkey and D. F. Pilkey, *Peterson's Stress-Concentration Factors,* 3rd ed. John Wiley & Sons, New York, 2008, Chart 3.11.

solid model was found to be −0.245 mm. This is 4.6 percent higher in magnitude than the −0.235 mm in deflection for the three-element beam model. For slopes, the brick element does not support rotational degrees of freedom, so the rotation at the ends has to be computed from the displacements of adjacent nodes at the ends. This results in the slopes at the ends of $\theta_A = -0.103°$ and $\theta_F = 0.0732°$; these are 6.7 and 6.6 percent higher in magnitude than the three-element beam model, respectively. However, the point of this exercise is, if deflections were the only result desired, which model would you use?

There are countless modeling situations which could be examined. The reader is urged to read the literature, and peruse the tutorials available from the software vendors.[14]

19-8   Thermal Stresses

A heat transfer analysis can be performed on a structural component including the effects of heat conduction, convection, and/or radiation. After the heat transfer analysis is completed, the same model can be used to determine the resulting thermal stresses. For a simple illustration, we will model a 254 mm × 102 mm, 6-mm-thick steel plate with a centered 25-mm-diameter hole. The plate is supported as shown in Fig. 19–9a, and the temperatures of the ends are maintained at temperatures of 38°C and −18°C. Other than at the walls, all surfaces are thermally insulated. Before placing the plate between the walls, the initial temperature of the plate was 0°F. The thermal coefficient of expansion for steel is $\alpha_s = 11.7 \times 10^{-6}$ °C^{-1}. The plate was meshed with 1312 two-dimensional elements, with the mesh refined along the border of the hole. Figure 19–9b shows the temperature contours of the steady-state temperature distribution obtained by the FEA. Using the same elements for a linear stress analysis, where the temperatures were transferred from the heat transfer analysis, Fig. 19–9c shows the resulting stress contours. As expected, the maximum compressive stresses occurred at the top and bottom of the hole, with a magnitude of 220 MPa.

19-9   Critical Buckling Load

Finite elements can be used to predict the *critical buckling load* for a thin-walled structure. An example was shown in Fig. 4–25 (p. 204). Another example can be seen in Fig. 19–10a, which is a thin-walled aluminum beverage can. A specific pressure was applied to the top surface. The bottom of the can was constrained in translation vertically, the center node of the bottom of the can was constrained in translation in all three directions, and one outer node on the can bottom was constrained in translation tangentially. This prevents rigid-body motion, and provides vertical support for the bottom of the can with unconstrained motion of the bottom of the can horizontally. The finite element software returns a value of the load multiplier, which, when multiplied with the total applied force, indicates the critical buckling load. Buckling analysis is an eigenvalue problem, and a reader who reviews a basic mechanics of materials textbook would find

---

[14]See, for example, R. D. Cook, *Finite Element Modeling for Stress Analysis,* Wiley & Sons, New York, 1995; and R. G. Budynas, *Advanced Strength and Applied Stress Analysis,* 2nd ed., McGraw-Hill, New York, 1999, Chap. 10.

(a)

(b)

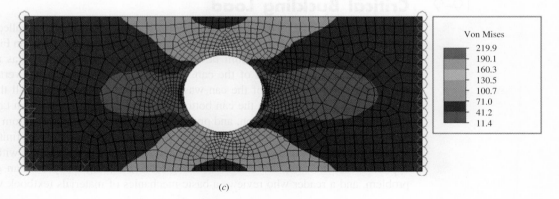

(c)

Figure 19–9

(a) Plate supported at ends and maintained at the temperatures shown; (b) steady-state temperature contours; (c) thermal von Mises stress contours where the initial temperature of the plate was −18°C.

Figure 19–10

(a) Thin-walled aluminum beverage container loaded vertically downward on the top surface; (b) isometric view of the buckled can (deflections greatly exaggerated).

there is a deflection mode shape associated with the critical load. The buckling mode shape for the buckled beverage can is shown in Fig. 19–10b.

19–10 Vibration Analysis

The design engineer may be concerned as to how a component behaves relative to dynamic input, which results in vibration. For vibration, most finite element packages start with a *modal analysis* of the component. This provides the natural frequencies and mode shapes that the component naturally vibrates at. These are called the eigenvalues and eigenvectors of the component. Next, this solution can be transferred (much the same as for thermal stresses) to solvers for forced vibration analyses, such as frequency response, transient impact, or random vibration, to see how the component's modes behave to dynamic input. The mode shape analysis is primarily based on stiffness and the resulting deflections. Thus, similar to static stress analysis, simpler models will suffice. However, if, when solving forced response problems, stresses are desired, a more detailed model is necessary (similar to the shaft illustration given in Sec. 19–7).

A modal analysis of the beam model without the bearing steps was performed for a 20-element beam model,[15] and the 56 384-element brick and tetrahedron model.

---

[15]For static deflection analysis, only three beam elements were necessary. However, because of mass distribution for the dynamics problem, more beam elements are necessary.

Figure 19-11

First free vibration mode of step beam. (*a*) Twenty-element beam model, $f_1 = 322$ Hz; (*b*) 56 384-element brick and tetrahedron model, $f_1 = 316$ Hz.

Needless to say, the beam model took less than 9 seconds to solve, whereas the solid model took *considerably* longer. The first (fundamental) vibration mode was bending and is shown in Fig. 19–11 for both models, together with the respective frequencies. The difference between the frequencies is about 1.9 percent. Further note that the mode shape is just that, a shape. The actual magnitudes of the deflections are unknown, only their relative values are known. Thus, any scale factor can be used to exaggerate the view of the deflection shape.

The convergence of the 20-element beam model was checked by doubling the number of elements. This resulted in no change.

Figure 19–12 provides the frequencies and shapes for the second mode.[16] Here, the difference between the models is 3.6 percent.

As stated earlier, once the mode shapes are obtained, the response of the structure to various dynamic loadings, such as harmonic, transient, or random input, can be obtained. This is accomplished by using the mode shapes together with modal super-position. The method is called *modal analysis*.[17]

19–11   Summary

As stated in Sec. 1–4, the mechanical design engineer has many powerful computational tools available today. Finite-element analysis is one of the most important and is easily integrated into the computer-aided engineering environment. Solid-modeling

---

[16]*Note:* Both models exhibited repeated frequencies and mode shapes for each bending mode. Since the beam and the bearing supports (boundary conditions) are axisymmetric, the bending modes are the same in all transverse planes. So, the second mode shown in Fig. 19–12 is the next unrepeated mode.

[17]See S. S. Rao, *Mechanical Vibrations,* 5th ed., Pearson Prentice Hall, Upper Saddle River, NJ, 2010, Sec. 6.14.

Figure 19–12

Second free-vibration mode of step beam. (*a*) Twenty-element beam model, $f_2 = 1296$ Hz; (*b*) 56 384-element brick and tetrahedron model, $f_2 = 1249$ Hz.

CAD software provides an excellent platform for the easy creation of FEA models. Several types of analysis have been described in this chapter, using some fairly simple illustrative problems. The purpose of this chapter, however, was to discuss some basic considerations of FEA element configurations, parameters, modeling considerations, and solvers, and not to necessarily describe complex geometric situations. Finite-element theory and applications is a vast subject, and will take years of experience before one becomes knowledgeable and skilled with the technique. There are many sources of information on the topic in various textbooks; FEA software suppliers (such as ANSYS, MSC/NASTRAN, and ALGOR) provide case studies, user's guides, user's group newsletters, tutorials, etc.; and the Internet provides many sources. Footnotes 11, 12, and 14 referenced some textbooks on FEA. Additional references are cited below.

Additional FEA References

K. J. Bathe, *Finite Element Procedures,* Prentice Hall, Englewood Cliffs, NJ, 1996.

R. D. Cook, D. S. Malkus, M. E. Plesha, and R. J. Witt, *Concepts and Applications of Finite Element Analysis,* 4th ed., Wiley, New York, 2001.

D. L. Logan, *A First Course in the Finite Element Method,* 4th ed., Nelson, a division of Thomson Canada Limited, Toronto, 2007.

J. N. Reddy, *An Introduction to the Finite Element Method,* 3rd ed., McGraw-Hill, New York, 2002.

O. C. Zienkiewicz and R. L. Taylor, *The Finite Element Method,* 4th ed., Vols. 1 and 2, McGraw-Hill, New York, 1989 and 1991.

PROBLEMS

The following problems are to be solved by FEA. It is recommended that you also solve the problems analytically, compare the two results, and explain any differences.

19–1 Solve Ex. 3–6.

19–2 For Ex. 3–10, apply a torque of 2700 N · m, and determine the maximum shear stress and angle of twist. Use 3-mm-thick plate elements.

19–3 The steel tube with the cross section shown is transmitting a torsional moment of 100 N · m. The tube wall thickness is 2.5 mm, all radii are $r = 6.25$ mm, and the tube is 500 mm long. For steel, let $E = 207$ GPa and $\nu = 0.29$. Determine the average shear stress in the wall and the angle of twist over the given length. Use 2.5-mm-thick plate elements.

Problem 19–3

19–4 For Fig. A–15–1, let $w = 50$ mm, $d = 7.5$ mm, and estimate K_t. Use 1/4 symmetry and 3-mm-thick 2-D elements.

19–5 For Fig. A–15–3, let $w = 38$ mm, $d = 25$ mm, $r = 2.5$ mm, and estimate K_t. Use 1/4 symmetry and 3-mm-thick 2-D elements.

19–6 For Fig. A–15–5, let $D = 75$ mm, $d = 50$ mm, $r = 6$ mm, and estimate K_t. Use 1/2 symmetry and 3-mm-thick 2-D elements.

19–7 Solve Prob. 3–122, using solid elements. *Note:* You may omit the top part of the eyebolt above the applied force.

19–8 Solve Prob. 3–132, using solid elements. *Note:* Since there is a plane of symmetry, a one-half model can be constructed. However, be very careful to constrain the plane of symmetry properly to assure symmetry without overconstraint.

19–9 Solve Ex. 4–11, with $F = 45$ N, $d = 3$ mm, $a = 12$ mm, $b = 25$ mm, $c = 50$ mm, $E = 207$ GPa, and $\nu = 0.29$, using beam elements.

19–10 Solve Ex. 4–13, modeling Fig. 4–14b with 2-D elements of 50-mm thickness. Since this example uses symmetry, be careful to constrain the boundary conditions of the bottom horizontal surface appropriately.

19–11   Solve Prob. 4–12, using beam elements.

19–12   Solve Prob. 4–47, using beam elements. Pick a diameter, and solve for the slopes. Then, use Eq. 7–18, p. 373, to readjust the diameter. Use the new diameter to verify.

19–13   Solve Prob. 4–63, using beam elements.

19–14   Solve Prob. 4–78, using solid elements. Use a one-half model with symmetry. Be very careful to constrain the plane of symmetry properly to assure symmetry without overconstraint.

19–15   Solve Prob. 4–79, using beam elements. Use a one-half model with symmetry. At the plane of symmetry, constrain translation and rotation.

19–16   Solve Prob. 4–80, using beam elements. Model the problem two ways: (*a*) Model the entire wire form, using, 200 elements. (*b*) Model half the entire wire form, using 100 elements and symmetry. That is, model the form from point *A* to point *C*. Apply half the force at the top, and constrain the top horizontally and in rotation in the plane.

19–17   Solve Prob. 4–88, using beam elements.

19–18   Solve Prob. 10–42, using beam elements.

19–19   An aluminum cylinder ($E_a = 70$ MPa, $\nu_a = 0.33$) with an outer diameter of 150 mm and inner diameter of 100 mm is to be press-fitted over a stainless-steel cylinder ($E_s = 190$ MPa, $\nu_s = 0.30$) with an outer diameter of 100.20 mm and inner diameter of 50 mm. Determine (*a*) the interface pressure *p* and (*b*) the maximum tangential stresses in the cylinders.

Note: Solve the press-fit problem, using the following procedure. Using the plane-stress two-dimensional element, utilizing symmetry, create a quarter model meshing elements in the radial and tangential directions. The elements for each cylinder should be assigned their unique material properties. The interface between the two cylinders should have common nodes. To simulate the press fit, the inner cylinder will be forced to expand thermally. Assign a coefficient of expansion and temperature increase, α and ΔT, respectively, for the inner cylinder. Do this according to the relation $\delta = \alpha \Delta T b$, where δ and *b* are the radial interference and the outer radius of the inner member, respectively. Nodes along the straight edges of the quarter model should be fixed in the tangential directions, and free to deflect in the radial direction.

20

Geometric Dimensioning and Tolerancing

Early in the design process the designer works largely from a macro point of view, in which components are nominally sized to satisfy the design requirements, typically including control of stresses and deflections. But when it comes to issues of manufacturability, fits between components, and assembly of components, the designer must take a much closer look at the precise specification of the parts. Section 1–14, p. 27, addressed some of the basic issues of dimensioning and tolerancing that a designer must consider. In this chapter, the focus is on a standardized method of defining part geometry that takes into consideration that no part is perfectly formed—nor should it need to be. The method known as *Geometric Dimensioning and Tolerancing* allows clarity for functionality, flexibility for manufacturing, and level of precision for inspection.

20–1   Dimensioning and Tolerancing Systems

The traditional method of dimensioning and tolerancing is referred to as the *coordinate dimensioning system.* In this system, every dimension is associated with a plus/minus tolerance, either directly specified immediately adjacent to the dimension, or implicitly specified with a general tolerance notation. This method of tolerancing has been used for many generations. It works reasonably well for parts that are not mass produced and that do not need to assemble with other parts. In general, this method is acceptable when a high level of precision is not needed. However, this system is lacking in many respects, particularly in its inability to address geometric issues of form and orientation.

As an example, consider the part shown in Fig. 20–1. This simple part is fully dimensioned and toleranced according to the traditional coordinate dimensioning system. In general, the designer's intent is clear, and most machine shops could manufacture such a part. But suppose during inspection of an actual part, it is found that the bar stock is not perfectly flat, the corners are not perfectly square, the hole is not perfectly perpendicular to the face of the bar stock, and the hole is not perfectly round. In fact, this will always be the case since manufacturing can never achieve perfect form. Figure 20–2 shows an exaggerated view of the imperfections of the manufactured part. The problem is that every dimension may be within its tolerance, but the part may be unusable for its application owing to one of the geometric imperfections. A more challenging problem is that it is not even clear how to measure

Figure 20–1

Part dimensioned with traditional coordinate dimensioning system.

All dimensions +/− 0.01

Figure 20-2

Exaggerated view of imperfections of a manufactured part.

some of the dimensions. For example, the center of the hole is to be 2 inches from the edge of the part. If the hole is not perfectly round, how is its center defined? If the corner of the part is not square, from where should the 2 inches be measured? Bottom corner? Top corner? Closest edge? There is no defined correct answer with this dimensioning system.

For many applications these issues can be overlooked because typical manufacturing methods are deemed to be good enough. However, mass production calls for the most efficient and inexpensive operation allowable. The manufacturer legitimately needs to be able to cut corners as much as possible. This requires the part specification to define precisely how good is good enough. In fact, the designer should always view the task of geometry specification as simultaneously restricting and freeing the manufacturer—restricting within the necessary limits for functional requirements, and freeing from unnecessary levels of perfectness. Proper balance of these two aspects provides for cost-effective *and* functional parts.

Clearly, to address issues of functionality, manufacturability, interchangeability, and quality control requires parts that can be uniquely defined and consistently measured. This requires a dimensioning and tolerancing method that takes into account not only size, but geometric location, orientation, and form as well. This system is known as Geometric Dimensioning and Tolerancing (GD&T).

20-2 Definition of Geometric Dimensioning and Tolerancing

Geometric Dimensioning and Tolerancing (GD&T) is a comprehensive system of symbols, rules, and definitions for defining the nominal (theoretically perfect) geometry of parts and assemblies, along with the allowable variation in size, location, orientation, and form of the features of a part. It serves as a means of accurately representing the part for the purposes of design, manufacture, and quality control. GD&T is not new. It has been developing as a standard within industry since the 1940s. Today, most of the major manufacturing companies utilize GD&T. The part previously considered in Fig. 20–1 is shown again in Fig. 20–3 using GD&T terminology. Unfortunately, many mechanical engineers are not able to interpret the drawing. During the time when the standard was becoming most prevalent in manufacturing, most engineering schools were phasing out comprehensive drafting courses in favor of computerized CAD instruction. Consequently, GD&T is often

Figure 20–3

Part dimensioned and
toleranced with GD&T
terminology.

missing from the engineering curriculum. A full understanding of GD&T is usually obtained through an intensive course or training program, readily available to practicing engineers. Some mechanical engineers will benefit from such a rigorous training. *All* mechanical engineers, however, should at least be familiar with the basic concepts and notation. The purpose of the coverage of GD&T in this chapter is to provide this foundational exposure that is essential for all machine designers. The coverage is not comprehensive. The focus is on the foundational concepts and the most commonly used notation. A first exposure to GD&T can feel overwhelming, as there are numerous concepts and terminology that all depend upon each other. This chapter is organized to help a beginner gradually build the most important concepts first, adding detail as needed, and culminating with a section on practical applications. Section 20–9 includes a glossary of some of the most important vocabulary terms used in GD&T, and should be used as a reference for clarification of terms while reading the rest of the chapter.

GD&T Standards

GD&T is defined and controlled by standards to provide uniformity and clarity on a global scale. One widely utilized standard is published by the American Society of Mechanical Engineers as *ASME Y14.5–2009 Dimensioning and Tolerancing*. It is part of the broader set of ASME Y14 standards that cover all aspects of engineering drawings and terminology. The International Organization of Standards (ISO) also publishes a series of standards that have been more commonly used in European countries. The two standards have developed in parallel and are mostly similar in concept and terminology. The ASME standard tends to put more emphasis on *design intent* while the ISO standards have a greater emphasis on *metrology*, or the measurement of the resulting part. According to the ASME approach, the parts are defined primarily in a way to ensure that they will perform the desired function, without specifying what equipment or processes should be used to manufacture or inspect the parts. The ASME Y14.5–2009 standard is utilized in this textbook.

Four Geometric Attributes of Features

A *feature* of a part is a general term referring to a clearly identifiable physical portion of a part. Examples include a hole, pin, slot, surface, or cylinder. There are four geometric attributes of every feature that must be considered in order to define the geometry of the feature. They are *size, location, orientation,* and *form*. These attributes are illustrated in Fig. 20–4. It is important in understanding GD&T to distinguish these four geometric attributes. They are each briefly described here, followed by further elaboration in the following sections.

The term *feature of size* refers to a feature that has a size that can be measured across two opposing points, such as a hole, cylinder, or slot. A helpful rule of thumb is that a feature of size can usually be measured with the head of a caliper tool, such as illustrated in Fig. 20–5. A dimension that is measured with the probe end of the caliper does not have opposed points, and is therefore not a feature of size. Such a dimension would be a *locating* dimension rather than a *size* dimension.

Location refers to the location of a feature with respect to some origin of measurement. *Orientation* refers to the angle of a feature or center-line of a feature with respect to some origin of measurement. It includes parallelism, perpendicularity, and angularity. *Form* refers to imperfections in the shape of a feature, and includes straightness, flatness, circularity, and cylindricity.

Figure 20–4

The four geometric attributes of a feature.
(*a*) Size; (*b*) location;
(*c*) orientation; (*d*) form.

(*a*) size   (*b*) location   (*c*) orientation   (*d*) form

Figure 20–5

Measuring opposing points of features of size with a caliper head.

Internal feature of size containing opposed points

$\varnothing 40 \pm 0.1$

40 ± 0.01

Not a feature of size since there are no opposed points

40 ± 0.1

External feature of size containing opposed points

According to the GD&T standard, plus/minus tolerancing should only be directly applied to the size dimension of a feature of size. The other attributes (location, orientation, and form) are controlled through *geometric tolerancing,* described in Sec. 20–4.

Symbolic Language

The ASME Y14.5–2009 standard utilizes an international language of symbols that minimizes the need for, and potential confusion from, written notes on machine drawings. The symbols are classified into two categories—geometric characteristic and modifying. For reference, a brief introductory summary of the symbols will be given here. The usage of the symbols will be clarified in later sections.

Table 20–1 shows the 14 geometric characteristics and their symbols. These geometric characteristics are refinements of the geometric attributes (size, location, orientation, and form), and are each used to control a geometric tolerance of a feature. Each geometric characteristic has a symbol that is used on the drawing to specify a tolerance zone associated with a geometric characteristic. The geometric characteristic symbols are directly associated with the geometry of a feature, not with a size dimension. This is why they are referred to as geometric tolerancing. As shown in the table, the geometric characteristics are further subdivided into the types of tolerance defined by GD&T (form, profile, orientation, location, and runout), as well as the broader geometric attribute being controlled (size, location, orientation, and form). The table also indicates whether the symbol can be associated with a datum reference and any material condition modifiers, which will be clarified in later sections.

Table 20–1

Geometric Characteristic Controls and Symbols

Type of Tolerance	Geometric Characteristic	Symbol	Geometric Attribute Controlled	Datum Referencing?	Material Condition Modifier Allowed
Form	Straightness	—	Form	No	Ⓜ Ⓛ or RFS
	Flatness	▱			Ⓜ Ⓛ or RFS
	Circularity	○			RFS
	Cylindricity	⌭			RFS
Profile	Profile of a line	⌒	Location, orientation, size, and form	Optional	Ⓜ Ⓛ or RFS
	Profile of a surface	⌓			
Orientation	Angularity	∠	Orientation	Required	Ⓜ Ⓛ or RFS
	Perpendicularity	⊥			
	Parallelism	∥			
Location	Position	⊕	Location and orientation of feature of size	Required	Ⓜ Ⓛ or RFS
	Concentricity	◎	Location of derived median points or planes		RFS
	Symmetry	⩶			RFS
Runout	Circular runout	↗	Location of cylinder	Required	RFS
	Total runout	↗↗			RFS

Tables 20–2 and 20–3 show the modifying symbols. Table 20–2 includes the dimensioning modifiers and their symbols. These are used to modify or clarify the meaning of a dimension on the drawing. Table 20–3 includes the tolerance modifiers and their symbols. These are used in a feature control frame (to be defined later) to modify or clarify the tolerance specification.

Table 20–2

Dimensioning Modifiers and Symbols

Description	Symbol
Basic Dimension	⊢ 98 ⊣
Diameter	⌀
Spherical Diameter	S⌀
Radius	R
Spherical Radius	SR
Controlled Radius	CR
Square	☐
Reference	( )
Arc Length	⌒
Dimension Origin	⊕→
All Around	⌀→
All Over	⌀→
Independency	Ⓘ
Continuous Feature	⟨CF⟩
Counterbore	⌴
Countersink	⌵
Spotface	⌴ SF
Depth	↧

Table 20–3

Tolerance Modifiers and Symbols

Description	Symbol
Maximum Material Condition (applied to tolerance) Maximum Material Boundary (applied to datum)	Ⓜ
Least Material Condition (applied to tolerance) Least Material Boundary (applied to datum)	Ⓛ
Translation	▷
Projected Tolerance Zone	Ⓟ
Free State	Ⓕ
Tangent Plane	Ⓣ
Statistical Tolerance	⟨ST⟩
Between	◄►
Unequally Disposed Profile	Ⓤ

20–3 Datums

Several concepts are foundational to the implementation of GD&T. In this section, the concept of datums will be explained, followed by specific methods of implementation in GD&T.

The geometric characteristics of features are defined and measured by relating them to clearly defined datums. A *datum* is an origin from which *location* or *orientation* of part features is established. Note that *size* dimensions and *form* control do not require an origin for measurement, and therefore do not need to be referenced to a datum. For a feature of a part to be manufactured or inspected, the entire part is located with respect to a datum reference frame. A *datum reference frame* is a set of up to three mutually perpendicular planes that are defined as the origin of measurement for locating the features of a part. The datum reference frame is idealized and geometrically perfect. It is necessary to consider its relationship to the nonideal physical part and the processing equipment. To do so, it is necessary to distinguish between several related terms, namely datum, datum reference frame, datum feature, and datum feature simulator.

A *datum feature* is a nonideal physical surface of the part that is specified in order to establish a theoretically exact datum. A datum feature is always a surface of the part that can be physically touched, not a centerline or other theoretical entity. Since the datum feature is not perfect, it is not directly used for measurements. Suppose a flat surface of a part is selected as a datum feature. The surface is an imperfect plane with localized hills and valleys, and is not perfectly flat. If the part is placed on a polished granite surface plate, a minimum of three high points on the surface of the datum feature will contact the nearly perfect plane of the polished surface plate. The surface plate serves as a *datum feature simulator* of the actual datum feature. A datum feature simulator is a precision embodiment, such as a surface plate, gauge pin, or machine tool bed, of the datum described by an imperfect datum feature. The datum feature simulator is often a physical gauging surface, but may also be simulated by "soft gauging" optical or probing methods. The *datum* itself is a theoretically exact point, axis, or plane derived from the datum feature simulator.

A recap of the relationship between the various datum terms may be helpful, using the example shown in Fig. 20–6. An imperfect actual surface of a part, such as the bottom surface, is designated as a datum feature. The datum feature (bottom surface) is placed in contact with a nearly perfect datum feature simulator (granite surface

Figure 20–6

Example demonstrating datum terminology.

Part

Datum

Datum feature simulator

Datum feature

Granite surface plate

Figure 20–7

Example demonstrating a three-plane datum reference frame.

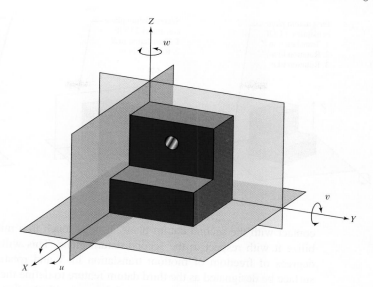

plate). A theoretical datum (true plane) is defined in association with the datum feature simulator. The process is repeated as necessary to define enough datums to obtain a three-plane datum reference frame. For example, if the back surface and one side surface are also selected as datum features, then the three-plane datum reference frame in Fig. 20–7 may be obtained.

How, then, are locations and orientations of features handled through the process of design, manufacture, and inspection? The designer specifies datum features that are best suited for the functionality, manufacture, and inspection of the part. Locations and orientations are defined by the designer on the drawing with respect to the datum reference frame. They are actually manufactured with respect to a datum feature simulator inherent in the manufacturing equipment, such as the table surface of a milling machine. They are measured for quality control with respect to a datum feature simulator, such as a granite surface plate. Note that measurements of location and orientation are not made with respect to the actual surface of the datum feature, but from the datum feature simulator.

Immobilization of the Part

The selection of datum features can be thought of as selecting which surfaces of the part will be put into contact with datum feature simulators in order to immobilize the part for manufacture and inspection. The part floating in space has six degrees of freedom (three translations and three rotations). Each datum constrains some of the degrees of freedom in order to immobilize the part in a precise, repeatable location. Consider the process of immobilizing a part with three datum planes, demonstrated in Fig. 20–8. First, let the bottom surface of the part be selected as a datum feature to be constrained by the first datum plane, as shown in Fig. 20–8a. Remember that the datum feature is imperfect, so it may touch the datum plane in only a few places. Specifically, a minimum of three points of contact are required to prevent the part from rocking on the datum plane. This contact with the datum plane will constrain three degrees of freedom of motion of the part: translation in Z, rotation in u, and rotation in v. Next, let the back surface of the part be designated as the second datum feature, from which the second datum plane is derived, as shown in Fig. 20–8b. Imagine holding the part in contact with the first datum plane and sliding it into

Figure 20–8

Immobilization of a part by sequential application of datum planes.

First datum plane constrains 3 DOF
1. Translation in Z
2. Rotation in u
3. Rotation in v

Second datum plane constrains 2 DOF
1. Translation in X
2. Rotation in w

Third datum plane constrains 1 DOF
1. Translation in Y

(a)　　　　　　　　(b)　　　　　　　　(c)

contact with the second datum plane. It must touch in a minimum of two points to stabilize it with respect to the second datum plane. This will constrain an additional two degrees of freedom of motion: translation in X and rotation in w. Finally, let a side surface be designated as the third datum feature to define the third datum plane, as shown in Fig. 20–8c. Maintaining contact of the part with the first two datum planes, and sliding it into contact with the third datum plane will result in a minimum of one point of contact with the third datum plane. This constrains the final degree of freedom: translation in Y. The part is now fully constrained in a precise, repeatable location.

Order of Datums

Notice that the order of application of the datum planes is important. Suppose the part from Fig. 20–8 is constrained by the first datum plane as before, but then the order of application of the second and third datum planes is reversed. Figure 20–9a shows a top view of the part that has been constrained by the YZ plane first, then by the XZ plane. Figure 20–9b shows the same part with the order of application of the two datum planes reversed. The final position of the two parts is not the same. Because measurements are made from the datum planes, not from the edges of the part itself, the measured locations of features on the part are clearly dependent on the choice of datum features and the order of application of the resulting datum planes. It is necessary for the part drawings to specify clearly, for each feature to be located on the part, the datum features as well as the order of application of the resulting datum planes. The features do not all have to use the same datums and order of application.

Non-planar Datum Features

So far, only planar datum features have been presented since they are the easiest to visualize the progression from the datum feature to the datum reference frame. Several

Figure 20–9

Comparison of order of application of datum planes.
(a) YZ plane constrained first;
(b) XZ plane constrained first.

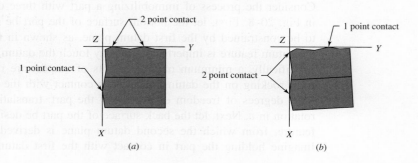

2 point contact

1 point contact

1 point contact

2 point contact

(a)　　　　　　　　(b)

other datum features are also provided for in the Y14.5 standard. In particular, cylindrical features such as shafts, bosses, and holes are often useful as datum features. Suppose in the part shown in Fig. 20–10, the hole is selected as a datum feature. The actual surface of the hole is the datum feature; the center axis of the hole is the datum. The center axis defines the intersection of two perpendicular datum planes. In conjunction with another datum feature, say the back surface, the part is constrained and a datum reference frame is defined.

Actual Mating Envelopes

In the previous paragraph, it was stated that the center axis of the hole is the datum. This is a simplified statement of a more detailed concept that warrants a better explanation. Since the hole feature is imperfect in form (that is, it does not have a perfectly circular cross section, or a perfectly straight center line, or a perfectly smooth surface), how is its theoretically perfect datum axis determined? To answer this question, a few GD&T terms will be introduced.

An *actual mating envelope* is a perfectly shaped counterpart of an imperfect feature of size, which can be contracted about an external feature, or expanded within an internal feature, so that it contacts the high points of the feature's surface. For example, Fig. 20–11a shows an imperfect dowel pin (the feature of size) that is circumscribed by the smallest possible perfect cylinder (the actual mating envelope). The imperfect pin does not technically have a center axis. Instead, it has a collection of *derived median points* that represent the centroids of each cross section. When referring to the center axis of an imperfect feature such as the pin, what is actually meant is the theoretically perfect center axis of the theoretically perfect actual mating envelope of the pin. The same concept can be applied to a feature of size with an internal surface, such as the hole feature shown in Fig. 20–11b.

Figure 20–10

An example of a hole as a datum feature.

Figure 20–11

Definition of terms for an actual mating envelope.
(a) External feature;
(b) internal feature.

(a)   (b)

Actual mating envelopes are categorized as *related* or *unrelated* to a datum. An *unrelated actual mating envelope* is sized to fit the feature without any constraint to any datum. In other words, it is free to float to find the best fit. A *related actual mating envelope* is sized to fit the feature while maintaining some constraint in orientation or location with respect to a datum. For example, for the hole feature in Fig. 20–10, the related actual mating envelope with respect to the back plane datum surface is the largest pin that can fit in the hole while being held perpendicular to the back datum plane.

Now, returning to the datum example in Fig. 20–10, the datum axis corresponding to the datum feature (the hole) is defined by the actual mating envelope of the hole, that is the largest cylinder that can fit within the hole. In practical implementation, this largest cylinder can be determined by physically inserting very precisely manufactured gauge pins of increasing size until the largest one is found. Alternatively, an expanding mandrel can be used. The largest gauge pin serves as the datum feature simulator (previously defined). In the case of an external datum feature, such as the surface of a shaft, the datum feature simulator is typically the jaws of a chuck or collet that is closed onto the surface. The center axis of the chuck is then the datum axis.

Datum Feature Symbol

On a drawing, a datum feature is defined symbolically by a capital letter enclosed in a square frame, attached to a leader line that terminates at the datum feature with a triangle. The triangle can be filled or unfilled. Any letter can be used, except I, O, and Q, which may be confused with numbers. The letters need not be assigned alphabetically as the precedence of the datums will be specified later as needed for each feature to be controlled.

The triangle may attach directly to the datum feature surface outline, point to it with a leader line, or attach to an extension line to the surface. All three of these methods are illustrated in Fig. 20–12, where the datum features correlate to those previously demonstrated for the part in Fig. 20–8.

If the datum is to be an axis or center plane of a feature of size, then the datum triangle is placed in-line with the dimension line of the feature of size. In the case of a cylinder it may be attached directly to the surface of the cylinder. The triangle may optionally replace one of the dimension arrowheads if both will not fit. Several examples are shown in Fig. 20–13. A datum triangle always indicates the datum feature (a physical surface) from which the datum (a theoretical axis or center plane) is derived. Accordingly, the triangle is never placed directly on an axis, center line, or center plane.

The datum triangle may also be attached to a feature control frame (to be defined in a later section) that controls the geometric tolerance of the datum feature. An example is included in Fig. 20–13b.

Note that in the case of a feature of size, the subtle difference in placement of the triangle in Figs. 20–14a and 20–14b has a significant difference in meaning. When the triangle is placed in-line with the dimension line, as in Fig. 20–14a, the datum is

Figure 20–12

Three methods of designating a datum feature.

Figure 20–13

Methods of designating an
axis or center plane of a
feature of size as a datum.
(*a*) Datums A and B are
defined as the center axes
of two different cylindrical
surface features.
(*b*) Datum symbols A
and B are attached to width
dimensions, defining the
center planes as the datums.
Datum symbols C, D, and E
are attached to features with
axes, defining the center axes
as the datums.

Figure 20–14

Different datum plane
designation due to placement
of datum symbol. (*a*) Symbol
in-line with dimension;
(*b*) symbol out-of-line with
dimension.

the center plane of the feature of size. When the triangle is placed away from the dimension line, as in Fig. 20–14*b*, the datum is the plane defined by the edge of the part. The center plane datum would likely be used if it was important that the hole be centered in the part, regardless of fluctuations in the overall width of the part. The edge datum has the advantage of simpler setup for manufacturing and inspecting, but because the hole location is controlled from one edge, the part may not be symmetric.

20–4   Controlling Geometric Tolerances

Tolerance Zones

GD&T generally only uses direct application of plus/minus tolerances when dimensioning a *feature of size*. A fundamental concept of GD&T is that in addition to any dimensional tolerance on a feature of size, the geometric shape and location of the surfaces must be controlled to stay within *tolerance zones*. Tolerance zones are defined in a variety of ways, such as by two parallel planes or concentric cylinders, to define limiting boundaries for the physical surfaces of the parts. The tolerance zones are defined relative to a theoretically exact location or shape. The actual shape and location of the part surfaces may vary from the theoretically exact location and shape, so long as they stay within the limiting boundaries of the tolerance zones.

There are special terms defined to represent the maximum and minimum boundaries of a feature of size. The *Maximum Material Condition (MMC)* is the condition in which a feature of size contains the maximum amount of material within the stated limits of size. For an external feature, such as the outer surface of a shaft, the MMC is when the shaft diameter is at its maximum value allowed by the tolerance. For an internal feature, such as a hole, the MMC is when the hole is at its smallest diameter allowed by the tolerance. Similarly, the *Least Material Condition (LMC)* is the condition in which a feature of size contains the least amount of material within the stated limits of size. This would correlate with the smallest allowed shaft diameter, or the largest allowed hole diameter. These terms are used extensively in GD&T.

The Y14.5 standard specifies a default tolerance zone for features of size through what is referred to as *Rule #1,* also known as the *envelope principle.* This rule states that when only a tolerance of size (i.e., a plus/minus tolerance) is specified for a feature of size, the limits of size prescribe the extent of which variation in its geometric form, as well as size, are allowed. Specifically, the envelope principle states that the surface of a feature of size may not extend beyond an envelope of perfect form at MMC. Consider a simple example of a dowel pin in Fig. 20–15a. The limiting envelope is a perfectly formed pin (e.g., perfectly straight, perfectly circular cross section, etc.) that is at its largest allowed size, as shown in Fig. 20–15b. An imperfect dowel pin would still meet the specifications so long as its diameter at any location is within the allowed tolerance, and its surface does not exceed the envelope. An implication of this is that if the pin is manufactured at its MMC, then it must have perfect form. As the diameter is reduced from the MMC, the pin may deviate from perfect form, as shown in Fig. 20–15c.

Other geometric controls (described in Sec. 20–5) may be specified when the default tolerance zone provided by Rule #1 is not sufficient to meet the requirements of the application.

Figure 20–15

The envelope principle (Rule #1). (*a*) Size and tolerance specification on the drawing; (*b*) the envelope of perfect form at MMC; (*c*) an acceptable imperfect pin within the envelope.

$\varnothing\ 20 \pm 0.5$

(*a*)

$\varnothing\ 20.5$

Envelope of perfect form at MMC

(*b*)

$\varnothing\ 20.5$

Pin with imperfect form, always within the MMC envelope

Envelope of perfect form at MMC

Pin diameter is between 19.5 and 20.5 at every cross-section

(*c*)

Basic Dimensions

A theoretically exact location is specified with a *basic dimension*. A basic dimension is a theoretically exact dimension which does not have a tolerance directly associated with it, but is instead associated with a geometric control of a tolerance zone. When a basic dimension is used to locate a part feature, the feature itself must include a geometric control that defines a tolerance zone specifying the permissible variation from perfect form and location.

Basic dimensions are indicated on a drawing by enclosing the dimension within a rectangular box, or by a general note indicating that all untoleranced dimensions are basic.

Feature Control Frames

A geometric control is specified on a drawing in a *feature control frame*. A feature control frame is a rectangular box attached to a feature on a drawing, containing the necessary information to define the tolerance zone of the specified feature. The frame is subdivided into compartments in a specific order, as shown in the example in Fig. 20–16.

The first compartment always contains one of the geometric control symbols from Table 20–1 to indicate what aspect of the feature is being controlled by the tolerancing information to follow.

The second compartment always contains a numerical value designating the total allowed tolerance. The tolerance value specified is always a total tolerance, that is, the entire range of the tolerance, not a plus/minus tolerance value from a midpoint. If the tolerance is circular or cylindrical, the diameter symbol will precede the specified tolerance.

The third and following compartments are used as needed to define the datum(s) necessary to immobilize the part. The order of the datum letters from left to right defines the precedence of the application of the datums. The number of datum letters may vary from zero to three, depending on the datum reference frame required for the particular tolerance being controlled. Tolerances of *form* affect only the designated feature, independent of any other feature or datum (as indicated in Table 20–1), and therefore never include datum reference letters in the feature control frame. Tolerances of location, orientation, and runout always relate the designated feature to some other feature or datum, and therefore always require the specification of datum reference letters.

Modifiers from Table 20–3 can be included in a compartment immediately following a tolerance value, or after a datum specification. The effect of these modifiers is discussed in Sec. 20–6.

Figure 20–16

Example of a feature control frame.

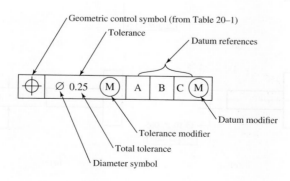

Geometric control symbol (from Table 20–1)

Tolerance

Datum references

Ø 0.25 (M) A B C (M)

Datum modifier

Tolerance modifier

Total tolerance

Diameter symbol

Figure 20–17

Reading a feature
control frame.

This feature must be
positioned within

a cylindrical
tolerance zone
of 0.25 total

when the feature is produced
at its maximum material condition

with respect to the datum reference frame
established by datum features A (primary),
B (secondary), and C (tertiary),

where datum feature C is at maximum
material condition.

The feature control frame is read from left to right, as illustrated in Fig. 20–17. If more than one geometric control is to be applied to the same feature, the feature control frames can be stacked, and are applied from top to bottom.

A feature control frame controls the feature to which it is attached. The four attachment methods are as follows:

1 A leader from the feature control frame points directly to the feature. See Fig. 20–18a.

2 The feature control frame is attached to an extension line from a planar feature. See Fig. 20–18b.

3 The feature control frame is placed below a leader-directed dimension or note pertaining to the feature. See Fig. 20–18c.

4 The feature control frame is attached to an extension of the dimension line pertaining to a feature of size. See Fig. 20–18d.

A feature control frame associated with a feature of size may control either the actual surface of the feature, or the axis or center line of the feature. To control the axis or center line of a feature of size, the feature control frame is associated with the dimension of the feature of size using either attachment method 3 or 4. The

Figure 20–18

Four methods of attaching
a feature control frame to a
feature.

dimension and feature control frame should be shown in a drawing view in which the axis or center line appears as a line. Further, if the feature dimension is for a diameter, the diameter symbol precedes the tolerance value in the feature control frame. As an example, in Fig. 20–18a, the straightness control applies to the surface of the small cylinder, whereas in Figs. 20–18c and 20–18d, the straightness control applies to the axis of the small cylinder. The significance of the difference between controlling the surface or the axis is described later.

20–5   **Geometric Characteristic Definitions**

Each of the geometric characteristic symbols in Table 20–1 is used to define a tolerance zone particular to a certain geometric characteristic. The geometric controls are categorized as controls of *form, orientation, profile, location,* and *runout.* Some of the controls are quite general and encompass most of the common needs, while some are very specific to a particular geometric need. For reference, a basic description will be given for each geometric control, followed by a broader discussion on practical implementation. The reader may find it helpful to only skim this section the first time through, then refer back to it as a reference when the details are needed for practical applications.

Form Controls

The four geometric characteristics that provide *form* control are *straightness, flatness, circularity,* and *cylindricity.* These control the form of an individual feature, independent of that feature's location or relationship to any other feature. Consequently, form controls never include reference to datums. Note that the form controls are a further refinement of any size tolerance, which must also be satisfied.

Straightness ——

The straightness control specifies a tolerance zone within which line elements of a surface or an axis must lie. When applied to the surface of a feature with either a leader line or an extension line to the surface, the straightness applies to all lines within the surface that appear as straight lines in the drawing view. See Fig. 20–19. When the feature control frame is applied beneath the size dimension of a feature of size, the straightness control is on the axis or derived median line of the feature. The

Figure 20–19

Straightness control applied to a surface. (*a*) Drawing specification; (*b*) Interpretation.

Translation: Line elements in this surface, that are seen as lines in this view, must be straight within 0.05.

0.05 tolerance zone for each line in the surface.

0.05

or

0.05

20 ± 0.1

(*a*)

(*b*)

diameter symbol is included with the straightness tolerance to apply a cylindrical tolerance zone to the axis of a cylinder. See Fig. 20–20.

Flatness ▱

The flatness control specifies a tolerance zone of a specified distance between two parallel planes within which all points of a surface (or derived median plane) must lie. The feature control frame is applied with a leader line or an extension line to the surface in a drawing view in which the surface appears as a line. See Fig. 20–21. If the feature control frame is applied beneath the size dimension of a feature of size, the flatness control is on the derived median plane rather than on the surface. The flatness control is often used to provide additional control of the primary datum feature to improve reproducibility of measurement.

Circularity ◯

The circularity control is used to control the periphery of the circular cross sections of a cylinder, cone, or sphere. The tolerance zone is the annulus between two concentric circles within which each circular element of the surface must lie. See Fig. 20–22. The

Figure 20–20

Straightness control applied to an axis of a feature.
(a) Drawing specification;
(b) Interpretation.

Translation: The derived median line of the feature must be straight within a cylindrical tolerance zone with a diameter of 0.05.

Figure 20–21

Flatness control. (a) Drawing specification; (b) Interpretation.

Translation: The entire surface seen as a line in this view, must be flat within 0.05, i.e. all points must lie between two parallel planes separated by 0.05.

Figure 20–22

Circularity control. (a) Drawing specification; (b) Interpretation.

Translation: Every circular element (slice) in this surface must lie between two concentric circles with a radial difference of 0.01.

circularity control should be used sparingly as it is difficult to inspect, since every circular cross section of the surface must be evaluated independently from one another and independently from any datum. The runout or profile controls provide alternate methods that are usually sufficient to ensure circularity with easier inspection methods.

Cylindricity ⌭

The cylindricity control is used to control a combination of circularity and straightness of a cylinder. The tolerance zone is the space bounded by two concentric cylinders with a difference of radius equal to the tolerance. See Fig. 20–23.

Orientation Controls ∠ ⊥ //

The three geometric characteristics that provide *orientation* control are *angularity, parallelism,* and *perpendicularity*. These control the orientation of a feature with respect to one or more datums, therefore mandating the inclusion of at least one datum reference in the feature control frame. The parallelism and perpendicularity controls are essentially convenient subsets of the angularity control in which the desired angle is 0° or 90°, respectively. The orientation controls may be applied to surfaces, axes, or center planes. In the case of surfaces and planes, the tolerance zone is defined by two parallel planes oriented at the specified basic angle from the datum reference. See Figs. 20–24 and 20–25. When controlling axes or center planes, the feature control frame is placed beneath the size dimension. When controlling axes, the diameter symbol is included before the tolerance, and the tolerance zone is cylindrical, as demonstrated in Fig 20–26.

Orientation tolerances are constrained only in rotational degrees of freedom with respect to the referenced datum. Since the translational degrees of freedom are not

Figure 20–23

Cylindricity control. (*a*) Drawing specification; (*b*) Interpretation.

> Translation: The entire surface of the cylinder must lie between two concentric cylinders with a radial difference of 0.01.

⌭ 0.01

∅ 30 ± 0.1

(*a*)   (*b*)

Figure 20–24

Angularity control. (*a*) Drawing specification; (*b*) Interpretation.

> Translation: The entire surface must lie between two parallel planes 0.2 apart which are inclined at a basic angle of 30° to datum plane A.

∠ 0.2 A

30°

A

Datum plane A

0.2

(*a*)   (*b*)

Figure 20–25

Perpendicularity control.
(a) Drawing specification;
(b) Interpretation.

Translation: The entire surface must lie between two parallel planes 0.2 apart which are perpendicular to datum plane A. Note that this orients datum plane B with respect to datum A.

Figure 20–26

An orientation control applied to the axis of a feature.
(a) Drawing specification;
(b) Interpretation.

Translation: The axis of the hole feature must lie within a cylindrical tolerance zone with 0.05 diameter and perpendicular to datum plane A.

constrained by orientation tolerances, an orientation tolerance zone cannot be used to locate a feature. It should only be used as a refinement of a tolerance that is doing the locating, such as position or profile of a surface. The most common use of the orientation controls is to orient a secondary or tertiary datum with respect to the primary datum plane.

Profile Controls ⌒ ⌓

Profile controls are used to define a tolerance zone around a desired true profile that is defined with basic dimensions. The two geometric characteristics that provide profile control are *profile of a line* and *profile of a surface*. Profile of a line is a two-dimensional tolerance zone that controls each line within the feature's surface, similar to straightness or circularity controls. Profile of a surface applies a three-dimensional control, similar to flatness or cylindricity. Profile controls are often used for irregularly shaped features and for castings, forgings, or stampings where it is desired to provide a tolerance zone for the overall surface of the part.

A profile tolerance is implied to be an overall tolerance that is centered on the true profile. A nonsymmetrical tolerance zone can be specified by following the overall tolerance value with the *unequally disposed profile* modifier (an uppercase U in a circle), followed by the amount of tolerance that is in the direction that would allow additional material to be added to the true profile.

Figure 20–27

Application of the profile control.
(*a*) Applied to a single surface; (*b*) applied between two designated points; (*c*) applied all around; (*d*) applied all over.

(*a*)

(*b*)

(*c*)

(*d*)

The feature control frame is attached with a leader line in a drawing view where the true profile is shown. The profile tolerance applies only to the individual feature surface, unless modified with the "between," "all around," or "all over" symbols, as shown in Fig. 20–27.

The profile controls are the only geometric characteristics that have the option of including or not including a datum reference. If the profile tolerance does not reference a datum, then the tolerance zone "floats" around the true profile, providing a form control of the surface, but not a location control. This option should be used sparingly, as it usually makes the inspection of the part more difficult.

If a datum is referenced, then the profile tolerance can simultaneously control size, form, orientation, and location of a feature. This general capability makes this control extremely useful as an overall default tolerance control. When the profile tolerance feature control frame is placed in a general note on the drawing, the tolerance applies to all features in the drawing unless otherwise specified. With this general note, other controls (e.g., flatness, perpendicularity, etc.) are needed only if a tighter control is desired than provided by the general profile tolerance.

Consider the example in Fig. 20–28. The bottom surface is the datum feature that defines datum A. The flatness control establishes that when this surface is in contact with datum plane A, all points on the surface must be within the 0.05 tolerance zone. Datum plane B is defined to be exactly perpendicular to datum plane A, but the datum feature B (the actual surface), may vary in form and orientation so long as it stays within the 0.2 tolerance zone. The profile control on the curved surface indicates that the part is first placed into contact with datum plane A, then into contact with datum plane B. Then the ideal curved surface is defined with the basic dimensions. Then a tolerance zone is defined as the space between two curved surfaces centered around the true surface with a space of 0.1 between them. In addition, owing to the profile tolerance of 0.3 in the general note at the bottom of the drawing, all other surfaces that do not have a tighter tolerance specified will be within a profile tolerance zone of 0.3 centered on the basic (ideal) shape. Any deviation of size, form, orientation, and location is allowable so long as the surfaces remain within these tolerance zones.

Figure 20–28

An example using a profile control note as an overall default tolerance control, with tighter controls added as needed. (*a*) Drawing specification; (*b*) Interpretation.

Translation: This surface must lie within a tolerance zone defined as the space between two surfaces separated by 0.1 and centered on the ideal surface. The ideal surface is defined with the basic dimensions, while the part is in contact with datum A and datum B.

0.2 tolerance zone

0.1 tolerance zone

Datum B

These surfaces are controlled by the default surface profile of 0.3 (tolerance zones not shown).

0.05 tolerance zone

90°

Datum A

R4

B

2

0.05

A

3

2

0.2 A

0.1 A B

UNLESS OTHERWISE SPECIFIED
UNTOLERANCED DIMENSIONS ARE BASIC

Default surface profile specification. 0.3 A B

(*a*)

(*b*)

Location Controls

There are three *location* controls: *position, concentricity,* and *symmetry.* The location controls are specified to control the location of a *feature of size,* such as a hole, slot, boss, or tab, with respect to a datum or another feature.

Position ⊕

The *position* control is one of the most effective and oft-used controls, as it incorporates most of the advantages of GD&T. The position control defines the allowed location (and orientation) of the axis, center line, or center plane of a feature of size. It does not control the size or form of the feature.

The application of the position control is interpreted in Fig. 20–29. The true position of a feature of size is first located with respect to datums by specifying basic dimensions to the axis, center line, or center plane of the feature of size. Then the size of the feature of size is directly dimensioned, along with a plus/minus tolerance. Finally, the position control is applied with a feature control frame placed beneath the feature's dimension. The position control specifies a tolerance zone centered around the theoretically exact location of the feature's axis, center line, or center plane. The tolerance zone is cylindrical if the diameter symbol precedes the tolerance; otherwise, it is the space between two parallel planes. The specification for the part shown in the machine drawing in Fig. 20–29*a* leads to the interpretation shown in Fig. 20–29*b*. Note that the hole's diameter may vary within its specified dimensional tolerance, while the hole's axis may be at any position and orientation within the cylindrical tolerance zone.

Figure 20–29

Position control. (*a*) Drawing specification; (*b*) Interpretation.

Translation: The axis of the hole may be located and oriented anywhere within a 0.15-diameter cylindrical tolerance zone which is ideally located by the basic dimensions, with the part located by datums A, B, and C.

(*a*)

(*b*)

The position control also provides an excellent means of controlling the location of a group of features of size. A group of features is indicated to share the same dimension, dimensional tolerance, and position control by including the number of intended features preceding the specification of the size dimension, such as shown by the 3× in Fig. 20–30.

Figure 20–30 also demonstrates the use of a composite control frame where there are two rows associated with the position control specification. This allows different tolerance specifications for the overall location of the pattern and the interrelation of features within the pattern. This is commonly needed when the feature pattern must mate with similarly spaced features on another part.

The first row in the composite control frame is referred to as a *Pattern-Locating Tolerance Zone Framework (PLTZF)*, pronounced "plahtz." The PLTZF applies to the overall location of the pattern as a group with respect to the datums. The PLTZF defines tolerance zones for the center line of each feature just like previously described for the position control.

The second row of the composite control frame adds additional constraint to the feature-to-feature relationship within the pattern, and is referred to as the *Feature-Relating Tolerance Zone Framework (FRTZF)*, pronounced "fritz." The FRTZF is applicable to the basic dimensions between the features, but not to the basic dimensions locating the features with respect to the datums. The FRTZF defines another smaller tolerance zone for each feature, centered around the exact locations as defined by the basic dimensions between the features. The FRTZF tolerance zones may float anywhere within the PLTZF tolerance zones, as long as they maintain their relative positions with respect to each other. The actual feature center lines must lie within the FRTZF tolerance zones, whose center lines must in turn lie within the PLTZF tolerance zones.

Consider the example specified in Fig. 20–30, where the tolerance specifications are larger than typical in order to allow the tolerance zones to be more easily visualized.

Figure 20–30

Drawing specification for position control applied to a group of features. The interpretation is shown in Fig. 20–31.

First row is PLTZF, controlling the overall location of the pattern.

Second row is FRTZF, controlling the feature-to-feature relationship within the pattern.

Position control applied to all 3 holes.

Figure 20–31

Interpretation of the PLTZF and FRTZF for a pattern of features located with a composite control frame, as specified in Fig. 20–30.

Feature relating tolerance zones, with axes constrained to a rigid pattern with separation of 20. The rigid pattern of the axes is free to rotate and translate within the PLTZF tolerance zones.

Pattern locating tolerance zones, with axes located by basic dimensions, with the part located by datums A, B, and C.

For clarity, the actual holes are not shown. The axes of the holes may be anywhere within the FRTZF (smaller) cylindrical tolerance zones.

Figure 20–31 shows the tolerance zones, without showing the actual holes, to minimize the clutter. The first row of the composite control frame specifies a PLTZF consisting of cylindrical tolerance zones each having a diameter of 5, centered around axes located by the basic dimensions from the datums. These are shown in Fig. 20–31 as the larger cylinders. The second row specifies a FRTZF consisting of cylindrical tolerance zones each having a diameter of 2. The FRTZF tolerance zones form a rigid pattern that must maintain the relative distances of 20 with respect to each other, but which is free to rotate and translate within the cylindrical zones defined by the PLTZF. Think of the FRTZF tolerance zones as being rigidly connected to each other by a frame that maintains the basic dimensions between them. The entire frame may be translated and rotated to any position that keeps all of the FRTZF center lines within the PLTZF tolerance zones.

When datum references are included in the FRTZF specification, they govern only the rotation of the FRTZF relative to the specified datums. In the example in Fig. 20–30, since the second row references datum A, the FRTZF must be aligned with datum plane A, that is, the tolerance zone axes will be perpendicular to datum plane A. Likewise, if datum B had been specified, the tolerance zone axes would be required to lie in a plane that was parallel to datum plane B.

Concentricity ◎

Nominally, *concentricity* is the condition of the center axis of a surface of revolutio such as a cylinder, being congruent with a datum axis. The Y14.5 standard defines more precisely as the condition where the median points of all diametrically oppose elements of a surface of revolution are within a cylindrical tolerance zone centere around a datum axis. This means the feature's center is not determined as a sing straight line, but rather as a collection of all points obtained by finding the median of all diameter measurements across the surface. This is extremely difficult and exper sive to measure, so it is recommended that the concentricity control be rarely used The preferred options for controlling concentric features are position, profile, an runout. The concentricity control might be warranted in cases where it is critical t control the axis rather than the surface of a feature, such as in dynamic balancing c a high-speed rotating part.

Symmetry ⹀

Symmetry is the condition where a feature has the same profile on either side of th center plane of a datum feature. In the Y14.5 standard, it is defined similarly to con centricity, except it applies to the center plane of a feature of size instead of the cente axis of a surface of revolution. Because it is based on controlling a collection of al median points measured across the feature, rather than a single center plane of th feature, symmetry suffers from the same difficulties of measurement as concentricity Consequently, the symmetry control should rarely be used. In most cases, symmetrica features are best controlled with position or profile controls.

Runout Controls ⟋ ⟋⟋

Runout controls the surface variation on a feature as it rotates around a datum axis. A simple example is the surface of a rotating shaft. There are two runout controls circular runout and total runout. *Circular runout* measures the radial variation of each circular section of a feature independently from each other. *Total runout* measures the runout of the entire surface of a cylindrical feature simultaneously.

Both types of runout are demonstrated in Fig. 20–32. The cylinder on the left is defined as datum feature A. Chuck jaws clamped onto the surface of this datum feature serve as the datum feature simulator to define the centerline as the actual datum. Thus, when the chuck rotates, the part necessarily rotates around the datum centerline. The circular runout specification on the tapered feature requires that an

Figure 20–32

Circular runout and total runout. (*a*) Drawing specification; (*b*) Interpretation.

Translation: An indicator must not move more than 0.01 units when placed at *any* location along the tapered feature, when the part is rotated around datum axis A.

⟋ | 0.01 | A

∅ 20 ± 0.1

Datum axis defined by chuck jaws

0.01

0.01 at each circular element

Rotate

∅ 15 ± 0.2

⟋⟋ | 0.02 | A

Chuck jaws

0.02 all elements together

Translation: An indicator must not move more than 0.02 units for *all* locations along the cylinder, measured in one setup, when the part is rotated around datum axis A.

(*a*)

(*b*)

indicator at any location along the feature must not move more than 0.01 units during a complete rotation of the part. Each point where the indicator is located should independently satisfy the runout tolerance control. The total runout specification on the right cylindrical feature requires that an indicator must not move more than 0.02 for all locations along the cylinder, measured in one setup. Another way of stating this is that the entire surface of the controlled feature must lie within the zone between two concentric cylinders that are separated radially by the stated tolerance of 0.02.

Circular runout can be applied to any surface of revolution since the measurements are made independently at each cross section. It is inherently controlling both concentricity and circularity. Total runout applies only to cylindrical features, since the diameter at each cross section must fit within the same tolerance zone. It is inherently controlling cylindricity, circularity, straightness, and surface profile. Total runout is a particularly useful control for rotating shafts that carry bearings or gears that are sensitive to misalignment. Total runout can also be used effectively to control the coaxiality of multiple cylindrical surfaces by relating each cylinder to the same datum centerline.

The runout controls can also be applied to features constructed at right angles to a datum axis, such as the end-face of a cylinder. In this case, runout controls variations of perpendicularity (such as wobble) and flatness, measured while rotating the feature about the datum axis.

20–6 Material Condition Modifiers

Maximum material condition (MMC) and least material condition (LMC) can be applied as modifiers to most of the geometric controls that deal with a feature of size. The symbols for the modifiers, Ⓜ and Ⓛ, can be included in a feature control frame directly following the geometric tolerance value, and/or immediately following a datum reference.

When a geometric tolerance is modified with the maximum material condition modifier, it indicates that the stated tolerance value applies when the feature is produced at its MMC. If the feature is produced at a size with less material than its MMC, the deviation from the MMC is added to the allowed geometric tolerance value. The implication is that the value of the geometric tolerance is not constant, as it depends on the size of the actual produced part.

As an example, consider the part in Fig. 20–33a, which contains both an external feature of size (the cylindrical boss) and an internal feature of size (the hole), each of which is located with a position control that is toleranced at MMC. The position control on the external feature is interpreted as follows: "The center axis of this external cylindrical feature must be in position within a cylindrical tolerance zone of 0.2 diameter if the feature is produced at its maximum material condition of 50.3, where the position is specified by the basic dimensions with respect to the datum reference frame established by datum features A, B, and C." If the cylinder is produced with a diameter less than its MMC of 50.3, say at 50.2, then the amount of the deviation from the MMC, that is 0.1, is added to the specified geometric tolerance, providing a realized tolerance of 0.3. This increase in tolerance is traditionally referred to as a "bonus" tolerance. The table in Fig. 20–33b shows how this bonus tolerance accumulates as the produced diameter of the cylinder is reduced from its MMC of 50.3 to its LMC of 50.0. Note that the bonus tolerance is applied to the tolerance associated with the geometric control (position, in this example), and not to the direct tolerance on the size of the feature.

Figure 20–33

Application of MMC to a
position tolerance.
(*a*) Drawing specification;
(*b*) summary of bonus
tolerance for external feature;
(*c*) summary of bonus
tolerance for internal feature.

(*a*)

External Feature ϕ	Additional Tolerance (Bonus)	Tolerance Zone Cylinder ϕ
50.3 (MMC)	0.0	0.2
50.2	0.1	0.3
50.1	0.2	0.4
50.0 (LMC)	0.3	0.5

(*b*)

Internal Feature ϕ	Additional Tolerance (Bonus)	Tolerance Zone Cylinder ϕ
30.3 (LMC)	0.3	0.5
30.2	0.2	0.4
30.1	0.1	0.3
30.0 (MMC)	0.0	0.2

(*c*)

This bonus tolerance is one of the significant advantages afforded by GD&T over the fixed tolerances provided by the traditional coordinate dimensioning. It is particularly useful for applications where mating parts need to fit with each other with a clearance for assembly. A minimum tolerance is guaranteed, but a bonus tolerance is made available to the manufacturer to reduce costs. This modifier would not be appropriate for an application where the fit between the mating parts is important, such as a press fit between a bearing and a shaft.

The same bonus tolerance concept is applicable for the hole in Fig. 20–33*a*, except that the MMC condition for the internal feature is the *smallest* hole size. The table in Fig. 20–33*c* shows how the bonus tolerance adds to the stated MMC tolerance as the hole size gets larger, moving from the MMC to the LMC.

The *least material condition* modifier works similarly to the MMC modifier, but in the opposite direction. When a geometric tolerance is modified with the LMC modifier, it indicates that the stated tolerance value applies when the feature is produced at its LMC. If the feature is produced at a size with more material than its LMC, the deviation from the LMC is added to the allowed geometric tolerance value. The LMC modifier is typically used for applications in which it is critical to maintain a minimum amount of material. Examples include the material between a hole and the edge of a part, or a wall thickness. LMC is also useful for specifying features on a casting that will later be machined, to ensure that enough material is left in the casting for the finish machining operation.

The MMC and LMC material condition modifiers can be similarly applied to most of the geometric tolerances, in particular to those that control features of size with a central axis or plane. If no material condition modifier is specified, the default material condition is known as *regardless of feature size (RFS)*. RFS means that the stated tolerance is applicable regardless of the feature size. In other words, no matter what size the feature is produced (within its tolerance), the geometric tolerance is fixed at the stated value. This is much more restrictive for manufacturing. It is warranted in applications where variable play between mating parts is not desirable, such as the press fit between a shaft and components like bearings and gears.

A fine detail to be noted is that in all the preceding discussion, the produced diameter of the feature is determined by the unrelated actual mating envelope, as defined in Sec. 20–3, as it is about the only practical way to determine a single value for the diameter of an imperfect feature.

The Ⓜ and Ⓛ symbols can also be applied immediately following a datum reference in the feature control frame, in particular when the datum is based on a feature of size. When applied to a datum, the symbols are referred to as *maximum material boundary (MMB)* and *least material boundary (LMB)*. A full explanation of the effect of applying MMB or LMB to a datum is beyond the scope of this chapter. In essence, it allows the part to float or shift relative to the datum reference frame as the produced datum feature of size deviates from its maximum or minimum material condition. Consequently, it does not change the tolerance on the considered feature, but just the relative position of the feature with respect to the datum reference frame.

20–7 Practical Implementation

The basic concept of GD&T is to define the ideal part, then specify the amount of variation that is acceptable. The allowable variation includes all four of the geometric attributes: size, location, orientation, and form. The size of a feature of size is directly dimensioned and toleranced with a plus/minus tolerance. The other three geometric attributes, broken down into more specific geometric characteristics, are controlled with geometric controls (Table 20–1). Some of the geometric controls are broad and inherently provide control of multiple characteristics. Anytime a boundary envelope is defined by *any* geometric control, it constrains *all* geometric variations from the ideal size and shape to fit within the envelope. For example, by applying profile of a surface to a feature, the bounding envelope that defines the allowable variation from the ideal profile also automatically controls orientation characteristics (e.g., parallelism and perpendicularity) and form characteristics (e.g., flatness and cylindricity). Consequently, most of the geometric characteristics can be controlled by a few controls, and refinements are only added as necessary.

A suggested general framework for implementation of GD&T comprises the following five steps:

1. Select the datum features.
2. Control the datum features.
3. Locate the features.
4. Size and locate the features of size.
5. Refine the orientation and the form of features, if needed.

Each step is elaborated in the following sections.

1. *Select the datum features.*

The datum features should be selected based on the functional use of the part first, rather than on the anticipated manufacturing method. The primary datum is usually the most critical to the part function, and of sufficient surface size to assure a stable setup for establishing the remaining datums. For mating parts, the corresponding interfacing features are usually selected as datum features. Remember that datum features identified with the datum symbol are only utilized when they are called out in the feature control frame of a feature being controlled. Though it is not unusual for every feature of a part to reference the same set of datums, it is not a requirement that they do so.

2. *Control the datum features.*

Though a datum is considered theoretically perfect, the physical datum feature is not. Consequently, the imperfect datum features need to be geometrically controlled just like any other feature. Sometimes the default controls are sufficient. However, since the datum features are used to stabilize the part for manufacturing and inspection, they may warrant additional consideration for that purpose, beyond what is needed just for the functionality of the part. If the primary datum feature is a flat surface, it may be helpful to consider a flatness or surface profile control on it. The secondary and tertiary datum features establish datum planes perpendicular to the primary datum plane. Consequently, it might be useful to use an orientation control (such as perpendicularity) on the secondary and tertiary datum features with respect to the primary datum. A default surface profile applied by a general note to the entire part might be sufficient for this.

When a feature of size is used as a datum feature, the size tolerance automatically provides form control through Rule #1 (see Sec. 20–4). Also, when the position control is used to locate a datum feature of size, it will automatically provide orientation control of the datum.

3. *Locate the features.*

All features have surfaces that need to be located with respect to appropriate datums. The best strategy for most situations is to use basic dimensions to locate the true position of each feature, accompanied by one or more appropriate geometric controls. A default surface profile control can be established with a general note.

4. *Size and locate the features of size.*

Features of size need to be sized as well as located. The following three steps are typical: (1) Locate the true position of the center axis or center plane of each feature of size with basic dimensions. (2) Directly dimension the feature's size, including a plus/minus tolerance. (3) Attach a position control to the feature's size callout to establish limits for location and orientation. For cylindrical features of size that are coaxial to a datum axis, the runout or surface profile controls may be used instead of the position control. The size tolerance on the feature of size automatically provides form control through Rule #1.

5. *Refine orientation and form of features, if needed.*

If any feature needs a tighter control in orientation or form than is provided by the previous steps, additional controls can be added.

The following example demonstrates this process.

EXAMPLE 20–1   Interpret and explain the GD&T notation for the part shown in Fig. 20–3.

Solution   Since this part has already been drawn, the five steps will be used to organize the explanation rather than to make the decisions.

1. *Select the datum features.*

The datum features that have been identified with the datum symbols are the back face and the left and bottom edges. The functional needs of the part are not specified, but this choice of datums is pretty common for this type of simple rectangular plate. Note that the callout for datum B is not placed in-line with the dimension, thus indicating the datum is at the edge of the part rather than the center plane of the part. Since the features are located with respect to the datum, the choice of datum indicates that it is more important to maintain the distance of the hole from the edge rather than to ensure that it is centered. The back face makes a good primary datum (as called out in the position control of the hole), as it is of sufficient size to stabilize the part with respect to three degrees of freedom while establishing the other datums. It is also likely that the back surface will be in contact with a mating part.

2. *Control the datum features.*

A flatness specification ensures that the surface of datum feature A does not vary more than 0.003. This is a common control for a primary datum that is a planar surface, especially if it must fit with a mating surface. Datum feature B, the left surface of the part, is required to be perpendicular to the back face, within a tolerance zone between two parallel planes separated by a distance of 0.005. Note that the envelope generated by this orientation control also constrains the form of this surface (e.g., flatness and straightness). Datum feature C must be perpendicular to datum features A and B, within a tolerance zone of 0.005.

3. *Locate the features.*

All features of this part are features of size, which are handled in the next step.

4. *Size and locate the features of size.*

There are four features of size: the hole, and the height, width, and thickness of the plate. The plate features do not need to be located, since each dimension starts at one of the datum features. Consequently, the three plate dimensions simply need a directly toleranced dimension.

The hole feature is located by basic dimensions from datums B and C. The hole diameter is specified to be within the range from 1.000 to 1.002. The position control on the hole stipulates that the center axis of the hole must be within a cylindrical tolerance zone that has a diameter of 0.003 if the hole is produced at its MMC of 1.000. The tolerance zone may increase to as much as 0.005 as the produced hole diameter increases from 1.000 to 1.002. The tolerance zone is determined with the part immobilized with respect to the three datums, applied in order of the back surface datum A, then the side edge datum B, then the lower edge datum C. Changing this order would change the location and orientation of the hole's tolerance zone. Note that the hole's tolerance zone, in addition to providing a tolerance on the location of the axis, also limits the orientation of the hole's axis. A separate orientation control could have been stipulated if the same tolerance zone was not suitable for both position and orientation.

5. *Refine the orientation and form of the features, if needed.*

No additional controls are specified, so apparently no further refinement is necessary. It is always a good idea to consider how far from the true geometry the part could be produced and still be within specifications. For example, consider how out-of-flat the front face (opposite of datum A) could be. Since the thickness of the plate is a feature of size, the front face is controlled by Rule #1, that is, the size tolerance also establishes an envelope for form control. The size tolerance allows the thickness to vary between 1.9 and 2.1. Since the back datum face may vary only by 0.003, the majority of the size tolerance can be realized by a front face that is curved, bent, or wavy. The plate could be 1.9 thick on one edge and 2.1 thick on the other. If this is not acceptable, the Rule #1 control should be overruled by another more specific refinement.

There are many concepts of GD&T that have not been addressed in this brief introductory chapter. Significant training is necessary to gain competence in defining appropriate geometric controls to achieve the desired function. However, a fundamental ability to "read" a drawing is not out of reach. The next example will provide an opportunity to practice.

EXAMPLE 20–2   Interpret and explain the GD&T notation for the part shown in Fig. 20–34.

Solution   Various aspects of the drawing are circled and annotated with note numbers that correlate with the explanations given here.

Note 1. The front face is defined as a datum feature and is used as the primary datum for several of the geometric controls on the drawing. Since the front face, rather than the back face, is selected as a datum, it is likely that this surface will fit against a mating surface. Functionally, this fit is apparently more important than the back face.

The flatness control is common for a surface that is a primary datum. The entire surface must fit within a tolerance zone defined by two parallel planes separated by a distance of 0.05.

| Figure 20–34

SECTION A–A

UNLESS OTHERWISE SPECIFIED:

Note 2. Consider the outer surface of the protruding cylinder, which is identified as a datum feature. Since it is a feature of size, the datum is the center axis. More precisely, the datum is the theoretically perfect center axis of the imperfect outer surface, as determined by the unrelated actual mating envelope. Note that if the bore through the center of the part had been chosen as the datum feature, the datum would also have been a center axis, but not precisely the same one. The fact that the outer surface is selected indicates it may have greater functional precedent over the bore. Perhaps the outer cylindrical surface will fit within the bore of a mating part. Even though the function of the part is not specified, the geometric requirements of the functional needs are clear, and the part should be manufactured and inspected accordingly.

Every feature of size must be controlled in all four geometric attributes: size, location, orientation, and form. Consider each in turn for the outer surface of the protruding cylinder.

The cylinder's size is directly specified. The diameter at any cross section must be within the stated range of 39.06 to 40.00.

The center axis of a feature of size must be located, usually by basic dimensions. In this case, since this is a datum feature, its own center axis becomes part of the definition of the origin of the datum reference frame. The feature's location is at the origin, so no further basic dimensions are needed to locate it.

The orientation of this cylinder feature is controlled by the perpendicularity control. This control requires the center axis of the feature to lie within a cylindrical tolerance zone that has a diameter of 0.05. The tolerance zone is exactly perpendicular to datum plane A; the center axis of the feature is allowed to tip within the tolerance zone.

No specific form controls are specified for this feature. The form of its surface is controlled by default through Rule #1 for features of size. Rule #1 requires that the surface of the cylinder must be within the cylindrical envelope of perfect form at the MMC diameter of 40.00.

This feature of size is controlled in all aspects: size, location, orientation, and form.

Note 3. Now consider the center bore. This is a feature of size, so its size is directly specified and toleranced. Its form is controlled by Rule #1. It is located by implication (since no basic dimensions are given) at the origin of the datum reference frame. Its orientation and location is controlled by the position control. The position control defines a cylindrical tolerance zone that is first held perpendicular to datum plane A, then centered at the datum axis defined from datum feature B. Since there is no material modifier specified on the tolerance, the default condition of *regardless of feature size (RFS)* applies. Therefore, the tolerance zone diameter is a constant 0.3, regardless of what size the feature is actually produced.

Note 4. The four holes are defined collectively as a pattern. They are features of size, so their diameters are directly dimensioned and toleranced, and their surfaces must not exceed the envelope of perfect form at the MMC diameter of 10 (Rule #1). Their location is specified by the basic dimension of 60 for the bolt circle, as well as by the implication that they are spaced at 90 degrees around the bolt circle.

Control of orientation and location is provided by the position control. Specifically, the position control requires the center axis of each hole to be within a cylindrical tolerance zone. The datum callouts require this tolerance zone to be first held perpendicular to datum A, then centered at the true location as measured with respect to datum axis B. The Ⓜ modifier that accompanies datum B allows a little more leeway on the overall shift of the hole pattern if datum feature B is produced at a diameter less than its MMC.

The zero tolerance value in the position control does not mean that no tolerance is allowed, as it is accompanied by the Ⓜ modifier. This is known as *zero tolerancing* at MMC. The meaning is that the tolerance zone diameter is zero if the hole is produced at its maximum material diameter of 10, but grows to a diameter of 0.1 as the hole's diameter increases to its least material diameter of 10.1. Consequently, the axis of the hole would need to be perfectly located and oriented if the hole is produced at its MMC of 10, but can deviate from perfection if the hole is produced at a larger size.

Note 5. This cylindrical feature is a feature of size, and could have been directly dimensioned and toleranced. Instead, it is specified with a basic dimension of 90 with no tolerance or geometric control in sight. This does not mean it has to be perfect. It means that it is controlled by the default profile tolerance that is stated at the bottom of the drawing. The ideal cylindrical surface is first sized at the basic diameter of 90, and located with respect to datums A and B. Then a tolerance zone is centered around this ideal surface with a total tolerance of 0.2. In this case, the tolerance zone is the space between two concentric cylinders with diameters of 89.8 and 90.2 (a radial difference of 0.2). This tolerance zone controls all four geometric attributes of size, location, orientation, and form.

Note 6. The basic dimension of 20 locates the ideal surface of the front face of the protruding cylinder. The surface profile control requires that the actual front face lie within a tolerance zone consisting of the space of 0.1 between two planes centered around the ideal surface, where each plane is parallel to datum A. This effectively establishes the location of the face (a distance of 19.95 to 20.05 from datum A), as well as the flatness of the face, and the parallelism of the face with respect to datum A.

20-8   GD&T in CAD Models

Many industries are utilizing 3D computer-aided design (CAD) data for some or all of the engineering, manufacturing, and inspection phases of the product life cycle. The prominent standard regulating this is ASME Y14.41-2003, Digital Production Definition Data Practices. The standard addresses many aspects of the practices, requirements, and interpretation of CAD data. The standard defines the use of GD&T in a digital environment where specifications are embedded directly into the data set—not just visually, but functionally as well.

Most of the concepts of GD&T apply directly to digital models. The significant difference in CAD models versus 2D drawings is that the CAD model represents the ideal geometry of the part. Any dimension can be queried from the model and the exact (ideal) dimension can be obtained. In fact, the ideal data can be directly utilized in computer-aided manufacturing operations. This leads to a misconception that a

Figure 20–35

An example of GD&T applied
to a 3D CAD model.

20 ± 0.1

4 × Ø10 +0.1 −0.0

⊕ | Ø 0 Ⓜ | A | B Ⓜ

Ø 20 ± 0.05

⊕ | Ø 0.3 | A | B

◠ | 0.1 | A

▱ | 0.05

A

Ø 40 +0.00 −0.04

⊥ | Ø 0.05 | A

B

part manufactured directly from the model on a computer numerical controlled
(CNC) machine will be perfect. In fact, CNC manufacturing has the same requirements as manual manufacturing methods with regard to the need to specify and
inspect geometric tolerances.

The Y14.41-2003 standard allows for *all* geometry data to be considered *basic,*
unless superseded by a toleranced dimension or defined as a reference dimension.
Geometric controls are applied to the 3D data models to control the features just
the same as on the 2D drawings. Direct tolerancing is recommended only for features of size. Essentially, GD&T works the same on 3D models as in 2D drawings,
except that in 3D models the basic dimensions do not have to be shown, since the
default is that all dimensions are basic. Figure 20–35 shows the 3D solid model
of the part used in Ex. 20–2, with appropriate GD&T elements embedded into the
CAD data.

The ongoing transition to 3D digital representation brings with it the possibilities
of tighter integration of the various processes of design, analysis, and manufacturing.
For example, the embedded geometric dimensioning and tolerancing information can
be directly accessed for tolerance analysis and process planning.

20–9 Glossary of GD&T Terms

Most of the concepts in GD&T are simple, but the vocabulary to describe them can
seem overwhelming at first. That is because the vocabulary needs to be precise enough
to be consistent and unambiguous. For convenient reference, some of the most commonly used terms in GD&T are summarized in this section.

Actual Mating Envelope—a perfectly shaped counterpart of an imperfect
feature of size, which can be contracted about an external feature, or
expanded within an internal feature, so that it contacts the high points
of the feature's surface.

Actual Mating Envelope, Related—an actual mating envelope that is sized to fit the feature while maintaining some constraint in orientation or location with respect to a datum.

Actual Mating Envelope, Unrelated—an actual mating envelope that is sized to fit the feature without any constraint to any datum.

Axis—a line defining the center of a cylindrical feature, established from the theoretical axis of the unrelated actual mating envelope of the cylindrical feature's extremities.

Basic Dimension—a theoretically exact dimension that ideally locates and/or orients the tolerance zone of a feature. It does not have a tolerance directly associated with it, but is instead associated with a geometric control of a tolerance zone. Basic dimensions are denoted by a box around the dimension, or by a general note.

Bonus Tolerance—additional tolerance that applies to a feature as its size shifts from a stated material condition of MMC or LMC.

Center Plane—the theoretical plane located at the center of a noncylindrical feature of size, established from the center plane of the unrelated actual mating envelope of the feature's extremities.

Combined Feature Control Frame—a feature control frame made of two or more feature control frames, each with a geometric characteristic symbol. The geometric controls are applied to the feature, in order from top to bottom.

Composite Feature Control Frame—a feature control frame made of two or more feature control frames sharing a common geometric characteristic symbol.

Datum—a theoretically exact point, axis, line, or plane derived from a datum feature simulator, used as an origin for repeatable measurements.

Datum Axis—the theoretical axis of a cylindrical datum feature, established from the axis of the unrelated actual mating envelope of the cylindrical feature's extremities.

Datum Feature—an actual physical surface of the part that is specified in order to establish a theoretically exact datum.

Datum Feature Simulator—a precision embodiment, such as a surface plate, gauge pin, or machine tool bed, of the datum described by an imperfect datum feature.

Datum of Size—a datum feature that is a feature of size, and therefore subject to size variation based on plus/minus tolerances.

Datum Reference Frame—a set of up to three mutually perpendicular planes that are defined as the origin of measurement for locating the features of a part.

Derived Median Line—an imperfect "line" formed by the center points of all cross sections of a feature, where the cross sections are normal to the axis of the Unrelated Actual Mating Envelope of the feature.

Derived Median Plane—an imperfect "plane" formed by the center points of all line segments bounded by a feature, where the line segments are normal to the center plane of the unrelated actual mating envelope of the feature.

Envelope Principle—See definition for Rule #1, given later in this section.

Feature—a general term referring to a clearly identifiable physical portion of a part, such as a hole, pin, slot, surface, or cylinder.

Feature Control Frame—a rectangular box attached to a feature on a drawing, containing the necessary information to define the tolerance zone of the specified feature.

Feature of Size, Irregular—a directly toleranced feature or collection of features that may contain or be contained by an actual mating envelope.

Feature of Size, Regular—a cylindrical surface, a spherical surface, a circular element, or a set of two opposed parallel elements or surfaces that are associated with a directly toleranced dimension. A regular feature of size has a size that can be measured across two opposing points, and has a reproducible center point, axis, or center plane.

Feature-Relating Tolerance Zone Framework (FRTZF)—a tolerance zone framework that governs the positional relationship from feature to feature within a pattern of features. It is specified in the bottom row of a composite feature control frame.

Geometric Attributes—the four broad attributes (size, location, orientation, and form) that must be considered to geometrically define a feature. This term is not strictly defined by GD&T.

Geometric Characteristics—the 14 characteristics defined in Table 20–1 that are available to control some aspect of a geometric tolerance of a feature. A geometric characteristic symbol is the first item in any feature control frame.

Least Material Boundary (LMB)—the limit defined by a tolerance or combination of tolerances that exists on or inside the material of a feature(s). When applied as a modifier to a datum reference in a feature control frame with the Ⓛ symbol, it establishes the datum feature simulator at a boundary determined by the combined effects of size (least material), and all applicable geometric tolerances.

Least Material Condition (LMC)—the condition in which a feature of size contains the least amount of material within the stated limits of size (e.g., maximum hole diameter or minimum shaft diameter). This condition may be specified as a tolerance modifier in a feature control frame with the Ⓛ symbol.

Material Condition Modifier—a modifier symbol, Ⓜ or Ⓛ, applied to a geometric tolerance to indicate that the tolerance applies at maximum material condition or least material condition, respectively. The absence of a material condition modifier indicates that the tolerance applies at all material conditions, that is, regardless of feature size (RFS).

Maximum Material Boundary (MMB)—the limit defined by a tolerance or combination of tolerances that exists on or outside the material of a feature(s). When applied as a modifier to a datum reference in a feature control frame with the Ⓜ symbol, it establishes the datum feature simulator at a boundary determined by the combined effects of size (maximum material), and all applicable geometric tolerances.

Maximum Material Condition (MMC)—the condition in which a feature of size contains the maximum amount of material within the stated limits of size (e.g., minimum hole diameter or maximum shaft diameter). This condition may be specified as a tolerance modifier in a feature control frame with the Ⓜ symbol.

Pattern-Locating Tolerance Zone Framework (PLTZF)—a tolerance zone framework that governs the positional relationship between a pattern of features and the datum features. It is specified in the top row of a composite feature control frame.

Regardless of Feature Size (RFS)—indicates that the stated tolerance applies, regardless of the actual size to which the feature is produced. This is the default condition for a tolerance with no modifier symbol (i.e., Ⓜ or Ⓛ).

Regardless of Material Boundary (RMB)—indicates that a datum feature simulator progresses from MMB toward LMB until it makes maximum contact with the extremities of a feature(s). This is the default condition for a datum reference with no modifier symbol (i.e., Ⓜ or Ⓛ).

Rule #1—when only a tolerance of size (i.e., a plus/minus tolerance) is specified for a feature of size, the limits of size prescribe the extent of which variation in its geometric form, as well as size, are allowed. The surface of a feature of size may not extend beyond an envelope of perfect form at MMC. This rule is also referred to as the envelope principle.

Tolerance—the total amount a specific dimension is permitted to vary, between the maximum and minimum limits.

Tolerance Zone—a limiting boundary in which the actual feature must be contained.

Virtual Condition—a constant "worst case" boundary defined by the collective effects of a feature's size, geometric tolerance, and material condition.

PROBLEMS

20–1   In the traditional coordinate dimensioning system, which of the following is true? (Select one.)
   i. Only "features of size" need to include a tolerance.
   ii. Only dimensions that are important need to include a tolerance.
   iii. Only dimensions that need to be tightly controlled need to include a tolerance.
   iv. All dimensions should include a tolerance.

20–2   In GD&T, which type of dimension should generally be directly toleranced with a plus/minus tolerance?

20–3   What underlying purpose is emphasized by the *ASME Y14.5–2009* standard in the dimensioning and tolerancing of a part? (Select one.)
   i. The method of manufacturing.
   ii. The design intent.
   iii. The inspection process.
   iv. Equal attention to all of the above.

20–4 What is the term that refers to a feature which has a size that can be measured across two opposing points?

20–5 What are the four geometric attributes that must be considered to define the geometry of a feature of a part?

20–6 What are the four geometric characteristics that provide form control?

20–7 What are the three geometric characteristics that provide orientation control?

20–8 What are the three geometric characteristics that provide location control? Which of the three is most prominently used?

20–9 How is a basic dimension toleranced? (Select one.)

    i. Basic dimensions receive the default tolerance specified in the title block.

    ii. Basic dimensions are not toleranced.

    iii. Basic dimensions receive the tolerance from an associated feature control frame.

20–10 For the part shown, identify all of the features of size.

Problems 20–10 to 20–14

20–11 For the part shown, clearly identify each of the following, with labels and sketches on the drawing.

    (*a*) Datum features A, B, and C.

    (*b*) Datums A, B, and C.

    (*c*) Datum reference frame based on datum features A, B, and C.

20–12 For the part shown, the ideal position of the cylindrical boss is located with the basic dimensions of 100 and 50. These basic dimensions are measured from which of the following? (Select one.)

    i. The physical edges of the part.

    ii. The high points on the physical edges of the part.

    iii. The low points on the physical edges of the part.

    iv. The theoretical datum planes B and C.

The basic dimensions are locating which of the following? (Select one.)

    i. The physical location of the center axis of the boss.

    ii. The center axis of the boss as determined by the actual mating envelope of the boss.

    iii. The ideal location of the center axis of the tolerance zone specified by the position control.

If the part is produced and is being inspected, the location of the boss will be measured from which of the following? (Select one.)

    i. The physical edges of the part.

    ii. The high points on the physical edges of the part.

    iii. The low points on the physical edges of the part.

    iv. The theoretical datum planes B and C.

    v. The datum feature simulators for datum features B and C.

20–13   For the part shown, answer the following questions with regard to the cylindrical boss.

    (*a*) What are the maximum and minimum diameters allowed for the boss?

    (*b*) What is the effect of the position tolerance of 0.2 on the diameters specified in part (*a*)?

    (*c*) The position control defines a tolerance zone. Specifically what must stay within that tolerance zone?

    (*d*) What is the diameter of the tolerance zone if the boss is produced with a diameter of 50.3?

    (*e*) What is the diameter of the tolerance zone if the boss is produced with a diameter of 49.7?

    (*f*) Describe the significance of the datum references to the determination of the position tolerance zone.

    (*g*) What is the perpendicularity tolerance with respect to datum A? (Select one.)

        i. Not defined.

        ii. Controlled by the position tolerance; 0.2 at MMC to 0.5 at LMC.

        iii. Controlled by the size tolerance; 0.3.

        iv. Must be perfectly perpendicular; 0.

    (*h*) What controls the cylindricity? (Select one.)

        i. There is no control on the cylindricity.

        ii. From Rule #1, the envelope of a perfect cylinder with diameter of 50.

        iii. From Rule #1, the envelope of a perfect cylinder with diameter of 50.3.

        iv. From the position control, the center axis of each cross section must be within the 0.2 cylindrical tolerance zone.

20–14   For the part shown, answer the following questions with regard to the hole.

    (*a*) What are the maximum and minimum diameters allowed for the hole?

    (*b*) What is the effect of the position tolerance of 0.3 on the diameters specified in part (*a*)?

    (*c*) The position control defines a tolerance zone. Specifically what must stay within that tolerance zone?

    (*d*) What is the diameter of the tolerance zone if the hole is produced with a diameter of 30.1?

    (*e*) What is the diameter of the tolerance zone if the hole is produced with a diameter of 29.9?

    (*f*) Describe the significance of the datum references to the determination of the position tolerance zone.

    (*g*) What is the perpendicularity tolerance with respect to datum A? (Select one.)

        i. Not defined.

        ii. Controlled by the position tolerance; 0.3.

        iii. Controlled by the size tolerance; 0.1.

        iv. Must be perfectly perpendicular; 0.

    (*h*) What controls the cylindricity? (Select one.)

        i. There is no control on the cylindricity.

        ii. From Rule #1, the envelope of a perfect cylinder with diameter of 30.

        iii. From Rule #1, the envelope of a perfect cylinder with diameter of 29.9.

        iv. From the position control, the center axis of each cross section must be within the 0.3 cylindrical tolerance zone.

20–15   Describe how a center axis is determined for a physical hole that is not perfectly formed.

20–16 According to the envelope principle (Rule #1), the size tolerance applied to a feature of size controls the size and _____ of the feature. (Select one.)
  i. location
  ii. orientation
  iii. form
  iv. runout
  v. All of the above.

20–17 If a shaft diameter is dimensioned 20 ± 0.2, determine the diameter of the shaft at MMC and at LMC.

20–18 If a hole diameter is dimensioned 20 ± 0.2, determine the diameter of the hole at MMC and at LMC.

20–19 A hole diameter is dimensioned 20 ± 0.2. According to the form control provided by the envelope principle (Rule #1), is the limiting envelope a perfect cylinder with diameter of 19.8, 20.2, or both? Explain your answer.

20–20 A shaft diameter is dimensioned 20 ± 0.2. According to the form control provided by the envelope principle (Rule #1), is the limiting envelope a perfect cylinder with diameter of 19.8, 20.2, or both? Explain your answer.

20–21 The diameter of a cylindrical boss is dimensioned 25 ± 0.2. A position control is used to control the basic location of the boss. Specify the diameters allowed for the position tolerance zone if the boss is produced with diameters of 24.8, 25.0, and 25.2, for each of the following position tolerance specifications:
  (*a*) ⌀0.1 (*b*) ⌀0.1Ⓜ (*c*) ⌀0.1Ⓛ

20–22 A hole diameter is dimensioned $32^{+0.4}_{-0.0}$. A position control is used to control the basic location of the hole. Specify the diameters allowed for the position tolerance zone if the hole is produced with diameters of 32.0, 32.2, and 32.4, for each of the following position tolerance specifications:
  (*a*) ⌀0.3 (*b*) ⌀0.3Ⓜ (*c*) ⌀0.3Ⓛ

20–23 What is the name of the geometric characteristic that effectively controls a combination of circularity and straightness of a cylinder?

20–24 What is the name of the geometric characteristic that can be specified in a note to provide a default tolerance zone to control size, form, orientation, and location of all features that are not otherwise controlled?

20–25 Which geometric characteristics never reference datums? Why?

20–26 Answer the following questions regarding material condition modifiers.
  (*a*) What are the three material condition modifiers?
  (*b*) Which one is the default if nothing is specified?
  (*c*) Which one(s) can provide "bonus" tolerance?
  (*d*) Which of the following is increased by a bonus tolerance? (Select one.)
    i. A size dimension.
    ii. A ± tolerance of a size dimension.
    iii. A basic dimension locating a feature.
    iv. A size of a tolerance zone controlling a feature.
  (*e*) To which of the following can a material condition modifier symbol be applied? (Select one.)
    i. A size dimension.
    ii. A ± tolerance of a size dimension.

iii. A tolerance of a geometric characteristic controlling a feature of size.

iv. A tolerance of a geometric characteristic controlling any feature.

(*f*) Which material condition modifier should be considered if the goal is to ensure a minimum clearance fit for a bolt in a hole, but to give greater manufacturing flexibility if the hole is produced with a greater clearance?

(*g*) Which material condition modifier should be considered if the goal is to provide a consistent press fit between interchangeable parts?

(*h*) Which material condition modifier should be considered if the goal is to ensure a minimum wall thickness for a casting, but to give greater manufacturing flexibility if the wall is produced with a greater thickness?

20–27 The drawing shown is of a mounting fixture to locate and orient a rod (not shown) through the large bore. The fixture will be bolted to a frame through the four bolt holes that are countersunk to recess the bolt heads. The bolt holes have too much clearance to properly align the rod, so the fixture will be aligned with two locating pins in the frame that will fit in the Ø6 hole and slot.

(*a*) Determine the minimum diameter allowed for the countersink.

(*b*) Determine the maximum depth allowed for the countersink.

(*c*) Determine the diameter of the bolt holes at MMC.

(*d*) Identify every feature that qualifies as a feature of size.

(*e*) The width of the base is specified with a basic dimension of 60, with no tolerance. (Note that as a feature of size, it could have had a tolerance directly specified.) What are the minimum and maximum allowed dimensions for the base width? Explain how they are determined.

Problem 20–27

(f) Describe the datum features A, B, and C. Describe their corresponding datums. Describe the datum reference frame that is defined by applying A, B, and C in that order. Describe how the part is stabilized by these datums. Explain why this is more appropriate for this application than using the edges of the base for datums B and C. (Notice that the basic dimensions are either measured from, or implied to be centered on, the datums of the datum reference frame.)

(g) If datum feature B is produced with a diameter of $\varnothing 6.00$, what is the diameter of the tolerance zone in which its axis must lie? What if it is produced at $\varnothing 6.05$?

(h) If the bolt holes are produced at $\varnothing 6.0$, what is the diameter of the tolerance zones locating the bolt hole pattern with respect to the true position specified by the basic dimensions? What if the bolt holes are produced at $\varnothing 6.1$?

(i) If the bolt holes are produced at $\varnothing 6.0$, what is the diameter of the tolerance zones locating the position of the bolt holes with respect to one another? What if the bolt holes are produced at $\varnothing 6.1$?

(j) Explain why the Ⓜ modifier is appropriate for the bolt hole position tolerance.

(k) For the large bore, explain what provides control of each of the following: orientation, straightness of its center axis, and cylindricity of its surface.

(l) Assume the part is cast, and the casting operation can provide a surface profile tolerance of less than 0.5. Which surfaces can likely be left in the as-cast condition without compromising any of the requirements of the drawing? How would this change if the drawing were modified to use the edges of the base as datum features B and C, while still maintaining the functional goals for the alignment of the rod?

Appendix Outline

Table A–1

Standard SI Prefixes*[†]

Name	Symbol	Factor
exa	E	$1\ 000\ 000\ 000\ 000\ 000\ 000 = 10^{18}$
peta	P	$1\ 000\ 000\ 000\ 000\ 000 = 10^{15}$
tera	T	$1\ 000\ 000\ 000\ 000 = 10^{12}$
giga	G	$1\ 000\ 000\ 000 = 10^{9}$
mega	M	$1\ 000\ 000 = 10^{6}$
kilo	k	$1\ 000 = 10^{3}$
hecto[‡]	h	$100 = 10^{2}$
deka[‡]	da	$10 = 10^{1}$
deci[‡]	d	$0.1 = 10^{-1}$
centi[‡]	c	$0.01 = 10^{-2}$
milli	m	$0.001 = 10^{-3}$
micro	μ	$0.000\ 001 = 10^{-6}$
nano	n	$0.000\ 000\ 001 = 10^{-9}$
pico	p	$0.000\ 000\ 000\ 001 = 10^{-12}$
femto	f	$0.000\ 000\ 000\ 000\ 001 = 10^{-15}$
atto	a	$0.000\ 000\ 000\ 000\ 000\ 001 = 10^{-18}$

*If possible use multiple and submultiple prefixes in steps of 1000.

[†]Spaces are used in SI instead of commas to group numbers to avoid confusion with the practice in some European countries of using commas for decimal points.

[‡]Not recommended but sometimes encountered.

Table A–2

Conversion Factors A to Convert Input X to Output Y Using the Formula $Y = AX^*$

Multiply Input X	By Factor A	To Get Output Y	Multiply Input X	By Factor A	To Get Output Y
British thermal unit, Btu	1055	joule, J	mile, mi	1.610	kilometer, km
			mile/hour, mi/h	1.61	kilometer/hour, km/h
Btu/second, Btu/s	1.05	kilowatt, kW	mile/hour, mi/h	0.447	meter/second, m/s
calorie	4.19	joule, J	moment of inertia, $\text{lbm} \cdot \text{ft}^2$	0.0421	kilogram-meter^2, $\text{kg} \cdot \text{m}^2$
centimeter of mercury (0°C)	1.333	kilopascal, kPa	moment of inertia, $\text{lbm} \cdot \text{in}^2$	293	kilogram-millimeter^2, $\text{kg} \cdot \text{mm}^2$
centipoise, cP	0.001	pascal-second, $\text{Pa} \cdot \text{s}$	moment of section (second moment of area), in^4	41.6	centimeter^4, cm^4
degree (angle)	0.0174	radian, rad	ounce-force, oz	0.278	newton, N
foot, ft	0.305	meter, m	ounce-mass	0.0311	kilogram, kg
foot^2, ft^2	0.0929	meter^2, m^2	pound, lbf†	4.45	newton, N
foot/minute, ft/min	0.0051	meter/second, m/s	pound-foot, lbf $\cdot$ ft	1.36	newton-meter, N $\cdot$ m
foot-pound, ft $\cdot$ lbf	1.35	joule, J	pound/foot^2, lbf/ft^2	47.9	pascal, Pa
foot-pound/second, ft $\cdot$ lbf/s	1.35	watt, W	pound-inch, lbf $\cdot$ in	0.113	joule, J
foot/second, ft/s	0.305	meter/second, m/s	pound-inch, lbf $\cdot$ in	0.113	newton-meter, N $\cdot$ m
gallon (U.S.), gal	3.785	liter, L	pound/inch, lbf/in	175	newton/meter, N/m
horsepower, hp	0.746	kilowatt, kW	pound/inch^2, psi (lbf/in^2)	6.89	kilopascal, kPa
inch, in	0.0254	meter, m			
inch, in	25.4	millimeter, mm	pound-mass, lbm	0.454	kilogram, kg
inch^2, in^2	645	millimeter^2, mm^2	pound-mass/second, lbm/s	0.454	kilogram/second, kg/s
inch of mercury (32°F)	3.386	kilopascal, kPa	quart (U.S. liquid), qt	946	milliliter, mL
kilopound, kip	4.45	kilonewton, kN	section modulus, in^3	16.4	centimeter^3, cm^3
kilopound/inch^2, kpsi (ksi)	6.89	megapascal, MPa (N/mm^2)	slug	14.6	kilogram, kg
			ton (short 2000 lbm)	907	kilogram, kg
mass, lbf $\cdot$ s^2/in	175	kilogram, kg	yard, yd	0.914	meter, m

*Approximate.

†The U.S. Customary system unit of the pound-force is often abbreviated as lbf to distinguish it from the pound-mass, which is abbreviated as lbm.

Table A–3

Optional SI Units for
Bending Stress
$\sigma = Mc/l$, Torsion Stress
$\tau = Tr/J$, Axial Stress
$\sigma = F/A$, and Direct
Shear Stress $\tau = F/A$

Bending and Torsion					Axial and Direct Shear		
M, T	**I, J**	**c, r**	**σ, τ**		**F**	**A**	**σ, τ**
N·m*	m⁴	m	Pa		N*	m²	Pa
N·m	cm⁴	cm	MPa (N/mm²)		N†	mm²	MPa (N/mm²)
N·m†	mm⁴	mm	GPa		kN	m²	kPa
kN·m	cm⁴	cm	GPa		kN†	mm²	GPa
N·mm†	mm⁴	mm	MPa (N/mm²)				

*Basic relation.
†Often preferred.

Table A–4

Optional SI Units for
Bending Deflection
$y = f(Fl^3/El)$ or
$y = f(wl^4/El)$ and
Torsional Deflection
$\theta = Tl/GJ$

Bending Deflection					Torsional Deflection				
F, wl	**l**	**I**	**E**	**y**	**T**	**l**	**J**	**G**	**θ**
N*	m	m⁴	Pa	m	N·m*	m	m⁴	Pa	rad
kN†	mm	mm⁴	GPa	mm	N·m†	mm	mm⁴	GPa	rad
kN	m	m⁴	GPa	μm	N·mm	mm	mm⁴	MPa (N/mm²)	rad
N	mm	mm⁴	kPa	m	N·m	cm	cm⁴	MPa (N/mm²)	rad

*Basic relation.
†Often preferred.

Table A–5

Physical Constants of Materials

Material	Modulus of Elasticity E		Modulus of Rigidity G		Poisson's Ratio ν	Unit Weight w		
	Mpsi	GPa	Mpsi	GPa		lbf/in³	lbf/ft³	kN/m³
Aluminum (all alloys)	10.4	71.7	3.9	26.9	0.333	0.098	169	26.6
Beryllium copper	18.0	124.0	7.0	48.3	0.285	0.297	513	80.6
Brass	15.4	106.0	5.82	40.1	0.324	0.309	534	83.8
Carbon steel	30.0	207.0	11.5	79.3	0.292	0.282	487	76.5
Cast iron (gray)	14.5	100.0	6.0	41.4	0.211	0.260	450	70.6
Copper	17.2	119.0	6.49	44.7	0.326	0.322	556	87.3
Douglas fir	1.6	11.0	0.6	4.1	0.33	0.016	28	4.3
Glass	6.7	46.2	2.7	18.6	0.245	0.094	162	25.4
Inconel	31.0	214.0	11.0	75.8	0.290	0.307	530	83.3
Lead	5.3	36.5	1.9	13.1	0.425	0.411	710	111.5
Magnesium	6.5	44.8	2.4	16.5	0.350	0.065	112	17.6
Molybdenum	48.0	331.0	17.0	117.0	0.307	0.368	636	100.0
Monel metal	26.0	179.0	9.5	65.5	0.320	0.319	551	86.6
Nickel silver	18.5	127.0	7.0	48.3	0.322	0.316	546	85.8
Nickel steel	30.0	207.0	11.5	79.3	0.291	0.280	484	76.0
Phosphor bronze	16.1	111.0	6.0	41.4	0.349	0.295	510	80.1
Stainless steel (18-8)	27.6	190.0	10.6	73.1	0.305	0.280	484	76.0
Titanium alloys	16.5	114.0	6.2	42.4	0.340	0.160	276	43.4

Table A–6

Properties of Structural-
Steel Equal Legs
Angles*†

w = weight per foot, lbf/ft
m = mass per meter, kg/m
A = area, in² (cm²)
I = second moment of area, in⁴ (cm⁴)
k = radius of gyration, in (cm)
y = centroidal distance, in (cm)
Z = section modulus, in³, (cm³)

Size, in	w	A	I_{1-1}	k_{1-1}	Z_{1-1}	y	k_{3-3}
$1 \times 1 \times \frac{1}{8}$	0.80	0.234	0.021	0.298	0.029	0.290	0.191
$\times \frac{1}{4}$	1.49	0.437	0.036	0.287	0.054	0.336	0.193
$1\frac{1}{2} \times 1\frac{1}{2} \times \frac{1}{8}$	1.23	0.36	0.074	0.45	0.068	0.41	0.29
$\times \frac{1}{4}$	2.34	0.69	0.135	0.44	0.130	0.46	0.29
$2 \times 2 \times \frac{1}{8}$	1.65	0.484	0.190	0.626	0.131	0.546	0.398
$\times \frac{1}{4}$	3.19	0.938	0.348	0.609	0.247	0.592	0.391
$\times \frac{3}{8}$	4.7	1.36	0.479	0.594	0.351	0.636	0.389
$2\frac{1}{2} \times 2\frac{1}{2} \times \frac{1}{4}$	4.1	1.19	0.703	0.769	0.394	0.717	0.491
$\times \frac{3}{8}$	5.9	1.73	0.984	0.753	0.566	0.762	0.487
$3 \times 3 \times \frac{1}{4}$	4.9	1.44	1.24	0.930	0.577	0.842	0.592
$\times \frac{3}{8}$	7.2	2.11	1.76	0.913	0.833	0.888	0.587
$\times \frac{1}{2}$	9.4	2.75	2.22	0.898	1.07	0.932	0.584
$3\frac{1}{2} \times 3\frac{1}{2} \times \frac{1}{4}$	5.8	1.69	2.01	1.09	0.794	0.968	0.694
$\times \frac{3}{8}$	8.5	2.48	2.87	1.07	1.15	1.01	0.687
$\times \frac{1}{2}$	11.1	3.25	3.64	1.06	1.49	1.06	0.683
$4 \times 4 \times \frac{1}{4}$	6.6	1.94	3.04	1.25	1.05	1.09	0.795
$\times \frac{3}{8}$	9.8	2.86	4.36	1.23	1.52	1.14	0.788
$\times \frac{1}{2}$	12.8	3.75	5.56	1.22	1.97	1.18	0.782
$\times \frac{5}{8}$	15.7	4.61	6.66	1.20	2.40	1.23	0.779
$6 \times 6 \times \frac{3}{8}$	14.9	4.36	15.4	1.88	3.53	1.64	1.19
$\times \frac{1}{2}$	19.6	5.75	19.9	1.86	4.61	1.68	1.18
$\times \frac{5}{8}$	24.2	7.11	24.2	1.84	5.66	1.73	1.18
$\times \frac{3}{4}$	28.7	8.44	28.2	1.83	6.66	1.78	1.17

Table A–6

Properties of Structural-Steel Equal Legs Angles*†

(*Continued*)

Size, mm	m	A	I_{1-1}	k_{1-1}	Z_{1-1}	y	k_{3-3}
25 × 25 × 3	1.11	1.42	0.80	0.75	0.45	0.72	0.48
× 4	1.45	1.85	1.01	0.74	0.58	0.76	0.48
× 5	1.77	2.26	1.20	0.73	0.71	0.80	0.48
40 × 40 × 4	2.42	3.08	4.47	1.21	1.55	1.12	0.78
× 5	2.97	3.79	5.43	1.20	1.91	1.16	0.77
× 6	3.52	4.48	6.31	1.19	2.26	1.20	0.77
50 × 50 × 5	3.77	4.80	11.0	1.51	3.05	1.40	0.97
× 6	4.47	5.59	12.8	1.50	3.61	1.45	0.97
× 8	5.82	7.41	16.3	1.48	4.68	1.52	0.96
60 × 60 × 5	4.57	5.82	19.4	1.82	4.45	1.64	1.17
× 6	5.42	6.91	22.8	1.82	5.29	1.69	1.17
× 8	7.09	9.03	29.2	1.80	6.89	1.77	1.16
× 10	8.69	11.1	34.9	1.78	8.41	1.85	1.16
80 × 80 × 6	7.34	9.35	55.8	2.44	9.57	2.17	1.57
× 8	9.63	12.3	72.2	2.43	12.6	2.26	1.56
× 10	11.9	15.1	87.5	2.41	15.4	2.34	1.55
100 × 100 × 8	12.2	15.5	145	3.06	19.9	2.74	1.96
× 12	17.8	22.7	207	3.02	29.1	2.90	1.94
× 15	21.9	27.9	249	2.98	35.6	3.02	1.93
150 × 150 × 10	23.0	29.3	624	4.62	56.9	4.03	2.97
× 12	27.3	34.8	737	4.60	67.7	4.12	2.95
× 15	33.8	43.0	898	4.57	83.5	4.25	2.93
× 18	40.1	51.0	1050	4.54	98.7	4.37	2.92

*Metric sizes also available in sizes of 45, 70, 90, 120, and 200 mm.

†These sizes are also available in aluminum alloy.

Table A–7

Properties of Structural-Steel Channels*

a, b = size, in (mm)
w = weight per foot, lbf/ft
m = mass per meter, kg/m
t = web thickness, in (mm)
A = area, in² (cm²)
I = second moment of area, in⁴ (cm⁴)
k = radius of gyration, in (cm)
x = centroidal distance, in (cm)
Z = section modulus, in³ (cm³)

a, in	b, in	t	A	w	I_{1-1}	k_{1-1}	Z_{1-1}	I_{2-2}	k_{2-2}	Z_{2-2}	x
3	1.410	0.170	1.21	4.1	1.66	1.17	1.10	0.197	0.404	0.202	0.436
3	1.498	0.258	1.47	5.0	1.85	1.12	1.24	0.247	0.410	0.233	0.438
3	1.596	0.356	1.76	6.0	2.07	1.08	1.38	0.305	0.416	0.268	0.455
4	1.580	0.180	1.57	5.4	3.85	1.56	1.93	0.319	0.449	0.283	0.457
4	1.720	0.321	2.13	7.25	4.59	1.47	2.29	0.433	0.450	0.343	0.459
5	1.750	0.190	1.97	6.7	7.49	1.95	3.00	0.479	0.493	0.378	0.484
5	1.885	0.325	2.64	9.0	8.90	1.83	3.56	0.632	0.489	0.450	0.478
6	1.920	0.200	2.40	8.2	13.1	2.34	4.38	0.693	0.537	0.492	0.511
6	2.034	0.314	3.09	10.5	15.2	2.22	5.06	0.866	0.529	0.564	0.499
6	2.157	0.437	3.83	13.0	17.4	2.13	5.80	1.05	0.525	0.642	0.514
7	2.090	0.210	2.87	9.8	21.3	2.72	6.08	0.968	0.581	0.625	0.540
7	2.194	0.314	3.60	12.25	24.2	2.60	6.93	1.17	0.571	0.703	0.525
7	2.299	0.419	4.33	14.75	27.2	2.51	7.78	1.38	0.564	0.779	0.532
8	2.260	0.220	3.36	11.5	32.3	3.10	8.10	1.30	0.625	0.781	0.571
8	2.343	0.303	4.04	13.75	36.2	2.99	9.03	1.53	0.615	0.854	0.553
8	2.527	0.487	5.51	18.75	44.0	2.82	11.0	1.98	0.599	1.01	0.565
9	2.430	0.230	3.91	13.4	47.7	3.49	10.6	1.75	0.669	0.962	0.601
9	2.485	0.285	4.41	15.0	51.0	3.40	11.3	1.93	0.661	1.01	0.586
9	2.648	0.448	5.88	20.0	60.9	3.22	13.5	2.42	0.647	1.17	0.583
10	2.600	0.240	4.49	15.3	67.4	3.87	13.5	2.28	0.713	1.16	0.634
10	2.739	0.379	5.88	20.0	78.9	3.66	15.8	2.81	0.693	1.32	0.606
10	2.886	0.526	7.35	25.0	91.2	3.52	18.2	3.36	0.676	1.48	0.617
10	3.033	0.673	8.82	30.0	103	3.43	20.7	3.95	0.669	1.66	0.649
12	3.047	0.387	7.35	25.0	144	4.43	24.1	4.47	0.780	1.89	0.674
12	3.170	0.510	8.82	30.0	162	4.29	27.0	5.14	0.763	2.06	0.674

Table A–7

Properties of Structural-Steel Channels*   (*Continued*)

$a \times b$, mm	m	t	A	I_{1-1}	k_{1-1}	Z_{1-1}	I_{2-2}	k_{2-2}	Z_{2-2}	x
76 × 38	6.70	5.1	8.53	74.14	2.95	19.46	10.66	1.12	4.07	1.19
102 × 51	10.42	6.1	13.28	207.7	3.95	40.89	29.10	1.48	8.16	1.51
127 × 64	14.90	6.4	18.98	482.5	5.04	75.99	67.23	1.88	15.25	1.94
152 × 76	17.88	6.4	22.77	851.5	6.12	111.8	113.8	2.24	21.05	2.21
152 × 89	23.84	7.1	30.36	1166	6.20	153.0	215.1	2.66	35.70	2.86
178 × 76	20.84	6.6	26.54	1337	7.10	150.4	134.0	2.25	24.72	2.20
178 × 89	26.81	7.6	34.15	1753	7.16	197.2	241.0	2.66	39.29	2.76
203 × 76	23.82	7.1	30.34	1950	8.02	192.0	151.3	2.23	27.59	2.13
203 × 89	29.78	8.1	37.94	2491	8.10	245.2	264.4	2.64	42.34	2.65
229 × 76	26.06	7.6	33.20	2610	8.87	228.3	158.7	2.19	28.22	2.00
229 × 89	32.76	8.6	41.73	3387	9.01	296.4	285.0	2.61	44.82	2.53
254 × 76	28.29	8.1	36.03	3367	9.67	265.1	162.6	2.12	28.21	1.86
254 × 89	35.74	9.1	45.42	4448	9.88	350.2	302.4	2.58	46.70	2.42
305 × 89	41.69	10.2	53.11	7061	11.5	463.3	325.4	2.48	48.49	2.18
305 × 102	46.18	10.2	58.83	8214	11.8	539.0	499.5	2.91	66.59	2.66

*These sizes are also available in aluminum alloy.

Table A–8

Properties of Round
Tubing

w_a = unit weight of aluminum tubing, lbf/ft
w_s = unit weight of steel tubing, lbf/ft
m = unit mass, kg/m
A = area, in^2 (cm^2)
I = second moment of area, in^4 (cm^4)
J = second polar moment of area, in^4 (cm^4)
k = radius of gyration, in (cm)
Z = section modulus, in^3 (cm^3)
$d \times t$ = size (OD) and thickness, in (mm)

Size, in	w_a	w_s	A	I	k	Z	J
$1 \times \frac{1}{8}$	0.416	1.128	0.344	0.034	0.313	0.067	0.067
$1 \times \frac{1}{4}$	0.713	2.003	0.589	0.046	0.280	0.092	0.092
$1\frac{1}{2} \times \frac{1}{8}$	0.653	1.769	0.540	0.129	0.488	0.172	0.257
$1\frac{1}{2} \times \frac{1}{4}$	1.188	3.338	0.982	0.199	0.451	0.266	0.399
$2 \times \frac{1}{8}$	0.891	2.670	0.736	0.325	0.664	0.325	0.650
$2 \times \frac{1}{4}$	1.663	4.673	1.374	0.537	0.625	0.537	1.074
$2\frac{1}{2} \times \frac{1}{8}$	1.129	3.050	0.933	0.660	0.841	0.528	1.319
$2\frac{1}{2} \times \frac{1}{4}$	2.138	6.008	1.767	1.132	0.800	0.906	2.276
$3 \times \frac{1}{4}$	2.614	7.343	2.160	2.059	0.976	1.373	4.117
$3 \times \frac{3}{8}$	3.742	10.51	3.093	2.718	0.938	1.812	5.436
$4 \times \frac{3}{16}$	2.717	7.654	2.246	4.090	1.350	2.045	8.180
$4 \times \frac{3}{8}$	5.167	14.52	4.271	7.090	1.289	3.544	14.180

Size, mm	m	A	I	k	Z	J
12 × 2	0.490	0.628	0.082	0.361	0.136	0.163
16 × 2	0.687	0.879	0.220	0.500	0.275	0.440
16 × 3	0.956	1.225	0.273	0.472	0.341	0.545
20 × 4	1.569	2.010	0.684	0.583	0.684	1.367
25 × 4	2.060	2.638	1.508	0.756	1.206	3.015
25 × 5	2.452	3.140	1.669	0.729	1.336	3.338
30 × 4	2.550	3.266	2.827	0.930	1.885	5.652
30 × 5	3.065	3.925	3.192	0.901	2.128	6.381
42 × 4	3.727	4.773	8.717	1.351	4.151	17.430
42 × 5	4.536	5.809	10.130	1.320	4.825	20.255
50 × 4	4.512	5.778	15.409	1.632	6.164	30.810
50 × 5	5.517	7.065	18.118	1.601	7.247	36.226

Table A–9

Shear, Moment, and
Deflection of Beams
(*Note:* Force and
moment reactions are
positive in the directions
shown; equations for
shear force V and
bending moment
M follow the sign
conventions given in
Sec. 3–2.)

1 Cantilever—end load

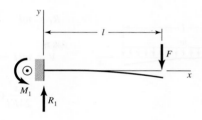

$$R_1 = V = F \qquad M_1 = Fl$$

$$M = F(x - l)$$

$$y = \frac{Fx^2}{6EI}(x - 3l)$$

$$y_{max} = -\frac{Fl^3}{3EI}$$

2 Cantilever—intermediate load

$$R_1 = V = F \qquad M_1 = Fa$$

$$M_{AB} = F(x - a) \qquad M_{BC} = 0$$

$$y_{AB} = \frac{Fx^2}{6EI}(x - 3a)$$

$$y_{BC} = \frac{Fa^2}{6EI}(a - 3x)$$

$$y_{max} = \frac{Fa^2}{6EI}(a - 3l)$$

(*Continued*)

Table A–9

Shear, Moment, and
Deflection of Beams
(*Note:* Force and
moment reactions are
positive in the directions
shown; equations for
shear force V and
bending moment
M follow the sign
conventions given in
Sec. 3–2.)
(*Continued*)

3 Cantilever—uniform load

$$R_1 = wl \qquad M_1 = \frac{wl^2}{2}$$

$$V = w(l - x) \qquad M = -\frac{w}{2}(l - x)^2$$

$$y = \frac{wx^2}{24EI}(4lx - x^2 - 6l^2)$$

$$y_{max} = -\frac{wl^4}{8EI}$$

4 Cantilever—moment load

$$R_1 = V = 0 \qquad M_1 = M = M_B$$

$$y = \frac{M_B x^2}{2EI} \qquad y_{max} = \frac{M_B l^2}{2EI}$$

Table A–9

Shear, Moment, and
Deflection of Beams
(*Note:* Force and
moment reactions are
positive in the directions
shown; equations for
shear force V and
bending moment
M follow the sign
conventions given in
Sec. 3–2.)
(*Continued*)

5 Simple supports—center load

$$R_1 = R_2 = \frac{F}{2}$$

$$V_{AB} = R_1 \qquad V_{BC} = -R_2$$

$$M_{AB} = \frac{Fx}{2} \qquad M_{BC} = \frac{F}{2}(l - x)$$

$$y_{AB} = \frac{Fx}{48EI}(4x^2 - 3l^2)$$

$$y_{max} = -\frac{Fl^3}{48EI}$$

6 Simple supports—intermediate load

$$R_1 = \frac{Fb}{l} \qquad R_2 = \frac{Fa}{l}$$

$$V_{AB} = R_1 \qquad V_{BC} = -R_2$$

$$M_{AB} = \frac{Fbx}{l} \qquad M_{BC} = \frac{Fa}{l}(l - x)$$

$$y_{AB} = \frac{Fbx}{6EIl}(x^2 + b^2 - l^2)$$

$$y_{BC} = \frac{Fa(l - x)}{6EIl}(x^2 + a^2 - 2lx)$$

(Continued)

Table A–9

Shear, Moment, and
Deflection of Beams
(*Note:* Force and
moment reactions are
positive in the directions
shown; equations for
shear force V and
bending moment
M follow the sign
conventions given in
Sec. 3–2.)
(*Continued*)

7 Simple supports—uniform load

$$R_1 = R_2 = \frac{wl}{2} \qquad V = \frac{wl}{2} - wx$$

$$M = \frac{wx}{2}(l - x)$$

$$y = \frac{wx}{24EI}(2lx^2 - x^3 - l^3)$$

$$y_{max} = -\frac{5wl^4}{384EI}$$

8 Simple supports—moment load

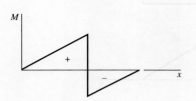

$$R_1 = R_2 = \frac{M_B}{l} \qquad V = \frac{M_B}{l}$$

$$M_{AB} = \frac{M_B x}{l} \qquad M_{BC} = \frac{M_B}{l}(x - l)$$

$$y_{AB} = \frac{M_B x}{6EIl}(x^2 + 3a^2 - 6al + 2l^2)$$

$$y_{BC} = \frac{M_B}{6EIl}[x^3 - 3lx^2 + x(2l^2 + 3a^2) - 3a^2l]$$

Table A–9

Shear, Moment, and
Deflection of Beams
(*Note:* Force and
moment reactions are
positive in the directions
shown; equations for
shear force *V* and
bending moment
M follow the sign
conventions given in
Sec. 3–2.)
(*Continued*)

9 Simple supports—twin loads

$$R_1 = R_2 = F \qquad V_{AB} = F \qquad V_{BC} = 0$$

$$V_{CD} = -F$$

$$M_{AB} = Fx \qquad M_{BC} = Fa \qquad M_{CD} = F(l - x)$$

$$y_{AB} = \frac{Fx}{6EI}(x^2 + 3a^2 - 3la)$$

$$y_{BC} = \frac{Fa}{6EI}(3x^2 + a^2 - 3lx)$$

$$y_{max} = \frac{Fa}{24EI}(4a^2 - 3l^2)$$

10 Simple supports—overhanging load

$$R_1 = \frac{Fa}{l} \qquad R_2 = \frac{F}{l}(l + a)$$

$$V_{AB} = -\frac{Fa}{l} \qquad V_{BC} = F$$

$$M_{AB} = -\frac{Fax}{l} \qquad M_{BC} = F(x - l - a)$$

$$y_{AB} = \frac{Fax}{6EIl}(l^2 - x^2)$$

$$y_{BC} = \frac{F(x - l)}{6EI}[(x - l)^2 - a(3x - l)]$$

$$y_c = -\frac{Fa^2}{3EI}(l + a)$$

(*Continued*)

Table A–9

Shear, Moment, and
Deflection of Beams
(*Note:* Force and
moment reactions are
positive in the directions
shown; equations for
shear force V and
bending moment
M follow the sign
conventions given in
Sec. 3–2.)
(*Continued*)

11 One fixed and one simple support—center load

$$R_1 = \frac{11F}{16} \qquad R_2 = \frac{5F}{16} \qquad M_1 = \frac{3Fl}{16}$$

$$V_{AB} = R_1 \qquad V_{BC} = -R_2$$

$$M_{AB} = \frac{F}{16}(11x - 3l) \qquad M_{BC} = \frac{5F}{16}(l - x)$$

$$y_{AB} = \frac{Fx^2}{96EI}(11x - 9l)$$

$$y_{BC} = \frac{F(l - x)}{96EI}(5x^2 + 2l^2 - 10lx)$$

12 One fixed and one simple support—intermediate load

$$R_1 = \frac{Fb}{2l^3}(3l^2 - b^2) \qquad R_2 = \frac{Fa^2}{2l^3}(3l - a)$$

$$M_1 = \frac{Fb}{2l^2}(l^2 - b^2)$$

$$V_{AB} = R_1 \qquad V_{BC} = -R_2$$

$$M_{AB} = \frac{Fb}{2l^3}[b^2l - l^3 + x(3l^2 - b^2)]$$

$$M_{BC} = \frac{Fa^2}{2l^3}(3l^2 - 3lx - al + ax)$$

$$y_{AB} = \frac{Fbx^2}{12EIl^3}[3l(b^2 - l^2) + x(3l^2 - b^2)]$$

$$y_{BC} = y_{AB} - \frac{F(x - a)^3}{6EI}$$

Table A–9

Shear, Moment, and
Deflection of Beams
(*Note:* Force and
moment reactions are
positive in the directions
shown; equations for
shear force V and
bending moment
M follow the sign
conventions given in
Sec. 3–2.)
(*Continued*)

13 One fixed and one simple support—uniform load

$$R_1 = \frac{5wl}{8} \qquad R_2 = \frac{3wl}{8} \qquad M_1 = \frac{wl^2}{8}$$

$$V = \frac{5wl}{8} - wx$$

$$M = -\frac{w}{8}(4x^2 - 5lx + l^2)$$

$$y = \frac{wx^2}{48EI}(l - x)(2x - 3l)$$

14 Fixed supports—center load

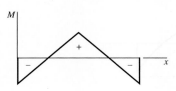

$$R_1 = R_2 = \frac{F}{2} \qquad M_1 = M_2 = \frac{Fl}{8}$$

$$V_{AB} = -V_{BC} = \frac{F}{2}$$

$$M_{AB} = \frac{F}{8}(4x - l) \qquad M_{BC} = \frac{F}{8}(3l - 4x)$$

$$y_{AB} = \frac{Fx^2}{48EI}(4x - 3l)$$

$$y_{max} = -\frac{Fl^3}{192EI}$$

(*Continued*)

Table A-9

Shear, Moment, and
Deflection of Beams
(*Note:* Force and
moment reactions are
positive in the directions
shown; equations for
shear force V and
bending moment
M follow the sign
conventions given in
Sec. 3–2.)
(*Continued*)

15 Fixed supports—intermediate load

$$R_1 = \frac{Fb^2}{l^3}(3a + b) \qquad R_2 = \frac{Fa^2}{l^3}(3b + a)$$

$$M_1 = \frac{Fab^2}{l^2} \qquad M_2 = \frac{Fa^2b}{l^2}$$

$$V_{AB} = R_1 \qquad V_{BC} = -R_2$$

$$M_{AB} = \frac{Fb^2}{l^3}[x(3a + b) - al]$$

$$M_{BC} = M_{AB} - F(x - a)$$

$$y_{AB} = \frac{Fb^2x^2}{6EIl^3}[x(3a + b) - 3al]$$

$$y_{BC} = \frac{Fa^2(l - x)^2}{6EIl^3}[(l - x)(3b + a) - 3bl]$$

16 Fixed supports—uniform load

$$R_1 = R_2 = \frac{wl}{2} \qquad M_1 = M_2 = \frac{wl^2}{12}$$

$$V = \frac{w}{2}(l - 2x)$$

$$M = \frac{w}{12}(6lx - 6x^2 - l^2)$$

$$y = -\frac{wx^2}{24EI}(l - x)^2$$

$$y_{max} = -\frac{wl^4}{384EI}$$

Table A–10

Cumulative Distribution Function of Normal (Gaussian) Distribution

$$\Phi(z_\alpha) = \int_{-\infty}^{z_\alpha} \frac{1}{\sqrt{2\pi}} \exp\left(-\frac{u^2}{2}\right) du$$

$$= \begin{cases} \alpha & z_\alpha \leq 0 \\ 1 - \alpha & z_\alpha > 0 \end{cases}$$

Z_α	0.00	0.01	0.02	0.03	0.04	0.05	0.06	0.07	0.08	0.09
0.0	0.5000	0.4960	0.4920	0.4880	0.4840	0.4801	0.4761	0.4721	0.4681	0.4641
0.1	0.4602	0.4562	0.4522	0.4483	0.4443	0.4404	0.4364	0.4325	0.4286	0.4247
0.2	0.4207	0.4168	0.4129	0.4090	0.4052	0.4013	0.3974	0.3936	0.3897	0.3859
0.3	0.3821	0.3783	0.3745	0.3707	0.3669	0.3632	0.3594	0.3557	0.3520	0.3483
0.4	0.3446	0.3409	0.3372	0.3336	0.3300	0.3264	0.3238	0.3192	0.3156	0.3121
0.5	0.3085	0.3050	0.3015	0.2981	0.2946	0.2912	0.2877	0.2843	0.2810	0.2776
0.6	0.2743	0.2709	0.2676	0.2643	0.2611	0.2578	0.2546	0.2514	0.2483	0.2451
0.7	0.2420	0.2389	0.2358	0.2327	0.2296	0.2266	0.2236	0.2206	0.2177	0.2148
0.8	0.2119	0.2090	0.2061	0.2033	0.2005	0.1977	0.1949	0.1922	0.1894	0.1867
0.9	0.1841	0.1814	0.1788	0.1762	0.1736	0.1711	0.1685	0.1660	0.1635	0.1611
1.0	0.1587	0.1562	0.1539	0.1515	0.1492	0.1469	0.1446	0.1423	0.1401	0.1379
1.1	0.1357	0.1335	0.1314	0.1292	0.1271	0.1251	0.1230	0.1210	0.1190	0.1170
1.2	0.1151	0.1131	0.1112	0.1093	0.1075	0.1056	0.1038	0.1020	0.1003	0.0985
1.3	0.0968	0.0951	0.0934	0.0918	0.0901	0.0885	0.0869	0.0853	0.0838	0.0823
1.4	0.0808	0.0793	0.0778	0.0764	0.0749	0.0735	0.0721	0.0708	0.0694	0.0681
1.5	0.0668	0.0655	0.0643	0.0630	0.0618	0.0606	0.0594	0.0582	0.0571	0.0559
1.6	0.0548	0.0537	0.0526	0.0516	0.0505	0.0495	0.0485	0.0475	0.0465	0.0455
1.7	0.0446	0.0436	0.0427	0.0418	0.0409	0.0401	0.0392	0.0384	0.0375	0.0367
1.8	0.0359	0.0351	0.0344	0.0336	0.0329	0.0322	0.0314	0.0307	0.0301	0.0294
1.9	0.0287	0.0281	0.0274	0.0268	0.0262	0.0256	0.0250	0.0244	0.0239	0.0233
2.0	0.0228	0.0222	0.0217	0.0212	0.0207	0.0202	0.0197	0.0192	0.0188	0.0183
2.1	0.0179	0.0174	0.0170	0.0166	0.0162	0.0158	0.0154	0.0150	0.0146	0.0143
2.2	0.0139	0.0136	0.0132	0.0129	0.0125	0.0122	0.0119	0.0116	0.0113	0.0110
2.3	0.0107	0.0104	0.0102	0.00990	0.00964	0.00939	0.00914	0.00889	0.00866	0.00842
2.4	0.00820	0.00798	0.00776	0.00755	0.00734	0.00714	0.00695	0.00676	0.00657	0.00639
2.5	0.00621	0.00604	0.00587	0.00570	0.00554	0.00539	0.00523	0.00508	0.00494	0.00480
2.6	0.00466	0.00453	0.00440	0.00427	0.00415	0.00402	0.00391	0.00379	0.00368	0.00357
2.7	0.00347	0.00336	0.00326	0.00317	0.00307	0.00298	0.00289	0.00280	0.00272	0.00264
2.8	0.00256	0.00248	0.00240	0.00233	0.00226	0.00219	0.00212	0.00205	0.00199	0.00193
2.9	0.00187	0.00181	0.00175	0.00169	0.00164	0.00159	0.00154	0.00149	0.00144	0.00139

(*Continued*)

Table A–10

Cumulative Distribution Function of Normal (Gaussian) Distribution* *(Continued)*

z_α	0.0	0.1	0.2	0.3	0.4	0.5	0.6	0.7	0.8	0.9
3	0.00135	0.0^3968	0.0^3687	0.0^3483	0.0^3337	0.0^3233	0.0^3159	0.0^3108	0.0^4723	0.0^4481
4	0.0^4317	0.0^4207	0.0^4133	0.0^5854	0.0^5541	0.0^5340	0.0^5211	0.0^5130	0.0^6793	0.0^6479
5	0.0^6287	0.0^6170	0.0^7996	0.0^7579	0.0^7333	0.0^7190	0.0^7107	0.0^8599	0.0^8332	0.0^8182
6	0.0^9987	0.0^9530	0.0^9282	0.0^9149	$0.0^{10}777$	$0.0^{10}402$	$0.0^{10}206$	$0.0^{10}104$	$0.0^{11}523$	$0.0^{11}260$

z_α	−1.282	−1.643	−1.960	−2.326	−2.576	−3.090	−3.291	−3.891	−4.417	
$F(z_\alpha)$	0.10	0.05	0.025	0.010	0.005	0.001	0.0005	0.0001	0.000005	
$R(z_\alpha)$	0.90	0.95	0.975	0.990	0.995	0.999	0.9995	0.9999	0.999995	

*The superscript on a zero after the decimal point indicates how many zeros there are after the decimal point. For example, $0.0^4481 = 0.000\ 048\ 1$.

Table A–11

A Selection of International Tolerance Grades—Metric Series (Size Ranges Are for *Over* the Lower Limit and *Including* the Upper Limit. All Values Are in Millimeters)

Source: Preferred Metric Limits and Fits, ANSI B4.2-1978. See also BSI 4500.

Basic Sizes	Tolerance Grades					
	IT6	IT7	IT8	IT9	IT10	IT11
0–3	0.006	0.010	0.014	0.025	0.040	0.060
3–6	0.008	0.012	0.018	0.030	0.048	0.075
6–10	0.009	0.015	0.022	0.036	0.058	0.090
10–18	0.011	0.018	0.027	0.043	0.070	0.110
18–30	0.013	0.021	0.033	0.052	0.084	0.130
30–50	0.016	0.025	0.039	0.062	0.100	0.160
50–80	0.019	0.030	0.046	0.074	0.120	0.190
80–120	0.022	0.035	0.054	0.087	0.140	0.220
120–180	0.025	0.040	0.063	0.100	0.160	0.250
180–250	0.029	0.046	0.072	0.115	0.185	0.290
250–315	0.032	0.052	0.081	0.130	0.210	0.320
315–400	0.036	0.057	0.089	0.140	0.230	0.360

Table A-12

Fundamental Deviations for Shafts—Metric Series
(Size Ranges Are for *Over* the Lower Limit and *Including* the Upper Limit. All Values Are in Millimeters)
Source: Preferred Metric Limits and Fits, ANSI B4.2-1978. See also BSI 4500.

Basic Sizes	Upper-Deviation Letter					Lower-Deviation Letter				
	c	d	f	g	h	k	n	p	s	u
0–3	−0.060	−0.020	−0.006	−0.002	0	0	+0.004	+0.006	+0.014	+0.018
3–6	−0.070	−0.030	−0.010	−0.004	0	+0.001	+0.008	+0.012	+0.019	+0.023
6–10	−0.080	−0.040	−0.013	−0.005	0	+0.001	+0.010	+0.015	+0.023	+0.028
10–14	−0.095	−0.050	−0.016	−0.006	0	+0.001	+0.012	+0.018	+0.028	+0.033
14–18	−0.095	−0.050	−0.016	−0.006	0	+0.001	+0.012	+0.018	+0.028	+0.033
18–24	−0.110	−0.065	−0.020	−0.007	0	+0.002	+0.015	+0.022	+0.035	+0.041
24–30	−0.110	−0.065	−0.020	−0.007	0	+0.002	+0.015	+0.022	+0.035	+0.048
30–40	−0.120	−0.080	−0.025	−0.009	0	+0.002	+0.017	+0.026	+0.043	+0.060
40–50	−0.130	−0.080	−0.025	−0.009	0	+0.002	+0.017	+0.026	+0.043	+0.070
50–65	−0.140	−0.100	−0.030	−0.010	0	+0.002	+0.020	+0.032	+0.053	+0.087
65–80	−0.150	−0.100	−0.030	−0.010	0	+0.002	+0.020	+0.032	+0.059	+0.102
80–100	−0.170	−0.120	−0.036	−0.012	0	+0.003	+0.023	+0.037	+0.071	+0.124
100–120	−0.180	−0.120	−0.036	−0.012	0	+0.003	+0.023	+0.037	+0.079	+0.144
120–140	−0.200	−0.145	−0.043	−0.014	0	+0.003	+0.027	+0.043	+0.092	+0.170
140–160	−0.210	−0.145	−0.043	−0.014	0	+0.003	+0.027	+0.043	+0.100	+0.190
160–180	−0.230	−0.145	−0.043	−0.014	0	+0.003	+0.027	+0.043	+0.108	+0.210
180–200	−0.240	−0.170	−0.050	−0.015	0	+0.004	+0.031	+0.050	+0.122	+0.236
200–225	−0.260	−0.170	−0.050	−0.015	0	+0.004	+0.031	+0.050	+0.130	+0.258
225–250	−0.280	−0.170	−0.050	−0.015	0	+0.004	+0.031	+0.050	+0.140	+0.284
250–280	−0.300	−0.190	−0.056	−0.017	0	+0.004	+0.034	+0.056	+0.158	+0.315
280–315	−0.330	−0.190	−0.056	−0.017	0	+0.004	+0.034	+0.056	+0.170	+0.350
315–355	−0.360	−0.210	−0.062	−0.018	0	+0.004	+0.037	+0.062	+0.190	+0.390
355–400	−0.400	−0.210	−0.062	−0.018	0	+0.004	+0.037	+0.062	+0.208	+0.435

Table A–13

A Selection of
International Tolerance
Grades—Inch Series
(Size Ranges Are for
Over the Lower Limit
and *Including* the Upper
Limit. All Values Are in
Inches, Converted from
Table A–11)

Basic Sizes	Tolerance Grades					
	IT6	IT7	IT8	IT9	IT10	IT11
0–0.12	0.0002	0.0004	0.0006	0.0010	0.0016	0.0024
0.12–0.24	0.0003	0.0005	0.0007	0.0012	0.0019	0.0030
0.24–0.40	0.0004	0.0006	0.0009	0.0014	0.0023	0.0035
0.40–0.72	0.0004	0.0007	0.0011	0.0017	0.0028	0.0043
0.72–1.20	0.0005	0.0008	0.0013	0.0020	0.0033	0.0051
1.20–2.00	0.0006	0.0010	0.0015	0.0024	0.0039	0.0063
2.00–3.20	0.0007	0.0012	0.0018	0.0029	0.0047	0.0075
3.20–4.80	0.0009	0.0014	0.0021	0.0034	0.0055	0.0087
4.80–7.20	0.0010	0.0016	0.0025	0.0039	0.0063	0.0098
7.20–10.00	0.0011	0.0018	0.0028	0.0045	0.0073	0.0114
10.00–12.60	0.0013	0.0020	0.0032	0.0051	0.0083	0.0126
12.60–16.00	0.0014	0.0022	0.0035	0.0055	0.0091	0.0142

Table A–14

Fundamental Deviations for Shafts—Inch Series (Size Ranges Are for *Over* the Lower Limit and *Including* the Upper Limit. All Values Are in Inches, Converted from Table A–12)

Basic Sizes	Upper-Deviation Letter						Lower-Deviation Letter			
	c	d	f	g	h	k	n	p	s	u
0–0.12	−0.0024	−0.0008	−0.0002	−0.0001	0	0	+0.0002	+0.0002	+0.0006	+0.0007
0.12–0.24	−0.0028	−0.0012	−0.0004	−0.0002	0	0	+0.0003	+0.0005	+0.0007	+0.0009
0.24–0.40	−0.0031	−0.0016	−0.0005	−0.0002	0	0	+0.0004	+0.0006	+0.0009	+0.0011
0.40–0.72	−0.0037	−0.0020	−0.0006	−0.0002	0	0	+0.0005	+0.0007	+0.0011	+0.0013
0.72–0.96	−0.0043	−0.0026	−0.0008	−0.0003	0	+0.0001	+0.0006	+0.0009	+0.0014	+0.0016
0.96–1.20	−0.0043	−0.0026	−0.0008	−0.0003	0	+0.0001	+0.0006	+0.0009	+0.0014	+0.0019
1.20–1.60	−0.0047	−0.0031	−0.0010	−0.0004	0	+0.0001	+0.0007	+0.0010	+0.0017	+0.0024
1.60–2.00	−0.0051	−0.0031	−0.0010	−0.0004	0	+0.0001	+0.0007	+0.0010	+0.0017	+0.0028
2.00–2.60	−0.0055	−0.0039	−0.0012	−0.0004	0	+0.0001	+0.0008	+0.0013	+0.0021	+0.0034
2.60–3.20	−0.0059	−0.0039	−0.0012	−0.0004	0	+0.0001	+0.0008	+0.0013	+0.0023	+0.0040
3.20–4.00	−0.0067	−0.0047	−0.0014	−0.0005	0	+0.0001	+0.0009	+0.0015	+0.0028	+0.0049
4.00–4.80	−0.0071	−0.0047	−0.0014	−0.0005	0	+0.0001	+0.0009	+0.0015	+0.0031	+0.0057
4.80–5.60	−0.0079	−0.0057	−0.0017	−0.0006	0	+0.0001	+0.0011	+0.0017	+0.0036	+0.0067
5.60–6.40	−0.0083	−0.0057	−0.0017	−0.0006	0	+0.0001	+0.0011	+0.0017	+0.0039	+0.0075
6.40–7.20	−0.0091	−0.0057	−0.0017	−0.0006	0	+0.0001	+0.0011	+0.0017	+0.0043	+0.0083
7.20–8.00	−0.0094	−0.0067	−0.0020	−0.0006	0	+0.0002	+0.0012	+0.0020	+0.0048	+0.0093
8.00–9.00	−0.0102	−0.0067	−0.0020	−0.0006	0	+0.0002	+0.0012	+0.0020	+0.0051	+0.0102
9.00–10.00	−0.0110	−0.0067	−0.0020	−0.0006	0	+0.0002	+0.0012	+0.0020	+0.0055	+0.0112
10.00–11.20	−0.0118	−0.0075	−0.0022	−0.0007	0	+0.0002	+0.0013	+0.0022	+0.0062	+0.0124
11.20–12.60	−0.0130	−0.0075	−0.0022	−0.0007	0	+0.0002	+0.0013	+0.0022	+0.0067	+0.0130
12.60–14.20	−0.0142	−0.0083	−0.0024	−0.0007	0	+0.0002	+0.0015	+0.0024	+0.0075	+0.0154
14.20–16.00	−0.0157	−0.0083	−0.0024	−0.0007	0	+0.0002	+0.0015	+0.0024	+0.0082	+0.0171

Table A–15

Charts of Theoretical Stress-Concentration Factors K_t^*

Figure A–15–1

Bar in tension or simple compression with a transverse hole. $\sigma_0 = F/A$, where $A = (w - d)t$ and t is the thickness.

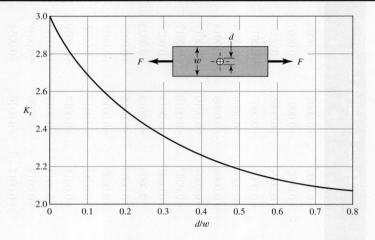

Figure A–15–2

Rectangular bar with a transverse hole in bending. $\sigma_0 = Mc/I$, where $I = (w - d)h^3/12$.

Figure A–15–3

Notched rectangular bar in tension or simple compression. $\sigma_0 = F/A$, where $A = dt$ and t is the thickness.

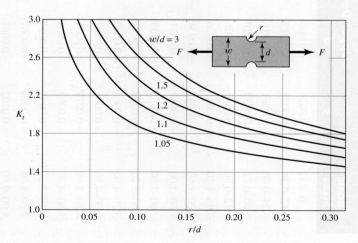

Table A–15

Charts of Theoretical Stress-Concentration Factors K_t^* (*Continued*)

Figure A–15–4

Notched rectangular bar in bending. $\sigma_0 = Mc/I$, where $c = d/2$, $I = td^3/12$, and t is the thickness.

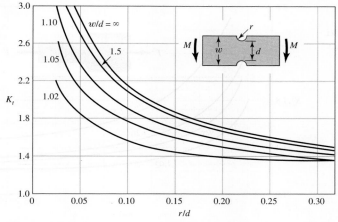

Figure A–15–5

Rectangular filleted bar in tension or simple compression. $\sigma_0 = F/A$, where $A = dt$ and t is the thickness.

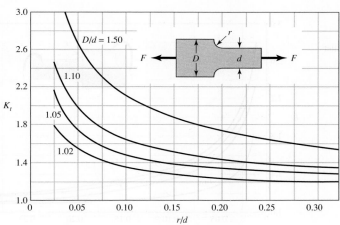

Figure A–15–6

Rectangular filleted bar in bending. $\sigma_0 = Mc/I$, where $c = d/2$, $I = td^3/12$, t is the thickness.

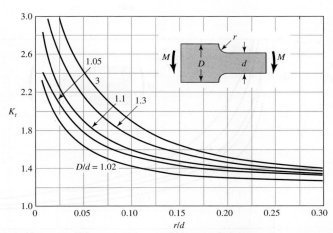

(*Continued*)

*Factors from R. E. Peterson, "Design Factors for Stress Concentration," *Machine Design*, vol. 23, no. 2, February 1951, p. 169; no. 3, March 1951, p. 161, no. 5, May 1951, p. 159; no. 6, June 1951, p. 173; no. 7, July 1951, p. 155. Reprinted with permission from *Machine Design*, a Penton Media Inc. publication.

Table A-15

Charts of Theoretical Stress-Concentration Factors K_t^* (*Continued*)

Figure A-15-7

Round shaft with shoulder fillet in tension. $\sigma_0 = F/A$, where $A = \pi d^2/4$.

Figure A-15-8

Round shaft with shoulder fillet in torsion. $\tau_0 = Tc/J$, where $c = d/2$ and $J = \pi d^4/32$.

Figure A-15-9

Round shaft with shoulder fillet in bending. $\sigma_0 = Mc/I$, where $c = d/2$ and $I = \pi d^4/64$.

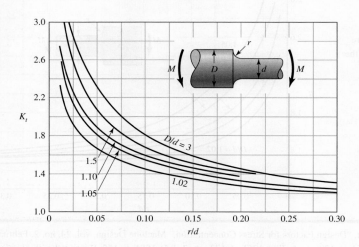

Table A–15

Charts of Theoretical Stress-Concentration Factors K_t^* (*Continued*)

Figure A–15–10

Round shaft in torsion with transverse hole.

$$\frac{J}{c} = \frac{\pi D^3}{16} - \frac{dD^2}{6} \text{ (approx)}$$

Figure A–15–11

Round shaft in bending with a transverse hole. $\sigma_0 = M/[(\pi D^3/32) - (dD^2/6)]$, approximately.

Figure A–15–12

Plate loaded in tension by a pin through a hole. $\sigma_0 = F/A$, where $A = (w - d)t$. When clearance exists, increase K_t 35 to 50 percent. (*M. M. Frocht and H. N. Hill, "Stress-Concentration Factors around a Central Circular Hole in a Plate Loaded through a Pin in Hole," J. Appl. Mechanics, vol. 7, no. 1, March 1940, p. A-5.*)

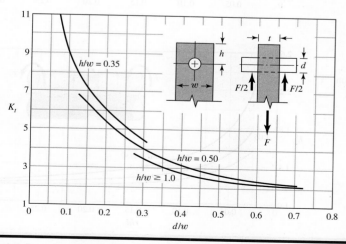

(*Continued*)

*Factors from R. E. Peterson, "Design Factors for Stress Concentration," Machine Design, vol. 23, no. 2, February 1951, p. 169; no. 3, March 1951, p. 161, no. 5, May 1951, p. 159; no. 6, June 1951, p. 173; no. 7, July 1951, p. 155. Reprinted with permission from Machine Design, a Penton Media Inc. publication.

Table A–15

Charts of Theoretical Stress-Concentration Factors K_t^* (*Continued*)

Figure A–15–13

Grooved round bar in tension.
$\sigma_0 = F/A$, where $A = \pi d^2/4$.

Figure A–15–14

Grooved round bar in bending.
$\sigma_0 = Mc/I$, where $c = d/2$
and $I = \pi d^4/64$.

Figure A–15–15

Grooved round bar in torsion.
$\tau_0 = Tc/J$, where $c = d/2$
and $J = \pi d^4/32$.

*Factors from R. E. Peterson, "Design Factors for Stress Concentration," *Machine Design*, vol. 23, no. 2, February 1951, p. 169; no. 3, March 1951, p. 161, no. 5, May 1951, p. 159; no. 6, June 1951, p. 173; no. 7, July 1951, p. 155. Reprinted with permission from *Machine Design*, a Penton Media Inc. publication.

Table A–15

Charts of Theoretical Stress-Concentration Factors K_t^* (*Continued*)

Figure A–15–16

Round shaft with flat-bottom groove in bending and/or tension.

$$\sigma_0 = \frac{4F}{\pi d^2} + \frac{32M}{\pi d^3}$$

Source: W. D. Pilkey and D. F. Pilkey, *Peterson's Stress-Concentration Factors,* 3rd ed. John Wiley & Sons, Hoboken, NJ, 2008, p. 115.

(*Continued*)

Table A–15

Charts of Theoretical Stress-Concentration Factors K_t^* (*Continued*)

Figure A–15–17

Round shaft with flat-bottom groove in torsion.

$$\tau_0 = \frac{16T}{\pi d^3}$$

Source: W. D. Pilkey and D. F. Pilkey, *Peterson's Stress-Concentration Factors,* 3rd ed. John Wiley & Sons, Hoboken, NJ, 2008, p. 133

Table A–16

Approximate Stress-Concentration Factor K_t of a Round Bar or Tube with a Transverse Round Hole and Loaded in Bending

Source: R. E. Peterson, *Stress-Concentration Factors,* Wiley, New York, 1974, pp. 146, 235.

The nominal bending stress is $\sigma_0 = M/Z_{net}$ where Z_{net} is a reduced value of the section modulus and is defined by

$$Z_{net} = \frac{\pi A}{32D}(D^4 - d^4)$$

Values of A are listed in the table. Use $d = 0$ for a solid bar

| | d/D | | | | | |
| | 0.9 | | 0.6 | | 0 | |
a/D	A	K_t	A	K_t	A	K_t
0.050	0.92	2.63	0.91	2.55	0.88	2.42
0.075	0.89	2.55	0.88	2.43	0.86	2.35
0.10	0.86	2.49	0.85	2.36	0.83	2.27
0.125	0.82	2.41	0.82	2.32	0.80	2.20
0.15	0.79	2.39	0.79	2.29	0.76	2.15
0.175	0.76	2.38	0.75	2.26	0.72	2.10
0.20	0.73	2.39	0.72	2.23	0.68	2.07
0.225	0.69	2.40	0.68	2.21	0.65	2.04
0.25	0.67	2.42	0.64	2.18	0.61	2.00
0.275	0.66	2.48	0.61	2.16	0.58	1.97
0.30	0.64	2.52	0.58	2.14	0.54	1.94

(Continued)

Table A–16 (*Continued*)

Approximate Stress-Concentration Factors K_{ts} for a Round Bar or Tube Having a Transverse Round Hole and Loaded in Torsion   *Source:* R. E. Peterson, *Stress-Concentration Factors,* Wiley, New York, 1974, pp. 148, 244.

The maximum stress occurs on the inside of the hole, slightly below the shaft surface. The nominal shear stress is $\tau_0 = TD/2J_{net}$, where J_{net} is a reduced value of the second polar moment of area and is defined by

$$J_{net} = \frac{\pi A(D^4 - d^4)}{32}$$

Values of A are listed in the table. Use $d = 0$ for a solid bar.

	d/D									
	0.9		**0.8**		**0.6**		**0.4**		**0**	
a/D	**A**	**K_{ts}**	**A**	**K_{ts}**	**A**	**K_{ts}**	**A**	**K_{ts}**	**A**	**K_{ts}**
0.05	0.96	1.78							0.95	1.77
0.075	0.95	1.82							0.93	1.71
0.10	0.94	1.76	0.93	1.74	0.92	1.72	0.92	1.70	0.92	1.68
0.125	0.91	1.76	0.91	1.74	0.90	1.70	0.90	1.67	0.89	1.64
0.15	0.90	1.77	0.89	1.75	0.87	1.69	0.87	1.65	0.87	1.62
0.175	0.89	1.81	0.88	1.76	0.87	1.69	0.86	1.64	0.85	1.60
0.20	0.88	1.96	0.86	1.79	0.85	1.70	0.84	1.63	0.83	1.58
0.25	0.87	2.00	0.82	1.86	0.81	1.72	0.80	1.63	0.79	1.54
0.30	0.80	2.18	0.78	1.97	0.77	1.76	0.75	1.63	0.74	1.51
0.35	0.77	2.41	0.75	2.09	0.72	1.81	0.69	1.63	0.68	1.47
0.40	0.72	2.67	0.71	2.25	0.68	1.89	0.64	1.63	0.63	1.44

Table A-17

Preferred Sizes and Renard (R-Series) Numbers

(When a choice can be made, use one of these sizes; however, not all parts or items are available in all the sizes shown in the table.)

Fraction of Inches

$\frac{1}{64}$, $\frac{1}{32}$, $\frac{1}{16}$, $\frac{3}{32}$, $\frac{1}{8}$, $\frac{5}{32}$, $\frac{3}{16}$, $\frac{1}{4}$, $\frac{5}{16}$, $\frac{3}{8}$, $\frac{7}{16}$, $\frac{1}{2}$, $\frac{9}{16}$, $\frac{5}{8}$, $\frac{11}{16}$, $\frac{3}{4}$, $\frac{7}{8}$, 1, $1\frac{1}{4}$, $1\frac{1}{2}$, $1\frac{3}{4}$, 2, $2\frac{1}{4}$, $2\frac{1}{2}$, $2\frac{3}{4}$, 3, $3\frac{1}{4}$, $3\frac{1}{2}$, $3\frac{3}{4}$, 4, $4\frac{1}{4}$, $4\frac{1}{2}$, $4\frac{3}{4}$, 5, $5\frac{1}{4}$, $5\frac{1}{2}$, $5\frac{3}{4}$, 6, $6\frac{1}{2}$, 7, $7\frac{1}{2}$, 8, $8\frac{1}{2}$, 9, $9\frac{1}{2}$, 10, $10\frac{1}{2}$, 11, $11\frac{1}{2}$, 12, $12\frac{1}{2}$, 13, $13\frac{1}{2}$, 14, $14\frac{1}{2}$, 15, $15\frac{1}{2}$, 16, $16\frac{1}{2}$, 17, $17\frac{1}{2}$, 18, $18\frac{1}{2}$, 19, $19\frac{1}{2}$, 20

Decimal Inches

0.010, 0.012, 0.016, 0.020, 0.025, 0.032, 0.040, 0.05, 0.06, 0.08, 0.10, 0.12, 0.16, 0.20, 0.24, 0.30, 0.40, 0.50, 0.60, 0.80, 1.00, 1.20, 1.40, 1.60, 1.80, 2.0, 2.4, 2.6, 2.8, 3.0, 3.2, 3.4, 3.6, 3.8, 4.0, 4.2, 4.4, 4.6, 4.8, 5.0, 5.2, 5.4, 5.6, 5.8, 6.0, 7.0, 7.5, 8.5, 9.0, 9.5, 10.0, 10.5, 11.0, 11.5, 12.0, 12.5, 13.0, 13.5, 14.0, 14.5, 15.0, 15.5, 16.0, 16.5, 17.0, 17.5, 18.0, 18.5, 19.0, 19.5, 20

Millimeters

0.05, 0.06, 0.08, 0.10, 0.12, 0.16, 0.20, 0.25, 0.30, 0.40, 0.50, 0.60, 0.70, 0.80, 0.90, 1.0, 1.1, 1.2, 1.4, 1.5, 1.6, 1.8, 2.0, 2.2, 2.5, 2.8, 3.0, 3.5, 4.0, 4.5, 5.0, 5.5, 6.0, 6.5, 7.0, 8.0, 9.0, 10, 11, 12, 14, 16, 18, 20, 22, 25, 28, 30, 32, 35, 40, 45, 50, 60, 80, 100, 120, 140, 160, 180, 200, 250, 300

Renard Numbers*

1st choice, R5: 1, 1.6, 2.5, 4, 6.3, 10

2d choice, R10: 1.25, 2, 3.15, 5, 8

3d choice, R20: 1.12, 1.4, 1.8, 2.24, 2.8, 3.55, 4.5, 5.6, 7.1, 9

4th choice, R40: 1.06, 1.18, 1.32, 1.5, 1.7, 1.9, 2.12, 2.36, 2.65, 3, 3.35, 3.75, 4.25, 4.75, 5.3, 6, 6.7, 7.5, 8.5, 9.5

*May be multiplied or divided by powers of 10.

Table A–18

Geometric Properties

Part 1 Properties of Sections

A = area

G = location of centroid

$I_x = \int y^2 \, dA$ = second moment of area about x axis

$I_y = \int x^2 \, dA$ = second moment of area about y axis

$I_{xy} = \int xy \, dA$ = mixed moment of area about x and y axes

$J_G = \int r^2 \, dA = \int (x^2 + y^2) \, dA = I_x + I_y$

= second polar moment of area about axis through G

$k_x^2 = I_x/A$ = squared radius of gyration about x axis

Rectangle

$A = bh \qquad I_x = \dfrac{bh^3}{12} \qquad I_y = \dfrac{b^3h}{12} \qquad I_{xy} = 0$

Circle

$A = \dfrac{\pi D^2}{4} \qquad I_x = I_y = \dfrac{\pi D^4}{64} \qquad I_{xy} = 0 \qquad J_G = \dfrac{\pi D^4}{32}$

Hollow circle

$A = \dfrac{\pi}{4}(D^2 - d^2) \qquad I_x = I_y = \dfrac{\pi}{64}(D^4 - d^4) \qquad I_{xy} = 0 \qquad J_G = \dfrac{\pi}{32}(D^4 - d^4)$

Table A-18

Geometric Properties
(*Continued*)

Right triangles

$$A = \frac{bh}{2} \qquad I_x = \frac{bh^3}{36} \qquad I_y = \frac{b^3h}{36} \qquad I_{xy} = \frac{-b^2h^2}{72}$$

Right triangles

$$A = \frac{bh}{2} \qquad I_x = \frac{bh^3}{36} \qquad I_y = \frac{b^3h}{36} \qquad I_{xy} = \frac{b^2h^2}{72}$$

Quarter-circles

$$A = \frac{\pi r^2}{4} \qquad I_x = I_y = r^4\left(\frac{\pi}{16} - \frac{4}{9\pi}\right) \qquad I_{xy} = r^4\left(\frac{1}{8} - \frac{4}{9\pi}\right)$$

Quarter-circles

$$A = \frac{\pi r^2}{4} \qquad I_x = I_y = r^4\left(\frac{\pi}{16} - \frac{4}{9\pi}\right) \qquad I_{xy} = r^4\left(\frac{4}{9\pi} - \frac{1}{8}\right)$$

(*Continued*)

Table A–18

Geometric Properties
(*Continued*)

Part 2 Properties of Solids (ρ = Mass Density, Mass per Unit Volume)

Rods

$$m = \frac{\pi d^2 l \rho}{4} \qquad I_y = I_z = \frac{ml^2}{12}$$

Round disks

$$m = \frac{\pi d^2 t \rho}{4} \qquad I_x = \frac{md^2}{8} \qquad I_y = I_z = \frac{md^2}{16}$$

Rectangular prisms

$$m = abc\rho \qquad I_x = \frac{m}{12}(a^2 + b^2) \qquad I_y = \frac{m}{12}(a^2 + c^2) \qquad I_z = \frac{m}{12}(b^2 + c^2)$$

Cylinders

$$m = \frac{\pi d^2 l \rho}{4} \qquad I_x = \frac{md^2}{8} \qquad I_y = I_z = \frac{m}{48}(3d^2 + 4l^2)$$

Hollow cylinders

$$m = \frac{\pi (d_o^2 - d_i^2) l \rho}{4} \qquad I_x = \frac{m}{8}(d_o^2 + d_i^2) \qquad I_y = I_z = \frac{m}{48}(3d_o^2 + 3d_i^2 + 4l^2)$$

Table A–19

American Standard Pipe

Nominal Size, in	Outside Diameter, in	Threads per inch	Wall Thickness, in		
			Standard No. 40	Extra Strong No. 80	Double Extra Strong
$\frac{1}{8}$	0.405	27	0.070	0.098	
$\frac{1}{4}$	0.540	18	0.090	0.122	
$\frac{3}{8}$	0.675	18	0.093	0.129	
$\frac{1}{2}$	0.840	14	0.111	0.151	0.307
$\frac{3}{4}$	1.050	14	0.115	0.157	0.318
1	1.315	$11\frac{1}{2}$	0.136	0.183	0.369
$1\frac{1}{4}$	1.660	$11\frac{1}{2}$	0.143	0.195	0.393
$1\frac{1}{2}$	1.900	$11\frac{1}{2}$	0.148	0.204	0.411
2	2.375	$11\frac{1}{2}$	0.158	0.223	0.447
$2\frac{1}{2}$	2.875	8	0.208	0.282	0.565
3	3.500	8	0.221	0.306	0.615
$3\frac{1}{2}$	4.000	8	0.231	0.325	
4	4.500	8	0.242	0.344	0.690
5	5.563	8	0.263	0.383	0.768
6	6.625	8	0.286	0.441	0.884
8	8.625	8	0.329	0.510	0.895

Table A–20

Deterministic ASTM Minimum Tensile and Yield Strengths for Some Hot-Rolled (HR) and Cold-Drawn (CD) Steels [The strengths listed are estimated ASTM minimum values in the size range 18 to 32 mm ($\frac{3}{4}$ to $1\frac{1}{4}$ in). These strengths are suitable for use with the design factor defined in Sec. 1–10, provided the materials conform to ASTM A6 or A568 requirements or are required in the purchase specifications. Remember that a numbering system is not a specification.]   *Source:* 1986 SAE Handbook, p. 2.15.

1 UNS No.	2 SAE and/or AISI No.	3 Process-ing	4 Tensile Strength, MPa (kpsi)	5 Yield Strength, MPa (kpsi)	6 Elongation in 2 in, %	7 Reduction in Area, %	8 Brinell Hardness
G10060	1006	HR	300 (43)	170 (24)	30	55	86
		CD	330 (48)	280 (41)	20	45	95
G10100	1010	HR	320 (47)	180 (26)	28	50	95
		CD	370 (53)	300 (44)	20	40	105
G10150	1015	HR	340 (50)	190 (27.5)	28	50	101
		CD	390 (56)	320 (47)	18	40	111
G10180	1018	HR	400 (58)	220 (32)	25	50	116
		CD	440 (64)	370 (54)	15	40	126
G10200	1020	HR	380 (55)	210 (30)	25	50	111
		CD	470 (68)	390 (57)	15	40	131
G10300	1030	HR	470 (68)	260 (37.5)	20	42	137
		CD	520 (76)	440 (64)	12	35	149
G10350	1035	HR	500 (72)	270 (39.5)	18	40	143
		CD	550 (80)	460 (67)	12	35	163
G10400	1040	HR	520 (76)	290 (42)	18	40	149
		CD	590 (85)	490 (71)	12	35	170
G10450	1045	HR	570 (82)	310 (45)	16	40	163
		CD	630 (91)	530 (77)	12	35	179
G10500	1050	HR	620 (90)	340 (49.5)	15	35	179
		CD	690 (100)	580 (84)	10	30	197
G10600	1060	HR	680 (98)	370 (54)	12	30	201
G10800	1080	HR	770 (112)	420 (61.5)	10	25	229
G10950	1095	HR	830 (120)	460 (66)	10	25	248

Table A–21

Mean Mechanical Properties of Some Heat-Treated Steels
[These are typical properties for materials normalized and annealed. The properties for quenched and tempered (Q&T) steels are from a single heat. Because of the many variables, the properties listed are global averages. In all cases, data were obtained from specimens of diameter 0.505 in, machined from 1-in rounds, and of gauge length 2 in. unless noted, all specimens were oil-quenched.] *Source: ASM Metals Reference Book,* 2d ed., American Society for Metals, Metals Park, Ohio, 1983.

1	2	3	4	5	6	7	8
AISI No.	Treatment	Temperature °C (°F)	Tensile Strength MPa (kpsi)	Yield Strength, MPa (kpsi)	Elongation, %	Reduction in Area, %	Brinell Hardness
1030	Q&T*	205 (400)	848 (123)	648 (94)	17	47	495
	Q&T*	315 (600)	800 (116)	621 (90)	19	53	401
	Q&T*	425 (800)	731 (106)	579 (84)	23	60	302
	Q&T*	540 (1000)	669 (97)	517 (75)	28	65	255
	Q&T*	650 (1200)	586 (85)	441 (64)	32	70	207
	Normalized	925 (1700)	521 (75)	345 (50)	32	61	149
	Annealed	870 (1600)	430 (62)	317 (46)	35	64	137
1040	Q&T	205 (400)	779 (113)	593 (86)	19	48	262
	Q&T	425 (800)	758 (110)	552 (80)	21	54	241
	Q&T	650 (1200)	634 (92)	434 (63)	29	65	192
	Normalized	900 (1650)	590 (86)	374 (54)	28	55	170
	Annealed	790 (1450)	519 (75)	353 (51)	30	57	149
1050	Q&T*	205 (400)	1120 (163)	807 (117)	9	27	514
	Q&T*	425 (800)	1090 (158)	793 (115)	13	36	444
	Q&T*	650 (1200)	717 (104)	538 (78)	28	65	235
	Normalized	900 (1650)	748 (108)	427 (62)	20	39	217
	Annealed	790 (1450)	636 (92)	365 (53)	24	40	187
1060	Q&T	425 (800)	1080 (156)	765 (111)	14	41	311
	Q&T	540 (1000)	965 (140)	669 (97)	17	45	277
	Q&T	650 (1200)	800 (116)	524 (76)	23	54	229
	Normalized	900 (1650)	776 (112)	421 (61)	18	37	229
	Annealed	790 (1450)	626 (91)	372 (54)	22	38	179
1095	Q&T	315 (600)	1260 (183)	813 (118)	10	30	375
	Q&T	425 (800)	1210 (176)	772 (112)	12	32	363
	Q&T	540 (1000)	1090 (158)	676 (98)	15	37	321
	Q&T	650 (1200)	896 (130)	552 (80)	21	47	269
	Normalized	900 (1650)	1010 (147)	500 (72)	9	13	293
	Annealed	790 (1450)	658 (95)	380 (55)	13	21	192
1141	Q&T	315 (600)	1460 (212)	1280 (186)	9	32	415
	Q&T	540 (1000)	896 (130)	765 (111)	18	57	262

(Continued)

Table A–21 (*Continued*)

Mean Mechanical Properties of Some Heat-Treated Steels
[These are typical properties for materials normalized and annealed. The properties for quenched and tempered (Q&T) steels are from a single heat. Because of the many variables, the properties listed are global averages. In all cases, data were obtained from specimens of diameter 0.505 in, machined from 1-in rounds, and of gauge length 2 in. unless noted, all specimens were oil-quenched.] *Source: ASM Metals Reference Book,* 2d ed., American Society for Metals, Metals Park, Ohio, 1983.

1 AISI No.	2 Treatment	3 Temperature °C (°F)	4 Tensile Strength MPa (kpsi)	5 Yield Strength, MPa (kpsi)	6 Elongation, %	7 Reduction in Area, %	8 Brinell Hardness
4130	Q&T*	205 (400)	1630 (236)	1460 (212)	10	41	467
	Q&T*	315 (600)	1500 (217)	1380 (200)	11	43	435
	Q&T*	425 (800)	1280 (186)	1190 (173)	13	49	380
	Q&T*	540 (1000)	1030 (150)	910 (132)	17	57	315
	Q&T*	650 (1200)	814 (118)	703 (102)	22	64	245
	Normalized	870 (1600)	670 (97)	436 (63)	25	59	197
	Annealed	865 (1585)	560 (81)	361 (52)	28	56	156
4140	Q&T	205 (400)	1770 (257)	1640 (238)	8	38	510
	Q&T	315 (600)	1550 (225)	1430 (208)	9	43	445
	Q&T	425 (800)	1250 (181)	1140 (165)	13	49	370
	Q&T	540 (1000)	951 (138)	834 (121)	18	58	285
	Q&T	650 (1200)	758 (110)	655 (95)	22	63	230
	Normalized	870 (1600)	1020 (148)	655 (95)	18	47	302
	Annealed	815 (1500)	655 (95)	417 (61)	26	57	197
4340	Q&T	315 (600)	1720 (250)	1590 (230)	10	40	486
	Q&T	425 (800)	1470 (213)	1360 (198)	10	44	430
	Q&T	540 (1000)	1170 (170)	1080 (156)	13	51	360
	Q&T	650 (1200)	965 (140)	855 (124)	19	60	280

*Water-quenched

Table A–22

Results of Tensile Tests of Some Metals* *Source:* J. Datsko, "Solid Materials," chap. 32 in Joseph E. Shigley, Charles R. Mischke, and Thomas H. Brown, Jr. (eds.-in-chief), *Standard Handbook of Machine Design*, 3rd ed., McGraw-Hill, New York, 2004, pp. 32.49–32.52.

Number	Material	Condition	Yield S_y, MPa (kpsi)	Ultimate S_u, MPa (kpsi)	Strength (Tensile) Fracture, σ_f, MPa (kpsi)	Coefficient σ_0, MPa (kpsi)	Strain Strength, Exponent m	Fracture Strain ϵ_f
1018	Steel	Annealed	220 (32.0)	341 (49.5)	628 (91.1)†	620 (90.0)†	0.25	1.05
1144	Steel	Annealed	358 (52.0)	646 (93.7)	898 (130)†	992 (144)	0.14	0.49
1212	Steel	HR	193 (28.0)	424 (61.5)	729 (106)†	758 (110)	0.24	0.85
1045	Steel	Q&T 600°F	1520 (220)	1580 (230)	2380 (345)	1880 (273)†	0.041	0.81
4142	Steel	Q&T 600°F	1720 (250)	1930 (280)	2340 (340)	1760 (255)†	0.048	0.43
303	Stainless steel	Annealed	241 (35.0)	601 (87.3)	1520 (221)†	1410 (205)	0.51	1.16
304	Stainless steel	Annealed	276 (40.0)	568 (82.4)	1600 (233)†	1270 (185)	0.45	1.67
2011	Aluminum alloy	T6	169 (24.5)	324 (47.0)	325 (47.2)†	620 (90)	0.28	0.10
2024	Aluminum alloy	T4	296 (43.0)	446 (64.8)	533 (77.3)†	689 (100)	0.15	0.18
7075	Aluminum alloy	T6	542 (78.6)	593 (86.0)	706 (102)†	882 (128)	0.13	0.18

*Values from one or two heats and believed to be attainable using proper purchase specifications. The fracture strain may vary as much as 100 percent.

†Derived value.

Table A–23

Mean Monotonic and Cyclic Stress-Strain Properties of Selected Steels   *Source: ASM Metals Reference Book*, 2nd ed., American Society for Metals, Metals Park, Ohio, 1983, p. 217.

Grade (a)	Orientation (e)	Description (f)	Hardness HB	Tensile Strength S_{ut} MPa	ksi	Reduction in Area %	True Strain at Fracture ε_f	Modulus of Elasticity E GPa	10^6 psi	Fatigue Strength Coefficient σ'_f MPa	ksi	Fatigue Strength Exponent b	Fatigue Ductility Coefficient ε'_f	Fatigue Ductility Exponent c
A538A (b)	L	STA	405	1515	220	67	1.10	185	27	1655	240	−0.065	0.30	−0.62
A538B (b)	L	STA	460	1860	270	56	0.82	185	27	2135	310	−0.071	0.80	−0.71
A538C (b)	L	STA	480	2000	290	55	0.81	180	26	2240	325	−0.07	0.60	−0.75
AM-350 (c)	L	HR, A		1315	191	52	0.74	195	28	2800	406	−0.14	0.33	−0.84
AM-350 (c)	L	CD	496	1905	276	20	0.23	180	26	2690	390	−0.102	0.10	−0.42
Gainex (c)	LT	HR sheet		530	77	58	0.86	200	29.2	805	117	−0.07	0.86	−0.65
Gainex (c)	L	HR sheet		510	74	64	1.02	200	29.2	805	117	−0.071	0.86	−0.68
H-11	L	Ausformed	660	2585	375	33	0.40	205	30	3170	460	−0.077	0.08	−0.74
RQC-100 (c)	LT	HR plate	290	940	136	43	0.56	205	30	1240	180	−0.07	0.66	−0.69
RQC-100 (c)	L	HR plate	290	930	135	67	1.02	205	30	1240	180	−0.07	0.66	−0.69
10B62	L	Q&T	430	1640	238	38	0.89	195	28	1780	258	−0.067	0.32	−0.56
1005-1009	LT	HR sheet	90	360	52	73	1.3	205	30	580	84	−0.09	0.15	−0.43
1005-1009	LT	CD sheet	125	470	68	66	1.09	205	30	515	75	−0.059	0.30	−0.51
1005-1009	L	CD sheet	125	415	60	64	1.02	200	29	540	78	−0.073	0.11	−0.41
1005-1009	L	HR sheet	90	345	50	80	1.6	200	29	640	93	−0.109	0.10	−0.39
1015	L	Normalized	80	415	60	68	1.14	205	30	825	120	−0.11	0.95	−0.64
1020	L	HR plate	108	440	64	62	0.96	205	29.5	895	130	−0.12	0.41	−0.51
1040	L	As forged	225	620	90	60	0.93	200	29	1540	223	−0.14	0.61	−0.57
1045	L	Q&T	225	725	105	65	1.04	200	29	1225	178	−0.095	1.00	−0.66
1045	L	Q&T	410	1450	210	51	0.72	200	29	1860	270	−0.073	0.60	−0.70
1045	L	Q&T	390	1345	195	59	0.89	205	30	1585	230	−0.074	0.45	−0.68
1045	L	Q&T	450	1585	230	55	0.81	205	30	1795	260	−0.07	0.35	−0.69
1045	L	Q&T	500	1825	265	51	0.71	205	30	2275	330	−0.08	0.25	−0.68
1045	L	Q&T	595	2240	325	41	0.52	205	30	2725	395	−0.081	0.07	−0.60
1144	L	CDSR	265	930	135	33	0.51	195	28.5	1000	145	−0.08	0.32	−0.58

Grade	Orient.	Condition											
1144	L	DAT	305	1035	150	25	0.29	200	28.8	1585	230	-0.09	0.27	-0.53
1541F	L	Q&T forging	290	950	138	49	0.68	205	29.9	1275	185	-0.076	0.68	-0.65
1541F	L	Q&T forging	260	890	129	60	0.93	205	29.9	1275	185	-0.071	0.93	-0.65
4130	L	Q&T	258	895	130	67	1.12	220	32	1275	185	-0.083	0.92	-0.63
4130	L	Q&T	365	1425	207	55	0.79	200	29	1695	246	-0.081	0.89	-0.69
4140	L	Q&T, DAT	310	1075	156	60	0.69	200	29.2	1825	265	-0.08	1.2	-0.59
4142	L	DAT	310	1060	154	29	0.35	200	29	1450	210	-0.10	0.22	-0.51
4142	L	DAT	335	1250	181	28	0.34	200	28.9	1250	181	-0.08	0.06	-0.62
4142	L	Q&T	380	1415	205	48	0.66	205	30	1825	265	-0.08	0.45	-0.75
4142	L	Q&T and deformed	400	1550	225	47	0.63	200	29	1895	275	-0.09	0.50	-0.75
4142	L	Q&T	450	1760	255	42	0.54	205	30	2000	290	-0.08	0.40	-0.73
4142	L	Q&T and deformed	475	2035	295	20	0.22	200	29	2070	300	-0.082	0.20	-0.77
4142	L	Q&T and deformed	450	1930	280	37	0.46	200	29	2105	305	-0.09	0.60	-0.76
4142	L	Q&T	475	1930	280	35	0.43	205	30	2170	315	-0.081	0.09	-0.61
4142	L	Q&T	560	2240	325	27	0.31	205	30	2655	385	-0.089	0.07	-0.76
4340	L	HR, A	243	825	120	43	0.57	195	28	1200	174	-0.095	0.45	-0.54
4340	L	Q&T	409	1470	213	38	0.48	200	29	2000	290	-0.091	0.48	-0.60
4340	L	Q&T	350	1240	180	57	0.84	195	28	1655	240	-0.076	0.73	-0.62
5160	L	Q&T	430	1670	242	42	0.87	195	28	1930	280	-0.071	0.40	-0.57
52100	L	SH, Q&T	518	2015	292	11	0.12	205	30	2585	375	-0.09	0.18	-0.56
9262	L	A	260	925	134	14	0.16	205	30	1040	151	-0.071	0.16	-0.47
9262	L	Q&T	280	1000	145	33	0.41	195	28	1220	177	-0.073	0.41	-0.60
9262	L	Q&T	410	565	227	32	0.38	200	29	1855	269	-0.057	0.38	-0.65
950C (d)	LT	HR plate	159	565	82	64	1.03	205	29.6	1170	170	-0.12	0.95	-0.61
950C (d)	L	HR bar	150	565	82	69	1.19	205	30	970	141	-0.11	0.85	-0.59
950X (d)	L	Plate channel	150	440	64	65	1.06	205	30	625	91	-0.075	0.35	-0.54
950X (d)	L	HR plate	156	530	77	72	1.24	205	29.5	1005	146	-0.10	0.85	-0.61
950X (d)	L	Plate channel	225	695	101	68	1.15	195	28.2	1055	153	-0.08	0.21	-0.53

Notes: (a) AISI/SAE grade, unless otherwise indicated. (b) ASTM designation. (c) Proprietary designation. (d) SAE HSLA grade. (e) Orientation of axis of specimen, relative to rolling direction; L is longitudinal (parallel to rolling direction); LT is long transverse (perpendicular to rolling direction). (f) STA, solution treated and aged; HR, hot rolled; CD, cold drawn; Q&T, quenched and tempered; CDSR, cold drawn strain relieved; DAT, drawn at temperature; A, annealed. From *ASM Metals Reference Book, 2nd edition,* 1983; ASM International, Materials Park, OH 44073-0002; table 217. Reprinted by permission of ASM International®, www.asminternational.org.

Table A–24

Mechanical Properties of Three Non-Steel Metals

(a) Typical Properties of Gray Cast Iron

[The American Society for Testing and Materials (ASTM) numbering system for gray cast iron is such that the numbers correspond to the *minimum tensile strength* in kpsi. Thus an ASTM No. 20 cast iron has a minimum tensile strength of 20 kpsi. Note particularly that the tabulations are *typical* of several heats.]

ASTM Number	Tensile Strength S_{ut}, kpsi	Compressive Strength S_{uc}, kpsi	Shear Modulus of Rupture S_{su}, kpsi	Modulus of Elasticity, Mpsi Tension†	Torsion	Endurance Limit* S_e, kpsi	Brinell Hardness H_B	Fatigue Stress-Concentration Factor K_f
20	22	83	26	9.6–14	3.9–5.6	10	156	1.00
25	26	97	32	11.5–14.8	4.6–6.0	11.5	174	1.05
30	31	109	40	13–16.4	5.2–6.6	14	201	1.10
35	36.5	124	48.5	14.5–17.2	5.8–6.9	16	212	1.15
40	42.5	140	57	16–20	6.4–7.8	18.5	235	1.25
50	52.5	164	73	18.8–22.8	7.2–8.0	21.5	262	1.35
60	62.5	187.5	88.5	20.4–23.5	7.8–8.5	24.5	302	1.50

*Polished or machined specimens.

†The modulus of elasticity of cast iron in compression corresponds closely to the upper value in the range given for tension and is a more constant value than that for tension.

Table A–24

Mechanical Properties of Three Non-Steel Metals   (*Continued*)

(b) Mechanical Properties of Some Aluminum Alloys

[These *are typical* properties for sizes of about $\frac{1}{2}$ in; similar properties can be obtained by using proper purchase specifications. The values given for fatigue strength correspond to $50(10^7)$ cycles of completely reversed stress. Alluminum alloys do not have an endurance limit. Yield strengths were obtained by the 0.2 percent offset method.]

Aluminum Association Number	Temper	Yield, S_y, MPa (kpsi)	Strength Tensile, S_u, MPa (kpsi)	Fatigue, S_f, MPa (kpsi)	Elongation in 2 in, %	Brinell Hardness H_B
Wrought:						
2017	O	70 (10)	179 (26)	90 (13)	22	45
2024	O	76 (11)	186 (27)	90 (13)	22	47
	T3	345 (50)	482 (70)	138 (20)	16	120
3003	H12	117 (17)	131 (19)	55 (8)	20	35
	H16	165 (24)	179 (26)	65 (9.5)	14	47
3004	H34	186 (27)	234 (34)	103 (15)	12	63
	H38	234 (34)	276 (40)	110 (16)	6	77
5052	H32	186 (27)	234 (34)	117 (17)	18	62
	H36	234 (34)	269 (39)	124 (18)	10	74
Cast:						
319.0*	T6	165 (24)	248 (36)	69 (10)	2.0	80
333.0†	T5	172 (25)	234 (34)	83 (12)	1.0	100
	T6	207 (30)	289 (42)	103 (15)	1.5	105
335.0*	T6	172 (25)	241 (35)	62 (9)	3.0	80
	T7	248 (36)	262 (38)	62 (9)	0.5	85

*Sand casting.

†Permanent-mold casting.

(c) Mechanical Properties of Some Titanium Alloys

Titanium Alloy	Condition	Yield, S_y (0.2% offset) MPa (kpsi)	Strength Tensile, S_{ut} MPa (kpsi)	Elongation in 2 in, %	Hardness (Brinell or Rockwell)
Ti-35A†	Annealed	210 (30)	275 (40)	30	135 HB
Ti-50A†	Annealed	310 (45)	380 (55)	25	215 HB
Ti-0.2 Pd	Annealed	280 (40)	340 (50)	28	200 HB
Ti-5 Al-2.5 Sn	Annealed	760 (110)	790 (115)	16	36 HRC
Ti-8 Al-1 Mo-1 V	Annealed	900 (130)	965 (140)	15	39 HRC
Ti-6 Al-6 V-2 Sn	Annealed	970 (140)	1030 (150)	14	38 HRC
Ti-6Al-4V	Annealed	830 (120)	900 (130)	14	36 HRC
Ti-13 V-11 Cr-3 Al	Sol. + aging	1207 (175)	1276 (185)	8	40 HRC

†Commercially pure alpha titanium.

Table A-25

Stochastic Yield and Ultimate Strengths for Selected Materials   *Source:* Data compiled from "Some Property Data and Corresponding Weibull Parameters for Stochastic Mechanical Design," Trans. *ASME Journal of Mechanical Design*, vol. 114 (March 1992), pp. 29–34.

Material	μ_{Sut}	$\hat{\sigma}_{Sut}$	x_0	θ	b	μ_{Sy}	$\hat{\sigma}_{Sy}$	x_0	θ	b	C_{Sut}	C_{Sy}
1018 CD	87.6	5.74	30.8	90.1	12	78.4	5.90	56	80.6	4.29	0.0655	0.0753
1035 HR	86.2	3.92	72.6	87.5	3.86	49.6	3.81	39.5	50.8	2.88	0.0455	0.0768
1045 CD	117.7	7.13	90.2	120.5	4.38	95.5	6.59	82.1	97.2	2.14	0.0606	0.0690
1117 CD	83.1	5.25	73.0	84.4	2.01	81.4	4.71	72.4	82.6	2.00	0.0632	0.0579
1137 CD	106.5	6.15	96.2	107.7	1.72	98.1	4.24	92.2	98.7	1.41	0.0577	0.0432
12L14 CD	79.6	6.92	70.3	80.4	1.36	78.1	8.27	64.3	78.8	1.72	0.0869	0.1059
1038 HT bolts	133.4	3.38	122.3	134.6	3.64						0.0253	
ASTM40	44.5	4.34	27.7	46.2	4.38						0.0975	
35018 Malleable	53.3	1.59	48.7	53.8	3.18	38.5	1.42	34.7	39.0	2.93	0.0298	0.0369
32510 Malleable	53.4	2.68	44.7	54.3	3.61	34.9	1.47	30.1	35.5	3.67	0.0502	0.0421
Pearlitic Malleable	93.9	3.83	80.1	95.3	4.04	60.2	2.78	50.2	61.2	4.02	0.0408	0.0462
604515 Nodular	64.8	3.77	53.7	66.1	3.23	49.0	4.20	33.8	50.5	4.06	0.0582	0.0857
100-70-04 Nodular	122.2	7.65	47.6	125.6	11.84	79.3	4.51	64.1	81.0	3.77	0.0626	0.0569
201SS CD	195.9	7.76	180.7	197.9	2.06						0.0396	
301SS CD	191.2	5.82	151.9	193.6	8.00	166.8	9.37	139.7	170.0	3.17	0.0304	0.0562
A	105.0	5.68	92.3	106.6	2.38	46.8	4.70	26.3	48.7	4.99	0.0541	0.1004
304SS A	85.0	4.14	66.6	86.6	5.11	37.9	3.76	30.2	38.9	2.17	0.0487	0.0992
310SS A	84.8	4.23	71.6	86.3	3.45						0.0499	
403SS	105.3	3.09	95.7	106.4	3.44	78.5	3.91	64.8	79.9	3.93	0.0293	0.0498
17-7PSS	198.8	9.51	163.3	202.3	4.21	189.4	11.49	144.0	193.8	4.48	0.0478	0.0607
AM350SS A	149.1	8.29	101.8	152.4	6.68	63.0	5.05	38.0	65.0	5.73	0.0556	0.0802
Ti-6AL-4V	175.4	7.91	141.8	178.5	4.85	163.7	9.03	101.5	167.4	8.18	0.0451	0.0552
2024 0	28.1	1.73	24.2	28.7	2.43						0.0616	
2024 T4	64.9	1.64	60.2	65.5	3.16	40.8	1.83	38.4	41.0	1.32	0.0253	0.0449
T6	67.5	1.50	55.9	68.1	9.26	53.4	1.17	51.2	53.6	1.91	0.0222	0.0219
7075 T6 .025"	75.5	2.10	68.8	76.2	3.53	63.7	1.98	58.9	64.3	2.63	0.0278	0.0311

Table A-26

Stochastic Parameters for Finite Life Fatigue Tests in Selected Metals   *Source:* E. B. Haugen, *Probabilistic Mechanical Design*, Wiley, New York, 1980, Appendix 10–B.

1	2	3	4	5		Stress Cycles to Failure			
Number	**Condition**	**TS MPa (kpsi)**	**YS MPa (kpsi)**	**Distribution**		10^4	10^5	10^6	10^7
1046	WQ&T, 1210°F	723 (105)	565 (82)	W	x_0	544 (79)	462 (67)	391 (56.7)	
					θ	594 (86.2)	503 (73.0)	425 (61.7)	
					b	2.60	2.75	2.85	
2340	OQ&T 1200°F	799 (116)	661 (96)	W	x_0	579 (84)	510 (74)	420 (61)	
					θ	699 (101.5)	588 (85.4)	496 (72.0)	
					b	4.3	3.4	4.1	
3140	OQ&T, 1300°F	744 (108)	599 (87)	W	x_0	510 (74)	455 (66)	393 (57)	
					θ	604 (87.7)	528 (76.7)	463 (67.2)	
					b	5.2	5.0	5.5	
2024 Aluminum	T-4	489 (71)	365 (53)	N	σ	26.3 (3.82)	21.4 (3.11)	17.4 (2.53)	14.0 (2.03)
					μ	143 (20.7)	116 (16.9)	95 (13.8)	77 (11.2)
Ti-6Al-4V	HT-46	1040 (151)	992 (144)	N	σ	39.6 (5.75)	38.1 (5.53)	36.6 (5.31)	35.1 (5.10)
					μ	712 (108)	684 (99.3)	657 (95.4)	493 (71.6)

Statistical parameters from a large number of fatigue tests are listed. Weibull distribution is denoted W and the parameters are x_0, "guaranteed" fatigue strength; θ, characteristic fatigue strength; and b, shape factor. Normal distribution is denoted N and the parameters are μ, mean fatigue strength; and σ, standard deviation of the fatigue strength. The life is in stress-cycles-to-failure. TS = tensile strength, YS = yield strength. All testing by rotating-beam specimen.

Table A–27

Finite Life Fatigue Strengths of Selected Plain Carbon Steels   *Source:* Compiled from Table 4 in H. J. Grover, S. A. Gordon, and L. R. Jackson, *Fatigue of Metals and Structures,* Bureau of Naval Weapons Document NAVWEPS 00-25-534, 1960.

Material	Condition	BHN*	Tensile Strength kpsi	Yield Strength kpsi	RA*	Stress Cycles to Failure							
						10^4	$4(10^4)$	10^5	$4(10^5)$	10^6	$4(10^6)$	10^7	10^8
1020	Furnace cooled		58	30	0.63			37	34	30	28	25	
1030	Air-cooled	135	80	45	0.62		51	47	42	38	38	38	
1035	Normal	132	72	35	0.54			44	40	37	34	33	33
	WQT	209	103	87	0.65		80	72	65	60	57	57	57
1040	Forged	195	92	53	0.23				40	47	33	33	
1045	HR, N		107	63	0.49	80	70	56	47	47	47	47	
1050	N, AC	164	92	47	0.40	50	48	46	40	38	34	34	
	WQT 1200	196	97	70	0.58		60	57	52	50	50	50	50
.56 MN	N	193	98	47	0.42	61	55	51	47	43	41	41	41
	WQT 1200	277	111	84	0.57	94	81	73	62	57	55	55	55
1060	As Rec.	67 Rb	134	65	0.20	65	60	55	50	48	48	48	
1095		162	84	33	0.37	50	43	40	34	31	30	30	30
	OQT 1200	227	115	65	0.40	77	68	64	57	56	56	56	56
10120	OQT 1200	224	117	59	0.12		60	56	51	50	50	50	56
	OQT 860	369	180	130	0.15		102	95	91	91	91	91	

*BHN = Brinell hardness number; RA = fractional reduction in area.

Table A-28

Decimal Equivalents of Wire and Sheet-Metal Gauges* (All Sizes Are Given in Inches)

Name of Gauge:	American or Brown & Sharpe	Birmingham or Stubs Iron Wire	United States Standard†	Manufacturers Standard	Steel Wire or Washburn & Moen	Music Wire	Stubs Steel Wire	Twist Drill
Principal Use:	Nonferrous Sheet, Wire, and Rod	Tubing, Ferrous Strip, Flat Wire, and Spring Steel	Ferrous Sheet and Plate, 480 lbf/ft³	Ferrous Sheet	Ferrous Wire Except Music Wire	Music Wire	Steel Drill Rod	Twist Drills and Drill Steel
7/0			0.500		0.490			
6/0	0.580 0		0.468 75		0.461 5	0.004		
5/0	0.516 5		0.437 5		0.430 5	0.005		
4/0	0.460 0	0.454	0.406 25		0.393 8	0.006		
3/0	0.409 6	0.425	0.375		0.362 5	0.007		
2/0	0.364 8	0.380	0.343 75		0.331 0	0.008		
0	0.324 9	0.340	0.312 5		0.306 5	0.009		
1	0.289 3	0.300	0.281 25		0.283 0	0.010	0.227	0.228 0
2	0.257 6	0.284	0.265 625		0.262 5	0.011	0.219	0.221 0
3	0.229 4	0.259	0.25	0.239 1	0.243 7	0.012	0.212	0.213 0
4	0.204 3	0.238	0.234 375	0.224 2	0.225 3	0.013	0.207	0.209 0
5	0.181 9	0.220	0.218 75	0.209 2	0.207 0	0.014	0.204	0.205 5
6	0.162 0	0.203	0.203 125	0.194 3	0.192 0	0.016	0.201	0.204 0
7	0.144 3	0.180	0.187 5	0.179 3	0.177 0	0.018	0.199	0.201 0
8	0.128 5	0.165	0.171 875	0.164 4	0.162 0	0.020	0.197	0.199 0
9	0.114 4	0.148	0.156 25	0.149 5	0.148 3	0.022	0.194	0.196 0
10	0.101 9	0.134	0.140 625	0.134 5	0.135 0	0.024	0.191	0.193 5
11	0.090 74	0.120	0.125	0.119 6	0.120 5	0.026	0.188	0.191 0
12	0.080 81	0.109	0.109 357	0.104 6	0.105 5	0.029	0.185	0.189 0
13	0.071 96	0.095	0.093 75	0.089 7	0.091 5	0.031	0.182	0.185 0
14	0.064 08	0.083	0.078 125	0.074 7	0.080 0	0.033	0.180	0.182 0
15	0.057 07	0.072	0.070 312 5	0.067 3	0.072 0	0.035	0.178	0.180 0
16	0.050 82	0.065	0.062 5	0.059 8	0.062 5	0.037	0.175	0.177 0
17	0.045 26	0.058	0.056 25	0.053 8	0.054 0	0.039	0.172	0.173 0

(Continued)

Table A–28
Decimal Equivalents of Wire and Sheet-Metal Gauges* (All Sizes Are Given in Inches)   *(Continued)*

Name of Gauge:	American or Brown & Sharpe	Birmingham or Stubs Iron Wire	United States Standard†	Manu-facturers Standard	Steel Wire or Washburn & Moen	Music Wire	Stubs Steel Wire	Twist Drill
Principal Use:	Nonferrous Sheet, Wire, and Rod	Tubing, Ferrous Strip, Flat Wire, and Spring Steel	Ferrous Sheet and Plate, 480 lbf/ft³	Ferrous Sheet	Ferrous Wire Except Music Wire	Music Wire	Steel Drill Rod	Twist Drills and Drill Steel
18	0.040 30	0.049	0.05	0.047 8	0.047 5	0.041	0.168	0.169 5
19	0.035 89	0.042	0.043 75	0.041 8	0.041 0	0.043	0.164	0.166 0
20	0.031 96	0.035	0.037 5	0.035 9	0.034 8	0.045	0.161	0.161 0
21	0.028 46	0.032	0.034 375	0.032 9	0.031 7	0.047	0.157	0.159 0
22	0.025 35	0.028	0.031 25	0.029 9	0.028 6	0.049	0.155	0.157 0
23	0.022 57	0.025	0.028 125	0.026 9	0.025 8	0.051	0.153	0.154 0
24	0.020 10	0.022	0.025	0.023 9	0.023 0	0.055	0.151	0.152 0
25	0.017 90	0.020	0.021 875	0.020 9	0.020 4	0.059	0.148	0.149 5
26	0.015 94	0.018	0.018 75	0.017 9	0.018 1	0.063	0.146	0.147 0
27	0.014 20	0.016	0.017 187 5	0.016 4	0.017 3	0.067	0.143	0.144 0
28	0.012 64	0.014	0.015 625	0.014 9	0.016 2	0.071	0.139	0.140 5
29	0.011 26	0.013	0.014 062 5	0.013 5	0.015 0	0.075	0.134	0.136 0
30	0.010 03	0.012	0.012 5	0.012 0	0.014 0	0.080	0.127	0.128 5
31	0.008 928	0.010	0.010 937 5	0.010 5	0.013 2	0.085	0.120	0.120 0
32	0.007 950	0.009	0.010 156 25	0.009 7	0.012 8	0.090	0.115	0.116 0
33	0.007 080	0.008	0.009 375	0.009 0	0.011 8	0.095	0.112	0.113 0
34	0.006 305	0.007	0.008 593 75	0.008 2	0.010 4		0.110	0.111 0
35	0.005 615	0.005	0.007 812 5	0.007 5	0.009 5		0.108	0.110 0
36	0.005 000	0.004	0.007 031 25	0.006 7	0.009 0		0.106	0.106 5
37	0.004 453		0.006 640 625	0.006 4	0.008 5		0.103	0.104 0
38	0.003 965		0.006 25	0.006 0	0.008 0		0.101	0.101 5
39	0.003 531				0.007 5		0.099	0.099 5
40	0.003 145				0.007 0		0.097	0.098 0

*Specify sheet, wire, and plate by stating the gauge number, the gauge name, and the decimal equivalent in parentheses.
†Reflects present average and weights of sheet steel.

Table A–29

Dimensions of Square and Hexagonal Bolts

Nominal Size, in	Square		Regular Hexagonal			Head Type Heavy Hexagonal			Structural Hexagonal		
	W	H	W	H	R_{min}	W	H	R_{min}	W	H	R_{min}
$\frac{1}{4}$	$\frac{3}{8}$	$\frac{11}{64}$	$\frac{7}{16}$	$\frac{11}{64}$	0.01						
$\frac{5}{16}$	$\frac{1}{2}$	$\frac{13}{64}$	$\frac{1}{2}$	$\frac{7}{32}$	0.01						
$\frac{3}{8}$	$\frac{9}{16}$	$\frac{1}{4}$	$\frac{9}{16}$	$\frac{1}{4}$	0.01						
$\frac{7}{16}$	$\frac{5}{8}$	$\frac{19}{64}$	$\frac{5}{8}$	$\frac{19}{64}$	0.01						
$\frac{1}{2}$	$\frac{3}{4}$	$\frac{21}{64}$	$\frac{3}{4}$	$\frac{11}{32}$	0.01	$\frac{7}{8}$	$\frac{11}{32}$	0.01	$\frac{7}{8}$	$\frac{5}{16}$	0.009
$\frac{5}{8}$	$\frac{15}{16}$	$\frac{27}{64}$	$\frac{15}{16}$	$\frac{27}{64}$	0.02	$1\frac{1}{16}$	$\frac{27}{64}$	0.02	$1\frac{1}{16}$	$\frac{25}{64}$	0.021
$\frac{3}{4}$	$1\frac{1}{8}$	$\frac{1}{2}$	$1\frac{1}{8}$	$\frac{1}{2}$	0.02	$1\frac{1}{4}$	$\frac{1}{2}$	0.02	$1\frac{1}{4}$	$\frac{15}{32}$	0.021
1	$1\frac{1}{2}$	$\frac{21}{32}$	$1\frac{1}{2}$	$\frac{43}{64}$	0.03	$1\frac{5}{8}$	$\frac{43}{64}$	0.03	$1\frac{5}{8}$	$\frac{39}{64}$	0.062
$1\frac{1}{8}$	$1\frac{11}{16}$	$\frac{3}{4}$	$1\frac{11}{16}$	$\frac{3}{4}$	0.03	$1\frac{13}{16}$	$\frac{3}{4}$	0.03	$1\frac{13}{16}$	$\frac{11}{16}$	0.062
$1\frac{1}{4}$	$1\frac{7}{8}$	$\frac{27}{32}$	$1\frac{7}{8}$	$\frac{27}{32}$	0.03	2	$\frac{27}{32}$	0.03	2	$\frac{25}{32}$	0.062
$1\frac{3}{8}$	$2\frac{1}{16}$	$\frac{29}{32}$	$2\frac{1}{16}$	$\frac{29}{32}$	0.03	$2\frac{3}{16}$	$\frac{29}{32}$	0.03	$3\frac{3}{16}$	$\frac{27}{32}$	0.062
$1\frac{1}{2}$	$2\frac{1}{4}$	1	$2\frac{1}{4}$	1	0.03	$2\frac{3}{8}$	1	0.03	$2\frac{3}{8}$	$\frac{15}{16}$	0.062

Nominal Size, mm											
M5	8	3.58	8	3.58	0.2						
M6			10	4.38	0.3						
M8			13	5.68	0.4						
M10			16	6.85	0.4						
M12			18	7.95	0.6	21	7.95	0.6			
M14			21	9.25	0.6	24	9.25	0.6			
M16			24	10.75	0.6	27	10.75	0.6	27	10.75	0.6
M20			30	13.40	0.8	34	13.40	0.8	34	13.40	0.8
M24			36	15.90	0.8	41	15.90	0.8	41	15.90	1.0
M30			46	19.75	1.0	50	19.75	1.0	50	19.75	1.2
M36			55	23.55	1.0	60	23.55	1.0	60	23.55	1.5

Table A–30

Dimensions of Hexagonal Cap Screws and Heavy Hexagonal Screws (W = Width across Flats; H = Height of Head; See Figure in Table A–29)

Nominal Size, in	Minimum Fillet Radius	Type of Screw Cap W	Type of Screw Heavy W	Height H
$\frac{1}{4}$	0.015	$\frac{7}{16}$		$\frac{5}{32}$
$\frac{5}{16}$	0.015	$\frac{1}{2}$		$\frac{13}{64}$
$\frac{3}{8}$	0.015	$\frac{9}{16}$		$\frac{15}{64}$
$\frac{7}{16}$	0.015	$\frac{5}{8}$		$\frac{9}{32}$
$\frac{1}{2}$	0.015	$\frac{3}{4}$	$\frac{7}{8}$	$\frac{5}{16}$
$\frac{5}{8}$	0.020	$\frac{15}{16}$	$1\frac{1}{16}$	$\frac{25}{64}$
$\frac{3}{4}$	0.020	$1\frac{1}{8}$	$1\frac{1}{4}$	$\frac{15}{32}$
$\frac{7}{8}$	0.040	$1\frac{5}{16}$	$1\frac{7}{16}$	$\frac{35}{64}$
1	0.060	$1\frac{1}{2}$	$1\frac{1}{8}$	$\frac{39}{64}$
$1\frac{1}{4}$	0.060	$1\frac{7}{8}$	2	$\frac{25}{32}$
$1\frac{3}{8}$	0.060	$2\frac{1}{16}$	$2\frac{3}{16}$	$\frac{27}{32}$
$1\frac{1}{2}$	0.060	$2\frac{1}{4}$	$2\frac{3}{8}$	$\frac{15}{16}$

Nominal Size, mm				
M5	0.2	8		3.65
M6	0.3	10		4.15
M8	0.4	13		5.50
M10	0.4	16		6.63
M12	0.6	18	21	7.76
M14	0.6	21	24	9.09
M16	0.6	24	27	10.32
M20	0.8	30	34	12.88
M24	0.8	36	41	15.44
M30	1.0	46	50	19.48
M36	1.0	55	60	23.38

Table A–31

Dimensions of
Hexagonal Nuts

Nominal Size, in	Width W	Height H		
		Regular Hexagonal	Thick or Slotted	JAM
$\frac{1}{4}$	$\frac{7}{16}$	$\frac{7}{32}$	$\frac{9}{32}$	$\frac{5}{32}$
$\frac{5}{16}$	$\frac{1}{2}$	$\frac{17}{64}$	$\frac{21}{64}$	$\frac{3}{16}$
$\frac{3}{8}$	$\frac{9}{16}$	$\frac{21}{64}$	$\frac{13}{32}$	$\frac{7}{32}$
$\frac{7}{16}$	$\frac{11}{16}$	$\frac{3}{8}$	$\frac{29}{64}$	$\frac{1}{4}$
$\frac{1}{2}$	$\frac{3}{4}$	$\frac{7}{16}$	$\frac{9}{16}$	$\frac{5}{16}$
$\frac{9}{16}$	$\frac{7}{8}$	$\frac{31}{64}$	$\frac{39}{64}$	$\frac{5}{16}$
$\frac{5}{8}$	$\frac{15}{16}$	$\frac{35}{64}$	$\frac{23}{32}$	$\frac{3}{8}$
$\frac{3}{4}$	$1\frac{1}{8}$	$\frac{41}{64}$	$\frac{13}{16}$	$\frac{27}{64}$
$\frac{7}{8}$	$1\frac{5}{16}$	$\frac{3}{4}$	$\frac{29}{32}$	$\frac{31}{64}$
1	$1\frac{1}{2}$	$\frac{55}{64}$	1	$\frac{35}{64}$
$1\frac{1}{8}$	$1\frac{11}{16}$	$\frac{31}{32}$	$1\frac{5}{32}$	$\frac{39}{64}$
$1\frac{1}{4}$	$1\frac{7}{8}$	$1\frac{1}{16}$	$1\frac{1}{4}$	$\frac{23}{32}$
$1\frac{3}{8}$	$2\frac{1}{16}$	$1\frac{11}{64}$	$1\frac{3}{8}$	$\frac{25}{32}$
$1\frac{1}{2}$	$2\frac{1}{4}$	$1\frac{9}{32}$	$1\frac{1}{2}$	$\frac{27}{32}$

Nominal Size, mm				
M5	8	4.7	5.1	2.7
M6	10	5.2	5.7	3.2
M8	13	6.8	7.5	4.0
M10	16	8.4	9.3	5.0
M12	18	10.8	12.0	6.0
M14	21	12.8	14.1	7.0
M16	24	14.8	16.4	8.0
M20	30	18.0	20.3	10.0
M24	36	21.5	23.9	12.0
M30	46	25.6	28.6	15.0
M36	55	31.0	34.7	18.0

Table A–32

Basic Dimensions of American Standard Plain Washers (All Dimensions in Inches)

Fastener Size	Washer Size	Diameter ID	Diameter OD	Thickness
#6	0.138	0.156	0.375	0.049
#8	0.164	0.188	0.438	0.049
#10	0.190	0.219	0.500	0.049
#12	0.216	0.250	0.562	0.065
$\frac{1}{4}$ N	0.250	0.281	0.625	0.065
$\frac{1}{4}$ W	0.250	0.312	0.734	0.065
$\frac{5}{16}$ N	0.312	0.344	0.688	0.065
$\frac{5}{16}$ W	0.312	0.375	0.875	0.083
$\frac{3}{8}$ N	0.375	0.406	0.812	0.065
$\frac{3}{8}$ W	0.375	0.438	1.000	0.083
$\frac{7}{16}$ N	0.438	0.469	0.922	0.065
$\frac{7}{16}$ W	0.438	0.500	1.250	0.083
$\frac{1}{2}$ N	0.500	0.531	1.062	0.095
$\frac{1}{2}$ W	0.500	0.562	1.375	0.109
$\frac{9}{16}$ N	0.562	0.594	1.156	0.095
$\frac{9}{16}$ W	0.562	0.625	1.469	0.109
$\frac{5}{8}$ N	0.625	0.656	1.312	0.095
$\frac{5}{8}$ W	0.625	0.688	1.750	0.134
$\frac{3}{4}$ N	0.750	0.812	1.469	0.134
$\frac{3}{4}$ W	0.750	0.812	2.000	0.148
$\frac{7}{8}$ N	0.875	0.938	1.750	0.134
$\frac{7}{8}$ W	0.875	0.938	2.250	0.165
1 N	1.000	1.062	2.000	0.134
1 W	1.000	1.062	2.500	0.165
$1\frac{1}{8}$ N	1.125	1.250	2.250	0.134
$1\frac{1}{8}$ W	1.125	1.250	2.750	0.165
$1\frac{1}{4}$ N	1.250	1.375	2.500	0.165
$1\frac{1}{4}$ W	1.250	1.375	3.000	0.165
$1\frac{3}{8}$ N	1.375	1.500	2.750	0.165
$1\frac{3}{8}$ W	1.375	1.500	3.250	0.180
$1\frac{1}{2}$ N	1.500	1.625	3.000	0.165
$1\frac{1}{2}$ W	1.500	1.625	3.500	0.180
$1\frac{5}{8}$	1.625	1.750	3.750	0.180
$1\frac{3}{4}$	1.750	1.875	4.000	0.180
$1\frac{7}{8}$	1.875	2.000	4.250	0.180
2	2.000	2.125	4.500	0.180
$2\frac{1}{4}$	2.250	2.375	4.750	0.220
$2\frac{1}{2}$	2.500	2.625	5.000	0.238
$2\frac{3}{4}$	2.750	2.875	5.250	0.259
3	3.000	3.125	5.500	0.284

N = narrow; W = wide; use W when not specified.

Table A–33

Dimensions of Metric Plain Washers (All Dimensions in Millimeters)

Washer Size*	Minimum ID	Maximum OD	Maximum Thickness	Washer Size*	Minimum ID	Maximum OD	Maximum Thickness
1.6 N	1.95	4.00	0.70	10 N	10.85	20.00	2.30
1.6 R	1.95	5.00	0.70	10 R	10.85	28.00	2.80
1.6 W	1.95	6.00	0.90	10 W	10.85	39.00	3.50
2 N	2.50	5.00	0.90	12 N	13.30	25.40	2.80
2 R	2.50	6.00	0.90	12 R	13.30	34.00	3.50
2 W	2.50	8.00	0.90	12 W	13.30	44.00	3.50
2.5 N	3.00	6.00	0.90	14 N	15.25	28.00	2.80
2.5 R	3.00	8.00	0.90	14 R	15.25	39.00	3.50
2.5 W	3.00	10.00	1.20	14 W	15.25	50.00	4.00
3 N	3.50	7.00	0.90	16 N	17.25	32.00	3.50
3 R	3.50	10.00	1.20	16 R	17.25	44.00	4.00
3 W	3.50	12.00	1.40	16 W	17.25	56.00	4.60
3.5 N	4.00	9.00	1.20	20 N	21.80	39.00	4.00
3.5 R	4.00	10.00	1.40	20 R	21.80	50.00	4.60
3.5 W	4.00	15.00	1.75	20 W	21.80	66.00	5.10
4 N	4.70	10.00	1.20	24 N	25.60	44.00	4.60
4 R	4.70	12.00	1.40	24 R	25.60	56.00	5.10
4 W	4.70	16.00	2.30	24 W	25.60	72.00	5.60
5 N	5.50	11.00	1.40	30 N	32.40	56.00	5.10
5 R	5.50	15.00	1.75	30 R	32.40	72.00	5.60
5 W	5.50	20.00	2.30	30 W	32.40	90.00	6.40
6 N	6.65	13.00	1.75	36 N	38.30	66.00	5.60
6 R	6.65	18.80	1.75	36 R	38.30	90.00	6.40
6 W	6.65	25.40	2.30	36 W	38.30	110.00	8.50
8 N	8.90	18.80	2.30				
8 R	8.90	25.40	2.30				
8 W	8.90	32.00	2.80				

N = narrow; R = regular; W = wide.

*Same as screw or bolt size.

Table A–34

Gamma Function*

Source: Reprinted with permission from William H. Beyer (ed.), *Handbook of Tables for Probability and Statistics,* 2nd ed., 1966. Copyright CRC Press, Boca Raton, Florida.

Values of $\Gamma(n) = \int_0^\infty e^{-x}x^{n-1}\,dx$

n	Γ(n)	n	Γ(n)	n	Γ(n)	n	Γ(n)
1.00	1.000 00	1.25	.906 40	1.50	.886 23	1.75	.919 06
1.01	.994 33	1.26	.904 40	1.51	.886 59	1.76	.921 37
1.02	.988 84	1.27	.902 50	1.52	.887 04	1.77	.923 76
1.03	.983 55	1.28	.900 72	1.53	.887 57	1.78	.926 23
1.04	.978 44	1.29	.899 04	1.54	.888 18	1.79	.928 77
1.05	.973 50	1.30	.897 47	1.55	.888 87	1.80	.931 38
1.06	.968 74	1.31	.896 00	1.56	.889 64	1.81	.934 08
1.07	.964 15	1.32	.894 64	1.57	.890 49	1.82	.936 85
1.08	.959 73	1.33	.893 38	1.58	.891 42	1.83	.939 69
1.09	.955 46	1.34	.892 22	1.59	.892 43	1.84	.942 61
1.10	.951 35	1.35	.891 15	1.60	.893 52	1.85	.945 61
1.11	.947 39	1.36	.890 18	1.61	.894 68	1.86	.948 69
1.12	.943 59	1.37	.889 31	1.62	.895 92	1.87	.951 84
1.13	.939 93	1.38	.888 54	1.63	.897 24	1.88	.955 07
1.14	.936 42	1.39	.887 85	1.64	.898 64	1.89	.958 38
1.15	.933 04	1.40	.887 26	1.65	.900 12	1.90	.961 77
1.16	.929 80	1.41	.886 76	1.66	.901 67	1.91	.965 23
1.17	.936 70	1.42	.886 36	1.67	.903 30	1.92	.968 78
1.18	.923 73	1.43	.886 04	1.68	.905 00	1.93	.972 40
1.19	.920 88	1.44	.885 80	1.69	.906 78	1.94	.976 10
1.20	.918 17	1.45	.885 65	1.70	.908 64	1.95	.979 88
1.21	.915 58	1.46	.885 60	1.71	.910 57	1.96	.983 74
1.22	.913 11	1.47	.885 63	1.72	.912 58	1.97	.987 68
1.23	.910 75	1.48	.885 75	1.73	.914 66	1.98	.991 71
1.24	.908 52	1.49	.885 95	1.74	.916 83	1.99	.995 81
						2.00	1.000 00

*For $n > 2$, use the recursive formula

$$\Gamma(n) = (n-1)\,\Gamma(n-1)$$

For example, $\Gamma(5.42) = 4.42(3.42)\,(2.42)\,\Gamma(1.42) = 4.42(3.42)\,(2.42)\,(0.886\,36) = 32.4245$

†For large positive values of x, $\Gamma(x)$ can be expressed by the asymptotic series based on Stirling's approximation

$$\Gamma(x) \approx x^x e^{-x}\sqrt{\frac{2\pi}{x}}\left(1 + \frac{1}{12x} + \frac{1}{288x^2} - \frac{139}{51\,840x^3} - \frac{571}{2\,488\,320x^4}\right)$$

B–1 **Chapter 1**

1–8 $P = 100$ units

1–11 (a) $e_1 = 0.005\ 751\ 311\ 1$, $e_2 = 0.008\ 427\ 124\ 7$, $e = 0.014\ 178\ 435\ 8$, (b) $e_1 = -0.004\ 248\ 688\ 9$, $e_2 = -0.001\ 572\ 875\ 3$, $e = -0.005\ 821\ 564\ 2$

1–15 $L_{10} = 84.1$ kcycles

1–18 $\bar{n} = 1.32$, $d = 31.9$ mm

1–20 $\bar{n} = 1.17$, $R = 94.9$ percent

1–21 (a) $w = 0.020 \pm 0.018$ in, (b) $\bar{d} = 6.528$ in

1–23 $a = 1.569 \pm 0.016$ in

1–24 $D_o = 4.012 \pm 0.036$ in

1–31 (a) $\sigma = 1.90$ kpsi, (b) $\sigma = 397$ psi, (c) $y = 0.609$ in, (d) $\theta = 4.95°$

B–2 **Chapter 2**

2–6 (b) $E = 30.5$ Mpsi, $S_y = 45.6$ kpsi, $S_{ut} = 85.6$ kpsi, area reduction = 45.8 percent

2–9 (a) Before: $S_y = 32$ kpsi, $S_u = 49.5$ kpsi, After: $S'_y = 61.8$ kpsi, 93% increase, $S'_u = 61.9$ kpsi, 25% increase, (b) Before: $S_u/S_y = 1.55$, After: $S'_u/S'_y = 1$

2–15 $\bar{S}_u = 177$ kpsi, $s_{Su} = 1.27$ kpsi

2–17 (a) $u_R \approx 34.7$ in · lbf/in³, (b) $u_T \approx 66.7\ (10^3)$ in · lbf/in³

2–26 Aluminum alloys have greatest potential followed closely by high carbon heat-treated steel. Warrants further discussion.

2–34 Steel, titanium, aluminum alloys, and composites

B–3 **Chapter 3**

3–1 $R_B = 33.3$ lbf, $R_O = 66.7$ lbf, $R_C = 33.3$ lbf

3–6 $R_O = 740$ lbf, $M_O = 8080$ lbf · in

3–14 (a) $M_{max} = 253$ lbf · in, (b) $a_{min} = 2.07$ in, $M_{min} = 214$ lbf · in

3–15 (a) $\sigma_1 = 22$ kpsi, $\sigma_2 = -12$ kpsi, $\sigma_3 = 0$ kpsi, $\phi_p = 14.0°$ cw, $\tau_1 = 17$ kpsi, $\sigma_{ave} = 5$ kpsi, $\phi_s = 31.0°$ ccw, (b) $\sigma_1 = 18.6$ kpsi, $\sigma_2 = 6.4$ kpsi, $\sigma_3 = 0$ kpsi, $\phi_p = 27.5°$ ccw, $\tau_1 = 6.10$ kpsi, $\sigma_{ave} = 12.5$ kpsi, $\phi_s = 17.5°$ cw, (c) $\sigma_1 = 26.2$ kpsi, $\sigma_2 = 7.78$ kpsi, $\sigma_3 = 0$ kpsi, $\phi_p = 69.7°$ ccw, $\tau_1 = 9.22$ kpsi, $\sigma_{ave} = 17$ kpsi, $\phi_s = 24.7°$ ccw, (d) $\sigma_1 = 25.8$ kpsi, $\sigma_2 = -15.8$ kpsi, $\sigma_3 = 0$ kpsi, $\phi_p = 72.4°$ cw, $\tau_1 = 20.8$ kpsi, $\sigma_{ave} = 5$ kpsi, $\phi_s = 27.4°$ ccw

3–20 $\sigma_1 = 24.0$ kpsi, $\sigma_2 = 0.819$ kpsi, $\sigma_3 = -24.8$ kpsi, $\tau_{max} = 24.4$ kpsi

3–23 $\sigma = 34.0$ kpsi, $\delta = 0.0679$ in, $\epsilon_1 = 1.13\ (10^{-3})$, $\epsilon_2 = -3.30(10^{-4})$, $\Delta d = -2.48(10^{-4})$ in

3–27 $\delta = 5.86$ mm

3–29 $\sigma_x = 382$ MPa, $\sigma_y = -37.4$ MPa

3–35 $\sigma_{max} = 84.3$ MPa, $\tau_{max} = 5.63$ MPa

3–40 model c: $\sigma = 17.8$ kpsi, $\tau = 3.4$ kpsi, model d: $\sigma = 25.5$ kpsi, $\tau = 3.4$ kpsi, model e: $\sigma = 17.8$ kpsi, $\tau = 3.4$ kpsi

3–51 (a) $T = 1318$ lbf · in, $\theta = 4.59°$, (b) $T = 1287$ lbf · in, $\theta = 4.37°$

3–53 (a) $T_1 = 1.47$ N · m, $T_2 = 7.45$ N · m, $T_3 = 0$ N · m, $T = 8.92$ N · m, (b) $\theta_1 = 0.348$ rad/m

3–59 $H = 55.5$ kW

3–66 $d_c = 1.4$ in

3–69 (a) $T_1 = 2880$ N, $T_2 = 432$ N, (b) $R_C = 1794$ N, $R_O = 3036$ N, (d) $\sigma = 263$ MPa, $\tau = 57.7$ MPa, (e) $\sigma_1 = 275$ MPa, $\sigma_2 = -12.1$ MPa, $\tau_{max} = 144$ MPa

3–72 (a) $F_B = 750$ lbf, (b) $R_{Cy} = 183.1$ lbf, $R_{Cz} = 861.5$ lbf, $R_{Oy} = 208.5$ lbf, $R_{Oz} = 259.3$ lbf, (d) $\sigma = 35.2$ kpsi, $\tau = 7.35$ kpsi, (e) $\sigma_1 = 36.7$ kpsi, $\sigma_2 = -1.47$ kpsi, $\tau_{max} = 19.1$ kpsi

3–80 (a) Critical at the wall at top or bottom of rod. (b) $\sigma_x = 16.3$ kpsi, $\tau_{xz} = 5.09$ kpsi,

(c) $\sigma_1 = 17.8$ kpsi, $\sigma_2 = -1.46$ kpsi,
$\tau_{max} = 9.61$ kpsi

3–84 (a) Critical at the top or bottom.
(b) $\sigma_x = 28.0$ kpsi, $\tau_{xz} = 15.3$ kpsi,
(c) $\sigma_1 = 34.7$ kpsi, $\sigma_2 = -6.7$ kpsi,
$\tau_{max} = 20.7$ kpsi

3–95 $x_{min} = 8.3$ mm

3–97 $x_{max} = 1.9$ kpsi

3–100 $p_o = 82.8$ MPa

3–104 $\sigma_l = -254$ psi, $\sigma_t = 5710$ psi, $\sigma_r = -23.8$ psi,
$\tau_{1/3} = 2980$ psi, $\tau_{1/2} = 2870$ psi, $\tau_{2/3} = 115$ psi

3–108 $\tau_{max} = 2.68$ kpsi

3–110 $\delta_{max} = 0.021$ mm, $\delta_{min} = 0.0005$ mm,
$p_{max} = 65.2$ MPa, $p_{min} = 1.55$ MPa

3–116 $\delta = 0.025$ mm, $p = 57.5$ MPa,
$(\sigma_t)_i = -57.5$ MPa, $(\sigma_t)_o = 149.5$ MPa

3–120 $\sigma_i = 300$ MPa, $\sigma_o = -195$ MPa

3–126 (a) $\sigma = \pm 8.02$ kpsi, (b) $\sigma_i = -10.1$ kpsi,
$\sigma_o = 6.62$ kpsi, (c) $K_i = 1.26$, $K_o = 0.825$

3–129 $\sigma_i = 64.6$ MPa, $\sigma_o = -21.7$ MPa

3–133 $\sigma_{max} = 352F^{1/3}$ MPa, $\tau_{max} = 106F^{1/3}$ MPa

3–138 $F = 117.4$ lbf

3–141 $\sigma_x = -35.0$ MPa, $\sigma_y = -22.9$ MPa,
$\sigma_z = -96.9$ MPa, $\tau_{max} = 37.0$ MPa

B–4 **Chapter 4**

4–3 (a) $k = \dfrac{\pi d^4 G}{32}\left(\dfrac{1}{x} + \dfrac{1}{l-x}\right)$,

$T_1 = 1500\dfrac{l-x}{l}$, $T_2 = 1500\dfrac{x}{l}$,

(b) $k = 28.2 \,(10^3)$ lbf · in/rad, $T_1 = T_2 = 750$ lbf · in,
$\tau_{max} = 30.6$ kpsi

4–7 $\delta = 5.262$ in, % elongation due to weight = 3.21%

4–10 $y_{max} = -25.4$ mm, $\sigma_{max} = -163$ MPa

4–13 $y_O = y_C = -3.72$ mm, $y|_{x=550mm} = 1.11$ mm

4–16 $d_{min} = 32.3$ mm

4–24 $y_A = -7.99$ mm, $\theta_A = -0.0304$ rad

4–27 $y_A = 0.0805$ in, $z_A = -0.1169$ in,
$(\theta_A)_y = 0.00115$ rad, $(\theta_A)_z = 8.06(10^{-5})$ rad

4–30 $(\theta_O)_z = 0.0131$ rad, $(\theta_C)_z = -0.0191$ rad

4–33 $(\theta_O)_y = 0.0104$ rad, $(\theta_O)_z = 0.00751$ rad,
$(\theta_C)_y = -0.0193$ rad, $(\theta_C)_z = -0.0109$ rad

4–36 $d = 62.0$ mm

4–39 $d = 2.68$ in

4–41 $y = 0.1041$ in

4–43 Stepped bar: $\theta = 0.026$ rad, simplified bar:
$\theta = 0.0345$ rad, 1.33 times greater, 0.847 in

4–46 $d = 38.1$ mm, $y_{max} = -0.0678$ mm

4–51 $y_B = -0.0155$ in

4–52 $k = 8.10$ N/mm

4–69 $\delta = 0.0102$ in

4–73 Stepped bar: $\delta = 0.706$ in, uniform bar:
$\delta = 0.848$ in, 1.20 times greater

4–76 $\delta = 0.0338$ mm

4–78 $\delta = 0.0226$ in

4–81 $\delta = 0.551$ in

4–85 $\delta = 6.067$ mm

4–90 (a) $\sigma_b = 48.8$ kpsi, $\sigma_c = -13.9$ kpsi,
(b) $\sigma_b = 50.5$ kpsi, $\sigma_c = -12.0$ kpsi

4–92 $R_B = 1.6$ kN, $R_O = 2.4$ kN, $\delta_A = 0.0223$ mm

4–97 $R_C = 1.33$ kips, $R_O = 4.67$ kips,
$\delta_A = 0.00622$ in, $\sigma_{AB} = -14.7$ kpsi

4–101 $\sigma_{BE} = 20.2$ kpsi, $\sigma_{DF} = 10.3$ kpsi,
$y_B = -0.0255$ in, $y_C = -0.0865$ in,
$y_D = -0.0131$ in

4–106 (a) $t = 11$ mm, (b) No

4–112 $F_{max} = 143.6$ lbf, $\delta_{max} = 1.436$ in

B–5 **Chapter 5**

5–1 (a) MSS: $n = 3.5$, DE: $n = 3.5$, (b) MSS: $n = 3.5$,
DE: $n = 4.04$, (c) MSS: $n = 1.94$, DE: $n = 2.13$,
(d) MSS: $n = 3.07$, DE: $n = 3.21$,
(e) MSS: $n = 3.34$, DE: $n = 3.57$

5–3 (a) MSS: $n = 1.5$, DE: $n = 1.72$,
(b) MSS: $n = 1.25$, DE: $n = 1.44$,
(c) MSS: $n = 1.33$, DE: $n = 1.42$,
(d) MSS: $n = 1.16$, DE: $n = 1.33$,
(e) MSS: $n = 0.96$, DE: $n = 1.06$

5–7 (a) $n = 3.03$

5–12 (a) $n = 2.40$, (b) $n = 2.22$, (c) $n = 2.19$,
(d) $n = 2.04$, (e) $n = 1.92$

5–17 (a) $n = 1.81$

5–19 (a) BCM: $n = 1.2$, MM: $n = 1.2$,
(b) BCM: $n = 1.5$, MM: $n = 2.0$, (c) BCM: $n = 1.18$,
MM: $n = 1.24$, (d) BCM: $n = 1.23$, MM: $n = 1.60$,
(e) BCM: $n = 2.57$, MM: $n = 2.57$

5–24 (a) BCM: $n = 3.63$, MM: $n = 3.63$

5–29 (a) $n = 1.54$

5–34 (a) $n = 1.54$

5–40 MSS: $n = 1.29$, DE: $n = 1.32$

5–48 MSS: $n = 13.9$, DE: $n = 14.3$

5–53 MSS: $n = 1.30$, DE: $n = 1.40$

5–58 For yielding: $p = 934$ psi,
For rupture: $p = 1.11$ kpsi

5–63 $d = 0.892$ in

5–65 Model c: $n = 1.80$, Model d: $n = 1.25$,
Model e: $n = 1.80$

5–67 $F_x = 2\pi f T/(0.2d)$

5–68 (a) $F_i = 16.7$ kN, (b) $p_i = 111.3$ MPa,
(c) $\sigma_t = 185.5$ MPa, $\sigma_r = -111.3$ MPa
(d) $\tau_{\max} = 148.4$ MPa, $\sigma' = 259.7$ MPa,
(e) MSS: $n = 1.52$, DE: $n = 1.73$

5–74 $n_o = 1.84$, $n_i = 1.80$

5–76 $n = 1.91$

5–84 (a) $F = 958$ kN, (b) $F = 329.4$ kN

B–6 Chapter 6

6–1 $S_e = 433$ MPa

6–3 $N = 116\ 700$ cycles

6–5 $S_f = 117.0$ kpsi

6–9 $(S_f)_{ax} = 162\ N^{-0.0851}$ kpsi for $10^3 \le N \le 10^6$

6–15 $n_f = 0.73$, $n_y = 1.51$

6–17 $n_f = 0.49$, $N = 4600$ cycles

6–20 $n_y = 1.66$, (a) $n_f = 1.05$, (b) $n_f = 1.31$,
(c) $n_f = 1.32$

6–24 $n_y = 2.0$, (a) $n_f = 1.19$, (b) $n_f = 1.43$, (c) $n_f = 1.44$

6–25 $n_y = 3.32$, using Goodman: $n_f = 0.64$,
$N = 34\ 000$ cycles

6–28 (a) $n_f = 0.94$, $N = 637\ 000$ cycles,
(b) $n_f = 1.16$ for infinite life

6–30 The design is controlled by fatigue at the hole,
$n_f = 1.48$

6–33 (a) $T = 23.0$ lbf · in, (b) $T = 28.2$ lbf · in,
(c) $n_y = 2.14$

6–35 $n_f = 1.21$, $n_y = 1.43$

6–38 $n_f = 0.56$

6–46 $n_f = 6.06$

6–47 $n_f = 1.40$

6–51 $n_f = 0.72$, $N = 7500$ cycles

6–57 $P = 4.12$ kips, $n_y = 5.29$

6–59 (a) $n_2 = 7\ 000$ cycles, (b) $n_2 = 10\ 000$ cycles

B–7 Chapter 7

7–1 (a) DE-Gerber: $d = 25.85$ mm, (b) DE-Elliptic:
$d = 25.77$ mm, (c) DE-Soderberg: $d = 27.70$ mm,
(d) DE-Goodman: $d = 27.27$ mm

7–2 Using DE-Elliptic, $d = 24$ mm, $D = 32$ mm,
$r = 1.6$ mm

7–6 These answers are a partial assessment of potential
failure. Deflections: $\theta_O = 5.47(10)^{-4}$ rad,
$\theta_A = 7.09(10)^{-4}$ rad, $\theta_B = 1.10(10)^{-3}$ rad. Compared
to Table 7–2 recommendations, θ_B is high for an
uncrowned gear. Strength: Using DE-Elliptic at the
shoulder at A, $n_f = 3.42$

7–18 (a) Fatigue strength using DE-Elliptic: Left
keyway $n_f = 3.8$, right bearing shoulder $n_f = 5.0$,
right keyway $n_f = 2.7$. Yielding: Left keyway
$n_y = 4.3$, right keyway $n_y = 2.7$, (b) Deflection
factors compared to minimum recommended in
Table 7–2: Left bearing $n = 1.38$, right bearing
$n = 0.72$, gear slope $n = 0.63$

7–28 (a) $\omega = 883$ rad/s (b) $d = 50$ mm
(c) $\omega = 1766$ rad/s (doubles)

7–30 (b) $\omega = 466$ rad/s $= 4450$ rev/min

7–34 $\frac{1}{4}$-in square key, $\frac{7}{8}$-in long, AISI 1020 CD

7–36 $d_{\min} = 14.989$ mm, $d_{\max} = 15.000$ mm,
$D_{\min} = 15.000$ mm, $D_{\max} = 15.018$ mm

7–42 (a) $d_{\min} = 35.043$ mm, $d_{\max} = 35.059$ mm,
$D_{\min} = 35.000$ mm, $D_{\max} = 35.025$ mm,
(b) $p_{\min} = 35.1$ MPa, $p_{\max} = 115$ MPa,
(c) Shaft: $n_y = 3.4$, hub: $n_y = 1.9$,
(d) Assuming $f = 0.8$, $T = 2700$ N · m

B–8 Chapter 8

8–1 (a) Thread depth 2.5 mm, thread width 2.5 mm,
$d_m = 22.5$ mm, $d_r = 20$ mm, $l = p = 5$ mm

8–4 $T_R = 15.85$ N · m, $T_L = 7.83$ N · m, $e = 0.251$

8–8 $F = 182$ lbf

8–11 (a) $L = 45$ mm, (b) $k_b = 874.6$ MN/m, (c) $k_m = 3\ 116.5$ MN/m

8–14 (a) $L = 3.5$ in, (b) $k_b = 1.79$ Mlbf/in, (c) $k_m = 7.67$ Mlbf/in

8–19 (a) $L = 60$ mm, (b) $k_b = 292.1$ MN/m, (c) $k_m = 692.5$ MN/m

8–25 From Eqs. (8–20) and (8–22), $k_m = 2\ 762$ MN/m. From Eq. (8–23), $k_m = 2\ 843$ MN/m

8–29 (a) $n_p = 1.10$, (b) $n_L = 1.60$, (c) $n_0 = 1.20$

8–33 $L = 55$ mm, $n_p = 1.29$, $n_L = 11.1$, $n_0 = 11.8$

8–37 $n_p = 1.29$, $n_L = 10.7$, $n_0 = 12.0$

8–41 Bolt sizes of diameters 8, 10, 12, and 14 mm were evaluated and all were found acceptable. For $d = 8$ mm, $k_m = 854$ MN/m, $L = 50$ mm, $k_b = 233.9$ MN/m, $C = 0.215$, $N = 20$ bolts, $F_i = 6.18$ kN, $P = 2.71$ kN/bolt, $n_p = 1.22$, $n_L = 3.53$, $n_0 = 2.90$

8–46 (a) $T = 823$ N · m, (b) $n_p = 1.10$, $n_L = 17.7$, $n_0 = 57.7$

8–51 (a) Goodman: $n_f = 7.55$, (b) Gerber: $n_f = 11.4$, (c) ASME-elliptic: $n_f = 9.73$

8–55 Goodman: $n_f = 11.9$

8–60 (a) $n_p = 1.16$, (b) $n_L = 2.96$, (c) $n_0 = 6.70$, (d) $n_f = 4.56$

8–63 $n_p = 1.24$, $n_L = 4.62$, $n_0 = 5.39$, $n_f = 4.75$

8–67 Bolt shear, $n = 2.30$; bolt bearing, $n = 4.06$; member bearing, $n = 1.31$; member tension, $n = 3.68$

8–70 Bolt shear, $n = 1.70$; bolt bearing, $n = 4.69$; member bearing, $n = 2.68$; member tension, $n = 6.68$

8–75 $F = 2.32$ kN based on channel bearing

8–77 Bolt shear, $n = 4.78$; bolt bearing, $n = 10.55$; member bearing, $n = 5.70$; member bending, $n = 4.13$

B–9 Chapter 9

9–1 $F = 49.5$ kN

9–5 $F = 51.2$ kN

9–9 $F = 31.1$ kN

9–14 $\tau = 123.7$ MPa

9–18 (a) $F = 9.75$ kN, (b) $F = 5.3$ kN

9–22 $F = 19.4$ kN

9–26 $F = 26$ kN

9–29 $F = 49.8$ kN

9–31 $F = 5.04$ kN

9–34 All-around square, four beads each $h = 6$ mm, 75 mm long, Electrode E6010

9–45 $\tau_{\max} = 184.9$ MPa

9–47 $\tau_{\max} = 45.3$ MPa

9–48 $n = 3.56$

9–51 $F = 61.2$ kN

B–10 Chapter 10

10–3 (a) $L_0 = 162.8$ mm, (b) $F_s = 167.9$ N, (c) $k = 1.314$ N/mm, (d) $(L_0)_{cr} = 149.9$ mm, spring needs to be supported

10–5 (a) $L_s = 65$ mm, (b) $F_s = 289.5$ N, (c) $n_s = 2.05$

10–7 (a) $L_0 = 47.7$ mm, (b) $p = 5.61$ mm, (c) $F_s = 81.12$ N, (d) $k = 2.643$ N/mm, (e) $(L_0)_{cr} = 105.2$ mm buckling is unlikely

10–11 Spring is not solid safe, $n_s = 0.56$ $L_0 \le 14.4$ mm

10–17 Spring is solid safe, $n_s = 1.04$, $L_0 \le 68.2$ mm

10–20 (a) $N_a = 12$ turns, $L_s = 44.2$ mm, $p = 10$ mm, (b) $k = 1.08$ N/mm, (c) $F_s = 81.9$ N, (d) $\tau_s = 271$ MPa

10–23 With $d = 2$ mm, $L_0 = 48$ mm, $k = 4.286$ N/mm, $D = 13.25$ mm, $N_a = 15.9$ coils, $n_s = 2.63 > 1.2$, ok. No other d works.

10–28 (a) $d = 6$ mm, (b) $D = 42$ mm, (c) $k = 24$ N/mm, (d) $N_t = 9.29$ turns, (e) $L_0 = 102.2$ mm

10–30 Use A313 stainless wire, $d = 2.3$ mm, OD $= 209$ mm, $N_t = 24.6$ turns, $L_0 = 118.9$ mm

10–39 $\Sigma = 31.3°$ (see Fig. 10–9), $F_{\max} = 87.3$ N

10–42 (a) $k = 12\ EI\{4l^3 + 3R[2\pi l^2 + 4(\pi - 2)\ lR + (3\pi - 8)\ R^2]\}^{-1}$, (b) $k = 36.3$ lbf/in, (c) $F = 3.25$ lbf

B–11 Chapter 11

11–1 $x_D = 525$, $F_D = 3.0$ kN, $C_{10} = 24.3$ kN, 02–35 mm deep-groove ball bearing, $R = 0.920$

11–6 $x_D = 456$, $C_{10} = 145$ kN

11–8 $C_{10} = 20$ kN

11–15 $C_{10} = 26.1$ kN

11–21 (a) $F_e = 5.34$ kN, (b) $\mathcal{L}_D = 444$ h

11–24 60 mm deep-groove

11–27 (a) $C_{10} = 12.8$ kips

11–33 $C_{10} = 5.7$ kN, 02–12 mm deep-groove ball bearing

11–34 $R_O = 112$ lbf, $R_C = 298$ lbf, deep-groove 02–17 mm at O, deep-groove 02–35 mm at C

11–38 $l_2 = 0.267(10^6)$ rev

11–43 $F_{RA} = 35.4$ kN, $F_{RB} = 17.0$ kN

B–12 **Chapter 12**

12–1 $c_{min} = 0.015$ mm, $r = 12.5$ mm, $r/c = 833$, $N_j = 18.3$ rev/s, $S = 0.182$, $h_0/c = 0.3$, $rf/c = 5.4$, $Q/(rcNl) = 5.1$, $Q_s/Q = 0.81$, $h_0 = 0.0045$ mm, $H_{loss} = 11.2$ W, $Q = 219$ mm³/s, $Q_s = 177$ mm³/s

12–3 SAE 10: $h_0 = 0.000\ 275$ in, $p_{max} = 847$ psi, $c_{min} = 0.0025$ in

12–7 $h_0 = 0.00069$ in, $f = 0.007\ 87$, $Q = 0.0833$ in³/s

12–9 $h_0 = 0.011$ mm, $H = 48.1$ W, $Q = 1426$ mm³/s, $Q_s = 1012$ mm³/s

12–11 $T_{av} = 154°F$, $h_0 = 0.00113$ in, $H_{loss} = 0.0750$ Btu/s, $Q_s = 0.0802$ in³/s

12–20 Approx: $\mu = 45.7$ mPa · s, Fig. 12–13: $\mu = 39$ mPa · s

B–13 **Chapter 13**

13–1 35 teeth, 3.25 in

13–2 400 rev/min, $p = 3\pi$ mm, $C = 112.5$ mm

13–4 $a = 0.3333$ in, $b = 0.4167$ in, $c = 0.0834$ in, $p = 1.047$ in, $t = 0.523$ in, $d_1 = 7$ in, $d_{1b} = 6.578$ in, $d_2 = 9.333$ in, $d_{2b} = 8.77$ in, $p_b = 0.984$ in, $m_c = 1.55$

13–5 $d_P = 2.333$ in, $d_G = 5.333$ in, $\gamma = 23.63°$, $\Gamma = 66.37°$, $A_0 = 2.910$ in, $F = 0.873$ in

13–10 (a) 13, (b) 15, 45, (c) 18

13–12 10:20 and higher

13–15 (a) $p_n = 3\pi$ mm, $p_t = 10.40$ mm, $p_x = 22.30$ mm, (b) $m_t = 3.310$ mm, $\phi_t = 21.88°$, (c) $d_P = 59.58$ mm, $d_G = 105.92$ mm

13–17 $e = 4/51$, $n_d = 47.06$ rev/min cw

13–24 $N_2 = N_4 = 15$ teeth, $N_3 = N_5 = 44$ teeth

13–29 $n_A = 68.57$ rev/min cw

13–36 (a) $d_2 = d_4 = 2.5$ in, $d_3 = d_5 = 7.33$ in, (b) $V_i = 1636$ ft/min, $V_o = 558$ ft/min,

(c) $W_{ti} = 504$ lbf, $W_{ri} = 184$ lbf, $W_i = 537$ lbf, $W_{to} = 1478$ lbf, $W_{ro} = 538$ lbf, $W_o = 1573$ lbf, (d) $T_i = 630$ lbf · in, $T_o = 5420$ lbf · in

13–38 (a) $N_{Pmin} = 15$ teeth, (b) $P = 1.875$ teeth/in, (c) $F_A = 311$ lbf, $F_B = 777.6$ lbf

13–41 (a) $N_F = 30$ teeth, $N_C = 15$ teeth, (b) $P = 3$ teeth/in, (c) $T = 900$ lbf · in, (d) $W_r = 65.5$ lbf, $W_t = 180$ lbf, $W = 192$ lbf

13–43 $\mathbf{F}_A = 71.5\ \mathbf{i} + 53.4\ \mathbf{j} + 350.5\ \mathbf{k}$ lbf, $\mathbf{F}_B = -178.4\ \mathbf{i} - 678.8\ \mathbf{k}$ lbf

13–50 $\mathbf{F}_C = 1565\ \mathbf{i} + 672\ \mathbf{j}$ lbf, $\mathbf{F}_D = 1610\ \mathbf{i} - 425\ \mathbf{j} + 154\ \mathbf{k}$ lbf

B–14 **Chapter 14**

14–1 $\sigma = 7.63$ kpsi

14–4 $\sigma = 32.6$ MPa

14–7 $F = 2.5$ in

14–10 $m = 2$ mm, $F = 25$ mm

14–14 $\sigma_c = -617$ MPa

14–17 $W^t = 16\ 390$ N, $H = 94.3$ kW (pinion bending); $W^t = 3469$ N, $H = 20.0$ kW (pinion and gear wear)

14–18 $W^t = 1283$ lbf, $H = 32.3$ hp (pinion bending); $W^t = 1633$ lbf, $H = 41.1$ hp (gear bending); $W^t = 265$ lbf, $H = 6.67$ hp (pinion and gear wear)

14–22 $W^t = 775$ lbf, $H = 19.5$ hp (pinion bending); $W^t = 300$ lbf, $H = 7.55$ hp (pinion wear), AGMA method accounts for more conditions

14–24 Rating power = min(157.5, 192.9, 53.0, 59.0) = 53 hp

14–28 Rating power = min(270, 335, 240, 267) = 240 hp

14–34 $H = 69.7$ hp

B–15 **Chapter 15**

15–1 $W_P^t = 690$ lbf, $H_1 = 16.4$ hp, $W_G^t = 620$ lbf, $H_2 = 14.8$ hp

15–2 $W_P^t = 464$ lbf, $H_3 = 11.0$ hp, $W_G^t = 531$ lbf, $H_4 = 12.6$ hp

15–8 Pinion core 300 Bhn, case, 373 Bhn; gear core 339 Bhn, case, 345 Bhn

15–9 All four $W^t = 690$ lbf

15–11 Pinion core 180 Bhn, case, 266 Bhn; gear core, 180 Bhn, case, 266 Bhn

B–16 Chapter 16

16–1 (a) Right shoe: $p_a = 734.5$ kPa cw rotation,
(b) Right shoe: $T = 277.6$ N · m; left shoe:
144.4 N · m; total $T = 422$ N · m, (c) RH shoe:
$R^x = -1.007$ kN, $R^y = 4.13$ kN, $R = 4.25$ kN,
LH shoe: $R^x = 570$ N, $R^y = 751$ N, $R = 959$ N

16–3 LH shoe: $T = 2.265$ kip · in, $p_a = 133.1$ psi,
RH shoe: $T = 0.816$ kip · in, $p_a = 47.93$ psi,
$T_{total} = 3.09$ kip · in

16–5 $p_a = 27.4$ psi, $T = 348.7$ lbf · in

16–8 $a' = 1.209r$, $a = 1.170r$

16–10 $P = 1.25$ kips, $T = 25.52$ kip · in

16–14 (a) $T = 8200$ lbf · in, $P = 504$ lbf, $H = 26$ hp,
(b) $R = 901$ lbf, (c) $p|_{\theta=0} = 70$ psi,
$p|_{\theta=270°} = 27.3$ psi

16–17 (a) $F = 1885$ lbf, $T = 7125$ lbf · in,
(c) torque capacity exhibits a stationary point
maximum

16–18 (a) $d* = D/\sqrt{3}$, (b) $d* = 3.75$ in, $T* = 7173$
lbf · in, (c) $(d/D)* = 1/\sqrt{3} = 0.577$

16–19 (a) Uniform wear: $p_a = 14.04$ psi,
$F = 243$ lbf, (b) Uniform pressure: $p_a = 13.42$ psi,
$F = 242$ lbf

16–23 $C_s = 0.08$, $t = 143$ mm

16–26 (b) $I_e = I_M + I_P + n^2 I_P + I_L/n^2$,
(c) $I_e = 10 + 1 + 10^2(1) + 100/10^2 = 112$

16–27 (c) $n* = 2.430$, $m* = 4.115$, which are
independent of I_L

B–17 Chapter 17

17–2 (a) $F_c = 4$ N, $F_i = 422$ N,
$F_{1a} = 630$ N, $F_c = 222$ N, (b) $H_a = 1.88$ kW,
$n_{fs} = 1.0$, (c) 4 mm

17–4 A-3 polyamide belt, $b = 6$ in, $F_c = 77.4$ lbf,
$T = 10\ 946$ lbf · in, $F_1 = 573.7$ lbf, $F_2 = 117.6$ lbf,
$F_i = 268.3$ lbf, dip = 0.562 in

17–6 (a) $T = 742.8$ lbf · in, $F_i = 148.1$ lbf,
(b) $b = 4.13$ in, (c) $(F_1)_a = 289.1$ lbf, $F_c = 17.7$ lbf,
$F_i = 147.6$ lbf, $F_2 = 41.5$ lbf, $H = 20.6$ hp,
$n_{fs} = 1.1$

17–8 $R^x = (F_1 + F_2)\{1 - 0.5[(D - d)/(2C)]^2\}$,
$R^y = (F_1 - F_2)(D - d)/(2C)$. From Ex. 17–2,
$R^x = 1214.4$ lbf, $R^y = 34.6$ lbf

17–14 With $d = 2$ in, $D = 4$ in, life of 10^6 passes,
$b = 4.5$ in, $n_{fs} = 1.05$

17–17 Select one B90 belt

17–20 Select nine C270 belts, life $> 10^9$ passes,
life $> 150\ 000$ h

17–24 (b) $n_1 = 1227$ rev/min. Table 17–20 confirms
this point occurs in the range 1200 ± 200 rev/min,
(c) Eq. (17–40) applicable at speeds exceeding
1227 rev/min for No. 60 chain

17–25 (a) $H_a = 7.91$ hp; (b) $C = 18$ in,
(c) $T = 1164$ lbf · in, $F = 744$ lbf

17–27 Four-strand No. 60 chain, $N_1 = 17$ teeth,
$N_2 = 84$ teeth, rounded $L = 100$ in, $C = 30.0$ in
$n_{fs} = 1.17$, life 15 000 h (pre-extreme)

B–20 Chapter 20

20–13 Partial answers: (a) 50.3, 49.7 (b) No effect
(c) the center axis of the boss, as determined by the
related actual mating envelope (d) 0.2 (e) 0.8

20–15 Hint: Read about actual mating envelopes.

20–17 20.2, 19.8

20–21 (a) 0.1, 0.1, 0.1, (b) 0.5, 0.3, 0.1,
(c) 0.1, 0.3, 0.5

Index